P9-CLB-097

1970
ANNUAL BOOK OF ASTM STANDARDS

PART 27

Plastics—General Methods of Testing, Nomenclature

ASTM 215-299-5400

Includes Standards of:

Committee D-20 on Plastics

Price $27.00

AMERICAN SOCIETY FOR TESTING AND MATERIALS

1916 Race St., Philadelphia, Pa. 19103

Library of Congress Catalog Card Number: 40-10712
ISBN 0-8031-3034-1 Coden: ASTSA

Printed in Easton, Md., U.S.A.

FOREWORD

The American Society for Testing and Materials

The American Society for Testing and Materials, founded in 1898, is an international, nonprofit, technical, scientific, and educational society devoted to ". . . . the promotion of knowledge of the materials of engineering, and the standardization of specifications and methods of testing."

The Society operates through more than 105 main technical committees. These committees function in prescribed fields under regulations that ensure balanced representation among producers, consumers, and general interest participants.

The Society currently has 16,000 active members and also 8500 other individuals serving as technical experts who represent corporate and institutional members on committees.

Membership in the Society is open to all concerned with the fields of work in which ASTM is active. A membership application with detailed information may be found at the back of this book. Additional information may be obtained from the Membership Development Department, American Society for Testing and Materials, 1916 Race St., Philadelphia, Pa. 19103.

1970 Annual Book of ASTM Standards

The *1970 Annual Book of ASTM Standards* consists of 33 parts, of which this part is one. It contains all current formally approved ASTM standard and tentative specifications, methods of test, recommended practices, definitions, and other related material, such as proposed methods.

Standards comprise those specifications, methods of test, and definitions that have been formally adopted by the Society, requiring a letter ballot approval by the entire membership. In accordance with Society regulations, standards that have been published for a period of five years or more without revision have been reviewed by the responsible technical committees prior to the appearance of the Book and have been reapproved in their present form.

Tentatives have been approved by the sponsoring committee as representing the latest thoughts and practices and have been accepted by the Society in accordance with established procedures, for use pending adoption as standard.

Tentative Revisions of standards appear at the end of the standards to which they apply. A list of the tentative revisions that apply to the standards in this part appears at the end of the Table of Contents.

A new edition of the Book of Standards is issued annually. Each part contains all actions accepted at the Annual Meeting preceding the issue date and all other actions accepted by the Society at least six months before the issue date. Later actions are included wherever this is possible without delaying the appearance of the part. The 1970 edition of the Book of Standards is comprised of over 30,000 pages and includes over 4300 ASTM standards and tentatives.

The combined *Index to ASTM Standards*, found in Part 33, will be helpful in locating any desired standard. All current standards and tentatives as well as new and revised items accepted by the Society between the appearance of any given part are made available as separate reprints.

An ASTM standard represents a common viewpoint of those parties concerned with its provisions, namely, producers, consumers, and general interest groups. It is intended to aid industry, government agencies, and the general public. The use of an ASTM standard is purely voluntary. It is recognized that, for certain work or in certain regions, ASTM specifications may be either more or less restrictive than needed. The existence of an ASTM standard does not preclude anyone from manufacturing, marketing, or purchasing products, or using products, processes, or procedures not conforming to the standard. Because ASTM standards are subject to periodic review and revision, those who use them are cautioned to obtain the latest revisions.

Using the Annual Book of ASTM Standards

The standards in each part of this Book are assembled in alpha-numerical sequence of their ASTM designation numbers. The table of contents in each part is presented in duplicate—one, a list of the standards in numerical sequence of their ASTM designations; the other, a list of the standards classified according to the subjects covered.

A subject index of the standards and tentatives in each part appears at the back of each volume. This index should be the most convenient means of locating any standard appearing in the volume if only the general subject matter covered is known.

Availability of Individual Standards

Each ASTM standard is available as a separate publication from the American Society for Testing and Materials at a price of $1.00 each. Special quantity prices are also available. When ordering, give ASTM standard designation, quantity desired, and shipping instructions.

Obsolete Editions

This new edition of the *Annual Book of ASTM Standards* makes last year's edition obsolete. Each part of the *Annual Book of ASTM Standards* is replaced annually because of additions of new standards and significant revisions in existing standards. On the average, about 30 percent of each part is new or revised. While old parts are obsolete, they are not completely without value. Many small colleges and other educational institutions in the United States and Canada, as well as underdeveloped countries, can put these books to good use. Donate it to a local small educational institution; contact your local ASTM district which may have an active book distribution program; contact Engineers and Scientists Committee Inc., a People-to-People Activity, Mr. E. J. Sparling, 124 Hilton Ave., Garden City, N.Y. 11530; contact the Office of Private Cooperation, United States Information Agency, Washington, D.C., or contact the Overseas Institute of Canada at Overseas Book Centre, 207 Queen's Quay West, Toronto, Ontario, Canada.

Legal counsel believes such contributions to be tax-deductible at approximately 50 percent of purchase price if donated to tax-exempt organizations.

Disclaimer

Nothing contained in any publication of the American Society for Testing and Materials is to be construed as granting any right, by implication or otherwise, for manufacture, sale, or use in connection with any method, apparatus, or product covered by Letters Patent, nor as insuring anyone against liability for infringement of Letters Patent.

1970 ANNUAL BOOK OF ASTM STANDARDS LISTED BY PARTS

Part 1—Steel Piping, Tubing, and Fittings(*April*)
Part 2—Ferrous Castings; Ferro-Alloys (*April*)
Part 3—Steel Sheet, Strip, Bar, Rod, Wire, Chain, and Spring; Wrought Iron Bar and Sheet; Metallic Coated Products; Ferrous Surgical Implants (*April*)
Part 4—Structural Steel; Concrete Reinforcing Steel; Pressure Vessel Plate; Steel Rails, Wheels, and Tires; Bearing Steel; Steel Forgings (*April*)
Part 5—Copper and Copper Alloys (Including Electrical Conductors) (*July*)
Part 6—Die-Case Metals; Light Metals and Alloys (Including Electrical Conductors) (*July*)
Part 7—Nonferrous Metals and Alloys (Including Corrosion Tests); Electrodeposited Metallic Coatings; Metal Powders; Nonferrous Surgical Implants (*July*)
Part 8—Magnetic Properties; Metallic Materials for Thermostats and for Electrical Resistance, Heating, and Contacts; Materials for Electron Devices and Microelectronics (*November*)
Part 9—Cement; Lime; Gypsum (*November*)
Part 10—Concrete and Mineral Aggregates (*November*)
Part 11—Bituminous Materials for Highway Construction, Waterproofing, and Roofing; Soils; Skid Resistance (*April*)
Part 12—Chemical-Resistant Nonmetallic Materials; Clay and Concrete Pipe and Tile; Masonry Mortars and Units; Asbestos-Cement Products; Natural Building Stones (*April*)
Part 13—Refractories, Glass, and Other Ceramic Materials; Manufactured Carbon and Graphite Products (*April*)
Part 14—Thermal and Cryogenic Insulating Materials; Acoustical Materials; Building Joint Sealants; Fire Tests; Building Constructions (*November*)
Part 15—Paper; Packaging; Cellulose; Casein; Flexible Barrier Materials; Carbon Paper; Leather (*April*)
Part 16—Structural Sandwich Constructions; Wood; Adhesives (*July*)
Part 17—Petroleum Products—Fuels, Solvents, Burner Fuel Oils, Lubricating Oils, Cutting Oils, Lubricating Greases, Hydraulic Fluids (*November*)
Part 18—Petroleum Products—Measurement and Sampling, Liquefied Petroleum Gases, Light Hydrocarbons, Plant Spray Oils, Aerospace Materials, Sulfonates, Crude Petroleum, Petrolatum, Wax, Graphite (*November*)
Part 19—Gaseous Fuels; Coal and Coke (*November*)
Part 20—Paint, Varnish, Lacquer, and Related Products—Materials Specifications and Tests; Naval Stores; Aromatic Hydrocarbons and Related Chemicals (*April*)
Part 21—Paint, Varnish, Lacquer, and Related Products—Tests for Formulated Products and Applied Coating (*April*)
Part 22—Sorptive Mineral Materials; Soap; Engine Antifreezes; Wax Polishes; Halogenated Organic Solvents; Activated Carbon; Industrial Chemicals (*July*)
Part 23—Water; Atmospheric Analysis (*November*)
Part 24—Textile Materials—Yarns, Fabrics, General Methods (*November*)
Part 25—Textile Materials—Fibers, Zippers; High Modulus Fibers (*November*)
Part 26—Plastics—Specifications; Methods of Testing Pipe, Film, Reinforced and Cellular Plastics (*July*)
Part 27—Plastics—General Methods of Testing, Nomenclature (*July*)
Part 28—Rubber; Carbon Black; Gaskets (*July*)
Part 29—Electrical Insulating Materials (*July*)
Part 30—General Test Methods; Radioisotopes and Radiation Effects; Statistical Methods; Appearance of Materials; Emission, Molecular, and Mass Spectroscopy; Sensory Evaluation of Materials and Products; Chromatography; Temperature Measurement; Space Simulation; Resinography; Microscopy; Deterioration of Nonmetallic Materials (*July*)
Part 31—Metals—Physical, Mechanical, Nondestructive, and Corrosion Tests, Metallography, Fatigue, Effect of Temperature (*July*)
Part 32—Chemical Analysis of Metals; Sampling and Analysis of Metal Bearing Ores (*April*)
Part 33—Index and Glossary of Definitions (*November*)

1970 ANNUAL BOOK OF ASTM STANDARDS LISTED BY SUBJECTS

STANDARDS LISTED BY SUBJECTS

CONTENTS

1970 ANNUAL BOOK OF ASTM STANDARDS, PARTS 27 AND 26

ASTM Standards and Tentatives Relating to Plastics— Specifications; Methods of Testing Pipe, Film, Reinforced and Cellular Plastics; General Methods of Testing; Nomenclature

A complete Subject Index appears on p. 811.

This contents lists only those standards included in these books and those standards in the area covered by Parts 26 and 27 that have been superseded or discontinued within the past five years. Since the standards in this book are arranged in numeric sequence, no page numbers are given in this contents.

In the serial designations prefixed to the following titles, the number following the dash indicates the year of original issue as tentative or of adoption as standard or, in the case of revision, the year of last revision. Thus, standards adopted or revised during the year 1970 have as their final number, 70. A letter following this number indicates more than one revision during that year, that is 69a indicates the second revision in 1969, 69b the third revision, etc. Tentatives are identified by the letter T. Standards that have been reapproved without change are indicated by the year of last reapproval in parentheses as part of the designation number, for example (1969).

* Approved as **American National Standard** by the American National Standards Institute.

RELATED MATERIAL

CONTENTS, *PART 27* AND PART 26

LIST BY SUBJECTS, PART 27

1970 ANNUAL BOOK OF ASTM STANDARDS, PART 27

PLASTICS—GENERAL METHODS OF TESTING, NOMENCLATURE

Since the standards in this book are arranged in numeric sequence, no page numbers are given in this list by subjects.

PLASTICS

Mechanical Properties

Methods of Test for:

D 1044 – 56 (1969) Abrasion Resistance, Surface, of Transparent Plastics
D 1242 – 56 (1969) Abrasion Resistance of Plastic Materials
D 1602 – 60 (1969) Bearing Load of Corrugated Plastics Panels (see Part 26)
D 953 – 54 (1969) Bearing Strength of Plastics
D 1893 – 67 Blocking of Plastic Film
D 952 – 51 (1969) Bond Strength of Plastics and Electrical Insulating Materials
D 1180 – 57 (1961) Bursting Strength of Round Rigid Plastic Tubing
*D 1894 – 63 Coefficients of Friction of Plastic Film
*D 695 – 69 Compressive Properties of Rigid Plastics
**D 1621 – 64* *Compressive Strength of Rigid Cellular Plastics (see Part 26)*
D 2143 – 69 Cyclic Pressure Strength of Reinforced, Thermosetting Plastic Pipe (see Part 26)
D 2236 – 67 T Dynamic Mechanical Properties of Plastics by Means of a Torsional Pendulum
**D 1604 – 63 (1969)* *Flatness of Plastics Sheet or Tubing, Measuring (see Part 26)*
D 790 – 66 Flexural Properties of Plastics
D 671 – 63 T Flexural Stress (Fatigue), Repeated, of Plastics
D 2176 – 69 Folding Endurance of Paper by the M.I.T. Tester
D 2583 – 67 Hardness, Indentation, of Plastics by Means of a Barcol Impressor
D 2240 – 68 Hardness, Indentation, of Rubber and Plastics by Means of a Durometer
*D 785 – 65 Hardness, Rockwell, of Plastics and Electrical Insulating Materials
*D 256 – 56 (1961) Impact Resistance of Plastics and Electrical Insulating Materials
D 1709 – 67 Impact Resistance of Polyethylene Film by the Free Falling Dart Method (see Part 26)
D 673 – 44 (1969) Mar Resistance of Plastics
D 1504 – 61 Orientation Release Stress of Plastic Sheeting (see Part 26)
D 2582 – 67 Resistance to Puncture-Propagation of Tear in Thin Plastic Sheeting
D 732 – 46 (1969) Shear Strength of Plastics
*D 1043 – 69 Stiffness Properties of Plastics as a Function of Temperature by Means of a Torsion Test

Methods of Test for:

*D 747 – 63 Stiffness of Plastics by Means of a Cantilever Beam
D 1938 – 67 Tear Propagation, Resistance to, in Plastic Film and Thin Sheeting by a Single-Tear Method
D 1004 – 66 Tear Resistance of Plastic Film and Sheeting
D 1922 – 67 Tear Resistance, Propagation, of Plastic Film and Thin Sheeting
D 1822 – 68 Tensile Impact Energy to Break Plastics and Electrical Insulating Materials
D 638 – 68 Tensile Properties of Plastics
D 2289 – 69 Tensile Properties of Plastics at High Speeds
D 1708 – 66 Tensile Properties of Plastics by Use of Microtensile Specimens

* Approved as **American National Standard** by the American National Standards Institute.

Methods of Test for:

**D 1623 – 64*	*Tensile Properties of Rigid Cellular Plastics (see Part 26)*
D 882 – 67	Tensile Properties of Thin Plastic Sheeting
D 1923 – 67	Tensile Properties of Thin Plastic Sheeting Using the Inclined Plane Testing Machine
D 651 – 48(1966)	*Tensile Strength of Molded Electrical Insulating Materials (see Part 29)*
*D 412 – 68	Tension Testing of Vulcanized Rubber
D 374 – 68	Thickness of Solid Electrical Insulation
*E 252 – 67	Thickness of Thin Foil and Film by Weighing
D 1502 – 60 (1961)	*Transverse Load of Corrugated Reinforced Plastic Panels (see Part 26)*
**D 1789 – 65*	*Welding Performance for Poly(Vinyl Chloride) Structures, Evaluation of (see Part 26)*
D 797 – 64	*Young's Modulus in Flexure of Natural and Synthetic Elastomers at Normal and Subnormal Temperatures (see Part 28)*

Recommended Practices for:

D 674 – 56 (1969)	Creep and Stress-Relaxation, Long-Time, of Plastics Under Tension or Compression Loads at Various Temperatures, Testing
D 759 – 66	Physical Properties of Plastics at Subnormal and Supernormal Temperatures, Determining
D 2093 – 62 T	*Preparation of Surfaces of Plastics Prior to Adhesive Bonding (see Part 16)*

Radiation Methods

Methods of Test for:

*D 1671 – 63 (1970)	Absorbed Gamma Radiation Dose in the Fricke Dosimeter
D 2365 – 68	Calculation of Neutron Dose to Polymeric Materials and Application of Threshold-Foil Measurements

Recommended Practices for:

D 2568 – 70	Calculation of Absorbed Dose from X or Gamma Radiation
*D 1672 – 66	Exposure of Polymeric Materials to High Energy Radiation

Thermal Properties

Methods of Test for:

D 746 – 64 T	Brittleness Temperature of Plastics and Elastomers by Impact
D 1790 – 62	Brittleness Temperature of Plastic Film by Impact
D 696 – 70	Coefficient of Linear Thermal Expansion of Plastics
D 864 – 52 (1961)	Coefficient of Cubical Thermal Expansion of Plastics
*D 621 – 64	Deformation of Plastics Under Load
D 648 – 56 (1961)	Deflection Temperature of Plastics Under Load
D 568 – 68	Flammability of Flexible Plastics
*D 635 – 68	Flammability of Self-Supporting Plastics
D 1692 – 68	Flammability of Plastic Sheeting and Cellular Plastics
D 757 – 65	Flammability of Plastics, Self-Extinguishing Type
D 2863 – 70	Flammability of Plastics Using the Oxygen Index Method
*D 1433 – 58 (1966)	Flammability of Flexible Thin Plastic Sheeting
D 569 – 59 (1961)	Flow Properties of Thermoplastic Molding Materials, Measuring
D 1238 – 65 T	Flow Rates of Thermoplastics by Extrusion Plastometer, Measuring
D 1929 – 68	Ignition Properties of Plastics
E 289 – 65 T	*Linear Thermal Expansion of Rigid Solids with Interferometry (see Part 31)*
D 2843 – 70	Measuring the Density of Smoke from the Burning or Decomposition of Plastics
*D 2117 – 64	Melting Point of Semicrystalline Polymers
D 955 – 51 (1969)	Shrinkage from Mold Dimensions of Molded Plastics, Measuring

Methods of Test for:

D 731 – 67	Molding Index of Thermosetting Molding Powder, Measuring
D 1637 – 61	Tensile Heat Distortion Temperature of Plastic Sheeting
D 1525 – 65 T	Vicat Softening Point of Plastics

Recommended Practices for:

D 2115 – 67	*Oven Heat Stability of Poly(Vinyl Chloride) Compositions (see Part 26)*
D 1703 – 62	Presentation of Capillary Flow Data on Molten Thermoplastics

Optical Properties

Methods of Test for:

D 881 – 48 (1965)	Deviation of Line of Sight Through Transparent Plastics
D 1494 – 60 (1969)	*Diffuse Light Transmission Factor of Reinforced Plastics Panels (see Part 26)*

Methods of Test for:

*D 1003 – 61	Haze and Luminous Transmittance of Transparent Plastics
*D 542 – 50 (1965)	Index of Refraction of Transparent Organic Plastics
*D 523 – 67	Specular Gloss
D 2457 – 65 T	Specular Gloss of Plastic Films
D 637 – 50 (1965)	Surface Irregularities of Flat Transparent Plastic Sheets
D 1746 – 70	Transparency of Plastic Sheeting
D 1729 – 64	*Visual Evaluation of Color Differences of Opaque Materials* (*see Part 21*)
D 1925 – 70	Yellowness Index of Plastics

Permanence Properties

Methods of Test for:

D 756 – 56 (1966)	Accelerated Service Conditions, Resistance of Plastics to
D 1042 – 51 (1966)	Changes in Linear Dimensions of Plastics, Measuring
D 1204 – 54 (1966)	Changes in Linear Dimensions of Nonrigid Thermoplastic Sheeting or Film, Measuring
D 543 – 67	Chemical Reagents, Resistance of Plastics to
D 1693 – 70	*Environmental Stress-Cracking of Type I Ethylene Plastics* (*see Part 26*)
**D 1604 – 63 (1969)*	*Flatness of Plastics Sheet or Tubing, Measuring* (*see Part 26*)
D 1434 – 66	Gas Transmission Rate of Plastic Film and Sheeting
D 1203 – 67	Loss of Plasticizer from Plastics (Activated Carbon Methods)
*D 1712 – 65	Resistance of Plastics to Sulfide Staining
D 1239 – 55 (1966)	Resistance of Plastic Films to Extraction by Chemicals
D 1299 – 55 (1966)	Shrinkage of Molded and Laminated Thermosetting Plastics at Elevated Temperature
D 2151 – 68	Staining of Poly(Vinyl Chloride) Compositions by Rubber Compounding Ingredients
D 1181 – 56 (1966)	Warpage of Sheet Plastics
*D 570 – 63	Water Absorption of Plastics

Recommended Practices for:

D 1501 – 65 T	Exposure of Plastics to Fluorescent Sunlamp
D 1672 – 66	Exposure of Polymeric Materials to High Energy Radiation
D 1920 – 69	Light Dosage in Carbon-Arc Light Aging Apparatus, Determining
D 822 – 60 (1968)	*Operating Light- and Water-Exposure Apparatus* (*Carbon-Arc Type*) *for Testing Paint, Varnish, Lacquer, and Related Products* (*see Part 21*)
E 42 – 69	Operating Light- and Water-Exposure Apparatus (Carbon-Arc Type) for Exposure of Nonmetallic Materials
*D 1499 – 64	Operating Light- and Water-Exposure Apparatus (Carbon-Arc Type) for Exposure of Plastics
D 2565 – 66 T	Operating Xenon Arc-Type (Water-Cooled) Light- and Water-Exposure Apparatus for Exposure of Plastics
D 794 – 68	Permanent Effect of Heat on Plastics, Determining
D 2299 – 68	Relative Stain Resistance of Plastics, Determining
*D 1924 – 63	Resistance of Synthetic Polymeric Materials to Fungi, Determining
D 2676 – 67 T	*Resistance of Plastics to Bacteria, Determining* (*see Part 26*)

Recommended Practices for:

D 2648 – 67 T	Time-to-Failure by Rupture of Plastics Under Tension in Various Environments, Measuring
D 1435 – 69	Weathering, Outdoor, of Plastics

Analytical Methods

Methods of Test for:

D 2238 – 68	Absorbance of Polyethylene Due to Methyl Groups at 1378 cm^{1}
*D 1671 – 63	Absorbed Gamma Radiation Dose in the Fricke Dosimeter
*D 494 – 46 (1967)	Acetone Extraction of Phenolic Molded or Laminated Products
*D 834 – 61 (1967)	Ammonia in Phenol-Formaldehyde Molded Materials
D 2124 – 70	Analysis of Components in Poly(Vinyl Chloride) Compounds Using an Infrared Spectrophotometric Technique
D 1603 – 68	Carbon Black in Ethylene Plastics
*D 817 – 65	Cellulose Acetate Propionate and Cellulose Acetate Butyrate
*D 1303 – 55 (1967)	Chlorine, Total, in Vinyl Chloride Polymers and Copolymers
D 1895 – 69	Density, Apparent, Bulk Factor, and Pourability of Plastic Materials
**D 1622 – 63*	*Density, Apparent, of Rigid Cellular Plastics* (*see Part 26*)
D 1505 – 68	Density of Plastics by the Density-Gradient Technique
D 2857 – 70	Dilute-Solution Viscosity of Polymers
D 1434 – 66	Gas Transmission Rate of Plastic Film and Sheeting

Methods of Test for:

D 1705 – 61 (1968) *Particle Size Analysis of Powdered Polymers and Copolymers of Vinyl Chloride (see Part 26)*
*D 1921 – 63 Particle Size (Sieve Analysis) of Plastic Materials
D 1045 – 58 (1967) Plasticizers Used in Plastics, Sampling and Testing
D 793 – 49 (1965) *Short-Time Stability at Elevated Temperatures of Plastics Containing Chlorine (see Part 26)*
*D 792 – 66 Specific Gravity and Density of Plastics by Displacement
D 1638 – 70 *Urethane Foam Raw Materials (see Part 26)*
D 1823 – 66 *Viscosity, Apparent, of Plastisols and Organosols at High Shear Rates by Castor Severs Viscometer (see Part 26)*
D 1824 – 66 *Viscosity, Apparent, of Plastisols and Organosols at Low Shear Rates by Brookfield Viscometer (see Part 26)*
*D 1243 – 66 Viscosity of Vinyl Chloride Polymers, Dilute Solution
D 1601 – 61 Viscosity of Ethylene Polymers, Dilute Solution

Specimen Preparation

Specifications for:

*D 647 – 68 Design of Molds for Test Specimens of Plastic Molding Materials

Methods of Test for:

*D 955 – 51 (1969) Shrinkage from Mold Dimensions of Molded Plastics, Measuring

Recommended Practices for:

D 2292 – 68 Compression Molding of Specimens of Styrene-Butadiene Molding and Extrusion Materials
D 1897 – 68 Injection Molding of Specimens of Thermoplastic Molding and Extrusion Materials
*D 1130 – 63 (1969) Injection Molding of Specimens of Thermoplastic Materials
*D 796 – 65 Molding Test Specimens of Phenolic Molding Materials
D 956 – 51 (1969) Molding Specimens of Amino Plastics
D 957 – 50 (1969) Mold Surface Temperature of Commercial Molds for Plastics, Determining
D 958 – 51 (1969) Temperatures of Standard ASTM Molds for Test Specimens of Plastics, Determining
*D 1896 – 65 Transfer Molding Test Specimens from Thermosetting Materials

Definitions

Definitions of Terms:

D 1600 – 69 Abbreviations of Terms Relating to Plastics
D 1009 – 55 (1961) Code for Designating Form of Material and Direction of Testing Plastics
*E 41 – 63 Conditioning, Relating to
D 883 – 69a Nomenclature Relating to Plastics

Statistical Techniques

Recommended Practices for:

D 1898 – 68 Sampling of Plastics
D 2188 – 69 Statistical Design in Interlaboratory Testing of Plastics

Conditioning

Specifications for:

*E 171 – 63 Standard Atmospheres for Conditioning and Testing Materials

Methods for:

*D 618 – 61 Conditioning Plastics and Electrical Insulating Materials for Testing

Recommended Practices for:

D 832 – 64 *Conditioning of Elastomeric Materials for Low-Temperature Testing (see Part 28)*

Definitions of Terms Relating to:

*E 41 – 63 Conditioning

LIST BY SUBJECTS, PART 27

Electrical Insulating Materials

Specifications for:

D 922 – 65	*Nonrigid Vinyl Chloride Polymer Tubing* (*see Part 29*)
D 1532 – 68	*Polyester Glass-Mat Sheet Laminate* (*see Part 29*)
**D 710 – 63*	*Vulcanized Fiber Sheets, Rods, and Tubes Used for Electrical Insulation* (*see Part 29*)

Methods of Test for:

D 150 – 68	A-C Loss Characteristics and Dielectric Constant (Permittivity) of Solid Electrical Insulating Materials
D 495 – 61	Arc Resistance, High-Voltage, Low-Current, of Solid Electrical Insulating Materials
*D 257 – 66	D-C Resistance or Conductance of Insulating Materials
*D 149 – 64	Dielectric Breakdown Voltage and Dielectric Strength of Electrical Insulating Materials at Commercial Power Frequencies
D 669 – 59 (1967)	*Dissipation Factor and Dielectric Constant Parallel with Laminations of Laminated Sheet and Plate Insulating Materials* (*see Part 29*)
D 876 – 65	*Nonrigid Vinyl Chloride Polymer Tubing* (*see Part 29*)
D 229 – 68	Sheet and Plate Materials, Rigid, Used for Electrical Insulation
D 651 – 48 (1966)	*Tensile Strength of Molded Electrical Insulating Materials* (*see Part 29*)
D 374 – 68	Thickness of Solid Electrical Insulation

Recommended Practice for:

D 1371 – 68	*Cleaning Plastic Specimens for Insulation Resistance Testing* (*see Part 29*)

Rubber and Rubber-Like Materials

Specifications for:

D 1565 – 66	*Flexible Foams Made from Polymers or Copolymers of Vinyl Chloride* (*see Part 28*)
D 1277 – 59 (1965)	*Nonrigid Thermoplastic Compounds for Automotive and Aeronautical Applications* (*see Part 28*)
D 1351 – 67	*Polyethylene Insulated Wire and Cable* (*see Part 28*)
D 1047 – 67	*Poly(Vinyl Chloride) Jacket Compound for Electrical Insulated Wire and Cable* (*see Part 28*)
D 2219 – 68	*Vinyl Chloride Plastic Insulation for Wire and Cable, 60 C Operation* (*see Part 28*)
D 2220 – 68	*Vinyl Chloride Plastic Insultion for Wire and Cable, 75 C Operation* (*see Part 28*)

Methods of Test for:

D 1565 – 66	*Flexible Foams Made from Polymers or Copolymers of Vinyl Chloride* (*see Part 28*)
D 470 – 68	*Rubber and Thermoplastic Insulated Wire and Cable* (*see Part 28*)
D 1564 – 64 T	*Slab Flexible Urethane Foam* (*see Part 28*)
*D 412 – 68	Tension Testing of Vulcanized Rubber
D 797 – 64	*Young's Modulus in Flexure of Natural and Synthetic Elastomers at Normal and Subnormal Temperatures* (*see Part 28*)

Recommended Practices for:

D 832 – 64	*Conditioning of Elastomeric Materials for Low-Temperature Testing* (*see Part 28*)
D 1371 – 68	*Cleaning Plastic Specimens for Insulation Resistance Testing* (*see Part 29*)
E 104 – 51	Maintaining Constant Relative Humidity by Means of Aqueous Solutions

Definitions of Terms Relating to:

D 1566 – 68a	*Rubber and Rubber-like Materials* (*see Part 28*)

Color Tests

Recommended Practices for:

D 1729 – 64	*Color Differences of Opaque Materials, Visual Evaluation of* (*see Part 21*)
E 308 – 66	*Spectrophotometry and Description of Color in CIE 1931 System* (*see Part 30*)

GENERAL METHODS OF TESTING

Specifications for:

E 145 – 68	*Gravity-Convection and Forced-Ventilation Ovens* (*see Part 30*)
*E 171 – 63	Standard Atmospheres for Conditioning and Testing Materials

LIST BY SUBJECTS, PART 27

Methods of Test for:

E 162 – 67	*Surface Flammability of Materials Using a Radiant Heat Energy Source (see Part 14)*
*C 177 – 63 (1968)	Thermal Conductivity of Materials by Means of the Guarded Hot Plate
*E 4 – 64	Verification of Testing Machines
E 96 – 66	Water Vapor Transmission of Materials in Sheet Form

Recommended Practices for:

E 187 – 66	*Conducting Natural Light (Sunlight and Daylight) Exposures Under Glass (see Part 30)*
E 29 – 67	Indicating Which Places of Figures Are to Be Considered Significant in Specified Limiting Values
E 166 – 63	*Goniophotometry of Transmitting Objects and Materials (see Part 30)*
E 167 – 63	*Goniophotometry of Reflecting Objects and Materials (see Part 30)*
E 104 – 51	Maintaining Constant Relative Humidity by Means of Aqueous Solutions
E 188 – 63 T	*Operating Enclosed Carbon-Arc Type Apparatus for Artificial Light Exposure Tests (see Part 30)*
E 42 – 69	Operating Light- and Water-Exposure Apparatus (Carbon-Arc Type) for Exposure of Nonmetallic Materials
E 122 – 58	*Sample Size, Choice of, to Estimate the Average Quality of a Lot or Process (see Part 30)*
E 177 – 68 T	Use of the Terms Precision and Accuracy as Applied to Measurement of a Property of a Material

Definitions of Terms Relating to:

*E 41 – 63	Conditioning
E 170 – 63 (1968)	*Dosimetry (see Part 30)*
*E 6 – 66	Methods of Mechanical Testing

METRIC PRACTICE GUIDE

Standard:

E 380 – 70	*Metric Practice Guide (Excerpts) (see Related Materials section)*

type to aim to establish details of construction and procedure to cover all contingencies that might offer difficulties to a person without technical knowledge concerning the theory of heat flow, temperature measurement, and general testing practices. Standardization of the method does not reduce the need for such technical knowledge. It is recognized also that it would be unwise, because of the standardization of this method, to restrict in any way the further development of improved or new methods or procedures by research workers.

2. Significance

2.1 The thermal conductivity of a material (*1*) may vary due to variability of the material or samples of it, (*2*) may be affected by moisture or other conditions, and (*3*) may change with time or high temperatures. It must be recognized therefore that the selection of a typical value of thermal conductivity representative for a material, or for particular applications, if it is practically feasible, must be based on a consideration of these factors and an adequate amount of test information.

3. Symbols and Definitions

3.1 *Symbols:*

k = thermal conductivity, Btu · in./h · ft^2 · deg F (W · cm/cm^2 · deg C),

C = thermal conductance, Btu/h · ft^2 · deg F (W/cm^2 · deg C),

R = thermal resistance, deg F · h · ft^2/Btu (deg C · cm^2/W),

q = time rate of heat flow, Btu/h (W),

A = area measured on a selected isothermal surface, ft^2 (cm^2),

L = thickness of specimen measured along a path normal to isothermal surfaces, in. (cm),

t_1 = temperature of hot surface, deg F (deg C), and

t_2 = temperature of cold surface, deg F (deg C).

3.2 *thermal conductivity*, k, of a homogeneous material (Note 2)—the time rate of heat flow, under steady-state conditions, through unit area, per unit temperature gradient in the direction perpendicular to an isothermal surface. For a flat slab, it is calculated as follows (Note 3):

$$k = qL/A(t_1 - t_2) \qquad (1)$$

NOTE 2—Materials are considered homogeneous for the purpose of this method when the value of the thermal conductivity is not affected by a change in thickness, L, or in area, A, within the range normally used. Some materials are not isotropic with respect to thermal conductivity. Care should be taken that the test method is suitable for the particular material and gives a value of conductivity applicable to the intended use.

NOTE 3—For convenience and in the interest of standardization, thermal conductivity is expressed in U.S. customary units. Conversion factors for thermal conductivity in other units in common use are given in Table 1. The values stated in U.S. customary units are to be regarded as the standard. The metric equivalents of U.S. customary units may be approximate.

3.3 *thermal conductance, average*, C, of a body between two definite surfaces—the time rate of heat flow between these surfaces, under steady-state conditions, divided by the difference of their average temperatures and by the area of one of the surfaces. The average temperature is one which adequately approximates that obtained by integrating the temperature of the entire surface. The thermal conductance of a flat slab is calculated as follows (Note 3):

$$C = q/[A\ (t_1 - T_2)] = k/L \qquad (2)$$

3.4 *thermal resistance*, R—the reciprocal of thermal conductance. For a flat slab, it is calculated as follows (Note 4):

$$R = 1/C = L/k \qquad (3)$$

4. Metal-Surfaced Hot Plate

4.1 The general features of the metal-surfaced guarded hot plate are shown schematically in Fig. 1. The plates are usually square, but round plates are sometimes used. The term "guarded hot plate" is applied to the entire assembled apparatus, including the heating unit, the cooling units, and the edge insulation. The heating unit consists of a central or metering section and a guard section. The central section consists of a central heater and central surface plates. The guard section consists of one or more guard heaters and the guard surface plates. The surface plates are usually made of noncorroding metal of high thermal conductivity. The working surfaces of the heating unit and cooling plates should be smoothly finished to conform to a true plane as closely as possible, and should be checked periodically. The maximum departure of the surface from a plane shall not exceed 0.003

Designation: C 177 – 63
(Reapproved 1968)

American National Standard Z98.1-1964 (R 1969)
American National Standards Institute

Standard Method of Test for

THERMAL CONDUCTIVITY OF MATERIALS BY MEANS OF THE GUARDED HOT PLATE[1]

This Standard is issued under the fixed designation C 177; the number immediately following the designation indicates the year of original adoption or, in the case of revision, the year of last revision. A number in parentheses indicates the year of last reapproval.

1. Scope

1.1 This method covers the determination of the existing thermal conductivity of dry specimens of insulating, building, and other materials within the limits specified in 1.2, and the coefficients obtained apply strictly only to the particular samples as tested.

1.2 For practical purposes, this method is limited to determinations on specimens having thermal conductances not in excess of 10 Btu/h · ft^2 · deg F (0.006 W/cm^2 · deg C) and thickness conforming to 6.2.

1.3 Two different types of guarded hot plate apparatus are described. They are similar in principle but differ enough in construction to warrant separate descriptions for each in regard to design. The metal-surface guarded hot plate apparatus is generally used for measurements at mean temperatures such that the temperature of the cold surface may be as low as −100 F (−75 C) or that of the hot surface as high as 500 F (260 C). It is described in Section 4. The refractory guarded hot plate apparatus is ordinarily used for measurements where the hot plate temperature is greater than 200 F (95 C) and less than 1300 F (700 C). It is made of a cast, or otherwise formed, electrically nonconducting refractory material. It is described in Section 5. In all other respects, the method is the same for both types of apparatus. It is intended, in presenting these descriptions, to indicate the essential elements and details which experience has shown to be necessary or important for reliable measurements by this method.

NOTE 1—For the convenience of new workers in the field, detailed drawings and descriptions for the construction, and some phases of the operation, of typical hot plates complying with the requirements of this method have been made available.[2] Two of these hot plates, known as the National Bureau of Standards plate and the National Research Council plate, are metal-surfaced plates; two are refractory plates.

1.4 The guarded hot plate is used for determining the thermal conductivity of homogeneous materials in the form of flat slabs, and this method covers the procedure for such tests. It is recognized, however, that frequently it is desirable to determine the thermal conductivity of some materials in the forms in which they are used, such as molded pipe coverings, etc., or the thermal transmittance of a composite wall construction or part thereof. For such purposes, ASTM Method C 335, Test for Thermal Conductivity of Pipe Insulation[3] and ASTM Method C 236, Test for Thermal Conductance and Transmittance of Built-Up Sections by Means of the Guarded Hot Box[3] are recommended.

1.5 For satisfactory results in conformance with this method, the principles governing the size, construction, and use of the apparatus described in this method should be followed. If the results are to be reported as having been obtained by this method, then all pertinent requirements prescribed in this method shall be met.

1.6 It is not practicable in a method of this

[1] This method is under the jurisdiction of ASTM Committee C-16 on Thermal and Cryogenic Insulating Materials. A list of committee members may be found in the ASTM Yearbook. This standard is the direct responsibility of Subcommittee C-16.30 on Thermal Conductance.

Current edition effective Sept. 30, 1963. Originally issued 1942. Replaces C 177 – 45.

[2] Detailed information on all of these hot plates is available at a nominal cost from the American Society for Testing and Materials, 1916 Race St., Philadelphia, Pa. 19103.

[3] *Annual Book of ASTM Standards*, Part 27.

in./ft (0.00025 cm/cm).

NOTE 4—The planeness of the surface can be checked with a steel straightedge held against the surface and viewed at grazing incidence with a light behind the straightedge. Departures as small as 0.001 in. (0.0025 cm) are readily visible, and larger departures can be measured using shim-stock or thin paper.

4.2 In the design of the guarded hot plate, consider the materials used in its construction with respect to their performance at the temperatures at which the plate will be operated. Consider also the electrical design of the heater and the design of the cooling plate to assure adequate capacity and suitable characteristics for the intended use. In all cases, design and construct the guarded hot plate so that in operation the two faces of the central section, and of the guard section, shall be substantially at the same uniform temperature, and that the heating units do not warp or depart from planeness at the operating temperatures.

4.3 Heating units shall have a definite separation or gap not greater than $^1/_8$ in. (0.3 cm) between the central surface plates and the guard surface plates (Note 5). The separation between the heating windings of the central section and the contiguous guard section shall not exceed $^3/_4$ in. (1.9 cm), and this separation is allowable only if the spacing bars on either side of the separation are of heavy copper, in order to distribute heat to the surface plates. In all other cases, the heater winding separation shall not exceed $^1/_8$ in. Establish the dimensions of the test area by measurements to the centers of the separations that surround this area. Paint or otherwise treat the surfaces of all plates to have a total emittance greater than 0.8 at operating temperatures.

NOTE 5—It is recommended that the area of the gap in the plane of the surface plate be not more than 6 percent of the metering section area on that side.

4.4 Provide the guarded hot plate with a suitable means of detecting temperature imbalance between the areas of the central and guard surface plates contiguous to the separation between them. Distribute the temperature-sensing elements to register adequately (Note 6) the temperature balance existing along the length of the central section periphery. The temperature-sensing elements may be read either individually to indicate any temperature difference that may exist, or they may be connected to be read differentially to indicate such temperature difference directly. Thermocouples are generally used for this purpose, with connections arranged so that they are read as a differentially connected thermopile. The detection system shall be sufficiently sensitive to assure that variation in conductivity due to gap temperature imbalance (Note 6) shall be restricted to not more than 0.5 percent.

NOTE 6—For information on determining this requirement, see the references at the end of this method.

4.5 The cooling units shall have surface dimensions at least as large as those of the heating unit. They shall consist of metal plates maintained at a uniform temperature lower than that of the heating unit, either by a constant-temperature fluid, or by electrical heating, or by thermal insulation of uniform conductance applied on the outermost surfaces, as appropriate for the cooling unit temperature desired.

4.6 For measuring the surface temperature of the central section of the heating unit, provide each of the central surface plates with permanently-installed thermocouples set in grooves or just under the working surface. The number of such thermocouples on each side shall be not less than $^1/_8 \sqrt{A}$, where A is the area in square inches (or square centimeters) of one side of the central surface plate. There shall be the same number of thermocouples permanently and similarly installed at corresponding positions in the facing cold plate. If the hot and cold plate thermocouples on each side are to be connected differentially, which is usual, they must be electrically insulated from the plates.

4.7 Provide means (*1*) for imposing a reproducible constant pressure of the plates against the specimens to promote good thermal contact (Note 7) and (*2*) for measuring the effective thickness of the specimen to within 0.5 percent (Note 8).

NOTE 7—A steady force thrusting the cold plates toward each other can be imposed by means of a calibrated compressed spring, or a system of levers and dead weights, or an equivalent method. It is unlikely that a pressure greater than 50 lb/ft^2 (2.5 kgf/cm^2) on the specimens would be required; for easily-compressible specimens, small stops interposed between the corners or edges of the cold plates, or some other positive means, may be used to limit the compression of the specimens, and a constant-pressure arrangement is not needed.

NOTE 8—Because of the changes of specimen thickness possible as a result of temperature, or compression by the plates, it is recommended that specimen thickness be measured in the apparatus, at the existing test temperature and compression conditions, when possible. Gaging-points, or measuring studs at the outer four corners of the cold plates or along the axis perpendicular to the plates at their centers, will serve for these measurements. The effective combined specimen thickness is determined by the difference in the micrometered distance, or average distance, between the gaging-points when the specimen is in place in the apparatus, and is not in place, and the same force is used to press the cold plates toward each other.

4.8 The best method of determining the temperature drop in the specimen depends upon its characteristics, and in some instances the choice of method is left to the judgment of the operator. For nonrigid specimens with flat uniform surfaces that conform well to the flat working surfaces of the plates, take the temperature drop in specimens of thermal conductance less than 2.0 Btu/h · ft^2 · deg F (0.001 W/cm^3 · deg C) as that indicated by the thermocouples permanently set in the hot and cold surface plates, and take the thickness of the specimen as the mean distance between the working surfaces of the hot and cold plates. For nonrigid specimens of conductance greater than 2.0, the operator's judgment should rule in accordance with the circumstances. Rigid specimens to be tested must, imperatively, have parallel surfaces flat to within 0.003 in./ft (Note 9). One method of testing rigid specimens is to install them in the apparatus with a sheet of suitable homogeneous material (for example, for temperatures up to 160 F (70 C), $^1/_8$-in. thick commercial gum rubber (30 to 45 Durometer floating stock, Federal Specification MIL R-880A)) interposed between the specimen and each plate surface. Determine the conductance of the composite sandwich (sheet-rigid specimen-sheet) using the temperature drop indicated by the permanent thermocouples in the hot and cold surface plates. If it is not already known, the conductance of the interposed sheets alone is similarly measured in a separate test made at the same mean temperature and with the same compressive force on the plates. The conductance of the rigid specimen is then calculated from the two conductances obtained. Another method of determining the conductance of a rigid specimen is to interpose resilient sheets between the specimen and plates as indicated above, and to determine the temperature drop across the rigid specimen by means of separate thermocouples mounted flush with, or interior to, the surface of the rigid specimen (Note 10). The number of separate thermocouples used on each side of the specimen shall be not less than $^1/_4 \sqrt{A}$, where A is the area in square inches (or square centimeters) of one side of the central surface plate. If separate thermocouples are used, take the effective thickness of the specimen as the average distance, perpendicular to the face of the specimen, between the centers of the thermocouples on the two sides.

NOTE 9—A rigid specimen is one of a material too hard and unyielding to be appreciably altered in shape by the pressure of the hot and cold plates, for example, a slab of glass or hard plastic.

NOTE 10—This method of measuring the specimen temperature drop may be subject to uncertainties difficult to evaluate, among them being the effects of (*1*) distortion of heat flow lines in the immediate vicinity of the thermocouple, due to its presence, (*2*) imprecision in ascertaining the exact position of the effective thermocouple junctions, and (*3*) local inhomogeneities in the surface of the specimen at the thermocouple junctions, such as pores, voids, or inclusions.

4.9 Thermocouples mounted in the surfaces of the plates shall be made of wire not larger than 0.0226 in. (0.0574 cm) in diameter (No. 23 B&S gage); specimen surface thermocouples shall be made of wire not larger than 0.0113 in. (0.0287 cm) in diameter (No. 29 B&S gage).

4.10 A potentiometer-galvanometer system having a sensitivity of 1 μV or better shall be used for all measurements of thermocouple emf.

4.11 Restrict heat losses from the outer edges of the guard section and the specimens by edge insulation, or by governing the surrounding air temperature, or by both methods. On the basis that the total of such heat losses shall not exceed $^1/_5$ of the heat flow through the two specimens, the required minimum thermal resistance against edge heat losses, in deg F · h · ft^2/Btu is given by the equation:

$$R = \frac{5x}{Sk}\left[\left(4x + 2y\left(\frac{T_m - T_a}{\Delta T}\right) + y\right]$$

where:

x = thickness of each specimen, in. (cm),
y = thickness of the heating unit, in. (cm)
S = length of the side (or diameter) of the guard section, in. (cm)
k = thermal conductivity of the specimen, in Btu · in./h · ft^2 · deg F (W · cm/cm^2 · deg C)
T_m = mean temperature of the specimen, in deg F (deg C),
T_a = temperature of the air surrounding the apparatus, in deg F (deg C), and
ΔT = temperature difference across the specimens, in deg F (deg C).

It should be noted that R depends on both $(T_m - T_a)$ and ΔT, and also that since it is desirable that the net heat transfer from the outer edges of the specimens should be kept nearly equal to zero, $(T_m - T_a)$ should be kept small.

4.12 A cabinet or enclosure surrounding the guarded hot plate, and equipped for maintaining the desired interior air temperature and dew point, is recommended for use in tests conducted at mean temperatures differing substantially from the laboratory air temperature.

5. Refractory Hot Plate

5.1 The general features of the refractory guarded hot plate are shown schematically in Fig. 2. The plates are usually round, but square plates are sometimes used. The term "guarded hot plate" is applied to the entire assembled apparatus, including the heating unit, the cooling units, and the edge insulation. The heating unit consists of a central or metering section and a guard section. The central section consists of a central heater and the guard section consists of one or more guard heaters. The cooling plates may be made of noncorroding metal of high thermal conductivity or, when their temperature is controlled by means of electrical heating, they may be made of refractory or of metal. The working surfaces of the heating unit and cooling plates should be smoothly finished to conform to a true plane as closely as possible, and should be checked periodically. The maximum departure of the surface from a plane shall not exceed 0.003 in./ft (0.00025 cm/cm) (Note 4).

5.2 In the design of the refractory guarded hot plate, give due consideration to obtaining satisfactory performance at the temperatures at which the plate will be operated. Consider also the electrical design of the heater and the design of the cooling plate to assure adequate capacity and suitable characteristics for the intended use. In all cases, design and construct the guarded hot plate so that in operation the two faces of the central section, and of the guard section, shall be substantially at the same uniform temperature, and that the heating units do not warp or depart from planeness at the operating temperatures.

5.3 In a refractory plate having a gap between the central and the guard areas, the gap shall not exceed 0.08 in. (0.2 cm) (Note 5). The effective metering area of the refractory plate is determined by the positions of the potential taps used to evaluate the power input to the metering area winding.

NOTE 11—For a double-spiral (bifilar) winding with the spacing between wires equal to b, and with the potential taps for the metering section in effect at points on the wires at the ends of a diameter $2a$ (or for a single-spiral winding of spacing b with the potential taps in effect at the center and at a radius a), the effective metering area is equal to

$$\pi a^2[1 + (1/12)(b/a)^2]$$

5.4 Provide the guarded hot plate with a suitable means for detecting temperature imbalance between the central and guard sections of the plate. Distribute the temperature-sensing elements to register adequately (Note 6) the temperature balance existing between the outer edge of the metering area and the inner edge of the guard section. Locate the thermocouple junctions used for detecting temperature imbalance at the edge of the metering area on the same radius, and distant from the edge of the metering area by not more than one-quarter of the guard width. The temperature-sensing elements may be read either individually to indicate any temperature difference that may exist, or they may be connected to read differentially to indicate such temperature difference directly. The detection system shall be sufficiently sensitive to assure that variation in conductivity due to temperature imbalance (Note 6) between the central and guard sections shall be restricted to not more than 0.5 percent.

5.5 The cooling units shall have surface dimensions at least as large as those of the heating unit. They shall consist of metal or refractory plates maintained at a uniform temperature lower than that of the heating unit, either by a constant-temperature fluid, or by electrical heating, or by thermal insulation of uniform conductance applied on the outermost surfaces, as appropriate for the cooling unit temperature desired.

5.6 Permanently installed thermocouples to be used in determining the temperature difference across the specimen shall be set flush with the working surfaces, and shall number not less than $^1/_8 \sqrt{A}$ on each working surface, where A is the area in square inches (or square centimeters) of the metering area of the hot plate on one side. However, permanently installed thermocouples are not mandatory if the temperature difference across the specimen is to be determined by means of separate thermocouples (see 4.8).

5.7 Same as 4.7.

5.8 The best method of determining the temperature drop in the specimen depends on the circumstances, and is therefore left to the best judgment of the operator. One method often used is to attach separate thermocouples on a sheet of asbestos paper or other suitable material, and to interpose the sheet between the specimen and the adjacent working surface of the apparatus, with the thermocouples in contact with the specimen. For rigid and hard specimens (Note 9), which should be flat to within 0.003 in./ft, it may be important to set the separate thermocouples in tight grooves in the faces of the specimens. The number of separate thermocouples used, at each face of the specimen, shall be not less than $^1/_4 \sqrt{A}$, where A is the metering area in square inches (or square centimeters) of one side of the hot plate. If separate thermocouples are used, take the effective thickness of the specimen as the average distance, perpendicular to the face of the specimen, between the centers of the thermocouples on the two sides. Another method, feasible if a suitable resilient sheet material is available for the test temperatures in question, is to use the composite sandwich (sheet-specimen-sheet) technique described in 4.8, in which permanently installed thermocouples in the plates are used.

NOTE 12—This method automatically compensates for any effective or virtual thermal resistance between the positions where the permanently installed plate thermocouples are located and the plate working surface. Such resistance may be appreciable in the case of a refractory plate.

5.9 Thermocouples mounted in the surfaces of the plates shall be made of wire not larger than 0.0253 in. (0.0643 cm) in diameter (No. 22 B&S gage); specimen surface thermocouples shall be made of wire not larger than 0.0179 in. (0.0455 cm) in diameter (No. 25 B&S gage).

5.10 Same as 4.10.

5.11 To reduce heat losses from the outer edges of the guard section, test specimens, and cooling plates, surround the assembly by a coaxial cylindrical nonmetallic container of inside diameter at least 12 in. (30 cm) greater than the diameter of the assembly. Fill the volume around the assembly with a suitable loose-fill insulation within at least the axial length between the outermost surfaces of the cooling plates.

6. Test Specimens

6.1 Prepare two specimens from each sample. They shall be as nearly identical as possible, of such size as to completely cover the heating unit surfaces, and shall be of sufficient thickness to give a true average representation of the material to be tested. The relationship between the thickness of the test specimen used and the dimensions of the guarded hot plate shall be as follows:

Maximum Thickness of Test Specimen, in. (cm)	Minimum Linear Surface Dimensions of Guarded Hot Plate (Square or Round), in. (cm)	
	Central Section of Heating Unit	Width of Guard Area Around Heating Unit
$1^1/_3$ (3.3)	4 (10)	2 (5)
2 (5)	6 (15)	3 (7.6)
$2^1/_2$ (6.4)	12 (30)	3 (7.6)
4 (10)	12 (30)	6 (15)

6.2 In testing all forms of homogeneous materials, make the surfaces of the test specimens as plane as possible, by sandpapering, face-cutting in a lathe, grinding, or otherwise, so that intimate contact between the specimens and the plates or interposed sheets can be effected. For rigid materials, make the faces of the specimens flat to within 0.003 in./ft. (0.00026 cm/cm).

6.3 When testing homogeneous solid or

blanket-type materials, prepare the specimens in accordance with 6.1 and 6.2. Determine their weight before and after they have been dried to constant weight in a ventilated oven at a temperature from 215 to 250 F (102 to 120 C) (Note 13). From these weights calculate the percentage as-received moisture. Promptly after drying and weighing, place the specimens in the apparatus taking care to prevent loss of material or moisture gain. Determine the as-tested thickness and volume by measurements made preferably at the end of the test under conditions of test-temperature equilibrium, and from these data and the dry weight, calculate the as-tested density. If it is feasible to do so with good thermal contact between the plates of the apparatus and the specimens, test blanket- or batt-type material at approximately the thickness and density determined by ASTM Methods C 167, Test for Thickness and Density of Blanket- or Batt-Type Thermal Insulating Materials[3] (Note 7).

NOTE 13—If the material may be adversely affected by heating to 215 F (102 C), it shall be dried in a desiccator at a temperature from 130 to 140 F (55 to 60 C).

6.4 When testing loose-fill materials, take a representative portion slightly greater than the amount needed for the test from the sample. Weigh this material before and after it has been dried to constant weight in a ventilated oven at a temperature from 215 to 250 F (102 to 120 C) (Note 13). From these weights, calculate the percentage as-received moisture. Then weigh out an amount of the dried material such that it will produce two specimens of the desired as-tested density using either Method A or B, given in the Appendix. As the volume of the specimen space is known, the required weight can be determined. During placement, take care to prevent loss of the weighed material, or significant moisture gain. When Method A is used, or Method B with covers of insignificant thermal resistance, take the specimen surface temperatures as equal to those of the surfaces of the hot and cold plates.

7. Procedure

7.1 For any test, adjust the temperature difference between the hot and cold plates to not less than 10 F (−12 C). It is recommended that for good insulators, the temperature gradient in the specimen be 40 F/in. (90 C/cm) or more.

7.2 Adjust the atmosphere surrounding the guarded hot plate during a test to a dew point temperature 10 F (−12 C) or more lower than the cold-plate temperature.

7.3 Supply the heating element of the central heater with electrical energy for heating it, providing for the measurement of the average rate of heat generation therein to an accuracy of not less than 0.5 percent. Automatic regulation of the input power is recommended; in any case, fluctuations or changes in input power shall not cause the temperature of the hot-plate surfaces to fluctuate, or to change in 1 h of a test period, by more than 0.5 percent of the temperature difference between the hot and cold plates. Measure the input power in such a way that the average power input during a test period can be determined; if the power input is of a fluctuating kind, make an integrated energy measurement. Adjust and maintain the power input to the guard section, preferably by automatic control, so as to effect the degree of temperature balance between the central and guard heater sections that is required for conformance to 4.4 (or 5.4).

7.4 Adjust the cooling units so that the temperature drops through the two specimens do not differ by more than 1 percent. The temperatures of the cooling plates during a test period shall not fluctuate, or change in 1 h, by more than 0.5 percent of the temperature difference between the hot and cold plates.

7.5 After steady-state conditions have been attained, continue the test at the steady state, making the necessary observations to determine the temperature difference across the specimens, the hot- and cold-plate temperatures, the center-to-guard temperature balance, and the electrical power input to the central section, at intervals of not less than 30 min, until four successive sets of observations give thermal conductivity values differing by not more than 1 percent.

7.6 Upon completion of the observations in 7.5, measure the thickness of the specimens (see Section 4.8 and Note 8) for use in calculating the final results, and determine the weight of the specimens or of the test material in the specimens immediately.

8. Calculations

8.1 Calculate the density of the dry specimen as tested, D, the as-received moisture content of the material, M, and the moisture regain of the specimen during test, R_w or R_v, as follows:

$$D = W_2/V$$
$$M = [(W_1 - W_2)/W_2] \times 100$$
$$R_w = [(W_4 - W_3)/W_3] \times 100$$
$$R_v = [(W_4 - W_3)/62.4v] \times 100$$

where:

D = density of the dry material as tested, lb/ft^3 (g/cm^3),
M = moisture content of the material as received, dry weight percent,
R_w = moisture regain of material during test, dry weight percent,
R_v = moisture regain of material during test, dry volume percent,
W_1 = weight of material in as-received condition, lb (g),
W_2 = weight of material after drying, lb (g),
W_3 = weight of dry material in specimens, lb (g),
W_4 = weight of material in specimens immediately after test, lb (g), and
V = volume occupied by material in specimens during test, ft^3 (cm^3).

8.2 Calculate thermal conductivity (or conductance) by means of Eq 1 (or 2) (3.2 and 3.3), using average values of the observed steady-state data.

9. Report

9.1 The report of the results of each test shall include the following (the numerical values reported shall represent the average values for the two specimens as tested):

9.1.1 Name and any other pertinent identification of the material,

9.1.2 Thickness of specimen (or material) as tested, in. (cm),

9.1.3 Method and temperature of drying,

9.1.4 Density of dry material as tested, lb/ft^3 (g/cm^3),

9.1.5 As-received moisture content of material, dry weight percent,

9.1.6 Moisture regain of material during test, dry weight percent or volume percent, or both,

9.1.7 Temperature gradient in material during test, deg F/in. (deg C/cm),

9.1.8 Mean temperature of test, deg F (deg C),

9.1.9 Heat flux through each specimen during test, Btu/h · ft^2 (W/cm^2),

9.1.10 Thermal conductivity, in Btu · in./h · ft^2 · deg F (W · cm/cm^2 deg C) or thermal conductance of specimen, in Btu/h · ft^2 · deg F (W/cm^2 deg C),

9.1.11 Orientation of plane of specimen: vertical or horizontal,

9.1.12 For tests made using sheet material interposed between the specimen and the plate surfaces, give information as to the nature, thickness, and thermal conductance of the sheet material. If separate thermocouples were used to determine the temperature drops in the specimens, also give information as to the size of the thermocouples, the method of affixing them to the specimen, and the measured distance between their centers, and

9.1.13 The approximate resistance of edge insulation, and the air temperature surrounding the guarded hot plate during the test.

9.2 Include a graphical representation of the results in the report when pertinent. This shall consist of a plot of each value of thermal conductivity obtained versus the corresponding test mean temperature, plotted as ordinates and abscissas, respectively.

REFERENCES

(1) Woodside, W., and Wilson, A.G., "Unbalance Errors in Guarded Hot Plate Measurements," *Symposium on Thermal Conductivity Measurements and Applications of Thermal Insulations, ASTM STP 217*, Am. Soc. Testing Mats., 1957, pp. 32–48.

(2) Gilbo, C. F., "Experiments with a Guarded Hot Plate Thermal Conductivity Set," *Symposium on Thermal Insulating Materials, ASTM STP 119*, Am. Soc. Testing Mats., 1951, pp. 45–57.

(3) Donaldson, I. G., "A Theory for the Square Guarded Hot Plate—A Solution of the Heat Conduction Equation for a Two Layer System," *Quarterly of Applied Mathematics*, Vol XIX, 1961, pp. 205–219.

TABLE 1 **Conversion Factors for Thermal Conductivity**

	cal · cm/s · cm² · deg C	W · cm/cm² · deg C	Btu · ft/h · ft² · deg F	Btu · in./h · ft² · deg F	kg-cal · m/h · m² · deg C
1 cal · cm/s · cm² · deg C	1	4.187	241.9	2903	360
1 W · cm/cm² · deg C	0.2389	1	57.78	693.3	85.98
1 Btu · ft/h ft² · deg F	0.004134	0.01731	1	12.00	1.488
1 Btu · in./h ft² · deg F	0.0003445	0.001442	0.08333	1	0.1240
1 kg-cal · m/h · m² · deg C	0.002778	0.01163	0.6720	8.064	1

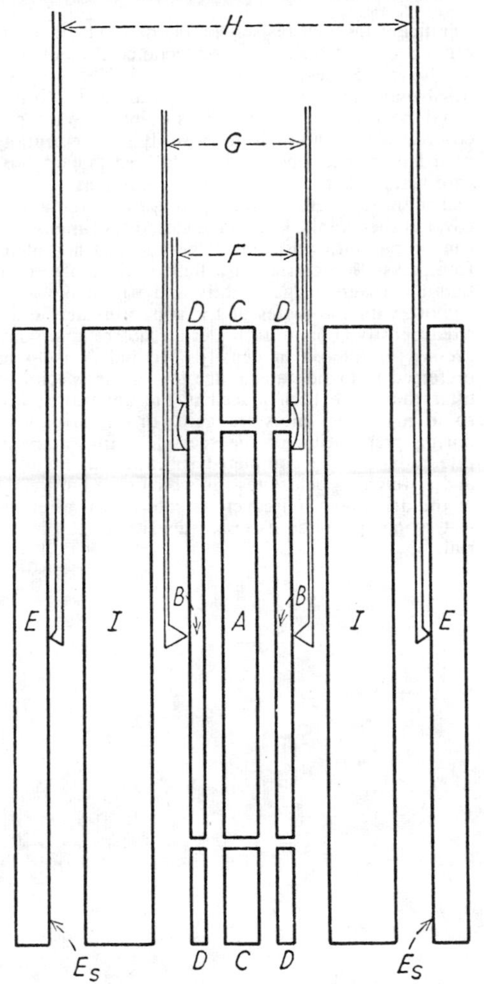

Heating Unit

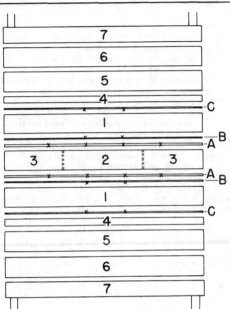

A—Central Heater
B—Central Surface Plates

Central Section of Heating Unit

C—Guard Heater
D—Guard Surface Plates

Guard Section of Heating Unit

E—Cooling Units
E_s—Cooling Unit Surface Plates
F—Differential Thermocouples
G—Heating Unit Surface Thermocouples
H—Cooling Unit Surface Thermocouples
I—Test Specimens

FIG. 1 General Features of the Metal-Surfaced Hot Plate.

7
6
5
4
C
1
B
A
3 2 3
A
B
1
C
4
5
6
7

1—Test Specimens
2—Center Test Section Heater
3—Guard Heaters
4—Nichrome Grounded Plates (if used)
5—Cold Surface Heaters (if used)
6—Insulation
7—Cold Water Plates
A—Differential Thermocouples Center to Guard Balance
B—Hot-Surface Thermocouples
C—Cold-Surface Thermocouples

FIG. 2 General Features of the Refractory Hot Plate.

APPENDIX

A1. Methods for Preparing Test Specimens of Loose Materials

A1.1 *Method A*—The guarded hot plate is set up with the required spacings between the heating unit and the cold plates. Low conductivity material that is suitable for confining the sample is placed around or between the outer edges of the guard section and the cold plates in such a manner that it forms two boxes, one on either side of the heating unit, and each open at the top. The weighed dried material mentioned in 6.4 is divided into eight equal portions, four for each specimen. Each portion is placed in turn in each of the two specimen spaces, each portion being vibrated, packed, or tamped in position until it occupies its appropriate one-quarter volume of the space, care being taken to produce specimens of uniform density.

A1.2 *Method B*—Two shallow boxes of thin-walled low-conductivity material are used, having outside dimensions the same as those of the heating unit. The box edges are of such width as to make the depth of the box equal to the thickness of the specimen to be tested. Covers for the open faces of the boxes are made, using (*1*) thin sheet plastic material not more than 0.002 in. thick or (*2*) blotting or asbestos paper or other suitable uniform sheet material, these to be glued or otherwise fastened to the edges of the boxes. If the covers have significant thermal resistance, the methods of determining the net specimen conductance presented in 4.8 for rigid specimens can be used. The weighed dried material mentioned in 6.4 is divided into two equal portions, one for each specimen. With one cover in place, and with the boxes lying horizontally on a flat surface, one portion is placed in each box, care being taken to produce two specimens of equal and uniform density throughout. The remaining cover is then applied, to make closed specimens that can be put into position in the guarded hot plate. Compressible materials are fluffed when placed so that the covers bulge slightly and will make good contact with the plates of the apparatus at the desired density. For loose materials not readily compressed or altered in density, Method A is to be preferred to avoid gaps at the top of the specimens after they have been placed in the apparatus. For some materials, moisture gain, or material loss, during preparation of the specimens may necessitate redrying the completed specimen in an oven and reweighing before test, in which case the weight of the dried box and covers is determined after the test to compute the as-tested density of the material.

Designation: C 581 - 68

American National Standard K66.10-1969
American National Standard Institute

Standard Method of Test for
CHEMICAL RESISTANCE OF THERMOSETTING RESINS USED IN GLASS FIBER REINFORCED STRUCTURES[1]

This Standard is issued under the fixed designation C 581; the number immediately following the designation indicates the year of original adoption or, in the case of revision, the year of last revision. A number in parentheses indicates the year of last reapproval.

1. Scope

1.1 This method is intended for use as a relatively rapid test to evaluate the chemical resistance of thermosetting resins used in the fabrication of glass fiber reinforced self-supporting structures under anticipated service conditions of use. This method provides for the determination of changes in the properties described in 1.1.1 through 1.1.4 of the test specimens and test reagent after exposure of the specimens to the reagent:

1.1.1 Hardness,

1.1.2 Appearance of specimen,

1.1.3 Appearance of immersion media, and

1.1.4 Flexural strength.

NOTE 1—The values stated in U.S. customary units are to be regarded as the standard. The metric equivalents of U.S. customary units may be approximate.

2. Significance

2.1 The results obtained by this method shall serve as a guide in, but not as the sole basis for, selection of a thermosetting resin used in a glass fiber reinforced plastic self-supporting structure. No attempt has been made to incorporate into the method all the various factors which may enter into the serviceability of a glass fiber reinforced resin structure when subjected to chemical environment.

3. Apparatus

3.1 *Micrometer*—A micrometer suitable for measurement to 0.001 in. (0.025 mm).

3.2 *Containers:*

3.2.1 Wide-mouth glass jars of sufficient capacity fitted with plastic-lined metal screw caps for low temperature tests involving solutions or solvents of low volatility.

3.2.2 Containers of sufficient capacity, each fitted with standard taper joints and reflux condenser attachment for use with volatile solutions or solvents.

3.2.3 Containers as described in 3.2.1 and 3.2.2 having an inert coating on their inner surfaces, or containers of a suitable inert material for use with solutions which attack glass.

3.3 *Constant*—A temperature oven or liquid bath capable of maintaining temperature within a range of ±4 F (±2.2 C).

3.4 *Flexural Properties Testing Apparatus*—This shall be in accordance with the testing machine described in ASTM Method D 790, Test for Flexural Properties of Plastics.[2]

3.5 *Impressor Type Instrument.*[3]

3.6 *Analytical Balance*—The balance shall have adequate capacity and be suitable for accurate weighing to 0.0001 g.

3.7 *Desiccator.*

3.8 *Heat-Resistant, Non-Reactive Crucibles.*

3.9 *Muffle Furnace*—The furnace shall be equipped with temperature controls.

---

[1] This method is under the jurisdiction of ASTM Committee C-3 on Chemical-Resistant Nonmetallic Materials. A list of members may be found in the ASTM Yearbook.
Current edition effective July 16, 1968. Originally issued 1965. Replaces C 581 - 65 T.

[2] *Annual Book of ASTM Standards*, Part 27.

[3] The Barcol Impressor GYZJ 934-1 has been found suitable for this purpose.

4. Reagents

4.1 The test media shall consist of the reagents, solutions, or products to which the resin system is to be exposed in service.

5. Test Specimens

5.1 Prepare laminates, using identical reinforcement in all of the specimens as outlined in 5.1.1 through 5.1.3.

5.1.1 *Standard Laminate*—The standard resin-glass laminate from which test specimens are cut shall be molded in accordance with instructions of 5.1. It shall have the general properties described in 5.1.1.1 through 5.1.1.7.

5.1.1.1 Glass content shall be 25 ± 2 weight percent, based on 1.1 sp gr cured resin, determined by weight loss on ignition of four 1 by 1-in. (or 25 by 25-mm) specimens. The glass content will be determined using the procedure described in 5.1.1.2 through 5.1.1.4.

5.1.1.2 *Weighing*[4]—Weigh the specimen on an analytical balance in a previously weighed, ignited crucible.

5.1.1.3 *Ignition*[4]—Place the specimen in the furnace at a temperature not greater than 650 F (343 C). Raise the temperature of the furnace to 1050 ± 50 F (565 ± 28 C), at a rate that will not cause blowing or loss of inorganic filler. The specimen and crucible shall be ignited at this maximum temperature to constant weight (2 to 6 h depending on the thickness), and allowed to cool in a desiccator. The loss in weight shall be determined by weighing the residue.

5.1.1.4 *Calculation*—The glass content shall be calculated as follows:

$$\text{Glass content, weight percent} = (w_o - w_l)/w_o \times 100$$

where:

w_l = loss in weight, and

w_o = original weight.

5.1.1.5 Thickness of laminate shall be 100 to 125 mils (2.5 to 3.2 mm).

5.1.1.6 Surfaces of laminate shall be uniform and glossy.

5.1.1.7 Hardness shall be at least that of fully cured, clear casting of the resin as defined by the manufacturer. Hardness shall be determined with an impressor type instrument.[3]

5.1.2 *Fabrication of Standard Laminate*—Dimensions of the standard laminate shall be 26 by 33 in. (660 by 840 mm). The sequence of lay-up (contact molding) shall be as described in 5.1.2.1 through 5.1.2.9.

5.1.2.1 Apply catalyzed resin and a 10-mil (0.25-mm) surfacing mat, type C, glass with a resin compatible finish on a flat surface covered with a 5-mil (0.13-mm) plastic film.[5]

5.1.2.2 Follow with two plies of $1^1/_2$ oz/ft^2 (457 g/m^2) chopped strand mat and resin. Mat shall be type E glass with resin compatible finish.

5.1.2.3 Follow with a 10-mil (0.25-mm) surfacing mat as in 5.1.2.1.

5.1.2.4 Densify the laminate and remove the air by rolling over the surface by means of a nonabsorbent roller, such as a short-fiber paint roller. Do not use pressure plates. Take precautions not to express enough resin to raise the glass content above permissible maximum.

5.1.2.5 After the lay-up is completed, cover laminate with a 5-mil (0.13-mm) plastic film.[5] Carefully smooth down to exclude entrapped air and pull the film taut, then fasten edges of the film to prevent wrinkling prior to gelation of resin.

5.1.2.6 Cure with heat or at room temperature as required to meet hardness requirements of the manufacturer.

5.1.2.7 Trim edges as required to obtain laminate 33 by 26 in. (840 by 660 mm). Check thickness.

5.1.2.8 Cut a strip 1 by 26 in. (25 by 660 mm) from the center of the laminate. From this strip, remove four specimens, each 1 in. (25 mm) long, at equal intervals along the length of the strip. Determine glass content in accordance with the procedure prescribed in 5.1.1.1.

5.1.2.9 If the laminate meets the requirements of this specification, retain the two 16 by 26-in. (410 by 660-mm) laminate sections for preparation of test specimens.

5.1.3 *Individual Test Specimens:*

5.1.3.1 Specimens for immersion in test solutions shall be 4 by 5 by $^1/_8$ in. (100 by 130 by 32 mm), cut from the standard laminate.

---

[4] Refer to Federal Test Method Standard No. 406-Method 7061.

[5] DuPont Mylar has been found suitable for this purpose.

5.1.3.2 Identity of specimens may be maintained by means of notches in the edge of specimen.

5.1.3.3 Cut edges and drilled holes for suspension, if used, shall be sanded smooth and coated with resin.

5.1.3.4 The number of specimens required is dependent upon the number of test solutions to be employed, the number of different temperatures at which testing is performed, and the frequency of test intervals. In addition, one 4 by 5-in. (100 by 130-mm) specimen shall be available for test immediately following the curing period and other 4 by 5-in. (100 by 130-mm) specimens, equivalent to the number of test temperatures, for test after aging in air at the test temperature for the total test period.

5.1.3.5 The total number of specimens required shall be calculated as follows:

$$N = n(S \times T \times I) + nT + n$$

where:

N = number of specimens,
n = number of specimens for a single test (min of one 4 by 5 by $^1/_8$-in. (100 by 130 by 32-mm) specimen),
S = number of solutions,
T = number of test temperatures, and
I = number of test intervals.

6. Test Conditions

6.1 The test conditions shall simulate the anticipated service conditions as closely as possible.

7. Procedure

7.1 *Measurement of Specimens*—Immediately following the curing period, measure the thickness of the specimens to the nearest 0.001 in. (0.025 mm) at the geometric center of each of the intended 1 by 4-in. (25 by 100-mm) specimens.

7.2 *Exposure, Visual Inspection of Test Specimens, and Hardness*—Following the conditioning period, as specified in 5.1.2.6, prior to immersion, record a brief description of the color and surface appearance of the coupons and the color and the clarity of the test solution. Also, determine the Barcol hardness by taking 10 readings and recording the average on the specimen described in 5.1.2.8. The total number of coupons per container is not limited except by the ability of the container to hold the coupons without touching each other or the container. The coupons must always be completely immersed. Add a minimum of 750 ml of test solution for each coupon and place the closed container in a constant temperature oven adjusted to the required temperature or in a suitably adjusted liquid bath. Examine the coupons if necessary after 30, 90, 180 days and 1 year of immersion or other time intervals as required to determine the rate of attack. The tests may be terminated if the coupon decomposes. If the coupons have not disintegrated, they are removed from the solution, rinsed with water, and wiped dry prior to measuring and taking Barcol hardness readings described in 7.2.1 through 7.2.4.

7.2.1 Clean the coupon by three quick rinses in running cold tap water and quick-dry by blotting with a paper towel between each rinse. After final blotting, allow the coupon to dry for $^1/_2$ h and measure the coupon to the nearest 0.001 in. (0.025 mm) for thickness in the geometric center of each intended 1 by 4-in. (25 by 100-mm) specimen. The Barcol hardness can then be checked taking an average of ten readings on each coupon, a maximum of 1 in. (25 mm) from the edge.

7.2.2 The 4 by 5-in. coupons, after the proper exposure period, are cut into three specimens measuring 1 by 4 in. (25 by 100 mm) (see Fig. 1). They are cut on a bandsaw and the edges are smoothed by routing, or by the use of a nonmetallic cut-off wheel. To be sure that there are no nicks on the edges, sand the coupons lightly by using a fine sandpaper.

7.2.3 Run flexural tests in accordance with Procedure A of ASTM Method D 790. The coupons can be forwarded to a testing laboratory for flexural tests; they may either be transported in the corroding environment or each clean and dried sample placed in a polyethylene air-tight bag for shipment. The elapsed time between the removal of the coupons from the corroding environment and the flexural tests should be uniform for all coupons.

7.2.4 Note any indication of surface attack on a coupon, any discoloration of the test solution, and the formation of any sediment. Determine the flexural strength for: (*1*) one set of three specimens immediately following the curing period, (*2*) one set of three specimens after each inspection, for each solution and each test temperature, and for (*3*) one set of three specimens after aging in air for the total test period at each test temperature. Test the specimens for flexural strengths following measurement and the Barcol hardness determination and record the maximum load.

7.3 *Changing of Immersion Medium*—Discard and replace the test solution with fresh material after each period. Solutions which are known to be unstable, for example aqueous sodium hypochlorite, shall be replaced as often as necessary in order to maintain original chemical composition and concentration.

8. Calculations

8.1 *Barcol Hardness Change*—Tabulate or construct a graph employing the actual hardness readings of the specimens broken at a given temperature, plotting the hardness as the ordinate and the test period, in days, as the abscissa.

8.2 *Change in Flexural Strength*—Calculate to the nearest 0.1 percent, the percentage decrease or increase in flexural strength of the specimen during immersion for each examination period, taking the flexural strength after curing as 100 percent.

$$\text{Change in flexural strength, percent} = [(S_2 - S_1)/S_1] \times 100$$

where:

S_1 = flexural strength of specimen after conditioning period, and

S_2 = flexural strength of specimen after test period.

8.2.1 Calculate flexural strength properties in accordance with Section 11 of Method D 790.

NOTE 2—A result showing a plus (+) sign will indicate a gain in flexural strength and a minus (−) sign will indicate a loss.

8.2.2 Construct a graph employing the average percentage of change in flexural strength of the specimens broken at a given examination period after immersion in a particular test solution at a given temperature, plotting the percentage of change in flexural strength as the ordinate and the test period, in days, as the abscissa.

9. Interpretation of Results

9.1 *Mechanical Properties of the Specimen*—Because of the chemical nature of certain types of plastic materials, the rate of change with time is of more significance than the actual value at any one time. A plot of the test results will indicate whether a particular specimen will approach constant flexural strength or hardness with time or will continue to change as the test progresses.

9.2 *Appearance of Specimen*—Visual inspection of the exposed specimen for surface cracks, loss of gloss, etching, pitting, softening, changes in thickness, etc. is very important, for these conditions indicate some degradation of the laminate by a chemical environment.

9.3 *Appearance of Immersion Medium*—Discoloration of the test solution and the formation of sediment are significant factors. An initial discoloration may indicate extraction of soluble components. Continuation of the test with fresh solution will indicate whether or not the attack is progressive.

10. Report

10.1 The report shall include the following:

10.1.1 Company and individual preparing standard laminates.

10.1.2 Complete identification of material tested, including resin, accelerator, catalyst and reinforcement.

10.1.3 Conditioning procedure.

10.1.4 Glass content of standard laminate.

10.1.5 Hardness, flexural strength of control coupons.

10.1.6 Color and surface appearance of specimen before testing.

10.1.7 Test conditions; immersion medium, temperature, etc.

10.1.8 Total duration of test in days, and examination periods, in days. For each examination period, the data listed in 10.1.8.1 through 10.1.8.4 are required.

10.1.8.1 Barcol hardness of the specimens.

10.1.8.2 Appearance of specimens after immersion (surface cracks, loss of gloss, etching, pitting, softening, etc.).

10.1.8.3 Appearance of immersion medium (discoloration, sediment, etc.).

10.1.8.4 Flexural strength of coupons and average percent change in flexural strength.

10.1.9 Graph showing percent change in flexural strength plotted against test period.

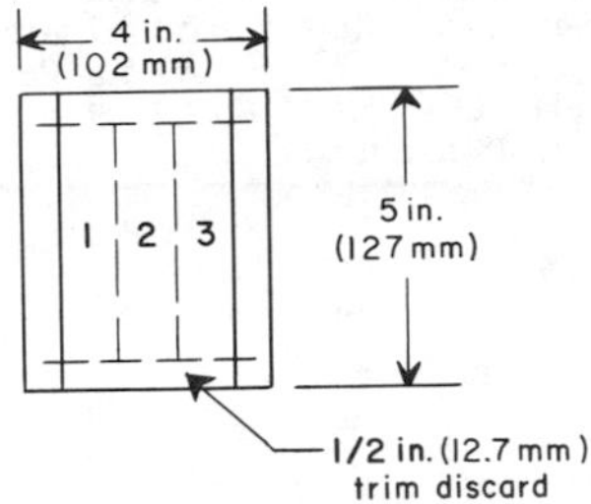

FIG. 1. Specimen Cutting Guide.

APPENDIXES

A1. Basis for Selection of Standard Test Specimen

A1.1 Chemical resistance of thermoset resins used in fabrication of reinforced plastic equipment for aggressive chemical service must be considered in terms of how the resin is employed in conjunction with reinforcement. Design of laminates for chemical process equipment follows certain established principles:

A1.1.1 The resin provides the primary chemical resistance.

A1.1.2 The reinforcement enhances mechanical properties.

A1.1.3 The reinforcement often detracts from effective chemical resistance, in varying degree, depending on the type of reinforcement, its form (mat, cloth, woven roving, roving strand), the reinforcement-resin ratio, and the location of the reinforcement in the cross section of the laminates.

A1.1.4 For maximum chemical resistance, reinforcement should not exceed approximately 15 percent by volume, and be in random short fiber form.

A1.2 In order to provide adequate chemical resistance and strength, the above principles dictate use of an unbalanced laminate, which is a laminate wherein the form and amount of reinforcement changes across the thickness of the laminate. The face of the laminate exposed to the aggressive environment is made resin-rich with enough random fiber reinforcement to stabilize the resin and impart reasonable strength without seriously impairing chemical resistance. This portion of the laminate is 60 to 125 mils thick and is termed the liner. The remainder of the laminate is the high-strength portion usually containing woven roving or roving filaments. This structural portion is relatively poor in chemical resistance as compared with the liner.

A1.3 Since the liner is the chemical resistance barrier, resins must be evaluated in terms of how they perform in liner-type laminates. This is the basis for this test method. The standard laminate for immersion testing represents the "liner" used in glass-reinforced laminates prepared by hand lay-up (contact molding). This is the predominant reinforcement and fabrication method in the Corrosion Resistant Structures Division of the reinforced plastics industry.

A1.4 Chemical resistance of other liner formulations and constructions may be determined by comparing their performance in this same test with the standard resin-glass laminate.

A2. Use of Method for General Characterization of Resins

A2.1 The method is applicable for broad characterization of chemical resistance of resin systems. This is normally done by resin manufacturers as a technical aid to customers in resin selection.

A2.2 Reagents recommended are those necessary to characterize the general chemical-resistant nature of a resin system. Chemicals used in this test shall be of technical grade or greater purity. All solutions shall be made with distilled or demineralized water.

A2.3 Tests are conducted at room temperature (20 to 30 C), and 50 C, 70 C, 100 C, or reflux.

A2.4 The standard reagents for basic evaluation of resin-glass systems are as follows. Solutions strengths are stated in terms of weight percent (absolute). Approximately 1000 ml are required for each exposure period.

A2.4.1 Sulfuric acid (H_2SO_4), 25 percent.

A2.4.2 Hydrochloric acid (HCl), 15 percent.

A2.4.3 Nitric acid (HNO_3), 5 percent.

A2.4.4 Acetic acid (CH_3COOH), 25 percent.

A2.4.5 Sodium hydroxide (NaOH), 5 percent.

A2.4.6 Ammonium hydroxide (NH_4OH, sp gr 0.90), concentrated.

A2.4.7 Toluene.

A2.4.8 Methylethyl ketone.

A2.4.9 Carbon tetrachloride.

A2.4.10 Distilled or demineralized water.

A2.5 Additional reagents for extended characterization should be chosen from the following:

A2.5.1 Phosphoric acid (H_3PO_4), 15 percent.

A2.5.2 Sodium chloride (NaCl), saturated solution.

A2.5.3 Sodium hypochlorite ($NaHClO_2$), 5¹/₄ percent (replace with fresh solution at 48-h intervals).

A2.5.4 Sodium carbonate (Na_2CO_2), 10 percent.

A2.5.5 Aluminum potassium sulfate ($KAl(SO_4)_2 \cdot 12H_2O$).

A2.5.6 Ethyl acetate.

A2.5.7 Monochlorbenzene.

A2.5.8 Perchlorethylene.

A2.5.9 Kerosine.

Designation: D 149 - 64 (Reapproved 1970)

USA Standard C59.48-1968
USA Standards Institute

Standard Methods of Test for DIELECTRIC BREAKDOWN VOLTAGE AND DIELECTRIC STRENGTH OF ELECTRICAL INSULATING MATERIALS AT COMMERCIAL POWER FREQUENCIES[1]

This Standard is issued under the fixed designation D 149; the number immediately following the designation indicates the year of original adoption or, in the case of revision, the year of last revision. A number in parentheses indicates the year of last reapproval.

1. Scope

1.1 These methods cover the determination of dielectric strength of electrical insulating materials at commercial power frequencies.

1.2 The procedures appear in the following order:

Procedure	Section
All Materials:	
Methods of Test	4
Electrical Apparatus	5
Voltage Application and Measurement	6
Criteria of Breakdown	7
Solids:	
Electrodes	8 and Table 1
Test Specimens	9
Thickness	10
Conditioning	11
Surrounding Medium	12
Test Chamber	13
Number of Tests	14
Solid Filling and Treating Compounds	15
Liquids	16 to 18
Report	19

1.3 Since some materials require special treatment, reference should also be made to the ASTM methods applicable to the material to be tested. These methods are listed in Appendix A2.

2. Significance

2.1 This method is intended for use as a control and acceptance test. It may be used also in the partial evaluation of materials for specific end uses and as a means for detecting changes in materials due to specific deteriorating causes. A more complete discussion of the significance of dielectric breakdown tests is given in Appendix A1.

3. Definitions

3.1 *dielectric breakdown voltage*—the voltage at which electrical breakdown of a specimen of electrical insulating material between two electrodes occurs under prescribed conditions of test.

3.2 *dielectric strength*—the ratio of the dielectric breakdown voltage to the thickness of an insulating material.

3.3 *flashover*—an electric discharge around the edge or over the surface of electrical insulation.

4. Types of Test

4.1 When determining the dielectric breakdown voltage of a material any one of three different methods may be employed for applying the test voltage. These are the Short-Time Test for quick determination, and the Step-by-Step Test and the Slow-Rate-of-Rise Test, where duration of voltage application is important. In choosing the type of test, reference should always be made to the ASTM method applicable to the material to be tested.

5. Electrical Apparatus

5.1 *Transformers*—The desired test voltage may be obtained from a step-up transformer energized from an adjustable low-voltage

[1] These methods are under the jurisdiction of ASTM Committee D-9 on Electrical Insulating Materials. A list of members may be found in the ASTM Yearbook.

Current edition effective Dec. 1, 1964. Originally issued 1922. Replaces D 149 – 61.

source. The transformer and its controlling equipment shall be of such size and design that, with the test specimen in the circuit, the crest factor (ratio of crest to effective) of the test voltage shall not differ by more than ±5 percent from that of a sinusoidal wave over the upper half of the range of test voltage. The crest factor may be checked by means of an oscillograph, an oscilloscope, a sphere gap, or a peak-reading voltmeter in conjunction with an rms voltmeter. For test specimens of small capacitance, a testing transformer as small as 500-V-A rating or a well-designed potential transformer may be adequate. For specimens of high capacitance or leakage, a transformer of ample kilovoltampere rating must be used. Where the wave form cannot be determined conveniently, a transformer having a rating of not less than 2 kVA has been found adequate for voltages not exceeding 50 kV and one having a rating of not less than 5 kVA for voltages exceeding 50 kV.

NOTE 1—The current drawn from the high-voltage winding should not exceed the full-load, full-voltage current except momentarily.

5.2 *Circuit-Interrupting Equipment*—The test transformer circuit shall be protected by an automatic circuit-breaking device capable of opening as quickly as practical on the current produced by breakdown of the specimen. Devices capable of opening the circuit in five cycles or less are available. A prolonged flow of current at the time of breakdown is undesirable as it causes unnecessary burning of the test specimen and pitting and heating of the electrodes, thereby increasing maintenance work and time of testing.

5.3 *Voltage-Control Equipment*—Control of voltage may be obtained by the use of a motor-driven, variable-ratio autotransformer or any device which gives at least as uniform rate of rise of the voltage without distortion of the wave form beyond the limits of 5.1. Preference should be given to equipment having an approximately straight-line voltage - time curve over the operating range. Motor drive with variable speed control should be preferred to manual drive because of the difficulty of maintaining a reasonably uniform rate of voltage rise with the latter.

6. Voltage Application and Measurement

6.1 *Method*—In making breakdown tests, the voltage shall be applied, as may be specified, by one or more of the following methods:

6.1.1 *Short-Time Test*—The voltage shall be increased from zero to breakdown at a uniform rate. The rate of rise shall be 100, 500, 1000, or 3000 V/s, depending on the total test time required and the voltage - time characteristic of the material. For the rate applicable to a given material, reference shall be made to the test method for that material.

6.1.2 *Slow-Rate-of-Rise Test*—An initial voltage shall be applied, approximately equal to 50 percent of the breakdown voltage in the short-time test, or as specified in the material specification. The voltage shall then be increased at a uniform rate, as stated in the material specifications, up to the point of breakdown. The specified rate should be such as to give approximately the same voltage - time exposure of the test specimen as provided in the step-by-step test.

6.1.3 *Step-by-Step Test*—Unless otherwise specified, an initial voltage equal to 50 percent of the breakdown voltage ±5 percent, as obtained on the short-time test, shall be applied to the test specimen. The voltage shall then be increased in equal increments and held for periods of time as specified in the various material specifications. The change from each step to the next higher shall be made as rapidly as possible and the time of change included in the succeeding test interval.

NOTE 2—The means used to obtain the various voltage steps shall not introduce voltage transients that increase the peak value more than 10 percent.

6.2 *Rate-of-Rise*—The rate-of-rise of voltage shall not, for the short-time test, vary more than ±20 percent from the specified rate at any point. It may be calculated from measurements of time required to raise the voltage between two selected values. Voltage control shall be obtained as indicated in 5.3.

6.3 *Measurement*—The voltage shall be measured in accordance with IEEE Standard No. 4, "Measurement of Voltage in Dielectric Tests." The response time of the voltmeter shall be such that its time lag shall not introduce an error greater than 1 percent of full scale at any rate-of-rise used. The overall accuracy of the voltmeter and the voltage measuring device used shall be such that measurement error will not exceed 5 percent.

7. Criteria of Breakdown

7.1 Observation of actual rupture or decomposition is positive evidence of voltage breakdown. When neither of these physical evidences of breakdown is apparent, it is common practice to reapply the voltage. In most cases this will give a positive indication.

7.2 Dielectric breakdown is generally accompanied by an increase in current in the circuit which may either trip a circuit breaker or blow a fuse. If the circuit breaker is well coordinated with the characteristics of the test equipment and the material under test, its operation may be a positive indication of breakdown. However, tripping of a circuit breaker may be influenced by flashover, leakage currents, corona currents, or equipment magnetizing currents. Failure of the circuit breaker to operate may not be a positive criterion of the absence of breakdown. A breaker may fail to trip because it is set for too great a current or because of malfunction.

SOLIDS

8. Electrodes

8.1 The dielectric strength of an insulating material varies with the thickness of the material, the area and geometry of the test electrodes, temperature, and humidity. Tests made with electrodes of different configuration at different temperatures and with different specimen thicknesses may not be comparable. The dimensions and geometry of the electrodes listed in Table 1 have been selected for the testing of various materials having different thicknesses.

8.2 The contact pressure of the electrodes shall be adequate to obtain good electrode contact with the test specimen. In some special cases it is desirable to paint the contact surface of the test specimen with conducting paint or spray with a liquid metal in order to assure good electrical contact.

NOTE 3—The results obtained with painted or sprayed electrodes may not be comparable with those obtained with other types of electrodes.

9. Test Specimens

9.1 The specimens shall be representative of the material to be tested. Sufficient material shall be available to permit making five satisfactory tests. In the preparation of test specimens from solid materials, care shall be taken that the surfaces in contact with the electrodes are parallel and as plane and smooth as the material permits.

9.2 *Thin Solid Materials* (*Sheets and Plates*)—The test specimens shall be only of sufficient area to prevent flashover under the conditions of test.

9.3 *Thick Solid Materials*—The breakdown of thick solid materials is generally so high that the specimen must be immersed in oil to prevent flashover and to minimize corona. Other techniques which may be used to prevent flashover are:

9.3.1 The machining of a well in the test specimen for a recessed electrode.[2]

9.3.2 The use of shrouds on the test specimen.[3]

9.3.3 The application of a sealing apparatus under pressure to the upper and lower faces of the test specimen.[4]

9.4 *Other Materials*—For a description of the test specimens of other materials and their preparation, reference shall be made to the ASTM methods applicable to the material to be tested.

10. Thickness

10.1 The thickness used in computing the dielectric strength shall be the thickness of the specimen measured as specified in the test method for the material involved. If not specified, the thickness measurements shall be made at a temperature of 25 ± 5 C in accordance with ASTM Methods D 374, Test for Thickness of Solid Electrical Insulation.[5]

---

[2] Chapman, J. J., and Frisco, L. J., "A Practical Interpretation of Dielectric Measurement Up to 100 Mc," published by U.S. Dept. of Commerce, Office of Technical Service, Dec. 31, 1958.

[3] See Section 27 of ASTM Methods D 295, Testing Varnished Cotton Fabrics and Varnished Cotton Fabric Tapes Used for Electrical Insulation, *Annual Book of ASTM Standards*, Part 29. See also Blanck, A. R., and Holman, B. T., "Method of Testing the Dielectric Breakdown Voltage and Dielectric Strength of Miniature Specimens," Technical Paper No. 44, Feltman Research and Engineering Laboratories, Picatinny Arsenal Dover, N. J., May 1960.

[4] Hand, W., and Winans, R. R., "An Apparatus for Measuring Dielectric Strength at Elevated Temperatures," *ASTM Bulletin*, No. 171, January 1951, pp. 63–65.

[5] *Annual Book of ASTM Standards*, Part 29.

11. Conditioning

11.1 The dielectric strength of most insulating materials varies with temperature and humidity. Materials that may be affected by variation of temperature and also by changes in relative humidity should be conditioned in a suitably controlled chamber. For information concerning the conditioning treatment, reference shall be made to the particular method for a given material. The test specimens shall be kept in the chamber long enough to reach a uniform temperature and humidity before tests are started. It may be desirable to determine the dielectric behavior of a material over a range of temperature and humidity to which it is likely to be subjected in use.

NOTE 4—If conditioning is performed under conditions where condensation will occur, it may be desirable to wipe off the surface of the test specimen carefully immediately before testing, as this will generally tend to minimize flashover.

11.2 Since some materials require a long time to attain equilibrium at normal conditions of temperature and humidity, conditioning should be specified when specimens are to be subjected to tests for evaluation of quality control and purchase specification requirements, in order to obtain reproducible results.

12. Surrounding Medium

12.1 In general, it is preferable to test materials in the medium in which they are to be used, whether air, other gas, or oil. Where conditions of use are not well-defined, materials should be tested in air up to the point where the breakdown is so high that an excessive amount of material is required to prevent flashover or excessive burning of the surface. For the medium to be employed with a particular material, reference should be made to the ASTM method applicable to that material.

12.2 Flashover must be avoided and the effects of corona minimized even during short-time tests. For specimens having a high breakdown voltage, the dielectric strength tests may be made under oil. However, it should be understood that breakdown values obtained under oil may not be comparable with those obtained in air.

When tests are to be made under oil, an oil bath of adequate size shall be provided. The oil should be of good grade of clean transformer oil or similar oil, having a viscosity not exceeding 100 SUS at 100 F, and dielectric breakdown voltage not less than 26 kV as determined in accordance with ASTM Method D 877, Test for Dielectric Breakdown Voltage of Insulating Liquids Using Disk Electrodes.[5]

13. Test Chamber

13.1 The test chamber or area shall be of adequate size to hold the test equipment conveniently and shall be provided with a safety interlock on its door so that the electrical circuit supplying voltage to the test specimen will be broken when the door is opened. For tests at elevated temperatures, an oven of suitable size, capable of meeting the requirements set forth in ASTM Specification D 2436, for Forced Convection Laboratory Ovens for Electrical Insulation,[5] may be found convenient.

14. Number of Tests

14.1 Unless otherwise specified, five tests shall be made.

SOLID FILLING AND TREATING COMPOUNDS

15. Procedure

15.1 Determine the dielectric strength of solid filling and treating compounds in accordance with Sections 50 to 54, incl, of ASTM Methods D 176, Testing Solid Filling and Treating Compounds Used for Electrical Insulation.[5]

LIQUIDS

16. Sampling

16.1 Obtain the sample in accordance with ASTM Method D 923, Sampling Electrical Insulating Liquids.[5]

17. Test Specimens

17.1 The preparation of the test specimens, the test cups to be used, and the care of the electrodes shall be as described in the respective test methods for dielectric strength.

18. Procedure

18.1 To determine the dielectric strength of an insulating liquid ASTM Method D 877,

Test for Dielectric Breakdown Voltage of Insulating Liquids Using Disk Electrodes,[5] may be used. However, very slightly contaminated oils may be evaluated more sensitively by using ASTM Method D 1816, Test for Dielectric Strength of Insulating Oil of Petroleum Origin Using VDE Electrodes.[5]

REPORT

19. Report

19.1 Unless otherwise specified, the report shall include the following:

19.1.1 Average thickness of the specimen,

19.1.2 Breakdown voltage at each puncture,

19.1.3 Average, maximum, and minimum breakdown voltage for each specimen,

19.1.4 Dielectric strength of each specimen,

19.1.5 Ambient temperature,

19.1.6 Ambient relative humidity in percent,

19.1.7 Conditioning,

19.1.8 Type of test employed:

19.1.8.1 When the short-time test is used, give the rate-of-rise of voltage;

19.1.8.2 When the step-by-step test is used, give the magnitude of the initially applied voltage, subsequent step voltages, and time period at each step;

19.1.8.3 When the slow-rate-of-rise test is used, give the magnitude of the initially applied voltage and the subsequent rate-of-rise;

19.1.9 Size, shape, and type of the electrodes, and

19.1.10 Surrounding medium.

TABLE 1 Typical Electrodes for Dielectric Strength Testing at Room Temperature of Various Types of Insulating Materials

NOTE—In selecting the electrode material reference shall be made to the procedure governing the material to be tested.

Insulating Material	Metal, Dimension, and Geometry of Electrodes
Sheet and film material, rubber and rubber products	Brass or stainless steel cylinders 2 in. in diameter and 1 in. in in thickness with edges rounded to ¼-in. radius.
Tapes and films	(*a*) Opposing cylindrical brass or stainless steel rods ¼-in. diameter with edges rounded to 1⁄32-in. radius. Upper movable electrode shall weigh 50 ± 2 g. (*b*) Flat brass or stainless steel plates ¼ in. wide and 4¼ in. long with edges square and ends rounded to ⅛-in. radius cylinders. Effective electrode length 4 in.
Other solid materials	Brass or stainless steel cylinders 1 in. in diameter and 1 in. in length with edges rounded to ⅛-in. radius.
Compounds	Hemispherical brass or stainless steel electrodes ½ in. in diameter.
Petroleum oils and askarels	(*a*) Polished brass square edged disks 1 in. in diameter (Method D 877).[5] (*b*) Polished brass spherically-capped disks 36 mm in diameter (Method D 1816).[5]
Research liquids (such as silicone oils, etc.)	Under development by Committee D-27.
Gases	Under development by Committee D-27.

APPENDIXES

A1. SIGNIFICANCE OF THE DIELECTRIC STRENGTH TEST

A1.1 Introduction

A1.1.1 A brief review of three postulated mechanisms of breakdown, namely: (*1*) the discharge or corona mechanism, (*2*) the thermal mechanism, and (*3*) the intrinsic mechanism, as well as a discussion of the principal factors affecting tests on practical dielectrics, are given here to aid in interpreting the data. The breakdown mechanisms usually operate in combination rather than singly. The following discussion applies only to solid, semi-solid, and liquid materials.

A1.2 Postulated Mechanisms of Dielectric Breakdown

A1.2.1 *Breakdown Caused by Electrical Discharges*—In many tests on commercial materials, breakdown is caused by electrical discharges, which produce high local fields. With solid materials the discharges usually occur in the surrounding medium, thus increasing the test area and producing failure at or beyond the electrode edge. Discharges may occur in any internal voids or bubbles that are present or may develop. These may cause local erosion or chemical decomposition. These processes may continue until a complete failure path is formed between the electrodes.

A1.2.2 *Thermal Breakdown*—Cumulative heating develops in local paths within many materials when they are subjected to high electric field intensities, causing dielectric and ionic conduction losses which generate heat more rapidly than can be dissipated. Breakdown may then occur because of thermal instability of the material.

A1.2.3 *Intrinsic Breakdown*—If electric discharges or thermal instability do not cause failure, breakdown will still occur when the field intensity becomes sufficient to accelerate electrons through the material. This critical field intensity is called the intrinsic dielectric strength. It cannot be determined by this test method, although the mechanism itself may be involved.

A1.3 Nature of Electrical Insulating Materials

A1.3.1 Solid commercial electrical insulating materials are generally nonhomogeneous and may contain dielectric defects of various kinds. Dielectric breakdown often occurs in an area of the test specimen other than that where the field intensity is greatest and sometimes in an area remote from the material directly between the electrodes. Weak spots within the volume under stress sometimes determine the test results.

A1.4 Influence of Test and Specimen Conditions

A1.4.1 *Electrodes*—In general, the breakdown voltage will tend to decrease with increasing electrode area, this area effect being more pronounced with thin test specimens. Test results are also affected by the electrode geometry. Results may be affected also by the material from which the electrodes are constructed, since the thermal and discharge mechanism may be influenced by the thermal conductivity and the work function, respectively, of the electrode material. Generally speaking, the effect of the electrode material is difficult to establish because of the scatter of experimental data.

A1.4.2 *Specimen Thickness*—The dielectric strength of solid commercial electrical insulating materials is greatly dependent upon the specimen thickness. Experience has shown that for solid and semi-solid materials, the dielectric strength varies inversely as a fractional power of the specimen thickness, and there is a substantial amount of evidence that for relatively homogeneous solids, the dielectric strength varies approximately as the reciprocal of the square root of the thickness. In the case of liquids, or of solids that can be melted and poured to solidify between fixed electrodes, the effect of electrode separation is less clearly defined. Since the electrode separation can be fixed at will in such cases, it is customary to perform dielectric strength tests on liquids and usually on fusible solids, with electrodes having a standardized fixed spacing. Since the dielectric strength is so dependent upon thickness it is meaningless to report dielectric strength data for a material without stating the thickness of the test specimens used.

A1.4.3 *Temperature*—The temperature of the test specimen and its surrounding medium influence the dielectric strength, although for most materials small variations of ambient temperature may have a negligible effect. In general, the dielectric strength will decrease with increasing temperatures, but the extent to which this is true depends upon the material under test. When it is known that a material will be required to function at other than normal room temperature, it is essential that the dielectric strength - temperature relationship for the material be determined over the range of expected operating temperatures.

A1.4.4 *Time*—Test results will be influenced by the rate of voltage application. In general, the breakdown voltage will tend to increase with increasing rate of voltage application. This is to be expected because the thermal breakdown mechanism is time-dependent and the discharge mechanism is usually time-dependent, although in some cases the latter mechanism may cause rapid failure by producing critically high local field intensities.

A1.4.5 *Wave Form*—In general, the dielectric strength is influenced by the wave form of the applied voltage. Within the limits specified in this method the influence of wave form is not significant.

A1.4.6 *Frequency*—The dielectric strength is not significantly influenced by frequency variations within the range of commercial power frequencies provided for in this method. However, inferences concerning dielectric strength behavior at other than commercial power frequencies (50 to 60 Hz) must not be made from results obtained by this method.

A1.4.7 *Surrounding Medium*—The surrounding medium can affect the heat transfer rate, external discharges, and field uniformity, thereby greatly influencing the test results. Results in one medium cannot be compared with those in a different medium.

A1.4.8 *Relative Humidity*—The relative humidity influences the dielectric strength to the extent that moisture absorbed by, or on the surface of, the material under test affects the dielectric loss and surface conductivity. Hence, its importance will depend to a large extent upon the nature of the material being tested. However, even materials that absorb little or no moisture may be affected because of greatly increased chemical effects of discharge in the presence of moisture. Except in cases where the effect of exposure on dielectric strength is being investigated, it is customary to control or limit the relative humidity effects by standard conditioning procedures.

A1.5 Evaluation

A1.5.1 A fundamental requirement of the insulation in electrical apparatus is that it withstand the voltage imposed on it in service. Therefore there is a great need for a test to evaluate the performance of particular materials at high voltage stress. The dielectric breakdown voltage test represents a convenient preliminary test to determine whether a material merits further consideration, but it falls short of a complete evaluation in two important respects. First, the condition of a material as installed in apparatus is much different from its condition in this test, particularly with regard to the configuration of the electric field and the area of material exposed to it, corona, mechanical stress, ambient medium, and association with other materials. Second, in service there are deteriorating influences, heat, mechanical stress, corona and its products, contaminants, etc., which may reduce the breakdown voltage far below its value as originally installed. Some of these effects can be incorporated in laboratory tests, and a better estimate of the material will result, but the final consideration must always be that of the performance of the material in actual service.

A1.5.2 The dielectric breakdown test may be used as a material inspection or quality control test, as a means of inferring other conditions such as variability, or to indicate deteriorating processes such as thermal aging. In these uses of the test it is the relative value of the breakdown voltage that is important rather than the absolute value.

A2. ASTM Methods and Specifications Governing Electrical Insulating Materials Which Provide for Dielectric Strength Tests

ASTM Serial Designation	Material
	Fabric, Fiber, Paper, and Tape
D 69	Friction Tape
D 119	Rubber Tape
D 202	Insulating Paper
D 295	Varnished Cotton Fabric Tape
D 373	Black Bias-Cut Varnished Cloth and Tape
D 619	Vulcanized Fiber
D 902	Resin-Coated Glass Fabrics and Tapes
D 1000	Pressure-Sensitive Adhesive Coated Tapes
D 1373	Ozone-Resistant Rubber Tape
D 1389	Thin Solid Electric Insulating Material
D 1458	Silicone Rubber-Coated Glass Fabric and Tapes
D 1459	Silicone Varnished Glass Cloth and Tape
D 1677	Untreated Mica Paper
D 1930	Kraft Dielectric Tissue
	Molding Compounds
D 700	Phenolic Molding Compounds
D 704	Melamine-Formaldehyde Molding Compounds
D 705	Urea-Formaldehyde Molding Compounds
D 729	Vinylidene Chloride Molding Compounds
D 1636	Allyl Molding Compounds
	Mica, Glass, and Porcelain
D 116	Vitrified Ceramic Materials
D 352	Pasted Mica
D 748	Natural Block Mica
D 1039	Glass-Bonded Mica
D 1677	Untreated Mica Paper
	Sleeving, Tubes, Sheets, and Rods
D 229	Rigid Sheet and Plate Materials
D 348	Laminated Tubes
D 349	Laminated Round Rods
D 350	Flexible Treated Sleeving
D 709	Laminated Thermosetting Materials
D 876	Nonrigid Vinyl Chloride Polymer Tubing
D 1202	Cellulose Acetate Sheet and Film
D 1675	TFE-Fluorocarbon Tubing
D 1710	TFE-Fluorocarbon Rod
	Varnishes
D 115	Varnishes
D 1346	Silicone Insulating Varnishes
D 1932	Thermal Endurances of Flexible Electrical Insulating Varnishes
	Oils and Askarels
D 877	Insulating Oil of Petroleum Origin (Metal Disk Electrodes)
D 901	Askarels
D 1816	Insulating Oils of Petroleum Origin (VDE Electrodes)
	Rubber and Rubber Products
D 120	Rubber Insulating Gloves
D 178	Rubber Matting
D 1048	Rubber Insulating Blankets
D 1049	Rubber Insulator Hoods
D 1050	Rubber Insulating Line Hose
D 1051	Rubber Insulating Sleeves
	Filling Compounds
D 176	Solid Filling and Treating Compounds
	Adhesives
D 1304	Adhesives Relative to Use as Electrical Insulation

Standard Methods of Test for
A-C LOSS CHARACTERISTICS AND DIELECTRIC CONSTANT (PERMITTIVITY) OF SOLID ELECTRICAL INSULATING MATERIALS[1]

This Standard is issued under the fixed designation D 150; the number immediately following the designation indicates the year of original adoption or, in the case of revision, the year of last revision. A number in parentheses indicates the year of last reapproval.

1. Scope

1.1 These methods cover the determination of dielectric constant, dissipation factor, loss index, power factor, phase angle, and loss angle of specimens of solid electrical insulating materials when the standards used are lumped impedances. The frequency range that can be covered extends from less than 1 Hz to several hundred megahertz.

1.2 The methods are presented in the following sequence:

	Section
Definitions	2
Significance	3
General Measurement Considerations	4
Electrode Systems	5
Choice of Apparatus and Method for Measuring Capacitance and A-C Loss	6
Procedure	7
Report	8
Appendix I—Corrections for Series Inductance and Resistance and Stray Capacitances	
Appendix II—Effective Area of Guarded Electrode	
Appendix III—Factors Affecting Dielectric Constant and Loss Characteristics	
Appendix IV—Circuit Diagrams of Typical Measuring Circuits	
References	

1.3 Since some materials require special treatment, reference should also be made to the ASTM methods applicable to the material to be tested.

2. Definitions

2.1 *capacitance*, C—that property of a system of conductors and dielectrics which permits the storage of electrically separated charges when potential differences exist between the conductors. Numerically it is the ratio of a quantity, Q, of electricity to a potential difference, V. A capacitance value is always positive. The units are farads when the charge is expressed in coulombs and the potential in volts:

$$C = Q/V \tag{1}$$

2.2 *dielectric constant*, (*permittivity, capacitivity, or specific inductive capacity*), κ'—the ratio of the capacitance, C_x, of a given configuration of electrodes with a material as the dielectric, to the capacitance, C_v, of the same electrode configuration with a vacuum (or air for most practical purposes) as the dielectric:

$$\kappa' = C_x/C_v \tag{2}$$

Experimentally, the vacuum must be replaced by the material at all points where it makes a significant change in the capacitance.

NOTE 1—C_x is taken to be C_p, the equivalent parallel capacitance as shown in Fig. 1.

NOTE 2—The series capacitance is larger than the parallel capacitance by less than 1 percent for a dissipation factor of 0.1, and by less than 0.1 percent for a dissipation factor of 0.03. If a measuring circuit yields results in terms of series components, the parallel capacitance must be calculated from Eq 5 before the corrections and dielectric constant are calculated.

NOTE 3—The dielectric constant of dry air at 23 C and 760 mm Hg pressure is 1.000536 (**1**).[2] Its divergence from unity, $\kappa' - 1$, is inversely proportional to absolute temperature and directly proportional to atmospheric pressure. The increase in di-

[1] These methods are under the jurisdiction of ASTM Committee D-9 on Electrical Insulating Materials. A list of members may be found in the ASTM Yearbook.

Current edition effective Sept. 9, 1968. Originally issued 1922. Replaces D 150 - 65 T.

[2] The boldface numbers in parentheses refer to the list of references appended to these methods.

electric constant when the space is saturated with water vapor at 23 C is 0.00025 (**2,3**), and varies approximately linearly with temperature expressed in degrees Celsius, from 10 to 27 C. For partial saturation the increase is proportional to the relative humidity.

2.3 *dielectric phase angle*, θ—the angular difference in the phase between the sinusoidal alternating potential difference applied to a dielectric and the component of the resulting alternating current having the same period as the potential difference.

2.4 *dielectric loss angle*, δ—the difference between 90 deg and the dielectric phase angle.

NOTE 4—The relation of dielectric phase angle and dielectric loss angle is shown in Figs. 2 and 4. Dielectric loss angle is sometimes called the phase defect angle.

2.5 *dielectric dissipation factor, loss tangent*, *D*—the tangent of the loss angle or the cotangent of the phase angle.

NOTE 5:

$$D = \tan \delta = \cot \theta = X_p / R_p = G/\omega C_p = 1/\omega C_p R_p \qquad (3)$$

where:
G = equivalent a-c conductance,
X_p = parallel reactance,
R_p = equivalent a-c parallel resistance,
C_p = parallel capacitance, and
ω = $2\pi f$ (sinusoidal wave shape assumed).

The reciprocal of the dissipation factor is the storage factor, *Q*, sometimes called the quality factor. The dissipation factor, *D*, of the capacitor is the same for both the series and parallel representations as follows:

$$D = \omega R_s C_s = 1/\omega R_p C_p \qquad (4)$$

The relationships between series and parallel components are as follows:

$$C_p = C_s/(1 + D^2) \qquad (5)$$

$$R_p/R_s = (1 + D^2)/D^2 = 1 + (1/D^2) = 1 + Q^2 \qquad (6)$$

NOTE 6: *Series Representation*—While the parallel representation of an insulating material having a dielectric loss (Fig. 1) is usually the proper representation, it is always possible and occasionally desirable to represent a capacitor at a single frequency by a capacitance, C_s, in series with a resistance, R_s (Figs. 3 and 4).

2.6 *dielectric power factor*, *PF*—the ratio of the power in watts, *W*, dissipated in a material to the product of the effective sinusoidal voltage, *V*, and current, *I*, in volt-amperes. Numerically, it may be expressed as the cosine of the dielectric phase angle θ (or the sine of the dielectric loss angle δ).

$$PF = W/VI = G/\sqrt{G^2 + (\omega C_p)^2} = \sin \delta = \cos \theta \qquad (7)$$

When the dissipation factor is less than 0.1, the power factor differs from the dissipation factor by less than 0.5 percent. Their exact relationship may be found either by reference to trigonometric tables or from the following relationships:

$$\left.\begin{aligned} PF &= D/\sqrt{1 + D^2} \\ D &= PF/\sqrt{1 - (PF)^2} \end{aligned}\right\} \qquad (8)$$

2.7 *dielectric loss index*, κ''—the product of the dielectric constant and the dissipation factor (tangent of the dielectric loss angle). It is a measure of the a-c loss. The loss index is proportional to the power loss per cycle per squared potential gradient per unit volume as follows:

$$\kappa'' = \frac{\text{power loss}}{E^2 \times f \times \text{volume} \times \text{constant}} \qquad (9)$$

When the units are watts, volts per centimeter, hertz, and cubic centimeters, the constant has the value 5.556×10^{-13}.

NOTE 7—Dielectric loss index is the term agreed upon internationally. In the U.S.A. κ'' was formerly called the loss factor.

3. Significance

3.1 *Dielectric Constant*—Insulating materials are used in general in two distinct ways, (*1*) to support and insulate components of an electrical network from each other and from ground, and (*2*) to function as the dielectric of a capacitor. For the first use, it is generally desirable to have the capacitance of the support as small as possible, consistent with acceptable mechanical, chemical, and heat-resisting properties. A low value of dielectric constant is thus desirable. For the second use, it is desirable to have a high value of dielectric constant, so that the capacitor may be physically as small as possible. Intermediate values of dielectric constant are sometimes used for grading stresses at the edge or end of a conductor to minimize a-c corona. Factors affecting dielectric constant are discussed in Appendix III.

3.2 *A-C Loss*—For both cases (as electrical insulation and as capacitor dielectric) the a-c loss generally should be small, both in order to reduce the heating of the material and to minimize its effect on the rest of the network. In high frequency applications, a low value of loss index is particularly desirable, since for a given value of loss index, the dielectric loss increases directly with frequency. In certain dielectric configurations such as are used in terminating bushings and cables for test, an increased loss, usually obtained from increased conductivity, is sometimes introduced to control the voltage gradient. In comparisons of materials having approximately the same dielectric constant or in the use of any material under such conditions that its dielectric constant remains essentially constant, the quantity considered may also be dissipation factor, power factor, phase angle, or loss angle. Factors affecting a-c loss are discussed in Appendix III.

3.3 *Correlation*—When adequate correlating data are available, dissipation factor or power factor may be used to indicate the characteristics of a material in other respects, such as dielectric breakdown, moisture content, degree of cure, and deterioration from any cause. However, deterioration due to thermal aging may not affect dissipation factor unless the material is subsequently exposed to moisture. While the initial value of dissipation factor is important, the change in dissipation factor with aging may be much more significant.

4. General Measurement Considerations

4.1 *Fringing and Stray Capacitance*—These methods are based upon placing a specimen of material in an electrode system with a vacuum capacitance that can be either calculated accurately or determined by a calibration in the absence of the solid material. The problem of determining the two capacitance values that are required, directly or indirectly, to determine κ_x' is best illustrated by reference to Figs. 5 and 6, showing two parallel plate electrodes between which the unknown material is to be placed for measurement. In addition to the desired direct interelectrode capacitance, C_v, the system as seen at terminals *a-a'* includes the following:

C_e = fringing or edge capacitance,
C_g = capacitance to ground of the outside face of each electrode,
C_L = capacitance between connecting leads,
C_{Lg} = capacitance of the leads to ground, and
C_{Le} = capacitance between the leads and the electrodes.

Only the desired capacitance, C_v, is independent of the outside environment, all the others being dependent to a degree on the proximity of other objects. It becomes immediately necessary to distinguish between two possible measuring conditions in order to discuss the effects of the undesired capacitances. If one measuring electrode is grounded, as is often the case, all of the capacitances described, except the ground capacitance of the grounded electrode and its lead, are in parallel with the desired C_v. If a guarded test cell is used the capacitance to ground no longer appears and the capacitance seen at *a-a'* includes C_v, C_e, and the lead capacitances only. The lead capacitance can usually be made negligibly small. The edge capacitance, C_e, in air, can be calculated with reasonable accuracy, but in the presence of the dielectric the value changes. Empirical corrections have been derived for various conditions, and these are given in Table 1 (for the case of thin electrodes such as foil). In routine work, where best accuracy is not required it may be convenient to use unshielded, two-electrode systems and make the approximate corrections. However, for exacting measurements it is necessary to use guarded electrodes.

4.2 *Guarded Electrodes*—The fringing and stray capacitance at the edge of the guarded electrode is practically eliminated by the addition of a guard electrode as shown in Fig. 7. If the test specimen and guard electrode extend beyond the guarded electrode by at least twice the thickness of the specimen and the guard gap is very small, the field distribution in the guarded area will be identical with that existing when vacuum is the dielectric, and the ratio of these two direct capacitances is the dielectric constant. Furthermore, the field between the active electrodes is defined and the vacuum capacitance can be calculated with the accuracy limited only by the accuracy with which the dimensions are known. For these reasons the guarded electrode (three-terminal) method is to be used as the referee method unless otherwise agreed upon. Figure 8 shows a schematic representation of a completely guarded and shielded electrode system. Although the guard is commonly grounded, the arrangement shown permits grounding either measuring electrode or none of the electrodes to accommodate the particular three-terminal measuring system being used. If the guard is connected to ground, or to a guard terminal on the measuring circuit, the measured capacitance is the direct capacitance between the two measuring electrodes. If, however, one of the measuring electrodes is grounded, the capacitance to ground of the ungrounded electrode and leads is in parallel with the desired direct capacitance. To eliminate this source of error, the ungrounded electrode should be surrounded by a shield connected to guard as shown in Fig. 8. In addition to guarded methods, which are not always convenient or practical and which are limited to frequencies less than a few megacycles, techniques using special cells and procedures have been devised that yield, with two-terminal measurements, accuracies comparable to those obtained with guarded measurements. Such methods described here include shielded micrometer electrodes (5.3.2) and fluid displacement methods (5.3.3).

4.3 *Geometry of Specimens*—For determining the dielectric constant and dissipation factor of a material, sheet specimens are preferable. Cylindrical specimens can also be used, but generally with lesser accuracy. The source of the greatest uncertainty in dielectric constant is in the determination of the dimensions of the specimen, and particularly that of its thickness which should, therefore, be large enough to allow its measurement with the required accuracy. The chosen thickness will depend on the method of producing the specimen and the likely variation from point to point. For 1 percent accuracy a thickness of 1.5 mm (0.06 in.) is usually sufficient, although for greater accuracy it may be desirable to use a thicker specimen. Another source of error, when foil or rigid electrodes are used, is in the unavoidable gap between the electrodes and the specimen. For thin specimens the error in dielectric constant can be as much as 25 percent. A similar error occurs in dissipation factor, although when foil electrodes are applied with a grease, the two errors may not have the same magnitude. For the most accurate measurements on thin specimens, the fluid displacement method should be used (5.3.3). This method reduces or completely eliminates the need for electrodes on the specimen. The thickness must be determined by measurements distributed systematically over the area of the specimen that is used in the electrical measurement and should be uniform within ± 1 percent of the average thickness. If the whole area of the specimen will be covered by the electrodes, and if the density of the material is known, the average thickness can be determined by weighing. The diameter chosen for the specimen should be such as to provide a specimen capacitance that can be measured to the desired accuracy. With well guarded and screened apparatus there need be no difficulty in measuring specimens having capacitances of 10 pf to a resolution of 1 part in 1000. A thick specimen of low dielectric constant may necessitate a diameter of 10 cm or more to obtain the desired capacitance accuracy. In the measurement of small values of dissipation factor, the essential points are that no appreciable dissipation factor shall be contributed by the series resistance of the electrodes and that in the measuring network no large capacitance shall be connected in parallel with that of the specimen. The first of these points favors thick specimens; the second suggests thin specimens of large area. Micrometer electrode methods (5.3.2) can be used to eliminate the effects of series resistance. A guarded specimen holder (Fig. 8) can be used to minimize extraneous capacitances.

4.4 *Calculation of Vacuum Capacitance*—The practical shapes for which capacitance can be most accurately calculated are flat parallel plates and coaxial cylinders, the equations for which are given in Table 1. These equations are based on a uniform field between the measuring electrodes, with no fringing at the edges. Capacitance calculated on this basis is known as the direct interelectrode capacitance.

4.5 *Edge, Ground, and Gap Corrections*—The equations for calculating edge capacitance, given in Table 1, are empirical, based on published work (**1**). They are expressed in terms of picofarads per centimeter of perimeter and are thus independent of the shape of the electrodes. It is recognized that they are dimensionally incorrect, but they are found to give better approximations to the true edge capacitance than any other equations that have been proposed. Ground capacitance cannot be calculated by any equations presently known. When measurements must be made that include capacitance to ground, it is recommended that the value be determined experimentally for the particular setup used. This can be done using a dielectric sample holder or a three-electrode method. The difference between the capacitance measured in the two-terminal arrangement and the capacitance calculated from the dielectric constant and the dimensions of the specimen is the ground capacitance plus the edge capacitance. The edge capacitance can be calculated using one of the equations of Table 1. As long as the same physical arrangement of leads and electrodes is maintained, the ground capacitance will remain constant, and the experimentally determined value can be used as a correction to subsequently measured values of capacitance. The effective area of a guarded electrode is greater than its actual area by approximately half the area of the guard gap (**2,3**). Thus, the diameter of a circular electrode, each dimension of a rectangular electrode, or the length of a cylindrical electrode is increased by the width of this gap. When the ratio of gap width, g, to specimen thickness, t, is appreciable, the increase in the effective dimension of the guarded electrode is somewhat less than the gap width. Details of computation for this case are given in Appendix II.

5. Electrode Systems

5.1 *Contacting Electrodes*—A specimen may be provided with its own electrodes, of one of the materials listed below. For two-terminal measurements the electrodes may extend either to the edge of the specimen or may be smaller than the specimen. In the latter case the two electrodes may be equal or unequal in size. If equal in size and smaller than the specimen, the edge of the specimen must extend beyond the electrodes by at least twice the specimen thickness. The choice between these three sizes of electrodes will depend on convenience of application of the electrodes, and on the type of measurement adopted. The edge correction (see Table 1) is smallest for the case of electrodes extending to the edge of the specimen and largest for unequal electrodes. When the electrodes extend to the edge of the specimen, these edges must be sharp. Such electrodes must be used, if attached electrodes are used at all, when a micrometer electrode system is employed. When equal-size electrodes smaller than the specimen are used, it is difficult to center them unless the specimen is translucent or an aligning fixture is employed. For three-terminal measurements, the width of the guard electrode shall be at least twice the thickness of the specimen (**3,4**). The gap width shall be as small as practical (0.5 mm (0.02 in.) is possible). For measurement of dissipation factor at the higher frequencies, electrodes of this type may be unsatisfactory because of their series resistance. Micrometer electrodes should be used for the measurements.

5.2 *Electrode Materials:*

5.2.1 *Metal Foil*—Lead or tin foil from 0.0075 to 0.025 mm (0.0003 to 0.001 in.) thick applied with a minimum quantity of refined petrolatum, silicone grease, silicone oil, or other suitable low-loss adhesive is generally used as the electrode material. Aluminum foil has also been used, but it is not recommended because of its stiffness and the probability of high contact resistance due to the oxidized surface. Lead foil also may give trouble because of its stiffness. Such electrodes should be applied under a smoothing pressure sufficient to eliminate all wrinkles and to work excess adhesive toward the edge of the foil. One very effective method is to use a narrow roller, and to roll outward on the surface until no visible imprint can be made on the foil. With care the adhesive film can be reduced to 0.0025 mm (0.0001 in.). As this film is in series with the specimen, it will always cause the measured dielectric constant to be too low and probably the dissipation factor to be too high. These errors usually become excessive for specimens of thickness less than 0.125 mm (0.005 in.). The error in dissipation factor is negligible for such thin specimens only when the dissipation factor of the film is nearly the same as that of the specimen. When the electrode is to extend to the edge, it should be made larger than the specimen and then cut to the edge with a small, finely ground blade. A guarded and guard electrode can be made from an electrode that covers the entire surface, by cutting out a narrow strip (0.5 mm

(0.02 in.) is possible) by means of a compass equipped with a narrow cutting edge.

5.2.2 *Conducting Paint*—Certain types of high-conductivity silver paints, either air-drying or low-temperature-baking varieties, are commercially available for use as electrode material. They are sufficiently porous to permit diffusion of moisture through them and thereby allow the test specimen to condition after application of the electrodes. This is particularly useful in studying humidity effects. The paint has the disadvantage of not being ready for use immediately after application. It usually requires an overnight air-drying or low-temperature baking to remove all traces of solvent, which otherwise may increase both dielectric constant and dissipation factor. It also may not be easy to obtain sharply defined electrode areas when the paint is brushed on, but this limitation usually can be overcome by spraying the paint and employing either clamp-on or pressure-sensitive masks. The conductivity of silver paint electrodes may be low enough to give trouble at the higher frequencies. It is essential that the solvent of the paint does not affect the specimen permanently.

5.2.3 *Fired-On Silver*—Fired-on silver electrodes are suitable only for glass and other ceramics that can withstand, without change, a firing temperature of about 350 C. Its high conductivity makes such an electrode material satisfactory for use on low-loss materials such as fused silica, even at the highest frequencies, and its ability to conform to a rough surface makes it satisfactory for use with high-dielectric-constant materials, such as the titanates.

5.2.4 *Sprayed Metal*—A low-melting-point metal applied with a spray gun provides a spongy film for use as electrode material which, because of its grainy structure, has roughly the same electrical conductivity and the same moisture porosity as conducting paints. Suitable masks must be used to obtain sharp edges. It conforms readily to a rough surface, such as cloth, but does not penetrate very small holes in a thin film and produce short circuits. Its adhesion to some surfaces is poor, especially after exposure to high humidity or water immersion. Advantages over conducting paint are freedom from effects of solvents, and readiness for use immediately after application.

5.2.5 *Evaporated Metal*—Evaporated metal used as an electrode material may have inadequate conductivity because of its extreme thinness, and must be backed with electroplated copper or sheet metal. Its adhesion is adequate, and by itself it is sufficiently porous to moisture. The necessity for using a vacuum system in evaporating the metal is a disadvantage.

5.2.6 *Liquid Metal*—Mercury electrodes may be used by floating the specimen on a pool of mercury and using confining rings with sharp edges for retaining the mercury for the guarded and guard electrodes, as shown in Fig. 9. A more convenient arrangement, when a considerable number of specimens must be tested, is the test fixture shown in Fig. 1 of ASTM Method D 1082, Test for Power Factor and Dielectric Constant of Natural Mica.[3] There is some health hazard present due to the toxicity of mercury vapor, especially at elevated temperatures, and suitable precautions should be taken during use. In measuring low-loss materials in the form of thin films such as mica splittings, contamination of the mercury may introduce considerable error, and it may be necessary to use clean mercury for each test. Wood's metal or other low-melting alloy can be used in a similar manner. Elevated temperature measurements can then be made without the health hazard of mercury.

5.2.7 *Rigid Metal*—For smooth, thick, or slightly compressible specimens, rigid electrodes under high pressure can sometimes be used, especially for routine work. Electrodes 1 cm (0.4 in.) in diameter, under a pressure of 176 kgf/cm^2 (2500 psi) have been found useful for measurements on plastic materials, even those as thin as 0.025 mm (0.001 in.). Electrodes 5 cm in diameter, under pressure, have also been used successfully for thicker materials. However, it is difficult to avoid an air film when using solid electrodes, and the effect of such a film becomes greater as the dielectric constant of the material being tested increases and its thickness decreases. The uncertainty in the determination of thickness also increases as the thickness decreases. The dimensions of a specimen may continue to change for as long as 24 h after the application of pressure.

[3] *Annual Book of ASTM Standards*, Part 29.

5.2.8 *Water*—Water can be used as one electrode for testing insulated wire and cable when the measurements are made at low frequency (up to 1000 Hz, approximately). Care must be taken to ensure that electrical leakage at the ends of the specimen is negligible.

5.3 *Non-Contacting Electrodes:*

5.3.1 *Fixed Electrodes*—Specimens of sufficiently low surface conductivity can be measured without applied electrodes by inserting them in a prefabricated electrode system, in which there is an intentional airgap on one or both sides of the specimen. The electrode system should be rigidly assembled and should preferably include a guard electrode. For the same accuracy, a more accurate determination of the electrode spacing and the thickness of the specimen is required than if direct contact electrodes are used. However, if the electrode system is filled with a liquid, these limitations may be removed (see 5.3.3).

5.3.2 *Micrometer Electrodes*—The micrometer-electrode system, as shown in Fig. 10, was developed (**5**) to eliminate the errors caused by the series inductance and resistance of the connecting leads and of the measuring capacitor at high frequencies. A built-in vernier capacitor is also provided for use in the susceptance variation method. It accomplishes this by maintaining these inductances and resistances relatively constant, regardless of whether the test specimen is in or out of the circuit. The specimen, which is either the same size as, or smaller than, the electrodes, is clamped between the electrodes. Unless the surfaces of the specimen are lapped or ground very flat, metal foil or its equivalent must be applied to the specimen before it is placed in the electrode system. If electrodes are applied, they also must be smooth and flat. Upon removal of the specimen, the electrode system can be made to have the same capacitance by moving the micrometer electrodes closer together. When the micrometer-electrode system is carefully calibrated for capacitance changes, its use eliminates the corrections for edge capacitance, ground capacitance, and connection capacitance. In this respect it is advantageous to use it over the entire frequency range. A disadvantage is that the capacitance calibration is not as accurate as that of a conventional multiplate variable capacitor, and also it is not direct reading. At frequencies below 1 MHz, where the effect of series inductance and resistance in the leads is negligible, the capacitance calibration of the micrometer electrodes can be replaced by that of a standard capacitor, either in parallel with the micrometer-electrode system or in the adjacent capacitance arm of the bridge. The change in capacitance with the specimen in and out is measured in terms of this capacitor. A source of minor error in a micrometer-electrode system is that the edge capacitance of the electrodes, which is included in their calibration, is slightly changed by the presence of a dielectric having the same diameter as the electrodes. This error can be practically eliminated by making the diameter of the specimen less than that of the electrodes by twice its thickness (**3**). When no electrodes are attached to the specimen, surface conductivity may cause serious errors in dissipation factor measurements of low loss material. Unless materials are to be tested "as received" they should be cleaned and dried in accordance with ASTM Recommended Practice D 1371, Cleaning Plastic Specimens for Insulation Resistance Testing,[3] and measured in a dry atmosphere. When the bridge used for measurement has a guard circuit, it is advantageous to use guarded micrometer electrodes. The effects of fringing, etc., are almost completely eliminated. When the electrodes and holder are well made, no capacitance calibration is necessary as the capacitance can be calculated from the electrode spacing and the diameter. The micrometer itself will require calibration, however. It is not practicable to use electrodes on the specimen when using guarded micrometer electrodes unless the specimen is smaller in diameter than the guarded electrode.

5.3.3 *Fluid Displacement Methods*—When the immersion medium is a liquid, and no guard is used, the parallel-plate system preferably shall be constructed so that the insulated high-potential plate is supported between, parallel to, and equidistant from two parallel low-potential or grounded plates, the latter being the opposite inside walls of the test cell designed to hold the liquid. This construction makes the electrode system essentially self-shielding, but normally requires duplicate test specimens. Provision must be made for precise temperature measurement of the liquid (**6,7**). Cells should be constructed of brass and gold plated. The high-potential electrode shall be removable for cleaning. The faces must be as nearly optically flat and plane parallel as possible. A suitable liquid cell for measurements up to 1 MHz is shown in Fig. 1 of ASTM Method D 1531, Test for Dielectric Constant and Dissipation Factor of Polyethylene by Liquid Displacement Procedure.[3] Changes in the dimensions of this cell may be made to provide for testing sheet specimens of various thicknesses or sizes, but, such changes should not reduce the capacitance of the cell filled with the standard liquid to less than 100 pF. For measurements at frequencies from 1 to about 50 MHz, the cell dimensions must be greatly reduced, and the leads must be as short and direct as possible. The capacitance of the cell with liquid shall not exceed 30 or 40 pF for measurements at 50 MHz. Experience has shown that a capacitance of 10 pF can be used up to 100 MHz without loss of accuracy. Guarded parallel-plate electrodes have the advantage that single specimens can be measured with full accuracy. Also a prior knowledge of the dielectric constant of the liquid is not required as it can be measured directly (**8**). If the cell is constructed with a micrometer electrode, specimens having widely different thicknesses can be measured with high accuracy since the electrodes can be adjusted to a spacing only slightly greater than the thickness of the specimen. If the dielectric constant of the fluid approximates that of the specimen the effect of errors in determination of specimen thicknesses are minimized. The use of a nearly matching liquid and a micrometer cell permits high accuracy in measuring even very thin film.

All necessity for determining specimen thickness and electrode spacing is eliminated if successive measurements are made in two fluids of known dielectric constant (**9**). This method is not restricted to any frequency range; however, it is best to limit use of liquid immersion methods to frequencies for which the dissipation factor of the liquid is less than 0.01 (preferably less than 0.0001 for low-loss specimens).

When using the two-fluid method it is important that both measurements be made on the same area of the specimen as the thickness may not be the same at all points. To ensure that the same area is tested both times and to facilitate the handling of thin films, specimen holders are convenient. The holder can be a U-shaped piece that will slide into grooves in the electrode cell. It is also necessary to control the temperature to at least 0.1 C. The cell may be provided with cooling coils for this purpose (**10**).

6. Choice of Apparatus and Methods for Measuring Capacitance and A-C Loss

6.1 *Frequency Range*—Methods for measuring capacitance and a-c loss can be divided into three groups: null methods, resonance methods, and deflection methods. The choice of a method for any particular case will depend primarily on the operating frequency. The resistive- or inductive-ratio-arm capacitance bridge in its various forms can be used over the frequency range from less than 1 Hz to a few megahertz. Parallel-T networks are used at the higher frequencies from 500 kHz to 30 MHz, since they partake of some of the characteristics of resonant circuits. Resonance methods are used over a frequency range from 50 kHz to several hundred megahertz. The deflection method, using commercial indicating meters, is employed only at power-line frequencies from 25 to 60 Hz, where the higher voltages required can easily be obtained.

6.2 *Direct and Substitution Methods*—In any direct method, the values of capacitance and a-c loss are in terms of all the circuit elements used in the method, and are therefore subject to all their errors. Much greater accuracy can be obtained by a substitution method in which readings are taken with the unknown capacitor both connected and disconnected. The errors in those circuit elements that are unchanged are in general eliminated; however, a connection error remains (Note 8).

6.3 *Two- and Three-Terminal Measurements*—The choice between three-terminal and two-terminal measurements is generally one between accuracy and convenience. The use of a guard electrode on the dielectric specimen nearly eliminates the effect of edge and ground capacitance, as explained in 4.2. The provision of a guard terminal eliminates some of the errors introduced by the circuit elements. On the other hand, the extra circuit elements and shielding usually required to provide the guard terminal add considerably to the size of the measuring equipment, and the number of adjustments required to obtain the final result may be increased many times. Guard circuits for resistive-ratio-arm capacitance bridges are rarely used at frequencies above 1 MHz. Inductive-ratio-arm bridges provide a guard terminal without requiring extra circuits or adjustments. Parallel-T networks and resonant circuits are not provided with guard circuits. In the deflection method a guard can be provided merely by extra shielding. The use of a two-terminal micrometer-electrode system provides many of the advantages of three-terminal measurements by nearly eliminating the effect of edge and ground capacitances but may increase the number of observations or balancing adjustments. Its use also eliminates the errors caused by series inductance and resistance in the connecting leads at the higher frequencies. It can be used over the entire frequency range to several hundred megahertz. When a guard is used, the possibility exists that the measured dissipation factor may be less than the true value. This is caused by resistance in the guard circuit at points between the guard point of the measuring circuit and the guard electrode. This may arise from high contact resistance, lead resistance, or from high resistance in the guard electrode itself. In extreme cases the dissipation factor may appear to be negative. This condition is most likely to exist when the dissipation factor without the guard is higher than normal due to surface leakage. Any point capacitively coupled to the measuring electrodes and resistively coupled to the guard point can be a source of difficulty. The common guard resistance produces an equivalent negative dissipation factor proportional to $C_h C_l R_g$, where C_h and C_l are guard-to-electrode capacitances and R_g is the guard resistance (**11**).

6.4 *Fluid Displacement Methods*—The fluid displacement method may be employed using either three-terminal or self-shielded, two-terminal cells. With the three-terminal cell the dielectric constants of the fluids used may be determined directly. The self-shielded, two-terminal cell provides many of the advantages of the three-terminal cell by nearly eliminating the effects of edge and ground capacitance, and it may be used with measuring circuits having no provision for a guard. If it is equipped with an integral micrometer electrode, the effects on the capacitance of series inductance in the connective leads at the higher frequencies may be eliminated.

6.5 *Accuracy*—The methods outlined in 6.1 contemplate an accuracy in the determination of dielectric constant of ±1 percent and of dissipation factor of ± (5 percent + 0.0005). These accuracies depend upon at least three factors: the accuracy of the observations for capacitance and dissipation factor, the accuracy of the corrections to these quantities caused by the electrode arrangement used, and the accuracy of the calculation of the direct interelectrode vacuum capacitance. Under favorable conditions and at the lower frequencies, capacitance can be measured with an accuracy of ± (0.1 percent + 0.02 pF) and dissipation factor with an accuracy of ± (2 percent + 0.00005). At the higher frequencies these limits may increase for capacitance to ± (0.5 percent + 0.1 pF) and for dissipation factor to ± (2 percent + 0.0002). Measurements of dielectric specimens provided with a guard electrode are subject only to the error in capacitance and in the calculation of the direct interelectrode vacuum capacitance. The error caused by too wide a gap between the guarded and the guard electrodes will generally amount to several tenths percent, and the correction can be calculated to a few percent. The error in measuring the thickness of the specimen can amount to a few tenths percent for an average thickness of 1.6 mm ($^1/_{16}$ in.), on the assumption that it can be measured to ±0.005 mm (0.0002 in.). The diameter of a circular specimen can be measured to an accuracy of ±0.1 percent, but enters as the square. Combining these errors, the direct interelectrode vacuum capacitance can be determined to an accuracy of ±0.5 percent. Specimens with contact electrodes, measured with micrometer electrodes, have no corrections other than that for direct interelectrode capacitance, provided they are sufficiently smaller in diameter than the micrometer electrodes. When two-terminal specimens are measured in any other manner, the calculation of edge capacitance and determination of ground capacitance will involve considerable error, since each may be from 2 to 40 percent of the specimen capacitance. With the present knowledge of these capacitances, there may be a 10 percent error in calculating the edge capacitance and a 25 percent error in evaluating the ground capacitance. Hence the total error involved may be from several tenths to 10 percent or more. However, when neither electrode is grounded, the ground capacitance error is minimized (4.1). With micrometer electrodes, it is possible to measure dissipation factor of the order of 0.03 to within ±0.0003 and dissipation factor of the order of 0.0002 to within ±0.00005, of the true values. The range of dissipation factor is normally 0.0001 to 0.1 but may be extended above 0.1. Between 10 and 20 MHz it is possible to detect a dissipation factor of 0.00002. Dielectric constants from 2 to 5 may be determined to ±2 percent. The accuracy is limited by the accuracy of the measurements required in the calculation of direct interelectrode vacuum capacitance and by errors in the micrometer-electrode system.

7. Procedure

7.1 *Preparation of Specimens:*

7.1.1 *General*—Cut or mold the test specimens to a suitable shape and thickness determined by the material specification being followed or by the accuracy of measurement required, the test method, and the frequency at which the measurements are to be made. Measure the thickness in accordance with the standard method required by the material being tested. If there is no standard for a particular material, then measure thickness in accordance with ASTM Methods D 374, Test for Thickness of Solid Electrical Insulation.[3] The actual points of measurement shall be uniformly distributed over the area to be covered by the measuring electrodes. Apply suitable measuring electrodes to the specimens (Section 5) (unless the fluid displacement method will be used), the choice as to size and number depending mainly on whether three-terminal or two-terminal measurements are to be made and, if the latter, whether or not a micrometer-electrode system will be used (5.3). The material chosen for the specimen electrodes will depend both on convenience of application and on whether or not the specimen must be conditioned at high temperature and high relative humidity (Section 5). Obtain the dimensions of the electrodes (of the smaller if they are unequal) preferably by a traveling microscope, or by measuring with a steel scale graduated to 0.25 mm (0.01 in.) and a microscope of sufficient power to allow the scale to be read to the nearest 0.05 mm (0.002 in.). Measure the diameter of a circular electrode, or the dimensions of a rectangular electrode, at several points to obtain an average.

7.1.2 *Micrometer Electrodes*—The area of the specimen may be equal to or less than the area of the electrodes, but no part of the specimen shall extend beyond the electrode edges. The edges of the specimens shall be smooth and perpendicular to the plane of the sheet and shall also be sharply defined so that the dimensions in the plane of the sheet may be determined to the nearest 0.025 mm (0.001 in.). The thickness may have any value from 0.025 mm (0.001 in.) or less to about 6 mm ($^1/_4$ in.) or greater, depending upon the maximum usable plate spacing of the parallel-plate electrode system. The specimens shall be as flat and uniform in thickness as possible, and free of voids, inclusions of foreign matter, wrinkles, and other defects. It has been found that very thin specimens may be tested more conveniently and accurately by using a composite of several or a large number of thicknesses. The average thickness of each specimen shall be determined as nearly as possible to within ±0.0025 mm (0.0001 in.). In certain cases, notably for thin films and the like but usually excluding porous materials, it may be preferable to determine the average thickness by calculation from the known or measured density of the material, the area of the specimen face, and the mass of the specimen (or specimens, when tested in multiple thicknesses of the sheet), obtained by accurate weighing on an analytical balance.

7.1.3 *Fluid Displacement*—When the immersion medium is a liquid, the specimen may be larger than the electrodes if the dielectric constant of the standard liquid is within about 1 percent of that of the specimen (see Method D 1531).[3] Also, duplicate specimens will normally be required for a cell of the type described in 5.3.3, although it is possible to test a single specimen at a time in such cells. In any case, the thickness of the specimen preferably should not be less than about 80 percent of the electrode spacing, this being particularly true when the dissipation factor of the material being tested is less than about 0.001.

7.1.4 *Cleaning*—Since it has been found that in the case of certain materials when tested without electrodes the results are affected erratically by the presence of conducting contaminants on the surfaces of the specimens, clean the test specimens by a suitable solvent or other means (as prescribed in the material specification) and allow to dry thoroughly before test (**12**). This is particularly important when tests are to be made in air at low frequencies (60 to 10,000 Hz), but is less important for measurements at radio frequencies. Cleaning of specimens will also reduce the tendency to contaminate the immersion medium in the case of tests performed using a liquid medium. Use Recommended Practice D 1371 as a guide to the choice of suitable cleaning procedures. After cleaning, handle the specimens only with tweezers and store in individual envelopes to preclude further contamination before testing.

7.2 *Measurement*—Place the test specimen with its attached electrodes in a suitable measuring cell, and measure its capacitance and a-c loss by a method having the required sensitivity and accuracy. For routine work when the highest accuracy is not required, or when neither terminal of the specimen is grounded, it is not necessary to place the solid specimen in a test cell.

NOTE 8—The method used to connect the specimen to the measuring circuit is very important, especially for two-terminal measurements. The connection method by critical spacing, formerly recommended in Methods D 150 for parallel substitution measurements can cause a negative error of 0.5 pF. A similar error occurs when two-terminal specimens are measured in a cell used as a guard. Since no method for eliminating this error is presently known, when an error of this magnitude must be avoided, an alternative method must be used, that is, micrometer electrodes, fluid immersion cell, or three-terminal specimen with guarded leads.

NOTE 9—Detailed instructions for making the measurements needed to obtain capacitance and dissipation factor and for making any necessary corrections due to the measuring circuit are given in the instruction books supplied with commercial equipment. The following paragraphs are intended to furnish the additional instruction required.

7.2.1 *Fixed Electrodes*—Adjust the plate spacing accurately to a value suitable for the specimen to be tested. For low-loss materials in particular, the plate spacing and specimen thickness should be such that the specimen will occupy not less than about 80 percent of the electrode gap. For tests in air, plate spacings less than about 0.1 mm (0.004 in.) are not recommended. When the electrode spacing is not adjustable to a suitable value, specimens of the proper thickness must be prepared. Measure the capacitance and dissipation factor of the cell, and then carefully insert and center the specimen between the electrodes of the micrometer electrodes or test cell. Repeat the measurements. For maximum accuracy determine ΔC and ΔD directly, if possible with the measuring equipment used. Record the test temperature.

7.2.2 *Micrometer Electrodes*—Micrometer electrodes are commonly used with the electrodes making contact with the specimen or its attached electrodes. To make a measurement first clamp the specimen between the micrometer electrodes, and balance or tune the network used for measurement. Then remove the specimen, and reset the electrodes to restore the total capacitance in the circuit or bridge arm to its original value by moving the micrometer electrodes closer together.

7.2.3 *Fluid Displacement Methods*—When a single liquid is used, fill the cell and measure the capacitance and dissipation factor. Carefully insert the specimen (or specimens if the two-specimen cell is used) and center it. Repeat the measurements. For maximum accuracy determine ΔC and ΔD directly, if possible with the measuring equipment used. Record the test temperature to the nearest 0.01 C. Remove specimens promptly from the liquid to prevent swelling, and refill the cell to the proper level before proceeding to test additional specimens. Equations for calculation of results are given in Table 3. Method D 1531 describes in detail the application of this method to the measurement of polyethylene. When a guarded cell, preferably with micrometer electrode, is available, greater accuracy can be obtained by measuring the specimen in two fluids. This method also eliminates the need to know the specimen dimensions. The procedure is the same as before except for the use of two fluids having different dielectric constants (**9,10**). It is convenient to use air as the first fluid since this avoids the necessity for cleaning the specimen between measurements. The use of a guarded cell permits the determination of the dielectric constant of the liquid or liquids used (2.2). When either the one- or two-fluid method is used, greatest accuracy is possible when the dielectric constant of one liquid most nearly matches that of the specimen.

NOTE 10—When the two-fluid method is used, the dissipation factor can be obtained from either set of readings (most accurately from the set with the higher κ_f').

7.3 *Calculation of Dielectric Constant, Dissipation Factor, and Loss Index*—The measuring circuits used will give, for the specimen being measured at a given frequency, a value of capacitance and of a-c loss expressed as *Q*, dissipation factor, or series or parallel resistance. When the dielectric constant is to be calculated from the observed capacitance values, these values must be converted to parallel capacitance, if not so expressed, by the use of Eq 5. The equations given in Table 2 can be used in calculating the capacitance of the specimen when micrometer electrodes are used. The equations given in Table 3 for the different electrode systems can be used in calculating dielectric constant and dissipation factor. When the parallel substitution method is used, the dissipation factor readings must be multiplied by the ratio of the total circuit capacitance to the capacitance of the specimen or cell. *Q* and series or parallel resistance also require calculation from the observed values. Dielectric constant is:

$$\kappa_x' = C_p/C_v \qquad (10)$$

Expressions for the vacuum capacitance (4.4) for flat parallel plates and coaxial cylinders are given in Table 1. When the a-c loss is expressed as series resistance or parallel resistance or conductance, the dissipation factor may be calculated using the relations given in Eqs 3 and 4. Loss index is the product of dissipation factor and dielectric constant 2.7.

4.4 *Corrections*—The leads used to connect the specimen to the measuring circuit have both inductance and resistance which, at high frequencies, increase the measured capacitance and dissipation factor. When extra capacitances have been included in the measurements, such as edge capacitance, and ground capacitance, which may occur in two-terminal measurements, the observed parallel capacitance will be increased and the observed dissipation factor will be decreased. Corrections for these effects are given in Appendix A1 and Table 1.

8. Report

8.1 The report shall include the following:

8.1.1 Description of the material tested, that is, the name, grade, color, manufacturer, and other pertinent data,

8.1.2 Shape and dimensions of the test specimen,

8.1.3 Type and dimensions of the electrodes and measuring cell,

8.1.4 Conditioning of the specimen, and test conditions,

8.1.5 Method of measurement and measurement circuit,

8.1.6 Applied voltage, effective voltage gradient, and frequency, and

8.1.7 Values of parallel capacitance, dissipation factor or power factor, dielectric constant, loss index, and estimated accuracy.

References

(1) Scott, A. H., and Curtis, H. L., "Edge Correction in the Determination of Dielectric Constant," *Journal of Research*, JNBAA, Nat. Bureau Standards, Vol 22, June, 1939, pp. 747–775.

(2) Amey, W. G., and Hamburger, F., Jr., "A Method for Evaluating the Surface and Volume Resistance Characteristics of Solid Dielectric Materials," *Proceedings*, ASTEA, Am. Soc. Testing Mats., Vol 49, 1949, pp. 1079–1091.

(3) Field, R. F., "Errors Occurring in the Measurement of Dielectric Constant," *Proceedings*, ASTEA, Am. Soc. Testing Mats., Vol 54, 1954, pp. 456–478.

(4) Moon, C., and Sparks, C. M., "Standards for Low Values of Direct Capacitance," *Journal of Research*, JNBAA, Nat. Bureau Standards, Vol 41, November, 1948, pp. 497–507.

(5) Hartshorn, L., and Ward, W. H., "The Measurement of the Permittivity and Power Factor of Dielectrics at Frequencies from 10^4 to 10^8 Cycles per Second," *Proceedings*, PIEEA, Institution of Electrical Engineers (London), Vol 79, 1936, pp. 597–609; or *Proceedings of the Wireless Section*, *Ibid*., Vol 12, March, 1937, pp. 6–18.

(6) Coutlee, K. G., "Liquid Displacement Test Cell for Dielectric Constant and Dissipation Factor up to 100 Mc," presented at Conference on Electrical Insulation (NRC), October, 1959, and reviewed in *Insulation*, INULA, November, 1959, p. 26.

(7) Maryott, A. A., and Smith, E. R., "Table of Dielectric Constants of Pure Liquids," NBS Circular No. 514, 1951.

(8) Hartshorn, L., Parry, J. V. L., and Essen, L., "The Measurement of the Dielectric Constant of Standard Liquids," *Proceedings*, PPSBA, Physical Soc. (London), Vol 68B, July, 1955, pp. 422–446.

(9) Harris, W. P., and Scott, A. H., "Precise Measurement of the Dielectric Constant by the Two-fluid Technique," 1962 Annual Report, Conference on Electrical Insulation, NAS-NAC, p. 51.

(10) Endicott, H. S., and McGowan, E. J., "Measurement of Permittivity and Dissipation Factor Without Attached Electrodes," 1960 Annual Report, Conference on Electrical Insulation, NAS-NAC.

(11) Harris, W. P., "Apparent Negative Impedances and their Effect on Three-terminal Dielectric Loss Measurements," 1965 Annual Report, Conference on Electrical Insulation, NAS-NRC Publication 1356.

(12) Field, R. F., "The Formation of Ionized Water Films on Dielectrics Under Conditions of High Humidity," *Journal Applied Physics*, JAPIA, Vol 17, May, 1946, pp. 318–325.

(13) Lauritzen, J. I., "The Effective Area of a Guarded Electrode," 1963 Annual Report, Conference on Electrical Insulation, NAS-NRC Publication 1141.

(14) Murphy, E. J., and Morgan, S. O., "The Dielectric Properties of Insulating Materials," *Bell System Technical Journal*, BSTJA, Vol 16, October, 1937, pp. 493–512.

TABLE 1 Calculation of Vacuum Capacitance and Edge Corrections

Type of Electrode (dimensions in centimeters)	Direct Inter-Electrode Capacitance in Vacuum, pF	Correction for Stray Field at an Edge, pF
Disk electrodes with guard-ring:	$C_v = \epsilon_0' \frac{A}{t} = \frac{1}{\mu_0 C^2} \frac{A}{t}$ $= 0.088542 \frac{A}{t}$ $A = \frac{\pi}{4}(d_1 + B^*g)^2$	$C_e = 0$
Disk electrodes without guard-ring: Diameter of the electrodes = diameter of the specimen:		where $a \ll t$, $\frac{C_e}{P} = 0.029 - 0.058 \log t$
Equal electrodes smaller than the specimen:	$C_v = 0.06954 \frac{d_1^2}{t}$	$\frac{C_e}{P} = 0.019\ \kappa_s' - 0.058 \log t + 0.010$ where: κ_s' = an approximate value of the specimen dielectric constant, and $a \ll t$.
Unequal electrodes:		$\frac{C_e}{P} = 0.041\ \kappa_s' - 0.077 \log t + 0.045$ where: κ_s' = an approximate value of the specimen dielectric constant, and $a \ll t$.
Cylindrical electrodes with guard-ring:	$C_v = \frac{0.24160\ (l_1 + B^* g)}{\log \frac{d_2}{d_1}}$	$C_e = 0$
Cylindrical electrodes without guard-ring: 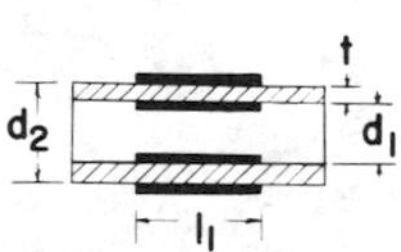	$C_v = \frac{0.24160\ l_1}{\log \frac{d_2}{d_1}}$	If $\frac{t}{t + d_1} < \frac{1}{10}$, $\frac{C_e}{2P} = 0.019\ \kappa_s' - 0.058 \log t + 0.010$ $P = \pi(d_1 + t)$ where κ_s' = an approximate value of the specimen dielectric constant.

* See Appendix A2 for corrections to guard gap.

TABLE 2 Calculation of Capacitance—Micrometer Electrodes

Parallel Capacitance	Definitions of Symbols
$C_p = C' - C_r + C_{vr}$	C' = calibration capacitance of the micrometer electrodes at the spacing to which the electrodes are reset, C_{vr} = vacuum capacitance for the area between the micrometer electrodes, which was occupied by the specimen, calculated using Table 1, C_r = calibration capacitance of the micrometer electrodes at the spacing r, and r = thickness of specimen and attached electrodes.

The true thickness and area of the specimen must be used in calculating the dielectric constant. This double calculation of the vacuum capacitance can be avoided with only small error (0.2 to 0.5 percent due to fringing at the electrode edge), when the specimen has the same diameter as the electrodes, by using the following procedure and equation:

$C_p = C' - C_t + C_{vt}$	C_t = calibration capacitance of the micrometer electrodes at the spacing t, C_{vt} = vacuum capacitance of the specimen area, and t = thickness of specimen.

TABLE 3 Calculation of Dielectric Constant and Dissipation Factor, Noncontacting Electrodes

Dielectric Constant	Dissipation Factor	Definitions of Symbols
Micrometer electrodes in air (with guard ring): $\kappa_x' = \dfrac{1}{1 - \dfrac{\Delta C}{C_1}\dfrac{t_0}{t}}$ or, if t_0 is adjusted to a new value, t_0', such that $\Delta C = 0$: $\kappa_x' = \dfrac{t}{t - (t_0 - t_0')}$	$D_x = D_c + M\kappa_x'\Delta D$	ΔC = capacitance change when specimen is inserted (+ when capacitance increases), C_1 = capacitance with specimen in place, ΔD = increase in dissipation factor when specimen is inserted, D_c = dissipation factor with specimen in place, t_0 = parallel-plate spacing, t = average thickness of specimen, M = $t_0/t - 1$, C_f = $\kappa_f'C_v$ = capacitance with fluid alone. κ_f' = dielectric constant of fluid at test temperature ($\cong$1.00 for air), C_v = vacuum capacitance for area considered (0.08854 A/t_0, pF), A = area of one face of specimen, cm^2 (or area of electrodes when specimen is exactly the same size as or larger than the electrodes), d_0 = OD of inner electrode, d_1 = ID of specimen, d_2 = OD of specimen, and d_3 = ID of outer electrode. NOTE 1—In the equation for the two-fluid method, subscripts 1 and 2 refer to the first and second fluids, respectively. These subscripts may be completely interchanged. NOTE 2—C and D in these equations are the values for the cell and may require calculations from the readings of the measuring circuit (as when using parallel substitution). Refer to Note 9.
Plane electrodes—fluid displacement: $\kappa_x' = \dfrac{\kappa_f'}{1 + D_x^2} \cdot \left[\dfrac{(C_f + \Delta C)(1 + D_c^2)}{C_f + M[C_f - (C_f + \Delta C)(1 + D_c^2)]}\right]$	$D_x = D_c + \Delta DM \cdot \left[\dfrac{(C_f + \Delta C)(1 + D_c^2)}{C_f + M[C_f - (C_f + \Delta C)(1 + D_c^2)]}\right]$	
When the dissipation factor of the specimen is less than about 0.1, the following equations can be used: $\kappa_x' = \dfrac{\kappa_f'}{1 - \dfrac{\Delta C}{\kappa_f'C_v + \Delta C}\dfrac{t_0}{t}}$	$D_x = D_c + M\dfrac{\kappa_x'}{\kappa_f'}\Delta D$	
Cylindrical electrodes (with guard rings)—fluid displacement (for D_x less than about 0.1): $\kappa_x' = \dfrac{\kappa_f'}{1 - \dfrac{\Delta C}{C_1}\dfrac{\log d_3/d_0}{\log d_2/d_1}}$	$D_x = D_c + \Delta D\dfrac{\kappa_x'}{\kappa_f'}\left\{\dfrac{\log \dfrac{d_3}{d_0}}{\log \dfrac{d_2}{d_1}} - 1\right\}$	
Two-fluid method—plane electrodes (with guard ring) (for D_x less than about 0.1): $\kappa_x' = \kappa_{f1}' + \dfrac{\Delta C_1 C_2(\kappa_{f2}' - \kappa_{f1}')}{\Delta C_1 C_2 - \Delta C_2 C_1}$	$D_x = D_{c2} + \left(\dfrac{\kappa_x'C_v - C_2}{\Delta C_2}\right)\Delta D_2$	

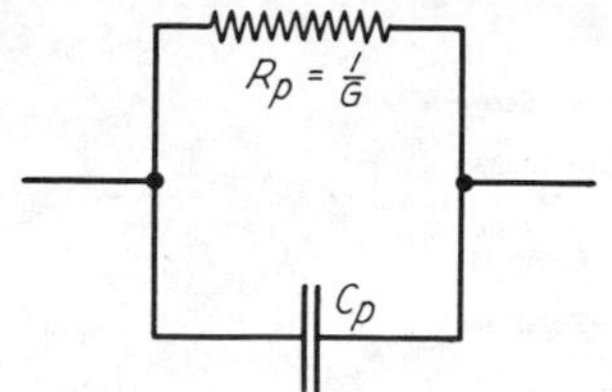

FIG. 1 Parallel Circuit.

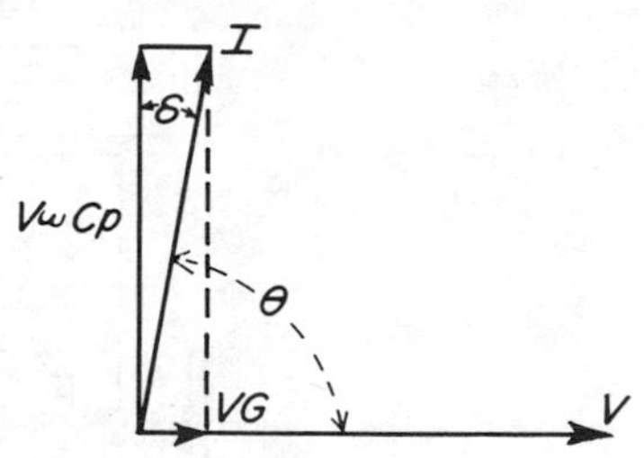

FIG. 2 Vector Diagram for Parallel Circuit.

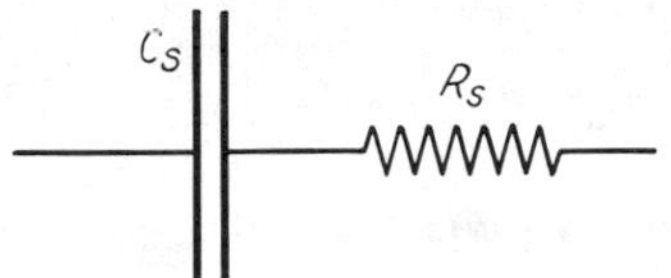

FIG. 3 Series Circuit.

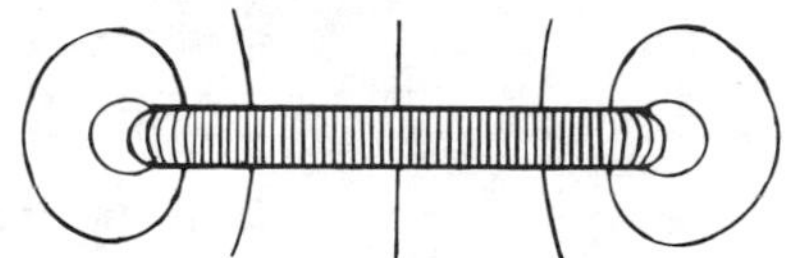

FIG. 5 Flux Lines Between Electrodes.

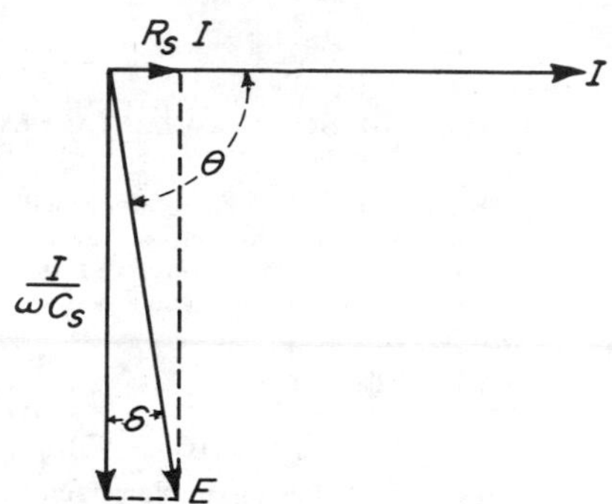

FIG. 4 Vector Diagram for Series Circuit.

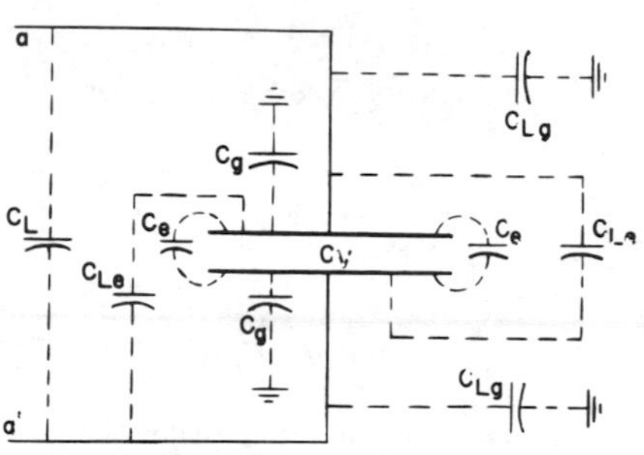

FIG. 6 Stray Capacitances.

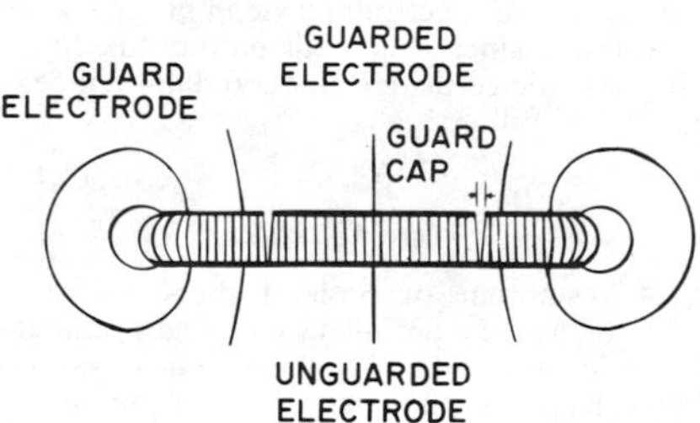

FIG. 7 Guarded Parallel-Plate Electrode System.

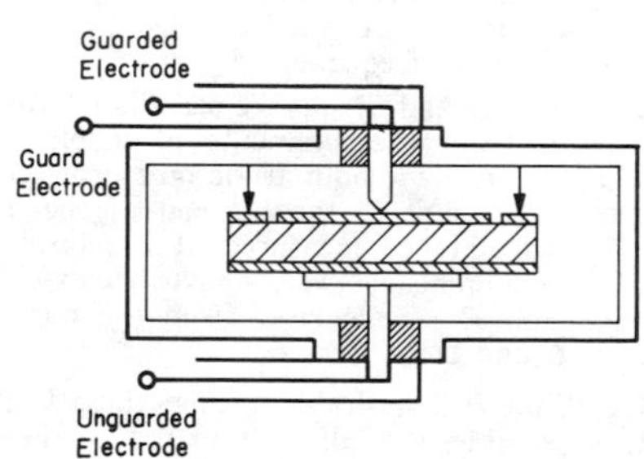

FIG. 8 Three-Terminal Cell for Solids.

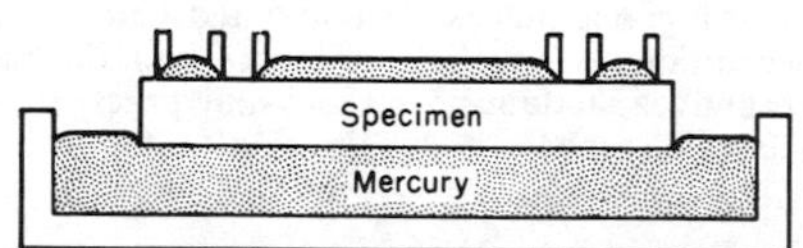

FIG. 9 Guarded Specimen with Mercury Electrodes.

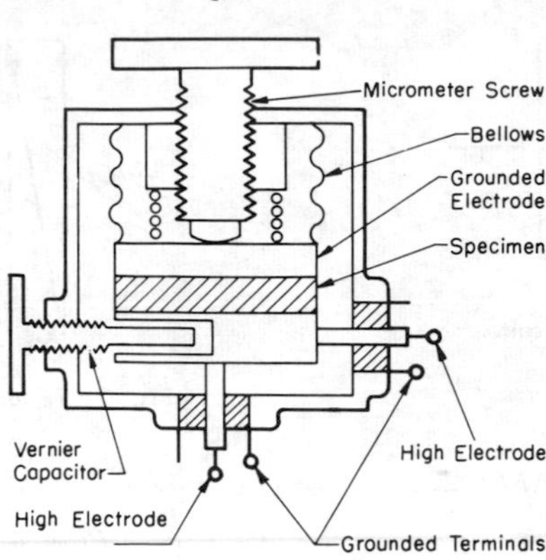

FIG. 10 Micrometer-Electrode System.

APPENDIXES

A1. Corrections for Series Inductance and Resistance and Stray Capacitances

A1.1 The increase in capacitance due to the inductance of the leads and of dissipation factor due to the resistance of the leads is calculated as follows:

$$\left.\begin{aligned}\Delta C &= \omega^2 L_s C_p^{\,2}\\ \Delta D &= R_s \omega C_p\end{aligned}\right\} \quad (11)$$

where:

C_p = true capacitance of the capacitor being measured,
L_s = series inductance of the leads,
R_s = series resistance of the leads, and
ω = 2π times the frequency, Hz.

Note A1—L and R can be calculated for the leads used, from measurements of a physically small capacitor, made both at the measuring equipment terminals and at the far end of the leads. C is the capacitance measured at the terminals, ΔC is the difference between the two capacitance readings, and R is calculated from the measured values of C and D.

A1.2 While it is desirable to have these leads as short as possible, it is difficult to reduce their inductance and resistance below 0.1 μH and 0.05 Ω at 1 MHz. The high-frequency resistance increases with the square root of the frequency. Hence these corrections become increasingly important above 1 MHz. When extra capacitances have been included in the measurements, such as edge capacitance, C_e, and ground capacitance, C_g, which may occur in two-terminal measurements, the observed parallel capacitance will be increased and the observed dissipation factor will be decreased. Designating these observed quantities by the subscript, m, the corrected values are calculated as follows:

$$\left.\begin{aligned}C_p &= C_m - (C_e + C_g)\\ D &= C_m D_m / C_p\\ &= C_m D_m / [C_m - (C_e + C_g)]\end{aligned}\right\} \quad (12)$$

A1.3 The expression for dissipation factor assumes that the extra capacitances are free from loss. This is essentially true for ground capacitance except at low frequencies, and also for edge capacitance when the electrodes extend to the edge of the specimen, since nearly all of the flux lines are in air. The dielectric constant and loss index are calculated as follows:

$$\left.\begin{aligned}\kappa' &= C_p/C_v = [C_m - (C_e + C_g)]/C_v\\ \kappa'' &= C_m D_m / C_v\end{aligned}\right\} \quad (13)$$

A1.4 When one or both of the electrodes are smaller than the specimen, the edge capacitance has two components. The capacitance associated with the flux lines that pass through the surrounding dielectric has a dissipation factor which, for isotropic materials, is the same as that of the body of the dielectric. There is no loss in the capacitance associated with the flux lines through the air. Since it is not practicable to separate the capacitances, the usual practice is to consider the measured dissipation factor to be the true dissipation factor.

A2. Effective Area of Guarded Electrode

A2.1 The effective area of a guarded electrode is greater than its actual area by approximately half the area of the guard gap. Generally, therefore, the diameter of a circular electrode, each dimension of a rectangular electrode, or the length of a cylindrical electrode is increased by the width of the gap. However, when the ratio of gap width, g, to electrode separation, t (usually approximately the specimen thickness), is appreciable, the increase in the effective dimension of the guarded electrode is less than the gap width by a quantity, 2δ, called the guard-gap correction, which is a function of: (*a*) the ratio g/t, (*b*) the ratio of the dielectric constant of the medium between the electrodes, κ', to the dielectric constant of the medium in the gap, κ_g', and (*c*) the thickness of the electrodes, *a* (**13**).

For $\kappa' = \kappa_g'$, and for relatively thick electrodes,

$$(2\delta/g) = (2/\pi)\tan^{-1}(g/2t) - (2t/\pi g)\ln[1 + (g/2t)^2] = (1/\pi)(g/2t) - (1/6\pi)(g/2t)^3 + \ldots \quad (14)$$

A2.2 The fraction of the guard gap to be added to the over-all electrode dimensions before calculating the effective electrode area is $B = 1 - 2\delta/g$.

A2.3 The guard-gap correction, 2δ, for thin electrodes is always equal to or greater than the guard-gap correction for thick electrodes. For thin electrodes, when $\kappa' = \kappa_g'$ and $g/t < 0.5$,

$$2\delta/g = (2t/\pi g)[1 - \cos(\pi g/4t)] = 1/8(\pi g/2t) - 1/384(\pi g/2t)^3 + \ldots \quad (15)$$

For $\kappa'' \gg \kappa_g'$, for thick or thin electrodes,

$$2\delta/g = (4t/\pi g)\ln\cosh(\pi g/4t) = 1/4(\pi g/2t) - 1/96(\pi g/2t)^3 + \ldots \quad (16)$$

If $g/t < 0.25$, the first term of each equation gives sufficient accuracy. Thus:

For $\kappa' = \kappa_g'$, thick electrodes,

$$2\delta/g = (1/\pi)(g/2t) \quad (17)$$

For $\kappa' = \kappa_g'$, thin electrodes,

$$2\delta/g = 1/8(\pi g/2t) \quad (18)$$

For $\kappa' \gg \kappa_g'$, thick or thin electrodes,

$$2\delta/g = 1/4(\pi g/2t) \quad (19)$$

It can be shown that Eq 19 represents an upper limit for the guard-gap correction, 2δ.

A2.4 For $g/t < 0.25$, a sufficiently accurate expression for showing the dependence of δ on κ'/κ_g' is:

For thin electrodes,

$$2\delta/g = [\kappa'/(\kappa' + \kappa'_g)](\pi/4)(g/2t) \quad (20)$$

For thick electrodes,

$$2\delta/g = [\kappa'/(\kappa' + 0.23\kappa'_g)][\kappa'/(\kappa' + \kappa'_g)](\pi/4)(g/2t) \quad (21)$$

A2.5 The ratio of the guard-gap correction, 2δ, to the guard gap, g, calculated from Eqs 14, 15, and 16 is given in Table A1.

A2.6 The difference between thick and thin electrodes is less important (in effect on $2\delta/g$) than the ratio κ'/κ_g'. The relation when $\kappa' = \kappa_g'$ is given in Eq 22. Values of f for various ratios of electrode thickness, a, to gap width, g, are given in Table A2.

$$2\delta/g = (2\delta/g_{\text{thick}}) + f(2\delta/g_{\text{thin}}) - (2\delta/g_{\text{thick}}) \quad (22)$$

A3. Factors Affecting Dielectric Constant and Loss Characteristics

A3.1 Frequency

A3.1.1 Insulating materials are used over the entire electromagnetic spectrum, from direct current to radar frequencies of at least 3×10^{10} Hz. There are only a very few materials, such as polystyrene, polyethylene, and fused silica, whose dielectric constant and loss index are even approximately constant over this frequency range. It is necessary either to measure dielectric constant and loss index at the frequency at which the material will be used or to measure them at several frequencies suitably placed, if the material is to be used over a frequency range.

A3.1.2 The changes in dielectric constant and loss index with frequency are produced by the dielectric polarizations which exist in the material. The two most important are dipole polarization due to polar molecules and interfacial polarization caused by inhomogeneities in the material. Dielectric constant and loss index vary with frequency in the manner shown in Fig. A1 (**14**). Starting at the highest frequency where the dielectric constant is determined by an atomic or electronic polarization, each succeeding polarization, dipole or interfacial, adds its contribution to dielectric constant with the result that the dielectric constant has its maximum value at zero frequency. Each polarization furnishes a maximum of both loss index and dissipation factor. The frequency at which loss index is a maximum is called the relaxation frequency for that polarization. It is also the frequency at which the dielectric constant is increasing at the greatest rate and at which half its change for that polarization has occurred. A knowledge of the effects of these polarizations will frequently help to determine the frequencies at which measurements should be made.

A3.1.3 Any d-c conductance in the dielectric caused by free ions or electrons, while having no direct effect on dielectric constant, will produce a dissipation factor that varies inversely with frequency, and that becomes infinite at zero frequency (dotted line in Fig. A1).

A3.2 Temperature

A3.2.1 The major electrical effect of temperature on an insulating material is to increase the relaxation frequencies of its polarizations. They increase exponentially with temperature at rates such that a tenfold increase in relaxation frequency may be produced by temperature increments ranging from 6 to 50 C. The temperature coefficient of dielectric constant at the lower frequencies would always be positive except for the fact that the temperature coefficients of dielectric constant resulting from many atomic and electronic polarizations are negative. The temperature coefficient will then be negative at high frequencies, become zero at some intermediate frequency and positive as the relaxation frequency of the dipole or interfacial polarization is approached.

A3.2.2 The temperature coefficient of loss index and dissipation factor may be either positive or negative, depending on the relation of the measuring to the relaxation frequency. It will be positive for frequencies higher than the relaxation frequency and negative for lower frequencies. Since the relaxation frequency of interfacial polarization is usually below 1 Hz, the corresponding temperature coefficient of loss index and dissipation factor will be positive at all usual measuring frequencies. Since the d-c conductance of a dielectric usually increases exponentially with decrease of the reciprocal of absolute temperature, the values of loss index and dissipation factor arising therefrom will increase in a similar manner and will produce a larger positive temperature coefficient.

A3.3 Voltage

A3.3.1 All dielectric polarizations except interfacial are nearly independent of the existing potential gradient until such a value is reached that ionization occurs in voids in the material or on its surface, or that breakdown occurs. In interfacial polarization the number of free ions may increase with voltage and change both the magnitude of the polarization and its relaxation frequency. The d-c conductance is similarly affected.

A3.4 Humidity

A3.4.1 The major electrical effect of humidity on an insulating material is to increase greatly the magnitude of its interfacial polarization, thus increasing both its dielectric constant and loss index and also its d-c conductance. These effects of humidity are caused by absorption of water into the volume of the material and by the formation of an ionized water film on its surface. The latter forms in a matter of minutes, while the former may require days and sometimes months to attain equilibrium, particularly for thick and relatively impervious materials (**12**).

A3.5 Water Immersion

A3.5.1 The effect of water immersion on an insulating material approximates that of exposure to 100 percent relative humidity. Water is absorbed into the volume of the material, usually at a greater rate than occurs under a relative humidity of 100 percent. However, the total amount of water absorbed when equilibrium is finally established is essentially the same under the two conditions. If there are water-soluble substances in the material, they will leach out much faster under water immersion that under 100 percent relative humidity without condensation. If the water used for immersion is not pure, its impurities may be carried into the material. When the material is removed from the water for measurement, the water film formed on its surface will be thicker and more conducting than that produced by a 100 percent relative humidity without condensation, and will require some time to attain equilibrium.

A3.6 Weathering

A3.6.1 Weathering, being a natural phenomenon, includes the effects of varying temperature and humidity, of falling rain, severe winds, impurities in the atmosphere, and the ultraviolet light and heat of the sun. Under such conditions the surface of an insulating material may be permanently changed, physically by roughening and cracking, and chemically by the loss of the more soluble components and by the reactions of the salts, acids, and other impurities deposited on the surface. Any water film formed on the surface will be thicker and more conducting, and water will penetrate more easily into the volume of the material.

A3.7 Deterioration

A3.7.1 Under operating conditions of voltage and temperature, an insulating material may deteriorate in electric strength because of the absorption of moisture, physical changes of its surface, chemical changes in its composition, and the effects of ionization both on its surface and on the surfaces of internal voids. In general, both its dielectric constant and its dissipation factor will be increased, and these increases will be greater the lower the measuring frequency. With a proper understanding of the effects outlined in A3.1 to A3.6, the observed changes in any electrical property, particularly dissipation factor, can be made a measure of deterioration and hence of decrease in dielectric strength.

A3.8 Conditioning

A3.8.1 The electrical characteristics of many insulating materials are so dependent on temperature, humidity, and water immersion, as indicated in the paragraphs above, that it is usually necessary to specify the past history of a specimen and its test conditions regarding these factors. Unless measurements are to be made at room temperature (20 to 30 C) and unspecified relative humidity, the specimen should be conditioned in accordance with ASTM Methods D 618, Conditioning Plastics and Electrical Insulating Materials for Testing.[3] The procedure chosen should be that which most nearly matches operating conditions. When data are required covering a wide range of temperature and relative humidity, it will be necessary to use intermediate values and possibly to condition to equilibrium.

A3.8.2 Methods of maintaining specified relative humidities are described in ASTM Recommended Practice E 104, Maintaining Constant Relative Humidity by Means of Aqueous Solutions.[3]

A3.8.3 Specifications for conditioning units are given in ASTM Specifications E 197, for Enclosures and Servicing Units for Tests Above and Below Room Temperature.[3]

A4. Circuit Diagrams of Typical Measuring Circuits

A4.1 The simplified circuits and equations presented in Figs. A2 to A9 are for general information only. The instruction book accompanying a particular piece of equipment should be consulted for the exact diagram, equations, and method of measurement to be used.

TABLE A1 Ratio of Guard-Gap Correction, 2δ, to Guard-Gap g

g/t	$\kappa' \gg \kappa_g'$	$\kappa' = \kappa_g'$	
	Thick or Thin Electrodes	Thin Electrodes	Thick Electrodes
0.01	0.0039	0.0020	0.0016
0.02	0.0079	0.0039	0.0032
0.03	0.0118	0.0059	0.0048
0.04	0.0156	0.0078	0.0064
0.05	0.0196	0.0098	0.0080
0.06	0.0236	0.0118	0.0096
0.08	0.0314	0.0157	0.0127
0.10	0.0392	0.0196	0.0159
0.12	0.0470	0.0236	0.0191
0.15	0.0588	0.0294	0.0239
0.20	0.0782	0.0393	0.0318
0.25	0.0975	0.0489	0.0398
0.3	0.1167	0.0586	0.0474
0.4	0.1546	0.0779	0.0634
0.5	0.1915	0.0969	0.0789
0.6	0.2274	0.1157	0.0942
0.8	0.2954	0.1519	0.1242
1.0	0.3579	0.1865	0.1531
1.2	0.4145	0.2187	0.1809
1.5	0.4885	0.2620	0.2203
2.0	0.5857	0.3183	0.2800

TABLE A2 Values of f for various a/g ratios

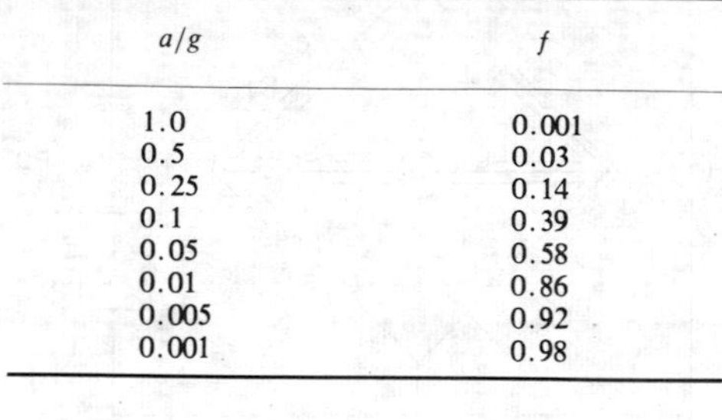

a/g	f
1.0	0.001
0.5	0.03
0.25	0.14
0.1	0.39
0.05	0.58
0.01	0.86
0.005	0.92
0.001	0.98

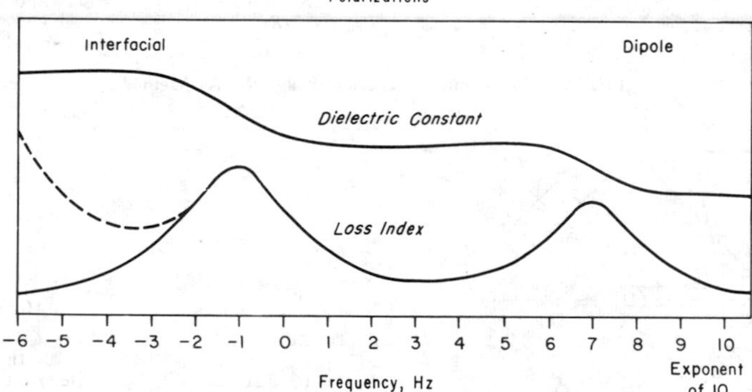

FIG. A1 Typical Polarizations (12).

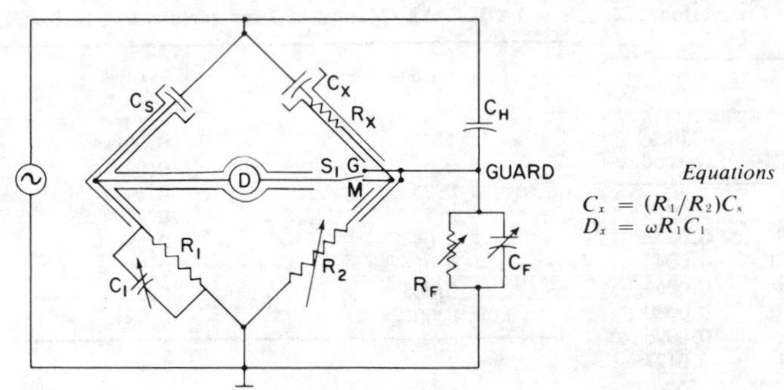

Equations

$$C_x = (R_1/R_2)C_s$$
$$D_x = \omega R_1 C_1$$

Method of Balance

Vary C_1 and R_2 with S_1 in position M to obtain minimum deflection in Detector D. Repeat with S_1 in position G by varying C_F and R_F. Repeat the above until the detector shows no change in balance by switching S_1 to M or G.

NOTE—This type of bridge is especially useful for high-voltage measurements at power frequencies as almost all of the applied voltage appears across the standard capacitor, C_s, and the specimen, C_x. The balancing circuits and detector are very nearly at ground potential.

FIG. A2 High-Voltage Schering Bridge.

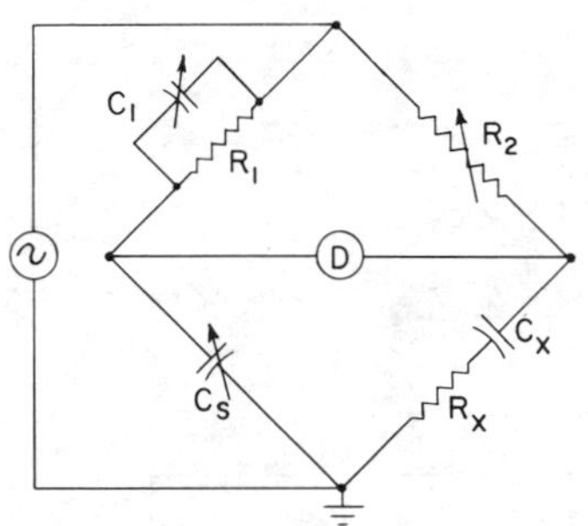

Equations

$$C_x = (R_1/R_2)C_s$$
$$D_x = \omega C_1 R_1$$

Method of Balance

Set ratio of R_1 to R_2 (range) and vary C_s and C_1 to obtain balance.

FIG. A3 Low-Voltage Schering Bridge, Direct Method.

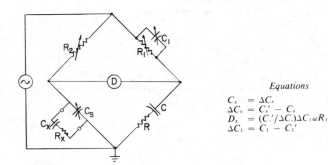

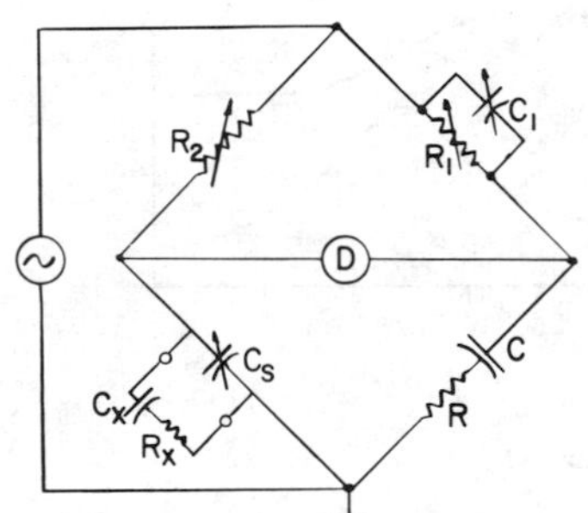

Equations

$$C_x = \Delta C_s$$
$$\Delta C_s = C_s' - C_s$$
$$D_x = (C_s'/\Delta C_s)\Delta C_1 \omega R_1$$
$$\Delta C_1 = C_1 - C_1'$$

Method of Balance

Vary C_1 and C_s, without and with the specimen connected, to obtain balance. Symbols used for the initial balance, with the ungrounded lead to the unknown disconnected, are primed.

FIG. A4 Low-Voltage Schering Bridge, Parallel Substitution Method.

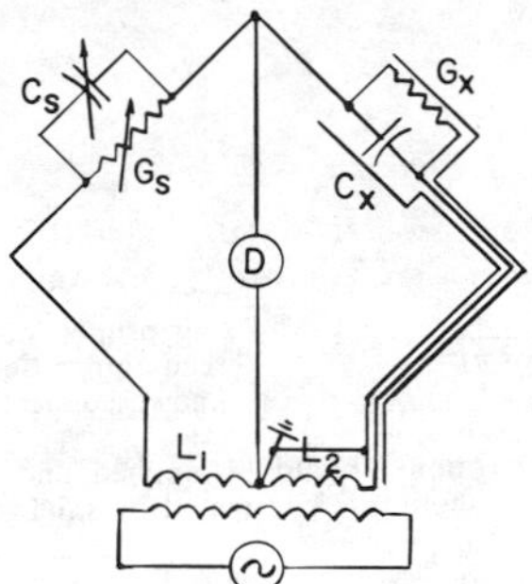

Equations	*Method of Balance*
$C_x = (L_1/L_2)C_s$ $G_x = (L_1/L_2)G_s$ $D_x = (G_s/\omega C_s)$	Set ratio of L_1 to L_2 (range) and adjust C_s and G_s to obtain balance.

FIG. A5 Inductive-Ratio-Arm (Transformer) Circuit.

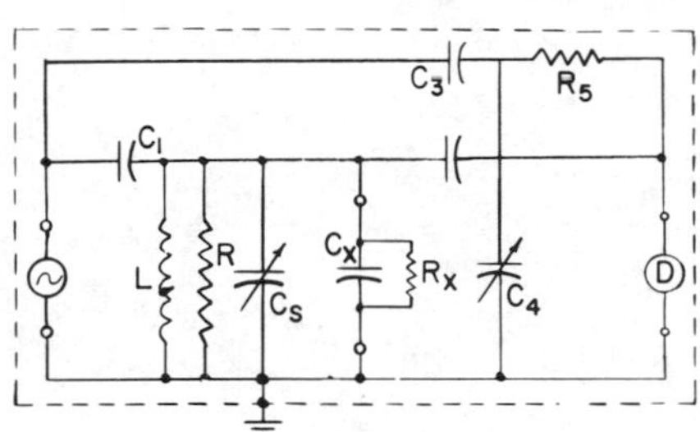

Equations	*Method of Balance*
$C_x = C_s' - C_s = \Delta C_s$ $G_x = (R_5\omega^2 C_1 C_2/C_s)(C_4 - C_4')$ $= \Delta G_x$ $D_x = G_x/\omega C_x = \Delta G_x/\omega\Delta C_s$	Balance without and rebalance with unknown connected, using C_s and C_4. Symbols used for the initial balance are primed.

FIG. A6 Parallel-T Network, Parallel Substitution Method.

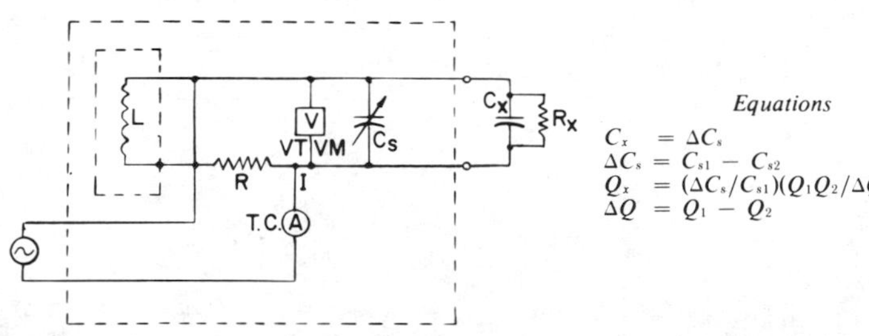

Equations	*Method of Balance*
$C_x = \Delta C_s$ $\Delta C_s = C_{s1} - C_{s2}$ $Q_x = (\Delta C_s/C_{s1})(Q_1 Q_2/\Delta Q)$ $\Delta Q = Q_1 - Q_2$	Adjust to resonance, without and with the specimen, noting I, maximum V and C_s. With a standard I, V on the $VTVM$ can be calibrated in terms of Q, since $Q = V/IR$. Subscripts 1 and 2 denote first and second balance respectively.

FIG. A7 Resonant-Rise (Q-Meter) Method.

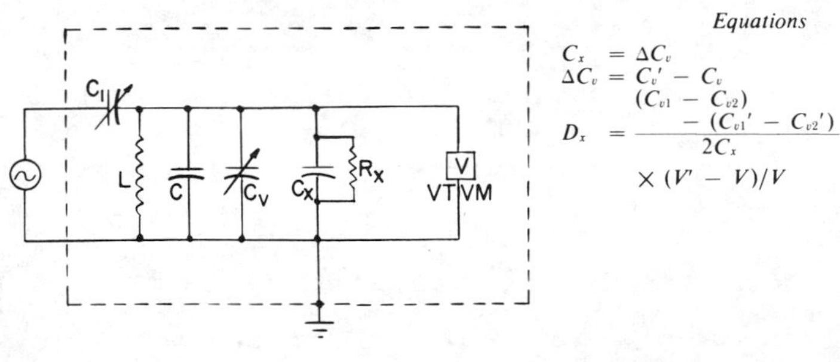

Equations	*Method of Balance*
$C_x = \Delta C_v$ $\Delta C_v = C_v' - C_v$ $D_x = \dfrac{(C_{v1} - C_{v2}) - (C_{v1}' - C_{v2}')}{2C_x} \times (V' - V)/V$	Adjust C_1 so maximum resonance V' is just under full scale. Note exact V' and capacitance C_v'. Adjust C_v so that V_a (=0.707 V') is first on one side of the resonance V' and then on the other. Record C_{v1} and C_{v2}. Repeat last process with C_x connected, noting C_{v1} and C_{v2x}, C_v and V.

FIG. A8 Susceptance-Variation Method.

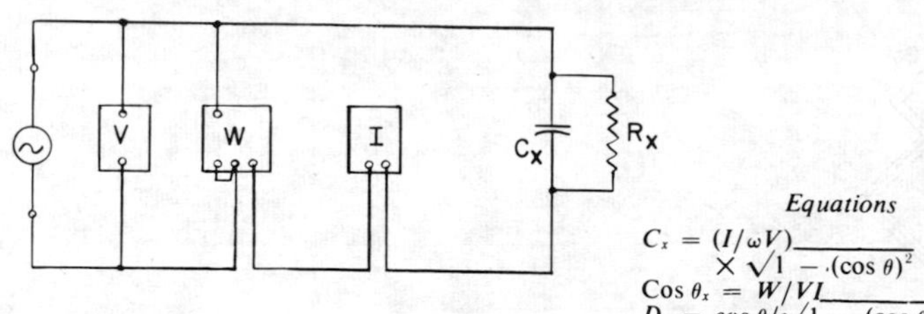

Equations

$$C_x = (I/\omega V) \times \sqrt{1 - (\cos\theta)^2}$$

$$\cos\theta_x = W/VI$$

$$D_x = \cos\theta/\sqrt{1 - (\cos\theta)^2}$$

Method

Using proper scale multiplier, read indications with unknown connected.

NOTE—This method is for use at power frequencies. Instrument corrections should be applied and an unusually sensitive wattmeter is required due to small losses. Errors from stray fields should be eliminated by shielding. Accuracy depends on combined instrument errors and is best at full scale.

FIG. A9 Voltmeter-Wattmeter-Ammeter Method.

Designation: D 229 – 69

Standard Method of
TESTING RIGID SHEET AND PLATE MATERIALS USED FOR ELECTRICAL INSULATION[1]

This Standard is issued under the fixed designation D 229; the number immediately following the designation indicates the year of original adoption or, in the case of revision, the year of last revision. A number in parentheses indicates the year of last reapproval.

1. Scope

1.1 These methods cover procedures for testing rigid electrical insulation normally manufactured in flat sheet or plate form. It is intended that these methods be used except where modified by individual methods or specifications for a particular material.

NOTE 1—The values stated in U.S. customary units are to be regarded as the standard. The metric equivalents of U.S. customary units given in the standard may be approximate.

NOTE 2—For tests applying to vulcanized fiber reference should be made to ASTM Methods D 619, Testing Vulcanized Fiber Used for Electrical Insulation.[2]

1.2 The methods appear on the following sections:

Test	Sections	ASTM Method[a]
Arc resistance	38	D 495
Ash	47 and 48	
Bonding strength	40 to 45	. . .
Compressive strength	20	D 695
Conditioning	3	D 618
Dielectric constant	27 to 32	D 150
Dissipation factor	27 to 32	D 669
Dielectric strength	23 to 26	D 149
Dust-and-fog tracking resistance	39	D 2132
Expansion (linear thermal)	61	D 696
Flammability and flame resistance	49 to 60	. . .
Flexural properties	9 to 19	D 790
Hardness (Rockwell)	46	D 785
Impact	21	D 256
Insulation resistance and resistivity	33 to 37	D 257
Tensile properties	5 to 8	D 638
Thickness	4	D 374
Warp or twist	62 to 65	. . .
Water absorption	22	D 570

[a] These designations refer to the following methods of the American Society for Testing and Materials:

D 149, Tests for Dielectric Breakdown Voltage and Dielectric Strength of Electrical Insulating Materials at Commercial Power Frequencies[3]

D 150, Tests for A-C Loss Characteristics and Dielectric Constant (Permittivity) of Solid Electrical Insulating Materials[3]

D 256, Tests for Impact Resistance of Plastics and Electrical Insulating Materials[3]

D 257, Tests for D-C Resistance or Conductance of Insulating Materials[3]

D 374, Tests for Thickness of Solid Electrical Insulation[3]

D 495, Test for High-Voltage, Low-Current Dry Arc Resistance of Solid Electrical Insulation[3]

D 570, Test for Water Absorption of Plastics[3]

D 618, Methods of Conditioning Plastics and Electrical Insulating Materials for Testing[3]

D 638, Test for Tensile Properties of Plastics[3]

D 669, Test for Dissipation Factor and Dielectric Constant Parallel with Laminations of Laminated Sheet and Plate Insulating Materials[2]

D 695, Test for Compressive Properties of Rigid Plastics[3]

D 696, Test for Coefficient of Linear Thermal Expansion of Plastics[3]

D 785, Test for Rockwell Hardness of Plastics and Electrical Insulating Materials[3]

D 790, Test for Flexural Properties of Plastics[3]

D 2132, Test for Dust-and-Fog Tracking and Erosion Resistance of Electrical Insulating Materials[2]

2. Definitions

2.1 Rigid electrical insulating materials are defined in these methods in accordance with ASTM Nomenclature D 883, Relating to Plastics.[2] They are generally used as terminal boards, spacers, voltage barriers, and circuit boards, and include materials such as laminated thermosetting plastics, hard rubber, and asbestos composition board.

2.2 In referring to the cutting of the specimens and the application of the specimens and the application of the load, the following definitions apply:

2.2.1 *flatwise*—load applied to the flat side of the original sheet or plate.

2.2.2 *edgewise*—load applied to the edge of

[1] These methods are under the jurisdiction of ASTM Committee D-9 on Electrical Insulating Materials. A list of members may be found in the ASTM Yearbook.

Current edition effective March 21, 1969. Originally issued 1925. Replaces D 229 – 68 T.

[2] *Annual Book of ASTM Standards*, Part 29.

[3] *Annual Book of ASTM Standards*, Part 27.

the original sheet or plate.

2.2.3 *lengthwise* (*LW*)—in the direction of the sheet known to be stronger in flexure.

2.2.4 *crosswise* (*CW*)—in the direction of the sheet known to be weaker in flexure and shall be 90 deg to the lengthwise direction.

3. Conditioning

3.1 The properties of the materials described in these methods are affected by the temperature and moisture exposure of the materials to a greater or lesser extent, depending on the particular material and the specific property. Controlled temperature and moisture exposure is undertaken to: (*1*) obtain satisfactory test precision, or (*2*) study the behavior of the material as influenced by specific temperature and humidity conditions.

3.2 Unless otherwise specified in these methods or by a specific ASTM material specification, or unless material behavior at a specific exposure is desired, test specimens shall be conditioned in accordance with Proceure A of ASTM Methods D 618, for Conditioning Plastics and Electrical Insulating Materials for Testing,[3] and tested in the Standard Laboratory Atmosphere (23 ± 1.1 C, 50 ± 2 percent relative humidity).

THICKNESS

4. Apparatus and Procedure

4.1 Measure thickness in accordance with ASTM Methods D 374, Test for Thickness of Solid Electrical Insulation.[3]

4.2 On test specimens, the use of a machinist's micrometer as specified in Method B is satisfactory for the determination of thickness for all of the methods that follow. Where it is convenient, the deadweight dial micrometer, Method C, may be used.

4.3 On large sheets, use Method B. Fit the micrometer with a yoke of sufficient size and rigidity to permit accurate measurements in the center of the sheet.

NOTE 3—Results of comparative tests in several factories, measuring 36-in. (91.4-cm) square sheets by a variety of such devices, indicate that the trade is able to measure sheets $^1/_{32}$ and $^1/_8$ in. (1 and 3 mm) in thickness to an accuracy of 0.0015 in. (0.0381 mm). (In the tests, σ, or root mean square deviations, of 0.0005 in. (0.0127 mm) were obtained.)

TENSILE PROPERTIES

5. Test Specimen

5.1 The test specimens shall be machined from sample material to conform to the dimensions of sheet and plate materials in Fig. 1.

5.2 Four LW and four CW specimens shall be tested.

6. Rate of Loading

6.1 The materials covered by these methods generally exhibit high elastic modulus. Any crosshead speed may be used provided that the load and strain indicators are capable of accurate measurement at the speed used, except that Speed A (0.05 in./min) shall be used in matters of dispute.

7. Procedure

7.1 Measure the tensile strength and elastic modulus in accordance with ASTM Methods D 638, Test for Tensile Properties of Plastics,[3] except as modified in the following paragraphs.

7.2 Measure the width and thickness of the specimen to the nearest 0.001 in. (0.254 mm) at several points along the length of the flat section. Record the minimum values of cross-sectional area so determined.

7.3 Place the specimen in the grips of the testing machine, taking care to align the long axis of the specimen and the grips with an imaginary line joining the points of attachment of the grips to the machine. Allow 0.25 in. (6.3 mm) between the ends of the gripping surfaces and the shoulders of the fillet of the flat test specimen; thus, the ends of the gripping surfaces should be 4.5 in. (114 mm) apart, as shown in Fig. 1, at the start of the test. Tighten the grips evenly and firmly to the degree necessary to prevent slippage of the specimen during the test but not to the point where the specimen would be crushed.

7.4 *Tensile Strength*—Set the rate of loading. Load the specimen at the indicated rate until the specimen ruptures. Record the maximum load (usually the load at rupture).

7.5 *Elastic Modulus*—When elastic modulus is desired, use a load-extension recorder with appropriate extension transmitter and proceed as in 7.3. Attach the extension transmitter, and proceed as in 7.4.

8. Report

8.1 The report shall include the following:

8.1.1 Complete identification of the material tested,

8.1.2 Type of test specimen (I or II),

8.1.3 Conditioning if other than specified,

8.1.4 Speed of testing if other than speed A,

8.1.5 Calculated tensile, strength, average, maximum, and minimum, for LW and CW specimens, respectively,

8.1.6 Calculated elastic modulus when applicable, average, maximum, and minimum, for LW and CW specimens, respectively, and

8.1.7 Any other tensile property calculated from the measurements obtained.

Flexural Properties

9. Test Specimens

9. Four LW and four CW specimens machined from sample material in accordance with Section 5.3 of ASTM Method D 790, Test for Flexural Properties of Plastics[3] shall be tested.

NOTE 4—Conventional flexture tests in a flatwise direction are not recommended for materials thinner than $^1/_{32}$ in. (1 mm) nor in the edgewise direction for materials thinner than $^1/_4$ in. (6 mm).

10. Rate of Loading

10.1 The materials covered by these methods generally rupture during flexural testing at small deflections. Therefore, Procedure A (strain rate of 0.01/min) is specified whenever it is desired to obtain the modulus of elasticity. Any crosshead speed that produces failure in no less than 1 min may be used when flexural strength only is desired, provided that the load indicator is capable of accurately indicating the load at the speed used, and except that in all matters of dispute a crosshead speed that produces the strain rate specified in Procedure A shall be considered to be the referee speed.

11. Procedure

11.1 Measure the flexural strength and modulus of elasticity in accordance with Procedure A of Method D 790, except that where modulus of elasticity is desired use a load-deflection recorder with appropriate deflection transmitter.

12. Report

12.1 The report shall include the following:

12.1.1 Complete identification of the material tested,

12.1.2 Conditioning if other than specified,

12.1.3 Speed of testing if other than Procedure A speed,

12.1.4 Calculated flexural strength, average, maximum, and minimum, for LW and CW specimens, respectively,

12.1.5 Calculated tangent modulus of elasticity when applicable, average, maximum, and minimum, for LW and CW specimens, respectively, and

12.1.6 Any other flexural property calculated from the measurements obtained.

Flexural Properties at Elevated Temperature

13. Scope

13.1 This method covers the determination of flexural properties at elevated temperature, and as a function of time of exposure to elevated temperature.

14. Significance

14.1 This method provides useful engineering information for evaluating the mechanical behavior of rigid electrical insulation at elevated temperature. When the proper exposure and test temperatures are chosen, depending on the material and end-use operating temperature, the method may be used as one means of indicating relative thermal degradation of rigid insulating materials.

15. Apparatus

15.1 *Testing Machine*—A universal testing machine and accessory equipment in accordance with Method D 790. Apparatus that is exposed to elevated temperature during the test shall be adjusted to function normally at the elevated temperature and, where necessary, accuracy shall be verified by calibration at the test temperature.

15.2 *Test Enclosure*—A test enclosure conforming to the Type I, Grade B, temperature requirements of ASTM Specifications E 197, for Enclosures and Servicing Units for Tests Above and Below Room Temperature.[2] The test enclosure may rest on the testing machine table, in which case the top shall have a hole of sufficient size so that adequate clearance is provided for the loading nose, or the test enclosure may rest on a dolly and contain a cradle which is supported by the loading members of the machine.

15.3 *Heat Aging Oven*—A heat aging oven for conditioning specimens at the test temperature for periods of more than 1 h. The oven shall conform to the requirements for Type I, Grade A, units of Specifications E 197, except with respect to the time constant.

15.4 *Specimen Transfer Device*—A means of transferring the test specimens from the heat-aging oven to the test enclosure when testing specimens exposed to elevated temperature for periods of more than 1 h. The specimens may be transferred without cooling either in a small mobile transfer oven or wrapped in previously heated asbestos cloth.

15.5 *Thermocouple*—Thermocouple made with No. 30 or 28 B & S gage thermocouple calibration wires to determine the temperature of the specimen. Any suitable indicating or recording device shall be used that provides an over-all (junction and instrument) accuracy of ±2 C.

16. Test Specimen

16.1 The specimen shall be tested flatwise and lengthwise and shall be machined from sample material in accordance with Section 9.

16.2 Where it is desired to evaluate relative thermal degradation, specimens shall be 1/8 in. (3 mm) in nominal thickness.

16.3 At least one specimen of each thickness for each sample material shall be fitted with a hole drilled into an edge that rests outside the support to a depth of at least 1/2 in. (13 mm). The thermocouple junction shall be inserted in this hole and cemented. This specimen shall be used to determine the temperature of the specimen on the support and the time required to reach the specified temperature for specimens that are tested after 15-min exposure or less.

16.4 Five specimens shall be tested at each temperature.

17. Conditioning

17.1 No special conditioning is required for specimens that are to be tested after more than 1-h exposure at elevated temperature.

18. Procedure

18.1 Adjust the rate of loading in accordance with Section 10 and test the specimen in accordance with Section 11.

18.2 Age in the flexural test enclosure the specimens that are to be tested 1 h or less after exposure to elevated temperature.

18.3 Exposures at elevated temperature for 15 min or less shall not include the time (previously determined from the specimen with the thermocouple) that is required for the specimen to reach the specified temperature. Rather, exposures for intervals of 15 min or less shall begin when the specimen reaches the specified temperature and end when the specified exposure period has expired.

18.4 Age in the heat-aging oven the specimens that are exposed to elevated temperature for more than 1 h. Do not allow the specimens to cool when removed from the heat-aging oven but rather transfer them in the mobile-transfer oven or wrap them in preiously heated asbestos cloth. Place them in the flexural test chamber which has been previously heated to the specified temperature.

18.5 The flexural test enclosure and accessory equipment inside shall be considered at equilibrium when a dummy specimen fitted with an internal thermocouple, and placed on the supports, has reached the specified temperature, as determined by the thermocouple measurement. Place test specimens in the flexural test enclosure only after equilibrium has been established.

19. Report

19.1 The report shall include the information required in Section 12 that is applicable, and shall in addition include the following:

19.1.1 Temperature at which the specimens were exposed and tested,

19.1.2 Time of exposure, and

19.1.3 Where sufficient measurements are made, the results should be plotted with flexural strength as ordinate and time at elevated temperature as abscissa for each temperature chosen.

COMPRESSIVE STRENGTH

20. Procedure

20.1 Determine the compressive strength in accordance with ASTM Method D 695, Test for Compressive Properties of Rigid Plastics,[3] except test four specimens.

RESISTANCE TO IMPACT

21. Procedure

21.1 Determine the resistance to impact in accordance with ASTM Methods D 256, Test for Impact Resistance of Plastics and Electrical Insulating Materials,[2] using Method A or C, whichever is applicable, except test four specimens conditioned in accordance with Section 3.2 of these methods.

WATER ABSORPTION

22. Procedure

22.1 Determine the water absorption in accordance with ASTM Method D 570, Test for Water Absorption of Plastics,[3] except test all sample material for water-soluble matter unless it has been previously demonstrated by test that there is negligible water-soluble matter in the sample. Test four specimens.

DIELECTRIC STRENGTH

23. Surrounding Medium

23.1 All tests shall be made in an electrical insulating oil medium that is equivalent to transformer oil. The temperature of the oil shall be essentially the same as that of the test specimen when voltage is first applied.

NOTE 5—A liquid medium is specified to obtain breakdown of a reasonable size test specimen rather than flashover in the medium. Except for the tapered-pin method the specified liquid medium effectively prevents flashover.

Transverse tests performed in an air medium will generally result in lower breakdown values than transverse tests performed in the liquid medium. This is particularly true when porous materials are tested. Tests performed in the liquid medium on specimens that have been thermally aged may produce misleading conclusions when change in dielectric strength is utilized as a criterion of thermal degradation.

Transverse tests in air for porous materials and thermally aged materials are encouraged. Various schemes may be utilized for potting or gasketing the electrodes to prevent flashover. Apparatus is being evaluated for use in a standard method for transverse tests in air.

23.2 *Oil Medium Breakdown*—If in a parallel-tapered-pin test the specimen should appar to fail below a specified minimum failure voltage and there is doubt as to whether the breakdown occurred in the oil or in the specimen, the oil itself shall be tested as specified in ASTM Method D 877, Test for Dielectric Breakdown Voltage of Insulating Liquids Using Disk Electrodes.[2] If the breakdown voltage of the oil, tested at room temperature, is found to be less than 22 kV, the oil shall be changed or reconditioned so that its breakdown voltage is no less than 22 kV.

NOTE 6—Breakdown of the oil above the specified value is not necessarily proof that actual specimen breakdown occurred during a parallel-tapered-pin test, since the specimen surface structure and its dielectric constant will influence the breakdown voltage of a given oil between the tapered pins with specimen in place.

24. Electrodes and Test Specimens

24.1 *Transverse Test*—Standard 51-mm (2-in.) diameter electrodes shall be used for voltage stress applied perpendicular to the flat side of the specimen. The test specimen shall be of such size that flashover in the oil medium does not occur before specimen breakdown. In general, a 4 in. (102 mm) square will be satisfactory.

24.2 *Parallel Test, Point-Plane Method*—The test specimens shall be $^1/_2$ in. (13 mm) in width by 1 in. (25 mm) in length by the thickness of the material. Minimum thickness of the material shall be $^1/_8$ in. (3 mm). Using a twist drill with a point angle of 60 to 90 deg, drill a hole in the approximate center of the 1-in. (25-mm) length in a direction parallel with the flat sides, to a depth of $^7/_{16}$ in. (11 mm), leaving a thickness of $^1/_{16}$ in. (1.6 mm) to be tested. Insert a snug-fitting metal pin electrode, with the end ground to conform with the shape of the drill used in the hole. Place the specimen on a flat metal plate that is at least 1 $^1/_2$ in. (38 mm) in diameter. This plate serves as the lower electrode. Thus, in effect, the material is tested parallel with the flat sides in a point-plane dielectric gap. The diameter of the hole shall be as shown in the following table:

Nominal Thickness of Sheets	Nominal Hole Diameter for Pin Electrode
$^1/_8$ to $^1/_4$ in. (3 to 6 mm)	$^1/_{16}$ in. (1.6 mm)
$^1/_4$ in. (6 mm)	$^1/_8$ in. (3 mm)

24.3 *Parallel Test, Tapered-Pin Method:*

24.3.1 *Significance*—Sheet and plate insulation, particularly laminated sheets, are frequently used in service in a manner such that the full thickness of the insulation is exposed to a voltage stress parallel to the flat sides between pin-type inserts. This method (employing tapered-pin electrodes) is recommended, rather than the method in 24.2, when it is desired to simulate the service condition described and when the need for obtaining quantitative dielectric breakdown data is secondary to acceptance and quality control needs.

24.3.2 *Nature of Test*—The tapered-pin electrodes extend beyond the test specimen on both flat sides. Therefore oil medium flashover or oil specimen interface failure may obscure specimen volume dielectric breakdown. This method is suited, consequently, for use primarily as a proof-type test, that is, to determine only that a material will withstand without failure a specified minimum electric stress applied in a prescribed manner under specified conditions. In some limited cases, however, (for example, specimens conditioned in water) it may be possible to employ the tapered-pin method to obtain quantitive specimen dielectric breakdown data. When numerous tests are made it may prove difficult to maintain the oil medium in such a condition as to obviate flashover (with specimen in place between pins spaced 1 in. (25 mm) apart) at voltage magnitudes above 60 kV. The practical limit, therefore, recommended for this test method is 60 kV.

24.3.3 *Test Specimens and Electrodes*—The test specimen shall be 2 by 3 in. (50 by 75 mm) by the thickness of the sheet. The electrodes shall be USA Standard tapered pins (such as Morse, Brown & Sharpe, or Pratt & Whitney) having a taper of $^1/_4$ in./ft (2 cm/m). For specimen thicknesses up to and including $^1/_2$ in. (13 mm), No. 3 USA Standard tapered pins[4] 3 in. (76 mm) long and having a diameter of $^7/_{32}$ in. (5.6 mm) at the large end shall be used. For specimen thicknesses over $^1/_2$ in. (13 mm) up to and including 2 in. (51 mm), No. 4 USA Standard Pins[4] 4 in. (102 mm) long having a diameter at the large end of $^1/_4$ in. (6 mm), shall be used. Drill two $^3/_{16}$-in. (5-mm) diameter holes, centrally located, 1 in. (25 mm) apart, center to center, and perpendicular to the faces of the specimen. Ream the holes to a sufficient depth to allow the pins to extend approximately 1 in. (25 mm) from the small ends of the holes. The electrodes shall be inserted from opposite sides of the specimen, after the conditioning period. Metal spheres of $^1/_2$-in. (13-mm) diameter placed on the extremities of the tapered pins may sometimes decrease the tendency to flashover in the oil.

25. Conditioning

25.1 Specimens shall be conditioned in accordance with Section 3 except that for the proof-type test no special conditioning is required.

26. Procedure

26.1 Determine the dielectric strength, dielectric breakdown voltage, and dielectric proof-type test in accordance with Methods D 149, except as follows: Make the tests perpendicular to or parallel with the flat sides, or both, depending upon whether the stress on the material when in use is to be perpendicular to or parallel with the flat sides, or both.

26.2 Make the tests by either the short-time method, the step-by-step method, or the slow-rate-of-rise method as follows:

26.2.1 *Short-Time Method*—Increase the voltage at the rate of 0.5 kV/s.

26.2.2 *Step-by-Step Method*—Apply the voltage at each step for 1 min and increase it in the following increments:

Breakdown Voltage by Short-Time Method, kV	Increment of Increase of Test Voltage, kV
25 or less	1.0
Over 25 to 50, incl	2.0
Over 50 to 100, incl	5.0
Over 100	10.0

26.2.3 *Slow-Rate-of-Rise Method*—Increase the voltage as follows:

Breakdown Voltage by Short-Time Method, kV	Rate of Test Voltage Rise, V/s
25 or less	17
Over 25 to 50, incl	33
Over 50 to 100, incl	83
Over 100	167

[4] For information on tapered pins, see Kent's Mechanical Engineers' Handbook, 12th edition, Design and Production Volume, Section 15, p. 14.

26.2.4 *Proof-Type Test*—Make the tests by either the step-by-step or the slow-rate-of-rise method as follows:

26.2.4.1 *Step-by-Step Method*—Starting at the prescribed percentage of the minimum failure voltage as specified in the appropriate material specification, increase the test voltage in 1-min steps. Use test voltage increments of 1.0 kV for starting voltage of 12.5 kV or less, 2.0 kV for starting voltages over 12.5 to 25 kV, inclusive, and 5.0 kV for starting voltage over 25 kV. Hold the test voltage for 1 min at the specified minimum failure voltage.

26.2.4.2 *Slow-Rate-of-Rise Method*—Starting at the prescribed percentage of the minimum failure voltages specified in the appropriate material specification, increase the test voltage at a uniform rate as indicated until the specified minimum failure voltage is reached. Calculate the slow rate-of-rise, in volts per second, as follows:

$$\text{Slow rate-of-rise, V/s} = (V_f - V_s)/(n \times 60)$$

where:

V_f = specified minimum failure voltage,
V_s = starting voltage, and
n = total number of 1-min steps that would be obtained using the step-by-step method of 26.2.4.1.

Dielectric Constant and Dissipation Factor

27. Apparatus

27.1 *Specimen Holder*—A well-designed specimen holder to support and shield the specimen and provide for connection of the electrodes to the terminals of the measuring apparatus is recommended. Two-terminal and three-terminal holders are described in Methods D 150. A specimen holder for use at elevated temperatures is described in ASTM Methods D 1674, Testing Polymerizable Embedding Compounds Used for Electrical Insulation.[2]

27.2 *Measuring Apparatus*—Suitable bridge or resonant-circuit apparatus conforming to the requirements of Methods D 150 shall be used. The choice of equipment will depend upon the frequency at which measurements are to be made, and in certain cases upon the applied voltage gradients when such are specified.

28. Electrodes (see Note 7)

28.1 Electrodes shall be applied to the specimens. Most of the electrode materials described in Methods D 150 are suitable except fired-on silver. Metal foil and conducting silver paint are generally recommended, but only the latter should be used for measurements at elevated temperatures. For laminated thermosetting materials to be tested at 1 MHz, either metal foil attached by a thin film of petrolatum or conducting silver paint shall be used, and the electrodes shall completely cover both sides of the specimen. For testing ultra-thin, that is, up to a thickness of about 0.03 in. (0.075 cm), glass-base laminated thermosetting materials, only conducting silver paint electrodes shall be used. When the same specimen is used for Condition A and for tests after immersion in water, metal foil electrodes must always be removed and the petrolatum cleaned off with a suitable solvent before immersion. Silver paint electrodes, on the other hand, are not removed prior to immersion of specimens in water.

NOTE 7—It has been found that satisfactory dielectric constant and dissipation factor measurements can be made on many sheet materials, particularly at radio frequencies, by the non-contacting electrode techniques (air-gap, liquid displacement, and two-fluid displacement) described in Methods D 150 when appropriate test cells and liquids are available. Such methods are permissible when agreed upon by the parties concerned. No electrodes of any kind are then applied directly to the test specimens.

29. Test Conditions

29.1 Unless otherwise specified, two specimens of each material shall be tested.

29.2 The thickness of the specimens preferably shall be the manufactured thickness of the sheet, but it may be necessary and is permissible to machine very thick specimens down to a usable thickness. The thickness shal be determined in accordance with Section 4, *except in the case of ultra-thin thermosetting glass-base laminates*, the mean effective thicknesses of which shall be calculated from the mass in grams and density in grams per cubic centimeter of accurately die-cut disks 5.08 cm (2.00 in.) in diameter, as follows:

$$\text{thickness} = (0.04933 \times \text{mass/density})\ \text{cm}$$
$$= (0.01942 \times \text{mass/density})\ \text{in.}$$

(The densities of the 5.08-cm disks shall be determined in accordance with ASTM Method D 792, Tests for Specific Gravity and Density of Plastics by Displacement.[3])

29.3 Generally, specimens shall be of such size as is practicable with the apparatus used. For measurements at frequencies up to about 1 MHz it is recommended that the specimens be of such size that the measured capacitances will be in the approximate range of 50 to 150 picofarads (pF). At higher frequencies, smaller specimens giving capacitances of 10 to 30 pF, approximately, will be required.

29.3.1 For laminated thermosetting materials, except as specified in 29.3.2 below, standard rectangular specimens for measurements at 1 MHz shall be sawed from sheets and shall have the following dimensions:

Thickness of Sheet	Size of Specimen
Up to $^3/_{64}$ in. (0.12 cm), incl	2 by 2 in. (5 by 5 cm)
Over $^3/_{64}$ in. (0.12 cm) to $^3/_{32}$ in. (0.24 cm)	3 by 3 in. (7.5 by 7.5 cm)
Over $^3/_{32}$ in. (0.24 cm) to $^1/_4$ in. (0.64 cm)	4 by 4 in. (10 by 10 cm)
Over $^1/_4$ in. (0.64 cm) to 2 in. (5 cm)	4 by 8 in. (10 by 20 cm)

29.3.2 *For ultra-thin thermosetting laminates*, particularly of the glass-base type, the specimens for measurements at 1 MHz shall be small disks accurately die-cut from larger 2-in. (5-cm) disks that have been coated previously on both sides with conducting silver paint first air-dried at room temperature, then heated in a circulating-air oven at 50 C for about 30 min, and finally cooled in a desiccator. The recommended specimen diameters are as follows:

Thickness of Sheet	Diameter of Specimen
Up to 0.003 in. (0.007 cm), approximately	0.50 in. (1.27 cm)
Over 0.003 in. (0.007 cm) to 0.010 in. (0.025 cm)	0.75 in. (1.90 cm)
Over 0.010 in. (0.025 cm) to 0.030 in. (0.075 cm)	1.00 in. (2.54 cm)

29.4 Unless otherwise specified, specimens shall be cleaned in accordance with ASTM Recommended Practice D 1371 for Cleaning Plastic Specimens for Insulation Resistance Testing[2] prior to application of electrodes and conditioning.

30. Conditioning

30.1 The dielectric constant and loss characteristics, especially at the lower frequencies, of the materials covered by these methods are significantly affected by conditioning.

30.2 Unless otherwise specified, specimens shall be conditioned for at least 40 h at 50 percent relative humidity, 23 C, immediately prior to performance of the electrical tests.

30.3 When water immersion conditions are specified, at the end of the conditioning period each specimen shall be removed separately, wiped or blotted with lint-free absorbent paper towels, and tested within approximately 2 or 3 min after removal from the water.

31. Procedure

31.1 Measure the dielectric constant and dissipation factor in accordance with Methods D 150, in the Standard Laboratory Atmosphere of 50 ± 2 percent relative humidity, 23 ± 1 C. Follow instructions given in manuals provided by manufacturers of testing apparatus employed.

31.2 In the case of the small disk specimens of ultra-thin laminates at 1 MHz, support the specimen directly on the high-measuring terminal of the apparatus and connect the specimen to the low or ground terminal by means of a small spring bronze clip attached to a banana plug. Place a small metal disk, such as a silver coin, smaller than the specimen between the free end of the clip and the low or ground electrode to improve contact and avoid damage to the specimen. In calculations of the dielectric constants of these small disk specimens, neglect the correction for edge capacitance.

NOTE 8—When measurements are made at commercial power frequencies, relatively high voltages may have to be used to obtain adequate sensitivity or to meet a requirement that tests be made at a specified voltage gradient on the specimen. The applied voltage shall not exceed the limitations of the instrument used, and must be below the corona starting voltage of the specimen-electrode system.

32. Report

32.1 The report shall include the following:

32.1.1 Description of the material tested, including the thickness,

32.1.2 The specimen size and type of electrodes employed, and

32.1.3 The dielectric constant and dissipation factor of each specimen, and the averages, for each test frequency and testing condition.

Insulation Resistance and Resistivity

33. Electrodes

33.1 *Electrodes for Volume and Surface Resistance*—Air drying or baking conductive silver paint shall be applied to the test specimen, approximately centered, in accordance with Fig. 4 of Method D 257, with the following dimensions:

D_1 = 2 in. (51 mm)
D_2 = 2 1/2 in. (63.5 mm)
D_3 = 3 in. (76 mm)

Note 9—Some materials may be metal clad. It may be desirable to utilize the metal foil clad to the insulating material for electrodes. In this event, specifications applicable to the specific material should be followed for etching the clad foil into a suitable electrode pattern.

33.2 *Electrodes for Insulation Resistance*—Metal electrodes in accordance with Fig. 3 of Methods D 257 for materials 1/32 in. (1 mm) or more in thickness, and in accordance with Fig. 1 of Methods D 257 for thinner materials, shall be used.

34. Conditioning Enclosure

34.1 A conditioning enclosure shall be used to provide the specified conditions, to support the specimens, and facilitate electrical connections for resistance measurements without introducing shunting resistances that interfere with the measurements.

34.2 *Humidity Test Enclosure*—The specified relative humidity at the specified temperature may be obtained by the use of solutions in accordance with ASTM Recommended Practice E 104, for Maintaining Constant Relative Humidity by Means of Aqueous Solutions.[2] The chamber containing the solution should be fitted with holders to support the specimen and make electrical connection for the resistance measurement. The chamber should be thermally insulated to prevent sudden temperature changes that can cause precipitation inside the chamber. The chamber shall be fitted with a small blower or propeller to circulate the air inside. The thermally insulated chamber shall be placed inside an oven maintained at the specified temperature. Figure 2 illustrates a suitable humidity test enclosure.

34.3 *Constant-Temperature Oven*—The oven used for elevated temperature resistance measurements shall conform to the Grade B requirements of ASTM Specifications E 197, for Enclosures and Servicing Units for Tests Above and Below Room Temperature,[2] except for the time constant. The oven should be fitted with holders to support the specimen and make electrical connection for the resistance measurements without introducing shunting resistances that interfere with the measurements. Figures 3 and 4 illustrate a suitable arrangement.

35. Test Specimen

35.1 The surface resistance, and therefore also insulation resistance, may be affected by the manner in which the specimen is prepared, cleaned, and handled. Before insertion or application of the electrodes, each specimen shall be cleaned (unless another procedure is specified) in accordance with ASTM Recommended Practice D 1371 for Cleaning Plastic Specimens for Insulation Resistance Testing.[2] Specimens should be handled by touching the edges only. Nylon, rayon, or surgical rubber gloves are recommended as a precaution against possible contamination of the specimens.

35.2 *Specimen for Volume and Surface Resistance Test*—The specimen shall be a 3 1/2-in. (89-mm) square or disk.

35.3 *Specimen for Insulation Resistance Test*—The specimen shall be a 3 by 2-in. (76 by 51-mm) rectangle for material 1/32 in. (1 mm) or more in thickness. For thinner materials a 2 1/2-in. (63.5-mm) wide strip, rectangular in shape, shall be used.

35.4 Four specimens shall be tested.

36. Conditioning

36.1 Resistance properties of materials covered by these methods are very sensitive to moisture and temperature conditions. Controlled conditioning is required.

36.2 Any controlled condition may be used to obtain the resistance information required. The resistance properties of the materials covered by these methods are generally so high at fairly dry and room temperature conditions that the resistance values have little, if any, practical engineering significance other than to establish quickly that they are high. The standard conditions recommended for obtaining useful engineering information are as follows:

36.2.1 Procedure C of Methods D 618, resistance to be measured while the specimen is in the conditioning atmosphere, and the conditioning to be accomplished in a forced-air circulated medium.

36.2.2 The volume resistance of the specimen shall be measured at the hottest-spot temperature at which the specimen is expected to be used, and 15 min after the specimen has reached and been maintained at this temperature, as determined by means of a thermocouple in the specimen so placed as to measure the temperature of the specimen without interfering with the resistance measurement.

37. Procedure

37.1 Determine the insulation resistance, volume resistance and resistivity, and surface resistance and resistivity in accordance with Methods D 257 and as further provided in the following paragraphs.

37.2 At the end of the conditioning period determine the presence of shunting resistances. If these cannot be effectively eliminated by guarding with the instrumentation used, make proper correction by calculation.

37.3 Measure the resistance of the specimen after applying 500 V of d-c potential difference for 1 min.

Arc Resistance

38. Procedure

38.1 Determine the arc resistance in accordance with ASTM Method D 495, Test for High-Voltage, Low-Current Dry Arc Resistance of Solid Electrical Insulation.[2]

Dust-and-Fog Tracking Resistance

39. Procedure

39.1 Determine the dust-and-fog tracking resistance in accordance with ASTM Method D 2132, Test for Dust-and-Fog Tracking and Erosion Resistance of Electrical Insulating Materials.[2]

Bonding Strength

40. Significance

40.1 The bonding strength is a measure of the adhesive strength of a heterogeneous material of the type covered by these methods. It is useful as a manufacturing control or acceptance test. It may serve to indicate whether or not a thermosetting laminated plastic is properly cured.

41. Definition

41.1 *bonding strength*—the force required to split a prescribed specimen under the test conditions specified herein.

42. Test Specimen

42.1 Any specimen $^3/_{16}$ in. (5 mm) or thicker may be tested. The bonding strength is dependent on specimen thickness, however, and therefore only specimens of the same thickness should be compared.

42.2 The standard specimen shall be 0.500 ± 0.005 in. (12.7 ± 0.127 mm) thick and 1 in. (25.4 mm) square. Two parallel edges shall be smooth within ±0.001 in. (±0.025 mm).

42.3 Four specimens shall be tested.

43. Apparatus

43.1 Any universal testing machine may be used, provided it is accurate to 1 percent of the lowest load to be applied. The machine shall be fitted with a head containing a 10-mm steel ball.

44. Procedure

44.1 Place the specimen with smooth edge on the testing machine table or a flat steel plate that rests on the testing machine table. Accurately center the steel ball between the edges and ends of the specimen.

44.2 Apply any constant load until the specimen splits. The load shall be accurately weighed and shall not exceed 0.050 in./min crosshead speed.

44.3 Record as the bonding strength the maximum force obtained.

45. Report

45.1 The report shall include the following:

45.1.1 The thickness of the material, and

45.1.2 The load, expressed in pounds or kilograms, required to split the specimen.

ROCKWELL HARDNESS

46. Procedure

46.1 Determine the Rockwell hardness in accordance with ASTM Method D 785, Test for Rockwell Hardness of Plastics and Electrical Insulating Materials,[3] except that under Method A use the M scale provided that the total indentation does not exceed the limits of the testing machine; in this case the L scale.

46.2 Test four specimens.

ASH

47. Test Specimen

47.1 The test specimen shall consist of 2 to 5 g of finely divided particles, such as millings or filings, of the material.

48. Procedure

48.1 Dry the test specimen for 2 h at 105 to 110 C, weigh, then ignite to constant weight in a crucible, and weigh. Calculate the percentage of ash, based on the weight of the dried specimen.

FLAMMABILITY AND FLAME RESISTANCE

49. Significance

49.1 Rigid electrical insulation is sometimes exposed to temperature sufficiently high to indicate a danger of ignition. This may occur due to malfunction of the apparatus of which the insulation is a part, due to failure of associated equipment in the system, or due to failure of the insulation to resist ignition in normal-usage exposure to electric arcs. It is therefore desirable to provide methods of test that allow the relative comparison of the ignition resistance of materials and the extent of burning if ignition does occur.

49.2 Two methods are provided: Flammability, Method I, is a relatively simple test that requires inexpensive apparatus. It is intended primarily as a control test and for screening quickly materials that are flame resistant from a population of various types. Flame Resistance, Method II, is intended primarily for use with materials that would be found non-burning or self-extinguishing, or both, by Method I. The equipment specified in Method II, which is relatively complex, allows more precise control of test conditions than Method I.

49.3 Neither method will directly produce information from which the performance of the insulating material in service can be quantitatively predicted, since the conditions of use in electrical apparatus are likely to be different than the test conditions. The methods do, however, provide means of comparing materials under controlled laboratory conditions.

49.4 Both methods provide for the measurement of resistance to ignition and resistance to continued burning. Method I simply distinguishes between specimens that will ignite (under conditions of the test) from those that will not. Resistance to burning is determined by the time the specimen burns. In Method II materials may be directly compared for resistance to ignition by determination of ignition time and for burning by the burning time. The comparison of burning, or the tendency of the material to contribute to the spread of fire, requires interpretation regardless of which method is used. Some materials may continue to burn for relatively long periods of time without the dissipation of much heat energy. Other materials, which may burn for relatively shorter periods, may do so with potentially damaging intensity. The determination of weight loss can aid in an interpretation of burning time test results on some materials and may be required by agreement between producer and consumer.

Method I—Flammability

50. Test Specimens

50.1 The test specimen shall be 5 by $^1/_2$ in. (127 by 13 mm) by the thickness of the sample except that this method shall be not applicable to specimens less than $^1/_{32}$ in. (1 mm) (nominal) thick.

50.2 At least four specimens in the as-received condition shall be tested. The test specimen shall have a scribe mark 1 in. from the end to be ignited. The test specimen shall have smooth edges.

50.3 Copper-clad specimens shall be tested with the copper removed by etching in accordance with ASTM Recommended Practice D 1825, for Etching and Cleaning Copper-Clad Thermosetting Laminates for Electrical Testing.[2]

51. Procedure

51.1 Determine the flammability in accordance with ASTM Method D 635, Test for Flammability of Self-Supporting Plastics,[3] except as follows:

51.2 Incline the bunsen burner illustrated in Fig. 1 of Method D 635 30 deg from a vertical plane toward the specimen and approximately parallel to the perpendicular vertical plane.

51.3 Apply the flame only once for a duration of 30 s.

51.4 After removal of the bunsen burner flame, measure the elapsed time (assuming that ignition was accomplished) until burning ceases or until the burning progresses to the scribe mark, whichever occurs first.

52. Report

52.1 The report shall include the following:

52.1.1 Description of the material tested including the thickness and whether the sample was copper-clad.

52.1.2 Whether the specimen ignites, and

52.1.3 Average and individual burning times in seconds, including a statement that informs whether burning progressed to the scribe mark.

Method II—Flame Resistance

53. Definitions

53.1 *ignition time* (*I*)—the elapsed time in seconds required to produce ignition under conditions of this method.

53.2 *burning time* (*B*)—the elapsed time that the specimen burns after removal of the ignition heat source under conditions of this method.

54. Apparatus

54.1 *Flame Cabinet*—A metal cabinet with heater coil, spark gaps, specimen holder, access door, and forced-air ventilation as illustrated in Fig. 5, or equipment that gives equivalent results.

54.2 *Control Cabinet*—A control assembly that provides adjustable, regulated power to the heater coils, ignition voltage to the spark gaps, and a timer or timers to indicate the required time intervals as illustrated in Fig. 6.

54.3 *Pyrometer*—An optical pyrometer calibrated to read directly for the emission of Nichrome V, or an optical pyrometer calibrated for black-body emission to which 6 C is added to the pyrometer reading to obtain the true temperature of the Nichrome V coil. The pyrometer shall include a scale for measurement of temperature near 860 C.

54.4 *Coil Form*—A grooved mandrel on which the Nichrome V resistance wire is wound into a heater coil as illustrated in Fig. 7(*a*).

54.5 *Coil Spacing Gage*—A spacing gage constructed of a sector of a coil form (54.4), as illustrated in Fig. 7(*b*) to check the coil turn-spacing.

55. Test Specimen

55.1 The specimen shall be $^1/_2 \pm 0.036$ in. (13 ± 0.8 mm) thick or nominal unmachined tolerance by $^1/_2 \pm 0.01$ in. (13 ± 0.25 mm) in width by $10 \pm {^1/_{16}}$ in. (254 ± 1.6 mm) in length. In cases of molded products the length of the specimen may be shorter.

NOTE 10—Methods of testing thinner thicknesses are being investigated.

55.2 The specimens shall be machined in a manner that produces a cut surface that is free from projecting fibers and ridges.

55.3 The test sample shall consist of five test specimens.

56. Calibration

56.1 Place a dummy specimen in the holder.

56.2 Adjust the heater coil so that the bottom turn is $1^1/_2$ in. (38 mm) above the top of the specimen holder, the coil is symmetrical about the specimen, and the coil height is $1^1/_2$ in. (38 mm). Use the coil spacing gage to adjust, if necessary, the individual coil turns for proper spacing.

56.3 Adjust the spark gap to $^3/_{16} \pm {^1/_{16}}$ in. (5 ± 1.6 mm) and determine that the arc is in an approximate horizontal plane. The total (in both electrodes) arc-current should be 20 ± 5 mA. The electrode tips should be approximately $^1/_8$ in. (3 mm) in a horizontal plane from the specimen and $^1/_2$ in. (13 mm) above the top turn of the heating coil.

56.4 Remove the dummy specimen. Close the door and energize the ventilating blower.

Energize the heating coil and adjust the heater current to approximately 55 A. Allow the coil to come to equilibrium temperature (approximately 120 s). If a new coil is being used, reduce the current to 50 A and allow to remain energized for 24 h to produce a stable oxide coating.

56.5 Open the peep-hole in the door; sight the optical pyrometer on the outside of the middle turn and adjust the heater current to obtain an equilibrium temperature of 860 ± 5 C. Keep the peep-hole closed during test.

NOTE 11—After the current has been adjusted, the variable-ratio autotransformer setting must not be disturbed during the test. In order to maintain the temperature within ±5 C, it is necessary that the average rms voltage across the heater remain constant within ±1.0 percent.

57. Conditioning

57.1 Specimens shall be conditioned for 168 h in the Standard Laboratory Atmosphere (23 C, 50 percent relative humidity) except that when it is demonstrated that test results for the specific type material are not significantly affected by conditioning, unconditioned specimens may be used.

57.2 Tests shall be conducted in a room that is controlled at the Standard Laboratory Atmosphere (Note 12) and is free of spurious drafts (Note 13).

NOTE 12—It is a well-established fact that the combustion process is influenced by the moisture content of the oxygen-providing atmosphere.

NOTE 13—Drafts, except those of unusual velocity, are not likely to disturb test conditions when the test is performed with properly constructed apparatus. However, changing drafts are likely to disturb the thermal equilibrium condition so that the heater coil temperature may change from the specified temperature even though constant input power is supplied.

58. Procedure

58.1 After calibration is completed, use an air jet to cool the coil to room temperature.

58.2 Insert the specimen in the holder with the cut side facing the spark gaps. (When laminates are tested, the plane of laminations shall be parallel to the plane of the front of the apparatus.) Close the peep-hole.

58.3 Move the arc electrodes to the horizontal position. Energize the ventilating blower.

58.4 Simultaneously energize the heater coil, arc gap, and timer circuit.

58.5 Record the elapsed time in seconds when the test specimen ignites as ignition time, *I*. (Gases released from the specimen may ignite before the specimen commences burning. Ignition time is determined from the instant that the specimen flames rather than from the instant of gas ignition.)

58.6 De-energize the heater and spark gaps 30 s after the specimen ignites; move the arc electrodes away from the specimen.

58.7 De-energize the timer circuit when the specimen ceases to burn (all flame has disappeared), and record the total elapsed, *T*, in seconds.

58.8 Before beginning the next test, cool the coil with an air jet, brush soot and contamination from the heater coil and arc gaps, and blow any debris from the test enclosure.

59. Calculations

59.1 *Burning Time*—Calculate the burning time, *B*, in seconds, as follows:

$$B = T - I - 30$$

where:

T = total elapsed time, and

I = ignition time.

Calculate the burning time by arranging the five values of burning time in increasing order of magnitude, as T_1, T_2, T_3, T_4, and T_5. Compute the following ratios:

$$(T_2 - T_1)/(T_5 - T_1)$$

and

$$(T_5 - T_4)/(T_5 - T_1)$$

If either of these ratios exceeds 0.642 then T_1 or T_5 is judged to be abnormal and is eliminated. The burning time reported shall be the average of the remaining four values.

59.2 *Average Ignition Time*—Calculate the average ignition time as the arithmetic mean of the five specimens.

60. Report

60.1 The report shall include the following:

60.1.1 Nominal thickness of the test specimen,

60.1.2 Average and individual burning times and ignition times, and

60.1.3 Description of how the specimen burns with particular attention to the intensity of the flame.

COEFFICIENT OF LINEAR THERMAL EXPANSION

61. Procedure

61.1 Test a minimum of two specimens in accordance with ASTM Method D 696, Test for Coefficient of Linear Thermal Expansion of Plastics.[3]

WARP OR TWIST

62. Significance

62.1 Warp and twist are expressions of deviation from flatness of a material. The extent of deviation is of interest primarily when it is intended to fabricate the sheet or plate material but may also affect the ability to use the full-size sheet in an assembly.

63. Conditioning

63.1 It is generally not necessary to condition the material. Where conditions of storage may cause warp or twist, the material shall be conditioned in a manner agreed by the purchaser and the supplier.

64. Procedure

64.1 Determine the warp or twist on the sheet in the as-received condition by holding a straightedge along the dimension to be measured. The concave side of the sheet shall be adjacent to the straightedge. Measure the greatest deviation of the concave surface from the straightedge by a metal scale.

64.2 *Warp*—Measure the warp by suspending the sheet freely from the center of one edge in a vertical position against a horizontal straightedge, then in succession by the other edges until the point of maximum warp is obtained.

64.3 *Twist*—Measure the twist by suspending the sheet in a vertical position from adjacent corners, singly and in succession, and then measuring the deviation along the diagonal from the straightedge connecting the corners opposite from the vertical. Report the maximum twist.

65. Calculations

65.1 Calculate the percentage warp or twist based on a 36-in. (914-mm) length as follows:

$$W_{914} = (914D/L^2) \times 100$$

or

$$W_{36} = (36D/L^2) \times 100$$

where:

W_{914} = percentage warp or twist calculated to a 914-mm length, or

W_{36} = percentage warp or twist calculated to a 36-in. length,

D = maximum deviation in millimeters or inches of the sheet from the straightedge, and

L = length in millimeters or inches of the dimension along which the warp or twist is measured.

65.2 When it is desired to compare the actual deviation for any length with the permissible deviation for that length, the following equation may be used:

$$D_x/D_{914} = L_x^2/(914)^2$$

or

$$D_x/D_{36} = L_x^2/(36)^2$$

where:

D_x = permissible deviation from straightedge in millimeters or inches for the given length,

D_{914} = permissible deviation in millimeters for 914-mm length, or

D_{36} = permissible deviation in inches for 36-in. length, and

L_x = given length in millimeters or inches.

NOTE 14—These requirements do not apply to cut pieces but only to sheet sizes as manufactured.

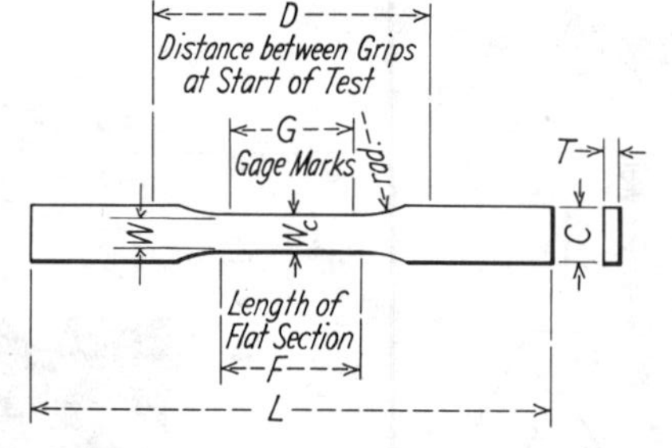

Dimension	Thickness,[a] T										Tolerance	
	1/4 in. (6 mm) or Under				Over 1/4 in. (6 mm) to 1/2 in. (13 mm), incl.				Over 1/2 in. (13 mm) to 1 in. (25 mm), incl.[b]			
	Type I		Type II[c]		Type I		Type II[c]		Type I			
	mm	in.	mm	in.	mm	in.	mm	in.	mm	in.	mm	in.
C—Width over-all	19.05	0.750	19.05	0.750	28.57	1.125	28.57	1.125	38.10	1.500	±0.40 −0.00	+0.016 −0.000
W—Width of flat section	12.70	0.500	6.35	0.250	19.05	0.750	9.52	0.375	25.40	1.000	+0.12	+0.005
F—Length of flat section	57.1	2.25	57.1	2.250	57.1	2.25	57.1	2.25	57.1	2.25	±0.40	±0.016
G—Gage length[d]	50.8	2.00	50.8	2.00	50.8	2.00	50.8	2.00	50.8	2.00	±0.40	±0.016
D—Distance between grips[a]	114	4 1/2	133	5 1/4	114	4 1/2	133	5 1/4	133	5 1/4	±3	±1/8
L—Length over-all	216	8 1/2	238	9 3/8	248	9 3/4	257	10 1/8	305	12	min	min
Rad.—Radius of fillet	76	3	76	3	76	3	76	3	76	3	min	min

[a] Tolerance on thickness shall be the standard for the grade of material tested.

[b] For sheets of a nominal thickness over 1 in. (25.4 mm) the specimens shall be machined to 1 in. (25.4 mm) ± 0.010 in. (0.25 mm) in thickness. For thicknesses between 1 in. (25.4 mm) and 2 in. (51 mm) approximately equal amounts shall be machined from each surface. For thicker sheets both surfaces shall be machined and the location of the specimen with reference to the original thickness shall be noted.

[c] The Type II specimen shall be used for material from which the Type I specimen does not give satisfactory failures in the gage length, such as for resin-impregnated compressed laminated wood.

[d] Test marks only.

FIG. 1 Tension Test Specimen for Sheet and Plate Insulating Materials.

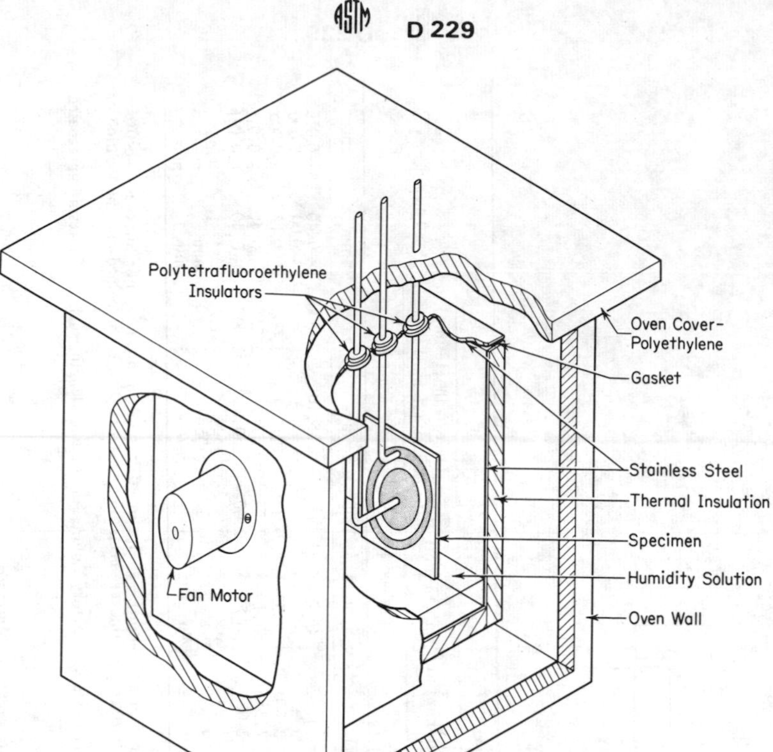

FIG. 2 Humidity Test Enclosure for Insulation Resistance and Resistivity Tests.

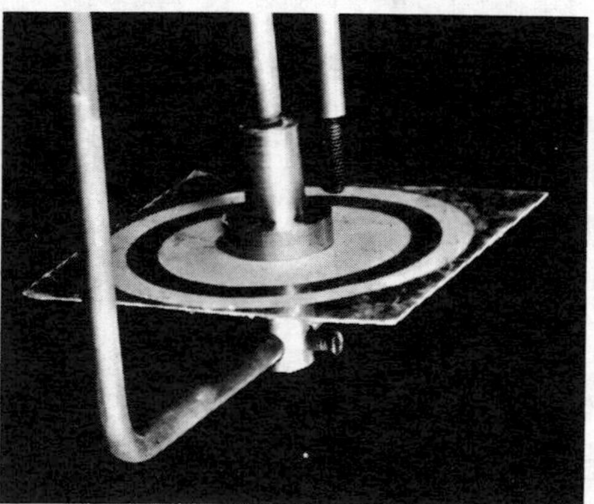

FIG. 3 Test Specimen for Insulation Resistance and Resistivity Tests Mounted in Specimen Holder.

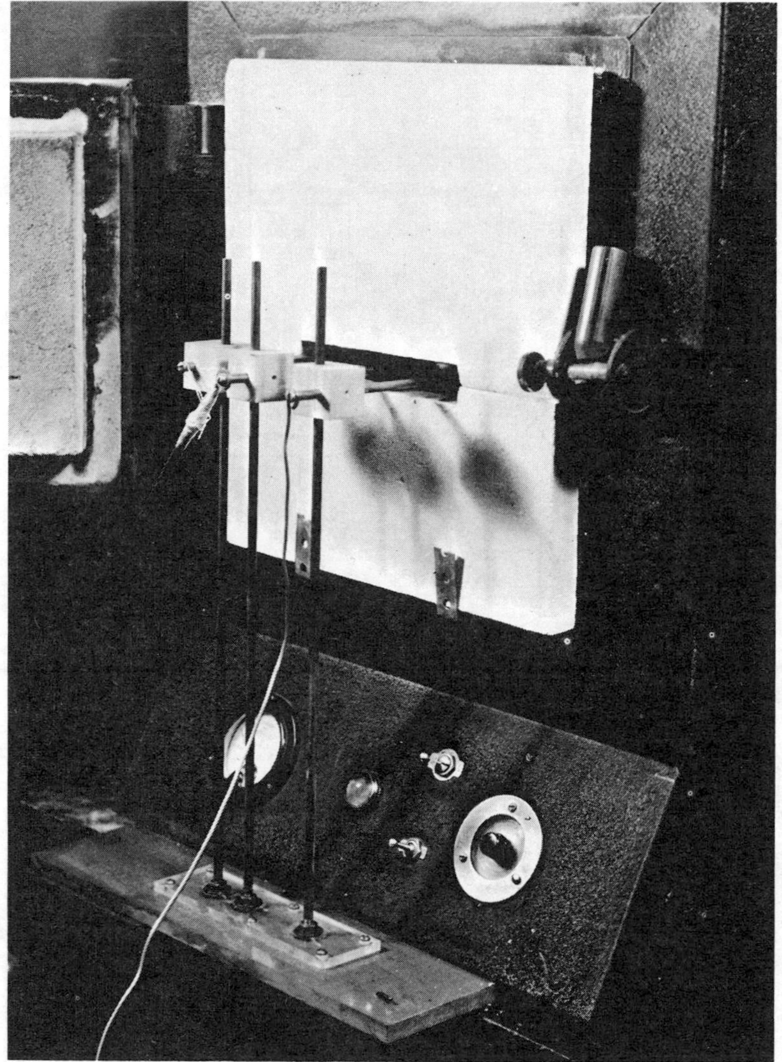

FIG. 4 Insulation Resistance and Resistivity Specimen Holder Brought Through a Split-Type Removable Oven Door.

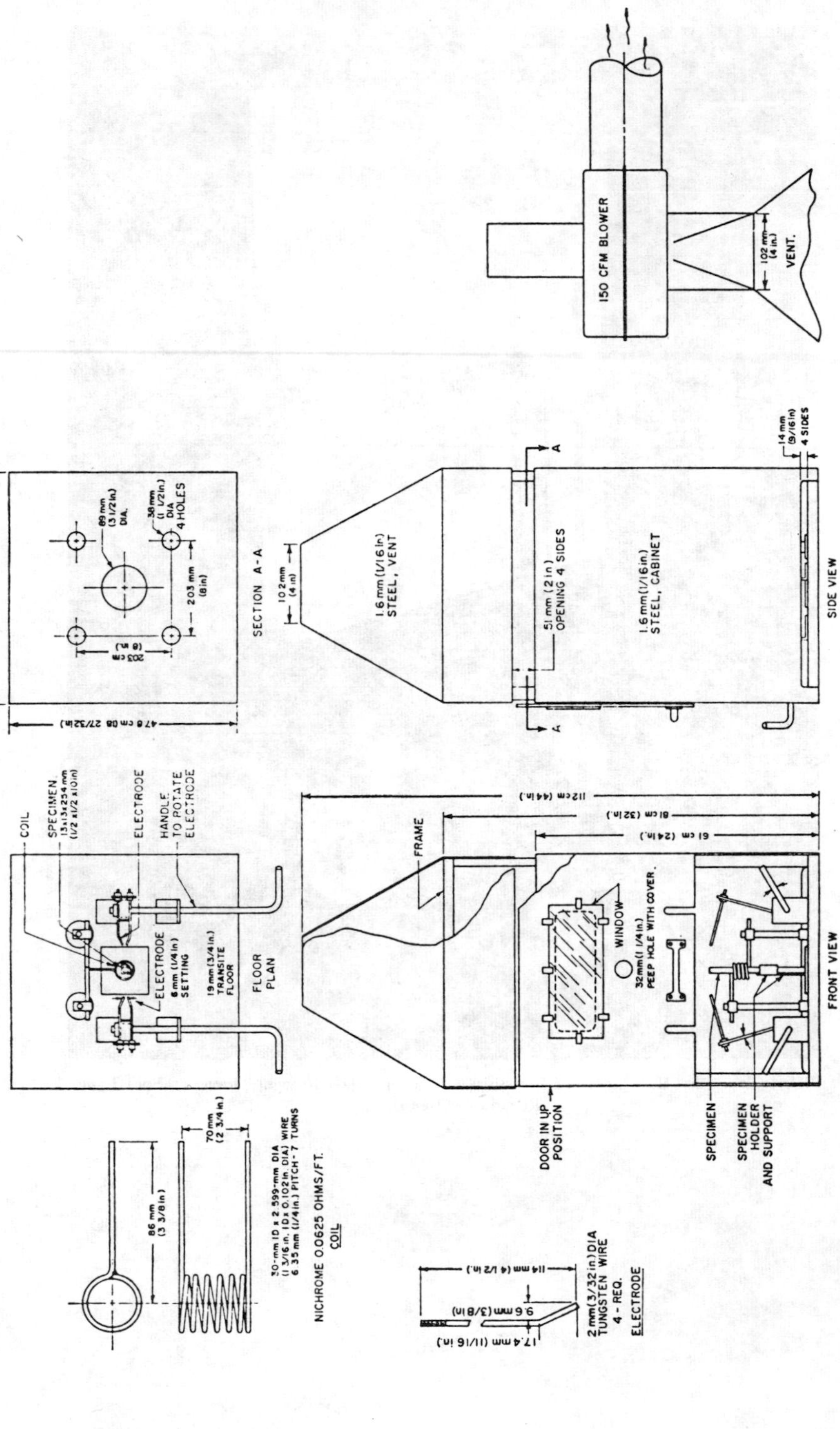

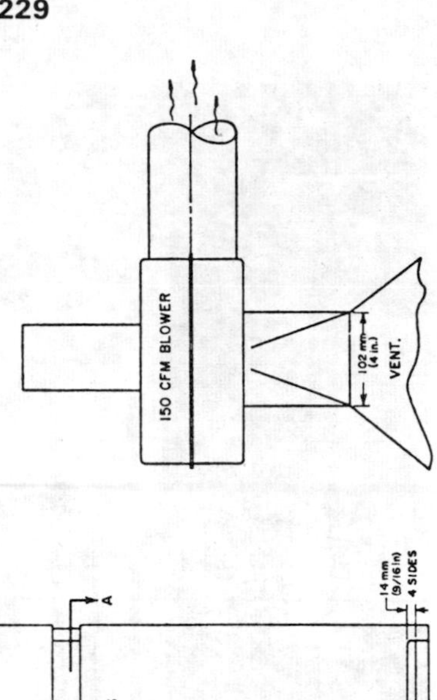

FIG. 5 Flame Cabinet.

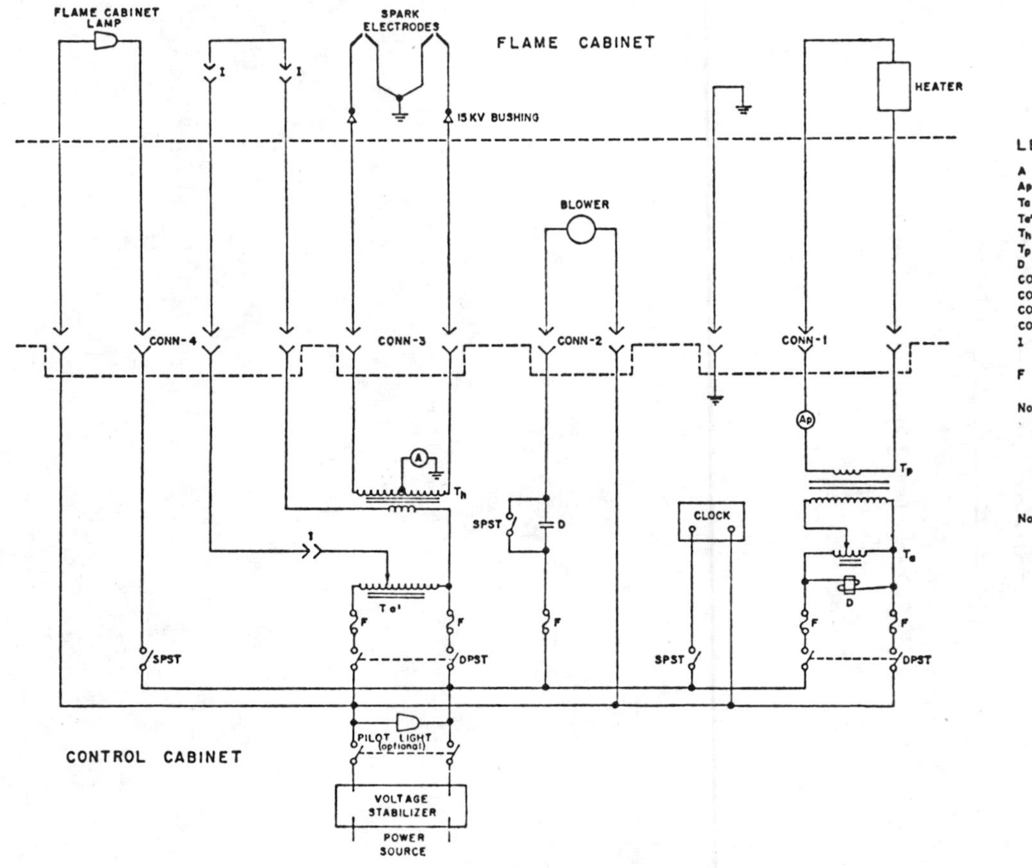

LEGEND:

A Ammeter, 0-50 ma
Ap Ammeter, 0-100 ma
Ta Variable ratio auto-transformer, 15 a
Ta' Variable ratio auto-transformer, 500 va
T_h HV Transformer, 12/6 kv - 250 va - See note 1
T_p LV Transformer, 12 v - 1.0 kva
D Voltage relay - SPST, NO
CONN-1 60 a - 3 pole
CONN-2 2 Pole
CONN-3 15 kv - 2 pole or use polystyrene board
CONN-4 4 Pole
I Mechanical interlock mounted on Flame Cabinet door and Control Cabinet door
F All fuses slow blow - See note 2

Note 1:
Alternatively, two 6 kv-125 va transformers may be used; or a single 6 kv-250 va transformer, or 12 kv-250 va transformer may be used. In any event, proper circuit connections should be made to obtain 20 ma arc-current, and the ammeter (A) should always be connected to ground.

Note 2:
Current rating of fuses and CONN-4 is determined by voltage rating of primary power source.

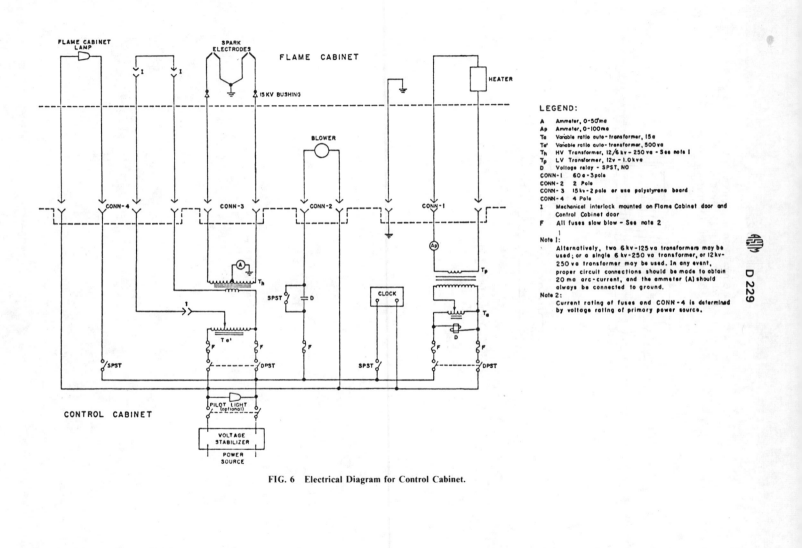

FIG. 6 Electrical Diagram for Control Cabinet.

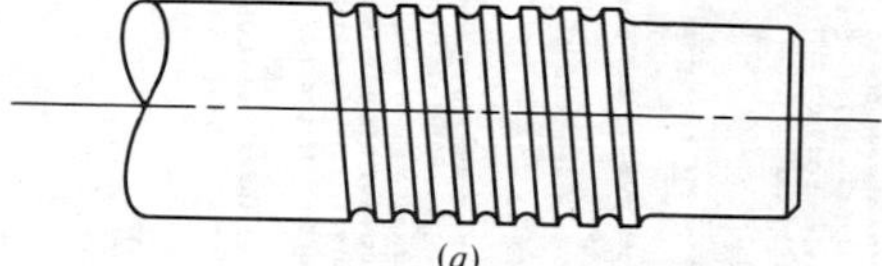

(*a*)

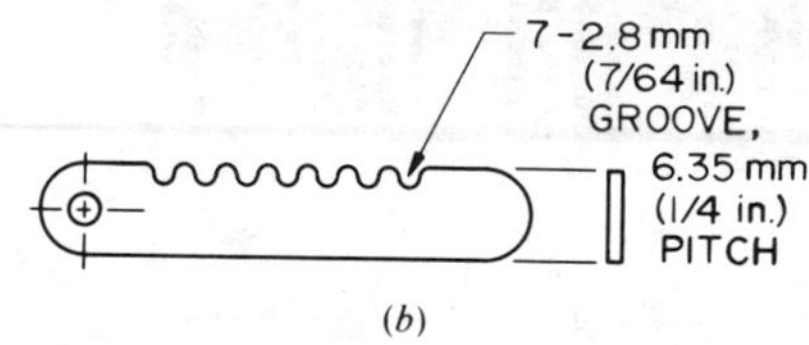

(*b*)

FIG. 7 Mandrel for Coil (*a*) **and Coil Spacing Gage** (*b*).

Designation: D 256 – 56
(Reapproved 1961)

American National Standard C59.11-1955 (R 1969)
American National Standards Institute

Standard Methods of Test for
IMPACT RESISTANCE OF PLASTICS AND ELECTRICAL INSULATING MATERIALS[1]

This Standard is issued under the fixed designation D 256; the number immediately following the designation indicates the year of original adoption or, in the case of revision, the year of last revision. A number in parentheses indicates the year of last reapproval.

NOTE—Note 3 was editorially deleted in March 1970.

1. Scope

1.1 These methods cover the determination of the relative susceptibility to fracture (Note 2) by shock of plastic materials and electrical insulating materials as indicated by the energy expended by a standard pendulum-type impact machine in breaking in one blow a standard specimen.

1.2 There are two types of pendulum impact machines and three related methods that use different specimens and differ in the method of holding and striking the specimen. Each specimen and method has characteristics that may dictate its use. Results by the different methods cannot be directly compared.

NOTE 1—The values stated in U.S. customary units are to be regarded as the standard. The metric equivalents of U.S. customary units may be approximate.

NOTE 2—The notch in the Izod specimen serves to concentrate the stress, minimize plastic deformation, and direct the fracture to the part of the specimen behind the notch. This reduces the data scatter. However, because of differences in the elastic and visco-elastic properties of plastics, response to a given notch varies among materials.

For materials which fracture in the unnotched condition, the ratio of the energy to break an unnotched Izod bar to the energy to break a bar with the standard notch is a measure of notch sensitivity. For such comparisons to be meaningful, the depth of the unnotched specimen must be reduced to equal the depth under the notch of the notched specimen.

Many materials cannot be broken in the unnotched condition, and in these instances a measure of notch sensitivity may be obtained by comparing the energy required to break specimens with identical notches except for the radius at the base of the notch. The standard 0.25-mm (0.010-in.) notch and supplementary 1.0-mm (0.040-in.) notch provide ratios of energy to break from 1 up to <6 for many common plastics and give guidance to designers as to the effect of stress concentration on the apparent toughness of molded parts.

2. Types of Tests

2.1 *Method A* is the cantilever beam or Izod-type test in which the specimen is held as a cantilever beam (usually vertical) and is broken by a blow delivered at a fixed distance from the edge of the specimen clamp. The test requires a notched specimen in all cases. The notch is intended to produce a standard degree of stress concentration.

2.2 *Method B* is the simple beam or Charpy-type test in which the specimen is supported as a simple beam (ususlly horizontal) and is broken by a blow delivered midway between the supports. In this test the specimen may be either plain or notched, as required by the characteristics of the material tested.

2.3 *Method C* is the same as Method A plus the inclusion of a method for determining the energy expended in tossing the specimen. The value reported is called the "estimated net Izod impact strength." This method is recommended in place of Method A for materials which have an Izod impact strength of less than 27 N · m/m (0.5 ft · lb/in.) of notch. It is not to be used for materials that have higher Izod impact strength values than 27 N · m/m (0.5 ft · lb/in.) of notch.

3. Significance

3.1 The excess energy pendulum impact

[1] These methods are under the jurisdiction of ASTM Committee D-20 on Plastics. A list of committee members may be found in the ASTM Yearbook. This standard is the direct responsibility of Subcommittee D-20.10 on Mechanical Properties.

Current edition effective Sept. 10, 1956. Originally issued 1926. Replaces D 256 – 54 T.

test indicates the energy to break standard test specimens of specified size under stipulated conditions of specimen mounting, notching (stress-concentration), and rate of loading.

3.2 The energy to break a standard test piece is actually the sum of several energies:

3.2.1 The energy to deform the specimen,

3.2.2 The energy to initiate fracture of the specimen,

3.2.3 The energy to propagate the fracture across the specimen,

3.2.4 The energy to throw the free end of the broken specimen ("toss factor"), and

3.2.5 The energy lost through friction and through vibration of the apparatus and its base.

3.3 The notch which is always present in the standard Izod test specimen, serves to concentrate the stress and largely prevents plastic deformation. Friction losses are largely eliminated by careful design and proper operation of testing machine. Energy losses due to vibration of the apparatus are generally assumed to be negligible for plastics, but may be considerable, if the machine is not correctly designed with sufficient mass for the specified range, and of rigid construction.

3.4 Thus, the indicated impact strength of a material, for all practical purposes, is based upon the sum of 3.2.2, 3.2.3, and 3.2.4 as indicated in 3.2. In the case of relatively brittle materials, the tearing energy is small compared to the fracture energy, whereas in the case of tough, ductile materials, or fiber-filled or cloth-laminated materials, the reverse is true. The toss factor may represent a very large fraction of the total energy absorbed when testing relatively brittle materials (less than 27 N · m/m (0.5 ft · lb/in.) of notch). A correction for the toss factor is specified in 18.4 of Method C. This correction is especially important when comparing materials of widely varying densities.

Method A. Cantilever Beam (Izod Type) Test

4. Apparatus

4.1 The machine shall be of the pendulum type, as illustrated in Fig. 1, of rigid construction, accurate to 13.6 N · m (0.01 ft · lb) for readings of less than 1355 N · m (1 ft · lb) and to 1 percent for higher values. Accurate correction shall be made for friction and windage losses.

4.2 The dimensions of the machine shall be such that the center of percussion of the striker (Note 3) is at the point of impact; that is, the center of the striking edge.

Note 3—The distance from the axis of support to the center of percussion may be determined experimentally from the period of oscillation of the pendulum through a small angle by means of the following equation:

$$l = 0.81\,P^2$$

where:

l = distance from the axis of support to the center of percussion, and

P = time in seconds of a complete swing (to and fro).

4.3 The pendulum shall be released from such a position that the linear velocity of the center of the striking edge (center of percussion) at the instant of impact shall be approximately 3.4 m/s (11 ft/s), which corresponds to an initial elevation of this point of 610 mm (2 ft).

4.4 The striking edge of the pendulum shall be a cylindrical surface of 0.79-mm ($^1/_{32}$-in.) radius, with its axis horizontal. The cylindrical surface shall be, when the pendulum is hanging free, tangent to the specimen in a line 22.0 mm (0.866 in.) above the top surface of the vise. The pendulum above the cylindrical portion of the striking edge shall be recessed or inclined at a suitable angle so that there is no chance of its coming in contact with the specimen during the break.

4.5 Means shall be provided for clamping the specimen rigidly in position with the edges of the supporting surfaces at 90-deg angles.

Note 4—Some plastics (such as polystyrene) are sensitive to clamping pressure; therefore, cooperating laboratories should agree on a means of standardizing the gripping force, such as by using a torque wrench on the screw of the specimen vise.

4.6 Means shall be provided for determining the impact value of the specimen, which is the energy expended by the machine in breaking the specimen. This value is equal to the difference between the energy in the pendulum blow and the energy remaining in the pendulum after breaking the specimen, after suitable correction has been made for windage and friction.

5. Test Specimen

5.1 The test specimen shall conform to the dimensions shown in Fig. 2, except as modified in accordance with 5.2, 5.3, and 5.4. To ensure the correct contour and condition of the specified notch, all specimens shall be notched as directed in Section 6.

5.2 For molded material (Note 5) the specimen shall be 12.5 mm (1/2 in.) by any dimension from 3 to 12.5 mm (1/8 to 1/2 in.) agreed upon as representative of the cross section in which the particular material is to be used. For all specimens having one dimension less than 12.5 mm (1/2 in.), the notch shall be cut in the narrower side (Note 6). For all compression-molded specimens, the notch shall be in the side parallel to the direction of application of the molding pressure (Note 6).

NOTE 5—The type of mold used to produce test specimens has an effect on the results obtained. Cooperating laboratories should, therefore, agree upon standard molds, conforming to ASTM Recommended Practice D 647, for Design of Molds for Test Specimens of Plastic Molding Materials[2] and upon molding procedure, to obtain concordant results.

NOTE 6—The indicated impact strength of some materials varies somewhat with the width (dimension along the notch) of the test specimen used. For interlaboratory comparisons on different compositions, molded specimens of a fixed width should be used.

NOTE 7—The impact strength may be different if the notch is perpentidular to the direction of molding instead of parallel.

5.3 For sheet material the specimens shall be cut from the sheet in both the lengthwise and crosswise directions, unless otherwise specified (Note 8). The thickness shall be the thickness of the sheet except that it shall not exceed 12.5 mm (1/2 in.). Sheet material thicker than 12.5 mm (1/2 in.) shall be machined down to 12.5 mm (1/2 in.). Such material may be tested either edgewise or flatwise, as specified. When specimens are tested flatwise, the notch shall be made on the machined surface, if the specimen is machined on one side only. When the specimen is cut from a thick sheet, notation shall be made from what portion of the thickness of the sheet the specimen is cut, for example, center, top surface, or bottom surface.

NOTE 8—In referring to the cutting of the specimens of laminated sheet materials and the applications of the load, the following descriptions of terms apply:

Flatwise—Load applied to the flat side of the original sheet or plate.

Edgewise—Load applied to the edge of the original sheet or plate.

Lengthwise—In the direction of the length of the sheet.

Crosswise—In the direction at right angles to the direction of the length of the sheet.

When the sheet has the same length and width, one dimension shall arbitrarily be designated as the "A" direction and the other as the "B" direction.

5.4 For sheet materials less than 12.5 mm (1/2 in.) in thickness, the test specimen may be a composite specimen consisting of a number of individual thin pieces aggregating 3 to 12.5 mm (1/8 to 1/2 in.) in thickness. The individual members of the test specimen shall all be accurately aligned with each other and shall be tested edgewise. Pieces less than 1.5 mm (1/16 in.) in thickness should be laminated together with a suitable adhesive (Note 9) to prevent buckling or twisting during impact. Single specimens less than 12.5 mm (1/2 in.) in thickness may be used, provided the width is sufficient for firm, accurate clamping and the impact value of the material is sufficiently high to be accurately determined by a machine of the capacity used.

NOTE 9—Care should be taken in selecting an adhesive that does not affect the impact values. If the adhesive contains solvents, a conditioning procedure should be followed which will allow the complete removal of the solvent. It is known that the presence of solvents with some materials (such as cellulosics) does affect the impact values.

6. Notching Test Specimens

6.1 Notching shall be done on a milling machine or lathe with a cutter which may be of the single-tooth or multi-tooth variety. The single tooth cutter is preferred because it is more readily ground to the desired contour. The cutting edge shall be carefully ground and honed to ensure sharpness and freedom from nicks and burrs. Tools with no rake and a work-relief angle of 15 to 20 deg have been found satisfactory.

6.2 The profile of the cutting tooth (teeth) shall be such as to produce in the test specimen at right angles to its principal axis a notch of the contour and depth specified in Fig. 2 (Note 10). The included angle of the notch shall be 45 ± 1 deg with a radius of

---

[2] *Annual Book of ASTM Standards*, Part 27.

curvature at the apex of 0.25 ± 0.025 mm (0.010 ± 0.001 in.) (Note 11). The plane bisecting the notch angle shall be perpendicular to the face of the test specimen with a tolerance of 2 deg.

NOTE 10—In the case of single-tooth cutters, the contour of the tip of the cutting tool may be checked instead of the contour of the notch in the specimen, if it can be shown that the two correspond, or that a definite relationship exists between them for the specific type of material being notched. There is some evidence to indicate that notches cut in materials of widely differing physical properties by the same cutter may differ in contour.

NOTE 11—For some materials, such as polystyrene, the acceptable tolerance range from 0.225 to 0.275 mm (0.009 to 0.011 in.) on the radius of curvature at the notch apex may be too broad to ensure reproducible impact test results. Whenever such notch sensitivity can be demonstrated, the permissible range should be reduced to 0.237 to 0.263 mm (0.0095 to 0.0105 in.).

6.3 Linear speeds of the tip of the cutting tool of about 90 to 185 m/min (300 to 600 ft/min) and feed rates in the range from 12 to 125 mm/min (0.5 to 5.0 in./min) have proved satisfactory for some materials.

NOTE 12—While the speed ranges given in 6.3 have produced satisfactory notches in some plastic and electrical insulating materials, it may be necessary to study the effect of variations in these conditions when unfamiliar materials are to be notched.

6.4 After each 500 notches, or more often, if hard abrasive materials are being notched, the notch in a specimen shall be inspected (Note 13) for sharpness, freedom from nicks, radius of tip, and angle. If the angle and radius do not fall within their specified limits, the cutter shall be replaced with a newly sharpened and honed one. A microscope with a camera lucida attachment (60× magnification) is suitable for checking the radius and angle of notch.

NOTE 13—Relatively close tolerances must be imposed upon the angle and radius of the notch for most materials, because these factors largely determine the degree of stress concentration at the base of the notch during the test. The maintenance of a sharp, clean-edged cutting tool is particularly important since minor defects at the base of the notch can cause large errors in test results.

7. Conditioning

7.1 *Conditioning*—Condition the test specimens at 23 ± 2 C (73.4 ± 3.6 F) and 50 ± 5 percent relative humidity for not less than 40 h prior to test in accordance with Procedure A of ASTM Methods D 618, Conditioning Plastics and Electrical Insulating Materials for Testing,[2] for those tests where conditioning is required. In cases of disagreement, the tolerances shall be ±1 C (±1.8 F) and ±2 percent relative humidity.

7.2 *Test Conditions*—Conduct tests in the Standard Laboratory Atmosphere of 23 ± 2 C (73.4 ± 3.6 F) and 50 ± 5 percent relative humidity, unless otherwise specified in the test methods or in this specification. In cases of disagreements, the tolerances shall be ±1 C (±1.8 F) and ±2 percent relative humidity.

8. Procedure

8.1 At least five and preferably ten individual determinations of impact value shall be made under the conditions prescribed in Section 7.

8.2 The test specimen shall be rigidly clamped with the center line of the notch on the level of the top of the clamping surface (Note 14) and the blow shall be struck on the notched side.

8.3 When a composite specimen is used, the individual members shall be held closely in contact and accurately aligned with each other when clamped.

NOTE 14—It is recommended that a jig or template be used to locate the specimen in the jaws, as specified.

9. Report

9.1 The report shall include the following:

9.1.1 A statement indicating the size of specimen, the method of test and type of preconditioning used, and, for sheet materials, the direction of testing and whether the specimens were cut lengthwise or crosswise from the sheet.

9.1.2 The value of energy expended in breaking each individual specimen expressed in newton-meters per meter (or foot-pounds per inch) of notch, determined by dividing the energy in newton-meters (or foot-pounds) expended in the individual test by the actual dimension in meters (or inches) along the notch of the specimen broken in each test, the range of these values for each material tested, and whether a single or a composite specimen was used and also the direction of cutting the specimen, and

9.1.3 The average of the values given in 9.1.2, reported as the "Izod Impact

Strength," the average thickness of the individual specimen, and the number of such specimens broken in each operation of the machine.

METHOD B. SIMPLE BEAM (CHARPY TYPE) TEST

10. Apparatus

10.1 The machine for Method B shall be of the pendulum type as shown in Fig. 3, of rigid construction, and accurate to 13.6 N · m (0.01 ft · lb) for readings of less than 1355 N · m (1 ft · lb) and to 1 percent for higher values. Accurate correction shall be made for friction and windage losses.

10.2 The dimensions of the machine shall be such that the center of percussion of the striker (Note 3, Section 4) is at the point of impact, that is, the center of the striking edge.

10.3 The pendulum shall be released from such a position that the linear velocity of the center of the striking edge (center of percussion) at the instant of impact shall be approximately 3.4 m/s (11 ft/s), which corresponds to an initial elevation of this point of 610 mm (2 ft).

10.4 The striking edge of the pendulum shall be tapered to have an included angle of 45 deg and shall be rounded to a radius of 3.17 mm (0.125 in.). It shall be so aligned that in the case of rectangular specimens it will make contact across the full width of the specimen.

10.5 The test specimen shall be supported against two rigid blocks in such a position that its center of gravity shall lie on a tangent to the arc of travel of the center of percussion of the pendulum drawn at the position of impact. The edges of the blocks shall be rounded to a radius of 3.17 mm (0.125 in.) and the points of support shall be 101.6 mm (4 in.) apart.

10.6 Means shall be provided for determining the impact value of the specimen, which is the energy expended by the machine in breaking the specimen. This value is equal to the difference between the energy in the pendulum blow and the energy remaining in the pendulum after breaking the specimen, after suitable correction has been made for friction and windage.

11. Test Specimen

11.1 For materials other than ceramic materials, the test specimen shall conform to the dimensions shown in Fig. 4, except as modified in accordance with 11.2, 11.3, and 11.4. To ensure the correct contour and condition of the specified notch, all specimens shall be notched as described in Section 6.

11.2 For molded material (Note 5, Section 5), the specimen shall be 12.5 mm ($^1/_2$ in.) by any dimension of 12.5 mm ($^1/_2$ in.) or less agreed upon as representative of the cross section in which the particular material is to be used. For all specimens having one dimension less than 12.5 mm ($^1/_2$ in.), the notch, when used, shall be cut in the narrower side. For all compression-molded specimens, the notch shall be in the side parallel to the direction of application of the molding pressure.

11.3 For sheet materials, the thickness shall be the thickness of the sheet, except that it shall not exceed 12.5 mm ($^1/_2$ in.). Sheet material thicker than 12.5 mm ($^1/_2$ in.) shall be machined down to 12.5 mm ($^1/_2$ in.). This material may be tested either edgewise or flatwise, as specified, and may be cut from the sheet either lengthwise or crosswise, as specified (Note 8, Section 5). When specimens are tested flatwise, the notch, when used, shall be made in the original surface.

11.4 When the individual specimen is less than 12.5 mm ($^1/_2$ in.) in thickness, the test specimen may be a composite specimen consisting of a number of individual thin pieces aggregating as nearly as possible 12.5 mm ($^1/_2$ in.) in thickness. The individual members of the test specimen shall all be accurately aligned with each other and shall be tested edgewise. Single specimens less than 12.5 mm ($^1/_2$ in. in thickness may be used, provided the width is sufficient to ensure stability during the test and the impact value of the material is sufficiently high to be accurately determined by a machine of the capacity used.

11.5 For ceramic material, the specimen shall be an unnotched cylinder 150 mm (6 in.) in length. The diameter shall be 12.5 mm (0.5 in.), 18.9 mm (0.75 in.), or 28.8 mm (1.125 in.), whichever value is comparable to that of the finished product.

12. Conditioning

12.1 Specimens shall be preconditioned and tested in accordance with Procedure A of ASTM Methods D 618.

13. Procedure

13.1 At least five and preferably ten individual determinations of impact value shall be made under the conditions prescribed in Section 12.

13.2 The test specimen shall be supported against the steel blocks so that the blow will be struck at the center of the specimen, and on the side opposite the notch for notched specimens.

13.3 When a composite specimen is used, the individual members shall be closely in contact and accurately aligned with each other.

14. Report

14.1 The report shall include the following:

14.1.1 A statement indicating the size of specimen, the method of test and type of preconditioning used, and, for sheet materials, the direction of testing and whether the specimens were cut lengthwise or crosswise from the sheet,

14.1.2 For ceramic materials, the energy expended in breaking each individual specimen expressed directly, and for other materials, the value of energy expended in breaking each individual specimen expressed in newton-meters per meter (or foot-pounds per inch) of notch, or per meter (or inch) of width of the face of the specimen against which the hammer strikes, determined by dividing the energy in newton-meters (or foot-pounds) expended in the individual test by the actual dimension in meters (or inches) along the notch or face of the specimen broken in each test; also, whether a single or a composite specimen was used, and

14.1.3 The average of the values given in 14.1.2, the average thickness or diameter of the individual specimen, and the number of such specimens broken in each operation of the machine.

METHOD C. CANTILEVER BEAM (IZOD TYPE) TEST FOR MATERIALS OF LESS THAN 680 N · M/M (0.5 FT · LB/IN.) OF NOTCH

15. Apparatus

15.1 The apparatus shall be the same as specified in Section 4.

16. Test Specimen

16.1 The test specimen shall be the same as specified in Section 5.

17. Conditioning

17.1 Specimens shall be preconditioned and tested in accordance with Methods D 618.

18. Procedure

18.1 At least five and preferably ten individual determinations of impact value shall be made under the conditions prescribed in Section 17.

18.2 The test specimen shall be rigidly clamped with the center line of the notch on the level of the top of the clamping surface (Note 14, Section 8) and the blow shall be struck on the notched side.

18.3 When a composite specimen is used, the individual members shall be held closely in contact and accurately aligned with each other when clamped (Note 14, Section 8).

18.4 An estimate of the magnitude of the energy to toss each different type of material and each different specimen size (width) shall be made. This is done by repositioning the free end of the broken specimen on the clamped portion and striking it a second time with the pendulum in such a way as to impart to it approximately the same velocity it attained during the test. This is done by releasing the pendulum from a height corresponding to that to which it rose following the breaking of the test specimen. The energy to toss is then considered to be the difference between the reading obtained as described above and the free swing reading obtained from this height.

19. Report

19.1 The report shall include the following:

19.1.1 A statement indicating the size of specimen, the method of test and type of preconditioning used, and, for sheet materials, the direction of testing and whether the specimens were cut lengthwise or crosswise from the sheet,

19.1.2 The value of energy expended in breaking each individual specimen expressed in newton-meters per meter (or foot-pounds per inch) of notch determined by dividing the energy in newton-meters (or foot-pounds)

expended in the individual test by the actual dimension in meters (or inches) along the notch of the specimen broken in each test, the range of these values for each material tested, and whether a single or a composite specimen was used and also the direction of cutting the specimen.

19.1.3 The average of the values given in 19.1.2, reported as the "Izod Impact Strength," the average thickness of the individual specimen, and the number of such specimens broken in each operation of the machine.

19.1.4 The estimated toss factor expressed in newton-meters per meter (or foot-pounds per inch) of notch, obtained by dividing the energy to toss in newton-meters (or foot-pounds) by specimen width (dimension along width), and

19.1.5 The difference between the "Izod Impact Strength" and the toss factor reported as the "Estimated Net Izod Impact Strength."

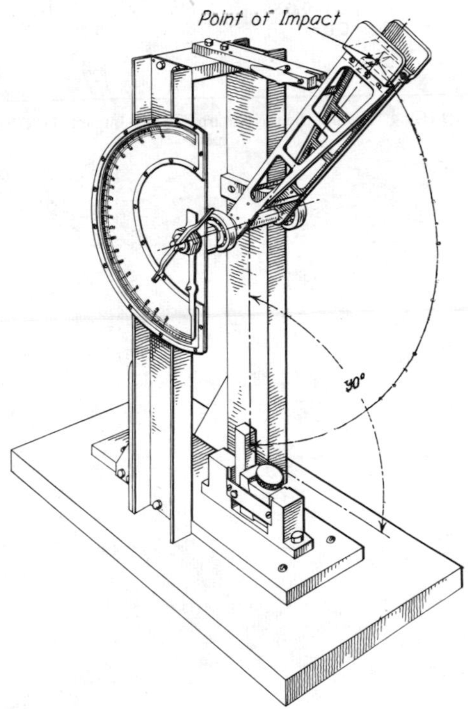

FIG. 1 Cantilever Beam (Izod-Type) Impact Machine.

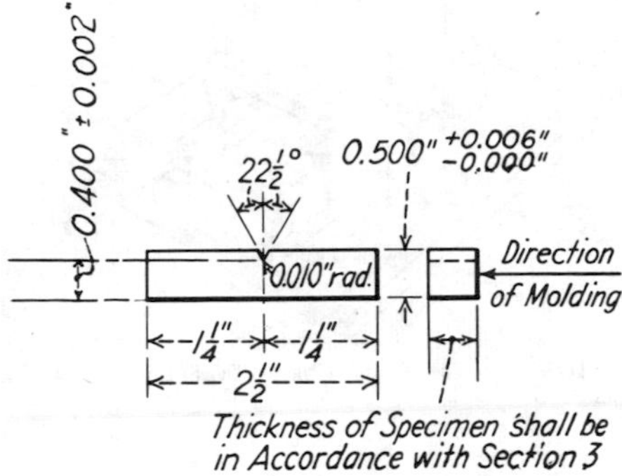

Metric Equivalents

in.	mm
0.400 ± 0.002	10.16 ± 0.05
0.500 + 0.006	12.70 + 0.15
−0.000	−0.00
0.010	0.25
1¼	31.75
2½	63.5

FIG. 2 Dimensions of Test Specimen.

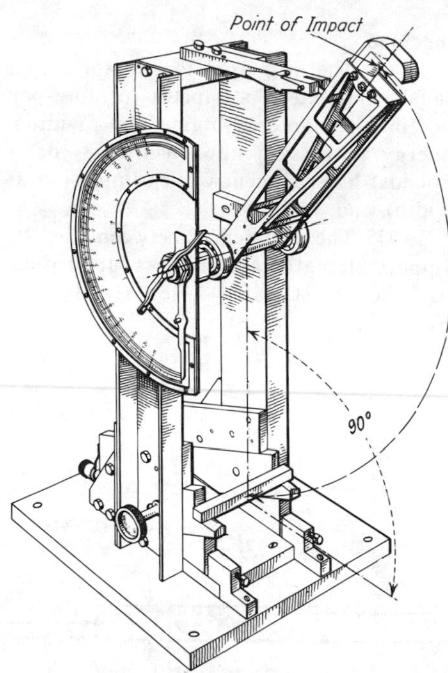

FIG. 3 Simple Beam (Charpy-Type) Impact Machine.

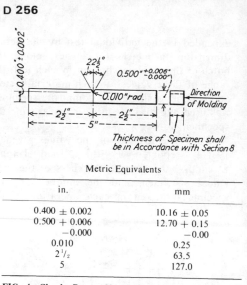

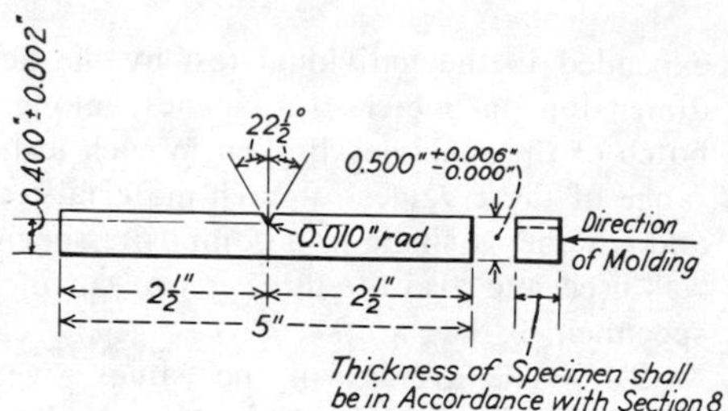

Metric Equivalents

in.	mm
0.400 ± 0.002	10.16 ± 0.05
0.500 + 0.006 −0.000	12.70 + 0.15 −0.00
0.010	0.25
2½	63.5
5	127.0

FIG. 4 Simple Beam (Charpy-Type) Impact Test Specimen.

Standard Methods of Test for
D-C RESISTANCE OR CONDUCTANCE OF INSULATING MATERIALS[1]

This Standard is issued under the fixed designation D 257; the number immediately following the designation indicates the year of original adoption or, in the case of revision, the year of last revision. A number in parentheses indicates the year of last reapproval.

NOTE—Editorial change in Table 2 was made in July 1967.

1. Scope

1.1 These methods cover procedures for the determination of d-c insulation resistance, volume resistance, volume resistivity, surface resistance, and surface resistivity of electrical insulating materials, or the corresponding conductances and conductivities.

NOTE 1—The values stated in U.S. customary units are to be regarded as the standard. The metric equivalents of U.S. customary units may be approximate.

1.2 The following ASTM standards are referenced:

Methods	Section	ASTM Designation References
Definitions	4	D 1711
Conditioning	9.1	D 618
Cleaning Solid Specimens	8.1	D 1371
Liquid Specimens and Cells	7.4	D 1169
Humidity Control	9.2	E 104

2. Summary of Methods

2.1 The resistance or conductance of a material specimen or of a capacitor is determined from a measurement of current or of voltage drop under specified conditions. By using the appropriate electrode systems, surface and volume resistance or conductance may be measured separately. The resistivity or conductivity can then be calculated when the required specimen and electrode dimensions are known.

3. Significance

3.1 Insulating materials are used in general to isolate and provide mechanical support for components of an electrical system from each other and from ground. For this purpose, it is generally desirable to have the insulation resistance as high as possible, consistent with acceptable mechanical, chemical- and heat-resisting properties. Since insulation resistance or conductance combines both volume and surface resistance or conductance, its measured value is most useful when the test specimen and electrodes have the same form as is required in actual use. Surface resistance or conductance changes very rapidly with humidity, while volume resistance or conductance changes only slowly although the final change may be greater.

3.2 Resistivity or conductivity may be used to predict, indirectly, the low-frequency dielectric breakdown and dissipation factor properties of some materials. Resistivity or conductivity is often used as an indirect measure of moisture content, degree of cure, mechanical continuity, and deterioration of various types. The usefulness of these indirect measurements is dependent on the degree of correlation established by supporting theoretical or experimental investigations. A decrease of surface resistance may result either in an increase of the dielectric breakdown voltage because the electric field intensity is reduced, or a decrease of the dielectric breakdown voltage because the area under stress is increased.

3.3 All the dielectric resistances or conductances may depend on the length of time

[1] These methods are under the jurisdiction of ASTM Committee D-9 on Electrical Insulating Materials. A list of members may be found in the ASTM Yearbook.

Current edition effective Nov. 9, 1966. Originally issued 1925. Replaces D 257 – 61.

of electrification and on the value of applied voltage (in addition to the usual environmental variables). These must be known to make the measured value of resistance or conductance meaningful.

3.4 Volume resistivity or conductivity can be used as an aid in designing an insulator for a specific application. The change of resistivity or conductivity with temperature and humidity may be great (**1,2,3,4**),[2] and must be known when designing for operating conditions. Volume resistivity or conductivity measurements are often used in checking the uniformity of an insulating material, either with regard to processing or to detect conductive impurities that affect the quality of the material and that may not be readily detectable by other methods.

3.5 Volume resistivities above 10^{21} $\Omega \cdot$cm, obtained on specimens under usual laboratory conditions, are of doubtful validity, when the limitations of commonly used measuring equipment are considered. Resistivities of 10^{22} $\Omega \cdot$cm require the measurement of the equivalent of a 2-μF, TFE-fluorocarbon capacitor having an apparent resistance of 10^{15} Ωs. This is a time constant of 2×10^9 s. If fully charged to 500 V, the voltage would decay about 25 μV/min.

3.6 Surface resistance or conductance cannot be measured accurately, only approximated, because more or less volume resistance or conductance is nearly always involved in the measurement. The measured value is largely a property of the contamination that happens to be on the specimen at the time. However, the permittivity of the specimen influences the deposition of contaminants and its surface characteristics affect the conductance of the contaminants. Thus surface resistivity or conductivity can be considered to be related to material properties when contamination is involved but is not a material property in the usual sense.

4. Definitions

4.1 *insulation resistance*, R_i—the insulation resistance between two electrodes that are in contact with, or embedded in, a specimen, is the ratio of the direct voltage applied to the electrodes to the total current between them. It is dependent upon both the volume and surface resistances of the specimen.

4.2 *volume resistance*, R_v—the volume resistance between two electrodes that are in contact with, or embedded in, a specimen, is the ratio of the direct voltage applied to the electrodes to that portion of the current between them that is distributed through the volume of the specimen.

4.3 *surface resistance*, R_s—the surface resistance between two electrodes that are on the surface of a specimen is the ratio of the direct voltage applied to the electrodes to that portion of the current between them which is primarily in a thin layer of moisture or other semi-conducting material that may be deposited on the surface.

4.4 *volume resistivity*, ρ_v—the volume resistivity of a material is the ratio of the potential gradient parallel to the current in the material to the current density.

NOTE 2—In the metric system, volume resistivity of an electrical insulating material in ohm-cm is numerically equal to the volume resistance in ohms between opposite faces of a 1-cm cube of the material.

4.5 *surface resistivity*, ρ_s—the surface resistivity of a material is the ratio of the potential gradient parallel to the current along its surface to the current per unit width of the surface.

NOTE 3—Surface resistivity of a material is numerically equal to the surface resistance between two electrodes forming opposite sides of a square. The size of the square is immaterial.

5. Electrode Systems

5.1 The electrodes for insulating materials should be of a material that is readily applied, allows intimate contact with the specimen surface, and introduces no appreciable error because of electrode resistance or contamination of the specimen (**8**). The electrode material should be corrosion-resistant under the conditions of test. For tests of fabricated specimens such as feed-through bushings, cables, etc., the electrodes employed are a part of the specimen or its mounting. Measurements of insulation resistance or conductance, then, include the contaminating effects of electrode or mounting materials and are generally related to the performance of the specimen in actual use.

[2] The bold face numbers in parentheses refer to the list of references appended to these methods.

5.1.1 *Binding-Post and Taper-Pin Electrodes*, Figs. 1 and 3, provide a means of applying voltage to rigid insulating materials to permit an evaluation of their resistive or conductive properties. These electrodes simulate to some degree the actual conditions of use, such as binding posts on instrument panels and terminal strips. In the case of laminated insulating materials having high-resin-content surfaces, somewhat lower insulation resistance values may be obtained with taper-pin than with binding posts, due to more intimate contact with the body of the insulating material. Resistance or conductance values obtained are highly influenced by the individual contact between each pin and the dielectric material, the surface roughness of the pins, and the smoothness of the hole in the dielectric material. Reproducibility of results on different specimens is difficult to obtain.

5.1.2 *Metal Bars* in the arrangement of Fig. 2 were primarily devised to evaluate the insulation resistance or conductance of flexible tapes and thin, solid specimens as a fairly simple and convenient means of electrical quality control. This arrangement is somewhat more satisfactory for obtaining approximate values of surface resistance or conductance when the width of the insulating material is much greater than its thickness.

5.1.3 *Silver Paint*, Figs. 4, 6, and 7, is available commercially with a high conductivity, either air-drying or low-temperature-baking varieties, which are sufficiently porous to permit diffusion of moisture through them and thereby allow the test specimen to be conditioned after the application of the electrodes. This is a particularly useful feature in studying resistance-humidity effects, as well as change with temperature. However, before conductive paint is used as an electrode material, it should be established that the solvent in the paint does not attack the material so as to change its electrical properties. Reasonably smooth edges of guard electrodes may be obtained with a fine-bristle brush. However, for circular electrodes, sharper edges can be obtained by the use of a ruling compass and silver paint for drawing the outline circles of the electrodes and filling in the enclosed areas by brush. A narrow strip of masking tape may be used, provided the pressure-sensitive adhesive used does not contaminate the surface of the specimen. Clamp-on masks also may be used if the electrode paint is sprayed on.

5.1.4 *Sprayed Metal*, Figs. 4, 6, and 7, may be used if satisfactory adhesion to the test specimen can be obtained. Thin sprayed electrodes may have certain advantages in that they are ready for use as soon as applied. They may be sufficiently porous to allow the specimen to be conditioned, but this should be verified. Narrow strips of masking tape or clamp-on masks must be used to produce a gap between the guarded and the guard electrodes. The tape shall be such as not to contaminate the gap surface.

5.1.5 *Evaporated Metal* may be used under the same conditions given in 5.1.4.

5.1.6 *Metal Foil*, Fig. 4, may be applied to specimen surfaces as electrodes. The usual thickness of metal foil used for resistance or conductance studies of dielectrics ranges from 0.006 to 0.08 mm (0.00025 to 0.003 in.). Lead or tin foil is in most common use, and is usually attached to the test specimen by a minimum quantity of petrolatum, silicone grease, oil, or other suitable material, as an adhesive. Such electrodes shall be applied under a smoothing pressure sufficient to eliminate all wrinkles, and to work excess adhesive toward the edge of the foil where it can be wiped off with a cleansing tissue. One very effective method is to use a hard narrow roller (1.0 to 1.5 cm wide), and to roll outward on the surface until no visible imprint can be made on the foil with the roller. This technique can be used satisfactorily only on specimens that have very flat surfaces. With care, the adhesive film can be reduced to 0.0025 mm (0.0001 in.). As this film is in series with the specimen, it will always cause the measured resistance to be too high. This error may become excessive for the lower-resistivity specimens of thickness less than 0.25 mm (0.010 in.). Also the hard roller can force sharp particles into or through thin films (0.05 mm (0.002 in.)). Foil electrodes are not porous and will not allow the test specimen to condition after the electrodes have been applied. The adhesive may lose its effectiveness at elevated temperatures necessitating the use of flat metal back-up plates under pressure. It is possible, with the aid of a suitable cutting device, to cut a proper width strip from one electrode to form a guarded and guard electrode.

Such a three-terminal specimen normally cannot be used for surface resistance or conductance measurements because of the grease remaining on the gap surface. It may be very difficult to clean the entire gap surface without disturbing the adjacent edges of the electrode.

5.1.7 *Colloidal Graphite*, Fig. 4, dispersed in water or other suitable vehicle, may be brushed on nonporous, sheet insulating materials to form an air-drying electrode. Masking tapes or clamp-on masks may be used (5.1.4). This electrode material is recommended only if all of the following conditions are met:

5.1.7.1 The material to be tested must accept a graphite coating that will not flake before testing,

5.1.7.2 The material being tested must not absorb water readily, and

5.1.7.3 Conditioning must be in a dry atmosphere (Procedure B of ASTM Methods D 618, Conditioning Plastics and Electrical Insulating Materials for Testing),[3] and measurements made in this same atmosphere.

5.1.8 Mercury or other liquid metal (indium) electrodes give satisfactory results. Mercury is not recommended for continuous use or at elevated temperatures due to possible toxic effects. The metal forming the upper electrodes should be confined by stainless steel rings, each of which should have its lower rim reduced to a sharp edge by beveling on the side away from the liquid metal. Figure 5 shows two electrode arrangements.

5.1.9 *Flat Metal Plates*, Fig. 4, (preferably guarded) may be used for testing flexible and compressible materials, both at room temperature and at elevated temperatures. They may be circular or rectangular (for tapes). To ensure intimate contact with the specimen, considerable pressure is usually required. Pressures of 20 to 100 psi have been found satisfactory (see material specifications).

5.1.10 *Conducting Rubber* has been used as electrode material, as in Fig. 4, without guard and has the advantage that it can quickly and easily be applied and removed from the specimen. As the electrodes are applied only during the time of measurement, they do not interfere with the conditioning of the specimen. The conductive-rubber material must be backed by proper plates and be soft enough so that effective contact with the specimen is obtained when a reasonable pressure is applied.

NOTE 4—There is evidence that values of conductivity obtained using conductive-rubber electrodes are always smaller (20 to 70 percent) than values obtained with tinfoil electrodes (**12**). When only order-of-magnitude accuracies are required, and these contact errors can be neglected, a properly designed set of conductive-rubber electrodes can provide a rapid means for making conductivity measurement.

5.1.11 *Water* is widely employed as one electrode in testing insulation on wires and cables. Both ends of the specimen must be out of the water and of such length that leakage along the insulation is negligible. Guard rings may be necessary at each end. It may be desirable to add a small amount of sodium chloride to the water to ensure high conductivity. Measurements may be performed at temperatures up to about 100 C.

6. Choice of Apparatus and Method

6.1 *Power Supply*—A source of very steady direct voltage is required (see A1.6.3). This may be either batteries or a properly regulated and rectified power supply.

6.2 *Direct Measurements*—The current through a specimen at a fixed voltage may be measured using any equipment that has the required sensitivity and accuracy (±10 percent is usually adequate). The current-measuring equipment may be a galvanometer, d-c amplifier, or electrometer. Typical methods and circuits are given in Appendix A3. When the measuring device scale is calibrated to read ohms directly no calculations are required.

6.3 *Comparison Methods*—A Wheatstone-bridge circuit may be used to compare the resistance of the specimen with that of a standard resistor.

6.4 *Precision and Accuracy:*

6.4.1 *General*—As a guide in the choice of apparatus, the pertinent considerations are summarized in Table 1, but it is not implied that the examples enumerated are the only ones applicable. This table is not intended to indicate the limits of sensitivity and error of the various methods *per se*, but rather is in-

[3] *Annual Book of ASTM Standards*, Part 29.

tended to indicate limits that are distinctly possible with modern apparatus. In any case, such limits can be achieved or exceeded only through careful selection and combination of the apparatus employed. It must be emphasized, however, that the errors considered are those of instrumentation only. Errors such as those discussed in Appendix A1 are an entirely different matter. In this latter connection, the last column of Table 1 lists the resistance that is shunted by the insulation resistance between the guarded electrode and the guard system for the various methods. In general, the lower such resistance, the less probability of error from undue shunting.

NOTE 5—No matter what measurement method is employed, the highest precisions are achieved only with careful evaluation of all sources of error. It is possible either to set up any of these methods from the component parts, or to acquire a completely integrated apparatus. In general, the methods using high-sensitivity galvanometers require a more permanent installation than those using indicating meters or recorders. The methods using indicating devices such as voltmeters, galvanometers, d-c amplifiers, and electrometers require the minimum of manual adjustment and are easy to read but the operator is required to make the reading at a particular time. The Wheatstone bridge (Fig. A4) and the potentiometer method (Fig. A2(*b*)) require the undivided attention of the operator in keeping a balance, but allow the setting at a particular time to be read at leisure.

6.4.2 *Direct Measurements:*

6.4.2.1 *Galvanometer*—With a galvanometer having a sensitivity of 500 × 10^{-12} A/scale division, 500 V applied to a resistance of 10^{12} Ω (conductance of 10^{-12} mho) will produce a deflection of one scale division. Since the Ayrton shunt errors are negligible, the maximum percentage error in the computed resistance or conductance is the sum of the percentage errors in galvanometer sensitivity, galvanometer deflection, and indicated voltage. If the galvanometer deflection is at least 25 scale divisions and is read to the nearest 0.5 division, the resultant galvanometer error will not exceed 4 percent. Hence 4 × 10^{10} Ω (2.5 × 10^{-11} mho) at 500 V can be measured with a maximum error of ±6 percent when the voltmeter reads full scale, and ±10 percent when it reads one-third full scale. The desirability of a good voltmeter is readily seen.

6.4.2.2 *Voltmeter-Ammeter*—The maximum percentage error in the computed value is the sum of the percentage errors in the voltages, V_x and V_s, and the resistance, R_s. The errors in V_s and R_s are generally dependent more on the characteristics of the apparatus used than on the particular method. The most significant factors that determine the errors in V_s are indicator errors, amplifier zero drift, and amplifier gain stability. With modern, well-designed amplifiers or electrometers, gain stability is usually not a matter of concern. With existing techniques, the zero drift of conductively coupled d-c amplifiers or electrometers cannot be eliminated but it can be made slow enough to be relatively insignificant for these measurements. The zero drift is virtually nonexistent for carefully designed converter-type amplifiers. Consequently, the null method of Fig. A2(*b*) is theoretically less subject to error than those methods employing an indicating instrument, provided, however, that the potentiometer voltage is accurately known. The error in R_s is to some extent dependent on the amplifier sensitivity. For measurement of a given current, the higher the amplifier sensitivity, the greater likelihood that lower valued, highly precise wire-wound standard resistors can be used. Such amplifiers can be obtained. Standard resistances of 10^{11} ohms known to ±2 percent, are a distinct possibility. If 10-mV input to the amplifier or electrometer gives full-scale deflection with an error not greater than 2 percent of full scale, with 500 V applied, a resistance of 5 × 10^{15} ohms can be measured with a maximum error of 6 percent when the voltmeter reads full scale, and 10 percent when it reads one-third scale.

6.4.2.3 *Comparison-Galvanometer*—The maximum percentage error in the computed resistance or conductance is given by the sum of the percentage errors in R_s, the galvanometer deflections or amplifier readings, and the assumption that the current sensitivities are independent of the deflections. The latter assumption is correct to well within ±2 percent over the useful range (above $^1/_{10}$ full-scale deflection) of a good, modern galvanometer (probably $^1/_3$ scale deflection for a d-c current amplifier). The error in R_s depends on the type of resistor used, but resistances of 1 MΩ with a limit of error ±0.5 percent are available. With a galvanometer or d-c current amplifier having a sensitivity of 10^{-8} A for full-scale deflection, 500 V applied to a resistance of 5×10^{12} Ω will produce a 1 percent deflection. At this voltage, with the above noted standard resistor, and with $F_s = 10^5$, d_s would be about half of full-scale deflection, with a readability error not more than ±1 percent. If d_x is approximately $^1/_4$ of full-scale deflection, the readability error would not exceed ± 4 percent, and a resistance of the order of 2×10^{11} Ω could be measured with a maximum error of $\pm 5^1/_2$ percent.

6.4.2.4 *Voltage Rate-of-Change*—The accuracy of the measurement is directly proportional to the accuracy of the measurement of applied voltage and time rate of change of the electrometer reading. The length of time that the electrometer switch is open and the scale used should be such that the time can be measured accurately and a full-scale reading obtained. Under these conditions, the accuracy will be comparable with that of the other methods of measuring current.

6.4.2.5 *Comparison Bridge*—When the detector has adequate sensitivity, the maximum percentage error in the computed resistance is the sum of the percentage errors in the arms *A*, *B*, and *N*. With a detector sensitivity of 1 mV/scale division, 500 V applied to the bridge, and $R_N = 10^9$ Ω, a resistance of 5×10^{14} Ω will produce a detector deflection of one scale division. Assuming negligible errors in R_A and R_B, with $R_N = 10^9$ Ω known to within ±2 percent, and with the bridge balanced to one detector-scale division, a resistance of 2×10^{15} Ω can be measured with a maximum error of ±6 percent.

7. Test Specimens

7.1 *Insulation Resistance or Conductance Determination:*

7.1.1 The measurement is of greatest value when the specimen has the form, electrodes, and mounting required in actual use. Bushings, cables, and capacitors are typical examples for which the test electrodes are a part of the specimen and its normal mounting means.

7.1.2 For materials, the test specimen may be of any practical form. The specimen forms most commonly used are flat plates, tapes, rods, and tubes. The electrode arrangements of Fig. 3 may be used for flat plates, rods, or rigid tubes whose inner diameter is about 20 mm ($^3/_4$ in.) or more. The electrode arrangement of Fig. 2 may be used for strips of sheet material or for flexible tape. For rigid strip specimens the metal support may not be required. The electrode arrangements of Fig. 7 may be used for flat plates, rods, or tubes. Comparison of materials when using different electrode arrangements is frequently inconclusive.

7.2 *Volume Resistance or Conductance Determination:*

7.2.1 The test specimen may have any practical form that allows the use of a third electrode to guard against error from surface effects when necessary. Test specimens may be in the form of flat plates, tapes, or tubes. Figures 4 and 5 illustrate the application and arrangement of electrodes for plate or sheet specimens. Figure 6 is a diametral cross section of three electrodes applied to a tubular specimen, in which electrode No. 1 is the guarded electrode, electrode No. 2 is a guard electrode and electrode No. 3 is the unguarded electrode (**6,7**). For materials that have negligible surface leakage, the guard ring may be omitted. Convenient and generally suitable dimensions applicable to Fig. 4 in the case of test specimens that are 0.3 cm ($^1/_8$ in.) in thickness are as follows: $D_3 = 10.0$ cm (4.0 in.), $D_2 = 8.8$ cm (3.5 in.), and $D_1 = 7.6$ cm (3.0 in.), or alternatively, $D_3 = 5.0$ cm (2.0 in.), $D_2 = 3.8$ cm (1.5 in.), and $D_1 = 2.5$ cm (1.0 in.). For a given sensitivity, the former allows measurements of higher resistance than the latter.

7.2.2 Measure the average thickness of the specimens in accordance with the method pertaining to the material being tested. The actual points of measurement shall be uniformly distributed over the area to be covered by the measuring electrodes.

7.2.3 It is not necessary that the electrodes have the circular symmetry shown in Fig. 4 although this is generally convenient. The guarded electrode (No. 1) may be circular, square, or rectangular, allowing ready computation of the guarded electrode area for volume resistivity or conductivity determination when such is desired. The diameter of a circular electrode, or the shortest side of a square or rectangular electrode, should be at least four times the specimen thickness. The gap width should be great enough so that the surface leakage between electrodes No. 1 and No. 2 does not cause an error in the measurement (this is particularly important for high-input-impedance instruments, such as electrometers). If the gap is made equal to twice the specimen thickness as suggested in 7.3.3 so that the specimen can also be used for surface resistance or conductance determinations, the effective area of electrode No. 1 can, usually with sufficient accuracy, be taken as extending to the center of the gap. If, under special conditions, it becomes desirable to determine a more accurate value for the effective area of electrode No. 1, the correction for the gap width can be obtained from Appendix A2. Electrode No. 3 may have any shape provided that it extends at all points beyond the inner edge of electrode No. 2 by at least twice the specimen thickness.

7.2.4 For tubular specimens, electrode No. 1 should encircle the outside of the specimen and its axial length should be at least four times the specimen wall thickness. Considerations regarding the gap width are the same as those given in 7.2.3. Electrode No. 2 consists of an encircling electrode at each end of the tube, the two parts being electrically connected by external means. The axial length of each of these parts should be at least twice the wall thickness of the specimen. Electrode No. 3 must cover the inside surface of the specimen for an axial length extending beyond the outside gap edges at least twice the wall thickness. The tubular specimen (Fig. 6) may take the form of an insulated wire or cable. If the length of electrode is more than 100 times the thickness of the insulation the effects of the ends of the guarded electrode become negligible, and careful spacing of the guard electrodes is not required. Thus, the gap between electrodes No. 1 and No. 2 may be several centimeters to permit sufficient surface resistance between these electrodes when water is used as electrode No. 1. In this case, no correction is made for the gap width.

7.3 *Surface Resistance or Conductance Determination:*

7.3.1 The test specimen may be of any practical form consistent with the particular objective, such as flat plates, tapes, or tubes.

7.3.2 The arrangements of Figs. 2 and 3 were devised for those cases where the volume resistance is known to be high relative to that of the surface (**2**). However, the combination of molded and machined surfaces makes the result obtained generally inconclusive for rigid strip specimens. The arrangement of Fig. 2 is somewhat more satisfactory when applied to specimens for which the width is much greater than the thickness, the cut edge effect thus tending to become relatively small. Hence, this arrangement is more suitable for testing thin specimens such as tape, than for testing relatively thicker specimens. The arrangements of Figs. 2 and 3 should never be used for resistance or conductance determinations without due considerations of the limitations noted above.

7.3.3 The three electrode arrangements of Figs. 4, 5, and 6 may be used for purposes of material comparison. The resistance or conductance of the surface gap between electrodes No. 1 and No. 2 is determined directly by using electrode No. 1 as the guarded electrode, electrode No. 3 as the guard electrode, and electrode No. 2 as the unguarded electrode (**6,7**). The resistance or conductance so determined is actually the resultant of the surface resistance or conductance between electrodes No. 1 and No. 2 in parallel with some volume resistance or conductance between the same two electrodes. For this arrangement the surface gap width, g, should be approximately twice the specimen thickness, t.

7.3.4 Special techniques and electrode dimensions may be required for very thin specimens having such a low volume resistivity that the resultant low resistance between the guarded electrode and the guard system would cause excessive error.

7.4 *Liquid Insulation Resistance*—The sampling of liquid insulating materials, the test cells employed, and the methods of cleaning the cells shall be in accordance with ASTM Method D 1169, Test for Specific Resistance (Resistivity) of Electrical Insulating Liquids.[3]

8. Specimen Mounting

8.1 In mounting the specimens for measurements, it is important that there shall be no conductive paths between the electrodes or between the measuring electrodes and ground that will have a significant effect on the reading of the measuring instrument (**9**). Insulating surfaces should not be handled with bare fingers (acetate rayon gloves are recommended). For referee tests of volume resistivity or conductance, the surfaces should be cleaned with a suitable solvent before conditioning, in accordance with ASTM Recommended Practice D 1371, for Cleaning Plastic Specimens for Insulation Resistance Testing.[3] When surface resistance is to be measured, the surfaces should be cleaned or not as specified or agreed upon.

9. Conditioning

9.1 The specimens shall be conditioned in accordance with ASTM Method D 618, Conditioning Plastics and Electrical Insulating Materials for Testing.[3]

9.2 Methods for controlling the relative humidity are described in ASTM Recommended Practice E 104, for Maintaining Constant Relative Humidity by Means of Aqueous Solutions.[3]

10. Procedure

10.1 *Insulation Resistance or Conductance*—Properly mount the specimen in the test chamber. If the test chamber and the conditioning chamber are the same (recommended procedure), the specimens should be mounted before the conditioning is started. Make the measurement with a suitable device having the required sensitivity and accuracy (see Appendix A3). Unless otherwise specified, the time of electrification shall be 1 min and the applied direct voltage shall be 500 ± 5 V.

10.2 *Volume Resistivity or Conductivity*—Measure the dimensions of the electrodes and determine the effective area of the measuring electrode and width of guard gap, *g*. Make the measurement with a suitable device having the required sensitivity and accuracy. Unless otherwise specified, the time of electrification shall be 1 min, and the applied direct voltage shall be 500 ± 5 V.

10.3 *Surface Resistance or Conductance:*

10.3.1 Measure the electrode dimensions and determine the effective perimeter, *P*, and the distance between the electrodes, *g*. Measure the surface resistance or conductance between electrodes No. 1 and 2 with a suitable device having the required sensitivity and accuracy. Unless otherwise specified, the time of electrification shall be 1 min, and the applied direct voltage shall be 500 ± 5 V.

10.3.2 When the electrode arrangement of Fig. 2 is used, *P* is taken as the perimeter of the cross section of the specimen. For thin specimens, such as tapes, this perimeter effectively reduces to twice the specimen width.

10.3.3 When the electrode arrangements of Fig. 7 are used (and the volume resistance is known to be high compared to the surface resistance), *P* is taken to be the length of the electrodes or circumference of the cylinder.

11. Calculation

11.1 Calculate the volume resistivity, ρ_v, and the volume conductivity, γ_v, using the equations in Table 2.

11.2 Calculate the surface resistivity, ρ_s, and the surface conductivity, γ_s, using the equations in Table 2.

12. Report

12.1 The report shall include at least the following:

12.1.1 A description and identification of the material (name, grade, color, manufacturer, etc.),

12.1.2 Shape and dimensions of the test specimen,

12.1.3 Type and dimensions of electrodes,

12.1.4 Conditioning of the specimen (cleaning, predrying, hours at humidity and temperature, etc.),

12.1.5 Test conditions (specimen temperature, relative humidity, etc., at time of measurement),

12.1.6 Method of measurement (see Appendix A3),

12.1.7 Applied voltage,

12.1.8 Time of electrification of measurement,

12.1.9 Measured values of the appropriate resistances in ohms or conductances in mhos, and

12.1.10 Computed values when required, of volume resistivity in ohm-cm units, volume conductance in mho-per-centimeter units, surface resistivity in ohms, or surface conductivity in mhos.

References

(**1**) Curtis, H. L., "Insulating Properties of Solid Dielectrics," *Bulletin*, Nat. Bureau Standards, Vol II, 1915, Scientific Paper No. 234, pp. 369–417.

(**2**) Field, R. F., "How Humidity Affects Insulation, Part I, D-C Phenomena," *General Radio Experimenter*, Vol 20, Nos. 2 and 3, July–August, 1945.

(**3**) Field, R. F., "The Formation of Ionized Water Films on Dielectrics Under Conditions of High Humidity," *Journal of Applied Physics*, Vol 5, May, 1946.

(**4**) Herou, R., and LaCoste, R., "Sur La Mésure Des Resistivities et L'Étude de Conditionnement des Isolantes en Feuilles," Report IEC 15-GT_2 (France) 4 April, 1963.

(**5**) Young, M. G., "New Approach to the Determination of Insulating Resistance," *Insulation*, August, 1963.

(**6**) Amey, W. G., and Hamberger, F., Jr., "A Method for Evaluating the Surface and Volume Resistance Characteristics of Solid Dielectric Materials," *Proceedings*, Am. Soc. Testing Mats., Vol 49, 1949, pp. 1079–1091.

(**7**) Witt, R. K., Chapman, J. J., and Raskin, B. L., "Measuring of Surface and Volume Resistance," *Modern Plastics*, Vol 24, No. 8, April, 1947, p. 152.

(**8**) Thompson, B. H., and Mathes, K. N., "Electrolytic Corrosion—Methods of Evaluating Materials Used in Tropical Service," *Transactions*, Am. Inst. Electrical Engrs., Vol 64, June, 1945, p. 287.

(**9**) Scott, A. H., "Insulation Resistance Measurements," Fourth Electrical Insulation Conference, Washington, D. C., Feb. 19–22, 1962.

(**10**) Kline, G. M., Martin, A. R., and Crouse, W. A., "Sorption of Water by Plastics," *Proceedings*, Am. Soc. Testing Mats., Vol 40, 1940, pp. 1273–1282.

(**11**) Greenfield, E. W., "Insulation Resistance Measurements," *Electrical Engineering*, Vol 66, July, 1947, pp. 698–703.

(**12**) Scott, A. H., "Anomalous Conductance Behavior in Polymers," Report of the 1965 Conference on Electrical Insulation, NRC-NAS.

(**13**) Turner, E. F., Brancato, E. L., and Price, W., "The Measurement of Insulation Conductivity," *NRL Report 5060*, Naval Research Laboratory, Feb. 25, 1958.

(**14**) Dorcas, D. S., and Scott, R. N., "Instrumentation for Measuring the D-C Conductivity of Very High Resistivity Materials," *Review of Scientific Instruments*, Vol 35, No. 9, Sept. 1964.

(**15**) Warner, A. J., and Nordlin, H. G., "Measurement of Insulation Resistance of High Quality Insulating Materials," Annual Report of the 1951 Conference on Electrical Insulation, NRC-NAS.

(**16**) Cole, K. S., and Cole, R. H., "Dispersion and Absorption in Dielectrics, II Direct Current Characteristics," *Journal of Chemical Physics*, Vol 10, 1942.

(**17**) Field, R. F., "Interpretation of Current-Time Curves as Applied to Insulation Testing," AIEE Boston District Meeting, April 19–20, 1944.

TABLE 1 Apparatus and Conditions for Use

Method	Reference Section	Reference Figure	Maximum Ohms Detectable at 500 V	Maximum Ohms Measurable to ±6 percent at 500 V	Type of Measurement	Ohms Shunted by Insulation Resistance from Guard to Guarded Electrode
Voltmeter-ammeter (galvanometer)	A3.1	A1	10^{12}	10^{11}	deflection	10 to 10^5
Comparison (galvanometer)	A3.4	A3	10^{12}	10^{11}	deflection	10 to 10^5
Voltmeter-ammeter (d-c amplification, electrometer)	A3.2	A2(*a*) (Position *1*)	10^{15}	10^{13}	deflection	10^2 to 10^9
		A2(*a*) (Position *2*)	10^{15}	10^{13}	deflection	10^2 to 10^3
		A2(*b*)	10^{17}	10^{15}	deflection	10^3 to 10^{11}
		A2(*b*)	10^{17}	10^{15}	null	0 (effective)
Comparison (Wheatstone bridge)	A3.5	A4	10^{15}	10^{14}	null	10^5 to 10^6
Voltage rate-of-change	A3.3	A5	$\sim 10^8$ megohm mfd		deflection	unguarded
Megohmmeter (typical)	commercial instruments		10^{15}	10^{14}	direct-reading	10^4 to 10^{10}

TABLE 2 Calculation of Resistivity or Conductivity

Type of Electrodes or Specimen	Volume Resistivity, Ω-cm	Volume Conductivity, mhos/cm
	$\rho_v = \frac{A}{h} R_v$	$\gamma_v = \frac{h}{A} G_v$
Circular (Fig. 4)	$A = \frac{\pi(D_1 + g)^2}{4}$	
Rectangular	$A = (a + g)\ (b + g)$	
Square	$A = (a + g)^2$	
Tubes (Fig. 6)	$A = \pi\ D_0\ (L + g)$	
Cables	$\rho_v = \frac{2\pi L R_v}{2.303 \log \frac{D_2}{D_1}}$	$\gamma_v = \frac{2.303 \log \frac{D_2}{D_1}}{2\pi L R_v}$
	Surface Resistivity, Ω	**Surface Conductivity, mhos**
	$\rho_s = \frac{P}{g} R_s$	$\gamma_s = \frac{g}{P} G_s$
Circular (Fig. 4)	$P = \pi D_0$	
Rectangular	$P = 2(a + b + 2g)$	
Square	$P = 4(a + g)$	
Tubes (Figs. 6 and 7)	$P = 2\ \pi\ D_2$	

Nomenclature:

A = the effective area of the measuring electrode for the particular arrangement employed,
P = the effective perimeter of the guarded electrode for the particular arrangement employed,
R_v = measured volume resistance in ohms,
G_v = measured volume conductance in mhos,
R_s = measured surface resistance in ohms,
G_s = measured surface conductance in mhos,
h = average thickness of the specimen,
D_0, D_1, D_2, g, L = dimensions indicated in Figs. 4 and 6 (see Appendix A2 for correction to g) and a, b, = lengths of the sides of rectangular electrodes.

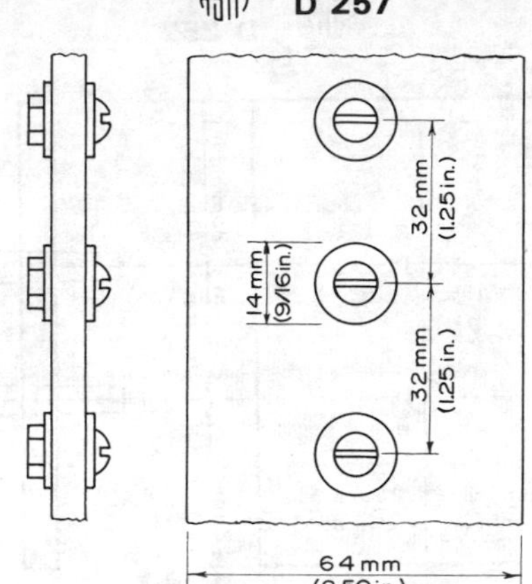

FIG. 1 Binding Post Electrodes for Flat, Solid Specimens.

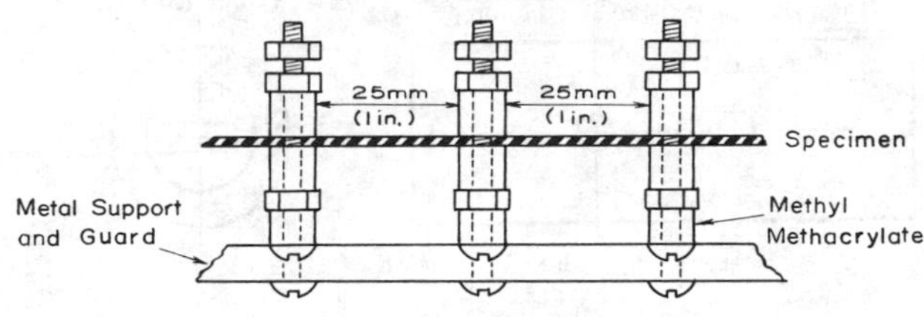

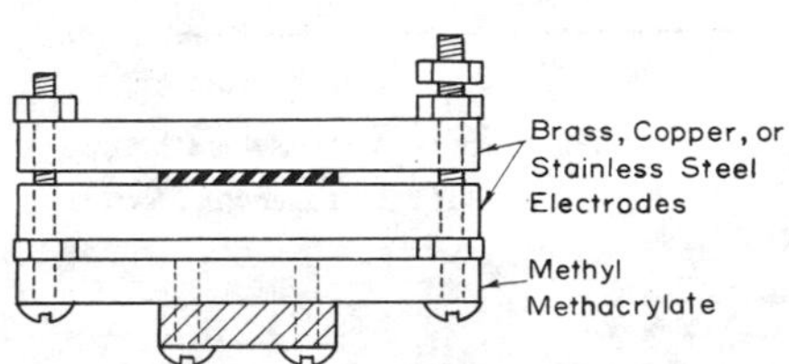

FIG. 2 Strip Electrodes for Tapes and Flat, Solid Specimens.

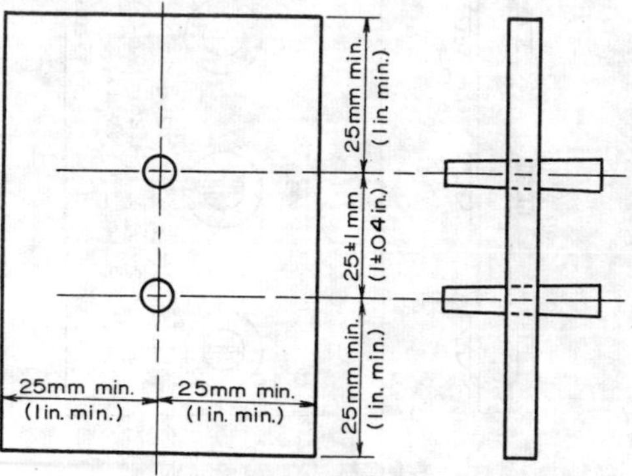

A. Plate Specimen

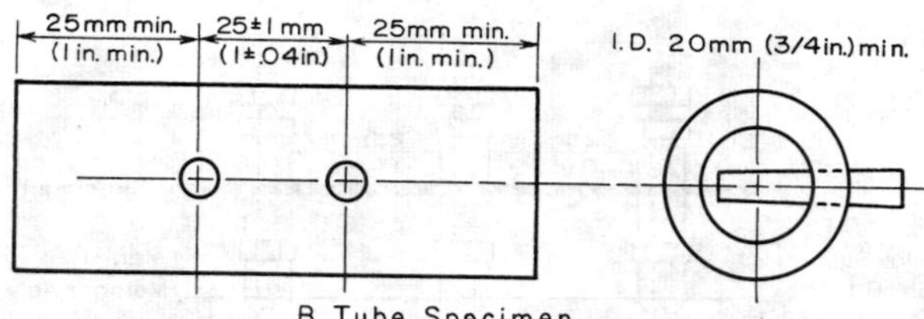

B. Tube Specimen

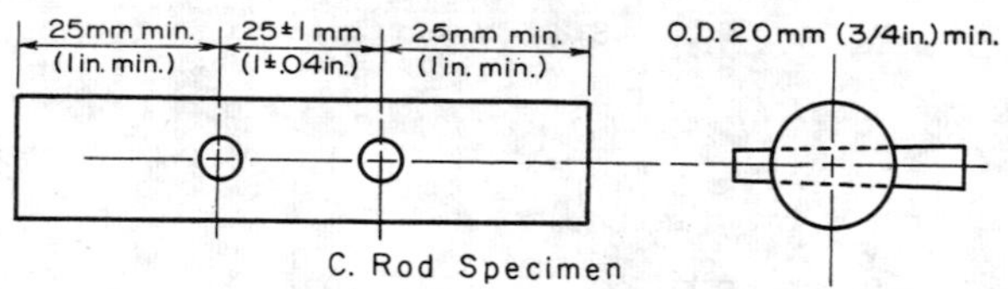

C. Rod Specimen

Use Pratt & Whitney No. 3 Taper Pins

FIG. 3 Taper-Pin Electrodes.

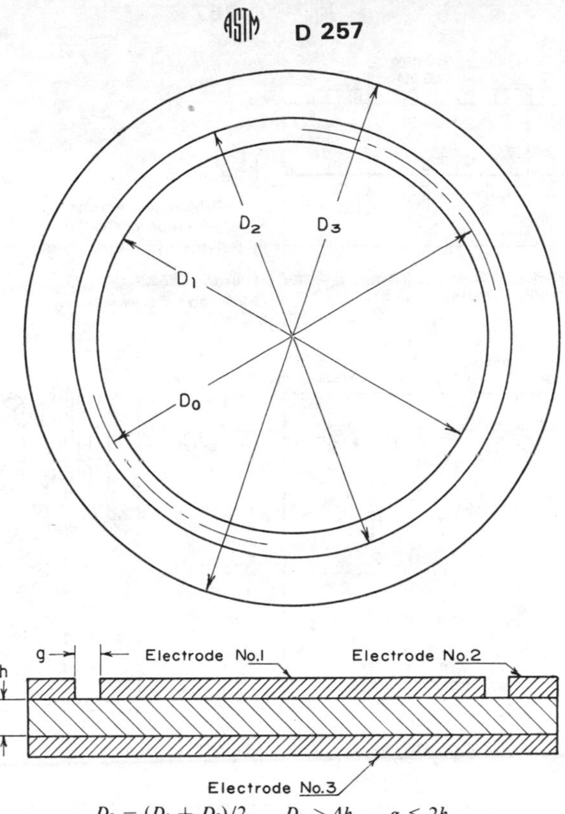

$$D_0 = (D_1 + D_2)/2 \qquad D_1 > 4h \qquad g \leqq 2h$$

FIG. 4 Flat Specimen for Measuring Volume and Surface Resistances or Conductances.

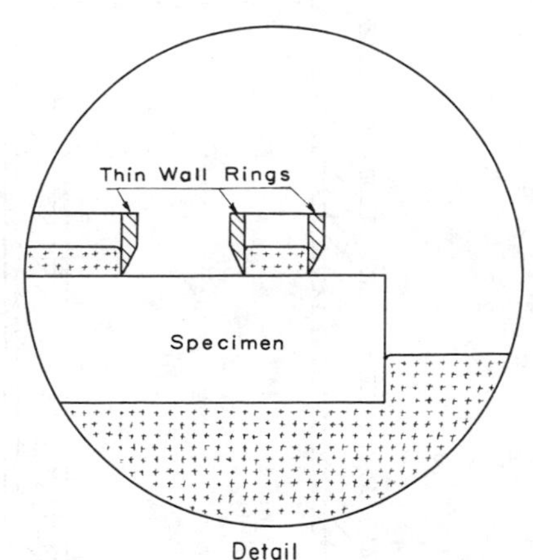

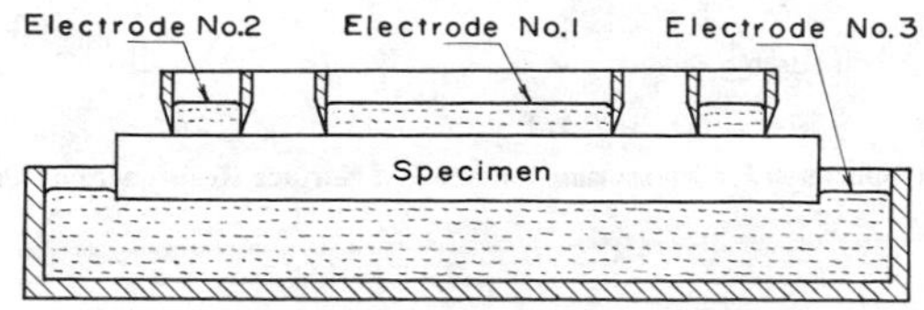

FIG. 5A Mercury Electrodes for Flat, Solid Specimens.

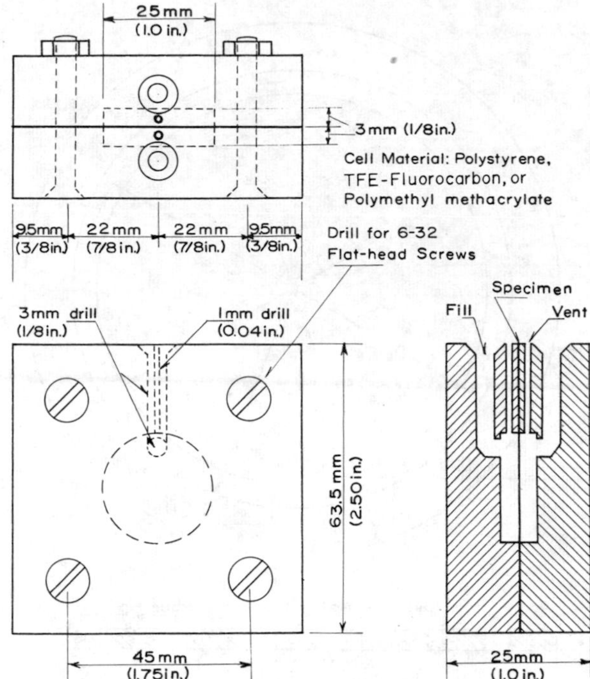

FIG. 5B Mercury Cell for Thin Sheet Material.

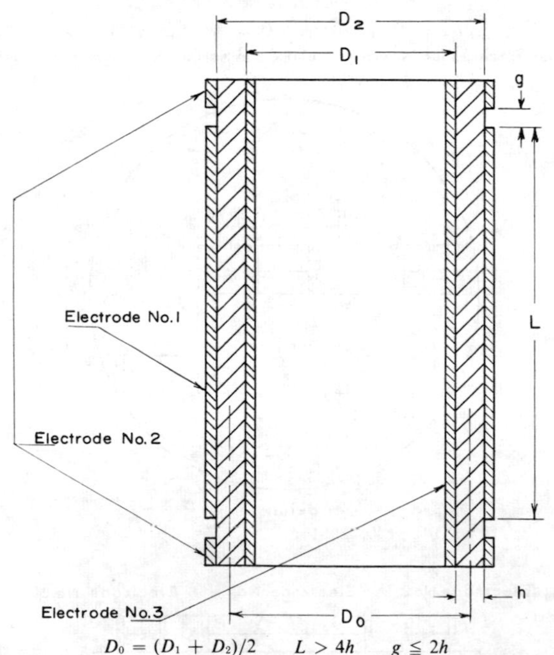

$D_0 = (D_1 + D_2)/2 \qquad L > 4h \qquad g \leqq 2h$

FIG. 6 Tubular Specimen for Measuring Volume and Surface Resistances or Conductances.

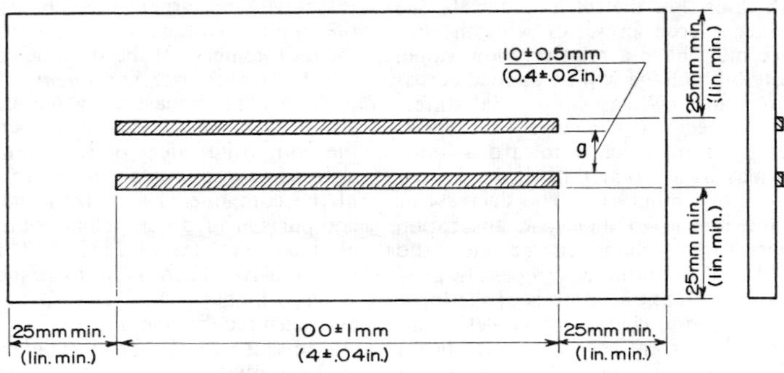

A — Plate Specimen

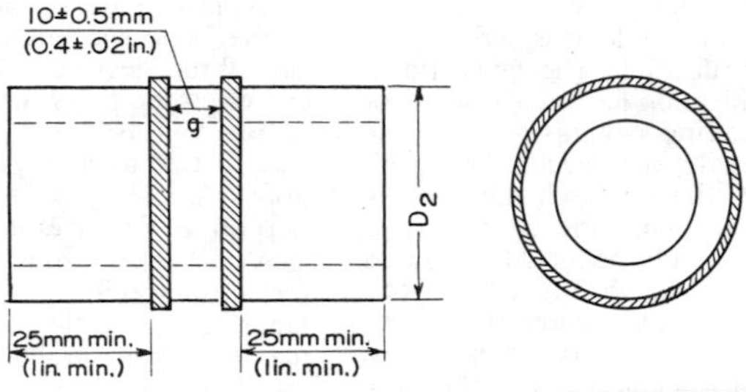

B—Tube or Rod Specimen

FIG. 7 Conducting-Paint Electrodes.

APPENDIXES

A1. Factors Affecting Insulation Resistance or Conductance Measurements

A1.1 *Inherent Variation in Materials*—Because of the variability of the resistance of a given specimen under similar test conditions and the nonuniformity of the same material from specimen to specimen, determinations are usually not reproducible to closer than 10 percent, and often are even more widely divergent (a range of values of 10 to 1 may be obtained under apparently identical conditions).

A1.2 *Temperature and Humidity*—The insulation resistance of solid dielectric materials decreases both with increasing temperature and with increasing humidity (**1,2,3,4,**). Volume resistivity is particularly sensitive to temperature changes, while surface resistance changes widely and very rapidly with humidity changes (**2,3**). In both cases the change is exponential. For some materials a change from 25 to 100 C may change insulation resistance or conductance by a factor of 100,000; a change from 25 to 90 percent relative humidity may change them by as much as a factor of 1,000,000 or even more. Insulation resistance or conductance being a function of both volume and surface resistance or conductance of the specimen, it is requisite to maintain both temperature and relative humidity within close limits when measuring insulation resistance or conductance. Surface resistance or conductance changes almost instantaneously with change of relative humidity. It is, therefore, necessary to make the measurements while the specimen is under specified conditions. In addition, when relative humidity of over 90 percent nominal is used, spurious values may result from surface condensation deposited at high instantaneous relative humidity portions of the control cycle. Extended periods of conditioning are required in determining the effect of humidity on volume resistance or conductance, since the absorption of water into the body of a dielectric is a relatively slow process (**10**); some specimens require months to come to equilibrium. This problem can be avoided by obtaining equivalent absolute humidity by use of a lower relative humidity at a higher temperature.

A1.3 *Time of Electrification*—Measurement of a dielectric material is not fundamentally different from that of a conductor except that an additional parameter, time of electrification, (and in some cases the voltage gradient) is involved. The relationship between the applied voltage and the current is involved in both cases. For dielectric materials, the standard resistance placed in series with the unknown resistance must have a relatively low value, so that essentially full voltage will be applied across the unknown resistance. When a potential difference is applied to a specimen, the current through it generally decreases asymptotically toward a limiting value which may be less than 0.01 of the current observed at the end of 1 min (**9,11**). This decrease of current with time is due to dielectric absorption (interfacial polarization, volume charge, etc.) and the sweep of mobile ions to the electrodes. In general, the relation of current and time is of the form $I = At^{-m}$, after the initial charge is completed and until the true leakage current becomes a significant factor. Depending upon the characteristics of the specimen material, the time required for the current to decrease to within 1 percent of this minimum value may be from a few seconds to many hours. Thus, in order to ensure that measurements on a given material will be comparable, it is necessary to specify the time of electrification. The conventional arbitrary time of electrification has been 1 min. For some materials, misleading conclusions may be drawn from the test results obtained at this arbitrary time. A resistance - time or conductance - time curve should be obtained under the conditions of test for a given material as a basis for selection of a suitable time of electrification, which must be specified in the test method for that material, or such curves should be used for comparative purposes. Occasionally, a material will be found for which the current increases with time. In this case either the time curves must be used or a special study undertaken, and arbitrary decisions made as to the time of electrification.

A1.4 *Magnitude of Voltage:*

A1.4.1 Both volume and surface resistance or conductance of a specimen may be voltage-sensitive (**4**). In that case, it is necessary that the same voltage gradient be used if measurements on similar specimens are to be comparable. Also, the applied voltage should be within at least 5 percent of the specified voltage. This is a separate requirement from that given in A1.6.3, which discusses voltage regulation and stability where appreciable specimen capacitance is involved.

A1.4.2 Commonly specified test voltages to be applied to the complete specimen are 100, 250, 500, 1000, 2500, 5000, 10,000 and 15,000 V. Of these, the most frequently used are 100 and 500 V. The higher voltages are used either to study the voltage-resistance or voltage-conductance characteristics of materials (to make tests at or near the operating voltage gradients), or to increase the sensitivity of measurement.

A1.4.3 Specimen resistance or conductance of some materials may, depending upon the moisture content, be affected by the polarity of the applied voltage. This effect, caused by electrolysis or ionic migration, or both, particularly in the presence of nonuniform fields, may be particularly noticeable in insulation configurations such as those found in cables where the test-voltage gradient is greater at the inner conductor than at the outer surface. Where electrolysis or ionic migration does exist in specimens, the electrical resistance will be lower when the smaller test electrode is made negative with respect to the larger. In such cases, the polarity of the applied voltage shall be specified according to the requirements of the specimen under test.

A1.5 *Contour of Specimen:*

A1.5.1 The measured value of the insulation resistance or conductance of a specimen results from the composite effect of its volume and surface resistances or conductances. Since the relative values of the components vary from material to material, comparison of different materials by the use of the electrode systems of Figs. 1, 2, and 3 is generally inconclusive. There is no assurance that, if material A has a higher insulation resistance than material B as measured by the use of one of these electrode systems, it will also have a higher resistance than B in the application for which it is intended.

A1.5.2 It is possible to devise specimen and electrode configurations suitable for the separate evaluation of the volume resistance or conductance and the approximate surface resistance or conductance of the same specimen. In general, this requires at least three electrodes so arranged that one may select electrode pairs for which the resistance or conductance measured is primarily that of either a volume current path or a surface current path, not both (**6**).

A1.6 *Deficiencies in the Measuring Circuit:*

A1.6.1 The insulation resistance of many solid dielectric specimens is extremely high at standard laboratory conditions, approaching or exceeding the maximum measurable limits given in Table 1. Unless extreme care is taken with the insulation of the measuring circuit, the values obtained are more a measure of apparatus limitations than of the material itself. Thus errors in the measurement of the specimen may arise from undue shunting of the specimen, reference resistors, or the current-measuring device, by leakage resistances or conductances of unknown, and possibly variable, magnitude.

A1.6.2 Electrolytic, contact, or thermal emf's may exist in the measuring circuit itself; or spurious emf's may be caused by leakage from external sources. Thermal emf's are normally insignificant except in the low resistance circuit of a galvanometer and shunt. When thermal emf's are present, random drifts in the galvanometer zero occur. Slow drifts due to air currents may be troublesome. Electrolytic emf's are usually associated with moist specimens and dissimilar metals, but emf's of 20 mV or more can be obtained in the guard circuit of a high-resistance detector when pieces of the same metal are in contact with moist specimens. If a voltage is applied between the guard and the guarded electrodes a polarization emf may remain after the voltage is removed. True contact emf's can be detected only with an electrometer and are not a source of error. The term "spurious emf" is sometimes applied to electrolytic emf's. To ensure the absence of spurious emf's of whatever origin, the deflection of the detecting device should be observed before the application of voltage to the specimen and after the voltage has been removed. If the two deflections are the same, or nearly the same, a correction can be made to the measured resistance or conductance, provided the correction is small. If the deflections differ widely, or approach the deflection of the measurement, it will be necessary to discover and remove the source of the spurious emf (**8**). Capacitance changes in the connecting shielded cables can cause serious difficulties.

A1.6.3 Where appreciable specimen capacitance is involved, both the regulation and transient stability of the applied voltage should be such that resistance or conductance measurements can be made to prescribed accuracy. Short-time transients, as well as relatively long-time drifts in the applied voltage may cause spurious capacitive charge and discharge currents which can significantly affect the accuracy of measurement. In the case of current-measuring methods particularly, this can be a serious problem. The current in the measuring instrument due to a voltage transient is $I_0 = C_x\, dV/dt$. The amplitude and rate of pointer excursions depend upon the following factors:

A1.6.3.1 The capacitance of the specimen,

A1.6.3.2 The magnitude of the current being measured,

A1.6.3.3 The magnitude and duration of the incoming voltage transient, and its rate of change,

A1.6.3.4 The ability of the stabilizing circuit used to provide a constant voltage with incoming transients of various characteristics, and

A1.6.3.5 The time-constant of the complete test circuit as compared to the period and damping of the current-measuring instrument.

A1.6.4 Changes of range of a current-measuring instrument may introduce a current transient. When $R_m \ll R_x$ and $C_m \ll C_x$, the equation of this transient is

$$I = (V_0/R_x)[I - e^{-t/RmCx}]$$

where:

V_0 = applied voltage,
R_x = apparent resistance of the specimen,
R_m = effective input resistance of the measuring instrument,
C_x = capacitance of the specimen at 1000 Hz,
C_m = input capacitance of the measuring instrument, and
t = time after R_m is switched into the circuit.

For not more than 5 percent error due to this transient,

$$R_m C_x \leq t/3$$

Microammeters employing feedback are usually free of this source of error as the actual input resistance is divided, effectively, by the amount of feedback, usually at least by 1000.

A1.7 *Residual Charge*—In A1.3 it was pointed out that the current continues for a long time after the application of a potential difference to the electrodes. Conversely, current will continue for a long time after the electrodes of a charged specimen are connected together. It should be established that the test specimen is completely discharged before attempting the first measurement, a repeat measurement, a measurement of volume resistance following a measurement of surface resistance, or a measurement with reversed voltage (**9**). The time of discharge before making a measurement should be at least four times any previous charging time. The specimen electrodes should be connected together until the measurement is to be made to prevent any build-up of charge from the surroundings.

A1.8 *Guarding:*

A1.8.1 Guarding depends on interposing, in all critical insulated paths, guard conductors which intercept all stray currents that might otherwise cause errors. The guard conductors are connected together, constituting the guard system and forming, with the measuring terminals, a three-terminal network. When suitable connections are made, stray currents from spurious external voltages are shunted away from the measuring circuit by the guard system.

A1.8.2 Proper use of the guard system for the methods involving current measurement is illustrated in Figs. A1 to A3, inclusive, where the guard system is shown connected to the junction of the voltage source and current-measuring instrument or standard resistor. In Fig. A4 for the Wheatstone-bridge method, the guard system is shown connected to the junction of the two lower-valued-resistance arms. In all cases, to be effective, guarding must be complete, and must include any controls operated by the observer in making the measurement. The guard system is generally maintained at a potential close to that of the guarded terminal, but insulated from it. This is because, among other things, the resistance of many insulating materials is voltage-dependent. Otherwise, the direct resistances or conductances of a three-terminal network are independent of the electrode potentials. It is usual to ground the guard system and hence one side of the voltage source and current-measuring device. This places both terminals of the specimen above ground. Sometimes, one terminal of the specimen is permanently grounded. The current-measuring device usually is then connected to this terminal, requiring that the voltage source be well insulated from ground.

A1.8.3 Errors in current measurements may result from the fact that the current-measuring device is shunted by the resistance or conductance between the guarded terminal and the guard system. This resistance should be at least 10 to 100 times the input resistance of the current measuring device. In some bridge techniques, the guard and measuring terminals are brought to nearly the same potentials, but a standard resistor in the bridge is shunted between the unguarded terminal and the guard system. This resistance should be at least 100 times that of the reference resistor.

A1.9 *General*—Calculation of volume resistivity from the measured volume resistance involves the quantity A, the effective area of the guarded electrode. Depending on the material properties and the electrode configuration, A differs from the actual area of the guarded electrode for either, or both, of the following reasons.

A1.9.1 Fringing of the lines of current in the region of the electrode edges may effectively increase the electrode dimensions.

A1.9.2 If, as in the arrangement of Fig. 4, the guarded and unguarded electrodes are not parallel planes, a correction may be required because the current density is not constant from one electrode to the other.

A1.10 *Fringing:*

A1.10.1 If the specimen material is homogeneous and isotropic, fringing effectively extends the guarded electrode edge by an amount (**4**):

$$g/2 - \delta$$

where:

$$\delta = t\{2/\pi \ln \cosh [(\pi/4)(g/h)]\},$$

and g and h are the dimensions indicated in Figs. 4 and 6. The correction may also be written

$$g[1 - (2\delta/g)] = Bg$$

where B is the fraction of the gap width to be added to the diameter of circular electrodes or to the dimensions of rectangular or cylindrical electrodes.

A1.10.2 Laminated materials, however, are somewhat anisotropic after volume absorption of moisture. Volume resistivity parallel to the laminations is then lower than that in the perpendicular direction, and the fringing effect is increased. With such moist laminates, δ approaches zero, and the guarded electrode effectively extends to the center of the gap between guarded and unguarded electrodes (**4**).

A1.10.3 The fraction of the gap width g to be added to the diameter of circular electrodes or to the electrode dimensions of rectangular or cylindrical electrodes, B, as determined by above equation for δ, is as follows:

$\frac{g}{h}$	B	$\frac{g}{h}$	B
0.10	0.96	1.0	0.64
0.20	0.92	1.2	0.59
0.3	0.88	1.5	0.51
0.4	0.85	2.0	0.41
0.5	0.81	2.5	0.34
0.6	0.77	3.0	0.29
0.8	0.71		

NOTE A1—The symbol "ln" designates logarithm to the base e = 2.718. . . . When g is approximately equal to $2h$, δ is determined with sufficient approximation by the equation:

$$\delta = 0.586h$$

NOTE A2—For tests on thin films when $h \ll g$, or when a guard electrode is not used and one electrode extends beyond the other by a distance which is large compared with h, $0.883h$ should be added to the diameter of circular electrodes or to the dimensions of rectangular electrodes.

NOTE A3—During the transition between complete dryness and subsequent relatively uniform volume distribution of moisture, a laminate is neither homogeneous nor isotropic. Volume resistivity is of questionable significance during this transition and accurate equations are neither possible nor justified, calculations within an order of magnitude being more than sufficient.

A1.11 *Voltmeter-Ammeter Method Using a Galvanometer:*

A1.11.1 A d-c voltmeter and a galvanometer with a suitable shunt are connected to the voltage source and to the test specimen as shown in Fig. A1. The applied voltage is measured by a d-c voltmeter, preferably having such a range and accuracy as to give minimum error in voltage indication. In no case shall a voltmeter be used that has an error greater than ±2 percent of full scale, nor a range such that the deflection is less than one third of full scale (for a pivot-type instrument). The current is measured by a galvanometer having a high current sensitivity (a scale length of 50 cm is assumed, as shorter scale lengths will lead to proportionately higher errors) and provided with a precision Ayrton universal shunt for so adjusting its deflection that the readability error does not, in general, exceed ±2 percent of the observed value. The galvanometer should be calibrated to within ±2 percent. The galvanometer can be made direct reading in current by providing a suitable additional fixed shunt.

A1.11.2 The unknown resistance, R_x, or conductance, G_x, is calculated as follows:

$$R_x = 1/G_x = V_x/I_x = V_x/KdF \quad (1)$$

where:

K = galvanometer sensitivity, in amperes per scale division,
d = deflection in scale divisions,
F = ratio of the total current, I_x, to the galvanometer current, and
V_x = applied voltage.

A1.12 *Voltmeter-Ammeter Method Using D-C Amplification or Electrometer:*

A1.12.1 The voltmeter-ammeter method can be extended to measure higher resistances by using d-c amplification or an electrometer to increase the sensitivity of the current measuring device (**12,13,14**). Generally, but not necessarily, this is achieved only with some sacrifice in precision, depending on the apparatus used. The d-c voltmeter and the d-c amplifier or electrometer are connected to the voltage source and the specimen as illustrated in Fig. A2. The applied voltage is measured by a d-c voltmeter having the same characteristics as prescribed in A1.11.1. This current is measured in terms of the voltage drop across a standard resistance, R_s, which may be included in the electrometer case.

A1.12.2 In the circuit shown in Fig. A2(*a*) the specimen current I_x, produces across the standard resistance, R_s, a voltage drop which is amplified by the d-c amplifier, and read on an indicating meter or galvanometer. The net gain of the amplifier usually is stabilized by means of a feedback resistance, R_f, from the output of the amplifier. The indicating meter can be calibrated to read directly in terms of the feedback voltage, V_s, which is determined from the known value of the resistance of R_f, and the feedback current passing through it. When the amplifier has sufficient intrinsic gain, the feedback voltage, V_s, differs from the voltage, I_xR_s, by a negligible amount. As shown in Fig. A2(*a*) the return lead from the voltage source, V_x, can be connected to either end of the feedback resistor, R_f. With the connection made to the junction of R_s and R_f (switch in dotted position *1*), the entire resistance of R_s is placed in the measuring circuit and any alternating voltage appearing across the specimen resistance is amplified only as much as the direct voltage, I_xR_s, across R_s. With the connection made to the other end of R_f (switch position *2*), the apparent resistance placed in the measuring circuit is R_s times the ratio of the degenerated gain to the intrinsic gain of the amplifier; any alternating voltage appearing across the specimen resistance is then amplified by the intrinsic amplifier gain.

A1.12.3 In the circuit shown in Fig. A2(*b*), the specimen current I_x, produces a voltage drop across the standard resistance, R_s, which may or may not be balanced out by adjustment of an opposing voltage, V_s, from a calibrated potentiometer. If no opposing voltage is used, the voltage drop across the standard resistance, R_s, is amplified by the d-c amplifier or electrometer and read on an indicating meter or galvanometer. This produces a voltage drop between the measuring electrode and the guard electrode which may cause an error in the current measurement unless the resistance between the measuring electrode and the guard electrode is at least 10 to 100 times that of R_s. If an opposing voltage, V_s, is used, the d-c amplifier or electrometer serves only as a very sensitive, high-resistance null detector. The return lead from the voltage source, V_x, is connected as shown, to include the potentiometer in the measuring circuit. When connections are made in this manner, no resistance is placed in the measuring circuit at balance and thus no voltage drop appears between the measuring electrode and the guard electrode. However, a steeply increasing fraction of R_s is included in the measuring circuit, as the potentiometer is moved off balance. Any alternating voltage appearing across the specimen resistance is amplified by the net amplifier gain. The d-c amplifier may be either conductively coupled or an a-c amplifier provided with input and output converters. Induced alternating voltages across the specimen often are sufficiently troublesome that a resistance-capacitance filter preceding the amplifier is required. The input resistance of this filter should be at least 100 times greater than the effective resistance that is placed in the measurement circuit by resistance R_s.

A1.12.4 The resistance, R_x, or the conductance, G_x, is calculated as follows:

$$R_x = 1/G_x = V_x/I_x = (V_x/V_s)R_s \quad (2)$$

where:
V_x = applied voltage,
I_x = specimen current,
R_s = standard resistance, and
V_s = voltage drop across R_s, indicated by the amplifier output meter, the electrometer or the calibrated potentiometer.

A1.13 *Voltage Rate-of-Change Method:*

A1.13.1 If the specimen capacitance is relatively large, or capacitors are to be measured, the apparent resistance, R_x, can be determined from the charging voltage, V_0, the geometric capacitance value, C_0 (capacitance of C_x at 1000 Hz), and the rate-of-change of voltage, dv/dt, using the circuit of Fig. A5. To make a measurement the specimen is charged by closing S_2, with the electrometer shorting switch S_1 closed. When S_1 is subsequently opened, the voltage across the specimen will fall because the leakage and absorption currents must then be supplied by the capacitance C_0 rather than by V_0. The drop in voltage across the specimen will be shown by the electrometer. If a recorder is connected to the output of the electrometer, the rate of change of voltage, dv/dt, can be read from the recorder trace at any desired time after S_2 is closed (1 min usually specified). Alternatively, the voltage, ΔV, appearing on the electrometer in a time, Δt, can be used. Since this gives an average of the rate-of-change of voltage during Δt, the time Δt should be centered at the specified electrification time (time since closing S_2).

A1.13.2 If the input resistance of the electrometer is greater than the apparent specimen resistance and the input capacitance is $1/100$ or less of that of the specimen, the apparent resistance at the time at which dv/dt or $\Delta v/\Delta t$ is determined

$$R_x = V_0/I_x = V_0 dt/C_0 dV_m \text{ or, } V_0\Delta t/C_0\Delta V_m \quad (3)$$

is depending on whether or not a recorder is used. When the electrometer input resistance or capacitance cannot be ignored or when V_m is more than a small fraction of V_0 the complete equation should be used.

$$R_x = \{V_0 - [(R_x + R_m)/R_m]V_m\}/(C_0 + C_m)dV_m/dt \quad (4)$$

where:
C_0 = capacitance of C_x at 1000 Hz,
R_m = input resistance of the electrometer,
C_m = input capacitance of the electrometer,
V_0 = applied voltage, and
V_m = electrometer reading = voltage decrease on C_x.

A1.14 *Comparison Method Using a Galvanometer or D-C Amplifier* (1):

A1.14.1 A standard resistance, R_s, and a galvanometer or d-c amplifier are connected to the voltage source and to the test specimen as shown in Fig. A3. The galvanometer and its associated Ayrton shunt is the same as described in A1.11.1. An amplifier of equivalent direct current sensitivity may be used in place of the galvanometer. It is convenient, but not necessary, and not desireable if batteries are used as the voltage source, to connect a voltmeter across the source for a continuous check of its voltage. The switch is provided for shorting the unknown resistance in the process of measurement. Sometimes provision is made to short either the unknown or standard resistance but not both at the same time.

A1.14.2 In general, it is preferable to leave the standard resistance in the circuit at all times to prevent damage to the current measuring instrument in case of specimen failure. With the shunt set to the least sensitive position and with the switch open, the voltage is applied. The Ayrton shunt is then adjusted to give as near maximum scale reading as possible. At the end of the electrification time the deflection, d_x, and the shunt ratio, F_x, are notex. The shunt is then set to the least sensitive position and the switch is closed to short the unknown resistance. Again the shunt is adjusted to give as near maximum scale reading as possible and the galvanometer or meter deflection, d_s, and the shunt ratio, F_s, are noted. It is assumed that the current sensitivities of the galvanometer or amplifier are equal for nearly equal deflections of d_x and d_s.

A1.14.3 The unknown resistance, R_x, or conductance, G_x, is calculated as follows:

$$R_x = 1/G_x = R_s[(d_sF_s/d_xF_x) - 1] \quad (5)$$

where:
F_x and F_s = ratios of the total current to the galvanometer or d-c amplifier with R_x in

the circuit, and shorted, respectively.

A1.14.4 In case R_s is shorted when R_x is in the circuit or the ratio of F_s to F_x is greater than 100, the value of R_x or G_x is computed as follows:

$$R_x = 1/G_x = R_s(d_s F_s/d_x F_x) \quad (6)$$

A1.15 *Comparison Methods Using a Wheatstone Bridge:*

A1.15.1 The test specimen is connected into one arm of a Wheatstone bridge as shown in Fig. A4. The three known arms shall be of as high resistance as practicable, limited by the errors inherent in such resistors. Usually, the lowest resistance, R_A, is used for convenient balance adjustment, with either R_B or R_N being changed in decade steps. The detector shall be a d-c amplifier, with an input resistance high compared to any of these arms.

A1.15.2 The unknown resistance, R_x, or conductance, G_x, is calculated as follows:

$$R_x = 1/G_x = R_B R_N/R_A \quad (7)$$

where R_A, R_B, and R_N are as shown in Fig. A4. When arm *A* is a rheostat, its dial can be calibrated to read directly in megohms after multiplying by the factor $R_B R_N$ which for convenience, can be varied in decade steps.

A1.16 *Recordings*—It is possible to record continuously against time the values of the unknown resistance or the corresponding values of current at a known voltage. Generally, this is accomplished by an adaption of the voltmeter-ammeter method, using d-c amplification (A1.12). The zero drift of conductively coupled d-c amplifiers, while slow enough for the measurements of A1.12, may be too fast for continuous recording. This situation can be met by periodic checks of the zero, or by using an a-c amplifier with the input and output converter. The indicating meter of Fig. A2(*a*) can be replaced by a recording milliammeter or millivoltmeter as appropriate for the amplifier used. The recorder may be of either the deflection type or the null-balance type, the latter usually having a smaller error. Null-balance-type recorders also can be employed to perform the function of automatically adjusting the potentiometer shown in Fig. A2(*b*) and thereby indicating and recording the quantity under measurement. The characteristics of amplifier, recorder balancing mechanism, and potentiometer can be made such as to constitute a well integrated, stable, electromechanical, feedback system of high sensitivity and low error. Such systems also can be arranged with the potentiometer fed from the same source of stable voltage as the specimen, thereby eliminating the voltmeter error, and allowing a sensitivity and precision comparable with those of the Wheatstone-bridge Method (A1.15).

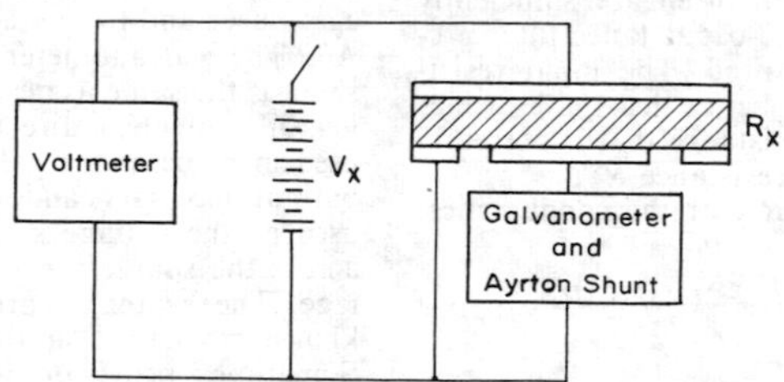

FIG. A1 Voltmeter-Ammeter Method Using a Galvanometer.

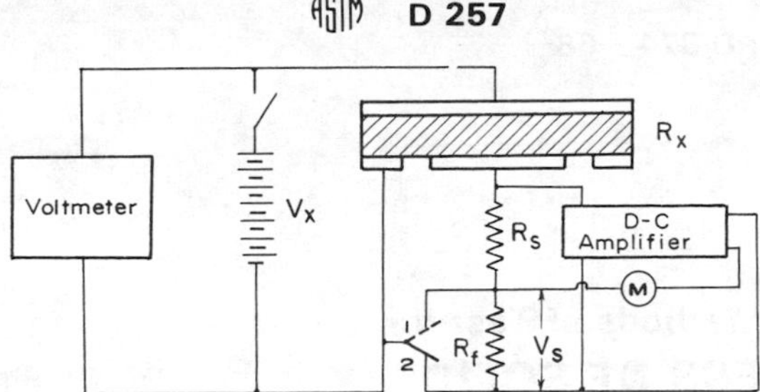

(a) Normal Use of Amplifier and Indicating Meter

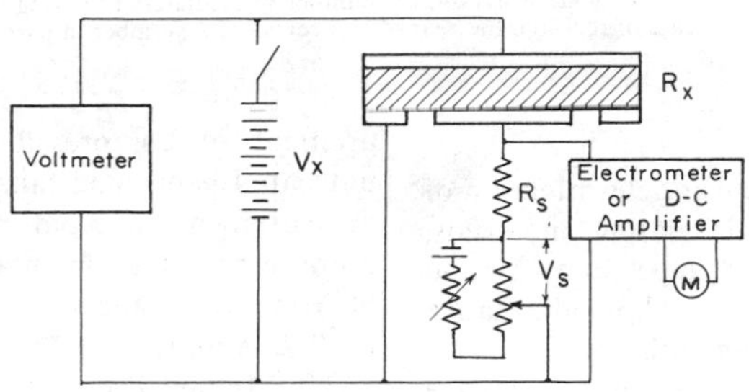

(b) Amplifier and Indicating Meter as Null Detector

FIG. A2 Voltmeter-Ammeter Method Using D-C Amplification.

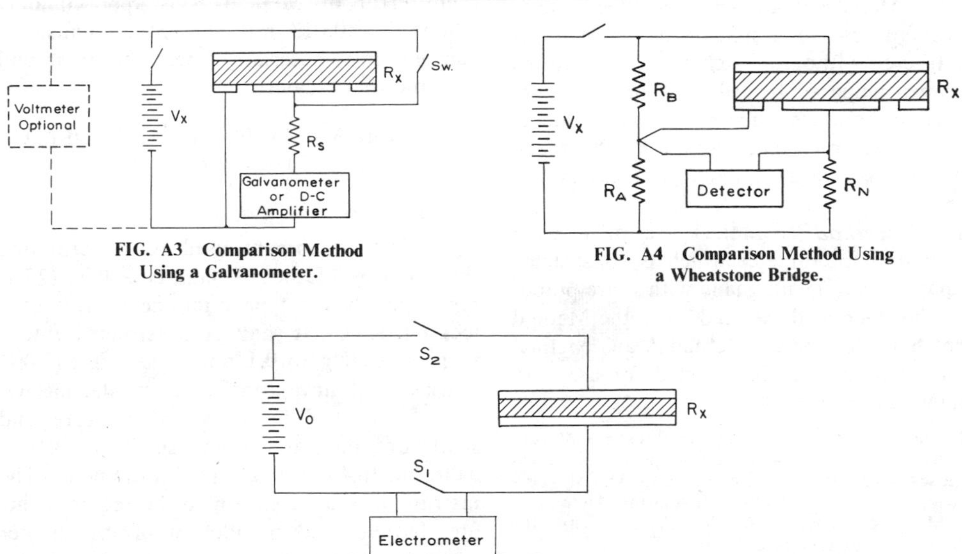

FIG. A3 Comparison Method Using a Galvanometer.

FIG. A4 Comparison Method Using a Wheatstone Bridge.

FIG. A5 Voltage Rate-of-Change Method.

Designation: D 374 - 68

Standard Methods of Test for
THICKNESS OF SOLID ELECTRICAL INSULATION[1]

This Standard is issued under the fixed designation D 374; the number immediately following the designation indicates the year of original adoption or, in the case of revision, the year of last revision. A number in parentheses indicates the year of last reapproval.

1. Scope

1.1 These methods[1a] cover determination of the thickness of solid insulating materials, except rubber insulating tape and friction tape, for electrical purposes. Three alternative procedures are described as follows:

1.1.1 *Method A* makes use of an adjusted ratchet micrometer together with a definite manipulative procedure by which the pressure exerted on the specimen is controlled.

1.1.2 *Method B*, known as the "feel" method, makes use of a machinist's micrometer constructed without a ratchet, pressure on the specimen being controlled by stopping closure of the micrometer when resistance to movement of the micrometer screw by the specimen between the instrument surfaces is first observed.

1.1.3 *Method C* makes use of a dead weight dial micrometer which is constructed so that measurements made with it are practically the same as those made with the Method A ratchet micrometer. Method A and Method C instruments may be selected for use with specific materials.

NOTE 1—For determining the thickness of rubber insulating tape and friction tape for electrical purposes, reference should be made to ASTM Specification D 119, for Rubber Insulating Tape[2] and ASTM Specification D 69, for Friction Tape for General Use for Electrical Purposes.[2]

NOTE 2—The values stated in U.S. customary units are to be regarded as the standard. The metric equivalents of U.S. customary units may be approximate.

2. Precision of Methods

2.1 Methods A and C are preferred as reference standard methods for use in cases of dispute. They are also preferred for the measurement of compressible materials such as untreated paper and fabrics, as well as for the measurement of rigid materials. The maximum error (instrumental plus manipulative) of Methods A and C is of the order of 0.0003 in. (0.008 mm).

2.2 Method B may be used on rigid materials or on yielding materials where it is necessary to measure the specimen with practically no compression or deformation. The maximum error of Method B is approximately 0.0005 in. (0.013 mm) except on unusually compressible materials, where the error will be somewhat greater.

METHOD A—MACHINIST'S MICROMETER WITH RATCHET

3. Apparatus

3.1 The instrument used for determining thickness by Method A shall be a 1-in. (25.4-mm) machinist's type micrometer without a locking device. It shall be constructed with a vernier reading to 0.1 mil (0.0001 in.) (0.003 mm); it shall have a ratchet or similar mechanism for controlling measuring pressure, and shall have anvil and spindle surfaces 0.250 ± 0.001 in. (6.45 ± 0.03 mm) in diameter. The instrument shall conform to the requirements for: flatness and parallelism of micrometer surfaces, zero reading, wear of micrometer screw, permissible micrometer screw error,

[1] These methods are under the jurisdiction of ASTM Committee D-9 on Electrical Insulating Materials. A list of members may be found in the ASTM Yearbook.

Current edition effective Nov. 6, 1968. Originally issued 1933. Replaces D 374 - 57 T.

[1a] This method was adopted without revision although revisions are presently in progress in Committee D-9.

[2] *Annual Book of ASTM Standards*, Part 28.

and ratchet pressure, as specified in Sections 11 to 16. The micrometer shall be tested and calibrated periodically for conformity to these requirements.

4. Procedure

4.1 Before starting measurements of thickness, close the micrometer on the specimen at a location outside the area to be measured. Then open the micrometer not more than 4 or 5 mils (0.004 or 0.005 in.) (0.102 or 0.13 mm) and move it into the area selected for measurement. Using the ratchet, close the micrometer surfaces so slowly on the specimen that the mil scale divisions may be counted easily as they move past the reference mark, or at rate of about 2 mils (0.002 in.) (0.051 mm)/s. Continue the closing motion at the same rate until the ratchet has clicked three times, and then read the thickness by means of the vernier.

4.2 In moving from one measurement location to another this operation shall be repeated, never opening the micrometer more than 4 or 5 mils (0.004 or 0.005 in.) more than the specimen thickness.

4.3 In making a measurement, all points on the peripheries of the micrometer surfaces shall be at least 1/4 in. (6.4 mm) from the edges of the specimen.

METHOD B—MACHINIST'S MICROMETER WITHOUT RATCHET

5. Apparatus

5.1 The instrument used for determining thickness by Method B shall be a 1-in. (25.4-mm) machinist's type micrometer without a locking device. It shall be constructed with a vernier reading to 0.1 mil (0.0001 in.) (0.003 mm), and with anvil and spindle surfaces 0.250 ± 0.001 in. (6.45 ± 0.03 mm) in diameter. The instrument shall conform to the requirements for: flatness and parallelism of micrometer surfaces, zero reading, and permissible micrometer screw error, as specified in Sections 11 to 13, and 15. The micrometer shall be tested and calibrated periodically for conformity to these requirements.

6. Procedure

6.1 For the determination of thickness, slowly close the micrometer on the specimen until contact is made without appreciable distortion of the specimen. The criterion of contact is the initial development of frictional resistance to movement of the micrometer screw by the specimen between the instrument surfaces. Then read the thickness by means of the vernier.

6.2 In moving from one measurement location to another, this operation shall be repeated, never opening the micrometer more than 4 or 5 mils (0.004 or 0.005 in.) (0.102 or 0.13 mm) more than the specimen thickness.

6.3 In making a measurement, all points on the peripheries of the micrometer surfaces shall be at least 1/4 in. (6.4 mm) from the edges of the specimens.

METHOD C—DEAD-WEIGHT DIAL MICROMETER

7. Apparatus

7.1 The micrometer shall be a dead-weight dial-type micrometer, having two ground and lapped circular surfaces, each 0.250 ± 0.001 in. (6.45 ± 0.03 mm) in diameter, and it shall have a capacity of not less than 0.170 in. (4.32 mm).

7.2 The surfaces shall be parallel to within 0.0001 in. (0.003 mm) and shall move on an axis perpendicular to themselves.

7.3 The presssure exerted on the specimen shall be within the limits of 23 and 27 psi (0.16 and 0.19 kgf/cm^2).

7.4 The dial spindle shall be vertical and the dial shall be at least 2 in. (50.8 mm) in diameter. It shall be continuously graduated to read directly to 0.0001 in. and shall be equipped with a telltale hand, recording the number of complete revolutions of the large hand. The dial indicator mechanism shall be full-jeweled.

7.5 The micrometer shall be capable of repeating its readings to 0.00005 in. (0.0012 mm) at zero setting or on a steel gage block.

7.6 Measurements made on standard steel gages shall be within the following tolerances:

Intervals, in. (mm)	Permissible Deviation of Reading from Actual Thickness of Standard Steel Gage, in. (mm)
0 to 0.01 (0.25)	0.0001 (0.003)
Over 0.01 (0.25)	0.0005 (0.012)

7.7 The deviations for the parts of the scale corresponding to the specimen thickness measured, shall be applied as corrections to the thickness reading.

7.8 The frame of the micrometer shall be of such rigidity that a load of 3 lb (1.36 kg) applied to the dial housing, out of contact with either the weight or the presser foot spindle, will produce a deflection of the frame not greater than 0.0001 in., as indicated on the micrometer dial.

8. Procedure

8.1 Place the micrometer on a solid, level table, free from excessive vibration. Place the specimen between the micrometer surfaces, and lower the presser foot onto the specimen at a location outside the area to be measured. Then raise the presser foot a distance of 0.3 to 0.4 mils, move the specimen to the measurement position, and drop the presser foot onto the specimen.

NOTE 3—The procedure described in 8.1 minimizes small errors present when the presser foot is lowered slowly on the specimen. Care should be taken not to raise the presser foot more than 0.4 mil above the position of rest on the specimen surface.

8.2 In making a measurement, all points on the peripheries of the micrometer surfaces shall be at least 1/4 in. from the edges of the specimen.

8.3 For each succeeding measurement, the presser foot shall be raised 0.3 to 0.4 mil, the specimen moved to the next measurement location, and the presser foot dropped.

8.4 When compressible papers or fabrics are measured, a slight settling of the presser foot occurs. To minimize the errors produced by this effect, the reading of the dial indicator should be deferred until the presser foot has been supported by the specimen for at least 2 s, or until the micrometer hand becomes stationary.

CALIBRATION OF MICROMETERS

9. Cleaning Surfaces of Machinist's and Dial Micrometers

9.1 Before and during instrument calibration and thickness measurements, the micrometer surfaces shall be maintained in a clean condition by lightly closing them on a clean sheet of smooth bond paper and moving the paper between the surfaces. To minimize the danger of the presence of lint, the edge of the paper should not be pulled between the surfaces.

10. Calibration of Machinist's Micrometers

10.1 *Method A Micrometer*—In calibrating controlled pressure micrometers used in Method A, close the micrometer on the gage or calibrating device and then open it 4 or 5 mils (0.004 or 0.005 in.) (0.102 or 0.13 mm). Using the ratchet, again close the micrometer so slowly on the calibrating device that the mil scale divisions may be easily counted as they move past the reference mark, or at the rate of about 2 mils (0.002 in.) (0.05 mm)/s. Continue the closing motion at the same rate until the ratchet has clicked three times, then take the reading.

10.2 *Method B Micrometer*—In calibrating the micrometers used in Method B, slowly close the micrometer on the gage or calibrating device until contact of the surfaces and gage is made. The criterion of contact is the initial development of frictional resistance to movement of the gage device between the micrometer surfaces.

11. Flatness of Surfaces of Machinist's Micrometers

11.1 The anvil and spindle surfaces of the machinist's micrometer shall be flat to within 0.00005 in. (0.0012 mm). The flatness may be determined by use of an optical flat. After cleaning the surfaces of the flat and the micrometer (Section 9) the latter shall be closed on the flat as described in 10.1 or 10.2. When illuminated by diffused daylight, interference bands are formed between the surfaces of the flat and those of the micrometer. The location, shape, and number of these bands indicates the deviation from flatness in increments of half the average of the wavelengths of white light, which is taken as 0.00001 in. (0.0003 mm).

11.1.1 A *flat surface* forms straight, parallel and equidistant fringes.

11.1.2 A *grooved surface* forms straight parallel fringes at unequal intervals. The estimated maximum displacement of any line from its normal position, where all lines would be equidistant, is a measure of deviation from flatness.

11.1.3 A *symmetrical concave or convex surface* forms concentric circular fringes, and their number is a measure of deviation from flatness.

11.1.4 An *unsymmetrical concave or convex surface* forms a series of curved fringes, cutting the periphery of the micrometer surface. The number of fringes cut by a straight line connecting the terminals of any fringe is a measure of the deviation from flatness.

12. Parallelism of Surfaces of Machinist's Micrometers

12.1 The anvil and spindle surfaces of the machinist's micrometer shall be parallel to each other to less than 0.0001 in. (0.003 mm) when tested with a pair of screw-thread-pitch wires or with a pair of $^1/_4$-in. nominal diameter plug gages. The diameters of the screw-thread-pitch wires or the plug gages, accurate to 0.00002 in. (0.005 mm) shall differ by an amount approximately equal to the axial movement of the spindle when rotated through 180 deg (12.5 mils). The micrometer shall be closed on the wires or on the plug gages according to the procedure described in 10.1 or 10.2. Observations made with either wire or with either plug gage placed at any location between the surfaces shall show differences of less than 0.001 in. (0.03 mm).

13. Zero Reading of Machinist's Micrometers

13.1 The position of the anvil shall be such that a zero reading is obtained when the machinist's micrometer is closed on the anvil as described in 10.1 or 10.2. Ten trials shall give ten readings of zero. The condition of zero reading is satisfied when examinations with a low-power magnifying glass show that at least two thirds of the widths of the zero graduation on the barrel and that of the reference mark coincide with each other.

14. Wear of Machinist's Micrometer Screws

14.1 The device for compensating for wear of the micrometer screw shall be adjusted so that the spindle has no perceptible lateral or longitudinal looseness, and yet may be rotated with a torque load of not more than $^1/_4$ oz · in. (0.0018 N · m).

15. Permissible Error in Machinist's Micrometer Screw

15.1 The micrometer screw error, after zero adjustment is made, shall be checked at 2, 5, and 10 mils (0.05, 0.13, and 0.25 mm), and at intervals of 100 mils (2.5 mm) over the remaining graduated scale. For checks up to and including a thickness of 10 mils, selected gage blades, the thicknesses of which are known to ±0.00002 in. (0.0005 mm), shall be used. Checks at values greater than 10 mils shall be made with standard gage blocks. At each value checked, ten readings shall be taken, and the arithmetic mean of these ten readings shall not differ from the thickness of the gage used by more than 0.1 mil (0.0001 in.) (0.003 mm). Manipulation of the instrument in these checks shall be in accordance with 10.1 or 10.2.

16. Calibration of Spindle Pressure on Ratchet Micrometers, Method A Only

16.1 *Apparatus:*

16.1.1 Triple-beam, single-plate balance, as shown in Fig. 1, graduated to 0.1 g, having a maximum capacity of about 2600 g using auxiliary weights. The balance should be equipped with an adjustable counterbalance.

16.1.2 An attachment[3] mounted on the plate of the balance, which supports a universal joint, one face of which is lapped flat.

16.1.3 Arm support, at right angles to the balance and universal joint, that will hold the micrometer for testing. This arm shall be adjustable up and down. The micrometer shall be held by this arm in such a way that the clamping pressure will not distort the micrometer frame.

16.2 *Procedure*—Place the micrometer in position on the supporting arm. Then adjust this arm to allow the balance pointer free travel between ±0.05 g, and lock it in position. Place a specimen, such as a pad of capacitor paper (ten layers), between the spindle foot of the micrometer and the lapped surface of the universal joint. Adjust the micrometer spindle so that the pointer of the balance reads −0.05 g. Apply the weights shown in Table 1 for the proper pressure, and turn

[3] This attachment can be adapted from a Starret Center Tester No. 65.

down on the micrometer with three clicks of the ratchet. This will indicate the pressure in pounds per square inch of the entire micrometer assembly. The micrometer ratchet should slip on 27 psi (186 kN/m^2) and bring the pointer up easily on 23 psi (159 kN/m^2). If the pressure is high, clip the spring until the proper range is obtained. If the pressure is low, start with a new spring.

NOTE 4—The micrometer spring should be made from 0.018-in. (0.46-mm) coil spring wire. The inside diameter of the spring coil should be 0.190 ± 0.003 in. (4.83 to 0.76 mm), with $^5/_{64}$-in. (1.98-mm) spacing between coils.

NOTE 5—Never pull a spring apart to increase pressure in pounds per square inch. Always use a new piece of spring of longer length.

NOTE 6—Be sure the ratchet is completely assembled before making a check.

NOTE 7—The ends of the spring should be smooth.

NOTE 8—Make certain that the spring is seated properly in the ratchet assembly.

NOTE 9—Do not use oil in the ratchet.

17. Calibration of Dial Micrometer

17.1 In making these calibration measurements, raise the presser foot from 0.3 to 0.4 mils (0.0076 to 0.0102 mm) above the position of contact with the steel ball or gage. After dropping the presser foot from the elevated position, observe the thickness reading.

17.2 *Parallelism of Surfaces*—A hardened steel ball about $^1/_{16}$-in. (1.6 mm) in diameter, fixed firmly in a flat metal handle about $^1/_{32}$ in. (0.8 mm) in thickness, shall be measured at several locations on the micrometer surfaces, and the maximum variations of readings noted. See 7.2.

17.3 *Accuracy of Scale Divisions*—The instrument shall be set at zero, and standard gages, the thickness of which is known to within 0.00001 in. (0.0003 mm) shall be measured. See also 7.5.

17.4 *Force*—The force applied to the presser foot spindle and weight necessary to move the pointer upward from the zero position shall be not greater than 650 g. The force applied to the presser foot spindle and weight necessary to just prevent movement of the pointer from a positive toward a lower reading shall be not less than 500 g.

TABLE 1 Applicable Weights and Corresponding Pressures

Pressure, psi (kN/m^2)	Weight, g
21 (145)	468
22 (152)	490
23 (158)	512
24 (165)	534
25 (172)	557
26 (179)	579
27 (186)	601
28 (193)	623
29 (200)	646

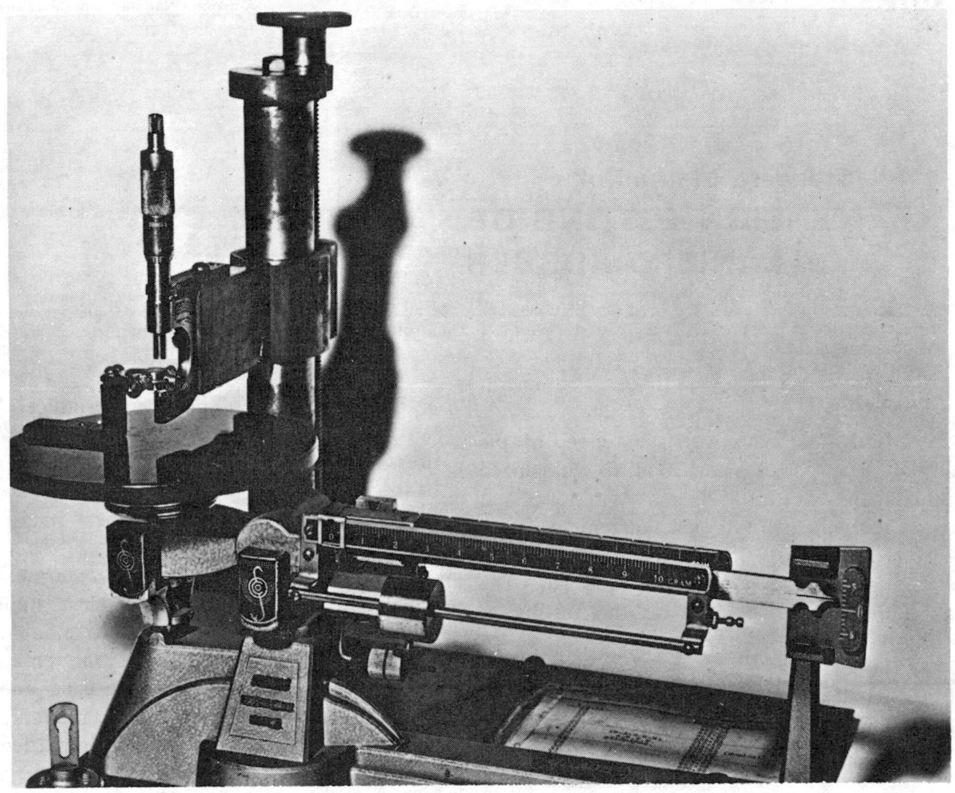

FIG. 1 A Triple-Beam, Single-Plate Balance for the Calibration of Spindle Pressue on Ratchet Micrometers.

ASTM Designation: D 412 - 68

American National Standard J2.1-1969
American National Standards Institute

Standard Method of
TENSION TESTING OF VULCANIZED RUBBER[1]

This Standard is issued under the fixed designation D 412; the number immediately following the designation indicates the year of original adoption or, in the case of revision, the year of last revision. A number in parentheses indicates the year of last reapproval.

1. Scope

1.1 This method covers the effect of the application of a tension load to vulcanized rubber and similar rubber-like materials at room temperature and elevated temperatures. Covered are tests for tensile stress, tensile strength, ultimate elongation, and set. The method is not applicable to the testing of material ordinarily classified as ebonite or hard rubber.

1.2 This method starts with a piece taken from the sample and covers (*1*) the preparation of the specimens for tension testing, and (*2*) the tension testing of the specimens.

NOTE 1—The values stated in U.S. customary units are to be regarded as the standard. The metric equivalents of U.S. customary units may be approximate.

2. Definitions

2.1 *sample*—a unit, collection of units, or a section of a unit taken from a sampling lot.

2.2 *piece*—the portion of the sample that is prepared for testing.

2.3 *specimen*—a piece of material appropriately shaped and prepared so that it is ready to use for a test. A specimen may be a complete article in the case of small rubber bands, gaskets, belts, and similar products that are uniform in cross section and are capable of being tested without reduction in length or cross sectional area.

2.4 *tensile stress*—The applied force per unit of original cross sectional area of specimen.

2.5 *tensile strength*—the maximum tensile stress applied during stretching a specimen to rupture.

2.6 *elongation or strain*—the extension of a uniform section of a specimen, produced by a tensile force applied to the specimen, expressed as a percentage of the original length of the section.

2.7 *ultimate elongation*—the maximum elongation prior to rupture.

2.8 *tensile stress at given elongation*—the tensile stress required to stretch a uniform section of a specimen to a given elongation.

2.9 *tension set*—the extension remaining after a specimen has been stretched and allowed to retract in a specified manner, expressed as a percentage of the original length.

2.10 *set after break*—the tension set of a specimen stretched to rupture.

3. Apparatus

3.1 *Dies and Cutters*—The shapes and dimensions of dumbbell dies for preparing dumbbell specimens shall conform with those shown in Fig. 1. The inside faces in the reduced section shall be polished and perpendicular to the plane formed by the cutting edges for a depth of at least 5 mm (0.2 in.). The dies shall be sharp and free of nicks in order to prevent ragged edges on the specimen. Dies and cutters for preparing ring specimens shall be designed to produce one of the standard ring specimens described in 4.4. A suggested cutter and holder are shown in Fig. 2 for preparing standard ring specimens from sheets prepared in accordance with ASTM Methods D 15, Sample Preparation

[1] This method is under the jurisdiction of ASTM Committee D-11 on Rubber and Rubber-Like Products. A list of committee members may be found in the ASTM Yearbook. This standard is the direct responsibility of Subcommittee D-11.10 on Physical Testing.

Current edition effective Sept. 13, 1968. Originally issued 1935. Replaces D 412 - 66.

for Physical Testing of Rubber Products.[2]

NOTE 2—Careful maintenance of die cutting edges is of extreme importance and can be obtained by light daily honing and touching up the cutting edges with jewelers' hard Arkansas honing stones. The condition of the die may be judged by investigating the rupture point on any series of broken specimens. When broken specimens are removed from the clamps of the testing machine it is advantageous to pile these specimens and note if there is any tendency to break at or near the same portion of each specimen. Rupture points consistently at the same place may be the indication that the die is dull, nicked, or bent at that particular position.

3.2 *Bench Marker*—The bench marker shall have two parallel straight marking surfaces ground smooth in the same plane. The surfaces shall be between 0.05 and 0.08 mm (0.002 and 0.003 in.) in width and at least 15 mm (0.6 in.) in length. The angles between the marking surfaces and the sides shall be at least 75 deg. The distance between the centers of the marking surfaces shall be within 0.08 mm (0.003 in.) of the required distance.

3.3 *Stamp Pad*—The stamp pad shall have plane unyielding surface (for example, hardwood, plate glass, or plastic). The ink shall have no deteriorating effect on the specimen and shall be of contrasting color to that of the specimen.

3.4 *Micrometers*—The dial micrometer used to measure the thickness of flat specimens shall be capable of exerting a pressure of 0.25 ± 0.05 kgf/cm^2 (3.6 ± 0.7 psi) on the specimens (Note 3) and measuring the thickness to within 0.025 mm (0.001 in.). The anvil of the micrometer shall be at least 35 mm (1.4 in.) in diameter and shall be parallel to the face of the contact foot. The dial micrometer used to measure the radial width of ring specimens having rectangular cross section shall be equipped with curved feet to fit the curvature of the ring. The screw micrometer used to measure ring specimens having a circular cross section shall be equipped with spherical tips either 5 mm or 0.25 in. in diameter and shall have graduations not exceeding 0.025 mm (0.001 in.).

NOTE 3—Dial micrometers exerting a force of 85 g on a circular foot 6.35 mm (0.25 in.) in diameter, or 20 g on a circular foot 3.2 mm (0.125 in.) in diameter conform to this pressure requirement. A micrometer should not be used to measure the thickness of specimens narrower in width than the diameter of the foot.

3.5 *Stepped Cone*—The cone or frustum of a cone used to measure the inside diameter of ring specimens shall have steps having diametric intervals not exceeding 2 percent of the diameter to be measured.

3.6 *Testing Machine*—Tension tests shall be made on a power-driven machine equipped with a suitable dynamometer and indicating or recording device for measuring the applied force within ±2 percent (Note 4). If the capacity range cannot be changed during a test, as in the case of the pendulum dynamometer, the applied force at break shall be measured within ±2 percent, and the smallest tensile force measured shall be accurate to within 10 percent. If the dynamometer is of the compensating type for measuring tensile stress directly, means shall be provided to adjust for the cross sectional area of the specimen. The response of either an indicator or recorder shall be sufficiently rapid that the applied force is measured with the requisite accuracy during the extension of the specimen to rupture. If the tester is not equipped with a recorder, a device shall be provided that indicates after rupture the maximum force applied during extension. Testers equipped with a device to measure elongation automatically shall be capable of determining extensions within 5 percent of the original length. If elongation is measured manually, a scale capable of measuring each 10 percent elongation shall be provided.

NOTE 4—An accuracy of 2 percent does not permit the use of the portion of the range below 50 times the smallest change in force that can be measured. In machines with close graduations the smallest change in force that can be measured may be the value of a graduation interval; with open graduations, or with magnifiers for reading, it may be an estimated fraction, rarely as fine as one tenth of a graduation interval; and with verniers it is customarily the difference between the scale and vernier graduations measured in terms of scale units. If the indicating mechanism includes a stepped detent, the detent action may determine the smallest change in force detectable.

3.7 *Grips*—The tester shall have two grips, one of which shall be connected to the dynamometer, and a mechanism for separating the grips at a uniform rate of 500 ± 50 mm (20 ± 2 in.)/min (Note 5) for a distance of at least 75 mm (30 in.).

[2] *Annual Book of ASTM Standards*, Part 28.

3.7.1 Grips for testing dumbbell specimens shall tighten automatically and exert a uniform pressure across the gripping surfaces, increasing as the tension increases in order to prevent uneven slipping and to favor failure of the specimen in its constricted section. At the end of each grip, a positioning device is recommended for inserting specimens to the same depth in the grip and aligning them with the direction of pull.

3.7.2 Grips for testing ring specimens shall consist of one or two rollers on each grip at least 9 mm (0.35 in.) in diameter and at least 10 mm (0.4 in.) in length. The surface of the rollers shall be lubricated with castor oil to facilitate equalizing the stress around the specimen. Smaller rollers may be provided for testing rings smaller than 25 mm (1 in.) in diameter.

3.7.3 Grips for testing straight specimens shall be either wedged or toggle type designed to transmit the applied force over a large surface area of the specimen.

NOTE 5—A rate of separation of 1 m (40 in.)/min may be used in routine work and notation of the speed used made on the report, but in case of dispute the rate shall be 500 ± 50 mm (20 ± 2 in.)/min.

3.8 *Calibration of Testing Machine*—The testing machine shall be calibrated in accordance with Procedure A of ASTM Methods E4, Verification of Testing Machines.[3] If the dynamometer is of the strain-gage type the tester shall be calibrated at one or more loads daily, in addition to the requirements in Sections 7 and 18 of Methods E4. Testers having pendulum dynamometers may be calibrated as follows: One end of a dumbbell specimen shall be placed in the upper grip of the testing machine. The lower grip shall be removed from the machine and attached to the specimen. To this lower grip shall be attached a hook suitable for holding weights. A weight shall be suspended from the hook on the specimen in such a way (Note 6) as to permit the weight assembly to rest on the machine grip holder. If the machine has a dynamometer head of the compensating type it shall be calibrated at two or more settings of the compensator. The motor shall be started and run as in normal testing until the weight assembly is freely suspended by the specimen. If the dial or scale (whichever is normally used in testing) does not indicate the weight applied (or its equivalent in stress for compensating tester) within the specified tolerance, the machine shall be thoroughly checked for excess friction in the bearings and all other moving parts. After eliminating as nearly as possible all the excess friction, the machine shall be recalibrated as just described. The machine shall be calibrated at a minimum of three points, using accurately known weight assemblies of approximately 10, 20, and 50 percent of capacity. The weight of the lower grip and hook shall be included as part of the calibration weight. If pawls and ratchet are used during test, they should also be used during the calibration. Friction in the head can be checked by calibrating with the pawls up.

NOTE 6—It is advisable to provide a means for preventing the weights from falling to the floor in case the dumbbell should break.

3.8.1 A rapid and frequent approximate check on accuracy of the tensile tester calibration may be obtained by using a spring calibration device.[4]

3.9 *Apparatus for Set Test*—The testing machine described under 3.6 or an apparatus similar to that shown in Fig. 3 may be used. A stop watch or other suitable timing device which will register the time in minutes for at least 30 min shall be provided. A scale or other device shall be provided for measuring set to within 1 percent.

3.10 *Test Chamber for Elevated Temperature*—The test chamber shall conform with the following requirements:

3.10.1 Air shall be circulated through the chamber with a speed of 60 to 120 m (200 to 400 ft)/min at the location of the grips and specimens, and maintained within 2 C of the specified temperature.

3.10.2 A calibrated sensing device shall be located near the grips for measuring the actual temperature.

3.10.3 The chamber shall be vented to an exhaust system or the outside atmosphere to remove any toxic fumes liberated at high temperatures.

---

[3] *Annual Book of ASTM Standards*, Part 30.

[4] Obtainable from Testing Machines Inc., 400 Bayview Ave., Amityville, L. I., N. Y. 11701.

3.10.4 Provision shall be made for suspending specimens vertically near the grips for conditioning prior to test. The specimens should not touch each other or the sides of the chamber except for momentary contact when agitated by the circulating air.

3.10.5 Suitable fast-acting grips for manipulation at high temperatures shall be provided to permit placing specimens in the grips without changing the temperature of the chamber.

3.10.6 The dynamometer shall be suitable for use at the temperature of test or thermally insulated from the chamber.

3.10.7 Provision shall be made for measuring elongation of specimens in the chamber. If a scale is used to measure the extension between bench marks, the scale shall be located parallel and close to the grip path during extension and shall be controlled from outside the chamber.

4. Test Specimens

4.1 *Dumbbell Specimens*—The piece of rubber to be tested shall, whenever possible, be flat, not less than 1.5 mm (0.06 in.) nor more than 3 mm (0.12 in.) in thickness and of a size which will permit cutting a dumbbell specimen by means of one of the standard dies. If obtained from a manufactured article the piece of rubber shall be freed of surface roughness, fabric layers, etc., in accordance with the procedure described in Methods D 15. Specimens from units made in long lengths such as dredging sleeves, hose tubing, or insulation, shall be taken in the length direction, except that specimens of belts wider than 300 mm (12 in.), of sheet packing, of hose, or of dredging sleeves more than 100 mm (4 in.) in inside diameter shall be taken in the transverse direction. In the case of specially cured sheets prepared according to Methods D 15 the specimen shall be died out in the direction of the grain. Dumbbell specimens shall conform in shape to those shown in Fig. 1. Unless otherwise specified, die *C*, Fig. 1, shall be used to prepare the specimens. In all cases, the cutting of test specimens shall be done with a single stroke of the cutting tool so as to assure obtaining smoothly cut surfaces.

4.2 *Marking Dumbbell Specimens*—Dumbbell specimens shall be marked with the bench marker described in 3.2. The sample shall be under no tension at the time it is marked. Marks shall be placed on the reduced section of the specimen equidistant from its center and perpendicular to its longitudinal axis. The centers of the marks shall be either 20.00 ± 0.08 mm, 25.00 ± 0.08 mm, or 1.000 ± 0.003 in. apart on specimens cut with dies *C* and *D*, and either 50.00 ± 0.08 mm or 2.000 ± 0.003 in. apart on specimens cut with the other dies shown in Fig. 1.

4.3 *Measuring Dumbbell Specimens*—Three measurements shall be made for thickness, one at the center and one at each end of the reduced section of the specimen. The median of the three measurements shall be used as the thickness in calculating the cross sectional area, except that specimens for which the differences between the maximum and minimum thickness exceeds 0.08 mm (0.003 in.), shall be discarded.

4.3.1 The width of the specimen shall be taken as the distance between the cutting edges of the die in the restricted section.

4.4 *Ring Specimens:*

4.4.1 The radial width should be less than 15 percent of the inside diameter of the ring, and the hardness should be less than 90, as determined by ASTM Method D 2240, Test for Indentation Hardness of Rubber and Plastics by Means of a Durometer,[2] or ASTM Method D 1415, Test for International Standard Hardness of Vulcanized Rubbers.[2] By means of suitable dies, ring specimens may be cut from manufactured articles that are flat or from thin-walled tubular articles that can be laid flat. In the case of heavy-walled tubing greater than 1.5 mm (0.06 in.) in thickness, the material shall be mounted on a mandrel, rotated in a lathe, and the ring specimen cut with a sharp tool. Ring specimens shall be of dimensions that permit the use of roller grips as described in 3.7.

4.4.2 A standard ring specimen may be prepared from flat sheets of certain thicknesses. From sheets 4 to 6 mm in thickness, the ISO standard ring specimen may be prepared with the following dimensions: Inside diameter 44.6 ± 0.1 mm, outside diameter 52.6 ± 0.1 mm, the radial width nowhere deviating by more than ±0.2 mm from the mean width. From sheets prepared in accordance with Methods D 15, one of the following standard ring specimens (Note 7) may be prepared: (*1*) inside diameter 29.5 ± 0.1 mm, outside diameter 33.5 ± 0.1 mm, for testing with an apparatus graduated in metric units, or (*2*) inside diameter 30 ± 0.1 mm, outside diameter 34 ± 0.1 mm, for testing with an apparatus graduated in U.S. customary units. The radial width shall be uniform within 0.02 mm.

4.4.3 The cross-sectional area of a ring specimen shall be calculated from its weight, density, and mean circumference. In control testing, the cross sectional area of a ring specimen having a rectangular section may be calculated from its axial thickness and radial width measured with the dial micrometer described in 3.4. The thickness of a standard ring specimen cut from a flat sheet may be assumed to be the thickness of the disk cut from inside the ring which can be measured with the micrometer (Note 3). Curved feet that fit the curvature of the ring shall be used in measuring the radial width. For rings having a circular cross section, the area may be calculated from the axial thickness measured with the ball-point micrometer described in 3.4, assuming radial width to be identical with axial thickness. In measuring the axial thickness, the ball-tips of the micrometer shall be closed with the specimen resting on them. Then the tips shall be slowly separated until the specimen falls through of its own weight. If the area is determined from axial thickness and radial width, three measurements shall be made at points distributed around the circumference of the ring, and the median of the three measurements shall be taken for the thickness or width of the specimen. The inside and outside diameters of ring specimens cut from a sheet may be assumed to be the same as that of the cutter determined from measurements of a ring of light cardboard cut by the cutter. The inside diameter of other ring specimens shall be measured by the stepped cone described in 3.5, to the nearest 2 percent interval, employing no stress in excess of that necessary to overcome any ellipticity of the ring. The inside circumference shall be obtained by multiplying the inside diameter by 3.14. The mean circumference shall be obtained by multiplying the sum of the inside diameter and the radial width by 3.14.

NOTE 7—The standard specimens cut from ASTM sheets are designed for autographic testing machines to give 100 percent elongation for each 20 mm on recorder charts graduated in metric units when the chart moves 200 mm/min while the grips separate at 500 mm/min, or for each inch on recorder charts graduated in U.S. customary units when the chart moves 10 in./min while the grips separate at 20 in./min. The size of these specimens gives the correct mean elongation within one percent when calculated in accordance with Note 9.

4.5 *Straight Specimens*—Straight specimens may be prepared where it is not practicable to cut either a dumbbell or a ring specimen, as in the case of narrow rubber strip, small tubing, or electrical insulation. These specimens shall be of sufficient length to permit their installation in the wedge or toggle grip used in the test. Bench marks shall be placed on the specimens as described for dumbbell specimens in 4.2. To determine the cross sectional area of straight specimens in the form of tubes, the weight, length, and density of the specimen shall be determined. The cross sectional area shall then be calculated from these measurements as follows:

$$A = W/DL$$

where:

A = cross-sectional area, cm^2,
W = weight in air, g,
D = density, g/cm^3, and
L = length, cm.

To determine the cross-sectional area in square inches, the area A in square centimeters shall be multiplied by 0.155.

5. Procedure

5.1 *Test Temperature*—Unless otherwise specified, the standard temperature (Note 8) for testing shall be 23 ± 1 C (73.4 ± 1.8 F). When testing at some other temperature is required, the temperature specified shall be one of those listed in ASTM Recommended Practice D 1349, for Standard Test Temperatures for Rubber and Rubber-Like Materials,[2] and the report shall include a statement of the temperature at which the test was made. Specimens shall be conditioned for at least 3 h if the test temperature is 23 C.

NOTE 8—This standard temperature is the same as prescribed for the Standard Laboratory Atmosphere in ASTM Specification E 171, for Standard Atmospheres for Conditioning and Testing Materials.[3]

5.2 *Determination of Tensile Stress, Tensile Strength, and Ultimate Elongation:*

5.2.1 Place dumbbell or straight specimens in the grips of the testing machine, using care to adjust it symmetrically in order that the tension will be distributed uniformly over the cross section. If tension is greater on one side of the specimen than on the other, the bench marks will not remain parallel and maximum strength of the rubber will not be developed. Start the machine and note continuously the distance between the center of the two bench marks, taking care to avoid parallax. Record the stress at the elongation specified for the materials under test and at the time of rupture, preferably by means of an autographic or spark recorder. At rupture measure and record the elongation to the nearest 10 percent on the scale.

5.2.2 In testing ring specimens, lubricate the surfaces of the grip rollers with castor oil. Place the specimen around the rollers with a minimum of tension. Start the machine and record or note continuously the distance between the centers of the rollers. If the stress and strain are not autographically recorded, predetermine the distance between the centers of the rollers for the elongation specified for the material under test by the following equation (Note 9):

$$D = 1/2[(EM/100) + C - G]$$

where:

D = distance between the roller centers of two grips,
E = specified elongation, percent,
C = inside circumference of the specimen,
M = mean circumference of the specimen, and
G = circumference of one grip roller (if each grip has two rollers, add twice the distance between the centers of the rollers on one grip).

NOTE 9—This equation is an approximation since changes in the cross sectional dimensions during extension have been neglected. For precision measurements, substitute $M - 3.14R\ (1 + 0.01E)^{-1/2}$ for C, where R is the radial width of the specimen (or the axial thickness if the specimen turns during extension so that the radial side is in contact with the grips).

5.2.3 Record the stress at the predetermined distances between the centers of the rollers and at the time of rupture, preferably by means of an autographic or spark recorder. At rupture measure the distance between the centers of rollers to within 2.5 mm (0.1 in.) and record.

5.3 For exposure at any temperature above the standard temperature (see 5.1), condition the specimen for 10 ± 2 min, and place each specimen in the test chamber at intervals ahead of testing so that all specimens of a series will be in the test chamber the same length of time, that is, if 1 min is required to run the test, the first specimen, placed in the test chamber 10 min prior to testing, would be followed by other specimens at 1-min intervals. The conditioning time at elevated temperatures must be limited to avoid additional curing or heat aging.

NOTE 10: **Caution**—Suitable heat-resistant gloves should be worn for hand and arm protection when testing at high temperature. An air mask is very desirable when the door of the chamber is opened to insert specimens; toxic fumes may be present and should not be inhaled by the operator conducting the test.

5.4 *Determination of Set*—Place the specimen in the grips of the testing apparatus and adjust symmetrically so as to distribute the tension uniformly over the cross section. Separate the grips at a rate of speed as uniformly as practicable, requiring about 15 s to reach the specified elongation. Then hold the specimen at the specified elongation for 10 min, release quickly without being allowed to snap back, and allow to rest for an additional 10 min. At the end of the 10-min rest period, measure the distance between the bench marks to the nearest 1 percent of original length. In stretching the specimen it has been found convenient to use a measured rod of a length equal to the exact distance required between the two bench marks. Holding the rod behind the test specimen while it is being stretched simplifies the operation and reduces the chance of stretching the specimen more than the required amount. Use a stop watch or equivalent timer for recording the time required for the various operations.

5.5 *Set at Break*—Set at break is the set determined on the specimen when stretched to rupture. Ten min after the specimen is broken, fit the two pieces carefully together so that they are in contact over the full area of the break. Measure the distance between the bench marks. The calculation is the same as that for tension set (see 5.4).

6. Calculation

6.1 *Dumbbell and Straight Specimens:*

6.1.1 Calculate the tensile stress as follows:

$$\text{Tensile stress} = F/A$$

where:

F = observed force, and

A = cross-sectional area of the unstretched specimen.

Calculate the tensile strength by letting F in the above equation for tensile stress be equal to the force required to break the specimen. Tensile stress and tensile strength are expressed in either kilograms-force per square centimeter or pounds per square inch. One kilogram-force per square centimeter is about 14.22 psi.

6.1.2 Calculate the elongation as follows:

$$\text{Elongation, percent} = [(L - L_o)/L_o] \times 100$$

where:

L = observed distance between the bench marks on the stretched specimen, and

L_o = original distance between the bench marks.

6.1.3 Calculate the ultimate elongation by letting L in the above equation for elongation be equal to the distance between the bench marks at the time of rupture. Calculate the tension set by substituting for L in the above equation, the distance between the bench marks after the 10-min retraction period.

6.2 *Ring Specimens:*

6.2.1 Calculate the tensile stress as follows:

$$\text{Tensile stress} = F/A$$

where

F = observed force, and

A = twice the cross-sectional area calculated from the axial thickness and radial width of the unstretched ring.

Calculate the tensile strength by letting F in the above equation for tensile stress be equal to the force required to break the specimen.

6.2.2 Determine the elongation or strain for an extension below rupture as described in 5.2.

6.2.3 Calculate the ultimate elongation (Note 11) as follows:

$$\text{Ultimate elongation, percent} = [(2D + G - C)/C] \times 100$$

where:

D = distance between the centers of the grip rollers at the time of rupture of the specimen,

G = circumference of one roller (if each grip has two rollers, add twice the distance between the centers of the rollers on one grip), and

C = inside circumference of the ring specimen

NOTE 11—The ultimate elongation of standard ring specimens cut from ASTM sheets may be determined by multiplying the highest elongation indicated on the recorder chart by 1.08. This factor gives the elongation based on inside circumference.

7. Characteristics of Piece Tested

7.1 The median of the values for three specimens shall be taken as the characteristics of the piece of rubber tested, except that under the following conditions the median of the values for five specimens shall be used:

7.1.1 If one or more values do not meet the specified requirements when testing for compliance with specifications.

7.1.2 If referee tests are being made.

8. Report

8.1 The report shall include the following:

8.1.1 Results calculated in accordance with 6. Calculation,

8.1.2 All observed and recorded data on which the calculations are based,

8.1.3 Date of vulcanization of the rubber, if known,

8.1.4 Date of Test,

8.1.5 Temperature of the test room shall be stated if it is other than as provided for in 5.1,

8.1.6 Type of testing machine used,

8.1.7 Type and dimensions of specimens used, and

8.1.8 When testing at elevated temperature, the temperature of the test chamber.

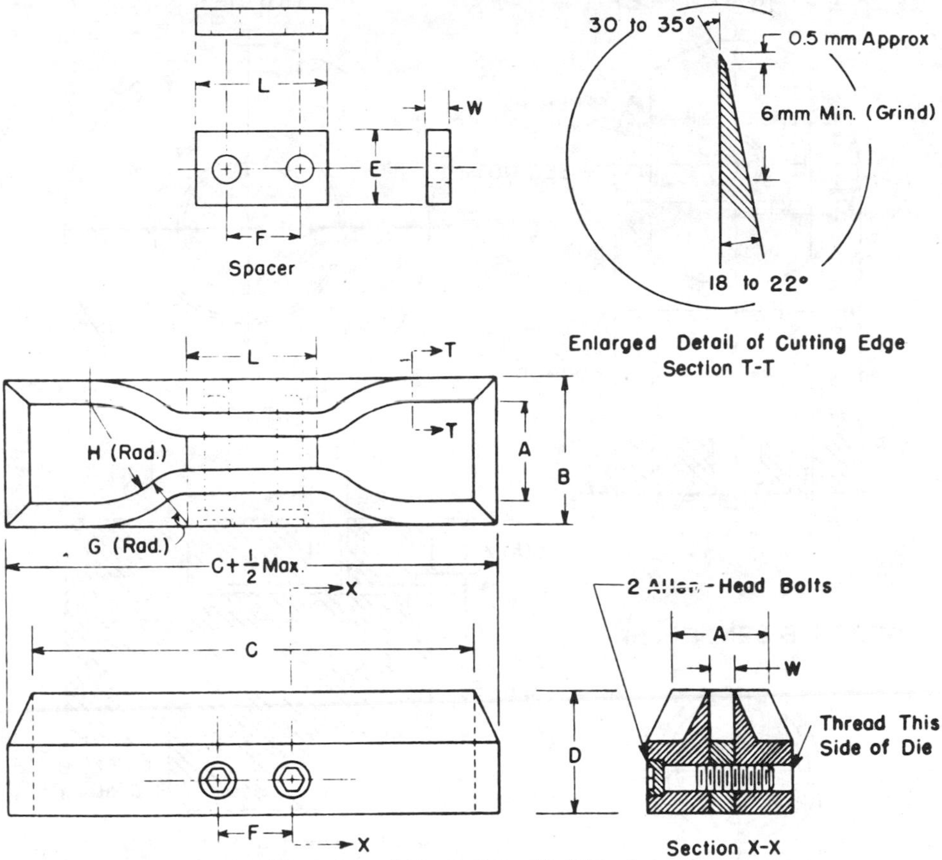

Dimensions of Standard Dumbbell Dies[a]

Dimension	Units	Tolerance	Die A	Die B	Die C	Die D	Die E	Die F
A	mm	±1	25	25	25	16	16	16
	in.	±0.04	1	1	1	0.62	0.62	0.62
B	mm	max	40	40	40	30	30	30
	in.	max	1.6	1.6	1.6	1.2	1.2	1.2
C	mm	min	140	140	115	100	125	125
	in.	min	5.5	5.5	4.5	4	5	5
D	mm	±6[b]	32	32	32	32	32	32
	in.	±0.25[b]	1.25	1.25	1.25	1.25	1.25	1.25
D-E	mm	±1	13	13	13	13	13	13
	in.	±0.04	0.5	0.5	0.5	0.5	0.5	0.5
F	mm	±2	38	38	19	19	38	38
	in.	±0.08	1.5	1.5	0.75	0.75	1.5	1.5
G	mm	±1	14	14	14	14	14	14
	in.	±0.04	0.56	0.56	0.56	0.56	0.56	0.56
H	mm	±2	25	25	25	16	16	16
	in.	±0.08	1	1	1	0.63	0.63	0.63
L	mm	±2	59	59	33	33	59	59
	in.	±0.08	2.32	2.32	1.31	1.31	2.32	2.32
W	mm	+0.05, −0.00	12	6	6	3	3	6
	in.	+0.002, −0.000	0.500	0.250	0.250	0.125	0.125	0.250

[a] Dies whose dimensions are expressed in metric units are not exactly the same as dies whose dimensions are expressed in U.S. customary units. However, equivalent results may be expected from either die. Dies dimensioned in metric units are intended for use with apparatus calibrated in metric units.

[b] For dies used in clicking machines it is preferable that this tolerance be ±0.5 mm or ±0.02 in.

FIG. 1 Standard Dies for Cutting Dumbbell Specimens.

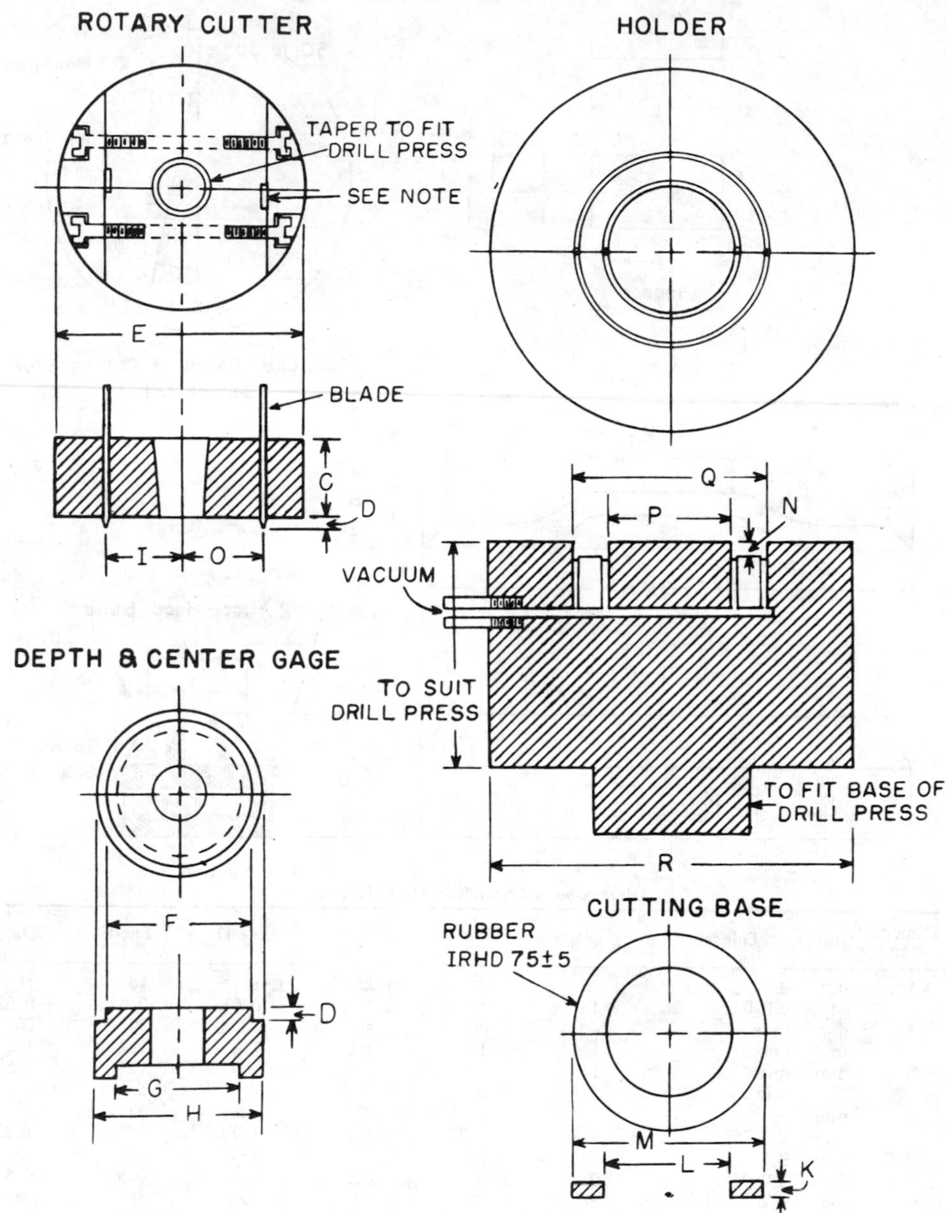

FIG. 2 Cutter and Holder for ASTM Standard Ring Specimens.
(Continued on next page.)

Dimensions of Ring Cutter and Holder

NOTE—Slot for cutting blade must be positioned so point is on diameter perpendicular to slot. For Bard-Parker stainless steel blade No. 15, slot is 5.4 ± 0.1 mm wide, 0.35 ± 0.02 mm deep, and offset from center line in direction of rotation by 1.7 mm. The run-out of the spindle to which the cutter is attached must not exceed 0.01 mm.

Dimension	mm	in.	Dimension	mm	in.
I[a]	14.75 ± 0.05	0.581 ± 0.002	H	39 max	1.54 max
I[b]	15.00 ± 0.05	0.591 ± 0.002	K	3 to 4	0.12 to 0.16
O[a]	16.75 ± 0.05	0.659 ± 0.002	L	$G^{+0.1}_{-0}$	$G^{+0.004}_{-0.000}$
O[b]	17.00 ± 0.05	0.669 ± 0.002			
C	15 min	0.6 min	M	39 ± 0.5	1.54 ± 0.02
D	3.0 to 3.5	0.12 to 0.14	N	K ± 0.05	K ± 0.002
E	50 min	2 min	P	$G^{+0.0}_{-0.1}$	$G^{+0.000}_{-0.004}$
F	30 max	1.2 max			
G	25 ± 0.5	1.00 ± 0.02	Q	40 ± 0.5	1.57 ± 0.02
			R	75 min	3 min

[a] Cutter for ring specimens tested with apparatus graduated in metric units.
[b] Cutter for ring specimens tested with apparatus graduated in English units.

FIG. 2—*Continued*

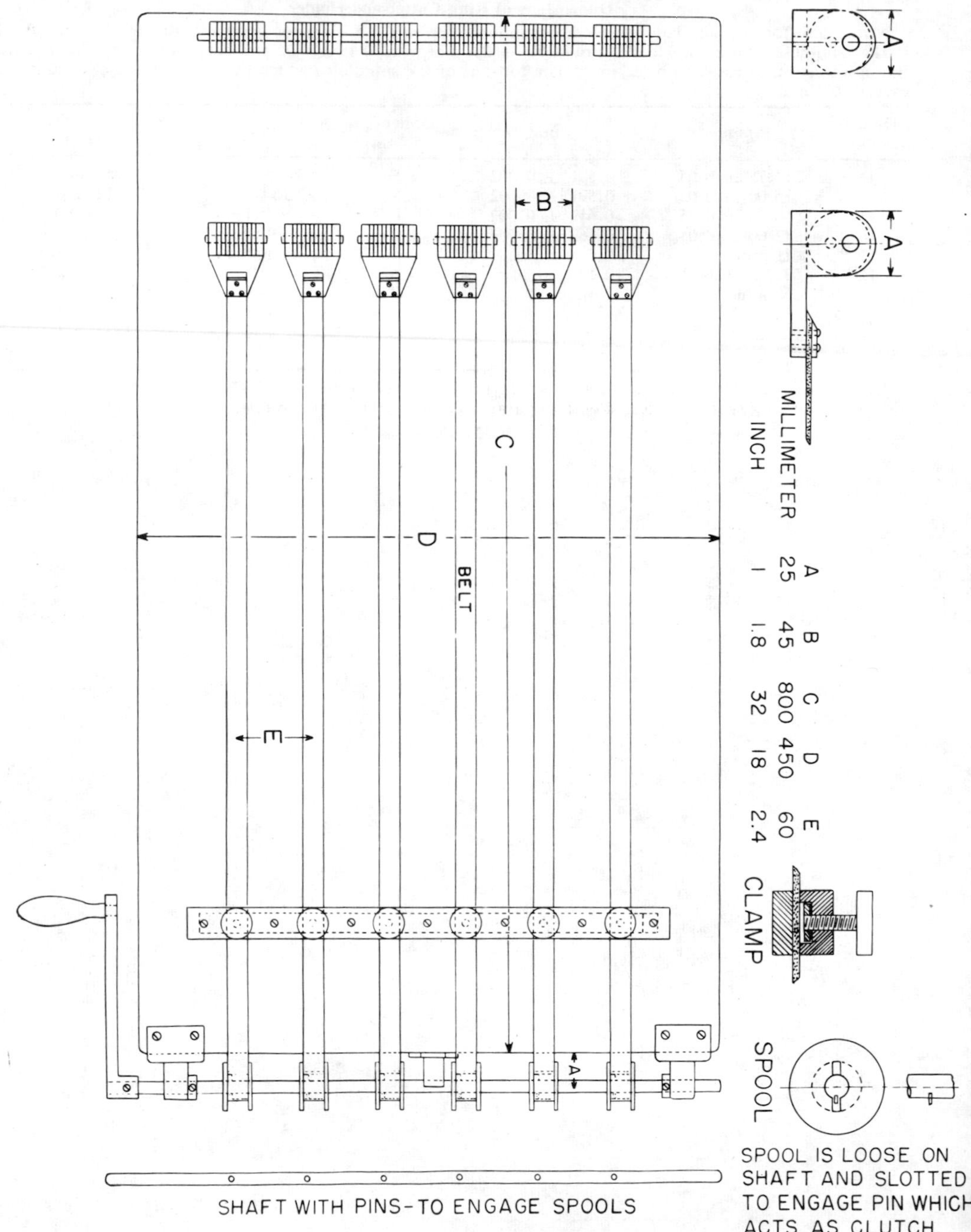

FIG. 3 Apparatus for Permanent Set Test.

Designation: D 494 – 46 (Reapproved 1967)

USA Standard K65.12-1959 (R 1969)
USA Standards Institute

Standard Method of Test for
ACETONE EXTRACTION OF PHENOLIC MOLDED OR LAMINATED PRODUCTS[1]

This Standard is issued under the fixed designation D 494; the number immediately following the designation indicates the year of original adoption or, in the case of revision, the year of last revision. A number in parentheses indicates the year of last reapproval.

1. Scope

1.1 This method covers the determination of the amount of acetone-soluble matter in molded or laminated phenolic products.

NOTE 1—The values stated in U.S. customary units are to be regarded as the standard. The metric equivalents of U.S. customary units may be approximate.

2. Significance

2.1 For molded phenolic products, acetone extraction should be considered solely as a quantitative expression of a property normally associated with degree of cure. There is no demonstrably rigorous relation between the optimum mechanical and electrical properties of a well-cured piece and the numerical value of the acetone test. The amount of acetone-soluble matter is affected by: (*1*) nature of resin and filler, (*2*) lubricant, (*3*) molding temperature, (*4*) length of cure, (*5*) thickness of the section from which sample is taken, (*6*) nature of molded piece, (*7*) technique used in molding, (*8*) distribution of fines in the material to be extracted, and (*9*) method of grinding the specimen. These variations under some conditions may cause a difference of 3 to 4 percent in acetone-extractable matter. For this reason, the method should be used only as a comparative test for measuring undercure.

2.2 For laminated phenolic products, acetone extraction indicates change in stage of cure, change in resin content, change in type of resin used, presence of plasticizers or other acetone-extractable addition agents, and is affected in general by the same factors as stated in 4.1.

3. Apparatus

3.1 *Sieves*—The set of sieves used shall consist of sieves Nos. 40 (420-μm) and 140 (105-μm), with a cover and receiving pan, conforming to the requirements of ASTM Specification E 11 for Wire-Cloth Sieves for Testing Purposes.[2]

3.2 *Extraction Apparatus*—The apparatus may be of the type shown in Fig. 1, or a Wiley-Richardson type, as shown in Fig. 2. The former type is more suitable for use with small electric hot plates, while the latter is more suitable for use with oil or water baths. In either case, it shall be possible to control the temperature so that the rate of extraction can be regulated accurately.

3.3 *Drying Dishes*—The drying dishes shall be lightweight dishes, approximately 63.5 mm (2 1/2 in.) in diameter and 38.1 mm (1 1/2 in.) in height.

4. Preparation of Sample

4.1 *Precautions*—It is of utmost importance that extreme care shall be taken during the preparation of the sample for extraction. The sample shall be drillings if possible; however, if not possible, other suitable means of producing particles equivalent to drillings may be used. Drillings taken from a large molded product shall be truly representative

[1] This method is under the jurisdiction of ASTM Committee D-20 on Plastics. A list of committee members may be found in the ASTM Yearbook. This standard is the direct responsibility of Subcommittee D-20.70 on Analytical Methods.

Current edition effective Sept. 16, 1946. Originally issued 1938. Replaces D 494 – 41.

[2] *Annual Book of ASTM Standards*, Part 30.

of all sections of the part in proper proportions. The drills for sampling shall be kept sharp and so operated that no undue heating of the material shall occur which will tend to precure the material.

4.2 If it is impracticable to obtain samples by drilling, the parts may be broken up with a lathe, planer, milling machine, or a suitable grinder. A mortar and pestle or a pebble mill is considered suitable as a grinder, provided no perceptible heating occurs during the grinding procedure. A sharp file or rasp may be used for procuring the sample where the size or shape of the part is such that no other method is suitable.

4.3 In any case, the particles of the sample shall be of the smallest size practicable, so that they will pass through the No. 40 sieve with the minimum of reworking or grinding. It is important in preparing the sample that the smallest possible volume shall be obtained for a unit weight of the material.

4.4 The sample shall be sieved through a No. 40 sieve and that part which will not pass through shall be reground and blended with the original material passing the sieve. After assembling the Nos. 40 and 140 sieves and the receiving pan, the sample shall be placed in the top sieve, the cover placed on, and the entire sample shall be resieved either by a mechanical sieve shaker or hand sieving. If the hand-sieving method is used, the sieve shall be rotated with slight tapping, the period of rotation being 5 min.

4.5 After sieving, the sample (that portion which has passed through the No. 40 sieve and has been retained on the No. 140 sieve) shall be placed immediately in an airtight container to prevent absorption of moisture by the powder and the consequent error in results.

5. Conditioning

5.1 *Conditioning*—Condition the test specimens at 23 ± 2 C (73.4 ± 3.6 F) and 50 ± 5 percent relative humidity for not less than 40 h prior to test in accordance with Procedure A of ASTM Methods D 618, Conditioning Plastics and Electrical Insulating Materials for Testing,[3] for those tests where conditioning is required. In cases of disagreement, the tolerances shall be 1 C (1.8 F) and ±2 percent relative humidity.

5.2 *Test Conditions*—Conduct tests in the Standard Laboratory Atmosphere of 23 ± 2 C (73.4 ± 3.6 F) and 50 ± 5 percent relative humidity, unless otherwise specified in the test methods or in this specification. In cases of disagreements, the tolerances shall be 1 C (1.8 F) and ±2 percent relative humidity.

6. Procedure

6.1 *Extraction*—The extraction procedure shall be carried out in triplicate. Accurately weigh a 3.000-g portion of the powdered sample into a tared, acid-hardened open-texture quantitative filter paper, 12.5 to 15 cm in diameter, or into a standard, single-thickness extraction thimble, 80 by 22 mm, trimmed if necessary. After folding over the thimble or filter paper containing the sample so that none of the powder can float out, place it in a desiccator until ready to insert in the siphon.

6.2 Press the filter paper or thimble containing the weighed sample into the siphon in such a way that the outlet of the bottom is not plugged. Place the condenser and the siphon in the extraction tube and add 50 ml of cp acetone. Start the water through the condenser and adjust the heat (Note 2) so that the siphon fills and empties between 15 and 20 times/h. This rate shall be carefully maintained, and the sample shall be extracted for 4 h. After the siphon empties, remove the flask and pour the contents into an individually weighed dish. Wash the flask three times with the smallest possible quantity of acetone, using a wash bottle, and add the washings to the extracted liquid in the dish.

NOTE 2—If an oil or water bath is used for heating, the height of the liquid in the bath should not come above 2.5 cm (1 in.) below the highest level of the acetone in the siphon before the siphon starts to discharge.

6.3 *Drying*—Place the dish in a well-ventilated drying chamber, maintained at 50 ± 2 C (Note 3), and dry the sample to constant weight. Between dryings, all dishes containing the residue shall be kept in a desiccator to prevent the absorption of moisture.

NOTE 3—It is very important that the specified temperature shall be maintained, otherwise consist-

[3] *Annual Book of ASTM Standards*, Part 27.

ent results can not be obtained between different laboratories. An electrically heated oven should not be used unless it is exceedingly well ventilated, as the acetone fumes are likely to come in contact with the heated coils and cause an explosion.

7. Calculation and Report

7.1 *Calculation*—Calculate the percentage of acetone-extractable matter in the specimen as follows:

$$\text{Acetone extractable matter, percent} = [(W - D)/S] \times 100$$

where:

W = weight of the dish and extract,
D = weight of the dish, and
S = weight of the original sample.

7.2 *Report*—The report shall include the percentage of acetone-extractable matter for each sample, and the average percentage of acetone-extractable matter for the three samples.

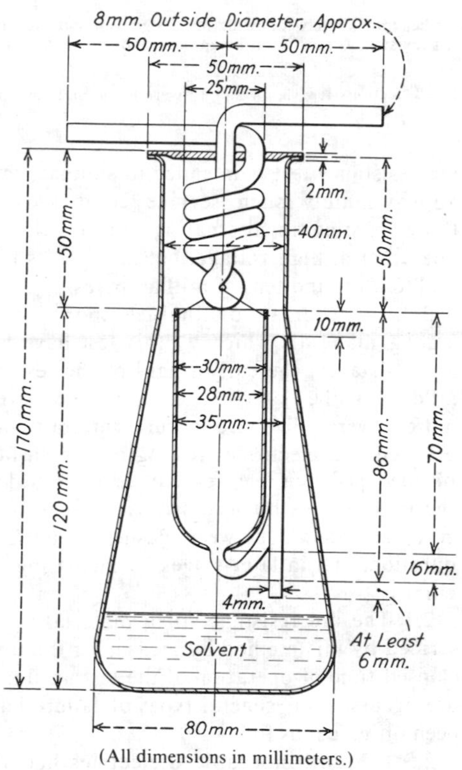

(All dimensions in millimeters.)

FIG. 1 Extraction Apparatus.

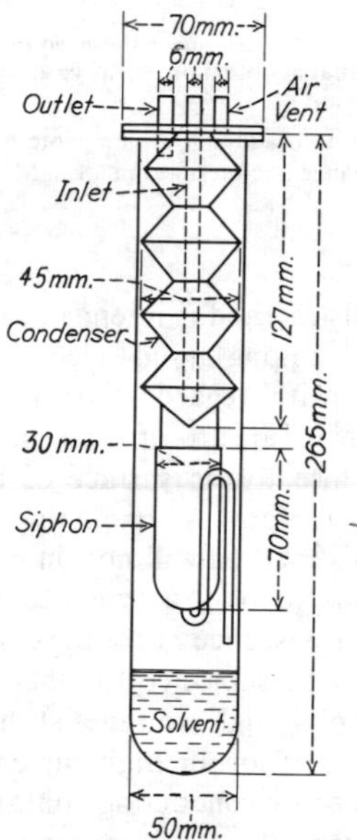

FIG. 2 Wiley-Richardson Type Extraction Apparatus.

 Designation: D 495 – 61

Standard Method of Test for
HIGH-VOLTAGE, LOW-CURRENT, ARC RESISTANCE OF SOLID ELECTRICAL INSULATING MATERIALS[1]

This Standard is issued under the fixed designation D 495; the number immediately following the designation indicates the year of original adoption or, in the case of revision, the year of last revision. A number in parentheses indicates the year of last reapproval.

NOTE—Editorial addition of new Note 6 and changes in Fig. 1 and Sections 1, 2, 3, 14 and 15 were made in June 1963. Editorial change in Note 6 was made in March 1964.

1. Scope

1.1 This method is intended to differentiate, relatively among solid electrical insulating materials with regard to their resistance to the action of an arc of high voltage and low current close to the surface of the insulation intending to form a conducting path therein.

1.2 This method will not, in general, permit conclusions to be drawn concerning the relative arc-resistance rankings of materials which may be subjected to other types of arcs, for example, high voltage at high currents, low voltage at low or high currents (promoted by surges or by conducting contaminants).

1.3 Although satisfactory for illustrating broadly the approximate arc-resistance rankings of insulators of different composition, this test must be used with caution in attempting close comparison (Note 2), except between specimens of almost the same composition, such as for control of quality, experimental development, and identity tests.

NOTE 1—The values stated in U.S. customary units are to be regarded as the standard. The metric equivalents of U.S. customary units may be approximate.

NOTE 2—By changing some of the circuit conditions described herein it has been found possible to rearrange markedly the order of arc resistance of a group of organic insulating materials consisting of vulcanized fiber and of molded phenolic and amino plastics, some containing organic, and some inorganic filler.

2. Significance

2.1 The high-voltage, low-current type of arc-resistance test is intended to simulate only approximately such service conditions as those existing in alternating current circuits operating at high voltage, but at currents limited to units and tens of milliamperes.

2.2 In order to distinguish more easily among materials which by this test have low arc resistance, the early stages of the test are mild and the later stages are successively more severe. The arc occurs intermittently between two electrodes resting on the surface of the specimen the severity is increased in the early stages by successively decreasing to zero the interval between flashes of uniform duration, and in later stages by increasing the current.

2.3 The arc resistance of a material is described by this method by measuring the total elapsed time of operation of the test until failure occurs. Four general types of failure have been observed, as follows:

2.3.1 Many inorganic dielectrics become incandescent, whereupon they are capable of conducting the current. Upon cooling, however, they return to their earlier insulating condition.

2.3.2 Some organic compounds burst into flame without the formation of a visible conducting path in the substance.

2.3.3 Others are seen to fail by "tracking,"

[1] This method is under the jurisdiction of ASTM Committee D-9 on Electrical Insulating Materials. A list of members may be found in the ASTM Yearbook.

Current edition effective Sept. 18, 1961. Originally issued 1938. Replaces D 495 – 58 T.

that is, a thin wiry line is formed between the electrodes.

2.3.4 The fourth type occurs by carbonization of the surface until sufficient carbon is present to carry the current.

2.4 Materials often fail within the first few seconds after a change in the severity stage. When comparing the arc resistance of materials, much more weight should be given to a few seconds that overlap two stages than to the same elapsed time within a stage. Thus there is a much greater difference in arc resistance between 178 and 182 s than between 174 and 178 s.

NOTE 3—Some investigators have reported attempts to characterize the remaining insulating value of the damaged area after failure as described above by allowing the specimen to cool to room temperature without disturbance of the original position of the electrodes and then either measuring the insulation resistance between the electrodes or determining the percentage of breakdown voltage remaining relative to that obtained on an undamaged area of the specimen. A recommended circuit arrangement and test procedure for carrying out the second of these two means of characterizing the remaining insulating value of the damaged area are described in an Appendix to this method. Still another, and obvious, method of re-evaluating the damaged area after failure would be simply to repeat the arc resistance test after the specimen had cooled, with the electrodes undisturbed in their original position. It should be clear, however, that none of these methods will be universally applicable because of the severe physical damage to the test area in many instances.

3. Arc Resistance

3.1 The end point in this test, that is, "failure," is usually quite definite. At a rather sharply defined time a conducting path is formed across the dielectric and the arc disappears into the material. At the same time a noticeable change in sound takes place. For some materials, however, the definition is more difficult. Definitions of failure of the following kinds are therefore listed:

3.1.1 The trend toward failure increases over a fairly broad interval of time before all parts of the arc have disappeared. In this instance the end point is considered to have been reached only when *all* parts of the surface between the electrodes are carrying the current, that is, no portion is apparent as an arc.

NOTE 4—In testing some materials, a persistent scintillation will be observed close to the electrodes after a tracking path has obviously been completed. The persistence of this remaining scintillation is not to be counted as part of the arc-resistance time.

3.1.2 Burning of the material obscures the arc. It may be preferable in such cases to report simply that failure occurred by burning.

3.1.3 In successive arc intervals the dielectric conducts the current only in the later part of the "on" time. The first such interval should be designated as failure.

3.2 Certain insulators are not readily classified by an arc-resistance test because of melting, gumming, fouling of electrodes, flaming, mechanical disintegration by cracking, etc. These characteristics should be recorded with no attempt at an arc-resistance measurement.

3.3 It has been found that changes in the timing of the intermittent arc, and in the current, as well as other circuit changes affecting the nature of the discharge, can not only affect the duration of a test of most samples in a group of dissimilar materials, but can drastically alter their positions in order of rank. It is therefore obvious that data obtained by this method, and the arbitrary conditions of test that have been assigned, should not be employed to infer that the materials examined occupy an unchanging quality relationship to each other, regardless of conditions of anticipated use.

3.4 This method permits the use of two electrode systems:

3.4.1 *Tungsten Rod Electrode System*—This system has a long history of use; a large amount of information concerning the arc resistance of materials has been obtained with these electrodes.

3.4.2 *Stainless Steel Strip Electrodes*—The use of this system is relatively recent. It offers the advantage of reducing the vagaries associated with tungsten rod electrodes when testing materials of relatively poor arc resistance (approximately 20 s). These materials, when tested with tungsten rod electrodes, exhibit sporadically higher arc resistance (as much as 150 s or more). The occurrence of this condition is much less frequent or may be completely absent when stainless steel strip electrodes are used. The stainless steel electrodes are not recommended for use beyond 180 s because of electrode distortion caused by electric loss heating.

3.5 Results obtained with the use of the tungsten rod electrode system may be different from those obtained with the use of the stainless steel strip electrode system. Experience will indicate whether it is possible to accept results interchangeably and for which materials. This method is referenced in numerous material specifications which contain requirements for arc resistance. When this method is being used in conformance with a specification, the publication date of which antedates the 1961 revision of this method, it is understood that the tungsten rod electrode system shall be used to determine compliance with arc-resistance requirements, unless the parties concerned have agreed otherwise.

4. Apparatus

4.1 The apparatus for tests with an arc current of up to 40 mA shall consist of the following (see Fig. 1):

4.1.1 *Transformer*, T_v—A self-regulating transformer (non-power factor corrected) with a rated primary potential of 115 V at 60 Hz a-c, a rated secondary potential, on open circuit, of 15,000 V and a related secondary current, on short circuit, of 0.060 A.

NOTE 5—Jefferson Electric type 721-411 luminous tube transformers have been found satisfactory. Any other type or make meeting the requirements of 4.1.1 may be used if it can be demonstrated that equivalent data are obtained.

4.1.2 *Variable Autotransformer*, T_a—An autotransformer rated at 7 A or more, and nominally adjustable up to 135 V.

NOTE 6—The Powerstat (Superior Electric Co.) or the Variac (General Radio Co.) of this or greater current-carrying capacity are both satisfactory. Other makes of the same general construction will probably also be adequate.

4.1.3 *Voltmeter*, V_1—One a-c voltmeter shall be permanently connected across the output of the autotransformer to indicate the voltage being supplied to the primary circuit. It shall be readable to 1 V in the range 90 to 130 V.

NOTE 7—It is recommended that a means be provided to maintain constant primary voltage. One of many commercially available line voltage stabilizers which does not distort the voltage wave form will provide a suitable means.

4.1.4 *Milliammeter*, *A*—The a-c milliammeter, *A*, of the moving vane type, shall be capable of reading from 10 to 40 mA with an error of not over ± 5 percent. Before use this meter should be calibrated in a test circuit containing no arc gap, to make certain of its readings. Since the milliammeter, *A*, is used only when setting up or making changes in the circuit, it may be shorted out by a by-pass switch when not in use.

NOTE 8—Although provision has been made for the suppression of radio-frequency components of current in the arc, it may be desirable to check for their presence when the apparatus is first constructed. This is done by use of a suitable thermocouple-type r-f milliameter temporarily inserted in series with milliammeter, *A*.

4.1.5 *Current Control Resistors*, R_{10}, R_{20}, R_{30}, R_{40}—Four resistors are required in series with the primary of T_v but in parallel with each other. These shall be adjustable to some extent to permit exact settings to be made during calibration of the circuit. R_{10} is always in the circuit, to obtain a current of 10 mA in the arc. Its value is approximately 60 Ω, with a current rating of at least 1 $^1/_4$ A. Closing switch S_{20} to add R_{20} in parallel with R_{10} shall result in an arc current of 20 mA. For this purpose R_{20} shall be about 50 Ω, with a current rating of at least 1 $^3/_4$ A. Similarly, R_{30} shall have values of about 30 Ω and at least 2 A, to provide an arc current of 30 mA and R_{40}, about 15 Ω and 5 A, for an arc current of 40 mA. (These figures were obtained with the transformer described in Note 6.)

4.1.6 *Suppressing Resistor*, R_s—Rates at 15,000 Ω and at least 24 W. The function of this resistor (as well as the inductors described in 4.1.7) is to suppress parasitic high frequency in the arc circuit.

4.1.7 *Air Core Inductors*—Inductance is required in the form of several coils of No. 30 cotton or enamel-covered wire, of 3000 to 5000 turns each, wound on insulating nonmetallic cores of about a $^1/_2$-in. (12.7-mm) diameter and $^5/_8$-in. (15.9-mm) inside length (Note 9). A sufficient number of these coils shall be used to total about 1.2 to 1.5 H. A single coil of this inductance is not satisfactory.

NOTE 9—About eight coils will be required. Similar coils may be obtained from Cutler-Hammer, Inc., Milwaukee, Wis., Catalog No. 70100-H-6.

4.1.8 *Interruptor, I*—A motor-driven device, to give the required cycles for the three lower steps of the test, shall be capable of opening and closing the primary circuit according to the schedule in Table 1 with an accuracy of $\pm {}^{1}/_{120}$ s or better. The interruptor can be made in the form of metal disks bearing suitable raised portions on their periphery to actuate contactor switches when rotated at an appropriate rate by a synchronous motor.

4.1.9 *Electrodes:*

4.1.9.1 *Tungsten rod electrodes* shall be made from ${}^{3}/_{32}$-in. (2.4-mm) diameter tungsten rod (Note 10) by sharpening one end as described in Section 7 (see Fig. 2).

4.1.9.2 *Stainless steel strip electrodes* shall be made from 0.002-in. (0.05-mm) thick stainless steel by cutting ${}^{1}/_{2}$ by 1 in. (12.7 by 25.4 mm) rectangles as described in Section 8 (see Fig. 5).

NOTE 10—Tungsten welding rod has been found suitable.

4.1.10 *Electrode Assemblies*—Means for holding the electrodes and specimen and applying the arc to the surface of the specimen as follows (see Figs. 4 and 5):

4.1.10.1 Specimens shall be supported in a uniform manner to assure ample air space below the specimens, and so that the upper surface of the specimens is level.

4.1.10.2 Adequate shielding from drafts shall be provided, with provision for venting of combustion products and allowing a clear view of the arc from a position slightly above the plane of the specimen, so as to distinguish readily between an arc and any flame of combustion.

4.1.10.3 Uniform and reproducible specified pressure of each electrode independently upon the specimen shall be maintained.

4.1.10.4 Inter-electrode spacing and required alignment of electrodes shall be maintained.

4.1.11 *Sharpening Jig for Tungsten Rod Electrodes*—A steel jig for securing the electrodes during sharpening to ensure finishing the pointed tips to the proper geometry (see Fig. 3).

4.1.12 *Alignment Jig for Stainless Steel Electrodes*—A metal jig to assure correct spacing and positioning of the electrodes (see Fig. 6).

4.1.13 *Time, TT*—A stop watch or electric interval timer operating at 115 V a-c, accurate to 1 s. The indicating member of the electric interval timer shall come to rest within ${}^{1}/_{2}$ s after the circuit is opened.

NOTE 11—Electric timers of either the dial or counter type can be obtained with a built-in brake that prevents coasting after the current is interrupted.

4.1.14 *Indicator Lamp, IL*—This pilot lamp indicates the interrupting cycle being used and permits the operator to start the first cycle of the intermittent portion of the test correctly by closing $S_{1/2}$ just after the lamp is extinguished. A 6-W, 115-V lamp with 2000 Ω, R_L, in series has been found satisfactory.

4.1.15 *Control Switches*—Toggle switches are convenient. All may be of the size rated at 3 A and 110 to 125 V, except S_1 and S_{40} which shall be rated at 10 A.

4.1.16 *Safety Interlocking Contactor, C_s*—For operator safety this contactor shall be so connected to the electrode assembly that raising the draft shield will cause it to open and thus remove high voltage from the electrodes. It shall be rated for a current of 10 A at 110 to 125 V.

4.1.17 *Interruptor Contactors, $C_{1/8}$, $C_{1/4}$, $C_{1/2}$*—Normally-open spring contactors are required (Note 12). These are caused to close and open by the rotary drive mechanism of the interruptor, thus closing and opening the primary circuit and providing the intermittent arc cycles listed in Table 1.

NOTE 12—Contactors similar in operation and current-carrying capacity to the microswitch are satisfactory. The contactor available from Micro Switch, a Division of Minneapolis Honeywell Regulator Co., Freeport, Ill., Catalog No. BZ-2RW2, is recommended, but having a stiff rather than flexible lever arm.

4.1.18 *High-Voltage Switch, S_4*—The single-pole, single-throw switch insulated for 15,000 V shall be isolated from the operator by a suitable enclosure through which projects a handle, made of insulating material, of sufficient length to ensure operator safety.

4.1.19 *Wiring*—All wiring in the arc circuit shall be of ignition wire rated at 15 kV or higher, and shall be so disposed that it and any of the circuit components are not readily accessible while energized.

5. Electrical Circuit

5.1 The electrical circuit has been closely specified to achieve a maximum of reproducibility among different test sets. Construction according to Fig. 1 with the apparatus described in Section 4 should result in a relatively quiet arc rather than the crackly blue spark characteristic of a condenser discharge.

5.2 Primary control shall by no means be attempted by using a variable inductance, because of its proved deleterious effect on the performance of the arc.

5.3 The high-voltage discharge from the transformer is dangerous. The circuit therefore shall be protectively interlocked (see 4.1.16).

6. Interruptor

6.1 Any apparatus that meets the requirement of 4.1.18 is satisfactory.

7. Tungsten Rod Electrodes

7.1 Tungsten rod with a diameter of $^3/_{32}$ in. (2.4 mm) and a length of about $^3/_4$ in. (19.1 mm) shall be selected to be free from cracks, pits, or rough spots. It shall be cleaned by heating to a dull red color, rubbing on a block of alkali nitrite and quenching immediately in water. It shall then be fastened into a shank, as shown in Fig. 2, by swaging, brazing, or silver solder. The shank ensures that the points when sharpened will be correctly oriented in the electrode assembly.

7.2 To sharpen the points of the electrodes the following technique shall be used:

7.2.1 The surface of the rod shall be polished by rotating it on its axis, using successive grades of emery paper from Nos. 1 to 000, held between the fingers.

7.2.2 The elliptical face is ground to the 30-deg angle to the axis shown in Fig. 2. Care shall be exercised to prevent cracking or chipping.

7.2.3 The electrode shall be mounted in the sharpening jig shown in Fig. 3 with its face flush with the surface of the jig. This surface shall be held against a rotating polishing disk or rubbed against a stationary machinist's flat, using Nos. 1 to 000 emery or emery paper to make successive cuts. The elliptical face should as a result be a highly polished surface inclined at 30 ± 1 deg to the axis of the electrode. No burrs or rough edges should be observable at 15× magnification.

7.2.4 The electrodes shall be placed in the support arms of the electrode assembly with the tips spaced 0.250 ± 0.002 in. (6.35 ± 0.05 mm) apart by means of an "over" and "under" block gage of 0.248 in. (6.29 mm) and 0.252 in. (6.45 mm).

NOTE 13—No attempt should be made to force the gage between the points; they may thus be distorted or chipped.

7.2.5 The points shall be resharpened when examination of the edge from which the arc originates shows that it has rounded to a diameter of 0.003 in. (0.08 mm) or that burrs or rough edges are observable at 15× magnification.

8. Stainless Steel Strip Electrodes

8.1 Stainless steel strip 0.002 in. (0.05 mm) thick shall be sheared into $^1/_2$ by 1-in. (12.7 by 25.4-mm) rectangles. The edges shall be cleanly cut so as to be free of burrs. The strip electrodes shall be bent slightly in the middle of the longer dimension so that an angle of approximately 160 deg will be formed when the electrodes are permitted to lie on a flat surface (see Fig. 6).

9. Tungsten Rod Electrode Assembly

9.1 The electrode assembly shall permit each electrode to rest on the surface of the specimen independently with a force of 50 ± 5 g as measured by raising and lowering with a spring balance. This shall be accomplished, as shown in Fig. 4, by holding the electrodes in supporting rods, electrically isolated by insulating blocks tnat rotate on pivots. A suitable weight shall be added, to increase or to counterbalance partially the force at the electrode, as needed.

9.2 For correct operation of the test, the electrodes shall lie in the same vertical plane and both be inclined at 35 deg from the horizontal, thus including 110 deg between their axes. Additionally, the minor axes of the polished elliptical surfaces of the tips shall be horizontal. These requirements are accomplished automatically if the following factors are controlled:

9.2.1 The axis of the tungsten rod is perpendicular to the axis of the support rod as a result of accurate machining.

9.2.2 The support rods are gripped in the pivot blocks in a position such that the axis of each electrode is inclined at 35 deg when the support rods are horizontal.

9.2.3 The electrodes, with square shank, are sharpened in a correctly made jig.

9.2.4 The spacing between points is adjusted with the support rods accurately horizontal.

9.2.5 The electrode assembly is checked for levelness before use, and

9.2.6 The surfaces to be tested are always brought to the same level height, namely, that at which the support rods will be horizontal.

9.3 Compliance with 9.2.3 depends upon obtaining a sharpening jig wherein the square hole for the shank of the electrode is aligned accurately at 30 deg to the working face and the sides of the hole are perpendicular to this face. The requirement of 9.2.5 can be met by means of a specimen holder of the correct height in which materials can be clamped against the under side of a frame. Thus samples of different thicknesses have their upper surfaces always brought to the correct height. Clamping shall be accomplished so that there is free access of air to the largest part of the under surface of the specimen and especially in the region directly below the test area.

9.4 The assembly shall be equipped with a transparent draft shield to prevent deflection of the arc by stray air currents. This shall completely surround the arc (not too closely) and shall be vented to permit convective flow to carry away combustion products (Note 14). It shall not hinder the self-alignment of the electrodes.

NOTE 14—Some electrical insulations give off toxic smoke and gases during this test. These should be vented away. When testing a material of unknown toxicity the fumes should also be removed.

9.5 For testing of larger insulating parts it may be necessary to remove the electrode assembly from the draft shield described in 9.4. Provision for screw legs for the assembly is desirable, to permit levelling at a height convenient for the part to be tested.

10. Stainless Steel Strip Electrode Assembly

10.1 Two stainless steel strip electrodes shall be placed on the test specimen surface with the corners down and touching the surface (see Fig. 5). The corners shall be 0.250 ± 0.003 in. (6.35 ± 0.08 mm) apart and the electrode edges shall make angles of 45 deg to a line joining the corners. An alignment jig (Fig. 6) should be used.

10.2 Each electrode shall be weighted at its center with a force of 50 ± 5 g. The tungsten electrode assembly can supply the required force. The tungsten electrodes must be retracted to a 1.25 in. (31.8 mm) spacing if this is to be done.

10.3 The electrode assembly shall be protected by a transparent draft shield as prescribed in 9.4 and 9.5.

11. Calibration

11.1 With the circuit connected as shown in Fig. 1, establish the correct conditions for operation by the following calibration procedure:

NOTE 15—In making calibrations the operator necessarily comes in contact with the high tension wiring of the secondary circuit. It is therefore important that the circuit be de-energized by opening S_1 before each change in connections is made.

11.1.1 *Open Circuit Operating Voltage*—It is necessary to establish a standard rate of increase of the secondary voltage wave in order that the duration of the arc for each reversal of current may be fixed. This can be done by specifying the value for the open circuit voltage. The standard value has been set at 12,500 V as measured by an electrostatic voltmeter.

11.1.2 *Adjustment of Secondary Current*—Attach the electrode assembly by closing S_4, and with the electrodes resting on a smooth lava block adjust the resistor R_{10} to give the required current of 10 mA as indicated by milliammeter A. If the apparatus is checked for r-f current in the secondary when the equipment is first set up, the r-f thermocouple milliammeter (See Note 8) should also read 10 mA to indicate proper suppression of r-f. Open S_4 again and adjust the autotransformer to give an open circuit secondary voltage of 12,500 V as indicated by an electrostatic voltmeter; then, with S_4 closed, adjust the secondary current again to 10 ma by means of R_{10}. Repeat this procedure until no further adjustment is necessary. Note the reading on voltmeter V at this time, while the arc is operating at 10 mA with the electrodes resting on the lava block and with the draft shield closed. This voltage shall be maintained for all subsequent calibration and testing (Note 16). Next close S_{20} and S_{10}, thus placing R_{20} in parallel with R_{10}. Adjust R_{20} to give an arc current of 20 mA. Close S_{30} and S_{40}, and adjust R_{30} and R_{40}, respectively, to obtain currents of 30 and of 40 mA.

NOTE 16—For the transformer described in Note 5, this reading of V has been found to be from about 110 to 115 V.

12. Test Specimens

12.1 For standard comparison of materials the test specimen shall be 0.125 ± 0.010 in. (3.17 ± 0.25 mm) in thickness, with a flat surface for testing such that no part of the arc is closer to the edge than 1/4 in. (6.4 mm) nor to a previously tested area than 1/2 in. (12.7 mm). Thin materials shall be tested by first clamping them together, such as by screws, to form a specimen as close as possible to 1/8 in. (3.2 mm) in thickness.

12.2 For testing molded parts the arc may be applied to a location deemed most significant. All comparison tests of parts shall be performed similarly.

12.3 Dust, moisture, or finger marks can affect the time required to track. If thus affected, specimens or parts shall be cleaned with a cloth containing water or other suitable solvent before conditioning and shall be wiped with a dry cloth immediately before testing.

NOTE 17—If the purpose of the test is to ascertain the effect of contamination, these instructions may be disregarded. However, the nature and quantity of contaminant present, if known, shall then be stated in the report.

13. Conditioning

13.1 Conditioning of all specimens prior to testing shall be carried out in accordance with Procedure A (Note 18) of D 618, Methods of Conditioning Plastics and Electrical Insulating Materials for Testing,[2] unless it can be shown that omission of this step has no effect.

NOTE 18—For some insulation the effect of exposure to humid atmosphere is of interest. The use of Procedure C of Methods D 618 is recommended for this purpose.

13.2 If specimens have been kept at a temperature close to or below the dew point of the surroundings in which the tests will be run, condensation of moisture must be prevented, such as by warming in a dry atmosphere.

NOTE 19—Insertion for 1/2 h in an oven operating at 50 C will be found suitable in most cases.

14. Procedure for Determination of Arc-Resistance Time

14.1 Place the specimen in a suitable holder to locate the test surface as specified in Sections 9 and 12 and insert under the tungsten rod electrodes. Alternatively place the specimen in a suitable holder and apply alignment jig and stainless steel strip electrodes as in Fig. 6; insert under the tungsten rods and remove the alignment jig as illustrated in Fig. 5.

14.2 Lower the draft shield, close the line switch, and adjust *Ta* if necessary to obtain the voltage specified in 11.2.

14.3 Begin the test sequence specified in Table 1 and simultaneously start the interval timer. At the end of each minute increase the severity in steps as shown in Table 1 until a failure occurs as defined in Section 3. Immediately interrupt the arc current and stop the interval timer. Record the seconds to failure.

14.4 The observer should watch the arc from a position nearly on a level with the surface of the specimen so as to be able to tell whether or not the arc is "normal" (lying close to the specimen with not over 1/16-in. (1.6-mm) lift at its middle). If the arc should "climb" or flare irregularly, it is an indication that the circuit constants are incorrect (Note 20). The low position of the observer's eye also enables him to decide when "tracking" occurs, with the extinguishment of the luminous sheath of gas in air which often continues for some time after a red-hot spot has formed upon the surface of the specimen.

NOTE 20—Gases are expelled forcibly from some insulators, thus causing the arc to flutter. The equipment should be checked for proper operation in these instances by examining the arc for quietness and maintenance of a position close to a surface such as glass or lava.

15. Cleaning Electrodes

15.1 *Tungsten Rod Electrodes*—After each test the tip of each tungsten rod electrode shall be freed from foreign matter. This shall be done by rubbing with rag-base letter paper backed with a strip of laminated plastic, after moving the specimen support aside. Often a gummy insulating substance boils out of the specimen because of the heat of the arc, and adheres to the electrode and changes the breakdown characteristics. In case the increased friction met while using the paper strip indicates the need of a slight abrasive, rub the electrode tips with a strip having crocus cloth mounted upon the surface. Keep the strip parallel with the elliptical flat surface while rubbing to prevent gouging. After the gummy substance is gone, rub the tips for a few strokes with the paper strip to be sure no abrasive particles adhere. Then hold the paper strip perpendicular to the flat surface and clean the working tip of the electrode, following the curve a little way. Application of a continuous 40-mA arc, without a specimen in place, for approximately 1 min often proves to be the most effective technique of cleaning the electrodes.

15.2 *Stainless Steel Strip Electrodes*—Two new corners of the stainless steel strip electrodes shall be used for each test.

16. Number of Tests

16.1 At least five tests of a material for arc-resistance time shall be made. Any larger number, as dictated by the behavior of the material, may be made. The average and the lowest single value of time shall be noted.

17. Report

17.1 The report shall include the following:

17.1.1 Type and trade name of material,

17.1.2 Conditions of fabrication,

17.1.3 Conditioning prior to test,

17.1.4 Thickness of specimen. If not a standard specimen, a description of the part,

17.1.5 Electrode system used,

17.1.6 Number of tests,

17.1.7 Averages and minimums of arc resistance time,

17.1.8 Special remarks, for instance burning, softening, and

17.1.9 Kind and amount of contaminant, if any.

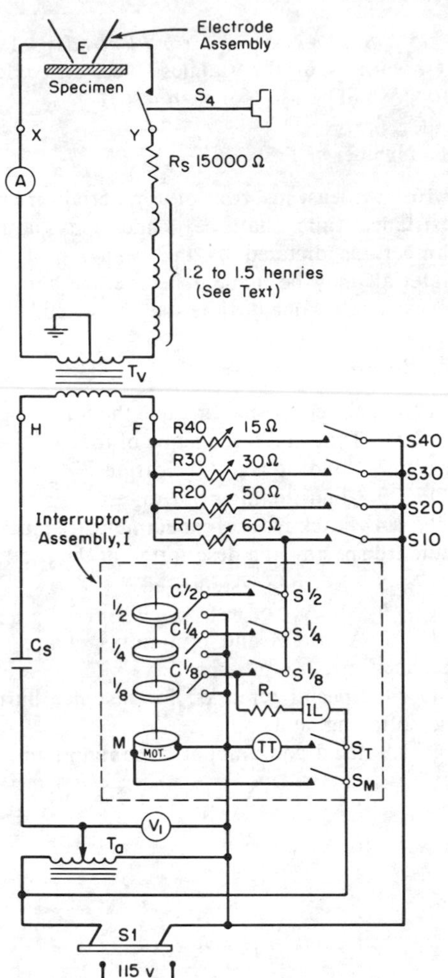

NOTE—Switches S_M to S_{40} are aligned in the sequence of their closing, from bottom to top, during a test.

FIG. 1 Arc-Resistance Test Circuit.

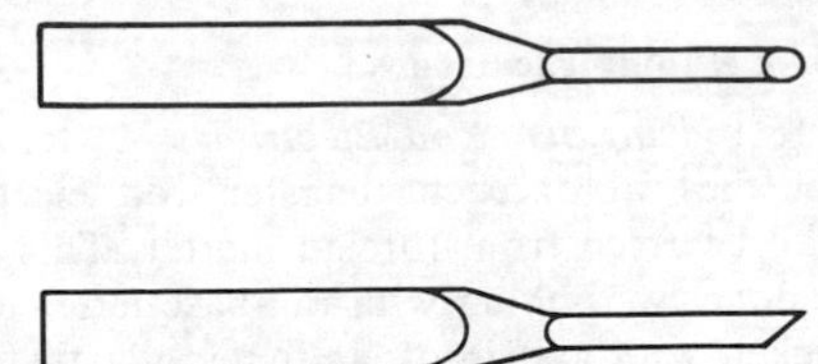

FIG. 2 Top and Side Views of Tungsten Rod Electrode

TABLE 1 Sequence of 1-min Current Steps.

Step	Current, mA	Time Cycle[a]	Approximate Rate of Heat Generation, W	Total Time, s
1/8–10	10	1/4 s on, 1 3/4 s off	3	60
1/4–10	10	1/4 s on, 3/4 s off	6	120
1/2–10	10	1/4 s on, 1/4 s off	12	180
10	10	continuous	24	240
20	20	continuous	34	300
30	30	continuous	45	360
40	40	continuous	56	420

[a] In the earlier steps an interrupted arc is used to obtain a less severe condition than the continuous arc; a current of less than 10 mA produces an unsteady (flaring) arc.

FIG. 3 Grinding and Polishing Block with Tungsten Rod Electrode in Place.

FIG. 4 Tungsten Rod Electrode Assembly.

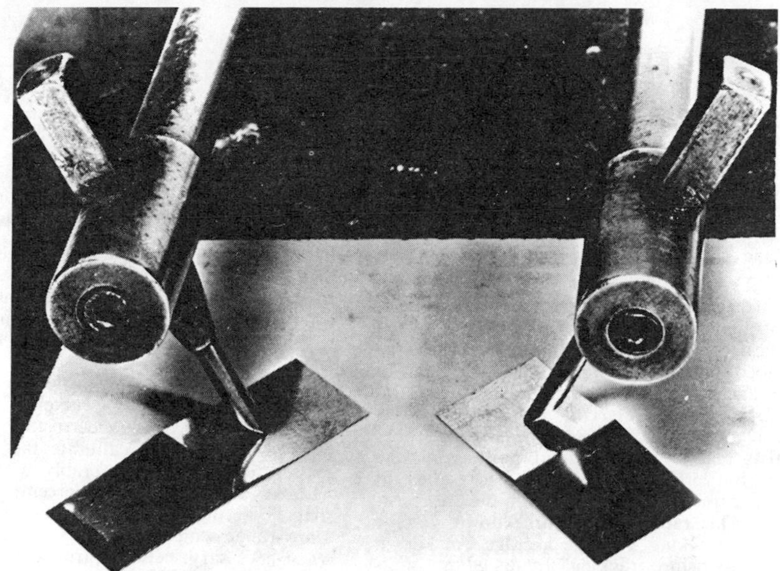

FIG. 5 Stainless Steel Strip Electrodes in Position on Test Specimen Surface.

FIG. 6 Alignment Jig for Stainless Steel Strip Electrodes.

APPENDIX

A1. VOLTS AFTER TRACKING TEST FOR DETERMINATION OF SURFACE BREAKDOWN VOLTAGE RATIO AFTER FAILURE OF SPECIMENS IN ARC-RESISTANCE METHOD D 495

A1.1 Scope

A1.1.1 This test is intended to determine the fraction of breakdown voltage remaining after completion of the track formed in a specimen of electrical insulating material when failure has occurred as described in ASTM Method D 495, Test for High-Voltage, Low-Current Arc Resistance of Solid Electrical Insulating Materials.[2] A breakdown voltage test is performed between the arc electrodes left undisturbed in their original position on the specimen, but after the specimen has cooled to room temperature following failure in the arc-resistance test. The ratio of this breakdown voltage ("volts after tracking") to the average breakdown voltage ("volts before tracking") obtained in a similar manner with the arc electrodes resting on undamaged areas of the specimen shall be known as the surface breakdown voltage ratio.

A1.1.2 This test should not be used when failure in the arc-resistance Method D 495 has resulted in such serious physical damage to the specimen that the performance of a voltage breakdown test at the damaged area would have little or no significance. The judgment of the operator shall be relied upon in such cases.

A1.2 Significance

A1.2.1 Certain insulating materials are known to recover a substantial measure of their original insulating value upon cooling after surface failure under electrical arcing. Plate glass, for example, will melt and become conducting in the area between the electrodes in an arc-resistance test, but will recover its insulating value completely upon stoppage of the arc and subsequent re-solidification. The degree to which a material will recover, after arcing is stopped, from the "failure" that may exist under arcing conditions is of great significance to the designer of electrical equipment. This characteristic of a material may determine whether or not insulation will have to be replaced following removal of the causes of a temporary or accidental arcing condition in service.

A1.2.2 Since the conditions under which breakdown voltage tests described herein are performed are highly arbitrary, caution should be exercised in applying the results indiscriminately to situations in which arc currents may not be limited to such low values as 10 mA.

A1.3 Apparatus

A1.3.1 The apparatus for this test is identical with that described in Section 4 of Method D 495 and the circuit arrangement shown in Fig. 1 is also identical, except that two additional voltmeters, V_2 and V_3, are required, as follows:

A1.3.1.1 Voltmeter V_2 shall be connected directly across the terminals H and F of the primary of the high-voltage transformer, T_v. It shall be accurately readable to $\pm^1/_2$ V or to ± 5 percent of the reading, whichever is larger. This can be met by a multiscale meter with full-scale readings of about 25, 75, and 125 V.

A1.3.1.2 Voltmeter V_3 shall be a high-voltage, root-mean-square reading meter of very high resistance, preferably of the electrostatic type. It is used only occasionally to calibrate the circuit. It shall read at least 5000 V, preferably to 6000 V or higher, with an accuracy of ± 5 percent. Below 500 V it shall be accurately readable to ± 10 V by means of a low-range scale, if necessary. It is to be connected, when necessary, between ground and position X or Y (Fig. 1) with switch S_4 open.

A1.4 Calibration of V_2 versus V_3

A1.4.1 In order to determine the breakdown voltage for specimens in the test for surface breakdown, it is necessary to relate the readings of the primary voltmeter, V_2, to the actual high voltage applied to the electrodes. To obtain this calibration, replace the electrode assembly, E, with the V_3 voltmeter, by opening S_4 and attaching the ungrounded side of V_3 to some convenient point electrically equivalent to X or Y (Fig. 1). With switches S_1 and S_{10} closed, raise the voltage from zero by means of the autotransformer, and take readings on the lowest appropriate scales of V_2 and V_3 for each increase that gives a total increment of about 100 V when the readings of V_3, when connected to X and then to Y, are added. Continue these steps up to about 1000 V, then take steps of about 500 V up to about 7000 V. Prepare a calibration curve from these data for use in describing the breakdown voltage of arc paths on specimens by converting from the reading of V_2.

A1.5 Test Specimens

A1.5.1 Use specimens as specified in Method D 495.

A1.6 Conditioning

A1.6.1 Condition the specimens as specified in Method D 495.

A1.7 Procedure for Determination of Arc-Resistance Breakdown Voltage

A1.7.1 The procedure for the determination of arc-resistance time is first carried out exactly as specified in Method D 495, as follows:

When failure occurs in the arc resistance test, stop the interval timer immediately by opening S_T. Then, if not already closed, close S_{10} to provide a steady arc for additional 3 s (estimated) before stopping the arc by opening all of $S_{1/8}$ to S_{40} that may have been closed. This is to ensure that a complete path has been formed between the electrodes.

A1.7.2 Without moving the electrodes or specimen from their positions during the procedure described in A1.7.1, permit them and the tested area to cool approximately to room temperature, assisted, if desired, by a mild air blast. Return T_a to zero, open S_4 and with S_{10} closed raise V_2 by means of T_a to a reading equivalent to a voltage of 200 at the electrodes. Close S_4 for an estimated 1 s, watching V_2 for a noticeable decrease. Continue raising the electrode voltage in steps of 100 by means of T_a, keeping S_4 open except for the 1-s test periods, until a voltage is reached where closing this switch produces a moderate but noticeable drop in the reading of V_2. The voltage at the electrodes for which this last step was set is the breakdown voltage of the arc path.

A1.8 Cleaning Electrodes

A1.8.1 The electrodes shall be cleaned as specified in Method D 495.

A1.9 Number of Tests

A1.9.1 At least five tests of a material for arc-resistance time and breakdown voltage of the exposed areas shall be made.

A1.10 Calculations

A1.10.1 The fraction of breakdown voltage (surface breakdown voltage ratio) remaining after performing an arc test in accordance with Method D 495 shall be calculated by conducting an equal number of similar tests of breakdown voltage on areas of the surface that have not been exposed to the action of the arc, and dividing the average of these into each of the values obtained for the exposed areas.

A1.11 Report

A1.11.1 The report shall include the following:

A1.11.1.1 Type and trade name of material,

A1.11.1.2 Conditions of fabrication,

A1.11.1.3 Conditioning prior to test,

A1.11.1.4 Thickness of specimen. If not a standard specimen, a description of the part.

A1.11.1.5 Number of tests,

A1.11.1.6 Averages and minimums of arc resistance times and surface breakdown voltage ratio,

A1.11.1.7 Special remarks, for instance burning, softening, and

A1.11.1.8 Kind and amount of contaminant, if any.

Designation: D 523 - 67

American National Standard Z131.1-1969
American National Standards Institute
Federation of Societies for
Paint Technology Standard No. Le-5-67

Standard Method of Test for SPECULAR GLOSS[1]

This Standard is issued under the fixed designation D 523; the number immediately following the designation indicates the year of original adoption or, in the case of revision, the year of last revision. A number in parentheses indicates the year of last reapproval.

1. Scope

1.1 This method covers the comparison of the specular gloss of nonmetallic specimens for glossmeter geometry of 60, 20, and 85 deg (**1**), (**2**), (**3**), (**5**), (**6**), (**7**).[2]

2. Definitions

2.1 *specular gloss*—the luminous fractional reflectance (**6**) of a specimen at the specular direction.

2.2 *luminous fractional reflectance*—the ratio of the luminous flux reflected from, to that incident on, a specimen for specified solid angles.

3. Summary of Method

3.1 Comparisons are made with 60, 20, or 85-deg geometry. The geometry of angles and apertures is chosen so that these procedures may be used as follows:

3.1.1 The 60-deg geometry for intercomparing most specimens, and for determining when either the 20-deg or the 85-deg geometry is applicable.

3.1.2 The 20-deg geometry for comparing specimens having 60-deg gloss higher than 70.

3.1.3 The 85-deg geometry for comparing specimens having 60-deg gloss lower than 30.

4. Apparatus[3]

4.1 *Instrumental Components*—The apparatus shall consist of an incandescent light source furnishing an incident beam, means for locating the surface of the specimen, and a receptor located to receive the required pyramid of rays reflected by the specimen. The receptor shall be a photosensitive device responding to visible radiation.

4.2 *Geometric Conditions*—The axis of the incident beam shall be at one of the specified angles from the perpendicular to the specimen surface. The axis of the receptor shall be at the mirror reflection of the axis of the incident beam. With a flat piece of polished black glass or other front-surface mirror in specimen position, an image of the source shall be formed at the center of the receptor field stop (receptor window). The length of the illuminated area of the specimen shall be equal to not more than one third of the distance from the center of this area to the receptor field stop. The axis of the incident beam and the axis of the receptor shall be within 0.1 deg of the nominal value indicated by the geometry. The dimensions and tolerances of the source and receptor shall be as indicated in Table 1. The angular dimensions of the receptor field stop are measured from the receptor lens in a collimated-beam type instrument, and from the test surface in a converging-beam type instrument. See Fig. 1 for a generalized illustration of the dimensions. The tolerances are chosen so that errors of no more than 1 gloss unit at any point on the scale will result from errors in the source and receptor apertures (**6**).

4.3 *Vignetting*—There shall be no vignetting of rays that lie within the field angles specified in 4.2.

4.4 *Spectral Conditions*—Results should not differ significantly from those obtained with a source-filter photocell combination that

[1] This method is under the jurisdiction of ASTM Committee D-1 on Paint, Varnish, Lacquer, and Related Products. A list of members may be found in the ASTM Yearbook.

Current edition effective Sept. 8, 1967. Originally issued 1939. Replaces D 523 - 66 T.

[2] The boldface numbers in parentheses refer to the list of references at the end of this method.

[3] List of manufacturers of glossmeters can be obtained from ASTM Headquarters, 1916 Race St., Philadelphia, Pa. 19103.

is spectrally corrected to yield CIE luminous efficiency with CIE Source C. Since specular reflection is, in general, spectrally nonselective, spectral corrections need be applied only to highly chromatic, low-gloss specimens upon agreement of users of this method.

4.5 *Measurement Mechanism*—The receptor-measurement mechanism shall give a numerical indication that is proportional to the light flux passing the receptor field stop within ±1 percent of full-scale reading.

5. Reference Standards

5.1 *Primary Working Standards* may be highly polished, plane, black-glass surfaces. Polished black glass with a refractive index of 1.567 shall be assigned a specular gloss value of 100 for each geometry. The gloss value for glass of another refractive index can be computed from the Fresnel equation (**6**). For small differences in refractive index, however, the gloss value will be a linear function of index, but the rate of change of gloss with index is different for each geometry. Each 0.001 increment in refractive index will produce a change of 0.27, 0.16, and 0.016 in the gloss value assigned to a polished standard for the 20, 60, and 85-deg geometries, respectively. For example, glass of index 1.527 would be assigned values of 89.2, 93.6, and 99.4 in order of increasing geometry.

5.2 *Secondary Working Standards*[4] of ceramic tile, depolished ground opaque glass, emery paper, and other semigloss materials having hard and uniform surfaces, are suitable when calibrated against a primary working standard on a glossmeter known to meet the requirements of this method. Such standards should be checked periodically for constancy, by comparing with primary standards.

6. Preparation and Selection of Test Specimens

6.1 This method does not cover preparation techniques. Whenever a test for gloss requires the preparation of a test specimen, specify the technique of specimen-preparation.

6.2 Use surfaces of good planarity, since surface warpage, waviness, or curvature may seriously affect test results. The directions of brush marks, or similar texture effects, should be parallel to the plane of the axes of the two beams.

NOTE 1—To determine the maximum gloss obtainable for a test specimen, such as a paint or a varnish film, use Methods C or D of ASTM Methods D 823, Producing Films of Uniform Thickness of Paint, Varnish, Lacquer, and Related Products on Test Panels.[5]

7. Procedure

7.1 Operate the glossmeter in accordance with the manufacturer's instruction.

7.2 Calibrate the instrument at the start and completion of every period of glossmeter operation, and during the operation at sufficiently frequent intervals to assure that the instrument response is practically constant. To calibrate, adjust the instrument to read correctly the gloss of a highly polished standard, and then read the gloss of a standard having poorer image-forming characteristics. If the instrument reading for the second standard does not agree within 1 percent of its assigned value, do not use the instrument without readjustment, preferably by the manufacturer.

7.3 Measure at least three portions of specimen surface to obtain an indication of uniformity.

8. Diffuse Correction

8.1 Apply diffuse corrections only upon agreement of buyer and seller. To apply the correction, subtract it from the glossmeter reading. To measure the correction, illuminate the specimen perpendicularly, and view at the incidence angle with the receiver aperture specified in 4.2 for the corresponding geometry. To compute the correction, multiply the 45-deg, 0-deg directional reflectance of the specimen by the effective fraction (Note 2) of the luminous flux reflected by magnesium oxide and accepted by the receiver aperture. Determine the 45-deg, 0-deg directional reflectance in accordance with ASTM Method E 97, Test for 45-deg, 0-deg Directional Reflectance of Opaque Specimens by Filter Photometry.[6] The effective fraction of the luminous flux from magnesium oxide entering the receiver aperture is listed as follows for each of the geometries:

[4] Gloss standards are available from the Gardner Laboratory, P. O. Box 5728, Bethesda, Md. 20014, and the Hunter Associates Laboratory, 9529 Lee Highway, Fairfax, Va. 22030.

[5] *Annual Book of ASTM Standards*, Part 21.

[6] *Annual Book of ASTM Standards*, Part 30.

Geometry, deg	Luminous Flux, parts per mil
60	2.1
20	1.3
85	0.002

NOTE 2—The effective fraction differs from the simple fraction because of the combined effects of surface reflectances, departure of practical diffusers from perfect, and polarization of the source (**6**).

9. Report

9.1 Report the average specular gloss reading and the geometry used.

9.2 Report the presence of any specimen, portions of the test surface of which differ in gloss from the average by more than 5 percent of the average.

9.3 Where preparation of the test specimen has been necessary, describe or otherwise identify the method of preparation.

9.4 Identify the glossmeter by the manufacturer's name and model designation.

9.5 Identify the working standard or standards of gloss used.

10. Precision

10.1 Readings obtained on the same instrument should be repeatable to within 1 percent of the magnitude of the readings. Readings obtained on different instruments should be reproducible to within 5 percent of the magnitude of the readings. Results obtained may be uncertain due to the cumulative effect of several sources of error, that is, difference between the geometric distribution of flux reflected from standards and specimens may bring about uncertainties in the measured gloss, even though the source and receiver apertures are within the tolerances specified in 4.2; inaccuracy of reading may result even though the precision of the measurement mechanism is held within the tolerance specified in 4.5; and lens arrangement and stray reflections from the interior walls of the instrument may cause errors in gloss readings.

REFERENCES

(**1**) Hunter, R. S., "Methods of Determining Gloss," *Proceedings*, Am. Soc. Testing Mats., Vol 36, 1936 Part II, p. 783; also *Journal of Research*, JRNBA Nat. Bureau Standards, Vol 18, January, 1937, No. 1, p. 19 (*Research Paper RP958*).
Six somewhat different appearance attributes are shown to be variously associated with gloss; therefore as many as six different photometric scales may be required to handle all gloss-measurement problems. (This paper out of print.)

(**2**) Hunter, R. S., and Judd, D. B., "Development of a Method of Classifying Paint According to Gloss," *ASTM Bulletin*, No. 97, March 1939, p. 11.
A comparison is made of several geometrically different photometric scales for separating paint finishes for gloss; the geometric conditions of test later incorporated in ASTM Method D 523 are recommended.

(**3**) Wetlaufer, L. A., and Scott, W. E., "The Measurement of Gloss," *Industrial and Engineering Chemistry*, Analytical Edition, IENAA, Vol 12, November 1940, p. 647.
A goniophotometric study of a number of paint finishes illuminated at 45 deg; a study of gloss readings affected by variation of aperture for 45 and 60-deg incidence.

(**4**) Horning, S. C., and Morse, M. P., "Measurement of the Gloss of Paint Panels," *Official Digest*, ODFPA Federation of Paint and Varnish Production Clubs, March 1947, p. 153.
A study of the effect of geometric conditions on results of gloss tests, with special attention to high-gloss panels.

(**5**) Hunter, R. S., "The Gloss Measurement of Paint Finishes," *ASTM Bulletin*, No. 150. January 1948, p. 72.
History of ASTM Method D 523.

(**6**) Hammond, H. K., III, and Nimeroff, I., "Measurement of Sixty-Degree Specular Gloss," *Journal of Research*, JRNBA Nat. Bureau of Standards, Vol 44, No. 6, June 1950, p. 585 (*Research Paper RP2105*); condensed account in *ASTM Bulletin*, No. 169 October 1950, p. 54.
A study of the effect of aperture variation on glossmeter readings; includes definitions of terms used in connection with specular gloss measurement, the Fresnel equation in a form readily usable for computation, and the derivation of diffuse correction formulas.

(7) Hunter, R. S., "Gloss Evaluation of Materials," *ASTM Bulletin*, No. 186, December 1952, p. 48.
A study of the history of gloss methods in the American Society for Testing and Materials and other societies; describes the background in the choice of geometry of these methods; contains photographs depicting gloss characteristics of a variety of materials.

(**8**) Huey, S., Hunter, R. S., Schreckendgust, J. G., and Hammond, H. K., III, Symposium on Gloss Measurement, *Official Digest*, Vol 36, No. 471, April, 1964, p. 343.
Contains discussion of industrial experience in measurement of 60-deg specular gloss (Huey), high-gloss measurement (Hunter), evaluation of low-gloss finishes with 85-deg sheen measurements (Schreckendgust), and gloss standards and glossmeter standardization (Hammond).

TABLE 1 Angles and Relative Dimensions of Source Image and Receptors

	In Plane of Measurement			Perpendicular to Plane of Measurement		
	θ, deg	2 tan $\theta/2$	Relative Dimension	θ, deg	2 tan $\theta/2$	Relative Dimension
Source image	0.75	0.0131	0.171	3.0[a]	0.0524	0.682
tolerance ±	0.25	0.0044	0.057			
60-deg receptor	4.4	0.0768	1.000	11.7	0.2049	2.668
tolerance ±	0.1	0.0018	0.023	0.2	0.0035	0.046
20-deg receptor	1.8	0.0314	0.409	3.6	0.0629	0.819
tolerance ±	0.05	0.0009	0.012	0.1	0.0018	0.023
85-deg receptor	4.0	0.0698	0.909	6.0	0.1048	1.365
tolerance ±	0.3	0.0052	0.068	0.3	0.0052	0.068

[a]Maximum; no minimum specification.

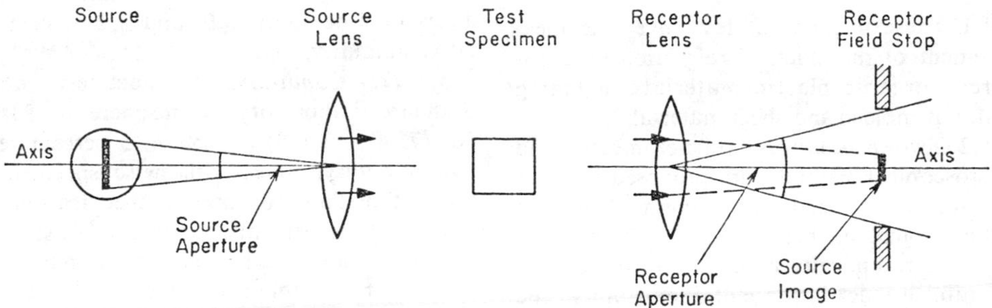

FIG. 1 Generalized Glossmeter Showing Apertures and Source Image Formation for a Collimated-Beam Type Instrument.

Designation: D 542 – 50 (Reapproved 1965)

USA Standard K65.7-1959 (R 1969)
USA Standards Institute

Standard Methods of Test for
INDEX OF REFRACTION OF TRANSPARENT ORGANIC PLASTICS[1]

This Standard is issued under the fixed designation D 542; the number immediately following the designation indicates the year of original adoption or, in the case of revision, the year of last revision. A number in parentheses indicates the year of last reapproval.

1. Scope

1.1 These methods of test cover the measurement of the index of refraction of transparent organic plastic materials including cast, hot-molded and sheet materials.

1.2 Two procedures, refractometric and microscopical methods, are proposed in order to cover satisfactorily the maximum range of indices found in these materials. Both methods require optically homogeneous specimens of uniform index for ease of duplicability. The refractometric method is to be preferred wherever it is applicable.

NOTE 1—The values stated in U.S. customary units are to be regarded as the standard. The metric equivalents of U.S. customary units may be approximate.

2. Significance

2.1 This test measures a fundamental property of matter useful for control of purity and composition, for simple identification purposes and for optical parts design. It is capable of much greater precision than ordinarily required. The refractometric method is accurate to four significant figures. The microscopical method, which is dependent upon the operator's skill in determining focus, is usually accurate to only three significant figures.

3. Conditioning

3.1 *Conditioning*—Condition the test specimens at 23 ± 2 C (73.4 ± 3.6 F) and 50 ± 5 percent relative humidity for not less than 40 h prior to test in accordance with Procedure A of ASTM Methods D 618, Conditioning Plastics and Electrical Insulating Materials for Testing,[2] for those tests where conditioning is required. In cases of disagreement, the tolerances shall be 1 C (1.8 F) and ±2 percent relative humidity.

3.2 *Test Conditions*—Conduct tests in the Standard Laboratory Atmosphere of 23 ± 2 C (73.4 ± 3.6 F) and 50 ± 5 percent relative humidity, unless otherwise specified in the test methods or in this specification. In cases of disagreements, the tolerances shall be 1 C (1.8 F) and ±2 percent relative humidity. If a material is found to have a high thermal coefficient of index, the temperature shall be accurately controlled to 23 C.

REFRACTOMETRIC METHOD

4. Apparatus

4.1 The apparatus for the preferred method shall consist of an Abbé refractometer (Note 2), a suitable source of white light, and a small quantity of a suitable contacting liquid (Notes 3 and 4).

NOTE 2—Other suitable refractometers can be used with appropriate modification of procedure as described in Section 6.

NOTE 3—A satisfactory contacting liquid is one which will not soften or otherwise attack the surface of the plastic within a period of 2 h of contact. The index of refraction of the liquid must be higher, by not less than one unit in the second decimal place, than the index of the plastic being measured.

The following liquids are suggested:

---

[1] These methods are under the jurisdiction of ASTM Committee D-20 on Plastics. A list of committee members may be found in the ASTM Yearbook. This standard is the direct responsibility of Subcommittee D-20.40 on Optical Properties.

Current edition effective Sept. 30, 1950. Originally issued 1939. Replaces D 542 – 42.

[2] *Annual Book of ASTM Standards*, Part 27.

Plastic	Contacting Liquid
Phenol formaldehyde resin	α-bromnaphthalene
Cellulose acetate and cellulose nitrate	{ cassia oil aniseed oil α-bromnaphthalene
Acrylic resins	saturated aqueous solution of zinc chloride made slightly acid
Vinyl resins[a]	α-monobromnaphthalene
Styrene resins	saturated aqueous solution of potassium mercuric iodide (Thoulet's or Sonstadt's solution)[b]

[a] No liquid has been found to date which is entirely satisfactory for vinyl resins.

[b] Avoid contact of this liquid with the skin or prolonged contact with the metal parts of the refractometer.

NOTE 4—In many cases a liquid as suggested above is not available and a second rapid reading must be made on a fresh specimen after preliminary trial.

5. Test Specimen

5.1 The test specimen shall be of a size such as will conveniently fit on the face of the fixed half of the refractometer prisms (Note 5). A specimen measuring 6.3 mm (0.25 in.) by 12.7 mm (0.5 in.) on one face is usually satisfactory.

NOTE 5—For maximum accuracy in the refractometer method the surface contacting the prism shall be quite flat. This surface can be judged for flatness, provided the specimen has been satisfactorily polished, by observing the sharpness of the dividing line between the light and dark field. A sharp, straight dividing line indicates a satisfactory contact of the specimen and prism surfaces.

5.2 The surface to be used in contact with the prism shall be flat and shall have a good polish. A second edge surface perpendicular to the first and on one end of the specimen shall be prepared with a fair polish (Note 6). The polished surfaces shall intersect without a beveled or rounded edge.

NOTE 6—It has been found possible to prepare a satisfactorily polished surface by hand polishing small specimens on an abrasive material backed by a piece of plate glass. Fine emery paper (for example, No. 000 Behr-Manning polishing paper) followed by a polishing rouge suspended in water on a piece of parchment paper have been used as the abrasive.

6. Procedure

6.1 Remove the hinged illuminating prism from the refractometer, if necessary. Place a source of diffuse light so that good illumination is obtained along the plane of the surface of contact between the specimen and the refractometer prism. Place a small drop of a suitable contacting liquid on the polished surface of the specimen and then place the specimen in firm contact with the surface of the prism and with the polished edge of the specimen toward the source of light. Determine the index of refraction in the same manner as for liquids. This shall be done by moving the index arm of the refractometer until the field seen through the eyepiece is one-half dark. Adjust the compensator (Amici prisms) drum to remove all color from the field. Adjust the index arm by means of the vernier until the dividing line between the light and dark portions of the field exactly coincides with the intersection of the cross hairs as seen in the eyepiece. Read the value of the index of refraction for the sodium D lines directly from the instrument. Determine the dispersion by reading the compensator drum and applying this figure, along with the index of refraction, to a chart or table supplied with the instrument.

NOTE 7—Sodium light from some type of a sodium burner is of use in increasing the accuracy and ease of setting of the refractometer.

MICROSCOPICAL METHOD

7. Apparatus

7.1 The apparatus for the auxiliary method shall consist of a microscope having a magnifying power of at least 200 diameters. This microscope shall be equipped with a means of measuring the longitudinal travel of the lens tube to within 0.025 mm (0.001 in.).

8. Test Specimens

8.1 The specimen described in Section 5 will be satisfactory for use with the microscopical method if it is not too thick to permit focusing the microscope through the polished surface (having a fair polish). A small specimen taken from a sheet of the material may be used if it is about 6.3 mm (0.25 in.) in thickness.

9. Procedure

9.1 Place a specimen having two parallel surfaces on the measuring microscope table with that surface having the best polish placed nearest to the objective. Focus the microscope through the specimen and on the bottom surface. Note the reading of the longitudinal displacement of the lens tube to the nearest 0.025 mm (0.001 in.) or less. Without moving the specimen, focus the microscope on the top surface of the specimen and again note the reading of the displacement. The difference between these two readings is the apparent thickness of the specimen. The index of refraction is found by dividing the actual thickness of the specimen by the apparent thickness.

10. Report

10.1 The report shall include the following:

10.1.1 The name of the method used.

10.1.2 The index of refraction to the nearest significant figure warranted by the accuracy and duplicability of the measurement. If the index is specified to more than three significant figures, the wave length of light for which the measurement was made shall be specified.

NOTE 8—In the case of nonisotropic materials, for example injection and compression molded materials, the index observed by the Abbé refractometer method will be the average value for a thin layer of small area at a point of contact near the center of the refractometer prism. For a complete and accurate determination of the variation of the index throughout the test specimen it is necessary to make the measurement at more than one point on the surface and within the body of the material. This can be done by preparing a contacting surface both perpendicular and parallel to the molding pressure or flow. After the test specimen is contacted to the prism it may be translated carefully for short distances along the prism surface in the direction of the light source while the variation of index is followed with the refractometer. This procedure should be repeated a sufficient number of times and for a sufficient number of specimens to determine the range of indices involved. The average value and the range of the index readings obtained from these specimens shall be reported if the range exceeds the accuracy of the measurement. If the variations in index are systematic with the orientation of the test specimen and if these variations exceed those found between specimens of the same material, the nature of these variations shall be reported with the average value.

Care should be taken to work rapidly to avoid changes in the refractive index of the plastic due to its absorption of the contacting liquid.

10.1.3 The temperature in degrees Celsius at which the index was measured.

10.1.4 If available, the dispersion shall be reported along with the index of refraction.

Designation: D 543 - 67

Standard Method of Test for
RESISTANCE OF PLASTICS TO CHEMICAL REAGENTS[1]

This Standard is issued under the fixed designation D 543; the number immediately following the designation indicates the year of original adoption or, in the case of revision, the year of last revision. A number in parentheses indicates the year of last reapproval.

1. Scope

1.1 This method covers the testing of all plastic materials, including cast, hot-molded, cold-molded, laminated resinous products, and sheet materials, for resistance to chemical reagents (Note 2). This method includes provisions for reporting changes in weight, dimensions, appearance, and strength properties (Note 3). Standard reagents are specified to establish results on a comparable basis. Provisions are made for various exposure times and exposure to reagents at elevated temperatures.

NOTE 1—The values stated in U.S. customary units are to be regarded as the standard. The metric equivalents of U.S. customary units may be approximate.

NOTE 2—The resistance of plastic films to extraction by chemicals shall be determined in accordance with ASTM Method D 1239, Test for Resistance of Plastic Films to Extraction by Chemicals.[2]

NOTE 3—The effect of chemical reagents on other properties shall be determined by making measurements on standard specimens for such tests before and after immersion in the reagents specified.

2. Significance

2.1 The limitations of the results obtained from this test should be recognized. The choice of types and concentrations of reagents, duration of immersion, temperature of the test, and properties to be reported is necessarily arbitrary. The specification of these conditions provides a basis for standardization and serves as a guide to investigators wishing to compare the relative resistance of various plastics to typical chemical reagents.

2.2 Correlation of test results with the actual performance or serviceability of plastics is necessarily dependent upon the similarity between the testing and end-use conditions. For applications involving continuous immersion, the data obtained in short time tests are of interest only in eliminating the most unsuitable materials or indicating a probable relative order of resistance to chemical reagents.

2.3 Evaluation of plastics for special applications involving corrosive conditions should be based on the particular reagents and concentrations to be encountered. The selection of test conditions should take into account the manner and duration of contact with reagents, the temperature of the system, and other performance factors involved in the particular application.

3. Apparatus

3.1 *Balance*—An accurate chemical balance.

3.2 *Micrometers*—Micrometers capable of measuring dimensions of test specimens to 0.025 mm (0.001 in.).

3.3 *Room*, or enclosed space capable of being maintained at the Standard Laboratory Atmosphere, as prescribed in ASTM Methods D 618, Conditioning Plastics and Electrical Insulating Materials for Testing.[2]

3.4 *Containers*—Suitable containers for immersing specimens in chemical reagents.

---

[1] This method is under the jurisdiction of ASTM Committee D-20 on Plastics. A list of committee members may be found in the ASTM Yearbook. This standard is the direct responsibility of Subcommittee D-20.50 on Permanence Properties.

Current edition effective Aug. 24, 1967. Originally issued 1939. Replaces D 543 – 65.

[2] *Annual Book of ASTM Standards*, Part 27.

They must be resistant to the corrosive effects of the reagents being used. Venting should be provided, especially when using volatile reagents at elevated temperatures.

3.5 *Oven or Constant-Temperature Bath*, capable of maintaining temperature within ± 2 C of the specified test temperatures.

3.6 *Testing Devices*—Testing devices for determining specific strength properties of specimens before and after immersion, conforming to the requirements prescribed in the ASTM methods of test for the specific properties being determined.

4. Standard Reagents

4.1 The following list of standard reagents is intended to be representative of the main categories of pure chemical compounds, solutions, and common industrial products. Chemicals used in this test shall be of technical grade or greater purity. All solutions shall be made with freshly prepared distilled water. Specified concentrations are on a weight percent or specific gravity basis. Mixing instructions are based on amounts of ingredients calculated to produce 1000 ml of solution of the specified concentration.

4.2 **Safety precautions should be taken to avoid personal contact, to eliminate toxic vapors, and to guard against explosion hazards in accordance with the hazardous nature of the particular reagents being used.**

4.3 The list of standard reagents is not intended to preclude the use of other reagents pertinent to particular chemical resistance requirements. It is intended to standardize on typical reagents, solution concentrations, and industrial products for general testing of the resistance of plastics to chemical reagents. Material specifications in which chemical resistance is indicated shall be based on reagents and conditions selected from those listed herein except by mutual agreement between the seller and the purchaser.

4.4 The standard reagents are as follows:

4.4.1 *Acetic Acid (sp gr 1.05)*—Glacial acetic acid.

4.4.2 *Acetic Acid (5 percent)*—Add 48 ml (50.5 g) of glacial acetic acid (sp gr 1.05) to 955 ml of water.

4.4.3 *Acetone.*

4.4.4 *Ammonium Hydroxide (sp gr 0.90)*—Concentrated ammonium hydroxide (NH_4OH).

4.4.5 *Ammonium Hydroxide (10 percent)*—Add 375 ml (336 g) of NH_4OH (sp gr 0.90) to 622 ml of water.

4.4.6 *Aniline.*

4.4.7 *Benzene.*

4.4.8 *Carbon Tetrachloride.*

4.4.9 *Chromic Acid (40 percent)*—Dissolve 549 g of chromic anhydride (CrO_3) in 822 ml of water.

4.4.10 *Citric Acid (10 percent)*—Dissolve 104 g of citric acid crystals in 935 ml of water.

4.4.11 *Cottonseed Oil*, edible grade.

4.4.12 *Detergent Solution, Heavy Duty (0.025 percent)*—Dissolve 0.05 g of alkyl aryl sulfonate and 0.20 g of trisodium phosphate in 1000 ml of water.

4.4.13 *Diethyl Ether.*

4.4.14 *Dimethyl Formamide.*

4.4.15 *Distilled Water*, freshly prepared.

4.4.16 *Ethyl Acetate.*

4.4.17 *Ethyl Alcohol (95 percent)*—Undenatured ethyl alcohol.

4.4.18 *Ethyl Alcohol (50 percent)*—Add 598 ml (482 g) of 95 percent undenatured ethyl alcohol to 435 ml of water.

4.4.19 *Ethylene Dichloride.*

4.4.20 *2-Ethylhexyl Sebacate.*

4.4.21 *Heptane*, commercial grade, boiling range 90 to 100 C.

4.4.22 *Hydrochloric Acid (sp gr 1.19)*—Concentrated hydrochloric acid (HCl).

4.4.23 *Hydrochloric Acid (10 percent)*—Add 239 ml (283 g) of HCl (sp gr 1.19) to 764 ml of water.

4.4.24 *Hydrofluoric Acid (40 percent)*—Slowly add 748 ml (866 g) of hydrofluoric acid (52 to 55 percent HF) to 293 ml of water.

4.4.25 *Hydrogen Peroxide Solution (28 percent, or USP 100 Volume).*

4.4.26 *Hydrogen Peroxide Solution (3 percent, or USP 10 Volume)*—Add 98 ml (108 g) of commercial grade (100 vol or 28 percent) hydrogen peroxide (H_2O_2) to 901 ml of water.

4.4.27 *Isooctane (2,2,4-Trimethyl Pentane).*

4.4.28 *Kerosine*—No. 2 fuel oil, ASTM Specification D 396 for Fuel Oils.[3]

4.4.29 *Methyl Alcohol.*

4.4.30 *Mineral Oil, White, USP (sp gr*

[3] *Annual Book of ASTM Standards*, Part 17.

0.830 to 0.860; Saybolt @ 100 F, 125 to 135 s).

4.4.31 *Nitric Acid (sp gr 1.42)*—Concentrated nitric acid (HNO_3).

4.4.32 *Nitric Acid (40 percent)*—Add 500 ml (710 g) of HNO_3 (sp gr 1.42) to 535 ml of water.

4.4.33 *Nitric Acid (10 percent)*—Add 108 ml (153 g) of HNO_3 (sp gr 1.42) to 901 ml of water.

4.4.34 *Oleic Acid*, cP.

4.4.35 *Olive Oil*, edible grade.

4.4.36 *Phenol Solution (5 percent)*—Dissolve 47 g of carbolic acid crystals, USP, in 950 ml of water.

4.4.37 *Soap Solution (1 percent)*—Dissolve dehydrated pure white soap flakes (dried 1 h at 105 C) in water.

4.4.38 *Sodium Carbonate Solution (20 percent)*—Add 660 g of sodium carbonate ($Na_2 \cdot CO_3 \cdot 10H_2O$) to 555 ml of water.

4.4.39 *Sodium Carbonate Solution (2 percent)*—Add 55 g of $Na_2CO_3 \cdot 10H_2O$ to 964 ml of water.

4.4.40 *Sodium Chloride Solution (10 percent)*—Add 107 g of sodium chloride (NaCl) to 964 ml of water.

4.4.41 *Sodium Hydroxide Solution (60 percent)*—Slowly dissolve 971 g of sodium hydroxide (NaOH) in 649 ml of water.

4.4.42 *Sodium Hydroxide Solution (10 percent)*—Dissolve 111 g of NaOH in 988 ml of water.

4.4.43 *Sodium Hydroxide Solution (1 percent)*—Dissolve 10.1 g of NaOH in 999 ml of water.

4.4.44 *Sodium Hypochlorite Solution, National Formulary (4 to 6 percent)*[4]—The concentration of this solution can be determined as follows: Weigh accurately in a glass-stoppered flask about 3 ml of the solution and dilute it with 50 ml of water. Add 2 g of potassium iodide (KI) and 10 ml of acetic acid, and titrate the liberated iodine with 0.1 *N* sodium thiosulfate ($Na_2S_2O_3$), adding starch solution as the indicator. Each ml of 0.1 *N* $Na_2S_2O_3$ solution is equivalent to 3.722 mg of sodium hypochlorite.

4.4.45 *Sulfuric Acid (sp gr 1.84)*—Concentrated sulfuric acid (H_2SO_4).

4.4.46 *Sulfuric Acid (30 percent)*—Slowly add 199 ml (366 g) of H_2SO_4 (sp gr 1.84) to 853 ml of water.

4.4.47 *Sulfuric Acid (3 percent)*—Slowly add 16.6 ml (30.6 g) of H_2SO_4 (sp gr 1.84) to 988 ml of water.

4.4.48 *Toluene.*

4.4.49 *Transformer Oil*, complying with the requirements of ASTM Specification D 1040, for Uninhibited Mineral Oil for Use in Transformers and in Oil Circuit Breakers.[5]

4.4.50 *Turpentine*—Gum spirits or steam distilled wood turpentine conforming to ASTM Specification D 13, for Spirits of Turpentine.[6]

5. Test Specimens

5.1 The type and dimensions of test specimens to be used depend upon the form of the material and the tests to be performed (Note 4). At least three specimens shall be used for each material being tested and for each reagent involved. The specimens shall be as follows:

5.1.1 *Molding and Extrusion Materials*—Specimens shall be molded to shape or cut from molded slabs as required below. The cut edges of specimens shall be made smooth by sharp cutting, machining, or by finishing with No. 0 or finer sandpaper or emery cloth. Molding shall conform to conditions recommended by the manufacturer of the material (Note 5). The shape and dimensions of specimens shall depend on the tests to be performed and shall conform to the following:

5.1.1.1 *Weight and Dimension Changes*—Standard specimens shall be in the form of disks 50.80 mm (2 in.) in diameter and 3.175 mm (0.125 in.) in thickness molded or cut from molded slabs. The nominal surface area of this standard disk is 45.60 cm^2 (7.068 $in.^2$).

5.1.1.2 *Mechanical Property Changes*—Standard tensile specimens shall be used according to the method of test prescribed in the appropriate specification for the material being tested, or by agreement among those concerned. Where the determination of other mechanical properties is agreed upon by the seller and the purchaser, standard specimens prescribed in the appropriate standard methods of test shall be used.

---

[4] Available from Fisher Scientific Co., Code No. SO-S-290.

[5] *Annual Book of ASTM Standards*, Part 29.

[6] *Annual Book of ASTM Standards*, Part 20.

5.1.2 *Sheet Materials*—Specimens from sheet materials shall be cut from a representative sample of the material (Note 6) in a manner depending on the tests to be performed and the thickness of the sheet as follows (see 5.1.1 regarding preparation of cut edges):

5.1.2.1 *Weight and Dimension Changes*—Standard specimens shall be in the form of bars 76.20 mm (3 in.) in length by 25.40 mm (1 in.) in width by the thickness of the material. The nominal surface area of the standard bar, having a thickness of 3.175 mm (0.125 in.), is 45.16 cm^2 (7.0 $in.^2$). Circular disk specimens 50.80 mm (2 in.) in diameter by the thickness of the material are permissible under mutual agreement between the seller and the purchaser. Permissible variations in thickness of both types of specimens are ±0.18 mm (0.007 in.) for hot molded and ±0.30 mm (0.012 in.) for cold molded or cast materials.

5.1.2.2 *Mechanical Property Changes*—Standard machined, sheared, or cut tensile specimens shall be used according to the methods of test prescribed in the appropriate specifications of the material being tested, or by agreement among those concerned (see 5.1.1.2).

NOTE 4—Specimen surface area greatly affects the weight change due to immersion in chemical reagents. Thickness influences percentage dimension change as well as percentage change in mechanical properties. In addition, molded specimens may not agree with specimens cut from molded or otherwise formed sheet of a given material. Consequently, comparison of materials should be made only on the basis of results obtained from specimens of identical dimensions and like methods of specimen preparation.

NOTE 5—Molding conditions can affect the resistance of plastics to chemical reagents. Compression moldings should be prepared in a manner that will disperse external lubricants and result in complete fusion of the particles. Injection molding should be accomplished in a manner that results in a minimum of molecular orientation and thermal stresses or a controlled level of both depending upon the conditions being simulated.

NOTE 6—For certain products, such as laminates, in which edge effects are pronounced, larger coupons may be exposed from which standard specimens can be cut after immersion for determining the effects of reagents on mechanical properties. This may be allowed in provisions of material specifications by mutual agreement between the seller and the purchaser and should be reported as such.

6. Conditioning

6.1 *Conditioning*—Condition the test specimens at 23 ± 2 C (73.4 ± 3.6 F) and 50 ± 5 percent relative humidity for not less than 40 h prior to test in accordance with Procedure A of ASTM Methods D 618, Conditioning Plastics and Electrical Insulating Materials for Testing,[2] for those tests where conditioning is required. In cases of disagreement, the tolerances shall be 1 C (1.8 F) and ±2 percent relative humidity.

6.2 *Test Conditions*—Conduct tests in the Standard Laboratory Atmosphere of 23 ± 2 C (73.4 ± 3.6 F) and 50 ± 5 percent relative humidity, unless otherwise specified in the test methods or in this specification. In cases of disagreements, the tolerances shall be 1 C (1.8 F) and ±2 percent relative humidity.

7. Procedure I. Weight and Dimension Changes (See Note 7 and 4.2.)

7.1 Weigh each conditioned specimen separately and measure its thickness at the center and its length and width, or two diameters at right angles to each other, to the nearest 0.025 mm (0.001 in.). In the case of laminates, edge swelling is not uncommon under certain conditions. Consequently, it may be necessary to measure thickness both at the center and at the edges and report the percentage change separately for each position.

7.2 Place the specimens in appropriate containers for the reagents being used and allow the specimens to be totally immersed in fresh reagent for 7 days in the Standard Laboratory Atmosphere. Suspend the specimens to avoid any contact with the walls or bottom of the container. For specimens of thin sheeting or those having a lower density than the reagent it may be necessary to attach small weights such as nichrome wire to prevent floating or curling. Several specimens of a given material may be immersed in the same container provided sufficient reagent is allowed for the total surface area exposed and the specimens do not touch each other. For specimens of nonextractable and relatively insoluble materials, the quantity of reagent shall be approximately 10 ml/$in.^2$ of specimen surface area. For specimens that tend to dissolve or which involve extraction of plasticizers, the quantity of reagent shall be approximately 40 ml/$in.^2$ of specimen surface area. Where there is any doubt in these matters, use the higher solvent ratio. For tests at other than room temperatures, it is recommended that the test temperature be 50 C, 70 C, or other temperatures recommended in Methods D 618. It is important that the reagent be at the elevated test temperature before the specimens are immersed.

7.3 Stir the reagents every 24 h by moderate manual rotation of the containers or other suitable means (Note 8).

7.4 After 7 days, or other agreed upon period of time, individually remove each specimen from the reagent, immediately weigh in a weighing bottle, and remeasure its dimensions. Wash with running water specimens removed from acid, alkali, or other aqueous solutions, wipe them dry with a cloth or tissue, and immediately weigh. Hygroscopic reagents such as concentrated sulfuric acid may remain adsorbed on the surface of the specimen even after rinsing, requiring immediate special handling to avoid moisture pickup before and during weighing. Rinse specimens removed from nonvolatile, nonwater-soluble organic liquids with a nonaggressive, but volatile solvent, such as ligroin, before wiping dry. Specimens removed from volatile solvents such as acetone, alcohol, etc., need no rinsing before wiping dry. Some specimens may become tacky due to dissolved material on the surface or solvent absorbed throughout the specimen. Take care in wiping such specimens not to disturb or contaminate the surface.

7.5 Observe the appearance of each specimen after exposure to chemical reagent. Observe and report appearance on the basis of examination for evidence of loss of gloss, developed texture, decomposition, discoloration, swelling, clouding, tackiness, rubberiness, crazing, bubbling, cracking, solubility, etc. See ASTM Nomenclature D 883, Relating to Plastics[2] for proper descriptive terminology.

NOTE 7—For some materials, absorption of the reagent over the 7-day immersion period is nearly balanced by the removal of soluble constituents from the plastic. This type of behavior may be revealed by comparing the initial conditioned weight of the specimen with its weight when dried for 7 days at 23 C and 50 percent relative humidity, after removal from the chemical reagent. A final weight lower than the initial weight may indicate removal of soluble constituents. However, only for particular combinations of reagent and test specimen can this weight difference be considered as due strictly to the removal of soluble constituents.

NOTE 8—In making tests for shorter or longer periods of time than 7 days, it is recommended that the tests be run at 1 and 3 days or for times that are made up of increments of 4 weeks, respectively. The containers should be stirred once each day during the first week, and once each week thereafter.

8. Procedure II. Mechanical Property Changes

8.1 Immerse and handle the mechanical test specimens in accordance with instructions given under Procedure I (Section 7).

8.2 Determine the mechanical properties of identical nonimmersed and immersed specimens in accordance with the standard methods for tensile tests prescribed in the specifications for the materials being tested (Note 9). Make mechanical property tests on nonimmersed and immersed specimens prepared from the same sample or lot of material in the same manner, and run under identical conditions. Test immersed specimens immediately after they are removed from the chemical reagent. Where specimens are exposed to reagents at elevated temperature, unless they are to be tested at the elevated temperature, they shall be placed in another container of the reagent at the Standard Laboratory Temperature for approximately 1 h to effect cooling prior to testing (Note 10).

NOTE 9—While tensile tests are more generally applicable and preferred for assessing mechanical property changes due to the effects of chemical reagents, other mechanical properties may be more significant in special cases. For example, flexural properties of rigid materials, which are not appreciably softened by the reagents under study, may be extremely sensitive to surface attack such as crazing. Consequently, in the use of this method for establishing chemical resistance levels in material or product specifications, consideration should be given to the choice of mechanical properties that properly characterize the effects of exposure to chemical reagents.

NOTE 10—To isolate the effects of certain chemical reagents on the mechanical properties of some plastics, it is necessary to test identical specimens that have been immersed in water. This is especially true of tests to determine the effects of aqueous solutions, where these may not differ greatly from the effects of immersion in water alone (see References **(3)** and **(4)** at the end of this method). When tests are run with a variety of aqueous solution reagents, the effects due to water alone should be established for better comparison of results. Similar behavior may result when tests are run at elevated temperatures, requiring knowledge of the effects of temperature alone to properly assess the effects due to the chemical reagents.

9. Report

9.1 The report shall include the following:

9.1.1 *Procedure I:*

9.1.1.1 Complete identification of the material tested including type, source, manufacturer's code, form, and previous history,

9.1.1.2 Method of preparing test specimens,

9.1.1.3 Conditioning procedure used,

9.1.1.4 Temperature of tests,

9.1.1.5 Chemical reagents,

9.1.1.6 Duration of immersion,

9.1.1.7 Initial length, width or two diameters, and thickness of each specimen, in inches, measured to the nearest 0.025 mm (0.001 in.) (see 7.1),

9.1.1.8 Initial weight of each specimen in grams to ±0.0005 g,

9.1.1.9 Length, width, and thickness after immersion,

9.1.1.10 Weight after immersion,

9.1.1.11 Average percentage increase or decrease in length and width, or in the two diameters, and in the thickness, taking the dimensions of the conditioned specimen as 100 percent,

9.1.1.12 Average percentage gain or loss in weight calculated to the nearest 0.01 percent, taking the conditioned weight as 100 percent, and

9.1.1.13 General appearance of specimens after immersion.

9.1.2 *Procedure II:*

9.1.2.1 Items 9.1.1.1 through 9.1.1.6 as for Procedure I, and the following:

9.1.2.2 Specimen type and dimensions,

9.1.2.3 Method of test,

9.1.2.4 Conditions of test,

9.1.2.5 Mechanical properties of identical nonimmersed and immersed specimens, and

9.1.2.6 Average percentage increase or decrease in mechanical properties, taking the properties of the conditioned nonimmersed specimens as 100 percent.

References

(1) Kline, G. M., Rinker, R. C., and Meindl, H. F., "Resistance of Plastics to Chemical Reagents," *Proceedings*, Am. Soc. Testing Mats., Vol 41, 1941, p. 1246; and *Modern Plastics*, Vol 19, December, 1941, p. 59.

(2) Delmonte, J., "Effect of Solvents Upon Solid Organic Plastics," *Industrial and Engineering Chemistry*, Vol 34, June, 1942, p. 764.

(3) Tyrie, W. R., "Chemical Resistance of Phenolic Laminates," *Transactions*, Am. Soc. Mechanical Engrs., Vol 69, October, 1947, p. 795.

(4) Crouse, W. A., Carickhoff, Margie, and Fischer, Margaret A., "Effect of Fuel Immersion on Laminated Plastics," *Transactions*, Am. Soc. Mechanical Engrs., Vol 72, February 1950, p. 175.

(5) Adams, W. H., and Lebach, H. H., "New Corrosion Test for Plastics," *Chemical Engineering*, Vol 56, July, 1949, pp. 98–101.

(6) Reinhart, Frank W., and Williams, Harry C., Jr., "Resistance of Representative Plastic Materials to Hydrofluoric Acid," ASTM Bulletin, No. 167, July 1950, p. 60.

(7) Meyerhans, Konrad, "Resistance of Ethoxyline Resins to Chemical Attacks. The Behavior of Araldite Casting Resin B towards Chemicals," *Kunststoffe*, Vol 44, 1954 pp. 135–42.

Designation: D 568 - 68

Standard Method of Test for
FLAMMABILITY OF FLEXIBLE PLASTICS[1]

This Standard is issued under the fixed designation D 568; the number immediately following the designation indicates the year of original adoption or, in the case of revision, the year of last revision. A number in parentheses indicates the year of last reapproval.

1. Scope

1.1 This method covers a procedure for comparison of plastics according to flammability when in the form of flexible, thin sheets or films.

NOTE 1—The values stated in U.S. customary units are to be regarded as the standard. The metric equivalents of U.S. customary units may be approximate.

NOTE 2—For tests of self-supporting plastics in the form of sheets, bars, plates or panels reference should be made to ASTM Method D 635, Test for Flammability of Self-Supporting Plastics.[2] Thin, flexible films may also be tested according to ASTM Method D 1433, Test for Flammability of Flexible Thin Plastic Sheeting.[2] Plastic foams and sheeting are referenced in ASTM Method D 1692, Test for Flammability of Plastic Foams and Sheeting.[2] ASTM Method E 84, Test for Surface Burning Characteristics of Building Materials[3] is utilized to evaluate plastics when applied or fastened to the surfaces of walls or ceilings as an interior finish.

2. Significance

2.1 This method provides a means of comparing the flammability of plastics in sheet form. Test data should be compared with data for a control material of known performance and of comparable thickness. Correlation with flammability under actual use conditions is not necessarily implied. Excessive shrinkage of specimens may invalidate the test.

3. Apparatus

3.1 *Shield*—A shield constructed from sheet metal or other fire-resistant material, having inside dimensions 30 cm (12 in.) in width by 30 cm (12 in.) in depth, by 76 cm (30 in.) in height, and open at the top. The shield shall be so constructed as to provide a ventilating opening approximately 2.54 cm (1 in.) in height around the bottom and shall have a viewing window in one side, of sufficient size and in such a position that the entire length of the specimen under test may be observed. Because of danger due to breaking glass, it may be necessary to use heat-resistant glass for the viewing window. One side of the shield shall be hinged (or some other suitable form of construction used) so that the shield may be readily opened and closed to facilitate the mounting and ignition of the test specimen.

3.2 *Clamp*—A spring type of paper clamp for holding the test specimen in position. The holding clamp shall be attached rigidly to the shield in such a manner that when the specimen is clamped therein, it is centered within the shield facing the viewing window.

3.3 *Burner*—Bunsen burner and gas supply.

3.4 *Timing Device*—Stop watch or other timer reading in seconds.

4. Test Specimens

4.1 Three test specimens 2.54 cm (1 in.) in width by 45.7 cm (18 in.) in length shall be cut in both machine and transverse directions from each of the materials being tested. The thickness of the specimens shall be determined in accordance with Method B of ASTM Methods D 374, Test for Thickness of Solid Electrical Insulation[2] or by any method at least as accurate.

4.2 Gage marks shall be drawn across the specimens 7.6 cm (3 in.) from each end, defining a 30.5-cm (12-in.) gage length over which

[1] This method is under the jurisdiction of ASTM Committee D-20 on Plastics. A list of committee members may be found in the ASTM Yearbook. This standard is the direct responsibility of Subcommittee D-20.30 on Thermal Properties.

Current edition effective Sept. 9, 1968. Originally issued 1940. Replaces D 568 - 61.

[2] *Annual Book of ASTM Standards*, Part 27.

[3] *Annual Book of ASTM Standards*, Part 14.

the burning rate is to be measured.

5. Conditioning

5.1 *Conditioning*—Condition the test specimens at 23 ± 2 C (73.4 ± 3.6 F) and 50 ± 5 percent relative humidity for not less than 40 h prior to test in accordance with Procedure A of ASTM Methods D 618, Conditioning Plastics and Electrical Insulating Materials for Testing,[2] for those tests where conditioning is required. In cases of disagreement, the tolerances shall be 1 C (1.8 F) and ±2 percent relative humidity.

5.2 *Test Conditions*—Conduct tests in the Standard Laboratory Atmosphere of 23 ± 2 C (73.4 ± 3.6 F) and 50 ± 5 percent relative humidity, unless otherwise specified in the test methods or in this specification. In cases of disagreements, the tolerances shall be 1 C (1.8 F) and ±2 percent relative humidity.

6. Procedure

6.1 Clamp the specimen vertically so that 43.2 cm (17 in.) of it is exposed below the clamp. Place the shield in a hood, with the ventilating fan turned off at the time of test. Adjust the bunsen burner to provide a flame about 2.54 cm (1 in.) in total height.

6.2 Apply the flame to the end of the specimen until it is ignited but not longer than 15 s. The flame may be momentarily withdrawn as needed to establish that the specimen is ignited.

6.3 If the specimen ignites, start a timer when the charred edge reaches the lower gage mark, and stop the timer when the charred edge reaches the upper gage mark. Record the elapsed time.

6.4 If the flame is extinguished before reaching the upper gage mark, stop the timer and note the apparent cause, such as melting and dripping, smothering, etc.

NOTE 3—Continued testing of highly flammable materials causes the holding clamp (see 3.2) to become heated. It is recommended that after each determination the holding clamp be cooled to room temperature by means of a cloth soaked in cool tap water.

NOTE 4—It should be noted that for some materials the products of burning are toxic, and care should be taken to guard the operator from the effects of these toxic gases. The ventilating fan in the hood under which the test is performed should be turned on immediately after the test is completed in order to remove any irritating products of the test.

7. Calculations

7.1 Calculate the burning rate in square centimeters (or square inches) per minute by dividing the area above the lower gage mark which is burned, charred, or melted off by the time in minutes as determined in 6.3.

8. Report

8.1 The report shall include the following:

8.1.1 Thickness of the material.

8.1.2 Classification as follows:

8.1.2.1 If the material does not ignite after 15 s of application of the flame, report it as "nonburning by this test."

8.1.2.2 If the flame is extinguished before reaching the upper gage mark, report the material as "self-extinguishing by this test" and report also the "apparent cause," such as melting, smothering.

8.1.2.3 If the specimen burns, melts off, or is charred up to the upper gage mark, calculate the burning rate in accordance with Section 7 and report the results.

Designation: D 569 – 59 (Reapproved 1961)

Standard Method of

MEASURING THE FLOW PROPERTIES OF THERMOPLASTIC MOLDING MATERIALS[1]

This Standard is issued under the fixed designation D 569; the number immediately following the designation indicates the year of original adoption or, in the case of revision, the year of last revision. A number in parentheses indicates the year of last reapproval.

1. Scope

1.1 This method covers the measurement of the following flow properties of thermoplastic molding materials:

1.1.1 *Procedure A*—The temperature at which a thermoplastic material attains a defined degree of flow when subjected to a prescribed pressure for a prescribed time in a specified extrusion mold, or

1.1.2 *Procedure B*—The degree of flow that a thermoplastic material attains when subjected to a prescribed pressure and temperature for a prescribed time in a specified extrusion mold.

NOTE 1—The values stated in U.S. customary units are to be regarded as the standard. The metric equivalents of U.S. customary units may be approximate.

2. Significance

2.1 The equilibrium state of the material in this test is not necessarily representative of the state at which the material is molded commercially. The indicated temperature of the heat-transfer medium in an injection or extrusion machine cylinder always is much in excess of that of the material itself and of the flow temperature of the material. Thus, the chief use of this test is for control and identification purposes rather than for prediction of molding conditions or physical properties.

3. Apparatus

3.1 The apparatus for the flow test (see Fig. 1) shall be a constant-force, vertical-orifice type machine[2] consisting essentially of the following:

3.1.1 *Orifice*—A vertical orifice 3.18 ± 0.013 mm (0.125 ± 0.0005 in.) in diameter and 38 mm ($1\frac{1}{2}$ in.) in length, *1*, Fig. 1, into which the material flows. The orifice is machined into a split cone 25.4 mm (1 in.) in diameter at the base which is clamped into a steam-heated block, *3*. A thermometer well 4.76 mm ($\frac{3}{16}$ in.) in diameter and 32 mm ($1\frac{1}{4}$ in.) in depth is drilled into the split cone. Temperature readings are taken at this point.

3.1.2 *Thermometer*—A $1\frac{1}{4}$-in. immersion mercury thermometer having a diameter just under $\frac{3}{16}$ in. and a temperature scale of not more than 20 C/in. of length.

3.1.3 *Charge Chamber*—Below the orifice and concentric with it is the charge chamber, *2*, 9.5 mm ($\frac{3}{8}$ in.) in diameter and 19 mm ($\frac{3}{4}$ in.) in length.

3.1.4 *Block*—A steam-heated block, *3*. Heat is supplied by steam at a line pressure of 620 to 1030 kN/m^2 (90 to 150 psi), passing through an accurate reducing valve; if the line pressure fluctuates too widely, two reducing valves in series may be used. The temperature is controlled by regulating the steam pressure.

3.1.5 *Ram*—A steam-heated ram, *4*, so arranged that it applies pressure to the charge chamber from the bottom, forcing the material into the orifice. The steam line is so arranged that the steam passes through the reducing valve, the ram, the block, and finally

[1] This method is under the jurisdiction of ASTM Committee D-20 on Plastics. A list of committee members may be found in the ASTM Yearbook. This standard is the direct responsibility of Subcommittee D-20.30 on Thermal Properties.

Current edition effective Sept. 10, 1959. Originally issued 1948. Includes former Method D 863 and replaces D 569 – 55.

[2] Rossi-Peakes flow tester as described in U. S. Patent 2,066,016.

through a suitable trap or small vent.

3.1.6 *Pressure System*—A mechanical system, *5*, for applying a net pressure of 10.3 MN/m^2 (1500 psi) to the ram. The pressure system illustrated in Fig. 1 is built so that any pressure up to 20.7 MN/m^2 (3000 psi) may be applied in increments of 690 kN/m^2 (100 psi).

3.1.7 *Flow Measuring Device*—Means for measuring the flow of material into the orifice. Measurements shall be accurate to ±0.25 mm (±0.01 in.). The following measuring systems are suitable:

3.1.7.1 A follower rod, guided by loose bearings in a swing arm, rests on the material in the orifice. A flexible chain attached to the upper part of the follower rod is passed part way around and fastened to the small diameter of a two-step pulley having a 3:1 ratio and mounted on the swing arm. A second flexible chain passing part way around and fastened to the larger diameter is attached to a sliding indicator and a counterweight. The indicator travels over a calibrated scale so that the amount of flow in the orifice is shown on the scale, magnified three times. The weight of the flow-indicating mechanism is such that a pressure of 275 kN/m^2 (40 psi) is exerted on the material in the orifice; this however, is counterbalanced by sufficient additional weight on the pressure system, *5*.

3.1.7.2 *Optional*—The flow measuring device illustrated in Fig. 1 may be equipped with a time clock, a recording pen, and suitable graph paper with the scale as one axis and time as the other so that the flow behavior over any given time period may be recorded.

4. Test Specimens

4.1 The test specimens shall be molded, machined, or preformed cylinders 9.5 mm ($^3/_8$ in.) in diameter and $^3/_8$ in. in height. They shall be free of air bubbles and they shall be as free as possible of strains.

5. Conditioning

5.1 *Preferred Method*—Test specimens shall be conditioned at room temperature in a desiccator over anhydrous calcium chloride for 72 h.

5.2 *Alternative Method*—Test specimens shall be conditioned in an oven at 50 C for 24 h.

PROCEDURE A

6. Procedure

6.1 Use a working pressure of 10.3 MN/m^2 (1500 psi).

6.2 The unit of flow time shall be 2 min ±1 s as measured with a stop watch or timer.

6.3 Insert the test specimens at room temperature in the hot charge chamber and test immediately.

6.4 If poisoning of the orifice occurs, clean it with acetone, benzene, or other suitable solvent before each test.

6.5 Test each material at three or more temperatures at which the flow will be within the interval of 12.7 to 38.1 mm (0.50 to 1.50 in.) with at least one measurement above and one below 25.4 mm (1 in.). Control temperatures within ±0.50 C. Make all temperature measurements at the split cone after it has come to equilibrium with the block.

NOTE 2—The split cone and the ram should be at the same temperature, and both temperatures should be held as near constant to any chosen temperature as possible. In order to assure this, it is essential that the condensate be bled from the steam lines in such a manner that a minimum fluctuation of steam pressure is introduced. This may be accomplished by the use of an orifice rather than a trap in the efflux steam line. Approximate orifice sizes suggested are:

Temperature, deg C	Orifice Size, mm (in.)
100 to 120	4.76 ($^3/_{16}$)
120 to 140	3.57 ($^9/_{64}$)
140 to 160	2.38 ($^3/_{32}$)
160 to 180	1.19 ($^3/_{64}$)

6.6 At the beginning of each series or at least once a day, check the flow tester against a sample of predetermined flow. At least two readings shall check within ±0.13 mm (±0.05 in.) or ±5 percent of the predetermined value, whichever is greater.

7. Plotting Results

7.1 The linear flow shall be plotted against temperature on semilogarithmic paper. With the flow measurements plotted on the logarithmic coordinate and temperature on the linear coordinate , a straight line will usually result. The temperature at which the flow is exactly 25.4 mm (1.0 in.) shall be read from the graph and reported as the flow temperature.

8. Report

8.1 The report shall include the following:

8.1.1 A statement indicating the nature of the material tested,

8.1.2 Curve and linear flow against temperature (Section 7),

8.1.3 Temperature at which flow is 25.4 mm (1 in.),

8.1.4 Report of any unusual behavior of the test specimens such as discoloration, sticking, etc., and

8.1.5 Details of conditioning.

8.2 In the case of cellulosic molding compounds, and molding compounds for which Procedure A is applicable, it is common practice to use the flow temperature as a basis of identifying or designating series of compositions covering the various base material-plasticizer ratios. This practice has been standardized in the following system of designation.[3] M flow shall be defined as 145 ± 5 C (293 ± 9 F), and all other flows shall extend from this temperature in multiples of 5 C (9 F) as follows:

Flow Designation	Flow Temperature Deg C	Deg F
H_n	for each 5 C (9 F) higher, add 1 flow (that is, H_3, H_4, H_5, etc.)	
H_2	160 ± 5	320 ± 9
H	155 ± 5	311 ± 9
MH	150 ± 5	302 ± 9
M	145 ± 5	293 ± 9
MS	140 ± 5	284 ± 9
S	135 ± 5	275 ± 9
S_2	130 ± 5	266 ± 9
S_n	for each 5 C (9 F) lower, add 1 flow (that is, S_3, S_4, S_5, etc.)	

Procedure B

9. Procedure

9.1 Use a working pressure of 3.45, 6.89, or 10.35 MN/m² (500 psi, 1000 psi, or 1500 psi).

Note 3—In the case of polystyrene molding compositions, whenever possible, a temperature of 135 C should be used with a pressure of either 1500 or 500 psi, 1500 psi being preferred.

9.2 The unit of flow time shall be 2 min ± 1 s as measured with a stop watch or timer.

9.3 Hold the temperature constant at some multiple of 5 C, for example 130, 135, 140, 145, 150, 155, 160, 165 C, etc., at which the flow will be within the interval of 12.7 to 38.1 mm (0.50 to 1.50 in.). Control temperatures within ±0.5 C. Make all temperature measurements at the split cone after it has come to equilibrium within the block (see Note 2).

9.4 Insert the test specimen at room temperature in the hot charge chamber and test immediately.

9.5 Run all tests in duplicate.

9.6 If poisoning of the orifice occurs, clean it with acetone, benzene, or other suitable solvent before each test.

9.7 At the beginning of each series of determinations or at least once a day, check the flow tester against a sample of predetermined flow. At least two readings shall check within ±1.3 mm (±0.05 in.) or ±5 percent of the predetermined value, whichever is greater.

10. Report

10.1 The report shall include the following information:

10.1.1 Statement indicating the nature of the material tested,

10.1.2 Temperature and pressure at which tests are run,

10.1.3 Degree of flow to the nearest 0.25 mm (0.01 in.),

10.1.4 Report of any unusual behavior of the test specimen such as discoloration, sticking, etc., and

10.1.5 Details of conditioning.

---

[3] Identical with the former Tentative Method of Designating the Flow Temperature of Thermoplastic Molding Materials (D 863 – 45 T), which was discontinued in 1948.

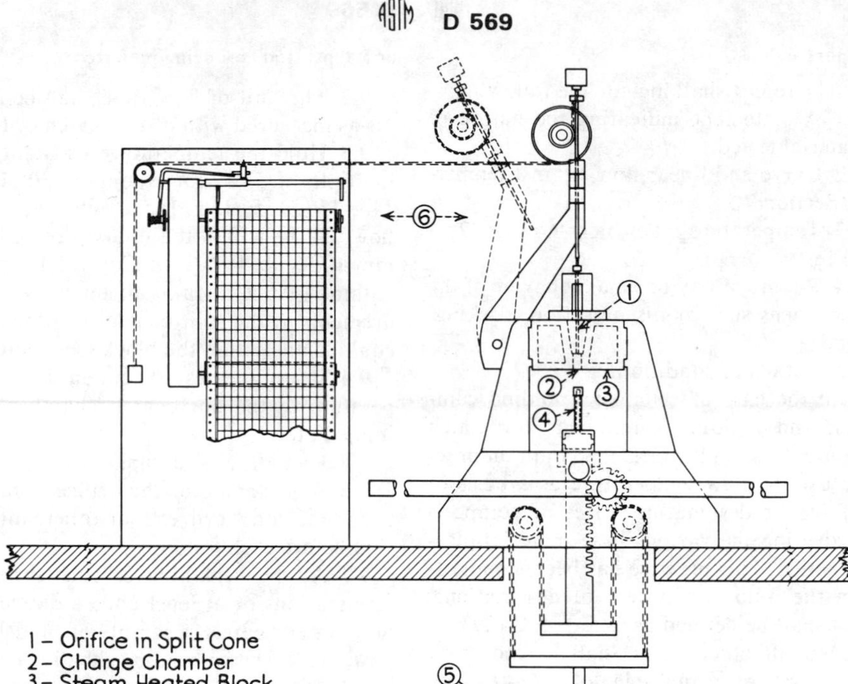

FIG. 1 Flow Test Apparatus.

Designation: D 570 – 63

American National Standard K65.16-1965
American National Standards Institute

Standard Method of Test for
WATER ABSORPTION OF PLASTICS[1]

This Standard is issued under the fixed designation D 570; the number immediately following the designation indicates the year of original adoption or, in the case of revision, the year of last revision. A number in parentheses indicates the year of last reapproval.

1. Scope

1.1 This method covers the determination of the relative rate of absorption of water by plastics when immersed. The method is intended to apply to the testing of all types of plastics, including cast, hot-molded, and cold-molded resinous products, and both homogeneous and laminated plastics in rod and tube form and in sheets 0.13 mm (0.005 in.) or greater in thickness.

NOTE 1—The values stated in U.S. customary units are to be regarded as the standard. The metric equivalents of U.S. customary units may be approximate.

2. Significance

2.1 The test for rate of water absorption has two chief functions: first, as a guide to the proportion of water absorbed by a material and consequently, in those cases where the relationships between moisture and electrical or mechanical properties, dimensions, or appearance have been determined, as a guide to the effects of exposure to water or humid conditions on such properties; and second, as a control test on the uniformity of a product. This second function is particularly applicable to sheet, rod, and tube arms when the test is made on the finished product.

2.2 Comparison of water absorption values of various plastics can be made on the basis of values obtained by Procedures A and D.

2.3 The moisture content of a plastic is very intimately related to such properties as electrical insulation resistance, dielectric losses, mechanical strength, appearance, and dimensions. The effect upon these properties of change in moisture content due to water absorption depends largely on the type of exposure (by immersion in water or by exposure to high humidity), shape of the part, and inherent properties of the plastic. With nonhomogeneous materials, such as laminated forms, the rate of water absorption may be widely different through each edge and surface. Even for otherwise homogeneous materials, it may be slightly greater through cut edges than through molded surfaces. Consequently, attempts to correlate water absorption with the surface area must generally be limited to closely related materials and to similarly shaped specimens. For materials of widely varying density, relation between water-absorption values on a volume as well as a weight basis may need to be considered.

3. Apparatus

3.1 *Balance*—An accurate chemical balance.

3.2 *Oven*, capable of maintaining uniform temperatures of 50 ± 3 C (122 ± 5.4 F) and of 105 to 110 C (221 to 230 F).

4. Test Specimen

4.1 The test specimen for molded plastics shall be in the form of a disk 50.8 mm (2 in.) in diameter and 3.2 mm ($^1/_8$ in.) in thickness (Note 2). Permissible variations in thickness are ±0.18 mm (±0.007 in.) for hot-molded and 0.30 mm (±0.012 in.) for cold-molded or cast materials.

NOTE 2—The disk mold prescribed in Section 3 of ASTM Recommended Practice D 647 for the

---

[1] This method is under the jurisdiction of ASTM Committee D-20 on Plastics. A list of committee members may be found in the ASTM Yearbook. This standard is the direct responsibility of Subcommittee D-20.50 on Permanence Properties.

Current edition effective Sept. 30, 1963. Originally issued 1940. Replaces D 570 – 59a T.

Design of Molds for Test Specimens of Plastic Molding Materials[2] is suitable for molding disk test specimens of thermosetting materials but not thermoplastic materials.

4.2 The test specimen for sheets shall be in the form of a bar 76.2 mm (3 in.) long by 25.4 mm (1 in.) wide by the thickness of the material. When comparison of absorption values with molded plastics is desired, specimens 3.2 mm ($^1/_8$ in.) thick should be used. Permissible variations in thickness of the 3.2-mm ($^1/_8$-in.) thick specimen shall be 0.20 mm (±0.008 in.) except for asbestos-fabric-base phenolic laminated materials or other materials which have greater standard commercial tolerances.

4.3 The test specimen for rods shall be 25.4 mm (1 in.) long for rods 25.4 mm (1 in.) in diameter or under, and 12.7 mm ($^1/_2$ in.) long for larger-diameter rods. The diameter of the specimen shall be the diameter of the finished rod.

4.4 The test specimen for tubes less than 76 mm (3 in.) in inside diameter shall be the full section of the tube and 25.4 mm (1 in.) long. For tubes 76 mm (3 in.) or more in inside diameter, a rectangular specimen shall be cut 76 mm (3 in.) in length in the circumferential direction of the tube and 25.4 mm (1 in.) in width lengthwise of the tube.

4.5 The test specimens for sheets, rods, and tubes shall be machined, sawed, or sheared from the sample so as to have smooth edges free from cracks. The cut edges shall be made smooth by finishing with No. 0 or finer sandpaper or emery cloth. Sawing, machining, and sandpapering operations shall be slow enough so that the material is not heated appreciably.

NOTE 3—If there is any oil on the surface of the specimen when received or as a result of machining operations, wash the specimen with a cloth wet with gasoline to remove oil, wipe with a dry cloth, and allow to stand in air for 2 h to permit evaporation of the gasoline. If gasoline attacks the plastic, use some suitable solvent or detergent that will evaporate within the 2-h period.

4.6 The dimensions listed in the following table for the various specimens shall be measured to the nearest 0.025 mm (0.001 in.). Dimensions not listed shall be measured within 0.8 mm (±$^1/_{32}$ in.).

Type of Specimen	Dimensions to be Measured to the Nearest 0.025 mm (0.001 in.)
Molded disk	thickness
Sheet	thickness
Rod	length and diameter
Tube	inside and outside diameter, and wall thickness

5. Conditioning

5.1 Three specimens shall be conditioned as follows:

5.1.1 Specimens of materials whose water-absorption value would be appreciably affected by temperatures in the neighborhood of 110 C, shall be dried in an oven for 24 h at 50 ± 3 C, cooled in a desiccator, and immediately weighed.

NOTE 4—If a static charge interferes with the weighing, lightly rub the surface of the specimens with a grounded conductor.

5.1.2 Specimens of materials, such as phenolic laminated plastics and other products whose water-absorption value has been shown not to be appreciably affected by temperatures up to 110 C, shall be dried in an oven for 1 h at 105 to 110 C.

5.1.3 When data for comparison with absorption values for other plastics are desired, the specimens shall be conditioned in accordance with 5.1.1.

6. Procedure

6.1 *24-h Immersion*—The conditioned specimens shall be placed in a container of distilled water maintained at a temperature of 23 ± 1 C (73.4 ± 1.8 F), and shall rest on edge and be entirely immersed. At the end of 24, +$^1/_2$, −0 h, the specimens shall be removed from the water one at a time, all surface water wiped off with a dry cloth, and weighed immediately. If the specimen is $^1/_{16}$ in. or less in thickness, it shall be put in a weighing bottle immediately after wiping and weighed in the bottle.

6.2 *2-h Immersion*—For all thicknesses of

[2] *Annual Book of ASTM Standards*, Part 27.

materials having a relatively high rate of absorption, and for thin specimens of other materials which may show a significant weight increase in 2 h, the specimens shall be tested as described in 6.1 except that the time of immersion shall be reduced to 120 ± 4 min.

6.3 *Repeated Immersion*—A specimen may be weighed after 2-h immersion, replaced in the water, and weighed again after 24 h.

NOTE 5—In using this method the amount of water absorbed in 24 h is likely to be less than it would have been had the immersion not been interrupted.

6.4 *Long-Term Immersion*—To determine the total water absorbed when substantially saturated, the conditioned specimens shall be tested as described in 6.1 except that at the end of 24 h they shall be removed from the water, wiped free of surface moisture with a dry cloth, weighed immediately, and then replaced in the water. The weighings shall be repeated at the end of the first week and every two weeks thereafter until the increase in weight per two-week period, as shown by three consecutive weighings, averages less than 1 percent of the total increase in weight, or 5 mg, whichever is greater; the specimen shall then be considered substantially saturated. The difference between the substantially saturated weight and the dry weight shall be considered as the water absorbed when substantially saturated.

6.5 *2-h Boiling Water Immersion*—The conditioned specimens shall be placed in a container of boiling distilled water, and shall be supported on edge and be entirely immersed. At the end of 120 ± 4 min, the specimens shall be removed from the water and cooled in distilled water maintained at room temperature. After 15 ± 1 min, the specimens shall be removed from the water, one at a time, all surface water removed with a dry cloth, and the specimens weighed immediately. If the specimen is $^1/_{16}$ in. or less in thickness, it shall be weighed in a weighing bottle.

6.6 *$^1/_2$-h Boiling Water Immersion*—For all thicknesses of materials having a relatively high rate of absorption, and for thin specimens of other materials which may show a significant weight increase in $^1/_2$ h the specimens shall be tested as described in 6.5, except that the time of immersion shall be reduced to 30 ± 1 min.

6.7 *Immersion at 50 C*—The conditioned specimens shall be tested as described in 6.5, except that the time and temperature of immersion shall be 48 ± 2 h and 50 ± 1 C (122.0 ± 1.8 F), respectively, and cooling in water before weighing shall be omitted.

6.8 When data for comparison with absorption values for other plastics are desired, the 24-h immersion procedure described in 6.1 and the equilibrium value determined in 6.4 shall be used.

7. Reconditioning

7.1 When materials are known or suspected to contain any appreciable amount of water-soluble ingredients, the specimens after immersion shall be reconditioned for the same time and temperature as used in the original drying period. They shall then be cooled in a desiccator and immediately reweighed. If the reconditioned weight is lower than the conditioned weight, the difference shall be considered as water-soluble matter lost during the immersion test. For such materials, the water-absorption value shall be taken as the sum of the increase in weight on immersion and of the weight of the water-soluble matter.

8. Calculation and Report

8.1 The report shall include the values for each specimen and the average for the three specimens as follows:

8.1.1 Dimensions of the specimens before test, measured in accordance with 4.6, and reported to the nearest 0.001 in.,

8.1.2 Conditioning time and temperature,

8.1.3 Immersion procedure used,

8.1.4 Time of immersion (long-term immersion procedure only),

8.1.5 Percentage increase in weight during immersion, calculated to the nearest 0.01 percent as follows:

$$\text{Increase in weight, percent} = \frac{\text{wet wt} - \text{conditioned wt}}{\text{conditioned wt}} \times 100$$

8.1.6 Percentage of soluble matter lost during immersion, if determined, calculated to the nearest 0.01 percent as follows (Note 6):

Soluble matter lost, percent

$$= \frac{\text{conditioned wt} - \text{reconditioned wt}}{\text{conditioned wt}} \times 100$$

NOTE 6—When the weight on reconditioning the specimen after immersion in water exceeds the conditioned weight prior to immersion, report "none" under 8.1.6.

8.1.7 The percentage of water absorbed, which is the sum of the values in 8.1.5 and 8.1.6, and

8.1.8 Any observations as to warping, cracking, or change in appearance of the specimens.

American National Standard C59.28-1965 (R 1969)
American National Standards Institute

Standard Methods of
CONDITIONING PLASTICS AND ELECTRICAL INSULATING MATERIALS FOR TESTING[1]

This Standard is issued under the fixed designation D 618; the number immediately following the designation indicates the year of original adoption or, in the case of revision, the year of last revision. A number in parentheses indicates the year of last reapproval.

1. Scope

1.1 In general, the physical and electrical properties of plastics and electrical insulating materials are influenced by temperature and relative humidity in a manner that materially affects test results. In order that reliable comparisons may be made of different materials and between different laboratories, it is necessary to standardize the humidity conditions, as well as the temperature, to which specimens of these materials are subjected prior to and during testing. These methods define procedures for conditioning plastics and electrical insulating materials (although not necessarily to equilibrium) prior to testing, and the conditions under which they shall be tested.

2. Significance

2.1 Conditioning of specimens may be undertaken: (*1*) for the purpose of bringing the material into equilibrium with normal or average room conditions, (*2*) simply to obtain reproducible results, regardless of previous history of exposure, or (*3*) to subject the material to abnormal conditions of temperature or humidity in order to predict its service behavior.

2.2 The conditioning procedures prescribed in these methods are designed to obtain reproducible results and may give physical values somewhat higher or somewhat lower than values under equilibrium at normal conditions, depending upon the particular material and test. To ensure substantial equilibrium under normal conditions of humidity and temperature, however, may require from 20 to 100 days or more depending upon thickness and type of material and its previous history. Consequently, conditioning for reproducibility must of necessity be used for general purchase specifications and product control tests.

NOTE 1—Conditioning of specimens for electrical tests is also necessary to obtain consistent results.

3. Definitions and Descriptions of Terms

3.1 *Standard Laboratory Temperature*—a temperature of 23 C (73.4 F) with standard tolerance as specified in 3.4 shall be the Standard Laboratory Temperature.

3.2 *Standard Laboratory Atmosphere*—an atmosphere having a temperature of 23 C (73.4 F) and a relative humidity of 50 percent with standard tolerances as specified in 3.4 shall be the Standard Laboratory Atmosphere.

3.3 *Room Temperature*—a temperature in the range from 20 to 30 C (68 to 85 F).

3.4 *Tolerances:*

3.4.1 *Temperature*—Standard tolerance shall be ±2 C (±3.6 F). Where the closer tolerance ±1 C (±1.8 F) is required, it may be specified and the tolerance used reported.

3.4.2 *Relative Humidity*—Standard tolerance shall be ±5 percent. Where the closer tolerance ±2 percent is required, it may be specified and the tolerance used reported.

3.4.3 *Temperature and Relative Humidity Measurement*—Measurements shall be made in the atmosphere as close as possible, to the

[1] These methods are under the joint jurisdiction of ASTM Committee D-9 on Electrical Insulating Materials and Committee D-20 on Plastics. A list of members may be found in the ASTM Yearbook. This standard is the direct responsibility of Subcommittee D-20.50 on Permanence Properties.

Current edition effective Sept. 18, 1961. Originally issued 1941. Replaces D 618 – 58.

specimen being conditioned or tested; but in no case shall the measurement be made more than 60 cm (2 ft) from such specimen.

NOTE 2—The temperature and relative humidity indicated at the control point, or on a recorder, may not be representative of a condition elsewhere in an enclosure or room because of local effects or deficiency in circulation of air. Tolerances at a controller must usually be smaller than those required at the specimen.

3.5 *Standard Test Temperatures Other than Standard Laboratory Temperature*—When data are to be obtained for comparison purposes at a specific temperature either above or below the Standard Laboratory Temperature, the temperature should be selected from the following:

Test Temperatures	Tolerance plus or minus
−70 C (−94 F)	2.0 C (3.6 F)
−55 C (−67 F)	2.0 C (3.6 F)
−40 C (−40 F)	2.0 C (3.6 F)
−25 C (−13 F)	2.0 C (3.6 F)
0 C (32 F)	2.0 C (3.6 F)
35 C (95 F)	1.0 C (1.8 F)
50 C (122 F)	2.0 C (3.6 F)
70 C (158 F)	2.0 C (3.6 F)
90 C (194 F)	2.0 C (3.6 F)
105 C (221 F)	2.0 C (3.6 F)
120 C (248 F)	2.0 C (3.6 F)
130 C (266 F)	2.0 C (3.6 F)
155 C (311 F)	2.0 C (3.6 F)
180 C (356 F)	2.0 C (3.6 F)
200 C (392 F)	3.0 C (5.4 F)
225 C (437 F)	3.0 C (5.4 F)
250 C (482 F)	3.0 C (5.4 F)
275 C (527 F)	3.0 C (5.4 F)
300 C (572 F)	3.0 C (5.4 F)
325 C (617 F)	4.0 C (7.2 F)
350 C (662 F)	5.0 C (9.0 F)
400 C (752 F)	6.0 C (10.8 F)
450 C (842 F)	8.0 C (14.4 F)
500 C (932 F)	10.0 C (18.0 F)
600 C (1112 F)	12.0 C (21.6 F)

4. Designations for Conditioning

4.1 *Designation for Conditioning Prior to Test:*

4.1.1 Conditioning of test specimens may be designated as follows:

4.1.1.1 A number indicating in hours the duration of the conditioning;

4.1.1.2 A number indicating in degrees Centigrade the conditioning temperature;

4.1.1.3 A number indicating relative humidity, whenever relative humidity is controlled, or a word to indicate immersion in a liquid.

4.1.2 The numbers shall be separated from each other by slant marks. A sequence of conditions shall be denoted by use of a plus (+) sign between successive conditions. "Des" shall be used to indicate desiccation over anhydrous calcium chloride. Temperature and relative humidity tolerances shall be those of Section 3, unless otherwise specified.

NOTE 3—*Examples:*

Condition 96/23/50—Condition 96 h at 23 C and 50 percent relative humidity.

Condition 48/50/water—Condition 48 h at 50 C in water.

Condition 48/50 + 96/23/50—Condition 48 h at 50 C; then condition 96 h at 23 C and 50 percent relative humidity.

Condition 48/50 + Des—Condition 48 h at 50 C followed by desiccation.

4.2 *Designation for Test Condition:*

4.2.1 Test condition may be designated as follows:

4.2.1.1 A capital letter "T" following the prior conditioning designation and separated therefrom by a colon.

4.2.1.2 A number indicating in degrees Centigrade the test temperature;

4.2.1.3 A number indicating the relative humidity in the test whenever relative humidity is controlled.

4.2.2 The numbers shall be separated from each other by a slant mark, and from the "T" by a dash. Temperature and relative humidity tolerances shall be those of Section 3, unless otherwise specified.

NOTE 4—*Examples:*

Condition 24/180: T—180—Condition 24 h at 180 C: Test at 180 C.

Condition 96/35/90: T—35/90—Condition 96 h at 35 C and 90 percent RH: Test at 35 C and 90 percent RH.

5. Standard Procedures for Conditioning Prior to Test

5.1 *Procedure A*—Condition 40/23/50 for specimens 7 mm (0.25 in.) or under in thickness, 88/23/50 for specimens over 7 mm—Condition test specimens 7 mm or under in thickness in the Standard Laboratory Atmosphere for a minimum of 40 h immediately prior to testing. Treat test specimens over 7 mm in thickness as above, except that the minimum time shall be 88 h. Provide adequate air circulation on all sides of the test specimens by placing them in suitable racks, hanging them from metal clips or laying them on wide-mesh, wire screen frames with at least 25 mm (1 in.) between the screen and the surface of the bench.

NOTE 5—Procedure A is generally satisfactory and is recommended unless other methods are specified. Note that Procedure A of Methods D 618 differs from Condition A of ASTM Methods D 709 and of the Armed Forces Specifications MIL-P designation in that Condition A means "as received, no special conditioning."

NOTE 6—If for any particular material or test, a specific longer time of conditioning is required, the time shall be agreed upon by the interested parties. Shorter conditioning times may be used for thin specimens provided equilibrium is substantially obtained.

5.2 *Procedure B*—Condition 48/50 + Des—Condition the specimens for a period of 48 h in a circulating-air oven at a temperature of 50 ± 2 C (122 ± 3.6 F). Remove the specimens from the oven and cool to the Room Temperature in a desiccator over anhydrous calcium chloride for a period of at least 5 h for specimens 7 mm (0.25 in.) or under in thickness, and at least 15 h for specimens over 7 mm in thickness, immediately prior to testing.

NOTE 7—Procedure B is commonly used for the purpose of obtaining reproducible test results on the thermosetting materials by means of a short-time conditioning period, or where the specific effects of moderate drying are to be determined. Other enclosures, desiccants, or desiccating techniques may be used that produce and maintain an atmosphere equivalent to that over anhydrous calcium chloride. Note that Procedure B of Methods D 618 is the same as Condition E-48/50 of Specifications D 709 and of the Armed Forces Specifications MIL-P designation.

5.3 *Procedure C*—Condition 96/35/90—Condition the specimens for a period of 96 h in an atmosphere of 90 percent relative humidity at a temperature of 35 C (95 F). The tolerances for this procedure shall be as follows:

Time	± 2 h
Temperature	± 1 C (1.8 F)
Humidity	± 2 percent

NOTE 8—Procedure C is recommended wherever the specific effects of exposure to severe atmospheric moisture are to be determined.

NOTE 9—It has been found that, for certain tests and materials, more reliable data are obtained in enclosures with circulating air rather than still air. In such cases enclosures with circulating air should be used.

5.4 *Procedure D*—Condition 24/23/water—Condition the specimens by immersion in distilled water for 24 ± ½ h at 23 ± 1 C (73.4 ± 1.8 F).

5.5 *Procedure E*—Condition 48/50/water + 1/23/water—Condition the specimens by immersion in distilled water for 48 ± ½ h at 50 ± 1 C (122 ± 1.8 F), and cool them by immersion in a sufficient quantity of distilled water to reduce the temperature to 23 C (73.4 F) within 1 h.

NOTE 10—*Procedures D* and *E* have been found useful in ASTM electrical and mechanical tests, and are used extensively in Armed Forces Specifications MIL-P designation.

5.6 *Procedure F*—Condition __/23/96 (time as specified in applicable materials specification)—Condition the specimens in an atmosphere of 96 ± 1 percent relative humidity at a temperature of 23 ± 1 C (73.4 ± 1.8 F) for a period of time as specified in the applicable materials specification.

NOTE 11—Constant relative humidity can be obtained only by careful temperature control. Procedures for maintaining close tolerances are outlined in 8.2 of ASTM Recommended Practice E 104, Maintaining Constant Relative Humidity by Means of Aqueous Solutions.[2]

NOTE 12—A considerable number of other procedures which might be considered as functional are outlined in ASTM Methods D 756, Test for Resistance of Plastics to Accelerated Service Conditions.[3]

NOTE 13—It has been found that, for certain tests and materials, more reliable data are obtained in enclosures with circulating air rather than still air. In such cases enclosures with circulating air should be used.

6. Tests at Normal Temperatures

6.1 Unless otherwise specified, test materials conditioned in the Standard Laboratory Atmosphere in the same atmosphere.

6.2 Unless otherwise specified, test materials conditioned according to Procedure B at Room Temperature conditions. Start the test as soon as possible, but do not allow more than ½ h to elapse between removal of the specimens from the desiccator and the start of the tests.

6.3 Unless otherwise specified, test materials conditioned according to Procedures C and F in the same atmosphere.

6.4 Unless otherwise specified, wipe materials conditioned according to Procedures D

[2] *Annual Book of ASTM Standards*, Part 30.
[3] *Annual Book of ASTM Standards*, Part 27.

and E immediately with a damp cloth, then with a dry cloth, and test them at Room Temperature. Specimens should only be removed from the water as the tests are ready to be conducted. Start the tests immediately and complete them as soon as possible.

7. Tests at Other Standard Test Temperatures

7.1 When tests are desired at Standard Test Temperatures prescribed in 3.5, transfer materials to the test conditions within $^1/_2$ h, preferably immediately, after completion of the preconditioning (according to Procedure A or B). Hold the specimens at the test temperature for no more than 5 h prior to test, and in no case for less than the time required to ensure thermal equilibrium.

8. Selection of Conditioning Procedure

8.1 In the case of materials covered by ASTM specifications, reference should be made thereto to determine the conditioning procedures to be used.

8.2 In the case of all other materials, the choice between procedures should preferably be based on the one that gives the most reproducible test results.

9. Report

9.1 The report of any test referencing this conditioning method shall state:

9.1.1 Conditioning procedure used,

9.1.2 Conditioning time used, to the nearest $^1/_2$ h, if not specified in the procedure,

9.1.3 Temperature, to the nearest degree Celsius, and the relative humidity, to the nearest percent, of the atmosphere in the vicinity of the specimen during the test, except that where the test extends longer than 30 min, the actual ranges of temperature and relative humidity shall be reported, and

9.1.4 Actual range of temperature and relative humidity if the standard tolerances are not used.

NOTE 14—The abbreviated nomenclature of Sections 4 and 5 should be used wherever practicable.

Designation: D 621 – 64

American National Standard K65.4-1964
American National Standards Institute

Standard Methods of Test for
DEFORMATION OF PLASTICS UNDER LOAD[1]

This Standard is issued under the fixed designation D 621; the number immediately following the designation indicates the year of original adoption as standard or, in the case of revision, the year of last revision. A number in parentheses indicates the year of last reapproval.

1. Scope

1.1 These methods cover the determination of the deformation under compression of nonmetallic sheet and molded plastic materials, of all classes and all commercial thicknesses, intended for structural and insulating purposes. Two methods are included, as follows:

Method A—For rigid plastics.

Method B—For nonrigid plastics.

1.2 The word deformation is used herein in the broad sense to cover (*1*) dimensional change due almost entirely to flow, and (*2*) dimensional change due to a combination of flow and shrinkage caused by loss of water or other volatile matter. The word flow as used in these methods may describe either plastic or elastic deformation or combinations thereof.

NOTE 1—The values stated in U.S. customary units are to be regarded as the standard. The metric equivalents of U.S. customary units may be approximate.

NOTE 2—Methyl methacrylate and polystyrene are examples of materials that deform almost entirely by flow. Cellulose acetate, cellulose acetate butyrate, phenolic laminated fiber, and vulcanized fiber are examples of materials that deform by a combination of flow and shrinkage.

2. Significance

2.1 Data obtained by Method A give a measure of the ability of a rigid plastic in assemblies of conductors and insulators that are held together by bolts, rivets, or similar fastening devices, to withstand compression without yielding and loosening the assembly with time. Method A also gives a measure of the rigidity of plastics at service temperatures and consequently can be used as an identification test for procurement purposes.

2.2 Data obtained by Method B give a measure of the ability of a nonrigid plastic to return to its original dimensions with time after having been deformed. Method B determines the extent to which the plastic will follow associated parts in applications requiring elastic properties.

METHOD A. RIGID PLASTICS

3. Nature of Test

3.1 The principle of Method A is essentially that of the parallel plate plastometer, namely, a constant-force system whereby a test specimen in conditioned, if necessary, and is then placed between the parallel plates of a constant-force device and the thickness observed over a required period at the stipulated temperature or temperatures.

4. Apparatus

4.1 *Testing Machine*—A machine capable of exerting a constant force of 113 kg (250 lb), 227 kg (500 lb), and 454 kg (1000 lb) ±1 percent between the parallel anvils of the machine, which shall be arranged so that they can be brought into contact with the test specimen before the load is applied. A machine suitable for this test is shown in Fig. 1. A recommended method for calibrating such a device is given in the Appendix. One of the anvils of the machine shall preferably be self-aligning and shall, in order that the load may be applied evenly over the face of the specimen, be arranged so that the specimen is ac-

[1] These methods are under the jurisdiction of ASTM Committee D-20 on Plastics. A list of committee members may be found in the ASTM Yearbook. This standard is the direct responsibility of Subcommittee D-20.30 on Thermal Properties.

Current edition effective Aug. 31, 1964. Originally issued 1941. Replaces D 621 – 59.

curately centered and the resultant of the load is through its center. The machine shall also be equipped with a dial gage or the equivalent capable of measuring the relative movement of the faces to 0.025 mm (0.001 in.) or less. A thermometer of the total immersion type shall be suspended in such a manner that the bulb is approximately level with the specimen and not more than 76 mm (3 in.) therefrom.

4.2 *Test Chamber*—A test chamber of suitable size and construction to enclose the testing machine and maintain it during the test within ±1 C (2 F) of the specified test temperature, except for the short period at the beginning when opening the door may cause a drop in temperature.

5. Test Specimens

5.1 The test specimen shall be a 12.7-mm (1/2-in.) cube, either solid or composite. Materials over 12.7 mm (1/2 in.) in thickness shall be reduced to 12.7 mm (1/2 in.), and thinner materials shall be piled up with the total height as near to 12.7 mm (1/2 in.) as possible. The squares of the composite specimen shall be accurately aligned in all cases. Surfaces of the test specimens shall be plane and parallel.

5.2 The specimens for materials made in large sheets, such as phenolic laminated fiber, where moisture absorption characteristics may vary over the entire surface, shall be prepared, unless otherwise specified, in accordance with the sampling method described in 5.3, which averages the effect over the sheet by selecting squares comprising the specimen in such a manner that each square represents the same proportional part of the entire area of the sheet.

5.3 *Specimen from Sheet Materials*—The specimen shall be selected from the whole sheet by cutting a strip 12.7 mm (1/2 in.) in width from the sheet, parallel to the two long edges if the sheet is not square or parallel to any two if it is square, midway between them, and extending from the edge to the center of the sheet. If quarter sheets are used, the strip shall be taken from the edge corresponding to the center-to-edge section of the original whole sheet. The strip shall be divided into eight equal parts numbered from 1 to 8, beginning with the piece corresponding to the edge of the sheet. The squares used to form the composite test specimen shall be taken from these pieces. If the whole piece is not used, the squares cut from it shall be taken from the end which was originally nearest the edge of the sheet. If any piece is insufficient for the number of squares required to be cut from a piece bearing that number, a second strip shall be cut adjacent to the first one, cut into pieces, and the required number of squares taken from the appropriate pieces. The number of squares to be cut from each piece is prescribed in Table 1.

NOTE 3—Substantial variations in test values may occur, particularly with composite specimens, if the opposite faces of the specimens are not plane and parallel or if sink marks or other similar imperfections are present. To minimize such variations, the squares used to form the composite specimens, if necessary, shall be rendered plane and parallel and their adjoining surfaces shall be freed of sink marks and other imperfections by milling, grinding, or other appropriate means. If solid cubes are used, their opposite faces in contact with the anvils of the testing machine, if necessary, shall be milled plane and parallel and shall be smooth and free of imperfections.

6. Conditioning

6.1 Where deformation under average room conditions is required, condition and test specimens in accordance with 6.1.1 and 6.1.2.

6.1.1 *Conditioning*—Condition the test specimens at 23 ± 2 C (73.4 ± 3.6 F) and 50 ± 5 percent relative humidity for not less than 40 h prior to test in accordance with Procedure A of ASTM Methods D 618, Conditioning Plastics and Electrical Insulating Materials for Testing,[2] for those tests where conditioning is required. In cases of disagreement, the tolerances shall be ±1 C (±1.8 F) and ±2 percent relative humidity.

6.1.2 *Test Conditions*—Conduct tests in the Standard Laboratory Atmosphere of 23 ± 2 C (73.4 ± 3.6 F) and 50 ± 5 percent relative humidity, unless otherwise specified in the test methods or in this specification. In cases of disagreement, the tolerances shall be ±1 C (±1.8 F) and ±2 percent relative humidity.

6.2 In those materials where shrinkage is a large part of total deformation, the specimens shall be preconditioned for 4 h at 65 ± 3 C (150 ± 5 F) and then conditioned for 68 h at a

[2] *Annual Book of ASTM Standards*, Part 27.

temperature of 35 ± 1 C (95 ± 1.8 F) and a relative humidity of 90 ± 2 percent, unless otherwise specified. The specimens shall be supported during conditioning upon a $^1/_8$-in. mesh wire screen or the equivalent in order to permit free access of the atmosphere to all surfaces.

6.3 Where a quick measurement of total deformation is desired, such as for procurement of materials wherein the flow is relatively great compared to shrinkage, the specimens shall be exposed for 16 to 18 h in a circulation air oven at 50 C (122 F). The specimens shall be removed from the oven and cooled to the Standard Room Temperature (Methods D 618) in a desiccator over anhydrous calcium chloride for a period of at least 3 h.

6.4 Where it is definitely known that the material is not moisture-responsive, conditioning of any kind may be omitted.

7. Test Temperatures

7.1 The test temperature, that is, the temperature of the chamber containing the testing apparatus shall be one or more of the following: 23 C (73.4 F), 50 C (122 F), and 70 C (158 F), each temperature being maintained within ±1 C (1.8 F).

8. Recommended Limiting Value of Deformation

8.1 It is recommended that, whenever practicable, deformation values should not exceed 25 percent since the engineering significance of substantially higher values is doubtful. In the event that the combination of force and temperature originally chosen as best representing probable service applications results in deformation in excess of 25 percent, it is suggested that the factor that is of lesser importance from a service standpoint be lowered to the next specified condition.

9. Procedure

9.1 Place the conditioned test specimens between the anvils of the testing machine immediately upon removal from the conditioning atmosphere. The specimens tested without conditioning shall be at room temperature when placed in the testing machine. Where composite specimens are used, take care to insure that the squares are well aligned.

9.2 Apply the load to the specimen without shock and take the initial reading 10 s after the full load is on the specimen. At the end of 24 h, take a second reading and record the total change in height in mils. Determine the original height in mils of the specimen by measuring the specimen after it is removed from the testing machine and adding to this the total change in height as read on the dial of the testing machine.

NOTE 4: *Composite Specimen Procedure*—The height of a composite specimen is difficult to determine accurately without disturbing the position of each square in the stack. Therefore it is recommended that the stack be removed as a unit by the use of a jig such as a small toolmaker's clamp (Fig. 2) and the height measured by a machinist micrometer to 0.0025 mm (0.0001 in.) by Method A of ASTM Method D 374, Test for Thickness of Solid Electrical Insulation.[2] In no case shall sufficient clamping pressure be used to mushroom the edges of the composite specimen. The initial and final height of the composite specimen in the test machine may also be measured by use of a long-range dial micrometer and calibrated slug.

10. Calculation

10.1 Calculate the deformation as the percentage change in height of the test specimens after 24 h, as follows:

$$\text{Deformation, percent} = (A/B) \times 100$$

where:

A = change in height in millimeters (mils) in 24 h, and

B = original height in millimeters (mils).

11. Report

11.1 The report shall include the following:

11.1.1 Original height of test specimen in mils,

11.1.2 Thickness of components in mils where a composite test specimen is used,

11.1.3 Conditioning procedure,

11.1.4 Temperature of test and force applied,

11.1.5 Change in height of the test specimen in 24 h in mils, and

11.1.6 Deformation (flow or combined flow and shrinkage) expressed as the percentage change in height of the test specimen calculated on the basis of its original height.

METHOD B. NONRIGID PLASTICS

12. Nature of Test

12.1 Method B is essentially the same as Method A except that the pressure is 0.69 MN/m^2 (100 psi), and the period of test used for deformation is 3 h. The recovery is based on removing the specimen from compression and allowing it to remain at the stated temperatures for a period of 1 h and at room temperature for $^1/_2$ h, after which the amount of recovery is measured.

13. Apparatus

13.1 *Testing Machine*—A testing machine similar in principle and operation to the one described in Method A (4.1), except that it shall be capable of exerting a constant force of 45 kg (100 lb) ± 1 percent. A machine suitable for this test is illustrated in Fig. 3. The machine shown in Fig. 1 may be used provided it is equipped with a locking device for the upper anvil so as to obtain a zero point. A lock nut added to the upper portion of the movable anvil has been found suitable for this purpose.

NOTE 5—In making tests in accordance with Method B, care should be taken to reduce machine friction to a minimum. This is particularly desirable when using the high-pressure machine shown in Fig. 1.

13.2 *Test Chamber*—A test chamber similar to that prescribed in Method A (4.2).

13.3 *Conditioning Chamber*—A chamber for drying, consisting of a desiccator provided with anhydrous calcium chloride.

14. Test Specimens

14.1 The test specimens shall be 28.67 mm (1.129 in.) in diameter by 12.7 mm ($^1/_2$ in.) in thickness, this size being suitable for the high-pressure testing machine shown in Fig. 1, whose arm without weights but including pan provides a load of 45 kg (100 lb) (Note 6). A single specimen shall be used. The specimen shall be molded or cut (Note 7) from a 12.7-mm ($^1/_2$-in.) sheet at least 25.4 mm (1 in.) from the edges and so selected as to be of uniform thickness, homogeneous, and free from bubbles.

NOTE 6—Unavoidable variations in diameter of the specimen may require slight adjustment of load to obtain 0.69 MN/m^2 (100 psi).

NOTE 7—A tool suitable for this purpose is described in ASTM Methods D 575, Test for Compression-Deflection Characteristics of Vulcanized Rubber.[3]

15. Conditioning

15.1 The specimen shall be conditioned by placing it in the desiccator for 24 h, unless it is definitely established that the material is not affected by moisture. The specimen shall be so supported and desiccated as to permit free access of the atmosphere to all faces and edges.

15.2 The test temperatures, that is, the temperatures of the chamber containing the testing apparatus, shall be one or more of the following: 23 C (73.4 F), 50 C (122 F), and 70 C (158 F), each temperature being maintained within ±1 C (1.8 F).

16. Procedure

16.1 The procedure for Method B shall be the same as that of Method A, as described in Section 9, except that the duration of the deformation test shall be 3 h. Also, obtain the zero point or initial reading for the beginning of the test by determining the dial reading with the anvils together under full load and calculating the zero point from the thickness of the specimen measured with a micrometer having a low-pressure ratchet attachment and capable of measuring to the nearest 0.2 mil. After determining the thickness of the specimen at the end of 3 h, remove the specimen from the machine and leave for 1 h in the chamber at the temperature at which it is being tested. Then remove it from the chamber and keep at room temperature for $^1/_2$ h, after which again determine the thickness with a micrometer.

17. Calculations

17.1 Calculate the deformation and recovery as follows:

$$\text{Deformation, percent} = [(H_0 - H_1)/H_0] \times 100$$

$$\text{Recovery, percent} = [(H_2 - H_1)/(H_0 - H_1)] \times 100$$

where:

H_0 = original specimen height or thickness,
H_1 = specimen thickness after 3 h under

[3] *Annual Book of ASTM Standards*, Part 28.

load of 0.69 MN/m^2 (100 psi), and

H_2 = specimen thickness $1\frac{1}{2}$ h after load removal.

18. Report

18.1 The report shall include the following:

18.1.1 Temperature of test,

18.1.2 Original thickness of test specimens in mils,

18.1.3 Total deformation in 3 h expressed as a percentage change in thickness calculated on the basis of its original thickness, and

18.1.4 Recovery, the total increase in specimen height which occurs during the $1\frac{1}{2}$-h period following load removal, expressed as a percentage of the original deformation.

TABLE 1 Number of Squares to be Cut from Each Piece

NOTE—In case of material of thicknesses not included in the table, the squares used in the composite test specimen shall be selected in accordance with the method for nearest thickness given in the table. In case the thickness is midway between two adjacent values in the table, the squares used in the composite test specimen shall be selected in accordance with the instructions for the thinner material.

Thickness of Material, mm (in.)	Total Number of Squares Required	Number of Piece							
		1	2	3	4	5	6	7	8
0.40 ($^1/_{64}$)	32	8	6	6	4	3	3	2	...
0.79 ($^1/_{32}$)	16	4	3	3	2	2	1	1	...
1.19 ($^3/_{64}$)	11	2	2	2	2	1	1	1	...
1.59 ($^1/_{16}$)	8	2	1	1	1	1	1	1	...
2.38 ($^3/_{32}$)	5	1	1	1	1	...	1	...	...
3.2 ($^1/_8$)	4	1	1	1	...	1	...	...	...
4.8 ($^3/_{16}$)	3	1	...	1	...	1	...	...	...
9.5 ($^3/_8$)	1	...	...	1	...	...	...	...	...
12.7 ($^1/_2$)	1	...	...	1	...	...	...	...	...

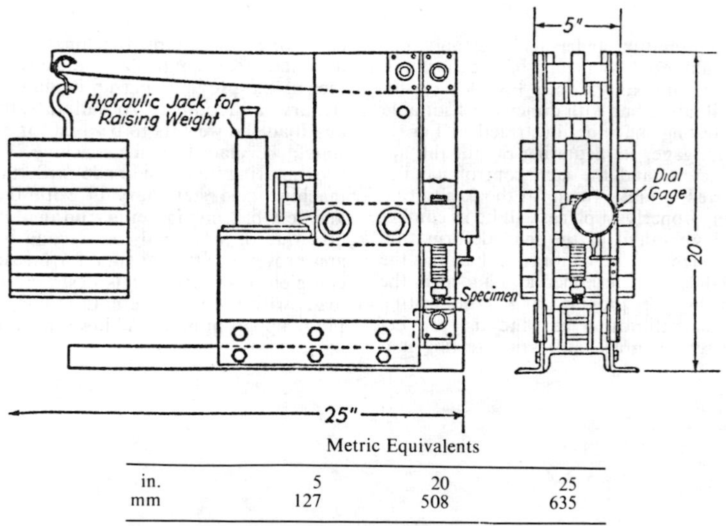

Metric Equivalents

in.	5	20	25
mm	127	508	635

FIG. 1 Deformation Testing Machine.

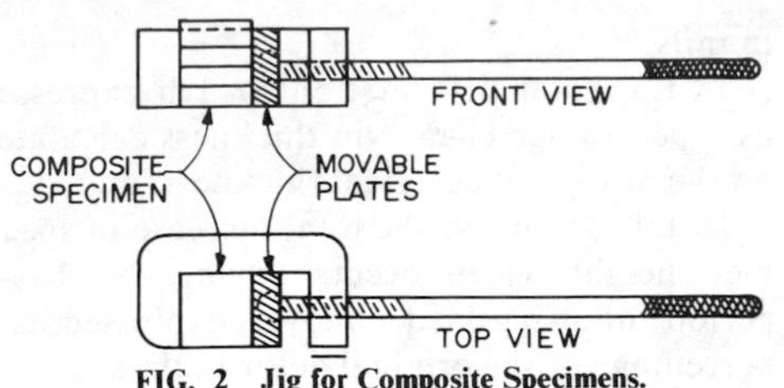

FIG. 2 Jig for Composite Specimens.

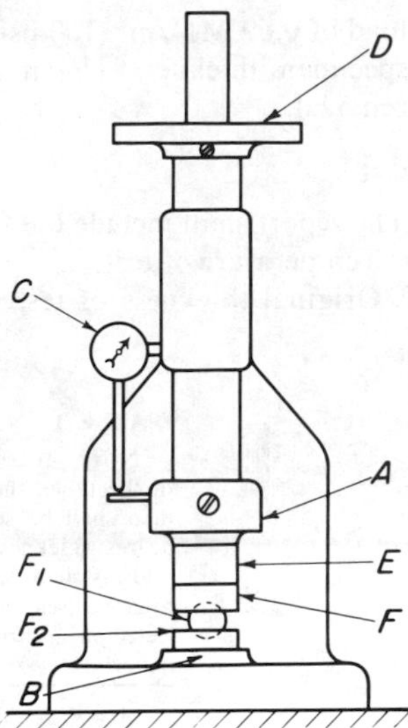

A—Movable anvil.
B—Stationary plate.
C—Thickness indicator.
D—Weight platform.
E—Test specimen.
F—Lower anvil mounted on ball support (F_1) and loose plate (F_2).

FIG. 3 Low-Pressure Deformation Tester.

APPENDIX

A1. Calibration of Deformation-Under-Load Testing Machine per Method A

A1.1 The deformation-under-load testing machine may be calibrated by means of a calibrating bar and a proving ring, as shown in Fig. A1. Figure A2 shows a calibrating bar with dimensions suitable for use in the testing machine illustrated in Fig. 1.

A1.2 The dial gage, its actuating detail, the upper anvil and screw, and the jack control handles should be removed to make room for the calibrating bar which, when properly in place, will be in contact with only the lower anvil of the load deformation machine. The weight of the calibrating bar and the proving ring should be determined and added to the load indicated by the proving ring. The testing machine with the calibrating bar and the proving ring should be set up in a compression testing machine or other suitable machine which has sufficient opening and is controllable to move the upper contact of the proving ring up or down or hold it stationary, as required. To calibrate the machine for any load the weights to be used for giving that load should be placed on the machine. Readings of the proving ring may be made with the compression machine held stationary or with the proving ring and test bar moving either up or down. Movement at a rate of a few mils per second is easy to attain and provides sufficient time for measurement. For a complete calibration it is desirable to measure the loads with the machine both moving and stationary in the upper, middle, and lower positions.

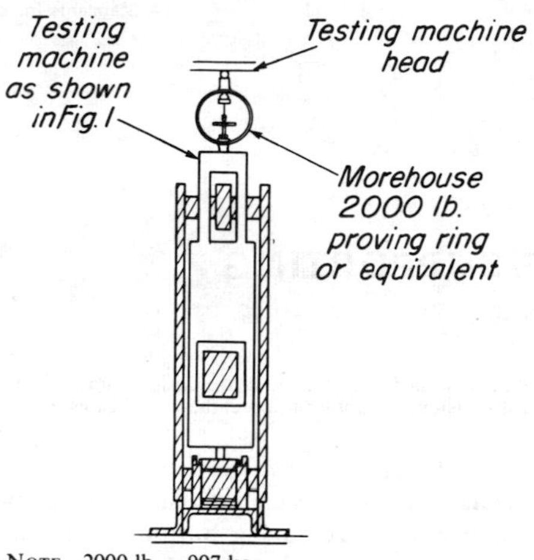

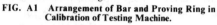

NOTE—2000 lb = 907 kg.

FIG. A1 Arrangement of Bar and Proving Ring in Calibration of Testing Machine.

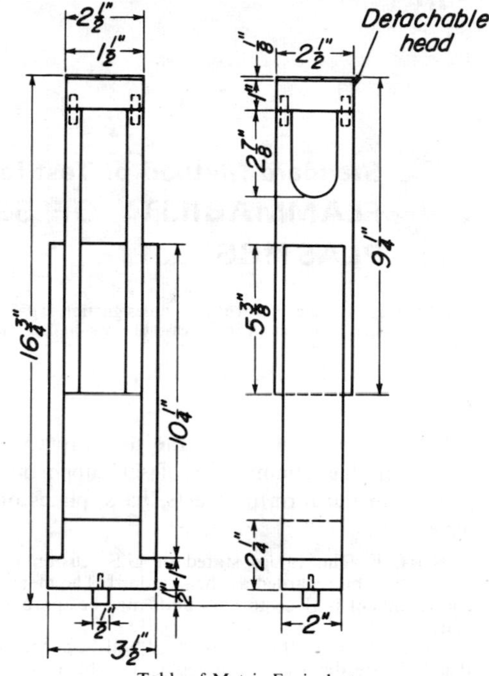

Table of Metric Equivalents

in.	mm
1/8	3.2
1/2	12.7
1	25.4
2	50.8
2 1/4	57
2 1/2	63
2 7/8	73
3 1/2	89
5 3/8	136
9 1/4	235
10 1/4	260
16 3/4	425

FIG. A2 Bar for Calibrating Deformation-Under-Load Testing Machine.

Designation: D 635 – 68

USA Standard K65.21-1969
USA Standards Institute

Standard Method of Test for
FLAMMABILITY OF SELF-SUPPORTING PLASTICS[1]

This Standard is issued under the fixed designation D 635; the number immediately following the designation indicates the year of original adoption or, in the case of revision, the year of last revision. A number in parentheses indicates the year of last reapproval.

1. Scope

1.1 This method covers the determination of the relative flammability of self-supporting plastics in the form of sheets, bars, plates or panels.

NOTE 1—The values stated in U.S. customary units are to be regarded as the standard. The metric equivalents of U.S. customary units may be approximate.

NOTE 2—For this method, "Self-Supporting Plastics" are defined as those plastics which, when mounted with the clamp end 1 cm ($^3/_8$ in.) above the horizontal screen, do not sag initially so that the end touches the screen. For tests of flexible plastics in the form of thin sheets and films, reference should be made to ASTM Method D 568, Test for Flammability of Flexible Plastics.[2] Plastic foams and sheeting are referenced in ASTM Method D 1692, Test for Flammability of Plastic Sheeting and Cellular Plastics.[2] ASTM Method E 84, Test for Surface Burning Characteristics of Building Materials,[3] is utilized to evaluate plastics when applied or fastened to the surfaces of walls or ceilings as an interior finish. Flammability tests on sheets and plate insulation are covered in ASTM Methods D 229, Testing Rigid Sheet and Plate Materials Used for Electrical Insulation.[2]

2. Significance

2.1 Tests made on a material under conditions herein prescribed can be of considerable value in comparing the flammability characteristics of different materials, in controlling manufacturing processes, or as a measure of deterioration or change in flammability rating prior to or during use. Correlation with flammability under actual use conditions is not necessarily implied.

2.2 The rate of burning will vary with thickness. Test data should be compared with data for a control material of known performance and of comparable thickness. Sheet materials that have been stretched during processing may relax during burning and give erratic results unless they are first heated above their heat distortion temperature for a time sufficient to permit complete relaxation.

3. Apparatus

3.1 *Test Chamber*—A laboratory hood, totally enclosed, with a heat-resistant glass window for observing the test. The exhaust fan is turned off during the test and turned on immediately following the test in order to remove products of combustion which may be toxic when testing some materials.

3.2 *Ring Stand*—A laboratory ring stand with two small clamps adjustable, by means of a check nut, to any angle.

3.3 *Burner*—Bunsen burner with gas supply.

3.4 *Wire Gauze*—A 20-mesh wire gauze, 10 cm (4 in.) square.

3.5 *Stop Watch.*

3.6 *Pan of Water.*

4. Test Specimens

4.1 At least ten test specimens, 12.7 cm (5 in.) in length by 1.27 cm (0.5 in.) in width and of the thickness of material normally supplied, shall be cut from sheet or molded from each of the samples to be tested. Ordinary

---

[1] This method is under the jurisdiction of ASTM Committee D-20 on Plastics. A list of committee members may be found in the ASTM Yearbook. This standard is the direct responsibility of Subcommittee D-20.30 on Thermal Properties.

Current edition effective Sept. 9, 1968. Originally issued 1941. Replaces D 635 – 63.

[2] *Annual Book of ASTM Standards*, Part 27.

[3] *Annual Book of ASTM Standards*, Part 14.

0.635 by 1.27 by 12.7-cm (0.25 by 0.5 by 5-in.) injection-molded or 1.27 by 1.27 by 12.7-cm (0.5 by 0.5 by 5-in.) compression-molded bars make satisfactory test specimens.

4.2 The specimens shall normally be tested in the as-received condition. Special preconditioning may be agreed upon by the purchaser and the seller.

4.3 Each test specimen shall be marked by scribing two lines, 2.54 cm and 10.2 cm (1 in. and 4 in.), from one end of the specimen.

4.4 The edges of the test specimens shall be smooth. In case of sawed edges, these should be fine sanded to a smooth finish.

5. Procedure

5.1 Clamp the specimen on one end in a support with its longitudinal axis horizontal and its transverse axis inclined at 45 deg to the horizontal. Under the test specimen clamp a piece of 20-mesh bunsen burner gauze about 10 cm (4 in.) square, in a horizontal position 1 cm ($^3/_8$ in.) below the edge of the specimen, and with about 1.3 cm ($^1/_2$ in.) of the specimen extending beyond the edge of the gauze (see Fig. 1). Any material remaining on the screen from the previous test must be burned off or a new screen used for each test. Place pan of water on the floor of the hood in position to catch any burning particles which may drop during the test.

5.2 Adjust a standard 1-cm ($^3/_8$-in.)-diameter bunsen burner with air ports open to produce a blue flame approximately 2.5 cm (1 in.) high. For each attempt to ignite the specimen, place the burner so that the tip of the flame contacts the end of the test specimen. At the end of 30 s remove the flame and place it at least 45 cm (18 in.) from the specimen to reduce the effects of drafts in the hood while the specimen is allowed to burn. In case the plastic does not continue to burn after the first ignition, place the burner in contact with the free end for a second period of 30 s immediately after the specimen ceases to burn. Extinguish the burner flame after the second application and close the hood door for the remainder of the test.

5.3 Measure the extent of burning along the lower edge of the test specimen. If the specimen does not ignite on two attempts, the result is judged to be "nonburning by this test." If the specimen continues to burn after the first or second ignition, start the stop watch when the flame along the lower edge reaches the mark 2.54 cm (1 in.) from the free end, and measure the time t (in seconds) until the flame reaches the mark 10.2 cm (4 in.) from the free end. A specimen that burns to the 10.2-cm (4-in.) mark is judged to be "burning by this test," and its burning rate is equal to $457/t$ cm/min ($180/t$ in./min). If a specimen does not burn to the 10.2-cm (4-in.) mark after the first or second ignition, it is judged to be "self-extinguishing by this test," and 10.2 cm (4 in.) minus the unburned length (in centimeters (or inches)) from the clamped end, measured along the lower edge, is its "extent of burning."

6. Judging Samples

6.1 Testing is discontinued as soon as three burning specimens have been found. In this case the sample is judged to be "burning by this test," and the "burning rate" of the sample is equal to the average of the burning rates of the burning specimens. The sample is also judged to be "burning by this test" if, of the ten specimens tested, two have been found to be burning. In this case, the burning rate of the sample is equal to the average of the burning rate of the two burning specimens.

6.2 If, among the ten specimens tested, one, and only one, has been found to be burning, and one or more have been found to be self-extinguishing, the sample is judged to be "self-extinguishing by this test." In this case, the extent of burning is equal to the average of the extent of burning of the self-extinguishing specimens, including the burning specimen which is assigned an extent of burning of 7.6 cm (3 in.).

6.3 If, among the ten specimens tested, one, and only one, has been found to be burning with the balance nonburning, start testing a second series of ten specimens. As soon as a burning or self-extinguishing specimen is found in the second set, testing shall be discontinued. In case of a burning specimen the sample is judged to be "burning by this test" with a rate of burning equal to the average of the rates of the burning specimens. In case of a self-extinguishing specimen, the sample is judged to be self-extinguishing. In this case, the burning specimen is assigned an extent of burning of 7.6 cm (3 in.), and this value is averaged with the extent of burning of the self-extinguishing specimen to give the extent of burning of the sample. In case all ten specimens of the second set are found to be nonburning, the sample shall be judged to be "nonburning by this test."

6.4 If, among the ten specimens tested, two or more are found to be self-extinguishing with the balance nonburning, the sample shall be judged to be "self-extinguishing by this test." The extent of burning of the sample is in this case equal to the average of the extent of burning of the self-extinguishing specimens.

6.5 If, among the ten specimens tested, one, and only one, has been found to be self-extinguishing with the balance nonburning, start with the testing of a second series of ten specimens. As soon as a burning or self-extinguishing specimen is found in the second set, testing shall be discontinued. In case of a burning or self-extinguishing specimen the sample shall be judged to be "self-extinguishing by this test." The burning sample, if any, shall be assigned an extent of burning of 7.6 cm (3 in.), and the extent of burning of the sample shall be the average of the two extents of burning. In case all ten specimens of the second set are found to be nonburning, the sample shall be judged to be "nonburning by this test."

6.6 If all ten specimens are found to be nonburning, the sample shall be judged to be "nonburning by this test."

7. Report

7.1 The report shall include the following:

7.1.1 A description of the material tested, including any treatment, such as relaxation of a stretched material, conditioning, etc., which the specimens received prior to testing,

7.1.2 A description of the method of specimen preparation,

7.1.3 The thickness of the specimens,

7.1.4 The length burned from the 2.54-cm (1-in.) mark when a material is reported as "self-extinguishing by this test,"

7.1.5 The burning rate in centimeters (or inches) per minute when a material is reported as "burning by this test,"

7.1.6 "Nonburning by this test" when the material does not burn as defined by this test,

7.1.7 The number of ignitions, and

7.1.8 The number of specimens tested.

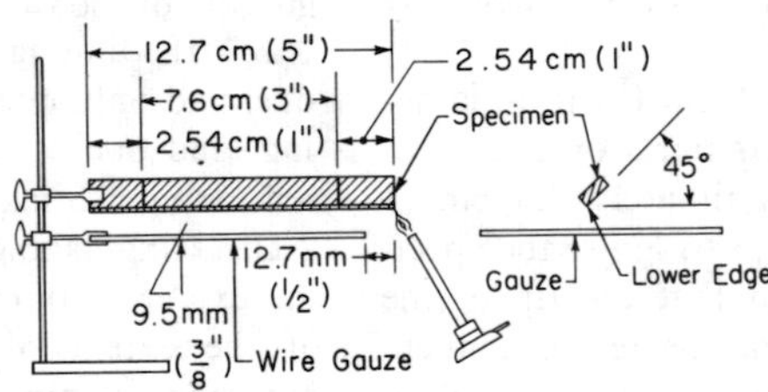

FIG. 1 Apparatus for Flammability Test.

Designation: D 637 – 50 (Reapproved 1965)

Standard Method of Test for
SURFACE IRREGULARITIES OF FLAT TRANSPARENT PLASTIC SHEETS[1]

This Standard is issued under the fixed designation D 637; the number immediately following the designation indicates the year of original adoption or, in the case of revision, the year of last revision. A number in parentheses indicates the year of last reapproval.

1. Scope

1.1 This method covers the measurement of the surface irregularities of flat transparent plastic sheets that are ordinarily used to cover openings through which visual and instrumental observations are made.

1.2 This method measures the distortion and the deviation of line of sight through flat sheets of transparent plastics. The method makes use of the prismatic or optical wedge deflection of a beam of light as it passes through a distortion spot or wave in the body of or on the surface of the material being inspected.

NOTE 1—The values stated in U.S. customary units are to be regarded as the standard. The metric equivalents of U.S. customary units may be approximate.

2. Significance

2.1 This test provides empirical results useful for control purposes and for correlation with service applications. The accuracy is much greater than required for most applications, but it is not expected to measure microscopic defects or the optical perfection of surfaces in terms of wave length of light.

3. Description of Terms

3.1 *Displacement Factor*—Displacement factor is the maximum movement (in inches) of the image of the cross, divided by the distance (in feet) from the projector to the screen, multiplied by 1000.

3.2 *Frequency of Image Movement*—Frequency of image movement shall be described as (*1*) irregular or wavy, (*2*) frequent, or (*3*) single shift.

3.3 *Pattern Distance*—Pattern distance is the maximum distance (in integer multiples of 127 mm (5 in.)) from the screen at which the specimen can be held without producing a sharply defined pattern of its minor surface irregularities.

4. Apparatus

4.1 *Projector*—A good quality lantern slide projector or similar assembly of light source and lenses that is capable of producing a sharply defined image on a screen at a distance of 7.62 m (25 ft). The objective lens of this system shall have an aperture approximately 50.8 mm (2 in.) in diameter and a focal length of approximately 12 in.

4.2 *Slide*—A transparent slide on which have been ruled two very fine black lines crossing at right angles at the center of the slide. This slide may be prepared by photographing a drawing or by directly ruling fine lines on some suitable transparent medium.

4.3 *Screen*—A square white screen measuring about 1.5 m (5 ft) on a side. This screen shall have a symmetrical cross ruled on it consisting of seven horizontal parallel fine lines spaced 12.7 mm (0.5 in.) apart and seven vertical parallel fine lines spaced 12.7 mm (0.5 in.) apart. These lines shall be as fine as it is possible to make them and still have them distinctly visible at a distance of 7.62 m (25 ft). A width of about 1.6 mm ($^1/_{16}$ in.) is suggested.

NOTE 2—An alternative form of target may be

[1] This method is under the jurisdiction of ASTM Committee D-20 on Plastics. A list of committee members may be found in the ASTM Yearbook. This standard is the direct responsibility of Subcommittee D-20.40 on Optical Properties.

Current edition effective Sept. 30, 1950. Originally issued 1941. Replaces D 637 – 43.

preferred, consisting of concentric circles having radii increasing by 25.4 mm (1 in.) from a 25.4-mm (1-in.) radius for the innermost circle, ruled in the center of a piece of white cardboard 0.3 m (1 ft) square. It is suggested that the first circle be ruled with a line thickness of 3.2 mm ($^1/_8$ in.), the second with a line thickness of 1.6 mm ($^1/_{16}$ in.), the third with a line thickness of 3.2 mm ($^1/_8$ in.), and so on.

4.4 *Supports*—A suitable means of rigidly supporting the projector and the screen.

4.5 *Dark Room*—A slightly darkened or dark room of sufficient length to accommodate the test setup. It has been found that an illumination level of not over 10 footcandles in the room will be satisfactorily dark when an ordinary lantern slide projector is being used.

5. Assembly of Apparatus

5.1 The screen shall be placed 7.62 m (25 ft) from the front lens of the projector and perpendicular to the direction of projection.

5.2 The cross-ruled slide shall be placed in the projector and the projector and its lens system adjusted to throw a sharply defined image of the cross directly on top of the center cross rules on the screen. The light source in the projector shall be adjusted to insure complete filling of the aperture in the projection lens.

NOTE 3—If the alternative form of target is used, the projector shall be adjusted so that the center of the image of the cross falls on the center of the circles.

6. Test Specimen

6.1 The test specimen shall consist of any flat sheet of a transparent plastic.

7. Conditioning

7.1 *Conditioning*—Condition the test specimens at 23 ± 2 C (73.4 ± 3.6 F) and 50 ± 5 percent relative humidity for not less than 40 h prior to test in accordance with Procedure A of ASTM Methods D 618, Conditioning Plastics and Electrical Insulating Materials for Testing[2] for those tests where conditioning is required. In cases of disagreement, the tolerances shall be ±1 C (±1.8 F) and ±2 percent relative humidity.

7.2 *Test Conditions*—Conduct tests in the Standard Laboratory Atmosphere of 23 ± 2 C (73.4 ± 3.6 F) and 50 ± 5 percent relative humidity, unless otherwise specified in the test methods or in this specification. In cases of disagreements, the tolerances shall be ±1 C (±1.8 F) and ±2 percent relative humidity.

8. Procedure

8.1 Hold the test specimen with its plane perpendicular to the direction of projection and in front of the projector at a distance of approximately 0 to 30 m (12 in.) from the front lens and move so that the entire area inside of a 25.4-mm (1-in.) border around the sheet is surveyed by the beam of light from the projector system. While the specimen is being moved about, observe the screen for movement of the projected image of the cross. Note the maximum amount and nature or frequency of movement of the image.

8.2 After the specimen has been completely surveyed in the position near the projector, move it parallel to and at a distance of 127 mm (5 in.) from the screen. Move the specimen back and forth parallel to the screen and observe any projected images of minor irregularities that are visible on the screen. Then move the specimen to positions 254, 381, or 508 mm (10, 15, or 20 in.) (integer multiples of 127 mm (5 in.)) from the screen until a pattern is observed. Note the maximum distance from the screen at which the specimen can be held without producing a pattern of its minor surface irregularities.

9. Report

9.1 The report shall include the following:

9.1.1 The displacement factor,

9.1.2 The frequency and nature of the shift of the image, and

9.1.3 The pattern distance in an integer multiple of 127 mm (5 in.).

[2] *Annual Book of ASTM Standards*, Part 27.

Designation: D 638 – 68

Standard Method of Test for
TENSILE PROPERTIES OF PLASTICS[1]

This Standard is issued under the fixed designation D 638; the number immediately following the designation indicates the year of original adoption or, in the case of revision, the year of last revision. A number in parentheses indicates the year of last reapproval.

1. Scope

1.1 This method covers the determination of the tensile properties of plastics in the form of standard test specimens and when tested under defined conditions of pretreatment, temperature, humidity, and testing machine speed. (Notes 2 to 7).

2. Definitions of Terms

2.1 Definitions of terms applying to this method appear in the Appendix.

3. Significance

3.1 This method is designed to produce tensile property data for the control and specification of plastic materials. These data may also be useful for material specification and qualitative characterization purposes and for research and development of plastics.

3.2 Tensile properties may vary with specimen preparation and speed and environment of testing. Consequently, where precise comparative results are desired, these factors must be carefully controlled.

3.3 Tensile properties may provide useful data for plastics engineering design purposes. However, due to the high degree of sensitivity exhibited by many plastics to rate of straining and environmental conditions, data obtained by this method cannot be considered valid for applications involving load-time scales or environments widely different from those of this method. In cases of such dissimilarity, no reliable estimation of the limit of usefulness can be made for most plastics. This sensitivity to rate of straining and environment necessitates testing over a broad load-time scale (including impact and creep) and range of environmental conditions if tensile properties are to suffice for engineering design purposes.

4. Apparatus

4.1 *Testing Machine*—A testing machine of the constant-rate-of-cross-head-movement type and comprising essentially the following:

4.1.1 *Fixed Member*—A fixed or essentially stationary member carrying one grip.

4.1.2 *Movable Member*—A movable member carrying a second grip.

4.1.3 *Grips*—Grips for holding the test specimen between the fixed member and the movable member. The grips shall be self-aligning, that is, they shall be attached to the fixed and movable member, respectively, in such a manner that they will move freely into alignment as soon as any load is applied, so that the long axis of the test specimen will coincide with the direction of the applied pull through the center line of the grip assembly. The test specimen shall be held in such a way that slippage relative to the grips is prevented insofar as possible. Grip surfaces that are deeply scored or serrated with a pattern similar to that of a coarse single-cut file, serrations about 2.4 mm ($^3/_{32}$ in.) apart and about 1.6 mm ($^1/_{16}$ in.) deep, have been found satisfactory for most thermoplastics. Finer serrations have been found to be more satisfactory for harder plastics such as the thermosetting materials. The serrations should be kept clean and sharp. Breaking in

[1] This method is under the jurisdiction of ASTM Committee D-20 on Plastics. A list of committee members may be found in the ASTM Yearbook. This standard is the direct responsibility of Subcommittee D-20.10 on Mechanical Properties.

Current edition effective July 16, 1968. Originally issued 1941. Replaces D 638 – 67 T.

the grips may occur at times when deep serrations or abraded specimen surfaces are used; other techniques must be used in these cases. Other techniques that have been found useful, particularly with smooth-faced grips, are abrading that portion of the surface of the specimen that will be in the grips, and interposing thin pieces of abrasive cloth, abrasive paper, or plastic or rubber-coated fabric, commonly called hospital sheeting, between the specimen and the grip surface. No. 80 double-sided abrasive paper has been found effective in many cases. An open-mesh fabric in which the threads are coated with abrasive has also been effective. Reducing the cross-sectional area of the specimen may also be effective. The specimens should be aligned as perfectly as possible with the direction of pull so that no rotary motion that may induce slippage will occur in the grips; there is a limit to the amount of misalignment self-aligning grips will accommodate. The use of special types of grips is sometimes necessary to eliminate slippage and breakage in the grips.

4.1.4 *Drive Mechanism*—A drive mechanism for imparting to the movable member a uniform, controlled velocity with respect to the stationary member, this velocity to be regulated as specified in Section 8.

4.1.5 *Load Indicator*—A suitable load-indicating mechanism capable of showing the total tensile load carried by the test specimen when held by the grips. This mechanism shall be essentially free from inertia-lag at the specified rate of testing and shall indicate the load with an accuracy of ± 1 percent of the indicated value, or better. The accuracy of the testing machine shall be verified in accordance with ASTM Methods E 4, Verification of Testing Machines[2] (Note 8).

4.1.6 The fixed member, movable member, drive mechanism, and grips shall be constructed of such materials and in such proportions that the total elastic longitudinal strain of the system constituted by these parts does not exceed 1 percent of the total longitudinal strain between the two gage marks on the test specimen at any time during the test and at any load up to the rated capacity of the machine.

4.2 *Extension Indicator*—A suitable instrument for determining the distance between two fixed points located within the gage length of the test specimen at any time during the test. It is desirable, but not essential, that this instrument automatically record this distance (or any change in it) as a function of the load on the test specimen or of the elapsed time from the start of the test, or both. If only the latter is obtained, load-time data must also be taken. This instrument shall be essentially free of inertia lag at the specified speed of testing and shall be accurate to ± 1 percent of strain or better (Note 10).

4.3 *Micrometers*—Suitable micrometers, reading to at least 0.025 $\pm$ 0.000 mm (0.001 $\pm$ 0.000 in.), for measuring the width and thickness of the test specimens. The thickness of nonrigid plastics should be measured with a dial micrometer that exerts a pressure of 0.25 $\pm$ 0.05 kgf/cm^2 (3.6 $\pm$ 0.7 psi) on the specimen and measures the thickness to within 0.025 mm (0.001 in.). The anvil of the micrometer shall be at least 35 mm (1.4 in.) in diameter and parallel to the face of the contact foot.

5. Test Specimens

5.1 *Sheet, Plate, and Molded Plastics*—The test specimen shall conform to the dimensions shown in Fig. 1. This specimen may be prepared by machining operations, or die cutting, from materials in sheet, plate, slab, or similar form, or it may be prepared by molding the material to be tested.

5.2 The test specimen for rigid tubes shall be as shown in Fig. 2. The length, *L*, shall be as shown in the table in Fig. 2. A groove shall be machined around the outside of the specimen at the center of its length so that the wall section after machining shall be 60 percent of the original nominal wall thickness. This groove shall consist of a straight section 5.72 cm (2$^1/_4$ in.) in length with a radius of 7.6 cm (3 in.) at each end joining it to the outside diameter. Steel or brass plugs having diameters such that they will fit snugly inside the tube and having a length equal to the full jaw length plus 2.5 cm (1 in.) shall be placed in the ends of the specimens

[2] *Annual Book of ASTM Standards*, Part 27.

to prevent crushing. They can be located in the tube conveniently by separating and supporting them on a threaded metal rod. Details of plugs and test assembly are shown in Fig. 2.

5.3 The test specimen for rigid rods shall be as shown in Fig. 3. The length, *L*, shall be as shown in the table in Fig. 3. A groove shall be machined around the specimen at the center of its length so that the diameter of the machined portion shall be 60 percent of the original nominal diameter. This groove shall consist of a straight section 5.72 cm (2¹/₄ in.) in length with a radius of 7.6 cm (3 in.) at each end joining it to the outside diameter.

5.4 All surfaces of the specimen shall be free from visible flaws, scratches, or imperfections. Marks left by coarse machining operations shall be carefully removed with a fine file or abrasive and the filed surfaces shall then be smoothed with abrasive paper (No. 00 or finer). The finishing sanding strokes shall be made in a direction parallel to the long axis of the test specimen. All flash shall be removed from a molded specimen, taking great care not to disturb the molded surfaces. In machining a specimen, undercuts that would exceed the dimensional tolerances shown in Fig. 1 shall be scrupulously avoided. Care shall also be taken to avoid other common machining errors.

5.5 If it is necessary to place gage marks on the specimen, this shall be done with a wax crayon or India ink that will not affect the material being tested. Gage marks shall not be scratched, punched, or impressed on the specimen.

5.6 When testing materials that may be suspected of anisotropy, duplicate sets of test specimens shall be prepared having their long axes respectively parallel with, and normal to the suspected direction of anisotropy. (Note 9.)

6. Conditioning

6.1 *Conditioning*—Condition the test specimens at 23 ± 2 C (73.4 ± 3.6 F) and 50 ± 5 percent relative humidity for not less than 40 h prior to test in accordance with Procedure A of ASTM Methods D 618 Conditioning Plastics and Electrical Insulating Materials for Testing,[2] for those tests where conditioning is required. In cases of disagreement, the tolerances shall be 1 C (1.8 F) and ±2 percent relative humidity.

6.2 *Test Conditions*—Conduct tests in the Standard Laboratory Atmosphere of 23 ± 2 C (73.4 ± 3.6 F) and 50 ± 5 percent relative humidity, unless otherwise specified in the test methods. In cases of disagreements, the tolerances shall be 1 C (1.8 F) and ±2 percent relative humidity.

7. Number of Test Specimens

7.1 At least five specimens shall be tested for each sample in the case of isotropic materials.

7.2 Ten specimens, five normal to, and five parallel with the principal axis of anisotropy, shall be tested for each sample in case of anisotropic materials.

7.3 Specimens that break at some obvious fortuitous flaw or that do not break between the predetermined gage marks shall be discarded and retests made, unless such flaws constitute a variable the effect of which it is desired to study.

8. Speed of Testing

8.1 Speed of testing shall be the relative rate of motion of the grips or test fixtures during the test. Rate of motion of the driven grip or fixture when the testing machine is running idle may be used if it can be shown that the resulting speed of testing is within the limits of variation allowed.

8.2 The speed of testing shall be chosen from one of the following:

Speed A	2.5 mm (0.10 in.) per min ± 25 percent
Speed B	5.1 mm (0.20 in.) per min ± 25 percent
Speed C	51 mm (2.0 in.) per min ± 10 percent
Speed D	510 mm (20 in.) per min ± 10 percent

The speed of testing shall be determined by the specification for the material being tested, or by agreement between those concerned. When the speed is not specified, then Speed B shall be used for rigid and semirigid plastics, and Speed C shall be used for nonrigid plastics (see ASTM Nomenclature D 883, Relating to Plastics,[2] for definitions). Modulus determinations may be made at the speed selected for the other tensile properties (Note 10).

9. Procedure

9.1 Measure the width and thickness of rigid flat specimens (Fig. 1) with a suitable micrometer to the nearest 0.025 mm (0.001 in.) at several points along their narrow sections. Measure the thickness of nonrigid specimens, produced by a Type IV die, in the same manner with the required dial micrometer. The width of this specimen shall be taken as the distance between the cutting edges of the die in the narrow section. Measure the diameter of rod specimens, and the inside and outside diameters of tube specimens to the nearest 0.025 mm (0.001 in.) at a minimum of two points 90 deg apart; make these measurements at the groove for specimens so constructed. Record the minimum values of cross-sectional area so determined. Use plugs in testing tube specimens as shown in Fig. 2.

9.2 Place the specimen in the grips of the testing machine, taking care to align the long axis of the specimen and the grips with an imaginary line joining the points of attachment of the grips to the machine. The distance between the ends of the gripping surfaces, when using flat specimens, shall be as indicated in Fig. 1. The location for the grips on tube and rod specimens shall be as shown in Figs. 2 and 3. Tighten the grips evenly and firmly to the degree necessary to prevent slippage of the specimen during the test but not to the point where the specimen would be crushed.

9.3 If modulus values are being determined, proceed as follows:

9.3.1 Attach the extension indicator.

9.3.2 Set the speed of testing at the proper rate (see 8.2) and start the machine.

9.3.3 Record loads and corresponding deformations at appropriate even intervals of strain.

9.3.4 Record the load carried by the specimen when the strain reaches 0.02 and the elapsed time from the start of the test until this point is reached. If rupture occurs before the strain reaches 0.02, record the elapsed time from the start of the test until the specimen breaks.

9.4 Determine the tensile strength and elongation by the following procedure:

9.4.1 Set the speed of testing at the proper rate as required in Section 8, and start the machine.

9.4.2 Record the load carried by the specimen when the strain reaches 0.02 and the elapsed time from the start of the test until this point is reached. If rupture occurs before the strain reaches 0.02, record the elapsed time from the start of the test until the specimen breaks.

9.4.3 Record the maximum load carried by the specimen during the test (also the load at the yield point or rupture, if different and applicable. (Note 12 and Appendix Note A1).

9.4.4 Record the extension at the moment maximum load is attained by the specimen (also the extension at the yield point or rupture, if different and applicable (Note 12 and Appendix Note A2)).

10. Calculations

10.1 *Tensile Strength*—Calculate the tensile strength by dividing the maximum load in kilograms (or pounds) by the original minimum cross-sectional area of the specimen in square centimeters (or square inches). Express the result in kilograms-force per square centimeter (or pounds per square inch) and report it to three significant figures as Tensile Strength at Yield or Tensile Strength at Break, whichever term is applicable. When a nominal yield or breaking load, less than the maximum, is present and applicable, it may be desirable also to calculate, in a similar manner, the corresponding Tensile Stress at Yield or Tensile Stress at Break and report it to three significant figures (Appendix Note A1).

10.2 *Percentage Elongation*—Calculate the percentage elongation by dividing the extension at the moment maximum load is attained by the specimen by the original gage length and multiply by 100. Report the percentage elongation to two significant figures as Percentage Elongation at Yield or Percentage Elongation at Break, whichever term is applicable. When a nominal yield or breaking load less than the maximum is present and applicable, it may be desirable to calculate both the Percentage Elongation at Yield and Percentage Elongation at Break and report them to two significant figures (Appendix Note A2).

10.3 *Mean Rate of Stressing*—Calculate the mean rate of stressing by dividing the tensile load carried by the specimen when the strain reaches 0.02 or at the moment of rupture, whichever occurs first, by the original minimum cross-sectional area of the specimen, and then dividing this result by the time in seconds, measured from the beginning of the test, required to attain this tensile load and strain. Express the result in kilograms-force per square centimeter (or pounds per square inch) per second and report it to three significant figures.

10.4 *Mean Rate of Straining*—Calculate the mean rate of straining from a strain-time curve, plotted for the purpose, by selecting any convenient point on the curve and dividing the strain represented by the point by the corresponding time. Express the result as a dimensionless ratio per second (units per second) and report it to three significant figures.

10.5 *Modulus of Elasticity*—Calculate the modulus of elasticity by extending the initial linear portion of the load-extension curve and dividing the difference in stress corresponding to a certain section on this straight line by the corresponding difference in strain. All elastic modulus values shall be computed using the average initial cross-sectional area of the test specimens in the calculations. The result shall be expressed in kilograms-force per square centimeter (or pounds per square inch) and reported to three significant figures.

10.6 For each series of tests calculate the arithmetic mean of all values obtained and report it as the "average value" for the particular property in question.

10.7 Calculate the standard deviation (estimated) as follows and report it to two significant figures:

$$s = \sqrt{(\Sigma X^2 - n\bar{X}^2)/(n - 1)}$$

where:

s = estimated standard deviation,
X = value of single observation,
n = number of observations, and
$\bar{X}$ = arithmetic mean of the set of observations.

11. Report

11.1 The report shall include the following:

11.1.1 Complete identification of the material tested, including type, source, manufacturer's code numbers, form, principal dimensions, previous history, etc.,

11.1.2 Method of preparing test specimens,

11.1.3 Type of test specimen and dimensions,

11.1.4 Conditioning procedure used,

11.1.5 Atmospheric conditions in test room,

11.1.6 Number of specimens tested,

11.1.7 Speed of testing,

11.1.8 Speed of testing for modulus determination,

11.1.9 Mean rate of stressing as calculated in 10.3,

11.1.10 Tensile Strength at Yield or Break, average value and standard deviation,

11.1.11 Tensile Stress at Yield or Break, if applicable, average value and standard deviation,

11.1.12 Percentage Elongation at Yield or Break, or both as applicable, average value and standard deviation,

11.1.13 Modulus of Elasticity, average value and standard deviation, and

11.1.14 Date of test.

Explanatory Notes

Note 1—The values stated in U.S. customary units are to be regarded as the standard. The metric equivalents of U.S. customary units may be approximate.

Note 2—This method is not well adapted to the testing of plastics in the form of thin sheets or films; a method for the testing of such forms is given in ASTM Method D 882, Test for Tensile Properties of Thin Plastic Sheeting.[2] Neither is this method particularly suitable for the testing of rubber or rubber-like polymers (elastomers), although the position of such materials as organic plastics is recognized.

Note 3—This method may be used for testing phenolic resin molded or laminated materials where comparative results are desired. However, where correlation with previous data is an object, such materials should be tested in accordance with the following methods of the American Society for Testing and Materials:

Methods D 229, Testing Rigid Sheet and Plate Materials Used for Electrical Insulation,[2] and
Method D 651, Test for Tensile Strength of Molded Electrical Insulating Materials.[3]

Note 4—It is realized that a material cannot be tested without also testing the method of preparation of that material. Hence, when comparative tests of materials *per se* are desired, the greatest care must be exercised to ensure that all samples are prepared in exactly the same way. Similarly, for referee or comparative tests of any given series of specimens, care must be taken to secure the maximum degree of uniformity in details of preparation, treatment, and handling.

---

[3] *Annual Book of ASTM Standards*, Part 29.

NOTE 5—This method is not intended to cover precise physical procedures. It is recognized that the constant-rate-of-crosshead-movement type of test leaves much to be desired from a theoretical standpoint, that wide differences may exist between rate of crosshead movement and rate of strain between gage marks on the specimen, and that the testing speeds specified disguise important effects characteristic of materials in the plastic state. Further, it is realized that variations in the thicknesses of test specimens, which are permitted by these procedures, produce variations in the surface-volume ratios of such specimens, and that these variations may influence the test results. Hence, where directly comparable results are desired, all samples should be of equal thickness. Special additional tests should be used where more precise physical data are needed.

NOTE 6—It is realized that "mean rate of stressing" as defined in these methods has only limited physical significance. It does, however, roughly describe the average rate at which most of the stress carried by the test specimen is applied and for that portion of the stress-strain curve in which principal stressing occurs. It is affected by the elasticity of the materials being tested but is fairly accurately determined by the method described. It can, if desired, be determined more precisely by calculation from load-time data, recorded especially for the purpose during a test.

NOTE 7—Since the existence of a true elastic limit in plastics, as in many other organic materials and in many metals, is debatable, the propriety of applying the term "elastic modulus" in its quoted generally accepted definition to describing the "stiffness" or "rigidity" of a plastic has been seriously questioned. The exact stress-strain characteristics of plastic materials are highly dependent on such factors as rate of application of stress, temperature, previous history of specimen, etc. However, stress-strain curves for plastics determined as described in this method almost always show a linear region at low stresses, and a straight line drawn tangent to this portion of the curve permits calculation of an elastic modulus of the usually defined type. Such a constant is useful if its arbitrary nature and dependence on time, temperature, and similar factors are realized.

NOTE 8—Experience has shown that many testing machines now in use are incapable of maintaining accuracy for as long as the periods between inspection recommended in ASTM Methods E 4. Hence, it is recommended that each machine be studied individually and verified as often as may be found necessary. It will frequently be necessary to perform this function daily.

NOTE 9—Before testing, all transparent specimens should be given a polariscopic inspection and those which show atypical or concentrated strain patterns should be rejected unless these "initial" strains constitute a variable the effect of which it is desired to study.

NOTE 10—Reference is made to ASTM Method E 83, Verification and Classification of Extensometers.[4]

NOTE 11—A speed of testing of 1.3 mm (0.05 in.)/min, formerly required by this method when determining modulus of elasticity on certain materials continues to appear in some specifications. When desirable this speed may be substituted for the recommended speed provided the moduli so obtained are identified as having been determined under this provision.

NOTE 12—If tensile properties at both yield and break are considered applicable, it may be necessary, in the case of highly extensible materials (polyethylene, etc.), to run two independent tests. The crosshead speed and the high magnification extensometer normally used to determine properties up to the yield point may not be suitable for tests involving high extensibility. A faster crosshead speed and a broad range incremental extensometer, or hand rule technique, may be needed when such materials are to be taken to rupture.

[4] *Annual Book of ASTM Standards*, Part 30.

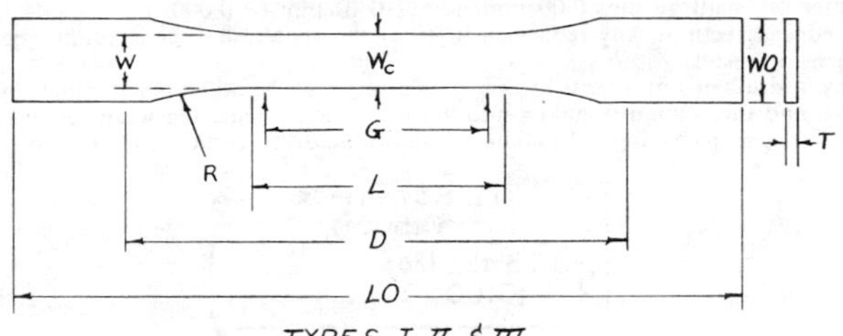

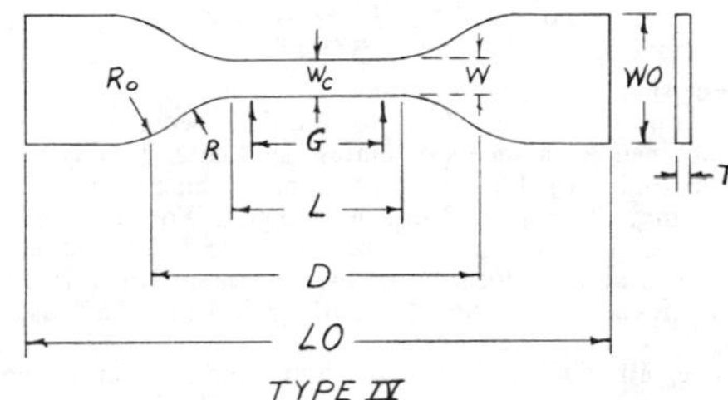

Specimen Dimensions for Thickness, T, mm[d]

Dimensions (see drawings)	7 or under		Over 7 to 14 incl.	4 or under	Tolerances
	Type I[e]	Type II[e]	Type III	Type IV[h]	
W—Width of narrow section[a,b]	13	6	19	6	±0.5[h]
L—Length of narrow section	57	57	57	33	±0.5
WO—Width over-all, min[f]	19	19	29	19	±6
LO—Length over-all, min[g]	165	183	246	115	no max
G—Gage length[c]	50	50	50	...	±0.25
G—Gage length[c]	...	...	...	25	±0.13
D—Distance between grips	115	135	115	64[i]	±5
R—Radius of fillet	76	76	76	14	±1
RO—Outer radius (Type IV)	...	...	...	25	±1

Specimen Dimensions for Thickness, T, in.[d]

Dimensions (see drawings)	0.28 or under		Over 0.28 to 0.55 incl.	0.16 or under	Tolerances
	Type I[e]	Type II[e]	Type III	Type IV[h]	
W—Width of narrow section[a,b]	0.50	0.25	0.75	0.25	±0.02[h]
L—Length of narrow section	2.25	2.25	2.25	1.30	±0.02
WO—Width over-all, min[f]	0.75	0.75	1.13	0.75	+0.25
LO—Length over-all, min[g]	6.5	7.2	9.7	4.5	no max
G—Gage length[c]	2.00	2.00	2.00	...	±0.010
G—Gage length[c]	...	...	...	1.00	±0.005
D—Distance between grips	4.5	5.3	4.5	2.5[i]	±0.2
R—Radius of fillet	3.00	3.00	3.00	0.56	±0.04
RO—Outer radius (Type IV)	...	...	...	1.00	±0.04

FIG. 1. Tension Test Specimens for Sheet, Plate, and Molded Plastics.

[a] The width at the center W_c shall be plus 0.00 mm, minus 0.10 mm (+0.000 in., −0.004 in.) compared with width W at other parts of the reduced section. Any reduction in W at the center shall be gradual, equally on each side so that no abrupt changes in dimension result.

[b] For molded specimens, a draft of not over 0.13 mm (0.005 in.) may be allowed for either Type I or II specimens 3.2 mm (0.13 in.) in thickness, and this should be taken into account when calculating width of the specimen. Thus a typical section of a molded Type I specimen, having the maximum allowable draft, could be as follows:

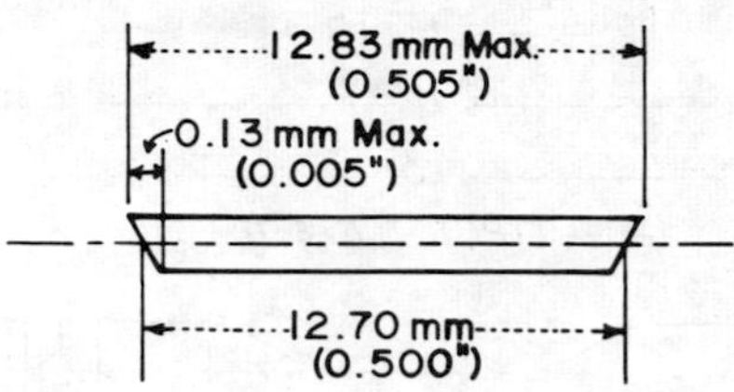

[c] Test marks or intial extensometer span.

[d] Thickness, T, shall be 3.2 ± 0.4 mm (0.13 ± 0.02 in.) for molded specimens and for other Types I and II specimens where possible. If specimens are machined from sheets or plates, thickness, T, may be the thickness of the sheet or plate provided this does not exceed 14 mm (0.55 in.). For sheets of nominal thickness greater than 14 mm (0.55 in.) the specimens shall be machined to 14 ± 0.4 mm (0.55 ± 0.02 in.) in thickness. For sheets of nominal thickness between 14 and 51 mm (0.55 and 2 in.) approximately equal amounts shall be machined from each surface. For thicker sheets both surfaces of the specimen shall be machined and the location of the specimen with reference to the original thickness of the sheet, shall be noted. Tolerances on thickness less than 14 mm (0.55 in.) shall be those standard for the grade of material tested.

[e] A Type I specimen, having an over-all width of 19 mm (0.75 in.) and an over-all length of 216 mm (8.5 in.) is the preferred specimen and shall be used whenever possible. The Type II specimen may be used when a material does not break in the narrow section with the preferred Type I specimen.

[f] Over-all widths greater than the minimum indicated may be desirable for some materials in order to avoid breaking in the grips.

[g] Over-all lengths greater than the minimum indicated may be desirable either to avoid breaking in the grips or to satisfy special test requirements.

[h] The Type IV specimen is intended for nonrigid plastics but may be used for rigid types where desirable. When used for nonrigid plastics, an over-all width of 25 mm (1.00 in.) is recommended and the dimensions given refer to the cutting die, when one is used. The internal width of the narrow section of the die shall be 6.00 ± 0.05 mm (0.250 ± 0.002 in.). The dimensions are essentially those of Die C in ASTM Method D 412, for Tension Testing of Vulcanized Rubber (*Annual Book of ASTM Standards*, Parts 27 and 28).

[i] When self-tightening grips are used, for highly extensible polymers, the distance between grips will depend upon the types of grips used and may not be critical if maintained uniform once chosen.

FIG. 1—*Continued.*

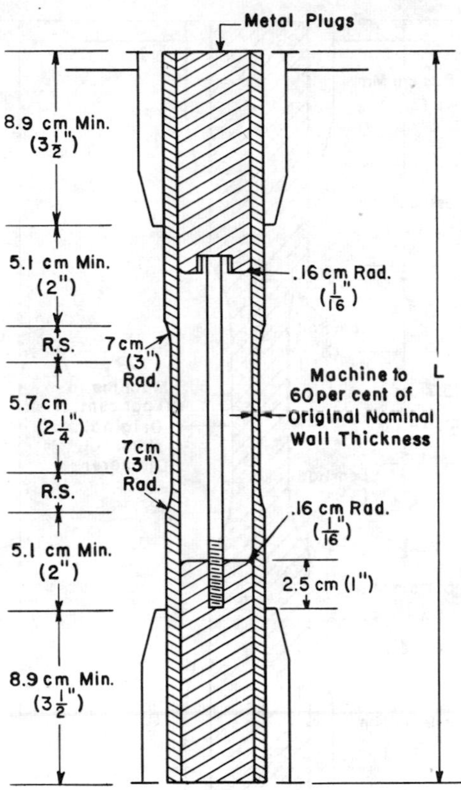

DIMENSIONS OF TUBE SPECIMENS

Nominal Wall Thickness	Length of Radial Sections, 2R.S.	Total Calculated Minimum Length of Specimen	Standard Length, *L*, of Specimen to be Used for 8.9-cm (3½-in.) Jaws
cm (in.)	cm (in.)	cm (in.)	cm (in.)
0.079 (1/32)	1.39 (0.547)	35.0 (13.80)	38.1 (15)
0.12 (3/64)	1.70 (0.670)	35.4 (13.92)	38.1 (15)
0.16 (1/16)	1.96 (0.773)	35.6 (14.02)	38.1 (15)
0.24 (3/32)	2.40 (0.946)	36.1 (14.20)	38.1 (15)
0.32 (1/8)	2.77 (1.091)	36.4 (14.34)	38.1 (15)
0.48 (3/16)	3.39 (1.333)	37.0 (14.58)	38.1 (15)
0.64 (1/4)	3.90 (1.536)	37.6 (14.79)	40.0 (15.75)
0.79 (5/16)	4.35 (1.714)	38.0 (14.96)	40.0 (15.75)
0.95 (3/8)	4.76 (1.873)	38.4 (15.12)	40.0 (15.75)
1.11 (7/16)	5.13 (2.019)	38.8 (15.27)	40.0 (15.75)
1.27 (1/2)	5.47 (2.154)	39.1 (15.40)	41.9 (16.5)

[a] For other jaws greater than 8.9 cm (3½ in.), the standard length shall be increased by twice the length of the jaws minus 17.8 cm (7 in.). The standard length permits a slippage of approximately 0.64 to 1.27 cm (¼ to ½ in.) in each jaw while maintaining maximum length of jaw grip.

FIG. 2. Diagram Showing Location of Tube Tension Test Specimens in Testing Machine.

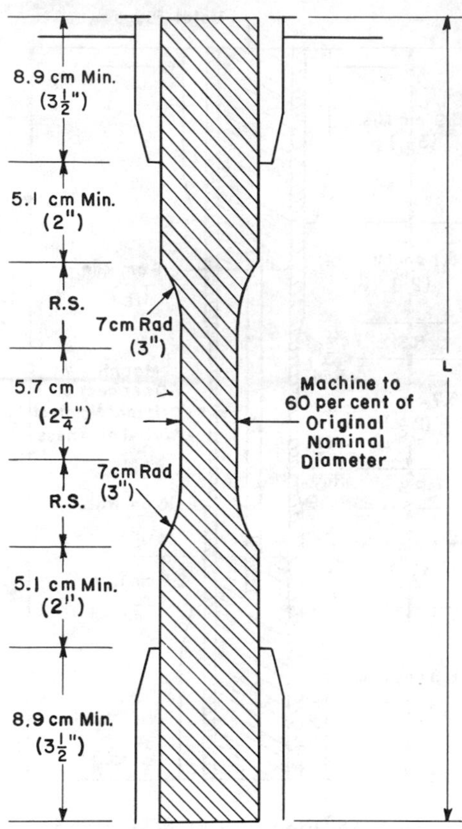

DIMENSIONS OF ROD SPECIMENS

Nominal Diameter	Length of Radial Sections, 2R.S.	Total Calculated Minimum Length of Specimen	Standard Length, L, of Specimen to be Used for 8.9-cm (3½-in.) Jaws
cm (in.)	cm (in.)	cm (in.)	cm (in.)
0.32 (⅛)	1.96 (0.773)	35.6 (14.02)	38.1 (15)
0.47 (³⁄₁₆)	2.40 (0.946)	36.1 (14.20)	38.1 (15)
0.64 (¼)	2.77 (1.091)	36.4 (14.34)	38.1 (15)
0.95 (⅜)	3.39 (1.333)	37.0 (14.58)	38.1 (15)
1.27 (½)	3.90 (1.536)	37.6 (14.79)	40.0 (15.75)
1.59 (⅝)	4.35 (1.714)	38.0 (14.96)	40.0 (15.75)
1.90 (¾)	4.76 (1.873)	38.4 (15.12)	40.0 (15.75)
2.22 (⅞)	5.13 (2.019)	38.8 (15.27)	40.0 (15.75)
2.54 (1)	5.47 (2.154)	39.1 (15.40)	41.9 (16.5)
3.18 (1¼)	6.09 (2.398)	39.8 (15.65)	41.9 (16.5)
3.81 (1½)	6.64 (2.615)	40.3 (15.87)	41.9 (16.5)
4.25 (1¾)	7.14 (2.812)	40.8 (16.06)	41.9 (16.5)
5.08 (2)	7.60 (2.993)	41.2 (16.24)	43.2 (17)

[a] For other jaws greater than 8.9 cm (3½ in.) the standard length shall be increased by twice the length of the jaw minus 17.8 cm (7 in.). The standard length permits a slippage of approximately 0.64 to 1.27 cm (¼ to ½ in.) in each jaw while maintaining length of jaw grip.

FIG. 3. Diagram Showing Location of Rod Tension Test Specimen in Testing Machine.

APPENDIX

A1. Definitions of Terms and Symbols Relating to Tension Testing of Plastics

A1.1 *Tensile Stress* (nominal) is the tensile load per unit area of minimum original cross-section, within the gage boundaries, carried by the test specimen at any given moment. It is expressed in force per unit area, usually kilograms-force per square centimeter (or pounds per square inch).

Note A1—The expression of tensile properties in terms of the minimum original cross-section is almost universally used in practice. In the case of materials exhibiting high extensibility, or "necking", or both, (A1.11) nominal stress calculations may not be meaningful beyond the yield point (A1.10) due to the extensive reduction in cross-sectional area that ensues. Under some circumstances it may be desirable to express the tensile properties per unit of minimum prevaling cross-section. These properties (that is, true tensile stress, etc.).

A1.2 *Tensile Strength* (nominal) is the maximum tensile stress (nominal) sustained by the specimen during a tension test. When the maximum stress occurs at the yield point (A1.10), it shall be designated Tensile Strength at Yield. When the maximum stress occurs at break, it shall be designated Tensile Strength at Break.

A1.3 *Gage Length* is the original length of that portion of the specimen over which strain or change in length is determined.

A1.4 *Elongation* is the increase in length produced in the gage length of the test specimen by a tensile load. It is expressed in units of length, usually centimeters (or inches). (Also known as *Extension*.)

Note A2—Elongation and strain values are valid only in cases where uniformity of specimen behavior within the gage length is present. In the case of materials exhibiting "necking phenomena," such values are only of qualitative utility after attainment of "yield" point. This is due to inability to assure that necking will encompass the entire length between the gage marks prior to specimen failure.

A1.5 *Percentage Elongation* is the elongation of a test specimen expressed as a percentage of the gage length.

A1.6 *Percentage Elongation at Yield and Break:*

A1.6.1 *Percentage Elongation at Yield* is the percentage elongation at the moment the yield point (A1.10) is attained in the test specimen.

A1.6.2 *Percentage Elongation at Break* is the percentage elongation at the moment of rupture of the test specimen.

A1.7 *Strain* is the ratio of the elongation to the gage length of the test specimen, that is, the change in length per unit of original length. It is expressed as a dimensionless ratio.

A1.8 *True Strain* (see Fig. A1) is defined by the following equation for ϵ_T:

$$\epsilon_T = \int_{L_o}^{L} dL/L = \ln L/L_o$$

where:

dL = the increment of elongation when the distance between the gage marks is L,
L_o = the original distance between gage marks, and
L = the distance between gage marks at any time.

A1.9 *Tensile Stress-Strain Curve* is a diagram in which values of tensile stress are plotted as ordinates against corresponding values of tensile strain as abscissas.

A1.10 *Yield Point* is the first point on the stress-strain curve at which an increase in strain occurs without an increase in stress (Fig. A2).

Note A3—Only materials whose stress-strain curves exhibit a point of zero slope may be considered as having a yield point.

Note A4—Some materials exhibit a distinct "break" or discontinuity in the stress-strain curve in the elastic region. This break is not a yield point by definition. However, this point may prove useful for material characterization in some cases.

A1.11 Necking is the localized reduction in cross-section which may occur in a material under tensile stress.

A1.12 *Yield Strength* is the stress at which a material exhibits a specified limiting deviation from the proportionality of stress to strain. Unless otherwise specified, this stress will be the stress at the yield point and when expressed in relation to the Tensile Strength shall be designated either Tensile Strength at Yield or Tensile Stress at Yield as required under A1.2 (Fig. A2). (See *Offset Yield Strength*.)

A1.13 *Offset Yield Strength* is the stress at which the strain exceeds by a specified amount (the offset) an extension of the initial proportional portion of the stress-strain curve. It is expressed in force per unit area, usually kilograms-force per square centimeter (or pounds per square inch).

Note A5—This measurement is useful for materials whose stress-strain curve in the yield range is of gradual curvature. The offset yield strength can be derived from a stress-strain curve as follows (Fig. A3):

On the strain axis lay off OM equal to the specified offset.

Draw OA tangent to the initial straight-line portion of the stress-strain curve.

Through M draw a line MN parallel to OA and locate the intersection of MN with the stress-strain curve.

The stress at the point of intersection *r* is the "offset yield strength." The specified value of the offset must be stated as a percentage of the original gage length in conjunction with the strength value. *Example:* 0.1 percent offset yield strength = ...kgf per sq cm (psi), or *yield strength* at 0.1 percent offset = ...kgf per sq cm (psi).

A1.14 *Proportional Limit* is the greatest stress which a material is capable of sustaining without any deviation from proportionality of stress to strain (Hooke's law). It is expressed in force per unit area, usually kilograms-force per square centimeter (or pounds per square inch).

A1.15 *Elastic Limit* is the greatest stress which a material is capable of sustaining without any permanent strain remaining upon complete release of the stress. It is expressed in force per unit area, usually kilograms-force per square centimeter (or pounds per square inch).

NOTE A6—Measured values of proportional limit and elastic limit vary greatly with the sensitivity and accuracy of the testing equipment, eccentricity of loading, the scale to which the stress-strain diagram is plotted, and other factors. Consequently, these values are usually replaced by yield strength.

A1.16 *Modulus of Elasticity* is the ratio of stress (nominal) to corresponding strain below the proportional limit of a material. It is expressed in force per unit area, usually kilograms-force per square centimeter (or pounds per square inch) (Also known as *Elastic Modulus* or *Young's Modulus*).

NOTE A7—The stress-strain relations of many plastics do not conform to Hooke's law throughout the elastic range but deviate therefrom even at stresses well below the elastic limit. For such materials the slope of the tangent to the stress-strain curve at a low stress is usually taken as the modulus of elasticity. Since the existence of a true proportional limit in plastics is debatable, the propriety of applying the term "modulus of elasticity" to describe the stiffness or rigidity of a plastic has been seriously questioned. The exact stress-strain characteristics of plastic materials are very dependent on such factors as rate of stressing, temperature, previous specimen history, etc. However, such a value is useful if its arbitrary nature and dependence on time, temperature, and other factors are realized.

A1.17 *Secant Modulus* is the ratio of stress (nominal) to corresponding strain at any specified point on the stress-strain curve. It is expressed in force per unit area, usually kilograms-force per square centimeter (or pounds per square inch), and reported together with the specified stress or strain.

NOTE A8—This measurement is usually employed in place of modulus of elasticity in the case of materials whose stress-strain diagram does not demonstrate proportionality of stress to strain.

A1.18 *Percentage Reduction of Area (Nominal)* is the difference between the original cross-sectional area measured at the point of rupture after breaking and after all retraction has ceased, expressed as a percentage of the original area.

A1.19 *Percentage Reduction of Area (True)* is the difference between the original cross-sectional area of the test specimen and the minimum cross-sectional area within the gage boundaries prevailing at the moment of rupture, expressed as a percentage of the original area.

A1.20 *Rate of Loading* is the change in tensile load carried by the specimen per unit time. It is expressed in force per unit time, usually kilograms (or pounds) per minute. The initial rate of loading can be calculated from the initial slope of the load versus time diagram.

A1.21 *Rate of Stressing* (nominal) is the change in tensile stress (nominal) per unit time. It is expressed in force per unit area per unit time, usually kilograms-force per square centimeter (or pounds per square inch) per minute. The initial rate of stressing can be calculated from the initial slope of the tensile stress (nominal) versus time diagram.

NOTE A9—The initial rate of stressing as determined in this manner has only limited physical significance. It does, however, roughly describe the average rate at which the initial stress (nominal) carried by the test specimen is applied. It is affected by the elasticity and flow characteristics of the materials being tested. At the yield point, the rate of stressing (nominal) may become zero, but the rate of stressing (true) may continue to have a positive value if the cross-sectional area is decreasing.

A1.22 *Rate of Straining* is the change in tensile strain per unit time. It is expressed either as strain per unit time, usually centimeters per centimeter (or inches per inch) per minute, or percentage elongation per unit time, usually percentage elongation per minute. The initial rate of straining can be calculated from the initial slope of the tensile strain versus time diagram.

NOTE A10—The initial rate of straining is synonymous with the rate of crosshead movement divided by the initial distance between crossheads only in a machine with constant-rate-of-crosshead movement and when the specimen has a uniform original cross-section, does not "neck down," and does not slip in the jaws.

A2. SYMBOLS

A2.23 The following symbols may be used for the above terms:

SYMBOL	TERM
W	Load
ΔW	Increment of load
L	Distance between gage marks at any time
L_o	Original distance between gage marks
L_u	Distance between gage marks at moment of rupture
ΔL	Increment of distance between gage marks = elongation
A	Minimum cross-sectional area at any time
A_o	Original cross-sectional area

ΔA Increment of cross-sectional area
A_u Cross-sectional area at point of rupture measured after breaking specimen
A_T Cross-sectional area at point of rupture, measured at the moment of rupture
t Time
Δt Increment of time
σ Tensile stress
$\Delta\sigma$ Increment of stress
σ_T True tensile stress
σ_U Tensile strength at break (nominal)
σ_{UT} Tensile strength at break (true)
ϵ Strain
$\Delta\epsilon$ Increment of strain
ϵ_U Total strain, at break
ϵ_T True strain
$\%El$ Percentage elongation
Y.P. Yield point
E Modulus of elasticity

A2.24 Relations between these various terms may be defined as follows:

$\sigma = W/A_o$

$\sigma_T = W/A$

$\sigma_U = W/A_o$ (where W is breaking load)

$\sigma_{UT} = W/A_T$ (where W is breaking load)

$\epsilon = \Delta L/L_o = (L - L_o)/L_o$

$\epsilon_U = (L_u - L_o)/L_o$

$\epsilon_T = \int_{L_o}^{L} dL/L = \ln L/L_o$

$\%El = (L - L_o)/L_o \times 100 = \epsilon \times 100$

Percentage reduction of area (nominal)

$$= (A_o - A_u)/A_o \times 100$$

Percentage reduction of area (true)

$$= (A_o - A_T)/A_o \times 100$$

Rate of loading $= \Delta W/\Delta t$

Rate of stressing (nominal)

$$= \Delta\sigma/\Delta t = (\Delta W/A_o)/\Delta t$$

Rate of straining $= \Delta\epsilon/\Delta t = (\Delta L/L_o)/\Delta t$

For the case where the volume of the test specimen does not change during the test, the following three relations hold:

$\sigma_T = \sigma(1 + \epsilon) = \sigma\, L/L_o$

$\sigma_{UT} = \sigma_U(1 + \epsilon_U) = \sigma_U\, L_u/L_o$

$A = A_o/(1 + \epsilon)$

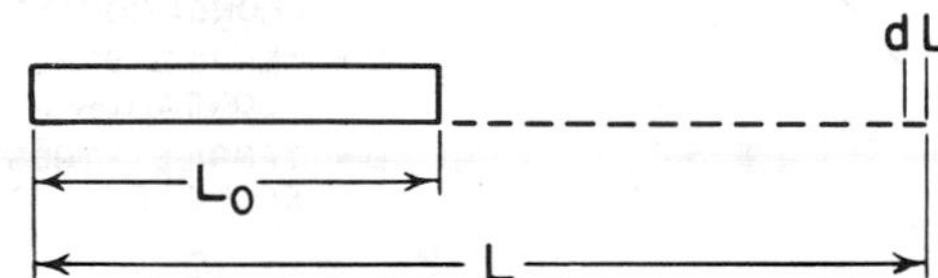

FIG. A1 Illustration of True Strain Equation

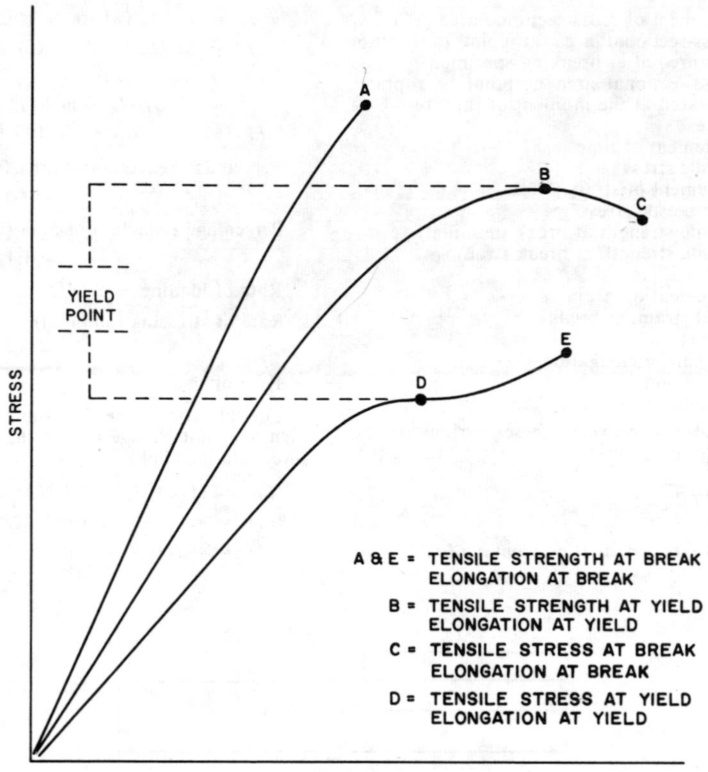

FIG. A2 Tensile Designations.

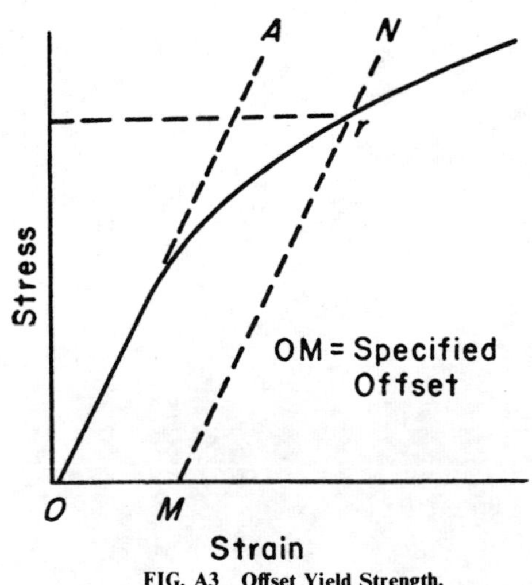

FIG. A3 Offset Yield Strength.

Designation: D 647 – 68

American National Standard K66.1-1969
American National Standards Institute

Recommended Practice for
DESIGN OF MOLDS FOR TEST SPECIMENS OF PLASTIC MOLDING MATERIALS[1]

This Recommended Practice is issued under the fixed designation D 647; the number immediately following the designation indicates the year of original adoption or, in the case of revision, the year of last revision. A number in parentheses indicates the year of last reapproval.

1. Scope

1.1 This recommended practice covers the design of molds for test specimens of plastic molding materials. The plastic and mold temperatures, pressure, and cycle timing used for any material should conform to the appropriate molding practice, if available, or material specification. If such instructions are not available, the recommendations of the material supplier should be followed.

1.2 Designs shown have been found suitable for a broad range of plastics, but may not be optimum for any given material. Only specimen dimensions are actually standardized.[2]

NOTE 1—Interlaboratory studies have shown that similar results were obtained with different mold designs when the same specimen geometry was used throughout. See ASTM Recommended Practice D 1897, for Injection Molding of Specimens of Thermoplastic Molding and Extrusion Materials.[3]

NOTE 2—The values stated in U.S. customary units are to be regarded as the standard. The metric equivalents of U.S. customary units may be approximate.

2. Molds for 12.7 by 12.7 by 127-mm (1/2 by 1/2 by 5-in.) and 6.4 by 12.7 by 127-mm (1/4 by 1/2 by 5-in.) Specimens

2.1 Molds for the 12.7 by 12.7 by 127-mm (1/2 by 1/2 by 5-in.) or 12.7 by 12.7 by 64-mm (1/2 by 1/2 by 2 1/2-in.) and 6.4 by 12.7 by 127-mm (1/4 by 1/2 by 5-in.) test specimens may be the single-bar, single-cavity positive mold type and shall conform to the design and dimensions shown in Fig. 1.[2]

3. Molds for Disk Test Specimens

3.1 Molds for the disk test specimens 51 mm (2 in.) or 102 mm (4 in.) in diameter shall be of the design and dimensions shown in Fig. 2.

4. Molds for Tension Test Specimen

4.1 Molds for the test specimen for determining the tensile strength of molded electrical insulating materials (Note 3) shall be of the design and dimensions shown in Fig. 3.[2]

NOTE 3—See ASTM Method D 651, Test for Tensile Strength of Molded Electrical Insulating Materials.[4]

4.2 Molds for the test specimen for determining the tensile properties of plastics (Note 4) shall be of the design and dimensions shown in Fig. 6.[2]

NOTE 4—See ASTM Method D 638, Test for Tensile Properties of Plastics.[3]

5. Mold for Injection Molding of Test Specimens

5.1 Layout—A 4-cavity mold has been designed as a suitable means of preparing frequently used test specimens with most injection molding machines of 55-g or greater capacity. The arrangement shown in Fig. 4 can be adapted to most mold bases 22.5 by 30.0 cm and larger.

5.1.1 The cavities for the 3.2 by 12.7 and

---

[1] This recommended practice is under the jurisdiction of ASTM Committee D-20 on Plastics. A list of committee members may be found in the ASTM Yearbook. This standard is the direct responsibility of Subcommittee D-20.90 on Specimen Preparation.

Current edition effective July 16, 1968. Originally issued 1941. Replaces D 647 – 65.

[2] Blueprints of detailed drawings for the construction of the molds shown in Figs. 1, 3, 6, and 7 are available at a nominal cost from the American Society for Testing and Materials, 1916 Race St., Philadelphia, Pa. 19103.

[3] *Annual Book of ASTM Standards*, Part 27.

[4] *Annual Book of ASTM Standards*, Part 29.

127 mm and 6.4 by 12.7 by 127 mm bars and the tension specimen are parted on the diagonal. This eliminates side draft in the test area and provides for easy ejection particularly of the thicker specimen. Alternatively, flat cavities, which are somewhat less expensive to machine, may be employed, with ejector pins. Locations for these pins are indicated in the recommended design. Ejector tabs are desirable for preventing damage to the test area of the specimen.

5.1.2 Tipping the 5-mm disk specimen cavity was not considered advantageous. Appropriate means of ejection have been provided.

5.2 *Cavity Dimensions*—Proper dimensions for each cavity should be obtained by considering the desired or specified dimensions and tolerances of the finished specimen and the mold shrinkage encountered with the material(s) being processed.

5.3 *Gates*—Rectangular gates were selected for simplicity of machining. All are large, to accommodate the great variety of plastics which can be injection molded. End gates are used throughout; side gates can be employed, providing they are not located within the test area. Gate land lengths of less than 0.8 mm are advised. Recommended gate sizes are given in Table 2.

5.4 *Runners:*

5.4.1 Full round runners 8 mm in diameter have been used for easy filling and ejection. Trapezoidal runners of the same cross-sectional area should also be satisfactory; ejector pins may be required, however. Separate runners for each cavity are provided.

5.4.2 Runner shut-offs have been incorporated. Critical specimen preparation may require that only one specimen be molded at a time. On other occasions it may be desirable to mold two or more simultaneously.

5.5 *Temperature Control*—Mold surface temperature is usually controlled by circulating a heat transfer fluid through the mold. The suggested coring should provide for a range of no more than 5 C over the entire mold surface.

6. Injection Mold for Weld Test Specimens

6.1 The injection mold for the weld test specimens shall be of the design and dimensions shown in Fig. 5.

NOTE 5—There are two cavities. Three specimens can be made as follows:

(*1*) *"Cold" or "Butt" Weld Specimen*—Separate streams of plastic are introduced in each end of a dumbbell specimen cavity. The streams meet in the center of the specimen as the final act of filling the cavity. The weld is at right angles to the length of the specimens.

(*2*) *"No Weld" Specimen*—The plastic is introduced in one end of the dumbbell specimen cavity. This is accomplished by blocking off one stream of plastic in the "cold" weld specimen.

(*3*) *"Hot" or "Streaming" Weld Specimen*—The plastic flow is interrupted at the gate to form two streams which are again united. The material to fill the cavity must all flow by this welded section. The weld is at right angles to the length of the specimen.

6.2 The test specimens are an alteration of the Type I specimen of Method D 638 and, except for the changes indicated in Table 3, all other dimensions are the same as shown in that method.

NOTE 6—It will be necessary to machine these specimens if other than tension tests are to be made.

7. Mold for Transfer Molding of Thermosetting Test Specimens

7.1 Mold—The cavities of this mold are designed to mold in one shot the following test specimens: a 6.4 by 12.7 by 127-mm ($^1/_4$ by $^1/_2$ by 5-in.) bar; a 12.7 by 12.7 by 127-mm ($^1/_2$ by $^1/_2$ by 5-in.) bar; a 3.2 by 51-mm ($^1/_8$ by 2-in.) diameter disk; a 3.2 by 102-mm ($^1/_8$ by 4-in.) diameter disk; and a 3.2 by 216-mm ($^1/_8$ by 8 $^1/_2$-in.) tension test specimen. The mold for these specimens shall be of the design shown in Fig. 7[2] and shall conform to the requirements prescribed in 7.2 to 7.5:

7.2 *Gate Size*—The gate size shall have a rectangular cross section 6.4 mm wide by 1.52 mm deep ($^1/_4$ by 0.060 in.).

7.3 *Mold for Tension Specimen*—The cavity of the mold for tension specimens for ASTM Method D 638, Test for Tensile Properties of Plastics,[3] shall have dimensions for the Type I specimen as shown in that method except for the following thickness:

$$3.175 \text{ mm} \quad {}^{+0.076}_{-0.000}$$

$$0.125 \text{ in.} \quad {}^{+0.003}_{-0.000}$$

7.4 *Mold for Bar Specimens*—The cavities of the mold for the bar specimens shall have the tolerances given in Table 4.

7.5 *Mold for Disk Specimens*—The cavities of the mold for the disk specimens shall have the dimensions given in Table 5.

TABLE 1 Specimen Sizes

Specimen	Approporiate ASTM Methods for Dimensions
Tension specimen, Type I	D 638[3]
3.2 by 12.7 by 127-mm specimen	D 256, D 648[3]
6.4 by 12.7 by 127-mm specimen	D 256, D 648, D 790[3]
3.2 by 5-mm diameter disk	D 570[3]

TABLE 2 Recommended Gate Sizes

Cavity	Recommended Gate Size
Tension specimen	3.2 by 9.6 mm
3.2 by 12.7 by 127-mm specimen	3.2 by 6.4 mm
6.4 by 12.7 by 127-mm specimen	6.4 by 6.4 mm
3.2 by 5-mm diameter disk	3.2 by 3.2 mm

TABLE 3 Comparison of Specimen Sizes D 647 versus D 638

Dimension	Method D 638 Type I Specimen	Specifications D 647
Length of flat section	57.2 mm (2.25 in.)	19.0 mm (0.75 in.)
Length over-all	216 mm (8½ in.)	127 mm (5 in.)
Length of grips	57 mm (2¼ in.) approx.	32 mm (1¼ in.) approx.

TABLE 4 Tolerances for Cavity Dimensions

Nominal Cavity Dimension	Specimen 12.7 by 12.7 mm (½ by ½ in.)	Specimen 6.4 by 12.7 mm (¼ by ½ in.)
12.7 mm	$12.70 \begin{cases} +0.13 \\ -0.00 \end{cases}$	$12.70 \begin{cases} +0.13 \\ -0.00 \end{cases}$
(½ in.)	$\left(0.500 \begin{cases} +0.005 \\ -0.000 \end{cases}\right)$	$\left(0.500 \begin{cases} +0.005 \\ -0.000 \end{cases}\right)$
6.4 mm	...	$6.350 \begin{cases} +0.076 \\ -0.000 \end{cases}$
(¼ in.)	...	$\left(0.250 \begin{cases} +0.003 \\ -0.000 \end{cases}\right)$
Length	127 mm (5 in.)	127 mm (5 in.)

TABLE 5 Dimensions of Mold for Disk Specimens

Maximum diameter	51 mm (2 in.)	102 mm (4 in.)
Depth	$3.175 \text{ mm} \begin{cases} +0.076 \\ -0.000 \end{cases}$	$3.175 \text{ mm} \begin{cases} +0.076 \\ -0.000 \end{cases}$
	$\left(0.125 \text{ in.} \begin{cases} +0.003 \\ -0.000 \end{cases}\right)$	$\left(0.125 \text{ in.} \begin{cases} +0.003 \\ -0.000 \end{cases}\right)$
Edge taper (to bottom of cavity)	10 deg	10 deg

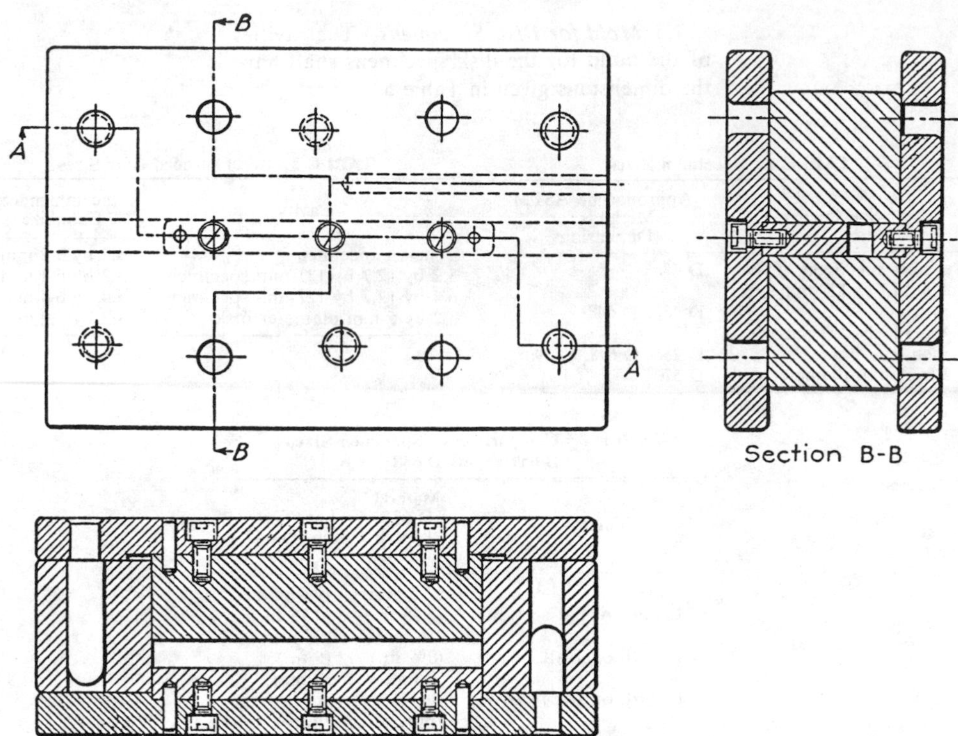

NOTE—Thermometer wells shall be 8 mm ($^5/_{16}$ in.) in diameter to permit use of a readily available thermometer.

FIG. 1 Single Bar, Single Cavity Positive Mold for 12.7 by 12.7 by 127-mm ($^1/_2$ by $^1/_2$ by 5-in.) and 6.4 by 12.7 by 127-mm ($^1/_4$ by $^1/_2$ by 5-in.) Test Specimens.[2]

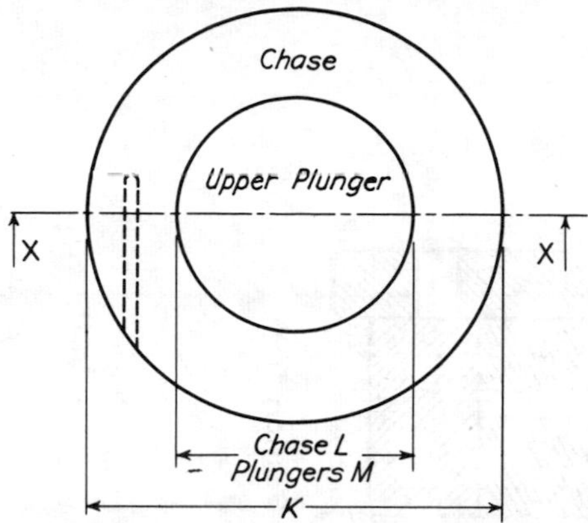

0.102 to 0.127-mm (0.004 to 0.005-in.) Clearance between Chase and Plungers.

Section XX

	Dimension K	Dimension L	Dimension M
Specimen for water absorption test	127 mm (5 in.)	51.61 mm (2.032 in.)	51.49 mm (2.027 in.)
Specimens for dielectric strength and power factor tests	178 mm (7 in.)	102.41 mm (4.032 in.)	102.29 mm (4.027 in.)

NOTE 1—A top plate may be added, if the material to be handled in the mold is considered to require breathing.

NOTE 2—Thermometer wells shall be 8 mm ($^{5}/_{16}$ in.) in diameter to permit use of a readily available thermometer.

FIG. 2 Mold for Disk Test Specimens.

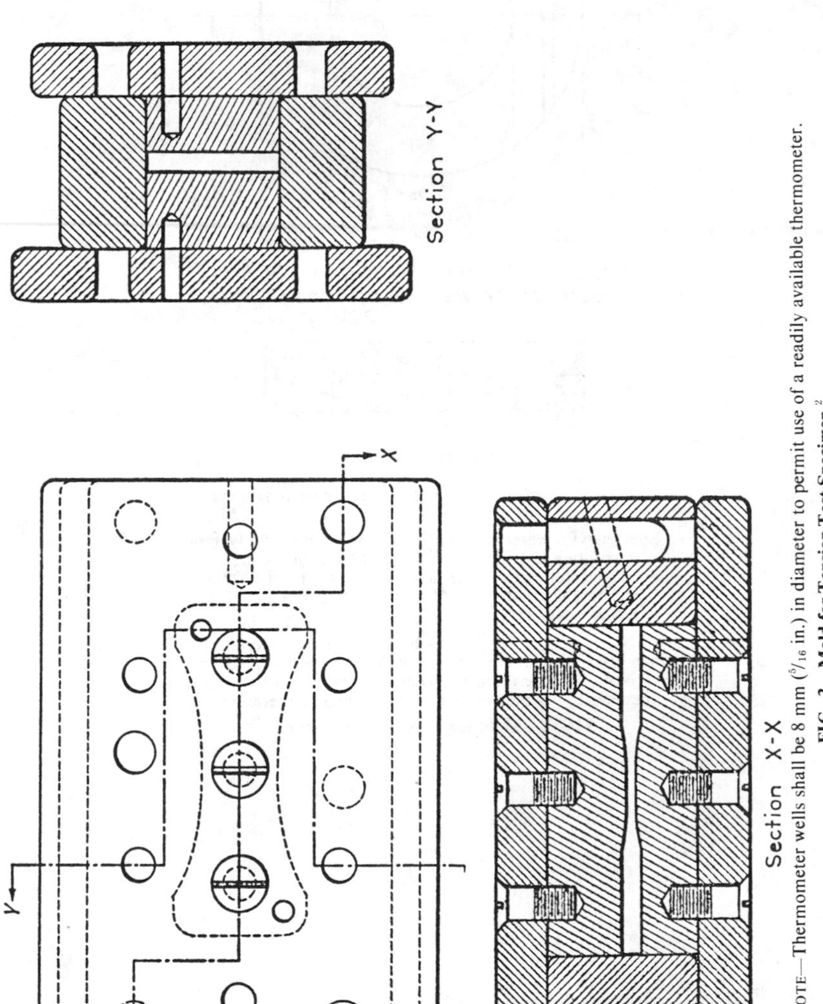

NOTE—Thermometer wells shall be 8 mm ($^5/_{16}$ in.) in diameter to permit use of a readily available thermometer.

FIG. 3 Mold for Tension Test Specimen.[2]

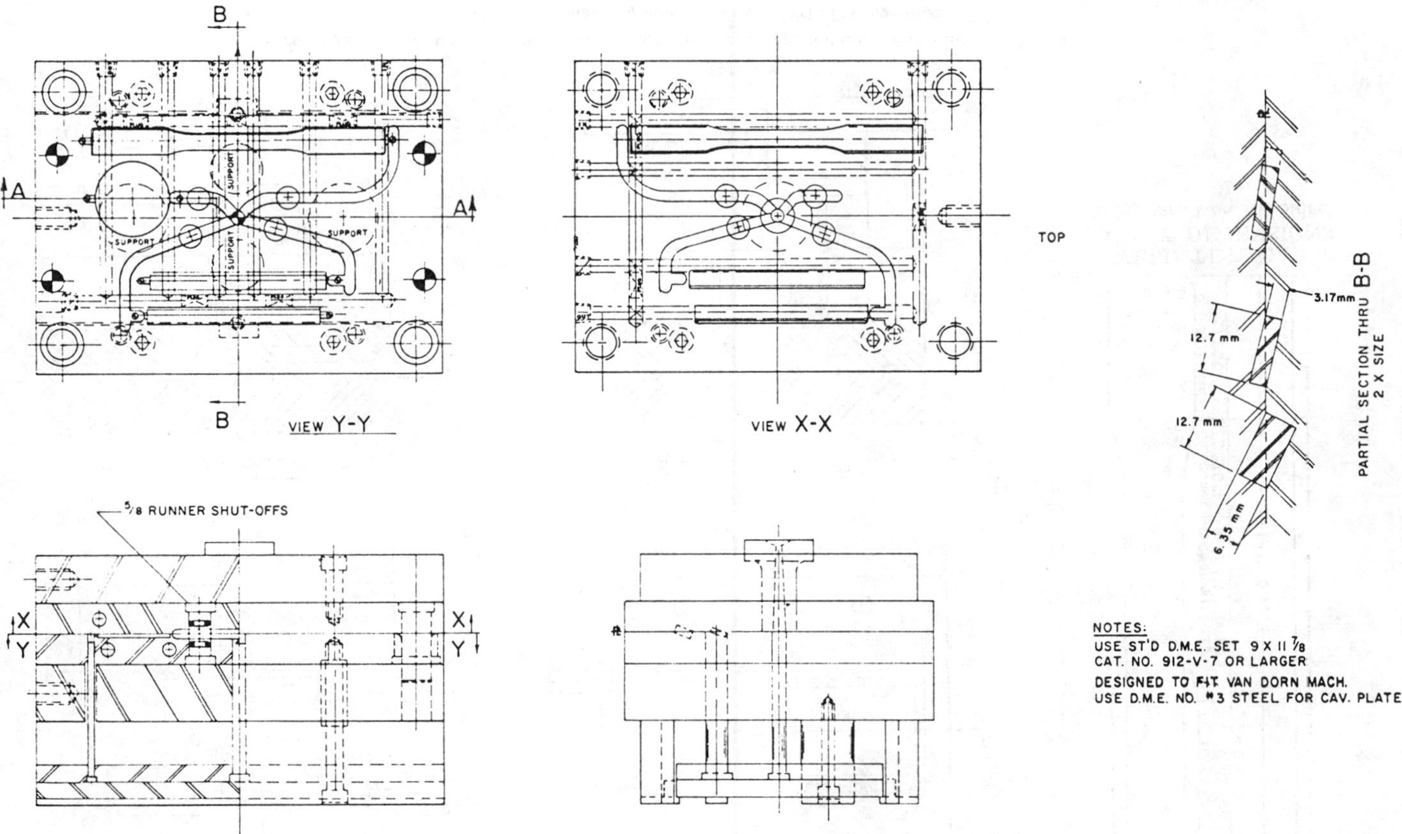

NOTE—Thermometer wells shall be 8 mm (5/16 in.) in diameter to permit use of a readily available thermometer.

FIG. 4 Mold for Injection Molding Test Specimens.

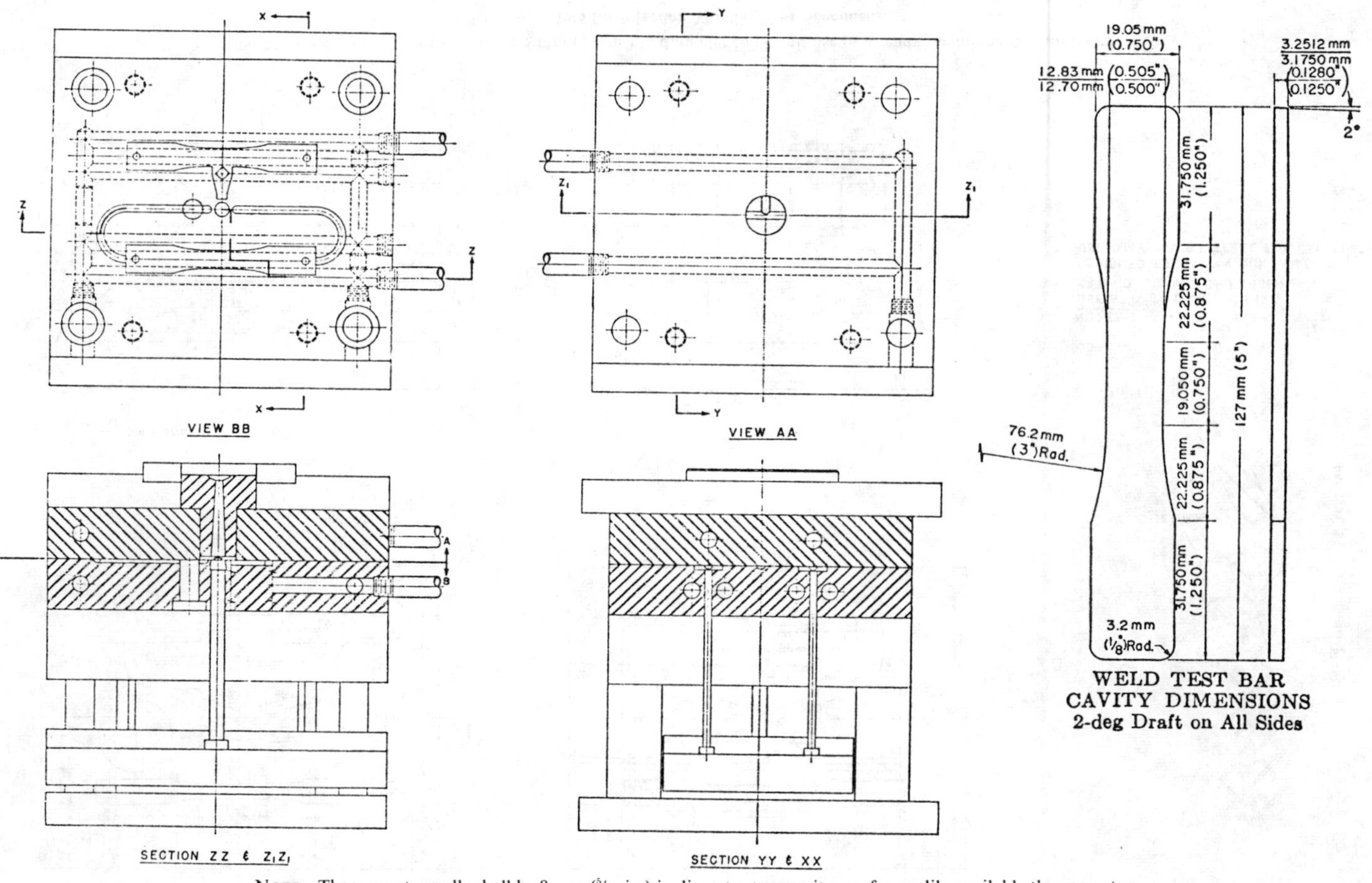

NOTE—Thermometer wells shall be 8 mm ($^5/_{16}$ in.) in diameter to permit use of a readily available thermometer.

FIG. 5 Mold for Injection Molding of Weld Test Specimen.

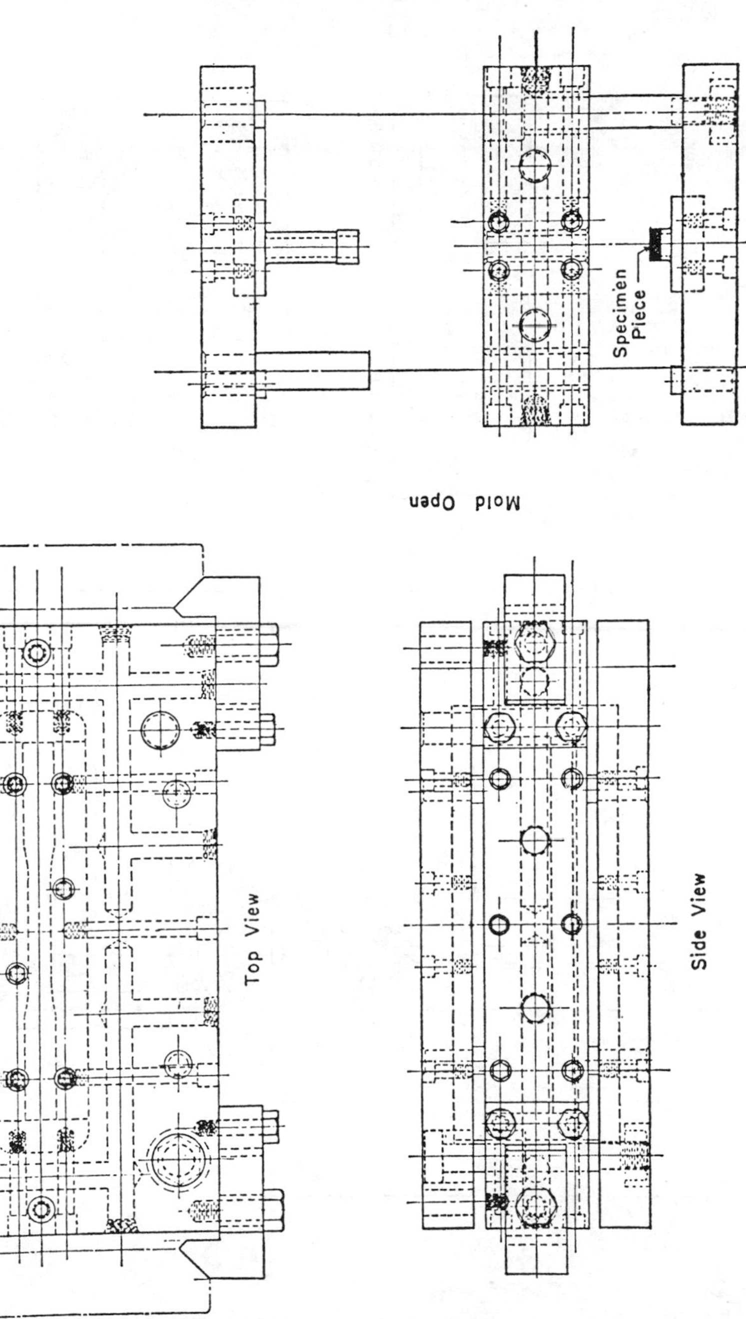

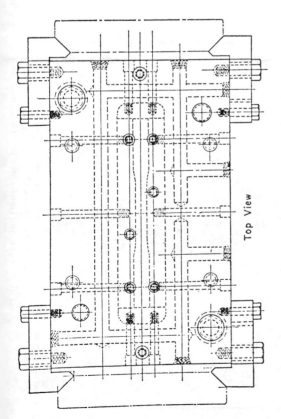

NOTE—Thermometer wells shall be 8 mm ($^5/_{16}$ in.) in diameter to permit use of a readily available thermometer.

FIG. 6 Semi-Automatic Compression Mold for Molding Tension Test Specimens.[2]

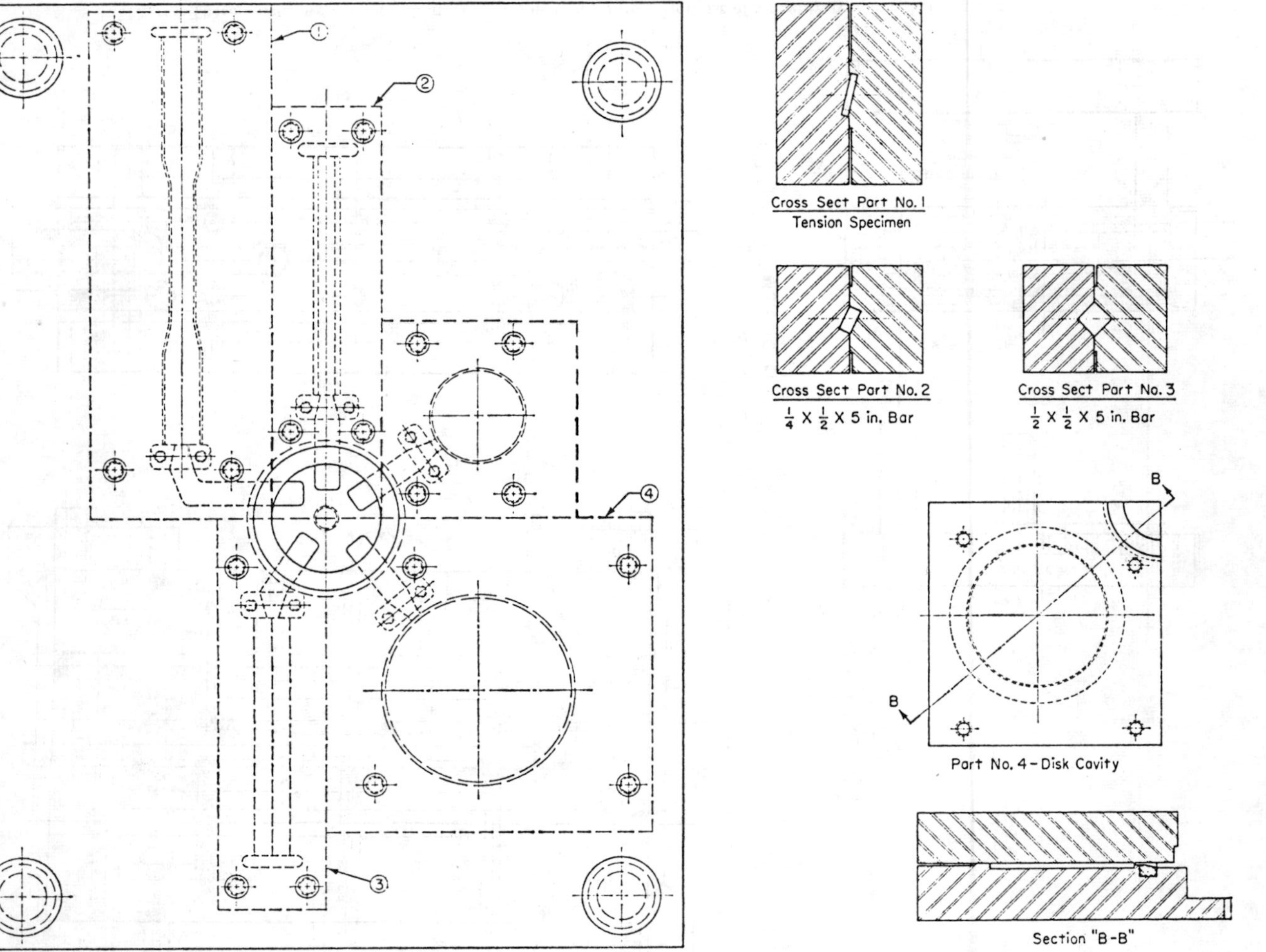

NOTE—Thermometer wells shall be 8 mm ($^5/_{16}$ in.) in diameter to permit use of readily available thermometers.

FIG. 7 Five-Cavity Transfer Mold for Thermosetting Plastic Test Specimens (Steam Cores Not Shown).[2]

 Designation: D 648 – 56 (Reapproved 1961)

Standard Method of Test for
DEFLECTION TEMPERATURE OF PLASTICS UNDER LOAD[1]

This Standard is issued under the fixed designation D 648; the number immediately following the designation indicates the year of original adoption or, in the case of revision, the year of last revision. A number in parentheses indicates the year of last reapproval.

Note—Note 4 was editorially changed in June 1965.

1. Scope

1.1 This method covers the determination of the temperature at which an arbitrary deformation occurs when specimens are subjected to an arbitrary set of testing conditions. Data obtained by this method may be used to predict the behavior of plastic materials at elevated temperatures only in applications in which the factors of time, temperature, method of loading, and fiber stress are similar to those specified in this test. The test is particularly suited to control and development work.

1.2 This method applies to molded and sheet materials available in thicknesses of 0.32 cm (1/8 in.) or greater and which are rigid at normal temperatures.

NOTE 1—Sheet stock less than 0.32 cm (0.125 in.) but more than 0.100 cm (0.040 in.) in thickness may be tested by use of a composite sample having a minimum thickness of 0.32 cm (0.125 in.). The laminae must be of uniform width in order to obtain a uniform stress distribution. One type of composite specimen has been prepared by cementing the ends of the laminae together and then smoothing the edges with sandpaper. The direction of loading shall be perpendicular to the edges of the individual laminae.

NOTE 2—The values stated in U.S. customary units are to be regarded as the standard. The metric equivalents of U.S. customary units may be approximate.

2. Apparatus

2.1 The apparatus shall be constructed essentially as shown in Fig. 1 and shall consist of the following:

2.1.1 *Specimen Supports*—Metal supports for the specimen which shall be 10 cm (4 in.) apart, allowing the load to be applied on top of the specimen vertically and midway between the supports. The contact edges of the supports and of the piece by which pressure is applied shall be rounded to a radius of 0.32 cm (1/8 in.). The vertical members which attach the specimen supports to the upper plate shall be made of a material having the same coefficient of linear thermal expansion as that used for the rod through which the load is applied.

NOTE 3—Unless these parts have the same coefficient of linear expansion, the differential change in length of these parts introduces an error in the reading of the apparent deformation of the specimen. A test should be made on each apparatus using a test bar made of material having a low coefficient of expansion.[2] The temperature ranges to be used should be covered and a correction factor determined for each temperature. If this factor is 0.013 mm (0.0005 in.) or greater, its algebraic sign should be noted and the factor should be applied to each test by adding it algebraically to the reading of apparent deflection of the test specimen.

2.1.2 *Immersion Bath*—A suitable liquid heat-transfer medium (Note 4) in which the specimen shall be immersed. It shall be well stirred during the test and shall be provided with a means of raising the temperature at 2 ± 0.2 C/min (Note 5).

NOTE 4—A liquid heat-transfer medium shall be chosen which will not affect the rigidity of the specimen. Mineral oil is considered safe from ignition to

---

[1] This method is under the jurisdiction of ASTM Committee D-20 on Plastics. A list of committee members may be found in the ASTM Yearbook. This standard is the direct responsibility of Subcommittee D-20.30 on Thermal Properties.

Current edition effective Sept. 10, 1956. Originally issued 1941. Replaces D 648 – 45.

[2] Invar or Pyrex glass has been found suitable for this purpose.

115 C. Silicone oils may be heated to about 260 C for short periods of time.

NOTE 5—The specified heating rate can be conveniently maintained by the use of an electric hot plate or immersion heater, if the current through these units is regulated by a variable transformer or rheostat.

2.1.3 *Weights*—A set of weights of suitable sizes so that the specimen can be loaded to a fiber stress of 18.5 kgf/cm^2 (264 psi) $\pm$ 2$^1/_2$ percent for the first part of the test described in 6.1, and 4.6 kgf/cm^2 (66 psi) $\pm$ 2$^1/_2$ percent for the procedure described in 6.2. The weight of the rod that applies the testing force shall be determined and included as part of the total load. If a dial gage is used, the force exerted by its spring shall be determined and shall be included as part of the load (Notes 6 and 7). The load shall be calculated as follows:

$$P = 2Sbd^2/3l$$

where:

P = load, kg (or lb),

S = maximum fiber stress in the specimen of 18.5 kgf/cm^2 (264 psi) when tested according to 6.1, 4.6 kgf/cm^2 (66 psi) when tested according to 6.2,

b = width of specimen, cm (or in.),

d = depth of specimen, cm (or in.), and

l = width of span between supports (10 cm (4 in.)).

NOTE 6—In some designs of this apparatus the force of the dial gage spring is directed upward and must be subtracted from the load, while in other designs this force acts downward and must be added to the load.

NOTE 7—Since the force exerted by the spring in certain dial gages varies considerably over the stroke, this force should be measured in that part of the stroke which is to be used.

2.1.4 *Thermometer*—The thermometer shall be one of the following, or its equivalent, as prescribed in ASTM Specifications E 1[3] for Thermometers: Thermometer 1C, 1F, 2C, or 2F, having ranges from −20 to +150 C, 0 to 302 F, −5 to +300C, or 20 to 580 F, respectively, whichever temperature range is most suitable for the material being tested.

3. Preparation of Apparatus

3.1 The apparatus shall be arranged so that the deflection of the midpoint of the specimen can be measured on a scale calibrated in 0.01 mm (0.0005 in.). The apparatus may be arranged to shut off the heat automatically and sound an alarm when the specified deflection has been reached. Sufficient heat-transfer liquid shall be used to cover the thermometers to the point specified in their calibration, or 7.6 cm (3 in.) in the case of the ASTM thermometers referred to in 2.1.4.

4. Test Specimens

4.1 At least two test specimens shall be used to test each sample at each fiber stress. The specimens shall be 12.7 by 1.27 cm (5 by $^1/_2$ in.) by any thickness from 0.3 to 1.3 cm ($^1/_8$ to $^1/_2$ in.).

NOTE 7—The types of mold and the molding process used to produce test specimens affect the results obtained in this test. Cooperating laboratories should therefore standardize mold and molding procedure to obtain concordant results.

5. Conditioning

5.1 The specimens shall be conditioned in accordance with the procedures specified for each material and in accordance with ASTM Methods D 618, Conditioning Plastics and Electrical Insulating Materials for Testing.[4]

6. Procedure

6.1 Place the test specimen in the apparatus with its 1.27-cm ($^1/_2$-in.) dimension vertical. In the case of 1.27 by 1.27-cm ($^1/_2$ by $^1/_2$-in.) bars compression-molded or cut from molded sheets, place the bars so that the direction of the testing force will be perpendicular to the direction of the molding pressure. The thermometers shall extend to within 0.32 cm ($^1/_8$ in.) of the specimen but shall not touch it. The temperature of the bath shall be 23 $\pm$ 1 C (73.4 $\pm$ 1.8 F) at the start of each test, unless previous tests have shown that for the particular material under test no error is introduced by starting at a higher temperature. Adjust the load so that the fiber stress is 18.5 kgf/cm^2 (264 psi) $\pm$ 2$^1/_2$ percent as calculated by the equation given in 2.1.3. Allow the load to act for 5 min (Note 9); take the zero reading or setting of the measuring device and start the heating. This waiting period may be omitted when testing materials that show no appreciable creep during the initial 5 min. Report the temperature at which the bar has

[3] *Annual Book of ASTM Standards*, Part 30.
[4] *Annual Book of ASTM Standards*, Part 27.

deflected 0.25 mm (0.010 in.) as the deflection temperature at 18.5 kgf/cm^2 (264 psi) fiber stress.

NOTE 9—The 5-min waiting period is provided to compensate partially for the creep exhibited by several materials at room temperature when subjected to the prescribed fiber stress. That part of the creep which occurs in the initial 5 min is usually a large fraction of that which occurs in the first 30 min.

6.2 The second part of the test shall be conducted exactly as prescribed in 6.1, except adjust the load to produce a fiber stress of 4.6 kgf/cm^2 (66 psi) ± 2 $^1/_2$ percent as calculated by the equation given in 2.1.3. Report the temperature at which a deflection of 0.25 mm (0.010 in.) is reached as the deflection temperature at 4.6 kgf/cm^2 (66 psi) fiber stress.

7. Report

7.1 The report shall include the following:

7.1.1 The width and depth of the specimen measured to 0.025 mm (0.001 in.),

7.1.2 The deflection temperature and corresponding fiber stress,

7.1.3 The immersion medium and the average heating rate, and

7.1.4 Any peculiar characteristics of the specimen noted during the test or after removal from the apparatus.

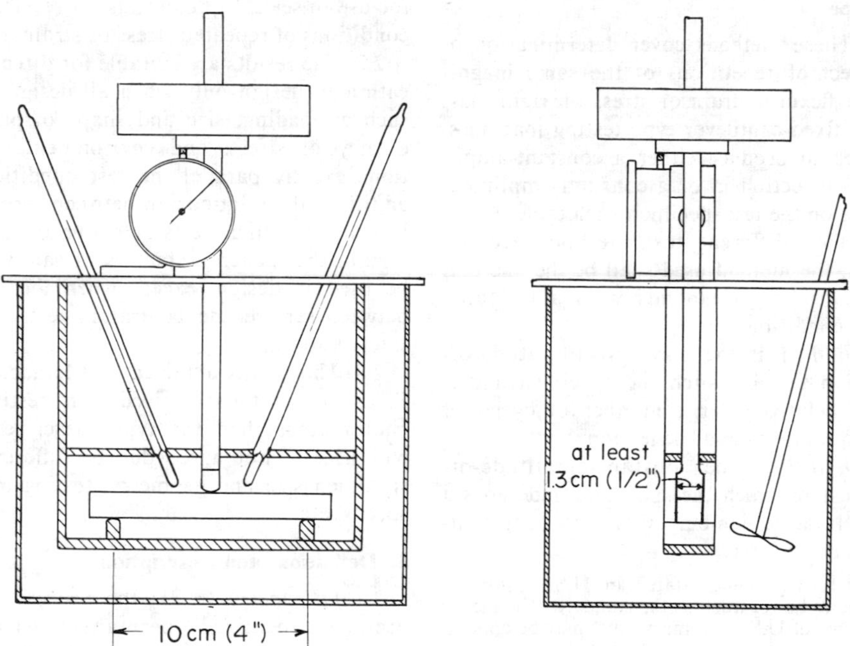

NOTE—The 1.27-cm ($^1/_2$-in.) dimension shown in the side elevation applies to the base of the loading rod.

FIG. 1 Apparatus for Deflection Temperature Test.

Designation: D 671 – 63 T

Tentative Methods of Test for
REPEATED FLEXURAL STRESS (FATIGUE) OF PLASTICS[1]

These Tentative Methods have been approved by the sponsoring committee and accepted by the Society in accordance with established procedures for use pending adoption as standard. Suggestions for revisions should be addressed to the Society at 1916 Race St., Philadelphia, Pa. 19103.

Pursuant to the warning note covering the 3-year limitation on tentatives, published in the previous edition, this tentative has received special permission from the Committee on Standards for continuation until July 1971.

1. Scope

1.1 These methods cover determination of the effect of repetitions of the same magnitude of flexural strain or stress on rigid plastics by fixed-cantilever type testing machines, designed to produce either a constant-amplitude-of-deflection in or a constant-amplitude-of-force on the test specimen each cycle.

1.2 Two different test methods are described; the method used shall be the one that more closely corresponds with anticipated service conditions:

Method A is the constant-amplitude-of-deflection test in which the specimen undergoes a relatively large number of cycles of constant-amplitude-of-deflection.

Method B is the constant-amplitude-of-force test in which the specimen undergoes a relatively large number of cycles of constant-amplitude-of-force.

NOTE 1—The values stated in U.S. customary units are to be regarded as the standard. The metric equivalents of U.S. customary units may be approximate.

2. Significance

2.1 The flexural fatigue test is expected to provide information on the ability of rigid plastics to resist the development of cracks or general mechanical deterioration of the material as a result of a relatively large number of cycles of constant amplitude of deflection or force.

2.2 These methods are useful for determining the effect of variations in material, stress concentrations, environmental conditions, etc., on the ability of a material to resist deterioration resulting from repeated stress and strain. They may also be used to provide data for use in selecting materials for service under conditions of repeated stress or strain.

2.3 The results are suitable for direct application in design only when all design factors such as loading, size and shape of part, frequency of stress, and environmental conditions exactly parallel the test conditions. In addition, the relationship between service and laboratory fatigue tests of materials is so uncertain that generous factors of safety should be used in design, except when the relation between service and laboratory tests is accurately known.

2.4 The results obtained by Methods A and B, as well as those derived from testing machines other than the types described here, will usually not agree due to differences in specimen size and geometry, testing machine speeds, etc.

3. Definitions and Descriptions

3.1 *fatigue* (Note 2)—the process of progressive localized permanent structural change occurring in a material subjected to conditions that produce fluctuating stresses and strains at some point or points and that may culminate in cracks or complete fraction after a sufficient number of fluctuations (Note 3).

NOTE 2—The term fatigue in the materials testing field, has, in at least one case, glass technology, been used for static tests of considerable duration, a type of test generally designated as stress-rupture.

---

[1] These methods are under the jurisdiction of ASTM Committee D-20 on Plastics. A list of committee members may be found in the ASTM Yearbook. This standard is the direct responsibility of Subcommittee D-20.10 on Mechanical Properties.

Current edition effective Sept. 30, 1963. Originally issued 1942. Replaces D 671 – 51.

NOTE 3—Fluctuations may occur both in stress and within time (frequency) as in the case of "random vibration."

3.2 *nominal stress, S, or strain,* ϵ—the stress or strain at a point calculated on the net cross section by simple elastic theory without taking into account the effect on stress or strain produced by geometric discontinuities such as holes, grooves, fillets, etc. Stress units shall be in kilograms-force per square centimeter or pounds per square inch, strain units in millimeters per millimeter or inches per inch.

3.3 *maximum stress,* S_{max}*, or strain,* ϵ_{max}—the stress or strain having the highest algebraic value in the stress or strain cycle, tensile stress or strain being considered positive and compressive stress or strain negative. In this definition, as well as in others that follow, the nominal stress or strain is used most commonly. Stress units shall be in kilograms-force per square centimeter or pounds per square inch, strain units in millimeters per millimeter or inches per inch.

3.4 *minimum stress,* S_{min}*, or strain,* ϵ_{min}—the stress or strain having the lowest algebraic value in the cycle, tensile stress or strain being considered positive and compressive stress or strain negative. Stress units shall be in kilograms-force per square centimeter or pounds per square inch, strain units in millimeters per millimeter or inches per inch.

3.5 *mean stress (or steady component of stress),* S_m—the algebraic average of the maximum and minimum stresses in one cycle, that is,

$$S_m = (S_{max} + S_{min})/2$$

The sign of the mean stress, whether tension (+) or compression (−), must be stated. Units shall be in kilograms force per square centimeter or pounds per square inch. See Fig. 1 and 3.22.

NOTE 4—The mean stress of a cycle is important because, in general, the fatigue strength of a given material will depend upon the magnitude of the mean stress. Thus it is necessary to specify both the fatigue strength and the mean stress.

3.6 *mean strain (or steady component of strain),* ϵ_m—the algebraic average of the maximum and minimum strains in one cycle, that is,

$$\epsilon_m = (\epsilon_{max} + \epsilon_{min})/2$$

The sign of the mean strain, whether tension (+) or compression (−) must be stated. Units shall be in millimeters per millimeter or inches per inch. This term is analogous to mean stress, as shown in Fig. 1.

3.7 *fatigue life, N*—the number of cycles of stress or strain of a specified character that a given specimen sustains before failure of a specified nature occurs.

3.8 *fatigue failure*—fatigue failure may usually be said to occur when a specimen has completely fractured into two parts. When fibrous materials, large specimens, residual stresses, etc., cause a large number of cycles to elapse from formation of macroscopic cracks to complete fracture, fatigue failure is arbitrarily defined as having occurred when the stiffness of the specimen has been reduced by a specified amount (one eighth for plastics) as a result of the cracks.

3.9 *stress ratio, A or R*—the algebraic ratio of two specified stress values in a stress cycle. Two commonly used stress ratios are: the ratio of the stress amplitude to the mean stress, that is,

$$A = S_a/S_m$$

and the ratio of the minimum stress to the maximum stress, that is,

$$R = S_{min}/S_{max}$$

3.10 *strain ratio,* A_ϵ *or* R_ϵ—the algebraic ratio of two specified strain values in a strain cycle. Two commonly used strain ratios are: the ratio of the strain amplitude to the mean strain, that is,

$$A_\epsilon = \epsilon_a/\epsilon_m$$

and the ratio of the minimum strain to the maximum strain, that is,

$$R_\epsilon = \epsilon_{min}/\epsilon_{max}$$

3.11 *range of stress,* S_r—the algebraic difference between the maximum and minimum nominal stresses in one cycle, that is,

$$S_r = S_{max} - S_{min}$$

3.12 *range of strain,* ϵ_r—the algebraic difference between the maximum and minimum nominal strains in one cycle, that is,

$$\epsilon_r = \epsilon_{max} - \epsilon_{min}$$

3.13 *stress amplitude (or variable component of stress),* S_a—one half the range of stress, that is,

$$S_a = S_r/2 = (S_{max} - S_{min})/2$$

Units shall be in kilograms-force per square centimeter or pounds per square inch. See Fig. 1.

3.14 *strain amplitude* (*or variable component of strain*), ϵ_a—one half the range of strain, that is,

$$\epsilon_a = \epsilon_r/2 = (\epsilon_{max} - \epsilon_{min})/2$$

Units shall be in millimeters per millimeter or inches per inch. This term is analogous to stress amplitude, as shown in Fig. 1.

3.15 *fatigue strength at N cycles*, S_N—a hypothetical value of stress for failure at exactly N cycles, determined from an S-N diagram. The value of S_N thus determined is subject to the same conditions as those which apply to the S-N diagram. Units shall be in kilograms-force per square centimeter or pounds per square inch.

NOTE 5—A value of S_N that is commonly found in the literature is the hypothetical value of S_{max}, S_{min}, or S_a at which 50 percent of the specimens of a given sample could survive N stress cycles in which $S_m = 0$. This is also known as the median fatigue strength for N cycles.

3.16 *fatigue strain at N cycles*, ϵ_N—a hypothetical value of strain for failure at exactly N cycles, determined from an ϵ-N diagram. The value of ϵ_N thus determined is subject to the same conditions as those which apply to the ϵ-N diagram. Units shall be in millimeters per millimeter or inches per inch.

NOTE 6—A value of ϵ_N that is commonly found in the literature is the hypothetic value of ϵ_{max}, ϵ_{min}, or ϵ_a at which 50 percent of the specimens of a given sample could survive N strain cycles in which $\epsilon_m = 0$. This is also known as the median fatigue strain for N cycles.

3.17 *fatigue limit*, S_f—the limiting value of S_N or ϵ_N as N becomes very large. Stress units shall be in kilograms-force per square centimeter or pounds per square inch, strain units in millimeters per millimeter or inches per inch.

NOTE 7—Certain materials and environments preclude the attainment of a fatigue limit. Values tabulated as "fatigue limits" in the literature are frequently (but not always) values of S_N or ϵ_N for 50 percent survival at N cycles of stress or strain in which $S_m = 0$ or $\epsilon_m = 0$.

3.18 *theoretical stress or strain concentration factor, or stress or strain concentration factor*, K_t *or* $K_{t\epsilon}$—the ratio of the greatest stress or strain in the region of a notch or other stress concentrator as determined by the theory of elasticity (or by experimental procedures that give equivalent values) to the corresponding nominal stress or strain.

NOTE 8—The theory of plasticity should not be used to determine K_t or $K_{t\epsilon}$.

3.19 *fatigure notch factor*, K_f—the ratio of the fatigue strength or strain of a specimen with no stress concentration to the fatigue strength or strain at the same number of cycles with stress concentration for the same conditions.

NOTE 9—In specifying K_f it is necessary to specify the geometry and the values of S_{max}, S_m, ϵ_{max}, ϵ_m, and N for which it is computed.

3.20 *S-N diagram*—a plot of S_{max}, S_{min}, or S_a against the number of cycles to failure for a specified value of S_m, A, or R and for a specified probability of survival. For N a log scale is almost always used. For S a linear scale is used most often, but a log scale is sometimes used.

3.21 *ϵ-N diagram*—a plot of ϵ_{max}, ϵ_{min}, or ϵ_a against the number of cycles to failure for a specified value of ϵ_m, A_ϵ, or R_ϵ and for a specified probability of survival. For N a log scale is almost always used. For ϵ either a linear or a log scale may be used.

3.22 In Fig. 1 stress versus time curves are shown. This graphical description of stresses applies to strains as well in the constant-amplitude-of-deflection test. These curves are shown for one cycle of stress with different combinations of mean stress, S_m, and superimposed alternating stress, S_a. In Fig. 1(*a*) the mean stress is zero so that the alternating stress is completely reversed in sign, that is, the maximum tension equals the maximum compression. In Fig. 1(*b*) the mean stress is of such value that the maximum tension is greater than the maximum compression. In Fig. 1(*c*) all values of stress in the cycle are tension. In Fig. 1(*d*) all values of stress in the cycle (including the mean stress) are compression.

METHOD A—CONSTANT-AMPLITUDE-OF-DEFLECTION TEST

4. Apparatus (Fig. 2)

4.1 *Testing Machine*—A fatigue testing

machine of the fixed-cantilever, repeated-constant-deflection type (see Appendix A1). In this machine the specimen, *A*, shall be held as a cantilever beam in a vise, *B*, at one end, *M*, and bent by a concentrated load applied to a holder, *J*, fastened to the other end. The bending shall be accomplished by a connecting rod, *C*, driven by a variable eccentric, *D*, mounted on a shaft. The shaft shall be rotated at constant speed by a motor. The vise may be set in the plane of the eccentric so that the beam will be deflected the same amount on either side of the neutral position (completely reversed stress) or the vise may be set so that the deflection is greater on one side than on the other (mean stress not zero).

4.2 *Counter*—A counter, *K*, geared to the shaft so as to record the number of cycles. A suitable mechanically or electrically operated cut-off switch, *L*, shall be provided to stop the machine when the specimen fractures.

4.3 *Holder*—A holder, *J*, for the test specimen proportioned in such a way as to place the center of percussion of the oscillating portion of the specimen and holder at the wrist pin, *F*. It shall have as little mass and as large rigidity as possible in order to minimize inertia effects. A diagram of the holder for the free end of the specimen is shown in Fig. 3. It is necessary, for some materials, to use rubber gaskets in the grips on both the holder and the dynamometer to prevent failure at the grip due to stress concentration and chafing.

4.4 *Dynamometer*—The vise, *B*, containing a dynamometer (Note 10) for measuring the load on the specimen. The dynamometer shall consist of a reduced section, *G*, which acts as a spring, and a dial gage, *H*, to measure the deflection of the spring when the specimen is deflected.

NOTE 10—Another form of this machine does not contain a separate dynamometer but uses the specimen as a dynamometer. In this case a dial is used to measure the deflection of the specimen under a known load, preferably the load which is repeatedly applied to the specimen. This requires a separate calibration for each specimen. The result obtained is very nearly identical to that obtained with a dynamometer.

5. Test Specimens

5.1 *Unnotched Specimen*—The unnotched specimen shall conform to the dimensions shown in Fig. 4(*a*). The cross-section at the center of the specimen shall be 7.6 mm (0.3 in.) square, except in the case of the specimen from thin sheet (Note 11) which shall be 7.6 mm (0.3 in.) by the thickness of the sheet. The 50.8 mm (2-in.) radius (Note 12) of the scalloped sides shall be formed by milling, using a very sharp cutter and such combination of speed and feed as will give a good finish with a minimum of heating of the specimen. The test specimen shall be polished with successively finer emery paper, finishing with No. 00 to remove all scratches and tool marks. The final polishing shall be lengthwise of the specimen, since even small scratches transverse to the direction of tensile stress tend to lower the fatigue strength. In order to avoid heating, all polishing shall be done either by hand or with light pressure on a slowly revolving sanding drum. Care shall be taken to avoid rounding the corners.

NOTE 11—For the purpose of this test a thin sheet shall be defined as a sheet less than 7.6 mm (0.3 in.) in thickness or a material for which the ratio of the modulus of elasticity to the endurance limit is less than 100. The reason for these restrictions is that thin sheets and materials having a low modulus of elasticity are bent so much under the required loads that the fatigue specimen cannot (in the deflected position) be considered a straight beam and hence the equation $S = Mc/I$ is not accurate. It should be noted that the grips (Fig. 3) may not be satisfactory for specimens of very thin sheet due to the inertia forces set up by the vibration.

NOTE 12—With a cantilever specimen the maximum stress does not occur exactly at the minimum section, but slightly farther from the point of loading, *F*, Fig. 2. However, with a moment arm of 127 mm (5 in.) and a specimen with a radius of 50.8 (2 in.) this discrepancy is small enough to be neglected.

5.2 *V-Notched Specimen*—The V-notched specimen for testing materials other than thin sheets shall conform to the dimensions shown in Fig. 4(*b*). The standard V-notch shall be at the center of the specimen on one side and shall be 0.25 mm (0.01 in.) in radius and 0.5 mm (0.02 in.) in depth. The notch shall be formed by milling or shaping in such a way that it will be free from scratches. After machining, all edges of the specimen shall be smoothed to remove burrs.

5.3 *Drilled-Hole Notched Specimen for Thin Sheets*—The notched specimen for testing thin sheets (Note 11) shall conform to the dimensions shown in Fig. 4(*c*). The notch shall consist of a transverse hole drilled with a sharp No. 31 drill 3 mm (0.120 in.) in diameter. The specimen shall be polished in the region of the hole before the hole is drilled and all burrs shall be removed after drilling. No attempt shall be made to polish the inside of the hole.

6. Possibilities and Limitations of Test

6.1 The fatigue test is inherently a time-consuming test and not convenient for acceptance testing. Its primary function is in determining the effect of variations in material, stress concentrations, environmental conditions, etc., on the ability of a material to resist deterioration from the cyclic fatigue.

6.2 The type of machine covered in this test method is suitable for determining the fatigue strength or strain of both notched and unnotched specimens for any value of mean stress or strain in flexure.

6.3 Specimens notched on one side (as in Fig. 4(*b*)) may be used to determine the notched fatigue strength or strain for any value of mean stress or strain in either tension or compression. It should be mentioned that for material (such as cellulose acetate), which creeps rapidly, the effect of a mean stress other than zero is to cause relaxation so that the stress cycle tends to approach the condition of complete reversal of stress.

6.4 The notched specimen is used to determine the effect of a notch of standard dimension upon the fatigue strength or strain of the material. The V-notch considered in this test as standard is shown in Fig. 4(*b*). The transverse-hole notch shown in Fig. 4(*c*) may be used for thin sheet and material such as plywood for which the continuity of the material would be altered by the V-notch.

6.5 In any plastic part there are likely to be changes in cross-section, such as fillets, holes, screw threads, and the like, at which the state of stress is complex (a combination of normal or shearing stresses, or both, in more than one direction) and at which the maximum value of stress is greater than that calculated by formulas such as $S = Mc/I$. It is known that the effect of a certain type of notch in reducing the fatigue strength of one material may be more severe than in another material; the use of a standard V-notch in a fatigue test specimen is therefore desirable as a means of evaluating the sensitivity of a material to the effect of a stress raiser.

6.6 Other notches have been proposed to replace those shown in Fig. 4(*b*) and (*c*) but experimental evidence of their suitability is at present lacking.

6.7 Tests of thin sheet (Note 11) yield results that vary with the thickness of the sheet even though the material is identical (Note 11). Because of this fact the thickness of the sheet shall be specified when reporting results of tests of thin sheet; and all comparisons of different materials, or selection of materials on the basis of fatigue strength shall be made from results of tests of standard specimens or tests in which the same thickness of sheet is used for all materials.

6.8 In the application of this test method it is assumed that test specimens of a given material are comparable and truly representative of the material. Departure from this assumption may introduce discrepancies as great as, if not greater than, those due to departure from details of procedure outlined in the test.

7. Conditioning

7.1 *Conditioning*—Condition the test specimens at 23 ± 2 C (73.4 ± 3.6 F) and 50 ± 5 percent relative humidity for not less than 40 h prior to test in accordance with Procedure A of ASTM Methods D 618, Conditioning Plastics and Electrical Insulating Materials for Testing,[2] for those tests where conditioning is required. In cases of disagreement, the tolerances shall be ±1 C (±1.8 F) and ±2 percent relative humidity.

7.2 *Test Conditions*—Conduct tests in the Standard Laboratory Atmosphere of 23 ± 2 C (73.4 ± 3.6 F) and 50 ± 5 percent relative humidity, unless otherwise specified in the test methods or in this specification. In cases of disagreement, the tolerances shall be ±1 C (±1.8 F) and ±2 percent relative humidity.

7.3 The mechanical properties of many plastics change rapidly with small changes in temperature. Since heat is generated as a result of the rapid flexing of the specimen in fatigue, all tests shall be conducted in still air to ensure uniformity of test conditions. The temperature at the region of the highest stress in the specimen shall be measured and recorded,

[2] *Annual Book of ASTM Standards*, Part 27.

if possible.

8. Procedure

8.1 *Calibration of Dynamometer*—Calibrate the dynamometer occasionally by using a dummy specimen, disconnecting the connecting rod at *F*, Fig. 2, hanging known weights from the wrist pin, *F*, and noting the corresponding reading of the dial, *H*. Plot the load in pounds against the corresponding dial reading. A straight line diagram should result. Calculate the load corresponding to one division of the dial from the slope of the line. This value is the calibration constant of the dynamometer. The calibration constant shall give results within 1 percent of the correct value of the load (Note 13). The dynamometer dial, *H*, shall read zero when a specimen, without grip, *J*, attached, is fastened to the dynamometer at *M*.

NOTE 13—The maximum allowable errors indicated for the measurements and calibration will ensure a maximum relative error, in the calculated stress, of less than 3 percent.

8.2 *Measurements*—Test three or more specimens at each of at least four different stress or strain amplitudes. Choose these amplitudes so as to yield a mean log of cycles to failure (log *N*) of about 4, 5, 6, and 7. Measure the minimum section of each specimen to the nearest 0.03 mm (0.001 in.). (For thin sheet, the relative error of measurement of the thickness of the sheet shall not exceed $^1/_3$ of 1 percent.) Clamp the specimen snugly in the grips, *J*, Fig. 2, and measure the distance *L*, from the center of the wrist pin to the minimum section of the specimen to the nearest 0.5 mm (0.02 in.). Clamp the other end of the specimen in the vise and affix the connecting rod to the wrist pin.

8.3 *Calculation of Test Load for Desired Alternating Stress*—Calculate the load to be applied to the specimen from the desired (known) alternating stress as follows:

$$P = Sbd^2/6L \quad (1)$$

where:

P = load be applied to the specimen, kg or lb,
S = desired alternating stress, or kgf/cm^2 or psi,
b = specimen width, cm or in.,
d = specimen thickness, cm or in., and
L = distance from the wrist pin to the minimum section of the specimen, cm or in.

For the unnotched specimen (Fig. 4(a)):

b = net width of the specimen, and
d = net depth of the minimum section of the specimen.

For the standard V-notched specimen (Fig. 4(b)):

b = net width of the specimen, and
d = thickness of the specimen at the root of the notch.

For the drilled-hole notched specimen (Fig. 4(c)):

b = width of the specimen minus the diameter of the hole, and
d = depth of the specimen.

Determine the reading of the dynamometer dial corresponding to load *P* from the dynamometer constant (8.1).

8.4 *Calculation of Test Load for Desired Alternating Strain*—Calculate the load to be applied to the specimen from the desired (known) alternating strain as follows:

$$P = \epsilon bd^2 E_s/6L \quad (2)$$

where *P*, *b*, *d*, and *L* are as described in 8.3, and ϵ = desired alternating strain in centimeters per centimeter or inches per inch, and E_s = secant modulus of elasticity in flexure in kilograms-force per square centimeter or pounds per square inch. Determine the secant modulus of elasticity in accordance with ASTM Method D 790, Test for Flexural Properties of Plastics.[2] The modulus value at a strain equal to ϵ is used in Eq. 2.

8.5 *Speed of Testing*—Conduct the test at 1720 ± 25 cpm of stress, except when it is desired to determine the effect of speed of testing. Record the speed of testing.

8.6 *Adjustments of Machine*—The initial load indicated shall be the same for both upward and downward deflection and shall be equal to that calculated in 8.3 or 8.4 for the desired stress or strain. To produce an alternating cycle of stress or strain in which the mean value is different from zero, first adjust the variable-throw crank to produce the desired amplitude of the alternating cycle of stress or strain, and then, with the variable-throw crank at mid-stroke, adjust the position of the vise so as to produce a stress or strain equal to the desired mean value of the cycle.

8.7 *Readings*—Read the dynamometer and revolution counter before starting the machine. Record dynamometer readings every 12 to 24 h during the testing of the specimen. After fracture of each specimen measure and record the cross-section at fracture and the distance from the fracture to the wrist pin before the specimen is removed from the grips. If the specimen fractures at a section more than 6.4 mm ($^1/_4$ in.) from the minimum section, mark the data for this specimen void.

8.8 *Determination of Number of Cycles for Fatigue Failure*—Plastics subject to repeated cycles of flexural stress as described herein may fail in one of three ways:

8.8.1 They may fail as a result of the formation of a spreading crack. For unfilled resins, the crack or cracks, when once formed, progress rapidly through the cross-section of the specimen, resulting in separation into two pieces. In such a case, the total number of cycles of the test is reported as the number of cycles for fatigue failure.

8.8.2 However, materials containing fibrous fillers (including wood, cotton, paper, etc.) do not always fail by complete separation under the conditions imposed by this test method. For these materials, the failure occurs as a single crack or more general cracking, which progresses into the material rapidly at first, then more slowly. Complete separation of the specimen into two pieces occurs only for tests at very high stresses. The following arbitrary definition of fatigue failure is used for materials for which a fatigue crack does not always result in complete separation.

Fatigue failure is said to have occurred when the stiffness of the specimen has been reduced by a specified amount as a result of a fatigue crack. Usually the load-deflection curve of a cracked fatigue specimen will not be a straight line owing to the opening and closing of the fatigue crack, and the curve may not have polar symmetry even for completely reversed bending tests. Also relaxation of stress in tests having a mean stress may cause both the maximum and minimum stress to decrease. For these reasons the decrease in minimum load is not a satisfactory measure of the decrease in stiffness.

Load-deflection curves for a cracked and an uncracked specimen in bending are illustrated in Fig. 5. The decrease in slope of the upper portion of the curve for the cracked specimen, compared to the slope of the curve before the specimen was cracked, is equal to the change in stiffness of the specimen.

This change is conveniently determined by measuring the loads at two different deflections, Δ and Δ_{max}, Fig. 5, before the fatigue test has started and after a crack has developed. The deflection Δ_{max} shall be taken as the maximum deflection, and the deflection Δ shall be a reproducible deflection large enough to be above the knee in the curve for the cracked specimen caused by the opening of the crack.

The ratio R_E, of the stiffness of the cracked specimen to the stiffness of the uncracked specimen is then given by the following:

$$R_E = (P'_{max} - P_1')/(P_{max} - P_1)$$

For tests of plastics, fatigue failure shall be defined as having occurred when $R_E = {}^7/_8$. A dynamometer is convenient for determining the values of loads involved in measuring R_E. Electric contacts between the dynamometer arm and connecting rod may be used to stop the test when the stiffness of the specimen is approaching the required value, instead of periodic manual checks.

NOTE 14—When testing materials whose stiffness is sensitive to small changes in temperature, the decrease in stiffness resulting from a rise in temperature due to internal friction in the specimen must be separated from the decrease in stiffness resulting from a fatigue crack.

8.8.3 Plastics may fail by overheating, or may have a failure which appears to be due to progressive fracture but which has been influenced by overheating, as a result of internal friction (damping). The overheating may result in softening of the specimen to a point where separation of the specimen into two pieces will result, as with certain unfilled thermoplastics. With some filled plastics, the over-heating may cause a weakening of the filler material, or of the resin, or both, so that failure may result at a smaller number of cycles than if no heating had occurred.

If tests on such material do not yield a progressive fracture (crack) type of failure unaccompanied by large temperature rise, or if there does not result a decrease of stiffness as described under Item (*2*) unaccompanied by large temperature rise, then this method of test is not applicable to the material being tested. Lower testing temperatures, lower speeds of testing, or provision for a faster removal of heat from the specimen (as by a current of air from a fan), or a combination of these methods, are necessary to produce failure by fatigue in such a case.

9. Plotting and Interpreting Results

9.1 For the following conditions, make no corrections to the value of stress or strain calculated at the minimum section of the specimen: (*1*) for fracture away from the minimum section, (*2*) for the position of the theoretical maximum stress or strain (see Note 12), or (*3*) for a change in dynamometer reading during the test.

9.2 *Plotting Results*—Plot an *S-N* (stress versus cycles-for-failure) or ϵ-*N* (strain versus cycles-for-failure) diagram with the alternating stress or strain amplitude as the ordinate against the common logarithm of the number of cycles required for fracture (see 8.8) as the abscissa. Plot all test data and use the geometric means of the number of cycles to failure (established for each stress or strain level) to define the *S-N* or ϵ-*N* diagram. Draw a smooth curve as closely as possible through the latter points. The use of semilogarithmic paper will facilitate the plotting. Indicate on the diagram specimens that did not fracture by an arrow directed away from the plotted point in the direction of increasing cycles.

9.3 *Interpretation of Results*—When a material shows a knee in the *S-N* or ϵ-*N* diagram such that the curve gives clear indications that it becomes asymptotic to a horizontal (constant stress or strain) line, it is sufficient to carry the number of cycles far enough beyond the knee to indicate with a good degree of accuracy that the curve becomes asymptotic to a constant stress or strain line. If the curve does not become asymptotic to a constant stress or strain line, continue the test until the number of cycles reached is greater than the number of cycles that the material will be expected to withstand in its life. Report the value of the alternating stress or strain amplitude corresponding to this number of cycles as either the "fatigue strength at __ cycles for a mean stress of __ kgf/cm^2 or psi," or the "fatigue strain at __ cycles for a mean strain of __ cm/cm or in./in." (Substitute the maximum number of cycles for which the stress-cycle or strain-cycle diagram has been well defined by the tests.)

10. Precautions Necessary in Applying Results

10.1 The relationship between service and laboratory fatigue tests of materials is so uncertain that generous factors of safety should be used in design except when the relation between service and laboratory tests is known.

10.2 It is necessary to make sure whether it is failure from fatigue that is the prime danger or failure from other causes such as creep, yielding, fracture under a single loading, or others.

10.3 For plastics that are subject to creep at the temperature of testing (this probably includes all plastics at room temperature), a fatigue test, in which the mean stress or strain is not zero, is accompanied by either creep or relaxation, depending on the type of testing machine used. As a result of this relaxation the range of stress tends to approach the condition of complete reversal of stress as time goes on. The magnitude of the effect that relaxation would have on the result obtained will depend upon the rate at which relaxation takes place and upon the time required for the endurance limit to be reached. The rate at which relaxation takes place will depend upon the material and the magnitude of the mean stress of the cycle.

10.4 Complete agreement between the results of tests performed on different types of fatigue machines will often not be found. Factors such as damping, thermal conductivity, specific heats, temperature sensitivity, and unknown factors may cause differences in results of tests of different shapes of specimens and different types of testing machines.

10.5 The nominal stress or strain, resulting from the applied load as determined in 8.3 or 8.4, does not always represent the actual magnitude of the applied stress or strain at the test section of the specimen. The elementary beam formula will not yield precise results for materials whose: (*1*) stress-strain relationship is not linear, (*2*) stress-strain curve in tension is not identical to that in compression, or (*3*) internal damping is large. Most plastics have one or more of these characteristics. No generally satisfactory method of taking these factors into account is yet available.

11. Report

11.1 The report (see Appendix A2) shall include the following:

11.1.1 Description of the material tested, including name, manufacturer, code number, date of manufacture, type of molding, and thickness of specimen or original material in case of thin sheet,

11.1.2 Dates of the test,

11.1.3 Type of testing machine,

11.1.4 Size and type of specimen used and whether notched or unnotched. If notched, the type of notch,

11.1.5 Thickness of the specimen in the case of thin sheets,

11.1.6 Preconditioning used (number of hours at the temperature in degrees Celsius or Fahrenheit and relative humidity),

11.1.7 Temperature of the testing room in degrees Celsius or Fahrenheit and the relative humidity of the testing room,

11.1.8 Temperature of the specimen in degrees Celsius or Fahrenheit when operating under the prevailing test conditions,

11.1.9 Speed (number of cycles per minute) of the testing machine,

11.1.10 Stress ratio, A or R, or strain ratio, A_ϵ or R_ϵ,

11.1.11 Maximum number of cycles used in the test,

11.1.12 Fatigue strength in kilograms-force per square centimeter or pounds per square inch at the number of cycles of stress, and the mean stress, or fatigue strain in millimeters per millimeter or inches per inch at the number of cycles of strain, and the mean strain, and

11.1.13 Mean stress in kilograms-force per square centimeter or pounds per square inch, or mean strain in millimeters per millimeter or inches per inch of the cycle, and whether in tension or compression.

METHOD B—CONSTANT-AMPLITUDE-OF-FORCE TEST

12. Apparatus (Fig. 6)

12.1 *Testing Machine*—A fatigue testing machine of the fixed-cantilever, repeated-constant-force type (see Appendix A1). In this machine the specimen, *A*, shall be held as a cantilever beam in a vise, *B*, at one end, and bent by a concentrated load applied through a yoke, *C*, fastened to the opposite end. The alternating force shall be produced by an unbalanced, variable eccentric, *D*, mounted on a shaft. The shaft shall be rotated at constant speed by a motor.

12.2 *Counter*—A counter, *E*, to record the number of cycles.

12.3 *Cut-off Switch*—A suitable mechanically or electrically operated cut-off switch, *F*, shall be provided to stop the machine when the specimen fractures.

13. Test Specimens

13.1 The test specimens shall conform to one of the two geometries (Type I or Type II) shown in Fig. 7. Selection of a particular specimen type will depend upon specimen thickness and the stress range over which the measurements are to be made. The triangular form of these specimen types provides for uniform stress distribution over their respective test spans. Machining of each specimen shall be accomplished with a very sharp cutting tool, using such combination or speed and feed as will give a good finish with minimum of heating of the specimen. The test specimen shall be polished with successively finer emery paper, finishing with No. 00 to remove all scratches and tool marks. The final polishing shall be lengthwise of the specimen, since even small scratches transverse to the direction of tensile stress tend to lower the fatigue strength. In order to avoid heating, all polishing shall be done either by hand or with light pressure on a slowly revolving sanding drum. Care shall be taken to avoid rounding the edges and corners of the specimen.

14. Possibilities and Limitations of Test

14.1 See Section 6 except that the type of machine covered in this test method is not suitable for determining fatigue strain (6.2).

15. Conditioning

15.1 See Section 7.

16. Procedure

16.1 *Tuning Machine*—Tune machine by using a thin metal specimen having a natural frequency of 1800 ± 4 cpm when vibrating as a free cantilever beam. (Tuning specimens and weights are furnished by the manufacturer of the test apparatus.) Tune by the following procedure.

16.1.1 Set the eccentric, *D*, Fig. 6, very near zero load, that is, at about 0.1 to 0.3 units. Do not change this setting during subsequent tuning runs (tests).

16.1.2 Make several runs of a minute or two each, using different total tuning weight, *G*, Fig. 6, and plot the total tuning weight versus the amplitude of vibration to the nearest 0.5 mm (0.02 in.). At low values of weight the amplitude will be very small, increasing to higher values as proper tuning is reached by the addition of more weights. Further addition of weight will then result in a decrease of amplitude.

16.1.3 Select the proper tuning weight (to the nearest 0.02 percent of the dynamic machine capacity) from the peak of the weight-versus-amplitude curve. This value is a constant for the machine and is called the "complementary weight."

16.2 *Measurements*—Test three or more specimens at each of at least four different stress amplitudes. Choose stress amplitudes so as to yield a mean log of cycles to failure (log *N*) of about 4, 5, 6, and 7. Measure the minimum thickness of the triangular portion of each specimen to the nearest 0.03 mm (0.001 in.). Clamp the specimen snugly in the vise, *B*, Fig. 6, with its smaller end screwed securely to the vibrating yoke, *C*. Measure the test span, *L*, to the nearest 0.5 mm (0.02 in.) from the leading edge of the vise, *B*, along the principal axis of the specimen to the line-of-centers of the mounting screws in the yoke, *C*. Measure the width, *b*, of the specimen, defined by the intersection of the leading edge of the vise with the sides (or projections thereof) of the triangular portion of the specimen, to the nearest 0.3 mm (0.01 in.).

16.3 *Calculation of Effective Weight of Test Specimen*—Calculate the effective weight (Note 15) of the standard specimen. Types I and II (Fig. 7) as follows:

For Type I Specimen:

$$W_e = kDd$$

For Type II Specimen:

$$W_e = k'Dd$$

where:

W_e = effective weight, g or lb,
D = density of the specimen (material), g/cm^3 or lb/in.^{3},
d = average specimen thickness, cm or in.,
k = 2.48 cm^2 (0.385 in.^{2}), and
k' = 3.23 cm^2 (0.50 in.^{2}).

NOTE 15—The "effective weight" of a given specimen is best understood as follows: Consider an actual cantilever specimen and then another "ideal" specimen of the same bending stiffness, but which is weightless along its stressed length. Let a concentrated weight be applied to the free end of the ideal specimen. If this concentrated (lumped) weight produces in this specimen the same transverse inertial force as the total transverse inertial forces associated with a test specimen vibrating at the same amplitude and frequency, then this concentrated weight is equal to the effective weight of the test specimen.

16.4 *Attachment of Tuning Weights*—Determine the proper amount of tuning weights, *G*, Fig. 6, by subtracting the *effective weight* of the test specimen, as determined in 16.3, from the *complementary weight* of the machine, determined in 16.1. Add these tuning weights to the vibrating assembly of the testing machine.

16.5 *Calculation of Test Load*—Calculate the load to be applied to the specimen from the desired (known) alternating stress as follows:

$$P = Sbd^2/6L$$

where:

P = load to be applied to the specimen,
S = desired alternating stress, kgf/cm^2 or psi,
b = specimen width, cm or in.,
d = specimen thickness, cm or in.,
L = test span, cm or in.

The load, *P*, required to produce the desired stress, *S*, in the specimen is the force setting on the eccentric, *D*, Fig. 6.

16.6 *Speed of Testing*—Conduct the test at 1800 ± 4 cpm of stress.

16.7 *Readings*—Set the revolution counter at zero before starting a test. Upon failure of the test specimen, read the counter to determine the number of cycles to failure.

16.8 *Determination of Number of Cycles for Fatigue Failure*—See 8.8.

17. Plotting and Interpreting Results

17.1 *Plotting Results*—Plot an *S-N* or

stress versus cycles-for-failure diagram with the alternating stress amplitude as the ordinate against the common logarithm of the number of cycles required for fracture (see 16.8) as the abscissa. Plot all test data and use the geometric means of the number of cycles to failure (established for each stress level) to define the *S-N* diagram. Draw a smooth curve as closely as possible through the latter points. The use of semilogarithmic paper will facilitate the plotting. Indicate on the diagram specimens that did not fracture by an arrow directed away from the plotted point in the direction of increasing cycles.

17.2 *Interpretation of Results*—When a material shows a knee in the *S-N* or stress-cycle diagram such that the curve gives clear indications that it becomes asymptotic to a horizontal (constant stress) line, it is sufficient to carry the number of cycles far enough beyond the knee to indicate with a good degree of accuracy that the curve becomes asymptotic to a constant stress line. If the curve does not become asymptotic to a constant stress line, continue the test until the number of cycles reached is greater than the number of cycles that the material will be expected to withstand in its life. Report the value of the alternating stress amplitude corresponding to this number of cycles as the "fatigue strength at __ cycles for a mean stress of __ kgf per cm² or psi." (Substitute the maximum number of cycles for which the stress-cycle diagram has been well defined by the tests.)

18. Precautions Necessary in Applying Results

18.1 See Section 10 except for references to nominal strain (10.5), which are not applicable.

19. Report

19.1 See Section 11 except for references to strain, which are not applicable.

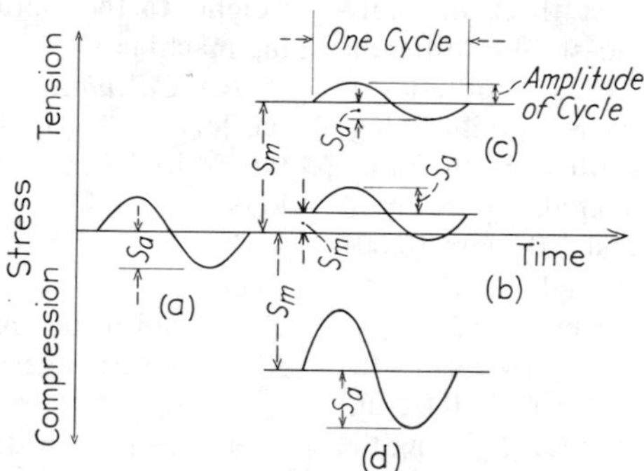

FIG. 1 Stress-Time Curves for Different Values Mean Stress.

FIG. 2 Fixed-Cantilever, Repeated-Constant-Deflection Type Fatigue Testing Machine.

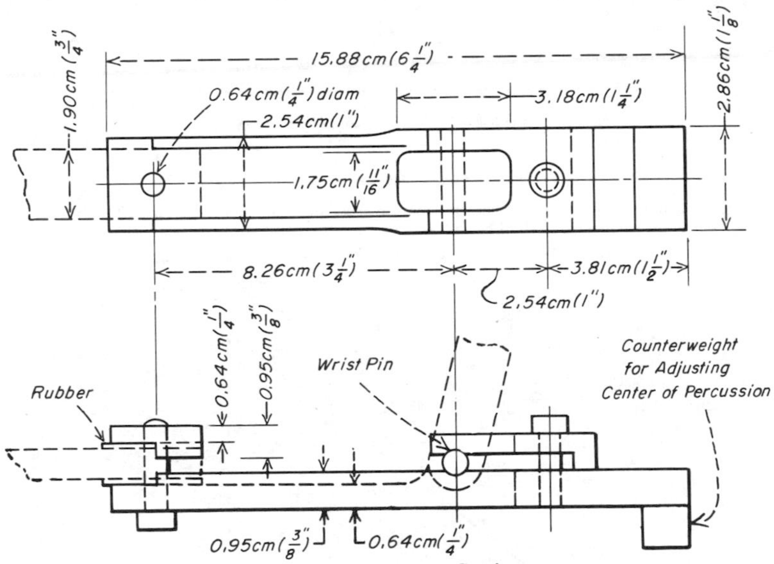

FIG. 3 Holder for Test Specimen.

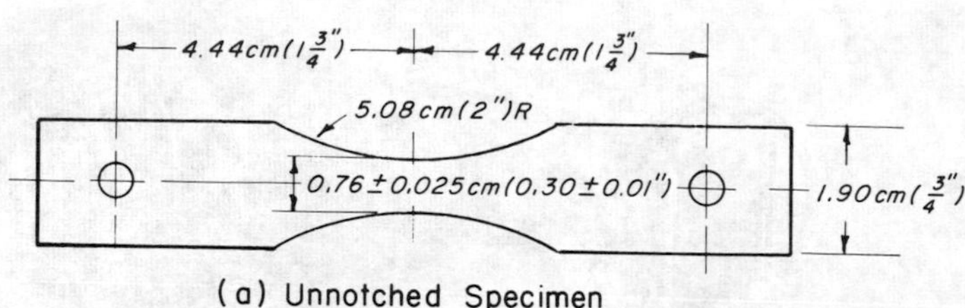

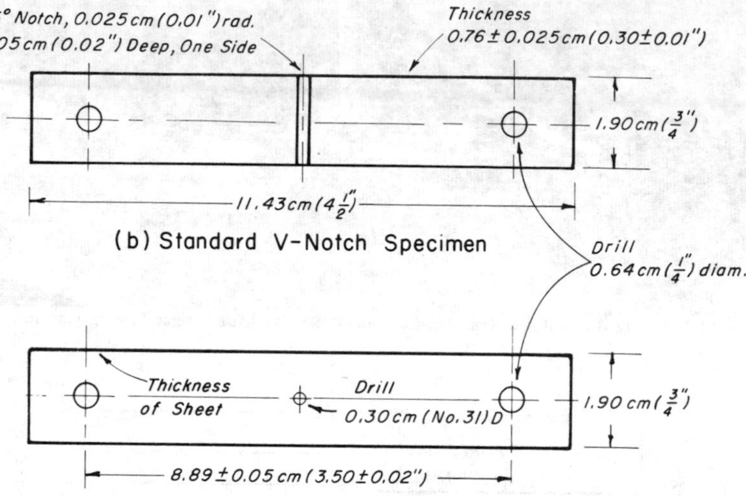

(c) Drilled Hole Specimen

NOTE—In the case of unnotched specimens the thickness of the specimen shall be 7.6 ± 0.25 mm (0.3 ± 0.01 in.) except unnotched specimens for sheets which shall be the thickness of the sheet.

FIG. 4 Dimensions of Constant-Deflection Fatigue Specimens.

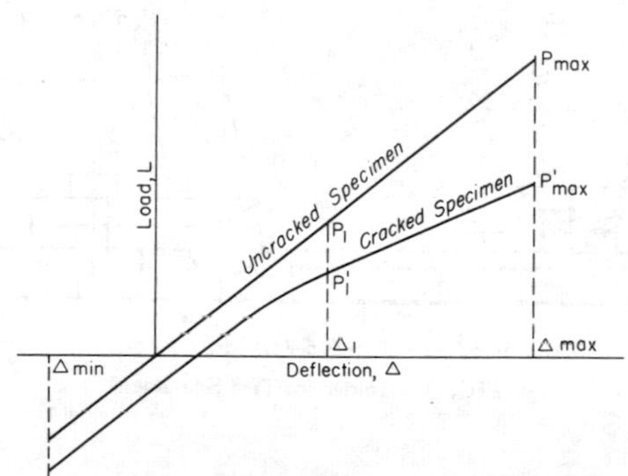

FIG. 5 Load - Deflection Diagram for an Uncracked and a Cracked Specimen in Bending.

GENERAL VIEW

INTERIOR VIEW

FIG. 6 Fixed-Cantilever, Repeated-Constant-Load Type Fatigue Testing Machine.

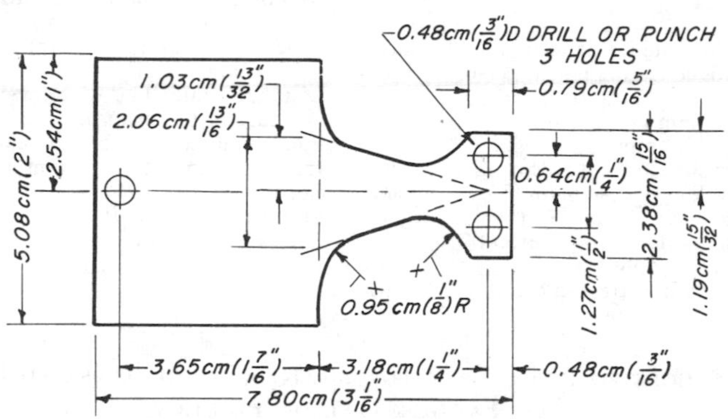

TYPE I

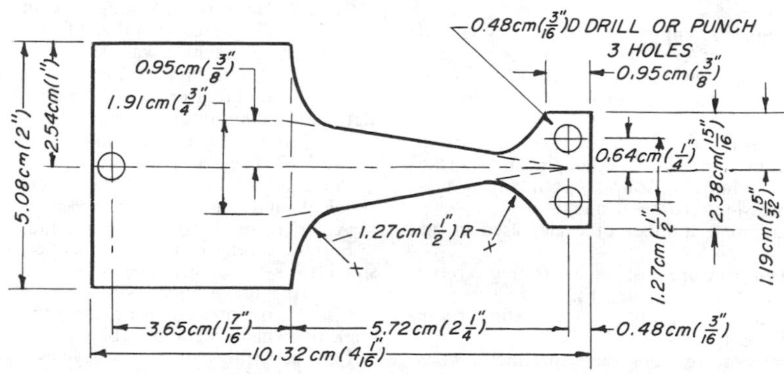

TYPE II

FIG. 7 Dimensions of Constant-Force Fatigue Specimens.

APPENDIX

A1. Classification of Fatigue Testing Machines

A1.1 Direct Stress (Tension of compression, or both)

A1.1.1 Repeated axial loading by alternating magnetic field or by inertia vibration.

A1.1.2 Repeated axial deformation by means of connecting rod or cam.

A1.2 Flexural Stress

A1.2.1 Repeated loading by:

A1.2.1.1 Rotating cantilever loaded by weight or spring.

A1.2.1.2 Fixed cantilever loaded by magnetic or inertia vibrator or by rotating spring.

A1.2.1.3 Rotating beam in pure bending loaded by spring or weights.

A1.2.2 Repeated deflection by:

A1.2.2.1 Fixed cantilever, repeated constant-deflection.

A1.3 Torsional Stress

A1.3.1 Repeated angular torque by:

A1.3.1.1 Inertia vibrator.

A1.3.2 Repeated angular twist by:

A1.3.2.1 Cam or connecting rod action on a torque arm.

A2. Suggested Form of Data Sheet for Repeated Flexural Fatigue Test of Plastics

Fatigue of Plastics

Material ....
Date ....
Test made by ....
Fixed-cantilever, repeated (-deflection ( ), -force ( )) type fatigue testing machine No. ....
Specimen:
Width, b ...... mm (...... in.)
Thickness, d ...... mm (...... in.)
Mean stress of cycle ...... kgf/cm^2 (...... psi)
Mean strain of cycle ...... mm/m (...... in./in.)
Room temperature ...... deg C (...... deg F)
Room relative humidity ...... percent
Speed of testing ...... cpm
Amplitude of alternating stress in cycle ...... kgf/cm^2 (...... psi)
Amplitude of alternating strain in cycle ...... cm/cm (...... in./in.)
Load ...... kg (...... lb) or ...... divisions on dynamometer
Distance center line of connecting rod pin to fracture or test span, L ...... cm (...... in.)
Final reading of revolution counter ....
Initial reading of revolution counter ....
Stress ratio, A or R ....
Strain ratio, A_ϵ or R_ϵ ....
Cycles for fracture ....
Remarks ....
....

A3. Suggested Form for Reporting Results of Repeated Flexural Fatigue Test of Plastics

A3.1 For evaluation of the fatigue strength of a material, the report shall contain the following information:

A3.1.1 Fatigue strength at ...... cycles of stress was ...... kgf/cm^2 or ...... psi for a mean stress of ...... kgf/cm^2 or ...... psi in tension or compression (cross out one), or (whichever applicable).

Fatigue strain at ...... cycles of strain was ...... mm/mm or ...... in./in. for a mean strain of ...... mm/mm or ...... in./in. in tension or compression (cross out one).

A3.1.2 Maximum number of cycles used in the test was ...... cycles.

A3.1.3 Temperature of the testing room was ...... deg C (...... deg F).

A3.1.4 Relative humidity of the testing room was ...... percent.

A3.1.5 Temperature of the specimen when operating at the fatigue strength was ...... deg C (...... deg F).

A3.1.6 Conditioning used was ...... h at ...... deg C (...... deg F) and ...... percent relative humidity.

A3.1.7 The type of testing machine was ....

A3.1.8 The speed (number of cycles per minute) of the testing machine was ....

A3.1.9 Description of material tested is: name ......, manufacturer ......, code number ......, date of manufacture ......, type of molding ......, and thickness of sample ......

A3.1.10 The dates of test were ....

A3.2 For evaluation of the fatigue strength of a notched material, the report shall contain Items A3.1.2 through A3.1.10 above, and:

Fatigue strength at ...... cycles of stress of a standard V-notch specimen was ...... kgf/cm^2 or ...... psi for a mean stress of ...... kgf/cm^2 or ...... psi in tension or compression (cross out one), or (whichever applicable).

Fatigue strain at ...... cycles of strain of a standard V-notch specimen was ...... mm/mm or ...... in./in. for a mean strain of ...... mm/mm or ...... in./in. in tension or compression (cross out one).

A3.3 For evaluation of the fatigue strength of thin sheet (Note 11, the report shall contain Items A3.1.2 through A3.1.10 above, and:

Fatigue strength at ...... cycles of stress of a sheet of ...... mm or ...... in. thickness was ...... kgf/cm^2 or ...... psi for a mean stress of ...... kgf/cm^2 or ...... psi in tension or compression (cross out one), or (whichever applicable).

Fatigue strain at ...... cycles of strain of a sheet of ...... mm or ...... in. thickness was ...... mm/mm or ...... in./in. for a mean strain of ...... mm/mm or ...... in./in. in tension or compression (cross out one).

A3.4 For evaluation of the fatigue strength of notched thin sheet, the report shall contain Items A3.1.2 through A3.1.10 above, and:

Fatigue strength at ...... cycles of stress of a sheet of ...... mm or ...... in. thickness with a drilled-hole notch was ...... kgf/cm^2 or ...... psi for a mean stress of ...... kgf/cm^2 or ...... psi in tension or compression (cross out one), or (whichever applicable).

Fatigue strain at ...... cycles of strain of a sheet of ...... -mm or ...... -in. thickness with a drilled-hole notch was ...... mm/mm or ...... in./in. for a mean strain of ...... mm/mm or ...... in./in. in tension or compression (cross out one).

Designation: D 673 – 44 (Reapproved 1969)

Standard Method of Test for

MAR RESISTANCE OF PLASTICS[1]

This Standard is issued under the fixed designation D 673; the number immediately following the designation indicates the year of original adoption or, in the case of revision, the year of last revision. A number in parentheses indicates the year of last reapproval.

NOTE—Section 5 was added editorially in December 1968.

1. Scope

1.1 This method covers the measurement of the resistance of glossy surfaces to abrasive action. Since measurement of progressive optical deterioration of a glossy surface is involved, such materials as molded and laminated plastics, paint and lacquer films, and plated coatings may be evaluated by this method. The test consists essentially of two steps as follows:

1.1.1 The production of a series of abraded spots on the surface of the test specimen, made by dropping increasing amounts of abrasive through a tube of fixed length, and

1.1.2 The measurement of the gloss of these abraded spots by an optical method and comparison with the original gloss.

NOTE 1—Since this method is intended for rating relative decrease in gloss, obviously only materials having originally high specular reflectance can be measured.

NOTE 2—The values stated in U.S. customary units are to be regarded as the standard. The metric equivalents of U.S. customary units may be approximate.

2. Apparatus[2]

2.1 *Abrader*—An abrader, shown in Fig. 1, consisting of the following:

2.1.1 *Tube*—A glass, metal, or plastic tube (*A*, Fig. 1) supported vertically through which the abrasive is dropped on a test specimen that is supported at an angle of 45 deg to the axis of the tube. The tube shall conform to the dimensions shown in Fig. 2.

2.1.2 *Hopper*—A hopper (*B*, Fig. 1) to distribute the feed of the abrasive in order to produce on the specimen an abraded spot nearly uniform in appearance over an area about 20 mm ($^3/_4$ in.) in diameter. The hopper shall be rotated at about 7 rpm and the abrasive fed from the hopper at a rate of about 200 to 250 g/min. The details of construction of the hopper are shown in Fig. 2. It may be formed from thin sheet metal with soldered seams. It shall have a double conical bottom with six holes 1.8 mm (0.070 in.) in diameter arranged in a circle 16 mm ($^5/_8$ in.) in diameter at the bottom edge, and one hole 1.8 mm (0.070 in.) in diameter at the apex of the middle cone. The inside of the hopper should be smooth and highly polished.

2.1.3 *Ball Bearing*—A ball bearing (*C*, Figs. 1 and 2) 57 mm (2 $^1/_4$ in.) in outside diameter and 31 mm (1 $^1/_4$ in.) in inside diameter, with sealed balls. The outside ring of the bearing shall be rigidly held in position.

2.1.4 *Brass or Copper Tube*—A brass or copper tube (see Fig. 2) about 38 mm (1 $^1/_2$ in.) in length fitted tightly into the bearing. The inside diameter shall be slightly over 25 mm (1 in.) to permit the top of glass tube *A* to extend up into it just below the lower edge of the hopper. To the tube shall be soldered four supports for holding the hopper in position. The projecting portion of the tube serves as a pulley to rotate the hopper assembly.

2.1.5 *Electric Motor*—A small electric motor[3] with built-in reducing gears for rotating the hopper assembly at a speed of about 7 rpm.

---

[1] This method is under the jurisdiction of the ASTM Committee D-20 on Plastics. A list of committee members may be found in the ASTM Yearbook. This standard is the direct responsibility of Subcommittee D-20.10 on Mechanical Properties.

Current edition effective Sept. 15, 1944. Originally issued 1942. Replaces D 673 – 42 T.

[2] Complete working drawings for this apparatus are obtainable at a nominal charge from the Headquarters of the American Society for Testing and Materials, 1916 Race St., Philadelphia, Pa. 19103.

[3] A suitable electric motor is No. 952W/11 rpm of the Speedway Mfg. Co., 1834 South 52nd Ave., Cicero, Ill.

2.1.6 *Receptacle for Abrasive*—A receptacle (*F*, Fig. 1) about 200 by 200 by 200 mm (8 by 8 by 8 in.) to receive the abrasive. A 5-gal can cut down to a height of 200 mm (8 in.) will be satisfactory. The receptacle shall be slotted on two sides at an angle of 45 deg as shown in Fig. 1 to permit sliding sheet specimens past the bottom of the glass tube.

2.1.7 *Support for Test Specimen*—A plywood board (*E*, Fig. 1), with spring clips to hold the test specimen, mounted in the receptacle at an angle of 45 deg to the axis of the tube and directly under the tube.

2.1.8 *Abrasive*—No. 80 carborundum or an equivalent abrasive.

NOTE 3—It has been found that continued use of the same batch of abrasive gradually improves the values of abrasion resistance of a given material, possible due to the breaking up of particles of the abrasive or by dulling their sharp cutting edges. An arbitrary limit of 50 tests is recommended for a given lot of abrasive after which it should be discarded. It is also advisable to keep the abrasive well screened through a No. 60 (250-μm) sieve to remove extraneous matter such as lint, hairs, wood splinters, etc., and on a No. 120 (125-μm) sieve to remove the fine, broken-down particles and dust.

2.2 *Glossmeter*—A light source and simplified form of glossmeter using a photoelectric cell as the active, element and a galvanometer, as shown in Fig. 3, conforming to the following requirements:

2.2.1 *Light Projector*—A model "M" SVE[4] projector to act as a light source. An extension tube made to permit the lens to form an image at a short distance shall be used. The size of the test spot used is determined by several circular holes ranging from 3 to 9 mm ($^1/_8$ to $^3/_8$ in.) in diameter punched in a piece of blackened 35-mm film loaded into the projector. A diaphragm may be added to the front of the projector lens to regulate the quantity of light independently of the size of the test spot. This may be a microscope condenser diaphragm with a 31-mm (1 $^1/_4$-in.) opening for fitting over the projector lens, or a series of caps with holes of various diameters in steps of $\sqrt{2}$ may be used. The projector may be permanently mounted at an angle of 45 deg and so that the distance from the lens surface to the perforated film is about 12 in. A voltage regulator on the line leading to the projector is desirable but not essential if the voltage fluctuations are not of too great an amplitude or if they are not of short duration.

2.2.2 *Receptor System*—A receptor system consisting of a lens, photoelectric cell, and galvanometer. A lens about 31 to 38 mm (1 $^1/_4$ to 1 $^1/_2$ in.) in diameter having a 100 to 125-mm (4 to 5-in.) focal length (a duplicate of the projector lens may be used) shall be used to pick up the reflected beam and converge it sufficiently to impinge on the sensitive surface of the photoelectric cell. The cell may be of any type having a linear current intensity relation.[5] The galvanometer[5,6] shall be used for indicating the current flow in the receptor circuit. The receptor system may be fastened to a hinged arm with its center in the plane of the test specimen and in line with the center of the illuminated spot, and shall have freedom of motion from the specular position at an angle of 45 deg to the plane of the specimen to the off-specular position at an angle of 60 deg to the plane of the specimen. When the glossmeter is assembled and a mirror placed in the position to be occupied by a test specimen, the circular incident light beam from the projector should be seen entering the receptor lens at the specular angle of 45 deg. When the receptor is tilted upward 15 deg to the off-specular angle of 60 deg, the specular beam shall just miss the receptor lens. The active surface of the photoelectric cell shall be masked to permit only the specularly reflected image of the spot to impinge on the cell.

NOTE 4—The glossmeter described may be used to measure gloss on any type of surface, transparent or opaque. For testing of transparent sheets, some form of hazemeter or densitometer may also be used.

3. Test Specimen

3.1 The test specimen shall consist of any material with a flat surface at least 50 by 50 mm (2 by 2 in.) in area for each determina-

[4] The Society for Visual Education, 10 E. Ohio St., Chicago, Ill.

[5] Any barrier layer cell having an active surface about 25 mm (1 in.) in diameter may be used. A suitable cell is the No. 732 Electrocell of Photovolt Corp., 95 Madison Ave., New York, N. Y., and other suitable cells are manufactured by the General Electric Co., Schenectady, N. Y., and the Westinghouse Electric and Mfg. Co., East Pittsburgh, Pa. The galvanometer may be the General Electric No. 32C 249G25 with built-in shunts with ranges of 10-5-2-1 or a similar instrument having an internal resistance of 300 Ω or less.

[6] Other suitable instruments that include both photoelectric cell and galvanometer are the Photrix Universal Photometer, model A, the Photovolt Corp.; and the Weston model 603 Illumination Meter with single No. 594 cell, the Weston Electrical Instrument Corp., Newark, N. J.

tion. A sheet 50 by 250 mm (2 by 10 in.) will be sufficient for five tests.

4. Procedure

4.1 Make five abraded spots on each test specimen (Note 5). Produce each succeeding spot by using a larger quantity of abrasive than that used for the preceding one. The following amounts of abrasive are recommended:

Spot	Abrasive, g
No. 1	200
No. 2	400
No. 3	800
No. 4	1200
No. 5	1600

NOTE 5—For a quick test at a mild degree of abrasive action, the results obtained using 200 or 400 g of abrasive may give sufficient information.

4.2 *Abrading Test Specimens*—Clip the test specimen to the supporting board, start the hopper rotating, and pour the proper amount of abrasive into the hopper. When all of the abrasive has passed through the glass tube, move the specimen to another position, and pour the next succeeding amount of abrasive into the hopper to produce another abraded spot. In this manner, produce five abraded spots, each by the respective amount of abrasive. It is recommended that a mild air blast be used to remove excess abrasive adhering to the surface of the specimen after the abraded spots have been produced.

4.3 *Measurement of Gloss*—Measure the gloss of the abraded test spots by means of the glossmeter and compare with the original gloss of the test specimen. The principle of operation of the glossmeter in measuring the gloss is based upon the following optical behavior of plane surfaces as shown in Fig. 4. A beam of light incident upon a mirror at an angle of 45 deg is reflected specularly at 45 deg; its polar distribution will have the shape of curve $AA'A_2$. If the mirror surface is disturbed by abrasion, progressively more light will be scattered on either side of the specular beam, and the light-distribution curve assumes the shape $BB'B_2$, approaching the uniform distribution shown by curve $CC'C_2$, which is characteristic of a nearly perfect diffusing surface such as magnesium carbonate. If a lens is placed in the path of the reflected beam and the image of the illuminated spot focused on a photoelectric cell and readings of the intensity taken at the specular angle I_1 and at an angle I_2, 15 deg from the specular angle, then the ratio of the readings is a measure of the sharpness of the peak of the light-distribution curve or a measure of the gloss. The angle of 15 deg off specular is selected because with the prescribed glossmeter the specular beam from a mirror surface just misses the opening of the collecting lens.

NOTE 6—For strict comparisons of mar resistance characteristics, materials of nearly similar reflectance values should be used, for example, a comparison of two types of resin coatings in which one was an opaque white and another an opaque black would not be as valid as if both materials had the same general color.

5. Conditioning

5.1 *Conditioning*—Condition the test specimens at 23 ± 2 C (73.4 ± 3.6 F) and 50 ± 5 percent relative humidity for not less than 40 h prior to test in accordance with Procedure A of ASTM Methods D 618, Conditioning Plastics and Electrical Insulating Materials for Testing,[7] for those tests where conditioning is required. In cases of disagreement, the tolerances shall be ±1 C (±1.8 F) and ±2 percent relative humidity.

5.2 *Test Conditions*—Conduct tests in the Standard Laboratory Atmosphere of 23 ± 2 C (73.4 ± 3.6 F) and 50 ± 5 percent relative humidity, unless otherwise specified in the test methods. In cases of disagreements, the tolerances shall be ±1 C (±1.8 F) and ±2 percent relative humidity.

6. Calculating Percentage Gloss

6.1 To avoid the complication of variability of photoelectric cell readings from surfaces of various original reflectivity and to minimize the effect of voltage variation on the light source, only the ratio of light intensity reflected at the two observation angles shall be used in the following equation for calculating percentage gloss:

$$\text{Gloss, percent} = 100\,[(I_1 - I_2)/I_1]$$

where:

I_1 = photoelectric cell reading at the specular angle, and

I_2 = photoelectric cell reading at the 60-deg angle (15 deg off specular).

(I_1/I_2 is a measure of the sharpness of the peak of light distribution, or a measure of the

[7] *Annual Book of ASTM Standards*, Part 27.

gloss.) From this equation, a reading of zero at the off-specular angle gives 100 percent gloss, and equal readings at specular and off-specular angles give 0 percent gloss.

6.2 In comparing abrasion characteristics of a series of materials that do not have the same original gloss, the initial gloss value of the unabraded surface shall be taken as 100 percent, and the progressive deterioration expressed in corresponding terms.

NOTE 7—Typical data are shown in the following table:

Abrasive, g	Readings at: 45 Deg, I_1	60 Deg, I_2	Gloss, percent	Percentage of Original Gloss
0	93.5	2.0	97.9	100
200	60.0	4.0	93.3	95.4
400	43.5	5.0	88.5	90.4
800	33.0	7.0	81.8	83.5
1200	25.0	6.0	74.0	75.5
1600	19.0	6.0	68.4	70.0

7. Plotting Results

7.1 The percentage gloss of the abraded spots shall be plotted against the respective amounts of abrasive used, to obtain a characteristic curve. Since such curves for different materials are often found to change slope irregularly, they may cross each other, and the rating of a series of different materials using a given amount of abrasive may not be representative of their relation at other amounts of abrasive. One way to arrive at an over-all performance is to average the percentage original gloss at the various amounts of abrasive, which would represent the area included between the curve and the coordinate representing the amount of abrasive (see Note 5).

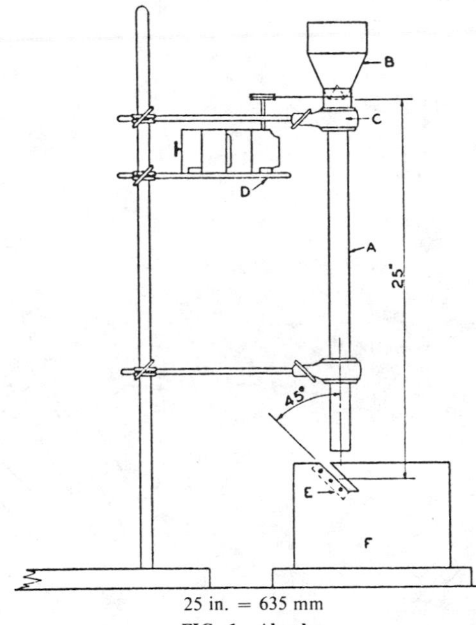

25 in. = 635 mm

FIG. 1 Abrader.

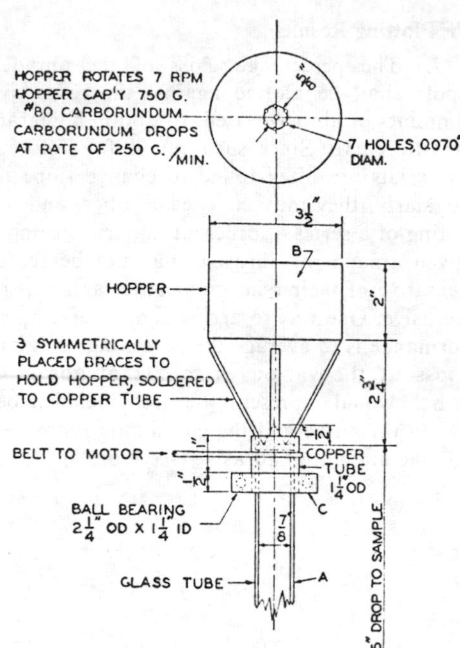

Table of Metric Equivalents

in.	mm
0.070	1.8
$^1/_2$	13
$^5/_8$	16
$^7/_8$	22
1	25
1 $^1/_4$	31
2	51
2 $^1/_4$	57
2 $^3/_4$	70
3 $^1/_2$	88
25	635

FIG. 2 Details of Hopper Assembly.

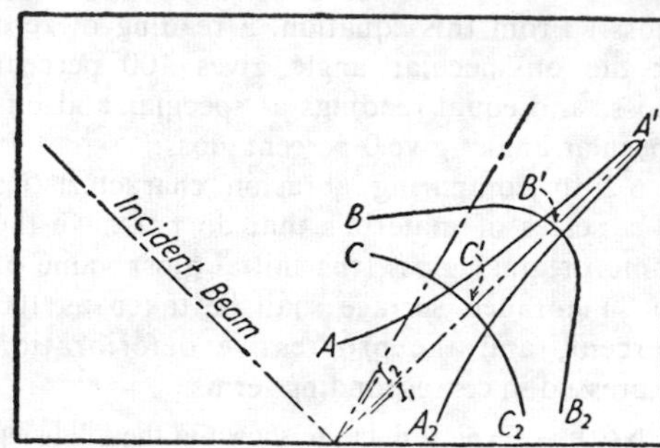

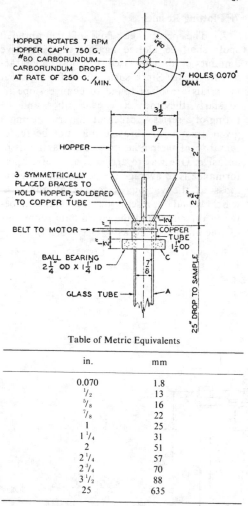

FIG. 4 Distribution of Reflected Light.

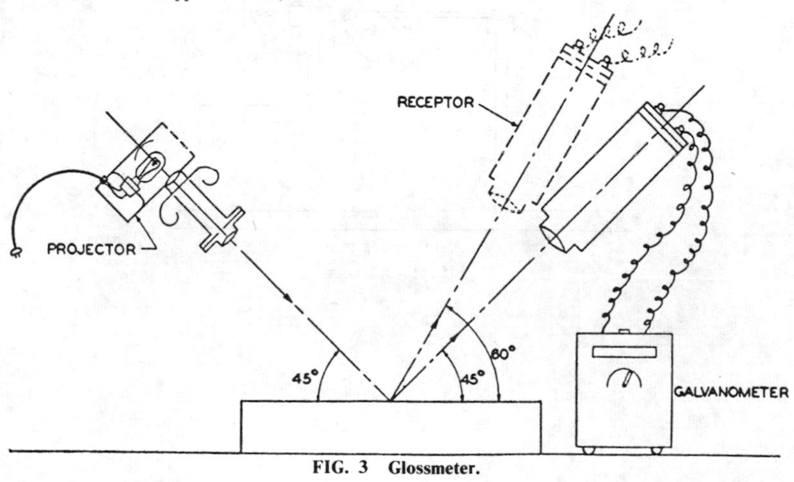

FIG. 3 Glossmeter.

 **Designation: D 674 – 56 (Reapproved 1969)**

Recommended Practices for

TESTING LONG-TIME CREEP AND STRESS-RELAXATION OF PLASTICS UNDER TENSION OR COMPRESSION LOADS AT VARIOUS TEMPERATURES[1]

These Recommended Practices are issued under the fixed designation D 674; the number immediately following the designation indicates the year of original adoption or, in the case of revision, the year of last revision. A number in parentheses indicates the year of last reapproval.

NOTE—Editorial change in title was made in December 1964.

1. Scope

1.1 These recommended practices describe procedures for determining the time dependence of the deformation and strength of plastics specimens resisting long-duration constant tension or compression loads, under conditions of constant temperature and relative humidity, and with negligible vibration.

1.2 They also describe a recommended practice for determining the time-dependence of stress (stress relaxation) of plastics resisting long-duration constant tension of compression strains at conditions of constant temperature and relative humidity and negligible vibration.

1.3 The subject is presented as recommended practices for guidance rather than a method for specification because the extremely time-consuming nature of the test makes it generally unsuited for routine testing or for specification in purchase of material. This test is therefore confined largely to research testing where standardization is generally undesirable because it tends to retard development of improved methods.

NOTE 1—The values stated in U.S. customary units are to be regarded as the standard. The metric equivalents of U.S. customary units may be approximate.

2. Significance

2.1 Data from creep tests are of considerable importance in predicting the strength of materials for resisting loads continuously applied for long times, and in predicting dimensional changes which may occur as a result of long-continued constant loads.

2.2 Data from stress-relaxation tests are useful in predicting the reduction of stress in materials subjected to constant deformation for long times, such as the decrease in tightness of bolted or riveted joints.

2.3 Both tests are highly sensitive to small changes in material composition and environmental conditions and hence are useful in evaluating the effect of such changes. However, the sensitivity requires such precise testing technique as to restrict its use to research purposes.

2.4 The reproducibility of the data is good when precise control is maintained over all testing conditions and material composition, including moisture content and effects resulting from aging.

3. Definitions

3.1 *creep*—the time-dependent part of the strain which results from the application of a constant stress (Note 2) to a solid at a constant temperature. That is, the creep at a given elapsed time is equal to the total strain at the given time minus the instantaneous strain (Note 3) on loading. The creep extension is usually expressed as a percentage of

[1] These recommended practices are under the jurisdiction of ASTM Committee D-20 on Plastics. A list of committee members may be found in the ASTM Yearbook. This recommended practice is the direct responsibility of Subcommittee D-20.10 on Mechanical Properties.

Current edition effective Sept. 10, 1956. Originally issued 1942. Replaces D 674 – 51.

the initial unstretched length.

NOTE 2—While constant-stress creep tests are desirable, the usual one is a constant load test. Where the total strain is small the difference is negligible, but for large strains corrections must be made in results of constant load tests, or the fact that the value reported is for constant load tests should be noted.

NOTE 3—Since time-dependent strain or stress occurs rapidly even during the application of the load, the instantaneous strain (or recovery or stress) is usually difficult to determine except in the arbitrary manner given in 3.2, 3.3, and 3.10.

3.2 *instantaneous strain*—the strain occurring immediately upon loading a creep specimen. Since it is nearly impossible to obtain strain readings at the instant of loading, the strain after a given small increment of time (such as 1 min) after loading is a more reproducible value. (The increment of time should be specified.)

3.3 *instantaneous recovery*—the decrease in strain occurring immediately upon unloading a specimen. As in instantaneous strain (3.2) a more reproducible value is obtained if the decrease in strain is measured after a given small increment of time (such as 1 min) following unloading. (The increment of time should be specified.)

3.4 *recovery*—the time-dependent portion of the decrease in strain following unloading of a specimen at the same constant temperature as the initial test. Recovery is equal to the total decrease in strain minus the instantaneous recovery (Note 3).

3.5 *time-for-fracture*—the duration of time in which a specimen of a material will withstand a given constant stress (Note 3), constant temperature, and relative humidity until fracture.

3.6 *creep strength*—the maximum stress which may be applied continuously for a specified time without causing fracture. The duration of time and temperature should always be stated.

3.7 *rate-of-creep*—the time rate at which straining occurs at a specified time during a creep test at a given stress and temperature. The stress and time (Note 4) and temperature should always be stated. The rate-of-creep may be determined as the slope of a strain-time curve plotted from creep test data.

NOTE 4—For materials having a creep curve which shows an inflection point the creep rate at the inflection point is sometimes used and is called the minimum creep rate.

3.8 *creepocity*—the percentage increase in strain during a creep test from one specified time (such as 1 min) to another specified time (such as 100,000 min). This quantity is a measure of the time-sensitivity of the material under creep at the given temperature. It is often independent of the stress at which the test was run. If the value reported is from tests at a single stress, the value of the stress should be stated. The times involved and temperature should always be stated.

3.9 *stress-relaxation*—the time-dependent change in the stress which results from the application of a constant total strain to a specimen at a constant temperature. The stress-relaxation at a given elapsed time is equal to the instantaneous stress (Note 3) resulting when the strain is applied minus the stress at the given time.

3.10 *instantaneous stress*—the stress occurring immediately upon straining a specimen in a relaxation test. Since it is nearly impossible to obtain stress readings at the instant of straining, the stress after a given small increment of time (such as 1 min) after straining is a more reproducible value. (The increment in time should be stated.)

4. Loading Device for Creep

4.1 For creep tests a means for applying a constant stress is desirable but a means for applying a constant load to the specimen is more usual. Direct application of dead weights or weights applied to a lever acting on the specimen is suitable if the laboratory is free of shock and vibration. The load on the specimen should be known within 1 percent.

5. Load Application and Measurement for Relaxation

5.1 For relaxation tests a suitable device for applying the load is a screw, driven by a worm-wheel and worm, whose action is controlled automatically by the strain in the specimen. A spring dynamometer may be used to weigh the load during the relaxation test. This dynamometer with its indicator should permit measurement of load within 1 percent, and the zero should not drift with time.

6. Grips

6.1 The grips and gripping technique should be designed to minimize eccentric loading of the specimen. To this end it is desirable to provide a swivel or universal joint near each end of the specimen wherever possible.

7. Temperature Control and Measurement

7.1 The test space (controlled temperature room, furnace or cold box) should be maintained at a constant temperature by a suitable automatic device. This is the most important single factor in a creep test, so that the closest possible control should be maintained. Where a furnace or cold box is used it is desirable to locate this unit in a constant-temperature room to aid in temperature control.

7.2 The determination of the temperature of the specimen during test is the most important single measurement in connection with creep testing, because small variations in temperature may produce large changes in creep rate.

7.3 Care must be taken to ensure accurate temperature measurements over the gage length of the specimen throughout the test.

7.4 Care should be taken to obtain reliable and accurate master temperature standards and to have these checked at suitable intervals. When thermocouples are used in the test, these should be checked by suitable means both before and after long-time tests.

7.5 Reliable temperature measurements should be made at sufficiently frequent intervals to ensure an accurate determination of the average test temperature and compliance with 7.6. At the conclusion of the test there should be at least 50 such observations of temperature, the average of which should be reported as the actual test temperature.

7.6 The maximum fluctuation of temperature throughout the duration of the test period should be kept as small as possible. The extent of this fluctuation should be stated in reporting the tests. The breadth of the normal control cycle should be made a matter of record for the apparatus.

7.7 The distribution of temperature over the gage length of the specimen should be kept to a minimum. For specimens having a gage length of less than 125 mm (5 in.) the maximum variation of temperature over the gage length can usually be confined to:

±0.5 C (±1 F) for tests between −45 to 150 C (−50 to +300 F),
±1 C (±2 F) for tests between 150 to 425 C (+300 to +800 F), and
±1.5 C (±3 F) for tests between 425 to 650 C (+800 to +1200 F).

The extent of variation in temperature should be reported for each test.

8. Humidity Control and Measurement

8.1 Since moisture (water) is a constituent of most plastics and affects the creep behavior, it is important to control the moisture content of specimens during the test. Procedures used for obtaining this objective are to condition the specimens as described in Section 13, so that their vapor pressure is in equilibrium with that of the test atmosphere, and then to maintain the relative humidity of the test atmosphere constant during the test.

8.2 The relative humidity should be controlled as closely as possible, 50 ± 2 percent is usual. The controlling and measuring instruments should be stable for long-time intervals and accurate within ±1 percent. Control of relative humidity is known to be difficult at temperatures much outside the range 10 to 40 C (50 to 100 F).

9. Vibration Control

9.1 Since creep tests especially are quite sensitive to shock and vibration, the location of the testing apparatus should be selected for a minimum of disturbance. When the possible locations are not free of vibration the test equipment and mounting should be designed so that the specimen is isolated from the vibration.

9.2 When the location is not free of shock or occasional vibration, an instrument for continuous recording of vibration should be used to permit differentiating between disturbances caused by humidity or temperature deviation, and shock or vibration.

10. Strain Measurement

10.1 The accuracy and sensitivity of extension-measuring equipment for determining the strain should be suitable for the purposes for which the materials under test are likely to be applied, and for the amount of creep to be expected. The sensitivity should never be less than 0.001 mm/mm (0.001 in./in.) and then only for tests wherein the expected creep will exceed 50 percent. In many instances sensitivity as high as one-millionth of an inch per inch is required.

10.2 In order to obtain sufficient sensitivity the gage length of the specimen should be chosen as long as consistent with other requirements, such as uniformity of temperature distribution along the gage length. It is always desirable and often necessary to use an averaging type extensometer or two extensometers acting on opposite faces of the specimen. The extensometers should be built of materials which are dimensionally stable at the test temperatures to be encountered. For example, the possibility of creep of the adhesive in the use of bonded wire strain gages makes their application for creep or relaxation tests questionable. It is also highly desirable that the extensometer be compensated for thermal expansion by proper selection of materials; so that the extensometer does not expand at all with temperature and hence indicates the total expansion or contraction of the specimen; or so that the thermal expansion of the extensometer is equal to that of the specimen, in which case the strain readings will be independent of any thermal changes. Extensometers which impose any appreciable elastic or frictional restraint on the extension of the specimen are undesirable.

10.3 The extensometers should be calibrated against a precision micrometer screw or other suitable standard under conditions as nearly identical with those encountered in service as possible.

11. Test Specimens and Preparation

11.1 Test specimens may have either circular, square or rectangular cross-sections and may be made by casting, injection or compression molding, or machining from sheets, plates, slabs or similar material. The shape specimen given in ASTM Method D 638, Test for Tensile Properties of Plastics[2] may be employed. The specimen should have a cross-sectional area which is uniform to within ±0.5 percent throughout the gage length. For round specimens, diameters of 12.82 mm, 9.06 mm, or 6.40 mm (0.505 in., 0.357 in., or 0.252 in.), and a gage length of not less than 50 mm (2 in.) are recommended. Longer gage lengths and the 12.82 mm (0.505-in.) diameter are preferred. Specimens should be straight and free from tool marks and scratches.

11.2 Specimens prepared from sheet or plate should be cut with their axes parallel to each other. When testing materials that may be suspected of anisotropy, duplicate sets of specimens may be prepared and tested: one set with their long axis parallel to the axis of greatest tensile strength, another set with their long axis at 45 deg to the axis of greatest tensile strength, and another set with the long axis normal to the direction of greatest tensile strength.

11.2 Specimens for compression-creep tests may suitably be prepared in the same manner described in ASTM Method D 695, Test for Compressive Properties of Rigid Plastics,[2] except that the length should be increased so that the slenderness ratio, l/r, lies between 11 and 15.

12. Number of Test Specimens

12.1 For creep tests it is desirable to test at least four to six specimens at each temperature. The stress values selected should preferably be so spaced as to provide a nearly uniform distribution of stress values from zero to the maximum value tested.

12.2 To determine curves for time for fracture under creep, it is usually desirable to test at least six specimens, starting at stress values high enough to produce failure in less than 10 h and decreasing the stress on succeeding specimens, until the longest duration reached is about 1000 h. However, special service conditions may make other ranges of the time-for-fracture more desirable.

13. Conditioning of Test Specimen

13.1 The problem of how to condition test specimens for creep testing is not easily resolved. It is complicated by the fact that not only do changes in moisture content, plasticizer content, and state of polymerization affect the creep behavior of the material, but these same factors produce dimensional changes (usually shrinkage at higher tempera-

[2] *Annual Book of ASTM Standards*, Part 27.

tures) which may be of such magnitude as to completely overshadow the creep phenomena.

13.2 The following procedures may be appropriate under certain circumstances: Subject the specimen to the preconditioning described in 4.1 of the ASTM Methods D 618, Conditioning Plastics and Electrical Insulating Materials for Testing.[2] This preconditioning may then be followed by one of the following procedures when the test temperature is different from the preconditioning temperature:

13.2.1 Subject the specimens previously fitted with extensometers to the test temperature and relative humidity to be used in the test for a period of time sufficient to obtain dimensional stability before starting the creep test;

13.2.2 Subject all specimens to be tested at all temperatures to the highest temperature to be used in the given series of tests for a period of time sufficient to obtain dimensional stability as indicated by shrinkage rate determined from extensometer readings; or

13.2.3 Start the creep test 48 h after the specimen has attained the test temperature.

13.3 Regardless of what procedure is used, a control specimen should be used for each set of tests at different temperatures. The control specimen should be identical to the test specimen and should be subjected to the same temperature and humidity history. It should be equipped with an extensometer, but should carry no load throughout the duration of the test.

14. Procedure

14.1 *Creep Tests*—In order to equalize the effects of aging, start all tests simultaneously, insofar as possible. Record readings of the extensometers prior to the application of the test temperature (if different from room temperature), immediately after the test temperature is reached, at intervals during any preshrinkage treatment, and immediately after the test load has been applied to the specimen. Take readings at the following intervals thereafter: 0.1, 0.2, 0.5, 0.7, 1, 2, 3, 5, 7, 10, 20 h; then every 24 h to 500 h; then every 48 h to 1000 h (Note 5). Apply the test load to the specimen quickly, but very gently. Make readings of temperature, relative humidity, and control specimen strain at the same time as strain readings for the test specimen. For time-to-fracture tests record the time at which fracture of the specimen occurs. Upon completion of the test interval without rupture remove the load quickly, but very gently, and read the extensometer immediately after removal of the load, and preferably also at the same intervals thereafter as used in the creep tests.

NOTE 5—If discontinuities in the creep curve are suspected or encountered, readings should be taken more frequently than every 24 or 48 h.

14.2 *Relaxation Tests*—The procedure for relaxation tests is essentially the same as for creep tests, except measure the force at intervals, instead of strain, as in the creep tests.

15. Plotting Results

15.1 It is usually desirable to plot curves of total creep or total strain as a function of time on linear rectangular coordinates, choosing a scale of ordinates large enough to be consistent with the accuracy of the strain measurement. The magnitude of the instantaneous strain should be indicated. It is preferable to use hours as the scale for the abscissa.

15.2 It is often more useful to plot the creep as a function of time on log-log coordinates. Such plottings often result in nearly straight-line diagrams, which facilitate interpolation and extrapolation, as well as interpretation.

15.3 Regardless of which method of plotting is used, the strain in the control specimen should be plotted or should be subtracted from corresponding readings of the creep test specimens, before values for these specimens are plotted.

15.4 A log-log plot or semi-log plot of stress against time for fracture is also desirable. From such diagrams at several temperatures the creep strength at a given number of hours duration can be computed; and a plot of creep strength at the given time *versus* the temperature constructed.

15.5 It is also desirable to determine the creep rate and creepocity for given values of time at each stress and temperature. The creep rate may then be plotted as a function of stress for each temperature on rectangular coordinates; and the creepocity values for each temperature may be averaged for all stresses at that temperature, and the average value plotted as a function of temperature.

15.6 Results of relaxation tests may be plotted in a manner similar to 15.1 and 15.2, except that stress is plotted as ordinates for the linear diagram, and decrease in stress is plotted as ordinates for the logarithmic diagram.

15.7 The relaxation rate may be measured for tests at each value of initial stress and each temperature at given values of time. The relaxation rate may then be plotted on rectangular coordinates as a function of initial stress for each temperature tested for each given value of time.

16. Precautions Necessary in Applying Results

16.1 Users of creep or relaxation data in design are particularly cautioned against assuming that the same rates of creep or of relaxation occur under conditions of combined stress (such as occur in piping under pressure or beams, etc.) as occur during a pure tension or compression test.

16.2 Since it is usually necessary to confine the period of creep or relaxation testing to a small fraction of the total service life of a part, the test data must often be extrapolated if quantitative values of creep or relaxation of stress are to be predicted. It should be remembered that the accuracy of such extrapolation is questionable, especially when the effects on the creep or relaxation of other environmental conditions than those which obtained during the tests are unknown. Thus generous factors of safety should be employed when designs are based on extrapolation substantially beyond the duration of the test data.

17. Form for Reporting Results

17.1 It is recommended that the following information be collected and used in reporting test results:

17.1.1 Description of the material tested, including all obtainable information on composition, preparation, manufacturer, trade name, code number, date of manufacture, type of molding, thickness or size of original stock or molding.

17.1.2 Dates of the test.

17.1.3 Type of testing equipment and the information tabulated on a form as shown in Appendix.

17.1.4 Dimensions of test specimen.

17.1.5 Preconditioning used.

17.1.6 Curves showing test data and derived values as described in Section 15.

APPENDIX

A1. Characteristics of Test Equipment

Note—Exact figures are not necessary but unless it is possible to furnish information which is approximately correct, the item should be left blank.

1. Gage length, mm (in.) .... ______
2. Specimen diameter, mm (in.) .... ______
3. Estimated friction in loading mechanism, percent of applied loads .... ______
4. Maximum temperature variation, deg C (deg F) .... ______
   (Over gage length, one normal condition: hottest point minus coldest point.)
5. Maximum temperature fluctuation, deg C (deg F) .... ______
   (During control cycle, one normal condition: hottest instant minus coldest instant.)
6. Number of temperature readings .... ______
   (Average of each specimen survey constitutes one reading.)
7. Number of standard temperature checks .... ______
   (How often furnace temperature is checked by independent standard temperature indicator.)
8. Temperature of reference rod .... ______
   (Answer "test" for gage rod inside furnace, or "room" for microscope standard outside.)
9. Expansion ratio of gage rod to specimen .... ______
   (For gage rod in furnace, the ratio of the coefficient of thermal expansion of gage rod to specimen.)
10. Are extension readings corrected for item 10? (State yes or no) .... ______
   (If ratio is near 1.0, no correction necessary.)

11. Thermal expansion of microscope standard .... ________
    (For microscope standard outside: coefficient of thermal expansion per degree Fahrenheit of microscope standard.)
12. Maximum room temperature variation, deg C (deg F) .... ________
    (Hotest room temperature while reading, minus coldest room temperature while reading.)
13. Are extension readings corrected for items 11 and 12? (State yes or no) .... ________
    (For fluctuation of temperature of specimen (item 5) and of standard (item 12).) (For small fluctuations and low thermal expansion, no correction necessary.)
14. Number of extension readings .... ________
15. Units of smallest scale division mm/mm (in./in.) .... ________
    (Reading, one full division mm/mm (in./in.) of gage length.) .... ________
16. Length of smallest scale division mm (in.) .... ________
    (Actual size.)
17. Magnification ratio .... ________
    (Item 16 divided by item 15.)
18. Usual reading, fraction .... ________
    (1/10, 1/5, or 1/2 of smallest scale division.)
19. Usual reading, mm/mm (in./in.) of gage length .... ________
20. Gage setting, automatic or manual .... ________
    (If dial or mirror is operated by gage rod, answer "automatic."
    If operator sets microscope, answer "manual.")
21. Width of cross-hair, mm/mm (in./in.) of gage length .... ________
22. Width of gage mark, mm/mm (in./in.) of gage length .... ________
23. Thermal capacity of furnace, kWh .... ________
24. Radiation rate, kW .... ________
25. Maximum power input, kW .... ________
26. Minimum power input, kW .... ________

Designation: D 695 – 69

American National Standard K65.1-1970
American National Standards Institute

Standard Method of Test for
COMPRESSIVE PROPERTIES OF RIGID PLASTICS[1]

This Standard is issued under the fixed designation D 695; the number immediately following the designation indicates the year of original adoption or, in the case of revision, the year of last revision. A number in parentheses indicates the year of last reapproval.

1. Scope

1.1 This method covers the determination of the mechanical properties of rigid plastics when loaded in compression at relatively low uniform rates of straining or loading. Test specimens of standard shape are employed.

NOTE 1—The values stated in U.S. customary units are to be regarded as the standard. The metric equivalents of U.S. customary units may be approximate.

2. Significance

2.1 Compression tests provide information about the compressive properties of plastics when employed under conditions approximating those under which the tests are made.

2.2 Compressive properties include modulus of elasticity, yield stress, deformation beyond yield point, and compressive strength (unless the material merely flattens but does not fracture). Materials possessing a low order of ductility may not exhibit a yield point. In the case of a material that fails in compression by a shattering fracture, the compressive strength has a very definite value. In the case of a material that does not fail in compression by a shattering fracture, the compressive strength is an arbitrary one depending upon the degree of distortion that is regarded as indicating complete failure of the material. Many plastic materials will continue to deform in compression until a flat disk is produced, the compressive stress (nominal) rising steadily in the process, without any well-defined fracture occurring. Compressive strength can have no real meaning in such cases.

2.3 Compression tests provide a standard method of obtaining data for research and development, quality control, acceptance or rejection under specifications, and special purposes. The tests cannot be considered significant for engineering design in applications differing widely from the load - time scale of the standard test. Such applications require additional tests such as impact, creep, and fatigue.

3. Definitions

3.1 *compressive stress* (nominal)—the compressive load per unit area of minimum original cross section within the gage boundaries, carried by the test specimen at any given moment. It is expressed in force per unit area.

NOTE 2—The expression of compressive properties in terms of the minimum original cross section is almost universally used. Under some circumstances the compressive properties have been expressed per unit of prevailing cross section. These properties are called "true" compressive properties.

3.2 *compressive strength*—the maximum compressive stress (nominal) carried by a test specimen during a compression test. It may or may not be the compressive stress (nominal) carried by the specimen at the moment of rupture.

3.3 *compressive strength at failure* (nominal)—the compressive stress (nominal) sustained at the moment of failure of the test specimen if shattering occurs.

3.4 *compressive deformation*—the decrease in length produced in the gage length of the

[1] This method is under the jurisdiction of ASTM Committee D-20 on Plastics. A list of committee members may be found in the ASTM Yearbook. This standard is the direct responsibility of Subcommittee D-20.10 on Mechanical Properties.

Current edition effective Dec. 19, 1969. Originally issued 1942. Replaces D 695 – 68 T.

test specimen by a compressive load. It is expressed in units of length.

3.5 *compressive strain*—the ratio of compressive deformation to the gage length of the test specimen, that is, the change in length per unit of original length along the longitudinal axis. It is expressed as a dimensionless ratio.

3.6 *percentage compressive strain*—the compressive deformation of a test specimen expressed as a percentage of the original gage length.

3.7 *compressive stress - strain diagram*—a diagram in which values of compressive stress are plotted as ordinates against corresponding values of compressive strain as abscissas.

3.8 *compressive yield point*—the first point on the stress - strain diagram at which an increase in strain occurs without an increase in stress.

3.9 *compressive yield strength*—normally the stress at the yield point (see also 3.10).

3.10 *offset compressive yield strength*—the stress at which the stress - strain curve departs from linearity by a specified percentage of deformation (offset).

3.11 *proportional limit*—the greatest stress which a material is capable of sustaining without any deviation from proportionality of stress to strain (Hooke's law). It is expressed in force per unit area.

3.12 *modulus of elasticity*—the ratio of stress (nominal) to corresponding strain below the proportional limit of a material. It is expressed in force per unit area based on the average unitial cross-sectional area.

3.13 *slenderness ratio*—the ratio of the length of a column of uniform cross-section to its least radius of gyration. For specimens of uniform rectangular cross-section, the radius of gyration is 0.289 times the smaller cross-sectional dimension. For specimens of uniform circular cross-section, the radius of gyration is 0.250 times the diameter.

3.14 *crushing load*—the maximum compressive force applied to the object, under the conditions of test, that produces a designated degree of failure.

4. Apparatus

4.1 *Testing Machine*—Any suitable testing machine capable of control of constant-rate-of-crosshead movement and comprising essentially the following:

4.1.1 *Drive Mechanism*—A drive mechanism for imparting to the cross-head movable member, a uniform, controlled velocity with respect to the base (fixed member), this velocity to be regulated as specified in Section 8.

4.1.2 *Load Indicator*—A load-indicating mechanism capable of showing the total compressive load carried by the test specimen. The mechanism shall be essentially free from inertia-lag at the specified rate of testing and shall indicate the load with an accuracy of ±1 percent of the maximum indicated value of the test (load). The accuracy of the testing machine shall be verified at least once a year in accordance with ASTM Methods E 4, Verification of Testing Machines.[2]

4.2 *Extensometer*—A suitable instrument for determining the distance between two fixed points on the test specimen at any time during the test. It is desirable that this instrument automatically record this distance (or any change in it) as a function of the load on the test specimen. The instrument shall be essentially free of inertia-lag at the specified rate of loading and shall conform to the requirements for a class B-2 extensometer as defined in ASTM Method E 83, Verification and Classification of Extensometers.[3]

4.3 *Compression Tool*—A compression tool for applying the load to the test specimen. This tool shall be so constructed that loading is axial within 1:1000 and applied through surfaces that are flat within 0.025 mm (0.001 in.) and parallel to each other in a plane normal to the vertical loading axis. Examples of suitable compression tools are shown in Figs. 1 and 2.

4.4 *Supporting Jig*—A supporting jig for thin specimens is shown in Figs. 3 and 4.

4.5 *Micrometers*—Suitable micrometers, reading to 0.01 mm or 0.001 in. for measuring the width, thickness, and length of the specimens.

5. Test Specimens

5.1 Unless otherwise specified in the materials specifications, the specimens described

[2] *Annual Book of ASTM Standards*, Part 27.
[3] *Annual Book of ASTM Standards*, Part 30.

in 5.2 to 5.7 shall be used. These specimens may be prepared by machining operations from materials in sheet, plate, rod, rube, or similar form, or they may be prepared by compression or injection molding of the material to be tested. All machining operations shall be done carefully so that smooth surfaces result. Great care shall be taken in machining the ends so that smooth, flat parallel surfaces and sharp, clean edges to within 0.025 mm (0.001 in.), perpendicular to the long axis of the specimen, result.

5.2 The standard test specimen, except as indicated in 5.3 to 5.7, shall be in the form of a right cylinder or prism whose length is twice its principal width or diameter. Preferred specimen sizes are 12.7 by 12.7 by 25.4 mm (0.50 by 0.50 by 1 in.) (prism), or 12.7 mm (0.50 in.) in diameter by 25.4 mm (1 in.) (cylinder). Where elastic modulus and offset yield-stress data are desired, the test specimen shall be of such dimensions that the slenderness ratio is in the range of 11 to 15:1.

5.3 For rod material, the test specimen shall have a diameter equal to the diameter of the rod and a sufficient length to allow a specimen slenderness ratio in the range of 11 to 15:1.

5.4 When testing tubes, the test specimen shall have a diameter equal to the diameter of the tube and a length of 25.4 mm (1 in.) (Note 3). For crushing-load determinations (at right angles to the longitudinal axis), the specimen size shall be the same, with the diameter becoming the height.

NOTE 3—This specimen can be used for tubes with a wall thickness of 1 mm (0.039 in.) or over, to inside diameters of 6.4 mm (0.25 in.) or over, and to outside diameters of 50.8 mm (2.0 in.) or less.

5.5 Where it is desired to test conventional high-pressure laminates in the form of sheets, the thickness of which is less than 25.4 mm (1 in.), a pile-up of sheets 25.4 mm (1 in.) square, with a sufficient number of layers to produce a height of at least 25.4 mm (1 in.), may be used.

5.6 When testing material that may be suspected of anisotropy, duplicate sets of test specimens shall be prepared having their long axis respectively parallel with and normal to the suspected direction of anisotropy.

5.7 The following specimens shall be used for glass-reinforced materials, and the specimens noted in 5.7.2 may be used for other materials when necessary both to comply with the slenderness ratio requirements and to permit attachment of a deformation-measuring device.

5.7.1 For materials 3.2 mm ($^1/_8$ in.) and over a specimen shall consist of a prism whose cross-section is 12.7 mm ($^1/_2$ in.) by the thickness of the material and whose length is of such dimensions that the slenderness ratio is in the range of 11 to 15:1 (Note 4).

5.7.2 For materials under 3.2 mm ($^1/_8$ in.) thick, a specimen conforming to that shown in Fig. 5 shall be used. The supporting jig shown in Figs. 3 and 4 shall be used to support the specimen during test (Note 5).

NOTE 4—If failure for materials in the thickness range of 3.2 mm ($^1/_8$ in.) is by delamination rather than by the desirable shear plane fracture, the material may be tested in accordance with 5.7.2.

NOTE 5—Round-robin tests have established that relatively satisfactory measurements of modulus of elasticity may be obtained by applying an extensometer to the edges of the jig-supported specimen.

5.8 When testing syntactic foam, the standard test specimen shall be in the form of a right cylinder 25.4 mm (1 in.) in diameter by 50.8 mm (2 in.) in length.

6. Conditioning

6.1 *Conditioning*—Condition the test specimens at 23 ± 2 C (73.4 ± 3.6 F) and 50 ± 5 percent relative humidity for not less than 40 h prior to test in accordance with Procedure A of ASTM Methods D 618, Conditioning Plastics and Electrical Insulating Materials for Testing,[2] for those tests where conditioning is required. In cases of disagreement, the tolerances shall be 1 C (1.8 F) and ±2 percent relative humidity.

6.2 *Test Conditions*—Conduct tests in the Standard Laboratory Atmosphere of 23 ± 2 C (73.4 ± 3.6 F) and 50 ± 5 percent relative humidity, unless otherwise specified in the test methods. In cases of disagreement, the tolerances shall be 1 C (1.8 F) and ±2 percent relative humidity.

7. Number of Test Specimens

7.1 At least five specimens shall be tested for each sample in the case of isotropic materials.

7.2 Ten specimens, five normal to and five parallel with the principal axis of anisotropy, shall be tested for each sample in the case of anisotropic materials.

7.3 Specimens that break at some obvious fortuitous flaw shall be discarded and retests made, unless such flaws constitute a variable, the effect of which it is desired to study.

8. Speed of Testing

8.1 Speed of testing shall be the relative rate of motion of the grips or test fixtures during the test. Rate of motion of the driven grip or fixture when the machine is running idle may be used if it can be shown that the resulting speed of testing is within the limits of variation allowed.

8.2 The standard speed of testing shall be 1.3 ± 0.3 mm (0.050 ± 0.010 in.) per min, except as noted in 9.6.4.

9. Procedure

9.1 Conduct the tests in the Standard Laboratory Atmosphere of 50 ± 2 percent relative humidity and 23 ± 1 C (73.4 ± 1.8 F) temperature.

9.2 Measure the width and thickness of the specimen to the nearest 0.01 mm (0.001 in.) at several points along its length. Calculate and record the minimum value of the cross-section area. Measure the length of the specimen and record the value.

9.3 Place the test specimen between the surfaces of the compression tool, taking care to align the center line of its long axis with the center line of the plunger and to ensure that the ends of the specimen are parallel with the surface of the compression tool. Adjust the crosshead of the testing machine until it just contacts the top of the compression tool plunger.

9.4 Place thin specimens in the jig (Figs. 3 and 4) so that they are centered and the ends project an even amount beyond the jig. The nuts or screws on the jig shall be finger tight (Note 6). Place the assembly in the compression tool as described in 5.2.

NOTE 6—A round robin on the effect of lateral pressure at the supporting jig has established that reproducible data can be obtained with the tightness of the jig controlled as indicated.

9.5 If compressive strength data only are desired, proceed as follows:

9.5.1 Set the speed control at 1.3 mm/min (0.050 in./min) and start the machine.

9.5.2 Record the maximum load carried by the specimen during the test (usually this will be the load at the moment of rupture).

9.6 If stress - strain data are desired, proceed as follows:

9.6.1 Attach extensometer.

9.6.2 Set the speed control at 1.3 mm/min (0.050 in./min) and start the machine.

9.6.3 Record loads and corresponding compressive strain at appropriate intervals of strain or, if the test machine is equipped with an automatic recording device, record the complete load - deformation curve.

9.6.4 After the yield point has been reached, it may be desirable to increase the speed to 5 to 6 mm/min (0.20 to 0.25 in./min) and allow the machine to run at this speed until the specimen breaks. This may be done only with relatively ductile materials and on a machine with a weighing system with response rapid enough to produce accurate results.

10. Calculations

10.1 *Compression Strength*—Calculate the compressive strength by dividing the maximum compressive load carried by the specimen during the test by the original minimum cross-sectional area of the specimen. Express the result in kilograms-force per square centimeter or pounds per square inch and report to three significant figures.

10.2 *Compressive Yield Strength*—Calculate the compressive yield strength by dividing the load carried by the specimen at the yield point by the original minimum cross-sectional area of the specimen. Express the result in kilograms-force per square centimeter or pounds per square inch and report to three significant figures.

10.3 *Offset Yield Strength*—Calculate the offset yield strength by the method referred to in 3.10.

10.4 *Modulus of Elasticity*—Calculate the modulus of elasticity by drawing a tangent to the initial linear portion of the load deformation on the stress - strain diagram, selecting any point on this straight line, and dividing the compressive stress represented by the point by the corresponding compressive strain. Express the result in kilograms-force per square centimeter or pounds per square inch and report to three significant figures.

10.5 For each series of tests, calculate to three significant figures the arithmetic mean of all values obtained and report as the "average value" for the particular property in question.

10.6 Calculate the standard deviation (estimated) as follows and report to two significant figures:

$$s = \sqrt{(\Sigma X^2 - n\bar{X}^2)/(n - 1)}$$

where:

s = estimated standard deviation,
X = value of single observation,
n = number of observations, and
$\bar{X}$ = arithmetic mean of the set of observations.

11. Report

11.1 The report shall include the following:

11.1.1 Complete identification of the material tested, including type, source, manufacturer's code number, form, principal dimensions, previous history, etc.,

11.1.2 Method of preparing test specimens,

11.1.3 Type of test specimen and dimensions,

11.1.4 Conditioning procedure used,

11.1.5 Atmospheric conditions in test room,

11.1.6 Number of specimens tested,

11.1.7 Speed of testing,

11.1.8 Compressive strength (Note 7), average value, and standard deviation,

NOTE 7—The method for determining the offset compressive yield stress is similar to that described in the Appendix of ASTM Method D 638, Test for Tensile Properties of Plastics.[2]

11.1.9 Compressive yield strength and offset yield strength average value, and standard deviation, when of interest,

11.1.10 Modulus of elasticity in compression, average value, standard deviation, and

11.1.11 Date of test.

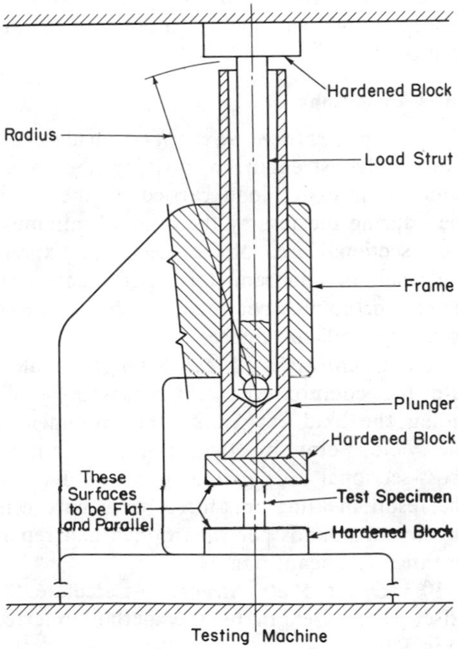

NOTE—Devices similar to the one illustrated have been successfully used in a number of different laboratories. Details of the device developed at the National Bureau of Standards are given in the paper by Aitchison, C. S., and Miller, J. A., "A Subpress for Compressive Tests," Natl. Advisory Committee for Aeronautics, Technical Note No. 912, 1943.

FIG. 1 Subpress for Compression Tests.

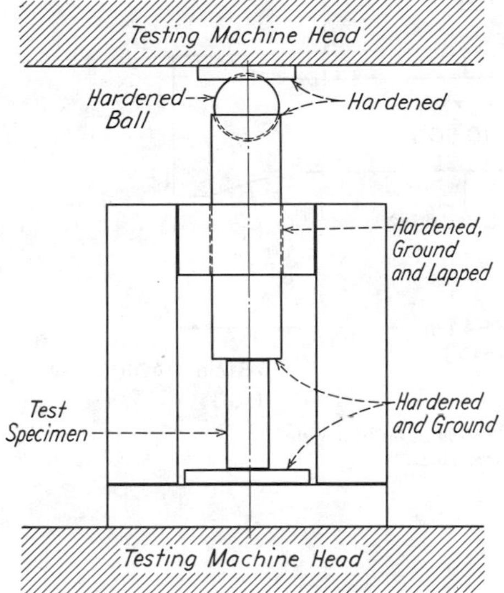

FIG. 2 Compression Tool.

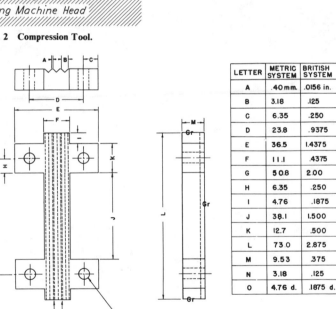

FIG. 3 Support Jig for Thin Specimens.

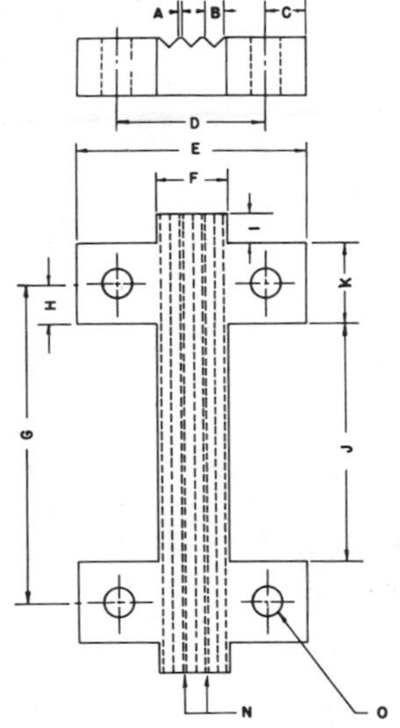

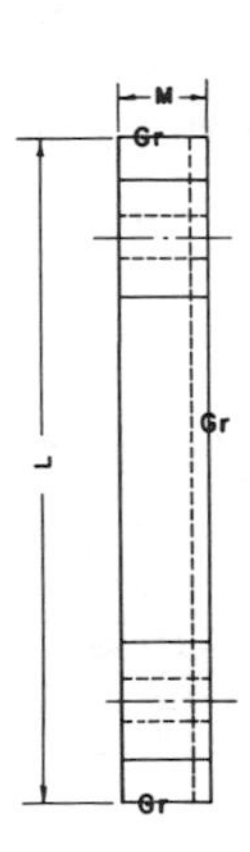

LETTER	METRIC SYSTEM	BRITISH SYSTEM
A	.40 mm.	.0156 in.
B	3.18	.125
C	6.35	.250
D	23.8	.9375
E	36.5	1.4375
F	11.1	.4375
G	50.8	2.00
H	6.35	.250
I	4.76	.1875
J	38.1	1.500
K	12.7	.500
L	73.0	2.875
M	9.53	.375
N	3.18	.125
O	4.76 d.	.1875 d.

NOTE 1—Cold rolled steel.
NOTE 2—Furnished four steel machine screws and nuts, round head, slotted, length 31.75 mm ($1^1/_4$ in.).
NOTE 3—Grind surfaces denoted "Gr."

FIG. 4 Support Jig, Details.

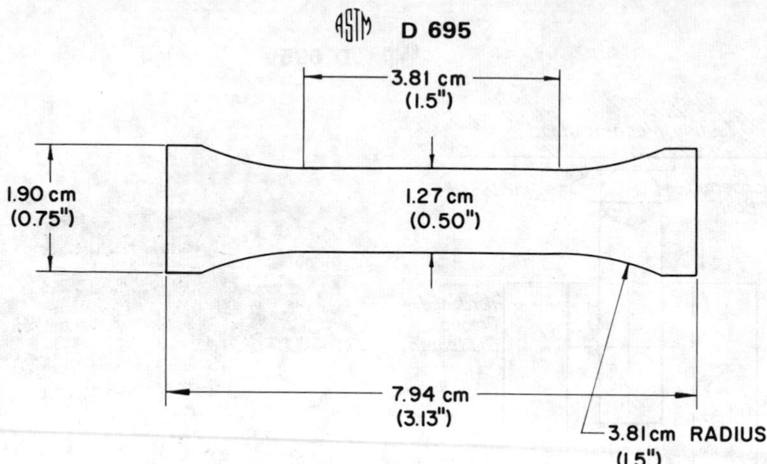

FIG. 5 Compression Test Specimen for Materials Less than 3.2 mm Thick.

ASTM Designation: D 696 - 70

Standard Method of Test for

COEFFICIENT OF LINEAR THERMAL EXPANSION OF PLASTICS[1]

This Standard is issued under the fixed designation D 696; the number immediately following the designation indicates the year of original adoption or, in the case of revision, the year of last revision. A number in parentheses indicates the year of last reapproval.

1. Scope

1.1 This method covers determination of the coefficient of linear thermal expansion for plastics by use of a vitreous silica dilatometer. At the test temperatures and under the stresses imposed, the plastic materials shall have a negligible creep or elastic strain rate or both, insofar as these properties would significantly affect the accuracy of the measurements.

NOTE 1—See also ASTM Method E 228, Test for Linear Thermal Expansion of Rigid Solids with a Vitreous Silica Dilatometer.[2]

1.2 The thermal expansion of a plastic is composed of a reversible component on which are superimposed changes in length due to changes in moisture content, curing, loss of plasticizer or solvents, release of stresses, phase changes and other factors. This method of test is intended for determining the coefficient of linear thermal expansion under the exclusion of these factors as far as possible. In general, it will not be possible to exclude the effect of these factors completely. For this reason, the method can be expected to give only an approximation to the true thermal expansion.

NOTE 2—The values stated in U.S. customary units are to be regarded as the standard. The metric equivalents of U.S. customary units may be approximate.

2. Significance

2.1 The coefficient of linear thermal expansion, α, between temperatures T_1 and T_2 for a specimen whose length is L_0 at the reference temperature, is given by the following equation:

$$\alpha = (L_2 - L_1)/[L_0(T_2 - T_1)] = \Delta L/L_0 \Delta T$$

where L_1 and L_2 are the specimen lengths at temperatures T_1 and T_2, respectively. α is, therefore, obtained by dividing the linear expansion per unit length by the change in temperature.

2.2 The nature of most plastics and the construction of the common type of dilatometer make -30 to $+30$ C a convenient temperature range for linear thermal expansion measurements of plastics. This range covers the temperatures in which plastics are very commonly used. When extending the range or when the measurement of a plastic whose characteristics are not known through the standard range is being made, particular attention shall be paid to the factors mentioned in 1.2 and special preliminary investigations by thermo-mechanical analysis, such as that prescribed in ASTM Method D 2236, Test for Dynamic Mechanical Properties of Plastics by Means of a Torsional Pendulum,[3] for the location of phase changes, may be required to avoid excessive error. Other ways of locating phase changes or transition temperatures using the dilatometer itself may be employed to cover the range of temperatures in question by using smaller steps than 30 C or by observing the rate of expansion during a steady rise in temperature of the specimen. Once such a transition point has been located, a separate coefficient of expansion for a temperature range below and above the transition point

[1] This method is under the jurisdiction of ASTM Committee D-20 on Plastics. A list of members may be found in the ASTM Yearbook. This method is the direct responsibility of Subcommittee D-20.30 on Thermal Properties.

Current edition effective Feb. 27, 1970. Originally issued 1942. Replaces D 696 - 44.

[2] *Annual Book of ASTM Standards*, Part 23.

[3] *Annual Book of ASTM Standards*, Part 27.

shall be determined. For specification and comparison purposes, the range of −30 C to +30 C (provided it is known that no transition point exists in this range) shall be used.

3. Summary of Method

3.1 This method is intended to provide a means of determining the coefficient of linear thermal expansion of plastics which are not distorted or indented by the thrust of the dilatometer on the specimen. The specimen is placed at the bottom of the outer dilatometer tube with the inner one resting on it. The measuring device which is firmly attached to the outer tube is in contact with the top of the inner tube and indicates variations in the length of the specimen with changes in temperature. Temperature changes are brought about by immersing the outer tube in a liquid bath at the desired temperature.

4. Apparatus

4.1 *Fused-Quartz-Tube Dilatometer* suitable for this method is illustrated in Fig. 1. A clearance of approximately 1 mm is allowed between the inner and outer tubes.

4.2 *Device* for measuring the changes in length (dial gage, LVDT, or the equivalent) shall be fixed on the mounting fixture so that its position may be adjusted to accommodate specimens of varying length (see 5.2). The accuracy shall be such that the error of indication will not exceed ±10 μm (4×10^{-5} in.) for any length change. The weight of the inner silica tube plus the measuring device reaction shall not exert a stress of more than 0.7 kgf/cm^2 (10 psi) on the specimen so that the specimen is not distorted or appreciably indented.

4.3 *Scale or Caliper* capable of measuring the initial length of the specimen with an accuracy of ±0.5 percent.

4.4 *Liquid Bath* to control the temperature of the specimen. The bath shall be so arranged that uniform temperature over the length of the specimen is assured. Means shall be provided for stirring the bath and for controlling its temperature within ±0.2 C (0.4 F) at the time of the temperature and measuring device readings.

NOTE 3—It is preferable and not difficult to avoid contact between the bath liquid and the test specimen. If such contact is unavoidable, care must be taken to select a liquid that will not affect the physical properties of the material under test.

4.5 *Thermometer or Thermocouple*—The bath temperature shall be measured by a thermometer or thermocouple capable of an accuracy of ±0.1 C (±0.2 F).

5. Test Specimens

5.1 The test specimens may be prepared by machining, molding, or casting operations under conditions which give a minimum of strain or anistropy. If specimens are cut from samples that are suspected of anisotropy, the specimens shall be cut along the principal axes of anisotropy and the coefficient of linear thermal expansion shall be measured on each set of specimens.

5.2 The specimen length shall be between 50 mm (2 in.) and 125 mm (5 in.).

NOTE 4—If specimens shorter than 50 mm are used, a loss in sensitivity results. If specimens greatly longer than 125 mm are used, the temperature gradient along the specimen becomes difficult to control within the prescribed limits. The length used will be governed by the sensitivity and range of the measuring device, the extension expected and the accuracy desired. Generally speaking, the longer the specimen and the more sensitive the measuring device, the more accurate will be the determination if the temperature is well controlled.

5.3 The cross section of the test specimen may be round, square, or rectangular and should fit easily into the outer tube of the dilatometer without excessive play on the one hand or friction on the other. The cross section of the specimen shall be large enough so that no bending or twisting of the specimen occurs. Convenient specimen cross sections are: 12.5 by 6.3 mm (1/2 in. by 1/4 in.), 12.5 by 3 mm (1/2 by 1/8 in.), 12.5 mm (1/2 in.) in diameter or 6.3 mm (1/4 in.) in diameter. If excessive play is found with some of the thinner specimen, guide sections may be cemented or otherwise attached to the sides of the specimen to fill out the space.

5.4 The ends of the specimens shall be reasonably flat and perpendicular to the length axis of the specimen. The ends shall be protected against indentation by means of flat, thin steel plates cemented or otherwise firmly attached to them before the specimen is placed in the dilatometer. These plates shall be 0.3 to 0.5 mm (0.012 to 0.020 in.) in thickness.

6. Conditioning

6.1 *Conditioning*—Condition the test specimens at 23 ± 2 C (73.4 ± 3.6 F) and 50 ± 5 percent relative humidity for not less than 40 h prior to test in accordance with Procedure A of ASTM Methods D 618, Conditioning Plastics and Electrical Insulating Materials for Testing,[3] for those tests where conditioning is required. In cases of disagreement, the tolerances shall be ±1 C (±1.8 F) and ±2 percent relative humidity.

6.2 *Test Conditions*—Conduct tests in the Standard Laboratory Atmosphere of 23 ± 2 C (73.4 ± 3.6 F) and 50 ± 5 percent relative humidity, unless otherwise specified in the test methods or in this specification. In cases of disagreement, the tolerances shall be ±1 C (±1.8 F) and ±2 percent relative humidity.

7. Procedure

7.1 Measure the length of the conditioned specimen at room temperature to the nearest 25 μm (0.001 in.) with the scale or caliper (see 4.3).

7.2 Cement or otherwise attach the steel plates to the ends of the specimen to prevent indentation (see 5.4).

7.3 Mount the specimen in the dilatometer. Carefully install the dilatometer in the −30 C (−22 F) bath so that the top of the specimen is at least 50 mm (2 in.) below the liquid level of the bath. Maintain the temperature of the bath in the range from −32 C to 28 C (−25 to −18 F) ±0.2 C (0.4 F) until the temperature of the specimen reaches the temperature of the bath as denoted by no further movement indicated by the measuring device over a period of 5 to 10 min. Record the actual temperature and the measuring device reading.

7.4 Without disturbing or jarring the dilatometer, change to the +30 C (86 F) bath, so that the top of the specimen is at least 50 mm (2 in.) below the liquid level of the bath. Maintain the temperature of the bath in the range from +32 to +28 C (+82 to 90 F) ±0.2 C (0.4 F) until the temperature of the specimen reaches that of the bath as denoted by no further changes in the measuring device reading over a period of 5 to 10 min. Record the actual temperature and the measuring device reading.

7.5 Without disturbing or jarring the dilatometer, change to the −30 C (−22 F) bath and repeat the procedure in 7.3.

NOTE 5—It is convenient to use alternately two baths at the proper temperatures. Great care should be taken not to disturb the apparatus during the transfer of baths. Tall Thermos bottles have been successfully used. The use of two baths is preferred because this will reduce the time required to bring the specimen to the desired temperature. The test should be conducted in as short a time as possible to avoid changes in physical properties during long exposures to high and low temperatures that might possibly take place.

7.6 If the change in length per degree of temperature difference due to heating does not agree with the change in length per degree due to cooling within 10 percent of their average, the cause of the discrepancy shall be investigated and, if possible, eliminated. The test shall then be repeated until agreement is reached.

8. Calculation

8.1 Calculate the coefficient of linear thermal expansion over the temperature range used as follows:

$$\alpha = \Delta L / L_0 \Delta T$$

where:

α = coefficient of linear thermal expansion per deg Celsius,
ΔL = change in length of test specimen due to heating or to cooling,
L_0 = length of test specimen at room temperature (ΔL and L_0 being measured in the same units), and
ΔT = temperature differences, deg C, over which the change in the length of the specimen is measured.

The values of α for heating and for cooling shall be averaged to give the value to be reported.

9. Report

9.1 The report shall include the following:

9.1.1 Designation of material, including name of manufacturer and information on composition when known,

9.1.2 Method of preparation of test specimen,

9.1.3 Form and dimensions of test specimen,

9.1.4 Type of apparatus used,

9.1.5 Temperatures between which the

coefficient of linear thermal expansion has been determined,

9.1.6 Average coefficient of linear thermal expansion per deg Celsius,

9.1.7 Location of phase change or transition point temperatures, if this is in the range of temperatures used, and

9.1.8 Complete description of any unusual behavior of the specimen, for example, differences of more than 10 percent in measured values of expansion and contraction.

10. Precision

10.1 The intralaboratory coefficient of variation by this method is 1.5 to 2 percent.

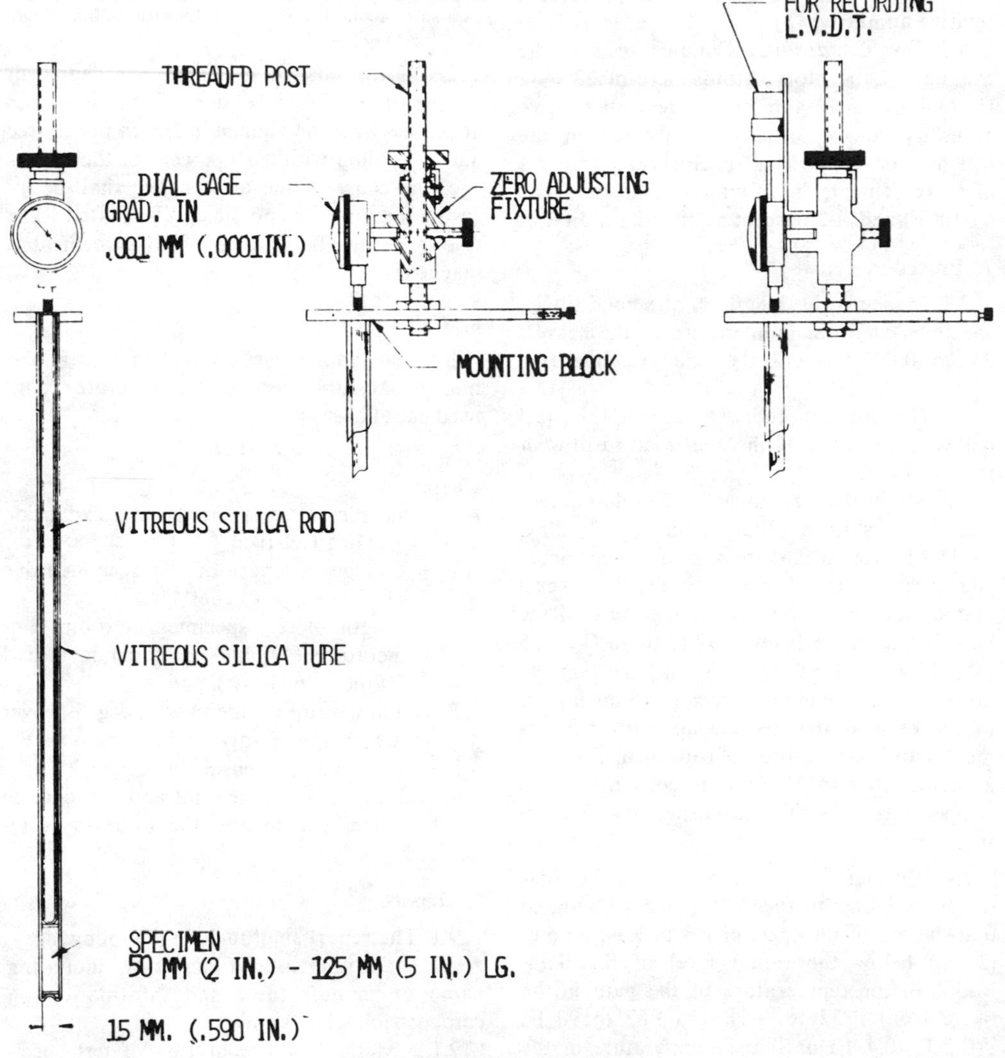

FIG. 1 Quartz-Tube Dilatometer.

Designation: D 731 - 67

Standard Method of
MEASURING THE MOLDING INDEX OF THERMOSETTING MOLDING POWDER[1]

This Standard is issued under the fixed designation D 731; the number immediately following the designation indicates the year of original adoption or, in the case of revision, the year of last revision. A number in parentheses indicates the year of last reapproval.

1. Scope

1.1 This method covers the measurement of the molding index of thermosetting plastics ranging in flow from soft to stiff by selection of appropriate molding pressures within the range from 46 to 463 kgf/cm^2 (600 to 6580 psi) (Note 1). A special mold is required and tests are made under specific conditions. This method does not necessarily denote that the molding behavior of different materials will be alike and trials may be necessary to establish the appropriate molding index for each material in question.

NOTE 1—The sensitivity of this test diminishes when the molding pressure is decreased below 67 kgf/cm^2 (950 psi) so pressures lower than this are not ordinarily recommended. This is due to the friction of moving parts and the insensitivity of the pressure switch actuating the timer at these low pressures.

NOTE 2—The values stated in U.S. customary units are to be regarded as the standard. The metric equivalent of U.S. customary units may be approximate.

2. Apparatus

2.1 *Mold*—A cup mold of the flash type, to produce a molded cup as shown in Fig. 1, operated under controlled pressure and temperatures and provided with stops so that flash or fin thickness cannot be less than 0.14 mm (0.0055 in.). The area of the mold casting creating the molded flash shall be located on top of the cup, flat, perpendicular to the axis of the cup, and in the form of an annular ring 3.17 mm (0.125 in.) in width.

2.2 *Thermometer*—A 32-mm (1$^1/_4$-in.) partial-immersion mercury thermometer having a diameter just under 4.8 mm ($^3/_{16}$ in.) and a temperature scale of not more than 20 C/in. of length. A pyrometer may be used to determine the temperature of the mold surfaces. For properly measuring mold temperatures, reference should be made to ASTM Recommended Practice D 958, Molds for Test Specimens of Plastics.[2]

2.3 *Heating System*—Any conventional means for heating the press platens, provided the heat source is constant enough to maintain the molding temperature within ±1 C of the specified temperature (see 5.2).

2.4 *Pressure System*—A semiautomatic press with a fixed mold and fully insulated to minimize heat losses shall be used. The use of hand molds is not recommended but may be used to give an estimate of the molding index. For the cup mold detail, see ASTM Recommended Practice D 647, for Design of Molds for Test Specimens of Plastic Molding Materials[2] (Note 3). The hydraulic system shall be provided with a means of pressure regulation so that the load on the mold shall differ by not more than ±113 kgf (250 lb) from the stated value. The capacity of the hydraulic system shall permit a ram travel of approximately 2.5 cm/s (1 in./s). It is recommended that the ram diameter not exceed 10 cm (4 in.).

NOTE 3—A detailed drawing of the mold design will be added to Recommended Practice D 647 at a later date.

---

[1] This method is under the jurisdiction of ASTM Committee D-20 on Plastics. A list of committee members may be found in the ASTM Yearbook. This standard is the direct responsibility of Subcommittee D-20.30 on Thermal Properties.

Current edition effective Nov. 1, 1967. Originally issued 1950. Replaces D 731 - 57.

[2] *Annual Book of ASTM Standards*, Part 27.

3. Conditioning

3.1 Materials are normally tested in the "as received" condition, except in referee tests, when they shall be conditioned in accordance with 3.2 (Note 4).

3.2 For referee testing, all materials shall be shipped and stored in sealed moisture barrier containers. These materials shall be stored for a minimum period of 48 h at standard laboratory temperature before breaking the seal on the carton. A representative sample shall be taken from this carton immediately after opening and tested within 3 min in order to preserve the original moisture content. Alternative methods of conditioning samples may be used provided they are mutually agreed upon between the manufacturer and the purchaser.

NOTE 4—Conditioning may alter the moisture content of most materials and thereby change their molding index or molding behavior.

4. Test Specimen

4.1 To determine the weight of the test specimen for materials having an Izod impact strength of 2.7 cm · kgf/cm (0.50 ft · lb/in.) of notch, or less, a cup having a flash or fin thickness of 0.15 to 0.20 mm (0.006 to 0.008 in.) shall be molded (Note 5). The adhering fin shall be removed and the cup weighed to the nearest 0.1 g. This weight multiplied by 1.1 shall be the weight of the test specimen used. For materials having an impact above 2.7 cm · kgf/cm (0.50 ft · lb/in.) of notch, the specimen weight is determined in a similar manner, except that cup flash shall not be more than 0.66 mm (0.026 in.) or less than 0.51 mm (0.020 in.) and the amount of material shall be 1.05 times the weight of this cup. The test specimen shall be in the form of loose powder unless preforming is necessary for materials of high bulk. Minimum pressure shall be employed in the preforming operation to minimize the increase in closing time resulting from the use of preforms.

NOTE 5—While the mold is provided with stops so that the flash or fin thickness cannot be less than 0.006 ± 0.0005 in., the molded cup itself may have a flash thickness of 0.006 to 0.008 in. as the micro switch controlling the closing time must have a tolerance in which to operate.

5. Procedure

5.1 Mount the mold in a press of the semiautomatic type. Past experience has shown that the rate of flow is sensitive to the condition of the mold surfaces; preceding materials may have deposited a film that influences the mold surfaces to the extent that erroneous results may be obtained unless properly conditioned prior to testing. A suggested procedure is to discard the first few cups molded and accept the flow time as correct when two successive cups molded under test do not differ by more than 1 s in time of flow.

5.2 The preferred mold temperature for testing the molding index for the following materials shall be:

	deg C	deg F
Phenolic	165 ± 1	330 ± 1.8
Melamine	155 ± 1	310 ± 1.8
Urea	150 ± 1	300 ± 1.8
Epoxy	150 ± 1	300 ± 1.8
Diallyl phthalate	150 ± 1	300 ± 1.8
Alkyd	150 ± 1	300 ± 1.8

Other temperatures may be used as agreed upon by the manufacturer and the purchaser.

5.3 First determine the proper weight to be used as outlined in 4.1. Then begin the test with a load sufficient to close the mold to the fin thickness specified for the type of material being tested as defined in 4.1. For example, if a 4540-kgf (10,000-lb) load is applied on the mold to make the initial cup and the required fin thickness is obtained, the next lower load, 3400 kgf (7500 lb) is applied as indicated in the following table. If the mold closes to the required thickness again, then the 2270-kgf (5000-lb) load is applied. If the mold then does not close, the "molding index" is the closing time obtained with the 3400-kgf (7500-lb) load. It is recommended that the mold loads used be selected from the following table:

Total Load		Molding Pressure	
kgf	lb	kgf/mm²	psi
1 130	2 500	0.46	660
1 630	3 600	0.66	950
2 270	5 000	0.93	1320
3 400	7 500	1.40	1970
4 540	10 000	1.85	2630
6 800	15 000	2.80	3950
9 070	20 000	3.70	5260
11 340	25 000	4.60	6580

5.4 The time of flow in seconds shall be measured from the instant that the hydraulic gauge indicates an applied load of 454 kg (1000 lb) to the instant that the fin has reached 0.20 mm (0.008 in.) in thickness for materials with an Izod impact strength of 2.7 cm · kgf/cm (0.50 ft · lb/in.) of notch, or less, and 0.66 mm (0.026 in.) for materials with an impact above 2.7 cm · kgf/cm (0.50 ft · lb/in.) of notch.

NOTE 6—The molding pressure is calculated by dividing the total load by the projected area of the mold cavity.

NOTE 7—A convenient method for determining fin thickness is to indicate the final movement of the platen with a dial micrometer.

6. Report

6.1 The report shall include the following:

6.1.1 A statement indicating the nature of the material tested and the manufacturer's material number and batch number,

6.1.2 The molding index expressed as the total minimum force in pounds required to close the mold, with the closing time in seconds as a subscript (for example, 6800_{18} kgf ($15,000_{18}$),

6.1.3 Temperature of the mold,

6.1.4 Weight of the test sample in grams, and form of sample (loose powder or preform),

6.1.5 Thickness of flash or fin measured to the nearest 0.025 mm (0.001 in.),

6.1.6 Description of any unusual characteristics of the test sample such as discoloration and sticking,

6.1.7 Details of conditioning,

6.1.8 Impact strength of the material expressed as Izod in foot-pounds per inch of notch in accordance with ASTM Methods D 256, for Impact Resistance of Plastics and Electrical Insulating Materials,[2] and

6.1.9 Indicate if preforms were necessary to run the test.

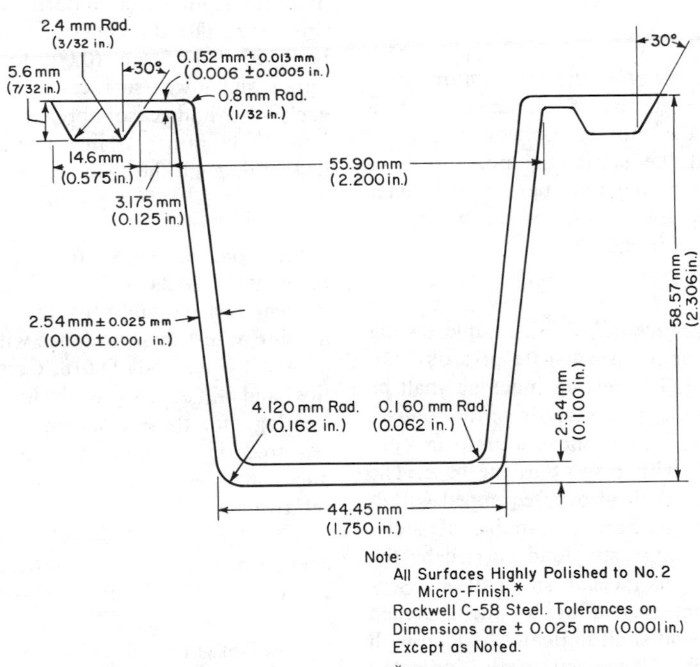

FIG. 1 Cup Mold.

Designation: D 732 – 46 (Reapproved 1969)

Standard Method of Test for
SHEAR STRENGTH OF PLASTICS[1]

This Standard is issued under the fixed designation D 732; the number immediately following the designation indicates the year of original adoption or, in the case of revision, the year of last revision. A number in parentheses indicates the year of last reapproval.

NOTE—Editorial changes in Section 5 were made in December 1968.

1. Scope

1.1 This method covers the punch-type of shear test and is intended for use in determining the shear strength of test specimens of organic plastics in the form of sheets and molded disks in thicknesses from 0.125 to 12.5 mm (0.005 to 0.500 in.).

NOTE 1—The values stated in U.S. customary units are to be regarded as the standard. The metric equivalents of U.S. customary units may be approximate.

2. Definition

2.1 *shear strength*—the maximum load required to shear the specimen in such a manner that the moving portion has completely cleared the stationary portion. It is expressed in meganewtons per square meter (pounds per square inch) based on the area of the sheared edge or edges.

3. Apparatus

3.1 *Testing Machine*—Any suitable testing machine of the constant-rate-of-crosshead movement type. The testing machine shall be equipped with the necessary drive mechanism for imparting to the crosshead a uniform, controlled velocity with respect to the base. The testing machine shall also be equipped with a load-indicating mechanism capable of showing the total compressive load carried by the test specimen. This mechanism shall be essentially free from inertia-lag at the specified rate of testing and shall indicate the load with an accuracy of ±1 percent of the indicated value or better. The accuracy of the testing machine shall be verified in accordance with ASTM Methods E 4, Verification of Testing Machines.[2]

3.2 *Shear Tool*—A shear tool of the punch type which is so constructed that the specimen is rigidly clamped both to the stationary block and movable block so that it cannot be deflected during the test. A suitable form of shear tool is shown in Fig. 1.

4. Test Specimen

4.1 The specimen shall consist of a 50-mm (2-in.) square or a 50-mm (2-in.) diameter disk cut from sheet material or molded into this form. The thickness of the specimen may be 0.125 to 12.5 mm (0.005 to 0.500 in.). The upper and lower surfaces shall be parallel to each other and reasonably flat. A hole approximately 11 mm ($^{7}/_{16}$ in.) in diameter shall be drilled through the specimen at its center.

5. Conditioning

5.1 *Conditioning*—Condition the test specimens at 23 ± 2 C (73.4 ± 3.6 F) and 50 ± 5 percent relative humidity for not less than 40 h prior to test in accordance with Procedure A of ASTM Methods D 618, Conditioning Plastics and Electrical Insulating Materials for Testing,[2] for those tests where conditioning is required. In cases of disagreement, the tolerances shall be ±1 C (±1.8 F) and ±2 percent relative humidity.

5.2 *Test Conditions*—Conduct tests in the Standard Laboratory Atmosphere of 23 ± 2 C (73.4 ± 3.6 F) and 50 ± 5 percent rela-

[1] This method is under the jurisdiction of ASTM Committee D-20 on Plastics. A list of committee members may be found in the ASTM Yearbook. This standard is the direct responsibility of Subcommittee D-20.10 on Mechanical Properties.

Current edition effective Sept. 16, 1946. Originally issued 1943. Replaces D 732 – 43.

[2] *Annual Book of ASTM Standards*, Part 27.

tive humidity, unless otherwise specified in the test methods or in this specification. In cases of disagreement, the tolerances shall be ±1 C (±1.8 F) and ±2 percent relative humidity.

6. Procedure

6.1 Make tests on materials conditioned in accordance with Procedure B of Methods D 618 at room temperature maintained at 23 ± 1 C (73.4 ± 1.8 F). Make tests on materials conditioned in accordance with Procedure A of Methods D 618 at 23 ± 1 C (73.4 ± 1.8 F) and at a relative humidity of 50 ± 2 percent.

6.2 Use five specimens.

6.3 Place the specimen over the 9.5 mm ($^3/_8$-in.) pin of the punch and fasten tightly to it by means of the washer and nut. Then assemble the tool jig and tighten the bolts.

6.4 Maintain the crosshead speed of the machine during test at 1.25 mm (0.05 in.)/min, measured when the machine is running idle.

6.5 Push down the punch far enough so that the shoulder clears the specimen proper. The specimen will then be adjacent to the necked-down portion of the punch and it should be possible to remove the specimen readily from the tool.

NOTE 2—For thick specimens of some materials the punched-out piece tends to stick in the die. If the test is continued only to the point where maximum load has been developed and starts to fall off rapidly, the specimen may be readily removed from both punch and die.

7. Report

7.1 The report shall include the following:

7.1.1 Complete identification of the material tested, including type, source, manufacturer's code number, form principal dimensions, previous history, etc.,

7.1.2 Method of test, type of test specimen, and dimensions,

7.1.3 Atmospheric conditions in the test room,

7.1.4 Conditioning procedure used,

7.1.5 Diameter of punch,

7.1.6 Load in newtons (pounds) required to shear each specimen, and

7.1.7 Calculated shear strength in meganewtons per square meter (pounds per square inch), determined by dividing the load by the area of the sheared edge, which shall be taken as the product of the thickness of the specimen by the circumference of the punch. Report individual and average values, also the average deviation.

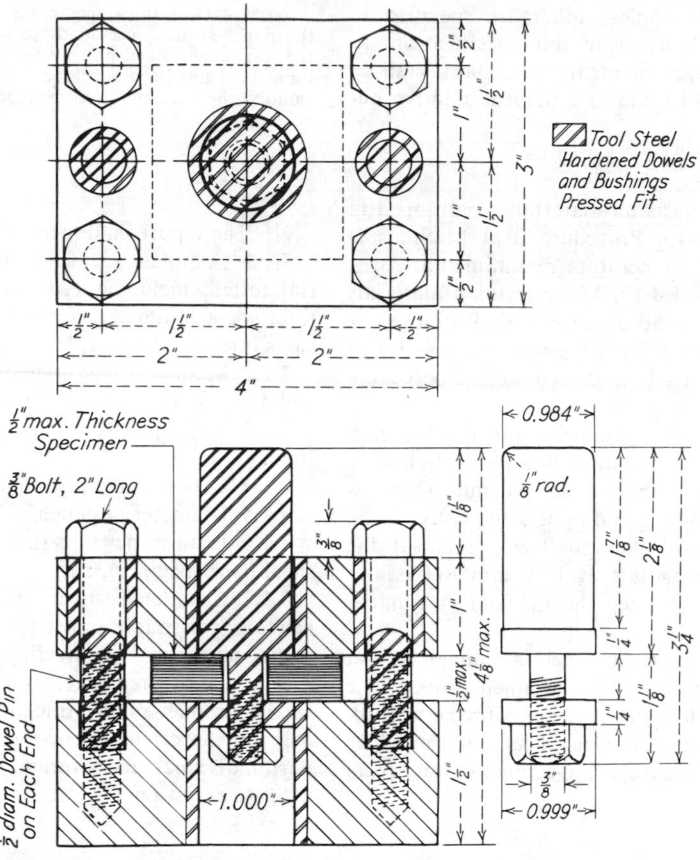

Cross-Sectional View

NOTE—In case of difficulty in obtaining hardened dowels and bushings, the entire shear tool may be made from a fairly good grade of steel, eliminating all of the bushings shown. The actual working surfaces will wear faster than when hardened tool steel is used. When they show signs of appreciable wear, the shear tool can then be bored out to take either hardened or unhardened bushings, depending upon which are available.

Table of Metric Equivalents

in.	1/8	1/4	3/8	1/2	0.984	0.999	1	1 1/8	1 1/2	1 7/8	2	2 1/8	3	3 1/4	4	4 1/8
mm	3.2	6.4	9.5	12.7	24.9	25.1	25.4	28	38	47	51	54	76	82	102	105

FIG. 1 Punch-Type Shear Tool for Testing Specimens 0.125 to 12.5 mm (0.005 to 0.500 in.) in Thickness.

Tentative Method of Test for
BRITTLENESS TEMPERATURE OF PLASTICS AND ELASTOMERS BY IMPACT[1]

This Tentative Method has been approved by the sponsoring committee and accepted by the Society in accordance with established procedures, for use pending adoption. Suggestions for revisions should be addressed to the Society at 1916 Race St., Philadelphia, Pa. 19103.

This tentative has been submitted to the Society for adoption as standard, but balloting was not complete by press time.

1. Scope

1.1 This method covers the determination of a temperature at which plastics and elastomers exhibit brittle failure under specified impact conditions. Two routine inspection and acceptance procedures are also provided.

NOTE 1—The values stated in U.S. customary units are to be regarded as the standard. The metric equivalents of U.S. customary units may be approximate.

2. Significance

2.1 This method establishes the temperature at which 50 percent of the specimens tested fail when subjected to the conditions herein specified. The test provides for the evaluation of long-time effects such as crystallization, or those that may be introduced by low-temperature incompatibility of plasticizers in the material under test. Plastics and elastomers are used in many applications requiring low-temperature flexing with or without impact. Data obtained by this method may be used to predict the behavior of plastic and elastomeric materials at low temperatures only in applications in which the conditions of deformation are similar to those specified in this method. This method has been found useful for specification purposes, but does not necessarily measure the lowest temperature at which the material may be used.

3. Definition

3.1 *brittleness temperature*—that temperature, estimated statistically, at which 50 percent of the specimens would fail in the specified test.

4. Apparatus

4.1 *Specimen Clamp and Striking Member*—The specimen clamp shall be designed to hold firmly the specimen or specimens as a cantilever beam. Each individual specimen shall be firmly and securely held in the clamp. The striking edge shall move relative to the specimens at a linear speed of 1.8 to 2.1 m/s (6.0 to 7.0 ft/s) at impact, and during at least the following 6.4 mm (0.25 in.) of travel. In order to maintain this speed it may be necessary to reduce the number of specimens tested at one time. The distance between the center line of the striking edge and the clamp shall be 7.87 ± 0.25 mm (0.31 ± 0.01 in.) at impact. The striking edge shall have a radius of 1.57 ± 0.13 mm (0.062 ± 0.005 in.). The striking arm and specimen clamp shall have a clearance of 6.35 ± 0.25 mm (0.25 ± 0.01 in.) at and immediately following impact. These dimensional requirements are illustrated in Fig. 1.

NOTE 2—Suitable apparatus is commercially available from several suppliers. The striking member may be motor-driven, solenoid-operated, gravity-actuated, or spring-loaded. The motor-driven tester should be equipped with a safety interlock to prevent striker arm motion when the cover is open.

4.2 *Thermocouple*, with associated temperature indicator graduated in 1 C divisions and having a range suitable for the temperature at

[1] This method is under the jurisdiction of ASTM Committee D-20 on Plastics. A list of committee members may be found in the ASTM Yearbook. This standard is the direct responsibility of Subcommittee D-20.30 on Thermal Properties.

Current edition effective Annual Meeting, 1964. Originally issued 1943. Replaces D 746 - 57 T.

which determinations are to be made. The thermocouple shall be constructed of copper-constantan wire, in the size range from No. 24 to No. 32 Awg, welded at the junction. It shall be located as near the specimens as possible.

NOTE 3—A thermometer may be used if it can be shown to agree with the specified thermocouple.

4.3 *Heat-Transfer Medium*—Any liquid heat transfer medium that remains fluid at the test temperature and will not appreciably affect the material tested may be used. Where a flammable or toxic solvent is used as the cooling medium, the customary precautions in handling such a material should be exercised. Methanol is the recommended heat transfer medium for rubber.

NOTE 4—The following materials have been used down to the indicated temperatures. When dichlorodifluoromethane refrigerant is used, it must be cooled below its boiling temperature of −29.8 C before being transferred from the cylinder to the tank of the testing machine.

Dow Corning 200 Fluids[2]	
5 cSt viscosity	−60 C
2 cSt viscosity	−76 C
Methyl alcohol	−90 C
Dichlorodifluoromethane	−120 C

NOTE 5—Specimens may be tested in a gaseous medium provided ample time has been allowed for them to reach equilibrium with the temperature of the medium.

4.4 *Temperature Control*—Suitable means (automatic or manual) shall be provided for controlling the temperature of the heat-transfer medium to within ±0.5 C of the desired value. Powdered solid carbon dioxide (dry ice) and liquid nitrogen are recommended for lowering the temperature, and an electric immersion heater for raising the temperature.

4.5 *Tank.*

4.6 *Stirrer*, to provide thorough circulation of the heat transfer medium.

5. Test Specimens

5.1 The test specimens shall be die punched as: (*1*) 6.35 ± 0.51 mm (0.25 ± 0.02 in.) wide pieces, or (*2*) modified T-50 specimens of the type illustrated in Fig. 2. The test specimens shall be 1.91 ± 0.25 mm (0.075 ± 0.010 in.) in thickness and shall be cut in lengths such that the length extending from the clamp is 2.5 ± 0.5 cm (1 ± 0.25 in.) with a minimum of 0.5 cm (0.25 in.) of material held in the clamp. The modified T-50 specimen shall be clamped so that the entire tab is inside the jaws for a minimum distance of 3.18 mm (0.125 in.). Other values of specimen thickness may be used, provided it can be shown that they give equivalent results for the material being tested.

NOTE 6—Sharp dies must be used in the preparation of specimens for this test if reliable results are to be achieved. Careful maintenance of die cutting edges is of extreme importance and can be obtained by daily lightly honing and touching up the cutting edges with jewelers' hard Arkansas honing stones. The condition of the die may be judged by investigating the rupture point on any series of broken specimens. When broken specimens are removed from the clamps of the testing machine it is advantageous to pile these specimens and note if there is any tendency to break at or near the same portion of each specimen. Rupture points consistently at the same place may be the indication that the die is dull, nicked, or bent at that particular position.

6. Conditioning

6.1 *Conditioning*—Condition the test specimens at 23 ± 2 C (73.4 ± 3.6 F) and 50 ± 5 percent relative humidity for not less than 40 h prior to test in accordance with Procedure A of ASTM Methods D 618, Conditioning Plastics and Electrical Insulating Materials for Testing,[3] for those tests where conditioning is required. In cases of disagreement, the tolerances shall be ±1 C (±1.8 F) and ±2 percent relative humidity.

6.2 *Test Conditions*—Conduct tests in the Standard Laboratory Atmosphere of 23 ± 2 C (73.4 ± 3.6 F) and 50 ± 5 percent relative humidity, unless otherwise specified in the test methods or in this specification. In cases of disagreement, the tolerances shall be ±1 C ±1.8 F) and ±2 percent relative humidity.

NOTE 7—Where long-time effects such as crystallization, incompatibility, etc., of materials are to be studied, the test specimens may be conditioned in accordance with ASTM Recommended Practice D 832, for Conditioning of Elastomeric Materials for Low-Temperature Testing.[4]

7. Procedure

7.1 Before running a test, prepare the bath and bring the apparatus to the starting temperature to be used. This may be accomplished by placing a suitable amount of powdered dry ice in the insulated tank and slowly adding the heat-transfer fluid until the tank is

[2] Available from the Dow Corning Corporation, Midland, Mich.

[3] *Annual Book of ASTM Standards*, Part 27.

[4] *Annual Book of ASTM Standards*, Part 28.

filled to a level approximately 30 to 50 mm (1 to 2 in.) from the top. During the test, the temperature of the bath may be maintained constant by the judicious addition of small quantities of coolant. Where temperatures below that obtainable with dry ice are required, liquid nitrogen may be used. Changing of temperature will require either the addition of more coolant or use of the electric immersion heater.

7.2 Mount the test specimen or specimens in the apparatus, and immerse them for 3.0 ± 0.5 min at the test temperature.

7.3 After immersion for the specified time at the test temperature, record the temperature and deliver a single impact blow to the specimens.

7.4 Examine each specimen to determine whether or not it has failed. Failure is defined as the division of a specimen into two or more completely separated pieces or as any crack in the specimen which is visible to the unaided eye. Where a specimen has not completely separated, it should be bent to an angle of 90 deg in the same direction as the bend caused by the impact. It should then be examined for cracks at the bend.

7.5 Use new specimens for each test or impact.

7.6 In establishing the brittleness temperature of a material, it is recommended that the test be started at a temperature at which 50 percent failure is expected. Test a minimum of ten specimens at this temperature and record the number of failures. In the event that all of the specimens fail or do not fail, increase or decrease, respectively, the temperature of the bath by 10 C, and repeat the test. When necessary, repeat this procedure until both failures and non-failures occur. Then change the temperature of the bath by uniform increments of 2 or 5 C, and test a minimum of ten specimens at each temperature until both the no-failure and all-failure temperatures are included.

8. Routine Inspection and Acceptance

8.1 *Procedure A*—Four routine inspection of materials received from an approved supplier, it shall be satisfactory to accept lots on the basis of testing a minimum of ten specimens at a specified temperature as stated in the relevant material specifications. Not more than five shall fail (see Section 7).

8.2 *Procedure B*—It shall be satisfactory to accept elastomeric compositions on a basis of testing five specimens at a specified temperature, as stated in the relevant material specifications. None shall fail.

9. Calculations

9.1 *Standard Method*—Using the number of specimens that failed, calculate the percentage of failures at each temperature. Calculate the brittleness temperature of the material as follows:

$$T_b = T_h + \Delta T[(S/100) - (1/2)]$$

where:

T_b = brittleness temperature, deg C,

T_h = highest temperature at which failure of all the specimens occurs (proper algebraic sign must be used), deg C,

ΔT = temperature increment, deg C, and

S = sum of the percentage of breaks at each temperature (from a temperature corresponding to no breaks down to and including T_h).

For the derivation of the above formula, see references (**4**) and (**5**).

NOTE 8: *Example*—The following example will illustrate application of this formula:

Material—Plasticized poly(vinyl chloride).

Number of Samples Tested at Each Temperature—Ten.

At −30 C, 0 specimens failed
At −32 C, 2 specimens failed
At −34 C, 3 specimens failed
At −36 C, 6 specimens failed
At −38 C, 8 specimens failed
At −40 C, 10 specimens failed
At −42 C, 10 specimens failed

Then: $T_h = -40$ C

$\Delta T = 2$

$S = 20 + 30 + 60 + 80 + 100 = 290$

Since: $T_b = T_h + \Delta T\,[(S/100) - (1/2)]$

Then: $T_b = -40 + 2\,[(290/100) - (1/2)] = -40 + 4.8 = -35.2$ C

Brittleness temperature, reported as −35 C.

9.2 *Alternative Graphic Method*—The value reported by this graphic method is essentially the same as that calculated by the Standard Method described in 9.1, but may be obtained without determining either the highest temperature at which all specimens fail or the lowest at which all pass the test. With those materials possessing a wide temperature range of brittleness transition, the graphic method also requires the testing of fewer sets of samples to determine the brittleness temperature. Select sets of ten specimens each in which both failures and non-failures occur at four or more temperatures. Choose temperatures above and below the estimated 50 percent failure point. Plot the data on probability graph paper with temperature on the linear scale and percent failure on the probability scale. Select the temperature scale so that it represents a minimum of two divisions for each degree. Draw the best fitting straight line through these points. The temperature indicated at the intersection of the data line with the 50 percent probability line shall be reported as the brittleness temperature, T_b. Where there is a question of conformance to the relevant material specification, the value obtained according to 9.1 shall be accepted as the T_b value.

NOTE 9—The graphic method applied to the data from the example in Note 8 is shown in Fig. 3. The brittleness temperature, T_b, estimated to the nearest degree is −35 C.

10. Report

10.1 The report shall include the following:

10.1.1 Brittleness temperature, to the nearest degree Celsius,

10.1.2 Complete identification of the material tested, including type, source, manufacturer's code designation, form, and previous history,

10.1.3 Method of calculation,

10.1.4 Type of apparatus used,

10.1.5 Thickness and width of test specimen,

10.1.6 Conditioning procedure followed, and

10.1.7 Date of test.

10.2 For routine inspection and acceptance testing only, the following shall be reported instead of 10.1.1 and 10.1.3:

10.2.1 Number of specimens tested,

10.2.2 Temperature of test, and

10.2.3 Number of failures.

REFERENCES

(1) Smith, E. F., and Dienes, G. J., *ASTM Bulletin*, ASTBA, No. 154, October 1948, p. 46.

(2) *Rubber Age*, RUBAA, Vol 66, No. 2, 1949, p. 182.

(3) Bimmerman, H. G., and Keen, W. N., *Industrial and Engineering Chemistry*, IECHA, Analytical Edition, Vol 16, 1944, p. 588.

(4) Cornfield, J., and Mantel, N., *Journal of the American Statistical Association*, JSTNA, Vol 45, 1950, p. 181.

(5) Finney, D. J., "Probit Analysis," Cambridge University Press, p. 39.

(6) Graves, F. L., *Rubber World*, RUBWA, Vol 113, No. 4, 1946, p. 521.

(7) Hoff, E. A. W., and Turner, S., "Study of Low Temperature Brittleness Testing of Polythene," *ASTM Bulletin*, ASTBA, No. 224, September 1957.

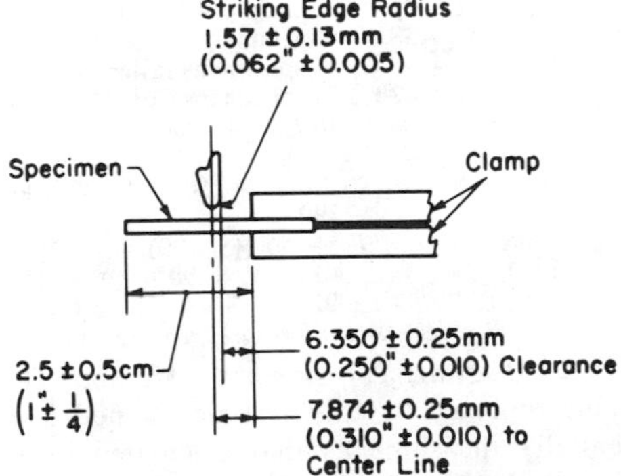

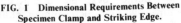

FIG. 1 Dimensional Requirements Between Specimen Clamp and Striking Edge.

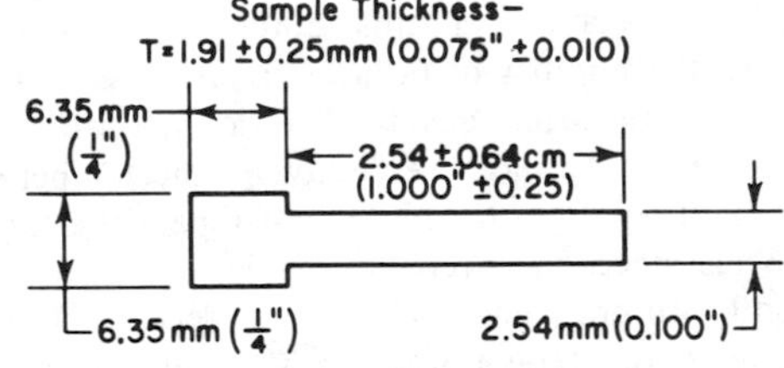

FIG. 2 Modified T-50 Specimen.

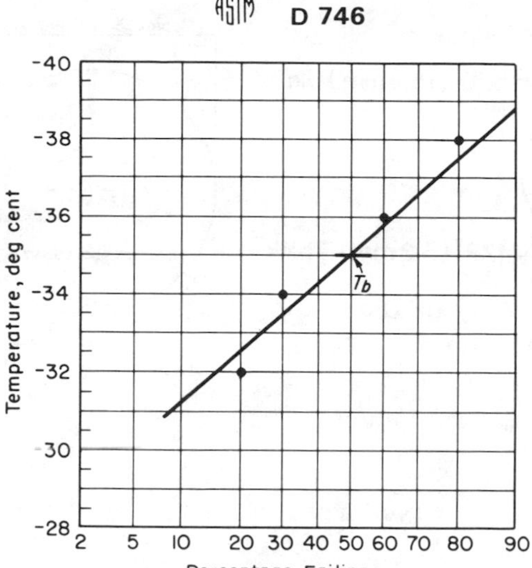

FIG. 3 Graphic Method for Determining the T_b Brittleness Temperature

APPENDIX

A1. Speed Calibration of the Solenoid-Actuated Brittleness Tester

A1.1. Calibration is accomplished by measuring the height, h, to which a steel ball, suspended on the striker mechanism of the tester, rises after the striker has had its upward motion halted by contact with a mechanical stop. The ball is accelerated in such a manner that the law governing a freely falling body applies. The velocity, v, of the striker is readily calculated from the following expression:

$$v = \sqrt{2gh}$$

A1.2 *Securing Ball Support*—Remove either one of the nuts that fasten the striking bar guide rods to the solenoid armature yoke. Place the small hole of the ball support (Fig. A1) over the guide rod and replace and secure the nut.

A1.3 *Adjusting Stroke of Striker*—Remove the metal guard from around the solenoid. Spread open the rubber bumper (Fig. A2) and insert it around the armature. Replace the solenoid guard. Insert a typical rubber or plastic specimen into the specimen holder of the tester. Raise the striking mechanism by hand until the end of the stroke is reached. It is essential that, with the striking mechanism raised to its maximum height, the striker bar of the tester be in contact with the specimen but that the bar not be in the plane of the specimen. If the striker bar is not in contact with the specimen, the rubber bumper must be removed and replaced by a thinner bumper. Conversely, if the striker bar moves into the plane of the specimen, the bumper must be replaced by a thicker one.

A1.3 *Placement of Ball and Measuring Tube*—Place a 19-mm ($^3/_4$-in.) diameter steel ball on the ball holder. (In theory, the upward flight of the ball is independent of the mass of the ball. However, if the mass is too large, the motion of the striker bar may be impeded.) Clamp a glass or clear plastic tube with a minimum inside diameter of 25.4 mm (1 in.) in a vertical position directly over the ball. The tube should contain a scale divided into 5-mm ($^1/_4$-in.) intervals. The zero position on the scale should be aligned with the top of the ball when the ball is at the top of the stroke of the striker mechanism.

A1.4 *Measurement and Calculation*—With the tester equipped as described above and devoid of test specimens and immersion medium, fire the solenoid and read the ball height to the closest 5 mm ($^1/_4$ in.) Make at least five measurements. Average all results and convert the average to meters (or feet). Determine the striker speed, v, from the following equation:

$$v = \sqrt{2\,gh}$$

where:
v = speed, m/s (or ft/s),
g = 9.8 m/s^2 (32.2 ft/s^2), and
h = average ball height, m (or ft).

NOTE A1—Calibration measurements should be made with the tester supported on a non-resilient surface, such as a laboratory bench or concrete floor. Resilient mountings tend to absorb some of the striker energy causing low ball height values.

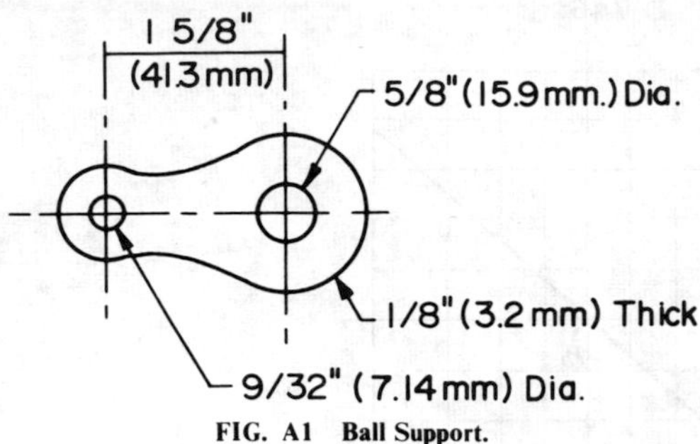

FIG. A1 Ball Support.

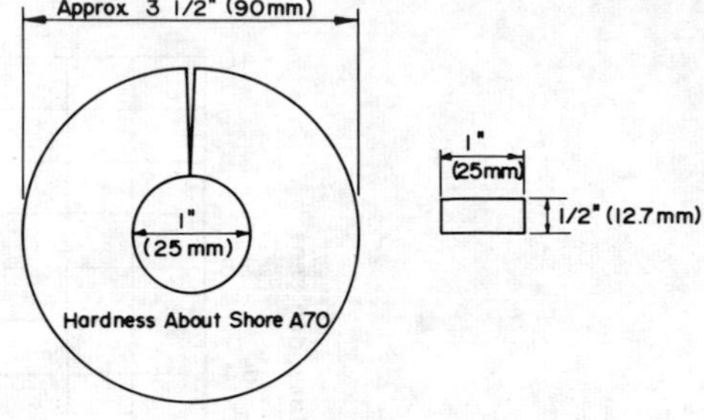

FIG. A2 Rubber Bumper.

Designation: D 747 – 63

American National Standard K65.3-1965
American National Standards Institute

Standard Method of Test for
STIFFNESS OF PLASTICS BY MEANS OF A CANTILEVER BEAM[1]

This Standard is issued under the fixed designation D 747; the number immediately following the designation indicates the year of original adoption or, in the case of revision, the year of last revision. A number in parentheses indicates the year of last reapproval.

1. Scope

1.1 This method covers determination of the stiffness of plastics by means of a cantilever beam. It is well suited for determining relative flexibility over a wide range.

NOTE 1—The values stated in U.S. customary units are to be regarded as the standard. The metric equivalents of U.S. customary units may be approximate.

2. Significance

2.1 This method provides a means of deriving an index of stiffness of a material by measuring force and angle of bend of a cantilever beam. The mathematical derivation assumes small deflections and elastic behavior. Under actual test conditions, the deformation has both elastic and plastic components. The method does not distinguish or separate these, hence a true elastic modulus is not calculable. Instead, an apparent value is obtained and is defined as the stiffness of the material.

2.2 Because of deviations from purely elastic behavior, changes in span length, width, and thickness of the specimen will affect the value of stiffness obtained; therefore, values obtained from specimens of different dimensions may not necessarily be comparable.

2.3 Rate of loading is controlled only to the extent that the rate of angular change of the rotating jaw is fixed at 58 to 66 deg/min. Actual rate of stressing will be affected by span length, width, thickness of the specimen, and weight of the pendulum.

NOTE 2—A discussion of the theory of the stiffness test will be found in Appendix A1.

3. Apparatus

3.1 The apparatus for the stiffness test, as shown in Fig. 1, shall be the cantilever beam bending type, consisting essentially of the following:

3.1.1 *Vise*—A specimen vise, V, to which the pointer indicator I_2 is attached, and which is capable of uniform clockwise rotation about the point O at a nominal rate of 60 deg of arc/min.

3.1.2 *Weighing System*—A pendulum weighing system, including an angular deflection scale, pointer indicator I_1, bending plate Q for contacting the free end of the specimen, and a series of detachable weights. This system shall be pivoted for nearly frictionless rotation about the point O. The total applied bending moment, M_W, consists of the effective moment of the pendulum and the bending plate, A_1, plus the moments of the added calibrated weights, A_2. Thus,

$$M_W = WL \sin \theta$$

where:

M_W = actual bending moment at the angle θ,
W = total applied load, kgf (or lb),
L = length of the pendulum arm, cm (or in.), and
θ = angle through which the pendulum rotates.

NOTE 3—Auxiliary weights for the Olsen stiffness tester are calibrated and marked directly with the values for M, the bending moment at a load reading of 100. Since M_W depends on the geometry of the testing machine, these weights are not interchangeable between machines of different capacities.

[1] This method is under the jurisdiction of ASTM Committee D-20 on Plastics. A list of committee members may be found in the ASTM Yearbook. This standard is the direct responsibility of Subcommittee D-20.10 on Mechanical Properties.

Current edition effective Sept. 30, 1963. Originally issued 1943. Replaces D 747 – 61.

3.1.3 *Load Scale*—A fixed scale that measures the load as a function of the deflection, θ, of the load pendulum system. It shall be calibrated such that

$$\text{Load scale reading} = 100\ WL \sin \theta / M$$

where:

M = bending moment at a load scale reading of 100.

Thus,

$$M_W = (M \times \text{load scale reading})/100$$

where:

M_W = actual bending moment.

3.1.4 *Angular Deflection Scale*—The angular deflection scale shall be calibrated in degrees of arc and shall indicate the angle through which the rotating vise has been turned relative to the pendulum system. This is the difference between the angle through which the vise has been turned and the angle through which the load pendulum has been deflected, and is designated as angle ϕ.

3.1.5 *Thickness Measuring Devices*—Suitable micrometers, or thickness gages, reading to 0.0025 mm (0.0001 in.) or less shall be used for measuring the thickness of the test specimens. The pressure exerted by the gage on the specimen being measured shall be between 160 and 185 kN/m^2 (1.5 and 2.0 kgf/cm^2) (23 and 27 psi) as prescribed in Method C of ASTM Methods D 374, Test for Thickness of Solid Electrical Insulation.[2]

3.1.6 *Width-Measuring Devices*—Suitable scales or other width measuring devices reading to 0.0025 cm (0.001 in.) or less shall be used for measuring the width of the test specimen.

4. Test Specimens

4.1 Test specimens may be molded or cut from molded, calendered, or cast sheets of the material to be tested. They shall have a rectangular cross-section and shall be cut with their longitudinal axes parallel to the direction of the principal axis of anisotropy, unless anisotropy effects are specifically to be evaluated. The length, width, and thickness of the specimen to be used will depend on the stiffness of the material and the capacity of the testing machine. Specimens shall have an even surface. If they exhibit a surface tackiness, they shall be dusted lightly with talc before being tested.

4.2 Specimen width shall be between 0.5 and 2.54 cm (0.25 and 1.00 in.) provided the material does not extend over the width of the anvil. Width shall be measured to the nearest 0.0025 cm (0.001 in.)

4.3 The minimum thickness shall be 0.05 cm (0.020 in.) and shall be measured to the nearest 0.00025 cm (0.0001 in.).

NOTE 4—A minimum thickness requirement is included since a large percentage error can result in the final stiffness value because of small errors in the thickness measurement. The reason for this large dependence of stiffness on thickness errors is because the thickness is to the third power in the formula.

4.4 The span-to-depth ratio shall be between 15 to 1 and 30 to 1.

4.5 The number of specimens tested shall be at least five.

5. Conditioning

5.1 *Conditioning*—Condition the test specimens at 23 ± 2 C (73.4 ± 3.6 F) and 50 ± 5 percent relative humidity for not less than 40 h prior to test in accordance with Procedure A of ASTM Methods D 618, Conditioning Plastics and Electrical Insulating Materials for Testing,[2] for those tests where conditioning is required. In cases of disagreement, the tolerances shall be ±1 C (±1.8 F) and ±2 percent relative humidity.

5.2 *Test Conditions*—Conduct tests in the Standard Laboratory Atmosphere of 23 ± 2 C (73.4 ± 3.6 F) and 50 ± 5 percent relative humidity, unless otherwise specified in the test methods or in this specification. In cases of disagreement, the tolerances shall be ±1 C (±1.8 F) and ±2 percent relative humidity.

5.3 Specimens to be tested at temperatures above or below normal shall be conditioned at the test temperature at least 2 h prior to testing, unless shorter equilibration time has been proven. The stiffness tester itself should be conditioned 2 h before testing.

5.4 *Lubrication of Test Apparatus*—For operations at temperatures below 0 C (32 F) it may be necessary to remove all the lubricant from the gear box, bearings, etc., of the apparatus and replace it with kerosine or silicone oil.

---

[2] *Annual Book of ASTM Standards*, Part 27.

6. Procedure

6.1 Place the machine on an approximately level surface. Add necessary weights to the pendulum and, if necessary, adjust the load scale to indicate zero. Set the bending pin or plate to the proper bending span as determined in 4.4. Start the motor and keep it running throughout the tests to minimize friction effects in the weighing system.

6.2 For maximum precision choose the value of *M* so that, at an angle of 3 deg, the load scale reading is between 5 and 10. If this value is not known, determine it by trial and error using the standard procedure. After obtaining *M*, test five specimens.

6.3 Firmly clamp the test specimen in the vise with the center line approximately parallel to the face of the dial plate. By turning the hand crank, apply sufficient load to the specimen to show a 1 percent load reading and then set the angle pointer to zero. Record this point and plot it as part of the data.

6.4 Hold down the motor engaging lever and take subsequent load scale readings at 3, 6, 9, 12, and 15 deg. Do not retest any specimen.

7. Calculation

7.1 Plot the data on coordinate paper with the load scale reading as ordinate and the angular deflection as abscissa.

7.2 Draw the steepest straight line through at least three consecutive points on the plot (see Figs. 2, 3, and 4). If this line does not pass through the origin, translate it parallel to itself until it passes through the origin. Use the data obtained from this line in the equation given in 7.3.

7.3 Calculate the stiffness to three significant figures, as follows:

$$E = (4S/wd^3) \times [(M \times \text{load scale reading})/100\,\phi]$$

where:

E = stiffness, kgf/cm² (or psi),
S = span length, cm (or in.),
w = specimen width, cm (or in.),
d = specimen thickness, cm (or in.),
M = total bending moment value of the pendulum system in kgf · cm (or lb · in.), based on the moment of the basic pendulum system, a_1, plus the moments indicated on the calibrated weight or weights, a_2, and
ϕ = reading on angular deflection scale converted to radians (Table 1).

8. Report

8.1 The report shall include the following:

8.1.1 Complete identification of the material tested, including type, source, manufacturer's code number, form, surface, width of the test specimens, span, and directionality,

8.1.2 Average stiffness and the nominal thickness used,

8.1.3 All observed and recorded data on which the calculations are based,

8.1.4 Test temperature, and

8.1.5 Date of test.

TABLE 1 Conversion Table: Degrees to Radians[a]

Degrees	Radians
3	0.0523
6	0.1047
9	0.1571
12	0.2094
15	0.2618
18	0.3141
20	0.3491
25	0.4363
30	0.5236
35	0.6109
40	0.6981
45	0.7854
50	0.8726
55	0.9599
60	1.0472

[a] 1 radian = 57 deg 18 min
1 deg = 0.01745 radians

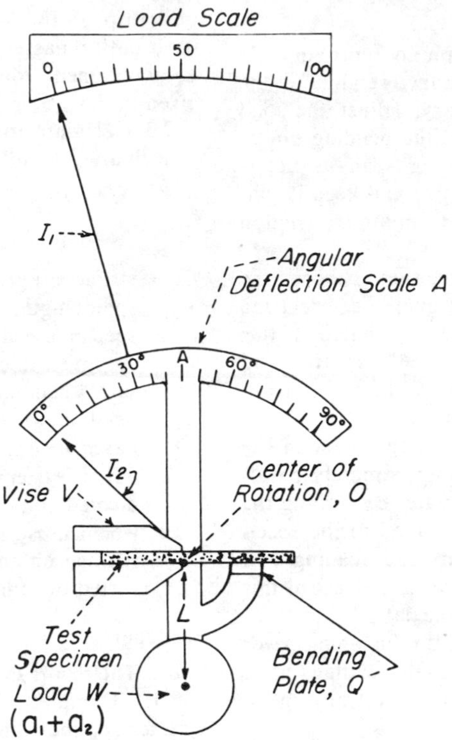

FIG. 1 Mechanical System of Stiffness Tester.

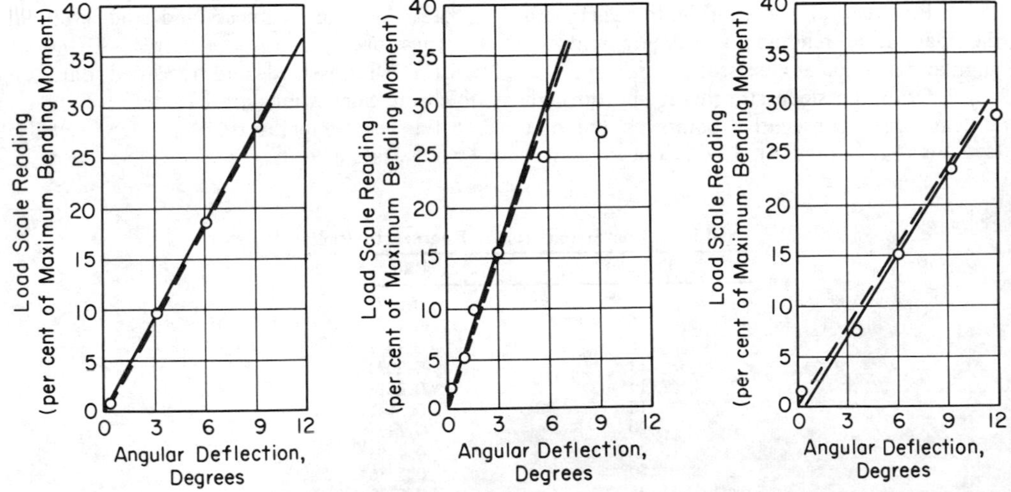

FIG. 2 Ideal Curve. FIG. 3 Typical Curve for Nonrigid Material. FIG. 4 Curve for Imperfect Specimen.

NOTE (Fig. 4)—If this type of curve is obtained, data should be taken at intervals of 5 deg until it is evident that the maximum slope has been obtained. (This type of curve may be obtained on a specimen that is warped or rough on the surface.)

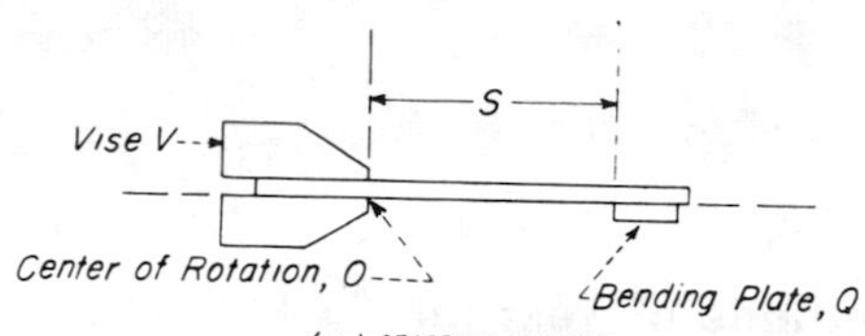

(a) STARTING POSITION

Vise V

S

P'

P

ϕ

Center of Rotation, O

Bending Plate, Q

(b) POSITION DURING TEST

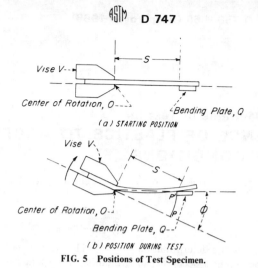

FIG. 5 Positions of Test Specimen.

APPENDIX

A1. THEORY OF OPERATION OF THE STIFFNESS TESTER

A1.1 The mechanical system of the stiffness tester is described in Section 3 of this method (Fig. 1).

A1.2 At the start of a test, the specimen is mounted as shown in Fig. 5(*a*). The load indicator I_1 reads zero on the load scale and the indicator I_2 reads zero on the angular deflection scale. During the test, the specimen vise, V, is rotated about the point O, bending the specimen through the angle ϕ against the plate Q as shown in Fig. 5(*b*). The point P on the specimen has been deflected to P' and the amount of deflection for small angles is given approximately by:

$$\overline{PP'} = M_W S^2/3EI = [(WL \sin \theta)S^2]/3EI \quad (1)$$

where:

$\overline{PP'}$ = deflection of point, P, cm (or in.),
W = applied load, kgf (or lb),
L = length of load pendulum arm to which weights are attached, cm (or in.),
θ = angular deflection of load pendulum system, deg,
S = span length of specimen, cm (or in.),
E = modulus of elasticity in flexure, kgf/cm^2 (or psi), and
I = moment of inertia of specimen cross section, which is $wd^3/12$ where w is the specimen width and d is the specimen thickness cm (or in.).

The angle, ϕ, through which the specimen bends, (Fig. 5(*b*)), which is registered by the angular deflection scale A and converted to radians, is given by:

$$\phi = \overline{PP'}/S = M_W S/3EI \quad (2)$$

Rearranging Eq 2 and substituting $wd^3/12$ for I gives:

$$E = M_W S/3I\phi = (4S/wd^3) \times (M_W/\phi) \quad (3)$$

Since the load scale is calibrated such that

$$M_W = (M \times \text{load scale reading})/100,$$

Eq 3 may be written:

$$E = (4S/wd^3) \times [(M \times \text{load scale reading})/100\,\phi] \quad (4)$$

 Designation: D 756 – 56 (Reapproved 1966)

Standard Methods of Test for
RESISTANCE OF PLASTICS TO ACCELERATED SERVICE CONDITIONS[1]

This Standard is issued under the fixed designation D 756; the number immediately following the designation indicates the year of original adoption or, in the case of revision, the year of last revision. A number in parentheses indicates the year of last reapproval.

1. Scope

1.1 These methods cover the determination of the weight and shape changes occurring in plastics under various conditions of use, not where exposure to direct sunlight, weathering, corrosive atmospheres, or heat alone is involved, but where changes in atmospheric temperature and humidity are encountered. This embraces the interior of buildings, and the interior of transport facilities such as motor vehicles, airplane cargo spaces or wing interiors, holds of ships, and railroad cars. Procedures are provided for exposing plastics to combinations of extreme humidity and temperature that will accelerate the changes taking place in the materials kept in sheltered spaces but subject to humidity and temperature variation. (See Explanatory Note).

1.2 Seven test procedures are provided which prescribe conditions for different types of exposure. Six of the procedures cover exposures at graduated levels of temperature and extremes of humidity; the seventh prescribes conditions involving alternate exposure to high and low temperatures. Insofar as weight and shape changes, embraced under the general term of dimensional stability, are concerned, these procedures provide a method of test. Further use of any of the conditions set up in the procedures is suggested, such as conditioning schedules prior to physical testing of the plastic either to test it at the particular condition involved, or to study changes resulting from exposure to that condition.

NOTE 1—The values stated in U.S. customary units are to be regarded as the standard. The metric equivalents of U.S. customary units may be approximate.

2. Significance

2.1 The test conditions covered in these methods represent a start towards organizing a group of test procedures for determining the effects of specified changes of atmospheric temperature and humidity upon plastic articles. These procedures have been used for testing both thermosetting and thermoplastic materials.

NOTE 2—The test procedures covered in these methods have been drawn from various tests used in branches of the plastics industry, and by government agencies in procurement of materials.

2.2 While several of the testing temperatures are not used in other ASTM methods, they have a background of several years experience in portions of the plastics industry. Especially in Procedures D and E (Sections 11 and 12) the accelerating effect of temperature in prompting changes caused by a rise in humidity leads to temperatures that are well above those encountered in normal service, and they are used simply to accelerate change.

2.3 Procedure A (Section 8) has been found to develop warping, weight change, and exudation in plastic parts. Procedure B (Section 9) is designed to reveal poorly cured plastics by developing cracks in them. Procedures C and D are more severe measures of the same tendencies developed in Procedure A; the conditions of Procedure D will produce noticeable chemical decomposition in many

[1] These methods are under the jurisdiction of ASTM Committee D-20 on Plastics. A list of committee members may be found in the ASTM Yearbook. This standard is the direct responsibility of Subcommittee D-20.50 on Permanance Properties.

Current edition effective Sept. 10, 1956. Originally issued 1944. Replaces D 756 – 50.

plastics. Procedure E (Section 12) is especially valuable in testing the behavior of plastic parts with metallic inserts and laminates, for cracking on exposure to temperature change. Procedures F and G are modifications of Procedure A, applying to impact-resistant and low heat-distortion temperature types of thermoplastics, respectively.

3. Apparatus

3.1 *Balance*—A balance capable of weighing accurately to 0.05 percent a test specimen weighing 100 g or less, and to 0.1 percent a test specimen weighing over 100 g.

3.2 *Oven*—A circulating-air oven capable of maintaining the required temperature of test within ±1 C (1.8 F).

3.3 *Containers*—Noncorroding containers with a shelf to support the test specimen above the solution used for maintaining the required humidity. The container shall be tightly sealed except for a small capillary which permits release of vapor pressure that might otherwise lift the top off the container. Each test specimen shall be tested preferably in a separate container.

3.4 *Desiccator*—A clean, dry, uncharged desiccator or equivalent closed container in which to bring test specimens to room temperature.

3.5 *Absorbent Cloth*—Clean, nonlinting absorbent cloth for use in wiping exudation or condensed moisture from test specimens.

3.6 *Micrometer*—A micrometer capable of measuring dimensions of test specimens to 0.025 mm (0.001 in.).

3.7 *Cold Box*—A cold box capable of maintaining the required temperature of test within ±3 C (5.4 F).

Test Specimens

4.1 The general term plastics is used to describe the samples tested, since it is the intent of these methods to provide testing procedures which apply with equal validity to finished articles comprising plastics components together with other materials, to plastics parts, to plastics articles, to test specimens molded from plastics, and, if shape and nature of the material permit, to sheets, rods, tubes, or other plastic shapes furnished in a finally processed form. The shape, size, and process of forming greatly influence the behavior of plastic objects, hence a standard size test specimen is not prescribed in these methods, but the type of test specimen to be used shall be specified by the purchaser. Each test shall be made in duplicate.

5. Conditioning

5.1 *Conditioning*—Condition the test specimens at 23 ± 2 C (73.4 ± 3.6 F) and 50 ± 5 percent relative humidity for not less than 40 h prior to test in accordance with Procedure A of ASTM Methods D 618, Conditioning Plastics and Electrical Insulating Materials for Testing,[2] for those tests where conditioning is required. In cases of disagreement, the tolerances shall be 1 C (1.8 F) and ±2 percent relative humidity.

5.2 *Test Conditions*—Conduct tests in the Standard Laboratory Atmosphere of 23 ± 2 C (73.4 ± 3.6 F) and 50 ± 5 percent relative humidity, unless otherwise specified in the test methods or in this specification. In cases of disagreement, the tolerances shall be 1 C (1.8 F) and ±2 percent relative humidity.

6. Measurements of Test Specimens

6.1 The following measurements shall be made on conditioned test specimens prior to testing, after reconditioning at the end of a test procedure, and at any intermediate stage as prescribed in the test procedures:

6.1.1 *Weight*—The weight within 0.05 percent if the specimen weighs 100 g or less, and within 0.1 percent if the specimen exceeds 100 g in weight.

6.1.2 *Dimensions*—The thickness to 0.025 mm (0.001 in.), the plane dimension in the direction of injection or transfer to 0.025 mm (0.001 in.), and the plane dimension across the direction of injection or transfer to 0.025 mm (0.001 in.).

6.1.3 *Dimensions of Compression Molded Specimen*—The thickness to 0.025 mm (0.001 in.), and the perpendicular dimensions in the plane at right angles to the direction of molding to 0.025 mm (0.001 in.).

6.2 Specimens shall be brought to room temperature in the uncharged desiccator, which will require 10 to 30 min. Then the specimen shall be weighed in less than 10 min after exposure to room conditions. The

[2] *Annual Book of ASTM Standards*, Part 27.

dimensions shall be measured immediately after weighing the specimen.

6.3 At the discretion of the purchaser, requirements for weights and measurements at intermediate stages given in the procedures may be omitted.

7. Visual Examination

7.1 Noticeable qualitative changes in surfaces, outline, and general appearance of the test specimen shall be recorded after each stage of the testing procedure. These changes include color, surface irregularities, odor, and splits, in accordance with ASTM Nomenclature D 883, Relating to Plastics.[2] Changes shall also be noted as they occur, especially those which alter the shape so that intended dimensions are no longer significant.

8. Procedure A

8.1 The test cycle for Procedure A shall be as follows:

24 h at 60 C (140 F) and 88 percent relative humidity, followed by
24 h at 60 C (140 F) in the oven.

8.2 Condition the specimen, weigh, and measure dimensions in accordance with Sections 5 and 6.

8.3 Expose the specimen for 24 h on the shelf of a container maintained at 60 ± 1 C (140 ± 1.8 F) in the oven, and containing a saturated solution of sodium sulfate to maintain a relative humidity of 85 to 89 percent.

8.4 Remove the specimen from the container, place it in the uncharged desiccator, and bring to room temperature in accordance with 6.2.

8.5 Wipe the specimen with the absorbent cloth, then weigh, measure dimensions, and examine visually in accordance with Sections 6 and 7.

8.6 Within 2 h after completion of the operation described in 8.3, expose the specimen for 24 h in the oven at 60 ± 1 C (140 ± 1.8 F).

8.7 Place the specimen in the uncharged desiccator, and bring to room temperature in accordance with 6.2.

8.8 Weigh the specimen, measure dimensions, and examine visually in accordance with Sections 6 and 7.

8.9 Recondition the specimen, weigh, and measure dimensions in accordance with Sections 5 and 6.

8.10 The specimen may be subjected to physical tests in accordance with Section 15.

9. Procedure B

9.1 The test cycle for Procedure B shall be as follows:

72 h at 60 C (140 F) in the oven.

9.2 Weigh and measure dimensions of the specimen in the as-received condition in accordance with Section 6.

9.3 Expose the specimen for 72 h in the oven at 60 ± 1 C (140 ± 1.8 F).

9.4 Place the specimen in the uncharged desiccator, and bring to room temperature in accordance with 6.2.

9.5 Weigh the specimen, measure dimensions, and examine visually in accordance with Sections 6 and 7.

10. Procedure C

10.1 The test cycle for Procedure C shall be as follows:

24 h at 70 C (158 F) and 70 to 75 percent relative humidity, followed by
24 h at 70 C (158 F) in the oven.

10.2 Condition the specimen, weigh, and measure its dimensions in accordance with Sections 5 and 6.

10.3 Expose the specimen for 24 h on the shelf of a container maintained at 70 ± 1 C (158 ± 1.8 F) in the oven, and containing a saturated solution of sodium chloride to maintain a relative humidity of 70 to 75 percent.

10.4 Remove the specimen from the container, place it in the uncharged desiccator, and bring to room temperature in accordance with 6.2.

10.5 Wipe the specimen with the absorbent cloth, then weigh, measure dimensions, and examine visually in accordance with Sections 6 and 7.

10.6 Within 2 h after the completion of the operation described in 10.3, expose the specimen for 24 h in the oven at 70 ± 1 C (158 ± 1.8 F).

10.7 Place the specimen in the uncharged desiccator and bring to room temperature in accordance with 6.2.

10.8 Weigh the specimen, measure dimensions, and examine visually in accordance with Sections 6 and 7.

10.9 Recondition the specimen, weigh, and measure dimensions in accordance with Sections 5 and 6.

10.10 The specimen may be subjected to physical tests in accordance with Section 15.

11. Procedure D

11.1 The test cycle for Procedure D shall be as follows:

24 h at 80 C (176 F) over water, followed by
24 h at 80 C (176 F) in the oven.

11.2 Condition the specimen, weigh, and measure dimensions in accordance with Sections 5 and 6.

11.3 Expose the specimen for 24 h on the shelf of a container maintained at 80 ± 1 C (176 ± 1.8 F) in the oven, and containing distilled water to maintain a humid atmosphere.

11.4 Remove the specimen from the container, place it in the uncharged desiccator, and bring to room temperature in accordance with 6.2.

11.5 Wipe the specimen with the absorbent cloth, then weigh, measure dimensions, and examine visually in accordance with Sections 6 and 7.

11.6 Within 2 h after the completion of the operation described in 11.3, expose the specimen for 24 h in the oven at 80 ± 1 C (176 ± 1.8 F).

11.7 Place the specimen in the uncharged desiccator, and bring to room temperature in accordance with 6.2.

11.8 Weigh the specimen, measure dimensions, and examine visually in accordance with Sections 6 and 7.

11.9 Recondition the specimen, weigh, and measure dimensions in accordance with Sections 5 and 6.

11.10 The specimen may be subjected to physical tests in accordance with Section 15.

12. Procedure E

12.1 The test cycle for Procedure E shall be as follows:

24 h at 80 C (176 F) and 70 to 75 percent relahumidity, followed by
24 h at −40 C (−40 F) or −57 C (−70.6 F) as specified.
24 h at 80 C (176 F) in the oven.
24 h at −40 C (−40 F) or −57 C(−70.6 F), as specified.

12.2 Condition the specimen, weigh, and measure its dimensions in accordance with Sections 5 and 6.

12.3 Expose the specimen for 24 h on the shelf of a container maintained at 80 ± 1 C (176 ± 1.8 F) in the oven and containing a saturated solution of sodium chloride.

12.4 Remove the specimen from the container, and wipe excess moisture from it with the absorbent cloth.

12.5 Within 30 min after the completion of the operation described in 12.3, expose the specimen for 24 h in a cold box maintained at −40 ± 2 C (−40 ± 3.6 F) or −57 ± 2 C (−70.6 ± 3.6 F), as specified.

12.6 Place the specimen in the uncharged desiccator and bring to room temperature in accordance with 12.2.

12.7 Wipe the specimen with the absorbent cloth, then weigh, measure dimensions, and examine visually in accordance with Sections 6 and 7.

12.8 Within 2 h of the completion of the operation described in 12.5, expose the specimen for 24 h in the oven at 80 ± 1 C (176 ± 1.8 F).

12.9 Place the specimen in the uncharged desiccator, and bring to room temperature in accordance with 6.2.

12.10 Within 30 min after the completion of the operation described in 12.6, expose the specimen for 24 h in the cold box at −40 ± 2 C (−40 ± 3.6 F) or −57 ± 2 C (−70.6 ± 3.6 F), as specified.

12.11 Place the specimen in the uncharged desiccator, and bring to room temperature in accordance with 6.2.

12.12 Wipe the specimen with the absorbent cloth, then weigh, measure dimensions, and examine visually in accordance with Sections 6 and 7.

12.13 Recondition the specimen, weigh, and measure dimensions in accordance with Sections 5 and 6.

12.14 The specimen may be subjected to physical tests in accordance with Section 15.

13. Procedure F

13.1 The test cycle for Procedure F shall be as follows:

24 h at 38 C (100.4 F) and 100 percent relative humidity, followed by
24 h at 60 C (140 F) in an oven.

13.2 Condition the specimen, weigh, and measure dimensions in accordance with Sections 5 and 6.

13.3 Expose the specimen for 24 h in a container maintained at 38 ± 1 C (100.4 ± 1.8 F) and 100 percent relative humidity (obtained by using distilled water as the humidifying medium).

13.4 Place the specimen in the uncharged desiccator, and bring to room temperature in accordance with 6.2.

13.5 Wipe the specimen with the absorbent cloth, then weigh, measure dimensions, and examine visually in accordance with Sections 6 and 7.

13.6 Within 2 h after the completion of the operations described in 13.3 and 13.4, expose the specimen for 24 h in the circulating air oven at 60 C (140 F).

13.7 Place the specimen in the uncharged desiccator and bring to room temperature in accordance with 6.2.

13.8 Weigh the specimen, measure dimensions, and examine visually in accordance with Sections 6 and 7.

13.9 Recondition the specimen, weigh, and measure dimensions in accordance with Sections 5 and 6.

13.10 The specimen may be subjected to physical tests in accordance with Section 15.

14. Procedure G

14.1 The test cycle for Procedure G shall be as follows:

24 h at 49 C (120.2 F) and 100 percent relative humidity, followed by
24 h at 49 C (120.2 F) in an oven.

14.2 Condition the specimen, weigh, and measure dimensions in accordance with Sections 5 and 6.

14.3 Expose the specimen for 24 h in a container maintained at 49 ± 1 C (120.2 ± 1.8 F) and 100 percent relative humidity (obtained by using distilled water as the humidifying medium).

14.4 Place the specimen in the uncharged desiccator, and bring to room temperature in accordance with 6.2.

14.5 Wipe the specimen with the absorbent cloth, then weigh, measure dimensions, and examine visually in accordance with Sections 6 and 7.

14.6 Within 2 h after the completion of the operations described in 14.3 and 14.4, expose the specimen for 24 h in the circulating air oven at 49 C (120.2 F).

14.7 Place the specimen in the uncharged desiccator and bring to room temperature in accordance with 6.2.

14.8 Weight the specimen, measure dimensions, and examine visually in accordance with Sections 6 and 7.

14.9 Recondition the specimen, weigh, and measure dimensions in accordance with Sections 5 and 6.

14.10 The specimen may be subjected to physical tests in accordance with Section 15.

15. Physical Tests

15.1 On appropriate test specimens, provided that prohibitive changes in dimension, structure, or shape have not taken place, physical tests, such as impact resistance and flexural strength may be conducted either at one of the test conditions, or after the final conditioning, for comparison with similar tests on specimens which have not been exposed to the conditions prescribed by these methods.

16. Report

16.1 The report shall include the following:

16.1.1 Description of test specimens, including material, shape, method of molding, and where obtained,

16.1.2 Test procedures used, and the number of cycles of each,

16.1.3 Average values of percentage changes in weight and dimensions found after (*1*) each stage of the testing procedure, and (*2*) reconditioning, when the entire cycle is finished. Notation shall be made whether all measurements prescribed in 6.1 were taken, or whether the simplified procedure of 6.3 was used, and

16.1.4 Description of qualitative changes in appearance that may have taken place during each stage of the test procedure.

EXPLANATORY NOTE

NOTE—To simulate the behavior of plastics subject to repeated exposure to heat and humidity, test procedures involving the use of repeated cycles of dry heat and humid heat may be selected from the single cycle test procedures described in these methods. In making such cycle tests, the initial conditioning shall be given followed by exposure to the test as stated. Subsequent test procedures shall be applied without reconditioning the specimen at the end of each unit of the cycle. For instance, in carrying out a three-cycle exposure to Procedure A, the first cycle shall be carried out in accordance with 8.1 to 8.8, the second cycle in accordance with 8.3 to 8.8, and the third cycle in accordance with 8.3 to 8.10. The reconditioning shall be always carried out at the end of the test cycle unless sample failure makes this step appear trivial. In reporting such tests, mention should be made of the various stages selected for the cycle, and the number of cycles to which the specimen is subjected.

ASTM Designation: D 757 - 65

Standard Method of Test for
FLAMMABILITY OF PLASTICS SELF-EXTINGUISHING TYPE[1]

This Standard is issued under the fixed designation D 757; the number immediately following the designation indicates the year of original adoption or, in the case of revision, the year of last revision. A number in parentheses indicates the year of last reapproval.

1. Scope

1.1 This method is designed for laboratory comparison evaluations of the flammability of rigid plastics in the form of sheets, plates, or molded bars having thicknesses greater than 1.3 mm (0.050 in.), when brought into contact with a surface at incandescence 950 ± 10 C (1742 ± 18 F) (Note 1). It is designed to be used for materials that have been judged self-extinguishing when tested by ASTM Method D 635, Test for Flammability of Self-Supporting Plastics[2] (Note 2).

NOTE 1—For tests of plastics in the form of thin sheets or films 1.3 mm (0.050 in.) and under in thickness, reference should be made to ASTM Method D 568, Test for Flammability of Flexible Plastics.[2] Flammability tests of sheet and plate insulation are covered in ASTM Methods D 229, Testing Sheet and Plate Materials Used for Electrical Insulation.[2]

NOTE 2—This test is not applicable to evaluating the flammability of plastic foams. The flammability test of plastic foams is referenced in ASTM Method D 1692, Test for Flammability of Plastic Sheeting and Cellular Plastics.[2]

NOTE 3—The values stated in U.S. customary units are to be regarded as the standard. The metric equivalents of U.S. customary units may be approximate.

2. Significance

2.1 This method provides comparative flammability data. Correlation with flammability under actual use conditions is not necessarily implied. It is useful in research and production control work when it is desirable to further evaluate plastics that have been judged "self-extinguishing" when tested in accordance with Method D 635.

2.2 The rate of burning will vary with thickness. Test data should be compared with data for a control material of known performance and of comparable thickness. Sheet materials that have been stretched during processing may relax during burning and give erratic results unless they are first heated above their heat distortion temperature for a time sufficient to permit complete relaxation.

3. Apparatus

3.1 *Test Chamber*—A laboratory hood, equipped with a door for complete enclosure and a window for observing the test. The exhaust fan used to remove products of combustion shall be either turned off during the test or run at a low speed that will not cause currents of air to affect the test.

3.2 *Test Unit*—A test unit consisting of parts shown in Figs. 1 and 2. The complete assembly is shown in Fig. 3. The igniting source, *A*, shall be a silicon carbide incandescent rod 8 mm ($^5/_{16}$ in.) in diameter and 27.9 cm (11 in.) in over-all length, supported in upright holders, *B*, with ceramic and asbestos bushings. The silicon carbide rod[3] with metallized contact ends for attachment of terminal straps, is mounted on upright supports, *C*.

3.2.1 *Variable Autotransformer* for alternating current or a rheostat for direct current that will provide 25 to 27 V at 14 to 15 A for heating the silicon carbide rod to 950 ± 10 C

[1] This method is under the jurisdiction of ASTM Committee D-20 on Plastics. A list of committee members may be found in the ASTM Yearbook. This standard is the direct responsibility of Subcommittee D-20.30 on Thermal Properties.

Current edition effective July 22, 1965. Originally issued 1944. Replaces D 757 - 49.

[2] *Annual Book of ASTM Standards*, Part 27.

[3] A suitable silicon rod is obtainable from the Globar Division, The Carborundum Co., Niagara Falls, N. Y., under designation of "Globar" R type "AT," 11 by 4 by $^5/_{16}$ in. The assembly requires two No. 2501 terminal straps.

(1742 ± 18 F). A variable autotransformer that will deliver up to 30 V and up to 15 A is satisfactory for a-c power sources. The voltmeter and ammeter shall show steady readings before starting the test.

3.2.2 *Wire Screen Guard, D*, with 6.4 by 6.4-mm ($^1/_4$ by $^1/_4$-in.) openings, and end plates, *E*, shield the silicon carbide rod and the terminal straps from accidental contact.

3.2.3 *Gauze Screen*, 20-mesh, for catching any molten material shall be located 6.4 mm (0.25 in.) below the lowest position of the lower edge of the specimen and shall be supported as shown in Fig. 2.

3.2.4 *Small Metal Tubes, F*, with ends restricted, direct jets of an inert gas to extinguish any flame at the end of the 3-min test time.

3.2.5 *Tubing, L*, through which gas is supplied.

3.3 *Optical Pyrometer*—A suitable optical pyrometer[4] to measure the temperature of the silicon carbide rod. The silicon carbide rod is incandescent (cherry red) at test temperature but color is not sufficient for good control of test temperature.

3.4 *Specimen Holder and Dash Pot*—Specimen holder, *H*, to clamp the test specimen with its length horizontal and at right angles to the axis of the silicon carbide rod, and its width in a vertical plane. The specimen clamp is welded to a bushing which is pinned to a shaft to permit rotation through 90 deg to position the test specimen. A flexible cable,[5] *G*, connected to the shaft and extending through the side of the hood, is for remote control without opening the hood during a test. A dash pot, *I*, with a loosely fitting plunger in water, controls the last 12.7 mm ($^1/_2$ in.) of travel of the test specimen before contact with the silicon carbide rod. The time of travel of this final distance shall be adjusted to within 2 to 5 s in order to prevent breakage of the silicon carbide rod due to impact of the test specimen.

3.5 *Counterweight and Adjustment Screw*—The 25.4 by 12.7-mm (1 by $^1/_2$-in.) thick counterweight, *J*, is adjusted to apply approximately 28 g (1 oz) of force on contacting the silicon carbide rod with the test specimen. By means of a 5.1-mm (0.200-in.) steel shim inserted between the base plate and the foot of the adjustment screw, set and lock the adjustment screw, *K*, then remove the shim before starting the test.

3.6 *Stop Watch or Timer Clock*.

4. Test Specimens

4.1 At least three test specimens of each thickness for any material shall be tested in 12.7-mm (0.50-in.) width. Specimen thickness shall exceed 1.3 mm (0.050 in.), and comparison of test results shall be on the same thicknesses. The standard test specimen shall be 12.1 cm ($4^3/_4$ in.) long by 3.17 mm ($^1/_8$ in.) thick unless special provisions are made to test other sizes. The tolerance on width shall be ±0.25 mm (±0.01 in.) and the test specimen shall be as nearly uniform in thickness as the type material permits. When specimens are sawed, cut, or milled from sheet stock or from bars, the surfaces shall be smoothed to remove fragments or burrs. Specimens thicker than 3.17 mm ($^1/_8$ in.) are inserted in the specimen holder by loosening the four socket head bolts and moving the side plate out to the desired width. Specimens shall be tested in the "as received" condition unless otherwise specified.

5. Conditioning

5.1 *Conditioning*—Condition the test specimens at 23 ± 2 C (73.4 ± 3.6 F) and 50 ± 5 percent relative humidity for not less than 40 h prior to test in accordance with Procedure A of ASTM Methods D 618, Conditioning Plastics and Electrical Insulating Materials for Testing,[2] for those tests where conditioning is required. In cases of disagreement, the tolerances shall be 1 C (1.8 F) and ±2 percent relative humidity.

5.2 *Test Conditions*—Conduct tests in the Standard Laboratory Atmosphere of 23 ± 2 C (73.4 ± 3.6 F) and 50 ± 5 percent relative humidity, unless otherwise specified in the test methods or in this specification. In cases of disagreements, the tolerances shall be 1 C (1.8 F) and ±2 percent relative humidity.

6. Procedure

6.1 Rotate the specimen holder away from the igniting bar and mount the specimen with one end flush with the rear face of the clamp.

[4] A portable instrument is obtainable from the Pyrometer Instrument Co. Inc., Bergenfield, N. J., under designation "Pyro Optical Pyrometer."

[5] Flexible cabel is available from automotive supply sources.

The length to be tested should be 10.2 cm (4 in.). Maintain the specimen in the vertical position until ready to start.

6.2 Heat the silicon carbide rod to 950 ± 10 C (1742 ± 18 F) by alternating or direct current. Adjust the electrical output to maintain this temperature as determined with the optical pyrometer, and wait until the voltmeter and ammeter show steady operation. Make any necessary adjustments to maintain temperature within the specified range before testing. Keep the hood closed with the fan either turned off or operating at very low speed to prevent drafts that will affect both temperature and burning rate.

6.3 Use the flexible cable, *G*, to rotate the specimen holder and bring the specimen to a point near the silicon carbide rod where the dash pot controls the final descent and the specimen rests upon igniting rod. At that instant start the stop watch or timer. As the specimen changes length due to burning, fusing, charring, or shrinking, its end face slides down the igniting bar until the adjustment foot contacts the base and prevents further movement.

6.4 After 3 min ± 3 s, rotate the specimen holder and any unburned length of test specimens away from the silicon carbide rod to a position between the inert gas jets, *F*, and extinguish any flame. This can be done with the hood closed. Then turn on the fan to normal capacity to remove combustion products formed during the test.

6.5 For convenience, the burned specimens may be placed beside an untested specimen and the length of the burned specimen measured to the nearest 1.0 mm (0.04 in.) at the center of the 12.7-cm ($^1/_2$-in.) side, to the point at which no charring or melting is visible.

7. Calculation

7.1 When the test specimens continue to burn for 3 min test time, calculate the rate of burning as inches per minute by dividing the length burned by three.

8. Report

8.1 The report shall include the following:

8.1.1 Identity of the sample.

8.1.2 Burning rate, in inches per minute, as the average of at least three tests.

8.1.3 When the specimen does not burn for the full 3 min, the burning time and the distance burned or charred.

8.1.4 In case the specimen melts without burning, it shall not be rated for flammability according to this test method.

8.1.5 In case the specimen melts with burning or shows any unusual behavior, this should be reported.

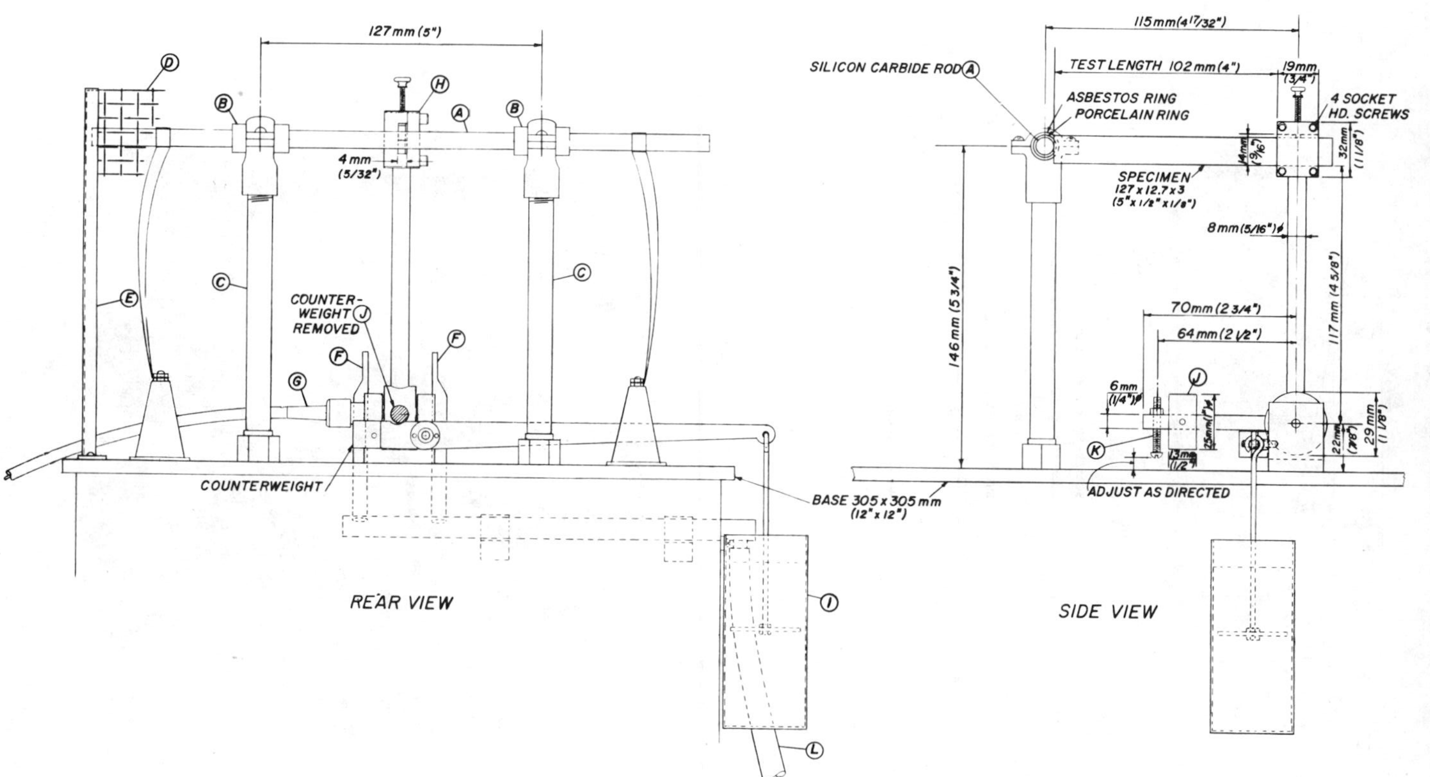

FIG. 1 Flammability Test Apparatus.

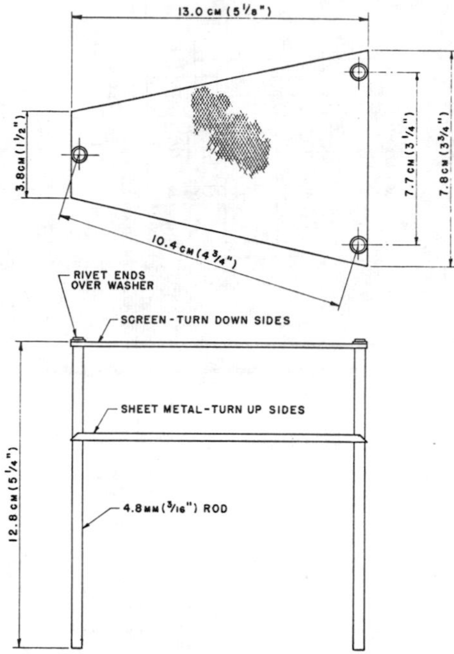

FIG. 2 Screen Ash Tray.

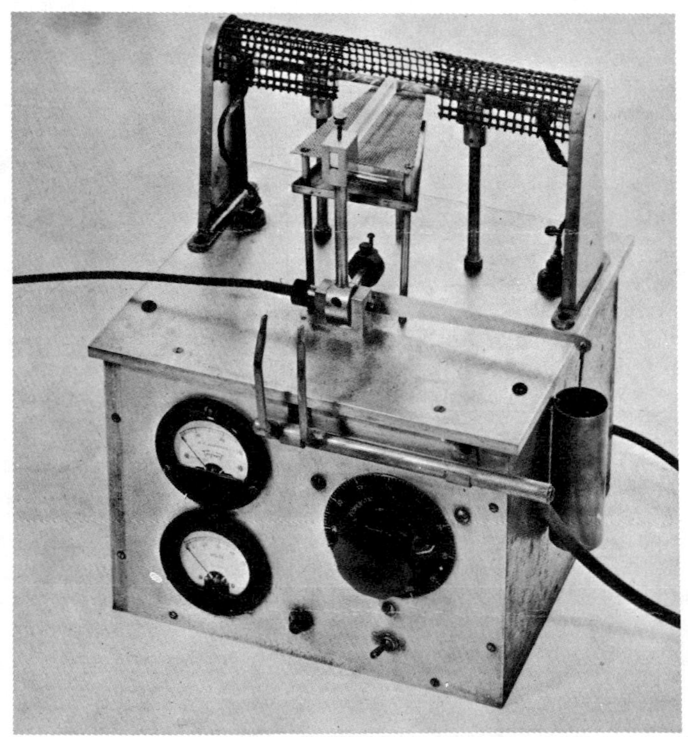

FIG. 3 Assembly of Test Unit

Designation: D 759 – 66

Recommended Practice for
DETERMINING THE PHYSICAL PROPERTIES OF PLASTICS AT SUBNORMAL AND SUPERNORMAL TEMPERATURES[1]

This Recommended Practice is issued under the fixed designation D 759; the number immediately following the designation indicates the year of original adoption or, in the case of revision, the year of last revision. A number in parentheses indicates the year of last reapproval.

1. Scope

1.1 This recommended practice is for use in determining the physical properties of plastics by means of appropriate ASTM test methods, at temperatures from −269 to +550 C (−452 to +1022 F) (Note 1), excluding the Standard Laboratory Temperature (23 C).

NOTE 1—Safeguards against the effects of extreme temperatures should be utilized to protect the operator.

1.2 These techniques are not intended for use in determining the effect of long continued exposure at elevated or reduced temperatures (see ASTM Recommended Practice D 794, for Determining Permanent Effect of Heat on Plastics[2]) or of cyclic exposures to ranges of temperature, although they may be used for such special tests if desired.

NOTE 2—The values stated in U.S. customary units are to be regarded as the standard. The metric equivalents of U.S. customary units may be approximate.

2. Apparatus

2.1 *Testing Equipment*—Any properly calibrated test equipment in accordance with the appropriate ASTM test method. All parts that are exposed to the high or low temperatures during the test shall be adjusted to function normally at these temperatures, and their calibration shall be verified by calibration at the chosen test temperature(s). Verification shall be repeated as often as necessary to ensure correct readings.

2.2 *Insulated Test Chamber*—An insulated test chamber, in which the heating or cooling medium (gas or liquid) is maintained at the desired temperature, to enclose the specimen during the test (Note 3). The test medium shall have no deleterious effect on the specimen. Where a medium of low heat transfer characteristics is used, adequate circulation or agitation shall also be used to ensure that conduction of heat through the components will not materially change the temperature of the specimen from that of the specified test temperature. The temperature of the medium (Note 4) in the test chamber may be regulated by any convenient means, within the following tolerances (unless otherwise specified):

±3 C (±5 F) from −70 to 300 C (−94 to 572 F), and

±2 percent over 300 C (572 F) and ±4 percent below −70 C (−94 F).

Where possible, the test temperature shall be selected from the test temperatures listed in ASTM Methods D 618 Conditioning Plastics and Electrical Insulating Materials for Testing.[2]

NOTE 3—Additional information regarding systems that may be used up to 300 C (572 F) and down to −70 C (−94 F) for maintaining the desired conditions is contained in ASTM Specifications E 197, for Enclosures and Servicing Units for Tests Above and Below Room Temperature.[3]

NOTE 4—Suggested cooling media and approximate temperatures are:

---

[1] This recommended practice is under the jurisdiction of ASTM Committee D-20 on Plastics. A list of committee members may be found in the ASTM Yearbook. This standard is the direct responsibility of Subcommittee D-20.10 on Mechanical Properties.

Current edition effective Sept. 7, 1966. Originally issued 1944. Replaces D 759 – 48.

[2] *Annual Book of ASTM Standards*, Part 27.

[3] *Annual Book of ASTM Standards*, Part 30.

Refrigerated air	−70 to +22 C (−94 to +72 F)
Carbon dioxide gas or dry ice (CO_2) in acetone	−79 C (−110 F)
Liquid nitrogen	−196 C (−320 F)
Liquid hydrogen	−253 C (−423 F)
Liquid helium	−269 C (−452 F)

Appropriate safety precautions should be observed when using flammable media, such as acetone or hydrogen.

2.3 *Temperature-Measuring Equipment*—Temperature-measuring device(s) capable of accuracy as required in 2.2 shall be placed in the medium surrounding the test specimen as close to the specimen as possible. In the case of large specimens, more than one temperature-sensing element may be needed, particularly if the medium is not stirred or circulated, and the temperature variation shall not exceed the limits stated in 2.2. Initially, a control specimen identical to the test specimen with the exception of a temperature-sensing element embedded in the center may be used to determine the time necessary to achieve thermal equilibrium at the test temperature. Thermocouple wire, 24-gage of finer, may be molded, or inserted in a drilled hole, in the control specimen.

2.4 *Preconditioning Oven*—A circulating-air oven, adjusted at 50 ± 3 C (122 ± 5 F) for preconditioning test specimens when required.

2.5 *Desiccators*—Desiccators containing anhydrous calcium chloride (or other suitable desiccant) for storage of the preconditioned specimens.

2.6 *Conditioning Chamber*—A chamber for conditioning the specimens at the desired test temperature just prior to testing and using the same medium as used in the test. This conditioning shall be done either in an insulated storage chamber of the circulating- or agitating-medium type or in the insulated test chamber. During the conditioning period care shall be exercised to allow free access of the conditioning medium to all of the test specimens.

3. Test Specimens

3.1 Test specimens shall conform to the applicable ASTM method or specification. In addition:

3.1.1 When machining is required, care should be taken that edges are smooth and square.

3.1.2 Standard methods shall be used for measuring dimensions of the specimen to the nearest 0.025 mm (0.001 in.) or better.

NOTE 5—Measurements on thin specimens (for example, films) require closer tolerances. Measurements on rough-surfaced specimens (for example, vacuum bag-molded reinforced plastics) should be obtained by a ball-point micrometer. Measurements of the specimen may change, especially when tests are conducted at extreme temperatures, and this change will have to be recognized in such cases.

3.13 The number of specimens shall conform to the applicable ASTM test method.

4. Conditioning

4.1 *Preconditioning*—All test specimens shall be preconditioned (Note 6) in accordance with Procedure B of Methods D 618, unless this condition is waived by agreement between the parties concerned.

NOTE 6—At the higher temperatures, tests will necessarily be made on dried specimens, regardless of the type of preconditioning. Consequently, in order that tests at all temperatures be made on the same basis as far as moisture content is concerned, a preconditioning treatment that removes all moisture is specified.

4.2 *Conditioning*—Test specimens shall be placed in the conditioning or test chamber in the test medium at the specified temperature and allowed to remain until thermal equilibrium is attained.

NOTE 7—Suggested minimum conditioning times are 1 h for specimens less than 6 mm (0.24 in.) in thickness, or 2 h for specimens greater than 6 mm (0.24 in.) in thickness (see also 5.2).

5. Procedure

5.1 If conditioned in a separate conditioning chamber, place such temperature-conditioned specimens in the test fixture of the test machine as rapidly as possible. Transfer from the temperature-conditioning chamber, if necessary, shall not require longer than 30 s. After thermal equilibrium (as defined below) has been restored, test the specimen in accordance with the applicable ASTM method.

5.2 Determine the time necessary for attainment of thermal equilibrium by appropriate measurements on the conditioned control specimens referred to in 2.3. Test the specimens after expsoure for a period 1.3 times the time found necessary for the control specimen, ±10 percent.

NOTE 8—The time requirement may be modified if it can be demonstrated that no appreciable error is introduced by so doing, and all concerned parties agree.

6. Calculations

6.1 Use standard equations for the applicable ASTM method in calculating the properties of the specimens.

6.2 For each series of tests, calculate the arithmetic means of all values obtained to three significant figures, and report as the "average value" for the particular property in question.

6.3 Calculate the standard deviation (estimated) as follows, and report to two significant figures.

$$s = \sqrt{\Sigma X^2 - n\bar{X}^2)/(n - 1)}$$

where:

s = estimated standard deviation,
X = value of a single observation,
n = number of observations, and
$\bar{X}$ = arithmetic mean of the set of observations.

7. Report

7.1 The report shall include information specified in the applicable ASTM method and shall include the following:

7.1.1 Complete identification of the material tested,

7.1.2 ASTM method used, with deviations, if any, from the method,

7.1.3 Preconditioning and conditioning procedures,

7.1.4 Temperature at the specimen during test and the duration of temperature exposure.

7.1.5 Temperature medium used,

7.1.6 Equilibrium temperature of control specimen as determined by 2.3 under chosen conditions of test, and

7.1.7 Calculated properties of materials tested and additional visible changes, if any.

Designation: D 785 - 65

USA Standard K65.3-1965
USA Standards Institute

Standard Method of Test for ROCKWELL HARDNESS OF PLASTICS AND ELECTRICAL INSULATING MATERIALS[1]

This Standard is issued under the fixed designation D 785; the number immediately following the designation indicates the year of original adoption or, in the case of revision, the year of last revision. A number in parentheses indicates the year of last reapproval.

1. Scope

1.1 This method covers two procedures for testing the indentation hardness of plastics and related plastic electrical insulating materials by means of the Rockwell hardness tester.

NOTE 1—The values stated in U.S. customary units are to be regarded as the standard. The metric equivalents of U.S. customary units may be approximate.

2. Significance

2.1 A Rockwell hardness number is a number derived from the net increase in depth of impression as the load on an indenter is increased from a fixed minor load to a major load and then returned to a minor load (Procedure A). A Rockwell alpha (α) hardness number represents the maximum possible remaining travel of a short-stroke machine from the net depth of impression, as the load on the indenter is increased from a fixed minor load to a major load (Procedure B). Indenters are round steel balls of specific diameters. Rockwell hardness numbers are always quoted with a scale symbol representing the indenter size, load, and dial scale used. This method is based on ASTM Methods E 18, Test for Rockwell Hardness and Rockwell Superficial Hardness of Metallic Materials.[2] Procedure A (Section 9) yields the indentation of the specimen remaining 15 s after a given major load is released to a standard 10-kg minor load for 15 s. Procedure B (Section 10) yields the indentation of the indenter into the specimen after a 15-s application of the major load while the load is still applied. Each Rockwell scale division represents 0.002-mm (0.00008-in.) vertical movement of the indenter.

2.2 A Rockwell hardness number is directly related to the indentation hardness of a plastic material, the higher the reading the harder the material. An α hardness number is equal to 150 minus the instrument reading. Due to a short overlap of Rockwell hardness scales by Procedure A, two different dial readings on different scales may be obtained on the same material, both of which may be technically correct.

2.3 For certain types of materials having creep and recovery, the time factors involved in applications of major and minor loads have a considerable effect on the results of the measurements.

2.4 The results obtained by this method are not generally considered a measure of the abrasion or wear resistance of the plastic materials in question.

2.5 Indentation hardness is used as an indication of cure of some thermosetting materials at room temperature. Generally, an undercured specimen has a hardness reading below normal.

2.6 Indentation hardness may be used as a control test for indicating the punching quality of laminated sheet stock at the processing temperature. Test methods are given in ASTM Method D 617, Test for Punching Quality of Phenolic Laminated Sheets.[3]

---

[1] This method is under the jurisdiction of ASTM Committee D-20 on Plastics. A list of committee members may be found in the ASTM Yearbook. This standard is the direct responsibility of Subcommittee D-20.10 on Mechanical Properties.

Current edition effective Aug. 31, 1965. Originally issued 1944. Replaces D 785 – 62.

[2] *Annual Book of ASTM Standards*, Part 31.

[3] *Annual Book of ASTM Standards*, Part 29.

2.7 Each Rockwell hardness scale in Table 1 is an extension of the preceding less severe scale, and while there is some overlap between adjacent scales, a correlation table is not desirable. Readings on one material may be satisfactory for such a table, but there is no guarantee that other plastic materials will give corresponding readings because of differences in elasticity, creep, and tear characteristics.

3. Factors Affecting Reproducibility and Accuracy

3.1 Rockwell hardness readings have been found reproducibile to ±2 divisons for certain homogeneous materials with a Young's modulus in compression over 35,000 kgf/cm^2 (5 × 10^5 psi). Softer plastics and coarse-filled plastic materials will have a wider range of variation. A large ball indenter will distribute the load more evenly and decrease the range of test results (Note 2). The sensitivity of the instrument decreases with an increase in the dial reading and becomes very poor for readings of 100 and over due to the shallow indention of the steel ball. It is desirable to use the smallest ball and highest load that is practical because of this loss of sensitivity. Rockwell hardness readings over 115 are not satisfactory and shall not be reported. Readings between zero and 100 are recommended, but readings to 115 are permissible. For comparison purposes, it may be desirable to take readings higher than 115 or lower than zero on any single scale. In such cases, Rockwell hardness readings may be reported, but the corresponding correct reading shall follow in parentheses, if possible. Such alternate readings are not always feasible when the specimen is subject to constantly changing conditions or irreversible reactions.

NOTE 2—Molded specimens containing coarse fiber fillers, such as woven glass fabric, will influence the penetration obtained. These variations in hardness may be reduced by testing with the largest ball indenter consistent with the overall hardness of the material.

3.2 If the bench or table on which a Rockwell hardness tester is mounted is subject to vibration, such as is experienced in the vicinity of other machines, the tester should be mounted on a metal plate with sponge rubber at least 2.5 cm (1 in.) thick, or on any other type of mounting that will effectually eliminate vibrations from the machine. Otherwise the indenter will indent farther into the material than when such vibrations are absent.

3.3 Dust, dirt, grease, and scale or rust should not be allowed to accumulate on the indenter, as this will affect the results. Steel ball indenters that have nicks, burrs, or are out of round shall not be used.

3.4 The condition of the test equipment is an important factor in the accuracy of the tests. Dust, dirt, or heavy oil act as a cushion to the load supporting members of the test equipment and cause erroneous readings of the instrument dial. The shoulders of the instrument housing, indenter chuck, ball seat in the instrument housing, capstan, capstan screw, and anvil shoulder seat should be kept clean and true. The capstan and screw should be lightly oiled.

3.5 Surface conditions of the specimen have a marked effect on the readings obtained in a test. Generally, a molded finish will give a higher Rockwell reading than a machined face due to the high resin content of filled materials or better orientation and lower plasticizer content of unfilled plastic materials. Tubular or unsupported curved specimens are not recommended for plastic hardness testing. Such curved surfaces have a tendency to yield with the load and produce an unsymmetrical indentation pattern.

3.6 Many plastic materials have anisotropic characteristics, which cause indentation hardness to vary with the direction of testing. In such cases, the hardest face is generally that one perpendicular to the molding pressure.

3.7 Rockwell hardness tests of the highest accuracy are made on pieces of sufficient thickness so that the Rockwell reading is not affected by the supporting anvil. A bulge, change in color, or other marking on the under surface of the test specimen closest to the anvil, is an indication that the specimen is not sufficiently thick for precision testing. Stacking of thin specimens is permitted provided they are flat, parallel, and free from dust or burrs. The precision of the test is reduced for stacked specimens, and results should not be compared to a test specimen of standard thickness.

4. Apparatus

4.1 *Rockwell Hardness Tester*, in accordance with the requirements of Section 6. A flat anvil at least 5 cm (2 in.) in diameter shall be used as a base plate for flat specimens.

4.2 For Rockwell hardness testing, it is necessary that the major load, when fully applied, be completely supported by the specimen and not held by other limiting elements of the machine. To determine whether this condition is satisfied, the major load should be applied to the test specimen. If an additional load is then applied, by means of hand pressure on the weights, the needle should indicate an additional indentation. If this is not indicated, the major load is not being applied to the specimen, and a long-stroke (PL) machine or less severe scale should be used. For the harder materials with a modulus around 55,000 kgf/cm^2 (8×10^5 psi) or over a stroke equivalent to 150 scale divisions, under major load application, may be adequate; but for softer materials the long-stroke (250 scale divisions under major load) machine is required.

4.3 A V-block anvil or double roller anvil shall be used if solid rods are tested.

5. Test Specimen

5.1 The standard test specimen shall have a minimum thickness of 0.6 cm ($^1/_4$ in.) unless it has been verified that, for the thickness used, the hardness values are not affected by the supporting surface and that no imprint shows on the under surface of the specimen after testing. The specimen may be a piece cut from a molding or sheet or it may be composed of a pile-up of several pieces of the same thickness, provided that precaution is taken that the surfaces of the pieces are in total contact and not held apart by sink marks, burrs from saw cuts, or other protrusions. Care shall be taken that the test specimen has parallel flat surfaces to ensure good seating on the anvil and thus avoid the deflection that may be caused by poor contact. The specimen shall be at least 2.5 cm (1 in.) square if cut from sheet stock, or at least 6 cm^2 (1 in.^{2}) in area if cut from other shapes. The minimum width shall be 1.2 cm ($^1/_2$ in.).

5.2 The diameter of solid rod specimens shall be not less than three times the diameter of the steel ball indenter. Tests parallel to the laminations of rods cut and ground from sheet stock are not recommended due to partial delamination of the specimen.

6. Calibration

6.1 Check the Rockwell hardness tester periodically with a small machinist's level along both horizontal axes from a flat anvil for correct positioning. Minor errors in leveling are not critical, but correct positioning is desirable.

6.2 The adjustment of speed-of-load application is of great importance. Adjust the dashpot on the Rockwell tester so that the operating handle completes its travel in 4 to 5 s with no specimen on the machine or no load applied by the indenter to the anvil. The major load shall be 100 kg for this calibration. When so adjusted, the period taken for the mechanism to come to a stop with the specimen in place will vary from 5 to 15 s, depending upon the particular specimen, the indenter, and the load used. The operator should check the instrument manual for this adjustment.

6.3 Select a test block in a range using ball type indenters (B or E, see Table 1 of Methods E 18), and make five impressions on the test surface of the block. Compare the average of these five tests against the hardness calibration of the block. If the error is more than ± 2 hardness numbers, bring the machine into adjustment as described in 6.4 or in 3.3 and 3.4.

NOTE 3—Standard test blocks for the other scales in Table 1 are not now available. If the equipment checks out on the E scale and K scale, it can be assumed that the basic Rockwell hardness tester is satisfactory. It may be assumed that other indenters are satisfactory if they meet the conditions of Section 3 and Table 1.

6.4 Check the index lever adjustment periodically and make adjustments if necessary. To adjust the index lever, place a specimen (plastic with no creep or soft metal) on the anvil and turn the knurled elevating ring to bring the specimen in contact with the indenter. Keep turning the ring to elevate the specimen until positive resistance to further turning is felt, which will be after the 10-kg minor load is encountered. When excessive power would have to be used to raise the specimen higher, set the dial so that the set position is at the top and take note of the position of the pointer on the dial. If the pointer is between B 50 and B 70 on the red scale, no adjustment is necessary; if the pointer is between B 45 and B 50, adjustment is advisable; and if the pointer position is anywhere else, adjustment is imperative. As the pointer revolves several times when the specimen is elevated, the readings mentioned above apply to that revolution of the pointer which occurs either as the reference mark on the gage stem disappears into the sleeve or as the auxiliary hand on the dial passes beyond the zero setting on the dial. The object of this adjustment is to determine if the elevation of the specimen to the minor load does not cause even a partial application of the major load. Apply the major load only through the release mechanism.

7. Conditioning

7.1 *Conditioning*—Condition the test specimens at 23 ± 2 C (73.4 ± 3.6 F) and 50 ± 5 percent relative humidity for not less than 40 h prior to test in accordance with Procedure A of ASTM Methods D 618, Conditioning Plastics and Electrical Insulating Materials for Testing,[4] for those tests where conditioning is required. In cases of disagreement, the tolerances shall be 1 C (1.8 F) and ±2 percent relative humidity.

7.2 *Test Conditions*—Conduct tests in the Standard Laboratory Atmosphere of 23 ± 2 C (73.4 ± 3.6 F) and 50 ± 5 percent relative humidity, unless otherwise specified in the test methods or in this specification. In cases of disagreements, the tolerances shall be 1 C (1.8 F) and ±2 percent relative humidity.

NOTE 4—Operation this test equipment above and below normal temperature is not recommended due to the change in viscosity of the dashpot oil and the close tolerance of the gage.

8. Number of Tests

8.1 At least five hardness tests shall be made on isotropic materials. For anisotropic materials, at least five hardness tests shall be made along each principal axis of anisotropy, provided the sample size permits.

9. Procedure A

9.1 Choose the correct scale for the specimen under test. Rockwell hardness values are reported by a letter to indicate the scale used and a number to indicate the reading. The choice of scales shall be governed by the considerations concerned with the total indentation readings and the final scale reading for a particular material and scale (see Table 1 and 3.1, 4.2 and 9.5). The Rockwell hardness scale used shall be selected from those listed in Table 1, unless otherwise noted in individual methods or specifications.

9.2 Discard the first reading after changing a ball indenter, as the indenter does not properly seat by hand adjustment in the housing chuck. The full pressure of the major load is required to seat the indenter shoulder into the chuck.

9.3 With the specimen in place on the anvil, turn the capstan screw until the small pointer is at a zero position and the large pointer is within ±5 divisions of B 30 or the "set" position on red scale. This adjustment applies without shock a minor load of 10 kg, which is built into the machine. Final adjustment of the gage to "set" is made by a knurled ring located on some machines just below the capstan hand wheel. If the operator should overshoot his "set" adjustment, another trial shall be made in a different test position of the specimen; under no circumstances should a reading be taken when the capstan is turned backward. Within 10 s after applying the minor load, and immediately after the "set" position is obtained, apply the major load by releasing the trip lever (Note 5). Remove the major load 15 (+1, −0) s after its application. Read the Rockwell hardness on the red scale to the nearest full-scale division 15 s after removing the major load.

NOTE 5—The application of the minor load starts when the gage pointer starts to move; this is not the point of final zero adjustment.

9.4 Record the readings as follows: Count the number of times the needle passes through zero on the red scale on the application of the major load. Subtract from this the number of times the needle passes through zero upon the removal of this load. If this difference is zero, record the value as the reading plus 100. If the difference is 1, record the reading without change, and, if the difference is 2, record the reading as the scale reading minus 100 (Note 6). Softer plastic materials, requiring a less severe scale than the R scale, shall be tested by ASTM Method D 2240, Test for Indentation Hardness of Rubber and Plastics by Means of a Durometer.[4]

---

[4] *Annual Book of ASTM Standards*, Part 27.

NOTE 6: *Example*—With a difference of two revolutions and a scale reading of 97, indentation hardness values should be reported as −3 = (97 −100).

9.5 If the total indentation, reading with the major load applied, for a particular scale exceeds the limits of the test machine used (150 divisions for regular machines and 250 divisions for PL machines), use the next less severe scale. Thus, if the M scale indentation (with major load) is 290 divisions, use L Rockwell scale. Determine the total indentation by the number of divisions the pointer travels from the zero set position during the 15 s from the time the lever is tripped.

9.6 Do not make the tests so near the edge of the specimen that the indenter will crush out the edge when the major load is applied. In no case shall the clearance be less than 0.6 cm ($^1/_4$ in.) to the edge of the specimen. Neither should tests be made too close to each other, as the plastic surface is damaged from the previous indentation. Never make duplicate tests on the opposite face of a specimen; if a specimen is turned over and retested on the opposite face, the ridges on the first face will contribute to a softer reading on the second face.

10. Procedure B

10.1 Use the R scale with a 1.27-cm ($^1/_2$-in.) indenter and 60-kg major load in this test.

NOTE 7—This is the only scale approved for plastics testing by Procedure B.

10.2 Determine the "spring constant" or correlation factor of the machine as follows: Place a soft copper block, of sufficient thickness and with plane parallel surfaces, on the anvil in the normal testing position. Raise the sample and anvil by the capstan screw to the 1.27-cm ($^1/_2$-in.) indenter until the small pointer is at the starting dot and the large pointer reads zero on the black scale. Apply the major load by tripping the load release lever. The dial gage will then indicate the vertical distance of indentation plus the spring of the machine frame and any other elastic compressive deformation of the indenter spindle and indenter. Repeat this operation several times without moving the block, but resetting the dial to zero after each test while under minor load, until the deflection of the dial gage becomes constant, that is, until no further indentation takes place and only the spring of the instrument remains. This value, in terms of dial divisions, is the correction factor to be used in 10.4.

10.3 Following the machine adjustment described in 6.2 and 6.4, place the test specimen in position on the anvil. With the specimen in place, apply the minor load of 10 kg and make the zero setting within 10 s; apply the major load *immediately* after the zero setting has been completed (Note 8). Observe and record the total number of divisions that the pointer of the dial gage passes through during 15 s of major load application. It is to be noted that the numbers shown on the standard dial gage are not to be used, but the actual scale divisions representing indentation shall be counted. For this reason, the black scale is recommended. A full revolution of the needle represents 100 divisions (Note 9). The limitations of 4.2 still apply to this procedure.

NOTE 8—Materials subject to excessive creep are not suitable for this procedure.

NOTE 9—The total indentation equals the number of revolutions × 100 + the reading on the black scale.

10.4 The total scale divisions that are indicated by the dial gage, as observed in 10.3, represent the indentation produced plus the spring constant correction of the test instrument. Subtract this correction factor (10.2) from the observed reading, and record the difference as the total indentation under load.

10.5 The hardness determined by this procedure shall be known as the alpha, α, Rockwell hardness number and shall be calculated as follows:

$$\alpha \text{ Rockwell hardness number} = 150 - \text{total indentation under load}$$

NOTE 10: *Examples*—With a total indentation of 30 divisions, obtained as described in 10.3 and 10.4, the value is α 120; or if the total indentation is 210, the value is $-\alpha$ 60.

11. Calculation

11.1 Calculate the arithmetic mean for each series of tests on the same material and at the same set of test conditions. Report the results as the "average value" rounded to the equivalent of one dial division.

11.2 Calculate the standard deviation (estimated) as follows, and report it to two significant figures:

$$s = \sqrt{(\Sigma X^2 - n\bar{X}^2)/(n - 1)}$$

where:

s = estimated standard deviation,
X = value of a single observation,
$\bar{X}$ = arithmetic mean of a set of observations, and
n = number of observations in the set.

12. Report

12.1 The report shall include the following:

12.1.1 Material identification,

12.1.2 Filler identification and particle size, if possible,

12.1.3 Total thickness of specimen,

12.1.4 The number of pieces in the specimen and the average thickness of each piece,

12.1.5 Surface conditions, for example, molded or machined, flat or round,

12.1.6 The procedure used (Procedure A or B),

12.1.7 The direction of testing (perpendicular or parallel to molding or anisotropy),

12.1.8 A letter indicating the Rockwell hardness scale used,

12.1.9 An average Rockwell hardness number calculated by Procedure A or B,

12.1.10 The standard deviation, and

12.1.11 The testing conditions.

TABLE 1 Rockwell Hardness Scales

Rockwell Hardness Scale (Red Dial Numbers)	Minor Load, kg	Major Load, kg[a]	Indenter Diameter	
			in.	cm
R	10	60	0.5000 ± 0.0001	1.27000 ± 0.00025
L	10	60	0.2500 ± 0.0001	0.63500 ± 0.00025
M	10	100	0.2500 ± 0.0001	0.63500 ± 0.00025
E	10	100	0.1250 ± 0.0001	0.31750 ± 0.00025
K	10	150	0.1250 ± 0.0001	0.31750 ± 0.00025

[a] This major load is not the sum of the actual weights at the back of the frame but is a ratio of this load, depending on the leverage arm of the machine. One make and model has a 25 to 1 leverage arm.

Standard Method of Test for FLEXURAL PROPERTIES OF PLASTICS[1]

This Standard is issued under the fixed designation D 790; the number immediately following the designation indicates the year of original adoption or, in the case of revision, the year of last revision. A number in parentheses indicates the year of last reapproval.

NOTE—Editorial correction to units of *m* in Eq 5 was made in September 1966.

1. Scope

1.1 This method covers the determination of flexural properties of plastics and electrical insulating materials in the form of rectangular bars molded directly or cut from sheets, plates, or molded shapes. The method is generally applicable to rigid and semirigid materials: however, flexural strength cannot be determined for those materials that do not break or that do not fail in the outer fibers. Two procedures are described, as follows:

1.1.1 *Procedure A* is designed principally for materials that break at comparatively small deflections.

1.1.2 *Procedure B* is designed particularly for those materials that undergo large deflections during testing.

1.2 Comparative tests may be run according to either procedure, provided that procedure is found satisfactory for the materials being tested. All specification tests, however, must be run according to Procedure A, unless otherwise stated in the material specifications.

NOTE 1—The values stated in U.S. customary units are to be regarded as the standard. The metric equivalents of U.S. customary units may be approximate.

2. Summary of Method

2.1 A bar of rectangular cross section is tested in flexure as a simple beam, the bar resting on two supports and the load applied by means of a loading nose midway between the supports. The specimen is deflected until rupture occurs, or until the maximum fiber strain (see 11.7) of 5 percent is reached, whichever occurs first.

3. Significance

3.1 Flexural properties determined by this method are especially useful for quality control and specification purposes. Reproducibility between specimens is approximately ±5 percent for homogeneous materials tested under comparable conditions. However, flexural properties may vary with specimen thickness, temperature, atmospheric conditions, and the differences in rate of straining specified in Procedures A and B (see also Note 8).

4. Apparatus

4.1 *Testing Machine*—A properly calibrated testing machine which can be operated at constant rates of crosshead motion over the range indicated, and in which the error in the load-measuring system shall not exceed ±1 percent. It shall be equipped with a deflection-measuring device. The stiffness of the testing machine shall be such that the total elastic deformation of the system does not exceed 1 percent of the total deflection of the test specimen during test, or appropriate corrections shall be made. The load-indicating mechanism shall be essentially free from inertia lag at the crosshead rate used. The accuracy of the testing machine shall be verified in accordance with ASTM Methods E 4, Verification of Testing Machines.[2]

---

[1] This method is under the jurisdiction of ASTM Committee D-20 on Plastics. A list of committee members may be found in the ASTM Yearbook. This standard is the direct responsibility of Subcommittee D-20.10 on Mechanical Properties.

Current edition effective March 31, 1966. Originally issued 1944. Replaces D 790 – 63.

[2] *Annual Book of ASTM Standards*, Part 27.

4.2 *Loading Nose and Supports*—The loading nose and supports shall have cylindrical surfaces. In order to avoid excessive indentation, or compressive failure, that is, nonrecoverable deformation or compressive failure due to stress concentration directly under the loading nose, the radius of nose and supports shall be at least 3.2 mm ($^1/_8$ in.) for all specimens. For specimens 3.2 mm ($^1/_8$ in.) in thickness or greater, the radius of the supports may be up to 1$^1/_2$ times the specimen depth, and the radius of the loading nose may be up to 4 times the specimen depth, and shall be this large if significant indentation or compressive failure occurs. The chord defining the arc of the loading nose in contact with the specimen shall be sufficiently large to prevent contact of the specimen with the sides of the nose. A minimum chord length of twice the specimen depth shall be used where possible (Fig. 1).

5. Test Specimens

5.1 The specimens may be cut from sheets, plates, or molded shapes, or may be molded to the desired finished dimensions.

NOTE 2—Any necessary polishing of specimens shall be done only in the lengthwise direction of the specimen.

5.2 *Sheet Materials* (*Except Laminated Thermosetting Materials and Certain Materials Used for Electrical Insulation, Including Vulcanized Fiber and Glass Bonded Mica*):

5.2.1 *For Materials 1.6 mm* ($^1/_{16}$ *in.*) *or Greater in Thickness*—For flatwise tests, the depth of the specimen shall be the thickness of the material. For edgewise tests, the width of the specimen shall be the thickness of the sheet and the depth shall not exceed the width (Notes 3 and 4). For all tests, the span shall be 16 (tolerance +4 or −2) times the depth of the beam. Specimen width shall not exceed one fourth of the span for specimens greater than 3.2 mm ($^1/_8$ in.) in thickness. Specimens 3.2 mm ($^1/_8$ in.) or less in thickness shall be 12.7 mm ($^1/_2$ in.) in width. The specimen shall be long enough to allow for overhanging on each end of at least 10 percent of the span, but in no case less than 6.4 mm ($^1/_4$ in.) on each end. Overhang shall be sufficient to prevent the specimen from slipping through the supports.

NOTE 3—Whenever possible, the original surfaces of the sheet shall be unaltered. However, where machine limitations make it impossible to follow the above criterion on the unaltered sheet, both surfaces shall be machined to the desired dimensions and the location of the specimens with reference to the total thickness shall be noted. The values obtained on specimens with machined surfaces may differ from those obtained on specimens with original surfaces. Consequently, any specifications for flexural properties on the thicker sheets must state whether the original surfaces are to be retained or not.

NOTE 4—Edgewise tests are not applicable for sheets that are so thin that specimens meeting these requirements cannot be cut. If specimen depth exceeds the width, buckling may occur.

5.2.2 *For Materials Less than 1.6 mm* ($^1/_{16}$ *in.*) *in Thickness*—The specimen shall be 50.8 mm (2 in.) long by 12.7 mm ($^1/_2$ in.) wide, tested flatwise on a 25.4-mm (1-in.) span.

NOTE 5—The formulas for simple beams used in this method for calculating results presuppose that the width is small in comparison with the span. Therefore, they do not apply rigorously to these dimensions.

NOTE 6—Where machine sensitivity is such that specimens of these dimensions cannot be measured, wider specimens or shorter spans, or both, may be used, provided span-to-depth ratio is at least 14 to 1. All dimensions must be stated in the report (see also Note 5).

5.3 *Laminated Thermosetting Materials and Sheet and Plate Materials Used for Electrical Insulation, Including Vulcanized Fiber and Glass-Bonded Mica*—Specimens shall be tested in accordance with Table 1. For paper-base and fabric-base grades over 25.4 mm (1 in.) in nominal thickness, the specimens shall be machined on both surfaces to a thickness of 25.4 mm (1 in.). For glass-base and nylon-base grades, specimens over 12.7 mm ($^1/_2$ in.) in nominal thickness shall be machined on both surfaces to a thickness of 12.7 mm ($^1/_2$ in.).

5.4 *Molding Materials* (*Including Phenolics, Polyesters, and Molding Materials Used for Electrical Insulation*)—The recommended specimen for molding materials is 127 by 12.7 by 6.4 mm (5 by $^1/_2$ by $^1/_4$ in.) tested flatwise on a 102-mm (4-in.) span.

5.5 *High-Strength Reinforced Plastic Composites for Structural and Semistructural Applications, Including Highly Orthotropic Laminates*—Specimens shall be tested in accordance with Table 1 or as described below. For flatwise tests, the depth of the specimen shall be the thickness of the laminate and the depth shall not exceed the width (Notes 3 and

4). Specimen width shall not exceed one fourth of the support span for specimens greater than 3.2 mm (1/8 in.) in thickness. Specimens 3.2 mm (1/8 in.) or less in thickness shall be at least 12.7 mm (1/2 in.) in width. The specimen shall be long enough to allow for overhanging on each end of at least 10 percent of the support span. Overhang shall be sufficient to prevent the specimen from slipping through the supports. For all tests the support span shall be a minimum of 16 and a maximum of 40 times the depth of the specimen (Note 7). The support span shall have a numerical value chosen so that the failures occur in the outer fibers of the specimens, due only to the bending moment. When laminate materials have low compression strength perpendicular to the laminations, they shall be loaded with a large-radius loading nose (up to maximum of 4 times specimen thickness) to prevent premature damage to the outer fibers. Three recommended span-to-depth ratios are 16, 32, and 40 to 1.

NOTE 7—As a general rule, span-to-depth ratios of 16 to 1 are satisfactory when the ratio of tensile strength to shear strength is less than 8 to 1, but the span-to-depth ratio must be increased for composite laminates having relatively low shear strength in the plane of the laminate and relatively high tensile strength parallel to the span.

6. Number of Test Specimens

6.1 At least five specimens shall be tested for each sample in the case of isotropic materials or molded specimens.

6.2 For each sample of anisotropic material in sheet form, at least five specimens shall be tested for each condition. Recommended conditions are flatwise and edgewise tests on specimens cut in lengthwise and crosswise directions of the sheet. For purposes of this test, "lengthwise" shall designate the principal axis of anisotropy, and shall be interpreted to mean the direction of the sheet known to be the stronger in flexure. "Crosswise" shall be the sheet direction known to be the weaker in flexure, and shall be at 90 deg to the lengthwise direction.

7. Conditioning

7.1 *Conditioning*—Condition the test specimens at 23 ± 2 C (73.4 ± 3.6 F) and 50 ± 5 percent relative humidity for not less than 40 h prior to test in accordance with Procedure A of ASTM Methods D 618, Conditioning Plastics and Electrical Insulating Materials for Testing,[2] for those tests where conditioning is required. In cases of disagreement, the tolerances shall be 1 C (1.8 F) and ±2 percent relative humidity.

7.2 *Test Conditions*—Conduct tests in the Standard Laboratory Atmosphere of 23 ± 2 C (73.4 ± 3.6 F) and 50 ± 5 percent relative humidity, unless otherwise specified in the test methods or in this specification. In cases of disagreements, the tolerances shall be 1 C (1.8 F) and ±2 percent relative humidity.

8. Procedure A

8.1 Use an untested specimen for each measurement. Measure the width and thickness of the specimen to the nearest 0.03 mm (0.001 in.) at the center of the span. For specimens less than 2.54 mm (0.100 in.) in thickness, measure the thickness to the nearest 0.003 mm (0.0001 in.).

8.2 Determine the span to be used as described in Section 5. After the span is set, measure the actual span length to the nearest 1 percent.

8.3 If Table 1 is used, set the machine or the specified rate of crosshead motion, or as near as possible to it. If Table 1 is not used, calculate the rate of crosshead motion as follows and set the machine for that calculated rate, or as near as possible to it:

$$N = ZL^2/6d \tag{1}$$

where:

N = rate of crosshead motion, mm (or in.)/min,
L = span, mm (or in.),
d = depth of beam, mm (or in.), and
Z = rate of straining of the outer fiber, mm/mm min (or in./in. min). Z shall equal 0.01.

8.4 Align the loading nose and supports so that the axes of the cylindrical surfaces are parallel and the loading nose is midway between the supports. This parallelism may be checked by means of a plate with parallel grooves into which the loading nose and supports will fit when properly aligned. Center the specimen on the supports, with the long axis of the specimen perpendicular to the loading nose and supports.

8.5 Apply the load to the specimen at the

specified crosshead rate, and take simultaneous load-deflection data. Measure deflection either by a gage under the specimen in contact with it at the center of the span, the gage being mounted stationary relative to the specimen support, or by measurement of the motion of the loading nose relative to the supports. In either case, make appropriate corrections for indentation in the specimens and deflections in the weighing system of the machine. Load-deflection curves may be plotted to determine the flexural yield strength, secant or tangent modulus of elasticity, and the total work measured by the area under the load-deflection curve.

8.6 Terminate the test if the maximum strain in the outer fiber has reached 0.05 mm/mm (in./in.) (Notes 8 and 9). The deflection at which this strain occurs may be calculated by letting r equal 0.05 mm/mm (or in./in.) as follows:

$$D = rL^2/6d \quad (2)$$

where:

D = deflection, mm (or in.),
r = strain, mm/mm (or in./in.),
L = span, mm (or in.), and
d = depth of beam, mm (or in.)

NOTE 8—For some materials the increase in strain rate provided under Procedure B may induce the specimen to yield or rupture, or both, within the required 5 percent strain limit.

NOTE 9—Beyond 5 percent strain, this test is not applicable, and some other method, such as ASTM Method D 638, Test for Tensile Properties of Plastics,[2] should be used.

9. Procedure B

9.1 Use an untested specimen for each measurement.

9.2 Testing conditions shall be identical to those described in Section 8, except that the rate of straining of the outer fiber shall be 0.10 mm/mm min (in./in. min).

10. Retests

10.1 Values for properties at rupture shall not be calculated for any specimen that breaks at some obvious, fortuitous flaw, unless such flaws constitute a variable being studied. Retests shall be made for any specimens on which values are not calculated.

11. Properties and Calculations

11.1 *Maximum Fiber Stress*—When a beam of homogeneous, elastic material is tested in flexure as a simple beam supported at two points and loaded at the midpoint, the maximum stress in the outer fiber occurs at midspan. This stress may be calculated for any point on the load-deflection curve by the following equation (Notes 10 and 11):

$$S = 3PL/2bd^2 \quad (3)$$

where:

S = stress in the outer fiber at midspan, kgf/cm^2 (or psi),
P = load at a given point on the load-deflection curve, kg (or lb),
L = span, cm (or in.),
b = width of beam tested, cm (or in.), and
d = depth of beam tested, cm (or in.).

NOTE 10—Equation 3 applies strictly to materials for which the stress is linearly proportional to strain up to the point of rupture and for which the strains are small. Since this is not always the case, a slight error will be introduced in the use of this equation. The equation will, however, be valid for comparison data and specification values up to maximum fiber strains of 5 percent for specimens tested by the procedures herein described.

NOTE 11—The above calculation is not valid if the specimen is slipping between the supports.

11.2 *Maximum Fiber Stress for Beams Tested at Large Spans*—If span-to-depth ratios greater than 16 to 1 are used so that large deflections occur, the maximum fiber stress of a simple beam can be reasonably approximated with the following equation (Note 12):

$$S = 3PL/2bd^2[1 + 6(D/L)^2 - 4(d/L)(D/L)] \quad (3a)$$

where S, P, L, b, and d are the same as for Eq 3 and D is the deflection of the centerline of the specimen at midspan with relation to the supports.

NOTE 12—When large span-to-depth ratios are used, significant end forces are developed which affect the moment in a simply supported beam. An approximate correction factor is given in Eq 3a to correct for this end force in large span-to-depth ratio beams where relatively large deflections exist.

11.3 *Flexural Strength* (*Modulus of Rupture*)—The flexural strength is equal to the maximum stress in the outer fiber at the moment of break. It is calculated in accordance with Eqs 3 and 3a by letting P equal the load at the moment of break. If the material does not break, this part of the test is not applicable. In this case, it is suggested that yield strength, if applicable, be calculated, and that the corresponding strain be reported also (see 11.4, 11.6, and 11.7).

11.4 *Flexural Yield Strength*—Some materials that do not break at outer fiber strains up to 5 percent may give load - deflection curves that show a point, *Y*, at which the load does not increase with an increase in deflection. In such cases, the flexural yield strength may be calculated in accordance with Eqs 3 and 3a by letting *P* equal the load at point *Y*.

11.5 *Flexural Offset Yield Strength*—Offset yield strength is the stress at which the stress - strain curve deviates by a given strain (offset) from the tangent to the initial straight-line portion of the stress - strain curve. The value of the offset must be given whenever this property is calculated.

NOTE 13—This value may differ from Flexural Yield Strength defined in 11.4. Both methods of calculation are described in the Appendix to Method D 638.

11.6 *Stress at a Given Strain*—The maximum fiber stress at any given strain may be calculated in accordance with Eqs 3 and 3a by letting *P* equal the load read from the load - deflection curve at the deflection corresponding to the desired strain.

11.7 *Maximum Strain*—The maximum strain in the outer fiber also occurs at mid-span, and may be calculated as follows:

$$r = 6Dd/L^2 \quad (4)$$

where:

r = maximum strain in the outer fiber, mm/mm (or in./in.),
D = maximum deflection of the center of the beam, mm (or in.),
L = span, mm (or in.), and
d = depth, mm (or in.).

11.8 *Modulus of Elasticity:*

11.8.1 *Tangent Modulus of Elasticity*—The tangent modulus of elasticity, often called "modulus of elasticity," is the ratio, within the elastic limit of stress to corresponding strain and shall be expressed in kilograms-force per square centimeter (or pounds per square inch). It is calculated by drawing a tangent to the steepest initial straight-line portion of the load - deformation curve and using Eq 5.

$$E_B = L^3m/4bd^3 \quad (5)$$

where:

E_B = modulus of elasticity in bending, kgf/cm^2 (or psi),
L = span, cm (or in.),
b = width of beam tested, cm (or in.),
d = depth of beam tested, cm (or in.), and
m = slope of the tangent to the initial straightline portion of the load - deflection curve, kg/cm (or lb/in.) of deflection.

11.8.2 *Secant Modulus of Elasticity*—The secant modulus of elasticity is the ratio of stress to corresponding strain at any given point on the stress - strain curve, or the slope of the straight line that joins the origin and the selected point on the actual stress - strain curve. It shall be expressed in kilograms-force per square centimeter (or pounds per square inch). The selected point is generally chosen at a specified stress or strain. It is calculated in accordance with Eq 5 by letting *m* equal the slope of the secant on the load - deflection curve.

11.9 *Arithmetic Mean*—For each series of tests, the arithmetic mean of all values obtained shall be calculated to three significant figures and reported as the "average value" for the particular property in question.

11.10 *Standard Deviation*—The standard deviation (estimated) shall be calculated as follows and reported to two significant figures:

$$s = \sqrt{(\Sigma X^2 - n\bar{X}^2)/(n - 1)}$$

where:

s = estimated standard deviation,
X = value of single observation,
n = number of observations, and
$\bar{X}$ = arithmetic mean of the set of observations.

12. Report

12.1 The report shall include the following:

12.1.1 Complete identification of the material tested, including type, source, manufacturer's code number, form, principal dimensions, and previous history,

12.1.2 Direction of cutting and loading specimens,

12.1.3 Conditioning procedure,

12.1.4 Depth and width of the specimen,

12.1.5 Span length,

12.1.6 Span-to-depth ratio,

12.1.7 Radius of supports and loading nose,

12.1.8 Rate of crosshead motion in millimeters (or inches) per minute,

12.1.9 Maximum strain in the outer fiber of the specimen,

12.1.10 Flexural strength (if applicable), average value and standard deviation,

12.1.11 Tangent or secant modulus of elasticity in bending, average value and standard deviation,

12.1.12 Flexural yield strength (if desired), average value and standard deviation,

12.1.13 Flexural offset yield strength (if desired), with offset or strain used, average value and standard deviation,

12.1.14 Stress at a given strain (if desired), with strain used, average value and standard deviation, and

12.1.15 Procedure used.

TABLE 1 Suggested Dimensions for Test Specimens of 5.3 and 5.5 (Note 7)

Nominal Specimen Thickness, mm (in.)	Width of Specimen, mm (in.)	Span-to-Depth Ratio								
		16 to 1			32 to 1			40 to 1		
		Length of Specimen, mm (in.)	Span, mm (in.)	Rate of Crosshead Motion (Procedure A), mm (in.)/min[a]	Length of Specimen, mm (in.)	Span, mm (in.)	Rate of Crosshead Motion (Procedure A), mm (in.)/min[a]	Length of Specimen, mm (in.)	Span, mm (in.)	Rate of Crosshead Motion (Procedure A), mm (in.)/min[a]
0.8 (1/32)	25 (1)	50 (2)	16 (5/8)[b]	0.5 (0.02)	50 (2)	25 (1)	1.3 (0.05)	60 (2¼)	30 (1¼)	2.0 (0.08)
1.6 (1/16)	25 (1)	50 (2)	25 (1)	0.8 (0.03)	80 (3)	50 (2)	2.8 (0.11)	90 (3½)	60 (2½)	4.3 (0.17)
2.4 (3/32)	25 (1)	60 (2½)	40 (1½)	1.0 (0.04)	100 (4)	80 (3)	4.1 (0.16)	120 (4¾)	95 (3¾)	6.4 (0.25)
3.2 (1/8)	25 (1)	80 (3)	50 (2)	1.3 (0.05)	130 (5)	100 (4)	5.3 (0.21)	180 (7)	130 (5)	8.4 (0.33)
4.8 (3/16)	13 (½)	100 (4)	80 (3)	2.0 (0.08)	191 (7½)	150 (6)	8.1 (0.32)	240 (9½)	191 (7½)	12.7 (0.50)
6.4 (¼)	13 (½)	130 (5)	100 (4)	2.8 (0.11)	250 (10)	200 (8)	10.9 (0.43)	330 (13)	250 (10)	17.0 (0.67)
9.6 (3/8)	13 (½)	191 (7½)	150 (6)	4.1 (0.16)	380 (15)	300 (12)	16.3 (0.64)	480 (19)	380 (15)	25.4 (1.00)
12.7 (½)	13 (½)	250 (10)	200 (8)	5.3 (0.21)	495 (19½)	410 (16)	21.6 (0.85)	640 (25)	510 (20)	34.0 (1.34)
19.1 (¾)	19 (¾)	380 (15)	300 (12)	8.1 (0.32)	740 (29)	610 (24)	32.5 (1.28)	940 (37)	760 (30)	50.8 (2.00)
25.4 (1)	25 (1)	495 (19½)	410 (16)	10.9 (0.43)	990 (39)	810 (32)	43.4 (1.71)	1240 (49)	1020 (40)	67.8 (2.67)

[a] Rates indicated are for Procedure A where strain rate is 0.01 mm/mm min (0.01 in./in. min). To obtain speeds for Procedure B where strain rate is 0.10 mm/mm min (0.10 in./in. min), multiply these values by 10. Procedure A is to be used for all specification purposes unless otherwise stated in specifications. See 8.3 for the method of calculation.

[b] This span-to-depth ratio is greater than 16 to 1 in order to give clearance between moving head and specimen support.

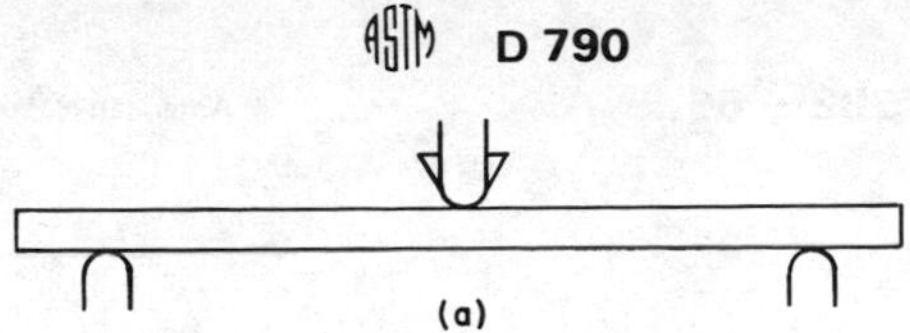

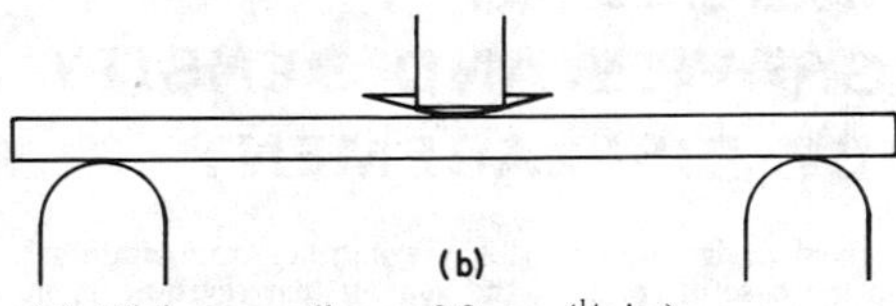

(*a*) Minimum radius = 3.2 mm ($^1/_8$ in.).
(*b*) Maximum radius supports = 1 $^1/_2$ times specimen depth, maximum radius loading nose = 4 times specimen depth (minimum length of chord defining arc in contact with the specimen = 2 times specimen depth).

FIG. 1 Allowable Range of Loading Nose and Support Radii for Specimen 6.4 mm ($^1/_4$ in.) Thick.

American National Standard K65.8-1965 (R1967)
American National Standards Institute

Standard Methods of Test for
SPECIFIC GRAVITY AND DENSITY OF PLASTICS BY DISPLACEMENT[1]

This Standard is issued under the fixed designation D 792; the number immediately following the designation indicates the year of original adoption or, in the case of revision, the year of last revision. A number in parentheses indicates the year of last reapproval.

1. Scope

1.1 These methods cover the determination of the specific gravity and density of solid plastics by displacement of liquid and determination of the change in weight.

1.2 These methods differ in applicability as follows:

1.2.1 *Methods A-1,-2,-3*—For plastics in forms such as sheets, rods, tubes, or molded items. Nondestructive testing of large items is covered. These methods also may be used to characterize a molding or casting compound by testing a specimen that has been prepared under controlled conditions.

1.2.2 *Method B*—For plastics in forms such as molding powder, flake, or pellets.

NOTE 1—Alternatively, ASTM Method D 1505, Test for Density of Plastics by the Density-Gradient Technique,[2] may be applied to many such forms, as well as to films and sheeting.

NOTE 2—The values stated in U.S. customary units are to be regarded as the standard. The metric equivalents of U.S. customary units may be approximate.

2. Significance

2.1 The specific gravity or density of a solid is a property that can be measured conveniently to identify a material, to follow physical changes in a sample, to indicate degree of uniformity among different sampling units or specimens, or to indicate the average density of a large item.

2.2 Changes in density of a single specimen may be due to changes in crystallinity, loss of plasticizer, absorption of solvent, or to other causes. Portions of a sample may differ in density because of difference in crystallinity, thermal history, porosity, and composition (types or proportions of resin, plasticizer, pigment, or filler).

NOTE 3—Reference is made to ASTM Method D 1622, Test for Apparent Density of Rigid Cellular Plastics.[3]

2.3 Density is useful for calculating strength - weight and cost - weight ratios.

3. Definitions

3.1 *specific gravity*—the ratio of the weight in air of a unit volume of the impermeable portion of the material at 23 C (73.4 F) to the weight in air of equal density of an equal volume of gas-free distilled water at the same temperature. The form of expression shall be:

Specific gravity 23/23 C (or sp gr 23/23 C)

NOTE 4—These definitions are essentially equivalent to the definitions of apparent specific gravity and apparent density in ASTM Definitions E 12, Terms Relating to Density and Specific Gravity of Solids, Liquids, and Gases,[4] because the small percentage difference introduced by not correcting for the buoyancy of air is insignificant for most purposes.

3.2 *density*—the weight in air in grams per cubic centimeter of impermeable portion of the material at 23 C. The form of expression shall be:

D^{23C}, g/cm^3 (Notes 4, 5, and 6)

NOTE 5—Densities in other units may be obtained, when desired, by multiplying by appropriate factors.

NOTE 6—Specific gravity 23/23 C can be converted to density 23 C, g/cm^3, by use of the follow-

[1] These methods are under the jurisdiction of ASTM Committee D-20 on Plastics. A list of committee members may be found in the ASTM Yearbook. This standard is the direct responsibility of Subcommittee D-20.70 on Analytical Methods.

Current edition effective Sept. 20, 1966. Originally issued 1944. Replaces D 792 – 64 T.

[2] *Annual Book of ASTM Standards*, Part 27.

[3] *Annual Book of ASTM Standards*, Part 26.

ing equation:

$$D^{23C}, g/cm^3 = \text{sp gr } 23/23 \text{ C} \times 0.9975$$

4. Sampling

4.1 The sampling units used for the determination of specific gravity shall be representative of the quantity of product for which the data are required, in accordance with ASTM Recommended Practice D 1898, Sampling of Plastics.[2] Particular attention shall be given to the following:

4.1.1 If it is known or suspected that the sample consists of two or more layers or sections having different specific gravities, either complete finished parts or complete cross sections of the parts or shapes shall be used as the specimens, or separate specimens shall be taken and tested from each layer.

4.1.2 In Method B, if it is known or suspected that fine and coarse particles may have different specific gravities, either the specimens shall contain the same proportions of each size as present in the bulk of the sample, or the sample shall be sieved into several size groups, each size group shall be tested separately, and the weighted average shall be calculated and reported.

5. Conditioning

5.1 *Conditioning*—Condition the test specimens at 23 ± 2 C (73.4 ± 3.6 F) and 50 ± 5 percent relative humidity for not less than 40 h prior to test in accordance with Procedure A of ASTM Methods D 618, Conditioning Plastics and Electrical Insulating Materials for Testing,[2] for those tests where conditioning is required. In cases of disagreement, the tolerances shall be 1 C (1.8 F) and ±2 percent relative humidity.

5.2 *Test Conditions*—Conduct tests in the Standard Laboratory Atmosphere of 23 ± 2 C (73.4 ± 3.6 F) and 50 ± 5 percent relative humidity, unless otherwise specified in the test methods or in this specification. In cases of disagreement, the tolerances shall be 1 C (1.8 F) and ±2 percent relative humidity.

NOTE 7—Certain ASTM specifications for plastic materials that achieve moisture equilibrium very slowly call for the testing of such materials in the "bone-dry" condition. Since molding powders or flake are rarely used in the Standard Laboratory Atmosphere of Methods D 618, the specific gravity or density of such materials may be determined under any condition considered desirable: as received, conditioned under standard conditioning procedures, or dry, etc., provided that the condition is stated. If conditioning includes immersion in a liquid, the time and temperature shall be stated.

METHOD A-1 FOR TESTING SOLID PLASTICS IN WATER (SPECIMENS 1 TO 50 G)

6. Scope

6.1 This method involves weighing a one-piece specimen of 1 to 50 g in water, using a sinker with plastics that are lighter than water. This method is suitable for plastics that are wet by, but otherwise not affected by water.

7. Apparatus

7.1 *Analytical Balance*—A balance with a precision within 0.1 mg, accuracy within 0.05 percent relative (that is, 0.05 percent of the weight of the specimen in air), and equipped with a stationary support for the immersion vessel above the balance pan ("pan straddle").

NOTE 8—Assurance that the balance meets the performance requirements should be provided by frequent checks on adjustments of zero point and sensitivity and by periodic calibration for absolute accuracy, using standard weights.

7.2 *Wire*—A corrosion-resistant wire for suspending the specimen.

7.3 *Sinker*—A sinker for use with specimens of plastics that have specific gravities less than 1.000. The sinker shall: (*1*) be corrosion-resistant; (*2*) have a specific gravity of not less than 7.0; (*3*) have smooth surfaces and a regular shape; and (*4*) be slightly heavier than necessary to sink the specimen. The sinker should have an opening to facilitate attachment to the specimen and wire.

7.4 *Immersion Vessel*—A beaker or other wide-mouthed vessel for holding the water and immersed specimen.

7.5 *Thermometer*—A thermometer with an accuracy of ±1 C (±2 F) is required if the test is not performed in the Standard Laboratory Atmosphere of Methods D 618, Conditioning Plastics and Electrical Insulating Materials for Testing[2] (compare 14.4).

8. Materials

8.1 *Water*—The water shall be substantially air-free, distilled, or demineralized water.

NOTE 9—Water may be rendered substantially air-free by boiling and cooling or by shaking under vacuum in a heavy-walled vacuum flask. (**Caution**—Use gloves and shielding.) If the water does not wet the specimen, a few drops of a wetting agent shall be added. If this solution does not wet the specimen, Method A-2 shall be used.

9. Test Specimens

9.1 The test specimen shall be a single piece of the material under test of any size and shape that can conveniently be prepared and tested, provided that its volume shall be not less than 1 cm^3 (0.06 $in.^3$), and its surface and edges shall be made smooth. The thickness of the specimen should be at least 1 mm (0.04 in.) for each 1 g of weight. A specimen weighing 1 to 5 g usually will be found convenient, but specimens up to approximately 50 g may be used (Note 10). Care should be taken in cutting specimens to avoid changes in density resulting from compressive stresses or frictional heating.

NOTE 10—Specifications for certain plastics require a particular method of specimen preparation and should be consulted if applicable.

9.2 The specimen shall be free from oil, grease, and other foreign matter.

10. Procedure

10.1 Weigh the specimen in air to the nearest 0.1 mg or 0.05 percent relative, whichever is greater.

10.2 Attach to the balance a piece of fine wire sufficiently long to reach from the hook above the pan to the support for the immersion vessel. Attach the specimen to the wire such that it is suspended about 2.5 cm (1 in.) above the vessel support.

NOTE 11—The specimen may be weighed in air after hanging from the wire. In this case, record the weight of the specimen, a = (weight of specimen + wire, in air) − (weight of wire in air).

10.3 Mount the immersion vessel on the support, and completely immerse the suspended specimen (and sinkers, if used) in water (8.1) at a temperature of 23 ± 2 C. The vessel must not touch wire or specimen. Remove any bubbles adhering to the specimen, wire, or sinker, paying particular attention to holes in the specimen and sinker. Usually these bubbles can be removed by rubbing them with another wire. If the bubbles cannot be removed by this method or if bubbles are continuously formed (as from dissolved gases), the use of vacuum is recommended (Note 13). Weigh the suspended specimen to the required precision (10.1) (Note 12). Record this weight as b (the weight of the specimen, sinker, if used, and the partially immersed wire in liquid). Unless otherwise specified, weigh rapidly in order to minimize absorption of water by the specimen.

NOTE 12—It may be necessary to change the sensitivity adjustment of the balance to overcome the damping effect of the immersed specimen.

NOTE 13—Some specimens may contain absorbed or dissolved gases, or irregularities which tend to trap air bubbles; any of these may affect the density values obtained. In such cases, the immersed specimen may be subjected to vacuum in a separate vessel until evolution of bubbles has substantially ceased before weighing (see Method B). It must also be demonstrated that the use of this technique leads to results of the required degree of precision.

10.4 Weigh the wire (and sinker, if used) in water with immersion to the same depth as used in the previous step (Notes 14 and 15). Record this weight as w (weight of the wire in liquid).

NOTE 14—It is convenient to mark the level of immersion by means of a shallow notch filed in the wire. The finer the wire, the greater the tolerance which may be permitted in adjusting the level of immersion between weighings. With wire Awg No. 36 or finer, disregard its degrees of immersion and, if no sinker is used, use the weight of the wire in air as w.

NOTE 15—If the wire is left attached to the balance arm during a series of determinations, the weight a may be determined either with the aid of a tare on the other arm of the balance or as in Note 13. In such cases, care must be taken that the change of weight of the wire (for example, from visible water) between readings does not exceed the desired precision.

10.5 Repeat the procedure for the required number of specimens.

11. Calculations

11.1 Calculate the specific gravity of the plastic as follows:

$$\text{Sp gr } 23/23 \text{ C} = a/(a + w - b)$$

where:

a = apparent weight of specimen, without wire or sinker, in air,

b = apparent weight of specimen (and of sinker, if used) completely immersed and of the wire partially immersed in liquid, and

w = apparent weight of totally immersed sinker (if used) and of partially immersed wire.

11.2 Calculate the density of the plastic as follows:

$$D^{23C}, g/cm^3 = \text{sp gr } 23/23\text{ C} \times 0.9975$$

12. Report

12.1 The report shall include the following:

12.1.1 Complete identification of the material or product tested, including method of specimen preparation and conditioning,

12.1.2 Average specific gravity for all specimens from a sampling unit, reported as sp gr 23/23 C = ———, or average density reported as $D^{23\text{ C}}$ = ——— g/cm³.

12.1.3 A measure of the degree of variation of specific gravity or density within the sampling unit such as the standard deviation and number of determinations on a homogeneous material or the averages plus these measures of dispersion on different layers or areas of a nonhomogeneous product,

12.1.4 Any evidence of porosity of the material or specimen,

12.1.5 The method of test (Method A-1 of Methods D 792), and

12.1.6 Date of test.

Method A-2 for Testing Solid Plastics in Liquids Other than Water (Specimens 1 to 50 g)

13. Scope

13.1 Method A-2 uses a liquid other than water for testing one-piece specimens, 1 to 50 g, of plastics that are affected by water or which are lighter than water.

14. Apparatus

14.1 The apparatus shall include the balance, wire, and immersion vessel of Section 7, and, optionally, the following:

14.2 *Pycnometer with Thermometer*—A 25-ml specific gravity bottle with thermometer, or

14.3 *Pycnometer*—A pycnometer of the Weld type, preferably with a capacity of about 25 ml and an external cap over the stopper.

14.4 *Thermometer*—A thermometer having not fewer than four divisions per deg C or two divisions per deg F over a temperature range of not less than 5 C or 10 F above and below the standard temperature, and having an ice point for calibration. A thermometer short enough to be handled inside the balance case will be found convenient. ASTM Thermometer 23C (see ASTM Specifications E 1, for ASTM Thermometers[4]) and Anschütz-type thermometers have been found satisfactory for this purpose.

14.5 *Constant-Temperature Bath*—An appropriate constant-temperature bath adjusted to maintain a temperature of 23 ± 0.1 C.

15. Materials

15.1 *Immersion Liquid*—The liquid used shall not dissolve, swell, or otherwise affect the specimen, but should wet it and should have a specific gravity less than that of the specimen. In addition, the immersion liquid should be nonhygroscopic, have a low vapor pressure, a low viscosity, and a high flash point, and should leave little or no waxy or tarry residue on evaporation. A narrow cut distilled from kerosine meets these requirements for many plastics. The specific gravity 23/23 C of the immersion liquid shall be determined shortly before and after each use in this method to a precision of at least 0.1 percent relative, unless it has been established experimentally in the particular application that a lesser frequency of determination can be used to assure the desired precision.

Note 16—For the determination of the specific gravity of the liquid, the use of a standard plummet of known volume (Note 17) or of Method A, C, or D of ASTM Methods D 891, Test for Specific Gravity of Industrial Aromatic Hydrocarbons and Related Materials,[5] using the modifications required to give specific gravity 23/23 C instead of specific gravity 60/60 F, is recommended. One suggested procedure is the following:

If a constant-temperature water bath is not available, determine the weight of the clean, dry pycnometer with thermometer to the nearest 0.1 mg on an analytical balance. Fill the pycnometer with water (8.1) cooler than 23 C. Insert the thermometer-stopper, causing excess water to be expelled through the side arm. Permit the filled bottle to warm in air until the thermometer reads 23.0 C. Remove the drop of water at the tip of the side arm with a bit of filter paper, taking care not to draw any liquid from within the capillary, place

[4] *Annual Book of ASTM Standards*, Part 30.

[5] *Annual Book of ASTM Standards*, Part 20.

the cap over the side arm, wipe the outside carefully, and weigh the filled bottle again to the nearest 0.2 mg. Empty the pycnometer, dry, and fill and weigh with the other liquid in the same manner as was done with the water. Calculate the specific gravity 23/23 C of the liquid, d, as follows:

$$d = (b - e)/(w - e)$$

where:
e = apparent weight of empty pycnometer,
w = apparent weight of pycnometer filled with water at 23.0 C, and
b = apparent weight of pycnometer filled with liquid at 23.0 C.

If a constant-temperature water bath is available, a pycnometer without a thermometer may be used (compare 29.2).

NOTE 17—One standard object which has been found satisfactory for this purpose is the Reimann Thermometer Plummet. These are normally supplied calibrated for measurements at temperatures other than 23/23 C, so that recalibration is necessary for the purposes of these methods. Calibration at intervals of 1 week is recommended.

16. Test Specimens

16.1 See Section 9.

17. Procedure

17.1 The procedure shall be similar to Section 10, except for the choice of immersion liquid, and the temperature during the immersed weighing (10.3) shall be 23 ± 0.5 C (73.4 ± 1.0 F).

18. Calculations

18.1 The calculations shall be similar to Section 11, except that d, the specific gravity 23/23 C of the liquid, shall be placed in the numerator:

$$\text{Sp gr } 23/23 \text{ C} = (a \times d)/(a + w - b)$$

19. Report

19.1 See Section 12.

METHOD A-3 FOR TESTING SOLID PLASTICS (SPECIMENS > 50 G)

20. Scope

20.1 This method is suitable for determining the specific gravity or density of items heavier than 50 g by immersion of the entire item, that is, nondestructively. Any appropriate liquid may be used.

21. Apparatus

21.1 *Balance*—A balance large enough to accommodate the desired specimen conveniently and to achieve a precision and accuracy within 0.05 percent relative (that is, of the weight of the specimen in air).

NOTE 18—A chainomatic balance has been found to meet these requirements. Alternatively, a torsion balance will meet these requirements and can be used for immersed weighings if modified as follows: (*1*) a hole is drilled in the base directly under the left hand pan; (*2*) the balance is mounted on a shelf, having a hole aligned with the hole in the base. This shelf is placed over working area so that the specimen can be attached to a wire suspended from the pan. A shield may be necessary to minimize the effects of air currents upon the weighings.

21.2 The apparatus shall also include the wire (7.2), sinker, if used (7.3), immersion vessel of appropriate size (7.4), thermometer (7.5 or 14.4), constant-temperature bath, if used, (14.5), and pycnometer, if used, (14.2).

22. Materials

22.1 See 8.1 or 15.1.

23. Test Specimens

23.1 The test specimen shall be a single item or piece of the material under test, of any shape that can conveniently be prepared and tested. The weight of the specimen in air should preferably be upwards of 50 g, to the capacity limit of the balance, but in any event shall be large enough so that the precision requirements (21.1) can easily be met.

24. Procedure

24.1 The procedure shall be similar to Section 10 or 17, depending upon the choice of immersion liquid. No support for the immersion vessel is necessary if a torsion balance is used and is mounted completely above the immersion vessel.

25. Calculations and Report

25.1 See Section 11 or 18, and Section 12.

METHOD B FOR TESTING MOLDING POWDERS, PELLETS, AND FLAKE

26. Scope

26.1 This method is suitable for determining the specific gravity or density of essentially nonporous plastic powders large enough to be retained on a 20-mesh (840-μm) sieve, flake, pellets, drillings, or small pieces cut from larger shapes (Note 19). The immersed specimen is weighed in a pycnometer. The immersion liquid must have a specific gravity less than that of the specimen.

NOTE 19—Immersion of powders finer than 20 mesh lay lead to incomplete wetting because of static charges and surface tension effects.

27. Apparatus

27.1 The apparatus shall include the balance, pycnometer, thermometer, and constant-temperature bath of Sections 7 and 14, and the following:

27.2 *Vacuum Pump*—A water aspirator or vacuum pump capable of producing a vacuum high enough to cause the liquid to boil at 23 C or to produce a vacuum of 3 mm Hg or better. The pump shall be provided with a trap, for preventing backflow of oil or water, and a stopcock on the inlet side.

27.3 *Vacuum Desiccator*—A vacuum desiccator or bell jar with plate constructed to withstand a vacuum of 1 atm and connections to the source of vacuum.

NOTE 20—Before a new desiccator is used for the first time, it should be wrapped in a towel, placed in a heavy shield, and tested under the vacuum to be used. The operator should be adequately protected by the proper safety devices, such as goggles. The same precautions should be observed under operating conditions.

28. Materials

28.1 *Water* (8.1) or other immersion liquid (15.1).

29. Procedure

29.1 Weigh the clean, dry pycnometer to the nearest 0.1 mg. Record this weight in grams as *e* (weight of empty pycnometer).

29.2 Fill the pycnometer with the liquid to be used at a temperature of approximately 23 C. Place the filled pycnometer without cap in the water bath and allow to remain until it has attained temperature equilibrium with the bath. If the liquid level in the capillary in the stopper has dropped, bring it level with the top of the stopper by adding more liquid. Wipe the outside of the pycnometer dry, cap, and weigh to the nearest 0.2 mg. Record this weight as *b* (weight of pycnometer filled with immersion liquid).

29.3 Clean and dry the pycnometer. Add 1 to 5 g of the material and weigh to the nearest 0.1 mg. Subtract *e* from this weight and record the difference (the weight of the specimen) as *a*.

29.4 Place sufficient liquid in the pycnometer to cover the specimen. Place the pycnometer without stopper in the vacuum desiccator and apply vacuum gradually until all the air has been removed from between the particles of the specimen. Take care not to permit any particles to be splashed out of the pycnometer or onto the ground portion of the neck, where they might prevent the stopper from seating properly.

29.5 Add additional liquid to fill the pycnometer. Insert the stopper, transfer to the bath, adjust to temperature, adjust the final volume, wipe, cap, and weigh. Record this weight as *m* (weight of pycnometer containing the specimen and filled with liquid).

NOTE 21—Since the precision of this method depends on the measurement of a small difference between two large weights, all weighings must be performed with unusual care. The temperature during the determinations of *b* and *m* must be exact in order that the volume of liquid displaced by the specimen can be accurately obtained. Other potential sources of error are retention of air by the specimen and absorption of the liquid.[6]

30. Calculations

30.1 Calculate the specific gravity 23/23 C of the specimen as follows (Note 22):

$$\text{Sp gr } 23/23\text{ C} = (a \times d)/(b + a - m)$$

NOTE 22—The weight of the liquid having a volume equal to that of the specimen is $(b + a - m)$ and the volume of this weight of liquid is $(b + a - m)/d$.

30.2 Calculate the density of the plastic as in 11.2.

31. Report

31.1 See Section 12.

32. Precision

32.1 In these methods, the following criterion should be used for judging the acceptability of results at a 95 percent confidence

[6] Determination of the specific gravity of powders is discussed in detail, together with methods of obtaining high precision in such measurements, in Hillebrand, W. F., and Lundell, G. E. F., *Applied Inorganic Analysis*, John Wiley & Sons, Inc., New York, N. Y., 1929, pp. 661–669.

level:

32.1 *Repeatability*—Two results (each the average of duplicate determinations) obtained by the same operator should be considered suspect if they differ by more than 0.1 percent, relative.

ASTM Designation: D 794 - 68

Recommended Practice for

DETERMINING PERMANENT EFFECT OF HEAT ON PLASTICS[1]

This Recommended Practice is issued under the fixed designation D 794; the number immediately following the designation indicates the year of original adoption or, in the case of revision, the year of last revision. A number in parentheses indicates the year of last reapproval.

1. Scope

1.1 This recommended practice is intended to define the conditions for testing the resistance of plastic sheet, plastic laminated materials, and molded plastics to changes in properties due to exposure at elevated temperatures. Only the procedure for heat exposure is specified, and not the test method or specimen. The effect of heat on any property may be determined by selection of an appropriate test method and specimen.

2. Significance

2.1 Plastic materials exposed to heat may be subject to many types of physical and chemical changes. The severity of the exposures in both time and temperature determines the extent and type of change that takes place. A plastic material is not necessarily degraded by exposure to elevated temperatures, but may be unchanged or improved. However, extended periods of exposure of plastics to elevated temperatures will generally cause some degradation, with progressive change in physical properties.

2.2 Generally, short exposures at elevated temperatures may drive out volatiles such as moisture, solvents, or plasticizers, relieve molding stresses, advance the cure of thermosets, and may cause some change in color of the plastic or coloring agent or both. Normally, additional shrinkage should be expected with loss of volatiles or advance in polymerization.

2.3 Some plastic materials may become brittle due to loss of plasticizers after exposure at elevated temperatures. Other types of plastics may become soft and sticky, either due to sorption of volatilized plasticiser or due to breakdown of the polymer.

2.4 The degree of change observed will depend on the property measured. Mechanical or electrical properties may not change at the same rate. For instance, the arc resistance[2] of thermosetting compounds improves up to the carbonization point of the material. Mechanical properties, such as flexural properties,[3] are sensitive to heat degradation and should be included in the test program.

2.5 Effects of exposure may be quite variable, especially when specimens are exposed for long intervals of time. Factors that affect the reproducibility of data are the degree of temperature control of the enclosure, the type of molding, cure, humidity of the oven room, air velocity over the specimen, and period of exposure. Errors in exposure are cumulative with time. Unless the effect of the molding procedure is a variable under investigation, transfer or injection molding is recommended as the trapped air and volatiles have a better chance to escape. Incompletely cured thermosets tend to blister or crack when exposed to high temperatures. Certain materials are susceptible to degradation due to the influence of humidity in long-term heat resistance tests. Materials susceptible to hydrolysis may undergo degradation when subjected to long-

---

[1] This recommended practice is under the jurisdiction of ASTM Committee D-20 on Plastics. A list of committee members may be found in the ASTM Yearbook. This standard is the direct responsibility of Subcommittee D-20.50 on Permanence Properties.

Current edition effective Sept. 9, 1968. Originally issued 1944. Replaces D 794 – 64 T.

[2] See ASTM Method D 495, Test for High-Voltage, Low-Current Arc Resistance of Solid Electrical Insulating Materials, which appears in this publication.

[3] See ASTM Method D 790, Test for Flexural Properties of Plastics, which appears in this publication.

term heat resistance tests.

2.6 Certain plastic materials require post-cure at or slightly below the expected use temperature. Manufacturer's recommendations in this regard should be considered in preparing specimens under this method.

3. Apparatus

3.1 *Oven*—A controlled horizontal or vertical air flow employing forced-draft circulation with substantial fresh air intake is recomended. When it is necessary to avoid contamination among specimens or materials, a tubular oven method such as ASTM Method D 1870, Test for Elevated Temperature Aging Using a Tubular Oven,[4] may be desirable. Oven apparatus shall be in accordance with ASTM Specifications E 145, for Gravity-Convection and Forced-Ventilation Ovens,[5] with the requirements for uniformity extended to include the range of test temperatures.

NOTE 1—Consideration should also be given for alleviating explosive hazards as well as disposal of any toxic gases that may be evolved.

3.2 *Specimen Rack*—A specimen rack of suitable design to allow ready air circulation around the specimens.

3.3 *Testing Apparatus*—Equipment for determining specific properties before and after exposure to heat shall conform to the requirements prescribed in the appropriate ASTM method of test.

4. Test Specimens

4.1. The number and type of test specimens required shall be in accordance with the ASTM test method for the specific property to be determined.

5. Conditioning

5.1 Conduct initial tests in the Standard Laboratory Atmosphere as specified in ASTM Methods D 618, Conditioning Plastics and Electrical Insulating Materials for Testing,[4] and conditioned in accordance with the requirements of the ASTM test method for determining the specific property or properties required.

5.2 Preconditioning of specimens prior to exposure at elevated temperatures should be such that the influence of any residual moisture, absorbed gas, or other contaminants are considered.

5.3 When required, conditioning of specimens following exposure at elevated temperatures and prior to testing, unless otherwise specified, shall be in accordance with Methods D 618.

6. Procedure

6.1 Suspend the test specimens in the circulating-air oven.

6.2 *Continuous Heat Tests*—Determine the temperature - time limit by the maximum temperature which the plastic withstands for a given time. This limit depends upon the type of plastic and the temperature - time requirements. Exposure periods may range from minutes to weeks. Start the exposure schedules at a temperature, in degrees Celsius divisible by 25, close to the maximum recommended use temperature. Should failure occur at the temperature first chosen, decrease the exposure temperature in steps of 25 C to that point at which failure does not occur. Conversely, if failure does not occur at the first test temperature, increase the temperature in steps of 25 C until failure occurs as defined in Section 7.

6.3 *Cycling Heat Tests*—Cycling tests may be run under conditions agreed upon between the purchaser and the seller.

NOTE 2—In view of continued interest in cycling heat tests, further investigation is being carried on in order to arrive at generally useful exposure procedures. Specific examples of cycling procedures are given in ASTM Methods D 756, Test for Resistance of Plastics to Accelerated Service Conditions.[4]

7. Failure

7.1 Failure due to heat is defined as a change in appearance, weight, dimension, or other properties that alter the plastic material to a degree that it is no longer acceptable for the service in question. Failure may result from blistering, cracking, loss of plasticizer or other volatile material that may cause embrittlement, shrinkage, or change in desirable mechanical or electrical properties.

8. Report

8.1 The report shall include the following:

8.1.1 Material and type of plastic subject

[4] *Annual Book of ASTM Standards*, Part 27.

[5] *Annual Book of ASTM Standards*, Part 30.

to exposure,

8.1.2 Preconditioning and postconditioning procedure followed, if any,

8.1.3 Maximum temperature, time and environmental atmosphere to which the plastic was exposed without failure,

8.1.4 Test method and results obtained before and after exposure to heat for each property at the temperature - time conditions of interest,

8.1.5 Observations of any visible changes in the test specimen, and

8.1.6 Type of oven used.

Designation: D 795 – 65 T

Tentative Recommended Practice for
EXPOSURE OF PLASTICS TO S-1 MERCURY ARC LAMP[1]

This Tentative Recommended Practice has been approved by the sponsoring committee and accepted by the Society in accordance with established procedures, for use pending adoption. Suggestions for revisions should be addressed to the Society at 1916 Race St., Philadelphia, Pa. 19103.

Pursuant to the warning note covering the 3-year limitation on tentatives, published in the previous edition, this tentative has received special permission from the Committee on Standards for continuation until June 1971.

1. Scope

1.1 This recommended practice[2] is intended to furnish a means for exposing plastics to the light of a mercury arc with or without water fog. A qualitative means for rating visual changes is provided.

2. Significance

2.1 This recommended practice is intended to simulate some of the aspects of outdoor weathering. However, direct correlation between this recommended practice and outdoor weathering should not be expected since outdoor conditions are variable.

3. Apparatus

3.1 The apparatus[3] shall consist of a lamp and turntable arranged as shown in Figs. 1 and 2.

3.2 *Lamp*—A General Electric S-1 lamp rated at 400 W. This lamp shall have been preaged at least 50 h and should be discarded after 300 h of use. The operating voltage shall be maintained at 110 ± 2 V, and because of voltage fluctuations it may be necessary to use a voltage stabilizer.[4] The reflector shall be of oxidized aluminum and have dimensions approximating those shown in Fig. 1.

3.3 *Turntable*—The turntable shall rotate at a speed of 30 to 40 rpm and shall be fitted with a disk of polished aluminum approximately 430 mm in diameter and 2.5 mm in thickness. Two sets of 21 corrosion-resistant machine screws and nuts shall be attached to the disk at holes equally spaced on concentric circles of approximately 100-mm and 200-mm radii, respectively. A third set of 21 machine screws and nuts shall be attached to the disk at holes spaced midway between the lines connecting the two previous sets of screws and on a circle of approximately 150-mm radius, concentric with the other circles. The screws may be American National Standard No. 10, 32 threads to the inch, 13 mm in length with fillister head. Corrosion-resistant washers about 160 mm in diameter and 1.5 mm in thickness shall be placed over the nuts to support the specimens about 5 mm above the surface of the disk (Figs. 1 and 2).

3.4 *Thermometer*—An ASTM Partial Immersion Thermometer having a range from −20 to +150 C and conforming to the requirements of Thermometer 1C as prescribed in ASTM Specification E 1, for ASTM Thermometers.[5]

3.5 *Fog Chamber*—A closed shallow chamber fitted with a unit for atomizing distilled water and a baffle to prevent the spray from impinging directly on the specimens. The distilled water mist should visibly fog the chamber with a continuous mist which deposits on

---

[1] This recommended practice is under the jurisdiction of ASTM Committee D-20 on Plastics. A list of members may be found in the ASTM Yearbook. This tentative is the direct responsibility of Subcommittee D-20.50 on Permanence Properties.

Current edition effective Dec. 21, 1965. Originally issued 1944. Replaces D 795 – 57 T.

[2] See R. Hirt et al, "Ultraviolet Spectral Energy Distributions of Natural Sunlight and Accelerated Test Light Sources," *Journal of the Optical Society of America*, JOSAA, Vol 50, 1960, pp. 706–713.

[3] Sunlamp 10-300 complete with turntable as manufactured by the American Instrument Co., conforms to these requirements.

[4] A suitable voltage stabilizing transformer is a General Electric No. 9T64Y4008.

[5] *Annual Book of ASTM Standards*, Part 30.

the specimens without a washing action. The temperature of the water feeding the atomizer shall be 21 to 27 C.

NOTE 1—A chamber and spray assembly constructed as shown in Figs. 3 and 4 is suitable.

4. Test Specimens

4.1 The specimens shall have the dimensions shown in Fig. 5 and should preferably not exceed 6 mm in thickness. Film specimens may be mounted in aluminum frames to add support so as to prevent the film from contacting the turntable.

5. Procedure A—Light Exposure Without Fog

5.1 Place the test specimens on the turntable as shown in Fig. 2 with the necessary air gap (see 3.3) and adjust their height so as to have their top surfaces in the same plane. Center the turntable under the S-1 lamp so that the plane of the top of the specimens is 180 mm from the bottom of the lamp.

5.2 In order to determine the extent of visual change during exposure, mask a portion of the specimen with an inert opaque plate 1 mm in thickness fitted over the outer screw to extend longitudinally along the specimen to cover one-third of the outer portion. If it is desired that changes other than visual be noted, the masking plate may be omitted.

NOTE 2—The cover plate may lead to an abnormal temperature situation which could produce a color change. Where this is suspected, it may be desirable to compare the exposed specimens with unexposed controls.

5.3 Conduct the test so that, with the thermometer placed on the turntable with its midpoint at the center of the disk, it will read 55 to 60 C. If, during the test, the thermometer reads higher than 60 C, provide a fan and place it so that it blows against the bottom of the turntable a volume of air sufficient to bring the temperature to 55 to 60 C.

NOTE 3—If the apparatus is not enclosed, and operates in a room in which the ambient temperature is 21 to 27 C, the thermometer reading will lie between 55 and 60 C. If the reading is below 55 C, the temperature can be raised by reducing somewhat the air circulation around the lamp.

5.4 Always expose the specimens throughout the test with the same surface up, and before making any tests, wipe the specimens with a damp cloth to remove any adhering dirt. Use the area exposed between the concentric circles of 100-mm and 150-mm radii, respectively, to determine the effects of the test conditions on the specimens.

5.5 Expose the specimens to the pre-agreed end point at which time remove the cover plate and compare the two areas.

6. Procedure B—Light Exposure with Fog

6.1 Expose the specimens to light as in Procedure A with interruptions for fog in the following order for each 24 h: 2 h in the fog chamber, 2 h of light, 2 h in the fog chamber, and 18 h of light.

6.2 The duration of this test shall be 240 h, unless otherwise specified.

7. Report

7.1 The report shall include the following:

7.1.1 Which procedure (A or B) was used,

7.1.2 Duration of exposure,

7.1.3 Color change, reported as none, slight, appreciable, or extreme (Note 4),

7.1.4 Observations regarding surface changes (such as dulling and chalking) and deep-seated changes (such as checking, crazing, warping, and discoloration), and

7.1.5 Results of any instrumental, physical, or chemical tests made to determine the extent of changes resulting from the exposure.

NOTE 4—A slight change is defined here as one that is perceptible with difficulty. An appreciable change is one that is readily perceptible without close examination but is insufficient to alter markedly the original color of the specimen. An extreme change is one that is very obvious and has resulted in a marked alteration of the original color of the specimen.

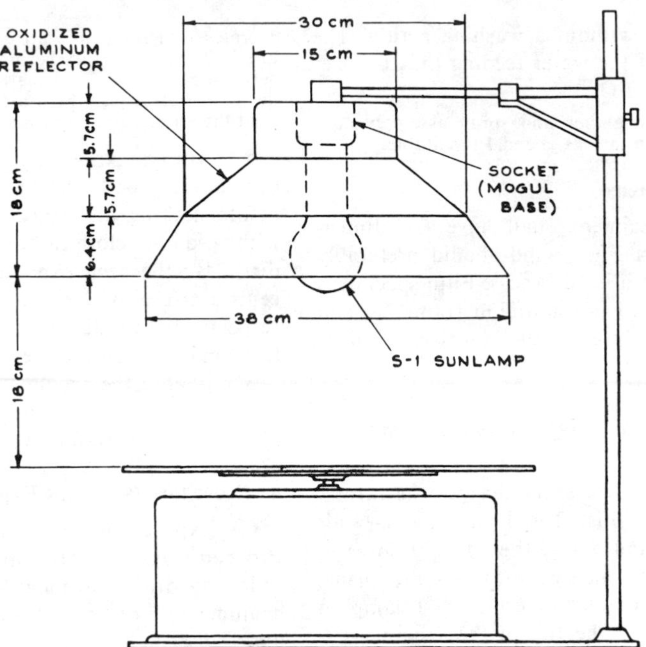

FIG. 1 Turntable with Disk Showing Assembly with Lamp.

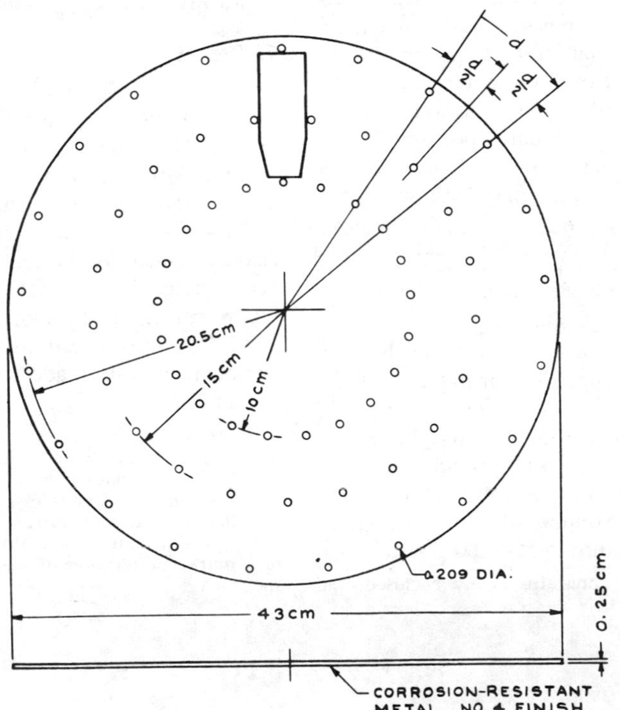

FIG. 2 Platform Disk and Test Specimen.

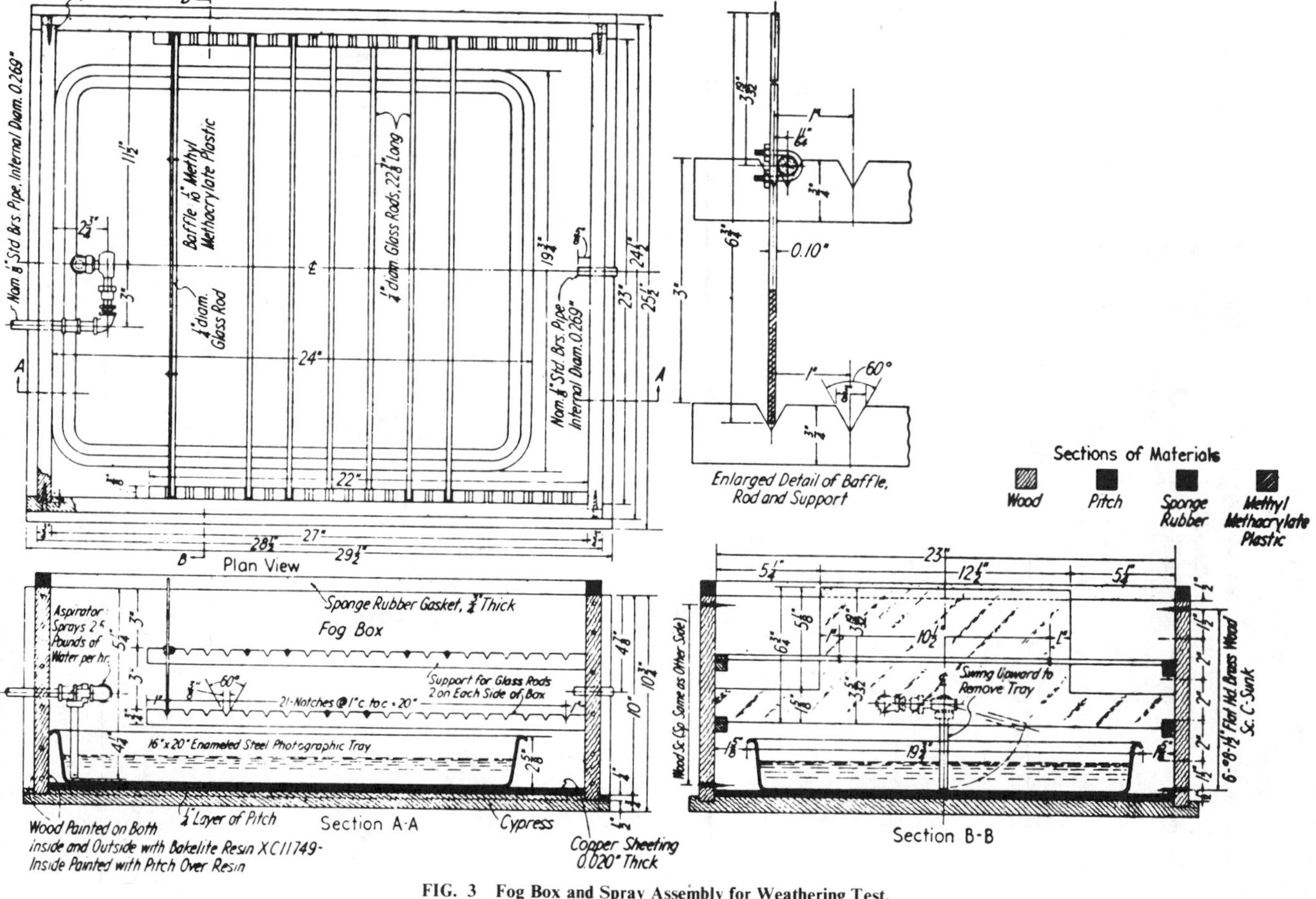

FIG. 3 Fog Box and Spray Assembly for Weathering Test.

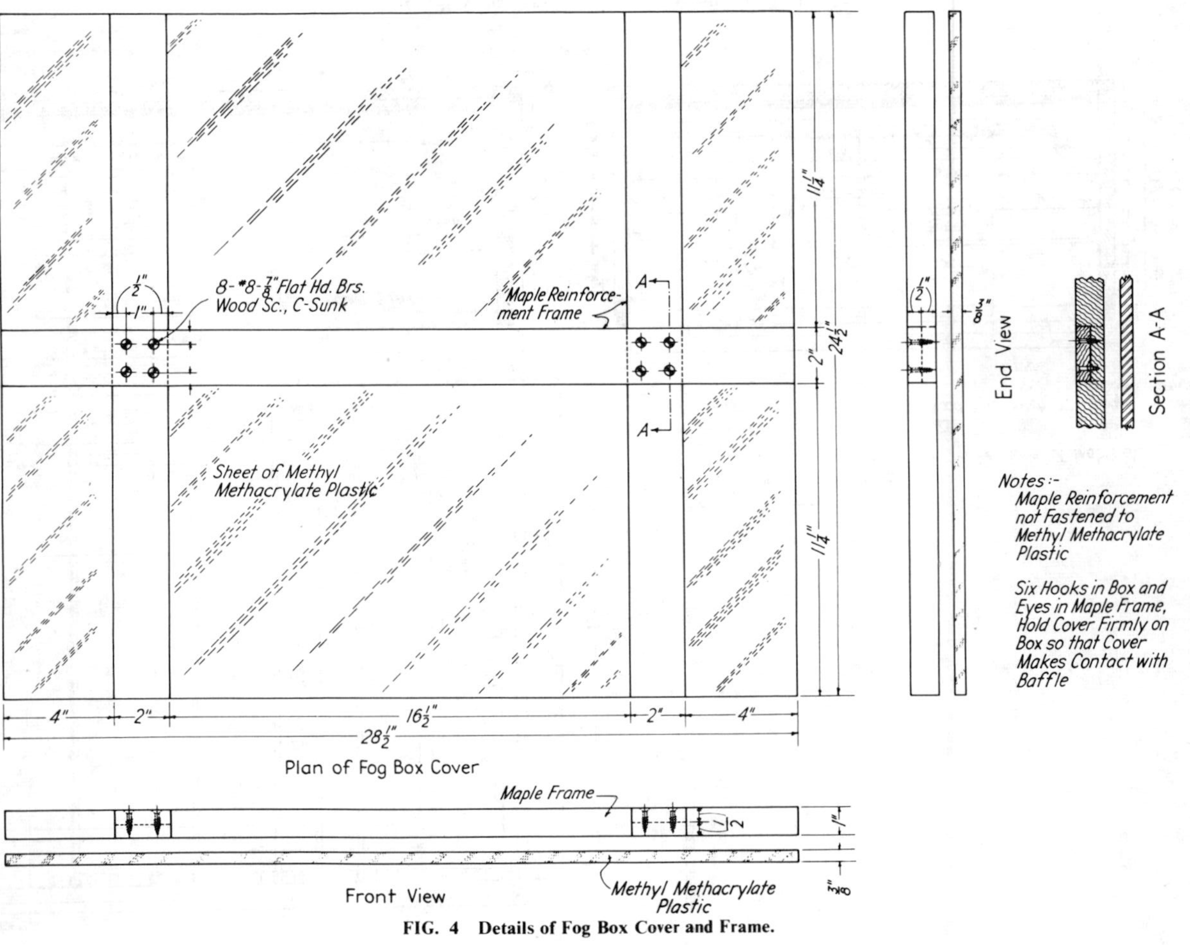

8-#8-7/8" Flat Hd. Brs. Wood Sc., C-Sunk
Maple Reinforce-ment Frame
Sheet of Methyl Methacrylate Plastic
Plan of Fog Box Cover
Maple Frame
Front View
Methyl Methacrylate Plastic
End View
Section A-A
Notes:-
Maple Reinforcement not Fastened to Methyl Methacrylate Plastic
Six Hooks in Box and Eyes in Maple Frame, Hold Cover Firmly on Box so that Cover Makes Contact with Baffle

FIG. 4 Details of Fog Box Cover and Frame.

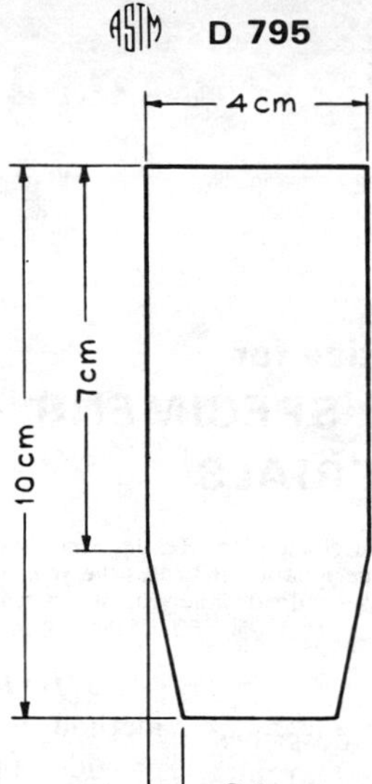

FIG. 5 Test Specimen.

Designation: D 796 – 65

USA Standard K66.8-1967
USA Standards Institute

Recommended Practice for

MOLDING TEST SPECIMENS OF PHENOLIC MOLDING MATERIALS[1]

This Recommended Practice of the American Society for Testing and Materials is issued under the fixed designation D 796; the number immediately following the designation indicates the year of original adoption or, in the case of revision, the year of last revision. A number in parentheses indicates the year of last reapproval.

1. Scope

1.1 This recommended practice describes procedures for compression molding of impact, flexure, tension, water absorption, and electrical test specimens of phenolic molding materials.

NOTE—The values stated in U.S. customary units are to be regarded as the standard. The metric equivalents of U.S. customary units may be approximate.

2. Apparatus

2.1 *Molds:*

2.1.1 For molding impact and flexure specimens, use the single bar, single cavity positive mold shown in Fig. 1 of ASTM Recommended Practice D 647 for Design of Molds for Test Specimens of Plastic Molding Materials.[2]

2.1.2 For molding tension specimens, use the positive mold shown in Fig. 3 of Specifications D 647 to make specimens required for ASTM Method D 651, Test for Tensile Strength of Molded Electrical Insulating Materials,[3] and the positive mold shown in Fig. 7 of Specifications D 647 to make specimens required for ASTM Method D 638, Test for Tensile Properties of Plastics.[2]

2.1.3 For molding water absorption specimens, use the 5.08-cm (2-in.) positive mold shown in Fig. 2 of Specifications D 647.

2.1.4 For molding specimens for electrical tests, use the 10.16-cm (4-in.) positive mold shown in Fig. 2 of Specifications D 647.

2.2 *Press*—The hydraulic press shall be such that the molding pressure on the specimen can be maintained at the value, and within the tolerances prescribed in Tables 1 and 2.

2.3 *Heating System*—Use any convenient method of heating the press platens or molds, provided the heat source is constant enough to maintain the mold temperature within ±3 C (±5 F).

2.4 *Thermometer*—Use a 7.62-cm (3-in.) immersion, mercury-filled thermometer having a temperature scale of not more than 25 C (45 F) per 2.54 cm (1 in.) of length, or a surface pyrometer graduated at 2 C intervals.

3. Conditioning

3.1 Except for referee tests, prior conditioning of phenolic molding materials is required only for electrical tests (see 3.2). For referee tests, condition a sufficient sample of the molding material, spread to a maximum depth of 2.54 cm (1 in.) in an open container, for 72 h in a Standard Laboratory Atmosphere (23 ± 1 C, 50 ± 2 percent relative humidity) for flexure, impact, tension and water absorption test specimens.

3.2 For molding electrical test specimens, condition a sufficient sample of the molding material in loose powder form for 30 min at 90 ± 3 C (194 ± 5 F) in a circulating-air oven and mold it immediately after conditioning. While being thus conditioned, the depth of the material in the container shall not exceed 1.27 cm (1/2 in.).

---

[1] This recommended practice is under the jurisdiction of ASTM Committee D-20 on Plastics. A list of committee members may be found in the ASTM Yearbook. This standard is the direct responsibility of Subcommittee D-20.90 on Specimen Preparation.

Current edition effective Aug. 31, 1965. Originally issued 1944. In 1949, incorporated Recommended Practice D 949. Replaces D 796 – 61 T.

[2] *Annual Book of ASTM Standards*, Part 27.

[3] *Annual Book of ASTM Standards*, Part 29.

4. Procedure

4.1 Mold test specimens under the conditions A, B, or C as prescribed in Tables 1 and 2. Transfer electronic preheated preforms to the mold immediately after preheating. Discard the initial preforms used to determine optimum electronic preheating after the proper preheating temperature has been determined.

5. Report

5.1 The report shall include the following:

5.1.1 Type and description of material used, and

5.1.2 Molding conditions (A, B, or C) used to mold the specimens.

TABLE 1 Molding Conditions for Impact, Flexure, Tension, and Water Absorption Test Specimens

Item	Molding Conditions	
	A	B
Charge	powder	preform[a]
Preheating	none	104 to 115 C (220 to 240 F) (electrical preheat)
Molding temperature	170 ± 3 C (338 ± 5 F)	170 ± 3 C (338 ± 5 F)
Molding pressure[b]	175 ± 35 kgf/cm² (2500 ± 500 psi)	175 ± 35 kgf/cm² (2500 ± 500 psi)
Molding time	5 min	2 min
Breathing	permitted	permitted

[a] A single preform preferably approximating the molded shape shall be used. If more than one preform is used, weld lines may affect the test value for the specimen.

[b] Highly plastic materials may flash excessively under the molding pressure specified. If that occurs, molding pressure may be reduced to a minimum of 106 kgf/cm² (1500 psi), in which case the pressure used shall be reported.

TABLE 2 Molding Conditions for Electrical Test Specimens

Item	Molding Condition C
Charge	powder
Preheating	30 min at 90 ± 3 C (194 ± 5 F) (3.2)
Molding temperature	170 ± 3 C (338 ± 5 F)
Molding pressure[a]	175 ± 35 kgf/cm² (2500 ± 500 psi)
Molding time	5 min
Breathing	permitted

[a] Highly plastic materials may flash excessively under the molding pressure specified. If that occurs, molding pressure may be reduced to a minimum of 106 kgf/cm² (1500 psi), in which case the pressure used shall be reported.

Designation: D 817 – 65

American National Standard K65.23-1967
American National Standards Institute

Standard Methods of
TESTING CELLULOSE ACETATE PROPIONATE AND CELLULOSE ACETATE BUTYRATE[1]

This Standard is issued under the fixed designation D 817; the number immediately following the designation indicates the year of original adoption or, in the case of revision, the year of last revision. A number in parentheses indicates the year of last reapproval.

1. Scope

1.1 These methods[2] cover procedures for the testing of cellulose acetate propionates and acetate butyrates. These esters may vary widely in composition and properties, so certain of the procedures can be used only in the ranges of composition where they are suitable.

1.2 The test procedures appear in the following sections:

	Sections
Acetyl and Propionyl or Butyryl Contents	21 to 28
Acetyl Content, Apparent	13 to 20
Acidity, Free	8 to 12
Ash	6 to 7
Color and Haze	59 to 62
Heat Stability	45 to 52
Hydroxyl Content	29 to 34
Hydroxyl Content, Primary	35 to 39
Moisture Content	4 to 5
Sulfur or Sulfate Content	40 to 44
Viscosity	57 to 58
Limiting Viscosity Number	53 to 56

2. Reagents

2.1 *Purity of Reagents*—Reagent grade chemicals or equivalent, as specified in ASTM Methods E 200, Preparation, Standardization, and Storage of Standard Solutions for Chemical Analysis,[3] shall be used in all tests.

2.2 Unless otherwise indicated, references to water shall be understood to mean reagent water conforming to ASTM Specification D 1193, for Reagent Water.[3]

3. Conditioning

3.1 *Conditioning*—Condition the test specimens at 23 ± 2 C (73.4 ± 3.6 F) and 50 ± 5 percent relative humidity for not less than 40 h prior to test in accordance with Procedure A of ASTM Methods D 618, Conditioning Plastics and Electrical Insulating Materials for Testing,[4] for those tests where conditioning is required. In cases of disagreement, the tolerances shall be ±1 C (±1.8 F) and ±2 percent relative humidity.

3.2 *Test Conditions*—Conduct tests in the Standard Laboratory Atmosphere of 23 ± 2 C (73.4 ± 3.6 F) and 50 ± 5 percent relative humidity, unless otherwise specified in the test methods or in this specification. In cases of disagreements, the tolerances shall be ±1 C (±1.8 F) and ±2 percent relative humidity.

MOISTURE CONTENT

4. Procedure

4.1 Transfer about 5 g of the sample to a tared, low, wide-form weighing bottle and weigh to the nearest 0.001 g. Dry in an oven for 2 h at 105 ± 3 C. Remove the bottle from the oven, cover, cool in a desiccator, and weigh.

---

[1] These methods are under the joint jurisdiction of ASTM Committee D-23 on Cellulose and Cellulose Derivatives and Committee D-20 on Plastics. A list of committee members may be found in the ASTM Yearbook. This standard is the direct responsibility of Subcommittee D-20.70 on Analytical Methods.

Current edition effective Aug. 31, 1965. Originally issued 1944. Replaces D 817 – 62 T.

[2] In 1962, the scope and the mixed ester procedure were extended to include cellulose acetate propionates. Methods for the determination of heat stability, hydroxyl, primary hydroxyl, sulfur or sulfate, intrinsic viscosity, and color and haze were added. An alternative method for apparent acetyl was substituted. Equations were added for calculation of mixed ester composition from hydroxyl instead of apparent acetyl.

[3] *Annual Book of ASTM Standards*, Part 23.

[4] *Annual Book of ASTM Standards*, Part 27.

5. Calculation

5.1 Calculate the percentage of moisture as follows:

$$\text{Moisture, percent} = (A/B) \times 100$$

where:
A = weight loss on heating, g, and
B = grams of sample used.

Ash

6. Procedure

6.1 Dry the sample for 2 h at 105 ± 3 C and weigh 10 to 50 g, to the nearest 0.01 to 0.1 g, depending on its ash content and the accuracy desired. Burn directly over a flame in a 100-ml tared platinum crucible that has been heated to constant weight and weighed to the nearest 0.1 mg. Add the sample in portions if more than 10 g is taken. The sample should burn gently and the portions should be added as the flame subsides. Continue heating with a burner only as long as the residue burns with a flame. Transfer the crucible to a muffle furnace and heat at 550 to 600 C for 3 h, or longer if required, to burn all the carbon. Allow the crucible to cool and then transfer it, while still warm, to a desiccator. When the crucible has cooled to room temperature, weigh accurately to the nearest 0.1 mg.

7. Calculation

7.1 Calculate the percentage of ash as follows:

$$\text{Ash, percent} = (A/B) \times 100$$

where:
A = grams of ash, and
B = grams of sample used.

Free Acidity

8. Reagents

8.1 *Acetone*, neutral.

8.2 *Methyl Red Indicator Solution (0.4 g/liter)*—Dissolve 0.1 g of methyl red in 3.72 ml of 0.1000 *N* NaOH solution and dilute to 250 ml with water. Filter if necessary.

8.3 *Phenolphthalein Indicator Solution (1 g/100 ml)*—Dissolve 1 g of phenolphthalein in 100 ml of ethyl alcohol (95 percent).

8.4 *Sodium Hydroxide, Standard Solution (0.01 N)*—Prepare and standardize a 0.01 *N* solution of sodium hydroxide (NaOH).

Method A—For Samples Containing Not More than About 30 percent Propionyl or Butyryl

9. Procedure

9.1 Shake 5 g of the sample, corrected for moisture content if necessary, in a 250-ml Erlenmeyer flask with 150 ml of freshly boiled, cold water. Stopper the flask and allow it to stand for 3 h. Filter off the cellulose ester and wash it with water. Titrate the combined filtrate and washings with 0.01 *N* NaOH solution, using phenolphthalein indicator solution.

9.2 Run a blank determination on the water, using the same volume as was used in extracting the sample.

10. Calculation

10.1 Calculate the percentage of acidity as free acetic acid as follows:

$$\text{Free acetic acid, percent} = \{[(A - B)C \times 0.06]/W\} \times 100$$

where:
A = milliliters of NaOH solution used to titrate the sample,
B = milliliters of NaOH solution used to titrate the blank,
C = normality of the NaOH solution, and
W = grams of sample used.

Method B—For Samples Containing More than About 7 percent Propionyl or Butyryl and Particularly Suitable for Samples Containing More than 30 percent Propionyl or Butyryl

11. Procedure

11.1 Dissolve 10.0 g of the sample, corrected for moisture content if necessary, in 200 ml of neutral acetone plus 20 ml of water. When completely dissolved, add 50 ml of water and shake well to precipitate the ester in a finely divided form. Add 3 drops of methyl red indicator solution and titrate to a lemon-yellow end point with 0.01 *N* NaOH solution.

11.2 Make a blank determination on the reagents.

12. Calculation

12.1 Calculate the free acid content as acetic acid as directed in Section 10.

Apparent Acetyl Content

13. Scope

13.1 The methods described in the following Sections 14 to 20 cover the determination of the saponification value of the sample calculated as percentage of apparent acetyl, equivalent weight 43. This value is required in the calculation of acetyl and propionyl or butyryl contents in 28.1.

13.2 The method used should be specified or agreed upon. The choice depends on the propionyl or butyryl content and the physical condition of the sample. Oridinarily, Method A is recommended for samples having less than about 35 percent propionyl or butyryl and Method B for samples having more than that amount.

Method A—For Samples Containing Less than About 35 percent Propionyl or Butyryl

14. Apparatus

14.1 *Weighing Bottle*, glass-stoppered, 15-ml capacity, 25-mm diameter by 50 mm high.

14.2 *Tray*, copper or aluminum, approximately 137 mm ($5^3/_8$ in.) square, containing 25 compartments 25 mm (1 in.) square. Each compartment shall have the correct dimensions to contain one weighing bottle. The entire tray shall fit inside a desiccator and should have a basket-type handle to facilitate the introduction and removal of the tray (convenient but not essential).

14.3 *Buret*, automatic zero, 35-ml, 25-ml bulb, stem graduated from 25 to 35 ml in 0.05-ml increments; or pipet, automatic zero, 30-ml for NaOH solution (40 g/liter).

14.4 *Buret*, automatic zero, 15-ml, 10-ml bulb, stem graduated from 10 to 15 ml in 0.05-ml increments, for 1 *N* H_2SO_4.

14.5 *Buret*, 5-ml, in 0.01 or 0.1-ml divisions, for back titration with 0.1 *N* NaOH solution.

14.6 *Magnetic Stirrer*, for single flask.

14.7 *Magnetic Stirrer*, capacity twelve or more flasks.

14.8 *Stirring Bars*, stainless steel Type 416, length 50 mm, diameter 5 to 6 mm or equivalent, dimensions not critical.

15. Reagents

15.1 *Acetone*—Add one 30-ml portion of 1.0 *N* NaOH solution to a mixture of 150 ml acetone and 100 ml hot water, allow to stand with frequent swirling for 30 min, and titrate with 1.0 *N* H_2SO_4. Add another 30-ml portion of 1.0 *N* NaOH solution to 100 ml of hot water, allow to stand for 30 min, and titrate as above. The difference between the two titrations shall not exceed 0.05 ml.

15.2 *Dimethyl Sulfoxide*.

15.3 *Pyridine*.

15.4 *Sodium Hydroxide Solution* (*40 g/liter*)—Dissolve 40 g of sodium hydroxide (NaOH) in water and dilute to 1 liter.

15.5 *Sodium Hydroxide, Standard Solution* (*0.1 N*)—Prepare and standardize a 0.1 *N* solution of NaOH.

15.6 *Sulfuric Acid, Standard* (*1.0 N*)—Prepare and standardize a 1.0 *N* solution of sulfuric acid (H_2SO_4).

15.7 *Phenolphthalein Indicator Solution* (*1 g/100 ml*)—Dissolve 1 g of phenolphthalein in 100 ml of ethyl alcohol (95 percent).

16. Procedure

16.1 Dry 1.9 ± 0.05 g of the ground well-mixed sample in weighing bottle for 2 h at 105 ± 3 C and weigh the dried sample by difference to the nearest 1 mg into a 500-ml Erlenmeyer flask. Prepare a blank by drying approximately 3.8 g of potassium acid phthalate and weighing it by difference into a flask as described above. Carry the blank through the entire procedure.

Note 1—Potassium acid phthalate is used so that the concentration of the NaOH in contact with the solvent in the blank will be approximately the same as that in contact with the sample and so that the titration of the blank will be approximately the same as the titration of the sample, thus avoiding errors caused by using a different buret for the titration of the blank and the sample or by refilling the 15-ml buret. If desired, however, the potassium acid phthalate may be omitted.

16.2 For acetone-soluble sample, put the sample into solution as follows: Add 150 ml of acetone and 5 to 10 ml of water and swirl to mix. Stopper the flask and allow it to stand with occasional swirling until solution is complete. Solution may be hastened by magnetic stirring or by any suitable mechanical shaking that will provide a gentle rocking type of agitation to avoid splashing the solution on the stopper. It is essential that complete solution be effected.

16.3 For acetone-insoluble samples of low propionyl or butyryl content, dissolve the sample by either of the following two methods:

16.3.1 Gently rotate the flask by hand to distribute and spread the sample in a thin layer over the bottom of the flask. Add 70 ml of acetone to the flask and swirl gently until the sample particles are completely wetted and evenly dispersed. Stopper the flask and allow it to stand undisturbed for 10 min. Carefully add 30 ml of dimethyl sulfoxide from a graduate to the flask, pouring the solvent down the sides of the flask to wash down any sample particles clinging to the side. Stopper the flask and allow it to stand with occasional swirling until solution is complete. Magnetic stirring or gentle mechanical agitation that will not splash the solution is recommended. When solution appears to be complete, add 50 ml of acetone and swirl or stir for 5 min. Proceed in accordance with 16.4.

16.3.2 Dimethyl sulfoxide is the preferred solvent, but if it is not available, spread the sample in a thin layer over the bottom of the flask, add 15 ml of acetone, swirl to wet the particles with acetone, stopper the flask, and allow the mixture to stand undisturbed for 20 min. Add 75 ml of pyridine without shaking or swirling and allow the mixture to stand for 10 min. Heat the solution just to boiling and swirl or stir for 5 min. Again heat to boiling and swirl or stir for 10 min. Continue to heat and stir until the mixture is homogeneous and all large gel masses are broken down into individual highly swollen particles. When these highly swollen gel particles are well dispersed and are not fused together in large gel masses, no further heating is necessary. Cool the flask, add 30 ml of acetone, and swirl or stir for 5 min.

16.4 Add 30 ml of NaOH solution (40 g/liter) with constant swirling or stirring to the solution of the sample and also to the blank. Use of a magnetic stirrer is recommended (Note 2). It is absolutely necessary that a finely divided precipitate of regenerated cellulose, free of lumps, be obtained. Stopper the flask and let the mixture stand with occasional swirling or stir on the magnetic stirring unit. Allow 30 min for saponification of lower acetyl samples, 2 h for high acetyl samples when dimethyl sulfoxide is the solvent, and 3 h when pyridine is the solvent. At the end of the saponification period, add 100 ml of hot water, washing down the sides of the flask, and stir for 1 or 2 min. Add 4 or 5 drops of phenolphthalein indicator solution and titrate the excess NaOH solution with 1.0 *N* H_2SO_4 (Note 3). Titrate rapidly with constant swirling or stirring until the end point is reached; then add an excess of 0.2 or 0.3 ml of H_2SO_4. Allow the mixture to stand with occasional stirring or preferably stir on the magnetic stirrer for a least 10 min. Then add 3 drops of phenolphthalein indicator solution to each flask and titrate the small excess of acid with 0.1 *N* NaOH solution to a persistent phenolphthalein end point. Take extreme care to locate this end point; after the sample is titrated to a faint pink end point, swirl the mixture vigorously or place it for a moment on the magnetic stirrer. If the end point fades because of acid soaking from the cellullose, continue the addition of 0.1 *N* NaOH solution until a faint persistent end point remains after vigorous swirling or stirring. Titrate the blank in the same manner as the sample.

NOTE 2—While the amount of magnetic stirring is somewhat optional, such stirring during the entire period of the determination is strongly recommended. Solution is more rapid, titrations are more rapid, and the end point can be approached directly and without a back titration.

NOTE 3—It is important to correct all 1.0 *N* H_2SO_4 buret readings for temperature and buret corrections.

17. Calculation

17.1 Calculate the percentage by weight of acetyl as follows:

$$\text{Acetyl, percent} = \{[(D - C)N_a - (B - A)N_b + P] \times 0.04305\}/W \times 100$$

$$P = (GH \times 1000)/204.2 \qquad \text{(Note 4)}$$

where:

A = milliliters of NaOH solution required for titration of the sample,

B = milliliters of NaOH solution required for titration of the blank,

N_b = normality of the NaOH solution,

C = milliliters of H_2SO_4 required for titration of the sample,

D = milliliters of H_2SO_4 required for titration of the blank,

N_a = normality of the H_2SO_4,

P = milliequivalents of potassium acid phthalate,
G = grams of potassium acid phthalate used,
H = purity factor for potassium acid phthalate, and
W = grams of sample used.

NOTE 4—When equal volumes of alkali or acid are added to samples and blank, these amounts cancel out. Thus only the amounts of each added in the titration enter into the calculations. Use of potassium acid phthalate in the blank is recommended. When it is not used, the term P drops out of the equation.

Method B—For Cellulose Esters Containing More than 30 percent Propionyl or Butyryl, by Varying the Reagents[5]

18. Reagents

18.1 *Acetone - Alcohol Mixture*—Mix equal volumes of acetone and methyl alcohol.

18.2 *Hydrochloric Acid, Standard (0.5 N)*—Prepare and standardize a 0.5 *N* solution of hydrochloric acid (HCl).

18.3 *Phenolphthalein Indicator Solution (1 g/100 ml)*—Dissolve 1 g of phenolphthalein in 100 ml of ethyl alcohol (95 percent).

18.4 *Pyridine - Alcohol Mixture*—Mix equal volumes of pyridine and methyl alcohol.

18.5 *Sodium Hydroxide, Aqueous Solution (20 g/liter)*—Dissolve 20 g of sodium hydroxide (NaOH) in water and dilute to 1 liter with water.

18.6 *Sodium Hydroxide, Methanol Solution (20 g/liter)*—Dissolve 20 g of NaOH in 20 ml of water and dilute to 1 liter with methyl alcohol.

19. Procedure

19.1 Dry the sample for 2 h at 105 ± 3 C and cool in a desiccator. Weigh 0.5-g portions of the sample to the nearest 0.005 g and transfer to 250-ml glass-stoppered Erlenmeyer flasks. Dissolve each sample in 100 ml of appropriate solvent (see 19.2 and 19.3) and prepare at least two blanks, which shall be carried through all steps of the procedure.

19.2 *Samples Containing 30 to 45 percent Propionyl or Butyryl*—Dissolve in 100 ml of the acetone - alcohol mixture. Add water and aqueous NaOH solution from a buret or pipet in the following order and swirl the contents of the flask vigorously during all additions: 10 ml of NaOH solution, 10 ml of water, 10 ml of NaOH solution, 5 ml of water, 20 ml of NaOH solution, and 5 ml of water. Stopper and allow to stand at room temperature for 16 to 24 h.

19.3 *Samples Containing More than 45 percent Propionyl or Butyryl*—Dissolve in 100 ml of the pyridine - alcohol mixture. Add 30 ml of the methanol solution of NaOH from a pipet or buret slowly, with swirling. Add 20 ml of water slowly in about 2-ml portions, with swirling, and swirl the flask until the solution becomes turbid. Stopper and allow to stand overnight at room temperature.

19.4 Back-titrate the excess NaOH with 0.5 *N* HCl just to the disappearance of color, using phenolphthalein indicator solution.

20. Calculation

20.1 Calculate the apparent acetyl content as follows:

$$\text{Apparent acetyl, percent} = \{[(A - B)N_a \times 0.04305]/W\} \times 100$$

where:

A = milliliters of HCl required for titration of the blank,
B = milliliters of HCl required for titration of the sample,
N_a = normality of the HCl, and
W = grams of sample used.

ACETYL AND PROPIONYL OR BUTYRYL CONTENTS

21. Scope

21.1 The method described in the following Sections 22 to 28 covers the determination of acetyl and propionyl or butyryl contents of cellulose mixed esters by calculation from the apparent acetyl content, determined in accordance with Sections 13 to 20, and the molar ratio of acetyl and propionyl or butyryl, determined in accordance with Sections 22 to 27. The molar ratio of acetyl and propionyl or butyryl is determined by saponifying, acidifying, vacuum distilling off the mixture of acids, and determining the distribution ratio of the acids between *n*-butyl acetate and wa-

[5] Malm, C. J., Genung, L. B., Williams, R. F., Jr., and Pile, M. A., "Analysis of Cellulose Derivatives: Total Acyl in Cellulose Organic Esters by Saponification in Solution," *Industrial and Engineering Chemistry*, Analytical Edition, IENAA, Vol 16, 1944, pp. 501–504.

ter. The distribution ratios are also determined for acetic, propionic, and butyric acids, using samples of known high purity, and the molar ratio of the acids in the sample is calculated from these values.[6]

21.2 The saponification conditions are varied depending on the propionyl or butyryl content of the sample. Use Procedure A (Section 24) for samples containing less than about 35 percent propionyl or butyryl, and use Procedure B (Section 25) for samples containing more than that amount.

22. Apparatus

22.1 *Vacuum Distillation Apparatus*—The vacuum distillation apparatus shown in Fig. 1 will be required. The 500-ml round-bottom flask, *A*, shall be fitted with a stopper carrying a very small capillary inlet tube, *B*, and a Kjeldahl distilling head, *C*. The Kjeldahl distilling head shall be connected to a vertical condenser, *D*, having an outlet tube long enough to reach within 3 in. of the bottom of the 500-ml distilling flask, *E*, used as a receiver. The Kjeldahl distilling head shall be equipped with a funnel or stoppered opening, *F*, for adding extra water during the distillation. A water bath, *G*, for heating the sample and a cooling bath, *H*, for cooling the receiver shall be provided.

23. Reagents

23.1 *Acetic, Propionic, and Butyric Acids*—Acetic, propionic, and butyric acids of tested purity.

23.2 *Bromcresol Green Indicator Solution (0.4 g/liter)*—Grind 0.1 g of tetrabromo-*m*-cresolsulfonphthalein in a mortar with 14.3 ml of 0.01 *N* NaOH solution and dilute to 250 ml.

23.3 *n-Butyl Acetate*—Prepare *n*-butyl acetate for use as an extraction solvent, free of acidity and water and containing not more than 2 percent butyl alcohol. Check for acidity by shaking 60 ml of the *n*-butyl acetate with 30 ml of water in a 125-ml separatory funnel for about 1 min. Allow to settle, draw off the water layer, and titrate with 0.1 *N* NaOH solution, using phenolphthalein as the indicator. If this requires more than 0.02 ml of 0.1 *N* NaOH solution, the butyl acetate should be purified or a correction for acidity applied to each titration.

23.4 *Ethyl Alcohol*, Formula 2B, 3A, or 30 (denatured).

23.5 *Phosphoric Acid (1 + 14)*—Dilute 68 ml of phosphoric acid (H_3PO_4, 85 percent) to 1 liter with water. Titrate the NaOH solution (20 g/liter) with this acid to a yellow end point, using bromcresol green indicator solution, and calculate the volume of the acid (approximately 50 ml) required for 100 ml of the NaOH solution.

23.6 *Sodium Hydroxide Solution (20 g/liter)*—Dissolve 20 g of sodium hydroxide (NaOH) in water and dilute to 1 liter.

23.7 *Sodium Hydroxide, Standard Solution (0.1 N)*—Prepare and standardize a 0.1 *N* solution of NaOH.

Isolation of the Mixed Acids

24. Procedure A—For Samples Containing Less than About 35 percent Propionyl or Butyryl

24.1 Heat duplicate 3-g portions of the sample, not especially dried nor accurately weighed, with 100 ml of NaOH solution (20 g/liter) in 500-ml, round-bottom, chemically resistant glass flasks in a water bath at 40 C for 48 to 72 h. At the end of this time add the required amount (approximately 50 ml) of H_3PO_4 (1 + 14) to each flask to form monosodium phosphate, whioh liberates the organic acids from their sodium salts.

24.2 Assemble the vacuum distillation apparatus as illustrated in Fig. 1. Heat the 500-ml round-bottom flask containing the sample in a water bath, and vacuum-distill the acid solutions to dryness, allowing a small stream of air bubbles to enter to avoid bumping. Keep the receiver cooled to 0 C. Add 25 ml of water to the residue in each flask and again distill to dryness. Repeat the distillation to dryness with a second 25-ml portion of water.

NOTE 5—In this operation it it not necessary to work with quantitative accuracy at all stages, but it is necessary to obtain water solutions of the acids in the same ratios as they occur in the esters. The volume of the distillate and rinsings is usually 200

[6] Malm, C. J., Nadeau, G. F., and Genung, L. B., "Analysis of Cellulose Derivatives: Analysis of Cellulose Mixed Esters by the Partition Method," *Industrial and Engineering Chemistry*, Analytical Edition, IENAA, Vol. 14, 1942, pp. 292–297. This reference may be consulted for application to other mixed esters and to three-component mixtures.

to 250 ml, which in the majority of cases automatically adjusts the acidity of the distillate to 0.06 to 0.12 N, the range desired for subsequent extractions.

24.3 Continue as directed in Section 26.

25. Procedure B—For Samples Containing More than About 35 percent Propionyl or Butyryl

25.1 Weigh duplicate 3-g samples, not especially dried nor accurately weighed, into 500-ml round-bottom flasks and add 100 ml of Formula 2B, 3A, or 30 denatured ethyl alcohol and 100 ml of NaOH solution (20 g/liter) to each flask. Allow the samples to stand stoppered at room temperature for 48 to 72 h. At the end of this period, filter off the regenerated cellulose, collecting the filtrates in 500-ml round-bottom flasks.

25.2 Assemble the vacuum-distillation apparatus as illustrated in Fig. 1. Heat the flasks in the water bath and vacuum-distill off all the alcohol. After distilling to dryness, release the vacuum, rinse out the distillation heads, condensers, and receivers, and discard the distillates and rinsings.

25.3 Add the required amount, about 50 ml, of H_3PO_4 (1 + 14) to form monosodium phosphate, which liberates the organic acids from their sodium salts. Also add 100 ml of water to each flask and reassemble the distillation apparatus. Vacuum-distill the volatile acids as described in 24.2.

25.4 Continue as directed in Section 26.

Determination of the Molar Ratios of the Acids

26. Procedure

26.1 Titrate a 25-ml portion of the distillate (24.2) with 0.1 N NaOH solution, using phenolphthalein as the indicator. Designate the volume of NaOH solution required as M. Shake 30 ml of the distillate in a small separatory funnel with 15 ml of n-butyl acetate. Measure these volumes accurately using pipets and burets. Shake the mixture thoroughly for 1 min, allow the layers to separate for 2 min, and draw off the aqueous (lower) layer. Pipet out 25 ml of the solution and titrate with 0.1 N NaOH solution (Note 6). Designate the volume of NaOH solution required as M_1. Calculate K, the percentage partition ratio of the acids in the distillate, as follows:

$$K = (M_1/M) \times 100$$

NOTE 6—It should be kept in mind that all these determination are ratios and not quantitative; however, accuracy of duplication is very important. All measurements must be made as exactly as those made by standardizations of the solutions and equipment.

26.2 In the same manner determine the distribution ratios for acetic, propionic, and butyric acids. Dilute a sample of each acid of tested purity with water to give an approximately 0.1 N solution. Titrate 25-ml portions and extract 30-ml portions, following exactly the same procedure as used for the mixtures (26.1). Calculate the partition ratios for the pure acids, as decimal fractions, as follows (Note 7):

$$k = M_1/M$$

where:

k_a = distribution ratio for acetic acid under the conditions described,

k_p = distribution ratio for propionic acid under the conditions described, and

k_b = distribution ratio for butyric acid under the conditions described.

NOTE 7—The constants must be checked occasionally and must be determined by each operator for each supply of butyl acetate. Blanks should be run on the butyl acetate, since it may develop acidity on standing, particularly if it contains a little water. All measurements should be made with good pipets or burets and extreme care and cleanliness observed during the whole operation. The accuracy of the procedure can be checked by testing an acid mixture of known composition.

27. Calculation

27.1 Calculate the molar ratios of acetic and propionic or butyric acids in the mixed acids as follows (Note 8):

$$P = (100k_a - K)/(k_a - k_p)$$

$$A = 100 - P$$

$$B = (100k_a - K)/(k_a - k_b)$$

$$A = 100 - B$$

where:

P = mole percentage of propionic acid,

B = mole percentage of butyric acid,

A = mole percentage of acetic acid,

K = percentage distribution ratio of the acids in the distillate (26.1),

k_a = distribution ratio of acetic acid (26.2),
k_p = distribution ratio of propionic acid (26.2), and
k_b = distribution ratio of butyric acid (26.2),

NOTE 8—In order to evaluate two unknowns, two simultaneous algebraic equations involving the two unknown quantities are necessary. In the case of a binary acid mixture, the sum of the mole percentages of the acids present represents the total acidity, or 100 percent. If A and B represent the mole percentages of acetic and butyric acids, respectively:

$$A + B = 100$$

$$Ak_a + Bk_b = K$$

The distribution ratios k_a and k_b are known and refer to the pure individual acids, whereas the distribution ratio K refers to the binary mixture. By solving these equations for B, the equations given in this section may be derived.

Calculation of Acetyl, Propionyl, and Butyryl Contents

28. Calculation

28.1 Calculate the percentages by weight of acetyl, propionyl, and butyryl as follows:

$$\text{Acetyl, percent} = AC/100$$

$$\text{Propionyl, percent} = (PC/100) \times (57/43)$$

$$\text{Butyryl, percent} = (BC/100) \times (71/43)$$

where:
A = mole percentage of acetic acid (Section 27),
P = mole percentage of propionic acid (Section 27),
B = mole percentage of butyric acid (Section 27), and
C = percentage by weight of apparent acetyl (Sections 17 and 20).

28.2 Hydroxyl can be measured precisely, particularly at high degrees of esterification (Sections 29 to 34). It is therefore sometimes advantageous to base the calculation of weight percentages of acetyl, propionyl, and butyryl on hydroxyl content rather than on apparent acetyl as in 28.1. The equations for this calculation are as follows:

For cellulose acetate propionates:

$$\text{Acetyl, percent} = 9.75A(31.5 - h)/(786 - A)$$

$$\text{Propionyl, percent} = 12.93P(31.5 - h)/(786 - A)$$

For cellulose acetate butyrates:

$$\text{Acetyl, percent} = 4.88A(31.5 - h)/(443 - A)$$

$$\text{Butyryl, percent} = 8.05B(31.5 - h)/(443 - A)$$

where, in addition to the definitions of terms in 28.1:
h = weight percentage of hydroxyl (Section 34).

NOTE 9—This calculation involves the assumption that there are exactly three hydroxyls, free plus esterified, for each anhydroglucose unit of cellulose.

HYDROXYL CONTENT

29. Scope

29.1 This method is applicable to pyridine-soluble cellulose esters and is especially useful when the hydroxyl content is low. (Samples containing plasticizer may be analyzed directly by this method because the plasticizer is removed during washing of the carbanilate).

30. Summary of Method

30.1 Hydroxyl in cellulose esters is determined by reaction with phenyl isocyanate in pyridine solution under anhydrous conditions to form the carbanilate derivative. The derivative is then analyzed for its carbanilate content by ultraviolet absorption.

31. Apparatus

31.1 *Spectrophotometer*,[7] complete with hydrogen light source and a set of four 1.00-cm silica cells or an equally suitable apparatus. The wavelength calibration, as checked against a mercury lamp, shall be within the manufacturer's tolerances. As a further check, measure the density of a potassium chromate (K_2CrO_4) solution prepared as follows. Dissolve 0.0400 g of K_2CrO_4 or 0.0303 g of potassium ($K_2Cr_2O_7$) in 0.05 N potassium hydroxide (KOH) solution and dilute to 1 liter in a volumetric flask with 0.05 N KOH solution. Using the hydrogen lamp measure the absorbance at 280 nm of a silica cell filled with the K_2CrO_4 solution and also of the same cell filled with water. The absorbance of the solution minus that of the blank shall be 0.723 ± 0.023.

31.2 *Bottles*, 4-oz, with screw caps, for washing the samples.

31.3 *Special Reflux Tubes* for the carbanilation, constructed as follows (see Fig. 2): Make a test tube approximately 20 by 150 mm from

[7] A Beckman Model DU Spectrophotometer has been found satisfactory for this purpose.

the outer part of a standard-taper 24/40 ground-glass joint by closing the open end in a blast lamp. Draw the tubing on the inner joint to a constriction just above the joint. Cut the glass at the point and seal on a short length of 8-mm tubing to provide a bearing for a glass stirrer. Make a stirrer of 4-mm glass rod with a semicircle at right angles to the shaft at the bottom and small enough to fit into the test tube. When properly constructed this unit acts as an air condenser, thus preventing the loss of solvent by evaporation.

31.4 *Pipet*, serological type, 5-ml capacity, graduated in 0.1-ml divisions.

31.5 *Büchner Funnel*, of a size accommodating 90-mm filter paper.

31.6 *Automatic Shaker*, with speed regulator mechanism.

31.7 *Electric Oven*, maintained at 105 ± 3 C.

31.8 *Oil Bath*, equipped with a rack to hold several of the special reflux tubes. This bath shall be kept between 115 and 120 C.

32. Reagents

32.1 *Acetone*.

32.2 *Ethyl Alcohol*, Formula 2B, 3A, or 30 (denatured).

32.3 *Methylene Chloride - Methyl Alcohol Mixture*—Mix 9 parts by weight of methylene chloride with 1 part of methyl alcohol. This mixture should have an absorbance of less than 0.2 at 280 nm in a 1.00-cm silica cell measured against air. Pure methylene chloride has an absorbance of about 0.05, but the commercial product may have an absorbance as high as 1.00. The methylene chloride and methyl alcohol should be selected to have low absorbance; otherwise, they should be redistilled.

32.4 *Phenyl Isocyanate*.[8]

32.5 *Pyridine*, redistilled, of low water content, preferably less than 0.05 percent.

33. Procedure

33.1 In the following procedure the phenyl isocyanate reagent shall be used under anhydrous conditions. Therefore, the sample, containers, pipet, and all other equipment shall be thoroughly dried.

33.2 Place a 0.5-g sample in a special reflux tube and dry in an electric oven at 105 ± 3 C for 2 h. Remove the tube from the oven, add 5 ml of pyridine, assemble the reflux apparatus complete with glass stirring rod, and place in the 115 to 120 C oil bath. Stir occasionally until the sample is completely dissolved. Add 0.5 ml of phenyl isocyanate, stir thoroughly, and reflux in the oil bath for $^1/_2$ h to complete the reaction. Use 0.1 ml of phenyl isocyanate for each 1 percent of estimated hydroxyl content, but never less than 0.5 ml.

33.3 At the end of the reaction time, remove the sample and dilute it with acetone to the proper viscosity for precipitation. The amount of acetone used to thin the solution is a critical factor in acquiring a good precipitate. Samples having low viscosity require little, if any, dilution. The average sample requires the addition of about an equal volume of acetone. Precipitate the carbanilate by pouring the solution into about 200 ml of ethyl alcohol, or if the ester contains more than 20 percent propionyl or butyryl, into the same volume of cold 80 percent alcohol. Stir the alcohol vigorously during the precipitation. The precipitate should be fluffy and white. Sticky precipitates indicate too little dilution. Filter off the precipitate using paper on a Büchner funnel, with suction applied only as long as is necessary to remove the bulk of the solvent; prolonged suction may cause undesirable clumping together of the precipitate.

33.4 Wash the precipitate with alcohol, unless the sample was precipitated in cold 80 percent alcohol. In this case, wash the precipitate in cold 90 percent alcohol. Washing is best accomplished by transferring the precipitate to a 4-oz screw cap bottle containing about 75 ml of alcohol and shaking for $^1/_2$ h on an automatic shaker. Filter, pressing out as much liquid as possible with a glass stopper. Repeat the washing and filtering operations twice more.

NOTE 10—Samples of high hydroxyl content and large amounts of propionyl or butyryl may give gummy precipitates when poured into cold 80 percent alcohol. Samples of this type give improved precipitates when precipitated in the reverse manner. Pour the diluted reaction solution

[8] Phenyl isocyanate available as Eastman No. 553 has been found satisfactory for this purpose.

into a 600-ml beaker, taking care to distribute the solution evenly on the bottom. Chill the beaker in a brine bath for 30 to 60 s. Pour about 200 ml of cold 80 percent alcohol onto the chilled liquid. Wash the resulting precipitate and filter in the usual manner using cold 90 percent alcohol.

33.5 Allow the precipitate to air-dry 1 to 2 h at room temperature with good ventilation or preferably overnight to ensure complete removal of the alcohol. (Samples wet with alcohol may sinter and stick to paper or glass when dried at 105 C.) Dry the sample at 105 C in the oven for 1 h and cool in a desiccator. Small manila envelopes are convenient for drying and cooling the samples.

33.6 Weigh 0.1231 g of the dry precipitate into a 100-ml volumetric flask fitted with a ground-glass stopper. Add 60 to 80 ml of methylene chloride-methyl alcohol mixture, and shake occasionally until complete solution occurs. Dilute to 100 ml and mix thoroughly. Using the spectrophotometer with a 1-cm silica cell, measure the absorbance of the solution at 280 nm against the solvent mixture as a reference.

34. Calculations

34.1 Calculate the percentage of carbanilate, c, for a sample weight of 0.1231 g as follows:[9]

$$\text{Carbanilate, percent} = A \times 17.1$$

where:
A = absorbance.

34.2 Calculate the percentage of hydroxyl as follows:

$$\text{Hydroxyl, percent} = 14.3c/(100 - c)$$

Primary Hydroxyl Content

35. Summary of Method

35.1 The primary hydroxyl content of cellulose ester is determined by formation of the triphenylmethyl (trityl) ether and measurement of the trityl group by ultraviolet absorbance.[9] Trityl chloride reacts preferentially with primary hydroxyls. Since there is also a slight reaction with secondary hydroxyls, standardized reaction conditions are important.[10]

36. Apparatus

36.1 See Section 31.

37. Reagents

37.1 *Acetone.*

37.2 *Ethyl Alcohol*, Formula 2B, 3A, or 30 (denatured).

37.3 *Methylene Chloride-Methyl Alcohol Mixture*—Mix 9 parts by weight of methylene chloride with 1 part of methyl alcohol. This mixture should have an absorbance of less than 0.2 at 259 nm in a 1-cm silica cell measured against air; otherwise, the solvents should be redistilled.

37.4 *Pyridine*, redistilled to a water content less than 0.05 percent. The water content may be reduced further by storing over a suitable drying agent, such as a molecular sieve.[11]

37.5 *Trityl Chloride (Chlorotriphenylmethane or Triphenylmethyl Chloride).*

38. Procedure

38.1 The reagents must be used under anhydrous conditions. It is imperative that the sample and all equipment be thoroughly dry.

38.2 Place a 0.5-g sample in the test tube of the special reflux apparatus and dry for 2 h at 105 ± 3 C. Add 5 ml of pyridine, insert the top of the reflux apparatus and the stirrer, and heat with stirring in a 115 to 120 C oil bath. After the sample has dissolved, add 0.5 g of trityl chloride. If the total hydroxyl content exceeds 3 percent, use an additional 0.075 g of trityl chloride for each additional 1 percent hydroxyl. Stir the mixture thoroughly and reflux in the oil bath for exactly 2 h at 115 to 120 C. Remove the tube and cool.

38.3 Dilute the sample with acetone to the proper viscosity for precipitation. The amount of acetone used to thin the solution is a critical factor in obtaining a good precipitate. Samples having low viscosity require little, if any, dilution. The average

[9] Malm, C. J., Tanghe, L. J., Laird, B. C., and Smith, G. D., "Determination of Total and Primary Hydroxyl in Cellulose Esters by Ultraviolet Absorption Methods," *Analytical Chemistry*, ANCHA, Vol 26, 1954, p. 189.

[10] Malm, C. J., Tanghe, L. J., and Laird, B. C., "Primary Hydroxyl Groups in Hydrolyzed Cellulose Acetate," *Journal of the American Chemical Society*, JACSA, Vol 72, 1950, p. 2674.

[11] Molecular Sieve Type 4A as manufactured by the Linde Air Products Co., has been found satisfactory for this purpose.

sample requires the addition of about an equal volume of acetone. Precipitate the trityl derivative by pouring the solution into about 200 ml of ethyl alcohol with vigorous stirring. The precipitate should be fluffy and white. Sticky precipitates indicate too little dilution. Separate the precipitate by filtering through paper on a Büchner funnel, with suction applied only as long as necessary to remove the bulk of the solvent; prolonged suction may evaporate the alcohol and cause the precipitate to partially redissolve in the remaining pyridine.

38.4 Wash the precipitate by transferring it to a 4-oz screw cap bottle containing 75 ml of ethyl alcohol, capping securely, and shaking for $^1/_2$ h on a shaker at medium speed. Again collect the precipitate on a Büchner funnel, pressing out as much liquid as possible with a glass stopper. Repeat this washing and filtering operation twice more, or until the absorbance of the filtrate at 259 nm is about the same as that of an alcohol blank. Allow the precipitate to air-dry on the filter paper for $^1/_2$ h at room temperature with good ventilation, or preferably overnight, to remove most of the alcohol. (Samples wet with alcohol may sinter or stick to paper or glass when dried at 105 C.) Transfer the sample to a manila envelope, dry it for 1 h at 105 C, and cool in a desiccator.

38.5 Weigh a 0.1231-g sample of the dry trityl ether derivative into a 100-ml volumetric flask fitted with a ground-glass stopper, and dissolve in the methylene chloride-methyl alcohol mixture. Dilute to 100 ml and mix thoroughly. Measure the abosrbance of this solution in a 1-cm silica cell using a spectrophotometer at 259 nm against the solvent as a reference.

39. Calculations

39.1 Calculate the trityl content, t, for this concentration of 0.1 g/100 g and with a correction of 0.015 for the absorbance of the cellulose acetate as follows:[9]

$$\text{Trityl, percent} = 25.25\,(A - 0.015)$$

where:

A = absorbance.

39.2 Calculate the weight percentage of primary hydroxyl as follows:

$$\text{Primary hydroxyl, percent} = 7.02t/(100.4 - t)$$

39.3 Calculate the percentage primary hydroxyl of the total hydroxyl as follows:

$$\text{Primary hydroxyl of total hydroxyl, percent} = (B/C) \times 100$$

where:

B = value of primary hydroxyl as determined in 39.2, and

C = value of total hydroxyl as determined in 34.2.

Sulfur or Sulfate Content

40. Summary of Method

40.1 The sulfur or sulfate content of cellulose acetate is measured by oxidizing the sample in a nitric acid-perchloric acid mixture and determining gravimetrically as barium sulfate. To determine combined sulfur the sample must first be reprecipitated into dilute acid to remove noncombined sulfur compounds.

41. Apparatus

41.1 *Funnel*, modified by cutting the stem off at the apex of the funnel and fire polishing.

41.2 *Crucibles*,[12] 30-ml, extra-fine porosity.

41.3 *Oven*, controlled at 120 to 125 C.

41.4 *Muffle Furnace*, controlled at 800 ± 50 C.

42. Reagents

42.1 *Acetone.*

42.2 *Acetic Acid (1+49)*—Mix 1 volume of glacial acetic acid with 49 volumes of water.

42.3 *Barium Chloride Solution (100 g/liter)*—Dissolve 100 g of barium chloride ($BaCl_2 \cdot 2H_2O$) in water and dilute to 1 liter.

42.4 *Hydrochloric Acid (1+1)*—Mix 1 volume of concentrated hydrochloric acid (sp gr 1.19) with 1 volume of water.

42.5 *Nitric Acid (sp gr 1.42)*—Concentrated nitric acid (HNO_3).

42.6 *Nitric Acid (2+3)*—Mix 2 volumes of concentrated nitric acid (sp gr 1.42) with 3 volumes of water.

42.7 *Nitric Acid-Perchloric Acid Mixture*—Mix 5 volumes of concentrated HNO_3 with

[12] Selas No. 3001 crucible has been found satisfactory for this purpose.

1 volume of concentrated perchloric acid ($HClO_4$, 70 percent).

42.8 *Phenolphthalein Indicator Solution (1 g/100 ml)*—Dissolve 1 g of phenolphthalein in 100 ml of ethyl alcohol (95 percent).

42.9 *Silver Nitrate Solution (50 g/liter)*—Dissolve 50 g of silver nitrate ($AgNO_3$) in water and dilute to 1 liter.

42.10 *Sodium Carbonate* (Na_2CO_3).

42.11 *Sodium Hydroxide Solution (400 g/liter)*—Dissolve 400 g of sodium hydroxide (NaOH) in water and dilute to 1 liter.

43. Procedure

Treatment Prior to Analysis

43.1 Remove uncombined sulfur as follows (Note 11): Dissolve 25 g of sample in approximately 300 ml of acetone, depending on the viscosity. If the sample is of too high acetyl content to be directly soluble in acetone, cool in a dry ice cabinet overnight; then allow to come to room temperature while tumbling or stirring. Filter the solution, if necessary, through felt or a coarse sintered-glass crucible. Precipitate with rapid stirring into a beaker or pail containing 2 to 3 liters of acetic acid (1+49). Filter through a cloth bag or a Büchner funnel and give two 15-min washes with water using mechanical agitation. A little Na_2CO_3 may be added to the last wash to stabilize samples of high sulfur content. Filter and dry overnight at 60 C.

NOTE 11—To analyze for total sulfur content omit this treatment.

Decomposition

43.2 Weigh 10 ± 0.1 g of cellulose acetate and transfer to a clean, wide-mouth, 500-ml Erlenmeyer flask. Add 50 ml of the HNO_3 - $HClO_4$ mixture to the flask, and swirl the flask gently to wet the sample thoroughly. Place the modified funnel in the mouth of the flask and heat the flask carefully on a hot plate in a fume hood.

NOTE 12: **Caution**—Use the utmost care in handling the HNO_3 - $HClO_4$ mixture. If a spill occurs, wash down with plenty of water. Wear safety glasses or a face shield.

43.3 After the mixture becomes hot and less viscous, increase the heat of the hot plate. Continue the digestion until all the sample has been oxidized and the thick reddish-brown fumes of nitric oxide have been expelled. At this point, white fumes will appear and a rather vigorous reaction will occur which is caused by the last traces of organic material being oxidized and the nitric acid fuming off.

43.4 When this reaction starts, remove the flask from the hot plate, swirl gently for a few seconds, and set it on the shelf in front of the hood until the reaction is complete. Place the flask back on the hot plate and continue the digestion until the $HClO_4$ refluxes about half way up the side of the Erlenmeyer flask and about 5 ml is left in the flask. The HNO_3 - $HClO_4$ mixture should be clear and colorless. If it is not, set the flask off the hot plate to cool and then add 3 to 5 ml of HNO_3 (sp gr 1.42). Replace the flask on the hot plate and continue heating until the $HClO_4$ refluxes half way up the flask. Remove the flask from the hot plate and allow the flask and its contents to cool.

Determination of Barium Sulfate

43.5 Wash the modified funnel top thoroughly with water, collecting the rinsings in the flask. Add 50 ml of water. Swirl the flask to mix the solution thoroughly. Add 2 drops of phenolphthalein indicator solution and neutralize the acid with the NaOH solution to a faint pink. Acidify immediately with HCl (1+1), dropwise, until the solution is just acid to phenolphthalein; then add 2 ml of HCl (1+1).

43.6 Filter through a 12.5-cm fine-porosity paper into a clean 400-ml beaker. Wash the flask thoroughly with water, filtering the washings through the paper. Finally wash the paper thoroughly with ten portions of hot water. Dilute the filtrate to approximately 200 ml. Place the beaker on the hot plate and heat almost to boiling. Slowly add 10 ml of $BaCl_2$ solution from a pipet, stirring the solution during the addition. Do not add the $BaCl_2$ solution rapidly, as from a graduate, since the rapid addition will produce an impure precipitate. Remove the stirring rod from the beaker and wash it with a stream of water from the wash bottle, collecting the washings in the beaker. Cover the beaker with a watch glass and keep the mixture near the boiling temperature for 6 h or overnight. Do not allow the liquid to evaporate to dryness.

43.7 Using suction, decant the supernatant liquid through an extra-fine porosity porcelain filter crucible that has been previously rinsed with acetone, ignited, and weighed to the nearest 0.1 mg. Transfer the precipitate with the aid of a stream of hot water. Always use a stirring rod in this transfer. Scrub the sides and bottom of the beaker with a rubber policeman to remove any adhering precipitate. The crucibles may be used to collect several precipitates one on top of the other. Close control of temperature and time of heating and cooling are necessary. Cleaning with hot water is generally sufficient; drastic attack with cleaning solution should be avoided.

43.8 Wash the precipitate on the filter until free of chlorides by the following test: To 5 ml of wash water, collected in a separate test tube or on a watch glass, add 1 ml of HNO_3 (2+3) and 1 ml of $AgNO_3$ solution. The appearance of a milky white precipitate indicates the presence of chlorides, and the washing should therefore continue until the test is negative. Do not attempt to get a completely negative test for chloride. Discontinue washing when no more than a faint opalescence is produced in the test.

43.9 Finally pour a few milliliters of pure acetone through the filter and suck it dry. Place the crucible in a larger crucible or in a metal tray with perforated sides and bottom for protection and place it in an oven at 120 to 125 C for 1 h. Do not handle the crucibles with the fingers between ignition and the completion of weighing; use forceps.

43.10 Remove the crucible from the oven and ignite it for 10 min in a muffle furnace at 800 ± 50 C. Cool in a desiccator for 75 ± 15 min and weigh to the nearest 0.0001 g. It is permissible to return the crucible to the oven for at least 15 min before transferring to the desiccator.

43.11 From time to time, and especially when using new reagents, run a blank in duplicate in the reagents. If the weight of the precipitate exceeds 0.0005 g, investigate and eliminate the cause. This is equivalent to an error of 0.002 percent on a 10-g sample.

44. Calculation

44.1 Calculate the percentage of sulfur and sulfate as follows:

$$\text{Sulfur, percent} = \{[(C - B) - (E - D)] \times 0.1374/A\} \times 100$$

$$\text{Sulfate, percent} = \{[(C - B) - (E - D)] \times 0.4115/A\} \times 100$$

where:

A = weight of sample,
B = weight of crucible for sample,
C = weight of crucible and $BaSO_4$ for sample,
$C - B$ = weight of $BaSO_4$ for sample,
D = weight of crucible for blank,
E = weight of crucible and $BaSO_4$ for blank, and
$E - D$ = weight of $BaSO_4$ for blank.

Heat Stability

45. Summary of Method

45.1 The heat stability of a cellulose ester is one indication of its quality. It is measured by heating the sample for a specified time and temperature, observing it for amount and uniformity of color developed, and possibly also measuring the loss of viscosity as a result of heating. Suggested times of heating are 8 h at 160 C, 8 h at 180 C, or 2 h at 190 C. The time and temperature of heating, method of grading, and limits are matters for agreement between the purchaser and the seller.

46. Apparatus

46.1 *Heater Block*—A metal block of suitable size heated electrically and maintained at the specified temperature within ±1 C. This is best accomplished by providing continuous heat to hold the temperature a few degrees below the specified temperature, and providing intermittent additional heat thermostatically controlled. Holes shall be drilled in the top of the block to hold test tubes, a thermoregulator, and a thermometer. The block should be insulated.

46.2 *Test Tubes*, either 18 by 150-mm or 20 by 150-mm, fitted with corks. The corks shall be fitted with glass tubes the length of the cork and 4 mm in inside diameter or shall have a small V-shaped notch of equivalent cross-section cut in a vertical position.

47. Solvent

47.1 *Methylene Chloride-Methyl Alcohol Mixture*—Mix 9 parts by weight of methylene chloride with 1 part of methyl alcohol.

48. Heat Treatment

48.1 Place the sample, ground to pass a No. 20 (841-μm) sieve, in a clean, dry test tube and pack it firmly and uniformly. Stopper with a cork having a notch or tube as described in 46.2. Heat the tube and contents for 8 h at 180 C or as otherwise specified.

49. Dry Color Evaluation

49.1 Examine the heated sample for uniformity of color and for the presence of charred or decomposed spots. Compare the color of the material at the bottom of the tube with standards prepared as follows: Heat portions of a check batch of similar particle size, representative quality and stability, and accepted by mutual agreement between the purchaser and the seller. Pack portions of this check batch firmly in each of twelve clean, dry test tubes and stopper with corks as described in 46.2. Heat the tubes at 180 C, or as otherwise specified, remove one tube each 2 h, and mark the time of heating in hours on each tube. This set of numbered tubes serves as the color standards. They should be checked and renewed if necessary every 6 months.

50. Solution Color Using Platinum - Cobalt Standards

50.1 Heat a 1-g sample for the specified time and temperature and, after cooling, examine for charred or decomposed spots. Dissolve the heated sample in 15 ml of the methylene chloride - methyl alcohol mixture. Compare the color of the solution (viewing transversely) with test tubes of platinum-cobalt color standards, prepared as described in Section 61. (It may be necessary to prepare standards having as much as 2000 ppm of platinum for this purpose or to dilute the sample solution before grading.)

51. Solution Color by Spectrophotometer

51.1 The color of the solution prepared as described in Section 50 may also be measured spectrophotometrically. Measure the absorbance at 400 nm against the solvent, using a suitable spectrophotometer with a 1-cm silica cell.

52. Viscosity Change

52.1 Measure the limiting viscosity number of the heated sample and of an unheated sample as described in Sections 53 to 56 of this test method. The percentage loss of viscosity as the result of heating is a measure of heat stability.

LIMITING VISCOSITY NUMBER

53. Summary of Method

53.1 Limiting viscosity number, expressed in milliliters of solution per gram of solute, is determined by measuring the flow times of a solution of known concentration and also of the solvent used and making a calculation by means of the modified Baker-Philippoff equation.

54. Apparatus

54.1 *Capillary Viscometer*, such as the Wagner apparatus (see Fig. 3) or an Ostwald-Fenske-Cannon pipet, that will give a flow time for the solvent of not less than 70 s.

54.2 *Water Bath*—A constant-temperature water bath controlled at 25.0 ± 0.1 C and with a pump for circulating the water through the viscometer jacket or tank.

54.3 *Stop Clock or Watch*, calibrated in tenths of a second.

55. Procedure

55.1 *Sample Preparation*—Dry about 0.26 g of sample in a weighing bottle at 105 ± 3 C for 2 h, stopper, and cool in a desiccator. Weigh the bottle containing the sample to the nearest 0.001 g, transfer the sample to a 250-ml flask, and reweigh the bottle. Pipet into the flask 100 ml of solvent at 25 ± 0.1 C. The solvent used should be mutually agreed upon by the purchaser and the seller. Suitable solvents are listed in Table 1. After the sample is completely dissolved, place it in the constant-temperature bath at 25 C along with a portion of the solvent used, and allow sufficient time for both to come to temperature before making the viscosity measurements.

During this conditioning period, water at 25 C should be circulating through the water jacket of the viscometer to allow ample time for the pipet to reach temperature equilibrium.

55.2 *Viscosity Measurements*—Rinse the reservoir and the outside of the capillary tube thoroughly with solvent. Rinse the inside of the capillary tube twice by alternately applying pressure at points *B* and *A* (see Fig. 3). Discard the wash portion of the solvent. Pour more solvent into the reservoir and allow several minutes for complete drainage and thermal equilibrium to be obtained. Adjust the outer meniscus to a reference point, *D*, that will give a flow time between 70 and 100 s. Apply air pressure at *B* to force the solvent up through the capillary past the upper timing mark, *C*, on the measuring bulb, *E*. Record the time in seconds required for the meniscus to fall between the timing marks, *C* and *F*. Take a minimum of two readings. Repeat these operations, substituting the solution for the solvent.

56. Calculation

56.1 Calculate the viscosity ratio, η/η_0 as follows:

$$\text{Viscosity ratio} = t_1/t_2$$

where:

t_1 = efflux time of solution, and

t_2 = efflux time of solvent.

NOTE 13—Strictly, the viscosity ratio is defined as η/η_0 where η and η_0 are the viscosities of the solution and solvent, respectively, and are related to the corresponding efflux times by:

$$\eta = Ct - E\rho/t^2$$

$$\eta_0 = Ct_0 - E\rho_0/t_0^2$$

where C and E are constants for the particular viscometer used. The equation in 56.1 follows if the second term in these relations, a kinetic energy correction, is negligible and the respective solvent and solution densities, ρ_0 and ρ, are substantially equal.

56.2 Calculate the limiting viscosity number, $[\eta]$, as follows:

$$[\eta] = (k/c)[\text{antilog}((\log \eta/\eta_0)/k) - 1]$$

where:

k = values from Table 1 (Note 14), and

c = concentration in g/ml.

NOTE 14—Different values may be used by agreement between the purchaser and the seller.

VISCOSITY

57. Procedure

57.1 *Solution*—Dry the sample for 1 to 2 h at 105 ± 3 C and cool in a desiccator. Prepare a solution of the dried sample in a solvent and at a concentration mutually agreed upon by the purchaser and the seller. Suitable solutions are listed in Table 2.

57.2 *Viscosity Determination*—Prepare the solution and measure the viscosity in accordance with ASTM Method D 1343, Test for Viscosity of Cellulose Derivatives by Ball-Drop Method[13] (see Note 15 in Section 62).

58. Report

58.1 Report the results in poises, unless otherwise specified. The viscosity value shall be prefixed with the letter A, B, C, etc., corresponding to the formula of the solution employed.

COLOR AND HAZE

59. Summary of Method

59.1 Color and haze determinations on cellulose ester solutions are made by comparison with standards. Simultaneous measurement of these properties is desirable because haze reduces the amount of color observed.

60. Apparatus

60.1 *Light Box*—A suitable light box (Fig. 4) is described as follows: The light source consists of a mercury vapor bulb[14] which requires an autotransformer[15] for the current source. The bulb is mounted horizontally across the lower front part of a plywood box 356 mm (14 in.) wide, 430 mm (17 in.) high, and 330 mm (13 in.) deep. This box is lined with Transite board for heat resistance and is painted black inside, except that the inside back surface toward the viewer is white. A bottle holder large enough to hold four bottles is built onto the front of the light box,

[13] *Annual Book of ASTM Standards*, Part 15.

[14] A General Electric mercury vapor bulb Type H-250, A5-1, or A5-81 (differing only in the shape of the glass globe) has been found satisfactory for this purpose.

[15] A General Electric Autotransformer Model 9T64Y21 has been found satisfactory for this purpose.

and a 64 by 150-mm (2 1/2 by 6-in) horizontal viewing hole is cut through the front of the box. This opening is covered with clear glass, and a 6-mm (1/4-in.) strip of black tape is fastened to the glass horizontally to aid in judging haze in the solution. Holes are cut in the bottom and top of the box for cooling by air convection. For continuous use, forced circulation of air would be desirable. A black metal baffle over the bulb prevents direct light on the viewing glass.[16]

60.2 *Sample Bottles*—The bottles used for the sample solutions are French square bottles, 470-ml (16-oz), with screw caps.[15] These same bottles may be used for the color and haze standards.[17]

NOTE 15—These bottles may also be used for determination of viscosity, as described in 57.2.

60.3 *Cap Liners*—Cap liners shall be of a composition not affected by the solvents used. Liners of fiber board covered with cellophane or aluminum foil are usually satisfactory, but vinyl resin or waxed liners may cause interference with viscosity, color, or haze measurements.

61. Reference Standards

61.1 *Color Standards*—A color standard containing 500 ppm of platinum may be purchased[18] or the solution may also be prepared as follows: Dissolve 1.245 g of potassium platinum chloride (K_2PtCl_6, containing 0.500 g of platinum, and 1.000 g of crystallized cobalt chloride ($CoCl_2 \cdot 6H_2O$), containing 0.248 g of cobalt, in water, add 100 ml of HCl (sp gr 1.19), and dilute to 1 liter with water. Prepare standards containing 50, 60, 70, 80, 90, 100, 125, 150, 175, 200, 250, 300, 350, 400 and 500 ppm of platinum by diluting suitable aliquots of the standard solution to 500 ml with water. Place these standards in the special bottles (see 60.2), taking care to select bottles with good clarity and free of flaws. Label and cap tightly.

61.2 *Haze Standards*—Prepare haze standards by diluting a stock solution having a turbidity of 1000 ppm.[19] Prepare bottles containing 10, 20, 30, 40, 50, 60, 70, 80, 90, 100, 125, 150, 175, 200, and 400 ppm of turbidity, label, and cap tightly.

62. Procedure

62.1 Prepare the solution to be graded by dissolving the cellulose ester in the specified amount and kind of solvent, in one of the square bottles. See Table 2 for suitable solutions. At least 350 ml are required. Tumble until a uniform solution is obtained. Allow the solution to stand until it is free of bubbles before grading it for color and haze.

62.2 Place the bottle containing the solution to be graded at the front of the shelf on the apparatus and place a similar bottle containing water behind it. Place the freshly shaken haze standard at the front of the shelf beside the bottle containing the solution to be tested and place the color standard behind it. Determine the amount of color and haze in the solution by changing the color and haze standards until as good a match as possible has been obtained. The haze standards settle out quickly so they must be reshaken at short intervals. Report results in parts per million for both color and haze.

NOTE 16—When viscosity, color, and haze determinations, and an observation of general appearance are to be made on a cellulose ester sample, a considerable saving in time can be made by using one solution in a square bottle for all three determinations. Dry the cellulose ester as required for the viscosity determination, prepare the solution carefully, and allow the bottle to stand long enough to form a thick solution before tumbling, to avoid solvent loss around the cap. Use a large enough sample to provide at least 350 ml of solution in the bottle. Measure the viscosity as described in ASTM Method D 1343.

[16] A detailed drawing of the color and haze apparatus may be obtained at a nominal cost from ASTM Headquarters, 1916 Race St., Philadelphia, Pa. 19103.

[17] Owens-Illinois French square bottles No. A6732 with No. 48-400 caps have been found satisfactory for the sample solutions, viscosity, and for the color and haze standards, although for the color and haze standards, No. A6720 bottles with No. 28-400 caps are more convenient.

[18] This color standard may be purchased as platinic cobalt chloride (APHA color standard) from the Hartman-Leddon Co., Philadelphia, Pa.

[19] A stock solution that meets this requirement may be purchased from Arthur H. Thomas Co., Philadelphia, Pa. This solution contains precipitated fuller's earth, water, and hydrochloric acid.

TABLE 1 Solvents for Limiting Viscosity Number Determination

Solvent[a]	Ingredients, percent by weight	Value of k for Calculation (Section 56.2)
A	90 percent acetone[b] 10 percent ethyl alcohol[c]	10
B	acetone[b]	10
C or D	90 percent methylene chloride[d] 10 percent ethyl alcohol[c]	3
E	96 percent acetone[b] 4 percent water	10
F	90 percent methylene chloride[d] 10 percent methanol[e]	3

[a] Solvent designations conform to those used in Table 2 for viscosity determinations.
[b] Acetone (99.4 ± 0.1 percent) containing 0.3 to 0.5 percent water and under 0.3 percent ethyl alcohol.
[c] Ethyl alcohol (95 percent by volume). Formula 2B or 3A denatured ethyl alcohol may be used.
[d] Methylene chloride having a boiling range of 39.2 to 40.0 C and less than 0.001 percent acidity calculated as HCl.
[e] Methyl alcohol (sp gr 20/20 C = 0.785 to 0.795).

TABLE 2 Solutions for Viscosity Determination

	Formula					
	A	B	C	D	E	F
	Ingredients, Weight percent					
Cellulose ester	20[a]	20[a]	20[b]	15[c]	20[a]	10[c]
Acetone[d]	72	80	...	...	...	...
Acetone, 96 percent Water, 4 percent	...	...	...	...	80	...
Ethyl alcohol[e]	8	...	8	8.5	...	...
Methyl alcohol[f]	...	...	...	...	...	9
Methylene chloride[g]	...	...	72	76.5	...	81
	Typical Solution Densities, g/ml at 25 C					
	0.85	0.86	1.25	1.23	0.86	1.24

[a] Suitable for most mixed esters having less than about 40 percent acetyl and more than about 8 percent propionyl or butyryl.
[b] Suitable for most of the commercial cellulose acetate propionates and acetate butyrates.
[c] Suitable for most of the commercial cellulose acetate propionates and acetate butyrates. Particularly good for esters containing more than 40 percent acetyl.
[d] Acetone (99.4 ± 0.1 percent) containing 0.3 to 0.5 percent water and under 0.3 percent ethyl alcohol.
[e] Ethyl alcohol (95 percent by volume). Formula 2B, 3A, or 30 denatured ethyl alcohol may be used.
[f] Methyl alcohol (sp gr 20/20 C = 0.785 to 0.795).
[g] Methylene chloride having a boiling range of 39.2 to 40.0 C and less than 0.001 percent acidity calculated as HCl.

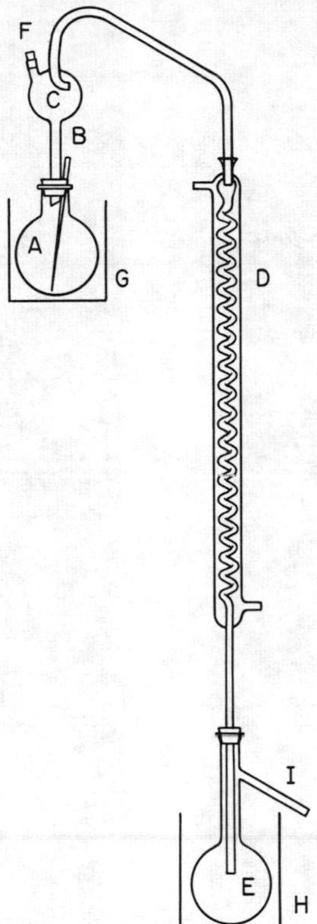

A—Flask containing sample (500-ml, round-bottom).
B—Capillary inlet tube.
C—Kjeldahl distilling head.
D—Condenser.
E—Receiver (500-ml distilling flask).
F—Opening for adding water.
G—Water bath for heating sample.
H—Cooling bath for receiver.
I—Side arm, connected to vacuum line.

FIG. 1 Vacuum Distillation Apparatus for Mixed-Ester Analysis.

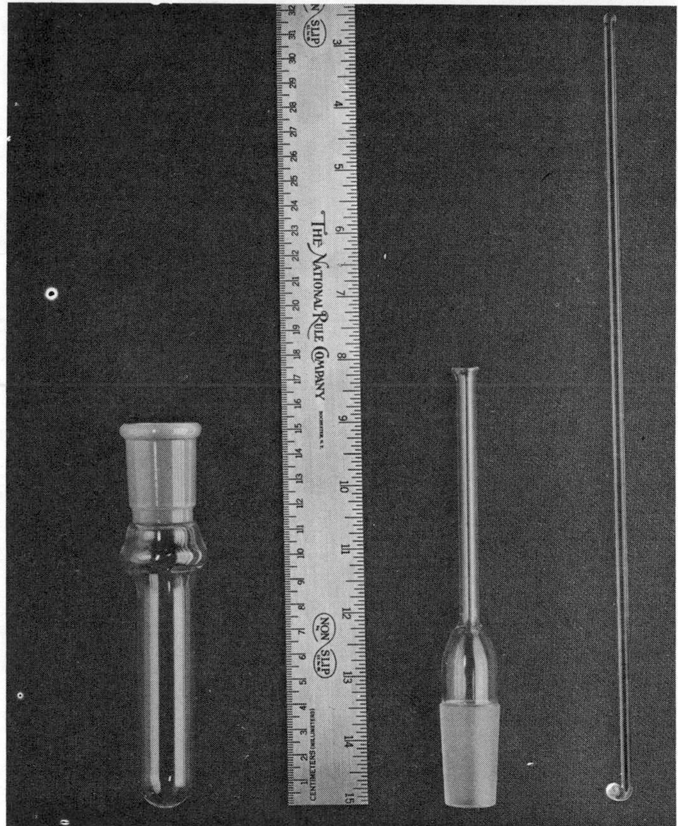

FIG. 2 Special Reflux Tube for Carbanilation.

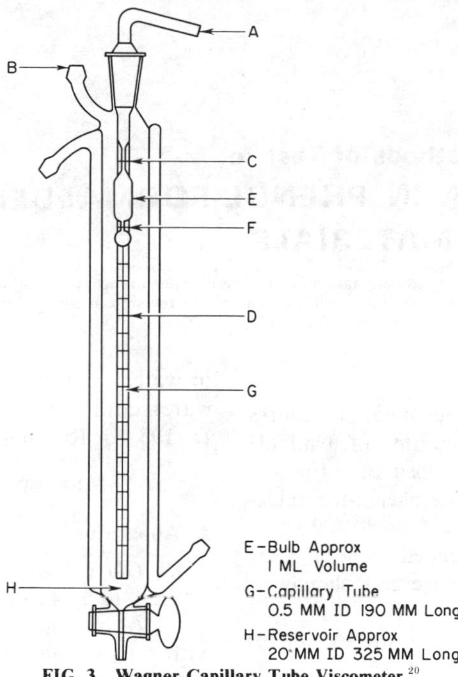

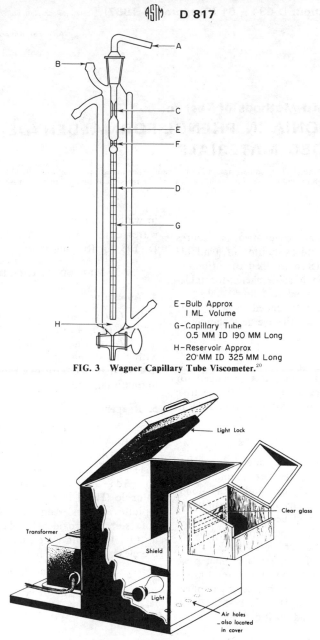

FIG. 3 Wagner Capillary Tube Viscometer.[20]

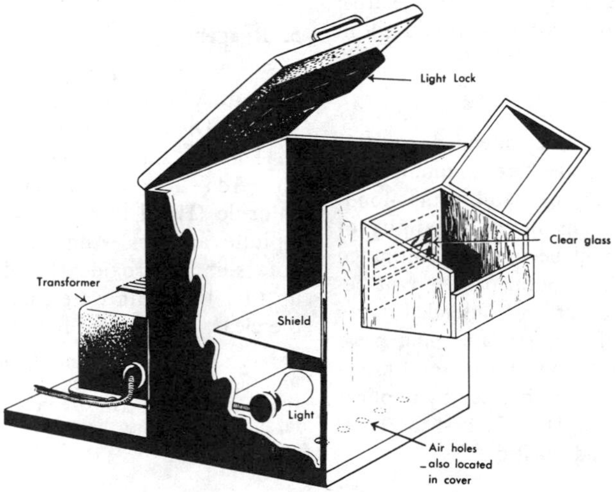

FIG. 4 Color and Haze Apparatus.

[20] Wagner, R. H., and Russell, John, "Capillary Tube Viscometer for Routine Measurement of Dilute High Polymer Solutions," *Analytical Chemistry*, ANCHA, Vol 20, 1948, pp. 151 – 7.

 **Designation: D 834 – 61 (Reapproved 1967)**

American National Standard K65.11-1960 (R 1969)
American National Standards Institute

Standard Methods of Test for
AMMONIA IN PHENOL-FORMALDEHYDE MOLDED MATERIALS[1]

This Standard is issued under the fixed designation D 834; the number immediately following the designation indicates the year of original adoption or, in the case of revision, the year of last revision. A number in parentheses indicates the year of last reapproval.

1. Scope

1.1 These methods cover two procedures for the detection and estimation of available free ammonia under prescribed conditions in parts molded of single-stage phenol-formaldehyde molding compound, as follows:

1.1.1 *Method A* is intended to be a guide as to whether a quantitative test should be applied.

1.1.2 *Method B* gives quantitative values under the conditions of the test.

1.2 The material under test is judged for the presence of free ammonia by the relative magnitude of the discoloration produced in the reagent.

2. Significance

2.1 These methods are intended to give information regarding the free ammonia content as an index of the probability that foodstuffs and similar items in the proximity of the molded part might become contaminated with ammonia or that metal parts might become corroded. They are not intended as an absolute measure of the free ammonia present. When an absolute measure of total nitrogen is desired, it may be obtained by application of ASTM Method D 1013, Test for Total Nitrogen in Resins and Plastics.[2]

3. Purity of Reagents

3.1 *Purity of Reagent*—Reagent grade chemicals or equivalent, as specified in ASTM Methods E 200, Preparation, Standardization, and Storage of Standard Solutions for Chemical Analysis,[3] shall be used in all tests.

3.2 Unless otherwise indicated, references to water shall be understood to mean reagent water conforming to ASTM Specifications D 1193, for Reagent Water.[4]

METHOD A. QUALITATIVE

4. Apparatus

4.1 *Glass Plate*, flat, approximately 100 by 100 mm (4 by 4 in.).

4.2 *Glass Vials*, 15 to 20 ml, flat bottom, with glass stoppers.

4.3 *Oven*, laboratory-type, capable of maintaining a temperature of 70 ± 2 C.

5. Reagent

5.1 *Nessler's Reagent*—Purchase from a laboratory supply house or prepare as follows:

5.1.1 Dissolve 50 g of potassium iodide (KI) in the smallest possible quantity of water. Add a saturated solution of mercuric chloride ($HgCl_2$) until a faint persistent precipitate appears. Add 400 ml of 50 percent potassium hydroxide (KOH) solution. Make up to 1 liter with water, let settle, and decant the clear, supernatant liquid for use.

NOTE 1—Keep container tightly closed and at the first sign of persistent cloudiness, discard the solution. The reagent is extremely sensitive to ammonia, and care must be taken to prevent access of ammonia from sources other than the sample under test.

[1] These methods are under the jurisdiction of the ASTM Committee D-20 on Plastics. A list of committee members may be found in the ASTM Yearbook. This standard is the direct responsibility of Subcommittee D-20.70 on Analytical Methods.

Current edition effective Sept. 18, 1961. Originally issued 1945. Replaces D 834 – 59 T.

[2] *Annual Book of ASTM Standards*, Part 20.

[3] *Annual Book of ASTM Standards*, Part 22.

[4] *Annual Book of ASTM Standards*, Part 23.

6. Test Specimen

6.1 The test specimen shall be a cup molded (Note 2) according to ASTM Method D 731, Measuring the Molding Index of Thermosetting Molding Powder.[5] Remove the fin and produce a smooth, flat surface on the open end by grinding lightly by elliptical motion of the cup on an emery base flat surface.

NOTE 2—Time, temperature, and pressure of 3 to 5 min, 160 to 168 C (320 to 334.4 F), and 210 to 350 kgf/cm^2 (3000 to 5000 psi) are suggested molding conditions. In any event, minimum conditions to produce a blister-free molding should be used.

7. Conditioning

7.1 Condition the cup by exposing it at 23 C (73.4 F) in freely circulating ammonia-free air for 4 hr, just prior to test.

8. Procedure

8.1 Adjust the temperature of the oven to 70 $\pm$ 2 C. Place the glass plate in the oven and allow it to prehcat for a minimum of 15 min.

8.2 Add 10 ml of Nessler's Reagent to each of two vials. Leave one stoppered for a blank and place it in the oven on the glass plate. Place the other, unstoppered, on the glass plate in the oven and cover with the molded cup so that the ground surface is on the glass.

8.3 After 30 min, remove the cup and vials and stopper the open vial. After cooling to room temperature, view the solutions through the full length of the vial against a white background and report evidence of discoloration in the test solution. Discoloration of the reagent shall be taken as positive indication of the liberation of ammonia, and no discoloration as a negative result.

NOTE 3—Convenient designations are the following according to McAlpine and Soule, "Prescott and Johnson's Qualitative Chemical Analysis," D. Van Nostrand Co., Inc., 2nd Edition, 1933.

No discoloration: Ammonia less than 1 ppm in 10 ml.

Slight discoloration: Ammonia 1 ppm up to about 30 ppm in 10 ml.

Light brown, precipitate: Ammonia about 30 ppm in 10 ml.

Dark brown: Ammonia greater than about 30 ppm in 10 ml.

9. Report

9.1 The report shall include the following:

9.1.1 Material identification,

9.1.2 Molding conditions,

9.1.3 Color of test solution, and

9.1.4 Positive or negative test for ammonia.

METHOD B. QUANTITATIVE

10. Apparatus

10.1 *Balance*, sensitive to at least 0.01 g.

10.2 *Condenser*.

10.3 *Crucible*, sintered-glass crucible of medium porosity (conveniently of 30-ml capacity), or equivalent.

10.4 *Distillation Flask*, 100-ml Claissen-type distillation flask.

10.5 *Nessler Tubes*, 50 ml capacity, either tall- or short-form.

10.6 *Nesslerimeter*.

11. Reagents

11.1 *Ammonia, Standard Solution (1 ml = 0.01 mg NH_3)*—Dissolve 3.141 g of ammonium chloride (NH_4Cl), which has been well dried at 100 C, in 1 liter of ammonia-free water in a 1-liter volumetric flask. Pipet 10 ml of the solution into a 1-liter volumetric flask and dilute to 1 liter with ammonia-free water. This standard is stable and, if kept in a well-stoppered flask, will keep indefinitely.

11.2 *Color Standards*—Pipet into Nessler tubes known volumes of the standard ammonia solution, such that the NH_3 content increases in 0.01-mg steps. Dilute each color standard to 50 ml, and then add to each standard 2 ml of Nessler's reagent.

NOTE 4: *Stability of Standards*—Since the color developed changes slowly and continuously, it is important to add the Nessler's reagent to sample and standards at about the same time. The maximum interval for best results is 30 min.

11.3 *Nessler's Reagent*—See Section 5.

11.4 *Potassium Permanganate*—Either solid potassium permanganate ($KMnO_4$) or a saturated aqueous solution.

11.5 *Sodium Hydroxide Solution (20 g/liter)*—Dissolve 20 g of sodium hydroxide (NaOH) in ammonia-free water and dilute to 1 liter.

11.6 *Water, Ammonia-Free*—Boil reagent water in a glass vessel for 2 min, and cool.

12. Preparation of Sample

12.1 Reduce the sample, if necessary, with

[5] *Annual Book of ASTM Standards*, Part 28.

a rasp or other means so as to pass a No. 60 (250-μm) sieve, taking care not to overheat the sample. Keep the sample in a tightly stoppered flask until analyzed. Due to the volatility of ammonia, carry out the analysis as rapidly as possible after reducing the sample.

13. Procedure

13.1 Weigh 10 g of the prepared sample on the balance to the nearest 0.01 g. Place in a 250-ml glass-stoppered flask and cover with 100 ml of ammonia-free water at 90 to 100 C. Stopper and allow to cool at room temperature for 1 h while shaking frequently.

13.2 Filter rapidly through a medium-porosity sintered-glass filter crucible. Pipet 10 ml of the filtrate into a 100-ml Claissen-type distillation flask. Add 1 g of solid $KMnO_4$ or 1.5 ml of saturated $KMnO_4$ solution. Add 10 ml of NaOH solution. Drop a few chips of porous plate into the flask to minimize "bumping." Stopper the flask with rubber stoppers. Mix well, and if necessary, add more $KMnO_4$ so that the purple color is maintained.

NOTE 5—Before any samples are tested, it is advisable to eliminate all traces of $NH_3$from the apparatus by running a blank determination using 10 ml of ammonia-free water. The apparatus is then ready for a sample determination.

13.3 Distill the sample slowly so as to prevent the entrainment of any $KMnO_4$, and collect the first 15 ml of the distillate in the 50-ml Nessler tube. Dilute the distillate to the 50-ml mark with ammonia-free water, and add 2 ml of Nessler's reagent.

13.4 Using the Nesslerimeter, compare the color developed in the Nessler tube containing the sample distillate against the prepared color standards (see 11.2). For good color definition, the highest concentration of the standard solution used should not exceed 0.06 mg of NH_3. If the color developed in the sample is darker than this, a 10-ml aliquot of the filtrate may be diluted to a known volume in a volumetric flask, and a 10-ml portion of this diluted solution may be used in the distillation (13.3).

14. Calculation

14.1 Calculate the percentage of ammonia in the sample as follows:

$$NH_3\text{, percent} = SD/10$$

where:

S = milligrams of NH_3 in the matching color standard, and

D = dilution factor (if sample is diluted before distillation).

Designation: D 864 – 52 (Reapproved 1961)

Standard Method of Test for
COEFFICIENT OF CUBICAL THERMAL EXPANSION OF PLASTICS[1]

This Standard is issued under the fixed designation D 864; the number immediately following the designation indicates the year of original adoption or, in the case of revision, the year of last revision. A number in parentheses indicates the year of last reapproval.

1. Scope

1.1 This method covers the determination of the coefficient of cubical thermal expansion of both rigid and nonrigid plastics.

1.2 The thermal expansion of a plastic is composed of a reversible component on which are superimposed changes in volume due to changes in moisture content, curing (degree of polymerization), loss of plasticizer or solvents, release of stresses, and other factors. This method covers determination of the reversible cubical thermal expansion under exclusion of these spurious factors. Some plastics show more or less abrupt changes in the coefficient of cubical thermal expansion in narrow temperature intervals, which indicate changes in molecular structure or second-order transition points. They can be investigated by careful measurements in the critical temperature range.

2. Apparatus

2.1 *Dilatometer*—A glass dilatometer as shown in Fig. 1 (Note 1). It shall consist of two pieces of borosilicate glass tubing sealed together as shown in Fig. 1(*a*). The lower or bulb member shall be approximately 20 cm in length, 10 mm in outside diameter, and 1 mm in wall thickness. The capillary part shall be approximately 1 m in length, 6 mm in outside diameter, and about 0.7 mm in inside diameter. The inside diameter shall be tested for uniformity and, if necessary corrections shall be made for irregularities in diameter.[2] A suitable scale shall be attached to the dilatometer as shown in Fig. 1(*b*). The scale shall extend from the top of the capillary to within a short distance of the stem mark and shall be graduated so that readings can be made to the nearest 0.05 cm.

NOTE 1—Glass dilatometers as described in this method have been used successfully. Dilatometer bulbs made from a metal with low thermal expansion may also be used, provided the expansion characteristics of the metal are known and are taken fully into consideration when calculating the coefficient of cubical thermal expansion of the specimen. It is hoped that a satisfactory design of a metal dilatometer can be shown in a future revision of this method.

2.2 *Temperature Bath*—A well-agitated liquid bath for controlling the temperature of the specimen.

NOTE 2—The following baths have been found useful:

Range	Bath Liquid
−35 to +10 C	solid carbon dioxide-acetone, controlled to ±0.2 C
+10 to +90 C	water controlled to ±0.2 C
+90 to +150 C	oil controlled to ±0.2 C

2.3 *Thermometer*—A thermometer to determine the temperature of the liquid bath with an accuracy of ±0.1 C.

3. Test Specimens

3.1 The samples from which test specimens are to be cut or shaped shall be prepared by methods and under conditions which ensure the absence of voids and gas bubbles, and which yield specimens that are substantially free of stresses. This requirement applies also

[1] This method is under the jurisdiction of ASTM Committee D-20 on Plastics. A list of committee members may be found in the ASTM Yearbook. This standard is the direct responsibility of Subcommittee D-20.30 on Thermal Properties.

Current edition effective Sept. 30, 1952. Originally issued 1945. Replaces D 864 – 45 T.

[2] For method of calibrating capillary see Joseph Reilly and W. N. Rae, *Physico-Chemical Methods*, Vol I, Van Nostrand, 1943, p. 362.

if the specimens are molded to their finished shape.

3.2 The test specimen shall be a rod approximately 15 cm by 6 by 6 mm with rounded edges. It may consist of two or more pieces although a single piece is preferred.

NOTE 3—Dimensions of specimen, bulb, and capillary may be changed to suit special conditions. In general, results will be more accurate if the ratio of mercury volume to specimen volume is small and if the ratio of meniscus travel to temperature difference is large.

4. Conditioning of Specimens

4.1 For those tests where conditioning is required, the test specimens shall be conditioned in accordance with Procedure B of ASTM Methods D 618, of Conditioning Plastics and Electrical Insulating Materials for Testing.[3] With materials whose water absorption is known to be negligible, the conditioning may be omitted.

5. Procedure

5.1 Determine the weight, W, in grams, of the specimen, and its volume, V, in cubic centimeters, by direct measurements or by means of specific gravity. Seal the test specimen into the bulb as shown in Fig. 1(*b*). Place a borosilicate glass bubble in the bottom of the bulb to prevent excessive heating of the specimen during sealing. Determine the weight, W_1, of the tube containing the sealed-in specimen.

5.2 Fill the tube with mercury by a vacuum technique, and adjust the level of the mercury meniscus by introducing a clean piano wire to remove excess mercury. Make this adjustment so that at the lowest and highest temperatures to be used the meniscus stays within the range of the scale attached to the dilatometer. Determine the weight, W_2, of the tube containing the specimen and the mercury.

5.3 Immerse the apparatus in the temperature bath up to the stem mark, and adjust the temperature of the bath to the lowest point of the range to be investigated maintaining the temperature at that point until the position of the mercury meniscus, X, has remained constant for 15 min. Increase the bath temperature in increments of 5 or 10 C, taking corresponding readings of the meniscus position.

5.4 An alternative heating schedule which is usually more rapid may be used, and consists of continuously varying the bath temperature at a maximum rate of 0.5 C per min (Note 4). Readings of the position of the mercury meniscus, X, shall be taken at increments of 5 C.

NOTE 4—If the bulb or the specimen are of a larger diameter than prescribed in 2.1 or 3.2 it may be necessary to reduce the rate of temperature rise in order to avoid excessive temperature lag between the bath and the specimen.

5.5 Make duplicate determinations with rising temperature, and in case of disagreement between these two determinations make a third or fourth determination until duplicate values are obtained.

NOTE 5—Values obtained on the first determination may disagree from values obtained on subsequent determinations because the specimen may have contained stresses that have been released at the temperatures reached during the first determination.

6. Calculation

6.1 Plot the position of the mercury meniscus against temperature on linear coordinate paper. In general, this curve will consist of two straight line portions which intersect at the transition temperature. Determine the slope of each straight-line portion; this shall form the basis for calculating the coefficient of cubical thermal expansion above and below the transition temperatures.

6.2 Determine the internal volume V_o, of the bulb below the stem mark as follows:

$$V_o = (W_2 - W_1)/S_m + V - (X_o + L)A$$

where:

$W_2 - W_1$ = total weight of mercury, in g,
S_m = density of mercury (13.53 g/cm^3),
V = volume of test specimen, cm^3,
$X_o + L$ = distance of mercury meniscus from stem mark, cm, at time of weighing, and
A = cross section of capillary bore, cm^2.

6.3 Calculate the coefficient of cubical thermal expansion, both above and below the transition temperature, as follows:

$$a = A\Delta X/V\Delta t - V_o/V\,(a_m - a_g) + a_m$$

where:

[3] *Annual Book of ASTM Standards*, Part 27.

a = coefficient of cubical thermal expansion of specimen,
A = cross section of capillary bore, cm^2,
V = volume of test specimen, cm^3,
V_o = internal volume of bulb below stem mark, cm^3,
a_m = coefficient of cubical thermal expansion of mercury (18.2×10^{-5}),
a_g = coefficient of cubical thermal expansion of borosilicate glass (1.0×10^{-5}), and
$\Delta X/\Delta t$ = slope of a straight line portion of the plot determined from two points (X_1, t_1) and (X_2, t_2) where $\Delta X = X_2 - X_1$ and $\Delta t = t_2 - t_1$.

7. Report

7.1 The report shall include the following:

7.1.1 Designation of material, including name of manufacturer and information on composition when known,

7.1.2 Method of preparation of test specimen,

7.1.3 Temperature range over which the measurements were made,

7.1.4 Coefficient of cubical thermal expansion per degree Celsius, and

7.1.5 Complete description of any unusual behavior of the specimen.

7.1.6 If a transition point is found, include the transition temperature and the coefficients of cubical thermal expansion below and above the transition temperature.

NOTE 6—For data to be published in tables of properties of plastics, it is recommended that the coefficient of cubical thermal expansion as derived from readings at −30 C and +30 C be given.

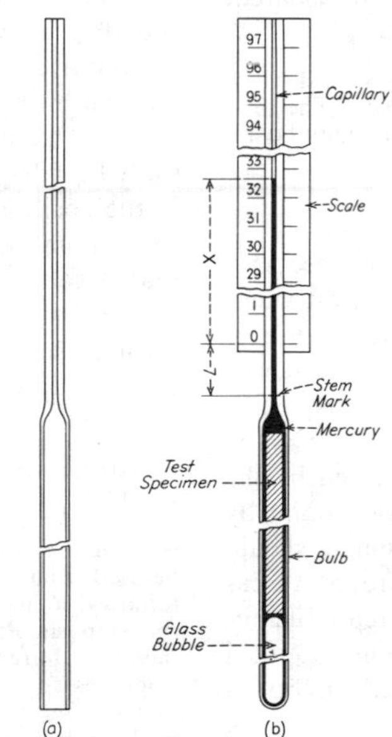

NOTE—The distance from the stem mark to the neck of the bulb shall be about 2 cm.

FIG. 1 Dilatometer.

Designation: D 881 – 48 (Reapproved 1965)

Standard Method of Test for
DEVIATION OF LINE OF SIGHT THROUGH TRANSPARENT PLASTICS[1]

This Standard is issued under the fixed designation D 881; the number immediately following the designation indicates the year of original adoption or, in the case of revision, the year of last revision. A number in parentheses indicates the year of last reapproval.

1. Scope

1.1 This method describes a procedure for measuring, by means of a telescope and a target, the deviation of the line of sight through flat or curved sections made of transparent plastics.

NOTE 1—The values stated in U.S. customary units are to be regarded as the standard. The metric equivalents of U.S. customary units may be approximate.

2. Significance

2.1 This is a functional test for sheets or formed articles, such as certain aircraft windshields and enclosures, the manner of use of which requires information about the deviation of the line of sight. The property measured is determined jointly by the prismatic power of the plastic at each point and by the angle made with the surface at that point by the line of sight. Thus the measurement should be considered as a specialized procedure capable of excellent precision and reproducibility but not applicable for determining optical distortions nor for defining optical quality in any general sense.

3. Summary of Method

3.1 A line of sight is established by focusing a telescope on a target. By placing the specimen to be tested in the line of sight, the apparent position of the cross hairs on the target is shifted. From the magnitude of the shift and the distance between the target and the specimen, the angular deviation of the line of sight by the specimen is calculated.

4. Apparatus

4.1 *Telescope*, rigidly mounted, equipped with cross hairs (Note 2). The telescope shall be sufficiently free of parallax so that when the eye of the observer is shifted, the corresponding movement of the image of the target relative to the cross hairs shall not exceed one tenth of the smallest scale division of the target. The magnifying power of the telescope shall be adequate to permit accurate reading of the position of the cross hairs on the target. The mounting of the telescope shall be such that it cannot be displaced readily after being clamped in position. To obtain clear images through some specimens, it may be necessary to stop down the objective of the telescope (Note 3).

NOTE 2—An engineer's transit is a very suitable instrument for measuring deviations of line of sight.

NOTE 3—The intended use of the article or material under test may determine the size of stop to be used. Thus, if the user will look through the material with his unaided eye, then the diameter of the stop should approximate that of the pupil of the eye. Larger apertures make for increase of brightness of the image, but frequently the definition is poorer and the measurements are more difficult or more fatiguing.

4.2 *Target*, with suitable scale. The markings shall be so made and the illumination shall be such that the divisions shall be clearly

[1] This method is under the jurisdiction of ASTM Committee D-20 on Plastics. A list of committee members may be found in the ASTM Yearbook. This standard is the direct responsibility of Subcommittee D-20.40 on Optical Properties.

Current edition effective Sept. 30, 1948. Originally issued 1946. Replaces D 881 – 46 T.

visible when viewed through the telescope (Note 4). Where deviation in one direction only is of interest, the target may consist of parallel straight lines; otherwise the most convenient form consists of concentric circles.

NOTE 4—Lines ruled in black on a white background, which may be either opaque with front lighting or translucent with back lighting, make the most suitable target for most purposes. The spacings of the lines can be calculated by a formula as follows:

$$s = \theta L/3440$$

where:

s = spacing of the lines, in.

θ = angular deviation, min, corresponding to an apparent shift of the cross hairs from one cross line to the next line, and

L = distance of the target from that part of the test specimen which is under examination, in.

Thus, if the distance of specimen to target is 40 ft, and if an apparent shift of the cross hairs from one line to the next is to correspond to one minute of angular deviation by the specimen, the spacing of the lines will be:

$$s = (L \times 40 \times 12)/3440 = 0.140 \text{ in.}$$

4.3 *Support for Specimen* A suitable fixture for support of the specimen (Note 5). The specimen shall be held firmly during each measurement, and means shall be provided for moving the specimen quickly to each specified position.

NOTE 5—Small flat sheets may be supported by hand while measurements are being made. Large flat sheets and curved articles require special fixtures, the design of which is dictated by the specific optical requirements of each article.

5. Conditioning

5.1 *Conditioning*—Condition the test specimens at 23 ± 2 C (73.4 ± 3.6 F) and 50 ± 5 percent relative humidity for not less than 40 h prior to test in accordance with Procedure A of ASTM Methods D 618, Conditioning Plastics and Electrical Insulating Materials for Testing,[2] for those tests where conditioning is required. In cases of disagreement, the tolerances shall be 1 C (1.8 F) and ±2 percent relative humidity.

5.2 *Test Conditions*—Conduct tests in the Standard Laboratory Atmosphere of 23 ± 2 C (73.4 ± 3.6 F) and 50 ± 5 percent relative humidity, unless otherwise specified in the test methods or in this specification. In cases of disagreement, the tolerances shall be 1 C (1.8 F) and ±2 percent relative humidity.

6. Procedure

6.1 Adjust the telescope so that the cross hairs coincide with a reference point or line on the target, and firmly lock or clamp in this position. Take great care not to displace the telescope or target during the measurements and recheck the initial setting frequently. Place the specimen in the prescribed position. Unless otherwise specified, place and orient the specimen so that the line of sight shall intersect its surface normally (Note 6) at a point 12 in. from the objective lens of the telescope. Record the position of the cross hairs on the image of the target if deviations at specific points are to be taken. If maximum deviations only are sought, the observer shall note continuously the apparent position of the cross hairs on the target while the sheet or article is moved through the line of sight until the entire specified area has been covered, and shall record the largest apparent displacement of the cross hairs from the reference point or line.

NOTE 6—The deviation of the line of sight is determined both by the angle of incidence and by any lack of parallelism of the surfaces. If the line of sight intersects the surface normally, any deviation is due to lack of parallelism, such as may be caused by "thinning out" during shaping operations. In actual use, however, the line of sight is rarely normal to the surface and the usual practice is to test aircraft enclosures and similar articles in such a way that the eye of the observer and his line of sight duplicate the position and direction of those of the ultimate user.

7. Calculations

7.1 The angular deviation of the line of sight, in minutes, produced by the specimen is given by the following formula:

$$a = N\theta$$

where:

N = number of rulings by which the apparent position of the cross hairs has been shifted, and

θ = angular deviation corresponding to the distance between adjacent rulings.

8. Report

8.1 Unless otherwise specified the report shall include the following:

[2] *Annual Book of ASTM Standards*, Part 27.

8.1.1 Nature of article tested,

8.1.2 Angle of intersection of line of sight with surface, if this is constant, or description of the arrangement used, and

8.1.3 Deviation of line of sight at specified points, with the direction of the deviation relative to a prescribed direction on the specimen, or maximum deviation of line of sight.

Designation: D 882 - 67

Standard Methods of Test for
TENSILE PROPERTIES OF THIN PLASTIC SHEETING[1]

This standard is issued under the fixed designation D 882; the number immediately following the designation indicates the year of original adoption or, in the case of revision, the year of last revision. A number in parentheses indicates the year of last reapproval.

These methods were prepared jointly by The Society of the Plastics Industry and the American Society for Testing and Materials.

NOTE—Section 6 was changed editorially in December 1968.

1. Scope

1.1 These methods cover the determination of tensile properties of plastics in the form of thin sheeting (less than 1.0 mm (0.04 in.) in thickness).

NOTE 1—Film has been arbitrarily defined as sheeting having nominal thickness not greater than 0.25 mm (0.010 in.).

NOTE 2—Tensile properties of plastics 1.0 mm (0.04 in.) or greater in thickness shall be determined according to ASTM Method D 638, Test for Tensile Properties of Plastics[2] if rigid, and according to ASTM Method D 412, Tension Testing of Vulcanized Rubber[2] if nonrigid. If the sample quality is limited, ASTM Method D 1708, Test for Tensile Properties of Plastics by Use of Microtensile Specimens[2] may be used.

1.2 Two types of tension tests are described in these methods (Note 3), differing basically only in manner of load application. These methods may be used to test all plastics within the thickness range described and the capacity of the machine employed.

1.2.1 *Method A. Static Weighing—Constant-Rate-of-Grip Separation Test*—This method employs a constant rate of separation of the grips holding the ends of the test specimen.

1.2.2 *Method B. Pendulum Weighing—Constant-Rate-of-Powered-Grip Motion Test*—This method employs a constant rate of motion of one grip and a variable rate of motion of the second grip. The variable-rate grip is attached to a pendulum weighing head, and its movement is dependent on the load-deformation behavior of the material under test.

NOTE 3—ASTM Method D 1923, Test for Tensile Properties of Thin Plastic Sheeting Using the Inclined Plane Testing Machine[2] describes a tension test using an inclined plane. This method was previously Method C of Methods D 882 - 56 T.

1.3 Specimen extension may be measured in these methods by grip separation, extension indicators, or displacement of gage marks.

1.4 A procedure for determining the tensile modulus of elasticity is included, using Method A at one strain rate.

NOTE 4—This modulus determination procedure is based on the use of grip separation as a measure of extension; however, the desirability of using extension indicators accurate to ±1.0 percent or better as specified in Method D 638 is recognized, and provision for the use of such instrumentation is incorporated in the procedure.

NOTE 5—The values stated in U.S. customary units are to be regarded as the standard. The metric equivalents of U.S. customary units may be approximate.

2. Significance

2.1 Tensile properties determined by these methods are of value for the identification and characterization of materials for control and specification purposes. Tensile properties may vary with specimen thickness, method of preparation, speed of testing, type of grips used, and manner of measuring extension.

[1] These methods are under the jurisdiction of ASTM Committee D-20 on Plastics. A list of committee members may be found in the ASTM Yearbook. This standard is the direct responsibility of Subcommittee D-20.10 on Mechanical Properties.

Current edition effective Dec. 12, 1967. Originally issued 1949. Replaces D 882 - 64 T.

[2] *Annual Book of ASTM Standards*, Part 27.

Consequently, where precise comparative results are desired, these factors must be carefully controlled. Since the actual loading rates vary between Methods A and B, the results obtained using these two methods cannot be directly compared. Method A is preferred and shall be used for referee purposes, unless otherwise indicated in particular material specifications.

2.2 Tensile properties may be utilized to provide data for research and development and engineering design as well as quality control and specification. However, data from such tests cannot be considered significant for applications differing widely from the load-time scale of the test employed.

2.3 The tensile modulus of elasticity is an index of the stiffness of thin plastic sheeting. The reproducibility of test results is good when precise control is maintained over all test conditions. When different materials are being compared for stiffness, specimens of identical dimensions must be employed.

2.4 Materials that fail by tearing give anomalous data which cannot be compared with those from normal failure.

3. Definitions

3.1 Definitions of terms and symbols relating to tension testing of plastics appear in the Appendix to Method D 638.

3.2 *Tear Failure*—A tensile failure characterized by fracture initiating at one edge of the specimen and progressing across the specimen at a rate slow enough to produce an anomalous load-deformation curve.

4. Apparatus

4.1 *Grips*—A gripping system that minimizes both slippage and uneven stress distribution.

NOTE 6—Grips lined with thin rubber, crocus cloth, or pressure-sensitive tape as well as file-faced or serrated grips have been successfully used for many materials. The choice of grip surface will depend on the material tested, thickness, etc. In cases where samples frequently fail at the edge of the grips, it may be advantageous to increase slightly the radius of curvature of the edges where the grips come in contact with the test area of the specimen. Air-actuated grips have been found advantageous, particularly in the case of materials that tend to "neck" into the grips, since pressure is maintained at all times.

4.2 *Thickness Gage*—A dead-weight dial micrometer as prescribed in Method C of ASTM Methods D 374, Test for Thickness of Solid Electrical Insulation,[2] reading to 0.0025 mm (0.0001 in.) or less.

4.3 *Width-Measuring Devices*—Suitable test scales or other width measuring devices capable of measuring 0.25 mm (0.010 in.) or less.

4.4 *Specimen Cutter*—Razor blades, fixtures incorporating razor blades, suitable paper cutters, or other devices capable of cutting the specimens to the proper width and producing straight, clean, parallel edges with no visible imperfections, shall be used. Devices that use razor blades have proved especially suitable for materials having an elongation-at-fracture above 10 to 20 percent. A device consisting of two parallel knives mounted firmly against a precision-ground base shear block (similar to a paper cutter) has also proved satisfactory. The use of striking dies is not recommended because of poor and inconsistent specimen edges which may be produced. It is imperative that the cutting edges be kept sharp and free from visible scratches or nicks.

4.5 *Conditioning Apparatus*—Apparatus for maintaining a room or enclosed space under atmospheric conditions of 23 ± 1 C (73.4 ± 1.8 F) and 50 ± 2 percent relative humidity for conditioning prior to and during testing. Procedure A of ASTM Methods D 618, Conditioning Plastics and Electrical Insulating Materials for Testing[2] shall be used unless otherwise specified.

4.6 *Extension Indicators* (if employed) shall conform to requirements specified in Method D 638. In addition, such apparatus shall be so designed as to minimize stress on the specimen at the contact points of the specimen and the indicator (see 7.3).

4.7 *Testing Machines:*

4.7.1 *For Method A*—A testing machine of the constant rate-of-jaw-separation type. The machine shall be equipped with a weighing system that moves a maximum distance of 2 percent of the specimen extension within the range being measured. The machine shall be equipped with a device for recording the tensile load and the amount of separation of the grips; both of these measuring systems shall be accurate to ±2 percent. The rate of separation of the jaws shall be uniform and capable of adjustment from approximately 1.3 to 500 mm (0.05 to 20 in.)/min in increments necessary to produce the strain rates specified in 8.3 and 8.4. This method (A) shall be used for tensile modulus of elasticity measurements (Note 7).

4.7.2 *For Method B*—A testing machine of the pendulum type. This machine shall be equipped with a pendulum weighing head to measure the load applied to the test specimen and a device for indicating or recording the tensile load carried by the specimen with an accuracy of ±2 percent. The rate of travel of the power-activated grip shall be uniform and capable of adjustment to 50.8 and 508 mm (2 and 20 in.)/min.

NOTE 7—A high response speed in the recording system is desirable, particularly when relatively high strain rates are employed for rigid materials. The speed of pen response for recorders is supplied by manufacturers of this equipment. Care must be taken to conduct tests at conditions such that response time (ability of recorder to follow actual load) will produce less than 2 percent error.

5. Test Specimens

5.1 The test specimens shall consist of strips of uniform width and thickness at least 50 mm (2 in.) longer than the grip separation used.

5.2 The nominal width of the specimens shall be not less than 5.0 mm (0.20 in.) or greater than 25.4 mm (1.0 in.).

5.3 A width-thickness ratio of at least eight shall be used. Narrow specimens magnify effects of edge strains or flaws, or both.

5.4 The utmost care shall be exercised in cutting specimens to prevent nicks and tears which are likely to cause premature failures (Note 8). The edges shall be parallel to within 5 percent of the width over the length of the specimen between the grips.

NOTE 8—Microscopical examination of specimens may be used to detect flaws due to sample or specimen preparation.

5.5 Wherever possible, the test specimens shall be selected so that thickness is uniform to within 10 percent of the thickness over the length of the specimen between the grips in the case of films 0.25 mm (0.010 in.) or less in thickness and to within 5 percent in the case of films greater than 0.25 mm (0.010 in.) in thickness but less than 1.00 mm (0.040 in.) in thickness.

NOTE 9—In cases where thickness variations are in excess of those recommended in 5.5, results may not be characteristic of the material under test.

5.6 If the material is suspected of being anisotropic, two sets of test specimens shall be prepared having their long axes respectively parallel with and normal to the suspected direction of anisotropy.

5.7 For tensile modulus of elasticity determinations, the length of the specimen between the grips shall be 250 mm (10 in.).

NOTE 10—A shorter test section length is permissible if extensometers are used.

NOTE 11—The 250-mm (10-in.) test sections are used in order to minimize the effects of grip slippage of test results. When this length is not feasible, shorter test sections may be used if it has been shown that results are not appreciably affected. The minimum test section shall be 100 mm (4 in.).

NOTE 12—Excessive jaw slippage becomes increasingly difficult to overcome in cases where high modulus materials are tested in thicknesses greater than 0.25 mm (0.010 in.).

6. Conditioning

6.1 *Conditioning*—Condition the test specimens at 23 ± 2 C (73.4 ± 3.6 F) and 50 ± 5 percent relative humidity for not less than 40 h prior to test in accordance with Procedure A of ASTM Methods D 618, Conditioning Plastics and Electrical Insulating Materials for Testing,[2] for those tests where conditioning is required. In cases of disagreement, the tolerances shall be 1 C (1.8 F) and ±2 percent relative humidity.

6.2 *Test Conditions*—Conduct tests in the Standard Laboratory Atmosphere of 23 ± 2 C (73.4 ± 3.6 F) and 50 ± 5 percent relative humidity, unless otherwise specified in the test methods or in this specification. In cases of disagreements, the tolerances shall be +1 C (±1.8 F) and ±2 percent relative humidity.

7. Number of Test Specimens

7.1 In the case of isotropic materials, at least five specimens shall be tested from each sample.

7.2 In the case of anisotropic materials, at least ten specimens, five normal and five parallel with the principal axis of anisotropy, shall be tested from each sample.

7.3 Specimens that fail at some obvious flaw, that do not fail within the gage length, or in the case of utilization of extension indicators, that fail at the contact point of the specimen and the indicator shall be discarded and retests made, unless such flaws or conditions constitute a variable whose effect is being studied.

NOTE 13—In the case of some materials, examination of specimens, prior to and following testing, under crossed "Polaroid" film provides a useful means of detecting flaws which may be, or are, responsible for premature failure.

8. Speed of Testing

8.1 The speed of testing is the rate of separation of the two members (or grips) of the testing machine when running idle (under no load). This rate of separation shall be maintained within 5 percent of no-load value when running under full-capacity load.

8.2 The speed of testing shall be calculated from the required initial strain rate as specified in 8.3. The rate of grip separation may be determined for the purpose of this test method from the initial strain rate as follows:

$$A = BC$$

where:

A = rate of grip separation, mm (or in.)/min,
B = initial distance between grips, mm (or in.), and
C = initial strain rate, mm/mm · min (or in./in. · min).

8.3 The initial strain rate shall be as follows unless otherwise indicated by the specification for the material being tested (Note 14):

Method	Percentage Elongation at Break	Initial Strain Rate, mm/mm · min (in./in. · min)
A	Less than 20	0.1
	20 to 100	0.5
	Greater than 100	10.0
B	Less than 100	0.5
	Greater than 100	10.0
Modulus of Elasticity Determination		
A		0.1

NOTE 14—Results obtained at different initial strain rates are not comparable; consequently, where direct comparisons between materials in various elongation classes are required, a single initial strain rate should be used. For some materials it may be advisable to select the strain rates on the basis of percentage elongation at yield.

8.4 In cases where conflicting material classification, as determined by percentage elongation at break values, results in a choice of strain rates, the lower rate shall be used.

8.5 If modulus values are being determined, separate specimens shall be used whenever strain rates and specimen dimensions are not the same as those employed in the test for other tensile properties.

9. Procedure

9.1 Select a load range such that specimen failure occurs within its upper two thirds. A few trial runs may be necessary to select a proper combination of load range and specimen width.

9.2 Measure the cross-sectional area of the specimen at several points along its length. Measure the width to an accuracy of 0.25 mm (0.010 in.) or better. Measure the thickness to an accuracy of 0.0025 mm (0.0001 in.) or better for films less than 0.25 mm (0.010 in.) in thickness and to an accuracy of 1 percent or better for films greater than 0.25 mm (0.010 in.) but less than 1.0 mm (0.040 in.) in thickness. Calculate the minimum cross-sectional area.

9.3 The initial grip separation shall be at least 50 mm (2 in.) for materials having a total elongation at break of 100 percent or more, and at least 100 mm (4 in.) for materials having a total elongation at break of less than 100 percent.

NOTE 15—Since slippage is a potential problem in these tests, as great an initial distance between grips as possible should be employed.

9.4 Set the rate of grip separation to give the desired strain rate based on the initial distance between the grips (Note 16). Balance, zero, and calibrate the load weighing and recording system.

NOTE 16—Suggested crosshead speeds and initial grip separation to give the desired initial strain rate described in 8.3 are as follows:

Method	Percentage Elongation at Break	Initial Strain Rate mm/mm·min (in./in.·min)	Initial Grip Separation		Rate of Grip Separation	
			mm	in.	mm/min	in./min
A	Less than 20	0.1	125	5	12.5	0.5
	20 to 100	0.5	100	4	50	2.0
	Greater than 100	10.0	50	2	500	20.0
B	Less than 100	0.5	100	4	50	2.0
	Greater than 100	10.0	50	2	500	20.0
Modulus of Elasticity Determination						
A		0.1–	250	10	25	1.0

9.5 In cases where it is desired to measure a test section other than the total length between the grips, mark the ends of the desired test section with a soft, fine wax crayon or with ink. Do not scratch these marks onto the surface since such scratches may act as stress raisers and cause premature specimen failure. Extensometers may be used if available; in this case, the test section will be defined by the contact points of the extensometer.

NOTE 17—Measurement of a specific test section is necessary with some materials having high elongation. As the specimen elongates, the accompanying reduction in area results in a loosening of material at the inside edge of the grips. This reduction and loosening moves back into the grips as further elongation and reduction in area takes place. In effect this causes problems similar to grip slippage, that is, exaggerates measured extension.

9.6 Place the test specimen in the grips of the testing machine, taking care to align the long axis of the specimen with an imaginary line joining the points of attachment of the grips to the machine. Tighten the grips evenly and firmly to the degree necessary to minimize slipping of the specimen during test.

9.7 Start the machine and record load versus extension.

9.7.1 When the total length between the grips is used as the test area, record load versus grip separation.

9.7.2 When a specific test area has been marked on the specimen, follow the displacement of the edge boundary lines with respect to each other with dividers or some other suitable device. If a load-extension curve is desired, plot various extensions versus corresponding loads sustained, as measured by the load indicator.

9.7.3 When an extensometer is used, record load versus extension of the test area measured by the extensometer.

9.8 If modulus values are being determined, select a load range and chart rate to produce a load - extension curve of between 30 and 60 deg to the *X* axis. For maximum accuracy, use the most sensitive load scale for which this condition can be met. The test may be discontinued when the load - extension curve deviates from linearity.

NOTE 18—In the case of materials being evaluated for secant modulus, the test may be discontinued when the specified extension has been reached.

10. Calculations

10.1 *Breaking Factor* (nominal) shall be calculated by dividing the maximum load by the original minimum width of the specimen. The result shall be expressed in force per unit of width, usually kilograms-force per centimeter (pounds per inch) of width, and reported to three significant figures. The thickness of the film shall always be stated to the nearest 0.0025 mm (0.0001 in.).

Example—Breaking Factor = 1.78 kgf/cm (10.0 lb/in.) of width for 0.13-mm (0.005-in.) thickness.

NOTE 19—This method of reporting is useful for very thin films (0.13 mm (0.005 in.) and less) for which breaking load may not be proportional to cross-sectional area and whose thickness may be difficult to determine with precision. Furthermore, films which are in effect laminar due to orientation, skin effects, nonuniform crystallinity, etc., have tensile properties disproportionate to cross-sectional area.

10.2 *Tensile Strength* (nominal) shall be calculated by dividing the maximum load by the original minimum cross-sectional area of the specimen. The result shall be expressed in force per unit area, usually kilograms-force per square centimeter (pounds per square inch). This value shall be reported to three significant figures.

NOTE 20—When tear failure occurs, so indicate and calculate results based on load and elongation at which tear initiates, as reflected in the load-deformation curve.

10.3 *Tensile Strength at Break* (nominal) shall be calculated in the same way as the tensile strength except that the load at break shall be used in place of the maximum load (Notes 20 and 21).

NOTE 21—In many cases tensile strength and tensile strength at break are identical.

10.4 *Percentage Elongation at Break* shall be calculated by dividing the elongation at the moment of rupture of the specimen by the initial gage length of the specimen and multiplying by 100. When gage marks or extensometers are used to define a specific test section, only this length shall be used in the calculation, otherwise the distance between the grips shall be used. The result shall be expressed in percent and reported to two significant figures (Note 20).

10.5 *Yield Strength*, where applicable, shall be calculated by dividing the load at the yield point by the original minimum cross-sectional area of the specimen. The result shall be expressed in force per unit area, usually kilograms-force per square centimeter (pounds per square inch). This value shall be reported to three significant figures.

NOTE 22—For materials that exhibit Hookean behavior but do not exhibit a yield point, an offset yield strength may be obtained as described in the Appendix of Method D 638. In this case the value should be given as "yield strength at — percent offset."

10.6 *Percentage Elongation at Yield* where applicable shall be calculated by dividing the elongation at the yield point by the initial gage length of specimen and multiplying by 100. When gage marks or extensometers are used to define a specific test section, only this length shall be used in the calculation. The results shall be expressed in percent and reported to two significant figures.

NOTE 23—For materials that exhibit Hookean behavior but do not exhibit a yield point, the elongation at the offset yield strength may be calculated.

10.7 *Elastic Modulus* shall be calculated by drawing a tangent (Note 24) to the initial linear portion of the load - extension curve, selecting any point on this tangent, and dividing the tensile stress by the corresponding strain. For purposes of this determination, the tensile stress shall be calculated by dividing the load by the average original cross section of the test section. The result shall be expressed in force per unit area, usually kilograms-force per square centimeter (pounds per square inch), and reported to three significant figures.

NOTE 24—For the correct method of drawing the tangent to the load-extension curve see the definitions in the Appendix to Method D 638.

10.8 *Secant Modulus* at a designated strain shall be calculated by dividing the corresponding stress (nominal) by the specified strain.

NOTE 25—Elastic modulus values are preferable and shall be calculated whenever possible. However, for materials where no proportionality is evident, the secant value shall be calculated.

10.9 For each series of tests, the arithmetic mean of all values obtained shall be calculated to the proper number of significant figures.

10.10 The standard deviation (estimated) shall be calculated as follows and reported to two significant figures:

$$s = \sqrt{(\Sigma X^2 - n\overline{X}^2/(n - 1)}$$

where:

s = estimated standard deviation,
X = value of a single observation,
n = number of observations, and
$\overline{X}$ = arithmetic mean of the set of observations.

11. Report

11.1 The report shall include the following:

11.1.1 Complete identification of the material tested, including type, source, manufacturer's code number, form, principal dimensions, previous history, and orientation of samples with respect to anisotropy (if any),

11.1.2 Method of preparing test specimens,

11.1.3 Thickness, width, and length of test specimens,

11.1.4 Number of specimens tested,

11.1.5 Strain rate employed,

11.1.6 Grip separation (initial),

11.1.7 Crosshead speed (rate of grip separation),

11.1.8 Gage length (if different from grip separation),

11.1.9 Type of grips used, including facing (if any),

11.1.10 Test method (A or B),

11.1.11 Conditioning procedure (test conditions, temperature and relative humidity if nonstandard),

11.1.12 Anomalous behavior such as tear failure,

11.1.13 Average breaking factor and standard deviation,

11.1.14 Average tensile strength (nominal) and standard deviation,

11.1.15 Average tensile strength at break (nominal) and standard deviation,

11.1.16 Average percentage elongation at break and standard deviation,

11.1.17 In the case of materials exhibiting "yield" phenomenon:

Average yield strength and standard deviation and

Average percentage elongation at yield and standard deviation,

11.1.18 For materials which do not exhibit a yield point:

Average__ percent offset yield strength and standard deviation and

Average percentage elongation at__ percent yield strength and standard deviation,

11.1.19 Average modulus of elasticity and standard deviation (if secant modulus is used, so indicate and report strain at which calculated), and

11.1.20 When an extensometer is employed, so indicate.

Designation: D 883 – 69a

Standard Nomenclature
RELATING TO PLASTICS[1]

This Standard is issued under the fixed designation D 883; the number immediately following the designation indicates the year of original adoption or, in the case of revision, the year of last revision. A number in parentheses indicates the year of last reapproval.

A-stage, *n*—an early stage in the reaction of certain thermosetting resins in which the material is still soluble in certain liquids and fusible. Sometimes referred to as Resol. (See also **B-stage** and **C-stage**.)

acetal plastics—plastics based on resins having a predominance of acetal linkages in the main chain (See also **polyoxymethylene**.) (1964)[2]

acrylic plastics—plastics based on resins made by the polymerization of acrylic monomers, such as ethyl acrylate and methyl methacrylate.

aging, *n*—the effect on materials of exposure to an environment for an interval of time. (*2*) the process of exposing materials to an environment for an interval of time. (1966)

air loss, *n*—the loss in weight by a plastic on exposure to air at room temperature.

alkyd plastics—plastics based on resins composed principally of polymeric esters, in which the recurring ester groups are an integral part of the main polymer chain, and in which ester groups occur in most cross-links that may be present between chains. (1961)

allyl plastics—plastics based on resins made by addition polymerization of monomers containing allyl groups.

amino plastics—plastics based on resins made by the condensation of amines, such as urea and melamine, with aldehydes.

apparent density—the weight per unit volume of a material including voids inherent in the material as tested. (1965)

NOTE—The term bulk density is commonly used for material such as molding powder.

artificial weathering—the exposure of plastics to cyclic laboratory conditions involving changes in temperature, relative humidity, and ultraviolet radiant energy, with or without direct water spray, in an attempt to produce changes in the material similar to those observed after long-term continuous outdoor exposure.

NOTE—The laboratory exposure conditions are usually intensified beyond those encountered in actual outdoor exposure in an attempt to achieve an accelerated effect. This definition does not involve exposure to special conditions such as ozone, salt spray, industrial gases, etc.

asbestos reinforcement—a variety of amphibole or serpentine in the form of fibers or fibrous masses that is used in the manufacture of reinforced plastic laminates and molding compounds. (See also **reinforced plastic**.)

autothermal extrusion—a method of extrusion in which the conversion of the drive energy, through friction, is the sole source of heat.

B-stage, *n*—an intermediate stage in the reaction of certain thermosetting resins in which the material swells when in contact with certain liquids and softens when heated, but may not entirely dissolve or fuse. The resin in an uncured thermosetting molding compound is usually in this stage. Sometimes referred to as Resitol. (See also **A-stage** and **C-stage**.)

blister—rounded elevation of the surface of a plastic, with boundaries that may be more or less sharply defined, somewhat resembling in shape a blister on the human skin. (1966)

---

[1] This nomenclature is under the jurisdiction of ASTM Committee D-20 on Plastics. A list of committee members may be found in the ASTM Yearbook. This standard is the direct responsibility of Subcommittee D-20.04 on Definitions.

Current edition effective Dec. 19, 1969. Originally issued 1946. Replaces D 883 – 69.

[2] Date indicates year of introduction or latest review or revision.

block copolymer—an essentially linear copolymer in which there are repeated sequences of polymeric segments of different chemical structure. (1965)

blocking, *n*—unintentional adhesion between plastic films or between a film and another surface. (1965)

bloom—a visible exudation or efflorescence on the surface of a plastic.

NOTE—Bloom can be caused by lubricant, plasticizer, etc.

bulk factor, *n*—the ratio of the volume of any given mass of loose plastic material to the volume of the same mass of the material after molding or forming. The bulk factor is also equal to the ratio of the density after molding or forming to the apparent density of the material as received. (1964)

burned—showing evidence of thermal decomposition through some discoloration, distortion, or destruction of the surface of the plastic. See also **discoloration.**

butylene plastics—plastics based on resins made by the polymerization of butene or copolymerization of butene with one or more unsaturated compounds, the butene being in greatest amount by weight. (1968)

C-stage, *n*—the final stage in the reaction of certain thermosetting resins in which the material is relatively insoluble and infusible. The resin in a fully cured thermoset molding is in this stage. Sometimes referred to as Resite. (See also **A-stage** and **B-stage**.)

cast film—a film made by depositing a layer of plastic, either molten, in solution, or in a dispersion, onto a surface, stabilizing this form, and removing the film from the surface. (1961)

cell, *n*—a single cavity formed intentionally by gaseous displacement in the making of a cellular plastic.

cellular plastic—a plastic the apparent density of which is decreased substantially by the presence of numerous cells disposed throughout its mass.

cellular striation—a layer of cells within a cellular plastic that differs greatly from the characteristic cell structure of the material.

cellulosic plastics—plastics based on cellulose compounds, such as esters (cellulose acetate) and ethers (ethyl cellulose).

chalking—dry, chalk-like appearance or deposit on the surface of a plastic. See also **frosting, lubricant-bloom.**

chemically foamed plastic—a cellular plastic produced by gases generated from chemical interaction of constituents.

chlorinated poly(vinyl chloride) plastics—plastics based on chlorinated poly(vinyl chloride) in which the chlorinated poly(vinyl chloride) is in the greatest amount by weight. (1969)

chlorofluorocarbon plastics—plastics based on resins made by the polymerization of monomers composed of chlorine, fluorine, and carbon only. (1964)

chlorofluorohydrocarbon plastics—plastics based on resins made by the polymerization of monomers composed of chlorine, fluorine, hydrogen, and carbon only. (1964)

circuit—in filament winding, the winding produced by a single revolution of mandrel or form. (1969)

clarity—clearness.

closed-cell foamed plastic—a cellular plastic in which there is a predominance of noninterconnecting cells.

cold flow—see **creep.**

collapse, *n*—inadvertent densification of a cellular plastic.

collimated roving—in filament winding roving characterized by essentially parallel strands. (1969)

cold molding—the shaping of compounds in a mold under pressure followed by transfer of the part to a heating medium for curing.

compound, *n*—the intimate admixture of a polymer or polymers with other ingredients such as fillers, softeners, plasticizers, catalysts, pigments, dyes, or curing agents. (1966)

condensation,[3] *n*—a chemical reaction in which two or more molecules combine with the separation of water or some other simple substance. If a polymer is formed, the process is called polycondensation. (See also **polymerization**.)

consistency, *n*—the resistance of a material to flow or permanent deformation when shear-

[3] These definitions are identical with those appearing in ASTM Definitions D 907, Terms Relating to Adhesives, which were prepared by ASTM Committee D-14 on Adhesives, and appear in the *Annual Book of ASTM Standards*, Part 16.

ing stresses are applied to it; the term is generally used with materials whose deformations are not proportional to applied stress.

NOTE—Viscosity is generally considered to be similar internal friction that results in flow in proportion to the stress applied. (See **viscosity** and **viscosity coefficient**.)

copolymer—see **polymer.**

copolymerization—see **polymerization.**

crater—small, shallow, crater-like, surface imperfection.

crazing—fine cracks at or under the surface of a plastic.

creep, *n*—the time-dependent part of strain resulting from stress. (1965)

cross laminated—see **laminated, cross.**

cure, *v*—to change the properties of a polymeric system into a final, more stable, usable condition by the use of heat, radiation, or reaction with chemical additives. (1966)

cure cycle—the schedule of time periods at specified conditions to which a reacting thermosetting plastic or rubber composition is subjected to reach a specified property level. (1967)

cure time—the period of time that a reacting thermosetting plastic is exposed to specific conditions to reach a specified property level. (1967)

cut-layers—as applied to laminated plastics, a condition of the surface of machined or ground rods and tubes and of sanded sheets in which cut edges of the surface layer or lower laminations are revealed.

degradation, *n*—a deleterious change in the chemical structure of a plastic. (See also **deterioration**.)

delamination, *n*—the separation of the layers of material in a laminate.

depth—in the case of a beam, the dimension parallel to the direction in which the load is applied. (1969)

deterioration, *n*—a permanent change in the physical properties of a plastic evidenced by impairment of these properties.

diffusion, *n*—the movement of a material in the body of a plastic.

discoloration—any change from an initial color possessed by a plastic; a lack of uniformity in color where color should be uniform over the whole area of a plastic object. In the latter sense, where they can be applied, use the more definite terms **mottle, segregation,** or **two-tone.**

dished—showing a symmetrical distortion of a flat or curved section of a plastic object, so that as normally viewed, it appears concave or more concave.

dispersion, *n*—a heterogeneous system in which a finely divided material is distributed in another material.

NOTE—In plastics technology, a dispersion is usually the distribution of a finely divided solid in a liquid or a solid. For example, (*1*) pigments or fillers in molded plastics, (*2*) plastisols, or (*3*) organosols. A dispersion of a solid in a liquid is a suspension. (See **plastisol** and **organosol.**)

dome—in reinforced plastics, an end of a filament-wound cylindrical container. (1969)

domed—showing a symmetrical distortion of a flat or curved section of a plastic object, so that as normally viewed, it appears convex or more convex.

dry-blend, *n*—a free-flowing dry compound prepared without fluxing or addition of solvent. (1962)

dry-spot—in reinforced plastics, an area of incomplete surface film where the reinforcement has not been wetted with resin. (1966)

elasticity, *n*—that property of plastics materials by virtue of which they tend to recover their original size and shape after deformation.

NOTE—If the strain is proportional to the applied stress, the material is said to exhibit Hookean or ideal elasticity.

elastomer,[3] *n*—a material which at room temperature can be stretched repeatedly to at least twice its original length and, upon immediate release of the stress, will return with force to its approximate original length.

envenomation, *n*—the process by which the surface of a plastic close to or in contact with another surface is deteriorated.

NOTE—Softening, discoloration, mottling, crazing, or other effects may occur.

epoxy plastics, *n*—thermoplastic or thermosetting plastics containing ether or hydroxy alkyl repeating units, or both, resulting from the ring-opening reactions of lower molecular weight polyfunctional oxirane resins, or compounds, with catalysts or with various polyfunctional acidic or basic coreactants. (1968)

NOTE—Epoxy plastics often are modified by the incorporation of diluents, plasticizers, fillers, thixotropic agents, or other materials. (1968)

ethylene plastics—plastics based on resins made by the polymerization of ethylene or copolymerization of ethylene with one or more other unsaturated compounds, the ethylene being in greatest amount by weight. (1964)

expandable plastic—a plastic in a form to be made cellular by thermal, chemical, or mechanical means.

expanded plastic—see **cellular plastic.**

extraction, *n*—the transfer of a material from a plastic to a surrounding liquid.

extrusion, *n*—a method whereby heated or unheated plastic forced through a shaping orifice becomes one continuously formed piece.

fabricating, *n*—the manufacture of plastic products from molded parts, rods, tubes, sheeting, extrusions, or other forms by appropriate operations such as punching, cutting, drilling, and tapping. Fabrication includes fastening plastic parts together or to other parts by mechanical devices, adhesives, heat sealing, or other means.

fading—any lightening of an initial color possessed by a plastic.

fiber show—strands or bundles of fibers not covered by resin at or above the surface of a reinforced plastic. (1962)

filler, *n*—a relatively inert material added to a plastic to modify its strength, permanence, working properties, or other qualities, or to lower costs. (See also **reinforced plastic**.)

filler-specks—visible specks of a filler used, such as woodflour or asbestos, which stand out in color contrast against a background of plastic binder. It should be stated whether the specks are visible before or only after removal of the surface film.

film, *n*—an optional term for sheeting having nominal thickness not greater than 0.25 mm (0.010 in.)

finishing, *n*—the removal of defects or the development of desired surface characteristics on plastics products.

fish-eye—small globular mass which has not blended completely into the surrounding material and is particularly evident in a transparent or translucent material.

fissure, *n*—a separation or split in a cellular plastic.

NOTE—For other forms of plastics the term "crack" is generally used.

fluorocarbon plastics—plastics based on resins made by the polymerization of monomers composed of fluorine and carbon only.

fluorocarbon resins—material made by the polymerization of monomers composed only of carbon and fluorine. (1964)

NOTE—When the monomer is essentially tetrafluoroethylene, the prefix TFE- may be used to designate these materials. When the resins are copolymers of tetrafluoroethylene and hexafluoropropylene, the resins may be designated with the prefix FEP-. Other prefixes may be adopted to designate other fluorocarbon resins.

fluorohydrocarbon plastics—plastics based on resins made by the polymerization of monomers composed of fluorine, hydrogen, and carbon only. (1964)

fluoroplastics, *n*—plastics based on resins made by the polymerization or copolymerization of monomers containing, on the average, one or more atoms of fluorine per monomer unit. For specific examples of fluoroplastics see **fluorocarbon plastics, chlorofluorocarbon plastics, fluorohydrocarbon plastics,** and **chlorofluorohydrocarbon plastics.** (1964)

foamed plastic—see **cellular plastic.**

foreign matter—particles of substance included in a plastic which seem foreign to its composition.

forming, *n*—a process in which the shape of plastic pieces such as sheets, rods, or tubes is changed to a desired configuration. (See also **thermoforming**.)

NOTE—The use of the term "forming" in plastics technology does not include such operations as molding, casting, or extrusion, in which shapes or pieces are made from molding materials or liquids.

frosting—a light-scattering surface resembling fine crystals. See also **chalking, haze, bloom.**

furan plastics—plastics based on resins in which the furan ring is an integral part of the polymer chain, made by the polymerization or polycondensation of furfural, furfuryl alcohol, or other compounds containing a furan ring, or by the reaction of these furan compounds with other compounds, the furan being in greatest amount by weight. (1964)

gel, *n*—(*1*) a semisolid system consisting of a network of solid aggregates in which liquid is held.

(*2*) the initial jelly-like solid phase that develops during the formation of a resin from a liquid.

NOTE—Both types of gel have very low strengths and do not flow like a liquid. They are soft, flexible and will rupture under their own weight unless supported externally.

gel point—the stage at which a liquid begins to exhibit pseudo-elastic properties.

NOTE—This stage may be conveniently observed from the inflection point on a viscosity-time plot. (See **gel.**)

gel time—the time from the initial mixing of the reactants of a plastic or rubber composition to the time when gelation occurs, as measured by a specific test.

NOTE—For a material that must be processed by exposure to some form of energy, the zero time is the start of exposure. (1967)

glass—an inorganic product of fusion which has cooled to a rigid condition without crystallizing.

NOTE—Glass is typically hard and relatively brittle, and has a conchoidal fracture (see ASTM Definitions C 162, Terms Relating to Glass and Glass Products.[4]

glass finish—a material applied to the surface of glass fibers used to reinforce plastics and intended to improve the physical properties of such reinforced plastics over that obtained using glass reinforcement without finish.

glass transition—the reversible change in an amorphous polymer or in amorphous regions of a partially crystalline polymer from (or to) a viscous or rubbery condition to (or from) a hard and relatively brittle one. (1969)

NOTE—The glass transition generally occurs over a relatively narrow temperature region and is similar to the solidification of a liquid to a glassy state; it is not a phase transition. Not only do hardness and brittleness undergo rapid changes in this temperature region but other properties, such as thermal expansibility and specific heat also change rapidly. This phenomenon has been called second order transition, rubber transition and rubbery transition. The word transformation has also been used instead of transition. Where more than one amorphous transition occurs in a polymer, the one associated with segmental motions of the polymer backbone chain or accompanied by the largest change in properties is usually considered to be the glass transition.

glass transition temperature (Tg)—the approximate midpoint of the temperature range over which the glass transition takes place.

NOTE—The glass transition temperature can be determined readily only by observing the temperature at which a significant change takes place in a specific electrical, mechanical, or other physical property. Moreover, the observed temperature can vary significantly depending on the specific property chosen for observation and on details of the experimental technique (for example, rate of heating, frequency). Therefore, the observed Tg should be considered only an estimate. The most reliable estimates are normally obtained from the loss peak observed in dynamic mechanical tests or from dialatometric data. (1969)

graft polymer—a copolymer in which polymeric side chains have been attached to the main chain of a polymer of different chemical structure. (1965)

granular structure—apparent incomplete fusion of, and at least partial retention of, their original form by the particles from which a plastic is formed, either within it or on its surface.

gusset, *n*—(*1*) a piece used to give additional size or strength in a particular location of an object.

(*2*) the folded-in portion of flattened tubular film. (1961)

halocarbon plastics—plastics based on resins made by the polymerization of monomers composed only of carbon and a halogen or halogens.

haze—the cloudy or turbid aspect of appearance of an otherwise transparent specimen caused by light scattered from within the specimen or from its surfaces. (1964)

NOTE—For the purpose of ASTM Method D 1003, Test for Haze and Luminous Transmittance of Transparent Plastics,[5] haze is the percentage of transmitted light which, in passing through the specimen, deviates from the incident

[4] *Annual Book of ASTM Standards*, Part 13.

[5] *Annual Book of ASTM Standards*, Part 27.

beam through forward scatter more than 2.5 deg on the average.

heat forming—see **thermoforming.**

heat mark—extremely shallow, depressicn or groove in the surface of a plastic having practically no depth (its area being very large compared with its depth) and visible because of a sharply defined rim or a roughened surface. See also **shrink-mark.**

hull—dark speck of foreign matter which appears to be in the fabric of fabric-base laminated sheet.

hydrocarbon plastics—plastics based on resins made by the polymerization of monomers composed of carbon and hydrogen only. (1964)

inhibitor, *n*—a substance which prevents or retards a chemical reaction.

insert, *n*—an object molded or cast into a plastic part for a definite purpose.

isocyanate plastics, *n*—plastics based on resins made by the condensation of organic isocyanates with other compounds.

isotactic, *adj*—pertaining to a type of polymeric molecular structure containing a sequence of regularly spaced asymmetric atoms arranged in like configuration in a polymer chain.

NOTE—Materials containing isotactic molecules may exist in highly crystalline form because of the high degree of order that may be imparted to such structures.

knuckle area—in reinforced plastics, the area of transition between sections of different geometry in a filament-wound part. (1969)

laminate,[3] *n*—a product made by bonding together two or more layers of material or materials. (See also **laminated, cross,** and **laminated, parallel.**) (1965)

NOTE—a single resin-impregnated sheet of paper, fabric, or glass mat, for example, is not considered a laminate. Such a single-sheet construction may be called a "lamina." (See also **reinforced plastic.**)

laminated, cross[3]—pertaining to a laminate in which some of the layers of material are oriented at right angles to the remaining layers with respect to the grain or strongest direction in tension. (See also **laminated, parallel.**)

NOTE—Balanced construction of the laminations about the center line of the thickness of the laminate is normally assumed.

laminated, parallel[3]—pertaining to a laminate in which all the layers of material are oriented approximately parallel with respect to the grain or strongest direction in tension. (See also **laminated cross.**)

lattice pattern—in reinforced plastics, a pattern of filament winding with a fixed arrangement of open voids. (1969)

lay, *n*—the length of twist produced by stranding singly or in groups filaments, such as fibers, wires or roving; or the angle that such filaments make with the axis of the strand during a stranding operation. (1969)

NOTE—Length of twist of a filament is usually measured as the distance parallel to the axis of the strand between successive turns of the filament. (1969)

lay up, *n*—in reinforced plastics, an assembly of layers of resin-impregnated material ready for processing. (1969)

lay up *v*—in reinforced plastics, to assemble layers of resin-impregnated material for processing. (1969)

let-go—an area in laminated glass over which an initial adhesion between interlayer and glass has been lost.

lignin plastics—plastics based on resins made by the treatment of lignin with heat or by reaction of lignin with chemicals or synthetic resins, the lignin being in greatest amount by weight. (1964)

lubricant-bloom—irregular, cloudy, greasy exudation on the surface, of a plastic.

luminous transmittance, *n*—the ratio of the luminous flux transmitted by a body to the flux incident upon it.

NOTE—Parallel definitions apply to spectral and radiant transmittance. See ASTM Recommended Practice E 308, for Spectrophotometry and Description of Color in CIE 1931 System,[6] for a detailed discussion of terminology. (1969)

mat, *n*—a fibrous material consisting of randomly oriented chopped or swirled filaments loosely held together with a binder. (1968)

mechanically foamed plastic—a cellular plastic in which the cells are formed by the physical incorporation of gases.

melamine plastics—plastics based on resins made by the condensation of melamine and aldehydes.

---

[6] *Annual Book of ASTM Standards*, Part 30.

metastable, *adj*—an unstable condition of a plastic evidenced by changes of physical properties not caused by changes in composition or in environment.

NOTE—Metastable refers, for example, to the temporarily more flexible condition of some plastics after molding. No physical tests should be made while the plastic is in a metastable condition unless data regarding this condition are desired.

migration, *n*—the transfer of a material from a plastic body to other contacting solids.

mold seam—line on a molded or laminated piece, differing in color or appearance from the general surface, caused by the parting line of the mold.

molding, bag—a method of molding or laminating which involves the application of fluid pressure, usually by means of air, steam, water or vacuum, to a flexible material which transmits the pressure to the material being molded or bonded.

molding, blow—a method of forming objects from plastic masses by inflating with compressed gas.

molding, compression—a method of forming objects from plastics by placing the material in a confining mold cavity and applying pressure and usually heat.

molding, contact pressure—a method of molding or laminating in which the pressure is only slightly more than necessary to hold the materials together during the molding operation. This pressure is usually less than 0.7 kgf/cm^2 (10 psi.)

molding, high-pressure—molding or laminating in which the pressure used is greater than 14 kgf/cm^2 (200 psi).

molding, injection—a method of forming plastic objects from granular or powdered plastics by fusing the plastic in a chamber with heat and pressure and then forcing part of the mass into a cooler chamber where it solidifies.

NOTE—This method is commonly used to form objects from thermoplastics.

molding, low-pressure—molding or laminating in which the pressure used is 14 kgf/cm^2 (200 psi) or less.

molding pressure, compression—the unit pressure applied to the molding material in the mold. The area is calculated from the projected area taken at right angles to the direction of applied force and includes all areas under pressure during complete closing of the mold. The unit pressure is calculated by dividing the total force applied by this projected area and is expressed in pounds per square inch.

NOTE—The land area in a flash or semi-positive mold shall be included in the projected area when material flows over this section and is under pressure when the mold is completely closed.

molding pressure, injection—the pressure applied to the cross-sectional area of the material cylinder, expressed in pounds per square inch.

molding pressure, transfer—the pressure applied to the cross-sectional area of the material pot or cylinder, expressed in pounds per square inch.

molding, transfer—a method of forming plastic objects from granular, powdered, or preformed plastics by fusing the plastic in a chamber with heat and then forcing essentially the whole mass into a hot chamber where it solidifies.

NOTE—The method is commonly used to form objects from thermosetting plastics.

monomer,[3] *n*—a relatively simple compound which can react to form a polymer. (See also **polymer**.)

mottle—an irregular distribution or mixture of colorants or colored materials giving a more or less distinct appearance of specks, spots, or streaks of color.

NOTE—Mottling is often purposely achieved although it may occur accidentally due to improper mixing.

necking, *n*—the localized reduction in cross-section which may occur in a material under tensile stress. (1961)

netting analysis, *n*—the stress analysis of filament-wound structures that neglects the strength of the resin and assumes that the filaments carry only axial tensile loads and possess no bending or shearing stiffness. (1968)

nonrigid plastic, *n*—for purposes of general classification, a plastic that has modulus of elasticity either in flexure or in tension of not over 700 kgf/cm^2 (10,000 psi) at 23 C and 50 percent relative humidity when tested in accordance with ASTM Method D 747, Test for Stiffness of Plastics by Means of a Cantilever Beam,[5] ASTM Method D 638, Test for Tensile Properties of Plastics,[5] or ASTM Methods D 882, Test for Tensile Properties of Thin Plastic Sheeting.[5] (1963)

novolak,[3] *n*—a phenolic-aldehydic resin which, unless a source of methylene groups is added, remains permanently thermoplastic. (See also **resinoid** and **thermoplastic**.)

nylon plastics—plastics based on resins composed principally of a long-chain synthetic polymeric amide which has recurring amide groups as an integral part of the main polymer chain.

olefin plastics—plastics based on resins made by the polymerization of olefins or copolymerization of olefins with other unsaturated compounds, the olefins being in greatest amount by weight. (1965)

oligomer, *n*—a polymer consisting of only a few monomer units such as a dimer, trimer, tetramer, etc., or their mixtures.

open-cell foamed plastic—a cellular plastic in which there is a predominance of interconnected cells.

optical distortion—any apparent alteration of the geometric pattern of an object when seen either through a plastic or as a reflection from a plastic surface.

orange-peel—uneven surface somewhat resembling an orange peel.

organosol—a suspension of a finely divided resin in a plasticizer, together with a volatile organic liquid.

NOTE—The volatile liquid evaporates at elevated temperatures, and the resulting residue is a homogeneous plastic mass, provided the temperature is high enough to accomplish mutual solution of the resin and plasticizer.

parallel laminated—see **laminated, parallel.**

permanence—the property of a plastic which describes its resistance to appreciable changes in characteristics with time and environment.

phenolic plastics—plastics based on resins made by the condensation of phenols, such as phenol and cresol, with aldehydes.

phenolic resin compound, single-stage—a phenolic resin compound in which the resin, by virtue of reactive groups, is capable of further polymerization by application of heat. (See also **phenolic resin compound, two-stage**).

phenolic resin compound, two-stage—a phenolic resin compound in which the resin is essentially not reactive at normal storage temperatures, and contains a reactive additive which causes further polymerization upon the application of heat.

NOTE—A two-stage resin is less sensitive to heat below a critical temperature and has a longer shelf life than does a single-stage resin.

pimple, *n*—small, sharp, or conical elevation on the surface of a plastic. (1966)

pinhole—very small hole through a plastic.

pit, *n*—small crater in the surface of thc plastic, with its width of approximately the same order of magnitude as its depth. (1966)

plastic, *n*—a material that contains as an essential ingredient an organic substance of large molecular weight, is solid in its finished state, and, at some stage in its manufacture or in its processing into finished articles, can be shaped by flow.

plastic, *adj*—the adjective plastic indicates that the noun modified is made of, consists of, or pertains to plastic.

NOTE 1—The above definition may be used as a separate meaning to the definitions contained in the dictionary for the adjective "plastic."

NOTE 2—The plural form may be used to refer to two or more plastic materials, for example, plastics industry. However, when the intent is to distinguish "plastic products," from "wood products" or "glass products," the singular form should be used. As a general rule, if the adjective is to restrict the noun modified with respect to type of material, "plastic" should be used; if the adjective is to indicate that more than one type of plastic material is or may be involved, "plastics" is permissible. (1964)

plastic foam—see **cellular plastic.**

plasticizer, *n*—a material incorporated in a plastic to increase its workability and its flexibility or distensibility.

NOTE—The addition of the plasticizer may lower the melt viscosity, the temperature of the second-order transition, or the elastic modulus of the plastic.

plastics pipe—a hollow cylinder of a plastic material in which the wall thicknesses are usually small when compared to the diameter and in which the inside and outside walls are essentially concentric.

NOTE—Much confusion exists in industry regarding use of the terms "pipe" and "tubing." Plastics pipe is usually made from rigid or semirigid plastics while plastics tubing is usually made from nonrigid plastics, or is of small diameter.

plastics tubing—see **plastics pipe.**

plastics welding—the joining of two or more pieces of plastic by fusion of the material in the pieces at adjoining or nearby areas either with or without the addition of plastic from another source.

plastisol, *n*—a suspension of a finely divided resin in a plasticizer.

NOTE—The resin does not dissolve appreciably in the plasticizer at room temperature but does at elevated temperatures to form a homogeneous plastic mass (plasticized resin).

plate-mark—any imperfection in a pressed plastic sheet resulting from the surface of the pressing plate.

polepiece—in reinforced plastics, the supporting part of the mandrel used in filament winding, usually on one of the axes of rotation. (1969)

polyamide plastics—see **nylon plastics.**

polybutylene, *n*—a polymer prepared by the polymerization of butene as the sole monomer. (See **polybutylene plastics** and **butylene plastics.**) (1968)

polybutylene plastics—plastics based on polymers made with butene as essentially the sole monomer. (1968)

polycarbonate—a polymer in which the repeating structural unit in the chain is of the carbonate type. (1966)

polycarbonate plastics—plastics based on polymers in which the repeating structural units in the chains are essentially all of the carbonate type. (1966)

polycondensation—see **condensation.**

polyester plastics—synonymous with **alkyd plastics.**

polyether, *n*—a polymer containing recurring ether groups as an integral part of the polymer chain. (1968)

polyethylene, *n*—a polymer prepared by the polymerization of ethylene as the sole monomer. (See **polyethylene plastics** and **ethylene plastics.**)

polyethylene plastics—plastics based on polymers made with ethylene as essentially the sole monomer.

NOTE—In common usage for this plastic, essentially means no less than 85 percent ethylene and no less than 95 percent total olefins. (1968)

polymer,[3] *n*—a compound formed by the reaction of simple molecules having functional groups that permit their combination to proceed to high molecular weights under suitable conditions. Polymers may be formed by polymerization (addition polymer) or polycondensation (condensation polymer). When two or more monomers are involved, the product is called a copolymer.

polymerization,[3] *n*—a chemical reaction in which the molecules of a monomer are linked together to form large molecules whose molecular weight is a multiple of that of the original substance. When two or more monomers are involved, the process is called copolymerization or heteropolymerization. (See also **condensation.**)

polymethylenic, *adj*—pertaining to a type of molecular structure consisting of a series of methylene groups.

polyolefin, *n*—a polymer prepared by the polymerization of an olefin(s) as the sole monomer(s). (See **polyolefin plastics** and **olefin plastics.**)

polyolefin plastics—plastics based on polymers made with an olefin(s) as essentially the sole monomer(s). (1968)

polyoxymethylene, *n*—a plastic or resin having a predominance of repeating oxymethylene units in the main chain. (1964)

polypropylene, *n*—a polymer prepared by the polymerization of propylene as the sole monomer. (See **polypropylene plastics** and **propylene plastics.**)

polypropylene plastics—plastics based on polymers made with propylene as essentially the sole monomer. (1968)

polystyrene, *n*—a plastic based on a resin made by polymerization of styrene as the sole monomer. (See **styrene plastics.**)

NOTE—Polystyrene may contain minor proportions of lubricants, stabilizers, fillers, pigments, and dyes.

poly(vinyl acetate)—a resin prepared by the polymerization of vinyl acetate alone.

poly(vinyl alcohol)—polymers prepared by the hydrolysis of polyvinyl esters.

poly(vinyl chloride)—a resin prepared by the

polymerization of vinyl chloride alone.

porosity—presence of numerous visible voids.

postcure—those additional operations to which a cured thermosetting plastic or rubber composition are subjected to enhance the level of one or more properties. (1966)

postforming, *n*—the forming of cured thermoset plastics. (See also **forming.**)

pot life—the period of time during which a reacting thermosetting plastic or rubber composition remains suitable for its intended processing after mixing with reaction-initiating agents. (1967)

powder blend—see **dry-blend.** (1962)

preform, *n*—a coherent, shaped mass of powdered, granular or fibrous plastic molding compound, or of fibrous filler material with or without resin.

NOTE—A preform may be made by compressing the material sufficiently to produce a coherent shaped mass for convenience in handling.

premix, *n*—in reinforced plastics, the admixture of resin, reinforcements, fillers, etc., not in web or filamentous form, usually prepared by the molder shortly before use. (1961)

prepolymer, *n*—a chemical structure intermediate between that of the monomer or monomers and the final polymer or resin.

prepreg, *n*—in reinforced plastics, the admixture of resin, reinforcements, fillers, etc., in web or filamentous form, ready for molding. (1961)

pressure-break—an apparent break in one or more outer sheets of the paper, fabric, or other base of a laminated plastic, visible through the surface layer of resin which covers it.

propylene plastics—plastics based on resins made by the polymerization of propylene or copolymerization of propylene with one or more other unsaturated compounds, the propylene being in greatest amount by weight. (1964)

pulled surface—imperfections in the surface of a laminated plastic ranging from a slight breaking or lifting of its surface in spots to pronounced separation of its surface from its body.

ream—layers of unhomogeneous material parallel to the surface in a transparent or translucent plastic.

reinforced plastic—a plastic with some strength properties greatly superior to those of the base resin, resulting from the presence of high strength fillers imbedded in the composition. (See also **filler.**)

NOTE—The reinforcing fillers are usually fibers, fabrics, or mats made of fibers. The plastic laminates are the most common and strongest type of reinforced plastics.

release agent, *n*—a material used to keep a molding material from adhering to a mold. (1968)

resin,[3] *n*—a solid, semisolid, or pseudosolid organic material which has an indefinite and often high molecular weight, exhibits a tendency to flow when subjected to stress, usually has a softening or melting range, and usually fractures conchoidally.

resin-pocket—an apparent accumulation of excess resin in a small localized area within a reinforced plastic. (1966)

resin-streak—streak of what appears to be excess resin on the surface of a laminated plastic.

resinoid,[3] *n*—any of the class of thermosetting synthetic resins, either in their initial temporarily fusible state or in their final infusible state. (See also **novolak** and **thermosetting.**)

reworked material (thermoplastic)—a plastic material that has been reprocessed, after having been previously processed by molding, extrusion, etc., in a fabricator's plant. (1965)

NOTE—In many specifications the use of reworked material is limited to clean plastic that has been generated from the processor's own production, meets the requirements specified for the virgin material, and yields a product essentially equal in quality to one made from only virgin material.

rigid plastic, *n*—for purposes of general classification, a plastic that has a modulus of elasticity either in flexure or in tension greater than 7000 kgf/cm^2 (100,000 psi) at 23 C and 50 percent relative humidity when tested in accordance with ASTM Methods D 747, Test for Stiffness of Plastics by Means of a Cantilever Beam,[5] ASTM Method D 790, Test for Flexural Properties of Plastics,[5] ASTM Method D 638, Test for Tensile Properties of Plastics,[5] or ASTM Methods D 882, Test for Tensile Properties of Thin Plastic Sheeting.[5] (1963)

rubber, n—term not defined by Committee D-20. Definition approved by Committee D-11 on Rubber and Rubber-Like Materials and included in ASTM Definitions D 1566, Terms Relating to Rubber and Rubber-Like Materials[7] is as follows:

"**rubber**—a material that is capable of recovering from large deformations quickly and forcibly, and can be, or already is, modified to a state in which it is essentially insoluble (but can swell) in boiling solvent, such as benzene, methylethylketone, and ethanol - toluene azeotrope.

"A rubber in its modified state, free of diluents, retracts within 1 min to less than 1.5 times its original length after being stretched at room temperature (20 to 27 C) to twice its length and held for 1 min before release."

runner, *n*—(*1*) the secondary feed channel that runs from the inner end of the sprue in an injection or transfer mold, to the cavity gate.

(*2*) the piece formed in a secondary feed channel or runner. (See first definition.)

sample, *n*—a small part or portion of a plastic material or product intended to be representative of the whole.

saran plastics—plastics based on resins made by the polymerization of vinylidene chloride or copolymerization of vinylidene chloride with other unsaturated compounds, the vinylidene chloride being in greatest amount by weight. (1964)

scaly—showing a flaked surface appearance.

segregation—a close succession of parallel, rather narrow and sharply defined, wavy lines of color on the surface of a plastic differing in shade from surrounding areas, and creating the impression that components of the plastic have separated.

semirigid plastic, *n*—for purposes of general classification, a plastic that has a modulus of elasticity either in flexure or in tension of between 700 and 7000 kgf/cm^2 (10,000 and 100,000 psi) at 23 C and 50 percent relative humidity when tested in accordance with ASTM Method D 747, Test for Stiffness of Plastics by Means of a Cantilever Beam,[5] ASTM Method D 790, Test for Flexural Properties of Plastics,[5] ASTM Method D 638, Test for Tensile Properties of Plastics,[5] or ASTM Methods D 882, Test for Tensile Properties of Thin Plastic Sheeting.[5] (1963)

sheet, *n*—a piece of plastic sheeting produced as an individual piece rather than in a continuous length, or cut as an individual piece from a continuous length. (See also **film.**)

sheeter lines—parallel scratches or projecting ridges distributed over considerable area of a plastic sheet such as might be produced during a slicing operation.

sheeting, *n*—a form of plastic in which the thickness is very small in proportion to length and width and in which the plastic is present as a continuous phase throughout, with or without filler. (See also **film.**)

sheeting, continuous—plastic sheeting made by a process which does not limit its length.

short, *n*—in a molded material, an incompletely filled out condition. (1966)

NOTE—This may be evident either through an absence of surface film in some areas, or as lighter unfused particles of material showing through a covering surface film, accompanied possibly by thin-skinned blisters.

shrink mark—depression in the surface of a molded material where it has retracted from the mold. (1966)

silicone plastics—plastics based on resins in which the main polymer chain consists of alternating silicone and oxygen atoms, with carbon-containing side groups.

skin, *n*—the relatively dense material that may form the surface of a cellular plastic.

slip-plane—plane within transparent plastic visible in reflected light because of poor welding and shrinkage on cooling.

softening range—the range of temperature in which a plastic changes from a rigid to a soft state.

NOTE—Actual values will depend on the method of test. Sometimes referred to as softening point.

specimen, *n*—an individual piece or portion of a sample used to make a specific test. Specific tests usually require specimens of specific shape and dimensions.

specular transmittance, *n*—the transmittance value obtained when the measured trans-

[7] *Annual Book of ASTM Standards*, Part 28.

mitted flux includes only that transmitted in essentially the same direction as the incident flux. (1969)

sprue, *n*—(*1*) the primary feed channel that runs from the outer face of an injection or transfer mold, to the mold gate in a single cavity mold or to the runners in a multiple cavity mold.

(*2*) the piece formed in a primary feed channel or sprue. (See first definition.)

stress-crack, *n*—external or internal cracks in a plastic caused by tensile stresses less than that of its short-time mechanical strength.

NOTE—The development of such cracks is frequently accelerated by the environment to which the plastic is exposed. The stresses which cause cracking may be present internally or externally or may be combinations of these stresses. The appearance of a network of fine cracks is called crazing.

striae—surface or internal thread-like inhomogeneities in transparent plastic.

styrene plastics—plastics based on resins made by the polymerization of styrene or copolymerization of styrene with other unsaturated compounds, the styrene being in greatest amount by weight. (1964)

styrene-rubber plastics—compositions based on rubbers and styrene plastics, the styrene plastics being in greatest amount by weight. (1964)

surface mat, *n*—a thin mat of fine fibers used primarily to produce a smooth surface on a reinforced plastic. (1968)

syneresis, *n*—the contraction of a gel accompanied by the separation of a liquid. (1969)

telomer, *n*—an addition polymer in which the growth of the chainlike molecules is intentionally terminated by radicals supplied by a chain transfer agent.

thermally foamed plastic—a cellular plastic produced by applying heat to effect gaseous decomposition or volatilization of a constituent.

thermoelasticity, *n*—rubber-like elasticity by a rigid plastic resulting from an increase in temperature.

NOTE—Retention of the desired shapes may be achieved by cooling in place after forming. In the case of thermosetting materials, prolonged heating may be necessary to effect cure in place.

thermoforming, *n*—forming with the aid of heat. (See also **forming.**)

thermoplastic, *n*—a plastic that repeatedly can be softened by heating and hardened by cooling through a temperature range characteristic of the plastic, and that in the softened state can be shaped by flow into articles by molding or extrusion. (1969)

thermoplastic, *adj*—capable of being repeatedly softened by heating and hardened by cooling through a temperature range characteristic of the plastic, and that in the softened state can be shaped by flow into articles by molding or extrusion. (1969)

NOTE—Thermoplastic applies to those materials whose change upon heating is substantially physical.

thermoset, *n*—a plastic which, when cured by application of heat or chemical means, changes into a substantially infusible and insoluble product.

thermoset,[3] *adj*—pertaining to the state of a resin in which it is relatively infusible.

thermosetting, *adj*—capable of being changed into a substantially infusible or insoluble product when cured under application of heat or chemical means.

thin-spot—an unpolished or poorly finished area on an otherwise polished sheet of plastic material.

translucent—allowing the passage of some light, but not a clear view of any object.

two-tone—two shades of its nominal color more or less entirely covering adjacent areas on a molding with a more or less sharp line of demarcation between.

urea plastics—plastics based on resins made by the condensation of urea and aldehydes.

urethane plastics—plastics based on resins made by the condensation of organic isocyanates with compounds or resins that contain hydroxol groups.

NOTE—Urethane plastics are a type of isocyanate plastics. (See **isocyanate plastics.**)

vacuum forming—a forming process in which a heated plastic sheet is changed to a desired shape by causing it to flow by reducing the air pressure on one side of the sheet.

vinyl acetate plastics—plastics based on resins made by the polymerization of vinyl acetate or copolymerization of vinyl acetate with other unsaturated compounds, the vinyl acetate being in greatest amount by weight. (1964)

vinyl alcohol plastics—plastics based on resins made by the hydrolysis of polyvinyl esters or copolymers of vinyl esters.

vinyl chloride plastics—plastics based on resins made by the polymerization of vinyl chloride or copolymerization of vinyl chloride with other unsaturated compounds, the vinyl chloride being in greatest amount by weight. (1964)

vinyl plastics—plastics based on resins made from monomers containing the vinyl group, $CH_2=CH—$. (1964)

NOTE—Typical vinyl plastics are vinyl chloride plastics, vinyl acetate plastics, and vinyl alcohol plastics. Some vinyl plastics are more commonly classified by names based on the more specific chemical designation of their basic monomers, for example, *acrylic plastics* derived from monomers based on acrylic acid, $CH_2=CHCOOH$, and *styrene plastics* derived from monomers based on styrene $CH_2=CHC_6H_5$.

vinylidene chloride plastics—see **saran plastics.**

virgin material—a plastic material in the form of pellets, granules, powder, floc, or liquid that has not been subjected to use or processing other than that required for its original manufacture.

viscosity—the property of resistance to flow exhibited within the body of a material. (1965)

NOTE—This property can be expressed in terms of the relationship between shear stress and corresponding rate of strain in shear. Viscosity is usually taken to mean "Newtonian Viscosity," in which case the ratio of shearing stress to the rate of shearing strain is constant. In non-Newtonian behavior, which is usual with plastic materials, the ratio varies with the parameters of the experiment. Such ratios are often called "apparent viscosities." (See **viscosity coefficient.**)

viscosity coefficient—the shearing stress necessary to induce a unit velocity flow gradient in a material. (1965)

NOTE—In actual measurement, the viscosity coefficient of a material is obtained from the ratio of shearing stress to shearing rate. This assumes the ratio to be constant and independent of the shearing stress, a condition which is satisfied only by Newtonian fluids. Consequently, in all other cases, values obtained are apparent and represent one point on the flow curve. In the cgs system, the viscosity coefficient is expressed in poises (dyne-seconds per square centimeter). (See **viscosity.**)

void, *n*—an unfilled space in a cellular plastic substantially larger than the characteristic individual cells. (See **bubble.**)

volatile loss—weight loss by vaporization.

volatilization, *n*—the vaporization of a material from the surface of a plastic or plasticizer.

vulcanization,[3] *n*—a chemical reaction in whcih the physical properties of a rubber are changed in the direction of decreased plastic flow, less surface tackiness, and increased tensile strength by reacting it with sulfur or other suitable agents.

warp, *n*—(*1*) the yarn running lengthwise in a woven fabric. (*2*) a group of yarns in long lengths and approximately parallel, put on beams or warp reels for further textile processing, such as weaving, etc. (1968)

waviness, surface—wave-like unevenness, or out-of-plane, in the surface of a plastic.

waviness, internal—an appearance of waviness seen *in* a transparent plastic.

weld- or knit-line—a mark on, or weakness in, a molded plastic formed by the union of two or more streams of plastic flowing together. (1966)

wet layup, *n*—a method of making a reinforced plastic in which the polymer compound is applied as a liquid and as the reinforcement is put in place. (1968)

wet winding, *n*—a method of making filament-wound reinforced plastics in which the fiber reinforcement is coated with a polymer compound as a liquid just prior to wrapping on a mandrel. (1968)

width—in the case of a beam, the shorter dimension perpendicular to the direction in which the load is applied. (1969)

window—a tiny, colorless, transparent area or speck in a sheet of colored or opaque plastic which looks like a hole when the sheet is held to the light.

wrinkle, *n*—in reinforced plastics, an imperfection that has the appearance of a wave molded into one or more plies of fabric or other reinforcement material. (1966)

A. GENERAL TERMS

A-stage, *n*
aging, *n*
apparent density
artifical weathering
asbestos reinforcement
B-stage, *n*

blocking, *n*
C-stage, *n*
compound, *n*
condensation, *n*
copolymerization
degradation, *n*
depth
deterioration, *n*
diffusion, *n*
dispersion, *n*
dry-blend, *n*
envenomation, *n*
extraction, *n*
isotactic, *adj*
mat
migration, *n*
netting analysis
plastic, *adj*
polycondensation
polymerization, *n*
polymethylenic, *adj*
reworked material
sample, *n*
specimen, *n*
specular transmittance, *n*
surface mat
thermoplastic, *adj*
thermoset, *adj*
thermosetting, *adj*
volatile loss
volatilization, *n*
vulcanization, *n*
warp
width

B. Materials

acetal plastics
acrylic plastics
alkyd plastics
allyl plastics
amino plastics
block copolymer
butylene plastics
cast film
cellulosic plastics
chlorinated poly(vinyl chloride) plastics
chlorofluorocarbon plastics
chlorofluorohydrocarbon plastics
collimated roving
copolymer
cross laminated
elastomer, *n*
epoxy plastics
ethylene plastics
filler, *n*
film, *n*
fluorocarbon plastics
fluorocarbon resins
fluorohydrocarbon plastics
fluoroplastics
furan plastics
glass, *n*
glass finish
graft polymer
halocarbon plastics
hydrocarbon plastics
inhibitor, *n*
isocyanate plastics
laminate, *n*
laminated, cross
laminated, parallel
lignin plastics
melamine plastics
monomer, *n*
nonrigid plastic
novolak, *n*
nylon plastics
olefin plastics
oligomer, *n*
organosol, *n*
parallel laminated
phenolic plastics
phenolic resin compound, single-stage
phenolic resin compound, two-stage
plastic, *n*
plasticizer, *n*
plastics pipe
plastics tubing
plastisol, *n*
polyamide plastics
polybutylene plastics
polycarbonate
polycarbonate plastics
polyester plastics
polyether
polyethylene, *n*
polyethylene plastics
polymer, *n*
polyolefin
polyolefin plastics
polyoxymethylene
polypropylene, *n*
polypropylene plastics
polystyrene, *n*

poly(vinyl acetate)
poly(vinyl alcohol)
poly(vinyl chloride)
preform, *n*
prepolymer, *n*
propylene plastics
reinforced plastics
resin, *n*
resinoid, *n*
rigid plastic
rubber, *n*
saran plastics
semirigid plastic
sheet, *n*
sheeting, *n*
sheeting, continuous
silicone plastics
styrene plastics
styrene-rubber plastics
telomer, *n*
thermoplastic, *n*
thermoset, *n*
urea plastics
urethane plastics
vinyl acetate plastics
vinyl alcohol plastics
vinyl chloride plastics
vinyl plastics
vinylidene chloride plastics
virgin material

C. Terms Relating to Processing

autothermal extrusion
circuit
bulk factor
cold molding
cure, *v*
cure cycle
cure time
delamination, *n*
dome
extrusion, *n*
fabricating, *n*
finishing, *n*
heat forming
insert, *n*
knuckle area
lattice pattern
lay, *n*
lay up, *n*
lay up, *v*
molding, bag
molding, blow
molding, compression
molding, contact pressure
molding, high-pressure
molding, injection
molding, low-pressure
molding pressure, compression
molding pressure, injection
molding pressure, transfer
molding, transfer
necking, *n*
plastics welding
pole piece
postcure
postforming, *n*
pot life
powder blend
release agent
runner, *n*
sprue, *n*
thermoforming, *n*
vacuum forming
wet layup
wet winding

D. Rheological and State Terms

cold flow
consistency, *n*
creep, *n*
elasticity, *n*
glass transition
glass transition temperature (Tg)
gel, *n*
gel point
gel time
metastable, *adj*
permanence, *n*
softening range
syneresis, *n*
thermoelasticity, *n*
viscosity, *n*
viscosity coefficient

E. Cellular Plastics

cell, *n*
cellular plastic
cellular striation
channel, *n*[8]

---

[8] Definition being developed.

chemically foamed plastic
closed cell foamed plastic
collapse, *n*
expandable plastics
expanded plastic
fissure, *n*
foam, *n*[8]
foamed plastic
mechanically foamed plastic
open-cell foamed plastic
plastic foam
skin, *n*
thermally foamed plastic
void, *n*

F. Terms Relating to Radiation

absorption, *n*[8]
activation, *n*[8]
barn, *n*[8]
beta particle[8]
capture, *n*[8]
contamination, *n*[8]
cooling, *n*[8]
cross section, nuclear[8]
curie, *n*[8]
decay, *n*[8]
decay constant[8]
decontamination, *n*[8]
dose, absorbed[8]
dose, exposure[8]
dose, rate[8]
flux, *n*[8]
G-value, *n*[8]
gamma ray[8]
geometry, nuclear[8]
half-life[8]
intensity, *n*[8]
ion-pair[8]
ion-pair yield[8]
irradiation, *n*[8]
isotope, *n*[8]
moderator, *n*[8]
monitor, *n*[8]
photoelectric effect[8]
rad, *n*[8]
radioactive, *adj*[8]
radioisotope, *n*[8]
roentgen, *n*[8]
survey instrument[8]

G. Descriptive Terms

burned, *adj*
clarity, *n*
cut layers
fading, *n*
filler specks
gusset, *n*
haze, *n*
mottle, *n*
stress-crack
translucent, *adj*
two-tone

H. Defects in Fabricated Articles

blister, *n*
bloom, *n*
chalking, *n*
crack, *n*
crater, *n*
crazing, *n*
discoloration, *n*
dished, *adj*
domed, *adj*
dry spot
fiber show
fish-eye, *n*
foreign matter
frosting, *n*
granular structure
heat mark
hull, *n*
let-go
lubricant bloom
mold seam
optical distortion
orange-peel
pimple, *n*
pinhole, *n*
pit, *n*
plate mark
porosity, *n*
pressure-break
pulled surface
ream, *n*
resin-pocket
resin-streak
scaly, *adj*
scratch, *n*
segregation, *n*
sheeter lines
short, *n*
shrink mark
slip-plane, *n*
striae, *n*

thin-spot
waviness, internal
waviness, surface
weld- or knit-line
window, *n*
wrinkle, *n*

I. Terms Not Recommended

blush—use **chalking, frosting,** or **lubricant-bloom.**
boil—use **bubble** and state size if desirable to do so.
bulging—use **blister** or **domed.**
cat's eye—use **fish-eye.**
chicken-skin—use **orange-peel.**
cold
cord—use **striae.**
crush—use **pressure-break.**
densification
dimple—use **shrink-mark.**
dog-skin—use **orange-peel.**
dry-area—use **dry-spot.**
flow-lines—use **striae,** or **weld-mark.**
fog—use **haze.**
gas-mark
gas-pocket—use **heat-mark.**
grease-mark—use **lubricant-bloom.**
inverted blister—use **shrink-mark.**
liquid resin—use **liquid polymer.**
low-spot—use **thin-spot.**
open bubble—use **pit.**
pebble—use **orange-peel.**
piping—use **shrink-mark.**
pock-mark—use **pit.**
precure—see **short.**
seed—use **bubble.**
sink-mark—use **shrink-mark.**
skip—use **thin-spot.**
string—use **striae.**
sun hour
tear-drop—use **fish-eye.**
ultraviolet sun hour
unconverted-spot—use **window.**

Designation: D 952 – 51 (Reapproved 1969)

Standard Method of Test for
BOND STRENGTH OF PLASTICS AND ELECTRICAL INSULATING MATERIALS[1]

This Standard is issued under the fixed designation D 952; the number immediately following the designation indicates the year of original adoption or, in the case of revision, the year of last revision. A number in parentheses indicates the year of last reapproval.

1. Scope

1.1 This method of test covers the determination of the bond strength or ply adhesion strength of sheet plastic (Note 2) and electrical insulating materials. It is applicable to such materials as laminated phenolics, laminated melamine with glass fabric base, and vulcanized fiber.

NOTE 1—The values stated in U.S. customary units are to be regarded as the standard. The metric equivalents of U.S. customary units may be approximate.

NOTE 2—The high-temperature cure characteristic of the adhesive employed limits the application of the method to those thermosetting plastics referred to in Section 1. It is planned to develop additional procedures in order that the method will be applicable to other thermosetting plastics as well as to the thermoplastic materials.

2. Significance

2.1 This test, when applied to laminated plastics, is a measure of the interlaminar or intralaminar strength, whichever is smaller. When applied to nonlaminated plastics, the test is a measure of the cohesive strength of the material. The property determined is of fundamental aspect and has not yet been correlated with the results of any other method for bond strength.

2.2 The test may be found to be useful as (*1*) a research test when studying the effects of changes in independent variables, (*2*) a specification test, or (*3*) a referee test. It is considered too time-consuming to be used effectively for quality control.

3. Apparatus and Materials

3.1 *Metal Blocks*—A pair of 51-mm (2-in.) square metal blocks of 24 ST aluminum alloy,[2] each having a maximum height of 51 mm (2 in.). Each block shall have in one end a hole (see Fig. 1) tapped 22.2 mm ($^7/_8$ in.) in accordance with the ANSI Standard for Unified Screw Threads (ANSI B1.1—1960), to accommodate threaded 22.2-mm ($^7/_8$-in.) studs of convenient length.

3.2 *Press*—A small press with temperature-controlled heated platens.

3.3 *Pyrometer*—A small contact pyrometer.

3.4 *Testing Machine*—A tension testing machine.

3.5 *Adhesive*—A suitable adhesive.

NOTE 3—Redux[3] has been found satisfactory for use in this test, and the instructions in Section 5 for preparation of the test assembly are based on the use of this material. Any adhesive that is found to perform satisfactorily under this test may be used, provided that the procedure for preparation of the test assembly is suitably modified to follow the manufacturer's recommendations for use of the adhesive.

4. Test Specimen

4.1 The test specimen shall consist of sheet material 51 mm (2 in.) square, prepared in such a manner as to produce smooth edges. The thickness of the specimen shall be the thickness of the material. Five specimens shall be tested.

---

[1] This method is under the jurisdiction of ASTM Committee D-20 on Plastics. A list of committee members may be found in the ASTM Yearbook. This standard is the direct responsibility of Subcommittee D-20.10 on Mechanical Properties.

Current edition effective Sept. 30, 1951. Originally issued 1948. Replaces D 952 – 48 T.

[2] This alloy is the same as alloy 2024 in ASTM Specification B 211 for Aluminum-Alloy Bars, Rods, and Wire. *Annual Book of ASTM Standards*, Part 6.

[3] Redux is available from the Shur-Lok Bonded Structures, Ltd., 1300 E. Normandy Place, Santa Ana, Calif.

5. Conditioning

5.1 *Conditioning*—Condition the test specimens at 23 ± 2 C (73.4 ± 3.6 F) and 50 ± 5 percent relative humidity for not less than 40 h prior to test in accordance with Procedure A of ASTM Methods D 618, Conditioning Plastics and Electrical Insulating Materials for Testing,[4] for those tests where conditioning is required. In cases of disagreement, the tolerance shall be ±1 C (±1.8 F) and ±2 percent relative humidity.

5.2 *Test Conditions*—Conduct tests in the Standard Laboratory Atmosphere of 23 ± 2 C (73.4 ± 3.6 F) and 50 ± 5 percent relative humidity, unless otherwise specified in the test methods or in this specification. In cases of disagreements, the tolerances shall be ±1 C (±1.8 F) and ±2 percent relative humidity.

6. Preparation of Test Assembly

6.1 Adjust the press platens to maintain the temperature at 300 F.

6.2 Determine the cross-sectional area of the test specimen in a plane parallel to the surface. Gently rub both sides of the specimen with 00 emery cloth. Do not rub the specimen on the cloth since the corners will then be abraded more than other parts of the surface. With a cloth saturated with acetone wipe clean both the surfaces of specimen and the surface of each block opposite the hole. Do not touch the cleaned surfaces with the hands. Apply a coating of Redux liquid (Note 3) to the cleaned surfaces of the blocks by the use of a brush. Sprinkle Redux powder on the wet liquid-coated surfaces of the blocks and remove excess powder by tapping each inclined block on a table or other suitable rigid working surface.

6.3 Place the specimen between the coated blocks (Note 4), being certain that the blocks are aligned (see Fig. 1); then insert the assembly into the hot press. Close the press and immediately apply and maintain a pressure of 3.45 MN/m^2 (500 psi) of specimen area (approximately 910 kgf (2000 lb) total pressure). By use of a small contact pyrometer, observe the temperature of the metal blocks near the surfaces adjacent to the specimen. Cure the assembly for 20 min at 300 F and a pressure of 3.45 MN/m^2 (500 psi) of specimen area (approximately 910 kgf (2000 lb) total pressure). Cool the assembly under the same pressure until the temperature of the metal blocks near the specimen reaches 100 C (212 F). Remove the assembly from the press without subjecting it to any shocks and place it in the Standard Laboratory Atmosphere, keeping it there until the test on the following day.

NOTE 4—In the development of this method, it was found that no delay before pressing the assembly was needed for the materials under consideration. In case failures occur at the adhesive interface, the strength of the adhesive bond may be improved by the use of open assembly times of 1/2 to 16 h.

7. Procedure

7.1 On the day following the preparation of the test assembly, insert the studs into the blocks and determine the tensile load at a speed of 1.3 mm (0.05 in.)/min required to break the specimen. Use self-aligning holders, and make the test in the Standard Laboratory Atmosphere. Calculate the unit stress in meganewtons per square meter (pounds per square inch) by dividing the load by the area of the test specimen.

NOTE 5—To prepare the metal blocks for reuse, grind the remaining portions of the test specimen and adhesive from the surfaces of the blocks by using a surface grinder. An abrasive wheel 178 mm (7 in.) in diameter and having a 12.7-mm (1/2-in.) face, running at 2900 rpm, has been found to be satisfactory.[5] In order to maintain a truly plane surface, it is recommended that the metal blocks be finished on a flat emery surface.

7.2 For testing the material specified, this method is so designed that failure between the adhesive and the metal should not occur. If failure does occur between the adhesive and the metal, discard the test and test another specimen.

8. Report

8.1 The report shall include the following:

8.1.1 Complete identification of the material tested, including type or grade, nominal thickness, source, principal dimensions, previous history, etc.,

---

[4] *Annual Book of ASTM Standards*, Part 27.

[5] Aloxite-AA-46-G6-V60 of the Carborundum Co., Niagara Falls, N. Y., has been found suitable for this purpose.

8.1.2 The adhesive used,

8.1.3 The atmospheric conditions in the test room,

8.1.4 The total load, in kilograms-force (pounds), required to break each specimen, and

8.1.5 The unit stress, in meganewtons per square meter (pounds per square inch).

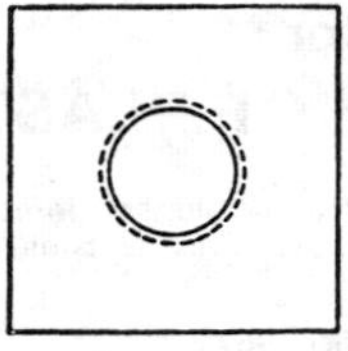

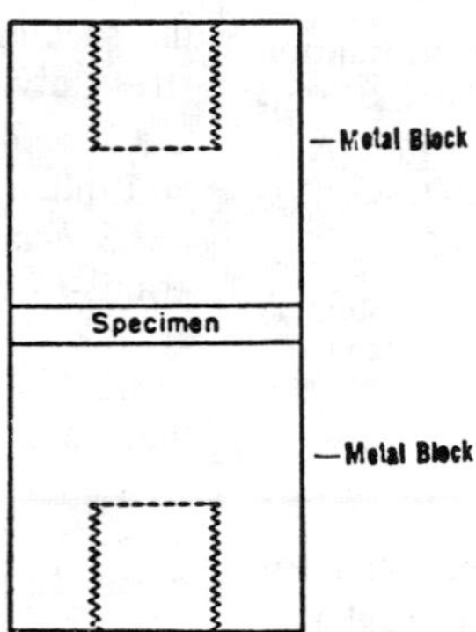

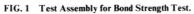

FIG. 1 Test Assembly for Bond Strength Test.

Standard Method of Test for

BEARING STRENGTH OF PLASTICS[1]

This Standard is issued under the fixed designation D 953; the number immediately following the designation indicates the year of original adoption or, in the case of revision, the year of last revision. A number in parentheses indicates the year of last reapproval.

NOTE—Editorial change was made in Section 7 in October 1969.

1. Scope

1.1 This method covers the determination of the bearing strength of rigid plastics in either sheet or molded form. Procedure A is applicable for tension loading and Procedure B for compression loading.

NOTE 1—The values stated in U.S. customary units are to be regarded as the standard. The metric equivalents of U.S. customary units may be approximate.

2. Significance

2.1 This bearing strength test for plastics is intended to apply in the specification of various thermoplastic or thermosetting products in sheet form where rivets, bolts, or similar fastenings are to be used in joining members or sections. It also is intended to apply wherever sheet materials of the classes indicated are required to sustain edgewise loads that are applied by means of pins or rods of circular cross section which pierce the sheet perpendicular to the surface.

2.2 The purpose of the test is to determine the bearing strength of the material and to show the bearing stress versus the deformation of the hole. The allowable deformation of the hole in the material should be such as to produce no looseness of joints.

2.3 While it is a known fact that higher strength materials will generally give higher bearing strengths, there is no satisfactory method by which bearing strength may be estimated from tensile or compressive properties of the material.

3. Definitions[2]

3.1 *edge distance ratio*—the distance from the center of the bearing hole to the edge of the specimen in the direction of the principle stress, divided by the diameter of the hole.

3.2 *bearing area*—the diameter of the hole multiplied by the thickness of the specimen.

3.3 *bearing stress*—the applied load in kilograms-force (pounds) divided by the bearing area.

3.4 *bearing strength*—the bearing stress at that point on the stress-strain curve where the tangent is equal to the bearing stress divided by 4 percent of the bearing hole diameter (see Appendix).

3.5 *maximum bearing stress*—the maximum load in kilograms-force (pounds) sustained by the specimen, divided by the bearing area.

4. Apparatus

4.1 *Testing Machine*—A properly calibrated universal testing machine that can be operated at a speed of 1.3 mm (0.05 in.)/min, except that in cases of certain types of material it may be necessary to operate at a slower speed. The percentage of error of the testing machine shall be no more than plus or minus 1 percent of the applied load as determined in accordance with ASTM Methods E 4, Verification of Testing Machines.[3]

---

[1] This method is under the jurisdiction of ASTM Committee D-20 on Plastics. A list of committee members may be found in the ASTM Yearbook. This standard is the direct responsibility of Subcommittee D-20.10 on Mechanical Properties.

Current edition effective Sept. 15, 1954. Originally issued 1948. Replaces D 953 - 48 T.

[2] Attention is also directed to ASTM Definitions E 6, Terms Relating to Methods of Mechanical Testing which appear in the *Annual Book of ASTM Standards*, Part 27.

[3] *Annual Book of ASTM Standards*, Part 27.

4.2 *Tension Loading Fixture*—A three-plate fixture of hardened steel similar to that shown in Fig. 1.

4.3 *Extension Indicator*—A suitable instrument reading to 0.0025 mm (0.0001 in.) for indicating the movement of the free end of the specimen with relation to the bearing pin in the tension loading fixture. A dial gage fitted with accessories for this purpose is shown in Fig. 2.

4.4 *Compression Loading Fixture*—A type of support with a suitable bearing pin similar to Fig. 3.

4.5 *Compression Indicator*—A suitable instrument reading to 0.0025 mm (0.0001 in.) for indicating the movement of the free end of the specimen with relation to the bearing pin in the compression loading fixture. A suggested dial gage assembly for this purpose is shown in Fig. 4.

5. Test Specimens

5.1 The test specimens shall conform to the dimensions shown in Fig. 5. A size of test specimen shall be chosen that most nearly conforms to production requirements of the material and the type of loading under consideration. The thicker specimens with the larger bearing hole are likely to give the more precise results, although it is advantageous to use the thinner specimens with the smaller bearing hole in testing certain relatively brittle plastics because they are less likely to fail prematurely. The specimen may be machined from sheet or molded to finished size. The bearing hole shall be located as shown in Fig. 5. It shall be drilled undersized and reamed to size as indicated. The hole shall be clean and smooth with sharp edges but not polished. It is suggested that the reaming operation be done in the drill press by hand without the use of a jig.

6. Number of Test Specimens

6.1 At least five specimens shall be tested for each sample in the case of isotropic materials.

6.2 Ten specimens, five perpendicular to and five parallel with the principal axis of anisotropy, shall be tested for each sample in case of anisotropic materials.

6.3 Specimens that break at some obvious fortuitous flaw shall be discarded and retests made, unless such flaws constitute a variable the effect of which it is desired to study.

7. Conditioning

7.1 *Conditioning*—Condition the test specimens at 23 ± 2 C (73.4 ± 3.6 F) and 50 ± 5 percent relative humidity for not less than 40 h prior to test in accordance with Procedure A of ASTM Methods D 618, Conditioning Plastics and Electrical Insulating Materials for Testing,[3] for those tests where conditioning is required. In cases of disagreement, the tolerances shall be ±1 C (±1.8 F) and ±2 percent relative humidity.

7.2 *Test Conditions*—Conduct tests in the Standard Laboratory Atmosphere of 23 ± 2 C (73.4 ± 3.6 F) and 50 ± 5 percent relative humidity, unless otherwise specified in the test methods or in this specification. In cases of disagreements, the tolerances shall be ±1 C (±1.8 F) and ±2 percent relative humidity.

8. Measurement of Dimensions

8.1 The width and thickness of the conditioned test specimen shall be measured to the nearest 0.025 mm (0.001 in.) at the bearing hole. The diameter of the bearing hole and the distance from the center of the bearing hole to the edge of the specimen in the direction of the principle stress shall also be measured to the nearest 0.025 mm (0.001 in.).

9. Procedure

9.1 The choice of either Procedure A for tension loading (9.2) or Procedure B for compression loading (9.3) is optional, but it should be recognized that Procedure B gives higher bearing strength values than Procedure A on the same material. Test specimens according to both Procedures A and B if a complete specification of bearing strength is required. Take care in aligning the long axis of the specimen with the center line of the testing fixture. Stress the specimen at the prescribed rate of crosshead travel and take deformation readings. In case autographic recording is not available, record the load sustained by the specimen for every 0.0127-mm (0.0005-in.) deformation up to a total strain of 4 percent of the bearing hole diameter. Continue the test until maximum stress is sustained and the corresponding deformation of the bearing hole recorded. Plot load – deflection curves to determine the bearing strength (see Appendix).

9.2 *Procedure A for Tension Loading*—Mount the specimen to be tested in the tension loading fixture and attach a deformation indicator. If a dial gage (Fig. 2) is used, adjust the yoke so that contact is made with the specimen at the level indicated in Fig. 1. Exception is taken in the case of those thermoplastic materials which exhibit extended plastic flow. Such specimens tend to "neck down" in the region of the bearing hole and it is necessary to mount the yoke 12.7 mm ($^1/_2$ in.) below the normal position indicated in *A*, Fig. 1, so that it will not slip during test.

9.3 *Procedure B for Compression Loading*—Mount the specimen to be tested in the compression loading fixture and load through a flat, hardened compression plate. Clamp the hardwood cheek blocks in place and attach the deformation indicator (see Fig. 4).

10. Speed of Testing

10.1 The mean rate of crosshead travel in the testing of specimens shall not exceed 1.3 mm (0.05 in.) min. In any case the crosshead movement shall be low enough so that deflection gage readings can be made accurately.

11. Retests

11.1 Results that deviate from the mean value of all tests shall be rejected if the deviation of the doubtful value is more than five times the average deviation from the mean obtained by excluding the doubtful value. Such doubtful values shall be discarded and retests made, unless the degree of variability is a factor that is being studied.

12. Report

12.1 The report shall include the following:

12.1.1 Complete identification of the material tested, including type, source, manufacturer's code number, form, principal dimensions, and previous history,

12.1.2 Direction of cutting and loading specimens,

12.1.3 Conditioning procedure,

12.1.4 Length, width, and thickness of specimen, in millimeters (inches),

12.1.5 Diameter of bearing hole,

12.1.6 Edge distance ratio,

12.1.7 Mean rate of crosshead motion in millimeters (inches) per minute,

12.1.8 Bearing strength in newtons per square millimeter (pounds per square inch), stating whether Procedure A for tension loading or Procedure B for compression loading was used, and

12.1.9 Maximum bearing stress in newtons per square millimeter (pounds per square inch), stating whether Procedure A or Procedure B was used.

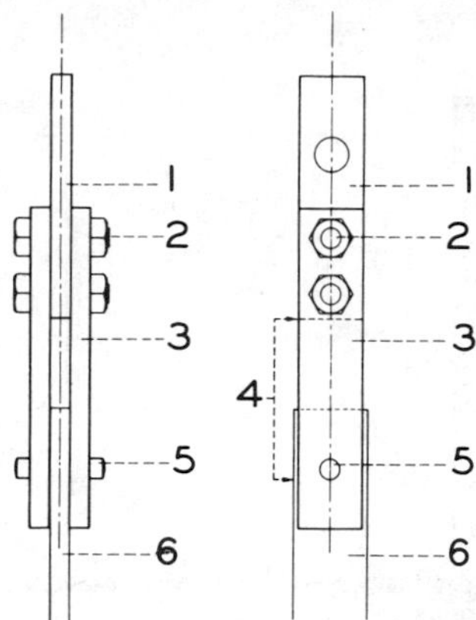

1—Hardened spacer plate.
2—6.3-mm (1/4-in.) steel bolts in reamed holes.
3—Hardened side plate.
4—Extensometer span.
5—Hardened steel pin in reamed hole.
6—Test specimen.

NOTE—Critical dimensions are as follows:

Type	Bearing Hole Diameter, mm (in.)	Bearing Pin Diameter, mm (in.)	Thickness of Spacer Plate, mm[a] (in.)
I	(3.175 +0.025 −0.000)	(3.150 +0.000 −0.025)	
	0.125 +0.0010 −0.0000	0.124 +0.0000 −0.0010	3.2 (1/8)
II	(6.350 +0.025 −0.000)	(6.325 +0.000 −0.025)	
	0.250 +0.0010 −0.0000	0.249 +0.0000 −0.0010	6.3 (1/4)

[a] The spacer plate shall be shimmed to a thickness of 0.025 to 0.125 mm (0.001 to 0.005 in.) greater than the specimen under test.

FIG. 1 Steel Tension Loading Fixture.

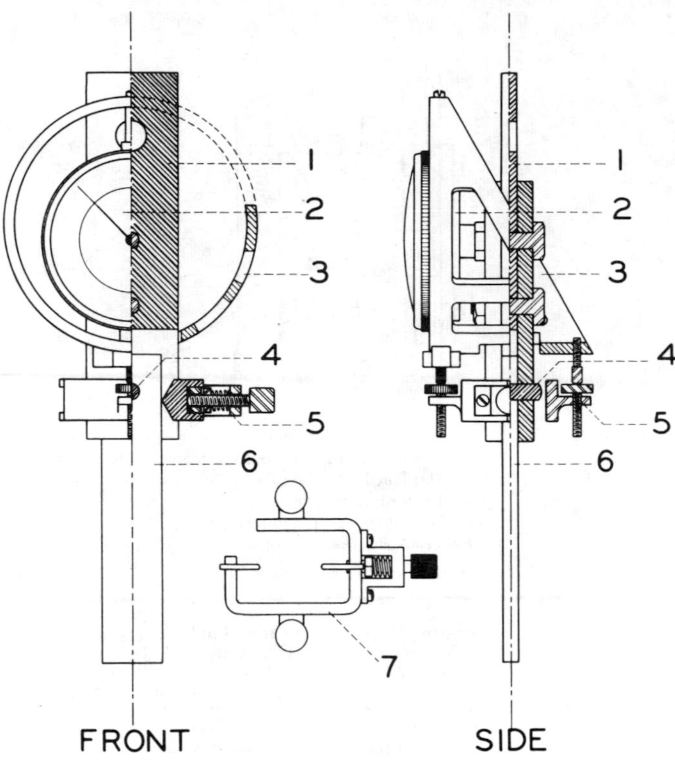

1—Spacer plate
2—Dial gage.
3—Double foot ring mounted on spindle of dial gage.
4—Bearing pin.
5—Yoke mounted on specimen.
6—Test specimen.
7—Plan view of yoke.

FIG. 2 Tension Loading Assembly.

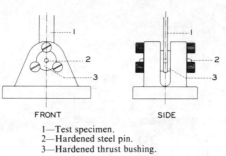

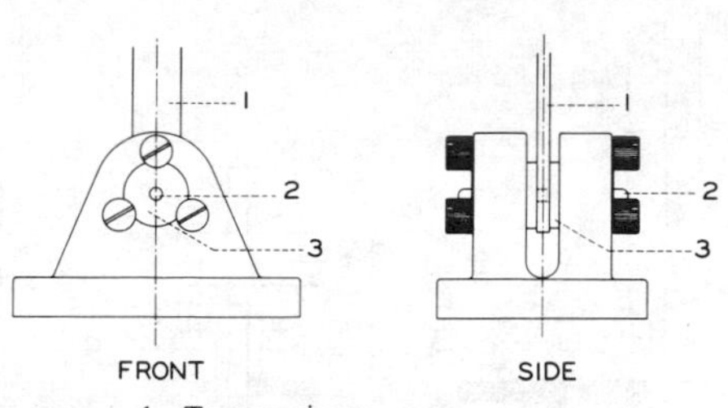

1—Test specimen.
2—Hardened steel pin.
3—Hardened thrust bushing.

NOTE—Critical dimensions are as follows:

Type	Bearing Hole Diameter, mm (in.)	Bearing Pin Diameter, mm (in.)	Minimum Clearance Between Bushings, mm (in.)
I	(3.175 +0.025 −0.000)	(3.150 +0.000 −0.025)	2.54
	0.125 +0.0010 −0.0000	0.124 +0.0000 −0.0010	0.10
II	(6.350 +0.025 −0.000)	(6.325 +0.000 −0.025)	5.58
	0.250 +0.0010 −0.0000	0.249 +0.0000 −0.0010	0.22

FIG. 3 Steel Compression Loading Fixture.

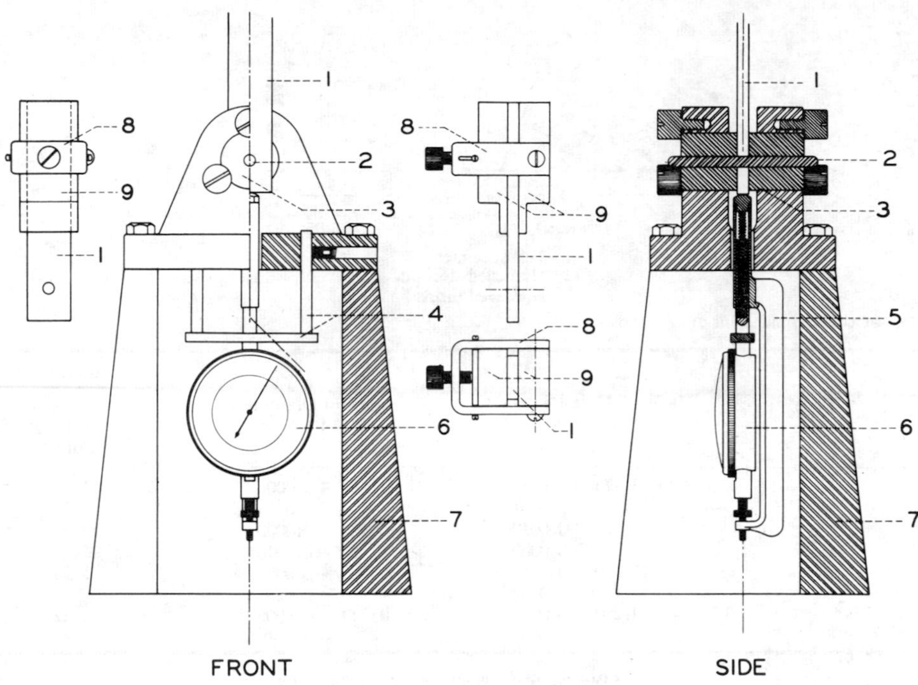

1—Test specimen.
2—Bearing pin.
3—Thrust bushings.
4—Hanger for dial gage.
5—Spring-supported thrust member to transfer deformation of specimen to dial gage foot.
6—Dial gage.
7—Pedestal support.
8—Clamp for cheek blocks.
9—Hard maple cheek blocks to stabilize test specimen.

FIG. 4 Compression Loading Assembly.

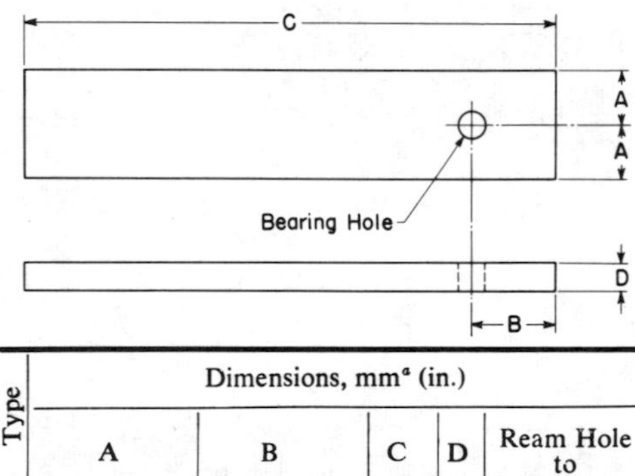

Type	Dimensions, mm[a] (in.)				
	A	B	C	D	Ream Hole to
I	11.913±0.127	19.050±0.127	120.6	3.2	3.200±0.025
	0.469±0.005	0.750±0.005	4¾	⅛	0.126±0.001
II	11.913±0.127	19.050±0.127	120.6	6.4	6.375±0.025
	0.469±0.005	0.750±0.005	4¾	¼	0.251±0.001

[a] All fractional dimensions shall be held to plus or minus 0.40 mm (1⁄64 in.) tolerance.

$$\text{Edge distance ratio} = \frac{\text{B}}{\text{hole diameter}}$$

FIG. 5 Dimensions of Bearing Strength Test Specimens.

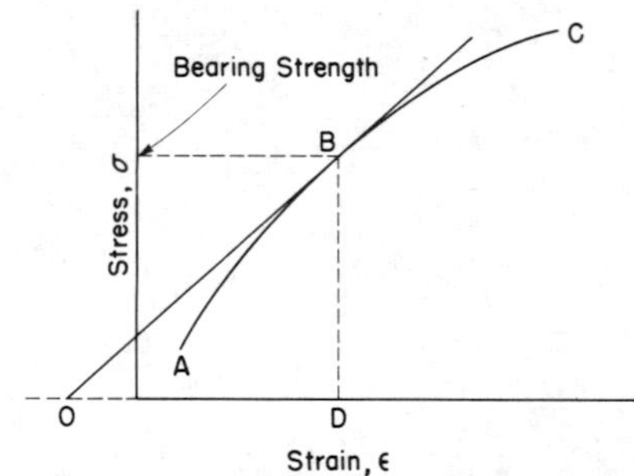

FIG. 6 Illustration of Method of Determining Bearing Strength from Bearing Stress-Deflection Curve.

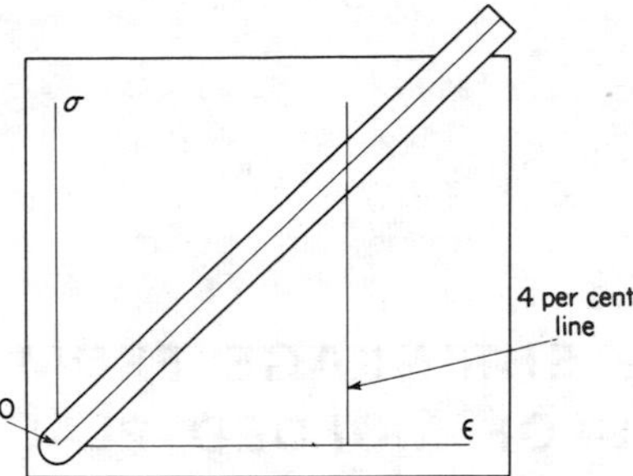

FIG. 7 Template for Determining Point *B* on Bearing Stress - Deflection Curve.

APPENDIX

A1. DETERMINATION OF BEARING STRENGTH

A1.1 Heretofore, bearing strength has been defined as the bearing stress at which the bearing hole is deformed a given percentage (4 percent) of the hole diameter, and is by definition, sensitive to zero load errors. The following procedure for determining bearing strength has been found to eliminate this ambiguity. It is insensitive to zero load errors, and as a result improves the precision of the bearing strength measurements.

A1.2 The method is best illustrated by Fig. 6. Given the bearing stress - deflection curve, *ABC*, the tangent is determined at a point, *B*, such that when the tangent is projected through the point, *O*, on the zero load axis, the distance between *O* and *D* shall be equal to 4 percent of the bearing hole diameter.

A1.3 A template to facilitate the determination of point *B* on the curve is shown in Fig. 7. It is designed to fit the coordinate paper upon which the stress-strain curve is drawn in such a way that a "4 percent line" on the template is established at a distance from the origin, *O*, equal to 4 percent of the bearing hole diameter. It consists of a thin rigid sheet of transparent plastic upon which the rectangular coordinates are ruled. A strip of transparent plastic is mounted on the sheet so as to rotate about point *O*. The strip is provided with a reference center line passing through the point of rotation.

A1.4 In practice, the ϵ axis of the template is superimposed on the zero load axis of the stress-strain curve. This can best be done with the aid of a drawing board and a parallel straight-edge. The template is then slid to right or left while in superposition until the rotating arm can be made tangent to the stress-strain curve at its intersection with the 4 percent line. The intersection is the point *B* on the curve in Fig. 6; *B* is projected horizontally to the load axis and the bearing strength is read.

A1.5 Bearing strength, obtained as described above (see Fig. 6), is directly related to the tangent modulus or bearing stiffness at the point indicated according to the general expression:

$$E_t = \Delta\sigma/\Delta\epsilon = (P/A)/(e/d)$$

Since the point *O* from which the deformation offset is measured is the point of intersection of the tangent to the curve at point *B* and the zero load axis,

$$E_t = \sigma/\epsilon = S_b/0.04 = 25\ S_b$$

where:

E_t = tangent modulus or bearing stiffness at point *B*,
$\Delta\sigma$ = unit bearing stress at point *B*,
$\Delta\epsilon$ = unit deformation at point *B*,
P = bearing load at point *B*,
A = bearing area,
e = deformation offset = 0.04 *d*,
d = bearing hole diameter, and
S_b = bearing strength.

Thus bearing strength, determined as prescribed herein, is related to the tangent modulus at the point specified by a factor that is dependent only on the offset distance, *OD*, Fig. 6.

Designation: D 955 – 51 (Reapproved 1969)

American National Standard K66.3-1959
(R 1969)
American National Standards Institute

Standard Method of

MEASURING SHRINKAGE FROM MOLD DIMENSIONS OF MOLDED PLASTICS[1]

This Standard is issued under the fixed designation D 955; the number immediately following the designation indicates the year of original adoption or, in the case of revision, the year of last revision. A number in parentheses indicates the year of last reapproval.

1. Scope

1.1 This method is intended to measure to batch-to-batch uniformity in initial shrinkage from mold to molded dimensions of either thermoplastic or thermosetting materials when molded by compression, injection, or transfer under specified conditions.

1.2 This method does not provide for the measurement of shrinkages that may occur as molded materials age, after the first 48 h out of the mold.

NOTE 1—The values stated in U.S. customary units are to be regarded as the standard. The metric equivalents of U.S. customary units may be approximate.

2. Factors Affecting Shrinkage

2.1 *Compression Molding*—In compression molding, the difference between the dimensions of a mold and of the molded article produced therein from a given material may vary according to the design and operation of the mold. It is probable that shrinkages will approach a minimum where design and operation are such that a maximum of material is forced solidly into the mold cavity or some part of it, or where the molded article is hardened to a maximum while still under pressure, particularly by cooling. In contrast, shrinkages may be much higher where the charge must flow in the mold cavity but does not receive and transmit enough pressure to be forced firmly into all its recesses, or where the molded article is not fully hardened when discharged. The plasticity of the material used may affect shrinkage in so far as it affects the retention and compression of the charge. The method described in Sections 3 to 6 measures minimum rather than maximum shrinkages and largely eliminates any effects that plasticity may have on shrinkage.

2.2 *Injection Molding*—In injection molding, also, the difference between the dimensions of the mold and of the molded article produced therein from a given material may vary according to the design and operation of the mold. It may vary with the type and size of molding machine, the thickness of molded sections, the degree of flow or movement of material in the mold, the size of the sprue and nozzle, the cycle on which the machine is operated, the temperature of the mold, and the length of time that follow-up pressure is maintained. As in the case of compression molding, shrinkages will approach a minimum where design and operation are such that a maximum of material is forced solidly into the mold cavity and where the molded article is hardened to a maximum while still under pressure as a result of the use of a runner, sprue, and nozzle of proper size, along with proper dwell. As in compression molding, shrinkages may be much higher where the charge must flow in the mold cavity but does not receive and transmit enough pressure to be forced firmly into all of the recesses of the mold. The plasticity of the material used may affect shrinkage indirectly, in that the more readily plasticized material will require a lower molding temperature.

---

[1] This method is under the jurisdiction of ASTM Committee D-20 on Plastics. A list of committee members may be found in the ASTM Yearbook. This standard is the direct responsibility of Subcommittee D-20.90 on Specimen Preparation.

Current edition effective Sept. 30, 1951. Originally issued 1948. Replaces D 955 – 48 T.

2.3 *Transfer Molding*—In transfer molding, also, the difference between the dimensions of the mold and of the molded article produced therein from a given material may vary according to the design and operation of the mold. It is affected by the size and temperature of the pot or cylinder and the pressure on it, as well as on mold temperature and molding cycle. Direction of flow is not as important a factor as would be expected, although it can have some bearing on results.

3. Sampling

3.1 Such samples as are necessary to represent the material, depending upon its uniformity, shall be kept at room temperature in nominally airtight containers until tested. If composite samples are desired, the method of sampling described in ASTM Recommended Practice D 1898, for Sampling of Plastics[2] shall be followed.

4. Apparatus

4.1 *Compression Mold*—A single bar, single-cavity positive mold and, for diametral shrinkage, a single cavity positive 102-mm (4-in.) disk mold, both conforming to ASTM Recommended Practice D 647, for Design of Molds for Test Specimens of Plastic Molding Materials.[2]

4.2 *Injection Mold*—For shrinkage parallel to flow, an impact bar mold having a cavity 12.7 by 127 mm ($^1/_2$ by 5 in.). The thickness shall be 3.2 mm ($^1/_8$ in.), unless otherwise agreed upon by the seller and the purchaser. The mold shall have at one end a gate 6.4 mm ($^1/_4$ in.) in width by 3.2 mm ($^1/_8$ in.) in depth (Note 2). For diametral shrinkage, the mold shall have a cavity 102 mm (4 in.) in diameter by 3.2 mm ($^1/_8$ in.) in thickness with a gate, placed radially at the edge, 12.7 mm ($^1/_2$ in.) in width by 3.2 mm ($^1/_8$ in.) in depth.

NOTE 2—If, for any reason, a test specimen of a different thickness is agreed upon, the depth of the gate should be equal to that thickness.

4.3 *Transfer Mold*—An impact bar mold having a cavity 12.7 by 12.7 by 127 mm ($^1/_2$ by $^1/_2$ by 5 in.) and having either an end gate or top gate at one end 6.4 mm ($^1/_4$ in.) in width by 6.4 ($^1/_4$ in.) in depth.

4.4 *Compression Press*—A suitable hydraulic press that will deliver 20 to 35 MN/m^2 (3000 to 5000 psi) to the material in the mold.

4.5 *Injection Press*—A suitable injection-molding machine that will fill the test molds when it is operated in the range from one half to three fourths of its rated shot capacity.

NOTE 3—If injection machines of appropriate capacity are not available, the requirement of 4.5 is met in machines of larger capacities by providing the test molds in multiple or by mounting the test molds with each other, or with molds for other pieces of similar thickness, to be filled from a common sprue, so that the total weight of the shot, including sprue and runners, will fall within the specified limits.

4.6 *Transfer Press*—A suitable hydraulic press that will deliver a pressure of 70 to 140 MN/m^2 (10,000 to 20,000 psi) on the material in the pot of the die or the cylinder of the press.

4.7 *Balance*—A balance for weighing compression-molding charges.

4.8 *Gages*—Gages accurate to 0.02 mm (0.001 in.) for measuring the molds and the test specimens.

5. Test Specimens

5.1 *Compression-Molding Materials*—For mold shrinkages of compression-molding materials, the test specimens shall be bars, 12.7 by 12.7 by 127 mm ($^1/_2$ by $^1/_2$ by 5 in.), or a disk 3.2 mm ($^1/_8$ in.) in thickness and 102 mm (4 in.) in diameter, made in a positive mold in such a way that there is no appreciable lateral movement of the plastic during the molding.

5.2 *Injection-Molding Materials*—For mold shrinkage of injection-molding materials, specimens of two types shall be used: (*1*) bars 12.7 by 3.2 by 127 mm ($^1/_2$ by $^1/_8$ by 5 in.), gated at the end to provide flow throughout the entire length, shall be used for measurements of shrinkage in the direction of flow, and (*2*) disks, 3.2 mm ($^1/_8$ in.) in thickness and 102 mm (4 in.) in diameter, gated radially at a single point in the edge, shall be used for measurements of shrinkages of diameters parallel and perpendicular to the flow.

5.3 *Transfer-Molding Materials*—For shrinkage of transfer-molding materials, specimens 12.7 by 12.7 by 127 mm ($^1/_2$ by $^1/_2$ by 5 in.), gated at the end or at the top near one

[2] *Annual Book of ASTM Standards*, Part 27.

end, so as to provide flow throughout their entire length, shall be used for measurement of shrinkage in the direction of flow.

6. Conditioning

6.1 *Conditioning*—Condition the test specimens at 23 ± 2 C (73.4 ± 3.6 F) and 50 ± 5 percent relative humidity for not less than 40 h prior to test in accordance with Procedure A of ASTM Methods D 618, Conditioning Plastics and Electrical Insulating Materials for Testing,[2] for those tests where conditioning is required. In cases of disagreement, the tolerances shall be ±1 C (±1.8 F) and ±2 percent relative humidity.

6.2 *Test Conditions*—Conduct tests in the Standard Laboratory Atmosphere of 23 ± 2 C (73.4 ± 3.6 F) and 50 ± 5 percent relative humidity, unless otherwise specified in the test methods or in this specification. In cases of disagreement, the tolerances shall be ±1 C (±1.8 F) and ±2 percent relative humidity.

7. Procedure

7.1 Measure the length of the cavity of the bar mold, or the diameter of the cavity of the disk mold parallel and perpendicular to the flow, to the nearest 0.02 mm (0.001 in.). Make the measurements at Standard Laboratory Temperature as defined in 3.1 of Methods D 618.

7.2 Mold at least three sound test specimens from the sample to be tested, under such conditions of pressure, temperature, time, etc., as the manufacturer and the purchaser may agree are suitable for the material. In the absence of other definite recommendations, the following are suggested as suitable molding procedures for plastics of various types:

7.2.1 *Thermoplastics Molded by Compression*—For thermoplastics, such as cellulose acetate, load the mold evenly at room temperature with the requisite quantity of granular material, also at room temperature, and place the mold in a hydraulic press such that a pressure of 20 to 35 MN/m^2 (3000 to 5000 psi) on the projected area of the mold cavity can be applied and the mold can be heated to 121 to 129 C (250 to 265 F) in 5 to 10 min. Apply heat and pressure, within these ranges, to mold the specimen. Cool the mold to room temperature under pressure and discharge the specimen. The rate of heating, the maximum temperature reached, and the rate of cooling do not seem critical, but the temperature of the mold at the time of discharge should be definitely controlled.

7.2.2 *Thermoplastics Molded by Injection*—The molding machine used should be such that it is operated under a pressure between 100 and 140 MN/m^2 (15,000 and 20,000 psi) without exceeding one half to three fourths of its rated shot capacity. The temperature of the heating cylinder should be maintained at a point midway between that at which the mold will fail to be filled and that at which the molded specimen is dull-surfaced or rough or shows blisters or bubbles. The mold should be maintained within 3 C (5 F) of the prescribed temperature, which depends upon the material being molded and ordinarily will lie between 49 and 93 C (120 and 200 F).

7.2.3 *Thermoplastics Molded by Transfer*—The press used for transfer molding thermoplastics should be of a size that will permit operation under a pressure on the stock pot or cylinder of from 70 to 140 MN/m^2 (10,000 to 20,000 psi). The temperature of the cylinder and of the die should be maintained at a point that will yield smooth pieces.

7.2.4 *Thermosetting Materials, Urea Type, Molded by Compression*—For thermosetting materials of the urea type, molding should be carried out in accordance with ASTM Recommended Practice D 956, for Molding Specimens of Amino Plastics.[2]

7.2.5 *Thermosetting Materials, Phenolic Type, Molded by Compression*—For thermosetting materials of the phenolic type, molding should be carried out in accordance with ASTM Recommended Practice D 796, for Molding Test Specimens of Phenolic Molding Materials.[2]

7.2.6 *Thermosetting Materials Molded by Transfer*—The press used should be such that pressure in the stock pot or cylinder can be maintained in the range between 55 and 140 MN/m^2 (8,000 and 20,000 psi). The temperature of the cylinder should be such that the entire charge can be forced into the die before premature curing can take place, which would be evidenced by roughness and poor finish.

7.3 After molding, allow the specimens to cool to approximately room temperature and then store in the Standard Laboratory Atmosphere prescribed in 3.2 of Methods D 618 before being measured. The period of storage for "initial molding shrinkage" for specimens 3.2 mm ($^1/_8$ in.) in thickness shall be from 1 to 2 h; for specimens 6.4 mm ($^1/_4$ in.) in thickness, 2 to 4 h; and for specimens 12.7 mm ($^1/_2$ in.) in thickness, from 4 to 6 h. Measure the length or diameter of each specimen at Standard Room Temperature to the nearest 0.02 mm (0.001 in.) and then return the specimens to storage in the Standard Laboratory Atmosphere. Measure the specimens again not less than 16 nor more than 24 h after molding, in order to obtain the "24-h shrinkage," and again not less than 40 nor more than 48 h after molding, in order to determine the "48-h" or "normal" mold shrinkage.

8. Calculation and Report

8.1 Calculate the shrinkage per millimeter (or inch, by subtracting the dimension of the specimen from the corresponding dimension of the mold cavity in which it was molded and dividing the difference by the latter.

8.2 The report shall include:

8.2.1 The molding procedure used,

8.2.2 For compression molding, the form and distribution of the mold charge and the molding pressure, temperature, and time,

8.2.3 For injection molding, the temperatures of cylinder and mold, the size of the nozzle, the molding pressure, the molding cycle, and the make, type, and the size of the machine used,

8.2.4 For transfer molding, the temperature of the pot or cylinder, the mold temperature, the pressure on the pot, and the molding cycle, and

8.2.5 The initial, 24-h, and 48-h shrinkages expressed in millimeters per millimeter (or inches per inch), and each representing the mean of values obtained on three or more specimens.

9. Precision

9.1 The probable limit of precision of this test is ±0.0005 mm/mm (in./in.). With this limitation in precision, it is possible for samples to appear to be the same in shrinkage and actually to differ by 0.001 mm/mm (in./in.), or for samples to appear to differ by 0.001 mm/mm (in./in.) and actually to be the same.

 **Designation: D 956 - 51 (Reapproved 1969)**

Recommended Practice for

MOLDING SPECIMENS OF AMINO PLASTICS[1]

This Recommended Practice is issued under the fixed designation D 956; the number immediately following the designation indicates the year of original adoption or, in the case of revision, the year of last revision. A number in parentheses indicates the year of last reapproval.

1. Scope

1.1 This recommended practice covers procedures for molding test specimens of amino plastics as follows:

1.1.1 *Cellulose-Filled Urea and Melamine Materials:*

Impact test specimens, 12.7 by 12.7 by 127 mm (1/2 by 1/2 by 5 in.)

Flexural test specimens, 6.4 by 12.7 by 127 mm (1/4 by 1/2 by 5 in.)

Water-absorption specimens, 51 by 3.2-mm (2 by 1/8-in.) disk.

1.1.2 *Shock-Resistant Melamine Materials:*

Impact test specimens, 12.7 by 12.7 by 127 mm (1/2 by 1/2 by 5 in.)

Flexural test specimens, 6.4 by 12.7 by 127 mm (1/4 by 1/2 by 5 in.)

Tension test specimens, "dog-bone" specimens shown in Fig. 2 of ASTM Method D 651, Test for Tensile Strength of Molded Electrical Insulating Materials.[2]

NOTE 1—The values stated in U.S. customary units are to be regarded as the standard. The metric equivalents of U.S. customary units may be approximate.

2. Apparatus

2.1 *Molds* (*Note 2*):

2.1.1 For molding impact specimens of cellulose-filled urea and melamine materials, a single bar, single cavity positive mold such as that shown in Fig. 1 of ASTM Recommended Practice D 647, for Design of Molds for Test Specimens of Plastic Molding Materials[3] shall be used.

2.1.2 For molding impact and flexure test specimens of shock-resistant urea and melamine materials, a single bar, single cavity positive mold such as that shown in Fig. 1 of Recommended Practice D 647 shall be used.

2.1.3 For molding water absorption specimens of urea and melamine materials, a mold such as the 51-mm (2-in.) positive mold shown in Fig. 2 of Recommended Practice D 647 shall be used.

2.1.4 For molding tension specimens, a single bar, single cavity positive mold such as that shown in Fig. 3 of Recommended Practice D 647 shall be used (Note 3).

NOTE 2—The molds shown in Recommended Practice D 647 are hand molds. It is recommended that they be made semi-automatic by bolting them to the press platens.

NOTE 3—The tension test specimen shown in Fig. 1 of ASTM Method D 638, Test for Tensile Properties of Plastics[3] will be included in this recommended practice when the mold for producing such specimens has been included in Recommended Practice D 647.

2.2 *Press*—A hydraulic press of sufficient sensitivity so that the molding pressure on the specimen can be controlled to ± 1.7 MN/m^2 (± 250 psi).

2.3 *Heating System*—Any convenient method of heating the press platens or molds may be used, provided the source of heat is constant enough to maintain the mold temperature within ± 3 C (5 F).

2.4 *Thermometer*—A 75-mm (3-in.) immersion mercury-filled thermometer having a temperature scale of not more than 25 C (45

[1] This recommended practice is under the jurisdiction of ASTM Committee D-20 on Plastics. A list of committee members may be found in the ASTM Yearbook. This standard is the direct responsibility of Subcommittee D-20.90 on Specimen Preparation.

Current edition effective Sept. 30, 1951. Originally issued 1948. Replaces D 956 – 50 T.

[2] *Annual Book of ASTM Standards*, Part 29.

[3] *Annual Book of ASTM Standards*, Part 27.

F) for each inch of length or a surface pyrometer graduated to read to within 2.5 C (5 F).

3. Conditioning

3.1 Materials shall be used as received. For specimens to be used in referee tests, the materials shall be conditioned for 72 h in a desiccator over anhydrous calcium chloride at room temperature. The material shall be spread to a depth of not more than 6 mm (1/4 in.).

4. Procedure

4.1 Mold test specimens under the conditions prescribed in Table 1.

TABLE 1 Molding Conditions for Specimens of Amino Plastics

Molding Conditions	Cellulose-Filled Urea Materials	Cellulose-Filled Melamine Materials	Shock-Resistant Melamine Materials
	Impact, Flexure, and Water-Absorption Specimens	Impact, Flexure, and Water-Absorption Specimens	Impact, Flexure, and Tension Specimens
Molding pressure			
MN/m^2	27.5 ± 1.7	27.5 ± 1.7	27.5 ± 1.7
psi	4000 ± 250	4000 ± 250	4000 ± 250
Molding temperature	145 ± 3 C (295 ± 5 F)	155 ± 3 C (310 ± 5 F)	155 ± 3 C (310 ± 5 F)
Charge	powder, granular or pill	powder, granular, or pill	as-received or preformed
Cure:			
51 by 3.2 mm (2 by 1/8-in.) disk	4 min	4 min	4 min
6.4 by 12.7 by 127-mm (1/4 by 1/2 by 5-in.) bar	5 min	5 min	5 min
12.7 by 12.7 by 127-mm (1/2 by 1/2 by 5-in.) bar	6 min	6 min	6 min
Breathing the mold	yes	yes	yes

Designation: D 957 – 50 (Reapproved 1969)

Recommended Practice for
DETERMINING MOLD SURFACE TEMPERATURE OF COMMERCIAL MOLDS FOR PLASTICS[1]

This Recommended Practice is issued under the fixed designation D 957; the number immediately following the designation indicates the year of original adoption or, in the case of revision, the year of last revision. A number in parentheses indicates the year of last reapproval.

NOTE—Metric equivalents were added and editorial changes made in August 1967.

1. Scope

1.1 This recommended practice is intended for determining the temperature at a specified point or points on the surface of the cavity and plunger of a commercial mold for plastics (Note 2). By determining the temperature at as many points as deemed necessary, the overall temperature condition of the surfaces can be determined.

NOTE 1—The values stated in U.S. customary units are to be regarded as the standard. The metric equivalents of U.S. customary units may be approximate.

NOTE 2—A commercial mold shall be considered any mold other than those designed to mold standard ASTM test specimens.

2. Apparatus

2.1 *Pyrometer*—A well-calibrated surface pyrometer (see Appendix) accurate within ± 1.5 C (2.5 F) shall be used.

3. Procedure

3.1 The size, shape, and contours of a commercial mold will determine the location and number of points at which temperature readings should be observed.

3.2 Place the thermocouple of the pyrometer at any specified point on the surface of the cavity and plunger of the mold, making certain that good contact is maintained between the two (Note 3). Maintain contact until the needle on the dial scale reaches equilibrium. Record the temperature reading on the dial scale at this point.

NOTE 3—Wax may be used to ensure intimate contact between the mold surface and the thermocouple of the pyrometer. This is accomplished by allowing a small portion of wax, or any suitable material with a low melting point, to melt at the point where the temperature reading is to be taken. The thermocouple is then placed in the puddle of molten material.

3.3 Take temperature readings at as many points as deemed necessary to determine the temperature condition of the mold surface.

---

[1] This recommended practice is under the jurisdiction of ASTM Committee D-20 on Plastics. A list of committee members may be found in the ASTM Yearbook. This standard is the direct responsibility of Subcommittee D-20.90 on Specimen Preparation.

Current edition effective Sept. 30, 1950. Originally issued 1948. Replaces D 957 – 48 T.

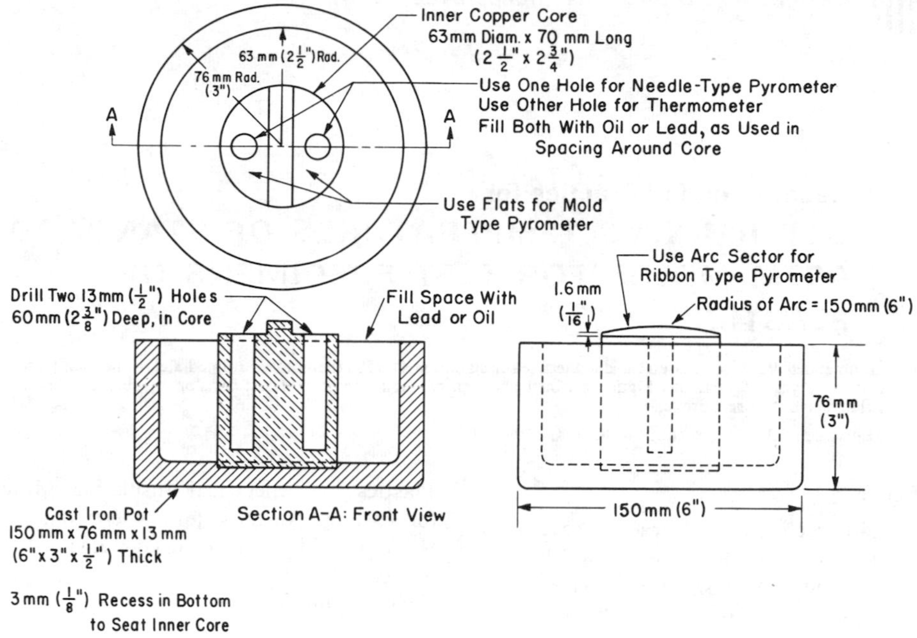

FIG. 1 Pyrometer Test Unit.

APPENDIX

A1. CALIBRATION OF PYROMETER

A1.1 Apparatus

A1.1.1 *Test Unit*—The apparatus shown in Fig. A1 shall be used to check the accuracy of the pyrometer. The apparatus shall consist of a copper core set in a heated bath contained in a cast iron pot. A suitable thermometer (A1.1.2) shall be used to obtain the temperature of the core. The core shall have two duplicate wells drilled in it, one for use as a thermometer well and the other for use to check the accuracy of needle thermocouples. If so desired, the needle thermocouple well may be used as a second thermometer well. The surface of the core shall contain a curved section for use in checking ribbon thermocouples and shall also have flat surfaces for thermocouples used on such surfaces. The latter type is generally used in determining temperatures of mold surfaces. The space in the pot surrounding the copper core shall be filled completely with a suitable oil as a heating medium. The wells in the core shall be filled with the same material. A suitable gas or electric heater can be used as a heat supply.

A1.1.2 *Thermometer*—The thermometer shall be an ASTM Partial Immersion Thermometer having a range, depending upon the temperature range being checked, from −5 to +110 C or +90 to 370 C (20 to 230 F or 200 to 700 F) and conforming to the requirements for Thermometer 9C or 10C (9F or 10F), respectively, as prescribed in ASTM Specifications E 1.[2]

A1.2 Procedure

A1.2.1 *Flat-Surface Thermocouples*—To check the accuracy of the pyrometer for thermocouples to be used on flat surfaces, place the thermocouple on a flat surface of the core, making certain that good contact is maintained between the two. Maintain contact until the needle on the dial scale reaches equilibrium. Record the temperature reading on the dial scale at this point and compare with the temperature reading of the thermometer in the well. To be considered accurate, the pyrometer reading should be within ±1.5 C (2.5 F) of the thermometer reading. Comparisons should be made at a minimum of three temperatures not closer than 55 C (100 F) to each other.

A1.2.2 *Needle and Ribbon Thermocouples*—In checking the accuracy of the needle and ribbon thermocouples, the same general procedure shall be used except that for contact the needle shall be immersed in the liquid in the well and the ribbon thermocouple shall be placed on the curved section.

[2] *Annual Book of ASTM Standards*, Part 30.

Designation: D 958 – 51 (Reapproved 1969)

Recommended Practice for
DETERMINING TEMPERATURES OF STANDARD ASTM MOLDS FOR TEST SPECIMENS OF PLASTICS[1]

This Recommended Practice is issued under the fixed designation D 958; the number immediately following the designation indicates the year of original adoption or, in the case of revision, the year of last revision. A number in parentheses indicates the year of last reapproval.

NOTE—Metric equivalents were added in February 1966.

1. Scope

1.1 This recommended practice covers two optional procedures for determining the temperature condition of standard ASTM molds for test specimens of plastics by means of either a thermometer or a pyrometer.

NOTE 1—The values stated in U.S. customary units are to be regarded as the standard. The metric equivalents of U.S. customary units may be approximate.

2. Molds

2. Requirements for the design of standard ASTM molds for test specimens of plastics are covered by ASTM Recommended Practice D 647, for Design of Molds for Test Specimens of Plastic Molding Materials.[2]

Procedure Using Thermometer

3. Apparatus

3.1 *Thermometer*—A 76-mm (3-in.) immersion, mercury-filled thermometer having a temperature scale of not more than 25 C (45 F) per 25 mm (1 in.) of length shall be used.

NOTE 2—The following ASTM standard thermometers (see Specifications E 1[3]) are considered satisfactory: 95C, 96C, 100C, and 101C.

4. Conditioning

4.1 *Conditioning*—Condition the test specimens at 23 ± 2 C (73.4 ± 3.6 F) and 50 ± 5 percent relative humidity for not less than 40 h prior to test in accordance with Procedure A of ASTM Methods D 618, Conditioning Plastics and Electrical Insulating Materials for Testing,[2] for those tests where conditioning is required. In cases of disagreement, the tolerances shall be 1 C (1.8 F) and ±2 percent relative humidity.

4.2 *Test Conditions*—Conduct tests in the Standard Laboratory Atmosphere of 23 ± 2 C (73.4 ± 3.6 F) and 50 ± 5 percent relative humidity, unless otherwise specified in the test methods or in this specification. In cases of disagreement, the tolerances shall be 1 C (1.8 F) and ±2 percent relative humidity.

5. Procedure

5.1 The chase of the mold shall be equipped with an immersion well 76 mm (3 in.) in depth. The base of the well shall be located at a point not more than 6.3 mm (1/4 in.) from the surface of the cavity. Insert the thermometer into the well as far as possible. Allow the thermometer to remain in the well until the mercury no longer moves in the column. The thermometer reading at this point shall be considered the mold temperature. In recording the temperature reading, state that the thermometer procedure was used.

[1] This recommended practice is under the jurisdiction of ASTM Committee D-20 on Plastics. A list of committee members may be found in the ASTM Yearbook. This standard is the direct responsibility of Subcommittee D-20.90 on Specimen Preparation.

Current edition effective Sept. 30, 1951. Originally issued 1948. Replaces D 958 – 50.

[2] *Annual Book of ASTM Standards*, Part 27.

[3] *Annual Book of ASTM Standards*, Part 30.

Procedure Using Pyrometer

6. Apparatus

6.1 *Pyrometer*—A calibrated surface pyrometer (Note 3) accurate within ±1.5 C (2.7 F) shall be used.

NOTE 3—For the procedure for calibration of surface pyrometers, see the Appendix to ASTM Recommended Practice D 957, for Determining Mold Surface Temperature of Commercial Molds for Plastics.[2]

7. Procedure

7.1 Place the thermocouple of the pyrometer at any point on the surface of the cavity and plunger of the mold, making certain good contact is maintained (Note 4). Maintain contact until the needle on the dial scale reaches equilibrium. Record the temperature reading on the dial scale at this point. Take temperature readings at at least four points each on the surfaces of the cavity and plunger of the mold. In recording temperature readings, state that the pyrometer procedure was used.

NOTE 4—Wax may be used to ensure intimate contact between the mold surface and the thermocouple of the pyrometer. This is accomplished by allowing a small portion of wax, or any suitable material with a low melting point, to melt at the point where the temperature reading is to be taken. The thermocouple is then placed in the puddle of molten material.

Designation: D 1003 – 61

American National Standard K65.5-1960
American National Standards Institute

Standard Method of Test for
HAZE AND LUMINOUS TRANSMITTANCE OF TRANSPARENT PLASTICS[1]

This Standard is issued under the fixed designation D 1003; the number immediately following the designation indicates the year of original adoption or, in the case of revision, the year of last revision. A number in parentheses indicates the year of last reapproval.

1. Scope

1.1 This method covers the measurement of the light-transmitting properties, and from these, the light-scattering properties, of planar sections of transparent plastics. Two procedures (Note 1) are covered as follows:

1.1.1 *Procedure A*. Using the Hazemeter.

1.1.2 *Procedure B*. Using the Recording Spectrophotometer.

NOTE 1—The method for measuring luminous transmittance in Procedure A is a simplified supplement, within stated limitations, to ASTM Method D 791, Test for Luminous Reflectance, Transmittance, and Color of Materials.[2] The corresponding method of Procedure B is a direct application of ASTM Method D 307, Test for Spectral Characteristics and Color of Objects and Materials.[2]

Results obtained should not be significantly different from those obtainable if the light source were CIE Source A or C (specified) and the receptor had a spectral response corresponding to the luminosity function $\bar{y}$ of the CIE standard observer.

1.2 Specimens with haze values greater than 30 percent obtained by these methods should be considered diffusing or translucent and should be tested in accordance with ASTM Recommended Practice E 166, for Goniophotometry of Transmitting Objects and Materials.[3]

NOTE 2—The values stated in U.S. customary units are to be regarded as the standard. The metric equivalents of U.S. customary units may be approximate.

2. Significance

2.1 The significance of the haze results obtained improves with the avoidance of heterogeneous surface and internal defects which can contribute to the diffusion or deviation of light. With this precaution and careful calibration of the apparatus, results may be obtained with a precision of ±0.1 percentage haze. Differences of this order are rarely detectable visually. In accordance with the definition as given, the accuracy of test results obtained by different operators and equipment is of the order of ±0.3 percentage haze. Although these results are empirical, they are of value for control purposes and may be related to more fundamental properties useful in research.

3. Definitions

3.1 *haze*—that percentage of transmitted light which in passing through the specimen deviates from the incident beam by forward scattering. For the purpose of this method only light flux deviating more than 2.5 deg on the average is considered to be haze.

3.2 *luminous transmittance* is defined in accordance with Method D 307 as the ratio of transmitted to incident light.

4. CIE Sources

4.1 CIE Sources A or C shall be provided by suitable filters if Procedure A is used, or by calculation if Procedure B is used. Use CIE Source C, unless there is reason to report the results for another CIE Source.

5. Test Specimen

5.1 The test specimen shall be large enough

[1] This method is under the jurisdiction of ASTM Committee D-20 on Plastics. A list of committee members may be found in the ASTM Yearbook. This standard is the direct responsibility of Subcommittee D-20.40 on Optical Properties.

Current edition effective Sept. 18, 1961. Originally issued 1949. Replaces D 1003 – 59 T.

[2] Discontinued, see *1967 Book of ASTM Standards*, Part 30.

[3] *Annual Book of ASTM Standards*, Part 30.

to cover a 2.5 by 2.5 cm (1 by 1-in.) aperture but small enough to be tangent to the sphere wall; a disk 3.5 cm (1 3/8 in.) in diameter is suggested. The specimen shall have substantially plane-parellel surfaces free of dust, grease, scratches, blemishes, and visibly distinct internal voids and particles, unless it is specifically desired to measure the contribution to haze due to these imperfections. Surface defects may be eliminated by immersing the specimen in a clear liquid of the same refractive index.

6. Conditioning

6.1 *Conditioning*—Condition the test specimens at 23 ± 2 C (73.4 ± 3.6 F) and 50 ± 5 percent relative humidity for not less than 40 h prior to test in accordance with Procedure A of ASTM Methods D 618, Conditioning Plastics and Electrical Insulating Materials for Testing,[4] for those tests where conditioning is required. In cases of disagreement, the tolerances shall be ±1 C (1.8 F) and ±2 percent relative humidity.

6.2 *Test Conditions*—Conduct tests in the Standard Laboratory Atmosphere of 23 ± 2 C (73.4 ± 3.6 F) and 50 ± 5 percent relative humidity, unless otherwise specified in the test methods or in this specification. In cases of disagreements, the tolerances shall be ±1 C (1.8 F) and ±2 percent relative humidity.

PROCEDURE A. USING THE HAZEMETER

7. Apparatus

7.1 The hazemeter[5] shall be constructed essentially as shown in Figs. 1 or 2 and shall conform to the requirements in 7.2 through 7.9.

7.2 An integrating sphere shall be used to collect transmitted flux; the sphere may be of any diameter so long as the total port area does not exceed 4.0 percent of the internal reflecting area of the sphere.

7.3 Figures 1 and 2 indicate possible arrangements of the apparatus. These contrast with that of Procedure B in that specimen and standard beams are the same beam. The entrance and exit ports shall be centered on the same great circle of the sphere and there shall be at least 170 deg of arc between centers. The exit port shall subtend an angle of 8 deg at the center of the entrance port. The axis of the irradiating beam shall pass through the centers of the entrance and exit ports. The photocell or photocells shall be positioned on the sphere 90 ± 10 deg from the entrance port. In the pivotable modification of this type (Fig. 2) designed to use the interior sphere wall adjacent to the exit port as the reflectance standard, the angle of rotation shall not exceed 10 deg.

7.4 The specimen shall be illuminated by a substantially unidirectional beam; the maximum angle which any ray of this beam makes with the direction of its axis shall not exceed 3 deg. This beam shall not be vignetted at either port of the sphere.

7.5 When the specimen is placed immediately against the integrating sphere at the entrance port, the angle between the normal to its surface and the axis of the beam shall not exceed 8 deg.

7.6 When the beam is unobstructed by a specimen, its cross-section at the exit port shall be approximately circular, sharply defined and concentric within the exit port, leaving an annulus of 1.3 ± 0.1 deg subtended at the entrance port.

7.7 The surfaces of the interior of the integrating sphere, baffles, and reflectance standards shall be of substantially equal reflectance, matte, and highly reflecting throughout the visible wave lengths.[6]

7.8 For some measurements the standard at the exit port is replaced by a light trap by actual removal of the reflectance standard or by pivoting the sphere (Fig. 2). The light trap shall absorb the beam completely when no specimen is present.[7]

7.9 The radiant flux within the sphere shall be measured by a photoelectric cell, the output measurements of which shall be propor-

[4] *Annual Book of ASTM Standards*, Part 27.

[5] The Pivotable-Sphere Hazemeter made by the H. A. Gardner Laboratories, Bethesda, Md., meets these requirements.

[6] Freshly smoked magnesium oxide is excellent for this purpose but highly reflecting matte sphere paints are more durable. See ASTM Recommended Practice E 259, for Preparation of Reference White Reflectance Standards. *Annual Book of ASTM Standards*, Part 30.

[7] Due to the absorbing annulus surrounding the unimpeded beam at the exit port, this trap will absorb slightly more than the undeviated portion of the total flux transmitted by the specimen.

tional within 1 percent to the incident flux over the range of intensity used. Spectral conditions for source and receiver must be constant throughout the test of each specimen. The design of the instrument shall be such that there shall be no galvanometer deflection when the sphere is dark.

8. Procedure

8.1 Determine the following four readings:

Reading Designation	Specimen in Position	Light Trap in Position	Reflectance Standard in Position	Quantity Represented
T_1	no	no	yes	incident light
T_2	yes	no	yes	total light transmitted by specimen
T_3	no	yes	no	light scattered by instrument
T_4	yes	yes	no	light scattered by instrument and specimen

8.2 Repeat readings for T_1, T_2, T_3 and T_4 with additional specified positions of the specimen to determine uniformity.

9. Calculation[8]

9.1 Calculate total transmittance, T_t (Note 3), equal to T_2/T_1.

9.2 Calculate diffuse transmittance T_d (Note 3), as follows:

$$T_d = [T_4 - T_3 (T_2/T_1)]/T_1$$

9.3 Calculate percentage haze as follows:

$$\text{Haze, percent} = T_d/T_t \times 100$$

NOTE 3—To obtain luminous transmittance values (total, T_t; diffuse, T_d) by this procedure, a proper source filter-photocell system must be employed. A Western Type III photocell, with Viscor Filter, approximates the response of the standard observer except at the extremes of the visible spectrum. The source may be operated at a color temperature approximating 2854 K (CIE Source A) by proper selection of voltage. CIE Source C may be achieved with CIE Source A and a filter described by Davis, R., and Gibson, K. S., *Miscellaneous Publication No. 114*, Nat. Bureau Standards, January 1931. An approximation of CIE Source C may be obtained with the lamp operating at 3100 K plus a Corning Type 5900 filter of standard thickness.

Commercially produced hazemeters[5] are equipped to provide luminous transmittance measurements, using either CIE Source, which closely duplicate those obtained by Method D 791.

To obtain greatest accuracy in luminous transmittance measurements, especially for dark or highly saturated colors, the photometer should be used as a comparison instrument with a standard of known transmittance similar to that of the specimen.

PROCEDURE B. USING THE RECORDING SPECTROPHOTOMETER

10. Apparatus

10.1 The recording spectrophotometer shall conform to the requirements in 10.2 through 10.10:

10.2 An integrating sphere shall be used to collect transmitted flux. The sphere may be of any diameter so long as the total area of the ports does not exceed 4.0 percent of the internal reflecting area of the sphere.

10.3 Emitted by one source, specimen and standard beams shall be deviated by a small angle so that they enter the sphere as two distinct beams through different entrance ports. These ports shall be located on the surface of the sphere, on a horizontal line and equidistant from a line from point of deviation to the center of the sphere. The angle of deviation and the planes of the exit ports shall be oriented so that the unobstructed beams at these ports are approximately circular.

10.4 The test specimen shall be illuminated by a substantially unidirectional beam; the maximum angle which any element of this beam makes with the direction of its axis shall not exceed 3 deg. This beam shall not be vignetted at either port of the sphere.

10.5 When the test specimen is placed immediately against the integrating sphere at the specimen entrance port, the angle between the normal to its surface and the axis of the sample beam shall not exceed 8 deg.

10.6 When the specimen beam is unobstructed by a specimen, its cross-section at the exit port must be approximately circular, sharply defined, and concentric within the port, leaving an annulus of 1.3 ± 0.1 deg.

10.7 The surfaces of the interior of the integrating sphere, baffles, and reflectance standards shall be substantially equal in reflectance, matte and highly reflecting throughout the visible wavelengths.[7]

10.8 When diffuse luminous transmittance T_d is determined, the standard at the speci-

[8] See Appendix for derivation of formulas.

men exit port is replaced by a light trap which shall absorb the specimen beam when unobstructed by a specimen.[8]

10.9 The radiant flux within the sphere shall be measured by a photoelectric cell as the flux transmitted by either of the following two acceptable methods:

10.9.1 The flux may be transmitted by a highly diffusing translucent window of opal glass so located in the sphere that it becomes a continuous portion of its interior surface. The photocell port is located at the lower pole of the vertical diameter. If the flux is of such low magnitude that insufficient sensitivity is obtained, improvement will result if the opal glass window is removed. If this is done, a baffle of similar translucency, for example, high quality bond paper, must be located on the interior surface of the sphere between photocell and exit port, to prevent direct reflection from the reflectance standards to the photoelectric cell.

10.9.2 The flux may be transmitted by a clear acrylic plastic rod extending up from the bottom of the sphere to just above the center between the two beams. This pipes the flux down to the phototube. It also prevents direct reflection from the reflectance standards to the photoelectric cell.

10.10 Values of spectral diffuse transmittance may be quite low; they must be recorded and integrated carefully. Improved accuracy will result from the use of a mechanically introduced 5× factor, applicable for curves that are less than 20 percent at all wavelengths. If a multiplying factor is not used, improved accuracy can be obtained by employing a correlation[9] between indicated photometric readings and transmittance values in this range. The operating instructions of the instrument manufacturer must be followed closely and particular attention must be given to the zero adjustment of the recording system.

NOTE 4—These requirements are met by a recording spectrophotometer of the Hardy type.[10,11] There are two different types of integrating spheres in general use with this instrument, as illustrated in Figs. 3 and 4. The principal difference is one of geometry. In the early or "diffuse (reflectance)" type the unobstructed light beams meet the reflectance standards normal to their surfaces, as shown in Fig. 3. In the later or "specular (reflectance)" type the axes of these beams meet the reflectance standard at an angle of 6 deg from the normal, as shown in Fig. 4.

11. Procedure

11.1 Check the wavelength and photometric scales of the instrument.[11,12]

NOTE 5—One method of making wavelength corrections is by running the curve of a calibrated didymium glass. On the General Electric Hardy type instrument the photometric scale can be checked by running 100 percent and zero curves and if desirable by running transmittance curves of standard glasses or solutions.

11.2 Determine whether or not specimens are birefringent by rotating them between crossed Polaroids or other polarizer and analyzer. Transmittance change with rotation indicates birefringence. Measure birefringent specimens with polarized light in such a manner that the results obtained are valid for unpolarized light.[13]

11.3 Place the sample directly in contact with the entrance port of the sphere for transmittance measurements.

NOTE 6—Direct contact with the entrance port is necessary if all of the scattered light is to enter

[9] Gibson, K. S., and Keegan, H. J., *Journal*, JOSAA, Optical Soc. of Am., Vol 28, 1938, pp.372 to 385.

[10] This type of instrument, manufactured by the General Electric Co., Schenectady, N. Y. exists in three models, Catalog Nos. 4980151, 5962004, and 7015E30. The model represented by Catalog No. 7015E30 may not meet the requirement given in 10.6. The manufacturer should be contacted for information. Construction details are given in the following publications:

Michaelson, J. L., "Construction of the General Electric Spectrophotometer," *Journal*, JOSAA, Optical Soc. of Am., Vol 28, 1938, p. 365.

Pritchard, B. S., and Holmwood, W. A., "New Recording Spectrophotometer," *Journal*, JOSAA, Optical Soc. of Am., Vol 45, 1955, p. 690.

[11] Pertinent directions for the operation and calibration of this instrument are given in the following publications:

Hardy, A. C., *Journal*, JOSAA, Optical Soc. of Am., Vol 28, No. 10, pp. 360 to 364, October 1938.

U.S. Patent No. 2,107,836; U.S. Patent No. 2,126,410; U.S. Patent No. 2,300,695.

Gibson, K. S., and Keegan, H. J., *Journal*, JOSAA, Optical Soc. of Am., Vol 28, 1938, pp. 372 to 385.

Pineo, O. W., *Journal*, JOSAA, Optical Soc. of Am., Vol 30, No. 7, July 1940, pp. 276 to 289.

[12] "Standards for Checking the Calibration of Spectrophotometers," *Letter Circular LC929*-Nat. Bureau Standards (sent free on request), contains instructions for obtaining standard materials for checking the wavelength and photometric scales of spectrophotometers.

[13] Buc, G. L., and Stearns, E. I., "Uses of Retardation Plates in Spectrophotometry—III. Measurement of Dichroic Samples," *Journal*, JOSAA, Optical Soc. of Am. Vol 35, August 1945, p. 521. This is a discussion of retardation plate modification applied to Hardy type spectrophotometers for the measurement of dichroic samples, whereby the results are valid and duplicate those which might be obtained using unpolarized light.

the sphere. To accomplish this with spheres having a connecting sleeve between sphere and sample compartment, a small specimen may be used, or a modification may be made in the instrument. One modification consists of removing a portion of the connecting sleeves on the specimen side, between the integrating sphere and the sample compartment. A portion of the specular cup and associated sphere sleeve must be removed to permit proper positioning of the specimen.

11.4 Record the following four curves:

Curve Designation	Specimen in Position	Light Trap in Position	Reflectance Standard in Position	Quantity Represented
C_1	no	no	yes	incident light
C_2	yes	no	yes	total light transmitted by specimen
C_3	no	yes	no	light scattered by instrument
C_4	yes	yes	no	light scattered by instrument and specimen

11.5 Repeat curves C_2 and C_4 above with different positions of specimen to determine uniformity. Recommended additional positions are:

11.5.1 Specimen rotated 90 deg in plane parallel to its surfaces,

11.5.2 Specimen reversed so that light beam is incident upon surface which was next to sphere, and

11.5.3 Specimen rotated 90 deg from position (11.5.2) in plane parallel to its surfaces.

12. Calculations[9]

12.1 Obtain the values of T_1, T_2, T_3, and T_4 by either of the following two acceptable methods:

12.1.1 For the recorded curves, compute T_1, T_2, T_3, and T_4 by integration for CIE Sources A or C and the luminosity function $\bar{y}$ (Method D 307). Values of $\bar{y}H$ required for this calculation may be computed from Tables 1 and 3 of Method D 307. A sufficient number of ordinates shall be used so that increasing their number produces a difference not exceeding 0.1 percent haze. An automatic digital read-out device,[14] connected directly to the recording spectrophotometer, may be used to obtain transmittances at stated wavelength intervals, eliminating the necessity of obtaining these values from the recorded curves.

12.1.2 Obtain the values T_1, T_2, T_3, and T_4 by the use of an automatic integrator[15] which is connected directly to the spectrophotometer, and which calculates the values accurately for the specified CIE Source.

12.2 Calculate total luminous transmittance, T_t, equal to T_2/T_1.

12.3 Calculate diffuse luminous transmittance as follows:

$$T_d = [T_4 - T_3(T_2/T_1)]/T_1$$

12.4 Calculate percentage haze as follows:

$$\text{Haze, percent} = T_d/T_t \times 100$$

13. Report

13.1 The report shall include the following data:

13.1.1 Source and identity of specimen,

13.1.2 Thickness of specimen in inches, to the nearest 0.025 mm (0.001 in.),

13.1.3 Total luminous transmittance T_t to the nearest 0.001 (indicate "average" when reporting average values and specify whether CIE Source A or C is used),

13.1.4 Diffuse luminous transmittance T_d to the nearest 0.001 (indicate "average" when reporting average values and specify whether CIE Source A or C is used),

13.1.5 Percentage haze, to nearest 0.1 percent (indicate "average" when reporting average values), and

13.1.6 Procedure used.

---

[14] A suitable automatic digital read-out device is manufactured by Librascope, Inc., Glendale, Calif.

[15] The Librascope Automatic Tristimulus Integrator, distributed by the General Electric Co., Schenectady, N.Y., Catalog No. 5469100, meets the requirements of this method. The principles of operation of this device are given in the following publication:

Davidson, H. R., and Imm, L. W., "Continuous Automatic Tristimulus Integrator for Use with the Recording Spectrophotometer," *Journal*, JOSAA, Optical Soc. of Am., Vol 39, 1949, p. 942.

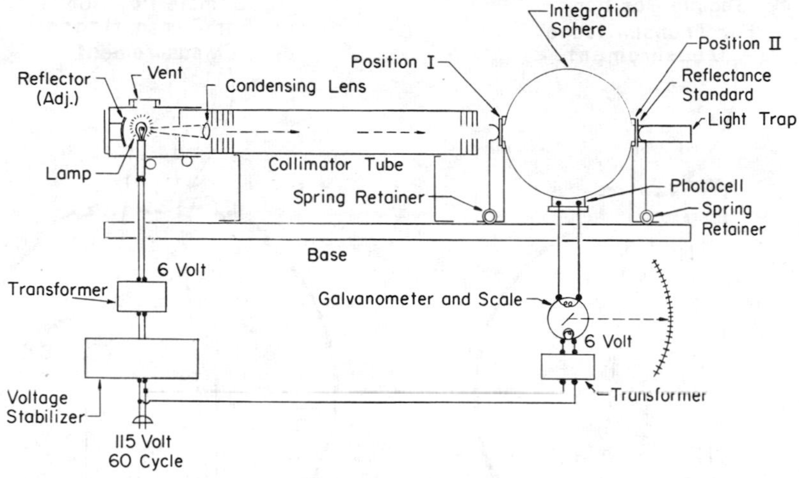

FIG. 1 Hazemeter.

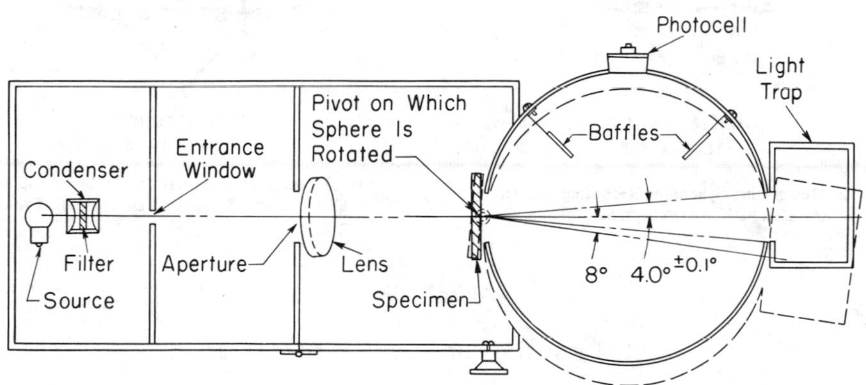

FIG. 2 Pivotable-Sphere Hazemeter.

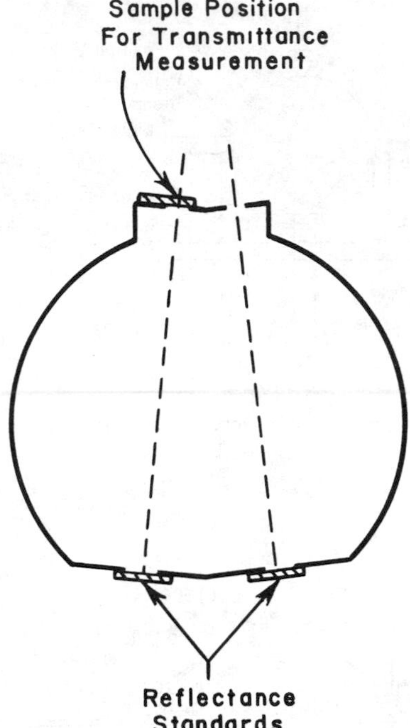

FIG. 3 Integrating Sphere for Recording Spectrophotometer, "Diffuse (Reflectance)" Type.

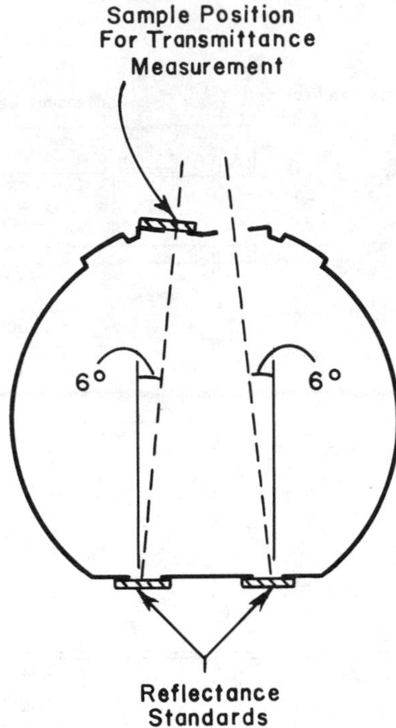

FIG. 4 Integrating Sphere for Recording Spectrophotometer, "Specular (Reflectance)" Type.

APPENDIX

A1. Derivation of Formulas for Haze

A1.1 The deviation of the formula for haze for both procedures is as follows:

A1.1.1 Total luminous transmittance, T_t, is calculated as follows:

$$T_t = T_2/T_1$$

where:

T_2 = total light transmitted by the specimen, and

T_1 = incident light.

A1.1.2 If T_3, the light scattered by the instrument, is zero, the diffuse luminous transmittance, T_d, is calculated as follows:

$$T_d = T_4/T_1$$

where:

T_4 = light scattered by the instrument and specimen.

A1.1.3 If T_3 is greater than zero due to light scattered by the instrument, the total scattered light, T_4, will be greater than the light scattered by the specimen by an amount proportional to T_3 and equal to T_3 times T_2/T_1. The corrected amount of light scattered by the specimen will then be:

$$T_4 - T_3\,(T_2/T_1)$$

A1.1.4 The diffuse luminous transmittance, T_d, is then calculated as follows:

$$T_d = [T_4 - T_3(T_2/T_1)/T_1$$

A1.1.5 Percentage haze is then calculated from the ratio of diffuse, T_d, to total luminous transmittance, T_t, as follows:

$$\text{Haze, percent} = T_d/T_t \times 100$$
$$= \{[T_4 - T_3(T_2/T_1)]/T_1 \div (T_2/T_1)\} \times 100$$
$$= [T_4 - T_3(T_2/T_1)]/T_2 \times 100$$
$$= [(T_4/T_2) - (T_3/T_1)] \times 100$$

Designation: D 1004 – 66

Standard Method of Test for
TEAR RESISTANCE OF PLASTIC FILM AND SHEETING[1]

This Standard is issued under the fixed designation D 1004; the number immediately following the designation indicates the year of original adoption or, in the case of revision, the year of last revision. A number in parentheses indicates the year of last reapproval.

This specification was prepared jointly by the American Society for Testing and Materials and The Society of the Plastics Industry.

NOTE—Section 5 was changed editorially in December 1968.

1. Scope

1.1 This method[2] covers the determination of the tear resistance of flexible plastic film and sheeting at very low rates of loading, 51 mm (2 in.)/min. The test is designed to measure the force to initiate tearing. The specimen geometry of this method produces a stress concentration in a small area of the specimen. The maximum stress, usually found near the outset of tearing, is recorded as the tear resistance in kilograms-force (pounds.)

NOTE 1—The values stated in U.S. customary units are to be regarded as the standard. The metric equivalents of U.S. customary units may be approximate.

2. Significance

2.1 Tear resistance of plastic film or sheeting is a complex function of its ultimate resistance to rupture. The specimen geometry and speed of testing in this method are controlled to produce tearing in a small area of stress concentration at rates far below those usually encountered in service.

2.2 The data of this method furnish comparative information for ranking the tearing resistance of plastic specimens of similar composition. Actual use performance in tearing of some plastics may not necessarily correlate with data from this method.

2.3 The resistance to tear of plastic film and sheeting, while partly dependent upon thickness, has no simple correlation with specimen thickness. Hence, tearing forces measured in kilograms-force (pounds) cannot be normalized over a wide range of specimen thickness without producing misleading data as to the actual tearing resistance of the material. Data from this method are comparable only from specimens which vary by no more than ±10 percent from the nominal or average thickness of all specimens tested. Therefore, the tearing resistance is expressed in maximum kilograms-force (pounds) of force to tear the specimen.

3. Apparatus

3.1 *Testing Machines*—A power driven machine of either of the two following types shall be used:

3.1.1 *Static Weighing—Constant Rate of Grip Separation Type*—Negligible movement of the upper jaw.

3.1.2 *Pendulum Weighing—Constant Rate of Powered Grip Motion Type*—Constant rate of lower jaw movement, variable upper jaw movement (Note 2). The machine shall conform to requirements given in the appropriate Method A or B of ASTM Methods D 882, Test for Tensile Properties of Thin Plastic Sheeting,[3] except that cross-head rates of other than 51 mm (2 in.)/min are not re-

---

[1] This method is under the jurisdiction of ASTM Committee D-20 on Plastics. A list of committee members may be found in the ASTM Yearbook. This standard is the direct responsibility of Subcommittee D-20.10 on Mechanical Properties.

Current edition effective March 31, 1966. Originally issued 1949. Replaces D 1004 – 61.

[2] The following reference may be of interest in connection with this method: Graves, F. L., "The Evaluation of Tear Resistance in Elastomers," *India Rubber World*, Vol 111, No. 3, December, 1944, pp. 305–308.

[3] *Annual Book of ASTM Standards*, Part 27.

quired. Either maximum load indicating devices or recorders are permissible in the testing machine. The accuracy of the testing machine shall be verified in accordance with ASTM Methods E 4, Verification of Testing Machines[3] (Note 3). The applied load, as indicated by a recorder, dial or scale, shall be accurate to within ±2 percent. If an indicating device is used, the indicator shall remain at the point of maximum load after rupture of the test specimen.

NOTE 2—If both jaws of the machine are power activated the rate of jaw separation shall be nominally 51 mm (2 in.)/min.

NOTE 3—Experience has shown that many testing machines now in use are incapable of maintaining accuracy for as long as the periods between inspection recommended in Methods E 4. Hence, it is recommended that each machine be studied individually and verified as often as may be found necessary. It will frequently be necessary to perform this function daily.

3.2 *Grips*—A gripping system that minimizes both slippage and uneven stress distribution on the specimen shall be used.

NOTE 4—Grips lined with thin rubber, No. 80 emery cloth, or crocus cloth have been successfully used. Grips may be of the self tightening type. In cases where specimens frequently fail at the edge of the grips, the radius of curvature of the edges of the grips may be increased slightly at the point where they come in contact with the specimen.

3.3 *Thickness Measuring Devices*—Suitable micrometers, or thickness gages, reading to 0.0025 mm (0.0001 in.) or less shall be used for measuring the thickness of the specimens. The pressure exerted by the gage on the specimen being measured shall be between 0.159 to 0.186 MN/m^2 (23 and 27 psi) as prescribed in Method C of ASTM Methods D 374, Test for Thickness of Solid Electrical Insulation.[3]

3.4 *Die*[4]—A die having the dimensions shown in Fig. 1 shall be used to cut all specimens. The 90-deg angle should be honed sharp with no radius or have a minimum practical radius. The cutting edge of the die shall have a 5-deg negative rake, and shall be kept sharp and free from nicks to avoid leaving ragged edges on the specimen. Cutting may be facilitated by wetting the surface of the sample and the cutting edges of the die with water. The sample shall rest on the smooth, slightly yielding surface that will not injure the die blade. Light weight cardboard or a piece of leather belting is suitable. Care should be taken that the cut edges of the specimen are perpendicular to its other surfaces and that the edges have a minimum of concavity.

3.5 *Conditioning Apparatus*—Apparatus for maintaining laboratory atmospheric conditions of 23 ± 1 C (73.4 ± 1.8 F) and 50 ± 2 percent RH for conditioning prior to and during testing shall be used as defined in Procedure A of ASTM Methods D 618, Conditioning Plastics and Electrical Insulating Materials for Testing,[3] unless otherwise specified.

4. Test Specimens

4.1 The test specimens shall conform to the dimensions shown in Fig. 1 and shall not vary by more than 0.5 percent from these dimensions.

4.2 Single specimens shall be used. Where orientation effects are significant and are to be evaluated, duplicate sets of specimens shall be prepared so that their direction of tear is respectively parallel with and normal to the direction of maximum orientation. The direction of maximum orientation may be established for transparent materials by examination of the film or sheeting between crossed Polaroids.

4.3 At least 10 specimens shall be tested for each sample, in the case of isotropic materials.

4.4 Twenty specimens, 10 parallel with and 10 normal to the principal axis of anisotropic or maximum orientation, shall be tested for each sample, in the case of anisotropic material.

4.5 Data from specimens which break at some obvious flaw or which break in or at the edges of the grips shall be discarded and retests made, unless such failures constitute a variable whose effect is being studied.

4.6 Data from specimens which deviate markedly from the mean value of all tests shall be rejected if the deviation of the doubtful value is more than 5 times the standard deviation from the mean value obtained by excluding the doubtful results.

NOTE 5—For certain materials whose properties vary considerably throughout film or sheeting, as

[4] Detail drawings of the die are available at a nominal cost from the American Society for Testing and Materials, 1916 Race St., Philadelphia Pa. 19103.

many as 50 specimens cut from random portions of the sheet must be tested if reliable tear resistance data is desired.

5. Conditioning

5.1 *Conditioning*—Condition the test specimens at 23 ± 2 C (73.4 ± 3.6 F) and 50 ± 5 percent relative humidity for not less than 40 h prior to test in accordance with Procedure A of ASTM Methods D 618, Conditioning Plastics and Electrical Insulating Materials for Testing,[3] for those tests where conditioning is required. In cases of disagreement, the tolerances shall be ±1 C (±1.8 F) and ±2 percent rclativc humidity.

5.2 *Test Conditions*—Conduct tests in the Standard Laboratory Atmosphere of 23 ± 2 C (73.4 ± 3.6 F) and 50 ± 5 percent relative humidity, unless otherwise specified in the test methods or in this specification. In cases of disagreements, the tolerances shall be ±1 C (±1.8 F) and ±2 percent relative humidity.

6. Speed of Testing

6.1 A jaw separation of 25.4 mm (1 in.) shall be used. The rate of travel of the power activated grip shall be 51 mm (2 in.)/min and shall be uniform at all times.

NOTE 6—In this method, resistance to tear is calculated from the maximum load recorded or indicated by the testing machine. In testing most plastics, this maximum load is generated at the outset of tearing across the 13 mm (0.5 in.) testing width of the specimen. Hence, the accuracy of the initial advance of the chart pen or dial from its origin is of utmost importance. The testing speed of 51 mm (2 in.)/min is sufficiently slow to make it virtually certain that the recorder pen or dial response is not exceeded. The response of the testing machine recorder pen shall be checked prior to testing specimens.

7. Procedure

7.1 Measure the width and thickness of the specimen at several points to the accuracy limits of the measuring devices specified in 3.4. Record the average thickness in inches.

7.2 Place the specimen in the grips of the testing machine so that the long axis of the enlarged ends of the specimen is in line with the points of attachment of the grips to the machine.

7.3 Apply the load at 51 mm (2 in.)/min rate of grip separation. After complete rupture of the specimen, the maximum tearing load in pounds shall be noted from the dial scale or recorder chart and recorded.

8. Calculation

8.1 The average resistance to tearing shall be calculated from all specimens tested in each principal direction of orientation. Data shall be recorded as pounds of tearing resistance.

NOTE 7—Resistance to tear may be expressed in pounds per mil of specimen thickness where correlation for the particular material being tested has been established. However, it should be realized that comparisons between films of dissimilar thickness may not be valid.

8.2 Standard deviation (estimated) shall be calculated as follows, or, by any mathematically equivalent expression. It shall be reported to two significant figures.

$$s = \sqrt{(\sum X^2 - n\bar{X}^9)/(n - 1)}$$

where:

s = estimated standard deviation,
X = value of a single observation,
n = number of observations, and
$\bar{X}$ = arithmetic mean of the set of observations.

9. Report

9.1 The report shall include the following:

9.1.1 Complete identification of the material tested, including type, source, manufacturer's code number, form, principal dimensions, previous history, and orientation of sample with respect to anisotropy, if any.

9.1.2 Average thickness of each test specimen and average thickness of all test specimens.

9.1.3 Type of testing machine used.

9.1.4 Number of specimens tested in each principal direction.

9.1.5 Average value of test strength calculated in kilograms-force (pounds).

9.1.6 The standard deviation from the averaged values obtained for specimens tested in each principal direction.

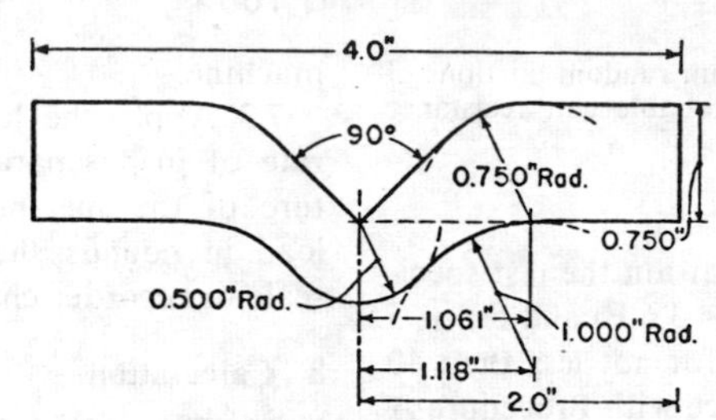

Tolerance ± 0.002″

Table of Metric Equivalents

in.	mm
4.0	101.60
0.750	19.05
1.061	26.95
1.000	25.40
1.118	28.40
2.0	50.80
0.002	0.051
0.500	12.70

FIG. 1 Die for Tear Test Specimen.

Designation: D 1009 – 55 (Reapproved 1961)

Recommended Code for

DESIGNATING FORM OF MATERIAL AND DIRECTION OF TESTING PLASTICS[1]

This Standard is issued under the fixed designation D 1009; the number immediately following the designation indicates the year of original adoption or, in the case of revision, the year of last revision. A number in parentheses indicates the year of last reapproval.

1. Scope

1.1 This code is recommended for use where the same plastic may be supplied in more than one form and where test values may depend upon the form of the material and the direction in which the plastic is tested. Its purpose is to make it possible, in reporting various values, to designate the form and the direction of testing with maximum brevity. Provision for the many forms in which plastics are supplied and the many directions in which they are tested of necessity makes the code more complex than otherwise would be desired. To be useful, codes should be simple and easily remembered. Ease of remembrance has been sought in this code through the use of initials and no more than a few simple rules of structure.

2. Form of Material

2.1 The letters shown in Table 1 are recommended for use in designating the form of material from which specimens are selected for testing.

3. Direction of Testing

3.1 One or more of the following letters, or letter combinations as may be necessary, should be added to those in Table 1, to designate the direction in which stress was applied to the test specimen. These letters may be used alone if there is no need to identify the form of material.

Letter[a]	Stress Applied in Direction of
—c	calendering
—d	diameter
—e	extrusion
—g	grain
—l	length
—pr	forming pressure
—s	slicing or sheeting
—ta	tangentially
—th	thickness
—w	width

[a] Replace the dash "—" with an "*x*" to indicate directions perpendicular to those listed.

4. Direction of Cutting Test Specimens

4.1 Where it is desirable to designate the direction in which the test specimen is cut from the material, enclose in parentheses the letter for direction of testing (Section 3).

NOTE—Applications of the above code are illustrated in the following examples:

CsS—th Cast sheet tested flatwise or in direction of thickness.

LSxth (l) Laminated sheet tested edgewise or perpendicular to direction of thickness, specimen cut lengthwise of sheet.

ClS(b)xth(xc) Composite specimen of calendered

[1] This recommended code is under the jurisdiction of ASTM Committee D-20 on Plastics. A list of committee members may be found in the ASTM Yearbook. This standard is the direct responsibility of Subcommittee D-20.04 on Definitions.

Current edition effective Sept. 12, 1955. Originally issued 1949. Replaces D 1009 – 49 T.

sheet tested edgewise or perpendicular to direction of thickness, specimen cut perpendicular to direction of calendering.

ClS(b)—cxth Composite specimen of calendered sheet tested in direction of calendering but edgewise or perpendicular to direction of thickness.

CM(m)—pr Specimen machined from a compression molded plate or block and tested in direction in which molding pressure was applied.

RTxthxl Specimen from wall of rolled tube tested edgewise or in direction perpendicular to its thickness and in direction perpendicular to length of tube. (To so use the code depends upon recognizing that tubes have thickness only in their walls.)

TABLE 1 Letters for Designating Form of Plastics

Letter	Form	Letter for Designating Form	
ClS	Calendered Sheet	Replace the letter *S* with *F* and *P*, respectively, to indicate film and plate.	Add (*b*) to indicate built-up specimens composed of more than one thinner specimen.
CsS	Cast Sheet		
LS	Laminated Sheet		
MS	Molded Sheet		
PrS	Pressed Sheet		
SS	Sliced Sheet		
US	Unidentified Sheet		
ER	Extruded Rod	Replace the letter *R* with *T* to indicate tubing.	
MR	Molded Rod		
UR	Unidentified Rod		
TR	Turned Rod		
RT	Rolled Tube		
C	Cast Specimen	Add (*m*) to indicate specimens machined from larger castings or moldings.	
CM	Compression Molded Specimen		
IM	Injection Molded Specimen		
JM	Jet Molded Specimen		
TM	Transfer Molded Specimen		
UM	Unidentified Molded Specimen		

Standard Method for

MEASURING CHANGES IN LINEAR DIMENSIONS OF PLASTICS[1]

This Standard is issued under the fixed designation D 1042; the number immediately following the designation indicates the year of original adoption or, in the case of revision, the year of last revision. A number in parentheses indicates the year of last reapproval.

1. Scope

1.1 This method is designed to provide a means for measuring in plastic specimens the dimensional changes resulting from exposure to service conditions. In particular, the method is suitable for measuring shrinkage or elongation developed in accordance with the procedures described in ASTM Methods D 756, Test for Resistance of Plastics to Accelerated Service Conditioning.[2]

NOTE—The values stated in U.S. customary units are to be regarded as the standard. The metric equivalent of U.S. customary units may be approximate.

2. Significance

2.1 This procedure is intended only as a convenient method for measurement of linear dimensional changes in plastics subjected to defined conditions of test as outlined in Methods D 756, and when all precautions are observed measurements are reproducible to ± 0.02 percent.

3. Apparatus

3.1 *Scriber*, so constructed that two sharp needle points are rigidly separated by 100 ± 0.2 mm. A very satisfactory scriber, as shown in Fig. 1, consists of two sharp steel phonograph needles (approximately 1.5 mm in diameter) inserted in holes drilled with their axes parallel to each other and perpendicular to and intersecting the long axis of a stainless steel rod, 7.9 mm ($^5/_{16}$ in.) in diameter by 125 mm (5 in.) in length. The needle points extend 6 mm beyond the supporting rod and are held in position by two set screws inserted through the ends of the rod. A stainless steel gage with reference points consisting of a center and two short concentric arcs (R_1 = 99.80 ± 0.02 mm, and R_2 = 100.20 ± 0.02 mm) is recommended for calibration of the scriber. Thickness of arc lines should not exceed 0.02 mm.

3.2 *Measuring Microscope*, graduated to read to 0.1 mm or better. (A Brinell microscope 20×, for measuring Brinell hardness, is very satisfactory). For more precise measurements, a micrometer microscope may be used.

4. Test Specimens

4.1 The test specimens shall be similar to those prescribed in Methods D 756 with the additional requirement that in the direction of test the length should be not less than 110 mm. In practice, specimens 150 by 25 mm and of the full thickness of the material are desirable. These specimens may be suspended individually in the specified environment by wire hooks inserted in a hole punched in one end of the specimen. When one type of plastic is to be compared with another type, a standard thickness of 3.2 mm (0.125 in.) is preferred.

5. Procedure

5.1 Immediately following the preconditioning period of the test, scribe an arc of 100-mm radius on the surface of the test specimen. Press one needle firmly into the specimen to form a center for this and subsequent measurements. The other needle scribes the

[1] This method is under the jurisdiction of ASTM Committee D-20 on Plastics. A list of committee members may be found in the ASTM Yearbook. This standard is the direct responsibility of Subcommittee D-20.50 on Permanence Properties.

Current edition effective Sept. 30, 1951. Originally issued 1949. Replaces D 1042 – 49 T.

[2] *Annual Book of ASTM Standards*, Part 27.

arc which is used as a reference for all subsequent measurements (see Fig. 2). Draw the arcs smoothly, using a pressure consistent with the surface hardness and test conditions to which the specimen is subjected. It is desirable to depress rather than scratch or tear the surface with the needle, although harder surfaced materials and those subjected to extreme test conditions may require deeper scratching of the surface.

5.2 At each phase of the test where it is desirable to measure linear changes, reinsert one needle in the original center and draw a short arc with the other. Measure the distance between the original arc and the new arc with the microscope. Measure the separation of the arcs between corresponding positions, for example, center to center.

5.3 If the test specimen is not flat, flatten it by pressing or clamping it against a plane surface before scribing the arcs and making the measurements. In no case shall the specimen be clamped or otherwise confined during the period of exposure to accelerated service conditions.

6. Calculation and Precision

6.1 Measure the distance between arcs to 0.1 mm and estimate to the nearest 0.01 mm (0.01 mm corresponds to a linear change of 0.01 percent). Measurements should be made preferably at constant temperature (±1 C) and relative humidity (±2 percent) and under these conditions measurement is reproducible to ±0.02 mm or 0.02 percent. Measurements made on materials not moisture sensitive at other than constant temperature, should be corrected for differential thermal expansion by the following amount:

$$C = (\alpha_p - \alpha_s)(T_f - T_i) \times 100$$

where:

C = correction to add to calculated percent shrinkage or subtract from calculated percent expansion,

α_p = coefficient of linear thermal expansion of the plastic specimen,

α_s = coefficient of linear thermal expansion of the steel,

T_f = temperature when final arc is scribed, and

T_i = temperature when initial arc is scribed.

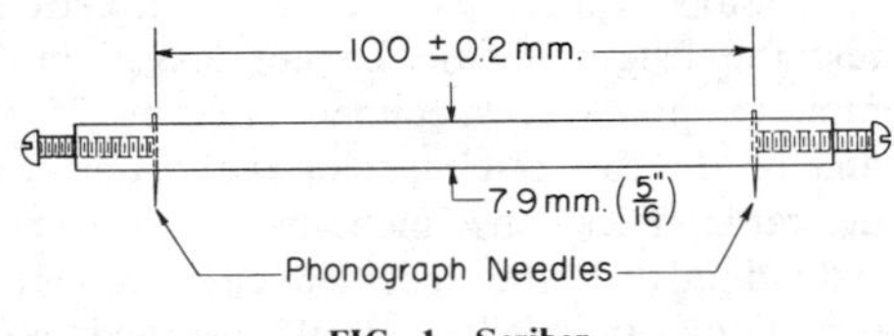

FIG. 1 Scriber.

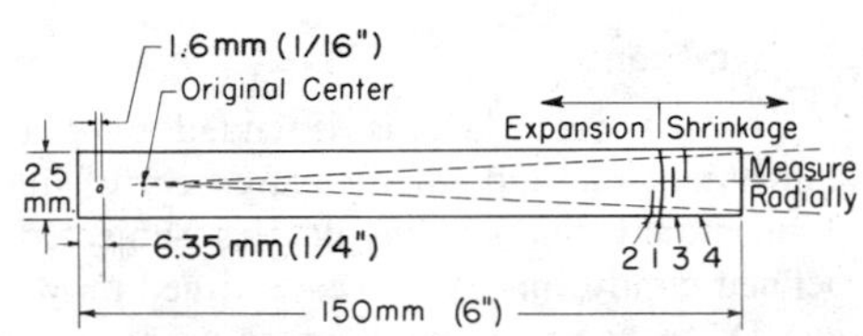

1—Original arc
2—After 37.7 C (100 F) and 100 percent humidity
3—After 60 C (140 F)
4—Final conditioning

NOTE—The arcs in this specimen illustrate the steps to use for Procedure VI of Method D 756.

FIG. 2 Scribing Test Specimens for Measurement.

Designation: D 1043 – 69

American National Standard K65.2-1970
American National Standards Institute

Standard Method of Test for

STIFFNESS PROPERTIES OF PLASTICS AS A FUNCTION OF TEMPERATURE BY MEANS OF A TORSION TEST[1]

This Standard is issued under the fixed designation D 1043; the number immediately following the designation indicates the year of original adoption or, in the case of revision, the year of last revision. A number in parentheses indicates the year of last reapproval.

1. Scope

1.1 This method covers the determination of the stiffness characteristics of plastics over a wide temperature range by direct measurement of the apparent modulus of rigidity.

NOTE 1—The values stated in U.S. customary units are to be regarded as the standard. The metric equivalents of U.S. customary units may be approximate.

2. Significance

2.1 The property actually measured by this method is the apparent modulus of rigidity, *G*, sometimes called the apparent *shear* modulus of elasticity. It is important to note that this property is not the same as the modulus of elasticity, *E*, measured in tension, flexure, or compression. The relationship between these properties is shown in the Appendix.

2.2 The measured modulus of rigidity is termed "apparent" since it is the value obtained by measuring the angular deflection occurring when the specimen is subjected to an applied torque. Since the specimen may be deflected beyond its elastic limit, the calculated value may not represent the true modulus of rigidity within the elastic limit of the material. In addition, the value obtained by this method will also be affected by the creep characteristics of the material, since the time of load application is arbitrarily fixed.

2.3 This method is useful for determining the relative changes in stiffness over a wide range of temperatures.

3. Apparatus

3.1 *Testing Machine*—A machine capable of exerting a torque sufficient to deflect a test specimen somewhere within the range from 5 to 10 deg of arc depending on the stiffness of the specimen and its span. A schematic diagram of a suitable machine is shown in Fig. 1.

NOTE 2—Two machines of different torque capacities are being used: one covers approximately 0.12 to 1.2 cm · kg (0.1 to 1.0 in. · lb) and the other approximately 1.2 to 19 cm · kg (1.0 to 16 in. · lb) or higher. Some machines also allow varying the span, especially important if shearing failures can occur (as in laminates at a span/width of 6).

NOTE 3—The amount of torque may be varied to suit the stiffness of the test specimen, and various weights should be available for this purpose. The actual amount of torque being applied by any given combination of weights, torque wheel radii, and shaft bearings should be determined by calibration.

NOTE 4—For operation at low temperatures the shaft of the machine must be provided with a heated collar next to the lower bearing to prevent the formation of ice.

3.2 *Temperature Control:*

3.2.1 *Flask*—A Dewar flask of suitable dimensions.

3.2.2 *Thermometer*—A thermometer graduated in 1 C divisions and having the necessary range. The bulb shall be located in close proximity to the test specimen.

[1] This method is under the jurisdiction of ASTM Committee D-20 on Plastics. A list of committee members may be found in the ASTM Yearbook. This standard is the direct responsibility of Subcommittee D-20.10 on Mechanical Properties.

Current edition effective April 25, 1969. Originally issued 1949. Replaces D 1043 – 61 T.

3.2.3 *Timer*—A timer accurate to 0.1 s.

3.2.4 *Heat Transfer Medium*—For normal laboratory purposes a substance that is liquid over the desired temperature range shall be used for the heat transfer medium, provided it has been shown that the liquid does not soften or otherwise affect the test specimen.

NOTE 5—Among the liquids found useful are acetone, ethanol, butanol, normal hexane, silicone oil, and a mixture of methyl phosphate and water in the ratio of 87 to 13 by volume. For temperatures to −70 C (−94 F) a mixture of 50 parts ethanol, 30 parts ethylene glycol, and 20 parts water may be found useful.

3.2.5 *Refrigeration*—Means shall be provided for cooling the heat transfer medium.

NOTE 6—For economical low-temperature use of the equipment, space for cooling enough containers of the heat-transfer medium for a day's work is necessary. Depending on the temperature ranges involved, mechanical refrigeration or a dry ice chest, or both, will be advantageous.

3.2.6 *Heater*—A controlled electric immersion heater in the Dewar flask when used in conjunction with an agitator shall be used to vary the temperature.

3.3 *Micrometer*—A dead-weight dial-type micrometer capable of measuring thicknesses accurately to 0.0025 mm (0.0001 in.) or better.

4. Test Specimens

4.1 *Geometry*—Test specimens shall be cut in the rectangular geometry shown in Fig. 2. They may be cut from compression-molded sheets. Care shall be taken to ensure that the test specimens are isotropic. Where the testing machine permits varying the span, the span to width (L/a) ratio should be 6 to 8. It is recommended that spans of 3.8 to 10 cm (1.5 to 4 in.) be used. The specimen shown in Fig. 3 is commonly used for nonrigid materials on the low range machine and has a span (L) of 38 mm (1.5 in.).

4.2 *Thickness*—The thickness of the specimen may vary between approximately 1 and 3 mm (0.040 and 0.125 in.). This range normally makes it possible to test materials of widely different stiffnesses.

4.3 Duplicate specimens of each material shall be tested. More replications are often needed, especially for nonhomogeneous materials.

5. Conditioning

5.1 *Conditioning*—Condition the test specimens at 23 ± 2 C (73.4 ± 3.6 F) and 50 ± 5 percent relative humidity for not less than 40 h prior to test in accordance with Procedure A of ASTM Methods D 618, Conditioning Plastics and Electrical Insulating Materials for Testing,[2] for those tests where conditioning is required. In cases of disagreement, the tolerances shall be ±1 C (±1.8 F) and ±2 percent relative humidity.

5.2 *Test Conditions*—Conduct tests in the Standard Laboratory Atmosphere of 23 ± 2 C (73.4 ± 3.6 F) and 50 ± 5 percent relative humidity, unless otherwise specified in the test methods or in this specification. In cases of disagreements, the tolerances shall be ±1 C (±1.8 F) and ±2 percent relative humidity.

6. Procedure

6.1 Measure the width and thickness of the specimen to three significant figures.

6.2 Carefully mount the specimen in the apparatus. Adjust the clamps so that the specimen has no slack or stress and is in complete contact with the clamps' internal surfaces.

6.3 Place the thermometer in position.

6.4 Fill the Dewar flask with the heat transfer medium, the temperature of which shall be somewhat below the lowest temperature of the range to be studied.

6.5 Place the flask in position on the tester, and start the agitator.

6.6 By intermittent use of the immersion heater, bring the bath to the desired test temperature.

6.7 Condition the specimen at the test temperature for 3 min, +15 s, −0 s.

6.8 Release the torque pulley. After 5 s note the angular deflection of the pulley and return the torque pulley to its initial position. If the reading thus obtained does not fall within the range from 5 to 100 deg of arc, vary the applied torque in such a way as to produce such a reading. For nonrigid materials, this reading should fall between 10 and 100 deg. After another waiting period of 3 min repeat the procedure at the same temperature.

---

[2] *Annual Book of ASTM Standards*, Part 27.

NOTE 7—In order to obtain measured values of apparent modulus of rigidity, G, that are comparable to the true value of G it is desirable that measurements be made within the elastic limit of the material being tested. Therefore, torques shall be chosen that will cause deflections that are as small as it is practical to measure accurately on the machine being used. It is often desirable to reduce the torque slightly before taking successive readings, particularly in the temperature range where the material is rapidly decreasing in rigidity.

NOTE 8—Better reproducibility is obtained if torques are chosen such that the deflection obtained at a given temperature is similar to or greater than that obtained at the previous lower temperature.

6.9 After each suitable reading is obtained, repeat the steps indicated in 6.6 to 6.8 for the next desired temperature. The torque may be lowered prior to each reading if desired (Notes 7 and 8).

7. Calculation

7.1 Calculate the apparent modulus of rigidity, G, for each temperature as follows:

$$G = 917TL/ab^3u\phi$$

where:

G = apparent modulus of rigidity, kgf/cm^2 (or psi),
T = applied torque, kg · cm (or lb · in.),
L = specimen length (span), cm (or in.),
a = specimen width (larger cross-sectional dimension), cm (or in.),
b = specimen thickness (smaller cross-sectional dimension), cm (or in.),
ϕ = angle of deflection of torque pulley, deg, and
u = value depending on ratio of a to b. Table 1 gives the values of u for various ratios of a to b. A third column gives thickness if the width is 6.350 mm (0.250 in.). If Table 1 is not adequate, u may be calculated by means of the equation given in Section A1 of the Appendix.

7.2 Plot the apparent modulus of rigidity values, calculated in accordance with 7.1, on a logarithmic scale versus temperature on a linear scale.

7.3 If desired, read from the graph the temperature at which the apparent modulus of rigidity is equal to 3160 kgf/cm^2 (45,000 psi). This value shall be designated T_F (see Note A1 in the Appendix).

8. Report

8.1 The report shall include the following:

8.1.1 Complete identification of the material, including name, stock or code number, date made, form, etc.,

8.1.2 Dimensions of the test specimen,

8.1.3 Details of conditioning the specimen prior to test, and identification of the heating medium used,

8.1.4 Table of data and results,

8.1.5 Plot of logarithm of apparent modulus of rigidity versus temperature,

8.1.6 Value of T_F if desired (see 7.3), and

8.1.7 Date of test.

TABLE 1 Values for u[a]

Ratio of Width, a, to Thickness. b	u	Thickness when Width is 6.350 mm (0.250 in.)
2.00	3.66	3.175 mm (0.125 in.)
2.25	3.84	2.819 mm (0.111 in.)
2.50	3.99	2.540 mm (0.100 in.)
2.75	4.11	2.311 mm (0.091 in.)
3.00	4.21	2.108 mm (0.083 in.)
3.50	4.37	1.829 mm (0.072 in.)
4.00	4.49	1.600 mm (0.063 in.)
4.50	4.59	1.422 mm (0.056 in.)
5.00	4.66	1.270 mm (0.050 in.)
6.00	4.77	1.067 mm (0.042 in.)
7.00	4.85	0.914 mm (0.036 in.)

[a] Taken from Trayer and March, National Advisory Committee for Aeronautics, Report 334 (1929).

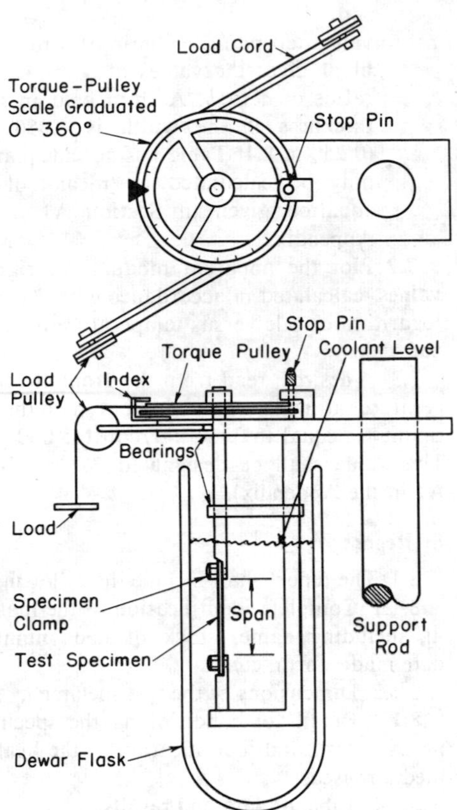

FIG. 1 Torsion Tester.

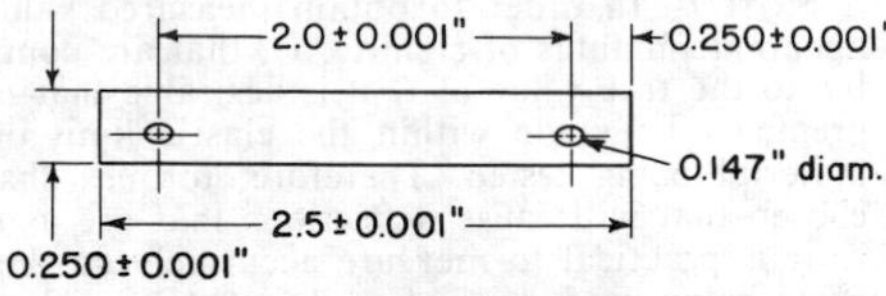

FIG. 2 Test Specimen, General.

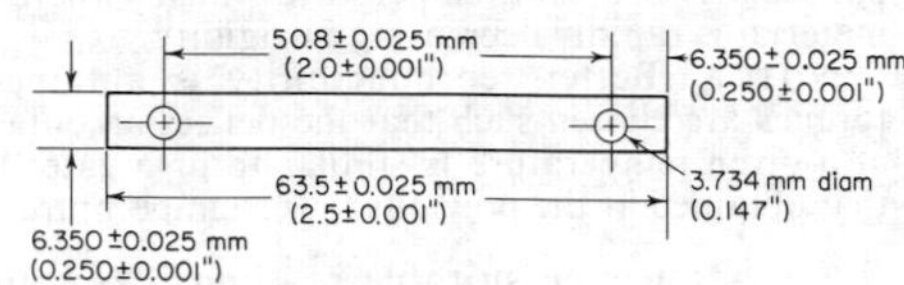

FIG. 3 Test Specimen.

APPENDIX

A1. Calculations

A1.1 Calculation of Factor *u*

A1.1.1 The factor u may be calculated by the following equation:

$$u = 5.33 - 3.36\, b/a(1 - b^4/12a^4)$$

where:
u = factor depending on the ratio of a to b,
a = specimen width, cm (or in.), and
b = specimen thickness, cm (or in.).

A1.2 Calculation of Apparent Modulus of Elasticity

A1.2.1 The relationship between modulus of rigidity, G, and modulus of elasticity, E, is expressed by the following equation (λ = Poisson's ratio):

$$E = 2G(1 + \lambda)$$

A1.2.2 It has been shown for some highly extensible plastics that the value for E calculated from the above equation agrees well with experimental values obtained by ASTM Method D 747, Test for Stiffness of Plastics by Means of a Cantilever Beam,[2] when Poisson's ratio is assumed to be 0.5, a suitable value for soft rubber at room temperature. It must be emphasized, however, that this is an experimental correlation only.

Note A1—Because of the correlation between E obtained from this method and stiffness from Method D 747, values in the literature have frequently been given in terms of E when this test method has been used. In this case, where Poisson's ratio is assumed to be 0.5, T_F indicates that temperature at which the material exhibits a modulus of elasticity, E, of 9490 kgf/cm^2 (135,000 psi). The term T_4 has been used to indicate the temperature at which the modulus of elasticity is 703 kgf/cm^2 (10,000 psi).

A1.2.3 In general, the calculation of modulus of elasticity from data obtained in this test is *not* recommended since there is evidence that suggests that Poisson's ratio varies from material to material, and may vary from temperature to temperature for the same material. In addition, data from this test may be obtained at deflections outside of the elastic limit of the material and thus would give misleading values of E. A value of modulus of elasticity may be obtained in tension as described in ASTM Method D 638, Test for Tensile Properties of Plastics.[2]

ASTM Designation: D 1044 – 56 (Reapproved 1969)

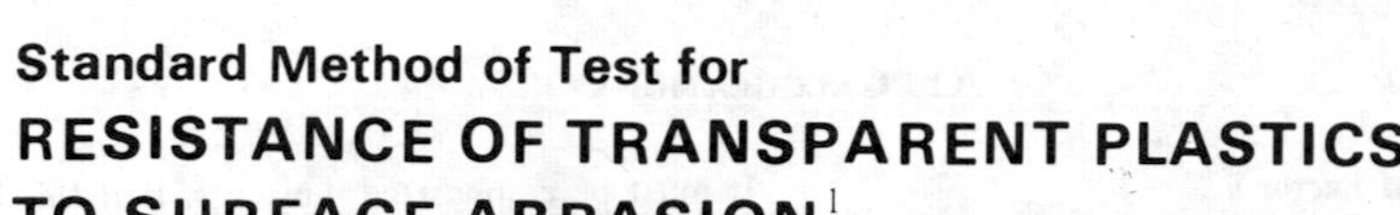

Standard Method of Test for

RESISTANCE OF TRANSPARENT PLASTICS TO SURFACE ABRASION[1]

This Standard is issued under the fixed designation D 1044; the number immediately following the designation indicates the year of original adoption or, in the case of revision, the year of last revision. A number in parentheses indicates the year of last reapproval.

1. Scope

1.1 This method of test describes a procedure for estimating the resistance of transparent plastics to one kind of surface abrasion by measurement of its optical effects.

NOTE 1—The values stated in U.S. customary units are to be regarded as the standard. The metric equivalents of U.S. customary units may be approximate.

NOTE 2—The abraser described herein was evaluated for weight and volume loss determinations with wheels more abrasive than that specified in this method. Meaningful results could not be obtained with two resilient materials because detached abrasive become embedded in them. With four other plastics, the interlaboratory reproducibility was unacceptable for an ASTM test. Sandpaper-wrapped wheels were found to be useful for a broad range of plastics. However, once again there was insufficient agreement among laboratories. Research report RR48: D-20, available from ASTM Headquarters, summarizes the results of this study.

NOTE 3—For determining resistance to abrasion of plastics by measurement of volume loss, reference should be made to ASTM Methods D1242, Test for Resistance to Abrasion of Plastic Materials.[2]

2. Significance

2.1 Transparent plastic materials, when used as windows or enclosures, are subject to wiping and cleaning, hence the maintenance of optical quality of a material after abrasion is of importance. It is the purpose of this test to provide a means of estimating the resistance of such materials to this type and degree of abrasion. The principal limitation of the test is its reproducibility, and the principal variable is the abradant. The intralaboratory precision of the method (95 percent confidence limits is about ±15 percent. The precision of interlaboratory measurements has been shown to be much poorer; therefore the instruments should be carefully checked against each other, if interlaboratory data are to be compared. Although the test does not provide fundamental data, it is suitable for grading materials relative to this type of abrasion in a manner which correlates with service.

3. Apparatus

3.1 *Abrader*—The Taber abraser or its equivalent, so constructed that wheels of several degrees of abrasiveness may be readily used. Loads of 250, 500, or 1000 g on the wheels may be obtained by use of changeable weights or counterweights.

3.2 *Refacing Stone*—An S-11 refacing disk shall be used for truing the abrasive wheels.

3.3 *Abrasive Wheels*—The grade of "Calibrase" wheel designated CS-10F shall be used. The "Calibrase" wheel shall meet the following requirements at the time of the test (Note 4):

3.3.1 The wheel shall not be used after the date stamped on it, and

3.3.2 The durometer hardness of the wheel shall be measured in accordance with ASTM Method D 2240, Test for Indentation Hardness of Rubber and Plastics by Means of a Durometer,[2] on at least four points equally spaced on the center of the abrading surface and one point on each side surface of the wheel. The test on the abrading surface shall

[1] This method is under the jurisdiction of ASTM Committee D-20 on Plastics. A list of committee members may be found in the ASTM Yearbook. This standard is the direct responsibility of Subcommittee D-20.10 on Mechanical Properties.

Current edition effective Sept. 10, 1956. Originally issued 1949. Replaces D 1044 – 54 T.

[2] *Annual Book of ASTM Standards*, Part 27.

be made with the pressure applied vertically along a diameter of the wheel, and the reading taken 10 s after full application of the pressure. Each wheel shall have a durometer hardness of 72 ± 5 at all measured points.

NOTE 4—The abrasive quality of the "Calibrase" wheels varies with hardness which changes with age.

3.4 *Abraser Turntable*—The turntable of the abraser shall rotate substantially in a plane with a deviation at a distance of 1.6 mm (1/16 in.) from its periphery of not greater than ±0.051 mm (±0.002 in.).

3.5 *Photometer*—An integrating sphere photoelectric photometer, described in ASTM Method D 1003, Test for Haze and Luminous Transmittance of Transparent Plastics,[2] shall be used to measure the light scattered by the abraded track.

3.6 *Diaphragm*—A circular diaphragm shall be centrally inserted in the apparatus to limit the light beam to the width of the abraded tracks. The light beam shall have a width of 7 mm ± 1 mm at the specimen, and shall be centered on the abraded area.

4. Test Specimens

4.1 The specimens shall be clean, transparent disks 102 mm (4 in.) in diameter or plates 102 mm (4 in.) square, having both surfaces substantially plane and parallel. They may be cut from sheets or molded in thicknesses up to 12.7 mm (1/2 in.). A 6.3-mm (1/4-in.) hole shall be centrally drilled in each specimen. Three such specimens shall be tested per sample, except for interlaboratory or specification tests when ten specimens shall be tested.

5. Conditioning

5.1 *Conditioning*—Condition the test specimens at 23 ± 2 C (73.4 ± 3.6 F) and 50 ± 5 percent relative humidity for not less than 40 h prior to test in accordance with Procedure A of ASTM Methods D 618, Conditioning Plastics and Electrical Insulating Materials for Testing,[2] for those tests where conditioning is required. In cases of disagreement, the tolerances shall be ±1 C (±1.8 F) and ±2 percent relative humidity.

5.2 *Test Conditions*—Conduct tests in the Standard Laboratory Atmosphere of 23 ± 2 C (73.4 ± 3.6 F) and 50 ± 5 percent relative humidity, unless otherwise specified in the test methods or in this specification. In cases of disagreements, the tolerances shall be ±1 C (±1.8 F) and ±2 percent relative humidity.

6. Procedure

6.1 Mount the pair of "Calibrase" wheels to be used on their respective flange holders, taking care not to handle them by their abrasive surfaces. Select the load to be used and affix it to the abraser. Mount a new S-11 refacing disk fine side up on the turntable and reface the wheels for 25 cycles, brushing the residue from the disk during the process.

CAUTION: Do not touch the surface of the wheels after they are refaced.

6.2 Mount the specimen on the specimen holder and subject it to abrasion for a selected number of cycles. Use an abrasion of 50 cycles with the 500-g load, unless otherwise specified. Wipe specimens after abrasion with dry, soft tissue.

NOTE 5—For plotting curves of light scattering *versus* cycles of abrasion, 10, 25, 50, and 100 cycles are recommended.

6.3 Where subjective data are adequate, compare abraded specimens visually with a measured, abraded standard. Where subjective data are not adequate, follow the procedure referred to in 6.4.

6.4 Using a photoelectric, integrating sphere photometer and Method D 1003, measure the percentage of the transmitted light which is diffused by the abraded track on at least four equally spaced intervals along the track.

NOTE 6—The specimen may be supported free to rotate on a 6.3-mm (1/4-in.) pin, so positioned as to center the light beam on the abraded area. The specimen may be automatically rotated and an integrated value thus obtained.

7. Report

7.1 The report shall include the following:

7.1.1 The percentage of the transmitted light which is diffused by the abraded specimens, averaged for the specimens tested.

7.1.2 The number of the specimens tested.

7.1.3 The load and the number of cycles used, if other than specified in 6.2.

7.1.4 A plot of the percentage of light scattered *versus* cycles abraded, if more than one number of cycles is used.

 Designation: D 1045 – 58 (Reapproved 1967)

Standard Methods of
SAMPLING AND TESTING PLASTICIZERS USED IN PLASTICS[1]

This Standard is issued under the fixed designation D 1045; the number immediately following the designation indicates the year of original adoption or, in the case of revision, the year of last revision. A number in parentheses indicates the year of last reapproval.

1. Scope

1.1 These methods cover the sampling and testing of liquid plasticizers used in the compounding of plastics.

NOTE 1—The values stated in U.S. customary units are to be regarded as the standard. The metric equivalents of U.S. customary units may be approximate.

2. Sampling

2.1 The method of sampling specified in 2.2 or 2.3 shall be used, according to the special conditions that exist.

2.2 *From Loaded Tank Car or Other Large Vessel*—The composite sample taken shall be not less than 1.9 liters ($^1/_2$ gal) and should consist of small samples of not more than 0.95 liter (1 qt) each, taken from near the top and bottom by means of a metal or glass container with removable stopper or top. This device, attached to a suitable pole, shall be lowered to the desired depth, when the stopper or top shall be removed and the container allowed to fill. A bomb sampler attached to a chain is convenient to use; the opening should be adjusted so that the bomb will fill on the way down.

2.3 *From Barrels and Drums*—At least 5 percent of the packages in any shipment shall be represented in the sample. The purchaser may increase the percentage of packages to be sampled at his discretion; in the case of plasticizers that are purchased in small quantity, each package may be sampled and analyzed, if desired. A portion shall be withdrawn from near the center of each package sampled by means of a "thief" or other sampling device and composited. The composite sample thus obtained shall be not less than 0.95 liter (1 qt) and shall consist of equal portions of not less than 0.24 liter ($^1/_2$ pt) from each package sampled.

3. Purity of Reagents

3.1 Reagent grade chemicals or equivalent, as specified in ASTM Methods E 200, Preparation, Standardization, and Storage of Standard Solutions for Chemical Analysis,[2] shall be used in all tests.

3.2 Unless otherwise indicated, references to water shall be understood to mean reagent water conforming to ASTM Specifications D 1193, for Reagent Water.[3]

ACIDITY

4. Reagents

4.1 *Alcohol*—Denatured alcohol, either Formula No. 3A or 2B of the U. S. Bureau of Internal Revenue.

4.2 *Alkali, Standard Solution (0.01 N)*—Prepare and standardize a 0.01 *N* aqueous solution of sodium hydroxide (NaOH) or a 0.01 *N* alcoholic solution of potassium hydroxide (KOH).

4.3 *Alkali, Standard Solution (0.1 N)*—Prepare and standardize a 0.1 *N* aqueous solution of sodium hydroxide (NaOH) or a 0.1 *N* alcoholic solution of potassium hydroxide (KOH).

4.4 *Benzene or Acetone.*

4.5 *Bromthymol Blue Indicator Solution.*

---

[1] These methods are under the jurisdiction of ASTM Committee D-20 on Plastics. A list of committee members may be found in the ASTM Yearbook. This standard is the direct responsibility of Subcommittee D-20.70 on Analytical Methods.

Current edition effective Sept. 22, 1958. Originally issued 1949. Replaces D 1045 – 55.

[2] *Annual Book of ASTM Standards*, Part 22.

[3] *Annual Book of ASTM Standards*, Part 23.

5. Procedure

5.1 Weigh 25 g of the sample into a 125-ml Erlenmeyer flask and dissolve in 50 ml of alcohol. If the sample is not completely soluble in alcohol, use 50 ml of a mixture of equal parts of alcohol and benzene or alcohol and acetone. With certain samples it may be necessary first to add 25 ml of benzene or acetone, warm to effect solution, and then add 25 ml of alcohol.

5.2 Add a few drops of bromthymol blue indicator solution and titrate with 0.01 *N* NaOH or KOH solution. If the titration exceeds 10 ml, repeat the determination using 0.1 *N* NaOH or KOH solution.

5.3 *Blank*—Make a blank titration on 50 ml of the solvent used to dissolve the sample.

6. Calculation

6.1 Calculate the acid number, expressed as milligrams of KOH per gram of sample, as follows:

$$\text{Acid number} = [(A - B)N \times 56.1]/C$$

where:

A = milliliters of NaOH or KOH solution required for titration of the sample,
B = milliliters of NaOH or KOH solution required for titration of the blank,
N = normality of the NaOH or KOH solution, and
C = grams of sample used.

6.2 If desired, in the case of esters, the results may be expressed as a percentage by weight of the appropriate acid, by using the proper factor in the equation in 6.1.

Ester Content

7. Reagents

7.1 *Bromthymol Blue Indicator Solution.*

7.2 *Hydrochloric Acid, Standard (0.5 N)*—Prepare and standardize a 0.5 *N* aqueous solution of hydrochloric acid (HCl).

7.3 *Potassium Hydroxide, Standard Solution (0.5 N)*—Prepare and standardize a 0.5 *N* alcoholic solution of potassium hydroxide (KOH).

8. Procedure

8.1 Weigh accurately about 2 g of the sample into a 250-ml Erlenmeyer flask with ground-glass joint. By means of a constant delivery pipet or buret, add 50 ml of 0.5 *N* KOH solution. Connect to a water-cooled condenser with ground-glass joint and reflux for a period of 1 to 4 h, depending on the ester being tested, or until saponification is complete.

8.2 After the apparatus has cooled, wash down the condenser with water and disconnect. Add a few drops of bromthymol blue indicator solution to contents of the flask and titrate with 0.5 *N* HCl.

8.3 *Blank*—Run a blank, containing 50 ml of the 0.5 *N* KOH solution, along with the sample.

9. Calculation

9.1 Calculate the ester content, expressed in milligrams of KOH per gram of sample, as follows:

$$\text{Ester content} = [(D - E)N \times 56.1]/G - F$$

where:

D = milliliters of HCl required for titration of the blank,
E = milliliters of HCl required for titration of the sample,
F = correction for acidity of sample (Sections 5 and 6),
N = normality of the HCl, and
G = grams of sample used.

9.2 If desired, the results may be expressed as a percentage of the appropriate ester by weight, by using the proper factor in the equation in 9.1.

Specific Gravity

10. Selection of Method

10.1 For materials that are not too viscous, the specific gravity may be measured by means of a hydrometer, as described under ASTM Method D 287, Test for API Gravity of Crude Petroleum and Petroleum Products, (Hydrometer Method),[4] or by a Westphal balance.

10.2 For the more viscous materials, a pycnometer or specific gravity bottle should be used, as described in Sections 12 and 13.

10.3 Certain samples may be too viscous at room temperature to seat the thermometer

---

[4] *Annual Book of ASTM Standards*, Parts 17, 18 and 29.

or capillary tube of a pycnometer properly. The specific gravity of such samples may be determined in accordance with ASTM Methods D 792, Test for Specific Gravity and Density of Plastics by Displacement,[5] or ASTM Method D 70, Test for Specific Gravity of Road Oils, Road Tars, Asphalt Cements, and Soft Tar Pitches.[6]

11. Thermometers

11.1 All temperature measurements shall be made with ASTM thermometers of suitable range, accurate to within 0.1 C and conforming to the requirements prescribed in ASTM Specification E 1, for ASTM Thermometers.[7]

12. Procedure Using Pycnometer

12.1 Determine the weight capacity of the pycnometer with water at 23 $\pm$ 1 C. Fill this standardized pycnometer with a portion of the sample that has been cooled to approximately 20 C. Insert the thermometer or capillary tube, taking care to avoid introduction of air bubbles. Set the pycnometer in a constant-temperature water bath maintained at 23 $\pm$ 1 C (73.4 $\pm$ 1.8 F) for a period of at least 30 min. Remove the droplet of sample from the overflow capillary and cover with the glass cap. Clean the outside of the pycnometer and weigh.

13. Calculation

13.1 Calculate the specific gravity as follows:

$$\text{Specific gravity at } 23/23\ C = A/B$$

where:

A = grams of sample used, and
B = water capacity of pycnometer in grams.

Color

14. Application

14.1 Useful comparisons between the color of relatively light colored plasticizers and platinum-cobalt standards may be made for colors in the range from 0 to 200, as defined in Section 15.

15. Preparation of Color Standards[8]

15.1 Dissolve 1.245 g of potassium chloroplatinate (K_2PtCl_6), containing 0.5 g of platinum, and 1 g of crystallized cobaltous chloride ($CoCl_2 \cdot 6H_2O$), containing about 0.248 g of cobalt, in water containing 100 ml of HCl (sp gr 1.19). Dilute to 1 liter with water to prepare a solution having a color of 500. This solution, already prepared, may be purchased from laboratory supply houses.

15.2 To prepare standards having colors of 5, 10, 15, etc., dilute 0.5, 1.0, 1.5 ml, etc., of the solution described in 15.1 with water to 50 ml in standard Nessler tubes. If 100-ml standard Nessler tubes are used, dilute 1.0, 2.0, 3.0 ml, etc., to 100 ml with water to obtain colors of 5, 10, 15, etc. Protect the tubes from evaporation and from dust when not in use. If desired, commercially available platinum-cobalt standards in Nessler tubes or colored glass disks with proper spectral transmission may be used, provided they are first checked against standards prepared from K_2PtCl_6 and $CoCl_2 \cdot 6H_2O$ as described above.

16. Procedure

16.1 Fill a 50 or 100-ml Nessler tube with the sample and compare with the standards by holding side by side and looking down through the columns of liquid at a matte white surface illuminated by northern daylight.

Refractive Index[9]

17. Apparatus

17.1 *Refractometer*—An Abbé refractometer with scale graduated directly in terms of refractive index of the D line of sodium at a temperature of 23 C.

17.2 *Water Supply*—A water supply, the temperature of which may be varied.

---

[5] *Annual Book of ASTM Standards*, Part 27.
[6] *Annual Book of ASTM Standards*, Part 11.
[7] *Annual Book of ASTM Standards*, Part 30.
[8] The preparation of these platinum-cobalt color standards was originally described by Hazen, A., *American Chemical Journal*, Vol 14, 1892, 300. The description given in these methods is substantially identical with that given in the *Standard Methods for the Examination of Water and Sewage*, Am. Public Health Assn., Ninth Edition, p. 14. A description of these standards is also given by Scott, W. W., *Standard Methods of Chemical Analyses*, Fifth Edition, Vol 2, p. 2048.
[9] This method is identical with the procedure in ASTM Methods D 1807, Test for Refractive Index and Specific Optical Dispersion of Electrical Insulating Liquids. *Annual Book of ASTM Standards*, Part 29.

18. Procedure

18.1 Place the refractometer in front of a suitable source of light (either daylight or electric light), insert the thermometer, and adjust the circulation of water so as to bring the prisms to the desired temperature (usually 23 C (73.4 F)). Clean with alcohol and wipe dry. Spread a drop of the liquid to be tested upon the lower prism and clamp the prisms together. Adjust the mirror so that the light enters the telescope. Focus the eyepiece on the cross-hairs and the reading lens of the scale. Upon moving the prism arm, a position can now be found where the lower part of the field is dark and the upper part light. In general, the borderline will be colored. Correct by turning the milled head on the right of the telescope so that a sharp black and white edge is obtained. Move the prism arm until this black edge just crosses the intersection of the cross-hairs. Read the refractive index from the scale, estimating the fourth decimal place.

18.2 The accuracy of the instrument may be checked by a small test plate of known refractive index, which is supplied with the refractometer. Attach this test plate to the upper prism with a liquid of high refractive index (usually monobromnaphthalene). Errors may be corrected by means of a small adjusting screw.

 **Designation: D 1130 – 63 (Reapproved 1969)**

American National Standard K66.7-1965
American National Standards Institute

Recommended Practice for

INJECTION MOLDING OF SPECIMENS OF THERMOPLASTIC MATERIALS[1]

This Recommended Practice is issued under the fixed designation D 1130; the number immediately following the designation indicates the year of original adoption or, in the case of revision, the year of last revision. A number in parentheses indicates the year of last reapproval.

1. Scope

1.1 This recommended practice covers only the general principles to be followed when injection molding test specimens of thermoplastics. This practice is to be used to obtain uniformity in methods of describing the various component parts of the molding operation, and also to set up uniform methods of reporting these conditions. The exact conditions required to prepare adequate specimens will vary for each plastic material. They should become a part of the specification for the material, or be agreed upon by the seller and the purchaser.

NOTE 1—There are many factors present in injection molding machines that may have an influence on the character of moldings and the numerical values of test results. Among those which may have some effect are the geometry and the temperature conditions of the heating chamber, the pressures applied, the size and shape of runners and gates, the mold temperature, and the time cycles used. It may be necessary to treat the molding granules in some manner prior to preparation of test specimens. Preheating, drying, etc., are often required, particularly with those plastics which absorb moisture.

NOTE 2—The values stated in U.S. customary units are to be regarded as the standard. The metric equivalents of U.S. customary units may be approximate.

2. Apparatus

2.1 The mold shall be of the design shown in Fig. 4 of ASTM Recommended Practice D 647, for Design of Molds for Test Specimens of Plastic Molding Materials.[2] If some other design of mold can be shown to give equivalent results, it may be used.

2.2 The mold shall be mounted in a suitable injection molding press.

2.3 There shall be provisions made to measure and control the following:

2.3.1 Feed of material to the injection cylinder,

2.3.2 Pressure applied to the injection ram,

2.3.3 Temperature of plastic heating chamber,

2.3.4 Temperature of the plastic,

2.3.5 Cavity and core surface temperatures, and

2.3.6 Cycle timing.

3. Conditioning

3.1 *Conditioning*—Condition the test specimens at 23 ± 2 C (73.4 ± 3.6 F) and 50 ± 5 percent relative humidity for not less than 40 h prior to test in accordance with Procedure A of ASTM Methods D 618, Conditioning Plastics and Electrical Insulating Materials for Testing,[2] for those tests where conditioning is required. In cases of disagreement, the tolerances shall be ±1 C (±1.8 F) and ±2 percent relative humidity.

3.2 *Test Conditions*—Conduct tests in the Standard Laboratory Atmosphere of 23 ± 2 C (73.4 ± 3.6 F) and 50 ± 5 percent relative humidity, unless otherwise specified in the test methods or in this specification. In cases of disagreements, the tolerances shall be 1 C (1.8 F) and ±2 percent relative humidity.

[1] This recommended practice is under the jurisdiction of ASTM Committee D-20 on Plastics. A list of committee members may be found in the ASTM Yearbook. This Standard is the direct responsibility of Subcommittee D-20.90 on Specimen Preparation.

Current edition effective Sept. 30, 1963. Originally issued 1950. Replaces D 1130 – 50.

[2] *Annual Book of ASTM Standards*, Part 27.

4. Procedure

4.1 *Feed*—Deliver an equal volume of granules to the injection chamber for each injection stroke.

4.2 *Injection Pressure*—Hold the injection pressure at a constant value for molding given sets of samples. Measure the hydraulic pressure on the injection ram by means of suitable gages. The injection pressure shall be the total pressure applied to the plastic (Note 3).

$$\text{Injection pressure} = \frac{\text{area of ram piston} \times \text{hydraulic fluid pressure}}{\text{area of injection ram}}$$

NOTE 3—The actual pressure applied on the plastic in the cavity will be less than this pressure. There will be pressure lost in compacting the granules and in moving plastic through the heating zone and through the sprues, runners, and gates. If the applied pressure is kept constant, the actual maximum mold pressure should be constant, provided the injection ram travels the same distance each stroke. This will hold for only one set of operating conditions of temperature, cycle time, etc., on a given molding machine.

4.3 *Cylinder Temperature*—Control the cylinder temperature by some suitable means. Generally what is actually measured and controlled is the temperature of some particular point on the metal wall of the cylinder. This temperature will often vary several degrees due to simple off-and-on types of controls. This may be a cause of considerable variation in the property being measured, even though all of the moldings will be of satisfactory appearance.

4.4 *Plastic Temperature*—Determine the actual plastic temperature by opening the press as for a "free" shot and inserting a needle thermocouple through the nozzle into the body of the plastic. Exercise care to keep the point of the needle away from any metal surface. Measure this temperature after the machine has been operated under a fixed set of conditions for a period of several cycles. (Two inventory changes of the heating cylinder.)

NOTE 4—Under a fixed set of operating conditions of cycle and shot-size, there will generally be a constant difference between the observed plastic temperature and the cylinder wall temperature. This difference will vary as conditions are changed.

4.5 *Cavity and Core Surface Temperature*—Control the cavity and core surface temperature by circulation of suitable coolant through the ports of the mold. Measure it in accordance with ASTM Recommended Practice D 957, for Determining Mold Surface Temperature of Commercial Molds for Plastics.[2]

4.6 *Cycle*—See Fig. 1. Various manufacturers of molding machines use different types of timers and time somewhat different portions of the molding cycle. To get uniformity in reporting of cycle components, set up some artificial concepts and definitions as follows:

4.6.1 *Starting Point or Zero Time* is the instant the plunger starts forward.

4.6.2 *Plunger Forward, D*, is the time from zero time until the instant the plunger begins to retract. It takes a measurable time for advance of the plunger, compression of the granules, and the build-up of pressure. If the machine is opened, the time from start of plunger forward to appearance of plastic flow from the nozzle can be measured. This is the approximate dead time, *F*. This time will vary from machine to machine and with the granular density, size of shot, etc. It also takes a measurable time to fill the cavity and to build up pressure. This will depend upon the plunger speed, gate and runner size, plastic temperature and pressure, etc. No simple method of estimation is available. After the cavity is filled, it is customarily retained under pressure for a period. This is, probably, the most important single phase of the molding cycle. It cannot be directly measured, but the time under pressure, *G*, can be approximated by subtracting the dead time from the total plunger forward time.

4.6.3 *Cooling Time, E*—After the plunger has retracted, it is generally necessary to leave the molds closed to give time for the plastic to cool.

4.6.4 *Mold Closed Time, B*, is the time from zero time on the cycle until the faces begin to open. This includes all the plunger forward time plus the cooling time.

4.6.5 *Mold Open Time, C*, is the time from when the mold faces start to part until zero time of the next cycle. This includes times necessary to open and close the mold.

4.6.6 *Total Cycle Time, A*, is the sum of mold closed time and mold open time. This is the total time from one point of one cycle to the same point of the next cycle.

NOTE 5—It is suggested that the total cycle time be set up as 60 s. A stop watch or sweep second hand of a clock can be used to check each component portion of every cycle. The periphery of the timer can be marked with crayon for this checking. Simple observation will suffice to check accuracy by this method. It is evident that the time to open the mold and the time to close the mold are included by definition in "mold open time." Time for plunger to advance is included in "plunger forward time." Time for the plunger to retract is included in "cooling time." In many machines, addition of timer settings will not add up to cycle times as defined here.

4.7 *Number of Moldings*—The total number of moldings to be made will depend upon the amount of material available, the number of test specimens required, and upon the need to reach molding equilibrium, so that test specimens produced are all truly made under one set of conditions. The injection process is not continuous, but is, instead, a series of repeated operations. Each operation should be performed so as to exactly duplicate the others. This can best be done with automatically controlled machines. The cycles must be repeated enough times so that steady conditions hold. Interruptions disturb this equilibrium so that a number of moldings must be discarded after each operation change. The exact number of moldings to discard before retaining samples must be agreed upon between buyer and seller, but it should probably represent at least one complete change of the material in the heating chamber.

5. Report

5.1 The report shall include the following:

5.1.1 Mold used,

5.1.2 Specimen description (such as tension specimen, $^1/_8$ by $^1/_2$ in. (3.2 by 12.7 mm) ASTM . . .),

5.1.3 Number of cavities molded per shot,

5.1.4 Gate size and description,

5.1.5 Make of machine used,

5.1.6 Automatic or hand control,

5.1.7 Rated capacity of heating chamber,

5.1.8 Weight of molding plus sprues (should be approximately 50 to 75 percent of rated heating chamber capacity),

5.1.9 Heater control indicated temperature,

5.1.10 Plastic temperature,

5.1.11 Mold surface temperature (cavity and core),

5.1.12 Injection pressure,

5.1.13 Length of cycle,

5.1.14 Mold open time,

5.1.15 Mold closed time,

5.1.16 Plunger forward time,

5.1.17 Dead time,

5.1.18 Machine timer settings to arrive at above cycle,

5.1.19 Number of moldings made,

5.1.20 Number of moldings discarded before selection of specimens,

5.1.21 Pretreatment used,

5.1.22 Material used,

5.1.23 Type of material, and

5.1.24 Designation

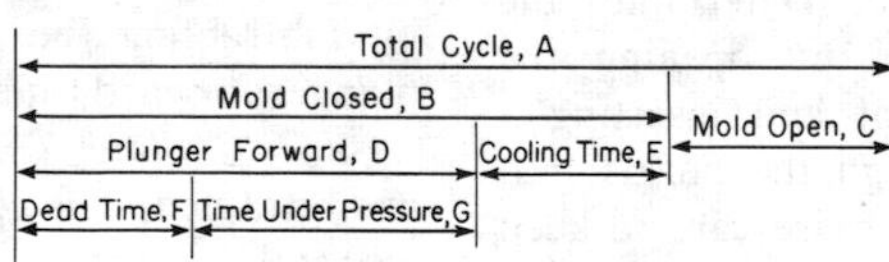

FIG. 1 Molding Cycle.

Designation: D 1180 – 57 (Reapproved 1961)

Standard Method of Test for
BURSTING STRENGTH OF ROUND RIGID PLASTIC TUBING[1]

This Standard is issued under the fixed designation D 1180; the number immediately following the designation indicates the year of original adoption or, in the case of revision, the year of last revision. A number in parentheses indicates the year of last reapproval.

1. Scope

1.1 This method covers measurement of the resistance offered by round rigid plastic tubing to rupture by internal hydrostatic pressure under the prescribed conditions of test, primarily involving hoop stresses. The method is applicable to rigid, round tubing, either molded or laminated, of thermosetting or thermoplastic material.

NOTE 1—The values stated in U.S. customary units are to be regarded as the standard. The metric equivalents of U.S. customary units may be approximate.

2. Apparatus

2.1 *Testing Machine*—Any suitable testing machine of the constant-rate-of-crosshead-movement type, with accuracy and sensitivity suitable for compression testing as prescribed in ASTM Method D 695, Test for Compressive Properties of Rigid Plastics.[2]

NOTE 2—In cases of questionable parallelism of test machine platens, recourse to a compression tool as described in Method D 695 may be necessary.

2.2 *Plugs*—Cylindrical plugs, 25, 50, and 75 mm (1, 2, and 3 in.) long, with parallel ends. The diameters of the plugs shall be 0.03 to 0.08 mm (0.001 to 0.003 in.) less than the inside diameter of the tube specimen, in order to provide a close sliding fit. The plugs may be of steel, thermosetting laminated rod, or other material having high compressive strength.

2.3 *Hydraulic Medium*—A very high-viscosity hydraulic medium—one that will not produce excessive flow between plug and specimen wall under conditions of the test.[3]

3. Test Specimens

3.1 The lengths of the test specimens and of the plugs to be used, together with the depth of insertion of the plug into the tube specimen, shall be as shown in Table 1.

4. Conditioning

4.1 *Conditioning*—Condition the test specimens at 23 ± 2 C (73.4 ± 3.6 F) and 50 ± 5 percent relative humidity for not less than 40 h prior to test in accordance with Procedure A of ASTM Methods D 618, Conditioning Plastics and Electrical Insulating Materials for Testing, for those tests where conditioning is required. In cases of disagreement, the tolerances shall be ±1 C (±1.8 F) and ±2 percent relative humidity.

4.2 *Test Conditions*—Conduct tests in the Standard Laboratory Atmosphere of 23 ± 2 C (73.4 ± 3.6 F) and 50 ± 5 percent relative humidity, unless otherwise specified in the test methods or in this specification. In cases of disagreements, the tolerances shall be ±1 C (±1.8 F) and ±2 percent relative humidity.

[1] This method is under the jurisdiction of ASTM Committee D-20 on Plastics. A list of committee members may be found in the ASTM Yearbook. This standard is the direct responsibility of Subcommittee D-20.10 on Mechanical Properties.

Current edition effective Sept. 30, 1957. Originally issued 1951. Replaces D 1180 – 51 T.

[2] *Annual Book of ASTM Standards*, Part 27.

[3] A suitable material for this purpose is General Electric's organosilicon compound G. E. SS-91, Silicone Bouncing Putty.

5. Test Temperature

5.1 All tests shall be made at a temperature of 23 ± 1 C (73.4 ± 1.8 F), unless otherwise specified.

6. Procedure

6.1 Test five specimens, filling the specimen tube with the hydraulic medium to conform with the depth of insertion specified in Section 3.

NOTE 3: *Example*—In testing a 25-mm (1-in.) diameter tube, the medium will be within 9 to 16 mm (3/8 to 5/8 in.) of each end of the specimen.

6.2 Insert plugs of the proper size, as specified in Section 3, and seat them firmly by hand. Place the prepared specimen in the testing machine and apply load to the ends of the plugs.

6.3 *Speed of Testing*—The crosshead speed of testing shall be approximately 2.5 mm (0.10 in.)/min until rupture occurs.

7. Report

7.1 The report shall include the following:

7.1.1 Inside and outside diameter of test specimen,

7.1.2 Type of tube, material and grade,

7.1.3 Load in pounds at first sign of rupture,

7.1.4 Ultimate bursting fluid pressure in pounds per square inch, calculated from the rupturing load and the area of the face of the plug,

7.1.5 Laboratory conditions, and

7.1.6 Presence of inside scores or other possible defects in the tube.

TABLE 1

Nominal Inside Diameter of Tube, mm (in.)	Length of Test Specimen, mm (in.)	Length of Plugs, mm (in.)	Depth of Insertion, mm (in.)
3.2 to 25.4 incl (1/8 to 1 incl)	51 (2)	25 (1)	12.7 ± 3.2 (1/2 ± 1/8)
Over 25.4 to 101.6 incl (Over 1 to 4 incl)	102 (4)	51 (2)	25.4 ± 6.4 (1 ± 1/4)
Over 101.6 to 152.4 incl (Over 4 to 6 incl)	153 (6)	76 (3)	50.8 ± 12.7 (2 ± 1/2)

Designation: D 1181 – 56 (Reapproved 1966)

Standard Method of Test for

WARPAGE OF SHEET PLASTICS[1]

This Standard is issued under the fixed designation D 1181; the number immediately following the designation indicates the year of original adoption or, in the case of revision, the year of last revision. A number in parentheses indicates the year of last reapproval.

1. Scope

1.1 This method covers the determination of the dimensional distortion of sheet plastic specimens resulting from exposure to service conditions. In particular, the method is suitable for measuring warpage such as that developed in accordance with the conditions described in ASTM Methods D 756, Test for Resistance of Plastics to Accelerated Service Conditions.[2]

NOTE 1—The values stated in U.S. customary units are to be regarded as the standard. The metric equivalents of U.S. customary units may be approximate.

2. Significance

2.1 This procedure is intended only as a convenient method for determination of dimensional distortion in sheet plastics, either when subjected to conditions of test such as are outlined in Method 5 D 756, or "as received" from a supplier. It is not applicable to specimens less than 1.3 mm (0.050 in.) in thickness, or where the span to be measured exceeds 21.5 cm (8.5 in.).

3. Apparatus

3.1 *Dial Indicator*—A 0-50-0 dial indicator[3] (Fig. 1) equipped with spindle extension (brass or aluminum) 3.2 mm (0.125 in.) in diameter, mounted in a small stainless steel block which shall be machined to slide laterally on a stainless steel bar. This bar shall be equipped with two supports, one stationary and the other movable. Machining shall be of such precision that movement of the indicator, while the jig is on the test plate surface, will not produce deviations in excess of ±0.013 mm (±0.0005 in.).

3.2 *Test Plate*—Stainless steel test plate, 18 by 18 by 1.3 cm (7 by 7 by 0.5 in.) with one 18 by 18-cm (7 by 7-in.) testing surface ground flat to ±0.013 mm (±0.0005 in.).

4. Test Specimens

4.1 The test specimens shall be not less than 10 cm (4 in.) nor more than 15 cm (6 in.) square, and shall be of uniform thickness. In practice, specimens 10 by 10 cm (4 by 4 in.) by the full thickness of the material are recommended, although the measuring device is capable of measuring a 15 by 15-cm (6 by 6-in.) specimen.

5. Conditioning

5.1 *Conditioning*—Condition the test specimens at 23 ± 2 C (73.4 ± 3.6 F) and 50 ± 5 percent relative humidity for not less than 40 h prior to test in accordance with Procedure A of ASTM Methods D 618, Conditioning Plastics and Electrical Insulating Materials for Testing,[2] for those tests where conditioning is required. In cases of disagreement, the tolerances shall be ±1 C (±1.8 F) and ±2 percent relative humidity.

5.2 *Test Conditions*—Conduct tests in the Standard Laboratory Atmosphere of 23 ± 2 C (73.4 ± 3.6 F) and 50 ± 5 percent relative humidity, unless otherwise specified in the test methods or in this specification. In cases of disagreements, the tolerances shall be ±1 C (±1.8 F) and ±2 percent relative

[1] This method is under the jurisdiction of ASTM Committee D-20 on Plastics. A list of committee members may be found in the ASTM Yearbook. This standard is the direct responsibility of Subcommittee D-20.50 on Permanence Properties.

Current edition effective Sept. 10, 1956. Originally issued 1951. Replaces D 1181 – 51 T.

[2] *Annual Book of ASTM Standards*, Part 27.

[3] Starrett No. 81C Dial Indicator (or equivalent), 0-50-0 by 0.001 in., center lug with top lift, jewel bearings. Tension springs removed to make the indicator the deadweight type. Spindle thrust not to exceed 10 g.

humidity.

6. Procedure

6.1 Zero the dial indicator by adjustment of its position in the sliding block while the jig is supported on the flat surface of the test plate.

6.2 Lay the specimen on the test plate (concave side up, if warped), and place over it the measuring jig, with the feet of the supporting bar resting on the test plate. Measure and record the height of the upper face of each corner, and the thickness of the specimen at each corner.

NOTE 2—Precise measurements can be made by fastening a small strip of metal foil 0.013 mm (0.0005 in.) thick, smoothly at the corners of the specimen with vaseline or stopcock grease. A suitable alarm circuit with this as one contact, and the indicator spindle as the other, will indicate immediately when the spindle contacts the specimen.

6.3 Suspend or support specimens vertically during exposure to a specified test environment. Determine warpage as in 6.2 following completion of each portion of the test cycle, or as specified.

7. Calculations

7.1 Calculate the warpage of the test specimens as follows:

$$\text{Initial warpage} = A_1 - B_1$$
$$\text{Final warpage} = A_2 - B_2$$
$$\text{Change in warpage} = (A_2 - B_2) - (A_1 - B_1)$$

where:

A_1 = average of initial corner heights,
B_1 = average initial thickness of specimen (measured at four corners),
A_2 = average of corner heights after test, and
B_2 = average thickness of specimen after test (measured at four corners).

NOTE 3—If the direction of warpage of the specimen has completely reversed, treat the final warpage ($A_2 - B_2$) as a negative quantity. The above expression then becomes $-(A_2 - B_2) - (A_1 - B_1)$.

8. Report

8.1 The report shall include the following:

8.1.1 Initial warpage, in millimeters (or inches),

8.1.2 Change in warpage, in millimeters (or inches) (a negative value indicates a change in the direction of warpage),

8.1.3 Maximum initial and final corner heights,

8.1.4 Specimen size, and

8.1.5 Test conditions.

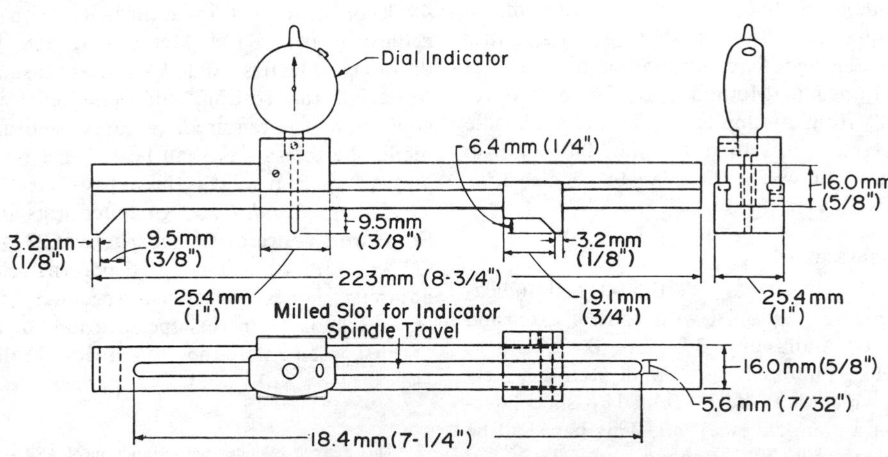

FIG. 1 Dial Indicator for Measuring Warpage of Sheet Plastics.

ASTM Designation: D 1203 – 67

Standard Methods of Test for

LOSS OF PLASTICIZER FROM PLASTICS (ACTIVATED CARBON METHODS)[1]

This Standard is issued under the fixed designation D 1203; the number immediately following the designation indicates the year of original adoption or, in the case of revision, the year of last revision. A number in parentheses indicates the year of last reapproval.

These methods were prepared jointly by the American Society for Testing and Materials and The Society of the Plastics Industry.

1. Scope

1.1 These methods cover the determination of the plasticizer loss from a plastic material under defined conditions of time and temperature, and using activated carbon as the immersion medium.

1.2 Two methods are covered as follows:

1.2.1 *Method A. Direct Contact with Activated Carbon*—In Method A the material is in direct contact with the carbon. This method is particularly useful in the rapid comparison of a large number of plastic specimens.

1.2.2 *Method B. Wire Cage*—Method B prescribes the use of a wire cage, which prevents the direct contact of the material with the carbon. By eliminating the direct contact of the specimen with the activated carbon the migration of the plasticizer or other volatile components to the surrounding carbon is minimized, and loss by volatilization is more specifically measured.

NOTE 1—The values stated in U.S. customary units are to be regarded as the standard. The metric equivalents of U.S. customary units may be approximate.

2. Significance

2.1 The methods are intended to be rapid empirical tests which may be useful in the relative comparison of materials having the same nominal thickness. The assumption is made that the material lost from the plastic is primarily plasticizer. This may or may not be valid, depending on the composition of the material being tested. The effect of moisture is considered to be negligible.

2.2 Correlation with ultimate application for various plastic materials should be determined by the user. To obtain accelerated tests that more nearly approach actual service conditions, reference should be made to ASTM Specification E 197, for Enclosures and Servicing Units for Tests Above and Below Room Temperature.[2]

3. Apparatus

3.1 *Balance*—An accurate analytical balance, equipped with class S weights or better.

3.2 *Oven or Bath*—A thermostatically controlled oven or bath capable of maintaining the temperature to within ± 1 C of the test temperature, which normally will be in the range of 50 to 150 C.

3.3 *Containers*—One-pint metal paint cans, 86.0 mm ($3^3/_8$ in.) in diameter, or 473-cm^3 (1-pt) Mason jars of approximately the same diameter.

3.4 *Micrometer*—A micrometer capable of measuring to the nearest 0.0025 mm (0.0001 in.) for measuring the thickness of the test specimens.

3.5 *Metal Cages (for Method B)*—Wire cages constructed from approximately 30-mesh bronze gauze, in cylindrical form, having a diameter of 60 mm and a height of 6

[1] These methods are under the jurisdiction of ASTM Committee D-20 on Plastics. A list of committee members may be found in the ASTM Yearbook. This standard is the direct responsibility of Subcommittee D-20.15 on Thermoplastic Materials.

Current edition effective April 11, 1967. Originally issued 1952. Replaces D 1203 – 61 T.

[2] *Annual Book of ASTM Standards*, Part 30.

mm, formed by soldering a strip of gauze at right angles to the periphery of a disk of bronze gauze. One of the bases acts as a lid.

4. Material

4.1 *Activated Carbon, $^6/_{14}$ Mesh*[3]—It has been found that different types and grades of activated carbon give differing results, thus making it necessary for the purchaser and the seller to agree on the same type and grade in order to obtain concordant results. Care should be taken that an airtight storage container is used for the activated carbon and that fresh material is used for each test, unless it can be shown that reuse does not affect the results.

5. Test Specimens

5.1 The test specimens shall be 2-in. diameter disks made of the plastic material to be tested. Three specimens of each formulation shall be tested.

5.2 Direct comparison of values between materials should not be made unless all specimens so compared do not vary by more than ±10 percent from a given nominal thickness. This precaution is necessary because of discrepancies that may arise due to edge effects, depletion of plasticizer, and the fact that the percentage weight loss is a direct function of thickness.

6. Conditioning

6.1 *Conditioning*—Condition the test specimens at 23 ± 2 C (73.4 ± 3.6 F) and 50 ± 5 percent relative humidity for not less than 40 h prior to test in accordance with Procedure A of ASTM Methods D 618, Conditioning Plastics and Electrical Insulating Materials for Testing,[4] for those tests where conditioning is required. In cases of disagreement, the tolerances shall be ±1 C (±1.8 F) and ±2 percent relative humidity.

6.2 *Test Conditions*—Conduct tests in the Standard Laboratory Atmosphere of 23 ± 2 C (73.4 ± 3.6 F) and 50 ± 5 percent relative humidity, unless otherwise specified in the test methods or in this specification. In cases of disagreements, the tolerances shall be ±1 C (±1.8 F) and ±2 percent relative humidity.

7. Procedure—Method A. Direct Contact with Activated Carbon

7.1 Weigh the conditioned specimens individually on the analytical balance and designate this weight as W_1.

NOTE 2—Only specimens of the same composition or formulation shall be tested in a single container, because of the possibility of cross-migration between compositions having varying plasticizer types or concentrations, or both.

7.2 Spread 120 cm³ of activated carbon evenly on the bottom of a container. Place one specimen on top of the activated carbon and cover it with 120 cm³ of activated carbon. Place a second specimen (Note 2) on top of the first and cover it with 120 cm³ of the carbon, followed by a third specimen and then 120 cm³ more of activated carbon. Place a cover on the container in such a manner that the container will be vented. This is necessary in order that possible pressure build-up in the container during heating be relieved. Take care that in no case shall the carbon be packed by pressure other than the weight of the composite sandwich in the container.

7.3 Place the container upright in the oven or bath. Unless otherwise specified, the temperature of the oven or bath shall be 70 ± 1 C and the duration of the test 24 h.

NOTE 3—If other conditions of test are desired due to purchase specifications or widely varying types of materials, these may be employed but shall be included as part of the report.

7.4 At the end of the 24-h period, remove the container from the oven or bath. Then, within 1 h, remove the specimens from the container, brush free of carbon, and recondition in accordance with Section 6.

7.5 After reconditioning, reweigh the specimens and designate this weight as W_2.

8. Procedure—Method B. Wire Cage[5]

8.1 Proceed as in Section 7 (Method A), except place every individual specimen in a small metal wire-mesh cage constructed as indicated in 3.5, and maintain the temperature at 100 ± 1 C (Note 3).

9. Calculations

9.1 Calculate the plasticizer loss, expressed

[3] Columbia activated carbon, grade AC, $^6/_{14}$ mesh, or grade ACC, $^6/_{14}$ mesh, as manufactured by the National Carbon Co., Division of Union Carbide Corp., 270 Park Ave., New York, N. Y. 10017, has been found satisfactory for this purpose.

[4] *Annual Book of ASTM Standards*, Part 27.

[5] This method is equivalent to Draft ISO Recommendation No. 191.

as percentage weight loss based on the original specimen weight, as follows:

$$\text{Weight loss, percent} = [(W_1 - W_2)/W_1] \times 100$$

where:

W_1 = initial weight of test specimen, and
W_2 = final weight of test specimen.

10. Report

10.1 The report shall include the following:

10.1.1 Complete identification of the material tested, including type, source, manufacturer's code number, and previous history,

10.1.2 Actual thickness to the nearest 0.0001 in. for each of the three specimens tested, and the average of the three,

10.1.3 Percentage weight loss of each of the three specimens, and the average of the three,

10.1.4 Any observations as to distortion or change in appearance of the specimens, and

10.1.5 Type of activated carbon used.

 Designation: D 1204 – 54 (Reapproved 1966)

Standard Method for
MEASURING CHANGES IN LINEAR DIMENSIONS OF NONRIGID THERMOPLASTIC SHEETING OR FILM[1]

This Standard is issued under the fixed designation D 1204; the number immediately following the designation indicates the year of original adoption or, in the case of revision, the year of last revision. A number in parentheses indicates the year of last reapproval.

This method was prepared jointly by the American Society for Testing and Materials and The Society of the Plastic Industry.

1. Scope

1.1 This method covers the measurement of changes in linear dimensions of nonrigid thermoplastic sheeting or film which result from exposure of the material to specified conditions of elevated temperature and time.

NOTE 1—The values stated in U.S. customary units are to be regarded as the standard. The metric equivalents of U.S. customary units may be approximate.

2. Significance

2.1 This method is particularly applicable to nonrigid thermoplastic sheeting or film made by the calender or extrusion process. The test gives an indication of lot-to-lot uniformity as regards the degree of internal strains introduced during processing.

3. Apparatus

3.1 *Oven*—A mechanical convection oven capable of maintaining a temperature of 100 ± 1 C.

3.2 *Scale* graduated in 0.25-mm (0.01-in.) divisions, 30 cm (12 in.) or more in length.

3.3 *Thermometer* graduated in 1 C divisions, with a range suitable for the test temperature used.

3.4 *Timer* graduated in minutes.

3.5 *Template* 25 by 25 cm (10 by 10 in.), for cutting test specimens.

3.6 *Heavy Paper Sheets*, approximately 40 by 40 cm (15 by 15 in.), with smooth, wrinkle- and crease-free surfaces.

3.7 *Talc*, finely ground.

4. Test Specimens

4.1 The test specimens shall be two pieces of the sheeting or film 25 by 25 cm (10 by 10 in.), cut with the aid of the template, one from either of the two transverse edges and one from the center of the sheet as shown in Fig. 1. Each specimen shall be marked to show the direction of calendering or extrusion. The midpoint of each edge shall be marked for use as a reference point when final measurements are made.

5. Conditioning

5.1 *Conditioning*—Condition the test specimens at 23 ± 2 C (73.4 ± 3.6 F) and 50 ± 5 percent relative humidity for not less than 40 h prior to test in accordance with Procedure A of ASTM Methods D 618, Conditioning Plastics and Electrical Insulating Materials for Testing,[2] for those tests where conditioning is required. In cases of disagreement, the tolerances shall be ±1 C (±1.8 F) and ±2 percent relative humidity.

5.2 *Test Conditions*—Conduct tests in the Standard Laboratory Atmosphere of 23 ± 2 C (73.4 ± 3.6 F) and 50 ± 5 percent relative humidity, unless otherwise specified in the

[1] This method is under the jurisdiction of ASTM Committee D-20 on Plastics. A list of committee members may be found in the ASTM Yearbook. This standard is the direct responsibility of Subcommittee D-20.50 on Permanence Properties.

Current edition effective Sept. 15, 1954. Originally issued 1952. Replaces D 1204 – 52.

[2] *Annual Book of ASTM Standards*, Part 27.

test methods or in this specification. In cases of disagreements, the tolerances shall be ±1 C (±1.8 F) and ±2 percent relative humidity.

6. Procedure

6.1 Place each specimen on the heavy paper which has been lightly dusted with talc, and cover with a second piece of dusted paper. Fasten the papers together with paper clips.

NOTE 2—The paper should be well dusted, and the specimens should not be restricted either by the paper or the clips. *It is imperative that the specimens be free to change shape as strains are relieved during the period of test.*

6.2 Place the paper-plastic sandwiches horizontally in the oven at the temperature and for the length of time applicable to the material being tested. Sandwiches must not be stacked, as this may restrict movement of the plastic between the papers.

6.3 At the end of the oven-exposure period, cool the specimens to room temperature and then remove the papers. Measure the distance between the opposite edges of the specimens at the reference marks to the nearest 0.25 mm (0.01 in.), under the conditions specified in. Section 5.

7. Calculations

7.1 Calculate the linear dimensional change as follows:

$$\text{Linear change, percent} = [(D_t - D_o)/D_o] \times 100$$

where:

D_f = final length (or width) of specimen, mm (or in.) after test, and

D_o = original length (or width) of specimen, mm (or in.).

A negative value denotes shrinkage, and a positive value indicates expansion.

7.2 Average the values obtained for each direction.

8. Report

8.1 The report shall include the following:

8.1.1 Identification of the material tested,

8.1.2 Test conditions (time and temperature), including conditioning of test specimens, and

8.1.3 Average percentage linear change in both the parallel and the transverse direction of processing.

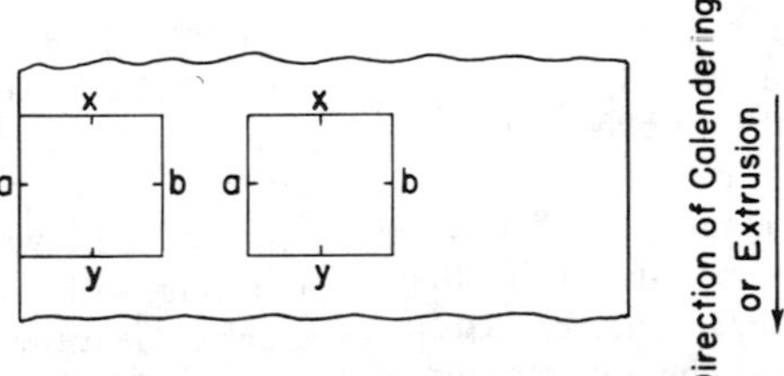

Points *a*, *b* and *x*, *y* are reference marks at midpoint of test specimen edges.

FIG. 1 Method of Cutting Test Specimens from Sample.

Designation: D 1238 – 65 T

Tentative Method of
MEASURING FLOW RATES OF THERMOPLASTICS BY EXTRUSION PLASTOMETER[1]

This Tentative Method has been approved by the sponsoring committee and accepted by the Society in accordance with established procedures, for use pending adoption as standard. Suggestions for revisions should be addressed to the Society at 1916 Race St., Philadelphia, Pa. 19103.

Pursuant to the warning note covering the 3-year limitation on tentatives, published in the previous edition, this tentative has received special permission from the Committee on Standards for continuation until June 1971.

1. Scope

1.1 This method covers measurement of the rate of extrusion of molten resins through an orifice of a specified length and diameter under prescribed conditions of temperature and pressure.

NOTE 1—The values stated in U.S. customary units are to be regarded as the standard. The metric equivalents of U.S. customary units may be approximate.

2. Significance

2.1 This method is particularly useful for quality control tests on thermoplastics having relatively low melt viscosities.

2.2 Procedure A is a manual cutoff operation used for materials having flow rates that fall between 0.15 and 25 g/10 min. Procedure B is an automatically timed flow rate measurement used for materials having flow rates from 0.15 to 300 g/10 min. Comparable flow rates have been obtained by these procedures in interlaboratory round-robin measurements of polyethylene and polypropylene.

2.3 This method serves to indicate the uniformity of the flow rate of the polymer as made by an individual process and, in this case, may be indicative of uniformity of other properties. However, uniformity of flow rate among various polymers as made by various processes does not, in the absence of other tests, indicate uniformity of other properties.

2.4 The flow rate obtained with the extrusion plastometer is not a fundamental polymer property. It is an empirically defined parameter critically influenced by the physical properties and molecular structure of the polymer and the conditions of measurement. The rheological characteristics of polymer melts depend on a number of variables. Since the values of these variables occurring in this test may differ substantially from those in large-scale processes, test results may not correlate directly with processing behavior.

3. Apparatus

3.1 *Plastometer:*

3.1.1 The apparatus shall be a deadweight piston plastometer consisting of a thermostatically controlled heated steel cylinder with an orifice at the lower end and a weighted piston operating within the cylinder (Note 2). The essential features of the plastometer, illustrated in Figs. 1 and 2, are described in 3.2 to 3.8.

NOTE 2—This instrument can be adapted for volumetric measurements by attaching a dial gage or recording equipment to measure piston travel. From simultaneous volume and weight measurements, melt density can be calculated. Flow rate measurements can also be made by measuring time of piston movement through a predetermined distance with a timer-actuating switch and an automatic timer. This latter method is particularly applicable to measuring the flow rates of the lower melt viscosity resins or

[1] This method is under the jurisdiction of ASTM Committee D-20 on Plastics. A list of committee members may be found in the ASTM Yearbook. This tentative is the direct responsibility of Subcommittee D-20.30 on Thermal Properties.

Current edition effective Oct. 13, 1965. Originally issued 1952. Replaces D 1238 – 62 T.

the faster flow rates resulting from the use of high testing pressures where the manual cutoff operation is most difficult to perform. See Procedure B.

3.1.2 Relatively minor changes in the design and arrangement of the component parts have been shown to cause differences in results between laboratories. It is important, therefore, for the best interlaboratory agreement that the design adhere closely to the description herein; otherwise, it should be determined that modifications do not influence the results.

3.2 *Cylinder*—The steel cylinder shall be 50.8 mm (2 in.) in diameter, 162 mm ($6^3/_8$ in.) in length with a smooth, straight hole 9.550 ± 0.007 mm (0.3760 ± 0.0003 in.) in diameter, displaced 4.8 mm ($^3/_{16}$ in.) from the cylinder axis. Wells for a thermoregulator and thermometer shall be provided as shown in Fig. 1. A 3.2-mm ($^1/_8$-in.) plate shall be attached to the bottom of the cylinder to retain the orifice. A hole in this plate, centered under the orifice and countersunk from below, allows free passage of the extrudate. At least two 10-mm ($^3/_8$-in.) diameter rods shall be screwed into the side of the cylinder for attaching to a vertical support. The essential dimensions of a satisfactory cylinder of this type are shown in Fig. 1 (Note 3). The cylinder bore should be finished by techniques known to produce approximately 12 rms or better in accordance with American National Standard B46.1, Surface Texture.

NOTE 3—Cylinders made of SAE 52100 steel heat-hardened to a Rockwell hardness, C scale, of 60 to 64 give good service when used at temperatures below 200 C. Cylinder liners of cobalt-chromium-tungsten alloy[2] are also satisfactory to 300 C.

3.3 *Orifice:*

3.3.1 The outside of the steel orifice shall be of such diameter that it will fall freely to the bottom of the 9.550-mm (0.376-in.) diameter hole in the cylinder (Note 4). The orifice shall have a smooth straight bore 2.095 ± 0.005 mm (0.0825 ± 0.0002 in.) in diameter and shall be 8.000 ± 0.025 mm (0.315 ± 0.001 in.) in length. The bore and its finish are critical. It shall have no visible drill or other tool marks and no detectable eccentricity. The orifice bore shall be finished by techniques known to produce approximately 12 rms or better in accordance with ANSI B46.1, Surface Texture.

NOTE 4—Orifice materials also found to be satisfactory are tungsten carbide,[3] synthetic sapphire, and cobalt-chromium-tungsten alloy.[2]

3.4 *Piston:*

3.4.1 The piston shall be made of steel with an insulating bushing at the top as a barrier to heat transfer from the piston to the weight. The land of the piston shall be 9.474 ± 0.007 mm (0.3730 ± 0.0003 in.) in diameter and 6.35 ± 0.13 mm (0.250 ± 0.005 in.) in length. Above the land, the piston shall be relieved to 8.890 ± 0.025 mm (0.350 ± 0.001 in.) in diameter. The finish of the piston foot shall be 12 rms in accordance with ANSI B46.1.

NOTE 5—Pistons of SAE 52100 steel with the bottom 25 mm (1 in.), including the foot, hardened to a Rockwell hardness, C scale, of 55 to 59 have been found to give good service when used at temperatures below 200 C.

3.4.2 A solid piston relieved to 7.900 ± 0.025 mm (0.311 ± 0.001 in.) in diameter above the land with a loose-fitting metal guiding sleeve at the top of the cylinder is desirable for use with the 21,600-g load. If wear or corrosion are a problem, the piston should be of stainless steel and equipped with a detachable foot for ease of replacement.

3.4.3 The piston shall be scribed with two reference marks 4 mm apart in such fashion that when the lower mark coincides with the top of the cylinder or other suitable reference point, the bottom of the piston is 48 mm (1.89 in.) above the top of the orifice. See Fig. 1.

3.4.4 The combined weight of piston and load shall be within a tolerance of ±0.5 percent of the selected load.

3.5 *Heater:*

3.5.1 Provision shall be made for heating the apparatus so that the temperature of the material can be maintained within ±0.2 C of the desired temperature during the test. At temperatures higher than 200 C, this degree of temperature control may be more difficult to obtain. The temperature specified shall be the equilibrium temperature of the material 12.7 mm ($^1/_2$ in.) above the orifice.

3.5.2 The temperature-indicating device shall be calibrated by means of a thermo-

[2] Haynes Stellite 6B, manufactured by the Haynes Stellite Co., Kokoma, Ind., has been found satisfactory for this purpose.

[3] Talide C-99, manufactured by the Metal Carbides Corp., Youngstown, Ohio, has been found satisfactory for this purpose.

couple inserted in the material 12.7 mm (1/2 in.) above the orifice. The thermocouple can be 28-gage iron - constantan insulated with TFE-fluorocarbon. About 150 mm (6 in.) of the thermocouple should be folded inside the cylinder to minimize heat loss by the wire. The thermocouple and junction may be folded several times, wrapped with thread, and inserted in an orifice so that the junction is 12.7 mm (1/2 in.) above the orifice. The entire assembly may be inserted in the cylinder and the lead wire brought out through the bottom of the cylinder to a potentiometer having a sensitivity of 0.005 mV. It is desirable to calibrate the thermocouple and mounting with reference to a platinum resistance thermometer.

3.5.3 Heat shall be supplied by an electric band heater which covers the entire length of the cylinder. The heater shall contain two 100-W concentric elements. As close to 90 percent as feasible of the power required to maintain the cylinder at the specified temperature should be supplied continuously by the outer of the two heating elements. Six tenths of this amount of power shall be applied intermittently by the inner element as required to maintain the specified temperature (Note 6). The cylinder with heater shall be lagged with 38 mm (1.5 in.) of foamed-glass insulation.[4] A TFE-fluorocarbon plate 3.2 mm (1/8 in.) in thickness shall be attached to the bottom of the cylinder to minimize heat loss at this point.

NOTE 6—Experience has shown that the correct adjustment of the power to the heaters is important in order that all instruments possess identical thermal characteristics. A convenient method for adjustment is obtained by measuring the voltage required to maintain the cylinder at the specified temperature (±1 C) using only the constant heater. Then the voltage to the constant heater is readjusted to the voltage obtained from the calculation:

$$E_c = \sqrt{0.9e^2} = 0.95e$$

where:

E_c = voltage required to maintain approximately 90 percent of the power required to keep the cylinder at the specified temperature, and

e = voltage required to maintain the power required to keep the cylinder at the specified temperature.

The intermittent heater voltage is then obtained by the calculation:

$$E_i = \sqrt{0.6e^2} = 0.78e$$

where:

E_i = voltage specified for the intermittent heater, and

e = same as above.

Measurement of the voltages is imperative, because variable autotransformer settings do not necessarily represent voltage.

3.6 *Thermoregulator:*

3.6.1 A mercury thermoregulator with an associated relay may be used for temperature control provided it will furnish the specified temperature in the material 12.7 mm (1/2 in.) above the orifice. These thermoregulators have a 90-deg bend with a 70-mm (2 3/4-in.) body, 155-mm (6 1/10-in.) stem to inside of bend, 28.6-mm (1 1/8-in.) bulb length, and bulb diameter not over 6.4 mm (1/4 in.). Other types of thermoregulators with a resistor as the sensitive element may be used for closer control and ease of changing temperature.

NOTE 7—A low-melting alloy such as Wood's metal or a suitable silicone fluid is necessary to improve the thermal contact for both the thermoregulator and thermometer. Care should be taken that it is suitable for the selected temperature. The thermoregulator and thermometer should be removed if the plastometer is cooled to room temperature or else breakage may result if Wood's metal is used.

3.7 *Thermometer:*

3.7.1 Thermometers having the same dimensions as the thermoregulators and having a range of 4 C, graduated in 0.2 C divisions may be used to indicate temperature. The temperature at this point may not necessarily be the temperature of the material 12.7 mm (1/2 in.) above the orifice. The thermometer may be used to monitor indirectly the temperature of the material 12.7 mm (1/2 in.) above the orifice and may be calibrated by reference to a thermocouple inserted in the material 12.7 mm (1/2 in.) above the orifice. See 3.5.2 for a description of the thermocouple and the method for measuring temperature.

3.8 *Accessory Equipment:*

3.8.1 Necessary accessories include equipment for charging samples to the cylinder, a funnel, a tool for cutting off the extruded sample, a timer or stop watch, cleaning equipment, and a balance accurate to ±0.001 g.

---

[4] "Foamglas" 54-mm (2 1/8-in.) diameter 1W pipe insulation as manufactured by the Pittsburgh Corning Corp., Pittsburgh, Pa., is recommended for this purpose.

4. Test Specimens

4.1 The test specimen may be in any form that can be introduced into the bore of the cylinder, for example, powder, granules, strips of film, or molded slugs. It may be desirable to preform or pelletize a powder.

5. Conditioning

5.1 Many thermoplastic materials do not require conditioning prior to testing. Materials that contain volatile components, that are chemically reactive or have other special characteristics most probably require appropriate conditioning procedures. Moisture not only affects reproducibility of flow rate measurement but, in some types of materials, degradation is accelerated by moisture at the high temperatures used in testing. If conditioning is necessary, see the applicable material specification and ASTM Methods D 618, Conditioning Plastics and Electrical Insulating Materials for Testing.[5]

6. Procedure A—Manual Operation

6.1 Select conditions of temperature and load from Table 1 in accordance with material specifications such that flow rates will fall between 0.15 to 25 g/10 min.

6.2 The following conditions have been found satisfactory for the materials listed:

	Condition
Acetals	E, M
Acrylics	H, I
Acrylonitrile-butadiene-styrene	G
Cellulose esters	D, E, F
Nylon	K, Q, R, S
Polychlorotrifluorethylene	J
Polyethylene	A, B, D, E, F, N
Polycarbonate	O
Polypropylene	L
Polystyrene	G, H, I, P
Vinyl acetal	C

6.3 The apparatus shall be clean. The parts are more easily cleaned while they are hot. The temperature of the cylinder with piston and orifice in place shall have been at the test temperature for at least 15 min before a test is begun. When the equipment is used repeatedly it should not be necessary to heat the piston and orifice for 15 min. See Note 6 for procedure for adjusting temperature. Take care that cleaning or previous use shall not have changed the dimensions. Make frequent checks to determine whether the orifice diameter is within the tolerances given in 3.3.1.

6.4 Experience has shown that for the best reproducibility the piston should operate within the same part of the cylinder each measurement. The piston is scribed so that the starting point for each extrusion is roughly the same. The charging weights shown in Table 2 are based on an amount such that it may be necessary to force some of the resin out of the cylinder to start the extrusion within the scribed marks. For convenience in running the test it may be desirable to force out some of the resin manually. This should be done during the first 4 min of preheat time and only when the operator is certain that this will not affect the results.

6.5 Charge the cylinder with $3\frac{1}{2}$ to 4 g of the sample, place the loaded piston in position, and start the timer.

NOTE 8—If flow rates of 10 to 25 g/10 min are being measured, the loss of sample while preheating will be appreciable. In this case a smaller weight on the piston may be used during the first 4 min of preheat time. It may be changed to the desired weight near the end of 4 min of preheat time. The 5-min cutoff is made after changing weights.

NOTE 9—In case a specimen has a flow rate at the borderline and slightly different values are obtained at different time intervals, the referee value shall be obtained at the longer time interval.

6.6 After 5 min cut off and discard the extruded portion. Make the cutoff flush with the bottom of the orifice. After one more minute, cut off and discard the extruded portion.

NOTE 10—There may be cases where 6 min of preheat time may not be sufficient. Six minutes is a minimum preheat time. Longer preheat periods are permissible when they can be shown to be necessary. If longer preheat times are used, the report shall so indicate. Care may be necessary to ensure that the sample so tested is adequately protected against changes in polymer structure, usually by appropriate antioxidant addition.

6.7 Start collecting the next extruded portion exactly according to the time interval given in Table 2, provided that the position of the piston is such that the top scribed mark is visible above the cylinder or index and the lower scribed mark is in the cylinder. Otherwise discard the charge and readjust the weight of charge. If the extrudate contains visible air bubbles, discard the complete

[5] *Annual Book of ASTM Standards*, Part 27.

charge and begin the test again.

NOTE 11—It is frequently helpful to take interim cuts of the extrudate at uniform time intervals during the specified extrusion time. Weights of these individual cuts give an indication of the presence of bubbles which may be masked due to their size or to opacity of the sample. This technique is particularly helpful in the case of highly pigmented materials. Forcing out some of the resin manually during the preheat period, as outlined in 6.4, often eliminates bubbles in the test extrudate.

6.8 Discharge the remainder of the specimen and push the orifice out through the top of the cylinder. Swab out the cylinder with cloth patches after the manner of cleaning a pistol barrel. The orifice may be cleaned by dissolving the residue in a solvent. A better method is pyrolytic decomposition of the residue in a nitrogen atmosphere. Place the orifice in a tubular combustion furnace or other device for heating to 550 ± 10 C and clean with a small nitrogen purge through the orifice. This method is preferable to flame or solvent cleaning, being faster than solvent cleaning and less detrimental to the orifice than an open flame. In certain cases where materials of a given class having similar flow characteristics are being tested consecutively, interim orifice cleaning may be unnecessary. In such cases, however, the effect of cleaning upon flow rate determination must be shown to be negligible if this step is avoided.

6.9 Weigh the extrudate to the nearest 1 mg when cool.

6.10 Multiply the weight by the factor shown in Table 2 to obtain the flow rate in g/10 min.

NOTE 12—Frequently, errors which defy all but the most careful scrutiny exist in test technique, apparatus geometry, or test conditions, causing discrepancy in flow rate determinations. The existence of such errors is readily determined by periodically measuring a reference sample of known flow rate. The flow rate value and range to be tolerated can be determined using a statistically correct test program composed of multiple determinations with various instruments. Standard samples are not available.

7. Precision

7.1 The following data should be used for judging the acceptability of results (95 percent confidence limits).

NOTE 13—These precision data are approximations based on limited data, but they provide a reasonable basis for judging the significance of results.

7.2 *Reproducibility*—The result reported by one laboratory should not be considered suspect unless it differs from that of another laboratory by more than that shown in Table 3.

8. Procedure B—Automatically Timed Flow Rate Measurement

8.1 *Apparatus:*

8.1.1 Extrusion plastometer and auxiliary equipment as detailed in Section 3 and below.

8.1.2 Device for electrically or mechanically timing piston movement within the specified travel range in the plastometer. The device should be designed so that it may be calibrated and corrected for its influence, if any, on the load being applied to the resin by the piston during test. It should be capable of measuring the time of travel over the specified distance to within ±0.1 s. Figure 3 shows the arrangement and essential features of one design of time-actuating switch and timer. In this device a switch operates a split-second electronic timer between two predetermined points set up by the switch. It is provided with an overriding mechanism which prevents damage to the follower arm or to the delicate contact bar assembly. The timer-actuating switch is so designed that it will actuate the timer either for piston travel of 6.35 mm (0.25 in.) for measurement of flow rates from 0.15 to 25 g/10 min or for piston travel of 25.4 mm (1 in.) for measurement of flow rates from 25 to 300 g/10 min. Figure 3 shows the essential features of the apparatus with the piston in position at the end of the timing operation. The switch is shown in position at the end of the timing operation. The switch is shown in position for measurement of flow rates from 25 to 300 g/10 min.

8.2 *Procedure:*

8.2.1 To ensure high interlaboratory reproducibility it is important that the timing device be adjusted so that the piston operates within a fixed part of the cylinder. Position the timer-actuating device so that, regardless of the distance the piston travels, the test will terminate with the piston foot 25.4 ± 1.5 mm (1 ± 0.06 in.) from the top of the orifice. For tests of flow rates within the range from 0.15 to 25 g/10 min, set or adjust the piston movement timing device so it will record the time for the piston to travel 6.35 ± 0.25 mm

(0.250 ± 0.010 in.). For flow rates greater than 25, set up the apparatus for a timed piston travel of 25.4 ± 0.25 mm (1.000 ± 0.010 in.). Determine the length of timed movement in either case to the nearest 0.025 mm (0.001 in.) for use in the calculation for melt flow rate.

8.2.2 For the timer illustrated by Fig. 3, the switch calibrating device shown in Fig. 4 is useful for adjusting the electrical contacts to open and close at the proper distances above the orifice.

8.2.3 Check orifice, cylinder, and piston dimensions for conformance to 3.2 through 3.4 and Figs. 1 and 2.

8.2.4 Refer to Table 1 for selection of conditions of temperature and load in accordance with the material specification.

8.2.5 Set the temperature of the plastometer as required for the material being tested and with voltage applied to the heaters as detailed in Note 6.

8.2.6 Clean the apparatus thoroughly prior to each test (see 6.8). Position the piston and orifice in the cylinder and seat firmly on the base plate. Maintain the temperature for at least 15 min before beginning a test. When the equipment is used repeatedly it should not be necessary to heat the piston and orifice for 15 min.

8.2.7 Charge from 2.5 to 7 g of sample depending on the expected flow rate of the material as shown in Table 4.

8.2.8 Place the loaded piston in position, and start the stop watch or preheat timer. For samples in the higher flow rate range, where loss of material during preheat is excessive and premature starting of the extrudate timer may occur, use a smaller weight on the piston during the first 4 min or support the piston 25.4 mm (1 in.) above the timer start point. If a smaller weight is used, change to the specified test weight before 4 min.

8.2.9 Within 30 s after completing 5 min of the preheat period, adjust the position of the piston in the cylinder by manually forcing out excess resin, if needed, so that upon subsequent free travel it will start the timer between the sixth and eighth minute of total preheat time. Sometimes no adjustment will be needed to achieve this timing, or it will be limited simply to a removal of the weight support, if used.

8.2.10 Observe the extrudate for voids during the flow rate measurement time. Also note the condition of the extrudate leaving the orifice for any indication of material leakage around the sides of the die.

8.2.11 Record the time to the nearest 0.1 s for the piston to complete the calibrated distance of travel. Discard any runs with voids occurring in the timed portion or leakage of extrudate around the die.

8.2.12 Discharge any remaining resin and clean the orifice and cylinder as detailed in 6.8.

9. Calculations

9.1 Calculate the flow rate in g/10 min as follows: (Note 14)

$$\text{Flow rate} = (427 \times L \times d)/t$$

where:

L = length of calibrated piston travel, cm,
d = density of resin at test temperature, g/cm^3 (see reference under Table 2),
t = time of piston travel for length L, s, and
427 = mean of areas of piston and cylinder × 600.

NOTE 14—The factors to be substituted in the following equation are given for some materials in Table 5.

$$\text{Flow rate, g/10 min} = F/t$$

where:

F = factor from Table 5, and
t = time of piston travel for length L, s.

10. Precision

10.1 The following data should be used for judging the acceptability of results (95 percent confidence limits).

NOTE 15—These precision data are approximations based on limited data, but they provide a reasonable basis for judging the significance of results.

10.2 *Reproducibility*—The result reported by one laboratory should not be considered suspect unless it differs from that of another laboratory by more than that shown in Table 6.

11. Report

11.1 The report shall include the following:

11.1.1 Statement indicating the nature and physical form of the material charged to the cylinder.

11.1.2 Temperature, load, and pressure at which the test is run may be reported. However, it is suggested that, since these are fixed by the condition under which the test was performed, the results and test conditions can be referred to as FR-A or FR-B, etc., where the last letter refers to the Condition in Table 1.

NOTE 16—It has become customary to refer to the flow rate of polyethylene as "melt index" when obtained under Condition E; however, the use of this term for designating the flow rate of other materials is discouraged.

11.1.3 Flow rate reported as the rate of extrusion in grams per 10 min.

11.1.4 Procedure used (A or B).

11.1.5 Any unusual behavior of the test specimen such as discoloration, sticking, etc.

11.1.6 Details of conditioning if any.

REFERENCES

(1) "Polyethylene Insulation and Sheathing for Electrical Cables," Government Department Electrical Specification No. 27, Great Britain, 1950.

(2) Tordella, J. P., and Jolly, R. E., "Melt Flow of Polyethylene," *Modern Plastics*, MOPLA, Vol 31, No. 2, 1953, p. 146.

(3) Dexter, F. D., "Plasticity Grading of Fluorothenes," *Modern Plastics*, MOPLA, Vol 30, No. 8, 1953, p. 125.

(4) Harban, A. A., and McGlamery, R. M., "Limitations on Measuring Melt Flow Rates of Polyethylene and Ethylene Copolymers by Extrusion Plastometer," *Materials Research and Standards*, MTRSA, Vol 3, No. 11, 1963, p. 906.

(5) Rudin, A., and Schreiber, H. P., "Factors in Melt Indexing of Polyolefins, "*SPE Journal*, SPEJA, Vol 20. No. 6, 1964, p. 533.

TABLE 1 Standard Test Conditions—Temperature and Load

Condition	Temperature, deg C	Load, Piston and Weight, g	Approximate Pressure	
			kgf/cm²	psi
A	125	325	0.46	6.5
B	125	2 160	3.04	43.25
C	150	2 160	3.04	43.25
D	190	325	0.46	6.5
E	190	2 160	3.04	43.25
F	190	21 600	30.40	432.5
G	200	5 000	7.03	100.0
H	230	1 200	1.69	24.0
I	230	3 800	5.34	76.0
J	265	12 500	17.58	250.0
K	275	325	0.46	6.5
L	230	2 160	3.04	43.25
M	190	1 050	1.48	21.0
N	190	10 000	14.06	200.0
O	300	1 200	1.69	24.0
P	190	5 000	7.03	100.0
Q	235	1 000	1.41	20.05
R	235	2 160	3.04	43.25
S	235	5 000	7.03	100.0

TABLE 2 Standard Test Conditions—Sample Weight and Testing Time—Procedure A

Flow Range, g/10 min	Suggested Weight[a] of Sample in Cylinder, g	Time Interval, min	Factor for Obtaining Flow Rate, g/10 min
0.15 to 1.0	3.5 to 4.0	6	1.67
1.0 to 3.5	3.5 to 4.0	3	3.33
3.5 to 10	3.5 to 4.0	1	10.00
10 to 25	3.5 to 4.0	0.5	20.00

[a] This is a suggested weight for materials with melt densities of about 0.7 g/cm³. Correspondingly greater quantities are suggested for materials of greater melt densities. Density of the molten resin (without filler) may be obtained using the procedure described by Terry, B. W., and Yang, K., "A New Method for Determining Melt Density as a Function of Pressure and Temperature," *SPE Journal*, SPEJA, Vol 20, No. 6, June 1964, p. 540. It may also be obtained from the weight of an extruded known volume of resin at the desired temperature. For example, 25.4 mm (1 in.) of piston movement extrudes 1.804 cm³ of resin. An estimate of the density of the material can be calculated from the following equation:

$$\text{Resin density at test temperature} = W/1.804$$

where W = weight of extruded resin.

TABLE 3 Expected Reproducibility for Some Polymers—Procedure A

Polymer	Nominal Flow Rate, g/10 min	Reproducibility, g/10 min
Polyethylene	0.25	±0.04
Polypropylene	0.4	±0.06
Polypropylene	6.0	±0.46
Polyethylene	7.0	±0.48
Polypropylene	12.5	±1.18
Polyethylene	20.0	±2.52

TABLE 5 Factors for Calculation of Flow Rate

Material (unpigmented)	Temperature, deg C	Piston Travel, *L*, mm (in.)	Factor for Calculation of Flow Rate, *F*
Polyethylene	190	25.4 (1)	822
Polyethylene	190	6.35 (0.25)	205
Polypropylene	230	25.4 (1)	799
Polypropylene	230	6.35 (0.25)	199

TABLE 4 Suggested Sample Charge Weight—Procedure B

Flow Rate, g/10 min	Suggested Weight of Sample, g
0.15 to 1.0	2.5
1.0 to 5.0	5.0
5.0 to 300	7.0

TABLE 6 Expected Reproducibility for Some Polymers—Procedure B

Polymer	Nominal Flow Rate, g/10 min	Reproducibility, g/10 min
Polyethylene	0.25	±0.04
Polypropylene	0.4	±0.06
Polypropylene	6.0	±0.54
Polyethylene	7.0	±0.34
Polypropylene	12.5	±1.02
Polyethylene	33	±2.2
Polypropylene	47	±5.6
Polypropylene	125	±14.6
Polypropylene	278	±37.8

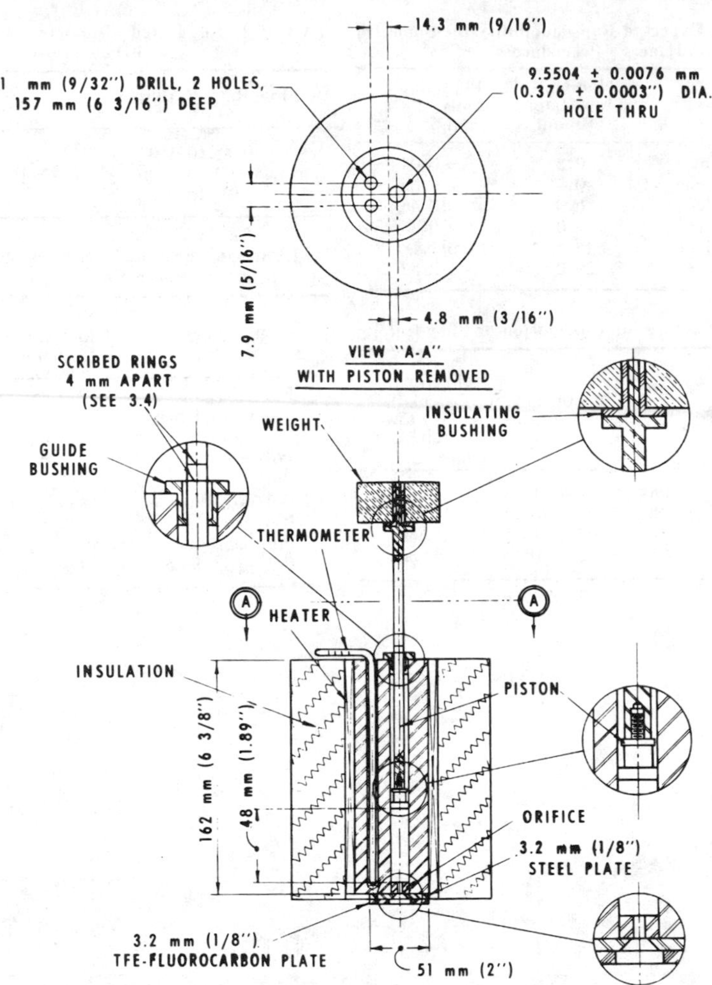

FIG. 1 General Arrangement of Extrusion Plastometer.

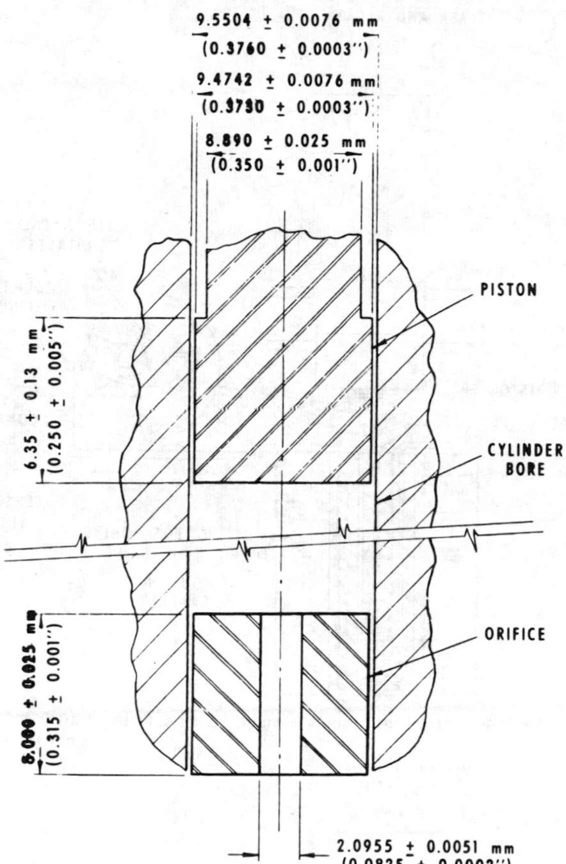

FIG. 2 Details of Extrusion Plastometer.

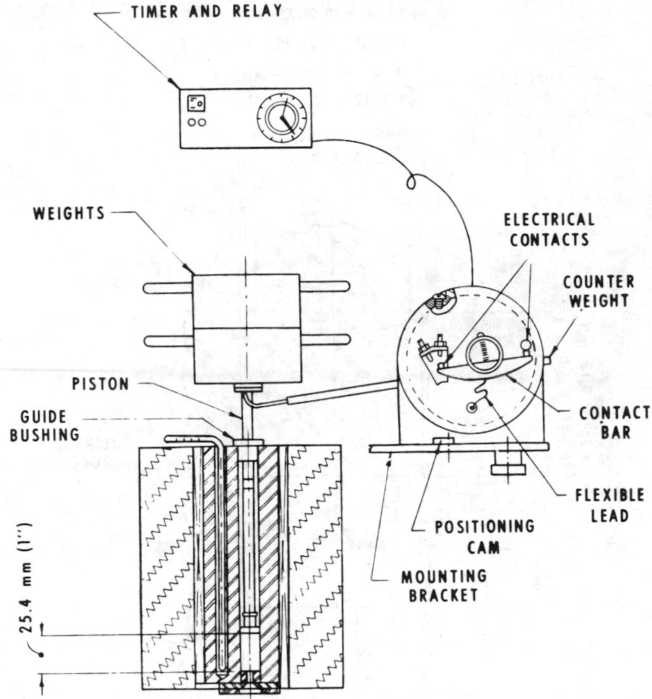

FIG. 3 Extrusion Plastometer with Automatic Timer-Actuating Switch and Timer.

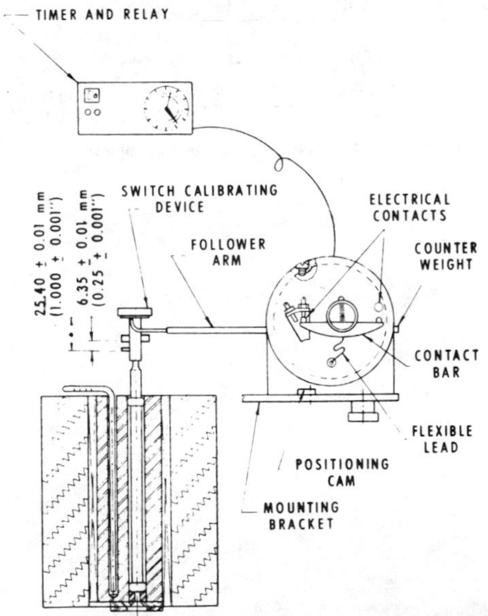

FIG. 4 Extrusion Plastometer with Automatic Timer-Actuating Switch, Switch Calibrating Device, and Timer.

Designation: D 1239 – 55 (Reapproved 1966)

Standard Method of Test for
RESISTANCE OF PLASTIC FILMS TO EXTRACTION BY CHEMICALS[1]

This Standard is issued under the fixed designation D 1239; the number immediately following the designation indicates the year of original adoption or, in the case of revision, the year of last revision. A number in parentheses indicates the year of last reapproval.

This method was prepared jointly by the American Society for Testing and Materials and The Society of the Plastics Industry.

1. Scope

1.1 This method for resistance of plastic films to chemicals covers the measurement of the weight loss of film after immersion in chemicals.

NOTE 1—Film is defined as sheeting having nominal thickness not greater than 0.25 mm (0.010 in.), in accordance with ASTM Nomenclature D 883, Relating to Plastics.[2]

NOTE 2—The values stated in U.S. customary units are to be regarded as the standard. The metric equivalents of U.S. customary units may be approximate.

2. Significance

2.1 The method is intended to be a rapid empirical test to determine the loss of the plasticizer or other extractable components from the plastic film when immersed in liquids commonly used in households.

3. Apparatus

3.1 *Balance*—An accurate analytical balance.

3.2 *Containers*—Pint jars or cans with a diameter of at least 6.5 cm (2.5 in.) (one container for each specimen).

4. Materials

4.1 *Distilled Water*—Freshly prepared distilled or deionized water.

4.1.2 *Soap Solution* (*1 percent*)—Dissolve 12 g of dehydrated pure white soap flakes (dried for 1 h at 105 C) in 1200 ml of distilled water. This is sufficient solution to test three specimens.

4.1.3 *Cottonseed Oil*—Household cooking grade.

4.1.4 *Mineral Oil, U.S.P.*—Heavy grade, sp gr 0.875 to 0.905.

4.1.5 *Kerosine.*

4.1.6 *Ethyl Alcohol* (*50 percent*) as described in ASTM Method D 543, Test for Resistance of Plastics to Chemical Reagents.[2]

4.1.7 Any other standard or supplementary reagent listed in Method D 543.

5. Test Specimens

5.1 The test specimens for plastic films shall be in the form of squares 50 ± 0.25 mm (2 in.) on each side. At least three specimens of each sample shall be tested with each chemical reagent.

5.2 Nothing in this test method precludes the use of specimens of other dimensions or the making of other tests on the same specimens after they have been exposed to the chemicals. Another acceptable specimen is a disk 50 ± 0.25 mm (2 in.) in diameter (total area 41.5 cm^2), or a tensile specimen 100 by 25 mm (4 by 1 in.) as prescribed in ASTM Methods D 882, Test for Tensile Properties of Thin Plastic Sheeting.[2] For such specimens, care should be taken that a proportionate amount of chemical be used and that the container is of appropriate dimensions so that the specimen can be immersed in a completely vertical position during the test. The

[1] This method is under the jurisdiction of ASTM Committee D-20 on Plastics. A list of committee members may be found in the ASTM Yearbook. This standard is the direct responsibility of Subcommittee D-20.50 on Permanence Properties.

Current edition effective Sept. 12, 1955. Originally issued 1952. Replaces D 1239 – 52 T.

[2] *Annual Book of ASTM Standards*, Part 27.

amount of chemical used shall be approximately 8 ml/cm^2, counting the area of both sides of the specimen.

NOTE 3—Direct comparison of values should not be made between samples of different thicknesses, since percentage weight loss is a function of thickness.

6. Conditioning

6.1 *Conditioning*—Condition the test specimens at 23 ± 2 C (73.4 ± 3.6 F) and 50 ± 5 percent relative humidity for not less than 40 h prior to test in accordance with Procedure A of ASTM Methods D 618, Conditioning Plastics and Electrical Insulating Materials for Testing,[2] for those tests where conditioning is required. In cases of disagreement, the tolerances shall be ± 1 C (±1.8 F) and ±2 percent relative humidity.

6.2 *Test Conditions*—Conduct tests in the Standard Laboratory Atmosphere of 23 ± 2 C (73.4 ± 3.6 F) and 50 ± 5 percent relative humidity, unless otherwise specified in the test methods or in this specification. In cases of disagreements, the tolerances shall be ±1 C (±1.8 F) and ±2 percent relative humidity.

7. Procedure

7.1 Maintain the chemical reagent at the test temperature for at least 4 h before the specimens are immersed in it.

7.2 After being weighed, immerse the specimens in the liquids, one specimen to each container. Each jar shall contain 400 ml of the liquid. Suspend the specimen freely in a vertical position (Note 4), but fully covered by the liquid.

NOTE 4—To prevent each specimen from floating or curling, it may be necessary to attach small weights, such as paper clips.

7.3 Cover the jars containing the specimens and keep at the test temperature for the specified time. The standard conditions of test shall be 24 h at 23 C. Alternative conditions suggested are 4 h at 23 C, or either 4 or 24 h at 40 C.

NOTE 5—The maximum weight loss by extraction is generally limited to approximately 50 percent of the plasticizer content. If, in a comparison of materials, one or several samples have a weight loss greater than 15 percent, tests should be made on all samples at a lower temperature or for less time.

7.4 Remove the specimens from the liquids and gently wipe with a soft cloth or absorbent tissue. Specimens taken from water or volatile solvents like acetone or gasoline require no rinsing, but simply wipe dry as directed; rinse specimens tested in salt solutions, soaps, acids, or alkalis, with water before wiping to dryness; and rinse specimens tested in nonvolatile oils with a poor but volatile solvent such as ligroine.

7.5 It is realized that if the immersion chemical is not volatile, has good adhesion to the film, and does not attack the film, there may be an increase in weight of the specimen at the end of the test. Determine a weight correction by conditioning another sample of the same film, before immersion, in the same manner as for the standard test, but immerse the specimens in the particular chemical for only 5 min and then rinse and wipe dry. Add the average percentage weight gain of this blank sample to the average percentage weight loss; if the blank sample has a weight loss by this procedure, do not make any correction. If a sample gains more weight than its blank, report the difference as a percentage weight gain.

8. Calculations

8.1 The percentage loss in weight from extraction, expressed as percentage weight loss compared to the original specimen weight, shall be calculated as follows:

$$\text{Weight loss, percent} = [(W_1 - W_2)/W_1] \times 100$$

where:

W_1 = weight of specimen after the conditioning period, and

W_2 = weight of specimen at the end of the test.

8.2 The values obtained for the three specimens for percentage weight loss shall be averaged and this value reported as the percentage weight loss of the sample being tested.

9. Report

9.1 The report shall include the following:

9.1.1 Complete identification of material tested, including type, thickness, source, manufacturer's code number, and previous history,

9.1.2 The length of time and temperature for each test,

9.1.3 The average and range of percentage weight loss or gain for all specimens from a given sample at each test condition, and

9.1.4 Any observation as to change in appearance of the specimens.

 Designation: D 1242 – 56 (Reapproved 1969)

Standard Methods of Test for RESISTANCE TO ABRASION OF PLASTIC MATERIALS[1]

This Standard is issued under the fixed designation D 1242; the number immediately following the designation indicates the year of original adoption or, in the case of revision, the year of last revision. A number in parentheses indicates the year of last reapproval.

1. Scope

1.1 These methods cover the determination of the resistance to abrasion of flat surfaces of plastic materials, measured in terms of volume loss, by two different types of abrasion-testing machines, as follows:

1.1.1 *Method A*—Loose Abrasive, and

1.1.2 *Method B*—Bonded Abrasive on Cloth or Paper (Abrasive Tape).

NOTE 1—The values stated in U.S. customary units are to be regarded as the standard. The metric equivalents of U.S. customary units may be approximate.

NOTE 2—For determining resistance of transparent plastics to surface abrasion by measuring the effect on its optical clarity, reference should be made to ASTM Method D 1044, Test for Resistance of Transparent Plastics to Surface Abrasion.[2]

2. Definition

2.1 *resistance to abrasion*—the ability of a material to withstand mechanical action such as rubbing, scraping, or erosion, that tends progressively to remove material from its surface.

3. Significance

3.1 The measurement of the resistance to abrasion of plastic materials is very complex. The resistance to abrasion is affected by many factors such as the physical properties of the polymeric material, particularly hardness, resilience, and the type and degree of added filler such as cellulose fiber or pigment. Resistance to abrasion is also affected by conditions of the test, such as the nature of the abradant, action of the abradant over the area of the specimen abraded, and development and dissipation of heat during the test cycle.

3.2 Although laboratory measurement of the resistance to abrasion of a given material may be closely allied, in some instances, to material quality, results should not be used as an absolute index of ultimate life, particularly in view of varied end-use applications. Correlation of laboratory abrasion data with end-use performance depends upon the ability to assess those actions in service which affect durability, and it is for this reason that many different types of testing machines have been employed. Results from machines such as those described in these methods are of value chiefly in the development of materials, and should not be used without qualification as a basis for commercial comparisons.

3.3 The procedures described in these methods have been prepared in an attempt to standardize methods of test used for measuring abrasion resistance of plastic materials. It is believed that the standardization of such methods, plus the addition of new methods as developed, and subsequent choice of that test method most closely simulating service conditions, will provide a means of measuring durability of plastic materials for a given application. These methods have been prepared in an attempt to attain these objectives.

---

[1] These methods are under the jurisdiction of ASTM Committee D-20 on Plastics. A list of committee members may be found in the ASTM Yearbook. This standard is the direct responsibility of Subcommittee D-20.10 on Mechanical Properties.

Current edition effective Sept. 10, 1956. Originally issued 1952. Replaces D 1242 – 52 T.

[2] *Annual Book of ASTM Standards*, Part 27.

Method A—Loose Abrasive[3]

4. Apparatus

4.1 The apparatus shown in Fig. 1[4] shall consist of the following: a rotating disk, a specimen-mounting plate holder, a cam arrangement for actuating the specimen holder, hopper and distributor for feeding fresh abrasive, hopper for collecting used abrasive, revolution counter, an added weight of 4.5 kg (10 lb), and suitable mechanism for driving the disk at 23 $^1/_2$ rpm and the specimen holder at 32 $^1/_2$ rpm.

4.2 *Metal Plate*, 63.5 by 88.9 by 1.6 mm (±0.05 mm) (2$^1/_2$ by 3$^1/_2$ by $^1/_{16}$ in. (±0.002 in.)), for adjusting height of drop.

4.3 *Specimen Mounting Plate*, made of aluminum in accordance with Fig. 2, for mounting test specimens.

4.4 *Analytical Balance*, for weighing specimen to a precision of 0.001 g.

4.5 *Aluminum Oxide Grit No. 80 TP*,[5] or other abradant as specified.

4.6 *Suitable Adhesive*,[6] for adhering test specimens to the specimen plate.

4.7 *Conditioning Room*, providing the Standard Laboratory Atmosphere of 50 ± 2 percent relative humidity at a temperature of 23 ± 1 C (73.4 ± 1.8 F).

5. Test Specimens

5.1 The test specimen shall measure 50.8 ± 0.4 by 76.2 ± 0.4 mm (2 ± $^1/_{64}$ by 3 ± $^1/_{64}$ in.), and shall be mounted face up on the raised 50.8 by 76.2-mm (2 by 3-in.) portion of the specimen mounting plate by means of a suitable adhesive. Good adhesion can be obtained by holding the mounted specimen for 24 h under a 2.3-kg (5-lb) weight in the conditioning room.

5.2 The required number of specimens for each test shall be indicated in the material specifications. In the event the material specifications omit this requirement, the material shall be sampled in triplicate, and duplicate measurements shall be made on each specimen. The average of the six measurements shall be taken as the abrasion loss for the material.

6. Conditioning

6.1 *Conditioning*—Condition the test specimens at 23 ± 2 C (73.4 ± 3.6 F) and 50 ± 5 percent relative humidity for not less than 40 h prior to test in accordance with Procedure A of ASTM Methods D 618, Conditioning Plastics and Electrical Insulating Materials for Testing,[2] for those tests where conditioning is required. In cases of disagreement, the tolerances shall be ±1 C (±1.8 F) and ±2 percent relative humidity.

6.2 *Test Conditions*—Conduct tests in the Standard Laboratory Atmosphere of 23 ± 2 C (73.4 ± 3.6 F) and 50 ± 5 percent relative humidity, unless otherwise specified in the test methods or in this specification. In cases of disagreements, the tolerances shall be ±1 C (±1.8 F) and ±2 percent relative humidity.

7. Procedure

7.1 Determine the specific gravity of the material to be tested in accordance with standard analytical procedures.

NOTE 3—If the specimen as received is not homogeneous, but possesses a surface that differs from the body or core, determine the specific gravity of the surface alone. If abrasion is to be carried beyond this surface into the body, then determine also the specific gravity of the latter and calculate and report the abrasion resistance of the two components separately.

7.2 Fill the abrasive container with the grit and adjust the rate of feed[7] to 44 ± 2 g/min.

---

[3] This method is based on the development described by A. W. Cizek, Jr., D. H. Kallas, and N. Nestlen in "A New Machine for Measuring Wear Resistance of Walkway Materials." *ASTM Bulletin*, No. 132, Jan. 1945.

[4] The Olsen Wearometer machine meets these requirements, and is manufactured by the Tinius Olsen Testing Machine Co., Willow Grove, Pa.

[5] No. 80 TP aluminum oxide grit is manufactured by the Carborundum Co., Niagara Falls, N.Y. Typical screen analysis on this grit has shown from a trace to 1.0 percent retained on the No. 60 (250-μm) sieve, and 5 to 15 percent passing the No. 100 (149-μm) sieve. Detailed requirements for these sieves are given in ASTM Specifications E 11 for Wire-Cloth Sieves for Testing Purposes, *Annual Book of ASTM Standards*, Part 30.

[6] Adhesive 3M, Type EC-711, as manufactured by the Minnesota Mining & Mfg. Co., Minneapolis, Minn.; Nairn Marine Deck Adhesive as manufactured by Congoleum-Nairn, Inc., Kearny, N. J.: or their equivalent, has been found satisfactory for this purpose.

[7] Feed rate may be measured by holding a tared Petri dish under the abrasive distributor for 1 min, while the machine is running, and weighing the amount of grit delivered. Feed rate may be adjusted by repositioning the abrasive feed impeller shaft upon which the abrasive hopper is mounted, either upward or downward. Fresh grit should always be used, and no attempt made to reclaim or run used grit.

7.3 Weigh the mounted test specimen to the nearest 0.01 g, and attach to the specimen plate holder by means of the end clamp provided.

7.4 Place the weight on the specimen shaft, brush off grit under the specimen, place the 1.6-mm ($^1/_{16}$-in.) spacer under the specimen with the larger dimension on the surface of the rotating disk, and adjust the cam follower to this maximum lift. Remove the spacer.

7.5 Set the counter for 1000 revolutions, and run the machine.

7.6 When the machine has stopped, check to see that the revolution counter has returned to zero, remove the mounted specimen, clean with a filtered air blast, and reweigh.

7.7 Inspect the abraded specimen for imbedded grit, unevenness on the abraded surface, peeling off of adhesive, chipped corners or edges, or any irregularities affecting loss of weight weighings.

7.8 For continuing runs, reattach the specimen mounting plate to the specimen plate holder, readjust to a 1.6-mm ($^1/_{16}$-in.) lift as described in 7.4, and repeat the steps given in 7.5, 7.6, and 7.7.

8. Calculation and Report

8.1 Report the resistance to abrasion or the abrasion loss as the loss in volume at 1000 revolutions, calculated as follows:

$$\text{Volume loss, cm}^3 = (W_1 - W_2)/S$$

where:
W_1 = initial weight, g,
W_2 = weight after abrasion, g, and
S = specific gravity of the material being abraded.

8.2 The average volume loss in cubic centimeters at 1000 revolutions for the three specimens tested in duplicate,

8.3 The 95 percent confidence limits, and

8.4 The name and grade of abrasive grit employed in making the test.

NOTE 4—It should be noted that abrasion loss can be expressed in other ways, using the data given above. For instance, the average loss in thickness of the abraded test specimen can be calculated by dividing the loss in volume by the area of the abraded specimen.

9. Reproducibility

9.1 The coefficient of variation[8] for the six measurements performed on a single machine shall not exceed 7 percent.

METHOD B—BONDED ABRASIVE ON CLOTH OR PAPER (ABRASIVE TAPE)[9]

10. Apparatus

10.1 The apparatus shown in Fig. 3,[10] with a schematic arrangement as shown in Fig. 4, shall consist of the following:

10.1.1 A continuous link belt with flat metal carrier plates upon which test specimens and zinc calibration standards may be mounted, and a constant-speed mechanism to provide a rate of travel of 2 m (80 in.)/min for the specimens.

10.1.2 A carriage assembly consisting of supply and take-up drums for the abrasive tape, including necessary guides and rollers for maintaining uniform tension, the entire carriage being mounted on rolls so as to move freely in either direction, with the abrasive kept in contact with the specimen by means of a dead weight attached to the carriage by a cable-spring-pulley arrangement. The constant-speed drive is designed to provide for continuous travel of the abrasive tape at a rate of 0.5 m (20 in.)/min. This motion in opposition to that of the specimens provides a total rate of slip of 2.5 m (100 in.)/min.

10.2 *Specimen Mounting Plates*, made of aluminum in accordance with Fig. 5, for mounting nonrigid or other specimens which require a rigid support.

10.3 *Suitable Adhesive*,[11] for adhering test specimens to the specimen mounting plates.

10.4 *Bonded Abrasive Cloth or Paper* (*abrasive tape*),[12] of agreed-upon specifica-

[8] The method of calculating the coefficient of variation may be found in the ASTM Manual on Quality Control of Materials. (*STP No. 15-C*), 1951.

[9] This method is based upon the development described by F. M. Gavan, S. W. Eby, Jr., and C. C. Schrader in "A New Sandpaper Abrasion Tester." *ASTM Bulletin*, No. 143, Dec. 1946.

[10] A machine meeting the requirements of this method is available from Custom Scientific Instruments, Inc., Box 170, Arlington, N.J.

[11] Adhesive N-111, produced by the Armstrong Cork Co., has been found satisfactory for most materials.

[12] OE Grade sandpaper in 50.8-mm wide rolls 50 m long (2-in.-wide rolls 50 yd long) is available from the Behr-Manning Div. of the Norton Abrasive Co., Troy, N. Y. 11/0-500 "snuffing paper" in similar dimensions is available from the Carborundum Co., Niagara Falls, N. Y. The width of the abrasive tape must be held to a tolerance of 50.8 + 0 − 0.8 mm (2 + 0 − $^1/_{32}$ in.), and all rolls in a package should be from the same production lot. It has been found advisable to have all rolls in a shipment made at the same time.

tions, in rolls 50.8 mm (2 in.) wide.

10.5 *Zinc Calibration Specimens*,[13] fabricated as shown in Fig. 6.

10.6 *Laboratory Balance*, for weighing specimens (including specimen mounting plates when used) to a precision of 0.01 g or better, as required by precision of the test.

10.7 *Conditioning Room*, providing the Standard Laboratory Atmosphere of 50 ± 2 percent relative humidity at a temperature of 23 ± 1 C (73.4 ± 1.8 F), as specified in Methods D 618.

10.8 *Die*, or other means for cutting nonrigid specimens to the required dimensions.

11. Test Specimens

11.1 The test specimens shall measure 44.5 ± 0.25 by 114.3 ± 0.25 mm (1.75 ± 0.01 by 4.50 ± 0.01 in.), and shall be prepared for attachment to the machine. Nonrigid or fragile specimens shall be mounted by means of an adhesive to the aluminum specimen mounting plates. Good adhesion can be obtained by holding the mounted specimen for 24 h under a 18-kg (40-lb) weight in the conditioning room. Rigid specimens of at least 1.6 mm ($^1/_{16}$ in.) in thickness can be directly mounted (without an adhesive) on the carrier plates of the link belt by means of two flat-headed machine screws (4–40) located on the long axis of each plate, along the center line, spaced 38.0 mm (1.5 in.) on each side of the center point of the plate, and countersunk below the thickness of the surface to be abraded.

11.2 At least one zinc calibration standard shall be abraded with each group of test specimens making up a "run" on the abrasion machine. The machine is provided with eight specimen-carrying plates.

11.3 The required number of specimens for each abrasion test shall be indicated in the material specifications. In the event the material specifications omit this requirement, three abrasion specimens shall be prepared from each lot or sub-lot, and the average of the three shall be taken as the abrasion loss for the lot or sub-lot.

12. Conditioning

12.1 All materials to be tested for abrasion shall be conditioned in accordance with 6.1.

12.2 The abrasion test shall be made in the Standard Laboratory Atmosphere in accordance with 6.2.

12.3 All abrasive tape, either cloth or paper, used in this test shall be stored until used in the Standard Laboratory Atmosphere prescribed in Methods D 618. No tape should be calibrated until it has been stored in the conditioned atmosphere for at least 1 week.

13. Procedure

13.1 Calibrate the abrasive tape to be used by selecting a roll at random from each package of 8 or 10 rolls, and make an abrasion test on this roll, using four standard zinc specimens that have been previously run with abrasive tape of known characteristics. Then adjust the number of cycles to be used for a "run" to bring the abrasive effect of the tape within the proper range of loss on the zinc specimens.

NOTE 5—The grade of the abrasive tape (type and grit size), the weight on the carriage, and the length of time of abrasion can be varied in this apparatus to fit the test to particular purposes. Two reference procedures are given below that have been found desirable for the use of coarse-grit and fine-grit abrasives, respectively.

(*1*) *Coarse Grit*—This procedure is based upon use of OE Flint abrasive paper 13.6-kg (30-lb) load on the carriage, and a number of abrasion cycles (complete revolutions of the link belt) which will produce an average loss of 0.525 ± 0.025 g on the zinc calibration specimens. The required number of cycles (approximately 60) must be varied, because of variations in the sandpaper. If it is desired to vary the test conditions, the proper loss on the zinc specimens for OE sandpaper can be calculated by referring to curves similar to those shown in Fig. 7 (zinc comparison standard, relation between abrasion loss and abrasion time) and in Fig. 8 (zinc comparison standard, relation between abrasion loss and force on the carriage). If other abrasive tapes are used, the proper loss on the zinc standard for the desired test characteristics must be determined by experiment and comparison.

(*2*) *Fine Grit*—This procedure is based upon use of 11/0-500 Flint abrasive paper (snuffing paper), 13.6-kg (30-lb) load on the carriage, and a number of abrasion cycles which will produce an average loss of 0.175 ± 0.010 g on the zinc calibration specimens. The required number of cycles is approximately 70. It is often desirable to vary the number of cycles to accommodate the amount of material that may be removed from the test specimens. The proper weight loss on the zinc standard for the de-

[13] Zinc plates 0.81 mm (0.032 in.) thick, with Scleroscope hardness of 15.5, may be secured from the New Jersey Zinc Co. (of Pa.), Palmerton, Pa. (New Jersey Zinc Co. Mix No. 103).

sired test shall be determined by experiment from which data similar to those given in Figs. 7 and 8 may be developed.

13.2 Cut two shims, each 127 mm long by 16 mm wide (5 in. long by $^5/_8$ in. wide), from the material to be tested, to accompany each abrasion specimen. If this is not feasible, use similar material of the same thickness.

13.3 Determine the specific gravity of the material to be abraded in accordance with the standard analytical procedure.

NOTE 6—If the specimen as received is not homogeneous, but possesses a surface which differs from the body or core, determine the specific gravity of the surface alone. If abrasion is to be carried beyond this surface into the body, then determine also the specific gravity of the latter, and calculate and report the abrasion resistance of the two components separately.

13.4 When the specimens have been prepared and conditioned (are ready for mounting on the abrader) weigh them and the zinc standard to the nearest 0.01 g for the coarse-grit test, or to the nearest 0.001 g for the fine-grit test. Mount the specimens on the carrier plates of the abrader with a shim at each end of each specimen. The purpose of these shims is to serve as bumper strips, leading the abrasive contact roll on and off the specimens without damaging its edges.

13.5 Place the weights on the yoke of the carriage. Unless otherwise specified, use a 11.4-kg (25-lb) weight to provide a total force of 13.6 kg (30 lb) acting on the carriage, since the hook, yoke, and platform of the machine weigh 2.3 kg (5 lb).

13.6 Set the counter of the machine to zero and run the abrasion test.

13.7 After the required number of cycles has been run, remove the mounted specimens and calibration standards, clean with a filtered air blast, and reweigh.

13.8 If the specimens are to be retested to determine the abrasion resistance below the surface or at various depths, they may be replaced on the machine and run for additional cycles as required.

13.9 In the case of very hard materials, such as zinc, a gasket of friction material may have to be used between the clamps and the specimen to avoid slippage. If different specimens are being tested, stagger the arrangement. No part of the shim shall protrude over the edge of the conveyor plate. Check the sandpaper carriage, springs, and pulleys for freedom of movement. After the test has been started, be sure that the abrasive tape is abrading all of the surface of the specimen.

NOTE 7—A single cycle of abrasion will usually reveal areas of the specimen not contacted by the abrasive. If only one side (more or less) is abraded, shimming can help in obtaining abrasions across the full width. When using the fine-grit test on rigid materials, it has been found helpful to employ a contact roll having a 1.6-mm ($^1/_{16}$-in.) thick cover of butyl rubber (with the same outside diameter as the standard steel roll). In this way general contact with surfaces having minor waviness ("orange peel") is improved.

13.10 Major surface irregularities are cause for rejection of the specimen for this test. Observe the test frequently to see that specimens and shims do not pull away from the specimen plate or clamps. Keep the specimens free from larger particles of dirt and grit.

13.11 If, at the end of the test, the loss on the zinc calibration standard is not in proportion to the limits specified, it should be assumed that the abrasive tape or the abrasion machine is out of calibration, and the test shall be repeated.

13.12 Be certain that the loss on the test specimen represents only material abraded, and that no loss is due to peeling off of adhesive, or broken corners and edges of the specimen. Cut test specimens from a portion of the sample that is representative of the material to be tested. Check the abrasion machine for level and alignment before each test.

14. Calculation and Report

14.1 Report the resistance to abrasion or the abrasion loss as the loss in volume for X cycles using Y grade of abrasive tape and a 13.6-kg (30-lb) load on the carriage, calculated as follows:

$$\text{Volume loss, cm}^3 = (W_1 - W_2)/S$$

where:

W_1 = initial weight, g,
W_2 = weight after abrasion, g, and
S = specific gravity of the material being abraded.

NOTE 8—In the case of nonrigid specimens W_1 and W_2, include the weight of the aluminum specimen-mounting plate and the adhesive used.

14.2 The report shall include the following:

14.2.1 The average volume loss in cubic centimeters for the three specimens tested,

14.2.2 The 95 percent confidence limits,

14.2.3 The number of cycles used,

14.2.4 The weight used, in pounds,

14.2.5 The name and grade of abrasive tape employed in making the test, and

14.2.6 The loss, in grams, obtained on the zinc calibration standards run at the same time.

NOTE 9—It should be noted that abrasion loss can be expressed in other ways, using the data given above. For instance, the average loss in thickness of the abraded test specimen can be calculated by dividing the loss in volume by the area of the abraded specimen.

15. Reproducibility

15.1 On highly reproducible substances such as the zinc standard plates, the coefficient of variation[8] has been found not to exceed 5 percent over many rolls of the same grade of abrasive.

NOTE—Distance from center line of camshaft to center line of rotating disk shaft shall be 5 in.

FIG. 1 Loose Abrasive Abrading Machine.

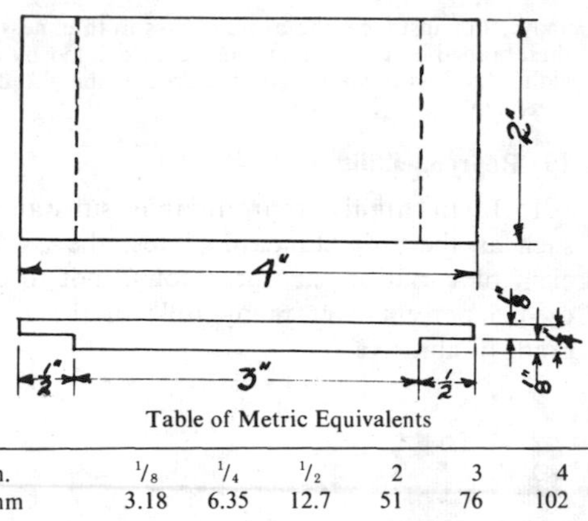

Table of Metric Equivalents

in.	1/8	1/4	1/2	2	3	4
mm	3.18	6.35	12.7	51	76	102

FIG. 2 Specimen Mounting Plates (Aluminum).

FIG. 3 Bonded Abrasive Abrading Machine.

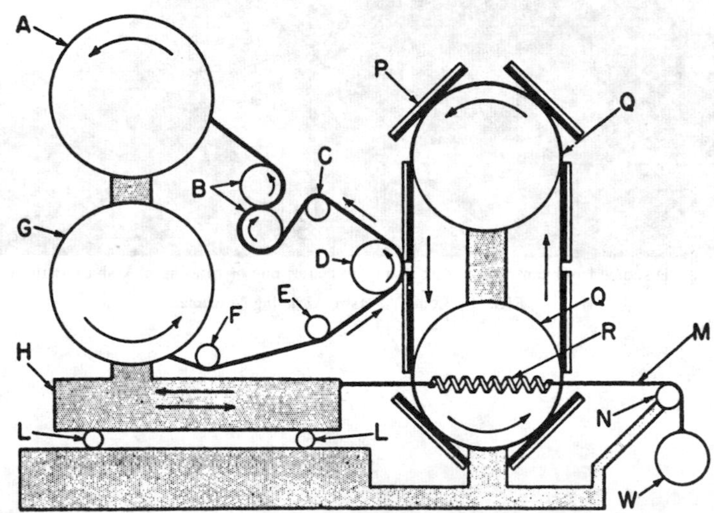

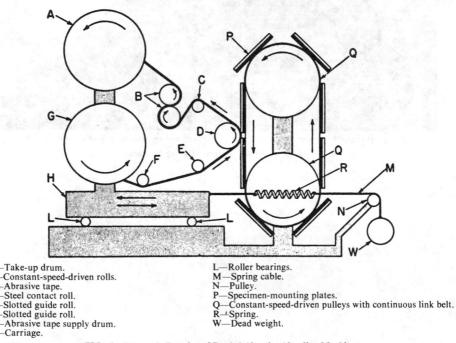

A—Take-up drum.
B—Constant-speed-driven rolls.
C—Abrasive tape.
D—Steel contact roll.
E—Slotted guide roll.
F—Slotted guide roll.
G—Abrasive tape supply drum.
H—Carriage.
L—Roller bearings.
M—Spring cable.
N—Pulley.
P—Specimen-mounting plates.
Q—Constant-speed-driven pulleys with continuous link belt.
R—Spring.
W—Dead weight.

FIG. 4 Schematic Drawing of Bonded Abrasive Abrading Machine.

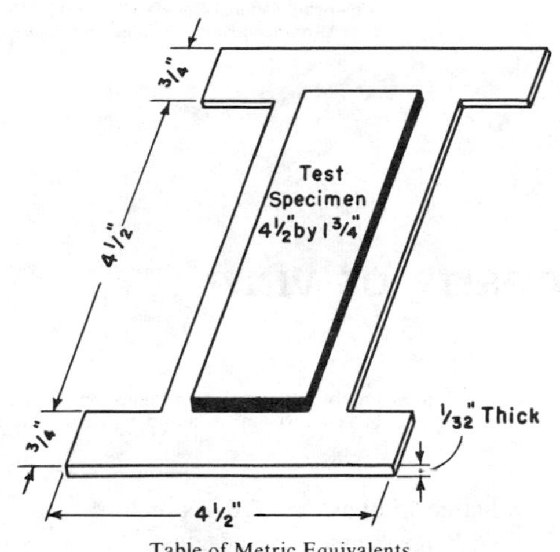

Table of Metric Equivalents

in.	1/32	3/4	1 3/4	4 1/2
mm	0.79	19	44	114

FIG. 5 Specimen Mounting Plate (Aluminum).

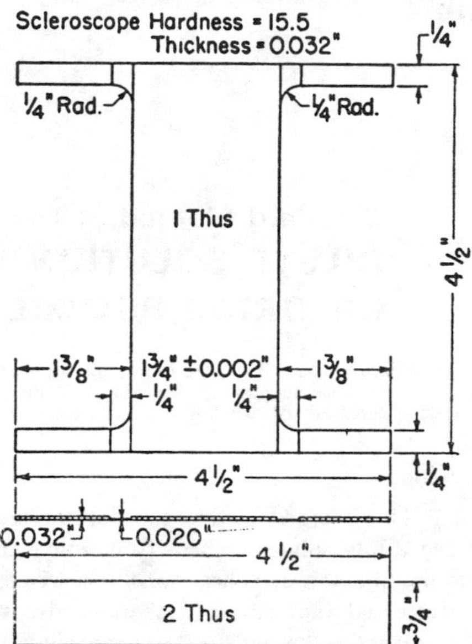

Table of Metric Equivalents

in.	mm
0.002	0.051
0.020	0.508
0.032	0.81
1/4	6.35
3/4	19.05
1 3/8	34.9
1 3/4	44.4
4 1/2	114
6	152

FIG. 6 Zinc Calibration Specimen.

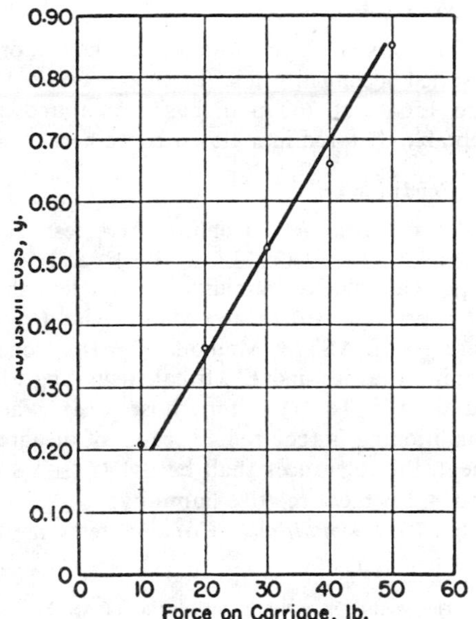

Table of Metric Equivalents

lb	10	20	30	40	50	60
kg	4.5	9.1	13.6	18.2	22.7	27.2

FIG. 8 Zinc Comparison Standard—Relation Between Abrasion Loss and Force on the Carriage.

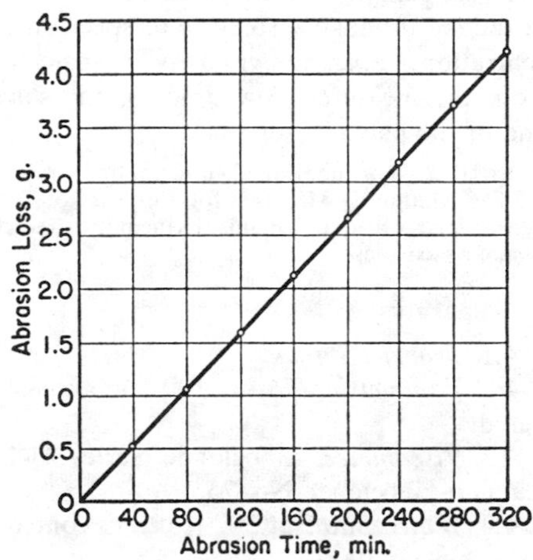

FIG. 7 Zinc Comparison Standard—Relation Between Abrasion Loss and Abrasion Time.

American National Standards K65.10-1967
American National Standards Institute

Standard Method of Test for

DILUTE SOLUTION VISCOSITY OF VINYL CHLORIDE POLYMERS[1]

This Standard is issued under the fixed designation D 1243; the number immediately following the designation indicates the year of original adoption as standard or, in the case of revision, the year of last revision. A number in parentheses indicates the year of last reapproval.

1. Scope

1.1 This method covers the determination of the dilute solution viscosity of vinyl chloride polymers in cyclohexanone. The viscosity is expressed in terms of inherent viscosity. The method is limited to those materials that give clear, uniform solutions at the test dilution.

NOTE 1—Other expressions for viscosity may be used as described in the Appendix, but any change from the method as specified shall be stated in the report.

2. Significance

2.1 Dilute solution viscosity values for vinyl chloride polymers are related to the average molecular size of that portion of the polymer which dissolves in the solvent.

3. Summary of Method

3.1 A sample of resin is dissolved in cyclohexanone to make a solution of specified concentration. Inherent viscosity is calculated from the measured flow times of the solvent and of the polymer solution.

NOTE 2—For additional information, refer to ASTM Method D 445, Test for Viscosity of Transparent and Opaque Liquids (Kinematic and Dynamic Viscosities).[2]

4. Apparatus

4.1 *Transfer Pipets.*

4.2 *Volumetric Flasks*, 100-ml, glass-stoppered.[3]

4.3 *Viscometer*, Ubbelohde Series U-1 or Cannon-Ubbelohde No. 75.

4.4 *Water Bath*, at 30 ± 0.5 C controlled to within ±0.01 C.

4.5 *Timer*, as specified in Method D 445, graduated in divisions of 0.1 s or less.

4.6 *Filter Funnel*, fritted-glass.[4]

4.7 *Thermometer*, standard, in accordance with requirements of ASTM Method E 77, Inspection, Test, and Standardization of Liquid-in-Glass Thermometers.[5]

5. Materials

5.1 *Solvent*—Cyclohexanone, laboratory-distilled technical grade, boiling between 155 and 156 C at 760 mm, has been found acceptable if stored in a closed container.

6. Conditioning

6.1 *Conditioning*—Condition the test specimens at 23 ± 2 C (73.4 ± 3.6 F) and 50 ± 5 percent relative humidity for not less than 40 h prior to test in accordance with Procedure A of ASTM Methods D 618, Conditioning Plastics and Electrical Insulating Materials for Testing,[6] for those tests where conditioning is required. In cases of disagreement, the tolerances shall be ±1 C (±1.8 F) and ±2 percent relative humidity.

6.2 *Test Conditions*—Conduct tests in the

[1] This method is under the jurisdiction of ASTM Committee D-20 on Plastics. A list of committee members may be found in the ASTM Yearbook. This standard is the direct responsibility of Subcommittee D-20.15 on Thermoplastic Materials.

Current edition effective March 31, 1966. Originally issued 1952. Replaces D 1243 – 60.

[2] *Annual Book of ASTM Standards*, Parts 18 and 29.

[3] Volumetric flasks should be tested in accordance with the procedures described in the National Bureau of Standards Circular No. C-434, "Testing of Glass Volumetric Apparatus," and should not exceed the limits of accuracy set forth in the circular.

[4] Filters may be obtained from Corning Glass, No. 36060 "Coarse" type.

[5] *Annual Book of ASTM Standards*, Part 30.

[6] *Annual Book of ASTM Standards*, Part 27.

Standard Laboratory Atmosphere of 23 ± 2 C (73.4 ± 3.6 F) and 50 ± 5 percent relative humidity, unless otherwise specified in the test methods or in this specification. In cases of disagreements, the tolerances shall be ±1 C (±1.8 F) and ±2 percent relative humidity.

7. Procedure

7.1 Weigh 0.2 ± 0.002 g of the sample (moisture content below 0.1 percent) and transfer it to a 100-ml glass-stoppered volumetric flask. Care should be taken to transfer all of the weighed resin into the flask. As an alternative method, the resin (0.2 ± 0.002 g) may be weighed directly into a tared, 100-ml glass-stoppered volumetric flask.

7.2 Add 50 to 70 ml of cyclohexanone to the flask, taking care to wet the resin so that lumps do not form.

7.3 Heat the flask at 85 ± 10 C until the resin is dissolved. Occasional shaking will reduce the time required for solution. Heating should not exceed 12 h and should preferably be less to minimize degradation. If any gel-like particles can be seen, prepare a new solution.

7.4 Cool the solution to the test temperature by immersing flask in the 30 C bath for a minimum time of 30 min and adjust to a solution volume of 100 ml. Filter through a fritted-glass filter directly into the viscometer.

7.5 Measure at 30 C the flow time of the prepared solution (6.4) and of the pure solvent (aged at 85 ± 10 C) in the viscometer. Allow 10 min for the viscometer to come to temperature equilibrium after placing it in the water bath. The flow time of the solution or the solvent should be within 0.1 percent on repeat runs on the same filling.

NOTE 3—Keep the Ubbelohde viscometer clean when not in use. Acetone may be used as a cleaning flush. The viscometer may be stored filled with pure solvent or it may be stored dry.

8. Calculation

8.1 Calculate the relative and inherent viscosity as follows:

$$\eta_{rel} = t/t_o$$

$$\eta_{inh} = (\ln \eta_{rel})/C$$

where:

η_{rel} = relative viscosity,
t = efflux time of the solution,
t_o = efflux time of the pure solvent,
C = weight of sample used (7.1) per 100 ml of solution,
η_{inh} = inherent viscosity, and
$\ln \eta_{rel}$ = natural logarithm of relative viscosity.

9. Report

9.1 Report the inherent viscosity to the nearest 0.001.

10. Precision

10.1 The following should be used for judging the acceptability of results (95 percent confidence limits):

10.1.1 *Repeatability*—Duplicate results by the same analyst should not be considered suspect unless they differ by more than 1 percent.

10.1.2 *Reproducibility*—The average result of two determinations reported by one laboratory should not be considered suspect unless it differs from that of another laboratory by more than 2 percent.

REFERENCES

(1) Billmeyer, F. W., Jr., *Journal of Polymer Science*, Vol 4, 1949, p. 83.
(2) Cragg, L. H., and Fern, C. R. H., *Journal of Polymer Science*, Vol 10, 1953, p. 185.
(3) Huggins, M. L., *Journal of the American Chemical Society*, Vol 64, 1942, p. 2716.
(4) International Union of Pure and Applied Chemistry, *Journal of Polymer Science*, Vol 8, 1952, p. 269.
(5) Streeter, D. J., and Boyer, R. F., *Industrial and Engineering Chemistry*, Vol 43, 1951, p. 1790.

APPENDIX

A1. OTHER EXPRESSIONS FOR VISCOSITY

A1.1 Definitions

A1.1.1 *relative viscosity*—ratio of the flow time of a specified solution of the polymer to the flow time of the pure solvent. The International Union of Pure and Applied Chemistry (IUPAC) term for relative viscosity is viscosity ratio.

A1.1.2 *specific viscosity*—relative viscosity minus one. Specific viscosity represents the increase in viscosity that may be attributed to the polymeric solute.

A1.1.3 *reduced viscosity*—ratio of the specific viscosity to the concentration. Reduced viscosity is a measure of the specific capacity of the polymer to increase the relative viscosity. The IUPAC term for reduced viscosity is viscosity number.

A1.1.4 *inherent viscosity*—ratio of the natural logarithm of the relative viscosity to the concentration. The IUPAC term for inherent viscosity is logarithmic viscosity number.

A1.1.5 *intrinsic viscosity*—limit of the reduced and inherent viscosities as the concentration of the polymeric solute approaches zero and represents the capacity of the polymer to increase viscosity. Interactions between solvent and polymer molecules have the affect of yielding different intrinsic viscosities for the same polymer in various solvents. The IUPAC term for intrinsic viscosity is limiting viscosity number.

A1.2 Determination of Intrinsic Viscosity

A1.2.1 To determine the intrinsic viscosity of a polymer from dilute solution viscosity data, the reduced and inherent viscosities of solutions of various concentrations of the polymer are determined at constant temperature and these values are then plotted against the respective concentrations. The two lines thus obtained converge to a point of zero concentration of the solute which represents the intrinsic viscosity of the polymer in that solvent at the temperature of the determination. Figure A1 illustrates this convergence.

A1.2.2 At higher concentrations the viscosity curves may deviate from linearity; therefore, the greatest accuracy is obtained at less than 0.5 g/ml of solution. Since extrapolation of either reduced viscosity or inherent viscosity curves to infinite dilution will give the same value for intrinsic viscosity, a plot of either type of viscosity will permit the calculation of valid intrinsic viscosity data.

A1.3 Estimation of Intrinsic Viscosity

A1.3.1 The mathematical method of Billmeyer permits a good approximation of intrinsic viscosity. This method makes use of equations derived from the power series expansion of viscosity versus concentration. Neglecting the higher order terms, equations may be written which can be used to estimate intrinsic viscosity. The following equation has been found suitable for poly(vinyl chloride) resins:

$$\eta = {}^{1}/_{4}\,[(\eta_{\rm rel} - 1)/C] + [({}^{3}/_{4}\,\ln \eta_{\rm rel})/C]$$

where C = concentration of polymer, g/100 ml.

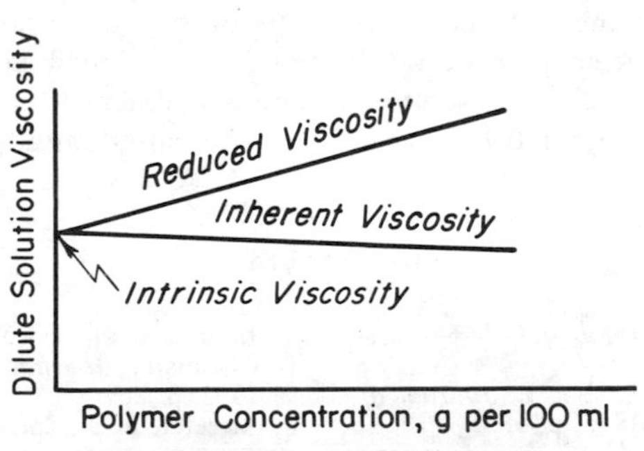

FIG. A1 Example of Plot to Determine Intrinsic Viscosity.

Designation: D 1299 – 55 (Reapproved 1966)

Standard Method of Test for
SHRINKAGE OF MOLDED AND LAMINATED THERMO-SETTING PLASTICS AT ELEVATED TEMPERATURE[1]

This Standard is issued under the fixed designation D 1299; the number immediately following the designation indicates the year of original adoption or, in the case of revision, the year of last revision. A number in parentheses indicates the year of last reapproval.

1. Scope

1.1 This method covers the determination of the shrinkage characteristics of various thermosetting molded and laminated plastics after exposure to certain prescribed temperatures.

1.2 The test is not expected to predict the behavior of materials in service, but rather provides a means of classifying plastics with respect to shrinkage on a relative basis.

NOTE 1—The values stated in U.S. customary units are to be regarded as the standard. The metric equivalents of U.S. customary units may be approximate.

2. Significance

2.1 Plastics experience dimensional change at elevated temperatures at varying rates, usually quite rapidly during the early stages and more slowly at the end of long periods of exposure. Different kinds of plastics also do not behave similarly. It is therefore frequently found that various plastic moldings will occupy different orders when rated according to shrinkage, depending on whether the exposure period has been short or long. For purposes of control work, short exposure periods will usually suffice to indicate shrinkage differences among different batches of the same material. Exposure periods of greater duration will more nearly reflect real differences in shrinkage phenomena, in comparing unlike materials for suitability in particular applications.

2.2 By the use of this method, it is possible to measure the change in dimensions of specimens at prescribed intervals during their exposure to circulating air at an elevated temperature. By this means, plastics can be compared on a relative basis for this shrinkage after any one of these intervals, or by examining their trends over a number of intervals. While certain industrial materials may be tested at higher temperatures, thus shortening the time required to produce significant shrinkage phenomena, such high temperatures may be inappropriate to other plastics, making it advisable to use lower temperatures and longer exposure periods. End use as well as the preceding stipulation should assist in the selection of the proper temperature. These data should be useful in making a choice of suitable materials for application where shrinkage is expected to result from atmospheric conditions that are approximately similar to those stipulated in this method.

3. Apparatus

3.1 *Jig*—A jig with micrometer dial gage attached, capable of measurements accurate to ±0.013 mm (±0.0005 in.). A schematic diagram of a fixture suitable for this test is shown in Fig. 1.

3.2 *Gage Block*—A stainless steel bar 13 by 13 by 100 ± 0.0025 mm (0.5 by 0.5 by 4.000 ± 0.0001 in.), with ends suitably marked for reference purposes.

3.3 *Oven*—An oven or enclosure meeting

[1] This method is under the jurisdiction of ASTM Committee D-20 on Plastics. A list of committee members may be found in the ASTM Yearbook. This standard is the direct responsibility of Subcommittee D-20.50 on Permanence Properties.

Current edition effective Sept. 12, 1955. Originally issued 1953. Replaces D 1299 – 53 T.

the requirements of Type I, Grade B, as specified in ASTM Specifications E 197, for Enclosures and Servicing Units for Tests Above and Below Room Temperature.[2]

3.4 *Desiccator*—A desiccator of the calcium chloride type.

4. Test Specimens

4.1 Test specimens shall be disks 100 ± 1.0 mm (4 in.) in diameter and 3.2 ± 0.13 mm (0.125 ± 0.005 in.) thick, or 100 ± 1.0 mm (4-in.) squares of the same thickness. Disks shall be molded according to standard technique and established practice—for thermosetting materials, compression molded, urea type, as described in ASTM Recommended Practice D 956, for Molding Specimens of Amino Plastics,[3] and for phenolic materials, as described in ASTM Recommended Practice D 796, for Molding Test Specimens of Phenolic Materials,[3] or as agreed upon between the purchaser and the seller. Specimens shall be marked with reference marks (either molded in or scratched in place by means of a scriber) at the edge, arranged diametrically opposite in pairs 100 mm (4 in.) apart, with the pairs at 90 deg to each other.

4.2 Three specimens of the material shall be tested, and the results averaged.

5. Conditioning

5.1 *Conditioning*—Condition the test specimens at 23 ± 2 C (73.4 ± 3.6 F) and 50 ± 5 percent relative humidity for not less than 40 h prior to test in accordance with Procedure A of ASTM Methods D 618, Conditioning Plastics and Electrical Insulating Materials for Testing,[3] for those tests where conditioning is required. In cases of disagreement, the tolerances shall be 1 C (1.8 F) and ±2 percent relative humidity.

5.2 *Test Conditions*—Conduct tests in the Standard Laboratory Atmosphere of 23 ± 2 C (73.4 ± 3.6 F) and 50 ± 5 percent relative humidity, unless otherwise specified in the test methods or in this specification. In cases of disagreements, the tolerances shall be 1 C (1.8 F) and ±2 percent relative humidity.

6. Procedure

6.1 Adjust the micrometer dial gage of the fixture to read zero after inserting the gage block as shown in Fig. 1. Take care to assure proper alignment of the gage block with the adjusting screw of the fixture and the travel stem of the dial gage. Calibrate the micrometer dial gage at the standard laboratory atmosphere prescribed in 3.1 of Method D 618.

6.2 Replace the gage block with a test specimen, and measure the specimen under the same atmospheric conditions as those employed in 6.1. Measure the specimen to the nearest 0.025 mm (0.001 in.) on the two marked reference axes at 90 deg to each other.

6.3 Place the specimens in metal racks, supported on edge and separated from each other, to permit free circulation of air.

6.4 Place the racked specimens in an oven maintained at the selected temperature, depending upon the type of material to be tested (see 2.2). For comparison of unlike plastics, the temperature chosen should be determined by the material with the lowest temperature limitation. Use one of the following test temperatures:

70 C
90 C
105 C
130 C
180 C
230 C

6.5 Heat the specimen at the temperature selected for 48, 96, 168, 504, or 1000 h, or longer, as required (see 2.2). The minimum permissible duration shall be that at which the shrinkage equals or exceeds 0.25 mm (0.010 in.). The schedules permitted here may be used either singly or in sequence.

6.6 At the end of an exposure period, remove the specimens from the oven and place them in the desiccator until they reach room temperature.

6.7 Measure again the two dimensions of each specimen, using the two marked reference axes. Calibrate the dial gage and make the measurements as prescribed in 6.1 and 6.2, respectively.

7. Report

7.1 The report shall include the following:

---

[2] *Annual Book of ASTM Standards*, Part 30.
[3] *Annual Book of ASTM Standards*, Part 27.

7.1.1 Complete identification of the material, including pertinent molding data.

7.1.2 Initial thickness of the test specimen measured to the nearest 0.025 mm (0.001 in.) at its center.

7.1.3 Test temperature and period or periods of exposure.

7.1.4 Change in dimensions for each reference axis, calculated in millimeters per millimeter or inches per inch, or in percent, based on the original dimensions of the conditioned specimen.

7.1.5 Observations on any visible changes in the test specimen, that is, warpage, crazing, etc.

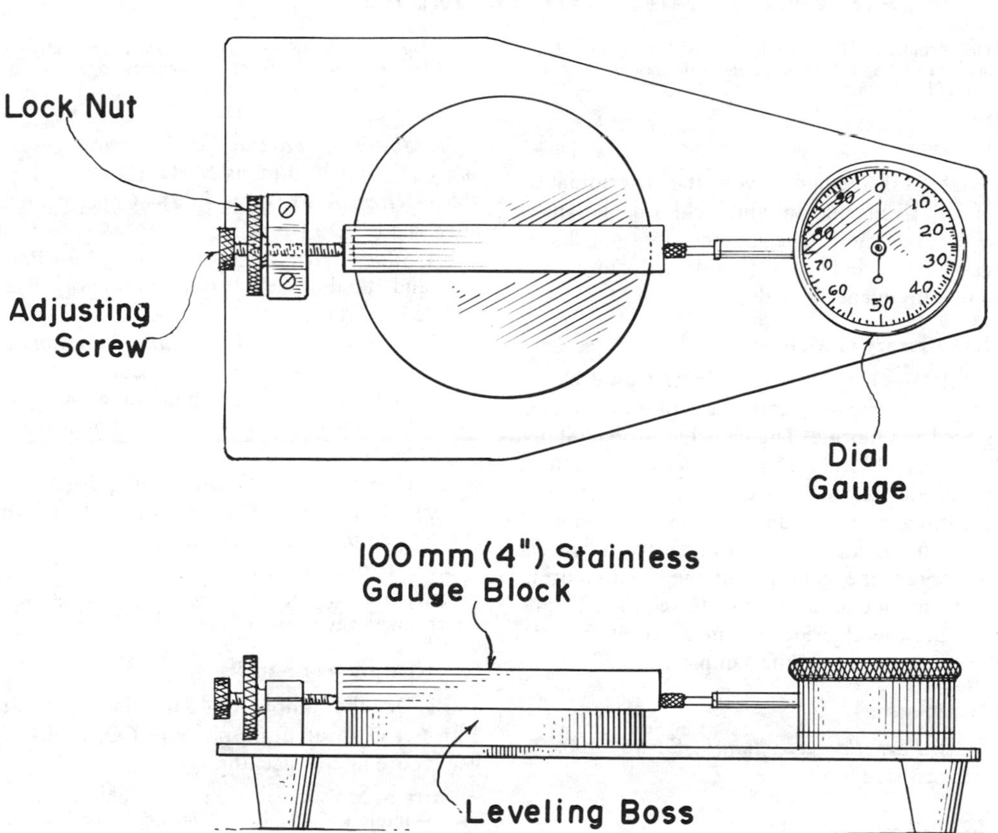

FIG. 1 Schematic Diagram of Jig with Micrometer Dial Gage Attached.

Designation: D 1303 – 55 (Reapproved 1967)

American National Standard K65.9-1959 (R 1969)
American National Standards Institute

Standard Method of Test for

TOTAL CHLORINE IN VINYL CHLORIDE POLYMERS AND COPOLYMERS[1]

This Standard is issued under the fixed designation D 1303; the number immediately following the designation indicates the year of original adoption or, in the case of revision, the year of last revision. A number in parentheses indicates the year of last reapproval.

1. Scope

1.1 This method covers the determination of total chlorine in vinyl chloride polymers and copolymers. The method does not distinguish between poly(vinyl chloride) and poly(vinylidene chloride).

2. Summary of Method

2.1 The organic matter is destroyed by hot sodium peroxide, and the chlorine is converted to chloride. The chloride is precipitated as silver chloride with an excess of standard silver nitrate, and this excess is determined by titration with standard thiocyanate solution using ferric nitrate indicator and nitrobenzene to depress the solubility of the silver chloride. The amount of silver nitrate required to precipitate the chloride is a measure of the total chlorine content of the sample.

3. Apparatus

3.1 *Parr Peroxide Bomb*, Series 2100, 22-ml, gas-fired.[2]

4. Purity of Reagents

4.1 *Purity of Reagents*—Reagent grade chemicals or equivalent, as specified in ASTM Methods E 200, Preparation, Standardization, and Storage of Standard Solutions for Chemical Analysis,[3] shall be used in all tests.

4.2 *Purity of Water*—Unless otherwise indicated references to water shall be understood to mean distilled water or water of equal purity.

5. Reagents and Materials

5.1 *Powdered Sugar* (preferably cp sucrose).

5.2 *Sodium Peroxide*, cp, calorific grade, assay 95 percent, min, as Na_2O_2 (Note 1).

5.3 *Nitric Acid (sp gr 1.42)*—Concentrated nitric acid (HNO_3).

5.4 *Silver Nitrate Solution (0.1 N)*—Prepare and standardize a 0.1 *N* solution of silver nitrate ($AgNO_3$).

5.5 *Ammonium or Potassium Thiocyanate Solution (0.1 N)*—Prepare and standardize a 0.1 *N* solution of either potassium thiocyanate (KCNS) or ammonium thiocyanate (NH_4CNS).

5.6 *Ferric Nitrate Solution (50 g $Fe(NO_3)_3$/liter)*—Dissolve 50 g of Ferric nitrate ($Fe(NO_3)_3$) in water and dilute to 1 liter.

5.7 *Nitrobenzene*, cp.

NOTE 1—Keep Na_2O_2 in a tightly closed container, and away from moisture.

6. Preparation of Sample

6.1 Dry the sample for 45 min in an oven at 100 ± 1 C, in an atmosphere of CO_2 or nitrogen. Store in a desiccator.

NOTE 2—Moisture pick-up is a problem, so the dried sample should be used promptly, and should not be stored for prolonged periods.

7. Conditioning

7.1 *Conditioning*—Condition the test speci-

[1] This method is under the jurisdiction of ASTM Committee D-20 on Plastics. A list of committee members may be found in the ASTM Yearbook. This standard is the direct responsibility of Subcommittee D-20.70 on Analytical Methods.

Current edition effective Sept. 12, 1955. Originally issued 1953. Replaces D 1303 – 55 (1961).

[2] The Parr Peroxide Bomb may be obtained from the Parr Instrument Co., Moline, Ill. Parr electrically fired bombs may be used in place of the gas fired peroxide bombs, provided tests of both apparatus have been conducted to demonstrate reproducible agreement of results.

[3] *Annual Book of ASTM Standards*, Part 30.

mens at 23 ± 2 C (73.4 ± 3.6 F) and 50 ± 5 percent relative humidity for not less than 40 h prior to test in accordance with Procedure A of ASTM Methods D 618, Conditioning Plastics and Electrical Insulating Materials for Testing,[4] for those tests where conditioning is required. In cases of disagreement, the tolerances shall be 1 C (1.8 F) and ±2 percent relative humidity.

7.2 *Test Conditions*—Conduct tests in the Standard Laboratory Atmosphere of 23 ± 2 C (73.4 ± 3.6 F) and 50 ± 5 percent relative humidity, unless otherwise specified in the test methods or in this specification. In cases of disagreements, the tolerances shall be 1 C (1.8 F) and ±2 percent relative humidity.

8. Procedure

8.1 To a clean, dry, 22-ml Parr bomb cup add 0.200 to 0.225 g of finely powdered sample. Add 0.5 g of dry powdered sugar and a standard measure (15 g) of Na_2O_2. Close the bomb tightly with the lid and the retaining rings, and shake vigorously to mix the contents. Tap the bomb lightly on the bench to dislodge particles sticking to the cap and to the upper portion of the bomb cup. Set the bomb over the hole in a steel housing supported on a tripod. A small steel chamber with an access door and a small sight hole built over the tripod is desirable for reasons of safety.

8.2 Add 2 ml of water to the depression in the bomb cover. Apply a very hot bunsen flame to the bottom of the bomb. Adjust the air to the burner to give a cone about 38 mm (1 ½ in.) high, the apex of which should be just short of contact with the bottom of the bomb. Allow to heat until the water in the lid begins to boil or until a dark red ring appears on the side of the cup, then quench in cold water. **Caution:** The heating of the bomb is of extreme importance, since underheating will cause imcomplete oxidation of the organic matter and low chlorine results, while overheating will cause possible damage to the bomb cup and even an explosion. These bombs are not intended to operate at a red heat, but there is no serious danger involved in this method of oxidation provided the proper charges are used in a bomb that is in a satisfactory condition, and provided the proper heating technique is used (Note 3). Discard the 22-ml bomb cups if the sides or bottom become visibly swollen, or if the interior surfaces become worn or corroded to an inside diameter of 27.4 mm (1.08 in.) at any point.

8.3 When the bomb is cold, remove it from the quenching bath, rinse if off with water, and open. Place the bomb cup on its side in a clean 600-ml beaker and add water until about half covered. Rinse off the bomb cover thoroughly with water, catching the rinsings in the same beaker. Cover the beaker with a watch glass, and warm slightly until the contents of the bomb are dissolved. Remove the bomb cup and rinse thoroughly with water, catching the rinsings in the same beaker. Transfer quantitatively to a 500-ml glass-stoppered flask, add 2 to 3 drops of phenolphthalein indicator solution, neutralize with HNO_3, using care to avoid loss by spattering and overheating, and add an excess of 5 ml of HNO_3 (Note 4). Add 2 ml of nitrobenzene (Note 5) and 50.0 ml of 0.1 *N* $AgNO_3$ solution, stopper, and shake until the precipitated AgCl becomes a spongy mass. Add 10 to 15 ml of $Fe(NO_3)_3$ indicator solution and titrate with 0.1 *N* thiocyanate solution to the first pink end point.

8.4 Run a blank determination at least every time a change in solutions or reagents is made, to ensure against high results caused by chlorides in the reagents.

NOTE 3—A careful reading of Parr Manual No. 121 on the method and precautions used in this method is suggested.

NOTE 4—When properly fired, the bomb leachings are free of black particles, and the acidified solution is substantially clear. If not, a new sample must be prepared and fired.

NOTE 5: **Caution**—Nitrobenzene is toxic, and care must be used to avoid contact with the skin.

9. Calculations

9.1 Calculate the chlorine and vinyl chloride content as follows:

$$\text{Total chlorine, percent} = [(V_1N_1 - V_2N_2)\,0.03546/W] \times 100$$

$$\text{Vinyl chloride, percent} = [(V_1N_1 - V_2N_2)\,0.06246/W] \times 100$$

where:

V_1 = milliliters of $AgNO_3$ solution added to

[4] *Annual Book of ASTM Standards*, Part 27.

the sample,
N_1 = normality of the $AgNO_3$ solution,
V_2 = milliliters of KCNS solution required for titration,
N_2 = normality of the KCNS solution, and
W = grams of sample used.

10. Reproducibility

10.1 Duplicate determinations should check within 0.3 percent.

Designation: D 1433 – 58 (Reapproved 1966)

American National Standard K65.28-1967
American National Standards Institute

Standard Method of Test for
FLAMMABILITY OF FLEXIBLE THIN PLASTIC SHEETING[1]

This Standard is issued under the fixed designation D 1433; the number immediately following the designation indicates the year of original adoption or, in the case of revision, the year of last revision. A number in parentheses indicates the year of last reapproval.

This method was prepared jointly by the American Society for Testing and Materials and The Society of the Plastics Industry.

1. Scope

1.1 This method covers the determination of the relative flammability of flexible plastics in film or thin sheeting form. The test measures the burning rate of plastic materials that support moderate combustion under the conditions of this test. Materials that shrink excessively upon ignition, or that melt to cause the flame to be carried away while dripping, cannot be evaluated by this method. Although the test is primarily intended for nonrigid plastic films, it may be used on rigid or nonrigid sheeting capable of being bent through the 45-deg angle specified in the test procedure.

NOTE 1—The values stated in U.S. customary units are to be regarded as the standard. The metric equivalents of U.S. customary units may be approximate.

2. Significance

2.1 This method is intended to provide data for comparing the relative burning rates of thin plastic sheeting under the conditions of this test. The rate of burning will vary with thickness. Test data should be compared with data for a control material of known performance and of comparable thickness. Correlation with flammability under actual use conditions is not necessarily implied.

2.2 Even though the test reproducibility within a laboratory will generally be within 5 to 10 percent, the reproducibility between various laboratories has been found to vary by as much as 40 to 50 percent. This must be recognized when making interlaboratory comparisons of data. The test is therefore not suitable as a precise general specification test. However, it is suitable for control testing or for making relative flammability evaluations within a laboratory.

3. Apparatus

3.1 The assembled apparatus[2] shall be essentially as shown in Fig. 1 and shall include the following:

3.1.1 *Specimen Holder*—A removable, flat specimen-holding rack, consisting of an upper and lower section constructed and hinged as shown in Fig. 2. The specimen holder shall be constructed in such a manner as to support the specimen by tight closure of the upper and lower sections around the sides of the specimens. The sections shall be held together at the bottom with two spring paper clips. The center of the holder contains an open U-shaped area in which the burning of the specimen occurs. At the open end of the holder the last inch of the forked sides is at an angle of 45 deg to the remainder of the holder. The switch actuators consist of suitable springs mounted on the side of the rack, one just beyond the curved portion at the open end, and the other near the closed end of the U-shaped

[1] This method is under the jurisdiction of ASTM Committee D-20 on Plastics. A list of committee members may be found in the ASTM Yearbook. This standard is the direct responsibility of Subcommittee D-20.30 on Thermal Properties.

Current edition effective Sept. 22, 1958. Originally issued 1956. Replaces D 1433 – 56 T.

[2] Available through Custom Scientific Instruments, Inc., Whippany, N. J.

holder. The springs are depressed and held in position prior to ignition by means of thread wound across the specimen and attached to the holder.

3.1.2 *Igniter Flame*—The igniter flame shall be produced at the tip of a No. 22 hypodermic needle jet (Fig. 3). The igniter shall be so located in the cabinet that the tip of the needle is $^1/_2$ in. from the surface of the specimen when the holder has been fully inserted on the track in the cabinet, as shown in Fig. 3. The forward point of the tapered hypodermic tip shall be up, and the burning end of the needle tilted slightly downward. The flame shall be adjusted to a 13-mm ($^1/_2$-in.) horizontal length, and the hypodermic needle adjusted to an angle which will permit the flame to just barely contact the specimen at the horizontal and vertical center of the 25-mm (1-in.) vertical section of the specimen holder. Precise adjustment of the flame is essential. This adjustment can be made preliminary to a test by using an asbestos sheet as a dummy specimen. The number of valves, connections, and length of tubing for the gas fuel shall be kept to a minimum.

3.1.3 *Cabinet*—The cabinet shall protect the igniter flame and specimen from air currents during test. It shall contain a hinged glass window for observation and fixed vents to supply sufficient air for normal combustion of the specimen. A specimen holder track shall be provided to position the specimen holder at a 45-deg angle, with the lower 1-in. section in a vertical position.

3.1.4 *Timing Mechanism*—The burning time shall be determined by an electric timer actuated by microswitches mounted on the specimen holder rack. As the first string burns apart, a timer is started by a microswitch which is actuated by the released spring. When the second string, which is 152 mm (6 in.) away from the first, burns apart a second microswitch is triggered and the timer is stopped to register the burning time.

3.1.5 *Butane, Commercial Grade*—Compressed butane gas shall be used as the igniter flame fuel.

3.1.6 *Thread*—White, nylon sewing thread, size D.[3]

3.1.7 *Hood*—A shielded hood with forced exhaust facilities shall be used to house the cabinet.

3.1.8 *Micrometer*—Suitable micrometer or other thickness gage, reading to 0.0025 mm (0.0001 in.) or less, for measuring the thickness of test specimens.

4. Test Specimens

4.1 Test specimens shall be 76 mm (3 in.) in width and 228 mm (9 in.) in length. They shall be free of folds or wrinkles. Five specimens from the machine direction and five specimens from the transverse direction of a material shall be tested. Handling of the specimens should be minimized.

5. Conditioning

5.1 *Conditioning*—Condition the test specimens at 23 ± 2 C (73.4 ± 3.6 F) and 50 ± 5 percent relative humidity for not less than 40 h prior to test in accordance with Procedure A of ASTM Methods D 618, Conditioning Plastics and Electrical Insulating Materials for Testing,[4] for those tests where conditioning is required. In cases of disagreement, the tolerances shall be ±1 C (±1.8 F) and ±2 percent relative humidity.

5.2 *Test Conditions*—Conduct tests in the Standard Laboratory Atmosphere of 23 ± 2 C (73.4 ± 3.6 F) and 50 ± 5 percent relative humidity, unless otherwise specified in the test methods or in this specification. In cases of disagreements, the tolerances shall be ±1 C (±1.8 F) and ±2 percent relative humidity.

6. Procedure

6.1 Thread the specimen holder to depress the switch actuator springs at least 6.4 mm ($^1/_4$ in.). Use a double string for each spring to avoid premature failures. Install the strings in such a manner as to lie flat over the specimen when the specimen and holder have been assembled.

6.2 Measure the specimen thickness at five places along the long center axis of the specimen and record the average thickness to the nearest 0.0025 mm (0.0001 in.).

6.3 Insert the specimen into the holder, making certain that it is clamped securely

[3] Nymo thread, available through Belding Hemingway Corticelli Co., 1407 Broadway, New York, N. Y. 10018, has been found suitable for this purpose.

[4] *Annual Book of ASTM Standards*, Part 27.

between the upper and lower U-frame, and between the two support wires at the open end of the burning channel. Make certain the specimen is free from wrinkles, stresses, or distortions when the holder is closed. Make certain the specimen does not extend beyond the lower end of the specimen holder, and the spring paper clips do not extend over the exposed specimen area.

6.4 Prior to inserting the specimen and holder into the track in the cabinet, set the automatic timer to zero and regulate the butane flame as prescribed in 3.1.2.

6.5 With the hood fan off, the clock set at zero, the flame adjusted, and the cabinet window closed, insert the specimen holder on the specimen holder rack in the cabinet. Slide the holder into place at a constant rate, covering the entire distance in approximately $^1/_2$ s. If there is any sticking or abnormal hesitation in this action make a retest.

6.6 Upon burning of the sample, turn on the hood blower to exhaust the smoke. Record the automatically timed burning time, in seconds, and remove the holder from the rack and clean.

6.7 Repeat the procedure described in 6.1 through 6.5 until a total of ten specimens, five in the machine direction and five in the transverse direction, have been tested. Designate any specimens which commence to burn through the lower string but extinguish before reaching the upper string as "self-extinguishing" by this test. Designate specimens which extinguish before the lower string is burned through as "nonburning" by this test.

7. Calculations

7.1 In calculating the average burning time, exclude any "nonburning" or "self-extinguishing" results and determine the average results on the basis of the remaining individual readings.

7.2 Calculate the average burning rate, in millimeters per second (or inches per second), by dividing 152 mm (6 in.) by the average burning time in seconds.

8. Report

8.1 The report shall include the following:

8.1.1 Complete sample identification,

8.1.2 The average burning rate in millimeters per second (or inches per second),

8.1.3 Any "self-extinguishing" or "nonburning" results, and

8.1.4 Specimen thickness in millimeters (or inches).

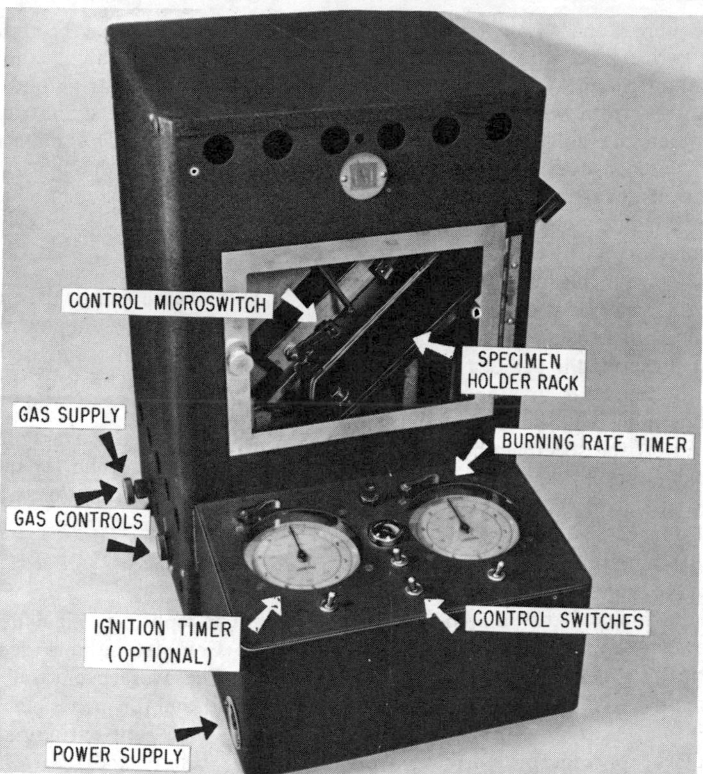

FIG. 1 Apparatus Assembled in Cabinet.

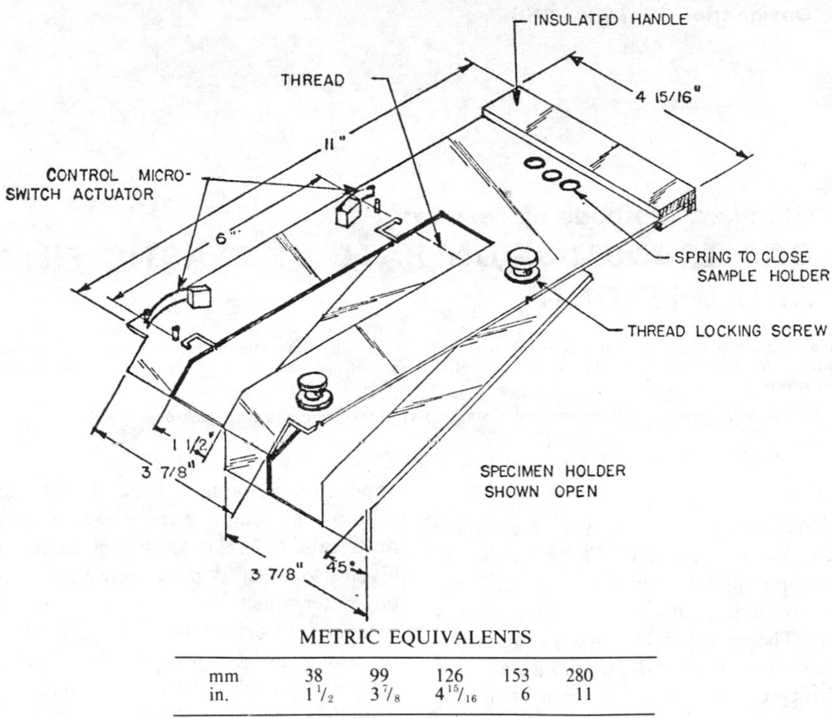

METRIC EQUIVALENTS

mm	38	99	126	153	280
in.	$1\frac{1}{2}$	$3\frac{7}{8}$	$4\frac{15}{16}$	6	11

FIG. 2 Specimen Holder Shown Open.

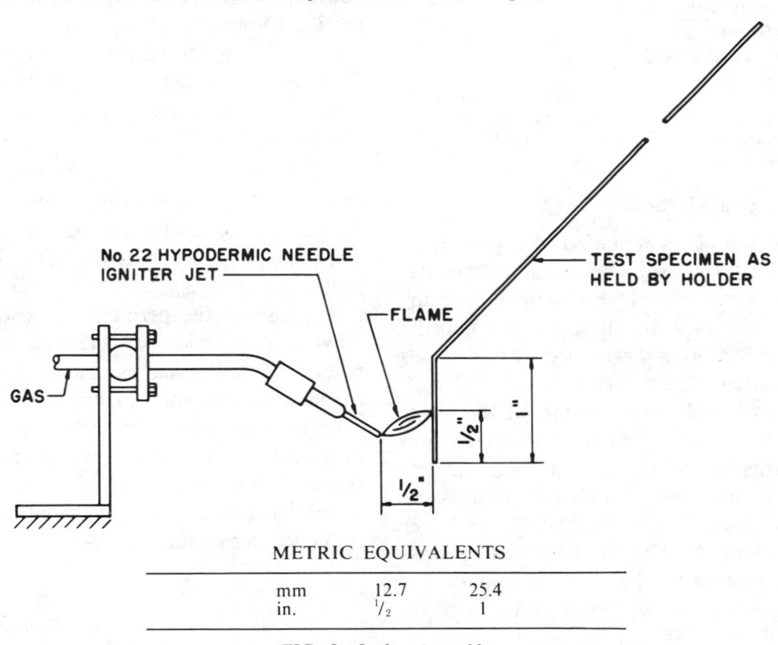

METRIC EQUIVALENTS

mm	12.7	25.4
in.	$\frac{1}{2}$	1

FIG. 3 Igniter Assembly.

ASTM Designation: D 1434 – 66

Standard Methods of Test for
GAS TRANSMISSION RATE OF PLASTIC FILM AND SHEETING[1]

This Standard is issued under the fixed designation D 1434; the number immediately following the designation indicates the year of original adoption or, in the case of revision, the year of last revision. A number in parentheses indicates the year of last reapproval.

These methods were prepared jointly by the American Society for Testing and Materials and The Society of the Plastics Industry.

1. Scope

1.1 These methods cover the determination of the steady-state rate of transmission of a gas through plastics in the form of film, sheeting, laminates, and plastic-coated papers or fabrics. They provide for the determination of (*1*) Gas Transmission Rate (GTR) and (*2*) Permeability Coefficient in the case of homogeneous materials.

1.2 Two methods are provided:

1.2.1 *Method M*—Manometric.

1.2.2 *Method V*—Volumetric.

NOTE 1—The values stated in U.S. customary units are to be regarded as the standard. The metric equivalents of U.S. customary units may be approximate.

2. Summary of Methods

2.1 The sample is mounted in a gas transmission cell forming a sealed semibarrier between two chambers. One chamber contains the test gas at a specific high pressure and the other chamber, at a lower pressure, receives the permeating gas.

2.1.1 *Method M*—In Method M the lower pressure chamber is initially evacuated and the transmission of the gas through the test specimen is indicated by a change in pressure.

2.1.2 *Method V*—In Method V the lower pressure chamber is maintained near atmospheric pressure and the transmission of the gas through the test specimen is indicated by a change in volume.

3. Significance

3.1 These tests give, by relatively simple apparatus, reliable values for the gas transmission of single pure gases through sheet materials. Correlation of test values with any given use, such as packaging protection, must be determined by experience. Interactions may affect performance in applications where several gases are present. The gas transmission is affected by conditions not specifically provided for in these tests, such as moisture content plasticizer content, and nonhomogeneities (Note 2). These tests do not include any provision for testing seals which may be involved in packing applications.

NOTE 2—The tests are run using gas with 0 percent relative humidity.

3.2 Experience has shown the tests to be capable of interlaboratory agreement within 10 percent or better for permeability coefficient ranging at least from 0.001 to 10 barrer units.

3.3 Use of the permeability coefficient (involving conversion of the gas transmission rate to a unit thickness basis) is not recommended unless the thickness-to-transmission rate relation has been predetermined. Even in essentially homogeneous structures, variations in morphology (for example, as indicated by density) and thermal history may influence permeability.

---

[1] These methods are under the jurisdiction of ASTM Committee D-20 on Plastics. A list of committee members may be found in the ASTM Yearbook. This Standard is the direct responsibility of Subcommittee D-20.70 on Analytical Methods.

Current edition effective Dec. 29, 1966. Originally issued 1956. Replaces D 1434 – 63.

4. Definitions

4.1 *gas transmission rate, GTR*—the volume of a gas flowing normal to two parallel surfaces, at steady-state conditions through unit area, under unit pressure differential and the conditions of test (temperature, specimen, thickness) (Note 2).

4.1.1 An accepted unit of 1 GTR is 1 cm^3 (at standard conditions)/24 h $\cdot m^2 \cdot$atm. The test specimen thickness and test temperature must also be stated.

$$GTR = cm^3/24h \cdot m^2 \cdot atm$$

4.2 *gas permeability coefficient,* $\bar{P}$—the volume of a gas flowing normal to two parallel surfaces at unit distance apart (thickness), under steady-state conditions, through unit area under unit pressure differential at a stated test temperature (Note 2).

4.2.1 An accepted unit of $\bar{P}$ is 1 cm^3 (at standard conditions)/s $\cdot cm^2 \cdot$cm Hg/cm of thickness at the stated temperature of the test (generally 23 C). A convenient unit is the barrer:

$$\bar{P}(\text{barrer}) = [10^{-10}\ cm^3\ (STP) \times cm]/(s \times cm^2 \times cm\ Hg)$$

The Gas Permeability Coefficient is a basic property of a material independent of specimen geometry. It is related to the diffusion rate and the solubility of a gas by the equation:

$$\bar{P} = DS$$

where:

D = diffusion rate, and

S = solubility.

NOTE 3—Additional units and conversions are shown in Appendix A2.

4.3 *steady-state*—the state attained when the volume of test gas transmitted as measured at STP (standard conditions of temperature and pressure) becomes linear with time.

5. Test Specimen

5.1 The test specimen shall be representative of the material, free of wrinkles, creases, pinholes, and other imperfections, and shall be of uniform thickness. The test specimen shall be cut to an appropriate size (generally circular) to fit the test cell. A minimum of three specimens shall be tested at each test condition.

5.2 The thickness of the specimen should be measured to the nearest 0.0025 mm (0.1 mil) with a calibrated dial gage (or equivalent) at a minimum of five points distributed over the entire test area. Maximum, minimum, and average values should be recorded. An alternative measure of thickness involving the weighing of a known area of specimens having a known density is also suitable for homogeneous materials.

6. Conditioning

6.1 *Conditioning*—Condition the test specimens at 23 ± 2 C (73.4 ± 3.6 F) and 50 ± 5 percent relative humidity for not less than 40 h prior to test in accordance with Procedure A of ASTM Methods D 618, Conditioning Plastics and Electrical Insulating Materials for Testing,[2] for those tests where conditioning is required. In cases of disagreement, the tolerances shall be ±1 C (±1.8 F) and ±2 percent relative humidity.

6.2 *Test Conditions*—Conduct tests in the Standard Laboratory Atmosphere of 23 ± 2 C (73.4 ± 3.6 F) and 50 ± 5 percent relative humidity, unless otherwise specified in the test methods or in this specification. In cases of disagreement, the tolerances shall be ±1 C (±1.8 F) and ±2 percent relative humidity.

METHOD M

(Pressure changes in the manometric cell may be determined by either visual or automatic recording.)

Manometric Visual Determination

7. Apparatus

7.1 The apparatus shown in Figs. 1 and 2 consists of the following items.[3]

7.1.1 *Cell Manometer System*—The calibrated cell manometer leg, which indicates the pressure of transmitted gas, shall consist of precision-bore glass capillary tubing at least 65 mm long with an inside diameter of 1.5 mm.

7.1.2 *Cell Reservoir System*, consisting of a glass reservoir of sufficient size to contain all the mercury required in the cell.

---

[2] *Annual Book of ASTM Standards*. Part 27.

[3] The Dow gas transmission cell supplied by Custom Scientific Instruments, Inc., Whippany, N. J., has been found satisfactory for this purpose.

7.1.3 *Adapters*—Solid and hollow adapters for measurement of the widely varying gas transmission rates. The solid adapter provides a minimum void volume for slow transmission rates. The hollow adapter increases the void volume by about a factor of eight for the faster transmission rates.

7.1.4 *Cell Vacuum Valve*, capable of maintaining a vacuum-tight seal.[4]

7.1.5 *Plate Surfaces* that contact the specimen and filter paper shall be smooth and flat.

7.1.6 *O-Ring*, for sealing the upper and lower plates.

7.1.7 *Manometer*—An open-end U-tube manometer or equivalent having a calibrated range from 0 to 2 atm.

7.1.8 *Barometer*, suitable for measuring the pressure of the atmosphere to the nearest 1 mm.

7.1.9 *Vacuum Gage*, to register the pressure during evacuation of the system to the nearest 0.1 mm.

7.1.10 *Vacuum Pump*, capable of reducing the pressure in the system to 0.2 mm or less.

7.1.11 *Needle Valve*, for slowly admitting and adjusting the pressure of the test gas.

7.1.12 *Cathetometer*, to measure accurately the height of mercury in the cell manometer leg. This instrument should be capable of measuring changes to the nearest 0.5 mm.

7.1.13 *Micrometer*, graduated to 0.0025 mm (0.1 mil) or better and accurate to ± 0.00125 mm (0.05 mil) to measure specimen thickness.

7.1.14 *Elevated - Temperature Fittings*—Special cell fittings are required for high-temperature testing.

8. Materials

8.1 *Test Gas*—The test gas shall be dry and of pure composition. The ratio of the volume of gas available for transmission to the volume of gas transmitted at the completion of the test shall be at least 100:1.

8.2 *Mercury*—Mercury used in the cell shall be triply distilled, checked regularly for purity, and replaced with clean mercury when necessary.

NOTE 4—Very low concentrations of mercury vapor in the air are known to be hazardous. Be sure to collect all spilled mercury in a closed container. Transfers of mercury should be made over a large plastic tray. Under normal daily laboratory use conditions, it is necessary to clean the cells about every 3 months. Dirty mercury is indicated when the drop of the capillary becomes erratic or when mercury clings to the side of the capillary, or both. Whenever such discontinuities occur, the mercury should be removed and the cell cleaned as follows:

(*1*) Wash with toluene (to remove greases and oils).

(*2*) Wash with acetone (to remove toluene).

(*3*) Wash with distilled water (to remove acetone).

(*4*) Wash with a 1+1 mixture of nitric acid and distilled water (to remove any mercury salts that may be present). This operation may be repeated if necessary in order to ensure complete cleaning of glassware.

(*5*) Wash with distilled water (to remove nitric acid).

(*6*) Wash with acetone (to remove water).

(*7*) Dry the cell at room temperature or by blowing a small amount of dry, clean air through it.

9. Calibration

9.1 Each cell shall be calibrated at the test temperature as follows (Fig. 3):

9.1.1 Determine the void volume of the filter paper from the absolute density of its fiber content (Note 5) and the weight and apparent volume of the filter paper (Note 6). Express the void volume so determined in cubic millimeters and designate as V_{CD}.

NOTE 5—Any high-grade, medium retention qualitative noncellulosic filter paper, 9.0 cm in diameter will be satisfactory for this purpose. Cellulose fiber has an approximate density of 1.45 g/cm^3.

NOTE 6—The apparent volume may be calculated from the thickness and diameter of the filter paper.

9.1.2 Determine the volume of the cell manometer leg from *B* to *C*, Fig. 3, by mercury displacement. (Since the void volume of the adapters is included in this part of the calibration, the volume from *B* to *C* should be determined twice, once with the solid adapter in place, and once with the hollow.) This volume is obtained by dividing the weight of the mercury displaced by its density (Note 7). Determine this volume to the nearest 1 mm^3 and designate as V_{BC}.

NOTE 7—The density of mercury at 23 C is 0.01354 g/mm.^{3}

9.1.3 Determine the volume, in cubic millimeters of the cell manometer leg from *A* to *B*, Fig. 3, by mercury displacement. Determine the average cross-sectional area of the capillary by dividing this volume by the length (expressed to the nearest 0.1 mm) from *A* to

[4] The Demi-G Valve ($^1/_4$-in. IPS) manufactured by G. W. Dahl Co., Inc., Bristol, R. I., or a precision-ground glass stopcock, meets this requirement.

B. Determine the area to the nearest 0.01 mm^2 and designate as *a*.

9.1.4 Determine the area of the filter paper cavity to the nearest 0.01 cm^2. Designate this area as A_2, the area of transmission.

9.1.5 Pour the mercury from the reservoir into the manometer system of the cell by carefully tipping the cell. The height of the mercury in the capillary leg shall be at approximately the same level as the upper calibration line *B* (Fig. 3). Record the distance from the datum plane to the top of the mercury meniscus in the capillary leg as h_B. Record the distance from the datum plane to the top of the mercury meniscus in the reservoir leg as h_L. Determine h_B and h_L to the nearest 1 mm.

10. Procedure

10.1 Transfer all the mercury into the reservoir of the cell manometer system by carefully tipping the cell such that the mercury pours into the reservoir.

10.2 Insert the appropriate adapter in the cell body.

10.3 Center a filter paper in the lower plate cavity.

10.4 Place the conditioned specimen smoothly on the lower plate so that it covers the filter paper and the entire exposed face of the lower plate.

10.5 Locate the O-ring on the upper plate; then carefully position this plate over the specimen and fix the plate with uniform pressure to ensure a vacuum-tight seal.

10.6 Connect the line in which the test gas will be subsequently admitted to the upper plate. (The entire cell is now directly connected to the test gas line.)

10.7 Connect the vacuum source to the nipple attached to the cell vacuum valve. Evacuate the bottom of the cell; then with the bottom still being evacuated, evacuate the top of the cell. Close off the vacuum line to the top of the cell; then close the line to the bottom (Fig. 4).

10.8 Flush the connecting line and the top of the chamber with test gas.

10.9 Re-evacuate the system in the same manner as in 10.7. The cell manometer system should be evacuated to a pressure of 0.2 mm or less, as indicated on the vacuum gage.

10.10 Pour mercury from the reservoir into the manometer system of the cell by carefully tipping the cell. The height of the mercury in the capillary leg shall be at approximately the same level as line *B* (Fig. 3) and relatively stationary.

NOTE 8—A leak is indicated if the height of the mercury does not remain stationary. If such a leak occurs, discontinue the test and repeat the entire procedure. (If a leak occurs on a second trial, this may indicate a mechanical failure of the equipment.)

10.11 Record the height of the mercury in the capillary leg, h_o, at the start of each test, that is, immediately before the test gas has been admitted to the top of the cell.

10.12 Slowly admit the test gas to the system to a pressure, *P*, of (*1*) approximately 1 atm or (*2*) any desired test pressure.

10.13 Record the height of the mercury, *h*, in the capillary leg to the nearest 0.5 mm versus time, *t*, in hours, to the nearest 0.05 h. Take several readings (at least six are recommended) during the test. Plot a straight line through the portion of the curve that represents steady-state conditions.

NOTE 9—If the calculations are not made at steady-state conditions, a false gas transmission rate will be obtained. If, after all the mercury has been displaced from the capillary, any doubt exists as to the attainment of steady-state, a good check is as follows:

(*1*) Return the mercury to the reservoir.

(*2*) Re-evacuate the bottom of the cell only, leaving the top pressurized with test gas.

(*3*) Repeat 10.10 and 10.11, omit 10.12, and repeat 10.13.

10.14 Return the mercury in the capillary leg to the reservoir by tipping the cell upon completion of the test and prior to opening the cell vacuum valve.

10.15 Remove the specimen from the cell and measure the thickness with micrometers (Note 10). Record the average of five determinations made uniformly throughout the specimen to the nearest 0.0025 mm (0.1 mil).

NOTE 10—If there is reason to believe that the specimen will expand or contract during transmission, the thickness should be measured prior to 10.4 as well as after transmission. If any change in thickness occurs, a note to this effect shall be included with the results.

10.16 Test three specimens with each gas.

10.17 If the requirements of 10.13 are not met in the 1-atm test, repeat the procedure at a higher (up to 3 atm) or lower (not less than 0.5 atm) test pressure.

11. Calculations

11.1 Calculate the gas transmission rate, GTR, in $cm^3/m^2 \cdot 24\ h \cdot atm$, from the following relationship:[5]

$$GTR = (4.09 \times 10^{12})/(P - P_t)A_t \times (dh/dt)[(2ah - a(h_L + h_B) - V_f)/RT]$$

where:

a = area of capillary $\bar{A}\bar{B}$, mm^2,
A_t = area of transmission, cm^2,
dh/dt = slope, at any given time, of the graph formed by plotting h in millimeters of mercury versus t in hours,
h = height of mercury in cell capillary leg at any given time, mm,
h_B = maximum height of mercury in the cell manometer leg from the datum plane to upper calibration line B, mm,
h_L = height of mercury in cell reservoir leg from datum plane to top of mercury meniscus, mm,
h_o = original height of mercury in the cell capillary immediately after mercury has been dumped from the reservoir at the start of the actual transmission run, mm,
P = pressure of gas to be transmitted, mm Hg,
P_t = $(h_L - h)$ = pressure of transmitted gas, mm Hg at time t,
R = universal gas constant, 62,360,000 $(mm^3 \times mm\ Hg)/(K \times g\text{-mols})$,
t = time, h,
T = absolute temperature, K,
V_{BC} = volume from B to C, mm^3,
V_{CD} = void volume of depression, mm^3, and
V_f = $(V_{BC} + V_{CD})$, mm^3.

Manometric Recording Determination

12. Apparatus

12.1 The description of the apparatus is identical to that in Section 7, with the omission of 7.1.12, which does not apply in this procedure, and the addition of the following apparatus:

12.2 *Resistance-Recording Instrument*—A resistance-recording instrument suitably connected to a uniform-diameter platinum wire (Note 11) which runs the calibrated length of the cell manometer leg shall be employed to measure changes in height of the mercury in the cell manometer leg versus time. This instrument shall be capable of measuring such changes to the nearest 0.5 mm.

NOTE 11—A recommended automatic recording device (Fig. 4 shows a simplified schematic of a setup utilizing an automatic recorder) consists of No. 44 platinum wire (with a resistance of 2 Ω/2.54 cm (1 in.)) with No. 30 tungsten leads to the glass. These are connected by means of No. 16 gage three-conductor copper wire to a suitable ten-turn potentiometer in series with a resistance recorder of 10 to 15 Ω resistance.[6]

13. Materials

13.1 Same as Section 8.

14. Calibration

14.1 Same as Section 9, but should also include the following:

14.2 The recording instrument with the cell, lead wires, and external resistance (Note 12) in series as used in the test shall be calibrated at test temperature initially and every time after a cell has been cleaned or repaired.

NOTE 12—The external resistance should be so chosen to permit complete traverse of the chart by the pen when a change in the height of mercury equal to the height of A to B occurs (Fig. 3).

14.3 The system shall be calibrated as follows:

14.3.1 Allow the cell to come to constant temperature at test temperature.

14.3.2 With the top of the cell removed and the vacuum valve open, pour the mercury into the cell manometer leg such that the mercury is approximately at the same level as line B (Fig. 3) and relatively stationary. Adjust the external resistance of the recorder so that a chart position of zero is indicated by the pen.

14.3.3 vary the height of the mercury column and note the position indicated by the chart pen so that a graph of chart position as abscissa versus mercury height as ordinate is

[5] The derivation of this equation is given in Appendix A3. Refer to Fig. 3 for location of symbols used in this equation.

[6] The Minneapolis-Honeywell Regulator Co. 60V Model 153X64W8-X-41 resistance recorder has been found satisfactory for this purpose. It is recommended that a quick-change variable speed chart drive, such as that supplied by Insco Co., Division of Barry Controls, Inc., Groton, Mass., be installed in the recorder.

obtained. A straight line should result.

14.3.4 Determine the slope of the graph in 14.3.3 in millimeters per 2.54 cm (1 in.) to the nearest 0.5 mm, and designate this slope as recorder travel constant (RTC).

14.3.5 Determine the rate of chart paper travel to the nearest 2.54 mm (0.1 in.)/h.

15. Procedure

15.1 Same as Section 10, with the following exceptions:

15.2 Adjust the pen of the resistance-recording instrument by means of the external resistance so that the pen position corresponds with the height of mercury in the capillary leg as determined in Section 14.

15.3 For best results, set the chart to run at a speed that will plot the gas transmission curve at a slope of about 45 deg (Note 13). Once experience is gained, the proper chart speed is easily selected.

NOTE 13—This applies only to charts that have a variable-speed drive.

16. Calculations

16.1 Same as Section 11, with the following exceptions:

16.2 The symbols are identical with 11.1. Furthermore, since h is always measured between the same two points, the term

$$[(2ah - a(h_L + h_B) - V_f)/RT]$$

becomes the cell constant.

16.3 For determinations in which measurements are made with an electrical resistance recorder, dh/dt is determined experimentally, and a, h_L, h_B, and V_f are constants for a given cell. The gas constant, R, and the temperature, T, are known at a given temperature. Thus, dn/dt (Eq 5 of Appendix A3) reduces the dh/dt multiplied by a constant. In this type of determination

$$dh/dt = (\text{RTC} \times \text{RPT})/\text{slope}$$

where:

RTC = recorder travel constant, mm/in. This value is calibrated as follows:

$$\text{RTC} = \frac{\text{difference in mercury levels from } h_B \text{ to any } h\text{, mm,}}{\text{record reading at } h\text{, in.}}$$

RPT = rate of paper travel, in./h, and

slope = slope of the points plotted by the recorder during a gas transmission run.

Calculate the gas transmission rate from the equation given in 11.1.

METHOD V

(Volumetric determinations may be made with several similar type apparatus. See Appendix A1 for a method using apparatus other than that described in this method.)

17. Apparatus

17.1 *Volumetric Gas Transmission Cell*[7] shown in Fig. 5.

17.2 *Precision Glass Capillaries* or manometers with various diameters (0.25, 0.50, and 1.0 mm are recommended). The glass capillaries should have a suitable U-bend to trap the manometer liquid and have a standard taper joint to fit into the cell.

17.3 *Cathetometer* or suitable scale for measuring changes in meniscus position to the nearest 0.5 mm.

17.4 *Temperature Control:*

17.4.1 A temperature-control liquid bath is recommended for controlling the temperature of the cell body to ±0.1 C.

17.4.2 The apparatus should be shielded to restrict the temperature variations of the capillary during the test to ±0.1 C.

17.5 *Micrometer*, graduated to 0.0025 mm (0.1 mil) or better and accurate to ±0.00125 mm (0.05 mil) to measure specimen thickness.

17.6 *Barometer*, suitable for measuring the pressure of the atmosphere to the nearest 1 mm.

17.7 *Manometer*—A U-tube mercury manometer suitable for measuring pressure differences up to 250 cm Hg.

18. Materials

18.1 *Cylinder of Compressed Gas* of high purity equipped with reducing valves and calibrated pressure gage(s) or manometer.

18.2 *Capillary Liquid*—4-Methyl-2-pentanone (methyl isobutyl ketone) (Note 14) or other appropriate liquid colored with a suitable dye[8] (Note 15).

---

[7] Suitable cells may be obtained from Custom Scientific Instruments, Whippany, N. J.

[8] Suitable dyes are Victoria Blue Base available from E. I. du Pont de Nemours & Co., Inc., or Sudan Red.

NOTE 14—4-Methyl-2-pentanone has a vapor pressure of 7 mm Hg at 23 C. Erroneous results may be obtained in some cases if the attainment of this equilibrium causes slug movement in the capillary. This may take an appreciable time, especially in small capillaries, and thereby cause an erroneous answer. Also, the vapor of 4-methyl-2-pentanone may swell some materials, causing a change of the permeation rate.

NOTE 15—Mercury is not recommended for the capillary liquid except for use in calibrating cross-sectional areas because of contact angle hysteresis and resulting pressure errors (about 3 cm Hg in a 0.5-mm capillary).

18.3 *Filter Paper*—Any high grade, medium-retention noncellulosic filter paper.

NOTE 16—Other porous filters such as sintered metal have been found to be satisfactory.

19. Calibration of Capillaries (Cross-Sectional Area)

NOTE 17: **Caution**—Very low concentrations of mercury vapor in the air are known to be hazardous. Be sure to collect all spilled mercury in a closed container. Transfers of mercury should be made over a large plastic tray.

19.1 Place a column of clean mercury, approximately 7 cm (3 in.) long, in the capillary and measure its length with a cathetometer.

19.2 Transfer all of the mercury to a tared beaker and obtain the weight of the mercury on an analytical balance. Discard the mercury to be cleaned.

19.3 Since the density and weight of the column of mercury are known, its volume, V_m, at room temperature (23 C) is given by the equation:

$$V_M = W/13.54$$

where:

W = weight of the mercury, g, and

13.54 g/cm^3 = density of mercury at 23 C.

Since for a cylinder:

$$V_M = al$$

where:

a = cross-sectional area, cm^2, and

l = length of mercury column, cm, then:

$$a = V_M/l$$

19.4 Calibrations shall be made at 23 C.

20. Procedure

20.1 Center a piece of filter paper in the upper portion of the test cell.

20.2 Place the conditioned specimen smoothly on the upper portion of the test cell.

20.3 Lightly grease[9] the rubber gasket, O-ring, or flat metal that the surface of the specimen will contact. Avoid excessive grease.

20.4 Place the upper half of the cell on the base and clamp it firmly to achieve a tight seal.

20.5 Apply positive test gas pressure to both sides of the cell, flushing out all air before closing the outlet vent. A recommended flushing time is at least 10 min at a flow rate of about 100 cm^3/min.

NOTE 18—The pressure differential is obtained by monitoring and adjusting the gage pressure on the high-pressure side of the cell so that it is the desired amount above the observed barometric pressure on the open (downstream) side. This is best measured with the mercury manometer. Changes in the barometric pressure during the test may cause significant errors. They may be obviated by closing the open end of the capillary with a reservoir of significantly large volume to minimize the pressure change resulting from movement of the capillary liquid. This action, of course, amplifies temperature-variation effects if the temperature is not closely controlled.

20.6 Introduce an approximately 2-cm liquid slug (not allowing it to separate into portions) at the top of the capillary (Note 19) and close the upper outlet vent after the slug rests on the bottom of the capillary (Note 20). The capillary shall be clean and free of obstructions.

NOTE 19—It is convenient to add the liquid slug to the capillary with a syringe fitted with a long thin needle to aid proper insertion.

NOTE 20—After lowering the liquid slug into the capillary, sufficient time must be allowed for drainage down the inner wall of the capillary before beginning to take a series of readings. Extreme care must be taken to avoid the entry of room air into the capillary or the test cell after flushing and prior to beginning the test.

20.7 Adjust the pressure across the specimen to maintain the exact pressure differential desired.

20.8 Check for leaks. Small leaks around connections and joints can often be detected with soap solutions, but in some cases it may be necessary to immerse the cell in water while applying gas pressure, so as to observe bubbles at leak sites. Leaks occurring on the high-pressure side of the cell should not be considered significant.

[9] DC Silicone has been found suitable.

20.9 After a suitable estimated time for attaining steady-state, begin measurements of the time-rate of motion of the slug, using a stop watch (or clock) and distance scale maintained on the capillary or cathetometer. Take measurements at the top of the meniscus.

20.10 On completion of the run, return the slug to its starting position by slightly opening the low-pressure vent.

20.11 Repeat the measurement as necessary to assure the attainment of a steady-state condition.

NOTE 21—The time required to reach steady-state will depend upon the nature of the specimen and its thickness. For specimens of low permeability, changes in barometric pressure may interfere, particularly if long periods of test and repeated measurements are required to obtain reliable results.

21. Calculations

21.1 Plot the capillary slug position versus the time on a graph paper and draw the best straight line through the points so obtained.

20.2 Calculate the volume-flow rate, V_r, in cubic centimeters per second of transmitted gas from the slope of this line as follows:

$$V_r = \text{slope} \times a$$

where:

slope = rate of rise of capillary slug, cm/s, and

a = cross-sectional area of capillary, cm^2.

21.3 The volume flow rate (V_r) is then used to calculate either the permeability coefficient ($\bar{P}$) or the gas transmission rate (GTR) as required.

NOTE 22—Correction of this volume to standard conditions (STP, 273 K and 76 cm Hg) may be achieved by the following relationship:

$$V_r(\text{STP} = V_r(\text{ambient}) \times (B/T) \times (273/76)$$

where:

B = measured barometric pressure during test, cm Hg, and

T = test temperature, deg (normally 296 K).

This correction to STP will be small enough to ignore in ordinary work.

21.3.1 Calculate the permeability coefficient, $\bar{P}$, in barrers, as follows:

$$\bar{P} = V_r \times (b/A\Delta p) \times 10^{10}$$

where:

A = transmitting area of specimen, cm^2,

b = specimen thickness, cm, and

Δp = test gas pressure differential, cm Hg.

$$1 \text{ barrer} = (10^{-10} \text{ cm}^3 \cdot \text{cm})/(\text{s} \cdot \text{cm}^2 \cdot \text{cm Hg})$$

21.3.2 Calculate the gas transmission rate, GTR, in $cm^3/24\ h \cdot m^2 \cdot atm$, as follows:

$$\text{GTR} = (V_r \times 6.566 \times 10^{10})/A\Delta p$$

22. Report[10]

22.1 The report shall include the following:

22.1.1 Procedure used,

22.1.2 Description of the sample, including identification of composition, presence of wrinkles, bubbles, or other imperfections, and manufacturer,

22.1.3 Test gas used, and test gas composition including purity and moisture content,

22.1.4 Test temperature in deg C,

22.1.5 Average results in barrers or GTR units as a permeability coefficient or a gas transmission rate and the unit of measure.

22.1.6 Average thickness, and

22.1.7 Maximum and minimum values or the standard deviation, or both, whenever the number of replicate determinations justifies (for example five or more).

[10] Supporting data for these methods have been filed at ASTM Headquarters as RR 49: D-20.

FIG. 1 Manometric Gas Transmission Cell.

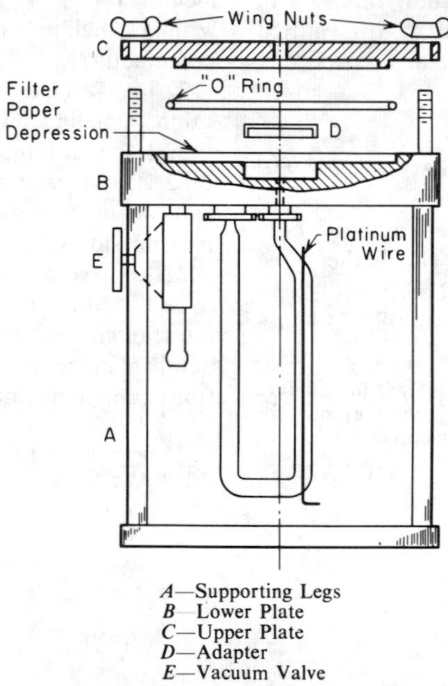

A—Supporting Legs
B—Lower Plate
C—Upper Plate
D—Adapter
E—Vacuum Valve

FIG. 2 Schematic View of Gas Transmission Cell.

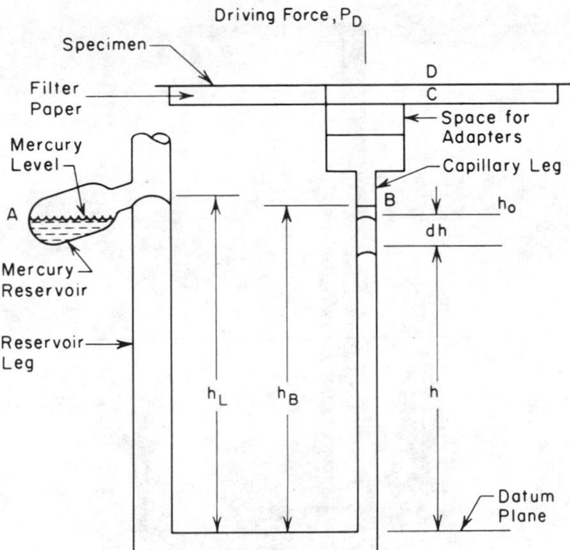

FIG. 3 Cell Manometer with Test Specimen in Place.

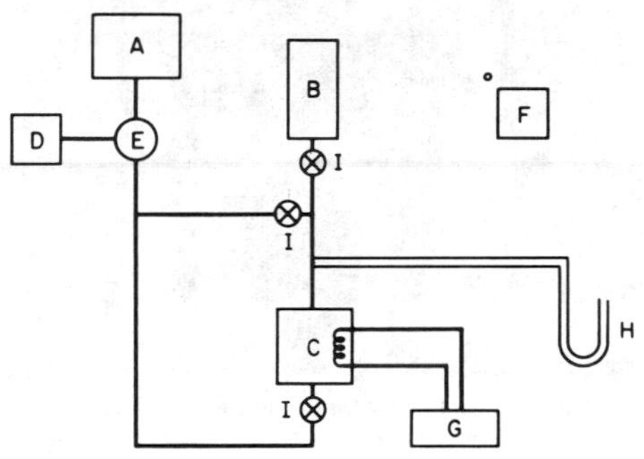

A—Vacuum Pump
B—Test Gas Cylinders
C—Gas Transmission Cell
D—Vacuum Gage
E—Trap
F—Barometer
G—Automatic Recorder (Optional)
H—Mercury Manometer
I—Needle Valve

FIG. 4 Component Arrangement of Gas Transmission Equipment.

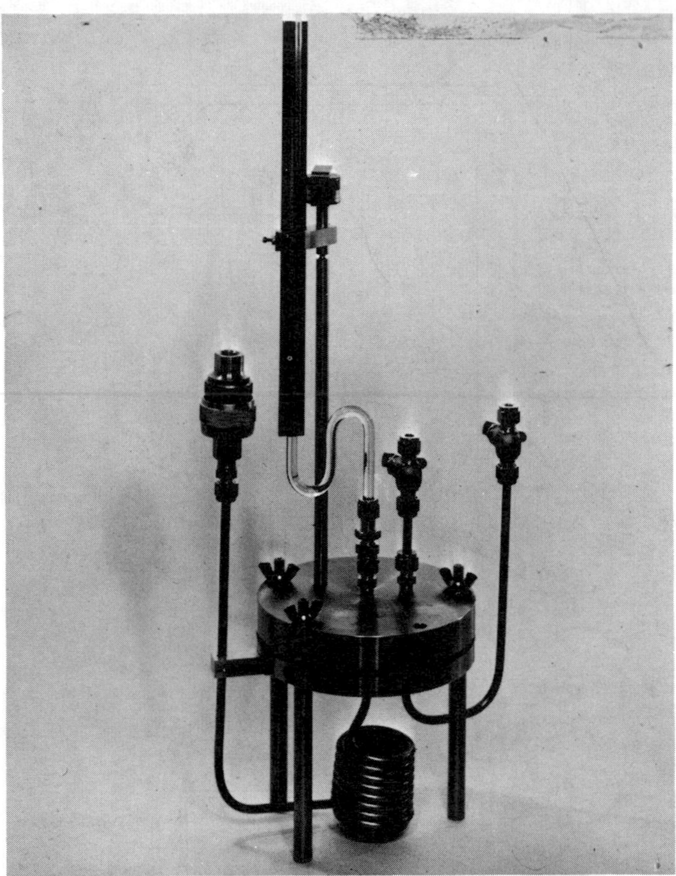

FIG. 5 Volumetric Gas Transmission Cell.

APPENDIXES

A1. Test Procedure for Aminco-Goodrich Volumetric Cell

A1.1 *Gas Transmission Measurements* can be made in accordance with Method V (volumetric) using the cell and procedure given below.

A1.2 *Gas Permeability Cell*[11] shown in Fig. A1:

A1.2.1 The arrangement differs in the following respects from the apparatus designated for Method V:

A1.2.1.1 The assembled cell is completely submerged into a rack within the controlled temperature bath with its inlet tube connected to a manifold, its outlet metal capillary to the manometer and its inlet and outlet, valved vent lines terminating above the liquid level.

A1.2.1.2 A separate 300-mm long capillary or manometer with a drain for releveling of both legs located at the bottom of the U-bend and a three-way ground-glass vent cock at the top of the inlet leg is connected to the outlet of the submerged cell by means of a flexible, detachable metal capillary.

A1.3 *Procedure:*

A1.3.1 The procedure differs only in the following respects from Method V, 17 through 22.1.7:

A1.3.1.1 The original as well as transmitted volume is recorded as the balanced or leveled meniscus height of both legs, one of which is open to atmosphere or to a controlled pressure, and the other leg to the pressure exerted by the transmitted gas.

A1.3.1.2 The manometer is filled prior to purging to the 300-mm graduation or sufficient height to permit necessary draining for releveling throughout a complete test while maintaining minimal transmitted volume space.

A1.3.1.3 The bottom manometer drain is used to withdraw liquid from the outlet leg to achieve releveling prior to recording the meniscus level, assuring isostatic conditions throughout the test.

A1.4 *Calculations*—Suggested calculations to conform with the requirements of this method are furnished with the cell.

---

[11] This cell is supplied by the American Instrument Co., Inc., Silver Spring, Md.

A2. Conversion Factors for Various Permeability Units

Note—To obtain unit in left-hand vertical column multiply unit in top horizontal column by figure opposite both units.

Multiply / To Obtain	$\frac{cm^3 \cdot cm}{cm^2 \cdot s \cdot cm\ Hg}$	$\frac{cm^3 \cdot mil}{m^2 \cdot 24\ h \cdot atm}$	$\frac{cm^3 \cdot mil}{100\ in.^2 \cdot 24\ h \cdot atm}$	$\frac{cm^3 \cdot mm}{m^2 \cdot 24\ h \cdot atm}$	$\frac{cm^3 \cdot mil}{100\ in.^2\ h \cdot atm}$	$\frac{in.^3 \cdot mil}{100\ in.^2 \cdot 24\ h \cdot atm}$
$\frac{cm^3 \cdot cm}{cm^2 \cdot s \cdot cm\ Hg}$	1.00	3.87×10^{-14}	6.00×10^{-13}	1.52×10^{-12}	9.26×10^{-15}	9.83×10^{-12}
$\frac{cm^3 \cdot mil}{m^2 \cdot 24\ h \cdot atm}$	2.58×10^{13}	1.00	15.5	39.4	0.24	2.54×10^{2}
$\frac{cm^3 \cdot mil}{100\ in.^2 \cdot 24\ h \cdot atm}$	1.67×10^{12}	6.45×10^{-2}	1.00	2.54	1.58×10^{-2}	16.4
$\frac{cm^3 \cdot mm}{m^2 \cdot 24\ h \cdot atm}$	6.57×10^{11}	2.54×10^{-2}	0.394	1.00	6.10×10^{-3}	6.46
$\frac{cm^3 \cdot mil}{100\ in.^2 \cdot h \cdot atm}$	1.08×10^{14}	4.16	64.6	1.64×10^{2}	1.0	1.06×10^{3}
$\frac{in.^3 \cdot mil}{100\ in.^2 \cdot 24\ h \cdot atm}$	1.02×10^{11}	3.94×10^{-3}	6.1×10^{-2}	1.55×10^{-1}	9.46×10^{-4}	1.0

A3. Derivation of Equation for Gas Transmission Rate Using Manometric Gas Cell

A3.1 The following symbols are used in the derivation of the equation used in Section 11 to calculate the gas transmission rate.

a = area of capillary $\bar{A}\bar{B}$, mm^2,
dP = dh = differential pressure change of transmitted gas, mm Hg,
dV = adh = differential volume change of transmitted gas, mm^3,
h = height of mercury in cell capillary leg at any given time, mm,
h_B = maximum height of mercury in the cell manometer leg from the datum plane to upper calibration line B, mm,
h_L = height of mercury in cell reservoir leg from datum plane to top of mercury meniscus, mm,
n = quantity of gas transmitted, mols,
$\bar{P}_t$ = $(h_L - h)$ = pressure of transmitted gas, mm Hg, at time t,
R = universal gas constant, 62,360,000 (mm^3 × mm Hg)/(K × g-mols),
t = time, h
T = absolute temperature, K,
V_{BC} = volume from B to C, mm^3,
V_{CD} = void volume of depression, mm^3,
V_f = $(V_{BC} + V_{CD})$, mm^3, and
V_t = $[V_f + a(h_B - h)]$ = volume of transmitted gas, mm^3 at time t.

A3.2 The number of mols of gas transmitted, n, may be determined from the ideal gas law (which is valid at the low pressure involved) as:

$$n = P_t V_t / RT \qquad (1)$$

For differential changes in P_t and V_t Eq 1 can be differentiated as follows:

$$nRT = P_t V_t$$

$$RT dn = d(P_t V_t)$$

$$RT dn = P_t dV + V_t dP \text{ (Note A1)} \qquad (2)$$

Note A1—This differentiation is of the form $d(uv) = udv + vdu$. Substitute into Eq 2 the values for dV and dP given in the definition of symbols:

$$RT dn = P_t a dh - V t dh, \quad \text{or}$$

$$dn = (-aP_t dh - V_t dh)/RT \qquad (3)$$

By definition,

$$P_t = (h_L - h), \text{ and}$$

$$V_t = [V_f + a(h_B - h)]$$

Therefore, substituting into Eq 3 and collecting terms,

$$dn = \{-a(h_L - h)dh - [V_f + a(h_B - h)]dh\}/RT$$

$$dn = [(-ah_L + ah)dh - (V_f + ah_B - ah)dh]/RT$$

$$dn = [dh(-ah_L + ah - V_f - ah_B + ah)]/RT$$

$$dn = \{dh[2ah - a(h_L + h_B) - V_f]\}/RT \qquad (4)$$

Since the rate of transmission is dn/dt, divide Eq 4 by the differential time, dt, to give the form:

$$dn/dt = (dh/dt)[(2ah - a(h_L + h_B) - V_f)/RT] \qquad (5)$$

A3.3 For determinations in which measurements are made visually, or with a cathetometer, dh/dt = the slope of the graph at any given time formed by plotting h in millimeters of mercury versus t in hours.

A3.4 By definition, dn/dt is expressed in mols per hour. To obtain the gas transmission rate (GTR) expressed in cm^3/m^2 · 24 h · atm, use the following conversion factors:

$$\text{GTR} = \frac{\frac{dn}{dt}\frac{\text{mol}}{\text{h}} \times 24 \frac{\text{h}}{24\text{ h}} \times 22{,}415 \frac{\text{cm}^3}{\text{mol}} \times 760 \frac{\text{mm}}{\text{atm}} \times 10{,}000 \frac{\text{cm}^2}{\text{m}^2}}{P_d(\text{mm}) \times A(\text{cm}^2)}$$

The term P_d, which is the driving force, and A, the area of transmission, must be included in this equation in order to obtain the GTR in the desired units. Or, collecting terms and substituting for

$$P_d = (P - p_t)$$

$$\text{GTR} = dn/dt \times (24 \times 22{,}415 \times 760 \times 10{,}000)/(P - P_t)A$$

or, substituting for dn/dt from Eq 5,

$$\text{GTR} = (22{,}415 \times 24 \times 760 \times 10{,}000)/(P - P_t)A \times (dh/dt)[(2ah - a(h_L + h_B) - V_f/RT], \quad \text{or}$$

$$\text{GTR, cm}^3\text{ m}^2\text{ 24 h atm} = (4.09 \times 10^{12})/(P - P_t)A \times (dh/dt)[2ah - a(h_L + h_B) - V_f/RT] \qquad (6)$$

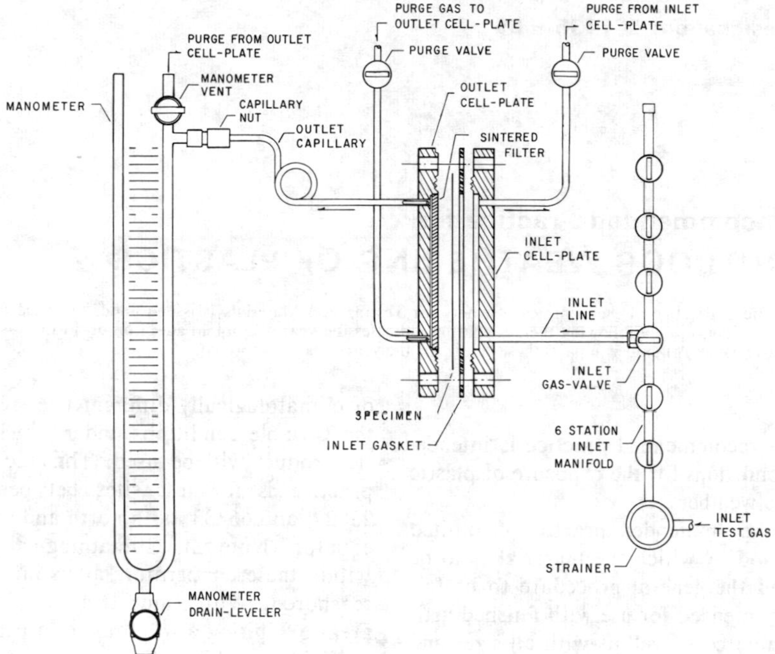

FIG. A1 Aminco-Goodrich Volumetric Cell.

Designation: D 1435 - 69

Recommended Practice for
OUTDOOR WEATHERING OF PLASTICS[1]

This Recommended Practice of the American Society for Testing and Materials is issued under the fixed designation D 1435; the number immediately following the designation indicates the year of original adoption or, in the case of revision, the year of last revision. A number in parentheses indicates the year of last reapproval.

1. Scope

1.1 This recommended practice is intended to define conditions for the exposure of plastic materials to weather.

1.2 This recommended practice is limited to the method by which the material is to be exposed and the general procedure to be followed. It is intended for use with finished articles of commerce as well as with all sizes and shapes of test specimens.

1.3 Means of evaluation of the effects of weathering will depend upon the intended usage of the test material.

NOTE 1—The values stated in U.S. customary units are to be regarded as the standard. The metric equivalents of U.S. customary units may be approximate.

2. Significance

2.1 Tests of the type described in this recommended practice may be used to evaluate the stability of plastic materials when exposed outdoors to the varied influences which comprise weather. Exposure conditions are complex and changeable. Important factors are climate, time of year, presence of industrial atmosphere, etc. Generally, because it is difficult, if not impossible, to define or measure precisely the factors directing degradation due to weathering, results of outdoor exposure tests must be taken as indicative only. Repeated exposure testing at different seasons and over a period of more than one year is required to confirm exposure tests at any one location. Control samples must always be utilized as weathering is a comparative test.

3. Test Sites

3.1 Weathering racks shall be located in cleared areas, preferably at a suitable number of climatologically different sites representing the variable conditions under which the plastic product will be used. This recommended practice is for test sites between latitudes 23°27′ and 66°33′ both north and south of the equator (Note 2). Climatological variations within these temperature areas include desert, seashore (salt air), industrial atmosphere, plus areas exhibiting a wide range in percentage of available sunshine. The area beneath and in the vicinity of the weathering racks should be characterized by low reflectance and be typical of the ground cover in that climatological area. In desert areas it should be sand whereas in most temperate areas it will be low cut grass.

NOTE 2—These are geographical limits of the Temperate Zones. The suitability of 45-deg mounting at sites in tropical or polar areas has not been established.

4. Apparatus

4.1 *Position of Racks*—The lowest samples in the rack shall be exposed at a minimum height of 760 mm (30 in.) above the cleared area. The rack shall be placed in a location so that no shadow from a neighboring obstruction, having an angle of elevation greater than 20 deg, shall fall on any sample.

4.1.1 *45-Deg Racks*—These racks shall be so positioned that the exposed surfaces of samples are at an angle of 45 deg to the horizontal and facing the equator.

4.1.2 *90-Deg Racks*—These racks shall be

---

[1] This recommended practice is under the jurisdiction of ASTM Committee D-20 on Plastics. A list of committee members may be found in the ASTM Yearbook. This standard is the direct responsibility of Subcommittee D-20.50 on Permanence Properties.

Current edition effective Nov. 14, 1969. Originally issued 1956. Replaces D 1435 - 65.

so positioned that the exposed surfaces of samples are at an angle of 90 deg to the horizontal and facing the equator.

NOTE 3—In some instances, exposures facing directly *away* from the equator may be desirable. Reports must contain specific reference to any such reverse direction exposures.

4.1.3 *Horizontal Racks*—These racks shall be so positioned that the exposed surfaces are horizontal.

4.2 *Materials of Construction*—Racks and hardware shall be constructed of materials that are noncorrodible without surface treatment. Untreated wood is applicable only in desert areas. Aluminum Alloy No. 6061-T6 has been found suitable for construction of racks and frames in all other areas. Steel and copper are unsuitable in all areas.

4.3 *Manner of Construction*—The racks shall be constructed to hold sample holders (or samples) of any convenient width and length. The structural members shall in no case constitute a backing to the samples under test. A typical aluminum rack of suitable design is illustrated in Fig. 1. A clamping device (Fig. 1), properly spaced slots, or aluminum alloy nails or screws shall be used for securely fastening the sample holders or samples without constraint.

4.4 *Sample Holders:*

4.4.1 Most samples under test will not be of exact size for mounting directly on the frame. Sample holders should be used to support the many sized specimens involved in this testing. In no case shall the sample holder constitute a backing for that portion of the material to be evaluated.

4.4.2 The sample holders shall be constructed of an inert material. (Aluminum extruded shapes have been found suitable.)

4.4.3 The design of the sample holders shall be such that each of the specimens or sheets contained cannot shift its position, yet is not constrained (that is, free to expand or contract with thermal changes, to swell because of moisture absorption, or to shrink because of plasticizer loss).

4.4.4 *Frame Holders*—These holders are in the shape of a frame which may be subdivided as necessary to provide proper spacing of samples. The exposure aperture of each frame shall be of sufficient size to expose the entire test area of each specimen when sufficient samples are contained.

4.4.5 *Plate Holders*—This type of holder is a universal panel consisting of a slotted aluminum plate on which electrical white glaze porcelain insulators[2] are mounted at proper positions to affix various sized specimens. The specimens are mounted in the grooves of the insulators at a fixed distance of 11 mm from the slotted back plate. The insulators provide inert mounting while the slotted plate permits free circulation of air behind the sample. This method of mounting is shown in Fig. 2.

5. Test Samples

5.1 Exposure test samples may be of any size or shape that can be mounted in a holder or directly applied to the racks. They may be specimens suited to the means of evaluating the effects of weathering on a specific physical property, or they may be larger samples from which samples for evaluation may be cut. Exposure test samples should be large enough so that mounting edges may be removed where evaluation test results would otherwise be affected.

5.2 Exposure test samples shall simulate as far as practical service conditions of an end use application. Normally all materials of unknown end use application will be run in an unbacked condition. When conditions of use are known, the sample exposed will consist of the plastic material being evaluated plus suitable backing materials to conform to projected practice. The effect of backing is highly significant and contributes to the degradation as a function of reflectance, heat absorption, etc. It shall always be used in relation to an end use system rather than as a standard mounting method.

5.3 The total number of samples will be determined by the removal schedule plus control samples for determining original and final control values. These control samples shall be retained at standard conditions of 23.0 ± 1 C (73.4 ± 1.8 F) and 50 ± 2 percent relative humidity. They shall be covered with inert wrapping to exclude light exposure during the aging period.

[2] A satisfactory insulator for this purpose is Catalog No. 615160 Special No. 6 Knob dry process porcelain insulator, Porcelain Products, Carey, Ohio.

6. Procedure

6.1 Inscribe the test samples to be exposed with an identifying number, letter, or symbol so that they may be readily identified after exposure. The marking shall be such that there is no interference with either the exposure or the subsequent testing. (Preferably, both sample and sample holder should be marked on the side not exposed to weather as advanced weathering can obscure even deeply scribed marks.)

6.2 Mount the test samples in the holder. It is convenient to group in one holder samples to be removed from exposure at the same time.

6.3 Record a diagram of the test holder specimen layout and record date of installation and length of exposure planned.

6.4 Record appearance and physical property data appropriate to the evaluation method used.

6.5 Mount the specimen holders on racks for the prescribed time.

6.6 Establish a fixed procedure of cleaning, visual examination, conditioning and testing of the specimens. This procedure will vary with material but must be uniform in a series of tests on one material to provide comparative results.

6.7 The face of the specimen shall not be masked for the purpose of showing on one panel the effects of various exposure times. Misleading results can be obtained by this method since the masked position of the specimen is still exposed to temperature and humidity which in many cases will affect results. The control samples shall be used for comparison with specimens exposed various exposure times.

6.8 Exposures and inspections may be planned to permit reporting:

6.8.1 Change after a specified exposure,

6.8.2 Exposure to a specified change, and

6.8.3 A record of a series of measurements versus exposure.

7. Report

7.1 The report shall include the following:

7.1.1 Angle of exposure (horizontal, 45 deg, or 90 deg) and direction of exposure.

7.1.2 Duration of exposure of each sample at each site and dates of exposure.

7.1.3 Sunlight energy data if available. Integrated incident energy expressed in langleys (gram-calories per square centimeter).

7.1.4 Description of climate at each site. Summary of pertinent U.S. Weather Bureau Climatological Data at each site for exposure period involved:

7.1.4.1 Rainfall.

7.1.4.2 Percentage of possible sunshine.

7.1.4.3 Temperature average and temperature extremes.

7.1.4.4 Humidity average and humidity extremes.

7.1.4.5 Geographical location of Weather Bureau relative to test site.

7.1.5 General appearance and properties of the control specimens both unaged and aged without exposure to light.

7.1.6 General appearance and properties of the exposed samples.

7.1.7 Data on reference samples exposed simultaneously.

7.1.8 Description or reference to tests used to evaluate weathering.

7.1.9 Suitably complete identification of sample.

FIG. 1 Typical Aluminum Exposure Rack.

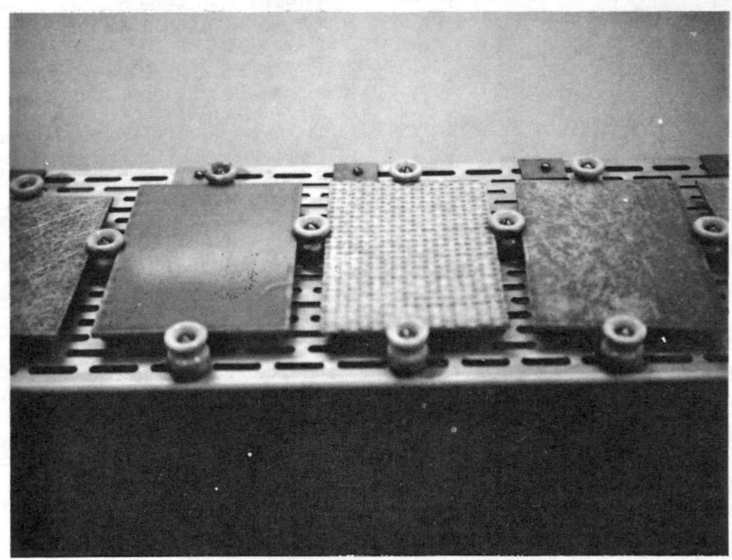
FIG. 2 Suitably Mounted Samples.

APPENDIX

A1. Evaluation of Weathering Effects

A1.1 Introduction

A1.1.1 The purpose of this appendix is to suggest test methods for evaluating the effect of outdoor exposure on plastic materials. A number of tests are listed by property classes. The selection of appropriate test methods for specific plastic materials, and the definition of failure criteria, are beyond the scope of this document. Outdoor weathering is known to affect various plastics in grossly different ways. The selection of test methods must be based on familiarity with the materials to be evaluated and with end use requirements. Through a knowledgeable selection from the test methods listed, and others when appropriate, a meaningful measure of weathering deterioration may be derived.

A1.2 Appearance Properties

A1.2.1 Appearance is principally judged by light reflection from or transmission through a specimen. Changes are usually in color or in geometric distribution of light, or both. These changes may be computed from measurements obtained prior to and after exposure, or they may be estimated by visual comparison with an unexposed sample. When such visual comparisons are made, it should be be recognized that the unexposed sample may also change with time.

Color

A1.2.1.1 *Color Difference:*

D 1535 Method of Specifying Color by the Munsell System[4]

D 1729 Method for Visual Evaluation of Color Differences of Opaque Materials[4]

D 2244 Method for Instrumental Evaluation of Color Differences of Opaque Materials[4]

E 308 Recommended Practice for Spectrophotometry and Description of Color in CIE 1931 System[4]

A1.2.1.2 *Yellowness:*

D 1925 Method of Test for Yellowness Index of Plastics[3]

E 313 Method of Test for Indexes of Whiteness and Yellowness of Near-White Opaque Materials[4]

Geometric Distribution

A1.2.1.3 *Gloss:*

D 523 Method of Test for Specular Gloss[3]

D 2457 Method of Test for Specular Gloss of Plastic Films[3]

A1.2.1.4 *Haze:*

D 1003 Method of Test for Haze and Luminous Transmittance of Transparent Plastics[3]

A1.2.1.5 *Transparency:*

D 1746 Method of Test for Transparency of Plastic Sheeting[3]

A1.3 Electrical Properties

A1.3.1 These tests should be run on samples as taken from the specimen holders before cleaning, as well as after cleaning, in accordance with 6.6.

A1.3.1.1 D 495 Method of Test for High-Voltage, Low-Current Arc Resistance of Solid Electrical Insulating Materials[3]

A1.3.1.2 D 149 Method of Test for Dielectric Breakdown Voltage and Dielectric Strength of Electrical Insulating Materials at Commercial Power Frequencies[3]

Note A1—It is suggested that this test be run at room temperature and at 100 C preferably in air, both by the short time and the step by step procedures. (Test temperatures are to be determined by recommended service temperature of material involved.)

A1.3.1.3 D 150 Test for A-C Loss Characteristics and Dielectric Constant (Permittivity) of Solid Electrical Insulating Materials[3]

A1.4 Mechanical Properties

A1.4.1 *Stiffness:*

D 747 Method of Test for Stiffness of Plastics by Means of a Cantilever Beam[3]

A1.4.2 *Indentation Hardness:*

D 2240 Method of Test for Indentation Hardness of Rubber and Plastics by Means of a Durometer[3]

D 785 Test for Rockwell Hardness of Plastics and Electrical Insulating Materials[3]

A1.4.3 *Impact:*

D 256 Methods of Test for Impact Resistance of Plastics and Electrical Insulating Materials[3]

A1.4.4 *Tear Resistance:*

D 1004 Method of Test for Tear Resistance of Plastic Film and Sheeting[3]

D 1938 Method of Test for Resistance to Tear Propagation in Plastic Film and Thin Sheeting by Single-Tear Method[3]

A1.4.5 *Dimensional Change:*

D 1042 Method of Measuring Changes in Linear Dimensions of Plastics[3]

D 1204 Method of Measuring Changes in Linear Dimensions of Nonrigid Thermoplastic Sheeting or Film[3]

A1.4.6 *Tensile and Elongation Properties:*

D 638 Method of Test for Tensile Properties of Plastics[3]

D 882 Methods of Test for Tensile Properties of Thin Plastic Sheeting[3]

D 1708 Method of Test for Tensile Properties of Plastics by Use of Microtensile Specimens[3]

D 1923 Method of Test for Tensile Properties of Thin Plastic Sheeting Using the Inclined Plane Testing Machine[3]

D 2289 Method of Test for Tensile Properties of Plastics at High Speeds[3]

---

[3] *Annual Book of ASTM Standards*, Part 27.

[4] *Annual Book of ASTM Standards*, Parts 21 and 30.

American National Standard K65.29-1967
American National Standards Institute

Recommended Practice for

OPERATING LIGHT- AND WATER-EXPOSURE APPARATUS (CARBON-ARC TYPE) FOR EXPOSURE OF PLASTICS[1]

This Recommended Practice is issued under the fixed designation D 1499; the number immediately following the designation indicates the year of original adoption or, in the case of revision, the year of last revision. A number in parentheses indicates the year of last reapproval.

1. Scope

1.1 This recommended practice covers specific variations in test conditions which shall be applicable when ASTM Recommended Practice E 42, for Operating Light-and Water-Exposure Apparatus (Carbon-Arc Type) for Exposure of Nonmetallic Materials[2] is employed for the exposure of plastics. Also covered are the preparation of test specimens, the test conditions best suited for plastics, and the evaluation of test results.

2. Significance

2.1 The ability of a plastic material to resist deterioration in its electrical, mechanical, and optical properties, when the plastic is exposed in light- and water-exposure apparatus can be very significant for many applications. This recommended practice is useful in comparing materials tested in different machines of the same type. The results of exposures in different types of machines should not be compared, unless correlation has been established for both the particular material and the machines. Because of the wide variation in weather with time and location, it is not possible to correlate results of outdoor exposure with data obtained in laboratory equipment by this method.

2.2 Accuracy of the test results will depend upon the care that is taken to operate the equipment according to this recommended practice. Factors that affect accuracy include regulation of line voltage, freedom from salt or other deposits from water, temperature control, and conditions of the electrodes. The results are empirical and provide a basis for laboratory comparisons.

3. Test Specimens

3.1 Specimens shall be cut to suit the requirements of the specimen holders in the apparatus used. Each specimen shall be mounted so that its back is open to the air and only the minimum area required for holding it touches the holder.

3.2 Since the thickness of a specimen may affect markedly the results of the test, the actual thickness shall be recorded at the start of the exposure period.

3.3 The face of a specimen shall not be masked for the purpose of showing on one panel the effects of various exposure times. Misleading results may be obtained by this method, since the masked portion of the specimen is still exposed to temperature and humidity cycles which in many cases will affect results. A separate control specimen obtained from the same lot as the specimen under test shall be used for comparison with specimens exposed for various exposure times.

4. Test Conditions

4.1 During the test cycle, the standard black panel thermometer shall reach an equilibrium temperature of 63 ± 5 C (145 ± 9 F). Lower black panel temperatures may be desirable when testing specimens that are particularly temperature sensitive. The black panel tem-

---

[1] This recommended practice is under the jurisdiction of ASTM Committee D-20 on Plastics. A list of committee members may be found in the ASTM Yearbook. This standard is the direct responsibility of Subcommittee D-20.50 on Permanence Properties.

Current edition effective Aug. 31, 1964. Originally issued 1959. Replaces D 1499 – 50 T.

[2] *Annual Book of ASTM Standards*, Part 27.

perature actually used shall be be reported.

NOTE 1—This equilibrium temperature is that attained during the dry portion of a cycle if a spray is included in the cycle.

4.2 The specimen water spray cycle described in 4.6 of Recommended Practice E 42 may be used in tests requiring specimen water spray.

NOTE 2—Spray cycles used shall permit the equilibrium temperature to be attained as specified in Note 1.

NOTE 3—Tests not requiring specimen water spray shall be conducted in accordance with ASTM Recommended Practice E 188, for Operating Enclosed Carbon-Arc Type Apparatus for Light Exposure of Nonmetallic Materials.[3]

4.3 If the specimen water spray leaves an objectionable deposit or stain on the specimens after continued exposure in the apparatus and there is disagreement between laboratories on the extent of this deposit, referee tests shall be run to resolve the disagreement using deionized or distilled water in the specimen spray containing not more than 20 ppm total solids content. If deionized or distilled water is employed the specimen spray piping from the demineralizer or still to the spray orifice shall be fabricated of an inert material such as aluminum, stainless steel, tin, saran, or polyethylene. Iron and copper are not satisfactory.

4.4 Specimens shall not be removed from the exposure apparatus for long periods of time and then returned for additional tests, since this does not produce the same results on all materials as tests run without this type of interruption.

NOTE 4—If it is necessary to remove specimens for observation during the test, the time they are off the exposure apparatus shall be reported.

5. Periods of Exposure and Evaluation of Test Results

5.1 The duration of the exposure under this recommended practice shall be one of the following:

5.1.1 A specified number of hours, as needed for the purpose of the test or as agreed upon by the parties concerned with the test.

5.1.2 The number of hours of exposure required, as agreed upon by the parties concerned with the test, to produce mimimum acceptable changes in either the test specimen or a standard sample.

5.1.3 The number of hours of exposure required, as agreed upon by the parties concerned with the test, to produce a minimum amount of change in the test specimen.

5.2 Changes in exposed test specimens shall be evaluated or rated by applicable ASTM specifications.

6. Report

6.1 In addition to the items specified in Section 5 of Recommended Practice E 42, the report shall include the following:

6.1.1 Light and water cycle employed,

6.1.2 Black panel temperature, and

6.1.3 Whether deionized, distilled, or tap water was used in the test.

[3] *Annual Book of ASTM Standards*, Part 30.

Designation: D 1501 – 65 T

Tentative Recommended Practice for

EXPOSURE OF PLASTICS TO FLUORESCENT SUNLAMP[1]

This Tentative Recommended Practice has been approved by the sponsoring committee and accepted by the Society in accordance with established procedures, for use pending adoption. Suggestion for revisions should be addressed to the Society at 1916 Race St., Philadelphia, Pa. 19103.

Pursuant to the warning note covering the 3-year limitation on tentatives, published in the previous edition, this tentative has received special permission from the Board of Directors for continuation until July 1971.

1. Scope

1.1 This recommended practice[2] defines conditions for the exposure of plastic materials to radiant energy from fluorescent sunlamps, heat, and fog.

1.2 Three alternative procedures may be used:

Procedure A for exposure to light at ambient temperature,

Procedure B, for exposure to heat and light, and

Procedure C, which includes moisture as well as heat and light.

2. Significance

2.1 The various procedures are designed to simulate some aspects of outdoor exposure conditions on a laboratory scale. However, direct correlation between this recommended practice and actual exposure outdoors should not be expected since outdoor conditions are variable.

3. Apparatus

3.1 *Light-Exposure Unit*—A turntable shall be used in conjunction with five fluorescent sunlamp bulbs which have been in use at least 50 and less than 1050 h. The bank of lamps shall be mounted so as to be centered directly above the turntable, and so that the distance from the plane of the top of the specimens on the turntable to the plane of the bottom of the lamps shall be 70 mm. It is recommended that a single lamp be replaced every 200 h so that the respective ages in the five lamps would be 50, 250, 450, 650, and 850 h.

3.2 *Lamp*—A Type FS-20-T12 Westinghouse fluorescent sunlamp. The lamp is 610 mm long, has a nominal input of 20 W, operates at 110 V ± 1 percent, and fits into standard fluorescent light fixtures. The bulbs shall be mounted parallel to one another on 60-mm centers. A polished flat aluminum reflector, 380 by 460 mm, shall be mounted above the lamps so that the polished surface shall be 25 mm from the plane of the top of the bulbs. The reflector shall be centered over the lamps with the long side parallel to their length.

3.3 *Turntable*—The turntable shall rotate at 30 to 40 rpm, and shall be fitted with a disk of polished aluminum approximately 430 mm in diameter and 2.5 mm thick. Means shall be provided to hold the specimens above the metal disk so that there is an air space of at least 5 mm between the bottom of the specimens and the top surface of the disk.

NOTE 1—Experience has shown that all areas of the top surface of the disk are equally effective so that the test specimens may be mounted anywhere on the disk.

3.4 *Heat Source*—A source of heated air for maintaining the desired chamber temperature.

3.5 *Thermometer*—An ASTM Partial

---

[1] This recommended practice is under the jurisdiction of ASTM Committee D-20 on Plastics. A list of committee members may be found in the ASTM Yearbook. This tentative is the direct responsibility of Subcommittee D-20.50 on Permanence Properties.

Current edition effective Dec. 21, 1965. Originally issued 1957. Replaces D 1501 – 57 T.

[2] See Hirt, R., et al., "Ultraviolet Spectral Energy Distributions of Natural Sumlight and Accelerated Test Light Sources," *Journal of the Optical Society of America*, JOSAA, Vol. 50, 1960, pp. 706–713.

Immersion Thermometer having a range from −20 to +150 C and conforming to the requirements of Thermometer 1C as prescribed in ASTM Specification E 1, for ASTM Thermometers.[3]

3.6 *Fog Chamber*—A closed shallow box fitted with a unit for atomizing distilled water and a baffle to prevent the spray from impinging directly on the specimens. The distilled water mist should visibly fog the chamber with a continuous mist which deposits on the specimens without a washing action. The temperature of the water feeding the atomizer shall be 21 to 27 C.

NOTE 2—The turntable and fog chamber shown in Figs. 3 and 4 of ASTM Recommended Practice D 795, for Exposure of Plastics to S-1 Mercury Arc Lamp,[4] are suitable.

4. Test Specimens

4.1 The type and number of test specimens to be used shall be specified in the particular material specification or by the purchaser.

NOTE 3—The test specimen shown in Fig. 5 of Recommended Practice D 795 is suitable and makes efficient use of the available disk space.

5. Procedure A—Light Exposure

5.1 Expose the specimens as prescribed (see 3.3) on the rotating turntable in an atmosphere maintained at an ambient temperature such that a thermometer placed with its midpoint at the center of the disk will read 21 to 25 C.

NOTE 4—Operating the apparatus in an atmosphere at the Standard Laboratory Temperature of 23 C will generally satisfy the temperature requirement.

5.2 Continue the exposure to a preagreed end point at which time examine the specimens.

6. Procedure B—Light and Heat Exposure

6.1 Expose the specimens as prescribed (see 3.3) on the rotating turntable. Blow heated air over the specimens at a temperature such that a thermometer placed with its midpoint at the center of the disk will read 55 to 60 C.

6.2 Continue the exposure to a preagreed end point at which time examine the specimens.

7. Procedure C—Light, Heat, and Fog Exposure

7.1 Expose the specimens to heat and light as in Procedure B with interruptions for fog in the following order for each 24 h: 2 h in the fog chamber, 2 h of heat and light, 2 h in the fog chamber, and 18 h of heat and light.

7.2 The duration of this test shall be 240 h, unless otherwise specified.

NOTE 5—Continuity of exposure under Procedures A, B or C may influence the results obtained, and deviation from continuity should be reported.

8. Report

8.1 The report shall include the following:

8.1.1 Which procedure (A, B, or C) was used,

8.1.2 Duration of exposure,

8.1.3 Color change, reported as none, slight, appreciable, or extreme (Note 6),

8.1.4 Observations regarding surface changes (such as dulling and chalking) and deep-seated changes (such as checking, crazing, warping, and discoloration), and

8.1.5 Results of any physical or chemical tests made to determine the extent of changes or degradation resulting from the exposure.

NOTE 6—A slight change is defined here as one that is perceptible with difficulty. An appreciable change is one that is readily perceptible without close examination but is insufficient to alter markedly the original color of the specimen. An extreme change is one that is very obvious and has resulted in a marked alteration of the original color of the specimen.

---

[3] *Annual Book of ASTM Standards*, Part 30.
[4] *Annual Book of ASTM Standards*, Part 27.

Standard Method of Test for
DENSITY OF PLASTICS BY THE DENSITY-GRADIENT TECHNIQUE[1]

This Standard is issued under the fixed designation D 1505; the number immediately following the designation indicates the year of original adoption or, in the case of revision, the year of last revision. A number in parentheses indicates the year of last reapproval.

NOTE—Editorial change in Section 6.1 was made in September 1969.

1. Scope

1.1 This method covers determination of the density of solid plastics.

1.2 The method is based on observing the level to which a test specimen sinks in a liquid column exhibiting a density gradient, in comparison with standards of known density.

NOTE 1—The values stated in U.S. customary units are to be regarded as the standard. The metric equivalents of U.S. customary units may be approximate.

2. Significance

2.1 The density of a solid is a conveniently measurable property which is frequently useful as a means of following physical changes in a sample, as an indication of uniformity among samples, and a means of identification.

2.2 This method is designed to yield results accurate to better than 0.05 percent.

NOTE 2—Where accuracy of 0.05 percent or better is desired, the gradient tube shall be constructed so that vertical distances of 1 mm shall represent density differences no greater than 0.0001 g/cm^3. The sensitivity of the column is then 0.0001 $g/cm^3 \cdot mm$. Where less accuracy is needed, the gradient tube shall be constructed to any required sensitivity.

3. Definition

3.1 *density of plastics*—the weight per unit volume of material at 23 C, expressed as follows:

$$D^{23\,C},\ g/cm^3$$

NOTE 3—Density is to be distinguished from specific gravity, which is the ratio of the weight of a given volume of the material to that of an equal volume of water at a stated temperature.

4. Apparatus

4.1 *Density-Gradient Tube*—A suitable graduate with ground-glass stopper.[2]

4.2 *Constant-Temperature Bath*—A means of controlling the temperature of the liquid in the tube at 23 ± 0.1 C. A thermostatted water jacket around the tube is a satisfactory and convenient method of achieving this.

4.3 *Glass Floats*—A number of calibrated glass floats covering the density range to be studied and approximately evenly distributed throughout this range.

4.4 *Pycnometer*, for use in determining the densities of the standard floats.

4.5 *Liquids*, suitable for the preparation of a density gradient (Table 1).

NOTE 3—It is very important that none of the liquids used in the tube exert a solvent or chemical effect upon the test specimens during the time of specimen immersion.

4.6 *Hydrometers*—A set of suitable hydrometers covering the range of densities to be measured. These hydrometers should have 0.001 density graduations.

4.7 *Analytical Balance*, with a sensitivity of 0.001 g.

4.8 *Siphon or Pipet Arrangement*, for filling the gradient tube. This piece of equipment

[1] This method is under the jurisdiction of ASTM Committee D-20 on Plastics. A list of committee members may be found in the ASTM Yearbook. This standard is the direct responsibility of Subcommittee D-20.70 on Analytical Methods.

Current edition effective Sept. 9, 1968. Originally issued 1957. Replaces D 1505 – 67.

[2] Tubes similar to those described in References (**6**) and (**12**) may also be used.

should be constructed so that the rate of flow of liquid may be regulated to 10 ± 5 ml/min.

5. Test Specimen

5.1 The test specimen shall consist of a piece of the material under test. The piece may be cut to any shape convenient for easy identification, but should have dimensions that permit the most accurate position measurement of the center of volume of the suspended specimen (Note 5). Care should be taken in cutting specimens to avoid change in density resulting from compressive stress.

NOTE 5—The equilibrium positions of film specimens in the thickness range from 0.025 to 0.051 mm (0.001 to 0.002 in.) may be affected by interfacial tension. If this aff-.t is suspected, films not less than 0.127 mm (0.005 in.) in thickness should be tested.

5.2 The specimen shall be free of foreign matter and voids and shall have no cavities or surface characteristics that will cause entrapment of bubbles.

6. Preparation of Density-Gradient Columns

6.1 *Preparation of Standard Glass Floats*[3]—Prepare glass floats by any convenient method such that they are fully annealed, approximately spherical, have a maximum diameter less than one quarter the inside diameter of the column and do not interfere with the test specimens. Prepare a solution (400 to 600 ml) of the liquids to be used in the gradient tube such that the density of the solution is approximately equal to the desired lowest density. When the floats are at room temperature, drop them gently into the solution. Save the floats that sink very slowly, and discard those that sink very fast or save them for another tube. If necessary to obtain a suitable range of floats, grind selected floats to the desired density by rubbing the head part of the float on a glass plate on which is spread a thin slurry of 400 or 500-mesh silicon carbide (Carborundum) or other appropriate abrasive. Progress may be followed by dropping the float in the test solution at intervals and noting its change in rate of sinking.

6.2 *Calibration of Standard Glass Floats* (See Appendix A1):

6.2.1 Place a tall cylinder in the constant-temperature bath maintained at 23 ± 0.1 C. Then fill the cylinder about two thirds full with a solution of two suitable liquids, the density of which can be varied over the desired range by the addition of either liquid to the mixture. After the cylinder and solution have attained temperature equilibrium, place the float in the solution, and if it sinks, add the denser liquid by suitable means with good stirring until the float reverses direction of movement. If the float rises, add the less dense liquid by suitable means with good stirring until the float reverses direction of movement.

6.2.2 When reversal of movement has been observed, reduce the amount of the liquid additions to that equivalent to 0.0001 g/cm^3 density. When an addition equivalent to 0.0001 g/cm^3 density causes a reversal of movement, or when the float remains completely stationary for at least 15 min, the float and liquid are in satisfactory balance. The cylinder must be covered whenever it is being observed for balance, and the liquid surface must be below the surface of the liquid in the constant-temperature bath. After vigorous stirring, the liquid may continue to move for a considerably length of time; make sure that the observed movement of the float is not due to liquid motion by waiting at least 15 min after stirring has stopped before observing the float.

6.2.3 When balance has been obtained, fill a freshly cleaned and dried pycnometer with the solution and place it in the 23 ± 0.1 C bath for sufficient time to allow temperature equilibrium of the glass. Determine the density of the solution by normal methods (ASTM Method D 941, Test for Density and Specific Gravity of Liquids by Lipkin Bicapillary Pycnometer[4]) and make "in vacuo" corrections for all weighings. Record this as the density of the float. Repeat the procedure for each float.

6.3 *Gradient Tube Preparation* (see Appendix for details):

6.3.1 *Method A*—Stepwise addition.

6.3.2 *Method B*—Continuous filling (liquid entering gradient tube becomes progressively less dense).

6.3.3 *Method C*—Continuous filling (liquid

[3] Glass floats may be purchased from the Scientific Glass Apparatus Co., Bloomfield, N. J.

[4] *Annual Book of ASTM Standards*, Part 17.

entering gradient tube becomes progressively more dense).

7. Conditioning

7.1 Test specimens whose change in density on conditioning may be greater than the accuracy required of the density determination shall be conditioned before testing in accordance with the method listed in the applicable ASTM material specification.

8. Procedure

8.1 Wet three representative test specimens with the less dense of the two liquids used in the tube and gently place them in the tube. Allow the tube and specimens to reach equilibrium, which will required 10 min or more. Thin films of 1 to 2 mils in thickness require approximately $1^1/_2$ h to settle, and rechecking after several hours is advisable (Note 5).

8.2 When a graduated tube is used, read the height of the floats and specimens by using a line through their center of volume. When a cathetometer is used, measure the height of the floats and specimens from an arbitrary level using a line through their center of volume. If equilibrium is not obtained, the specimen may be imbibing the liquid.

8.3 Old samples can be removed without destroying the gradient by slowly withdrawing a wire screen basket attached to a long wire (Note 6). This can be conveniently done by means of a clock motor. Withdraw the basket from the bottom of the tube and, after cleaning, return it to the bottom of the tube. It is essential that this procedure be performed at a slow enough rate (approximately 30 min/300-mm length of column) so as not to disturb the density gradient.

NOTE 6—Whenever it is observed that air bubbles are collecting on samples in the column, a vacuum applied to the column will correct this.

9. Calculations

9.1 The densities of the samples may be determined graphically or by calculation from the levels to which the samples settle by either of the following methods:

9.1.1 *Graphical Calculation*—Plot float position versus float density on a chart large enough to be read accurately to ± 1 mm and the desired precision of density. Plot the positions of the unknown specimens on the chart and read their corresponding densities.

9.1.2 *Numerical Calculation*—Calculate the density by interpolation as follows:

$$\text{Density at } x = a + [(x - y)(b - a)/(z - y)]$$

where:

a and b = densities of the two standard floats,

y and z = distances of the two standards, a and b, respectively, bracketing the unknown measured from an arbitrary level, and

x = distance of unknown above the same arbitrary level.

10. Report

10.1 The report shall include the following:

10.1.1 Density reported as D^{23C}, in grams per cubic centimeter, as the average for three representative test specimens,

10.1.2 Number of specimens tested if different than three,

10.1.3 Sensitivity of density gradient in grams per cubic centimeter per millimeter,

10.1.4 Complete identification of the material tested, and

10.1.5 Date of the test.

REFERENCES

(1) Linderstrøm-Lang, K., "Dilatometric Ultra-Micro-Estimation of Peptidase Activity," *Nature*, NATRA, Vol 139, 1937, p. 713.

(2) Linderstrøm-Lang, K., and Lanz, H., "Enzymic Histochemistry XXIX Dilatometric Micro-Determination of Peptidase Activity," *Comptes rendus des gravaus de laboratorie Carlsberg, Serie Chimique*, Vol 21, 1938, p. 315.

(3) Linderstrøm-Lang, K., Jacobsen, O., and Johansen, G., "Measurement of the Deuterium Content in Mixtures of H_2O and D_2O," *ibid.*, Vol 23, 1938, p. 17.

(4) Jacobsen, C. F., and Linderstrøm-Lang, K., "Method for Rapid Determination of Specific Gravity," *Acta Physiologica Scandinavica*, APSCA, Vol 1, 1940, p. 149.

(5) Boyer, R. F., Spencer, R. S., and Wiley, R. M., "Use of Density-Gradient Tube in the Study of High Polymers," *Journal of Polymer Science*, JPSCA, Vol 1, 1946, p. 249.

(6) Anfinsen, C., "Preparation and Measurement

of Isotopic Tracers: A Symposium Prepared for the Isotope Research Group," Edwards, J. W., Publishers, Ann Arbor, Mich., 1946, p. 61.

(7) Tessler, S., Woodberry, N. T., and Mark H., "Application of the Density-Gradient Tube in Fiber Research," *Journal of Polymer Science*, JPSCA, Vol 1, 1946, p. 437.

(8) Low, B. W., and Richards, F. M., "The Use of the Gradient Tube for the Determination of Crystal Densities," *Journal of the American Chemical Society*, JACSA, Vol 74, 1952, p. 1660.

(9) Sperati, C. A., Franta, W. A., and Starkweather, H. W., Jr., "The Molecular Structure of Polyethylene V, the Effect of Chain Branching and Molecular Weight on Physical Properties," *Journal of the American Chemical Society*, JACSA, Vol 75, 1953, p. 6127.

(10) Tung, L. H., and Taylor, W. C., "An Improved Method of Preparing Density Gradient Tubes," *Journal of Polymer Science*, JPSCA, Vol 21, 1956, p. 144.

(11) Mills, J. M., "A Rapid Method of Construction Linear Density Gradient Columns," *Journal of Polymer Science*, Vol 19, 1956, p. 585.

(12) Wiley, R. E., "Setting Up a Density Gradient Laboratory," *Plastics Technology*, PLTEA, Vol 8, No. 3, 1962, p. 31.

TABLE 1 Liquid Systems for Density-Gradient Tubes

System	Density Range, g/cm³
Methanol - benzyl alcohol	0.80 to 0.92
Isopropanol - water	0.79 to 1.00
Isopropanol - diethylene glycol	0.79 to 1.11
Ethanol - carbon tetrachloride	0.79 to 1.59
Toluene - carbon tetrachloride	0.87 to 1.59
Water - sodium bromide	1.00 to 1.41
Water - calcium nitrate	1.00 to 1.60
Carbon tetrachloride - trimethylene dibromide	1.60 to 1.99
Trimethylene dibromide - ethylene bromide	1.99 to 2.18
Ethylene bromide - bromoform	2.18 to 2.89

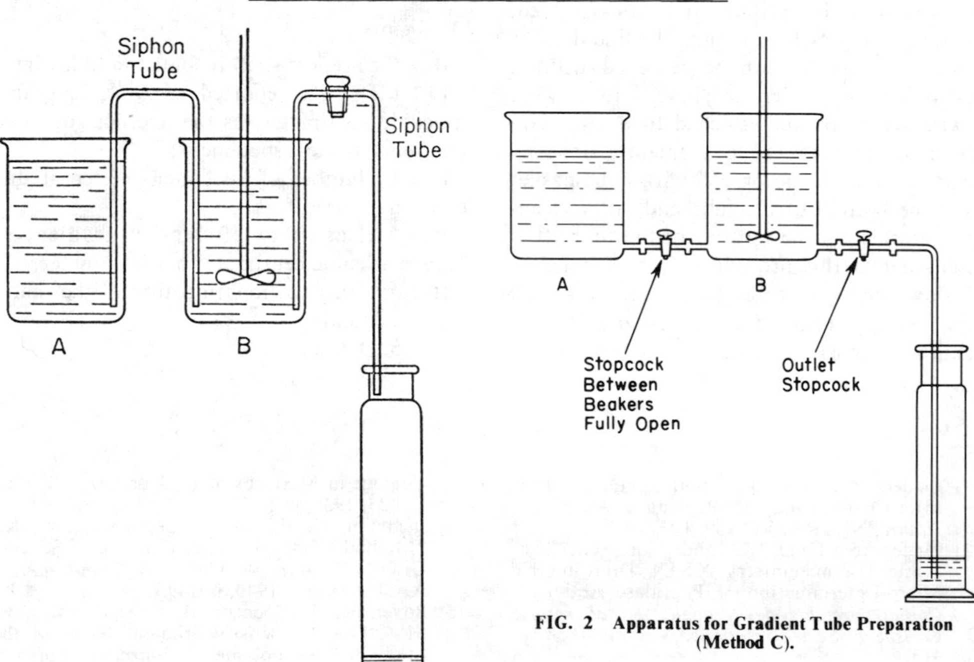

FIG. 1 Apparatus for Gradient Tube Preparation (Method B).

FIG. 2 Apparatus for Gradient Tube Preparation (Method C).

APPENDIXES

A1. FLOAT CALIBRATION—ALTERNATIVE METHOD

A1.1 This method of float calibration has been found by one laboratory to save time and give the same accuracy as the standard method. Its reliability has not been demonstrated by round-robin data.

A1.1.1 Prepare a homogeneous solution whose density is fairly close to that of the float in question.

A1.1.2 Fill a graduate about $^3/_4$ full with the solution, drop in the float, stopper, and place in a thermostatted water bath near 23 C. Fill a tared two-arm pycnometer (ASTM Method D 941,[4] or equivalent) with the solution. Place the pycnometer in the bath.

A1.1.3 Vary the bath temperature until the solution density is very near to that of the float. (If the float was initially on the bottom of the graduate, lower the bath temperature until the float rises; if the float floated initially, raise the bath temperature until the float sinks to the bottom.)

A1.1.4 Change the bath temperature in the appropriate direction in increments corresponding to solution density increments of about 0.0001 g/cm^3 until the float reverses direction of movement as a result of the last change. This must be done slowly (at least 15-min intervals between incremental changes on the temperature controller). Read the volume of liquid in the pycnometer.

A1.1.5 Change the bath temperature in increments in the opposite direction, as above, until a change in the float position again occurs. Read the volume of liquid in the pycnometer.

NOTE A1—The float should rise off the bottom of its own volition. As a precaution against surface tension effects when the float is floating, the float should be pushed about half-way down in the liquid column and then observed as to whether it rises or falls. For this purpose, a length of Nichrome wire, with a small loop on the lower end and an inch or so of length extending above the liquid surface, is kept within the graduate throughout the course of the run. To push a floating float down, the cylinder is unstoppered and the upper wire end grasped with tweezers for the manipulations. The cylinder is then quickly restoppered.

A1.1.6 Remove the pycnometer from the bath, dry the outside, and set aside until the temperature reaches ambient temperature. Weigh and calculate the "in vacuo" mass of solution to 0.0001 g. Using the average of the two observed solution volumes, calculate the density of the solution to 0.0001 g/cm^3. This solution density is also the float density.

A1.1.7 The pycnometer used should be calibrated for volume from the 23 C calibration, although the reading is taken at a different temperature. The alternative method is based on a number of unsupported assumptions but generally gives the same results as that described in 6.2 within the accuracy required. In case of disagreement, the method described in 6.2 shall be the referee method.

A2. GRADIENT TUBE PREPARATION

A2.1 *Method A—Stepwise Addition:*

A2.1.1 Using the two liquids that will give the desired density range, and sensitivity (S) in grams per cubic centimeter per millimeter, prepare four or more solutions such that each differs from the next heavier by 80 S g/cm^3. The number of solutions will depend upon the desired density range of the column and shall be determined as follows:

Number of solutions to prepare density-gradient column (Note A2) $= (1 + D_2 - D_1)/80\,S$

where:

D_2 = upper limit of density range desired,
D_1 = lower limit of density range desired, and
S = sensitivity, in grams per cubic centimeter per millimeter.

NOTE A2—Correct the value of $(1 + D_2 - D_1)/80\,S$ to the nearest whole number. To prepare these solutions, proceed as follows: Using the hydrometers, mix the two liquids in the proportions necessary to obtain the desired solutions. Remove the dissolved air from the solutions by gentle heating or an applied vacuum. Then check the density of the solutions at 23 ± 0.1 C by means of the hydrometers and, if necessary, add the appropriate air-free liquid until the desired density is obtained.

NOTE A3—Where aqueous mixtures are used, 0.5 percent aqueous sodium acetate should be used to prepare the mixture. This reduces the formation of bubbles from dissolution.

NOTE A4—In order to obtain a linear gradient in the tube, it is very important that the solutions be homogeneous and at the same temperature when their densities are determined. It is also important that the density difference between the solutions consecutively introduced into the tube be equal.

A2.1.2 By means of a siphon or pipet, fill the gradient tube with an equal volume of each liquid starting with the heaviest, taking appropriate measures to prevent air from being dissolved in the liquid. After the addition of the heaviest liquid, very carefully and slowly pour an equal volume of the second heaviest liquid down the side of the column by holding the siphon or pipet against the side of the tube at a slight angle. Avoid excess agitation and turbulence. In this manner, the "building" of the tube shall be completed.

NOTE A5—Density gradients may also be prepared by reversing the procedure described in A2.1.1 and A2.1.2. When this procedure is used, the lightest solution is placed in the tube and the next lightest solution is very carefully and slowly "placed" in the bottom of the tube by means of a pipet or siphon which just touches the bottom of the tube. In this manner the "building" of the tube shall be completed.

A2.1.3 If the tube is not already in a constant-temperature bath, transfer the tube, with as little agitation as possible, to the constant-temperature bath maintained at 23 ± 0.1 C. The bath level should approximately equal that of the solution in the tube, and provision should be made for vibrationless mounting of the tube.

A2.1.4 For every 254 mm of length of tube, dip a minimum of five clean calibrated floats, spanning the effective range of the column, into the less dense solvent used in the preparation of the gradient tube and add them to the tube. By means of a stirrer (for example, a small coiled wire or other appropriate stirring device) mix the different layers of the tube gently by stirring horizontally until the least dense and most dense floats span the required range of the gradient tube. If at this time it is observed that the floats are "bunched" together and not spread out evenly in the tube, discard the solution and repeat the procedure. Then cap the tube and keep it in the constant-temperature bath for a minimum of 24 h.

A2.1.5 At the end of this time, plot the density of floats versus the height of floats to observe whether or not a fairly smooth and nearly linear curve is obtained. Some small irregularities may be seen, but they should be slight. Whenever an irregular curve is obtained, the solution in the tube shall be discarded and a new gradient prepared.

NOTE A6—Gradient systems may remain stable for several months.

A2.2 *Method B—Continuous Filling with Liquid Entering Gradient Tube Becoming Progressively Less Dense:*

A2.2.1 Assemble the apparatus as shown in Fig. 1, using beakers of the same diameter. Then select an appropriate amount of two suitable liquids which previously have been carefully deaerated by gentle heating or an applied vacuum. Typical liquid systems for density-gradient tubes are listed in Table 1. The volume of the more dense liquid used in the mixer (beaker *B* shown in Fig. 1) must be equal to at least one half of the total volume desired in the gradient tube. An estimate of the volume of the less dense liquid required in beaker *A* to establish flow from *A* to *B* can be obtained from the following inequality:

$$V_A > d_B V_B / d_A$$

where:
V_A = starting liquid volume in beaker *A*,
V_B = starting liquid volume in beaker *B*,
d_A = density of the starting liquid in beaker *A*, and
d_B = density of the starting liquid in beaker *B*.

A small excess (not exceeding 5 percent) over the amount indicated by the above equality will induce the required flow from *A* to *B* and yield a very nearly linear gradient column.

A2.2.2 Place an appropriate volume of the denser liquid into beaker *B* of suitable size. Prime the siphon between beaker *B* and the gradient tube with liquid from beaker *B* and then close the stopcock. The delivery end of this siphon should be equipped with a capillary tip for flow control.

A2.2.3 Place an appropriate volume of the less dense liquid into beaker *A*. Prime the siphon between beakers *A* and *B* with the liquid from beaker *A* and close the stopcock. Start the highspeed, propeller-type stirrer in beaker *B* and adjust the speed of stirring such that the surface of the liquid does not fluctuate greatly.

A2.2.4 Start the delivery of the liquid to the gradient tube by opening the two siphon-tube stopcocks simultaneously. Adjust the flow of liquid into the gradient tube at a very slow rate, permitting the liquid to flow down the side of the tube. Fill the tube to the desired level.

NOTE A7—Preparation of a suitable gradient tube may require 1 to 1½ h or longer, depending upon the volume required in the gradient tube.

A2.3 *Method C—Continuous Filling with Liquid Entering Gradient Tube Becoming Progressively More Dense:*

A2.3.1 This method is essentially the same as Method B with the following exceptions:

A2.3.2 The lighter of the two liquids is placed in beaker *B*.

A2.3.3 The liquid introduced into the gradient column is introduced at the bottom of the column. The first liquid introduced is the lighter end of the gradient and is constantly pushed up in the tube as the liquid being introduced becomes progressively heavier.

A2.3.4 The liquid from beaker *A* must be introduced into beaker *B* by direct flow from the bottom of beaker *A* to the bottom of beaker *B*, rather than being siphoned over as it is in Method B. Filling the tube by this method may be done more rapidly than by Methods A or B. The stopcock between containers *A* and *B* should be of equal or larger bore than the outlet stopcock. A schematic drawing of the apparatus for Method C is shown in Fig. 2.

Designation: D 1525 - 65 T

Tentative Method of Test for
VICAT SOFTENING POINT OF PLASTICS[1]

This Tentative Method has been approved by the sponsoring committee and accepted by the Society in accordance with established procedures, for use pending adoption as standard. Suggestions for revisions should be addressed to the Society at 1916 Race St., Philadelphia, Pa. 19103.

This tentative has been submitted to the Society for adoption as standard, but balloting was not complete by press time.

1. Scope

1.1 This method covers determination of the temperature at which a specified needle penetration occurs when specimens are subjected to specified test conditions. This method is useful for many thermoplastic materials.

1.2 This test is not recommended for ethyl cellulose, nonrigid poly(vinyl chloride), poly(vinylidene chloride), or other materials having a wide Vicat softening range.

NOTE 1—The values stated in U.S. customary units are to be regarded as the standard. The metric equivalents of U.S. customary units may be approximate.

2. Significance

2.1 Data obtained by this method may be used to compare the heat-softening qualities of thermoplastic materials. Results obtained in different laboratories using the same rate of temperature rise are generally within a range of 4 C. Precision in a given laboratory is usually within 2 C.

2.1.1 When used with many thermoplastics, within-laboratory precision at the same rate of temperature rise has been observed to remain usually within the 2 C. Agreement between laboratories may be found to be somewhat poorer for some materials, for example, polyamides, with differences sometimes as high as 5 to 8 C.

2.2 Two rates of temperature rise are permissible:

Rate A—50 ± 5 C/h.
Rate B—120 ± 12 C/h.

Rate A may be used to conform to both the ISO Recommended Procedure R306 and to the previous editions of this method.

2.3 This test is useful in the areas of quality control, development, and characterization of materials.

3. Definition

3.1 *Vicat softening point*—the temperature at which a flat-ended needle of 1-mm^2 circular cross section will penetrate a thermoplastic specimen to a depth of 1 mm under a specified load using a selected uniform rate of temperature rise.

4. Apparatus

4.1 The equipment shall be constructed essentially as shown in Fig. 1 and shall consist of the following:

4.1.1 *Immersion Bath*—Bath for heat-transfer liquid with stirrer, thermometer, and heaters. The heaters should have controllers to permit manual or automatic control of the selected bath temperature rise rate (2.2).

4.1.2 *Heat-Transfer Liquid*—Several liquids, such as silicone oils, glycerin, ethylene glycol, and mineral oil, have been used successfully for various plastics.[2] The liquid used shall be known to have no short-time effect at elevated temperatures on the materials being tested, and shall be of low viscosity at room temperature.

---

[1] This method is under the jurisdiction of ASTM Committee D-20 on Plastics. A list of committee members may be found in the ASTM Yearbook. This tentative is the direct responsibility of Subcommittee D-20.30 on Thermal Properties.

Current edition effective July 22, 1965. Originally issued 1958. Replaces D 1525 – 58 T.

[2] Silicone oils having a room temperature viscosity of 100 cP have been found satisfactory and safe for short-term heat cycles up to 260 C.

4.1.3 *Specimen Support*—A suitable stand or support for the specimen to be placed in the bath. The vertical members that attach the specimen support to the upper plate shall be made of a material having the same coefficient of expansion as that used for the rod through which the load is applied in order that the dial gage reading caused by differential expansion over the intended temperature range does not exceed 0.02 mm when the specimen is replaced by a piece of heat-resistant material.[3]

4.1.4 *Weights and Indicators*—A stand to hold the loaded penetrating needle and penetration indicator whose smallest graduations shall be 0.01 mm. The net load on the needle point shall be equal to 1000, +40, −0 g, when the apparatus is assembled, and shall consist of the needle rod weight, the load (if any) caused by the spring of the penetration indicator, and the extra weight to be applied to the needle rod assembly to balance it if required. The required extra weight may be calculated as follows:

$$\text{Required weight, g} = 1000 - (R - T)$$

where:

R = weight of needle rod assembly, g, and

T = positive or negative thrust of penetration indicator spring (if any), g.

4.1.5 *Temperature Indicator*—The thermometer or thermocouple shall be adequate to cover the range being tested and shall be accurate within ±0.5 C.

4.1.6 *Needle*—Flat-ended, circular cross section of 1, +0.05, −0.02 mm^2 with the flat tip free from burrs and square to the axis of the needle rod. Stainless steel wire, diameter 1.143 ± 0.025 mm (0.045 ± 0.001 in.), is available and suggested for needles. The length shall be 5 to 12.5 mm.

5. Test Specimens

5.1 At least two specimens shall be used to test each sample. The specimen shall be flat and shall have a minimum width of 12.7 mm (0.5 in.) and a minimum thickness of 3.2, + 1.0, −0.3 mm (0.125, +0.040, −0.010 in.). When necessary, two specimens of suitable thickness may be used with one on top of the other. The specimens may be cut from sheet or molded material.

6. Procedure

6.1 Start the test at room temperature.

NOTE 2—It is desirable to have a cooling coil in the liquid bath in order to reduce the time required to lower the temperature from previous determinations. This should be removed or drained before starting test as boiling of the coolant can affect the rate of temperature rise.

6.2 Place the specimen, which is at room temperature, on the specimen support so that it is approximately centered under the needle, and gently lower the needle rod (without the additional weight) so that the needle rests on the surface of the specimen and holds it in position. The bulb of the thermometer or the thermocouple should be as close as practical to the specimen. The bath should be well stirred.

6.3 Carefully lower the assembly into the bath, taking care not to jar it in any way that would damage the specimen. Apply the load, set the penetration indicator at zero, and start the temperature rise. Rate of temperature increase shall be either 50 ± 5 C/h (Rate A) or 120 ± 12 C/h (Rate B) and uniform throughout the test. Selection of rate shall be as agreed upon by the interested parties.

NOTE 3—If manual control of the rate of temperature rise is used, it is advisable to check this rate at 2-min intervals and to make any necessary adjustments to the heater controls. Maintenance of proper rate of temperature rise is important.

6.4 Record the temperature when the indicator reads 1-mm penetration. Care should be taken with the 1-mm reading since the penetration will be increasing rapidly at this point.

NOTE 4—If a permanent record is desired, read and record the penetration for each 5 C rise in temperature until the penetration reaches 0.4 mm and at 5.0 C intervals thereafter.

7. Retests

7.1 Retests shall be made if the difference between the two specimens is greater than 2 C.

8. Report

8.1 The report shall include the following:

[3] "Invar" or "Pyrex" glass has been found satisfactory for this purpose.

8.1.1 Complete identification of the material tested,

8.1.2 Method of preparing test specimens,

8.1.3 Rate of temperature rise, Rate A, 50 C/h, or Rate B, 120 C/h.

8.1.4 Softening point, deg C, average of two values, and

8.1.5 Heat transfer liquid used.

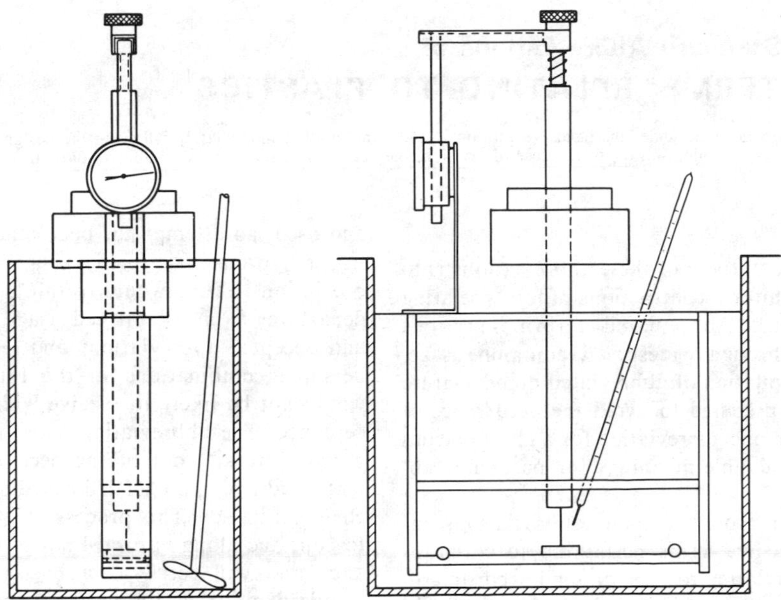

FIG. 1 Apparatus for Softening Point Determination.

Designation: D 1600 – 69

Standard Abbreviations of
TERMS RELATING TO PLASTICS[1]

This Standard is issued under the fixed designation D 1600; the number immediately following the designation indicates the year of original adoption or, in the case of revision, the year of last revision. A number in parentheses indicates the year of last reapproval.

1. Scope

1.1 The purpose of these abbreviations is to provide uniform contractions of terms relating to plastics. Abbreviated terminology has evolved through widespread common usage. This compilation of abbreviated nomenclature has been prepared to avoid the occurrence of more than one abbreviation for a given plastics term and double meanings for particular abbreviations.

1.2 The scope of these abbreviations includes plastics terms pertaining to composition and relating to type or kind according to mode of preparation or principal distinguishing characteristics. Also included are abbreviations for additives such as plasticizers, fillers, etc.

NOTE 1—Codes relating to forms of plastics, particularly as they pertain to method of forming, construction and direction of testing are given in ASTM Recommended Code D 1009, for Designating Form of Material and Direction of Testing Plastics.[2]

NOTE 2—A code relating to the composition of rubbers is given in ASTM Recommended Practice D 1418, for Nomenclature for Solid Elastomers and Latices.[3]

1.3 No attempt is made here to systematize formally a shorthand terminology for polymers. Terminology, symbols, and formulae designations for scientific literature in the field of natural and synthetic polymers are being standardized through intensive study by the International Union of Pure and Applied Chemistry.[4]

1.4 These abbreviations are by no means all inclusive of plastics terminology. They represent, in general, those abbreviations which have come into established use. Since it is recognized that abbreviations serve no useful purpose unless they are generally accepted and used, no attempt has been made to establish a rigorous code for devising standard abbreviations. This would result in awkward departures from established usage of existing and accepted abbreviations and lead to cumbersome combinations in the future, which would not be likely to receive widespread acceptance. The abbreviations now in use have grown naturally out of the need for convenient, readily comprehended shorthand for long chemical names. This process can be expected to continue along natural lines of least resistance and will serve as a basis for further standardization as the need arises. A general guide for the preparation of abbreviations appears desirable, however, to facilitate more organized and uniform standardization in the future. An Appendix is attached, which suggests a uniform way to prepare abbreviations.

2. Recommended Abbreviations

2.1 *Plastics and Resins:*

Term	Abbreviation
Acrylonitrile - butadiene - styrene plastics	ABS
Carboxymethyl cellulose	CMC
Casein	CS
Cellulose acetate	CA
Cellulose acetate-butyrate	CAB
Cellulose acetate propionate	CAP
Cellulose nitrate	CN

[1] These abbreviations are under the jurisdiction of ASTM Committee D-20 on Plastics. A list of committee members may be found in the ASTM Yearbook. This standard is the direct responsibility of Subcommittee D-20.04 on Definitions.

Current edition effective April 4, 1969. Originally issued 1958. Replaces D 1600 – 68 T.

[2] *Annual Book of ASTM Standards*, Part 27.

[3] *Annual Book of ASTM Standards*, Part 28.

[4] "Report on Nomenclature in the Field of Macromolecules," *Journal of Polymer Science*, JPSCA, Vol. VIII, 1952, pp. 257–277.

Cellulose propionate	CP
Chorinated poly(vinyl chloride)	CPVC
Cresol-formaldehyde	CF
Diallyl phthalate	PDAP
Epoxy, epoxide	EP
Ethyl cellulose	EC
Melamine-formaldehyde	MF
Perfluoro(ethylene-propylene) copolymer	FEP
Phenol-formaldehyde	PF
Poly(acrylic acid)	PAA
Polyacrylonitrile	PAN
Polyamide (nylon)	PA
Polybutadiene acrylonitrile	PBAN
Polybutadiene-styrene	PBS
Polycarbonate	PC
Poly(diallyl phthalate)	PDAP
Polyethylene	PE
Polyethylene terephthalate	PETP
Poly(methyl-α-chloroacrylate)	PMCA
Poly(methyl methacrylate)	PMMA
Polymonochlorotrifluoroethylene	PCTFE
Polyoxymethylene, polyacetal	POM
Polypropylene	PP
Polystyrene	PS
Polytetrafluoroethylene	PTFE
Poly(vinyl acetate)	PVAc
Poly(vinyl alcohol)	PVAL
Poly(vinyl butyral)	PVB
Poly(vinyl chloride)	PVC
Poly(vinyl chloride-acetate)	PVCAc
Poly(vinyl fluoride)	PVF
Poly(vinyl formal)	PVFM
Silicone plastics	SI
Styrene-acrylonitrile	SAN
Styrene-butadiene plastics	SBP
Styrene-rubber plastics	SRP
Urea-formaldehyde	UF
Urethane plastics	UP

2.2 *Plastic and Resin Additives:*

Term	Abbreviation
Dibutyl phthalate	DBP
Dicapryl phthalate	DCP
Diisodecyl adipate	DIDA
Diisodecyl phthalate	DIDP
Diisooctyl adipate	DIOA
Diisooctyl phthalate	DIOP
Dinonyl phthalate	DNP
Di-*n*-octyl *n*-decyl phthalate	DNODP
Dioctyl adipate	DOA
Dioctyl azelate	DOZ
Dioctyl phthalate	DOP
Dioctyl sebacate	DOS
Tricresyl phosphate	TCP
Trioctyl phosphate	TOF
Triphenyl phosphate	TPP

2.3 *Monomers:*

Term	Abbreviation
Diallyl chlorendate (diallyl ester of 1,4,5,6,7,7-hexachlorobicyclo-(2,2,1)-5-heptene-2,3-dicarboxylic acid)	DAC
Diallyl fumarate	DAF
Diallyl isophthalate	DAIP
Diallyl maleate	DAM
Diallyl orthophthalate	DAP
Methyl methacrylate	MMA
Monochlorotrifluoroethylene	CTFE
Tetrafluoroethylene	TFE
Triallyl cyanurate	TAC

2.4 *Miscellaneous Plastics Terms:*

Term	Abbreviation
General purpose	GP
Single stage	SS
Solvent welded plastics pipe	SWP

3. Use of Abbreviations

3.1 When abbreviations are used in publications or other written matter, their first occurrence in the text should be enclosed in parentheses and preceded by the written word or words being abbreviated. Subsequent references to such words in the article can then be by the appropriate abbreviations, as such.

APPENDIX

A1. A Suggested System for Preparing Abbreviations for Plastics and Related Terms[5]

NOTE—The use of the following guide in establishing abbreviated designations for plastics and related terms will promote uniform and systematic standardization.

A1.1 Use capital letters for the main components in the order in which they occur in the plastics term being abbreviated, for example:

Poly(vinyl chloride) = PVC

A1.2 Use small letters following capital letters in cases where duplications are likely to occur or where confusion may otherwise result, for example:

Poly(vinyl alcohol) has generally been abbreviated PVAL. The abbreviation for Poly(vinyl acetate) would then be PVAc.

A1.3 Use generic terms for family designations, for example:

saran for vinylidene chloride plastics.

A1.4 Use figures to distinguish between plastics prepared from various condensation units in a homologous series, for example:

Poly(hexamethylene sebacamide) = PA 6–10 where the letters refer to polyamide, the first number refers to the carbon atoms in the amine, and the second number refers to the carbon atoms in the acid.

[5] Based in part on Proposals under Consideration by the Division of Plastics and High Polymers, IUPAC.

Designation: D 1601 – 61

Standard Method of Test for
DILUTE SOLUTION VISCOSITY OF ETHYLENE POLYMERS[1]

This Standard is issued under the fixed designation D 1601; the number immediately following the designation indicates the year of original adoption or, in the case of revision, the year of last revision. A number in parentheses indicates the year of last reapproval.

1. Scope

1.1 This method covers the determination of the dilute solution viscosity of ethylene polymers at 130 C. It is applicable to a reasonably wide spectrum of ethylene polymers having densities from 0.910 to 0.970 g/cm^3. Directions are given for the determination of relative viscosity, inherent viscosity, and intrinsic viscosity.

2. Significance

2.1 The knowledge of dilute solution viscosity serves as an additional tool in characterizing ethylene polymers. Viscosity data alone may be of limited value in predicting the processing behavior of the polymer. However, when used in conjunction with other flow and physical property values, the solution viscosity of ethylene polymers may contribute to their characterizations.

2.2 Satisfactory correlation between solution viscosity and certain other properties is possible from polymers of a single manufacturing process. The solution viscosity test is not sensitive to some molecular configurational patterns that may occur among polymers from different manufacturing processes. Hence, its correlation with other properties of polymers produced by different processes, by even one manufacturer, may be limited.

2.3 The viscosity of polymer solutions may be drastically affected by the presence of known or unknown additives in the sample. The use of solution viscosity may be of questionable value where ethylene polymers are known or suspected to contain colorants, carbon black, low molecular weight hydrocarbons, fillers, or other additives.

2.4 The measurement of dilute solution viscosity of ethylene polymers presents problems not ordinarily encountered in viscosimetry. Ethylene polymers are not soluble at room temperature in any known solvent. Some of the higher density materials are insoluble below 100 C. Extreme care must be exercised in transferring the solution to the viscometer for the test if the correct solution concentration is to be maintained. This test has no significance unless the sample is completely soluble.

2.5 The solution viscosity is a function of the root-mean-square size of the polymer molecules in solution. It is known that the solvent selected and the temperature of the determination have an effect on the root-mean-square size of the particles. Hence, where a viscometer, solvent, or temperature other than specified is used, data may not be comparable to that obtained by this procedure.

3. Materials

3.1 *Solvent—Decahydronaphthalene*, practical grade, purified and redistilled, as follows:

3.1.1 The solvent shall be purified by percolation through 100 to 200 mesh commercial grade silica gel. This treatment removes naphthalene, tetrahydronaphthalene, and oxy compounds, particularly peroxides.

3.1.2 The redistilled product shall conform to the following requirements when tested in accordance with ASTM Method D 86, Test

[1] This method is under the jurisdiction of ASTM Committee D-20 on Plastics and is the direct responsibility of Subcommittee D-20.70 on Analytical Methods. A list of committee members may be found in the ASTM Yearbook.

Current edition effective Sept. 18, 1961. Originally issued 1958. Replaces D 1601 – 59 T.

for Distillation of Petroleum Products:[2]

Standard Distillation	ASTM Method D 86
Initial boiling point	190 C min
10 ml	191 C min
20 ml	192 C min
80 ml	194 C max
90 ml	195 C max
Dry point	196 C max

NOTE 1—While use of other solvents, such as tetrahydronaphthalene or xylene, may sometimes be advantageous, they will generally yield different values for solution viscosities.

3.1.3 Immediately after redistillation of the decahydronaphthalene, 0.05 to 0.1 percent by weight of an antioxidant may be added to inhibit oxidation during the viscosity determination. Phenyl betanaphthylamine and 4-methyl-2,6-ditertiary-butylphenol have been found satisfactory. There is evidence to show that stabilization of the solvent with one of these antioxidants renders its solutions with most ethylene polymers resistant to oxidation at 130 C for periods of many hours.

3.2 *Heat Transfer Medium*—Any liquid heat transfer medium that will not appreciably affect the accuracy of the test may be used. Care should be exercised in using fluids that discolor or smoke with prolonged heating.

NOTE 2—The following materials have been used: (*1*) Silicone fluid,[3] that has been found to be free of discoloration at 130 C for sustained periods. Some silicone fluids are not miscible with others, (*2*) Peanut oil, (*3*) Primol D, a mineral oil. Peanut oil and Primol D oil may darken on prolonged heating; however, this darkening may be inhibited through the use of antioxidants.

4. Apparatus

4.1 *Volumetric Flasks*, 100-ml, grade EXAX or better.[4]

4.2 *Transfer Pipets*, grade EXAX or better.[4]

4.3 *Constant Temperature Bath*, capable of maintaining 130 ± 0.1 C.

4.4 *Viscometer*—Modified Ubbelohde viscometer[5] as shown in Figs. A1, A2, and A3 and described in Appendix A1.

NOTE 3—Other types of viscometers may be used provided they can be shown to agree with the type specified.

4.5 *Oven*, maintained at 140 ± 5 C.

4.6 *Timer*, as specified in 4.5 of ASTM Method D 445, Test for Viscosity of Transparent and Opaque Liquids (Kinematic and Dynamic Viscosities).[2]

4.7 *Thermometer*—An ASTM High Softening Point Thermometer having a range from 30 to 200 C, and conforming to the requirements for Thermometer 16C in ASTM Specification E 1, for ASTM Thermometers.[6]

5. Sample

5.1 Ordinarily, ethylene polymers are readily soluble in decalin at 130 C. Where solubility is desired in minimum time the sample shall be pulverized to pass a 20-mesh screen.

NOTE 4—The sample can be conveniently pulverized in a rotary cutting mill with a 20-mesh screen fitted to its pulverizing chamber. The Wiley mill, containing blades on its rotor which impact the material against stationary blades fitted into the pulverizing chamber walls, has been found satisfactory. It is sold by a number of scientific supply houses. Care must be taken to prevent overheating unusually hard or tough polymer samples during pulverizing. Such overheating can possibly cause thermal degradation.

6. Conditioning

6.1 The sample need not be predried or conditioned, unless previous specific knowledge of the sample makes it advisable. Ordinarily, ethylene polymers containing no additives are not hygroscopic.

7. Measurement of Solvent Viscosity

7.1 Determine the efflux time of the solvent in the viscometer in accordance with 8.6 through 8.9 until three consecutive efflux times agree within 0.2 s. The average of these readings shall be used in calculating the relative viscosity of the solution, as described in Section 9.

NOTE 5—The solvent viscosity may be calculated from the efflux time in accordance with the following equation:

$$V = Ct - B/t$$

[2] *Annual Book of ASTM Standards*, Part 17.

[3] The silicone fluids available from the Dow Corning Corp., Midland, Mich., or from the Union Carbide Corp., Linde Silicones Div., New York, N. Y., have been found satisfactory for this purpose.

[4] Glassware used in this method should be tested in accordance with the procedures described in the National Bureau of Standards Circular No. C-434, "Testing of Glass Volumetric Apparatus," and should not exceed the limits of accuracy set forth in the circular.

[5] The modified Ubbelohde viscometer is available from J. V. Stabin, 84-21 Midland Parkway, Jamaica, Queens, New York, N. Y., 11431, and Cannon Instrument Co., P.O. Box 812, State College, Pa.

[6] *Annual Book of ASTM Standards*, Part 30.

where:
V = kinematic viscosity, cSt,
C = calibration constant for the viscometer,
t = efflux time, s, and
B = kinetic energy and end effect correction constant. (This constant is dependent on the design of the viscometer and may be determined experimentally, along with the constant C, by use of suitable calibrating fluids. In general, with viscometers designed so that B is relatively small, the correction B/t is negligible for relatively long efflux times.)

NOTE 6—Although the kinetic energy effect inherent in the design of the viscometer may be determined by Method D 445, that method is believed to be tedious for liquids with viscosities in the range of decalin at 130 C. The following equations provide a more convenient and precise means of calculating the relative viscosity, η_r, corrected for the kinetic energy contribution.

$$\eta_r = \frac{t_s + \Delta t_s}{t_o + \Delta t_o} = \frac{t_s - D/t_s}{t_o - D/t_o}$$

where:
t_s and t_o = flow times of the solution and solvent, respectively,
Δt = kinetic energy correction term for a particular liquid, and
D = a constant for the viscometer.

D and Δt may be calculated as follows:

$$-D/t = (-Vm\rho)/L\eta 8\pi = \Delta t$$

where:
t = flow time of the liquid in the viscometer,
V = efflux volume of the liquid,
m = kinetic energy correction coefficient. (If the viscometer capillary has trumpet-shaped ends that break away rather sharply, and Δt is no larger than 3 percent of t, it is probably safe to assume m equal to unity.)
ρ = density of the liquid,
L = length of the capillary, and
η = viscosity of the liquid.

Using this equation, the viscometer constant D, may be calculated after the following measurements are made:

(*1*) Flow time of a liquid (water for example) of known viscosity and density in the viscometer,

(*2*) Capillary length, and

(*3*) Volume of the viscometer bulb. This may be done by filling the bulb with water and measuring the volume of the water in a graduated cylinder. An accuracy of ±3 percent is adequate for all measurements, if Δt is no larger than 3 percent of t. This will provide a precision of better than 0.1 percent.

8. Procedure

8.1 If it is desired to determine only the relative viscosity or the inherent viscosity, weigh one specimen of 0.18 to 0.22 ± 0.0002 g and transfer it quantitatively to a 100-ml volumetric flask which has been purged with nitrogen.

NOTE 7—Usually the determination of relative or inherent viscosity may suffice.

8.2 To determine the intrinsic viscosity, weigh four specimens of the following approximate weights to ±0.0002 g:

Specimen 1	0.09 to 0.11 g
Specimen 2	0.18 to 0.22 g
Specimen 3	0.28 to 0.32 g
Specimen 4	0.38 to 0.42 g

NOTE 8—The intrinsic viscosity may also be measured using successive dilutions of only one, or duplicate, solutions. This has the advantage of requiring that only one polymer specimen be weighed. Dilutions to the solution of this specimen in the viscometer are a time-saving device. The technique is as follows: Weigh one, or a duplicate, specimen of 0.38 to 0.42 g to an accuracy of ± 0.0002 g and transfer it quantitatively to a 100-ml volumetric flask, which has been purged with nitrogen, using a funnel and washing the watch glass and funnel down with solvent. Place the specimen into solution in accordance with 8.3 through 8.5 and measure the relative viscosity in accordance with 8.6 through 8.9.

Separate dilutions of the original solution to desired concentrations can be made by adding solvent at 130 C in precisely measured quantities to the original solution in the viscometer and the relative viscosity at each dilution tested before the next dilution is made. For example, the following technique might be used.

First Dilution—Add *exactly* 5 ml of solvent at 130 C to the filter stick of the viscometer and calculate the solution concentration from the known weight of solute and total volume of solvent in the viscometer. (This concentration should be approximately 0.3 g/100 ml.) Mix the solution in the viscometer thoroughly by pulling it into the viscometer bulb and back again three times. Determine the relative viscosity at this concentration.

Second Dilution—Add *exactly* 10 ml of solvent at 130 C and calculate the concentration, as above. (This concentration should be approximately 0.2 g/100 ml.) Determine the relative viscosity at this concentration.

Volumetric limitations of the modified Ubbelohde viscometer require that approximately 15 ml of solution be added to the instrument for the first determination of relative viscosity. Also, the viscometer will not hold and properly mix liquid volumes very much in excess of 30 ml. Hence, use of the above procedure will provide only three values of relative viscosity from which intrinsic viscosity can be determined. Other dilution techniques may be used to obtain four or more values of relative viscosity, but the range of solute concentrations is limited unless the reverse procedure is used—that of introducing a minimum volume of solvent into the viscometer to obtain solvent viscosity, then adding aliquots of a concentrated stock solution to the viscometer to obtain successive values of relative viscosity at higher concentrations.

8.3 Add approximately 50 ml of solvent to the specimen flasks, purge them again with nitrogen, stopper them loosely and place them in an oven or bath maintained at 140 C. Vent the flasks at the end of 10 min. Shake the flasks once every 10 min until the solution is complete (Notes 9 and 11).

NOTE 9—Most ethylene polymers will go into complete solution at the solute concentrations used in this method within 2 h. Some polymer samples may require up to 4 h. If the solute has not dissolved in 6 h, this procedure may not be applicable, due to possibly incomplete solution of the sample not detectable to the eye. If incomplete solution of the sample is suspected, the viscometer filter stick may be removed from the instrument after the sample solution has been transferred and the filter stick dried under vacuum. Any weight difference after drying should show evidence of undissolved residue. Even with the use of an antioxidant in the solvent, significant oxidation of the polymer sample may occur in unusually long dissolving period.

NOTE 10: **Caution**—When the solution is considered complete, examine visually to be certain that no undissolved particles, gels, or particles of foreign matter are present.

8.4 Place the solutions in the volumetric flasks in the constant temperature bath maintained at 130 ± 0.1 C. Dilute the solutions to approximately 99 ml with the solvent maintained at the bath temperature, and thoroughly mix the solutions by hand. Replace the flasks in the bath.

8.5 After $^1/_2$ h, complete the dilution to the 100-ml mark with the solvent maintained at 130 C by means of a transfer pipet and shake the flasks again.

NOTE 11—Be certain the solutions are uniformly mixed. Do not allow excessive cooling while shaking the flasks. The error due to expansion of the flask at 130 C is believed not to be significant in this test.

8.6 Cut about 0.5 in. from the lower tip of a transfer pipet to permit faster transfer of its contents to the viscometer. Place the pipet in a 140 C oven for 30 min. It is preferable that the pipet be fitted with a suitable heating mantel to retard precipitation of the polymer on its walls during transfer of the solution (Note 12). Purge the viscometer thoroughly with nitrogen gas using a slow, gentle stream. Fill the pipet from the sample solution flask and transfer approximately 15 ml of the solution into the filter stick of the modified Ubbelohde viscometer which is permanently positioned in the bath (Note 13). (See Appendix A1 for directions for use and cleaning of the viscometer.)

NOTE 12—A graduate cylinder previously heated in the viscosity bath may be used instead of a pipet. The hot solution can be poured into the graduate positioned in the bath, the graduate removed after about 3 min, and its contents poured into the viscometer filter stick.

NOTE 13—If the dilution method described in Note 8 is used, *exactly* 15 ml of solution must be transferred to the viscometer, using the transfer pipet described in 8.6.

8.7 Bring the liquid level to approximately 10 mm above the upper graduation mark on the viscometer capillary.

8.8 Allow the solution to drain to the upper mark on the capillary. As the meniscus passes this point, start the timer and time the interval for the solution to drain to the lower mark on the capillary.

8.9 Measure the efflux time of the solution at least three times. Three consecutive readings shall agree within 0.2 s.

9. Calculations

9.1 *Viscosity Ratio* (*Relative Solution Viscosity*)—Calculate the viscosity ratio for each concentration measured, from the average efflux times as follows:

$$\eta_r = t/t_o$$

where:

η_r = viscosity ratio,
t = average efflux time of solution, and
t_o = average efflux time of pure solvent.

9.2 *Logarithmic Viscosity Number* (*Inherent Viscosity*)—Calculate the logarithmic viscosity number for each concentration measured, as follows:

$$\eta_{inh} = \ln \eta_r/C$$

where:

η_{inh} = logarithmic viscosity number at concentration C, and
$\ln \eta_r$ = natural logarithm of the viscosity ratio, and
C = concentration in grams/100 ml of solution.

9.3 *Limiting Viscosity Number* (*Intrinsic Viscosity*)[7]—Plot the four logarithmic viscosity numbers *versus* their respective concentrations on rectilinear graph paper. Draw the best straight line through the points and extrapolate it to zero concentration. The limiting viscosity number, $[\eta]$, is the intercept of the line at zero concentration.

[7] Schulken, R. M., and Sparks, M. L., *Journal of Polymer Science*, JPSCA, Vol XXVI, 1957, p. 227.

NOTE 14—If desired, the specific viscosity (η_{sp}) and the viscosity number (reduced viscosity) (η_{red}) may be calculated as follows:

$$\eta_{sp} = \eta_r - 1$$

$$\eta_{red} = \eta_{sp}/C$$

NOTE 15—As a check on the plot of the limiting viscosity numbers *versus* their concentrations, the specific viscosity (η_{sp}) divided by their concentrations for the four specimens may be plotted *versus* their concentrations on the same graph. The slopes of these two lines will not be the same, but they should converge to the same value at zero concentration. This additional line serves to more accurately position the zero concentration point.

10. Report

10.1 The report shall include the following:

10.1.1 Complete identification of the material tested including type, source, manufacturer's code numbers, and trade name.

10.1.2 Conditioning procedure used, if any.

10.1.3 The viscosity ratio of one or more concentrations, depending on whether it is desired to obtain relative, inherent, or intrinsic viscosity.

10.1.4 The intrinsic viscosity, when desired, to three significant figures.

APPENDIX

A1. DIRECTIONS FOR USE AND CLEANING OF THE MODIFIED UBBELOHDE VISCOMETER

A1.1 Using Viscometer

A1.1.1 Pour the hot solution immediately into the viscometer through tube 3 containing the filter stick, as shown in Figs. A1 and A2. Slip a rubber tube with one end closed off by a stopcock over tube 2 of the viscometer. Cautiously apply air or nitrogen pressure, using a pressure controller, in filling tube 3 in order to force the solution through the glass filter. Remove the pressure controller. Remove excess solution from the pores of the filter by applying the end of the intake tube of a suction flask to the filter surface; this is necessary in order to prevent the formation of an air trap in the system. Care must be taken in the use of suction in cleaning the filter to avoid loss of sample.

A1.1.2 Allow the solution to stand in the bath for at least 5 min to ensure temperature equilibration. Keep the stopcock on tube 2 closed, and cautiously apply pressure to tube 3, forcing the liquid into viscometer tube 1 until it fills the small upper bulb. Then release the pressure and open the stopcock on tube 2 to the atmosphere allowing the solution to flow freely out of the capillary tube. Measure the time, in seconds, required for the meniscus to pass from the upper to the lower mark on tube 1. Repeat the determinations until three successive efflux times agree within 0.2 s; the average of these three readings is used for calculating the viscosity.

A1.2 Cleaning Viscometer

A1.2.1 Empty the viscometer after each determination without removing it from the 130 C bath. Remove the filter stick (Fig. A3) from the viscometer and insert the intake tube of the suction flask into tube 3 so that its lower end touches the bottom of the viscometer tube. Apply a vacuum to the suction flask and withdraw the solution. Rinse the viscometer in place with fresh hot solvent. Measure the efflux time of pure solvent after every determination. If the value obtained differs by more than ±0.3 s from the value previously obtained in the clean viscometer, further cleaning is necessary.

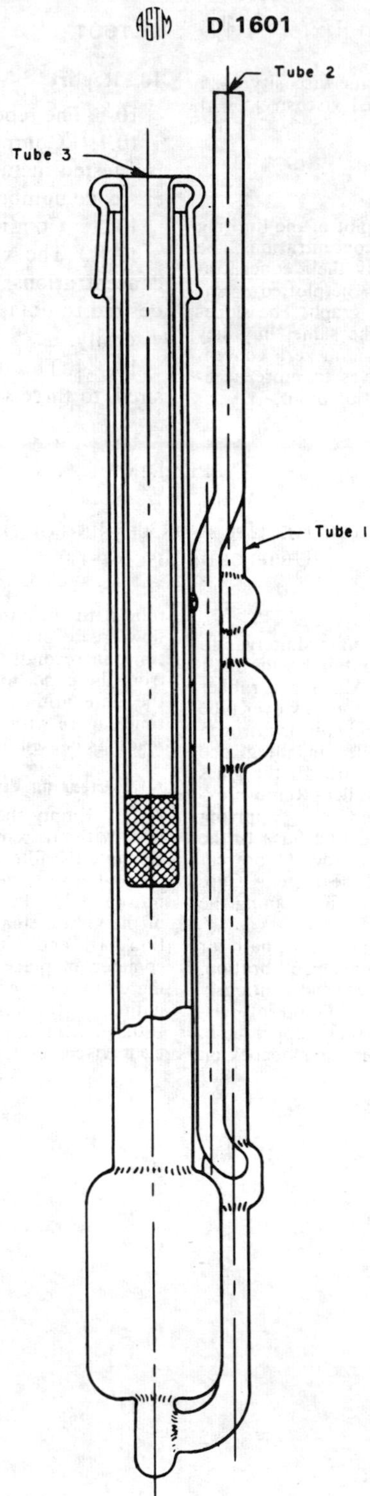

FIG. A1 Modified Ubbelohde Viscometer.

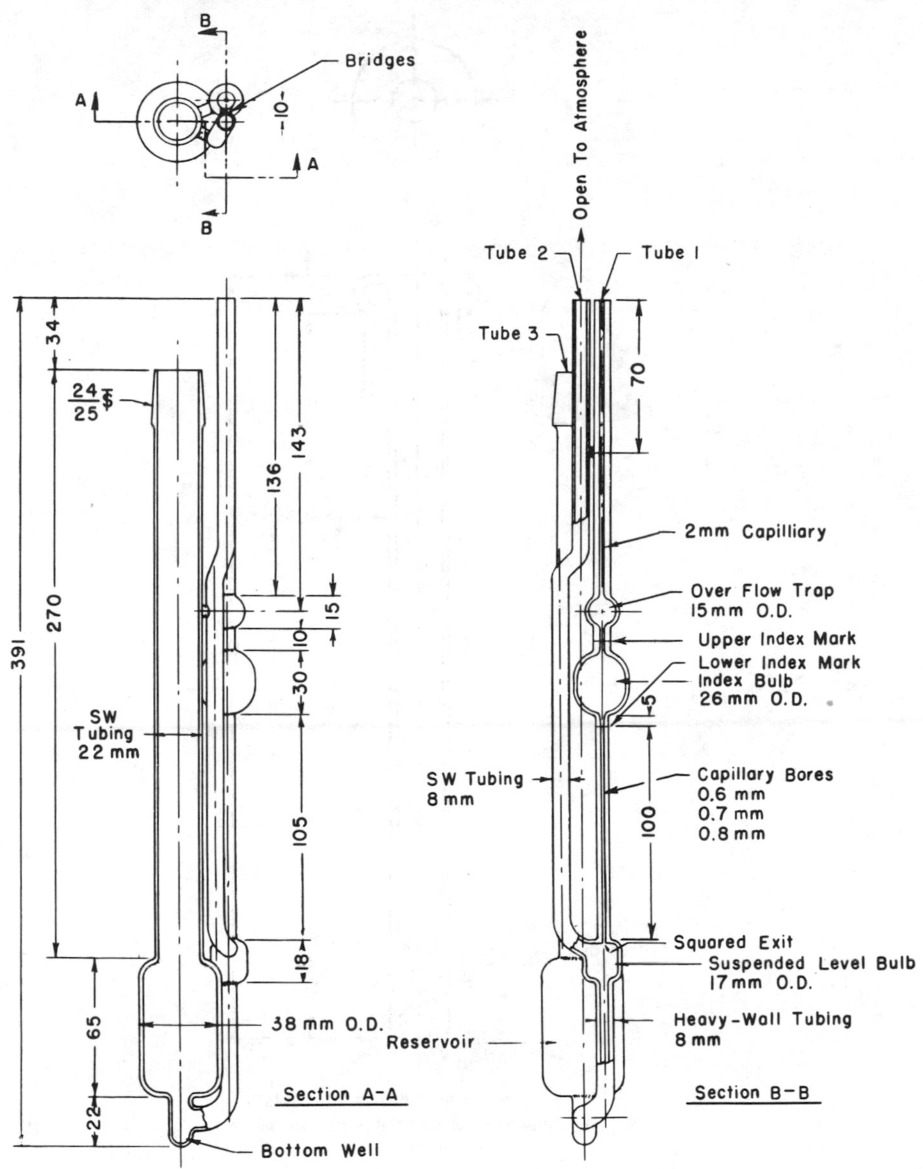

FIG. A2 Section Views of the Modified Ubbelohde Viscometer.

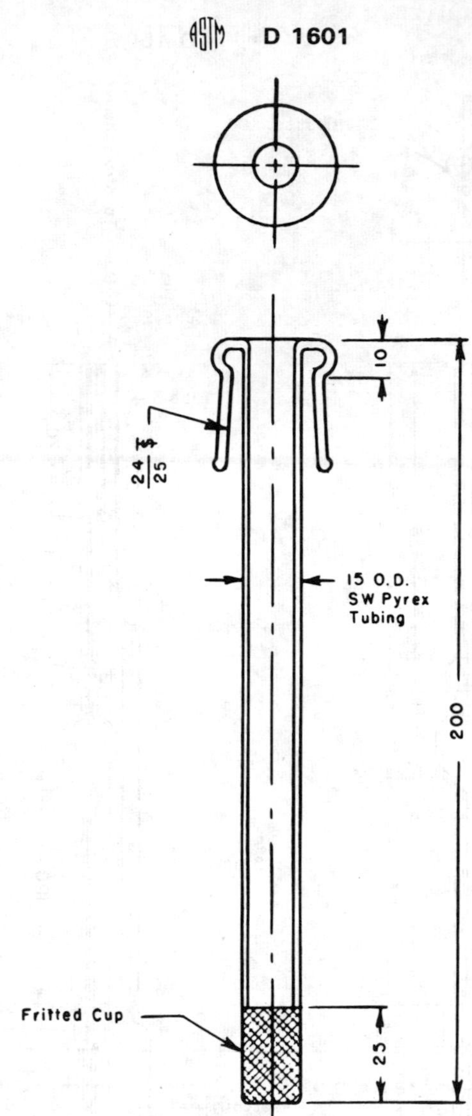

All Dimensions in Millimeters

FIG. A3 Filter Stick of the Modified Ubbelohde Viscometer

ASTM Designation: D 1603 - 68

Standard Method of Test for
CARBON BLACK IN ETHYLENE PLASTICS[1]

This Standard is issued under the fixed designation D 1603; the number immediately following the designation indicates the year of original adoption or, in the case of revision, the year of last revision. A number in parentheses indicates the year of last reapproval.

1. Scope

1.1 This method covers the determination of the carbon black content of ethylene plastics. Determinations of carbon black are made gravimetrically after pyrolysis of the sample under nitrogen. This method is not applicable to compositions that contain nonvolatile pigments or fillers other than carbon black.

NOTE 1—The values stated in U.S. customary units are to be regarded as the standard. The metric equivalents of U.S. customary units may be approximate.

2. Significance

2.1 The information provided by this method is useful for control purposes and is required for calculation of optical absorptivity.

3. Apparatus

3.1 *Electric Furnace*, at least 20 cm (8 in.) long suitable for use with the tubing described in 3.2.

3.2 *Tubing*,[2] 2.9 cm ($1^1/_8$ in.) in diameter, approximately twice as long as the furnace described in 3.1.

3.3 *Stoppers*—Two rubber or neoprene stoppers, to fit the tube described in 3.2, unless the tube is fitted with ground joints and mating connectors.

3.4 *Ten-mm Glass Tubing*, of sufficient amount, and matching rubber or plastic tubing for connections.

3.5 *Combustion Boat*, approximately 8 cm by 1.9 cm by 1.3 cm ($3^1/_2$ by $^3/_4$ by $^1/_2$ in.). Glazed porcelain, quartz high-silica glass, or platinum is suitable.

NOTE 2—A loose-fitting cover for the combustion boat is optional. If used, it shall be considered a part of the boat and handled and weighed with it.

3.6 *Iron-Constantan Thermocouple*, and a potentiometer or millivoltmeter suitable for determining temperatures in the range 300 to 700 C.

3.7 *Flow Meter*,[3] suitable for measuring gas flow at rates of 1 to 10 liters/min.

3.8 *Traps*, three glass traps with removable ground-glass connected heads and 10-mm diameter inner and connecting tubes.

NOTE 3—Only one trap is required if the entire apparatus train is placed in a fume hood. None is required if in addition, nitrogen of sufficient purity is used and produced by the alternative means provided in Section 4.

3.9 *Drying Tube*—A U-shaped drying tube, having an inside diameter of 20 mm or larger, fitted with ground glass or neoprene stoppers.

3.10 *Glass Wool*.

3.11 *Desiccator*, with desiccant.

3.12 *Bunsen Burner*.

3.13 *Balance*—An analytical balance having a sensitivity of 0.0001 g.

3.14 *Weights*—A set of Class S weights for use with the balance.

4. Reagents and Materials

4.1 *Carbon Dioxide, Solid* (Dry Ice).

NOTE 4—The solid carbon dioxide and the trichloroethylene are not required if the entire apparatus train is placed in a fume hood.

4.2 *Desiccant*, such as anhydrous calcium chloride ($CaCl_2$).

[1] This method is under the jurisdiction of ASTM Committee D-20 on Plastics. A list of committee members may be found in the ASTM Yearbook. This standard is the direct responsibility of Subcommittee D-20.70 on Analytical Methods.

Current edition effective Sept. 9, 1968. Originally issued 1958. Replaces D 1603 - 62.

[2] Pyrex, Vycor, or equivalent tubing has been found satisfactory for this purpose.

[3] Precision Bore Rotometer Tube No. 2B-25, available from the Fischer & Porter Co. has been found satisfactory.

4.3 *Nitrogen*, prepurified, having oxygen content below 0.01 percent. As a safeguard against accidental leakage, contamination, or inadequate purity, the gas shall be further purified by one of the following procedures:

4.3.1 Passage of the nitrogen through a glass trap inserted ahead of the drying tube (see Fig. 1), filled approximately one third full of potassium hydroxide - pyrogallol solution made to contain 5 g of pyrogallol and 50 g of KOH in 100 ml of water. Technical grade, or better, reagents are satisfactory.

4.3.2 Insertion of a plug, or roll, of clean copper tinsel, foil, or wire 7.5 to 10 cm (3 to 4 in.) long into the combustion tube ahead of the sample (see Fig. 1) so that it is completely within the heated region of the furnace. Care should be taken that channeling of the nitrogen through the plug does not occur. The extent of blackening of the copper may be taken as a guide for determining when the plug should be renewed.

4.3.3 Passage of the nitrogen through a combustion tube filled to a length of 15 cm (6 in.) or greater with clean copper tinsel, foil, or wire, and maintained in a furnace at a temperature around 500 C.

4.3.4 The need for the procedures described is eliminated if gas having an oxygen content of less than 0.002 percent (20 ppm) is used.

4.4 *Trichloroethylene*, technical grade (Note 3).

5. Procedure

5.1 Assemble the apparatus as shown in Fig. 1. Both cold traps following the combustion tube shall contain trichloroethylene, but only the first need be cooled with solid carbon dioxide. Alternatively, the entire apparatus may be placed in a fume hood and the two traps following the combustion tube omitted. Fill the drying tube with anhydrous $CaCl_2$ or other suitable desiccant. Hold between loose plugs of glass wool.

5.2 Heat a clean combustion boat to red heat in a bunsen flame; then transfer the boat to the desiccator and allow it to cool over fresh desiccant for not less than 30 min.

5.3 Remove the boat from the desiccator and weigh it to nearest 0.0001 g. Immediately place 1.0 ± 0.1 g of the ethylene plastic under test in the boat and quickly weigh to the nearest 0.0001 g.

5.4 Heat the furnace to a constant temperature of 600 C. Adjust the rate of nitrogen flow to 1.7 ± 0.3 liters/min. Open the inlet end of the 2.9-cm diameter tube, quickly place the combustion boat with the sample into the tube at the center of the furnace, and adjust the thermocouple so that the weld is touching the boat. Insert copper plug, if this is used (see 4.3.2). Quickly close the furnace and allow heating to proceed for at least 15 min.

NOTE 5—The exact temperature of heating is not critical in the range 500 to 700 C, although a heating time as long as 30 min is desirable at the lower temperature. If desired the sample may be put in the furnace at 300 C or less and the temperature of the furnace then programmed for sample heating to 350 C in 10 min, 450 C in another 10 min, and 500 C after a total of 30 min, finally heating at 500 C for an additional 15 min.

5.5 Move the tube or furnace so that the boat is no longer in the heated zone of the furnace and allow 5 min for cooling, while maintaining the flow of nitrogen. Remove the copper plug, if present, and the boat through the inlet end of the tube and allow it to cool in the desiccator for at least 30 min. Take care that the boat does not become contaminated from any deposits on the walls of the tube. Then quickly reweigh the boat and its contents to the nearest 0.0001 g.

5.6 Make all determinations in duplicate.

6. Calculation

6.1 Calculate the carbon black content as follows:

$$\text{Carbon black, percent} = (W_r/W_s) \times 100$$

$$\text{Concentration of carbon black, g/cm}^3 = (W_r \times D_s)/W_s$$

where:

W_r = weight of residue, g
W_s = weight of sample, g, and
D_s = density of sample.

7. Report

7.1 The report shall include the following:

7.1.1 Identity of the sample,

7.1.2 The individual values calculated as described in Section 6, and

7.1.3 The average of the values reported in 7.1.2.

8. Precision

8.1 The interlaboratory precision, exclusive

of sampling error, was determined by an interlaboratory study with seven laboratories to be ±0.156 percent (3S) as defined in ASTM Recommended Practice E 177, for Use of the Terms Precision and Accuracy as Applied to Measurement of a Property of a Material,[4] over the range 2.5 to 3.0 percent carbon black, and ±0.309 percent (3S) over the range 35 to 55 percent carbon black.

[4] *Annual Book of ASTM Standards*, Part 30.

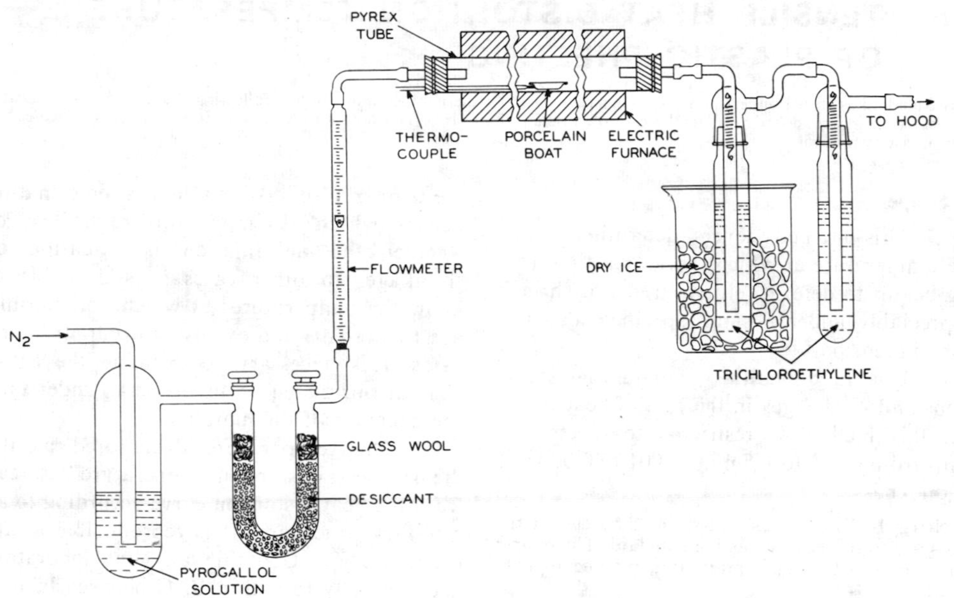

FIG. 1 Assembly of Apparatus.

ASTM Designation: D 1637 – 61

Standard Method of Test for
TENSILE HEAT DISTORTION TEMPERATURE OF PLASTIC SHEETING[1]

This Standard is issued under the fixed designation D 1637; the number immediately following the designation indicates the year of original adoption or, in the case of revision, the year of last revision. A number in parentheses indicates the year of last reapproval.

1. Scope

1.1 This method covers determination of the temperature at which thermoplastic sheeting begins to deform (either stretch or shrink) appreciably under a small specified load applied in tension.[2]

1.2 Due to difficulties of mounting specimens and to changes in the rate of heat transfer, this method is restricted to sheets and films from 0.025 to 1.5 mm (0.001 to 0.060 in.) in thickness.

NOTE 1—The values stated in U.S. customary units are to be regarded as the standard. The metric equivalents of U.S. customary units may be approximate.

NOTE 2—For plastic sheets over 1.5 mm (0.060 in.) in thickness, see ASTM Method D 648, Test for Deflection Temperature of Plastics Under Load.[3]

1.3 The test method is applicable only to materials that have a modulus greater than 700 kgf/cm^2 (10,000 psi) at 23 C (73.4 F).

2. Significance

2.1 This method is empirical and does not measure any fundamental property. It is, however, useful for obtaining a picture of the heat distortion or softening characteristics of plastic sheeting, and is especially valuable for comparing different materials.

2.2 Some plastic sheeting, when heated under a small tensile load, shows largely viscous behavior rather than elastic extension. Other materials that have been stretched during manufacture, when heated under a small tensile load, first relax or shrink due to release of "frozen-in" strain and then, as the temperature is raised further, begin to elongate under the applied load. In some applications it is the temperature at which the first deformation occurs which is important, regardless of whether the deformation is extension or shrinkage. In other cases it is desirable to know the temperatures at which both shrinkage (if present) and extension occur. In many cases it is necessary to examine the entire heat distortion curve to properly understand the behavior of the material.

2.3 The temperature designated as the "tensile heat distortion temperature" is read from the heat distortion curve according to an arbitrary procedure. It is reproducible to approximately 2 C within any one laboratory and generally to within 10 C between laboratories.

3. Apparatus

3.1 *Fixed Member*—A fixed or essentially stationary member carrying one grip.

3.2 *Movable Member*—A movable member carrying a second grip.

3.3 *Grips*—Grips for holding the test specimen between the fixed member and the movable member. The test specimen shall be held in such a way that slippage relative to the

---

[1] This method is under the jurisdiction of ASTM Committee D-20 on Plastics. A list of committee members may be found in the ASTM Yearbook. This standard is the direct responsibility of Subcommittee D-20.30 on Thermal Properties.

Current edition effective Sept. 18, 1961. Originally issued 1959. Replaces D 1637 – 59 T.

[2] For additional information, refer to the following:

Moelter, G. M., and Schweizer, E., "Heat Softening of Cellulose Acetate," *Industrial and Engineering Chemistry*, Analytical Edition, IENAA, Vol 41, 1949, p. 684.

Watson, M. T., Armstrong, G. M., and Kennedy, W. D., "An Autographic Apparatus for the Study of Thermal Distortion," *Modern Plastics*, MOPLA, Vol 34, 1956, p. 169.

[3] *Annual Book of ASTM Standards*, Part 27.

grips is prevented insofar as possible.

NOTE 3—If specimens are tested in the form of loops instead of strips (4.4) the grips are replaced by metal pins and yokes from which the tensile stress is applied. These metal pins should not be over 0.75 mm (0.030 in.) in diameter. Large diameter pins introduce errors in the effective sample length since the specimen is not free to elongate over the area in contact with the pins.

3.4 *Extension Indicator*—An instrument for determining the change in specimen length at any time during the test. This instrument shall be essentially free of inertia lag at the occurring rate of dimensional change. It shall permit measurement of the strain to ±0.05 mm/mm (0.05 in./in.).

NOTE 4—Calibrated dial indicators, cathetometers, and linear variable differential transformers have been used successfully.

3.5 *Heating Bath*—An oven or bath for raising the temperature of the specimen at a constant rate of 2 ± 0.2 C/min (Note 5). The heating medium shall be well stirred during the test to reduce the temperature lag between the medium and the specimen (Note 6).

NOTE 5—The specified heating rate can be conveniently maintained by the use of an immersion heater, if the current is regulated by a variable transformer or rheostat. This may be done either manually or by automatic controls. Any other apparatus designed to maintain this heating rate is satisfactory, such as a synchronous motor-driven bimetallic regulatory which energizes the immersion heater.

NOTE 6—A liquid heat-transfer medium such as mineral oil has been found to affect the shape of the heat distortion curve for some plastic materials. For this reason an air bath is recommended. One satisfactory approach is to have the specimen in an air shaft which is immersed in an oil bath such as silicone fluid. The oil bath facilitates uniform temperature distribution. Such an arrangement is shown in Fig. 1. A cooling coil may be inserted in the oil bath to reduce the delay between tests. An air oven may also be used as shown in Fig. 2.

3.6 *Weights*—Weights for applying the specified stress of 3.52 kgf/cm^2 (50 psi). The stress is applied to the specimen through the movable member.

3.7 *Micrometer*—A micrometer or other thickness gage, reading to the nearest 0.002 mm (0.0001 in.), for measuring the thickness of the test specimen.

3.8 *Thermometer*—The thermometer shall be one of the following, or its equivalent, as prescribed in ASTM Specification E 1, for ASTM Thermometers:[4] ASTM Partial Immersion Thermometer 1C, 1F, 2C, or 2F, having ranges from −20 to +150 C, 0 to 302 F, −5 to +300 C, or 20 to 580 F, respectively, whichever temperature range is most suitable for the material being tested.

3.9 *Desiccator*—A desiccator containing anhydrous calcium chloride (or other suitable desiccant).

4. Test Specimens

4.1 In the case of isotropic materials at least two specimens shall be tested from each sample.

4.2 In testing materials that are suspected of being anisotropic, duplicate sets of test specimens shall be tested in each of the two principal axes of anisotropy. When viewed between crossed polarizers, anisotropic samples often show positions of darkening or extinction every 90 deg as the sample is rotated. These extinction directions usually coincide with the principal axes of anisotropy but may or may not coincide with the machine and cross directions.

4.3 Materials shall be tested in the form of strips 6 to 25 mm (0.25 to 1.0 in.) wide and 50 to 180 mm (2 to 7 in.) long. The initial distance between grips shall be between 25 and 125 mm (1 and 5 in.). The length-to-width ratio shall be at least 2.

4.4 Flexible materials less than 0.075 mm (0.003 in.) in thickness, but which can be readily cemented or otherwise sealed together without changing their properties, may be tested in the form of loops. Strips 6 to 25 mm (0.25 to 1.0 in.) in width and 50 to 250 mm (2 to 10 in.) long shall be cut and the ends lapped 6 mm ($^1/_4$ in.) and sealed (Note 7), thus forming approximately 25 to 125-mm (1 to 5-in.) loops.

NOTE 7—The degree of overlap should not exceed 6 mm ($^1/_4$ in.), since the specimen has double thickness in this area. If a solvent cement is used, the specimen loop must be seasoned by heating at not over 60 C (140 F) until the residual solvent evaporates. Solvents which are difficult to remove should not be used. Heat sealing or high frequency sealing may be used if the properties of the specimen are not affected.

5. Conditioning

5.1 Specimens shall be conditioned by desiccating for 24 h over anhydrous calcium

[4] *Annual Book of ASTM Standards*, Part 30.

chloride or other suitable desiccant. A shorter desiccating time may be used for thin films if it can be shown that equilibrium is reached. This conditioning procedure may be omitted when the tensile heat distortion temperature has been shown to be unaffected by the desiccation treatment.

6. Procedure

6.1 Measure the width of the specimen to the nearest 0.25 mm (0.01 in.) and the thickness to the nearest 0.002 mm (0.0001 in.) (or to 1 percent of the thickness for specimens over 0.25 mm (0.010 in.)) at several points along its gage length. Record the average cross section found.

6.2 Condition the specimen in accordance with Section 5 before placing it in the apparatus.

6.3 Place the specimen in the grips attached to the fixed and movable members. Tighten the grips evenly and firmly to the degree necessary to prevent slippage of the specimen during the test, but not to the point where the specimen would be crushed. Where specimen loops are used instead of strips, the loop is held in position by pins attached to the fixed and movable members.

6.4 Apply weights to the movable member so that the test specimen is under a tensile stress of 3.52 kgf/cm^2 (50 $\pm$ 5 psi), calculated on the average cross-sectional area of the specimen between the grips. The weight required to give the same unit stress for a loop specimen is twice that required for a strip specimen.

NOTE 8—If this test is carried out using other stress values, the tensile stress used must be indicated.

6.5 Place the specimen in the immersion bath at a temperature approximately 20 C (68 F) below the temperature at which the specimen is expected to begin to deform.

NOTE 9—This latter temperature can be found for any new material from an initial determination of the tensile heat distortion curve with a starting temperature of 23 C (73.4 F).

6.6 Raise the temperature of the bath at a rate of 2 $\pm$ 0.2 C/min. Record the reading of the extension indicator and the bath temperature at appropriate intervals (more frequently when the length of the specimen is changing rapidly) until the specimen has either elongated approximately 50 percent or the temperature has reached 250 C (482 F).

6.7 Apply a temperature correction for the small lag between the temperature of the specimen and that of the immersion bath. This can be determined once for each apparatus by attaching a thermocouple to a test specimen and comparing the heating curve from the thermocouple with that from the thermometer in the immersion bath. The correction shall not exceed 3 C.

7. Calculation

7.1 The extension indicator readings shall be converted to percentage change in length. A smooth curve on rectangular coordinate paper shall be plotted of percentage change in length as ordinate *versus* temperature in degrees Celsius as abscissa.

7.2 The tensile head distortion temperature shall be reported as the temperature at the following points (Notes 10 and 11):

7.2.1 At 2 percent shrinkage (if the specimen shrinks), and

7.2.2 At 2 percent extension. If the specimen shrinks prior to extension, the 2 percent extension is calculated above the maximum shrinkage level, not above the zero percent deformation line (see Fig. 3).

NOTE 10—It has been found useful with certain materials to draw a straight line through the steeply rising portion of the tensile distortion temperature curve and to read the temperature where this line extrapolates to intercept the abscissa. However, this method is not applicable to all types of materials.

NOTE 11—It is impossible to completely characterize a tensile heat distortion temperature curve by means of one or two points. It is advisable to examine the entire curve when comparing the behavior of different types of materials (see Fig. 3).

8. Report

8.1 The report shall include the following:

8.1.1 Complete identification of the material tested, including type, source, manufacturer's code numbers, form, direction of specimen in original sheet, and dimensions of the test specimen,

8.1.2 Tensile heat distortion temperatures for shrinkage and extension, calculated in accordance with 7.2,

8.1.3 Maximum percentage of shrinkage or

percentage of extension, or both, if such maximums occur,

8.1.4 Tensile stress used, and

8.1.5 Any peculiar sample characteristics noted during the test. Where materials with different types of tensile heat distortion behavior are being compared, a plot of the complete curve for each material shall be included in the report.

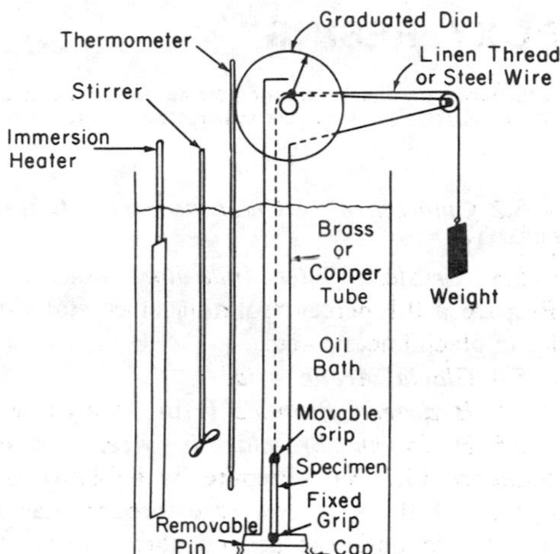

FIG. 1 Tensile Heat Distortion Apparatus.

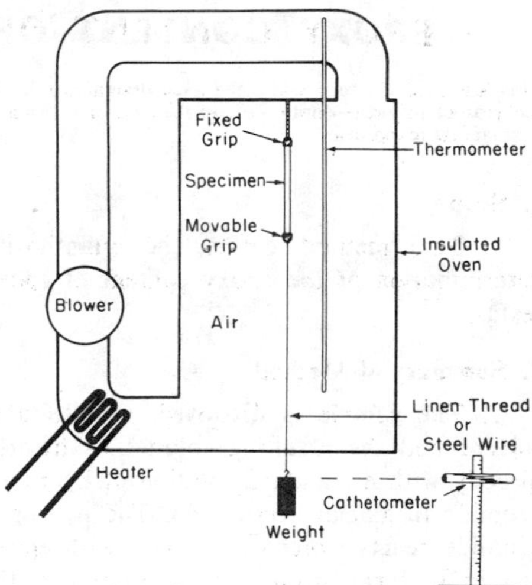

FIG. 2 Air Oven in Tensile Heat Distortion Apparatus.

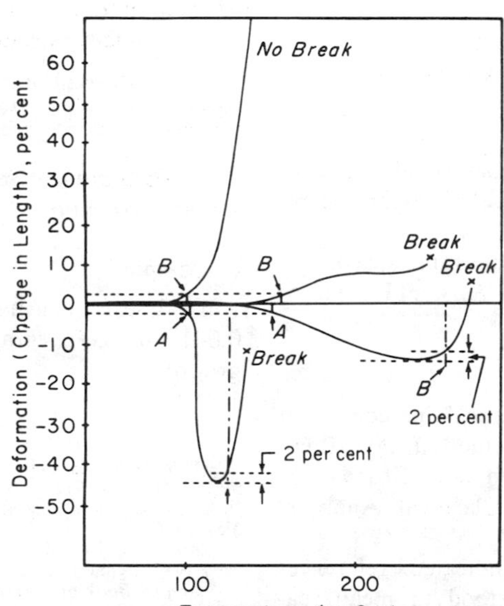

A—Tensile heat distortion temperature for 2 percent shrinkage.

B—Tensile heat distortion temperature for 2 percent extension.

FIG. 3 Typical Tensile Heat Distortion Curves for Different Types of Thermoplastics.

Designation: D 1652 - 67

Federation of Societies for Paint Technology Standard No. Ct-9-67

Standard Method of Test for
EPOXY CONTENT OF EPOXY RESINS[1]

This Standard is issued under the fixed designation D 1652; the number immediately following the designation indicates the year of original adoption or, in the case of revision, the year of last revision. A number in parentheses indicates the year of last reapproval.

1. Scope

1.1 This method covers the quantitative determination of the epoxy content of epoxy resins.

2. Summary of Method

2.1 The sample is dissolved in a suitable solvent and the resulting solution is titrated directly with a standard solution of hydrogen bromide in glacial acetic acid. The hydrogen bromide reacts stoichiometrically with epoxy groups to form bromohydrins; therefore, the quantity of acid consumed is a measure of the epoxy content.

3. Apparatus

3.1 *Buret*, closed-reservoir type. The buret tip should be fitted with a rubber stopper of proper size to fit the neck of the Erlenmeyer flask and the stopper should have an additional small hole to permit escape of replaced air during titration.

3.2 *Magnetic Stirrer*, adjustable speed.

3.3 *Magnetic Stirring Bars*, TFE-fluorocarbon-covered.

4. Purity of Reagents

4.1 Reagent grade chemicals or equivalent as specified in ASTM Methods E 200, Preparation, Standardization, and Storage of Standard Solutions for Chemical Analysis[2] shall be used in all tests.

4.2 Unless otherwise indicated, references to water shall be understood to mean reagent water conforming to ASTM Specification D 1193, for Reagent Water.[3]

5. Reagents and Materials

5.1 *Chlorobenzene*.

5.2 *Chloroform - Chlorobenzene Mixture* (1 + 1).

5.3 *Crystal Violet Indicator Solution*—Prepare a 0.1 percent solution of crystal violet in glacial acetic acid.

5.4 *Glacial Acetic Acid*.

5.5 *Hydrogen Bromide* (HBr), anhydrous.

5.6 *Hydrogen Bromide in Acetic Acid, Standard (0.1 N)*[4]—Prepare by bubbling anhydrous HBr at a slow rate through glacial acetic acid until the desired normality is attained (approximately 8 g of HBr/liter). Standardize daily against 0.4 g of potassium acid phthalate ($KHC_8H_4O_4$) accurately weighed and dissolved by gently heating in 10 ml of glacial acetic acid.

NOTE—Reagent of 0.1 *N* concentration has been specified, because as solutions exceed this concentration they became progressively less stable.

5.7 *Potassium Acid Phthalate*, ($KHC_8H_4O_4$) *Primary Standard*.[5]

6. Sample

6.1 Use a quantity of sample that contains 0.001 to 0.002 gram-equivalents of epoxy groups.

---

[1] This method is under the jurisdiction of ASTM Committee D-1 on Paint, Varnish, Lacquer, and Related Products. A list of members may be found in the ASTM Yearbook.

Current edition effective Nov. 30, 1967. Originally issued 1959. Replaces D 1652 - 62 T.

[2] *Annual Book of ASTM Standards*, Part 22.

[3] *Annual Book of ASTM Standards*, Part 23.

[4] Hydrobromic acid, 30 to 32 percent concentration in acetic acid, available from Distillation Products Industry, a division of Eastman Co. (Catalog No. 1161), has been found satisfactory for the preparation of 0.1 *N* HBr solution.

[5] National Bureau of Standards sample or Malinckrodt No. 6704 has been found satisfactory for this purpose.

7. Procedure

7.1 Weigh the desired amount of sample, to the nearest 1 mg, into an Erlenmeyer flask. Use a 50-ml flask for low-molecular-weight resins (liquid grades) and a 125-ml flask for high-molecular-weight resins (solid grades).

7.2 Dissolve the sample in a solvent at room temperature. Use 10 ml of chlorobenzene for liquid grade resins and 25 ml of a 1+1 mixture of chloroform and chlorobenzene for solid grade resins. Place a TFE-fluorocarbon-sealed magnetic stirring bar into the flask and allow mixture to swirl on magnetic stirrer to effect solution.

7.3 Add 4 to 6 drops of crystal violet indicator solution and attach the flask to the rubber stopper on the buret tip. Lower the buret tip to a point just above the solution and titrate with the hydrogen bromide in acetic acid to a blue-green end point with the magnetic stirrer rotating at a moderate speed to avoid splashing. Slow down the titration near the end point to allow ample time for the reaction to take place. Titrate to as nearly as possible the same color at the end point as that obtained during standardization of the reagent.

7.4 Make a blank determination on the reagents in an identical manner.

8. Calculations

8.1 Calculate the normality of the HBr in acetic acid as follows:

$$N = (A \times 1000)/204.2B$$

where:

N = normality of the HBr in acetic acid,
A = grams of $KHC_8H_4O_4$ used, and
B = milliliters of HBr used.

8.2 Calculate the epoxy content in gram equivalents of epoxy groups per 100 g of resin as follows:

$$\text{Epoxy content} = N(A - B)/10C$$

where:

N = normality of the HBr in acetic acid,
A = milliliters of HBr used for titration of the sample,
B = milliliters of HBr used for titration of the blank, and
C = grams of sample used.

8.3 Calculate the percentage of oxirane oxygen as follows:

$$\text{Oxirane oxygen, percent} = 1.6N(A - B)/C$$

8.4 Calculate the weight per epoxy equivalent (*WPE*), that is, grams of resin containing one gram equivalent of epoxy groups, as follows:

$$WPE = 1000C/N(A - B)$$

9. Precision

9.1 *Repeatability*—The average difference between duplicate runs performed by the same analyst may approximate 0.7 percent of the epoxy content of the resin tested. Two such values should be considered suspect if they differ by more than 2 percent.

9.2 *Reproducibility*—The average difference between two determinations performed by different analysts in different laboratories may approximate 2 percent of the epoxy content of the resin tested. Two such values should be considered suspect if they differ by more than 6 percent.

Designation: D 1671 – 63 (Reapproved 1970) American National Standard K65.15-1965
American National Standards Institute

Standard Method of Test for
ABSORBED GAMMA RADIATION DOSE IN THE FRICKE DOSIMETER[1]

This Standard is issued under the fixed designation D 1671; the number immediately following the designation indicates the year of original adoption or, in the case of revision, the year of last revision. A number in parentheses indicates the year of last reapproval.

1. Scope

1.1 This method, known as the Fricke dosimeter method, covers the measurement of gamma radiation in the range of 0.2×10^4 to 4×10^4 rads by oxidation of ferrous ammonium sulfate and determination of the ferric ion concentration spectrophotometrically.

2. Significance

2.1 *Dose Rate*—No effect on accuracy of measurement of radiation dose is observed up to rates of 10^7 rads/h.

2.2 *Temperature*—The accuracy of this method is not significantly changed by varying temperature of the system between 0 to 50 C during irradiation.

2.3 *Energy Dependence*—This method has been shown to be independent of energy in the range of 0.1 to 2 MeV.

3. Definitions

3.1 *rad*—The recommended unit for reporting absorbed radiation dose, defined as 100 ergs of energy absorbed per gram of specimen.

4. Apparatus

4.1 *Spectrophotometer*—Beckman Model DU, hydrogen lamp attachment, and 1-cm matched quartz cells, or equivalent apparatus.

4.2 *Containers:*

4.2.1 Chemically-resistant borosilicate glass shall be used to hold the reagent.

4.2.2 Containers should be of approximately the same dimensions, geometry, and materials as those employed to hold specimens for irradiation. A thin-walled glass liner, 1.0 to 1.2 mm maximum thickness, may be used inside the specimen container where necessary. No internal dimension of the containers should be less than 8 mm in diameter.

4.2.3 All containers should be cleaned with chromic-sulfuric acid cleaning solution, then rinsed thoroughly with distilled water. Care should be taken that the last three rinsings are water clear.

5. Purity of Reagents

5.1 *Purity of Reagents*—Reagent grade chemicals or equivalent, as specified in ASTM Methods E 200, for Preparation, Standardization, and Storage of Standard Solutions for Chemical Analysis,[2] shall be used in all tests.

5.2 *Purity of Water*—Unless otherwise indicated, references to water shall be understood to mean water conforming to ASTM Specifications D 1193, for Reagent Water.[2]

6. Reagent

6.1 *Dosimetric Solution*—Prepare a solution that is 0.001 *M* in ferrous ammonium sulfate, 0.001 *M* in sodium chloride, and 0.4 *M* in sulfuric acid, using distilled water. A convenient way is to make up a stock solution 0.5 *M* in Fe^{++} (ferrous ammonium sulfate) and 0.5 *M* in sodium chloride. This stock solution

---

[1] This method is under the joint jurisdiction of ASTM Committee D-9 on Electrical Insulating Materials and Committee D-20 on Plastics. A list of committee members may be found in the ASTM Yearbook. This standard is the direct responsibility of Subcommittee D-20.20 on Radiation Methods.

Current edition effective Sept. 30, 1963. Originally issued 1959. Replaces D 1671 – 59 T.

[2] *Annual Book of ASTM Standards*, Part 22.

may be stored for up to 3 months.

NOTE 1—The 0.5 *M* ferrous ammonium sulfate stock solution may be tested occasionally for depletion of ferrous iron by the following procedure. Dilute 2 ml of the stock solution to 1 liter with 0.4 *M* sulfuric acid solution. If desired, the dosimetric solution itself may be used. In either case, the optical density as read at 305 nm (slit width of 0.5 mm, quartz cells) with distilled water as a blank, must be less than 0.4. Otherwise, the stock and dosimetric solutions must be discarded.

6.2 When a test is to be run, add 2 ml of this solution to 1 liter of a solution of 0.4 *M* sulfuric acid that has previously been saturated with oxygen during the same day. To introduce the oxygen, use only clean, dust-free glass tubing in contact with the solution. Clean plastic tubing may be used to connect the glass sparging tube to the oxygen supply. This dosimetric solution should be made up fresh each day. All solutions should be stored in amber bottles out of direct sunlight when not in use.

NOTE 2—The use of rubber stoppers and rubber tubing for sealing specimen tubes or transferring the solutions should be avoided since contact between the dosimetric solution and any kind of organic material can lead to erratic results. Extreme care should be taken to insure cleanliness of all glassware.

7. Calibration Curve

7.1 *Preparation of Solutions*—Add a weighed quantity of ferric sulfate to the calculated amount of 0.4 *M* sulfuric acid required to make up a solution approximately 0.1 *M* in ferric ion. Place this mixture (with cap loose on the bottle) in an oven at 90 to 95 C overnight to obtain a solution. Analyze a sample of the cooled ferric solution for iron (Note 3). Dilute one, two, three, four, five, six, and seven portions of the stock solution to 1 liter with 0.4 *M* sulfuric acid solution.

NOTE 3—Methods such as those given in Scott's "Standard Methods of Chemical Analysis," Fifth Edition, 1939, pp. 474–476 and 482, may be used for the analysis of the iron content of the ferric sulfate calibration solution.

7.2 *Spectrophotometric Measurements*—Read the optical densities (absorbances) of the diluted solutions in the Beckman Spectrophotometer set at 305 nm, slit width of 0.5 mm, using quartz cells and using a portion of the 0.4 *M* sulfuric acid solution as the blank in the spectrophotometer.

7.3 *Preparation of Curve*—Convert the concentrations of ferric ion solutions whose optical densities were measured to equivalent radiation dosages as follows (Notes 4 and 5):

$$\text{Rads} = \text{micromoles of } Fe^{+++} \text{ per liter} \times 60.9$$

Plot the calculated radiation doses against the corresponding optical densities to obtain the calibration curve.

NOTE 4—For a development of this conversion factor, see the Appendix.

NOTE 5—All units of radiation dose have been expressed in rads, the unit recommended in Section 3 for expressing absorbed doses.

8. Procedure

8.1 Place a specimen of dosimetric solution in the radiation field for a carefully measured length of time. After removal from the radiation field, read the optical density of the specimen in the spectrophotometer on the same day. Use a portion of unirradiated ferrous solution as a blank in the spectrophotometer. From the optical density of the irradiated sample, read the radiation dose in rads directly from the calibration curve.

8.2 The measurement of optical densities with the spectrophotometer shall be carried out at as nearly constant temperature as possible, since the extinction coefficient has a rather large temperature coefficient of +0.7 percent/deg C. Alternatively, a correction for temperature may be made according to the following equation:

$$\text{Dose (corrected)} = \text{dose measured at } T_2/[1 + 0.007\,(T_2 - T_1)]$$

where:

T_1 = temperature at which the calibration curve was prepared, deg C, and

T_2 = temperature at which the unknown sample is read, deg C.

APPENDIX

A1. Development of Equation for Converting Ferric Ion Concentration to Radiation Exposure Units

A1.1 By definition, G = number of molecules reacting per 100 electron volts (eV) absorbed.

$$G \text{ molecules} \approx 100 \text{ eV}$$

$$1 \text{ molecule} \approx 100/G \text{ eV}$$

A1.2 Using Avogadro's number:

$$1 \text{ mole} \approx 100/G\ (6.02 \times 10^{23}) = 6.02/G \times 10^{25} \text{ eV}$$

$$1 \text{ micromole} \approx 6.02/G \times 10^{19} \text{ eV}$$

A1.3 Since,

$$1 \text{ eV} = 1.60 \times 10^{-12} \text{ erg},$$

$$1 \text{ micromole} \approx (6.02/G \times 10^{19})\ (1.60 \times 10^{-12}) \approx 9.63/G \times 10^{7} \text{ erg}$$

A1.4 Dividing both sides by weight:

$$\text{micromole/g} \approx 9.63/G \times 10^{7} \text{ ergs/g}$$

A1.5 The density of the dosimetric solution is 1.02; therefore:

$$(1 \text{ micromole/g})\ (1.02 \text{ g/ml}) \approx 9.63/G \times 10^{7} \text{ ergs/g}$$

$$1 \text{ micromole/ml} \approx 9.63/(1.02 \times G) \times 10^{7} \text{ ergs/g}$$

$$1 \text{ micromole/ml} \approx 9.44/G \times 10^{7} \text{ ergs/g}$$

$$1 \text{ micromole/liter} \approx 9.44/G \times 10^{4} \text{ ergs/g}$$

A1.6 Since 1 rad = 100 ergs/g

$$1 \text{ micromole/liter} \approx 944/G \text{ rads}$$

A1.7 To convert from micromoles of Fe^{+++}/liter to rads, multiply by $944/G$.

To convert from rads to micromoles of Fe^{+++}/liter, multiply by $G/944$.

A1.8 Where G = 15.5:

To convert from rads to micromoles of Fe^{+++}/liter, multiply by $G/944$.

A1.8 Where G = 15.5:

To convert from micromoles of Fe^{+++}/liter to rads, multiply by 60.9.

To convert from rads to micromoles of Fe^{+++}/liter, multiply by 0.0164.

American National Standard C59.83-1968
American National Standards Institute

Recommended Practice for

EXPOSURE OF POLYMERIC MATERIALS TO HIGH ENERGY RADIATION[1]

This Recommended Practice is issued under the fixed designation D 1672; the number immediately following the designation indicates the year of original adoption or, in the case of revision, the year of last revision. A number in parentheses indicates the year of last reapproval.

1. Scope

1.1 This recommended practice covers conditions for the exposure of plastics and elastomers to high energy radiation prior to determination of radiation induced changes in physical or chemical properties. The problems of measuring the properties of materials during irradiation are too varied and complex for consideration in this recommended practice. Only the conditions of irradiation are specified and not the specimens, so that the effect of radiation on any property may be determined by the use of the appropriate specimens and test methods. The method describes procedures for five types of exposure, as follows:

1.1.1 *Procedure A*—Exposure at Ambient Conditions.

1.1.2 *Procedure B*—Exposure at Controlled Temperature.

1.1.3 *Procedure C*—Exposure in a Medium other than Air.

1.1.4 *Procedure D*—Exposure Under Load.

1.1.5 *Procedure E*—Exposure Combining Two or More of the Variables Listed in Procedures A through D.

1.2 The method specifically covers the following kinds of radiation: gamma or X-radiation, electrons or beta radiation, neutrons, and mixtures of these such as reactor radiation.

2. Significance

2.1 The procedures outlined here are designed to standardize the exposure of organic polymers for the purpose of studying the effects of high energy radiation, but have been made flexible enough so that a large variety of conditions may be met within the scope of this one irradiation method. Because of this flexibility in the procedures it is important that the experimenter have some idea of the kind of changes that will occur, and of the conditions that will affect these changes.

3. Effects of Irradiation

3.1 The purpose of the following information on "Effects of Irradiation" is to provide a minimum background. Exposure to radiation results in extensive changes in the nature of high polymers, which owe their unique properties to chemical linking into giant molecules of chain or net structure. These chain or net structures may be crosslinked by radiation into a rigid three dimensional network or in other cases, may be cleaved into smaller molecules to produce a weaker material. In all cases some low molecular weight fragments are produced and, if exposures are large enough, general decomposition results.

3.2 One of the first results of the reaction of high energy radiation with polymers is the formation of free radicals or excited molecular fragments. The rate at which these molecular fragments are formed may be much greater than their annihilation rate and this leads to the accumulation of reactive species within the irradiated material and to the possibility of continuing reactions for days or weeks after the specimen has been removed from the radiation field. Because of these

[1] This recommended practice is under the joint jurisdiction of ASTM Committee D-9 on Electrical Insulating Materials and Committee D-20 on Plastics. A list of committee members may be found in the ASTM Yearbook. This standard is the direct responsibility of Subcommittee D-20.20 on Radiation Methods.

Current edition effective Sept. 20, 1966. Originally issued 1959. Replaces D 1672 – 61 T.

post-irradiation reactions it has been necessary to standardize the times and conditions of storage between irradiation and testing of specimens.

3.3 The resultant changes in the molecular structure of polymeric materials by radiation are dependent on the respective rates of recombination, cross-linking, or cleavage of the molecular fragments. These rates are affected by the mobility of the molecular fragments (which is strongly influenced by temperature) and by the concentration of the reactants.

3.4 The concentration of reactive species will vary with the rate of absorption of radiation. Either radiation intensity or dose rate is therefore specified in reporting the results of tests, even though a dose-rate effect is not often observed. The effect of dose rate and specimen thickness is observed when irradiations are carried out in the presence of oxygen, where oxygen reacts with radicals produced in the irradiated material. This oxygen reaction will be diffusion controlled. The reactivity of irradiated specimens toward oxygen makes it necessary to specify whether irradiation or storage has been carried out in the presence of air. When irradiations are carried out in air, the accessibility to an air supply undepleted in oxygen should be assured if possible.

3.5 The localized concentration of reactive species during irradiation will vary locally or microscopically depending on the type of radiation employed. The proton and carbon recoils from neutron bombardment produce densely ionized tracks in the specimen compared to the diffuse ionization in the wake of photons or electrons. The effects of different types of radiation may therefore be different. It is required that the type of radiation to which the specimen has been exposed be reported as well as the irradiation dose in energy absorbed units.

3.6 A wide variation in the stability of the various chemical structures toward radiation makes it difficult to select specific exposure levels for testing. Polystyrene requires about fifty times as much energy for one crosslink as polyethylene does. At the other end of the scale, poly(methyl methacrylate) and polytetrafluoroethylene show changes in engineering properties at about one-twentieth the exposure required for changes in polyethylene. An aromatic ring attached to the main chain at frequent regular intervals has been found to confer marked stability toward radiation, while a quaternary carbon atom in the polymer chain leads to cleavage under radiation and a loss of strength at fairly low exposures. The exposure levels should therefore be those which will produce significant changes in a stipulated property rather than a specified fixed irradiation dose. Furthermore, the change in property may progress at different rates, with some materials changing rapidly once a change has been initiated, while others may change quite slowly. *It is necessary therefore to irradiate to several fixed levels of property change in order to establish the rate of change* (see 11.4).

3.7 Materials that have been exposed to reactor radiation will become radioactive. For pure hydrocarbons the amount of induced radioactivity is not large, but fillers and small amounts of impurities may become highly radioactive, and thus create a handling problem. The other common radiation sources to which polymeric materials will be exposed will not normally produce significant amounts of induced radioactivity.

4. Test Specimens

4.1 Wherever possible, the types of specimens employed shall be in accordance with the ASTM test methods for the specific properties to be measured.

4.2 Where it is not possible to use standard test specimens, irradiated and non-irradiated specimens shall be of the same size and shape.

5. Conditioning

5.1 Specimens to be exposed in air shall be conditioned in accordance with Procedure A, of ASTM Methods D 618, Conditioning Plastics and Electrical Insulating Materials for Testing.[2]

5.2 Specimens to be exposed in a gas other than air shall be conditioned at the temperature of exposure in an appropriate container at a pressure of 10^{-3} mm Hg or less for at least 8 h followed by three flushes with the gas to be present during exposure. After flushing, the container shall be filled with the exposure gas and sealed.

5.3 Specimens to be exposed in a vacuum

[2] *Annual Book of ASTM Standards*, Part 29.

shall be conditioned at the temperature of exposure in an appropriate container at a pressure of 10^{-3} mm Hg or less for at least 48 h. The container shall then be sealed.

5.4 Specimens to be exposed in a liquid medium shall be conditioned in accordance with 5.1 before placing in the liquid medium. The specimens shall be completely immersed in the liquid during the entire period of irradiation.

6. Procedure A, Exposure at Ambient Conditions

6.1 After conditioning in accordance with 5.1, arrange the specimens to be exposed on suitable racks or in containers such that free access to air is assured on all slides.

6.2 When the nature of the radiation source requires that the specimens be enclosed in a container, package the specimens in the standard laboratory atmosphere described in 5.1.

NOTE 1—It is likely that the composition of the atmosphere in the container will be changed by radiation-induced reactions. Therefore it should be clearly stated in the report that the irradiation was made in a closed container.

6.3 Place the specimens in the radiation field and expose to only one total dose. If irradiation is performed with a beam-emitting machine, convey the specimens in some manner such that they traverse the radiation beam; maintain the ratio of exposure time to nonexposure time constant throughout this procedure. For each new total dose, expose additional properly conditioned specimens. Exposure in nuclear reactors or other sources having uniform radiation fields will not require traversing the radiation flux.

6.4 After the required period of time, remove the specimens from the field and condition prior to test in the standard laboratory atmosphere, 5.1, for no less than 16 h and no more than 32 h, unless it is necessary to store the specimens for longer periods of time because of radioactivity or other reasons. The time and condition of such storage should be reported.

6.5 Condition nonirradiated control specimens in accordance with 5.1 prior to test in the standard laboratory atmosphere.

7. Procedure B, Exposure at Controlled Temperature

7.1 Follow the procedure described in 6.1 and 6.2.

7.2 Irradiate the specimens as described in 7.3 at the desired temperature. Place a dummy specimen containing a grounded thermocouple in the radiation field at the same conditions as the test specimens to determine the temperature. If the temperature varies by more than ± 2 C this should be reported.

7.3 After irradiating, condition the specimens in accordance with 7.4.

7.4 After conditioning in accordance with 5.1, expose nonirradiated control specimens to the same temperature employed in 7.2 for the same period of time as the irradiated specimens.

7.5 After treatment, condition the control specimens along with the irradiated specimens in accordance with 5.1 prior to test.

8. Procedure C, Exposure in Medium Other than Air

8.1 After conditioning in accordance with 5.2, 5.3, or 5.4 irradiate the specimens as described in 6.3.

8.2 After removal from the medium, condition the specimens as described in 6.4.

8.3 Allow the nonirradiated control specimens that have been conditioned in accordance with 5.2, 5.3, or 5.4 to remain in the selected medium for the same period of time as the irradiated specimens.

8.4 After treatment, condition the control specimens along with the irradiated specimens in accordance with 6.4 prior to test.

9. Procedure D, Exposure Under Load

9.1 After conditioning in accordance with 5.1, arrange the specimens on a suitable fixture such that they may be subjected to a load during irradiation and have maximum access to air.

9.2 Follow the packaging and irradiating procedures outlined in 6.1 or 6.2 and 6.3.

9.3 After removal from the radiation field release the load, and condition the specimens prior to test in the standard laboratory atmosphere according to 6.4.

9.4 After conditioning in accordance with 5.1, load nonirradiated control specimens in the same manner and for the same period of time as the irradiated specimens. Remove the load and condition the control specimens along with the irradiated specimens in accordance with 6.4 prior to test.

10. Procedure E, Exposure Modifications

10.1 When any combination of two or more of the variables listed in Procedures A through D is used, follow a combined procedure and designate this as Procedure E. The combined procedure must incorporate all of the features of the separate procedures used.

11. Radiation Field and Irradiation Schedule

11.1 Three categories of high-energy radiation are specifically included in this recommended practice. It should be recognized that the radiation effect may be different for different kinds of radiation or for large differences in irradiation dose rate. The dose rates following show the range for a given kind of radiation within which past experience indicates that approximately equal effects will result for equal total exposure:

Radiation	Rads/h
Gamma radiation, X-radiation	10^5 to 10^7
Electrons, beta radiation	
Radioisotopes	10^5 to 10^7
Accelerators	10^8 to 10^{10}
Reactor radiation (neutrons and gamma radiation)	10^5 to 10^7

11.2 The energy of the photons or particles should be such that in passing through the specimen the intensity of the radiation is not reduced by more than 30 percent. The reductions in radiation intensity may be checked for accelerators by the beam current reading and for other radiation sources by one of the methods suggested in 12.1.

11.3 It may be desirable to inject the radiation beam from both sides of the specimen. If so, the specimen may be sufficiently thick so that the minimum dose at any point is no less than 70 percent of the maximum dose. In any case a uniform exposure should be given the entire specimen by traversing the beam if necessary.

NOTE 2—Experience is lacking for the heavy charged particles (alphas, protons, etc.) and therefore these kinds of radiation are not included in this method.

11.4 Two or more irradiation exposures shall be made. The irradiation dose shall be selected to produce a significant change in a stipulated property. To establish a significant trend a minimum of two changes in value is required for a particular property in a given material. Additional exposures are recommended.

NOTE 3—For the effect of irradiation on tensile strength, significant changes might be reduction to 80 percent and 20 percent of initial value. Significant change in density, however, might be 2 to 5 percent.

12. Determination of Irradiation Exposure

12.1 For engineering purposes the best correlation of radiation effects in polymers is made on the basis of energy absorbed in the specimen. The preferred unit for this purpose is the rad (Note 4). The monitoring method is left to the discretion of the experimenter. The determination of the irradiation exposure may be made by the use of a chemical dosimeter (ASTM Method D 1671, Test for Absorbed Gamma Radiation Dose in the Fricke Dosimeter),[2] by an ionization chamber, by change in physical property of a material, or by calorimetry.

NOTE 4—One rad is defined as 100 ergs of absorbed energy per gram of material.

12.2 The kind of radiation and the energy or energy spectrum of the radiation must be specified. The energy or energy spectrum may be specified by giving the source of radiation.

12.3 For reactor radiation the neutron flux and neutron energy spectrum must be specified as well as the energy absorbed dose rate. The minimum description of the neutron flux will be the thermal neutron flux, the epithermal neutron flux, and the fast neutron flux (above 1 MeV). Procedures for these measurements have not yet been standardized.

13. Report

13.1 The report shall include the exposure procedure used, including pertinent details such as temperature medium, stress on specimen, post-irradiation storage, description of containers.

13.2 Irradiation conditions shall be reported as follows:

13.2.1 Type of radiation source and kind of radiation, including energy spectrum, if pertinent.

13.2.2 Irradiation dose rate, rads per hour. For accelerators list pulse repetition rate, duty cycle, and pulse peak energy; also list traverse cycle of specimen and "in-time" and "out-time".

13.2.3 Irradiation time.

13.2.4 Total dose in rads.

13.2.5 For reactors or other neutron sources report neutron exposure as neutrons per square centimeter for thermal, epithermal, and fast neutrons.

13.2.6 Reference to or description of irradiation dose measurement procedure.

13.2.7 Dimensional description of the test specimen size, shape, thickness, etc. shall be reported.

13.2.8 Description of the material tested shall be reported, including as much of the following pertinent information as is available:

13.2.1.1 Type and description of polymer. Nonirradiated properties: density, melting point, crystallinity, orientation, solubility, etc.

13.2.1.2 Formulation and compounding data: fillers, plasticizers, stabilizing agent, light absorbers, etc., if available.

13.2.1.3 Manufacturer, manufacturer's designation, trade name.

13.2.1.4 History of material at time of exposure: age, storage conditions, etc.

Designation: D 1692 – 68

Standard Method of Test for

FLAMMABILITY OF PLASTIC SHEETING AND CELLULAR PLASTICS[1]

This Standard is issued under the fixed designation D 1692; the number immediately following the designation indicates the year of original adoption or, in the case of revision, the year of last revision. A number in parentheses indicates the year of last reapproval.

1. Scope

1.1 This method covers a small scale laboratory screening procedure for comparing the relative flammability of plastic sheeting and cellular plastics. This method should be used solely to establish relative burning characteristics and should not be considered or used as a fire hazard classification.

1.2 Materials that exhibit pronounced shrinking, curling, or melting away upon heating cannot be evaluated by this test, unless they are preconditioned to prevent distortion of the specimen while under test.

1.3 This method is not applicable to materials that cannot be ignited under the conditions of this test, or to materials that exhibit progressive combustion without flame (continued glowing or charring).

NOTE —The values stated in U.S. customary units are to be regarded as the standard. The metric equivalents of U.S. customary units may be approximate.

2. Significance

2.1 Tests made on a material under conditions herein prescribed can be of considerable value in comparing the flammability characteristics of different materials, in controlling manufacturing processes, or as a measure of deterioration or change in flammability prior to or during use. Correlation with flammability under actual use conditions is not implied.

2.2 This method is not intended to be a criterion for fire hazard. Fire hazard more properly includes such other factors in addition to burning rate and flame spread as ease of ignition, fuel contribution, intensity of burning, products of combustion, and others.

2.2.1 Burning characteristics are affected by such factors as density and direction of rise in the case of cellular plastics, and the thickness of the specimen. These factors must be considered in order to compare materials on the same basis.

2.2.2 Round-robin results between five laboratories indicate that burning extent can be determined with a precision of ±2 mm (±0.1 in.).

3. Apparatus

3.1 *Test Chamber*—Any enclosure is satisfactory that is large enough to provide quiet draft-free air around the specimen during test, yet will permit normal thermal circulation of air past the specimen during burning. A hood is recommended in order to remove the sometimes noxious products of combustion. If a test chamber is used, it should be of such a design that it can be used in a hood. For referee purposes, test results with the chamber should be the same whether or not the hood exhaust is on. In cases of discrepancy, values obtained with the damper closed or the hood fan off, or both, will be considered valid.

3.1.1 The recommended test chamber should be constructed of sheet metal or other fire-resistant material, having inside dimensions 300 mm (12 in.) wide, 600 mm (24 in.) long, and 760 mm (30 in.) high, open at the top, with a ventilating opening approximately 25 mm (1 in.) high around the bottom. A viewing window of heat-resistant glass should be

[1] This method is under the jurisdiction of ASTM Committee D-20 on Plastics. A list of committee members may be found in the ASTM Yearbook. This standard is the direct responsibility of Subcommittee D-20.30 on Thermal Properties.

Current edition effective Sept. 9, 1968. Originally issued 1959. Replaces D 1692 – 67 T.

of sufficient size and in such a position that the entire length of the specimen under test may be observed. The chamber should be readily opened and closed to facilitate mounting and ignition of the test specimen.

3.2 *Burner*—A standard 9.5-mm (3/8-in.) barrel bunsen or Tirrill burner fitted with a 48 ± 1-mm (1 7/8 ± 1/16-in.) wide wing top.

3.2.1 The wing top opening may have to be adjusted to approximately 3.2 mm (1/8 in.) to provide the flame required in 6.3.

3.3 *Fuel Supply*—Propane gas of at least 85 percent purity.

3.4 *Specimen Support*—Hardware cloth of 6.4-mm (1/4-in.) mesh using 0.8-mm (1/32-in.) diameter steel wire. Cloth sample 76 by 216 mm (3 by 8 1/12 in.) shall have 13 mm (1/2 in.) of length bent to form a right angle. This will form the specimen support as shown in Fig. 1.

3.5 *Specimen Support Holders*—Any holding devices which will clamp the hardware-cloth specimen support horizontally so that the bottom of the bent-up portion is 13 mm (1/2 in.) above the top of the burner wing top, as shown in Fig. 1. A typical arrangement consists of two laboratory ring stands with two adjustable flat-surface clamps which may be locked in place by set screw or lock nut.

3.6 *Timing Device*, accurate to ±1 s.

3.7 *Measuring Scale.*

3.8 A device to ensure correct relative positioning of burner and specimen.

4. Test Specimens

4.1 Ten specimens 51 by 152 mm (2 by 6 in.) are needed.

4.1.1 Cellular plastic specimens shall be cut from uniform density material. Materials supplied in thicknesses over 13 mm (1/2 in.) shall be cut to 13 mm (1/2 in.) thickness and any skin removed. Materials foamed in thicknesses of 13 mm (1/2 in.) or less shall be tested at the thickness supplied. Care should be taken to remove all dust and cut particles from the foam surfaces.

4.1.2 Sheet samples shall be cut from a thickness of sheet normally supplied or compression molded to a desired thickness. Edges of the cut specimens shall be filed smooth.

4.2 Each test specimen shall be marked across its width by one line 25 mm (1 in.) from one end.

5. Conditioning

5.1 Unless otherwise agreed upon between the purchaser and the seller, materials shall be conditioned prior to test in an atmosphere having a temperature of 23 ± 2 C (73.4 ± 3.6 F) and a relative humidity of 50 ± 5 percent, so that successive weighings at 24-h intervals do not differ by more than 1 percent of the sample weight. Tests shall be made in this atmosphere or immediately after removal therefrom.

6. Procedure

6.1 Clamp the hardware-cloth specimen support horizontally so that the bottom of the hardware-cloth support is 13 mm (1/2 in.) above the top of the burner wing top, as shown in Fig. 1. Burn off any material remaining on the hardware cloth from previous tests, or use a new screen for each test. If a new screen is not used for each test, the old screen should be cool to the touch before being used.

6.2 Place the specimen on the support with one end touching the 13 mm (1/2-in.) bent-up portion of the support and with the longitudinal axis parallel to that of the support. The end of the specimen nearest the gage mark should be away from the bent-up end of the specimen support, so that the gage mark is 127 mm (5 in.) away from the bent-up end, as shown in Fig. 1.

6.3 Adjust the burner with the wing top to provide a blue flame whose visible portion is 38 mm (1 1/2 in.) high with a clearly defined inner cone 6 mm (1/4 in.) high. Place the burner under the bent-up end of the specimen support so that one edge of the flame is in line with the vertical section of hardware cloth and the other edge of the flame extends into the front edge of the specimen, as shown in Fig. 2. The center of the wing top shall be directly under the center line of the specimen.

6.4 Start the timing device when the flame is first applied to the specimen. After 60 s, remove the burner at least 150 mm (6 in.) away from the test specimen. Note the time, in seconds, when the flame reaches the gage mark, if this occurs, or when the flame is extinguished before reaching the gage mark, if this occurs. The time at which the flame front reaches the gage mark is designated t_g, and the time at which the flame is extinguished before reaching the gage mark is designated t_e, or "extinguishment time." If the flame is extinguished before reaching the gage mark, the "burning extent" is equal to 152 mm (6 in.) minus the distance from the unburned end to the nearest evidence of the flame front, such as charring, along the upper surface of the specimen. Note burning characteristics such as intumescence, melting, dripping, or smothering.

6.4.1 In some cases, extinguishment may occur in the first 60 s. In such cases, the disappearance of the yellow or characteristic flame shall be construed as extinguishment.

6.5 Discontinue testing if three specimens out of the ten have been found which burn past the gage mark.

7. Calculations

7.1 If the flame front passes the gage mark, the burning rate is equal to $762/t_g$ cm/min ($300/t_g$ in./min). If the flame is extinguished before reaching the gage mark, the burning rate is equal to 60 times the burning extent divided by the extinguishment time t_e.

7.2 If in all ten specimens the flame is extinguished before reaching the gage mark, the average burning rate, burning extent, and extinguishment time of all ten specimens shall be calculated.

7.3 If one or more specimens have been found which burn past the gage mark and others do not, the burning rate of the sample is equal to the average of the burning rates of the specimens which burn past the gage mark, and burning extent and extinguishment time are not calculated for the sample.

8. Report

8.1 Unless otherwise agreed between the purchaser and the seller, the report shall include the following:

8.1.1 A description of the material tested, including any prior treatment or conditioning of the material before testing, other than cutting and trimming.

8.1.2 Density and direction of rise of cellular materials: direction of rise can be reported as 13-mm (1/2-in.) parallel, 50-mm (2-in.) parallel, or 150-mm (6-in.) parallel, depending on the specimen dimension parallel to the foam rise direction.

8.1.3 Thickness of the specimens to the nearest 1 mm (1/32 in.), and presence or absence of skins.

8.1.4 Average, high, and low values of burning rate, in millimeters per minute (or inches per minute), to the nearest 2.5 mm/min (0.1 in./min).

8.1.5 If no specimens burn past the gage mark, average, high, and low values of burning extent in millimeters (inches) and extinguishment time in seconds. Report the sample as "self-extinguishing by this test" (SE) with subscript data specifying average burning extent to the nearest 2.5 mm (0.1 in.) and average extinguishment time to the nearest 1.0 s.

Examples:

$SE_{2.0 \text{ in., } 65 \text{ s}}$ and $SE_{0.9 \text{ in., } 38 \text{ s}}$

8.1.6 Number of specimens tested, and number of specimens in which the flame front did not reach the gage mark.

8.1.7 A description of burning characteristics such as warping, intumescence, melting, dripping, charring, or smothering.

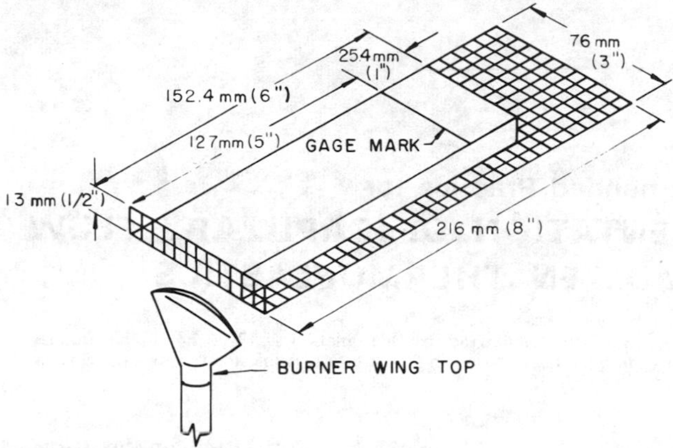

FIG. 1 Relative Positions of Burner Wing Top, Specimen, and Specimen Support.

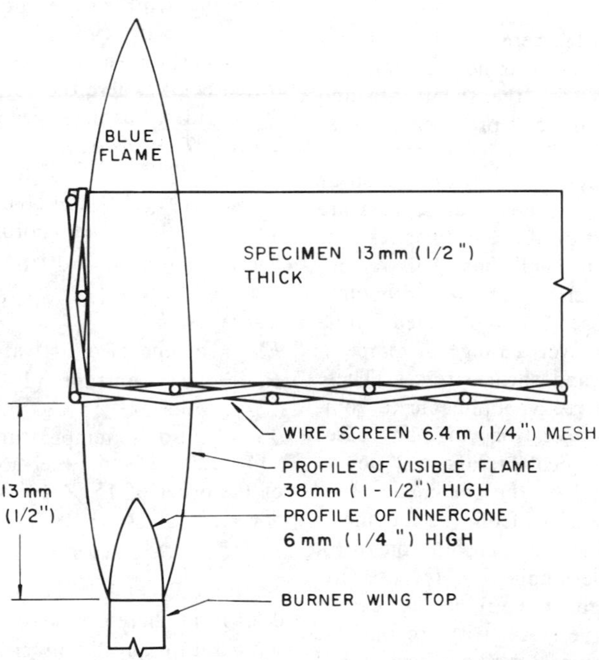

FIG. 2 Relative Positions of Burner Wing Top, Flame, Specimen, and Specimen Support.

Designation: D 1703 – 62

Recommended Practice for

PRESENTATION OF CAPILLARY FLOW DATA ON MOLTEN THERMOPLASTICS[1]

This Recommended Practice is issued under the fixed designation D 1703; the number immediately following the designation indicates the year of original adoption or, in the case of revision, the year of last revision. A number in parentheses indicates the year of last reapproval.

1. Scope

1.1 This recommended practice describes a preferred manner of reporting data on molten thermoplastics obtained from a capillary rheometer. Certain possible sources of experimental error are outlined.

2. Significance

2.1 At any given temperature the viscosity of a molten thermoplastic is not constant but varies considerably with the shear rate and shear stress. In commercial processing operations a thermoplastic is subjected to a wide range of temperatures, shear rates, and shear stresses. Therefore, it is desirable to measure the flow characteristics of thermoplastics over a range of extrusion conditions. This recommended practice describes a preferred manner of presenting rheological data obtained from a capillary rheometer over a range of temperatures, shear rates, and shear stresses. These data may not be directly applicable to commercial processing operations in many cases because some rheological factors may not be adequately simulated by the rheometer. Empirical correlations drawn from the rheometer data can be valuable, however, in control of the synthesis of thermoplastics, for user acceptance tests, and in engineering design problems. Past experience with rheometers that meet the requirements outlined below indicates that the rheological data can be obtained with a standard deviation, between instruments, of less than 10 percent.

3. Apparatus

3.1 Any rheometer in which molten thermoplastic can be forced from a reservoir through a capillary die and in which temperature, pressure, output rate, and die dimensions can be measured may be satisfactory.

4. Measurements and Tolerances

4.1 It is necessary to measure accurately at least the temperature, pressure, die dimensions, and output rate. The necessity for accurate measurement of these variables is illustrated below where the output rate is arbitrarily considered as the dependent variable.

4.1.1 *Temperature:*

4.1.1.1 The effect of temperature variation on output rate, Q, the other variables remaining constant, is given approximately by:

$$\text{Percent error in } Q = (dQ/Q) \times 100 = (E^*/R_g T_A^2) dT_A \times 100$$

where:

E^* = an energy of activation,
R_g = gas constant (2 cal/deg C · mole), and
T_A = absolute temperature, K.

4.1.1.2 For some thermoplastics E^* may be of the order of 15×10^3 cal/mole and T_A = 500 K. In this case:

$$\text{Percent error in } Q = 3\, dT_A$$

So for $dT_A = 0.2$ the percent error in Q is 0.6. It is therefore necessary that both the temperature and the precision of the tempera-

[1] This recommended practice is under the jurisdiction of ASTM Committee D-20 on Plastics. A list of committee members may be found in the ASTM Yearbook. This standard is the direct responsibility of Subcommittee D-20.30 on Thermal Properties.

Current edition effective Sept. 28, 1962. Originally issued 1959. Replaces D 1703 – 59 T.

ture measurement be known and reported.

NOTE—Temperature control to ±0.2 C at 190 C has been achieved (see ASTM Method D 1238, Test for Measuring Flow Rates of Thermoplastics by Extrusion Plastometer).[2] At higher temperatures this degree of control may be more difficult to obtain.

4.1.2 *Pressure:*

4.1.2.1 In non-Newtonian materials, such as thermoplastic melts, the output rate varies approximately as the pressure, P, raised to some power, b, greater than unity. Over a range of output rates b may not be constant. The effect of pressure variation on output rate, the other variables remaining constant, is given by:

$$\text{Percent error in } Q = (dQ/Q) \times 100 = b(dP/P) \times 100$$

4.1.2.2 Thus a 0.5 percent error in pressure measurement implies an error of $(b/2)$ percent in output rate. As the value of b can range from 1 to 3 a corresponding error in Q of 0.5 to 1.5 percent could result from this 0.5 percent error in P. It is therefore necessary that both the pressure and the precision of the pressure measurement be known and reported. With certain instruments it has been found possible to measure the pressure with a precision of ±0.5 percent or better.

4.1.3 *Capillary Dimensions:*

4.1.3.1 The b being defined as above, the output rate varies with the product $R^3 + b_L$ $-$ b, where R is the die radius and L the length of land. The error which arises in Q due to variations only in R and L is given by:

$$\text{Percent error in } Q = (dQ/Q) \times 100 = (3 + b)(dR/R) \times 100 - b(dL/L) \times 100$$

4.1.3.2 As the value of b can range from 1 to 3, the resultant error in Q due to a variation in R of ±0.5 percent can be ±2 to 3 percent, and the resultant error in Q due to variation in L of ±0.5 percent can be ±0.5 to 1.5 percent.

4.1.3.3 The precision with which die dimensions can be measured is dependent upon both the die radius and die length. In the case of Method D 1238, the die diameter is specified to ±0.25 percent and the die length to ±0.25 percent. With dies of much smaller diameter such precision is difficult. Due to the extreme sensitivity of flow data to die dimensions, it is most important that both the die dimensions and the precision with which the die dimensions are measured be known and reported.

4.1.4 *Output Rate*—The output rate and the precision with which the output rate is measured shall be known and reported.

5. Definitions

5.1 At a given temperature, or series of temperatures, values of output rates, Q (in cubic centimeters per second), and the corresponding values of applied pressure, P (in dynes per square centimeter), are obtained. These values of Q and P will depend on the capillary radius, R (in centimeters), and the capillary length, L(in centimeters). From these the quantities in Table 1 can be defined.

6. Treatment of Data

6.1 In some cases data may be reported in different terms than given in Section 5. For example, true shear rates, corrected for non-Newtonian flow behavior and true shear stresses, corrected for end effects or kinetic energy losses may be calculated. In such cases the exact details of the mode of correction must be reported.

7. Report

7.1 The report shall include the following information:

7.1.1 *Information Other than Flow Data:*

7.1.1.1 A description of the material being tested,

7.1.1.2 A description of the rheometer used,

7.1.1.3 The temperature at which the data were obtained and the precision of the temperature measurement,

7.1.1.4 The radius R, and the length to radius ratio, L/R, of the die land and the precision of these measurements,

7.1.1.5 A complete description of any die in which the entry or exit deviates from planes perpendicular to the axis of the die parallel, and

7.1.1.6 A statement as to any preconditioning which the sample has undergone.

7.1.2 *Flow Data*, which can be reported

---

[2] *Annual Book of ASTM Standards*, Part 27.

either in tabular or graphical form. The following graphical forms are acceptable:

7.1.2.1 Log shear stress *versus* log shear rate,

7.1.2.2 Log viscosity *versus* log shear stress or log shear rate,

7.1.2.3 Log viscosity *versus* the reciprocal of the absolute temperature,

7.1.2.4 Log viscosity *versus* the temperature in degrees Celsius,

7.1.2.5 Log critical shear stress or log critical shear rate *versus* the reciprocal of the absolute temperature, and

7.1.2.6 Log critical shear stress or log critical shear rate *versus* the temperature in degrees Celsius.

7.1.3 *Visual Observation*—In cases where observation is possible, distortion of the monofilament may be noted at or above a certain shear rate or stress. These values may or may not correspond to the critical shear rate and stress defined in Table 1. The data should be reported separately as "visual" critical shear rate and stress.

TABLE 1 Definitions

Quantity	Definition	Relation	Units
Shear Stress	Maximum apparent shear stress at the wall of the capillary.	$PR/2L$	dynes/cm^2
Shear Rate	Maximum apparent Newtonian shear rate at the wall of the capillary.	$4Q/\pi R^3$	s^{-1}
Viscosity	Apparent viscosity (ratio of shear stress to shear rate) at a given shear rate or shear stress.	$\pi PR^4/8LQ$	poise
Critical Shear Stress	The value of the shear stress at which there is a discontinuity in the slope of a log shear stress *versus* log shear rate plot.	From the flow curve of log shear stress *versus* log shear rate	dynes/cm^2
Critical Shear Rate	The shear rate corresponding to critical shear stress.	From the flow curve of log shear stress *versus* log shear rate	s^{-1}

Designation: D 1708 – 66

Standard Method of Test for
TENSILE PROPERTIES OF PLASTICS BY USE OF MICROTENSILE SPECIMENS[1]

This Standard is issued under the fixed designation D 1708; the number immediately following the designation indicates the year of original adoption or, in the case of revision, the year of last revision. A number in parentheses indicates the year of last reapproval.

1. Scope

1.1 This method covers the determination of the comparative tensile strength and elongation properties of plastics in the form of standard microtensile test specimens when tested under defined conditions of pretreatment, temperature, humidity, and testing machine speed. It can be used for specimens of any thickness up to 3.2 mm (1/8 in.), including thin films, and is designed specifically for use where only limited amounts of material are available. It is not intended to replace ASTM Methods D 882, Test for Tensile Properties of Thin Plastic Sheeting,[2] or ASTM Method D 638, Test for Tensile Properties of Plastics.[2] Where sufficient material is available, the latter shall be used.

1.2 This method cannot be used for the determination of modulus of elasticity. For the determination of modulus, see Method D 638 or D 882.

NOTE 1—The values stated in U.S. customary units are to be regarded as the standard. The metric equivalents of U.S. customary units may be approximate.

2. Significance

2.1 Tension tests provide data for research and development, engineering design, quality control, acceptance or rejection under specifications, and for special purposes. The tests cannot be considered significant for applications where the rate of straining differs widely from the rate used in this test. Such applications require suitable tests such as impact, creep, and fatigue. Data obtained at several levels of temperature, humidity, time, or other variables, however, usually furnish reasonably accurate indications of the behavior of the material.

2.2 Results obtained by this method are essentially comparable to results obtained on sheet materials by Methods D 638 and D 882, except in the case of brittle or notch-sensitive materials where imperfections caused by preparation techniques may cause premature failures (Note 2). For direct comparisons, however, all results should be obtained by a single procedure on specimens of comparable thickness and prepared in a comparable manner, unless such variables have been shown to be negligible for the materials in question.

NOTE 2—This has been proved only for thermoplastic materials where the microtensile specimens were prepared by milling or die-cutting from the sheet. It has not been proved for molded microtensile specimens or for thermosetting materials or reinforced laminates.

3. Definitions

3.1 Definitions of terms applying to this method appear in the Appendix to Method D 638.

4. Apparatus

4.1 The apparatus shall be as specified in Method D 638, with the following exceptions:

4.1.1 *Grips*—Serrated grips should be used with care, since yielding or tearing at the grips may interfere with measurement of

[1] This method is under the jurisdiction of ASTM Committee D-20 on Plastics. A list of committee members may be found in the ASTM Yearbook. This standard is the direct responsibility of Subcommittee D-20.10 on Mechanical Properties.

Current edition effective Sept. 20, 1966. Originally issued 1959. Replaces D 1708 – 59 T.

[2] *Annual Book of ASTM Standards*, Part 27.

elongation even when the specimen breaks in the reduced section. Rubber-faced grips are recommended for thin specimens. Self-tightening grips of the "V"-design type are not satisfactory for this test because of the change in grip separation that occurs as they bite on the specimen. If the specimen tab is not long enough to prevent the grip faces from cocking, shims should be inserted to provide more uniform clamping.

4.1.2 *Drive Mechanism*—The velocity of the drive mechanism shall be regulated as specified in Section 8.

4.1.3 The fixed and movable members, drive mechanism, and grips should be constructed of such materials and in such proportions that, after grip slack is taken up, the total elastic longitudinal deformation of the system constituted by these parts does not exceed 1 percent of the total longitudinal deformation between the grips at any time during the test. If this is not possible, appropriate corrections shall be made in the calculation of strain values.

4.1.4 *Extension Indicator*—The extension indicator shall be capable of determining the distance between grips at any time during the test. The instrument shall be essentially free of inertia lag at the specified speed of testing, and shall be accurate to ±1 percent of extension or better.

NOTE 3—It is desirable that the load indicator and the extension indicator be combined into one instrument, which automatically records the load as a function of the extension or as a function of time. In the latter case, the conversion to a load-extension record can readily be made because extension is proportional to time after the take-up of the initial grip slack.

NOTE 4—Extension may also be measured by timing the test with a stop watch and calculating the distance of crosshead movement during that time. Time shall be taken from the instant that the machine records a load on the specimen to the instant the specimen breaks.

4.1.5 *Micrometers*—Micrometers shall read to 0.0025 mm (0.0001 in.) or less.

5. Test Specimens

5.1 Microtensile test specimens shall conform to the dimensions shown in Fig. 1. This specimen shall be prepared by die-cutting or machining from sheet, plate, slab, or finished article. Dimensions of a die suitable for preparing die-cut specimens are shown in Fig. 2.

5.2 All surfaces of the specimen shall be free from visible flaws, scratches, or imperfections. Marks left by coarse machining operations shall be carefully removed with a fine file or abrasive, and the filed surfaces shall then be smoothed with abrasive paper (No. 00 or finer). The finishing sanding strokes shall be made in the direction parallel to the long axis of the test specimen.

NOTE 5—Tabs shown in Fig. 1 are minimum size for adequate gripping. Shims may be required with thicker specimens to keep grips from cocking. Handling is facilitated and gripping improved by the use of larger tabs wherever possible.

6. Conditioning

6.1 *Conditioning*—Condition the test specimens at 23 ± 2 C (73.4 ± 3.6 F) and 50 ± 5 percent relative humidity for not less than 40 h prior to test in accordance with Procedure A of ASTM Methods D 618, Conditioning Plastics and Electrical Insulating Materials for Testing,[2] for those tests where conditioning is required. In cases of disagreement, the tolerances shall be ±1 C (±1.8 F) and ±2 percent relative humidity.

6.2 *Test Conditions*—Conduct tests in the Standard Laboratory Atmosphere of 23 ± 2 C (73.4 ± 3.6 F) and 50 ± 5 percent relative humidity, unless otherwise specified in the test methods or in this specification. In cases of disagreements, the tolerances shall be ±1 C (±1.8 F) and ±2 percent relative humidity.

7. Number of Test Specimens

7.1 At least five specimens shall be tested for each sample in the case of isotropic materials.

7.2 Ten specimens, five normal to and five parallel to the principal axis of anisotropy, shall be tested for each sample in the case of anisotropic materials.

7.3 Results obtained on specimens that break at some obvious fortuitous flaw or at the edge of the grips shall be discarded and retests made, unless such flaws constitute a variable, the effect of which it is desired to study.

8. Speed of Testing

8.1 Speed of testing is the velocity of separation of the two members (or grips) of the testing machine when running idle (under no load).

8.2 The speed of testing shall be chosen such that the rate of straining shall be approximately the same as the rate of straining obtained when the material is tested at the designated speed according to Method D 638. Speeds giving rates of straining approximating those given in Methods D 638 are as follows:

Speed A	0.25 mm (0.01 in.)/min
Speed B	1 to 1.3 mm (0.04 to 0.05 in.)/min
Speed C	10 to 13 mm (0.4 to 0.5 in.)/min
Speed D	100 to 130 mm (4 to 5 in.)/min

These speeds are 0.20 to 0.25 times the speeds designated in Method D 638, since the effective gage length of bars specified in the latter method is 4 to 5 times that of the microtensile specimens. When the speed of testing is not specified, Speed B shall be used.

9. Procedure

9.1 Specimens shall be tested at the Standard Laboratory Atmosphere as defined in Methods D 618.

9.2 Measure and record the minimum value of the cross-sectional area of each specimen. Measure the width to the nearest 0.025 mm (0.001 in.), and the thickness to the nearest 0.0025 mm (0.0001 in.) for specimens less than 2.5 mm (0.1 in.) thick, or to the nearest 0.025 mm (0.001 in.) for specimens 2.5 mm (0.1 in.) or greater in thickness.

9.3 Set the testing machine so that the distance between the grips is 22.9 ± 0.25 mm (0.900 ± 0.010 in.).

NOTE 6—This may easily be checked by the use of a 22.86-mm (0.900-in.) gage block or a pair of inside calipers.

9.4 Place the specimen in the grips of the testing machine with the inside edge of each tab visible at the edge of the grip, taking care to align the long axis of the specimen and the grip with an imaginary line joining the points of attachment of the grips to the machine. Tighten the grips evenly and firmly to the degree necessary to prevent slippage of the specimen during the test, but not to the point where the specimen would be crushed.

9.5 Set the speed control at the speed desired (8.2) and start the machine.

9.6 Record the load at the yield point (if one exists), the maximum load carried by the specimen during the test, the load at rupture, and the elongation (extension between grips) at the moment of rupture.

10. Calculations

10.1 *Yield Strength, Tensile Strength, and Tensile Strength at Break*—Calculate the yield strength, tensile strength, and tensile strength at break in accordance with Method D 638.

10.2 *Percentage Elongation at Break*—Calculate the percentage elongation at break by dividing the elongation (extension) at the moment of rupture of the specimen by the original distance between tabs 22.250 ± 0.051 mm (0.876 ± 0.002 in.), and multiplying by one hundred. Report the percentage elongation to two significant figures.

10.3 *Percentage Elongation at the Yield Point*—Calculate the percentage elongation at the yield point, if desired, by dividing the elongation (extension) at the yield point by the original distance between tabs 22.250 ± 0.051 mm (0.876 ± 0.002 in.), and multiplying by one hundred.

10.4 Calculate the "average value" and standard deviation for each property in accordance with Method D 638.

11. Report

11.1 The report shall include the following:

11.1.1 Complete identification of the material tested, including type, source, manufacturer's code numbers, form, principal dimensions, previous history, and other pertinent information.

11.1.2 Method of preparing test specimens,

11.1.3 Specimen thickness,

11.1.4 Conditioning procedure used,

11.1.5 Atmospheric conditions in test room,

11.1.6 Number of specimens tested,

11.1.7 Speed of testing,

11.1.8 Yield strength (if any), average value and standard deviation,

11.1.9 Tensile strength, average value and standard deviation,

11.1.10 Tensile strength at break, average value and standard deviation,

11.1.11 Percentage elongation at break, average value and standard deviation,

11.1.12 Percentage elongation at the yield point, average value and standard deviation (if desired), and

11.1.13 Date of test.

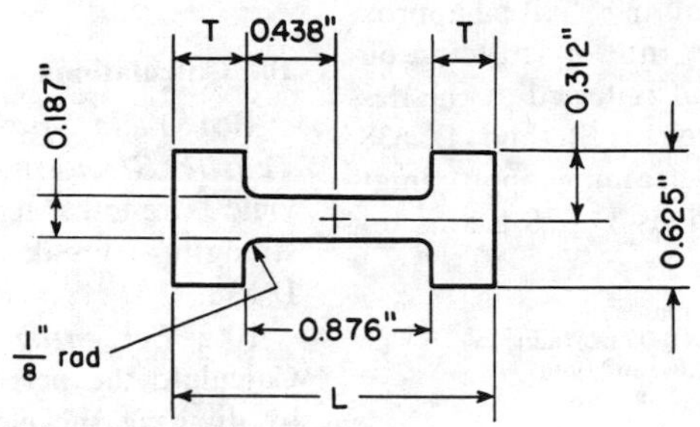

Tolerance: ±0.051 mm (±0.002 in.)
Thickness = thickness of sheet.
Minimum tab length, T = 7.9 mm (0.312 in.) (Larger tabs shall be used wherever possible.)
Minimum length, L = 38.1 mm (1.50 in.)

Table of Metric Equivalents

in.	mm
1/8	3.2
0.187	4.75
0.438	11.13
0.312	7.92
0.625	15.88
0.876	22.25

FIG. 1 Microtensile Test Specimen.

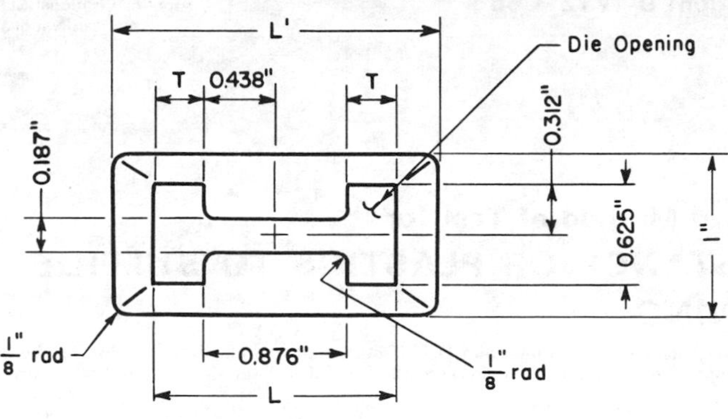

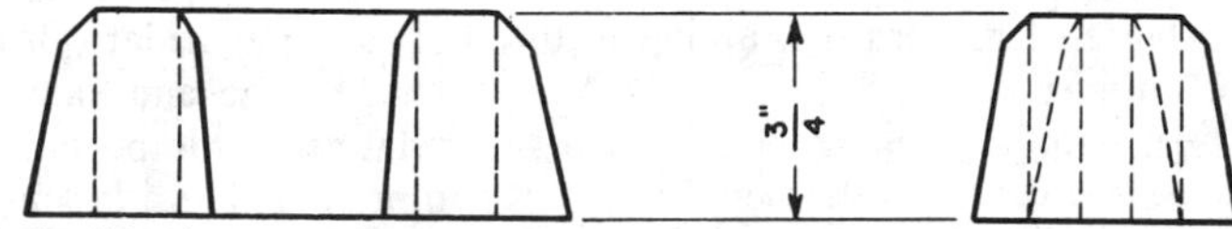

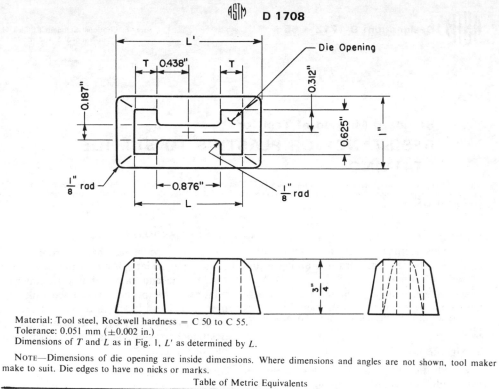

Material: Tool steel, Rockwell hardness = C 50 to C 55.
Tolerance: 0.051 mm (±0.002 in.)
Dimensions of T and L as in Fig. 1, L' as determined by L.

NOTE—Dimensions of die opening are inside dimensions. Where dimensions and angles are not shown, tool maker make to suit. Die edges to have no nicks or marks.

Table of Metric Equivalents

in.	1/8	0.187	0.312	0.438	0.625	3/4	0.876	1
mm	3.2	4.75	7.92	11.13	15.88	19.05	22.25	25.4

FIG. 2 Die for Die-Cutting Microtensile Specimens.

Designation; D 1712 - 65

American National Standard K65.33-1967
American National Standards Institute

Standard Method of Test for RESISTANCE OF PLASTICS TO SULFIDE STAINING[1]

This Standard is issued under the fixed designation D 1712; the number immediately following the designation indicates the year of original adoption or, in the case of revision, the year of last revision. A number in parentheses indicates the year of last reapproval.

1. Scope

1.1 This method covers the determination of the resistance of plastics to staining in the presence of sulfides.

NOTE 1—The values stated in U.S. customary units are to be regarded as the standard. The metric equivalents of U.S. customary units may be approximate.

2. Significance

2.1 Plastic compositions containing salts of lead, cadmium, copper, antimony, and certain other metals (as stabilizers, pigments, driers, or fillers) may stain due to the formation of a metallic sulfide when in contact with external materials that contain sulfide. The external sulfide source may be liquid, solid, or gas. Examples of materials that may cause sulfide stains are rubber, industrial fumes, foods, kraft paper, etc. This method provides a means of estimating the relative susceptibility of plastic composition to sulfide staining.

3. Reagent

3.1 *Hydrogen Sulfide Solution*—A freshly prepared saturated solution of hydrogen sulfide (**Caution,** see Note 2). Prepare the solution by rapidly bubbling hydrogen sulfide gas (Note 3) through water. Five minutes of bubbling is sufficient for 100 to 150 ml of water at room temperature (approximately 23 C).

NOTE 2: **Caution**—Hydrogen sulfide is a highly toxic gas and must be handled only in suitably ventilated area such as a hood. Avoid breathing of vapors.

NOTE 3—Hydrogen sulfide gas may be obtained commercially as compressed gas in cylinders, or may be conveniently prepared in the laboratory by the reaction of an acid with a sulfide or by heating a tube containing sulfur and paraffin wax.[2]

4. Test Specimens

4.1 Test specimens shall be representative of the particular plastic composition being tested. Size and shape of test specimens are relatively unimportant. Specimens 100 ± 25 mm (4 ± 1 in.) in length by 13 ± 6 mm (0.5 ± 0.25) in width by the thickness of the composition being tested, have been found suitable for this test.

5. Conditioning

5.1 *Conditioning*—Condition the test specimens at 23 ± 2 C (73.4 ± 3.6 F) and 50 ± 5 percent relative humidity for not less than 40 h prior to test in accordance with Procedure A of ASTM Methods D 618, Conditioning Plastics and Electrical Insulating Materials for Testing,[3] for those tests where conditioning is required. In cases of disagreement, the tolerances shall be ±1 C (±1.8 F) and ±2 percent relative humidity.

5.2 *Test Conditions*—Conduct tests in the Standard Laboratory Atmosphere of 23 ± 2 C (73.4 ± 3.6 F) and 50 ± 5 percent relative humidity, unless otherwise specified in the test methods or in this specification. In cases of disagreements, the tolerances shall be ±1 C (±1.8 F) and ±2 percent relative humidity.

---

[1] This method is under the jurisdiction of ASTM Committee D-20 on Plastics. A list of committee members may be found in the ASTM Yearbook. This standard is the direct responsibility of Subcommittee D-20.50 on Permanence Properties.

Current edition effective Aug. 31, 1965. Originally issued 1960. Replaces D 1712 - 60 T.

[2] Such tubes are commercially available under the name of Aitch-Tu-Ess from the Hengar Co., 1711 Spruce St., Philadelphia, Pa. 19103.

[3] *Annual Book of ASTM Standards*, Part 27.

6. Procedure

6.1 Using a 250-ml beaker or equivalent, immerse approximately half of each specimen (Note 3) in the saturated hydrogen sulfide solution for 15 min (Note 4). Cover the test container with a watch glass or aluminum foil during the test.

NOTE 4—It is recommended that a control material, whose tendency to sulfide stain is known, be included with each test series. This provides a reference point from series to series. Staining may be rated as more, less, or equal to that of the control.

NOTE 5—If desired, additional specimens may be tested for different periods of time. Suggested periods of immersion are 5, 15, 30, 60, and 120 min. If immersion times greater than 30 min are desired, remove the specimens from the reagent each 30 min and again bubble hydrogen sulfide gas through the solution for 3 min. Reimmerse the specimens and continue the test. Do not dry the specimens during the time they are removed for reconcentration of the solution. Then time compensation will not be necessary for the period the specimens are not immersed.

6.2 After immersion for 15 min, remove the specimens, wipe dry, and examine for discoloration of the immersed section compared to a sample of the identical plastic composition not exposed to hydrogen sulfide solution.

6.3 Compare the relative degree of staining for each material being tested in a series, and establish their relative order of sulfide stain resistance.

7. Report

7.1 The report shall include the following:

7.1.1 Complete identification of the material tested, including type, source, manufacturer's code numbers, form, previous history, and other pertinent information.

7.1.2 Duration of exposure, and

7.1.3 Effects of exposure, including whether or not staining occurred, color of stain, and severity of staining in relation to other materials in the series or in relation to a control material, if included.

Designation: D 1746 - 70

Standard Method of Test for
TRANSPARENCY OF PLASTIC SHEETING[1]

This Standard is issued under the fixed designation D 1746; the number immediately following the designation indicates the year of original adoption or, in the case of revision, the year of last revision. A number in parentheses indicates the year of last reapproval.

1. Scope

1.1 This method covers the measurement of the transparency of plastic sheeting in terms of specular transmittance (T_s). Although generally applicable to any translucent or transparent material, it is principally intended for use with nominally clear and colorless thin sheeting.

NOTE 1—The values stated in U.S. customary units are to be regarded as the standard. The metric equivalents of U.S. customary units may be approximate.

2. Significance

2.1 The attribute of clarity of a sheet, measured by its ability to transmit image-forming light, correlates with its specular transmittance. Sensitivity to differences improves with decreasing incident beam- and receptor-angle. If the angular width of the incident beam and of the receptor aperture (as seen from the specimen position) are of the order of 0.1 deg or less, films of commercial interest have a range of transparency of about 1 to 90 percent as measured by this test. Results obtained by the use of this method are greatly influenced by the design parameters of the instruments, for example, the resolution is largely determined by the angular width of the receptor aperture. Caution should therefore be exercised in comparing results obtained from different instruments, especially for samples with low specular transmittance.

2.2 Specular transmittance data according to this method correlate with the property commonly known as "see-through," which is rated subjectively by the effect of a hand-held specimen on an observer's ability to distinguish clearly a relatively distant target. This correlation is poor for highly diffusing materials because of interference of scattered light in the visual test.

3. Definitions

3.1 *transmittance*—the ratio of the radiant flux transmitted by a specimen to the radiant flux incident on the specimen.

3.2 *specular transmittance*—the transmittance value obtained when the measured transmitted radiant flux includes only that transmitted in essentially the same direction as that of the incident flux. The specular transmittance may equal the total transmittance but cannot exceed it. (For this method, limitations on the geometry of the optical system are specified in Section 4.)

4. Apparatus

4.1 The apparatus shall consist of a light source, source aperture, lens system, specimen holder, receptor aperture, photoelectric detector, and an indicating or recording system, arranged to measure specular transmittance. The system shall meet the following requirements:

4.1.1 An incandescent or vapor-arc lamp, with a regulated power supply such that fluctuations in light intensity shall be less than ± 1 percent. If an arc lamp is used, an appropriate filter shall be used to limit light only to the spectral range from 540 to 560 nm.

4.1.2 A system of apertures and lenses shall be used that will provide a symmetrical incident beam. When measured with the indicat-

[1] This method is under the jurisdiction of the ASTM Committee D-20 on Plastics. A list of committee members may be found in the ASTM Yearbook. This standard is the direct responsibility of Subcommittee D-20.40 on Optical Properties.

Current edition effective Jan. 22, 1970. Originally issued 1960. Replaces D 1746 - 62 T.

ing or recording system of the apparatus, using a receptor aperture having a width or diameter subtending an angle of 0.025 ± 0.005 deg at the plane of the specimen, the incident beam shall meet the following requirements:

Angle, deg	Maximum Relative Intensity
0	100
0.05	10
0.1	1
0.3	0.1

The source aperture may be circular or a rectangular slit having a length-to-width ratio of at least 10.

4.1.3 A holder shall be provided that will secure the specimen so that its plane is normal to the axis of the incident beam at a fixed distance from the receptor aperture. Provision must be made for rotating the specimen if slit optics are used. Provision for transverse motion may be provided to facilitate replication of measurements.

4.1.4 An aperture shall be provided over the receptor so that its diameter or width subtends an angle, at the plane of the specimen, of 0.1 ± 0.025 deg. The image of the source aperture with no specimen in place shall be the same shape as the receptor aperture, centered on and entirely within it.

4.1.5 A photoelectric detector shall be provided such that the indicated or recorded response to incident light shall be substantially a linear function and uniform over the entire range from the unobstructed beam (I_o) to 0.01 I_o or less.

4.1.6 Means shall be provided for relatively displacing the receptor or the image of the source aperture continuously (in the plane of the receptor aperture) by at least 1 deg from the optical axis of the undeviated incident beam; for circular apertures, in two directions at right angles to each other; for slit optics, in the direction of the short dimension of the slit.

NOTE 2—This provision is necessary for checking the geometry of the incident beam (4.1.2) and for readjusting for maximum light intensity in the event that the beam is deviated by a specimen with non-parallel surfaces.

NOTE 3—Apparatus meeting these requirements has been described in the literature,[2] and commercial versions are available.

5. Standards

5.1 Two or more clear glass filters of known spectral and total luminous transmittance, CIE Source C, measured in accordance with ASTM Method D 791, Test for Luminous Reflectance, Transmittance, and Color of Materials,[3] shall be used as standards for checking the photometric accuracy of the instrument. Transmittances shall be chosen so that check points are about evenly spaced between 0 and 90 percent. If a monochromatizing filter is used, the standard transmittance shall be selected from the spectral plot at the proper wavelength.

5.2 Two or more specimens of known specular transmittance (T_s) shall be provided as standards to check for changes in mechanical alignment.

NOTE 4—Optical misalignment of an instrument may not be detected by a photometric check as in 5.1. *Carefully protected* plastic films or glass slides with one rough surface have been used as standards.[4]

6. Test Specimens

6.1 All specimens should be nominally colorless (Note 5) and transparent to translucent, have essentially plane parallel surfaces, and be free of surface or internal contamination.

NOTE 5—Transparency of colored or highly reflective materials may be measured by the ratio of T_s/T_t, where T_t is the total luminous transmittance (see Method D 791 or ASTM Method D 1003, Test for Haze and Luminous Transmittance of Transparent Plastics.[5]

6.2 Nonrigid specimens must be held in a suitable holder so that they are flat and free from wrinkles.

7. Conditioning

7.1 *Conditioning*—Condition the test specimens at 23 ± 2 C (73.4 ± 3.6 F) and 50 ± 5 percent relative humidity for not less than 40 h prior to test in accordance with Procedure A of ASTM Methods D 618, Conditioning Plas-

[2] Webber, Alfred C., "Method for the Measurement of Transparency of Sheet Materials," *Journal of the Optical Society of America*, JOSAA, Vol 47, No. 9, September 1957, pp. 785–789.

[3] Discontinued, see *1967 Book of ASTM Standards*, Part 30.

[4] "Perrocolor" 2 by 2-in. slide binders distributed by E. Leitz, Inc., 468 Park Ave., New York, N. Y. 10016 and available in most photo supply stores, have been found suitable.

tics and Electrical Insulating Materials for Testing,[5] for those tests where conditioning is required. In cases of disagreement, the tolerances shall be 1 C (1.8 F) and ±2 percent relative humidity.

7.2 *Test Conditions*—Conduct tests in the Standard Laboratory Atmosphere of 23 ± 2 C (73.4 ± 3.6 F) and 50 ± 5 percent relative humidity, unless otherwise specified in the test methods or in this specification. In cases of disagreements, the tolerances shall be 1 C (1.8 F) and ±2 percent relative humidity.

8. Procedure

8.1 Check the instrument for:

8.1.1 Zero reading, no light reaching detector.

8.1.2 100 percent reading, light on, receptor centered.

8.1.3 Accuracy of photometric readings, using calibrated standards.

8.1.4 Accuracy of alignment, using standard films.

8.2 Adjust the instrument to read a maximum by carefully moving the receptor so that it receives the maximum flux from the light beam. Set the indicator or recorder to 100 or read the intensity with no specimen in the beam (I_o).

8.3 Mount the test specimen in the holder so that it is neither wrinkled nor stretched, and place the holder in the instrument in the desired position, centered and normal to the light beam.

8.4 Readjust the receptor position so that a maximum value is indicated or recorded. If slit optics are used, the specimen must also be rotated in its own plane to a maximum value of transmitted light. Note the reading (I_s).

8.5 Repeat with other specimens or other positions of specimen as desired.

9. Calculation

9.1 Calculate the specular transmittance, T_s, as follows:

$$T_s = 100 I_s / I_o$$

where:

I_s = light intensity with the specimen in the beam, and

I_o = light intensity with no specimen in the beam.

NOTE 6—No calculation is needed if I_o is set to 100 percent or a conversion chart or special scale is used to interpret the instrument reading.

10. Report

10.1 The report shall include the following:

10.1.1 Sample designation,

10.1.2 Instrument used,

10.1.3 Average specular transmittance,

10.1.4 Number of specimens tested,

10.1.5 Standard deviation, and

10.1.6 Any measured anisotropy.

---

[5] *Annual Book of ASTM Standards*, Part 27.

ASTM Designation: D 1790 – 62

Standard Method of Test for
BRITTLENESS TEMPERATURE OF PLASTIC FILM BY IMPACT[1]

This Standard is issued under the fixed designation D 1790; the number immediately following the designation indicates the year of original adoption or, in the case of revision, the year of last revision. A number in parentheses indicates the year of last reapproval.

1. Scope

1.1 This method covers the determination of that temperature at which plastic film 0.25 mm (10 mils) or less in thickness exhibits a brittle failure under specified impact conditions.

NOTE 1—The values stated in U.S. customary units are to be regarded as the standard. The metric equivalents of U.S. customary units may be approximate.

NOTE 2—This method was developed jointly with the Society of the Plastics Industry, primarily for use with plasticized vinyl films. Its applicability to other plastic films must be verified by the user.

2. Significance

2.1 Data obtained by this method may be used to predict the behavior of plastic film at low temperatures only if the conditions of deformation are similar to those specified in the method. This method has been useful for specification testing and for comparative purposes, but does not necessarily measure the lowest temperature at which the plastic film may be used.

2.2 It has been demonstrated that on calendered film the brittleness temperature is sensitive to the direction of fold. It has also been demonstrated that the brittleness temperature is sensitive to the direction of sampling. Therefore, it is imperative that the procedure for taking specimens and folding be explicitly followed.

3. Definition

3.1 *brittleness temperature*—that temperature statistically calculated where 50 percent of the specimens would probably fail 95 percent of the time when a stated minimum number are tested by this method. The 50 percent failure temperature may be determined by calculation, as described in 8.1, or by plotting the data on probability graph paper, as described in 8.2.

4. Apparatus

4.1 *Conditioning Equipment* conforming to Grade B requirements, or better, of ASTM Specifications E 197, for Enclosures and Servicing Units for Tests Above and Below Room Temperature,[2] shall be considered standard equipment for this test.

4.2 *Impact Tester*[3] shown in Fig. 1 shall be constructed of cold-rolled steel except for the bolts, screws, and rubber stopper. All structural parts (that is, base, anvil, arm, arm supports, and shaft) may be chromium plated. The arm, including the rubber stopper and bolt, shall weigh 3.09 kg ± 28 g (6 lb 13 ± 1 oz).

4.3 *Die*, 50.8 ± 0.4 mm by 146.1 ± 0.4 mm (2 ± $^1/_{64}$ in. by 5$^3/_4$ ± $^1/_{64}$ in.) to cut the test specimens from the sample.

4.4 *Stapler*, desk-type, with a stop mounted on the base exactly 12.7 mm ($^1/_2$ in.) back of the center of the groover that turns the staple.

4.5 *Index Card Stock*, 50 by 127 mm (2 by 5 in.).

---

[1] This method is under the jurisdiction of ASTM Committee D-20 on Plastics. A list of committee members may be found in the ASTM Yearbook. This standard is the direct responsibility of Subcommittee D-20.30 on Thermal Properties.

Current edition effective Sept. 28, 1962. Originally issued 1960. Replaces D 1790 – 60 T.

[2] *Annual Book of ASTM Standards*, Part 30.

[3] The tester is available from the United States Testing Co., 1415 Park Ave., Hoboken, N.J.

5. Test Specimens

5.1 The test specimens shall be cut with the 50.8 by 146.1-mm (2 by 5 3/4-in.) die. The long dimension of the test specimen shall be parallel with the transverse direction of the film. The 50-mm (2-in.) ends of each specimen shall be so collected that a gradual, closed loop is formed at room temperatures. The collected ends of the specimen and one end of the 50 by 127-mm (2 by 5-in.) card shall be matched exactly with the body of the loop lying on the card. Two staples shall be crimped through the matched ends of the specimen and card. These staples shall be 13 mm (1/2 in.) from, and parallel to, the 50-mm (2-in.) end of the stack.

5.2 Since it has been demonstrated that this test is sensitive to the direction of fold on calendered film, it is necessary to stipulate that all specimens be folded in the same direction. It is recommended that the specimen be folded so that the surface last in contact with the roll shall be inside the loop.

6. Conditioning

6.1 *Conditioning*—Condition the test specimens at 23 ± 2 C (73.4 ± 3.6 F) and 50 ± 5 percent relative humidity for not less than 40 h prior to test in accordance with Procedure A of ASTM Methods D 618, Conditioning Plastics and Electrical Insulating Materials for Testing,[4] for those tests where conditioning is required. In cases of disagreement, the tolerances shall be ±1 C (±1.8 F) and ±2 percent relative humidity.

6.2 *Test Conditions*—Conduct tests in the Standard Laboratory Atmosphere of 23 ± 2 C (73.4 ± 3.6 F) and 50 ± 5 percent relative humidity, unless otherwise specified in the test methods or in this specification. In cases of disagreements, the tolerances shall be ±1 C (±1.8 F) and ±2 percent relative humidity.

7. Procedure

7.1 Place the impact tester, with its base horizontal, in the cold conditioning chamber at the desired testing temperature, for a conditioning period of 1 h. Expose the specimens to the test temperature for 15 ± 5 min.

7.2 Place each of ten mounted test specimens individually on the anvil with the crimped ends of the staples on the back of the card fitted into the groove in the anvil. Handle the card, not the sample. Locate the bent loop of the film on the anvil with the loop centered on the anvil and facing away from the pivotal end of the impact arm.

7.3 Allow the impact arm to fall free from a position within 5 deg of perpendicular to the base onto the bent loop. Repeat until each of the ten conditioned specimens has been tested.

7.4 Remove the specimens and examine for failure. Partial fracture shall be construed as failure as well as complete division into two or more pieces.

7.5 Adjust the temperature of the cold conditioning chamber and condition a second group of ten specimens for 15 ± 5 min at a suitable temperature as indicated by the results of testing the first group of ten specimens. Repeat the testing procedure and examination.

7.6 When using the standard method of calculation, it is necessary to repeat this procedure of testing ten specimens at a series of temperatures differing by uniform increments of 2 or 5 C. If the alternate graphic method is used, it is not necessary that the increments be uniform. When using the calculation described in 8.1 it shall be necessary to obtain the lowest no-failure temperature and the highest all-failure temperature, and it is desirable to obtain two additional temperatures, one on either side of the 50 percent failure temperature. When using the probability graph paper in 8.2, it is not necessary to obtain the lowest no-failure temperatures or the highest failure temperature.

7.7 When testing for conformance to a specification, it shall be satisfactory to accept materials on the basis of testing a minimum of ten test specimens at the specified temperature. Not more than half of the specified number of test specimens shall fail at the specified temperature.

8. Calculations

8.1 *Standard Method*—Using the number of specimen failures, calculate the percentage of failures at each temperature. Calculate the 50 percent failure temperature, T_b in degrees

[4] *Annual Book of ASTM Standards*, Part 27.

Celsius as follows:

$$T_b = T_h + \Delta T\,[(S/100) - (1/2)]$$

where:

T_h = highest temperature at which failure of all specimens occurs, deg C (correct algebraic sign must be used),

ΔT = temperature increment, deg C, and

S = sum of the percentage of breaks at each temperature (over the range of temperatures from the lowest no-failure temperature down to and including T_h).

8.2 *Alternative Graphic Method*—Plot the data on probability graph paper with the temperature on the linear scale and the percentage failure on the probability scale. Select the temperature scale so that it represents a minimum of two divisions per degree. Draw the best straight line through a minimum of four points, two above and two below the 50 percent point. The temperature indicated at the intersection of the data line with the 50 percent probability line shall be reported as T_b, the 50 percent failure temperature.

9. Report

9.1 The report shall include the following:

9.1.1 Complete identification of the material tested, including type, source, manufacturer's code designation, form, and previous history,

9.1.2 Fifty percent failure temperature, to the nearest degree Celsius,

9.1.3 Method of calculation,

9.1.4 Thickness of test specimen,

9.1.5 Direction of fold,

9.1.6 Conditioning procedure, and

9.1.7 Date of test.

9.2 For routine acceptance testing, report the following instead of 9.1.1, 9.1.2, and 9.1.3:

9.2.1 Number of specimens tested,

9.2.2 Temperature of test, and

9.2.3 Number of failures.

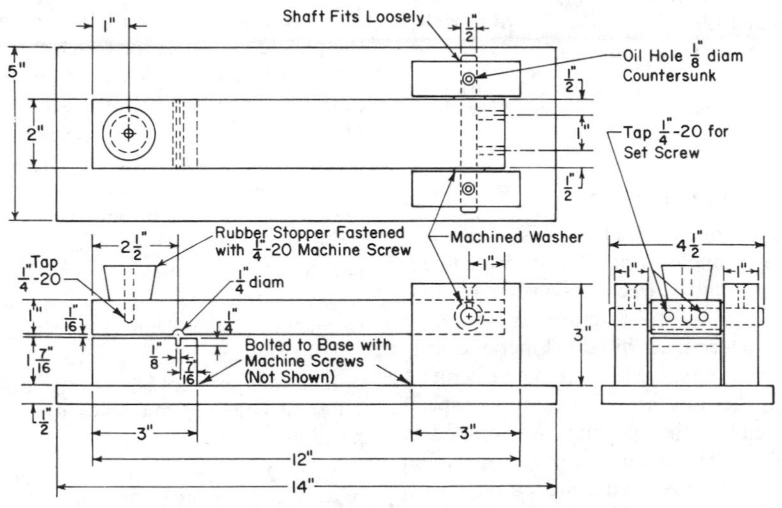

FIG. 1 **Brittleness Temperature Impact Testing Machine.**

Metric Equivalents

in.	1/8	1/4	7/16	1/2	1	2 1/2	3	4 1/2	12	14
mm	3.2	6.4	11.1	12.7	25.4	64	76	114	305	356

ASTM Designation: D 1822 - 68

Standard Method of Test for
TENSILE-IMPACT ENERGY TO BREAK PLASTICS AND ELECTRICAL INSULATING MATERIALS[1]

This Standard is issued under the fixed designation D 1822; the number immediately following the designation indicates the year of original adoption or, in the case of revision, the year of last revision. A number in parentheses indicates the year of last reapproval.

1. Scope

1.1 This method covers the determination of the energy required to rupture standard tension-impact specimens of plastic or electrical insulating materials. Materials that can be tested by this method are those too flexible or too thin to be tested in accordance with ASTM Methods D 256, Test for Impact Resistance of Plastics and Electrical Insulating Materials,[2] as well as more rigid materials.

NOTE 1—The values stated in U.S. customary units are to be regarded as the standard. The metric equivalents of U.S. customary units may be approximate.

2. Summary of Method

2.1 The energy utilized in this method is delivered by a single swing of a calibrated pendulum of a standardized tension-impact machine. The energy to fracture by shock in tension is determined by the kinetic energy extracted from the pendulum of an impact machine in the process of breaking the specimen. One end of the specimen is mounted in the pendulum. The other end of the specimen is gripped by a crosshead which travels with the pendulum until the instant of impact and instant of maximum pendulum kinetic energy, when the crosshead is arrested.

3. Significance

3.1 Tensile-impact energy is the energy required to break a standard tension-impact specimen in tension by a single swing of a standard calibrated pendulum under a set of standard conditions (Note 2). In order to compensate for the minor differences in cross-sectional area of the specimens as they will occur in the preparation of the specimens, the energy to break is normalized to units of kilogram-centimeters per square centimeter (foot-pounds per square inch) of minimum cross-sectional area. For a perfectly elastic material the impact energy might be reported per unit volume of material undergoing deformation. However, since much of the energy to break the plastic materials for which this test method is written is dissipated in drawing of only a portion of the test region, such normalization on a volume basis is not feasible. The method permits two specimen geometries so that the effect of elongation or rate of extension or both upon the result can be observed. With the Type S (short) specimen the extension is comparatively low, while with the Type L (long) specimen the extension is comparatively high. In general, the Type S specimen (with its greater reproducibility, but less differentiation among materials. Results obtained with different capacity machines may not be comparable.

NOTE 2—Friction losses are largely eliminated by careful design and proper operation of the testing machine. Attention is drawn to ASTM Methods E 23, for Notched Bar Impact Testing of Metallic Materials,[3] for a general discussion of impact equipment and procedures.

---

[1] This method is under the jurisdiction of ASTM Committee D-20 on Plastics. A list of committee members may be found in the ASTM Yearbook. This standard is the direct responsibility of Subcommittee D-20.10 on Mechanical Properties.

Current edition effective Aug. 15, 1968. Originally issued 1961. Replaces D 1822 - 61 T.

[2] *Annual Book of ASTM Standards*, Part 27.

[3] *Annual Book of ASTM Standards*, Part 31.

3.2 The scatter of data may be due to different failure mechanisms within a group of specimens. Some materials may exhibit a transition between different failure mechanisms; if so, the elongation will be critically dependent on the rate of extension encountered in the test. The impact energy values for a group of such specimens will have an abnormally large dispersion. Some materials retract at failure with insignificant permanent set. With such materials it may not be possible to determine the type of failure, ductile or brittle, by examining the broken pieces. A set of specimens may sometimes be sorted into two groups by observing the broken pieces to ascertain whether or not there was necking during the test. Qualitatively, the strain rates encountered here are intermediate between the high rate of the Izod test of Method D 256 and the low rate of usual tension testing in accordance with ASTM Method D 638, Test for Tensile Properties of Plastics.[2]

3.3 The energy for fracture is a function of the force times the distance through which the force operates. Thus, two materials may have properties that result in equal tensile-impact energies on the same specimen geometry, arising in one case from a large force associated with a small elogation and in the other from a small force associated with a large elongation. It cannot be assumed that this test will correlate with other tests or end uses unless such a correlation has been established by experiment.

3.4 Comparisons among specimens from different sources can be made with confidence only to the extent that specimen preparation, for example, molding history, has been precisely duplicated. Comparisons between molded and machined specimens must not be made without first establishing quantitatively the differences inherent between the two methods of preparation.

3.5 Only results from specimens of nominally equal thickness and tab width shall be compared unless it has been shown that the tensile-impact energy normalized to kilogram-centimeters per square centimeter (foot-pounds per square inch) of cross-sectional area is independent of the thickness over the range of thicknesses under consideration.

3.6 Slippage of specimens results in erroneously high values. The tabs of broken specimens should be examined for an undistorted image of the jaw faces optically, preferably under magnification, and compared against a specimen which has been similarly clamped but not tested. Because slippage has been shown to be present in many cases and suspected in others, the use of bolted specimens is mandatory. The function of the bolt is to assure good alignment and to improve the tightening of the jaw face plates.

3.7 The bounce of the crosshead supplies part of the energy to fracture the tests specimen (see Appendix A1).

4. Apparatus

4.1 The machine shall be of the pendulum type shown schematically in Figs. 1 and 2. The base and suspending frame shall be of sufficiently rigid and massive construction to prevent or minimize energy losses to or through the base and frame. The pendulum should be released from such a position that the linear velocity of the center of impact (center of percussion) at the instant of impact shall be approximately 3.444 m/s (11.3 ft/s), which corresponds to an initial elevation of this point of 609.6 mm (2 ft).

4.2 The pendulum shall be constructed of a single- or multiple-membered arm holding the head, in which the greatest mass is concentrated. A rigid pendulum is essential to maintain the proper clearances and geometric relationships between related parts and to minimize energy losses, which always are included in the measured impact energy value. It is imperative that the center of percussion of the pendulum system and the point of impact can be demonstrated to be coincident within ± 2.54 mm (± 0.100 in.) and that the point of contact occur in the neutral (free hanging) position of the pendulum within 2.54 mm (0.100 in.), both with and without the crosshead in place.

NOTE 3—The distance from the axis of support to the center of percussion may be determined experimentally from the period of small amplitude oscillations of the pendulum by means of the following equation:

$$L = (g/4\pi^2)\, p^2$$

where:

L = distance from the axis of support to the center of percussion, mm (ft),

g = local gravitational acceleration (known to an

accuracy of one part in one thousand), in mm/s[2] (ft/s[2]),

π = 3.14159, and

p = period, s, of a single complete swing (to and fro) determined from at least 50 consecutive and uninterrupted swings (known to one part in two thousand). The angle of swing shall be less than 5 deg each side of the center.

4.3 The positions of the rigid pendulum and crosshead clamps on the specimen are shown in Fig. 2. The crosshead should be rigid and light in weight. The crosshead shall be supported by the pendulum so that the test region of the specimen is not under stress until the moment of impact, when the specimen shall be subjected to a pure tensile force. The clamps shall have serrated jaws to prevent slipping. Jaws should have file-like serrations and the size of serrations should be selected according to experience with hard and tough materials and with the thickness of the specimen. The edge of the serrated jaws in close proximity to the test region shall have a 0.397 mm ($^1/_{64}$-in.) radius to break the edge of the first serrations.

4.4 A means shall be provided for measuring the energy absorbed from the pendulum by measuring on a suitable scale the height of the pendulum swing after the break.

4.5 Accurate means shall be available to determine and minimize energy losses due to windage and friction.

4.6 Setup and calibration procedures for tension-impact machines shall be followed as described in Appendix A2.

4.7 A ball-type micrometer shall be used for measuring the width of the restricted area of the Type S specimen. Either a ball-type or ordinary machinist's micrometer may be used to measure the thickness of the Type S specimen and the thickness and width of the Type L specimen. These measurements shall be made to an accuracy of 0.013 mm (0.0005 in.).

5. Test Specimen

5.1 At least five and preferably ten specimens from each sample shall be prepared for testing. For sheet materials that are suspected of anisotropy, duplicate sets of test specimens shall be prepared having their long axis respectively parallel with, and normal to, the suspected directions of anisotropy.

5.2 The test specimen shall be sanded, machined, or die cut to the dimensions of one of the specimen geometries shown in Fig. 3, or molded in a mold whose cavity has these dimensions. Figure 4A shows bolt holes and bolt hole location and Fig. 4B shows a slot as an alternative method of bolting for easy insertion of the specimens into the grips. The No. 8-32 bolt size is recommended for the 9.53-mm (0.375-in.) wide tab and No. 8-32 or No. 10-32 bolt size is suggested for the 12.70-mm (0.500-in.) wide tabs. Final machined, cut, or molded specimen dimensions cannot be precisely maintained because of shrinkage and other variables in sample preparation.

5.3 A nominal thickness of 3.175 mm ($^1/_8$ in.) is optimum for most materials being considered and for commercially available machines. Thicknesses other than 3.175 mm ($^1/_8$ in.) are nonstandard and they should be reported with the tension-impact value.

NOTE 4—Cooperating laboratories should agree upon standard molds and upon specimen preparation procedures and conditions.

6. Conditioning

6.1 *Conditioning*—Condition the test specimens at 23 ± 2 C (73.4 ± 3.6 F) and 50 ± 5 percent relative humidity for not less than 40 h prior to test in accordance with Procedure A of ASTM Methods D 618, Conditioning Plastics and Electrical Insulating Materials for Testing,[2] for those tests where conditioning is required. In cases of disagreement, the tolerances shall be ±1 C (±1.8 F) and ±2 percent relative humidity.

6.2 *Test Conditions*—Conduct tests in the Standard Laboratory Atmosphere of 23 ± 2 C (73.4 ± 3.6 F) and 50 ± 5 percent relative humidity, unless otherwise specified in the test methods or in this specification. In cases of disagreement, the tolerances shall be ±1 C (±1.8 F) and ±2 percent relative humidity.

7. Procedure

7.1 Measure the thickness and width of each specimen with a micrometer. Record these measurements along with the identifying markings of the respective specimens.

7.2 Bolt the specimen securely with a torque wrench in accordance with 4.3. Clamp the specimen to the crosshead while the crosshead is out of the pendulum. A jig may be necessary for some machines to position the specimen properly with respect to the crosshead during the bolting operation. With the crosshead properly positioned in the elevated pendulum, bolt the specimen at its other end to the pendulum itself, as shown in Fig. 1.

7.3 Use the lowest capacity pendulum available, unless the impact values go beyond the 85 percent scale reading. If this occurs, use a higher capacity pendulum.

NOTE 5—In changing pendulums, the tensile-impact energy will decrease as the mass of the pendulum is increased.

7.4 Measure the tension-impact energy of each specimen and record its value, and comment on the appearance of the specimen regarding permanent set or necking, and the location of the fracture.

8. Calculations

8.1 Calculate the corrected impact energy to break as follows:

$$X = E = Y + e$$

where:

X = corrected impact energy to break, in kg · cm (ft · lb),
E = scale reading of energy of break, in kg · cm (ft · lb),
Y = friction and windage correction in kg · cm (ft · lb) (Fig. 5).
e = bounce correction factor, in kg · cm (ft · lb) (Fig. 5).

NOTE 6—Figure 5 is a sample curve. A curve must be calculated in accordance with Appendix A1 for the crosshead and pendulum used before applying any bounce correction factors.

NOTE 7—*Examples:*

Case A—*Low-Energy Specimen:*

Scale reading of energy to break 5.947 kg · cm (0.43 ft · lb)

Friction and windage correction = −0.277 kg · cm (−0.02 ft ·lb)
Bounce correction factor, *e*, (from Fig. 5 in Appendix A1) +2.489 kg · cm (+0.18 ft · lb)
= +2.213 kg · cm (+0.16 ft · lb)
} +2.213 kg · cm (0.16 ft · lb)

Corrected impact energy to break 8.160 kg · cm (0.59 ft · lb)

Case B—*High-Energy Specimen:*

Scale reading of energy to break 23.788 kg · cm (1.72 ft · lb)

Friction and windage correction = −0.138 kg · cm (−0.01 ft · lb)
Bounce correction factor, *e* (from Fig. 5 in Appendix A1) +3.458 kg · cm
+3.319 kg · cm (+0.24 ft · lb)
} +3.319 kg · cm (0.24 ft · lb)

Corrected impact energy to break 27.107 kg · cm (1.96 ft · lb)

8.2 Calculate the standard deviation (estimated) as follows and report to two significant figures:

$$s = \sqrt{(\Sigma X^2 - n\bar{X}^2)/(n - 1)}$$

where:

s = estimated standard deviation,
X = value of single observation,
n = number of observations, and
$\bar{X}$ = arithmetic mean of the set of observations.

9. Report

9.1 The report shall include the following:

9.1.1 Complete identification of the material tested, including type, source, manufacturer's code number, form, principal dimensions, and previous history,

9.1.2 Specimen type (S or L), and tab width.

9.1.3 A statement of how the specimens were prepared, the testing conditions, including the size of the bolts and torque used, thickness range, and direction of testing with respect to anisotropy, if any.

9.1.4 The capacity of the pendulum in kilogram-centimeters (foot-pounds or inch-pounds).

9.1.5 The average and the standard deviation of the tensile-impact energy of specimens in the sample. If the ratio of the minimum value is less than 0.75, report average and maximum and minimum values. If there is an apparent difference in the residual elongation observed due to some of the sample necking, report the number of specimens displaying necking.

9.1.6 Number of specimens tested per sample or lot of material (that is, five or ten or more).

10. Precision[4]

10.1 In round-robin tests of triplicate bolted specimens the nine participating laboratories averaged the standard deviations shown in Table 1.

[4] Supporting data for this method have been filed at ASTM Headquarters as RR 47: D-20.

TABLE 1 Round-Robin Calibration Tests[4]

Kind of Specimen	Tensile-Impact Strength		Approximate Standard Deviation	
	kg·cm/cm²	(ft·lb/in.²)	kg·cm/cm²	(ft·lb/in.²)
Poly(methyl methacrylate) 3/8-in. tab 1/8-in. neck	42.6	20	5.1	2.4
Poly(methyl methacrylate) 1/2-in. tab 1/4-in. neck[a]	46.9	22	2.8	1.3

[a] Nonstandard geometry.

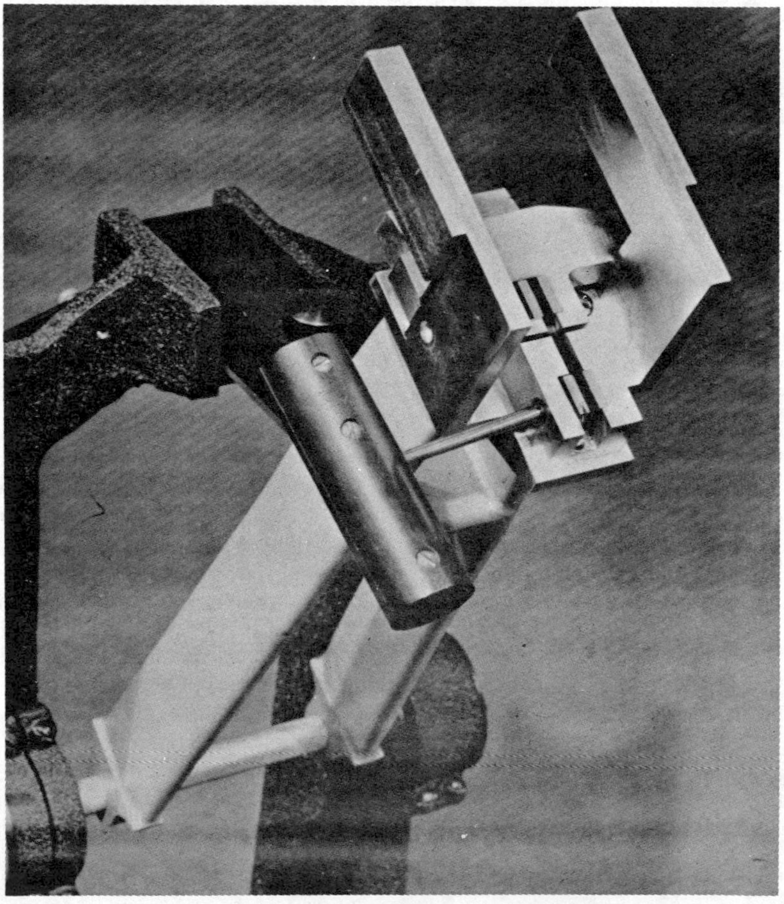

FIG. 1 Specimen-in-Head Tension-Impact Machine.

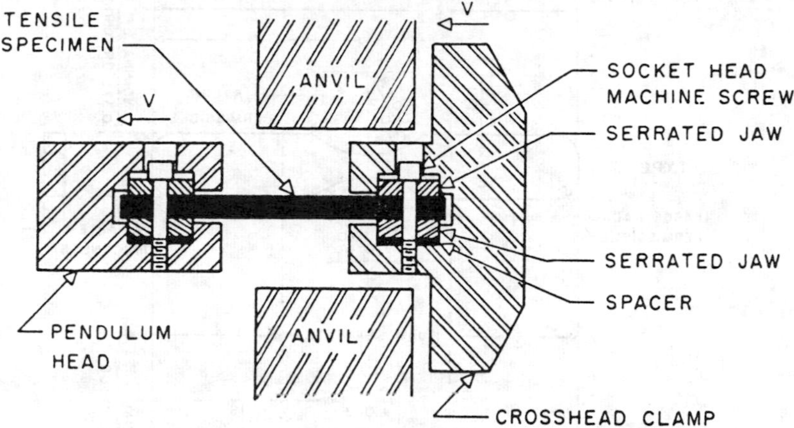

FIG. 2 Specimen-in-Head Tension-Impact Machine (Schematic).

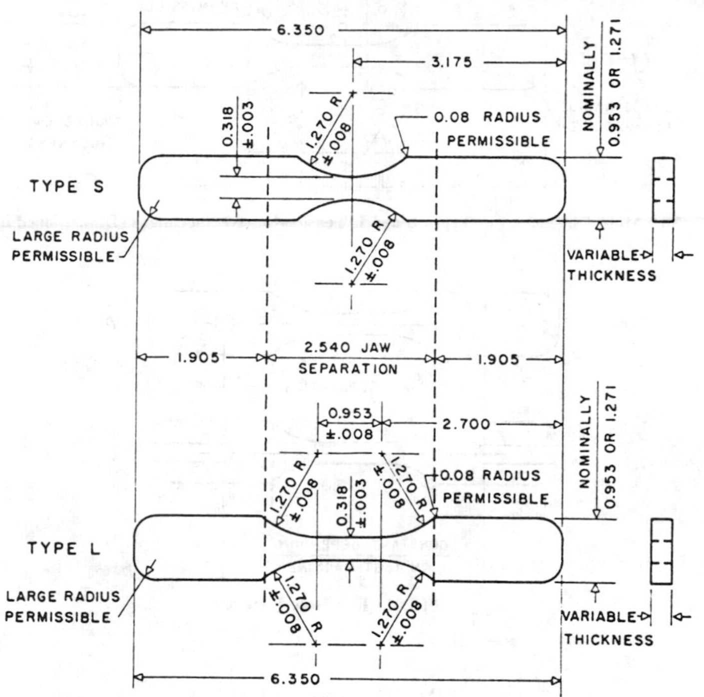

FIG. 3A Mold Dimensions of Types S and L Tension-Impact Specimens (Dimensioned in Centimeters).

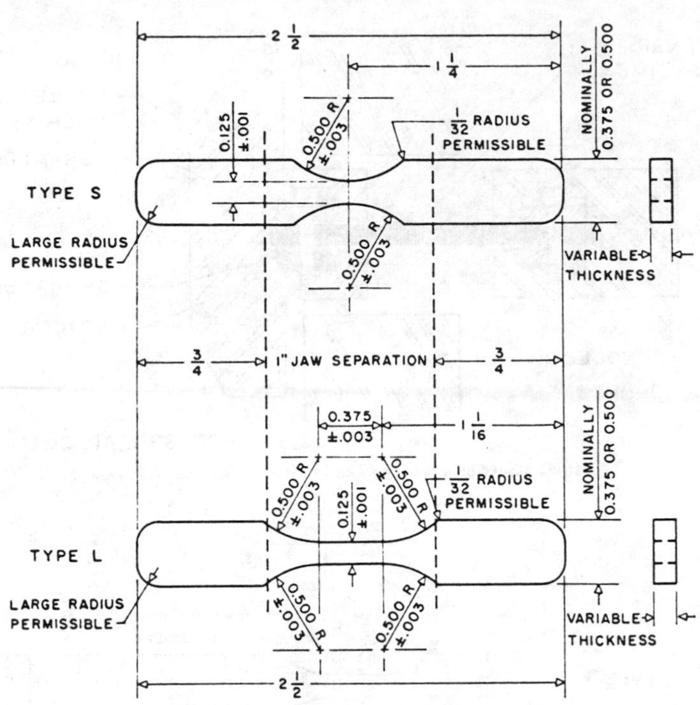

FIG. 3B Mold Dimensions of Types S and L Tension-Impact Specimens (Dimensioned in Inches).

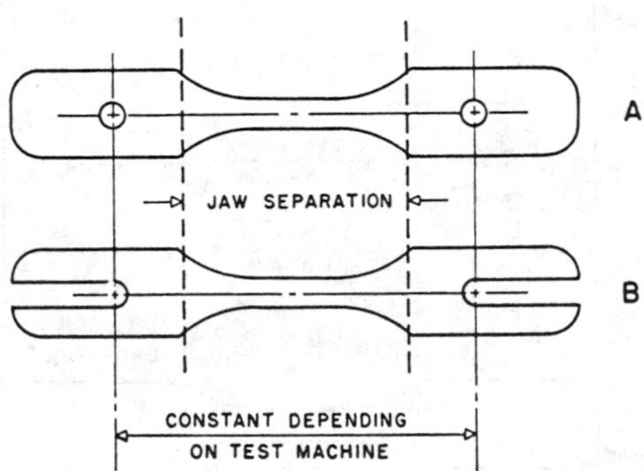

FIG. 4 Bolt Hole Location.

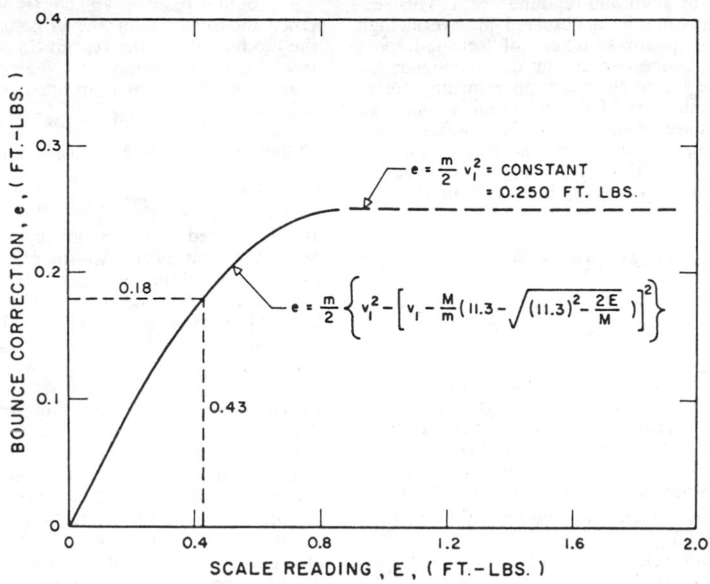

$$e = \frac{m}{2}v_1^2 = \text{CONSTANT} = 0.250 \text{ FT. LBS.}$$

$$e = \frac{m}{2}\left\{v_1^2 - \left[v_1 - \frac{M}{m}\left(11.3 - \sqrt{(11.3)^2 - \frac{2E}{M}}\right)\right]^2\right\}$$

FIG. 5 Typical Correction Factor Curve for Single Bounce of Crosshead for Specimen-in-Head. Tension-Impact Machine, 5 ft · lb Hammer, 0.428 lb Steel Crosshead (see Appendix A1).

APPENDIXES

A1. Determination of Bounce Velocity and Correction Factor

A1.1 General

A1.1.1 Upon contacting the anvil at the bottom of the swing of the pendulum, the crosshead bounces away with an initial velocity dependent upon the degree of elasticity of the contacting surface. The elastic compression and expansion of the metallic crosshead, both of which take place prior to the separation of the crosshead from the anvil, occur in a time interval given approximately by twice the crosshead thickness divided by the speed of sound in the metal of which the crosshead is made. This is usually of the order of 1 in. divided by 200,000 in. (5080 m)/s, or about 5×10^{-6} s. During this time the crosshead, moving at about 135 in. (3.4 m/s, moves along about 7×10^{-4} in. For a test specimen with a modulus of 500,000 psi and a specific gravity of 1.0 the sound speed in the sample would be only 70,000 in. (1778 m)/s and a stress wave would move only about 0.4 in. (10.160 mm) in 5×10^{-6} s. Thus in this short time a stress wave would not have traveled through the plastic specimen to the end of the specimen clamped to the pendulum, and so the specimen would exert no retarding force on the crosshead at the instant it bounced away. Since this is the case, one can assume that the initial rebound velocity of the crosshead, v_1, is the same as that which is measured with no specimen in the pendulum.

A1.2 Determination of Bounce Velocity

A1.2.1 The determination of bounce velocity, v_1, of the free crosshead can be made by photographic analysis (high-speed movies or stroboscopic techniques) or by the coefficient of restitution method.

A1.2.2 It has been observed that in several cases the rebound velocity of the crosshead is about 6.2 ft/s. It has also been noted that under certain geometrical conditions, the coefficient of restitution of steel on steel is about 0.55 (Eshbach's *Handbook of Engineering Fundamentals*). Since 0.55×11.3 ft (3.44 m)/s = 6.2 ft (1.88 m)/s, it appears that an approximate value of the rebound velocity might be taken as 6.2 ft/s for steel crossheads, if the use of high speed moving picutres is impractical. However, the preferred method of determining crosshead rebound velocity is by photographic analysis.

A1.3 Determination of Correction Factor

A1.3.1 After impact and rebound of the crosshead the specimen is pulled by two moving bodies, the pendulum with an energy of $MV^2/2$, and the crosshead with an energy of $mv^2/2$. When the specimen breaks, only that energy is recorded on the pendulum dial which is lost by the pendulum. Therefore, one must add the incremental energy contributed by the crosshead to determine the true energy used to break the specimen. Consider once again the moving crosshead before the specimen breaks. As the crosshead moves away from the anvil it is slowed down by the specimen, which is being stretched. If the specimen does not break very quickly, the crosshead velocity will diminish to zero and theoretically the crosshead could be brought back against the anvil and rebound again. This second bounce has not been observed in several high speed moving pictures taken of tensile-impact breaks, but if it does occur one can no longer assume that the specimen exerts no retarding force, and the determination of the crosshead velocity on the second bounce becomes relatively complex.

A1.3.2 If only a single bounce occurs, one can calculate the correction (that is, the incremental energy contributed by the crosshead) as follows:

By definition

$$E = (M/2)(V_2 - V_2^2) \quad (1)$$

and by definition:

$$e = (m/2)\,(v_1^2 - v_2^2) \quad (2)$$

where:

M = mass of pendulum, kg · s²/cm (lb · s²/ft),
m = mass of crosshead, kg · s²/cm (lb · s²/ft),
V = maximum velocity of center of percussion of of crosshead of pendulum, cm/s (ft/s),
V_2 = velocity of center of percussion of pendulum at time when specimen breaks, cm/s (ft/s),
v_1 = crosshead velocity immediately after bounce, cm/s (ft/s),
v_2 = crosshead velocity at time when specimen breaks, cm/s (ft/s),
E = energy read on pendulum dial, kg · cm (ft · lb), and
e = energy contribution of crosshead, that is, bounce correction factor to be added to pendulum reading, kg · cm (ft · lb).

Once the rebound of the crosshead has occurred, the momentum of the system (in a horizontal direction) must remain constant. Neglecting vertical components of the momentum one can write:

$$MV - mv_1 = MV_2 - mv_2 \quad (3)$$

Equations 1, 2, and 3 can be combined to give:

$$e = m/2\,\{v_1^2 - [v_1 - (M/m) \cdot (V - \sqrt{(V)^2 - (2E/M)})]^2\} \quad (4)$$

If e is plotted as a function of E (for fixed values of V, M, m, and v_1), e will increase from zero, pass through a maximum (equal to $mv^2/2$), and decrease, passing again through zero and becoming negative. The only part of this curve for which a reasonably accurate analysis has been made is the initial portion where the curve lies between values of zero and $mv_1^2/2$. Once the crosshead reverses the direction of its travel, the correction becomes less clearly defined, and after a second contact with the anvil has been made, the correction becomes much more difficult to evaluate. It is assumed, therefore, for the sake of simplicity, that once e has reached its maximum value, the correction factor will remain constant at a value of $mv_1^2/2$. It should be clearly realized that the use of that portion of the curve in Fig. 4 where e is constant does not give an accurate correction. However, as E grows larger, the correction factor becomes relatively less important and no great sacrifice of over-all accuracy results from the assumption that the maximum correction is $mv_1^2/2$.

A2. Set-Up and Calibration Procedure for Low Capacity 14 to 220 Kg · Cm (1 to 16 Ft · Lb), Tension-Impact Machines for Use with Plastic Specimens

A2.1 Locate impact machined on a sturdy bench. It shall not "walk" on the bench and the bench shall not vibrate appreciably. Loss on energy from vibrations will give high readings. It is recommended that the impact tester be bolted to the bench weighing at least 23 kg (50 lb) if it is used at capacities higher than 28 kg · cm (2 ft · lb).

A2.2 Check the levelness of the machine in both directions in the plane of the base with spirit levels mounted in the base, by a machinist's level if a satisfactory reference surface is available, or with a plumb bob. The machine should be level to within $\tan^{-1}$ 0.001 in the plane of swing and to within $\tan^{-1}$ 0.002 in the plane perpendicular to the swing.

A2.3 Check for signs of rubbing or interference between the pendulum head and the anvil and check the side clearance between the pendulum head and anvil while the pendulum hangs freely. Unequal side clearances may indicate a bent pendulum arm, bent shaft, or faulty bearings. Excessive side play may also indicate faulty bearings. If free-hanging side clearances are equal, but there are signs of interference the pendulum may not have sufficient rigidity. Readjust the shaft bearings, relocate the anvil, or straighten the pendulum shaft as necessary to attain the proper relationship between the pendulum head and the anvil.

A2.4 Check the pendulum arm for straightness within 1.2 mm (0.05 in.) with a straightedge or by sighting down the shaft. This arm is sometimes bent by allowing the pendulum to slam against the catch when high-capacity weights are on the pendulum.

A2.5 Swing the pendulum to a horizontal position and support it by a string clamped in the vise of the pendulum head so that the string is positioned exactly on the longitudinal centerline of the specimen. Attach the other end of the string to a suitable load-measuring device. The pendulum weight should be within 0.4 percent of the required weight for that pendulum capacity. If weight must be added or removed, take care to balance the added or removed weight about the center of percussion. It is not advisable to add weight to the opposite side of the bearing axis from the head to increase the effective length of the pendulum since the distributed mass will lead to large energy losses from vibration of the pendulum.

A2.6 Calculate the effective length of the pendulum arm, or the distance to the center of percussion from the axis of rotation, by the procedure of Note 3. The effective length must be within 1 percent of the distance from the center of rotation to the striking edge.

A2.7 Measure the vertical distance of fall of the pendulum center of percussion from the trip height to its lowest point. This distance should be 610 ± 2 mm (24 ± 0.1 in.). The vertical falling distance may be adjusted by varying the position of the pendulum latch.

A2.8 When the pendulum is in the position where the crosshead just touches the anvil the pointer should be at the full scale index within 0.2 percent of scale.

A2.9 The pointer friction should be adjusted so that the pointer will just maintain its position anywhere on the scale. The striking pin of the pointer should be securely fastened to the pointer. The friction device should be adjusted in accordance with the recommendations of the manufacturer.

A2.10 The free swing reading of a 28-kg · cm (2-ft · lb) pendulum (without specimen) from the tripping height should be less than 2.5 percent of scale on the first swing. If the reading is higher than this then the pointer friction is excessive or the bearings are dirty. To clean the bearings dip them in grease solvent and spin dry in an air jet. Clean the bearings until they spin freely, or replace them. Oil very lightly with instrument oil before replacing. A reproducible method of starting the pendulum from the proper height must be devised.

A2.11 The position of the pointer after three swings of the pendulum, each from the starting position, without manual readjustment of the pointer should be between $^1/_2$ and 1 percent of the scale. If the readings differ from this, then the machine is not level, the calibration dial is out of alignment, or the pendulum finger is out of calibration position.

A2.12 The shaft about which the pendulum rotates shall have no detectable radial play (less than 0.05 mm (0.002 in.)). An end play of 0.25 mm (0.010 in.) is permissible when a 1-kg (2.2-lb) axial force is applied in alternate directions. This shaft shall be horizontal within $\tan^{-1}$ 0.003 as checked with a level.

A2.13 The center of the anvil faces shall lie in a plane parallel to the shaft axis of the pendulum within $\tan^{-1}$ 0.001. The anvil faces shall be parallel to the transverse and vertical axes of the pendulum within $\tan^{-1}$ 0.001. One side of the crosshead shall not make contact with the anvil later than 0.05 mm (0.002 in.) after the other side has made contact. This measurement can be made by holding the crosshead in place in the pendulum head with the hand and feeling click of the sides of the crosshead against the anvil while thin shims are inserted between the side of the crosshead and the anvil. If the crosshead is not being contacted evenly check the crosshead for nicks and burrs and then the pendulum for twisting.

A2.14 The top of the machine base and the approach to the anvil should be surfaced with a soft rubber or plastic material having a low coefficient of friction with the crosshead so that the crosshead can slide or bounce free of the anvil after impact. This is to ensure that the crosshead does not come in contact with the pendulum on the backswing. Otherwise considerable damage to the crosshead and pendulum may result.

A2.15 The machine should not be used to indicate more than 85 percent of the energy capacity of the pendulum.

A2.16 A jig shall be provided for locating the specimen so that it will be parallel to the base of the machine at the instant of strike within $\tan^{-1}$ 0.01 and coincident with the center of strike within 0.25 mm (0.01 in.).

A2.17 For checking the accuracy and reliability of the tension-impact machines use standard specimens made from 5052 H-32 aluminum. For low-capacity tension-impact machines 0.458-mm (0.020-in.) thick specimens may be used and for high-capacity tension-impact machines 1.27-mm (0.050-in.) specimens are recommended. In round-robin calibration tests on ten standard "L" specimens the six participating laboratories average the standard deviation shown in Table A1.

A2.18 In converting tension-impact units use Table A2 to multiply the quantity in the units on the left by the number in the center to convert to the units on the right.

TABLE A1 Round-Robin Calibration Tests[4]

Thickness of 5052 Aluminum, in.	Machine Capacity		Tensile-Impact Strength		Approximate Standard Deviation	
	kg·cm	(ft·lb)	kg·cm/cm²	(ft·lb/in.²)	kg·cm/cm²	(ft·lb/in.²)
0.020	8.5 (1)[a]	4 (1)	630	296	55	26
	10.7 (4)	5 (4)	567	266	55	26
	32.0 (2)	5 (2)	550	258	110	52
0.050	8.5 (1)	4 (1)	758	356	72	34
	10.7 (4)	5 (4)	741	348	53	25
	32.0 (2)	15 (2)	675	317	45	21

[a] Numbers in parentheses show the number of machines used in obtaining the average values.

TABLE A2 Converting Tensile-Impact Units

Multiply	by	To Obtain
Inch-pounds	1.152	Kilogram-centimeters
Foot-pounds	13.83	Kilogram-centimeters
Kilogram-centimeter	0.8680	Inch-pounds
Kilogram-centimeter	0.07233	Foot-pounds
Foot-pounds/inch²	2.143	Kilogram-centimeters/centimeter²
Kilogram-centimeters/centimeter²	0.4666	Foot-pounds/inch²

Designation: D 1870 – 68

Standard Method of Test for
ELEVATED TEMPERATURE AGING USING A TUBULAR OVEN[1]

This Standard is issued under the fixed designation D 1870; the number immediately following the designation indicates the year of original adoption or, in the case of revision, the year of last revision. A number in parentheses indicates the year of last reapproval.

1. Scope

1.1 This method[2] covers a procedure for estimating the relative resistance of polymeric compounds to elevated temperature aging under controlled conditions of air circulation and freedom from contamination.

2. Summary of Method

2.1 This method consists of subjecting test specimens having previously determined properties to controlled deteriorating influences for known periods, after which the physical properties are again measured and the changes noted. In this method, the test involves exposure of specimens to a carefully controlled flow of air at an elevated temperature and at atmospheric pressure.

3. Significance

3.1 Any correlation between this test and natural life of these materials must be determined for the particular application in which the materials are to be used.

4. Apparatus

4.1 *Forced Circulating-Air Oven*, provided with tubes and a means for calibrating and controlling air velocity within the tubes. The air velocity variation shall not exceed 10 percent of the specified velocity. The temperature within the testing chamber shall be uniform within the tolerances given in Table 1 of ASTM Specification E 145, for Gravity-Convection and Forced-Ventilation Ovens,[3] for Type II Grade A ovens. The heating medium for the aging chamber shall be preheated air, circulated within it at atmospheric pressure. The heated air in both the chamber and in the tubes shall be thoroughly circulated by means of mechanical agitation. The design of the oven shall be such that heated fresh air enters one end of the tube and is exhausted from the system without being recirculated, eliminating the use of contaminated air and cross migration of volatile constituents contained in the specimens being tested. An apparatus meeting these requirements is described in the Appendix.

4.2 *Recording Temperature - Indicating Device*, to measure the temperature near the center of the chamber.

4.3 *Velometer*,[4] for checking the air velocity.

4.4 *Specimen Holders*—Suitable specimen holders shall be provided, which will position the specimens within the tubes, separated from each other and the tube walls, and permit free flow of air around the specimens.

5. Preparation and Calibration of Apparatus

5.1 Adjust the air within the tubes to the required test temperature within the range of 50 to 150 C.

5.2 Adjust the air flow in each tube to the required velocity.

---

[1] This method is under the jurisdiction of ASTM Committee D-20 on Plastics. A list of committee members may be found in the ASTM Yearbook. This standard is the direct responsibility of Subcommittee D-20.50 on Permanence Properties.

Current edition effective Jan. 12, 1968. Originally issued 1961. Replaces D 1870 – 61 T.

[2] For further information on this method the following reference may be consulted: Markus Royen, "A Study of the Accelerated Aging of Vinyl Plastic Compounds in a Modified Testing Oven," ASTM Bulletin. No. 243, Jan., 1961, p. 43 (TP 1).

[3] *Annual Book of ASTM Standards*, Part 30.

[4] Alnor Velometer Type 3002 has been found satisfactory for this purpose.

NOTE—Velocities of 100 to 250 m/min have been used.

6. Conditioning

6.1 *Conditioning*—Condition the test specimens at 23 ± 2 C (73.4 ± 3.6 F) and 50 ± 5 percent relative humidity for not less than 40 h prior to test in accordance with Procedure A of ASTM Methods D 618, Conditioning Plastics and Electrical Insulating Materials for Testing,[5] for those tests where conditioning is required. In cases of disagreement, the tolerances shall be ±1 C (±1.8 F) and ±2 percent relative humidity.

6.2 *Test Conditions*—Conduct tests in the Standard Laboratory Atmosphere of 23 ± 2 C (73.4 ± 3.6 F) and 50 ± 5 percent relative humidity, unless otherwise specified in the test methods or in this specification. In cases of disagreement, the tolerances shall be ±1 C (±1.8 F) and ±2 percent relative humidity.

7. Procedure

7.1 Mount the specimens in holders. Only specimens of identical compositions shall be permitted in any individual tube.

7.2 Place each loaded holder in a tube, place the tube in position in the oven, and age for the specified time and temperature.

7.3 At the termination of the aging interval, remove the holders from the tubes and allow them to cool to room temperature.

7.4 Measure the air velocity within each tube at the end of the test period. Any variation greater than ±10 percent of the required air velocity will be cause for rejection of the specimens from these tubes.

8. Calculations

8.1 Express the results of the aging test as a percentage of the change in each physical property, calculated as follows:

$$\text{Change, percent} = (O - A)/O \times 100$$

where:

O = original value, and
A = value after aging.

9. Report

9.1 The report shall include the following:

9.1.1 Results calculated in accordance with Section 7,

9.1.2 Air velocity in the tubes,

9.1.3 Aging interval,

9.1.4 Aging temperature,

9.1.5 Dates of original and final determinations of physical properties, and

9.1.6 Form and thickness of the test specimens.

[5] *Annual Book of ASTM Standards*, Part 27.

APPENDIX

A1. DESCRIPTION AND CALIBRATION OF TUBULAR OVEN

A1.1 Description

A1.1.1 *Tubular Oven*—The objectives of this method can be achieved with the tubular oven shown in Fig. A1. The unit (Fig. A2) is designed for insertion in standard ovens that are provided with removable side walls. The most widely used ovens today have this feature, and are adaptable to modification. The unit consists of two replacement walls of rigid metal (approximately 6 mm in thickness) provided with machined holes and adaptors to accommodate the specimen tubes. The air entrance wall is provided with air valves for each tube position. These valves are individually adjustable from within the oven for a predetermined setting of the air velocity, as measured in meters per minute. Both walls are provided with openings to permit passage of air outside the tubes for over-all temperature uniformity. Figures A1 and A3 show an oven under actual operating conditions. Holders are provided for positioning the specimens in the tube. The tube arrangement shown permits convenient removal of any single tube at any time. With this design, the standard 480 by 480 by 480-mm oven cavity will hold 16 tubes with 38 mm inside diameter.

A1.1.2 *Calibration Tube*—Auxiliary to the unit itself is the apparatus used in presetting the specified rate of air flow in the tubes. This consists of a metal calibration tube (Fig. A4) of dimensions equal to the specimen tube, and a velometer. The tube is provided with a slot to permit insertion of the probe of the velometer.

A1.2 Calibration

A1.2.1 Before using the oven for testing, adjust the air flow through each tube to the desired value. To adjust the rate of air flow, insert the calibration tube in place of one of the oven tubes. With the oven at the test temperature, with the oven intake and exhaust valves fully open, and with the oven damper preset, set the air velocity through the tubes to the desired rate in meters per minute by adjusting the valves. To eliminate seepage of outside air and also to assist in positioning, use a flexible rubber gasket on the probe of the velometer. Continue the calibration procedure with each tube.

NOTE A1—Provided that the setting of air valves is not changed, the air velocity within the tubes, once set, will usually remain constant for extended periods.

FIG. A1 Tubular Oven.

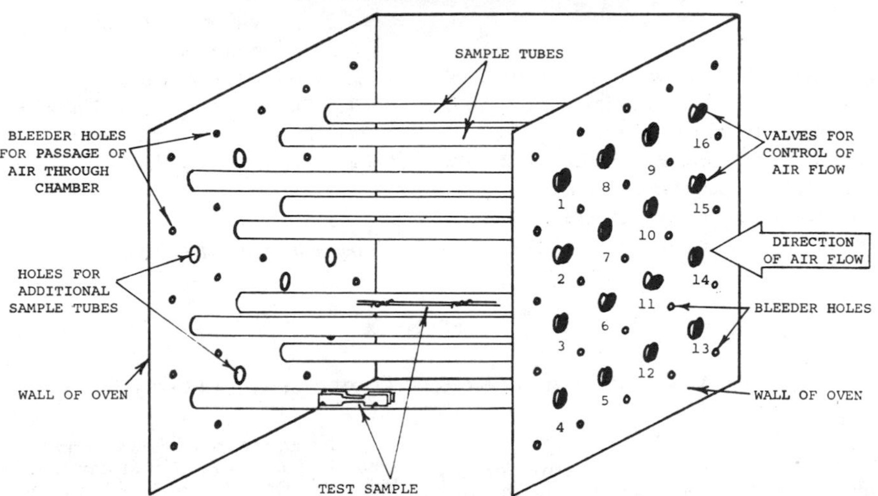

FIG. A2 Details of Tubular Oven for Insertion in Standard Ovens.

FIG. A3 Tubular Oven Under Actual Operating Conditions.

FIG. A4 Apparatus Used in Presetting the Specified Rate of Air Flow in the Tubes.

Designation: D 1893 - 67

Standard Method of Test for
BLOCKING OF PLASTIC FILM[1]

This Standard is issued under the fixed designation D 1893; the number immediately following the designation indicates the year of original adoption or, in the case of revision, the year of last revision. A number in parentheses indicates the year of last reapproval.

1. Scope

1.1 This method yields quantitative information regarding the degree of blocking existing between layers of plastic film (Note 2). It is not intended to measure susceptibility to blocking.

NOTE 1—The values stated in U.S. customary units are to be regarded as the standard. The metric equivalents of U.S. customary units may be approximate.

NOTE 2—Development of this test has been done with films of 0.025 to 0.10 mm (0.001 to 0.004 in.) in thickness.

1.2 This method is applicable to the comparison of films of different materials. It may be used to study films prepared by different processes, or under varied processing conditions. The method may also be used for quality control.

2. Significance

2.1 Blocking develops under a variety of temperatures and pressures, and may arise from processing, use, or storage. In nearly all cases, blocking results from either or both of the following:

2.1.1 Extremely smooth film surfaces, allowing intimate contact and nearly complete exclusion of air.

2.1.2 Pressure- or temperature-induced fusion of the surfaces in contact, or both.

3. Definitions

3.1 *blocking*—an adhesion between touching layers of plastic film.

3.2 *blocking force*—the average force (Note 3) per unit width of blocked surface required to separate progressively two layers of plastic film one from another by a rod 6.35 mm (0.250 in.) in diameter moving perpendicularly to its axis at a uniform rate of 125 mm (5 in.)/min. This force is expressed in grams per centimeter of width (or pounds per foot).

NOTE 3—Frictional forces may comprise a portion of the forces involved in this method, but they have not been observed in actual practice.

4. Apparatus

4.1 *Testing Machine*—A power-driven machine, preferably of the load-cell type, having a 125 mm (5 in.)/min crosshead speed. The machine shall be of the continuous recording type, and shall be accurate to within 1 percent of full scale.

NOTE 4—A more complete description of this machine may be found in ASTM Method D 638, Test for Tensile Properties of Plastics.[2]

4.2 *Rod and Frame*—A smoothly finished 6.35-mm (0.250-in.) diameter aluminum rod and rigid supporting frame, as shown in Figs. 1 and 2. Dimensions and method of construction of the frame are not critical.

NOTE 5—Should the optional clamp (4.4) be used, it will be found advantageous to modify this equipment such that the aluminum rod may be conveniently inserted into the assembled frame from one end at the start of each test cycle.

4.3 *Cutting Instrument*—A device for cleanly cutting and trimming test specimens. A precision paper cutter with a sharp knife-edge is suitable.

4.4 *Clamp (Optional)*—A device, attachable to the stationary jaws of the testing machine,

[1] This method is under the jurisdiction of ASTM Committee D-20 on Plastics. A list of committee members may be found in the ASTM Yearbook. This standard is the direct responsibility of Subcommittee D-20.10 on Mechanical Properties.

Current edition effective March 11, 1967. Originally issued 1961. Replaces D 1893 - 61 T.

[2] *Annual Book of ASTM Standards*, Part 27.

in which the specimen may be clamped uniformly over its width.

5. Test Specimens

5.1 Four specimens from each sample shall be prepared for testing as follows:

5.1.1 Each specimen shall consist of two blocked layers of film.

5.1.2 Specimens shall be rectangular. The following are recommended size limits for optimum test results:

	Min	Max
Length[a]	250 mm (10 in.)	300 mm (12 in.)
Width	200 mm (8 in.)	250 mm (10 in.)

[a] Length shall be in the machine direction for continuously manufactured film.

NOTE 6—Narrow widths may be dictated by the size of available samples. If such is the case, it must be realized that the measured forces will be small, film irregularities more significant, and that reproducibility, in general, will suffer. The foregoing also applies to shorter length specimens, but with less severity.

5.1.3 The sides shall be cut in such a manner as to ensure tear- and snag-free edges.

NOTE 7—Film samples exhibiting very little blocking shall be handled with extreme care during specimen preparation, otherwise premature separation may occur, causing large errors in the measured blocking force.

5.1.4 A small mark shall be made on one end, locating the longitudinal center line of the specimen.

6. Conditioning

6.1 *Conditioning*—Condition the test specimens at 23 ± 2 C (73.4 ± 3.6 F) and 50 ± 5 percent relative humidity for not less than 40 h prior to test in accordance with Procedure A of ASTM Methods D 618, Conditioning Plastics and Electrical Insulating Materials for Testing,[2] for those tests where conditioning is required. In cases of disagreement, the tolerances shall be ±1 C (1.8 F) and ±2 percent relative humidity.

6.2 *Test Conditions*—Conduct tests in the Standard Laboratory Atmosphere of 23 ± 2 C (73.4 ± 3.6 F) and 50 ± 5 percent relative humidity, unless otherwise specified in the test methods or in this specification. In cases of disagreement, the tolerances shall be ±1 C (±1.8 F) and ±2 percent relative humidity.

NOTE 8—Electrostatic charges are often encountered in handling plastic films of high resistivity. In order to reduce the possibility of interference from such charges, the testing should be conducted in an atmosphere of 50 ± 2 percent relative humidity.

7. Procedure

7.1 Attach the aluminum rod and supporting frame to the testing machine crosshead (Fig. 2). The rod shall be within 40 mm (1½ in.) of the stationary clamp at the start of each test cycle.

NOTE 9—When testing severely blocked or very light gage films, the standard clamping method described herein and shown in Fig. 2 may produce buckling and stretching of the film layers during test. An optional full-width clamp, (4.4), attached to the stationary jaws, may be used to overcome this difficulty. The foregoing procedure still applies, except that insertion of the aluminum rod (through the separated layers) into the assembled frame, after clamping of each specimen, will expedite handling.

7.2 Separate the surfaces of each test specimen along the marked end for free passage of 50 to 80 mm (2 to 3 in.) of each single film layer around the aluminum rod.

NOTE 10—In the case of severely blocked film, separation is most conveniently done as follows: Attach two small pieces of pressure-sensitive tape to opposite faces of one corner of the specimen in such fashion that they extend beyond the edge of the specimen and adhere also to each other. Separate the two pieces of tape, starting at the outermost edge. The film corners will follow the tape and separate in very clean fashion. Once this is done, a pencil or similar implement may be used to separate the area specified in the procedure.

7.3 Mount the test specimen in the machine and clamp in place, using the center mark as a guide (Fig. 2). Then make a zero adjustment to account for the weight of the specimen.

7.4 Start the cycle at a separation rate of 125 mm (5 in.)/min and a recorder chart speed of at least 50 mm (2 in.)/min, and continue until the specimen is completely separated. Initially, manual assistance may be needed to ensure an even, unwrinkled separation of the films. The load recorded during this interval should be ignored.

7.5 Determine the width of each test specimen to the nearest 1 mm (0.05 in.).

8. Calculations

8.1 Determine the actual blocking force

from the best average load line. Such a load line may be established by eye, by averaging selected, uniformly spaced ordinates, or by any conventional method, including graphical integration. Report the average load, expressed in grams, divided by the specimen width expressed in millimeters, as the blocking force in grams per millimeter for the particular specimen under test. In cases where there are appreciable differences in the blocking force from one location to another, it is recommended that the maximum and minimum blocking force also be reported.

8.2 Results may be converted to pounds per foot by use of the multiplier 0.672.

8.3 For each sample, calculate and report the arithmetic mean of the four test specimen results as the "average blocking force."

9. Report

9.1 The report shall include the following:

9.1.1 Complete identification of the specimens tested, including type, source, form, process of manufacture, age, and any other pertinent facts,

9.1.2 Conditioning and testing environments, if different from those recommended in Section 6.

9.1.3 Film thickness (single layer), and

9.1.4 Individual and average blocking force values for each specimen tested. Maximum and minimum values may also be recorded.

10. Precision

10.1 *Repeatability*—Duplicate results by the same operator should not be considered suspect unless they differ by more than $0.014 \times 2\sqrt{2}$ g/mm.

10.2 *Reproducibility*—The result submitted by each of two laboratories should not be considered suspect unless the two results differ by more than $0.019 \times 2\sqrt{2}$ g/mm.

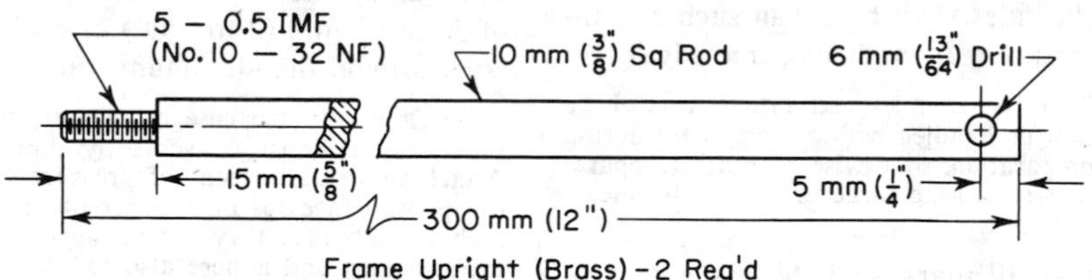

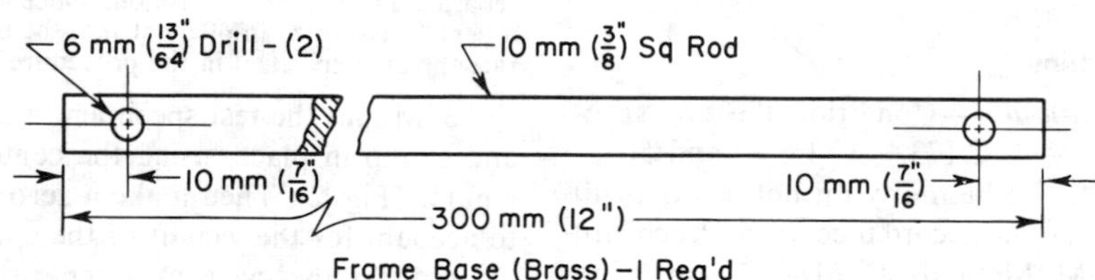

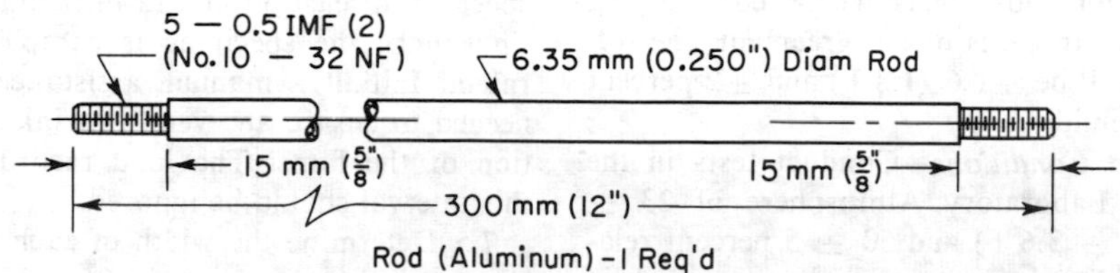

FIG. 1 Rod and Frame Components.

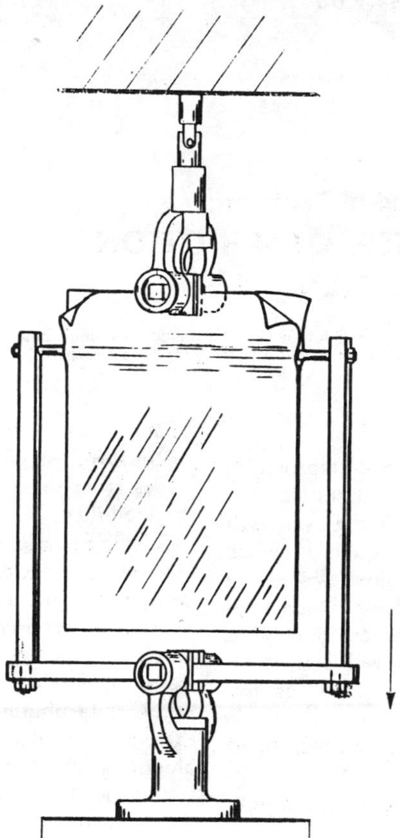

FIG. 2 Specimen Under Test.

Designation: D 1894 – 63

American National Standard K65.14-1965
American National Standards Institute

Standard Method of Test for
COEFFICIENTS OF FRICTION OF PLASTIC FILM[1]

This Standard is issued under the fixed designation D 1894; the number immediately following the designation indicates the year of original adoption or, in the case of revision, the year of last revision. A number in parentheses indicates the year of last reapproval.

1. Scope

1.1 This method covers the determination of the coefficients of starting and sliding friction of plastic sheeting when sliding over itself or other substances. The method is intended to be used for plastic film defined as sheeting having a nominal thickness of not greater than 0.0254 cm (0.010 in.) as indicated in ASTM Nomenclature D 883, Relating to Plastics.[2]

1.2 Two procedures are included, as follows:

1.2.1 *Procedure A*. Stationary Sled, Moving Plane Procedure, and

1.2.2 *Procedure B*. Moving Sled, Stationary Plane Procedure.

Both procedures yield the same coefficients of friction values for a given sample.

NOTE 1—The values stated in U.S. customary units are to be regarded as the standard. The metric equivalents of U.S. customary units may be approximate.

2. Significance

2.1 Measurements of frictional properties may be made on film-to-film surfaces, or film-to-surfaces of other substances. The coefficients of friction are related to the slip properties of plastic films that are of wide interest in packaging applications. These methods yield empirical data for control purposes in film production. Correlation of test results with actual performance can usually be established.

2.2 Slip properties are generated by additives in some plastic films, for example, polyethylene. These additives have varying degrees of compatibility with the film matrix. Some of them bloom, or exude to the surface, lubricating it and making it more slippery. Because this blooming action may not always be uniform on all areas of the film surface, values from these tests may be limited in reproducibility.

2.3 The frictional properties of plastic film may be dependent on the uniformity of the rate of motion between the two surfaces. Care should be exercised to ensure that the rate of motion of the equipment is as carefully controlled as possible.

2.4 Data obtained by these procedures are extremely sensitive to the age of the film. The blooming action of many slip additives is time-dependent. For this reason, it is sometimes meaningless to compare slip and friction properties of films produced at dissimilar times, unless it is desired to study this effect.

2.5 Frictional and slip properties of plastic film are based on measurements of surface phenomena. Where these products have been made by different processes, or even on different machines by the same process, their surfaces may be dependent on the equipment or its running conditions. Such factors must be weighed in evaluating data from these methods.

2.6 The measurement of the static coefficient of friction is highly dependent on the rate of loading and on the amount of blocking occurring between the loaded sled and the platform due to variation in time before mo-

[1] This method is under the jurisdiction of ASTM Committee D-20 on Plastics. A list of committee members may be found in the ASTM Yearbook. This standard is the direct responsibility of Subcommittee D-20.10 on Mechanical Properties.

Current edition effective Sept. 30, 1963. Originally issued 1961. Replaces D 1894 – 61 T.

[2] *Annual Book of ASTM Standards*, Part 27.

tion is initiated.

2.7 Care should be exercised to make certain that the speed of response of the recorder, either electronic or mechanical, is not exceeded.

3. Definitions

3.1 *slip*—a term denoting lubricity of two surfaces sliding in contact with each other. In a sense, it is the antithesis of friction in that high coefficient of friction denotes poor slip and low coefficient of friction good slip.

3.2 *friction*—the resisting force that arises when a surface of one substance slides, or tends to slide, over an adjoining surface of itself or another substance. Between surfaces of solids in contact there may be two kinds of friction: (*1*) the resistance opposing the force required to start to move one surface over another, and (*2*) the resistance opposing the force required to move one surface over another at a variable, fixed, or predetermined speed.

3.3 *nip*—the line of tangency between two parallel rolls in contact with one another, or between either of the rolls and the surface of an object passing between them.

3.4 *coefficient of friction*—the ratio of the frictional force, 3.2, to the force, usually gravitational, acting perpendicular to the two surfaces in contact. This coefficient is a measure of the relative difficulty with which the surface of one material will slide over an adjoining surface of itself, or of another material. The static or starting coefficient of friction (μs) is related to the force measured to begin movement of the surfaces relative to each other. The kinetic or sliding coefficient of friction (μk) is related to the force measured in sustaining this movement.

4. Test Specimens

4.1 The test specimens shall be of two sizes; one specimen shall be cut approximately 250 mm (10 in.) in the machine direction and 130 mm (5 in.) in the transverse direction, and the other shall be cut approximately 120 mm (4 $^1/_2$ in.) square.

4.2 Five specimens shall be tested for each sample unless otherwise specified.

NOTE 2—Plastic films may exhibit different frictional properties in their respective principal directions due to anisotropy or extrusion effects. Specimens may be tested with their long dimension in either the machine or transverse direction of the sample, but it is more common practice to test the specimen as described in 4.1 with its long dimension parallel to the machine direction.

NOTE 3: **Caution**—Extreme care must be taken in handling the specimens. The test surface must be kept free of all dust, lint, finger prints, or any foreign matter that might change the surface characteristics of the specimens.

5. Conditioning

5.1 *Conditioning*—Condition the test specimens at 23 ± 2 C (73.4 ± 3.6 F) and 50 ± 5 percent relative humidity for not less than 40 h prior to test in accordance with Procedure A of ASTM Methods D 618, Conditioning Plastics and Electrical Insulating Materials for Testing,[2] for those tests where conditioning is required. In cases of disagreement, the tolerances shall be ±1 C (±1.8 F) and ±2 percent relative humidity.

5.2 *Test Conditions*—Conduct tests in the Standard Laboratory Atmosphere of 23 ± 2 C (73.4 ± 3.6 F) and 50 ± 5 percent relative humidity, unless otherwise specified in the test methods or in this specification. In cases of disagreements, the tolerances shall be ±1 C (±1.8 F) and ±2 percent relative humidity. In specific cases, such as control testing, where the conditioning requirements cannot be met and the data still may be of direct assistance to the operation, other conditioning procedures may be used and recorded in the report. Frictional properties should be measured only after sufficient time has been allowed for the specimens to reach essential equilibrium with the ambient atmosphere.

6. Apparatus

Common to Both Procedures:

6.1 *Sled*—A metal block 63.5 mm (2 $^1/_2$ in.) square by approximately 6 mm (0.25 in.) thick with a suitable eye screw fastened in one end. The block shall be wrapped with 3.2-mm ($^1/_8$-in.) thick, medium-density foam rubber. The rubber shall be wrapped snugly around the sled end to end and fastened in position. The wrapped sled should weigh 200 ± 5 g.

6.2 *Scissors or Cutter*, suitable for cutting film specimens to the desired dimensions.

6.3 *Adhesive Tape*, cellophane.

6.4 *Adhesive Tape*, double-faced.[3]

6.5 *Nylon Monofilament*, having a 0.33 ± 0.05-mm (0.013 ± 0.002-in.) diameter and capable of supporting a 3.6-kg (8-lb) load.

6.6 *Low-Friction Pulleys*[4]—A phenolic-type pulley mounted in hardened steel cone bearings on a metal fork.

6.7 *Force-Measuring Device*, capable of measuring the frictional force to ±5 percent of its value. A spring gage[5] (Note 4), universal testing machine, or strain gage may be used.

NOTE 4—The capacity of the spring gage (Fig. 1 (*a*)) needed will depend upon the range of values to be measured. For most plastic film, a 500-g capacity gage with 10-g or smaller subdivisions will be satisfactory. This spring will measure coefficients of friction up to and including 2.5.

Procedure A. Stationary Sled, Moving Plane:

6.8 *Moving Plane*—A polished plastic, wood, or metal sheet,[6] approximately 150 by 300 by 1 mm (6 by 12 by 0.040 in.).

6.9 *Driving Device for Moving Plane*—The plane may be pulled by a driven pair of rubber-coated rolls not less than 203 mm (8 in.) long, capable of maintaining a uniform surface speed 152 ± 30 mm/min (0.5 ± 0.1 ft/min) (Fig. 1(*b*)) or by the crosshead of a universal testing machine (Fig. 1(*d*)) (Note 5). A constant-speed chain drive system has also been found satisfactory (Fig. 1(*a*)).

NOTE 5—Where the moving crosshead of a universal testing machine is used to pull the moving plane through a pulley system (Fig. 1(*d*)), the strain gage load cell, or other load-sensing instrument in the testing machine, acts as the force-measuring device.

6.10 *Supporting Base*—A smooth wood or metal base approximately 200 by 380 mm (8 by 15 in.) is necessary to support the moving plane. The supporting base may be a simple rectangular box. If a universal testing machine is used to pull the moving plane, a supporting base of sufficient structural strength and rigidity to maintain a firm position between the moving crosshead and the force-measuring device will be necessary.

Procedure B. Moving Sled, Horizontal Plane:

6.11 *Horizontal Plane*—A polished plastic, wood, or metal sheet,[6] approximately 150 by 300 by 1 mm (6 by 12 by 0.040 in.). This plane shall be securely mounted and level. If the apparatus of Fig. 1(*c*) is used, the plane may be constructed of smooth metal supported by a diagonal brace from its underside. It may be attached to the constant rate-of-motion, lower crosshead of a universal testing machine. A smooth, flat piece of glass may cover the upper side of the horizontal plane. This provides a smooth support for the specimen. A low-friction pulley shall be attached to the end of, or at the edge of this plane and centered on the crosshead directly beneath the load cell of the testing machine. The base shall be tested for levelness.

6.12 *Sled Pulling Device*—A power-operated source for pulling the sled over the horizontally-mounted specimen at a uniform speed of 152 ± 30 mm/min (0.5 ± 0.1 ft/min). A universal testing machine equipped with a load cell in its upper crosshead and a constant rate-of-motion lower crosshead has been found satisfactory (see Fig. 1(*c*)).

7. Preparation of Apparatus

7.1 Figure 1 shows four ways in which the apparatus may be assembled.

7.2 If the moving sled procedure is used, calibrate the sled-pulling device to operate at a speed of 152 ± 30 mm/min (0.5 ± 0.1 ft/min). If a universal testing machine is used, select the proper speed setting for crosshead motion at 152 ± 30 mm/min (0.5 ± 0.1ft/min). A similar speed for the stress-strain recorder is desirable. However, the speed of the recorder can be adjusted to give the desired accuracy in reading the pen trace.

7.3 If the apparatus of Fig. 1(*a*) or (*b*) is used, calibrate the scale of the spring gage as follows:

7.3.1 Mount the low-friction pulley in front of the spring gage.

7.3.2 Fasten one end of the nylon filament to the spring gage, bring the filament over the pulley, and suspend a known weight on the lower end of the filament to act downward.

NOTE 6—The reading on the scale shall correspond to the known weight within ±5 percent. The

[3] Available from Minnesota Mining and Manufacturing Co., Inc., or Permacel Tape Corp.

[4] Suitable pulleys are available from scientific supply houses.

[5] Model L-500, available from Hunter Spring Co., Lansdale, Pa., has been found satisfactory for this purpose.

[6] Acrylic or rigid poly(vinyl chloride) sheeting has been found satisfactory for this purpose.

weight used for this calibration shall be between 50 and 75 percent of the scale range on the gage.

7.4 If the moving plane procedure is used, start the drive and adjust its speed to 152 ± 30 mm/min (0.5 ± 0.1 ft/min). This speed may be checked by marking off a 152 mm (0.5-ft) section beside the polished plastic sheet and determining the time required for the sheet to travel 152 mm (0.5 ft).

7.5 Wipe the base free of foreign matter.

7.6 Lay down two strips of double-faced adhesive tape along the length of the supporting base so that they are approximately 100 mm (4 in.) between centers.

8. Procedure

8.1 Tape the 150 by 600-mm (6 by 24-in.) film specimen to the moving plane with the machine direction of the film corresponding to the length of the sheet. Smooth the specimen to eliminate wrinkes.

NOTE 7—For some film samples it has been found necessary to tape only the leading edge of the film specimen to the base. In some cases the film has been pulled through the nip rolls without the plate. However, should any dispute arise, taping of all four edges will be the referee method.

NOTE 8—For the sake of uniformity and later comparison, unless it is desired to measure other effects, the specimens shall be mounted so that the same side of the specimen shall be used as the contact surface for both the moving and staionary specimens.

NOTE 9—Coefficient of friction measurements may be made between film-to-film or film-to-other-substance surfaces wherein the movement is made in the transverse direction of the specimen. However, the methods described here will be confined to movements in the machine direction of the specimens.

8.2 Tape the edges of the 120-mm (4 $^{1}/_{2}$-in.) square specimen to the back of the sled, using cellophane tape and pulling the specimen tight to eliminate wrinkles without stretching it. Keep the machine direction of the film parallel to the length of the sled.

8.3 Put the sled lightly in position at the rear end of the horizontal specimen (Note 10). Attach the specimen-covered sled through its eye screw to the nylon filament. If a universal testing machine is used, pass the filament beneath the pulley and upward to the bottom of the load-sensing device. Tie it tightly to the load cell. If a spring gage is used, attach the specimen-covered sled to it. In all cases, align the eye screw and pulley so that the force on the sled is exactly parallel to the surface of the horizontal plane.

NOTE 10—The sled must be placed very lightly and gently on the platform to prevent any unnatural bond from developing. A high starting coefficient of friction may be caused by undue pressure on the sled when mounting it onto the platform.

8.4 Start the nip rolls or the moving crosshead (which have been adjusted previously to provide a speed of 152 ± 30 mm/min (0.5 ± 0.1 ft/min)). As a result of the frictional force between the contacting surfaces, no immediate relative motion may take place between the sled and the moving plane until the pull on the sled is equal to, or exceeds, the static frictional force acting at the contact surfaces. Record this initial, maximum reading as the force component of the static coefficient of friction.

8.5 Record the visual average reading during a run of approximately 130 mm (5 in.) while the surfaces are sliding uniformly over one another. This is equivalent to the kinetic force required to sustain motion between the surfaces and normally is lower than the static force required to initiate motion. After the sled has traveled over (5 in.) stop the apparatus and return to starting position.

8.6 If a strain gage and stress-strain recorder are used, draw the best straight line midway between the maximum points and minimum points shown on the chart while the sled was in motion. The mean load represented by this line is the kinetic friction force required to sustain motion on the sled.

8.7 Remove the film specimen from the sled and the horizontal plane. The apparatus is now ready for the next set of specimens. A new set of specimens shall be used for each run. No specimen surface shall be tested more than once.

NOTE 11—The maximum point at which initial motion takes place between the sled and the horizontal plane should be carefully examined with reference to the rate of loading and the speed of response of the sensing device. Failure to consider this factor can lead to meaningless results for the value of the static coefficient of friction.

9. Calculations

9.1 Calculate the static coefficient of friction μs, as follows:

$$\mu s = A/B$$

where:

A = initial motion scale reading, g, and
B = sled weight, g.

9.2 Calculate the kinetic coefficient of friction, μk, as follows:

$$\mu k = A/B$$

where:
A = average scale reading obtained during uniform sliding of the film surfaces, g, and
B = sled weight, g.

9.3 Calculate the standard deviation (estimated to be ±15 percent of the value of the coefficient of friction) as follows, and report it to two significant figures:

$$s = \sqrt{(\Sigma X^2 - n\bar{X}^2)/(n - 1)}$$

where:
s = estimated standard deviation,
X = value of a single observation,
n = number of observations, and
$\bar{X}$ = arithmetic mean of the set of observations.

10. Report

10.1 The report shall include the following:

10.1.1 Complete description of the plastic film sample, including manufacturer's code designation, thickness, method of production, surfaces tested, principal directions tested, and approximate age of sample after extrusion,

10.1.2 Procedure used (either A or B),

10.1.3 Apparatus used,

10.1.4 Average static and kinetic coefficients of friction, together with the standard deviation, and

10.1.5 Number of specimens tested for each coefficient of friction.

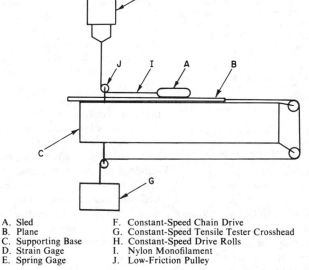

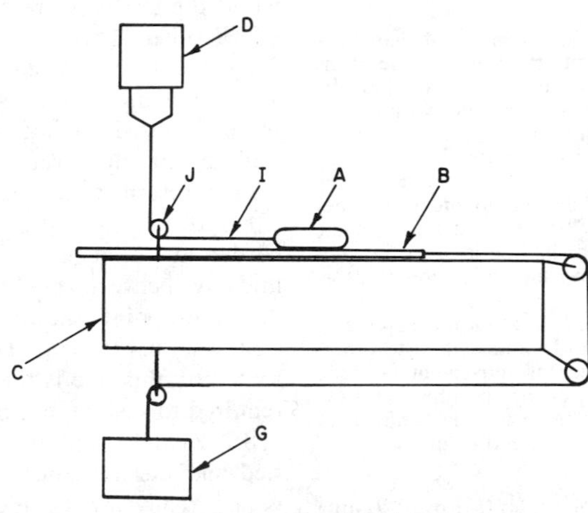

A. Sled
B. Plane
C. Supporting Base
D. Strain Gage
E. Spring Gage
F. Constant-Speed Chain Drive
G. Constant-Speed Tensile Tester Crosshead
H. Constant-Speed Drive Rolls
I. Nylon Monofilament
J. Low-Friction Pulley

FIG. 1 Typical Assembly of Apparatus for Determination of Coefficients of Friction of Plastic Film.

Designation: D 1895 - 69

Standard Methods of Test for
APPARENT DENSITY, BULK FACTOR, AND POURABILITY OF PLASTIC MATERIALS[1]

This Standard is issued under the fixed designation D 1895; the number immediately following the designation indicates the year of original adoption or, in the case of revision, the year of last revision. A number in parentheses indicates the year of last reapproval.

1. Scope

1.1 These methods cover the measurement of apparent density, bulk factor, and where applicable, the pourability of plastic materials such as molding powders. Different procedures are given for application to the various forms of these materials that are commonly encountered, from fine powders and granules to large flakes and cut fibers.

NOTE 1—The values stated in U.S. customary units are to be regarded as the standard. The metric equivalents of U.S. customary units may be approximate.

2. Significance

2.1 These methods provide useful indexes of performance of plastic materials such as powders and granules with respect to their handling in packaging and fabrication.

2.2 Apparent density is a measure of the fluffiness of a material.

2.3 Bulk factor is a measure of volume change that may be expected in fabrication.

2.4 Pourability characterizes the handling properties of a finely divided plastic material. It is a measure of the readiness with which such materials will flow through hoppers and feed tubes and deliver uniform weights of material.

3. Definitions

3.1 *apparent density*—the weight per unit volume of a material, including voids inherent in the material as tested.

NOTE 2—The term *bulk density* is commonly used for materials such as molding powder.

3.2 *bulk factor*—the ratio of the volume of any given quantity of the loose plastic material to the volume of the same quantity of the material after molding or forming. The bulk factor is also equal to the ratio of the density after molding or forming to the apparent density of the material as received.

3.3 *pourability*—a measure of the time required for a standard quantity of material to flow through a funnel of specified dimensions.

4. Conditioning

4.1 *Conditioning*—Condition the test specimens at 23 ± 2 C (73.4 ± 3.6 F) and 50 ± 5 percent relative humidity for not less than 40 h prior to test in accordance with Procedure A of ASTM Methods D 618, Conditioning Plastics and Electrical Insulating Materials for Testing,[2] for those tests where conditioning, is required. In cases of disagreement, the tolerances shall be ±1 C (±1.8 F) and ±2 percent relative humidity.

4.2 *Test Conditions*—Conduct tests in the Standard Laboratory Atmosphere of 23 ± 2 C (73.4 ±3.6 F) and 50 ± 5 percent relative humidity, unless otherwise specified in the test methods or in this specification. In cases of disagreement, the tolerances shall be ±1 C (±1.8 F) and ±2 percent relative humidity.

[1] These methods are under the jurisdiction of ASTM Committee D-20 on Plastics. A list of committee members may be found in the ASTM Yearbook. This standard is the direct responsibility of Subcommittee D-20.70 on Analytical Methods.

Current edition effective Nov. 14, 1969. Originally issued 1961. Replaces D 1895 - 67.

[2] *Annual Book of ASTM Standards*, Part 27.

Apparent Density

Method A

5. Scope

5.1 Method A[3] covers the measurement of the apparent density of fine granules and powders that can be poured readily through a small funnel.

6. Apparatus

6.1 *Measuring Cup*—A cylindrical cup of 100 ± 0.5-cm^3 capacity, having a diameter equal to half the height, for example, 39.9 mm (1.572 in.) inside diameter by 79.8 mm (3.144 in.) inside height, as shown in Fig. 1.

6.2 *Funnel*, having a 9.5-mm diameter opening at the bottom, and mounted at a height 38 mm above the measuring cup, as shown in Fig. 1.

7. Procedure

7.1 With the apparatus assembled as shown in Fig. 1, close the small end of the funnel with the hand or with a suitable flat strip and pour a 115 ± 5-cm^3 sample into the funnel. Open the bottom of the funnel quickly and allow the material to flow freely into the cup. If caking occurs in the funnel, the material may be loosened with a small glass rod.

7.2 After all the material has passed through the funnel, immediately scrape off the excess on the top of the cup with a straightedge without shaking the cup. Weigh the material in the cup to the nearest 0.1 g. Calculate the weight in grams of 1 cm^3 of the material.

NOTE 3—To convert grams per cubic centimeter to ounces per cubic inch, multiply by 0.578. To convert to grams per cubic inch multiply by 16.39. To convert grams per cubic centimeter to pounds per cubic foot, multiply by 62.43.

NOTE 4—Apparent density figures are not comparable except on materials having the same specific gravity after molding or forming.

7.3 Very fine materials that will bridge and not flow through the funnel may be poured lightly from a paper held approximately 38 mm above the opening of the measuring cup.

Method B

8. Scope

8.1 Method B covers the measurement of the apparent density of coarse, granular materials, including dice and pellets, that either cannot be poured or that pour with difficulty through the funnel described in Method A.

9. Apparatus

9.1 *Measuring Cup*—A cylindrical cup of 400-cm^3 capacity, as shown in Fig. 2.

9.2 *Funnel*, having a 25.4-mm diameter opening at the bottom, and mounted at a height 38 mm above the measuring cup, as shown in Fig. 2.

NOTE 5—Finely divided powders may collect electrostatic charges which, if present at the time of measurement, may result in variable apparent density values. Carbon black is a material that may be premixed with the sample at concentrations of 0.05 to 0.2 weight percent to reduce this variability (Superba Black with a bulk density of about 8 lb/ft^3 has been found satisfactory for some materials).

10. Procedure

10.1 With the apparatus assembled as shown in Fig. 2, close the small end of the funnel with the hand or with a suitable flat strip and pour a 500 ± 20-cm^3 sample into the funnel. Open the bottom of the funnel quickly and allow the material to flow freely into the cup.

NOTE 6—The funnel described in Method B is also used for more finely divided powders such as vinyl resins. While these powders usually will pour through this funnel, they may bridge in the 400-cm^3 cup shown in Fig. 2. To avoid this, the 100-cm^3 cup shown in Fig. 1 may be substituted, and the sample reduced to 115 ± 5 cm^3.

10.2 After all the material has passed through the funnel, immediately scrape off the excess on the top of the cup with a straightedge without shaking the cup. Weigh the material in the cup to the nearest 0.1 g; then calculate the weight in grams of 1 cm^3 of the material (Note 3). Make three determinations of the apparent density on each sample and average the results (Note 4).

Method C

11. Scope

11.1 Method C[3] covers the measurement

[3] Method A differs from the similar ISO method, R 60, Determination of Apparent Density of Molding Material that Can be Poured from a Specified Funnel, as described in the Appendix, Method C is substantially identical with ISO Method R 61, Determination of Apparent Density of Molding Material that Cannot be Poured from a Specified Funnel.

of the apparent density of materials supplied in the form of coarse flakes, chips, cut fibers, or strands. Such materials cannot be poured through the funnels described in Methods A and B. Also, since they ordinarily are very bulky when loosely poured and since they usually are compressible to a lesser bulk, even by hand, a measure of their density under a small load is appropriate and useful.

12. Apparatus

12.1 *Measuring Cylinder*—A cylinder of 1000-cm^3 capacity, as shown in Fig. 3.

12.2 *Weight Plunger*—A cylinder closed at one end and having an outside diameter slightly smaller than the inside diameter of the measuring cylinder, as shown in Fig. 3. The plunger shall have a scale cut on the outside surface, graduated at intervals of 1 mm.

13. Procedure

13.1 Place the measuring cylinder on a piece of paper. Loosely drop 60 ± 0.2 g of the material to be tested into the measuring cylinder from a height approximately that of the cylinder, taking care to lose none of the material and to drop it as uniformly as practicable. Before applying the weight, level the material in the measuring cylinder. Measure the height of the loose material in centimeters and identify this measurement as H_1.

13.2 Fill the weight plunger with lead shot to obtain a total weight of 2300 ± 20 g (5.07 ± 0.04 lb), including the plunger. Lower this weight gradually into the measuring cylinder until it is entirely supported by the material. Allow the weight to settle for 1 min and take the reading from the scale to the nearest 0.1 cm. This reading will give directly in centimeters the height, H_2, of the material.

14. Calculation

14.1 Calculate the apparent density of the material before and after application of the load as follows, making separate calculations for both H_1 and H_2:

$$V = H \times A$$

$$\text{Apparent density} = W/V$$

where:

V = volume occupied by the material in the measuring cylinder, cm^3,
H = height of the material in the measuring cylinder, cm,
A = cross-sectional area of the measuring cylinder, (61.89 cm^2), and
W = weight of the material in the cylinder (60 g).

14.2 Make three determinations of the apparent density on each sample and average these results separately.

15. Report

15.1 The report shall include each value of the apparent density and of the average density, both before and after loading.

BULK FACTOR

16. Procedure

16.1 *Apparent Density*—Measure the apparent density of the material in accordance with the applicable method as defined in Sections 5 to 15.

16.2 *Density After Molding or Forming*—Measure the density of the molded or formed plastic material in accordance with either ASTM Method D 1505, Test for Density of Plastics by the Density-Gradient Technique,[2] or Method A of ASTM Methods, D 792, Test for Specific Gravity and Density of Plastics by Displacement,[2] on two or more samples formed of the material under examination. When the latter method is used, the average specific gravity, in general, shall be assumed to be numerically equal to the average density in grams per cubic centimeter. If the shape of the formed specimen is such that its volume may be readily and accurately calculated from its dimensions, the density may be determined by dividing the weight of the specimen by its volume.

17. Calculation

17.1 Calculate the bulk factor of the plastic material as follows:

$$\text{Bulk factor} = D_2/D_1$$

where:

D_2 = average density of the molded or formed specimen, and
D_1 = average apparent density of the plastic material prior to forming.

18. Report

18.1 The report shall include the following:

18.1.1 Average apparent density of the plastic material and the method employed,

18.1.2 Average density of the molded or formed specimen, and

18.1.3 Bulk factors (Notes 7 and 8) calculated from them.

NOTE 7—Because bulk factor is a dimensionless ratio, it permits ranking of plastic materials, with respect to volume change upon fabrication, regardless of their molded or formed density (or specific gravity).

NOTE 8—For measurements made in accordance with Method C, bulk factor before and after loading shall be reported. These shall be clearly identified.

POURABILITY

19. Apparatus

19.1 The apparatus shall consist of the funnel described in Method A (6.2 and Fig. 1), mounted as shown, and either a stop watch or an electric timer of comparable accuracy.

NOTE 9—Pourability may be measured with the funnel described in Method B (9.2); however, the flow of material through this funnel is frequently too rapid to yield meaningful results. Method C does not permit measurement of pourability.

20. Procedure

20.1 Take a sample of the plastic material weighing, in grams, 100 times its specific gravity (or density) after molding or forming. Work this sample on a paper until the there is no tendency for the material to pack or cake. With the apparatus assembled as shown in Fig. 1, close the small end of the funnel with the hand or with a suitable flat strip and pour the sample lightly into the funnel, avoiding any tendency to pack it. Then quickly open the bottom of the funnel and start the stop watch or timer at the same instant. Allow the material to run from the funnel as freely as it will, and stop the watch or timer at the instant the last of it leaves the funnel.

21. Report

21.1 The report shall state the time in seconds required for the funnel to discharge, to the nearest 0.2 s; or, if so found, that the material will not run through the funnel.

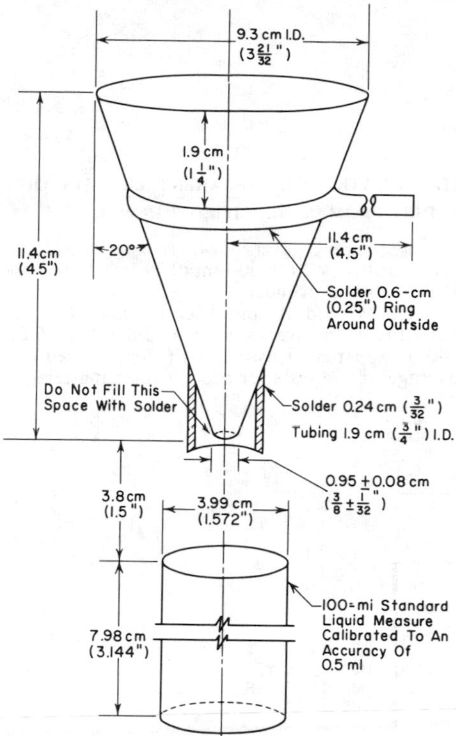

9.3 cm I.D.
(3 21/32")
1.9 cm
(1 1/4")
11.4cm
(4.5")
20°
11.4 cm
(4.5")
Solder 0.6-cm
(0.25") Ring
Around Outside
Do Not Fill This
Space With Solder
Solder 0.24 cm (3/32")
Tubing 1.9 cm (3/4") I.D.
0.95 ± 0.08 cm
(3/8 ± 1/32")
3.8cm
(1.5")
3.99 cm
(1.572")
100-ml Standard
Liquid Measure
Calibrated To An
Accuracy Of
0.5 ml
7.98 cm
(3.144")

FIG. 1 Apparatus for Apparent Density Test, Method A.

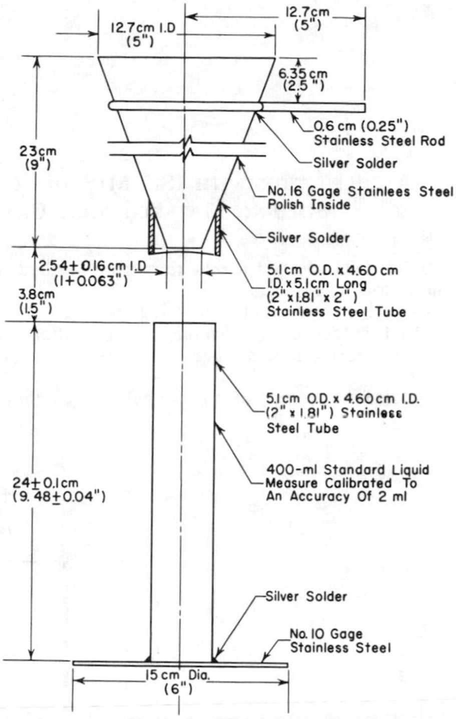

12.7cm I.D
(5")
12.7cm
(5")
6.35 cm
(2.5")
0.6 cm (0.25")
Stainless Steel Rod
23cm
(9")
Silver Solder
No. 16 Gage Stainlees Steel
Polish Inside
Silver Solder
2.54 ± 0.16 cm I.D
(1 ± 0.063")
3.8cm
(1.5")
5.1cm O.D. x 4.60 cm
I.D. x 5.1cm Long
(2" x 1.81" x 2")
Stainless Steel Tube
5.1cm O.D. x 4.60 cm I.D.
(2" x 1.81") Stainless
Steel Tube
400-ml Standard Liquid
Measure Calibrated To
An Accuracy Of 2 ml
24 ± 0.1cm
(9.48 ± 0.04")
Silver Solder
No. 10 Gage
Stainless Steel
15 cm Dia.
(6")

FIG. 2 Apparatus for Apparent Density Test, Method B.

Cut Millimeter Scale as Shown
Plunger
1
2
3
4
5
6
7
8
9
10
11
12
13
14
15
16
8.810 ± 0.008cm
(3.470 ± 0.003")
Machine From
Standard
3" Pipe
Plunger
16.15 ± 0.005 cm
(6.360 ± 0.002")
Lead Shot,
Vary Weight
as Desired
Weld
3/16" Steel Plate
8.890 ± 0.013cm
(3.500 ± 0.005")
Machine From
Extra Strong
3½" Pipe
16.15 ± 0.005 cm
(6.360 ± 0.002")
Measuring Cylinder

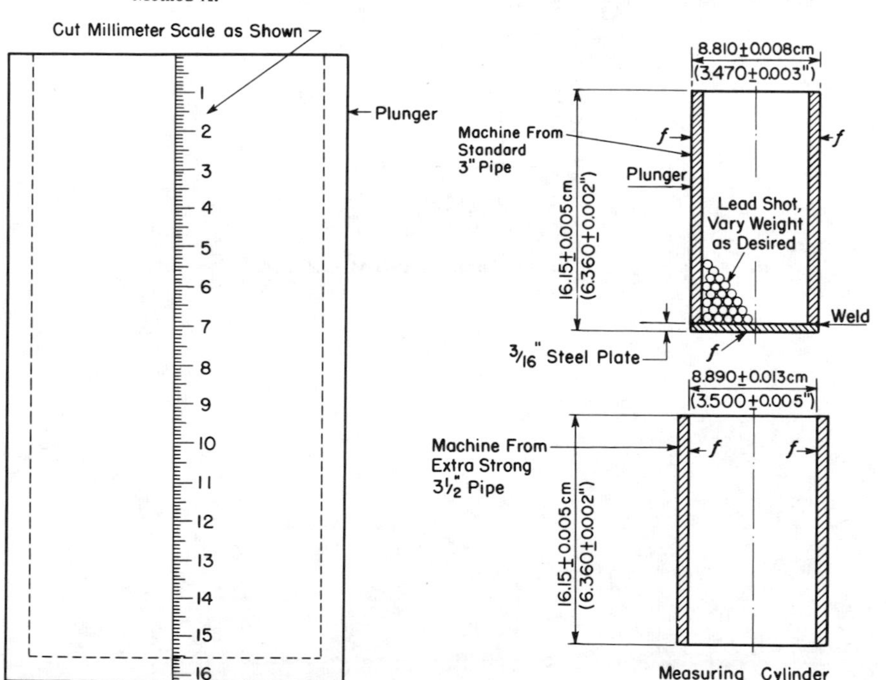

FIG. 3 Apparatus for Apparent Density Test, Method C.

APPENDIX

A1. Relation with ISO Method R 60, Determination of Apparent Density of Molding Material that Can be Poured from a Specified Funnel

A1.1 ISO Method R 60 differs from Method A in the following respects:

A1.1.1 *Funnel* as shown in Fig. A1.

A1.1.2 *Measuring Cylinder*, which differs only in that the internal diameter is 45 ± 5 mm (1.77 ± 0.20 in.).

A1.1.3 Mounting of the funnel is such that its lower orifice is 20 to 30 mm (0.79 to 1.18 in.) above the measuring cylinder.

A1.2 Method A and ISO Method R 60 have been found to give values that differ by −0.01 to +0.03 apparent density units (g/cm^3) based on the average of five tests for each of three materials.

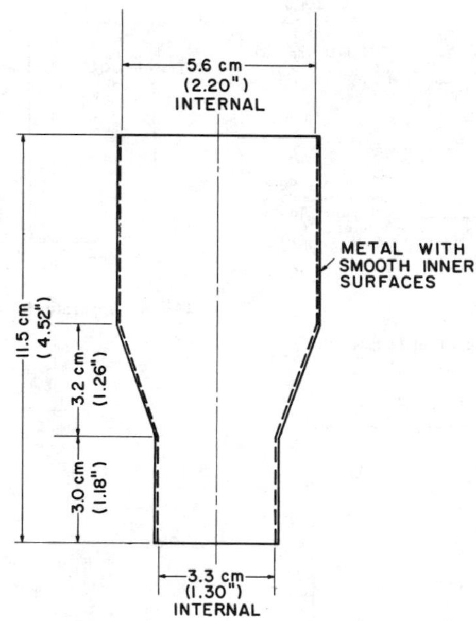

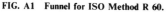

FIG. A1 Funnel for ISO Method R 60.

Designation: D 1896 – 65

American National Standard K66.9-1967
American National Standards Institute

Recommended Practice for
TRANSFER MOLDING TEST SPECIMENS FROM THERMOSETTING MATERIALS[1]

This Recommended Practice is issued under the fixed designation D 1896; the number immediately following the designation indicates the year of original adoption or, in the case of revision, the year of last revision. A number in parentheses indicates the year of last reapproval.

1. Scope

1.1 This recommended practice covers a general procedure for transfer molding test specimens from thermosetting materials for Izod or Charpy impact, flexure, tension, compression, water absorption, and electrical tests (Note 2).

NOTE 1—The values stated in U.S. customary units are to be regarded as the standard. The metric equivalents of U.S. customary units may be approximate.

NOTE 2—This method has been found useful for transfer molding of phenolics, polyesters, alkyds, and diallyl phthalate thermosetting molding compounds over the plasticity ranges specified in ASTM Method D 731, Measuring the Molding Index of Thermosetting Molding Powder.[2] Until more experience is obtained with this method, it cannot be recommended for thermosetting molding compounds not listed above, or lower plasticities than those given in Method D 731.

2. Significance

2.1 This recommended practice is subject to the definition of transfer molding as given in ASTM Nomenclature D 883 Relating to Plastics.[2]

2.2 Transfer molding by this recommended practice has the advantage of molding in one shot a set of test specimens for the tests mentioned above or any part thereof, as the occasion demands. Any runner may be blocked at will by reinserting part of the sprue of that runner.

2.3 Transfer molded specimens, in many cases, have an improved surface finish due to the mixing action of the gates and runners of the mold.

2.4 Transfer molding is particularly suited to materials of very soft plasticity with a long cure time. Very little resin is lost in flash at satisfactory molding pressures.

2.5 Bumping of the mold is not required to release trapped volatile matter as the gas is free to flow from the vent end of the mold. This is a particular advantage for heat-resistant compounds and lowers the tendency to blister at high exposure temperatures.

2.6 Long fiber filler, such as glass roving, chopped cloth, long fiber asbestos, or shavings, is not recommended for transfer molding. These filler materials tend to break, tear, or ball in passing through the gates of the mold and, consequently, lose their original strength.

2.7 Izod impact strength is generally lower by transfer molding than by compression molding for short-fiber molding compounds. Quite often this impact strength will vary along the axis of the bar due to molding conditions.

NOTE 3—If the temperature of the mold is held constant and the plunger pressure varied for a designated thermosetting molding compound, two extreme characteristic conditions can be obtained. If the pressure is low, then the vent end of the cavity will not fully fill and weld lines will form by incomplete knitting of the material. If the pressure is too high, the mold cavity will fill fast; the outside of the specimen will case harden while the pressure is still forcing material out the vent, and a ball-and-socket grain structure will be obtained. Thus, a ball-and-socket grain structure will be obtained for both over-pressure and under-pressure

---

[1] This recommended practice is under the jurisdiction of the ASTM Committee D-20 on Plastics. A list of committee members may be found in the ASTM Yearbook. This standard is the direct responsibility of Subcommittee D-20.90 on Specimen Preparation.

Current edition effective Aug. 31, 1965. Originally issued 1961. Replaces D 1896 – 61 T.

[2] *Annual Book of ASTM Standards*, Part 27.

on the plunger cavity of the mold. A ball-and-socket structure is an indication of poor molding conditions and a bar weak in impact.

2.8 Flow and weld lines in a molded piece are often positions of mechanical or electrical weakness. Weld lines may be found in some degree of severity throughout the molded piece. The semisolid molding compound passing through the gate, is subject to non-Newtonian flow and, consequently, wrinkles and folds as it travels down the mold cavity. Fibers and other reinforcements in the molding compound align with the flow pattern and, consequently, mold perpendicular to the axis of the bar at the center and parallel at the surface of the bar. Mold temperature, thermal conductivity, and plasticity of the molding compound, radio frequency preheating, and plunger pressure are factors that influence the time to fill the mold cavities and the knitting of the weld lines.

2.9 Flexural and tensile strength tests with transfer-molded specimens quite often yield higher values than tests made with compression-molded specimens of the same material. Flexural tests are particularly sensitive to transfer molding due to the thin resin skin formed at the surface of the bar during the final filling of the cavity and pressure build-up.

2.10 Fast-cure molding compounds that have good thermal conductivity are not recommended for transfer molding due to precure, more susceptibility to weld lines, and lower strengths.

2.11 Fixed plunger pressure cannot be specified for all types of materials. Molding compounds of the same type often come in many different plasticities as measured in accordance with Method D 731. One material may mold properly under a fixed set of conditions while another with a stiffer plasticity may not mold at all; while a third, having a softer plasticity, may even flash excessively through the vent and not take the molding pressure before curing.

3. Apparatus

3.1 *Press*—A hydraulic press designed to develop and maintain accurately any desired pressure between 211 and 844 kgf/cm^2 (3000 and 12,000 psi) on the plunger to ±17.6 kgf/cm^2 (±250 psi) and have a minimum plunger loading capacity of 230 cm^3 (14 in.^{3}) (Note 4). The ram pressure shall be at least 10 percent higher than the plunger pressure.

NOTE 4—Plunger molding pressure under actual molding conditions is a variable which is difficult to control. Pressure standardization should be done on an empty cavity with the plunger against the mold stop plate. The speed of the ram is not important as the mold is closed before the plunger operates. A ram speed of 3.6 m/min (140 in./min) and a plunger speed of 2.2 m/min (85 in./min) have been found satisfactory when the mold is not loaded. The plunger speed is subject to the flow properties of the molding material when the plunger cavity is loaded with molding compound.

3.2 *Mold*—The mold shall conform to the plunger cavity mold shown in Fig. 8 of ASTM Recommended Practice D 647, for Design of Molds for Test Specimens of Plastic Molding Materials.[2]

3.3 *Heating System*—Any convenient method of heating the press platens and plunger cavity may be used provided the heat source is constant enough to maintain the mold and plunger temperature within ±3 C (5 F).

3.4 *Temperature Indicator*—Either a thermometer or pyrometer as specified in ASTM Recommended Practice D 958, for Determining Temperatures of Standard ASTM Molds for Test Specimens of Plastics[2] may be used.

3.5 *Preforming*—Any preforming equipment or press may be used that will make satisfactory preform material for the plunger cavity and ease in handling in the radio frequency preheater.

3.6 *Gates*—A gate size 1.52 by 6.35 mm (0.060 by 0.250 in.) has been found satisfactory for most materials. Other gate sizes may be used if the molding conditions warrant.

4. Conditioning

4.1 Molding compounds are generally preformed, radio frequency preheated, and molded from the compound in the as-received condition.

4.2 Molding compounds, known to contain a high percentage of moisture, may be conditioned for 30 min at 90 ± 3 C (194 ± 5 F) in a forced draft oven and preformed immediately afterward. The preformed material shall be stored in a desiccator over anhydrous calcium chloride at room temperature until radio frequency preheated and molded. The depth of the molding compound in the over tray shall not exceed 13 mm (1/2 in.) while subject to oven preheating.

4.3 In the case of a referee test, the preform material shall be prepared as indicated in 4.2.

5. Procedure

5.1 Uniformly preheat by radio frequency sufficient preformed material of a given molding compound to 105 ± 5 C (221 ± 9 F) within a heating period of 1/2 min.

NOTE 5—The temperature of the preformed material after radio frequency preheating may be determined by a needle-type pyrometer of low thermal capacity.

5.2 Immediately remove the preheated preformed material from the radio frequency preheater, place it in the plunger cavity, and apply molding pressure within a period of 15 s after completion of radio frequency preheating.

5.3 Adjust the plunger molding pressure to the plasticity of the material, after first determining the minimum plunger pressure to fill each cavity used in the mold. Then increase the plunger pressure by 70.3 ± 17.6 kgf/cm^2 (1000 ± 250 psi) for a correct molding pressure.

NOTE 6—It has been found that a satisfactory plunger pressure of 337 kgf/cm^2 (4800 psi) is satisfactory for most general-purpose materials with a molding index of 3.32 kg (7500 lb), as measured in accordance with Method D 731 when the material is properly radio frequency preheated.

5.4 The recommended temperature of the mold and plunger cavity is 175 ± 3 C (347 ± 5 F). Other molding temperatures may be used when mutually accepted by the purchaser and the seller.

5.5 The minimum cure time shall be 3 min as measured from the time the pressure on the plunger is within 70.3 kgf/cm^2 (1000 psi) of the preset pressure.

5.6 No knockouts are required to remove the molded specimens, sprue, and cull from the mold. In fact, the whole molded spider can be removed as a unit from the mold with the aid of compressed air from a hose.

Designation: D 1897 - 68

Recommended Practice for

INJECTION MOLDING OF SPECIMENS OF THERMOPLASTIC MOLDING AND EXTRUSION MATERIALS[1, 2]

This Recommended Practice is issued under the fixed designation D 1897; the number immediately following the designation indicates the year of original adoption or, in the case of revision, the year of last revision. A number in parentheses indicates the year of last reapproval.

1. Scope

1.1 This recommended practice covers the injection molding of test specimens of thermoplastic molding and extrusion materials.

2. Summary

2.1 The injection molding of plastic test specimens for the characterization of materials involves many variables that may be categorized broadly under (*1*) type of molding machine, (*2*) design of molds, and (*3*) molding conditions. This recommended practice attempts to control some of these variables and nullify others. Definite molding temperatures are specified. By a sequence of operations pressure and cycle are established on the basis of their effects upon the molded part itself rather than upon external gage and timer settings.

3. Significance

3.1 It is well known that plastic test specimens molded under different conditions can exhibit significantly different properties. This has led to disputes, as between supplier and molder. This recommended practice is designed to minimize these differences without being overly restrictive. No attempt has been made to specify one type of molding machine or restrict the mold to one design.

3.2 The molding temperatures recommended are considered suitable to yield representative properties. It was not desired to maximize any particular property.

3.3 When molding conditions have been established, they should be recorded carefully. They may then be used as often as desired for that test specimen molded of that material on that machine with that mold.

3.4 This recommended practice would probably not be suitable for experimental materials or small samples. A relatively large amount of material is required for best results, depending upon the heating capacity of the molding machine.

3.5 This recommended practice is believed to be applicable to all thermoplastic molding and extrusion materials. However, interlaboratory studies have been limited to polystyrene, styrene-butadiene and polycarbonate plastics thus far.

4. Apparatus

4.1 *Molds*—The mold shall be of the design shown in Fig. 4 of ASTM Recommended Practice D 647, for Design of Molds for Test Specimens of Plastic Molding Materials.[3] Other mold designs may be used if they produce specimens within the specified dimensional tolerances.

4.2 *Injection Molding Machine*, in which the mold is mounted. There shall be provi-

[1] This recommended practice is under the jurisdiction of ASTM Committee D-20 on Plastics. A list of committee members may be found in the ASTM Yearbook. This Standard is the direct responsibility of Subcommittee D-20.90 on Specimen Preparation.

Current edition effective May 7, 1968. Originally issued 1961. Replaces D 1897 – 66 T.

[2] In 1968, the title and scope were broadened to encompass all thermoplastic materials, and specific provisions for polycarbonate materials were added.

[3] *Annual Book of ASTM Standards*, Part 27.

sions made to measure and control the following:

4.2.1 Feed of material by weight or volume to the injection cylinder,

4.2.2 Pressure applied to the injection ram,

4.2.3 Temperature of the plastic heating chamber,

4.2.4 Temperature of the plastic,

4.2.5 Temperatures of the cavity and core surfaces of the mold, and

4.2.6 Cycle timing.[4]

5. Procedure

5.1 Follow the molding procedure in the sequence given in the following paragraphs. Repeat the procedure in its entirety for each test specimen molded. Mold only one specimen per cycle unless in a multicavity mold the cavities are identical and symmetrically located. In multicavity molds in which the cavities are not identical, apply the molding procedure separately to each cavity, disregarding all others in those molds in which cavities cannot be blocked.

5.2 *Molding Temperatures:*

5.2.1 Based upon past experience, recommendations by the supplier, or other available information, select a preliminary molding cycle for the test specimen, the type of material, and the molding machine being used. All times should be accurate to within ± 0.2 s.

5.2.2 The mold surface temperature should be adjusted to the appropriate value as shown in Table 1. The temperature should be measured with a surface-type pyrometer to an accuracy of ± 2.0 C (3.6 F).

5.2.3 Adjust the heating cylinder temperature to give the average stock temperature shown in Table 1. Measure this temperature on cycle by actually taking the temperature of a free shot with a needle-type pyrometer to an accuracy of ± 2.0 C (3.6 F). Move the needle about constantly in the plastic mass. Make a sufficient number of measurements to establish a reliable average, and perform frequent checks. For a new material not covered by Table 1, establish stock and mold temperatures by agreement between the purchaser and the seller until such time as generally accepted values can be added to Table 1.

NOTE—If not established, the stock temperature for a material can be determined by holding all other conditions fixed and gradually reducing the temperature until satisfactory pieces cannot be produced. The stock temperature for molding is then established at a selected level above the minimum fill temperature, the level to be agreed upon by the molder and the material supplier.

5.3 *Injection Pressure:*

5.3.1 While filling the cavity at maximum ram speed, and with stock and mold temperatures set as described in 5.2, reduce the injection pressure to the point at which visibly defective specimens are obtained. Record the gage pressure at which this occurs as P_s. Read this pressure as precisely as possible.

5.3.2 Obtain the gage pressure for molding specimens P_m by increasing P_s by 30 to 35 percent, or

$$P_m = 1.30P_s \text{ to } 1.35P_s$$

5.3.3 Determine P_m separately for each type of specimen molded.

5.3.4 In multicavity molds in which two or more nonidentical cavities must fill simultaneously, determine the pressure with respect to the particular specimen desired, disregarding all others.

5.4 *Molding Cycle:*

5.4.1 Modify the initially selected cycle by reducing the plunger-forward time until visually defective pieces appear, but keep the total cycle time constant by keeping the mold-closed time constant. Designate the point at which defective pieces appear as the "minimum plunger-forward time." Starting at this point, increase the plunger-forward time in 3.0 ± 0.2-s increments until visually satisfactory pieces are obtained, keeping total cycle constant. Allow an increase in plunger-forward time no greater than absolutely necessary to produce good specimens.

5.4.2 Reduce the mold-closed time and thus total cycle time progressively until defective pieces again begin to appear. Designate this point as the "minimum mold-closed time." Starting at this point establish the final total cycle time by increasing the mold-closed time in 3.0 ± 0.2-s increments until the specimens become satisfactory, but keep the increase to the minimum required. Record the final molding cycle.

[4] See ASTM Recommended Practice D 1130, for Injection Molding of Specimens of Thermoplastic Materials, which appears in this publication.

5.4.3 Keep the mold-open time consistent within ±1.0 s.

5.5 *Readjustments:*

5.5.1 Check the stock or plastic temperature while the apparatus is running with the established pressure and cycle. This may have changed somewhat because the cycle established may differ from the cycle initially selected.

5.5.2 If necessary, readjust the cylinder temperature to return the stock temperature to the value specified in Table 1. If the readjustment is of considerable magnitude, as indicated by flashing or underfilling, repeat the procedures of 5.3 and 5.4 to make compensatory changes in injection pressure and cycle.

5.6 *Number of Moldings*—The total number of moldings to be made will depend upon the amount of material available, the number of test specimens required, and the need to reach molding equilibrium.[4]

6. Report

6.1 The report shall include the following:

6.1.1 Type and description of material used,

6.1.2 Pretreatment of material,

6.1.3 Mold used,

6.1.4 Description of specimen,

6.1.5 Number of specimens molded per shot,

6.1.6 Gate size and description,

6.1.7 Make of machine used,

6.1.8 Type of machine control, automatic or hand,

6.1.9 Rated capacity of heating chamber, (should be approximately 50 to 75 percent of rated heating chamber capacity),

6.1.10 Weight of molding plus sprues,

6.1.11 Heater control indicated temperature(s),

6.1.12 Stock (plastic) temperature,

6.1.13 Mold surface temperature (cavity and core),

6.1.14 Minimum molding pressure, P_s,

6.1.15 Gage pressure for molding P_m,

6.1.16 Preliminary molding cycle,

6.1.17 Mold-open time,

6.1.18 Minimum plunger-forward time,

6.1.19 Plunger-forward time for molding,

6.1.20 Minimum mold-closed time,

6.1.21 Mold-closed time for molding,

6.1.22 Screw speed (if applicable),

6.1.23 Back pressure of screw (if applicable),

6.1.24 Temperature of pressure readjustments required after establishment of final cycle,

6.1.25 Number of moldings discarded at final conditions before selection of specimens, and

6.1.26 Number of moldings made.

7. Precision

7.1 Round robins were conducted in which selected materials were molded by various laboratories with their equipment in accordance with this recommended practice. Precision estimates obtained from data generated by these studies are listed in Table 2 in terms of the estimated standard deviation (s) and the coefficient of variation ($v = 100s/\bar{x}$).

[5] *Annual Book of ASTM Standards*, Part 26.

TABLE 1 Stock and Mold Temperatures for Various Thermoplastic Materials

Material	Temperature, deg C	
	Stock	Mold
Polystyrene, Types 1, 2, 4[a], Styrene-butadiene Grades 1, 3, 5, 7[b]	220 ± 5	65 ± 5
Polystyrene, Type 3[a], Styrene-butadiene Grades 2, 4, 6, 8[b]	220 ± 5	80 ± 5
Polycarbonate, Type 1, Grade 1[c]	290 ± 5	90 ± 5
Polycarbonate, Type 1, Grade 2[c]	300 ± 5	90 ± 5
Polycarbonate, Type 1, Grade 3[c]	310 ± 5	90 ± 5

[a] The polystyrene molding and extrusion materials shall conform to ASTM Specification D 703 – 64, for Polystyrene Molding and Extrusion Materials.[5]

[b] The styrene-butadiene molding and extrusion materials shall conform to ASTM Specification D 1892 – 66, Styrene-Butadiene Molding and Extrusion Materials.[5]

[c] The polycarbonate molding and extrusion materials shall conform to ASTM Specification D 2473 – 66, for Polycarbonate Plastic Molding, Extrusion, and Casting Materials.[3]

TABLE 2 Precision Estimates from Round-Robin Data

Material	Molders	Testers	Tensile Strength[a]		Elongation[a]		Molders	Testers	D.T.L.[b]	
			s, kgf/cm²	*v*, percent	*s*, percent	*v*, percent			*s*, deg C	*v*, percent
Polystyrene, Type 3 (Material A)	6	1	27	4.9	0.46	14.0	5	1	0.50	0.58
Polystyrene, Type 3 (Material B)	6	1	24	4.2	0.29	14.0	5	1	0.74	0.81
Styrene-butadiene, Grade 6	6	1	25	6.2	4.0	16.0	3	2	0.97	1.3
Styrene-butadiene, Grade 7	6	1	14	5.4	13.0	33.0	3	2	2.4	2.8
									Izod Impact[c]	
									s, cm·kgf/cm	*v*, percent
Polycarbonate, Type 1, Grade 2	2	1	34	4.7	7.1	6.3	2	1	2.3	2.5
Polycarbonate Type 1, Grade 2	2	1	30	4.3	9.2	8.2	2	1	2.7	2.9

[a] Tensile strength and elongation were obtained in accordance with ASTM Method D 638, Test for Tensile Properties of Plastics.[3] Type 1 specimens were tested at Speed B (polystyrene and styrene-butadiene) or at Speed C (polycarbonate).

[b] Deflection temperature under load (D.T.L.) was determined on 1.27 by 0.64 by 12.7-cm (½ by ¼ by 5-in.) specimens at 18.5 kgf/cm² (264 psi) fiber stress in accordance with ASTM Method D 648, Test for Deflection Temperature of Plastics Under Load.[4] Specimens were conditioned in accordance with Procedure A of ASTM Methods D 618, Conditioning Plastics and Electrical Insulating Materials for Testing.[3]

[c] Izod impact strength (notched) was obtained in accordance with ASTM Method D 256, Test for Impact Resistance of Plastics and Electrical Insulating Materials.[4] Test specimens were 0.64 cm (¼ in.) thick.

Designation: D 1898 - 68

Recommended Practice for
SAMPLING OF PLASTICS[1]

This Recommended Practice is issued under the fixed designation D 1898; the number immediately following the designation indicates the year of original adoption or, in the case of revision, the year of last revision. A number in parentheses indicates the year of last reapproval.

1. Scope

1.1 This recommended practice is primarily a statement of principles to guide purchasers of plastic materials purchased under specifications to prepare sampling plans that will describe sampling procedures and that will enable them to determine within practical limits whether or not the products meet the specifications.

1.2 The same principles may be used to guide the preparation of specifications in quantitative terms.

1.3 Some of the same principles may be used to determine the actual quality of a product, including its average and variation above and below that average, with respect to a particular property.

1.4 Since the design of probability sampling depends upon the ultimate use of the samples and the resulting data, consideration must be given to the intended inspection of the samples. Hence, the design of plans both for examination and testing of samples for their attributes and variables is included.

1.5 This recommended practice is intended for general guidance when little information is available on the variability of the material and of the method of inspection. In some cases, quality control chart methods may be substituted for the procedures herein. In any event, a statistician should be consulted if difficulty is encountered in applying the recommendations herein to the design of a specific sampling problem.

1.6 There is no intent to guide the *disposition* of material that is found to be off-specification by sampling and inspection; disposition is a contractual matter.

1.7 The following outline is presented to facilitate use of this recommended practice:

NOTE 1—The values stated in U.S. customary units are to be regarded as the standard. The metric equivalents of U.S. customary units may be approximate.

2. Significance

2.1 The purpose of the sample may be as follows:

2.1.1 To estimate properties of a lot or shipment, such as the percentage of some constituent, the fraction of the items that fail to meet a specific requirement, the average weight or property of an item, or the quality or total weight of the shipment, or simple identity.

2.1.2 To dispose of the lot or shipment rationally, without the intermediate step of the formation of an estimate.

[1] This recommended practice is under the jurisdiction of ASTM Committee D-20 on plastics. A list of committee members may be found in the ASTM Yearbook. This recommended practice is the direct responsibility of Subcommittee D-20.13 on Statistical Techniques.

Current edition effective Jan. 10, 1968. Originally issued 1961. Replaces D 1898 - 61 T.

2.1.3 To define new materials in terms of their properties.

2.1.4 To provide material for evaluation of a test method.

3. General Philosophy

3.1 Sampling is a means employed to meet the problem of estimating the quality of a lot from the inspection of only part of the lot. When 100 percent inspection ("screening") of a lot is performed, sampling does not apply.

3.2 The inspection of the sample or portion according to a sampling plan results in an estimation of properties, hence, an uncertainty which can be expressed for a given sampling plan in terms of an operating characteristic curve. Each such curve, that is, for a given acceptable quality level, can be used to obtain two numerical quantities: α, the seller's risk, and β, the purchaser's risk. An example is given in Appendix A2.

3.3 The acceptable quality level (AQL) provides assurance to the producer that good material will seldom be rejected. There is a large probability that quality that is close to and slightly better than the AQL will be accepted (A2.2.1, Note A9). Consequently the seller is well protected against rejection of submitted lots of material from a process that is at the AQL or slightly above it. The operating characteristic curve is a graph of the risks that are taken in sampling inspection plotted as a function of incoming lot quality. Any of several units such as average, median, or coefficient of variation, can be used to establish the seller's and purchaser's risks. The unit most frequently used for this purpose is the arithmetic average.

3.4 An immediate question which arises in approaching the sampling of a given lot is that of selecting the number of units to be sampled, that is, defining the sampling plan. A sample of relatively small size will generally be adequate for a large lot. For a small lot, however, a relatively large sample is required for reliable estimation of properties. Such a sample is often uneconomical, and a compromise must then be found between the cost of an adequate sample and its inspection, and the risk entailed in the less certain information from a smaller sample. Actually, such a balance should always be sought in statistical practice so that the value of the information to be obtained from a given program is commensurate with the cost of obtaining the information. This recommended practice includes in Appendixes A2 and A3 some specific directions for choosing the number of units to be sampled. This has proved useful in extensive use in certain seller-purchaser relationships.

NOTE 2—Some detailed procedures for setting up the numerical features of a sampling plan for a particular characteristic of a material will also be found in ASTM Recommended Practice E 122, for Choice of Sample Size to Estimate the Average Quality of a Lot or Process.[2]

3.5 In the early stages of experience with a material, the required sample quantity for an isolated lot is generally calculated while considering the lot as an entity. Considerable advantage is often gained in evaluating a series of shipments of similar material by regarding their manufacture as a whole. This is generally done after there is some assurance that the manufacturing process is under control, as by the application of control chart methods.[3]

3.6 If the sampling plan is to be designed to help establish whether a lot meets the requirements of a material specification, it is imperative that the specification be formulated carefully.[4]

4. Sampling for Attributes

4.1 The quality of plastic materials or products is judged in part by the frequency of occurrence of their attributes, that is, properties or conditions which may or may not be present in a given sample. Examples include visual defects, contamination, and success or failure in meeting dimensional or test specifications. Probability sampling should be employed to assess the actual frequency, provided that a small number of undesirable attributes can be tolerated.

NOTE 3—If the presence of a given attribute is unacceptable, all the lot must be inspected, and statistical techniques do not apply.

4.2 The quality of a lot with respect to attributes is evaluated by inspection. The sampling plans for such an evalution are conveniently

[2] *Annual Book of ASTM Standards*, Part 30.

[3] See the *ASTM Manual on Quality Control of Materials, STP 15-C*, Am. Soc. Testing Mats. (1951).

[4] See the Recommendations on Form of ASTM Standards, availiable from ASTM Headquarters.

expressed in tabular and graphic forms (Appendix A2). The selection of AQL's for a given material is traditionally a contractual matter. Factors in determining the number of samples to be taken from a lot include the size of the lot and the degree and level of inspection.

4.3 Attributes may be evaluated by examination or by testing.

4.4 *Sampling for Examination* presents fewer problems. Examination generally is nondestructive and hence less costly. Contamination of molding powder, bubbles in sheets, fisheyes in film, warp and twist in extruded shapes, and dimensions by go and no-go gages fall in this category. Because of their subjective nature, it is difficult to put firm numbers on visual defects and contamination for a plastics material specification. Once the specification has been established, however, a sufficient number of samples for a good statistical evaluation of a lot can generally be examined at reasonable cost.

4.5 *Sampling for Physical and Chemical Tests* must be planned while keeping in mind that the tests are often time-consuming and destructive of the sample. The cost tends to limit the number of tests that will be performed. Tests include the determination of particle size and of moisture of molding powder, measurement of properties such as tensile strength, hardness, and deflection temperature, and measurements of color and transparency. For many of these tests, specimens must be cut or molded. The manipulations and calculations may require considerable time for each sample.

4.6 The definition of the sampling unit is important, since it affects the size of the sample. The conception of the sampling unit should also take into account the intended use of the material. For instance, a single bubble may seriously impair the usefulness of a large sheet of glazing material that is to be used in its entirety in military aircraft. If the sheet is to be cut into 152-mm (6-in.) portholes, however, more defects could presumably be tolerated because they can be marked and then avoided in the cutting.

5. Sampling for Variables

5.1 The ultimate object of sampling for variables is to estimate the average value of a property of the material that can be measured on a continuous scale and the variability of the property, with a desired degree of confidence. Such a procedure can be used to obtain data for establishing a specification, as well as for comparing a new lot with a specification.

5.2 In working against a specification for a variable, the concept of AQL is useful because it reduces to tabular and graphic forms the choice of sample size (that is, the number of units from the lot to be drawn into the sample) that is ncessary in order to minimize the risk of accepting substandard material.

5.3 In both inspection for attributes by means of tests, and in inspection for variables, values measured in tests are the ultimate bases for judgment of quality, that is, both types of inspection are intended to ensure that the percent defective is less than a certain value. The difference lies in that, in the first instance, the individual test results, or perhaps an average for each sampling unit, are used in deciding acceptance or rejection. In inspection for variables, however, the range (spread) of the group of measurements is also important. The estimated standard deviation, derived from the measurements, is multiplied by a factor from tables, and the resulting product is then subtracted from the average. The corrected average is used for comparison with a minimum specification value. (When the specification is a maximum limit, the product is added to the average.) This means, as one result, that a lot of product having a relatively poor average value, but a narrow range of values for a certain property, may be accepted as meeting a specification, while another lot having a better average but greater variation may be unacceptable. This puts a premium on consistent performance, which is probably more economical in the long run for everyone concerned.

5.4 In setting up a plan for sampling for variables, it must be remembered that it is generally easier to attain a desired degree of assurance for the average value than for its standard deviation. This can easily be demonstrated by an experiment which involves drawing numbers truly at random from a bowl. In general, it takes a very few trials in order to establish a reasonable value for the average number. It takes many more samples to arrive at a good estimate of the standard deviation of the average. This would be equally true for a plastic molding powder that happened to be quite heterogeneous with respect to some property, for example, if the moisture content varied from one particle to another. If random sampling were achieved, a good idea of the average moisture could be obtained from a few samples. It would require considerably more to establish the actual variability.

5.5 Detailed plans for sampling for variables, suggested for wide applicability, will be found in Appendix A3.

6. Means and Standard Deviations

6.1 In evaluating the results of measurements of a variable property of a material, several calculations should be made. It is common practice to perform a series of measurements on the material and to calculate the arithmetic mean or average of the series, $\bar{X}$, for comparison with the specification.

6.2 Even in the most careful measurements, a certain amount of variability or dispersion will be encountered. This must also be reported in order to allow meaningful comparisons of the values. For the purpose of this recommended practice, the most useful numerical expression of this variability is an estimate of the standard deviation, s, of the individual values, calculated as follows:

$$s = \sqrt{(\Sigma(X_i - \bar{X})^2/(n - 1)}$$

where:

X_i = value of a single observation,
$\bar{X}$ = arithmetic means of a set of observations, and
n = number of observations in the set.

The reliability of the mean can now be estimated as $s/\sqrt{n}$, the standard deviation of the mean.

6.3 The condition of the average and the standard deviation can also lead to an estimation of the "confidence limits" of the average, provided that only random errors influence the measurements. In order to achieve narrow confidence limits, a large number of determinations must be made. This is seldom realized in practice because of economic considerations of the cost of repeated measurements. Hence, it becomes necessary to report also the numbers of measurements made in arriving at the stated average. Then, confidence limits can be calculated from tables of Student's "t" values.

6.4 Sampling may be designed simply to obtain such information as the above on a new material or test method. This information is then applied to make decisions on the quality of subsequent lots of the material.

7. Comparison of Sampling Plans

7.1 Table 1 shows the sample sizes (number of units of product or specimens) and the qualities of lots (in percent defective) that would be accepted 90 and 10 percent of the time under three methods of inspection. Inspection Method A is inspection by means of attributes under Military Standard MIL-STD-105 at Inspection Level II. Method B is inspection by means of variables, with variability unknown, under Military Standard MIL-STD-414 at Inspection Level IV. Method C is inspection by means of variables, with variability known, under Military Standard MIL-STD-414 at Inspection Level IV. The plans shown are for lot sizes of 500 and 10,000 units subjected to "normal" single inspection under AQL's of 1.0 and 6.5 percent defective. Table 1 is presented to show the following:

7.1.1 Smaller sample sizes are needed for the same degree of discrimination for Method B than for Method A, and for Method C (variability known) than for Method B (variability unknown).

7.1.2 Because of larger sample size, greater discrimination (the power to separate good lots from bad) can be obtained by inspecting large lots than by inspecting small lots, provided that the large lots are reasonably homogeneous. However, increased discrimination in inspecting small lots may be obtained by using a higher inspection level, that is, larger sample if the increased cost of inspection is considered justified. This change is made by substituting a code letter later in the alphabet.

7.1.3 Inspection Level IV of Military Standard MIL-STD-414 is roughly equivalent in discrimination to Inspection Level II of Military Standard MIL-STD-105.

7.1.4 Under these plans the seller's risk (the risk of having good lots rejected) varies little between the several sampling plans. The chief difference in the plans is the difference in the purchaser's risk (the risk of accepting bad lots).

GENERAL SAMPLING PROCEDURES

8. Scope

8.1 A general sampling procedure may be outlined, both for random and stratified sampling, followed by specific directions for handling the physical nature of various plastic materials. Once the precise purpose of the sampling plan has been decided in terms of attributes or variables and the level of quality it is desired to attain, then one may proceed to formulate a specific procedure, possibly as follows:

8.1.1 Decide on the size of the sampling unit. This decision may be obvious from the nature of the lot, for example, in a lot of ten drums of a plastic molding powder of a single grade and color, known to be uniform within drums, the drums are the sampling units. This decision may have already been made in arriving at the desired purpose and AQL.

8.1.2 Count the number of units in the lot.

8.1.3 Designate, as by labeling, the sequential units in the lot.

8.1.4 Deduce, from the number of units and the sampling plan, the number of units to be sampled (demonstration graphs and tables in Appendixes A2 and A3).

8.1.5 Designate the actual units that are to be samples. For choice between random and stratified sampling, see 9.1 and 10.1.

8.1.6 Perform the sampling of these units.

8.1.7 Deliver the samples, properly identified, for inspection where the calculation will be performed and the report made.

8.1.8 Ascertain the quality of the data or lot, from the report, for further action.

9. Random Sampling (Unstratified Material)

9.1 Randomness is achieved when every part of the lot has an equal chance of being drawn into the sample in the first and in successive sampling operations. To permit random sampling, all units in the lot shall be designated.

NOTE 4—Random does not have the connotation of haphazard. A random sample ensures that repeat tests will provide the same result, within the sampling error of a normal distribution, if the distribution is normal.

9.1.1 Choose numbers in sequence or other serial code so that sampling by random numbers may be employed.

9.1.2 If possible, this sequence shall be in direct relation to order of manufacture, packaging, or filling of the package, for example, later units and strata in units shall have higher numbers, extruded shapes shall be numbered in unit areas, or lengths from end to end. Therefore, where possible, the manufacturer should code every package consecutively.

9.1.3 When a large quantity of material is contemporary, for example, the transverse direction of a wide extruded sheet, also make designations of subdivisions of this width, in sequence from one side to the other for each time interval in the machine direction.

9.2 If serial codes other than sequential numerals are normally applied to the units, the user can readily construct an equivalent table of random items, once the code is related to numerals, for example, the code sequence 1A, 2A, 3A, 1B, 2B, 3B,—could become 1, 2, 3, 4, 5, 6,—before randomizing. Although subsequent sampling and testing of the numbered units will be in random fashion, recording of such a "location designation" of each unit will still make it possible to reconstruct time and position trends of unit properties for process studies if desired.

9.3 Random sampling of the numbered units is accomplished by use of a table of random numbers. If the total number of units in the lot from which samples are to be taken is exactly 10, or 100, or 1000, use random numbers from 1 to 10, 1 to 100, or 1 to 1000, respectively. In the more likely cases of other totals, use the actual total to prepare a new table of random numbers from the next higher table; for example, for a total of 15, calculate from a table of random numbers from 1 to 100. Books of tables of random numbers are available (**4**).[5]

9.4 In practice, application of random numbers may be simplified. For example, where the actual use of random number tables is inconvenient, random numbers may be selected in advance and provided in envelopes for use as needed. In the selection of material from sheets, templates with random cutouts can be used. Random numbers may also be found by successive throws of a 20-sided die.

9.5 Since it is vital that every significant

[5] The boldface numbers in parentheses refer to the list of references at the end of this recommended practice.

unit in the lot have an equal chance to be represented among the final samples, it is important that no unit (or part of it) be omitted from the numbering scheme unless it is also to be omitted from ultimate use of the material, for example, because of visible contamination or other obvious reasons for off quality. Thus, the edge of a sheet may be omitted from the sampling pattern only if it is also to be trimmed off before use.

9.6 With regard to material in the central parts of sheets, lengths of pipe, and the like, the precaution against deliberate omission from the sampling pattern applies primarily to the preliminary or occasional experiments which are designed to establish the variability of the material. Economy generally dictates against sampling such central portions, but a correction must then be applied to the data if the material is known to be nonuniform.

9.7 In evaluation of the over-all quality of a lot, of course, it is imperative that all the samples of random choice be used, regardless of individual quality, unless they are to be discarded from the lot because of off-color or contamination.

9.8 Samples from a lot of molding powder should not be blended together before testing, if information about the variability within the lot is desired. Only the average quality of the lot can be deduced from a blend. If such blending is performed, the presence in the lot of portions of poor quality may not be detected.

NOTE 5—Systematic sampling, by sampling every *n*th unit, may be justified if it has been shown that the period of significant variation is very short or very long compared to the sampling interval.

10. Sampling Stratified Materials

10.1 It may be known or suspected that the value of a property of the material varies in a nonrandom fashion throughout the lot, for example, because of stratification or size segregation. Stratified sampling should then be employed. Divide the lot into a number of real or imaginary sections or strata, and sample every such division. In the ultimate calculation, the average property for the lot must take into account the relative sizes of the strata, if they vary in size, by appropriate weighting of the data. Exercise great caution in the application of the MIL-STD plans to stratified sampling.

10.2 Stratified sampling is generally employed in studies of sampling to determine the presence of any persistent bias in the property with time or direction (for example, size segregation within a drum of molding powder) which might preclude the application of random sampling.

10.3 Stratified sampling is recommended when compositing, since few materials are completely homogeneous. If a composite is prepared from a stratified material, each stratum should be sampled in an amount proportionate to its size.

SPECIFIC SAMPLING PROCEDURES

11. Scope

11.1 Depending on the physical form of a plastic material, the degree of uniformity that may have been imposed upon a lot before shipment (for example, blending of 4500 kg (10,000 lb) of molding powder before packaging in drums), and the size of the lot, various specific approaches to the sequential numbering of the sampling units in the lots and the subsequent sampling may be employed.

12. Sampling Molding Powder (Fluff, Pellets, Cubes)

12.1 *In Small Packages:*

12.1.1 The packages shall be the sampling units. Number the packages in the lot serially.

12.1.2 Choose the total number of packages to be used in sampling from considerations given above or in Recommended Practice E 122 for the most critical test to be run on the material.

12.1.3 Using random numbers, obtain numbers to indicate all the packages that are to be sampled.

12.1.4 Take a sample from each package designated in 12.1.3.

NOTE 6: *Quantity of Sample*—If more than one type of inspection is to be run on the sample, be sure to take enough sample for all such inspections from the package and make it uniform by blending. The gross quantity of sample taken from one unit should preferably be approximately twice that estimated to be required for the tests that are to be made. This allowance is made in order to permit repeat tests without resampling, in cases where the terms of retesting do not require fresh samples or samples from units not previously sampled.

NOTE 7: *Sample Handling*—In opening packages

for purposes of sampling, make certain that no contamination enters from scale, paint, shattered heads, torn liners, or other causes. Transfer each sample to a clean, dry container, capable of being tightly closed. The container must not react with nor otherwise contaminate the sample. Constructions of sheet metal, glass, or many plastic materials are generally suitable, but do not assume inertness of the container in the absence of certain knowledge. Cap liners may be attacked. Do not subject the container subsequently to any conditions that may alter the properties of the sample before testing.

12.2 *In Drums:*

12.2.1 Number the drums in the lot serially.

12.2.2 Choose the total number of drums to be sampled from considerations given above or in Recommended Practice E 122, for the most critical test to be run on the material.

12.2.3 Using random numbers, obtain numbers to designate all the drums that are to be sampled to achieve the total number designated in 12.2.2.

12.2.4 If nonuniformity of material within the drum is unlikely to occur, or to affect any of the test results if it does occur, take a grab sample 50 to 100 mm (2 to 4 in.) below the surface from each package designated in 12.2.3 (Notes 6 and 7).

12.2.5 If nonuniformity of material within the drum is likely to affect the test result, number ten strata of equal height from bottom to top of each drum to be sampled. Choose the total number of strata to be sampled within a drum as in 12.2.2. Obtain the numbers of the strata to be sampled using random numbers. Take a grab sample from each stratum to be sampled (Notes 6, 7, and 8).

NOTE 8: *Stratification and Size Segregation of Molding Powder*—Drums of molding powder in transit, and frequently during packing at the manufacturer's plant, become shaken down. Frequently dense compaction occurs, so that only a few ounces of material would fall out if the container were inverted. The few coarse particles not bound in the mass show plainly on the top surface, giving the impression that the fines have sunk and the coarse particles risen. Sieve analysis of samples taken just under the surface, and throughout the drum, ordinarily fails to disclose segregation.

12.3 *In Bulk* (packages containing more than 100 kg (220 lb), balloons, carloads wherein access can be had to all parts of the container):

12.3.1 Devise a sampling unit such that there will be approximately 100 equal such units in a bulk package or lot. This will give approximately 45 kg (100 lb) per unit in a 4500-kg (10,000-lb) lot. Each sampling unit should have approximately the same height, width, and depth dimensions. The units should be numbered serially in a logical fashion throughout the bulk package, for example, in a row from left to right across the top at the back, then similarly in the next top row toward the front, and finally repeated in similar rows for the next row down from the top.

12.3.2 Choose the total number of units to be used in sampling as described in 12.1.2.

12.3.3 Obtain the numbers of the unit to be sampled, using random numbers.

12.3.4 Take a grab sample from each unit designated in 12.3.3 (Notes 6, 7, and 8).

12.4 *In Bulk*, when access can be had to all contents of the container only as it is emptied and total time of emptying is known:

12.4.1 Devise a "sampling time interval" such that there will be approximately 100 such equal intervals numbered in sequence, in the process of emptying the bulk package at a constant rate of moving the material.

12.4.2 Choose the total number of intervals to be used in sampling as described in 12.1.2.

12.4.3 Obtain the numbers of intervals to be sampled, using random numbers. Arrange these numbers in increasing order.

12.4.4 As the bulk package is emptied, take a grab sample of material passing a fixed point during each time interval determined in 12.4.3 (Notes 6, 7, and 8).

12.5 *In Bulk*, when access can be had to all contents of container only as it is emptied and total time of emptying is unknown or prolonged. Take samples at ten random intervals selected for the first 40 h of use, then every such interval + 40 h thereafter, until the lot is consumed.

13. Sampling Fabricated Stock Shapes

13.1 *Sheets, Flat Film, Slit Tubing:*

13.1.1 Lay out on the sheet a rectangular pattern of unit areas of a size chosen so that uniformity within the unit is likely.

13.1.2 Number these units serially, first back to front (in the strip of areas at the extreme left) across the width of the sheet, then across the next width and similarly down the length of the sheet, moving from left to right.

13.1.3 Choose the total number of units to be sampled (12.1.2).

13.1.4 Obtain the numbers of the units to

be sampled using random numbers or by the use of a latin square (**5**).

13.1.5 Remove the units of area indicated by number in 13.1.4 for later sample cutting, or cut samples from the units directly. If the substrate is very likely to be uniform, use only units near one or both ends.

13.2 *Blocks*—Lay out unit volumes in rectangular coordinates in three dimensions, then apply the random sampling procedure in 13.1.

13.3 *Blown Film, Rods, Tubes, and Pipe of Large Circumference:*

13.3.1 Lay out unit areas around the circumference.

13.3.2 Number, choose, and sample the unit areas as in 13.1.

13.4 *Rods and Tubes of Small Circumference, Tapes, Strips, Beams, Monofilament, Rope, Insulated Wire and Cable, and Other Extruded Shapes:*

13.4.1 Number units of length from one end to the other.

13.4.2 Choose and sample the unit lengths as in 13.1.

14. Molded Items and Groups of Test Specimens (as prepared in large quantities for developmental and interlaboratory testing)

14.1 Number each specimen as a unit, preferably in order of preparation (molding).

14.2 Choose and sample the numbered specimens as described in 13.1.

REFERENCES

(**1**) ASTM Recommended Practice E 105, for Probability Sampling of Materials, *Annual Book of ASTM Standards*, Part 30.

(**2**) Sampling Procedures and Tables for Inspection by attributes, Military Standard MIL-STD-105B, U. S. Government Printing Office, Washington 25, D. C., 27 April 1954.

(**3**) Sampling Procedures and Tables for Inspection by Variables for Percent Defective, Military Standard MIL-STD-414, U. S. Government Printing Office, Washington 25, D. C., June 11, 1957.

(**4**) Rand Corp., *A Million Random Digits*, The Free Press, Glencoe, Ill. (1955).

(**5**) Dixon, W. J., and Massey, F. J., *Introduction to Statistical Analysis*, McGraw-Hill Book Co. (1951).

(**6**) Freeman, H. A., et al., *Sampling Inspection*, McGraw-Hill Book Co. (1948).

TABLE 1 Comparison of Sampling Plans
(Sample size and quality of lots accepted.)

Lot Size Units	Code Letter	Number of Units Sampled								
		Method A (II)			Method B (IV)			Method C (IV)		
		(a)	*(b)*	*(c)*	*(a)*	*(b)*	*(c)*	*(a)*	*(b)*	*(c)*
AQL = 1.0 percent Defective										
500	I	50	(1.0)	(7.7)	25	(1.2)	(8.0)	9	(1.2)	(8.0)
10 000	N	300	(1.5)	(3.8)	75	(1.3)	(4.2)	25	(1.3)	(4.2)
AQL = 6.5 percent Defective										
500	I	50	(7.5)	(22.0)	25	(7.2)	(21.0)	15	(7.2)	(21.0)
10 000	N	300	(9.0)	(13.5)	75	(7.5)	(14.7)	42	(7.5)	(14.7)

Example—Using Method A at AQL = 1.0 on a lot size of 500, a sample size (*a*) of 50 is required. Under this plan, 90 percent of lots containing 1.0 percent (*b*) defective material would be accepted, but only 10 percent of lots containing 7.7 percent (*c*) defective material would be accepted.

APPENDIXES

A1. Glossary

Sampling Plans make use of the theory of probability to combine a suitable method for deciding upon the sample size with an appropriate procedure for summarizing the test results so that inferences may be drawn and risks calculated from the test results by the theory of probability. For any given set of conditions there will usually be several possible plans, all valid, but differing in speed, simplicity, and cost.

Shipment, *n*—The total material of given type and grade obtained from one manufacturer in a delivery or some convenient time interval; it may represent one or more batches.

Batch, *n*—A unit of manufacture or "run" as produced.

Lot, *n*—A specific quantity of similar material or collection of similar units from a common source; the quantity offered for inspection and acceptance at any one time. A lot might comprise a shipment, batch, or smaller quantity.

Unit (Sampling Unit), *n*—One of the number of similar articles, parts, specimens, lengths, areas (package, drum), etc., of a material or product.

AQL (Acceptable Quality Level)—A quality of product (expressed as percent defective), such that a lot having this percent defective will have a probability of rejection equal to α. An ideal sampling and inspection plan would accept all lots of this or better quality and reject all lots of lower quality. Any practical plan can only approach this ideal.

$\boldsymbol{\alpha}$—The seller's risk, that is, the probability of rejecting a lot that is actually acceptable.

$\boldsymbol{\beta}$—The purchaser's risk, that is, the probability of accepting a lot that is actually unacceptable.

A2. Sampling Procedures and Tables for Inspection by Attributes (Adapted from MIL-STD-105B)

A2.1 Scope

A2.1.1 Table A1 shows the sample size code letters corresponding to various lot sizes under three different levels of inspection, the sample sizes corresponding to these sample size code letters, and the acceptance numbers of these samples at two different acceptable quality levels. Figure 1 shows the operating characteristic curves of sampling plans for two different lot size code letters, at each of two AQL's.

A2.2 Explanatory Notes for Table A1

A2.2.1 Explanatory notes for Table A1 are as follows:

Note A1—The sample sizes and acceptance numbers shown on the right of the table apply only to the code letters shown opposite for Inspection Level II. The size of the sample depends on the sample size code letter, not on the numerical size of the lot. A lot of 1020 units, when inspected under Level I, has a code letter "I" which calls for a sample size of 50 (13 percent) and an acceptance number of 1 for an AQL of 1.0. Unless otherwise specified, Inspection Level II shall be used.

Note A2—In single sampling plans, the rejection number is one larger than the acceptance number. Therefore, for the normal inspection plan of code letter "I" (sample size 50), the acceptance and rejection numbers for an AQL of 6.5 are 6 and 7, respectively. If 6 or fewer defectives are found in the sample of 50, the lot is accepted. If 7 or more defectives are found, the lot is rejected.

Note A3—When the sample size equals or exceeds the lot size, 100 percent inspection is required.

Note A4—When a downward-pointing arrow appears in the acceptance number column opposite the sample size, the first sampling plan appearing below the arrow shall be used. It may be noted that the smallest sample which may be inspected against an AQL of 1.0 is 15.

Note A5—When an upward-pointing arrow is found, the first sampling plan above the arrow shall be used.

Note A6—Eight additional lower inspection levels, "L1" through "L8," are available for use when small sample sizes are needed. These might be used for inspection by means of expensive or destructive tests. They involve the use of a new sample size code letter "AA" which calls for a sample size of 1 and an acceptance number of 0.

Note A7—The lot and sample sizes refer to the number of units or product contained in them.

Note A8—Double and multiple sampling plans are also available. These are designed to reduce the amount of inspection on very good or very bad lots. They work as follows: A first sample is sected and inspected. This sample is smaller than that used in the corresponding single sampling plan. The acceptance number for this first sample is smaller than that used in a single sampling plan having the same sample size and AQL, while the rejection number is larger. The rejection number is two or more digits larger than the acceptance number. If the number of defects in this first sample is equal to or fewer than the acceptance number, the lot is accepted. If it is equal to or greater than the rejection number, the lot is rejected. If it is between these two limits, an additional sample is inspected and the combined number of defects in the two samples is measured by the narrower acceptance and rejection numbers of the sampling plan for the combined sample. This is continued until the lot either passes or fails. In the last step of these double or multiple sampling plans, the rejection number is one more than the acceptance number.

NOTE A9: *AQL, Inspection Level, Degree of Inspection*—The selection of an AQL is a contractual matter, since it determines the minimum acceptable quality of a lot and hence its value. The actual AQL's illustrated herin are for example only.

When the purchaser does the inspection, the inspection level may be established or changed by the purchaser. This is done to balance the cost of inspection with the degree of risk he is willing to assume of accepting material that is appreciably worse than the AQL. The seller's risk has been made nearly the same for all inspection levels, and only the purchaser's risk changes with the inspection level.

When the supplier performs the inspection, "tightened" or "reduced" inspection may be employed in certain cases. During periods of high quality, less inspection is performed under the "reduced" plan. During periods of poor quality, additional inspection is performed under the "tightened" plan. This "previous quality history" is determined by the "process average," the determination of which also is explained in Military Standard MIL-STD-105.

The sampling plans for tightened inspection under AQL's of 1.5 and 10.0 are the same as those shown for normal inspection under AQL's of 1.00 and 6.50.

Reduced inspection plans are obtained from different tables in Military Standard MIL-STD-105. They involve smaller sample sizes for the same lot sizes and larger acceptance numbers than are specified for these small sample sizes in Table IV-A.

A2.3 Discussion of Fig. A1

A2.3.1 The operating characteristic curves of sampling plans show the efficiences of these plans, as measured by their abilities to distinguish between good and bad lots. The quality of submitted lots, in percent defective or defects per hundred units, is plotted on the horizontal axis, while the percentage of these lots that are expected to be accepted under the plans is plotted on the vertical axis.

A2.3.2 It may be seen that increasing the AQL merely moves the curves to the right, without changing their shapes or slopes (except that it permits the use of smaller samples from small lots), while increasing the sample size steepens the slopes and makes the plan more discriminatory. Although increasing the sample size reduces the seller's risk to some extent, the chief advantage is in reducing the purchaser's risk. It may be noted that at an AQL of 6.5, the risk of accepting lots that have a quality of 6.5 percent defective (or 6.5 defects per hundred units) is about the same for both plans (97 to 98 percent). However, the risk of accepting lots that are 15 percent defective is quite different under the two plans. The plan for code letter "I" permits an average of 35 percent of such lots to be accepted, while that for code letter "N" accepts only about 2 percent of such lots. The plan for code letter "I" would accept about 4 percent of lots that are 25 percent defective.

A3. Sampling Procedures and Table for Inspection by Variables for Percent Defective (Adapted from MIL-STD-414)

A3.1 Scope

A3.1.1 Table A2 shows the sample size code letters corresponding to various lot sizes under five different levels of inspection, the sample sizes corresponding to these sample size code letters, and the acceptance constants of these samples at two different acceptable quality levels. The table applies to situations wherein the variability of the material with respect to the property of interest is unknown, so that the standard deviation of the property throughout the sample must be estimated from the values measured on the various units of the sample. The AQL's are given for example only.

A3.2 Procedure for Sampling and Inspection

A3.2.1 The operation of a typical sampling and inspection procedure is as follows:

A3.2.1.1 Determine the sample size code letter from the size of the lot, in accordance with Table A2 and 8.1.1 of this recommended practice. Unless otherwise specified, use Inspection Level IV. The sample sizes and acceptance constants shown on the right of the table apply only to the code letters shown for Inspection Level IV.

A3.2.1.2 Derive the numerical sample size in units from the sample size code letter.

A3.2.1.3 Choose the sample units, withdraw them from the lot, and test in accordance with 8.1.2, 8.1.3, and 8.1.5 through 8.1.7 of this recommended practice.

A3.2.1.4 From the test values, calculate the arithmetic means for the lot, $\bar{X}$ (6.1) and the estimated standard deviations, s (6.2).

A3.2.1.5 Assume that the test is being run against a lower specification limit, then calculate the quality index, Q_L, as follows:

$$Q_L = (\bar{X} - L)/s,$$

where:

L = (lower) specification limit value.

If Q_L is negative, the lot is automatically rejected.

A3.2.1.6 Compare Q_L with the acceptance constant, k, for the associated sample size and AQL values.

If $Q_L \geq k$, accept the lot.
If $Q_L < k$, reject the lot.

A3.3 Alternative Procedures

A3.3.1 Procedures are also provided in Military Standard MIL-STD-414 for situations in which (*1*) an upper specification limit must be met, (*2*) both upper and lower specification limits must be met, (*3*) the average range of the test values is used as an estimate of lot variability, (*4*) the variability of the material is known, and (*5*) "tightened" or "reduced" inspection is practiced (see A2.2.1, Note A9).

A4. Example of Use of Previous Data for Judging the Qualities of Lots and for Setting Specification Limits

A4.1 When the variability of a material is known, the quality of a new lot of the material can be determined more efficiently and specification requirements can be set more equitably than when variability is unknown. An example may be cited in the data obtained by testing a sample of 270 paving bricks for transverse strength described in Example (*a*) of Part I; Tables II(*a*), III, VII(*a*), and Figs. 2, 3, and 4 of the ASTM Manual on Quality Control of Materials. The average transverse strength of the sample is 1000 lb and the standard deviation is 202 lb. The skewness factor of the distribution, 0.51, while not truly "normal" indicates that the distribution is close enough to normal that factors based on the normal curve may safely be used to calculate the properties of the lot of bricks from which this sample was taken.

A4.2 From the data, using appropriate principles, it is possible to calculate that: (*1*) if the sample is representative of the lot of bricks from which it was taken, approximately 99 percent of the bricks in the lot have transverse strengths greater than 540 lb; (*2*) 98 percent have strengths between $1000 - (2.326)(202) = 540$ lb and $1000 + (2.326)(202) = 1460$ lb; and (*3*) 95 percent of them have strengths above 668 lb. In the actual sample of 270 bricks, 2 bricks (0.7 percent) had strengths less than 540 lb, and 7 bricks (2.6 percent) had strengths less than 668 lb.

A4.3 From a strictly theoretical standpoint, these conclusions are not quite accurate, since the true average and standard deviation of the lot of bricks are unknown. Because of the large size of the sample, however, these calculations are sufficiently accurate to be useable. Similar conclusions based on a small sample would not have been reliable.

A4.4 These data, if considered to be representative of normal production, could be used to prepare specification requirements for future procurement of these bricks. It appears possible to specify that each brick in the lot shall have a transverse strength of not less than 500 lb, with an AQL of 1 percent defective, or a strength of 650 lb with an AQL of 5 percent defective.

A4.5 The requirement for 500-lb strength, 1 percent defective, written in the form of a statistical sampling plan for variables, could state that the average of a large sample minus 2.33 times its standard deviation shall not be less than 500 lb. This method of specifying quality has the advantage that it permits a seller furnishing a product of uniform quality (small standard deviation) to submit a somewhat lower average quality than that required of another seller whose product is more variable.

TABLE A1 Sample Size Code Letters for Lot Sizes under Three Inspection Levels, and Sample Sizes (Number and Percent of Mean Lot Size) and Acceptance Numbers of Normal Single Inspection at AQL's of 1.0 and 6.5[a,b]

Lot Size		Sample Size Code Letter for Inspection Level			Inspection Level II, Single Sampling			
					Sample Size		Acceptance Numbers, Normal Inspection	
Limits	Geometric Mean	I	II	III	No.	Percent	AQL = 1.00	AQL = 6.5
2 to 8	4	A	A	C	2	50	|	↓
9 to 15	11.6	A	B	D	3	26	|	0
16 to 25	20	B	C	E	5	25	|	↑
26 to 40	32	B	D	F	7	22	|	↓
41 to 65	52	C	E	G	10	19	↓	1
66 to 110	85	D	F	H	15	18	0	2
111 to 180	140	E	G	I	25	18	↑	3
181 to 300	230	F	H	J	35	15	↓	5
301 to 500	390	G	I	K	50	13	1	6
501 to 800	630	H	J	L	75	12	2	9
801 to 1 300	1 020	I	K	L	110	11	3	12
1 301 to 3 200	2 040	J	L	M	150	7.3	4	17
3 201 to 8 000	5 040	L	M	N	225	4.5	5	24
8 001 to 22 000	13 300	M	N	O	300	2.3	7	32
22 001 to 110 000	49 000	N	O	P	450	0.92	10	43
110 000 to 550 000	250 000	O	P	Q	750	0.30	15	68
550 001 and over	over 550 001	P	Q	Q	1 500	0.27	25	124

[a] Table based on MIL-STD-105B, Tables III and IV-A.

[b] See Explanatory Notes in A2.2.1.

TABLE A2 Master Table for Inspection for Plans Based on Variability Unknown, Standard Deviation Method (Single Specification Limit, Form 1)[a]

Lot Size in Units	Sample Size Code Letters[b]					Sample Size in Units	Acceptance Constant,[k]	
	Inspection Levels						AQL = 0.40[c]	AQL = 1.50[c]
	I	II	III	IV	V			
3 to 15	B	B	B	B		3	↓[d]	↓[d]
16 to 25	B	B	B	C	E	4	↓	1.34
26 to 40	B	B	B	D	F	5	↓	1.40
41 to 65	B	B	C	E	G	7	1.88	1.50
66 to 110	B	B	D	F	H	10	1.98	1.58
111 to 180	B	C	E	G	I	15	2.06	1.65
181 to 300	B	D	F	H	J	20	2.11	1.69
301 to 500	C	E	G	I	K	25	2.14	1.72
501 to 800	D	F	H	J	L	30	2.15	1.73
801 to 1 300	E	G	I	K	L	35	2.18	1.76
1 301 to 3 200	F	H	J	L	M	40	2.18	1.76
3 201 to 8 000	G	I	L	M	N	50	2.22	1.80
8 001 to 22 000	H	J	M	N	O	75	2.27	1.84
22 001 to 110 000	I	K	N	O	P	100	2.29	1.86
110 001 to 550 000	I	K	O	P	Q	150	2.33	1.89
550 001 and over	I	K	P	Q	Q	200	2.33	1.89

[a] Table based on MIL-STD-414, Tables A-2, B-1.

[b] Sample size code letters given in body of table are applicable when the indicated inspection levels are to be used.

[c] All AQL values are in percent defective.

[d] Use first sampling plan below arrow, that is, both sample size as well as *k* value. When sample size equals or exceeds lot size, every item in the lot must be inspected.

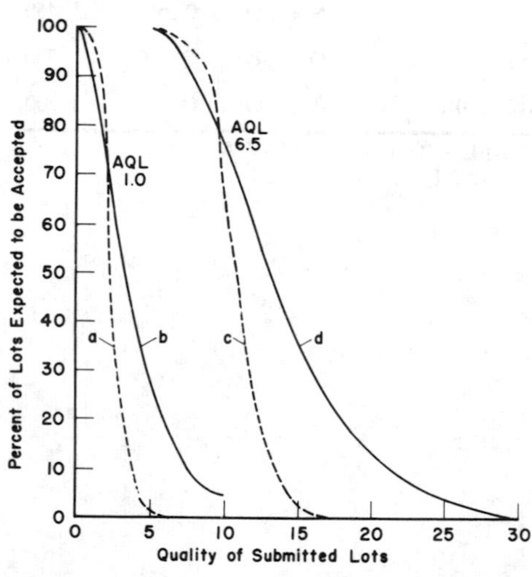

Curve	Lot Size	Code Letter	Sample Size	Acceptance Number
a	8 001 to 22 000	N	300	7
b	301 to 500	I	50	1
c	8 001 to 22 000	N	300	32
d	201 to 500	I	50	6

NOTE—Quality of submitted lots is expressed in percent defective.

FIG. A1 Operating Characteristic Curves for Single Sampling Plans from Military Standard. MIL-STD-105B (Curves for double and multiple sampling are essentially equivalent).

ASTM Designation: D 1920 - 69

Recommended Practice for

DETERMINING LIGHT DOSAGE IN CARBON-ARC LIGHT AGING APPARATUS[1]

This Recommended Practice is issued under the fixed designation D 1920; the number immediately following the designation indicates the year of original adoption or, in the case of revision, the year of last revision. A number in parentheses indicates the year of last reapproval.

1. Scope

1.1 This recommended practice covers the use of a yellow poly(methyl methacrylate) molded standard specimen for the evaluation of light dosage during tests conducted in carbon-arc light aging apparatus of the types described in ASTM Recommended Practice E 42, for Operating Light- and Water-Exposure Apparatus (Carbon-Arc Type) for Exposure of Nonmetallic Materials,[2] and ASTM Recommended Practice E 188, for Operating Enclosed Carbon-Arc Type Apparatus for Light Exposure of Nonmetallic Materials.[3]

NOTE 1—The values stated in U.S. customary units are to be regarded as the standard. The metric equivalents of U.S. customary units may be approximate.

2. Summary of Method

2.1 Data from periodic measurements at 420-nm wavelength on the spectral transmittance of the yellow plastic molded standard specimen exposed in the carbon-arc type apparatus, are plotted on log-log paper. The light dosage conditions are determined for a given type of carbon-arc apparatus by the rate of change in light transmittance with time of exposure of the yellow plastic specimen.

3. Significance

3.1 The light exposure conditions within light aging apparatus of the various carbon-arc types produce different rates of fading of the yellow plastic standard specimen. This recommended practice is useful in determining whether the carbon-arc apparatus described in Recommended Practice E 42 and light-exposure apparatus, carbon-arc type without water, as specified in Recommended Practice E 188, when operated under the prescribed conditions for any one type of apparatus, have given the same relative light dosage with time. Only apparatus of the same type should be compared by means of the graphs, and the carbon arc shall be equipped with a suitable filter as described in Recommended Practice E 42 or Recommended Practice E 188.

3.2 Equivalence of degree of fading of the yellow plastic standard specimen under different types of radiation does not indicate equivalence of weathering effect on the specimen being tested. Comparison of the slopes of the log-log plot of hours of radiation exposure as the abscissa and the spectral transmittance of the yellow plastic specimen as the ordinate, will serve as a guide to determine whether carbon-arc apparatus of the same type have different light dosage.

3.3 Contaminants such as salt or other deposits on the surface of the yellow plastic specimen can affect the fading rate.

4. Apparatus

4.1 *Holders*—The holders used to mount the 51 by 51-mm (2 by 2-in.) yellow plastic specimen shall have not less than 41-mm ($1^5/_8$-in.) opening both in height and width. The holders may be either aluminum or the stainless steel type furnished with the appara-

[1] This recommended practice is under the jurisdiction of ASTM Committee D-20 on Plastics. A list of committee members may be found in the ASTM Yearbook. This standard is the direct responsibility of Subcommittee D-20.50 on Permanence Properties.

Current edition effective June 26, 1969. Originally issued 1964. Replaces D 1920 – 66 T.

[2] *Annual Book of ASTM Standards*, Part 27.

[3] *Annual Book of ASTM Standards*, Part 30.

tus. The adapter shown in Fig. 1, Form 1, can be used to mount the 51 by 51-mm (2 by 2-in.) yellow plastic standard specimen in the holders for Types A, B, C, and D apparatus in Recommended Practice E 42. The adapter shown in Fig. 1, Form 2, can be used in Types E, F, and G apparatus.

4.2 *Transmittance Measuring Instrument*—An instrument capable of measuring the transmittance of the yellow plastic specimen at 420-nm wavelength before and after exposure. Different makes of spectrophotometers do not all have the same slit-width at 420-nm, and this will affect the readings. The lowest readings will be obtained with the instrument having the smallest slit-width. For any one test the same instrument and slit-width shall be used.

4.2.1 Spectrophotometers of the types described in ASTM Recommended Practice E 308, Spectrophotometry and Description of Color in CIE 1931 System,[3] and ASTM Method D 1003, Test for Haze and Luminous Transmittance of Transparent Plastics,[2] are recommended for referee tests. Abridged types of spectrophotometers can be used for other tests for comparison purposes. When none of these instruments is available, a band pass photometer can be constructed according to the schematic diagram illustrated in Fig. 2. Data obtained with this instrument will be useful for day-to-day observation of operation of the apparatus. They will not be directly comparable with data obtained with the recording spectrophotometer.

5. Yellow Plastic Standard Specimens

5.1 The yellow poly(methyl methacrylate) standard specimens shall be injection molded of material meeting the requirements of ASTM Specification D 788, for Methacrylate Molding and Extrusion Compounds,[4] Grade 8, with no ultraviolet absorber added, and containing Solvent Yellow 33 C. I. 47,000. The dye concentration shall be that which will produce 17 ± 1.5 percent light transmittance measured at 420-nm wavelength on a specimen 3.15 ± 0.04 mm (0.124 ± 0.0015 in.) thick using a GE recording spectrophotometer, or equivalent instrument. The molded specimens shall conform to the dimensions shown in Fig. 3.

5.2 Yellow plastic specimens produced from a single lot or batch of molding powder shall be marked with the same identification number, measured for thickness, and light transmittance determined at 420-nm wavelength, using the GE recording spectrophotometer or equivalent instrument. Representative specimens from each lot shall be tested in the National Bureau of Standards Master Lamp.[5] Graphs showing spectral transmittance versus exposure time shall be provided for each lot of the yellow plastic specimens as a control measurement of lot-to-lot variation.

5.3 The yellow plastic specimen shall be preconditioned for at least 48 h at 50 percent relative humidity, and room temperature as defined in ASTM Definitions E 41, Terms Relating to Conditioning.[2]

6. General Considerations

6.1 The moisture content of the yellow plastic specimen has an effect on the rate of change in light transmittance during the first part of the exposure period of approximately 50 h. Very dry conditioning will depress the graph below that for a normal moisture content. This will not affect the slope of the graph after longer hours of exposure but the level of light transmittance for these longer hours will be lower.

6.2 Shut-down of apparatus for as little as 2 days' time can affect the shapes of the fading graph. The change in shape is not apparent until the end of the exposure period following the shut-down. This usually results in an upward shift in the curve which is not entirely erased with longer exposure.

7. Procedure

7.1 Break the yellow plastic specimen in half along the center depression, and mount one of the 51 by 51-mm (2 by 2-in.) pieces in a holder suitable for the type carbon-arc apparatus used. Use the other 51 by 51-mm piece as a reference to check the original light

[4] *Annual Book of ASTM Standards*, Part 26.

[5] The Master Lamp consists of an Atlas Electric Devices Co. glass-enclosed type SMC-R lamp which meets requirements of Recommended Practice E 188, having a 20-in. diameter drum operating in a test chamber larger than a Model FDA-R machine, and relative humidity of the air where it leaves the test chamber at approximately 30 percent.

transmittance at 420-nm and each time the exposed specimen is measured. Between readings the reference specimen should be stored in the blue envelope in which it was received. Measure and record the thickness of the specimen to the nearest 0.025 mm (0.001 in.).

7.2 Mount the yellow plastic specimen in position *A* of the holder as shown in Fig. 1. If it is replaced in the holder after each reading, the location shall be the same. With the holder mounted in the upper part of the specimen rack in the test apparatus, the yellow plastic standard specimen will be in the same plane as the test specimens, and just above center in the vertical plane of the test rack. When it is mounted at different heights on the specimen rack, there will be some differences in fading rate of the specimen. For specific tests it may be desirable to have the yellow plastic specimen in another location on the test rack, in which case agreement should be reached among laboratories concerned with the test. The area directly behind the specimen shall be either open or have a nonreflective surface.

7.3 Prepare a convenient schedule at the start of the test to show times at which spectral transmittance will be measured at 420-nm wavelength. When specimens are removed after periods specified in the test schedule, they should be measured within a few hours. It may be advantageous to mount several specimens in the apparatus and remove them one at a time during continuous operation of a test.

7.4 Ordinary water may deposit salts on the yellow plastic specimen that need to be removed because of their effect on the light transmittance test. These are removed by immersing the yellow plastic specimen in 1 percent HCl solution, then rinsing with water that is free of contaminants. Some contaminants that affect light transmittance measurements are nearly impossible to remove except by careful buffing of the yellow plastic specimen (Note 2). Differences in quality of water between laboratories can affect these results.

NOTE 2—A satisfactory buff consists of a 100-mm (4-in.) wide soft cloth wheel operating at 240 surface m/min (800 surface ft/min).[6] Care should be taken to avoid burning or heating the surface of the yellow plastic standard sufficiently to cause the material to flow. Careful buffing will not harm the yellow plastic standard in such a way as to affect the spectral transmittance measurements.

7.5 Plot on log-log paper the values of light transmittance at 420-nm wavelength as ordinate and the exposure times as the abscissa. Compare slopes of these plots for tests run under the same time schedules on the same lot of yellow plastic specimens tested in a given type of carbon-arc apparatus.

8. Report

8.1 The report shall include the following:

8.1.1 Description of the carbon-arc apparatus used and of the instrument used to measure the spectral transmittance,

8.1.2 Location of the yellow plastic specimen in the test rack, and

8.1.3 The graph showing light transmittance versus hours of exposure.

[6] A satisfactory buffing compound is Matchless Z-102 cut and color compound, available from Matchless Metal Products, 726 Bloomfield Ave., Glen Ridge, N.J.

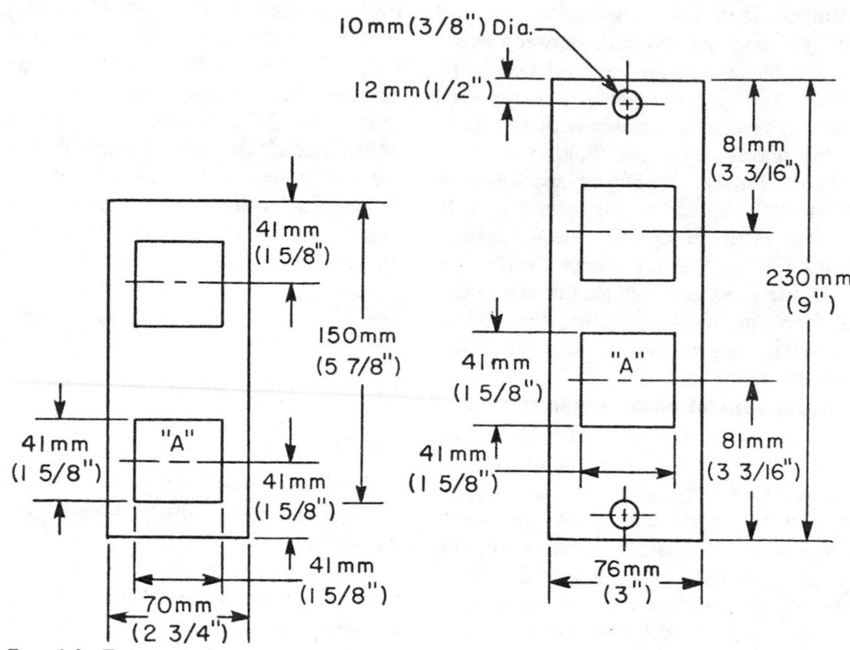

Form 1 for Types A, B, C, and D apparatus (Recommended Practice E 42)

Form 2 for Types E, F, and G apparatus (Recommended Practice E 42)

FIG. 1 Adapters for Mounting Standards in Holders.

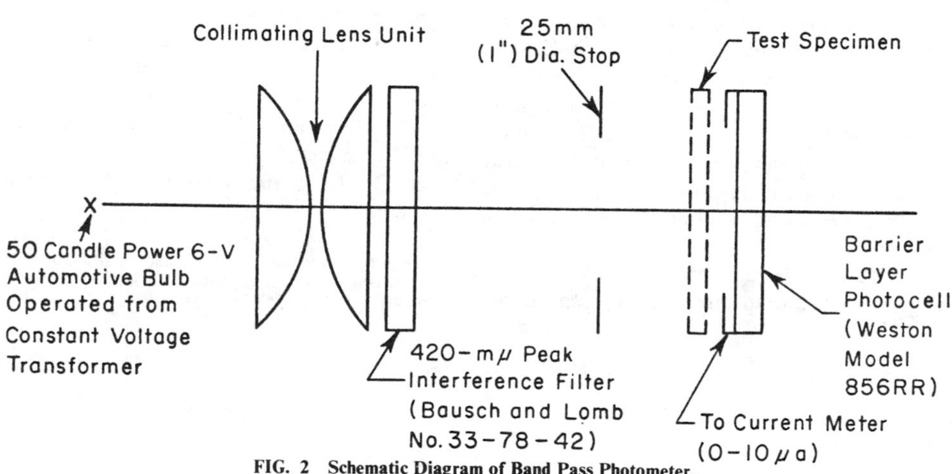

FIG. 2 Schematic Diagram of Band Pass Photometer.

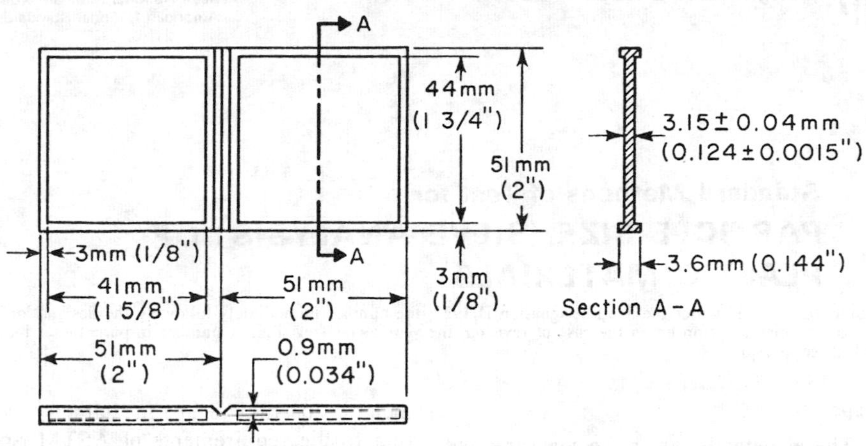

FIG. 3 Dimensions of Molded Standard Specimen.

Standard Methods of Test for
PARTICLE SIZE (SIEVE ANALYSIS) OF PLASTIC MATERIALS[1]

This Standard is issued under the fixed designation D 1921; the number immediately following the designation indicates the year of original adoption or, in the case of revision, the year of last revision. A number in parentheses indicates the year of last reapproval.

1. Scope

1.1 These methods cover the measurement of the particle size of plastic materials in the powdered, granular, or pelleted forms in which they commonly are supplied. As only dry sieving methods are described, the lower limit of measurement is considered to be about 37 μm (No. 400 sieve). For smaller particle sizes sedimentation methods are recommended.

1.2 Four methods are covered, as follows:

1.2.1 *Method A* describes analyses conducted with multiple sieves. Particle sizes and their distribution are determined by this method. Standard 203-mm (8-in.) diameter sieves and a mechanical shaker are employed. With appropriate screens and proper duration of shaking either fine or coarse materials can be tested.

1.2.2 *Method B* is an abbreviated version of Method A.

1.2.3 *Method C* describes sieve analyses conducted with the aid of vacuum. Small samples and special sieves of small diameter are employed. This method ordinarily is used only for finely powdered materials.

1.2.4 *Method D* is a special version of Method C, for use when complete investigation of particle size is not necessary or a supplementary analysis of a large-size sample is desired to detect occasional coarse particles.

NOTE 1—The values stated in U.S. customary units are to be regarded as the standard. The metric equivalents of U.S. customary units may be approximate.

METHOD A

2. Apparatus

2.1 *Sieves*—The sieves used shall be half-height, 203 mm (8 in.) in diameter, conforming to the requirements of ASTM Specification E 11, for Wire-Cloth Sieves for Testing Purposes.[2] A selection of sieves encompassing the expected range of particle sizes together with a cover and a bottom pan will be required.

NOTE 2—Handle the sieves carefully and protect them against damage in use or during storage. After each analysis, clean each sieve thoroughly with a suitable brush or with compressed air, or both. Periodically, wash the sieves with either mild soap and water or a suitable solvent.

2.2 *Mechanical Sieving Device and Time Switch*—A mechanical sieve-shaking device equipped with an automatic time switch. This device shall be capable of imparting uniform rotary motion and a tapping action at the rate of 150 ± 10 taps/min.

2.3 *Balance*—A laboratory balance of 500-g capacity, sensitive to 0.1 g, shall be used for weighing the sample and the residues retained on the sieves.

2.4 *Accessories* required are suitable brushes for cleaning the sieves, paper for collecting samples, and any other items of equipment ordinarily used in making sieve analyses.

3. Procedure

3.1 Select sieves sufficient in number to cover the expected range of particle sizes, and nest them together in order of diminishing

[1] These methods are under the jurisdiction of ASTM Committee D-20 on Plastics. A list of committee members may be found in the ASTM Yearbook. This standard is the direct responsibility of Subcommittee D-20.70 on Analytical Methods.

Current edition effective Sept. 30, 1963. Originally issued 1961. Replaces D 1921 – 61 T.

[2] *Annual Book of ASTM Standards*, Part 30.

openings with the coarsest sieve on top and the pan on the bottom. Weigh out a 100 ± 0.1-g sample (Note 3) and transfer it to the top sieve of the stack. For those finely divided powders which tend to clog the sieves, add 1.0 ± 0.1 g of carbon black to the sample (Note 4).

NOTE 3—If necessary this test may be made on a sample of any size from 50 to 200 g. The weight of sample used shall be stated in the report.

NOTE 4—The weight of the added carbon black has not been accounted for in this method. In most cases this weight is insignificant in relation to the sample weights. However if it is felt that the weight of the carbon black contributes to the results of the analysis, a separate sieve analysis of the carbon black may be made or its weight accounted for in some other manner.

3.2 Cover the stack and place it in the mechanical sieve shaker. Start this device and shake the sample for 10 min ± 15 s. For coarse powders, 5 min of shaking may be sufficient and will be permitted.

3.3 After shaking, carefully separate the stack of sieves, beginning at the top, and weigh the quantity of powder retained on each sieve and that contained in the pan to the nearest 0.1 g. This may be accomplished either by transferring the fractions to the balance or by weighing either the sieve or the pan and its contents and subtracting the tare weight from the total. If the powder is transferred to the balance, the sieve shall be carefully brushed on both sides to ensure that adhering particles are transferred.

NOTE 5—Ordinarily there is a small loss of dust indicated by the cumulative total of actual weight being less than 100 percent. If this loss is not over 2 percent, the amount reported through the finest sieve shall be increased until the total of all portions of the sample equals 100 percent. If the cumulative total of actual weight is less than 98 percent, the weights and operations shall be carefully checked and the work repeated, if necessary.

4. Analysis of Particle Distribution

4.1 To permit analysis of particle distribution, plot the values for cumulative percentage by weight on probability graph paper. Join these points by a smooth curve and investigate any marked departure from it. From this curve, determine and record the particle size in microns corresponding to the 16th, 50th, and 84th percentiles of cumulative weight through.

4.2 Determine the median, dispersion, and skew as follows:

$$\text{Median} = P_{50}$$

$$\text{Dispersion} = (P_{84} - P_{16})/2$$

$$\text{Skew} = [(P_{84} - P_{50}) - (P_{50} - P_{16})]/(P_{84} - P_{16})$$

where:

P_{16} = 16th percentile,
P_{50} = 50th percentile, and
P_{84} = 84th percentile.

5. Report

5.1 The report shall include the following:

5.1.1 Percentage of material retained on each sieve,

5.1.2 If required, the total cumulative percentage of material retained on each sieve and in the pan,

5.1.3 Median, dispersion, and skew of particle distribution, if calculated, and

5.1.4 Shaking time if less than 10 min.

METHOD B

6. Apparatus

6.1 *Sieves*—The sieves used shall be half-height, 203 mm (8 in.) in diameter, conforming to the requirements of Specification E 11. Either one selected sieve or two (one coarse and one fine) may be used. A cover and a bottom pan also will be required. By using a bottom pan with an extended rim, several samples can be tested simultaneously (Note 2).

6.2 *Mechanical Sieving Device and Time Switch*, as specified in 2.2.

6.3 *Balance*, as specified in 2.3.

6.4 *Accessories*, as specified in 2.4.

7. Procedure

7.1 If a single sieve is being used, stack it on the pan and transfer a sample weighing 100 ± 0.1 g to that sieve. If two sieves are to be used in the analysis of the sample, stack the coarse sieve over the fine sieve and transfer the weighed sample to the coarse sieve. For those finely divided powders which tend to clog the sieves, add 1.0 ± 0.1 g of carbon black to the sample (Note 4). If several samples are to be tested simultaneously, stack the sieves and add the samples successively in accordance with the foregoing instructions until the entire stack has been assembled.

7.2 Cover the stack and place it in the mechanical sieve shaker. Start this device and shake the sample for 10 min ± 15 s. For coarse powders, 5 min of shaking may be sufficient and will be permitted.

7.3 After shaking, carefully separate the stack of sieves, beginning at the top, and weigh the quantity of powder retained on each sieve to the nearest 0.1 g. If the analysis has been conducted on a single sample, the quantity of powder contained in the pan also may be weighed. This may be accomplished either by transferring the fractions to the balance or by weighing the sieve and its contents and subtracting the tare weight from the total. If the powder is transferred to the balance, the sieve shall be carefully brushed on both sides to ensure that adhering particles are transferred (Note 5).

8. Report

8.1 The report shall include the following:

8.1.1 Percentage of material (A) retained on the sieve and the percentage passing through it ($100 - A$),

8.1.2 If two sieves are used in the analysis of a sample, the cumulative percentage of material retained on the finer sieve (B) and the percentage passing through it ($100 - B$),

8.1.3 Presence of any particle of unusual size or character, and

8.1.4 Shaking time if less than 10 min.

METHOD C

9. Apparatus

9.1 *Sieves*—The sieves used shall be half-height, 76 mm (3 in.) in diameter, conforming to the requirements of Specification E 11. A selection of sieves encompassing the expected range of particle sizes will be required.

9.2 *Vacuum System*—A sheet metal adapter or funnel with its large diameter designed to permit nesting of the sieves and with its small diameter rigidly attached to a vertical standpipe, 25 mm (1 in.) in diameter, connected to either a central vacuum system or a household vacuum cleaner of the tank type. The standpipe shall be securely supported in the vertical position, and it shall be fitted with either a gate valve or other quick-opening adjustable valve capable of regulating the flow of air through the sieves at a velocity that will not cause their distortion.

9.3 *Vibrator*—A vibrator attached to the standpipe and adjusted to strike the upper edge of the adapter. A heavy-duty electric (110-V) bell with the bell removed from the striker has been found satisfactory for this purpose.

9.4 *Balance*—An analytical balance with Class S weights and with a sensitivity of 0.001 g shall be used for weighing the samples and the residues on the sieves.

9.5 *Accessories* required include watch glasses, 125-ml Erlenmeyer flasks, Griffin low-form beakers, a desiccator containing 2.4-mm (8-mesh) $CaCl_2$ or other active desiccant, spatulas, and a 13-mm (½-in.) wide, flat, soft camel's-hair brush.

10. Procedure

10.1 Prior to use, wash the sieves in clean acetone (containing less than 0.001 g of residue upon evaporation of 100 ml), dry thoroughly in a stream of air, and store in a desiccator for at least 15 min prior to use.

10.2 Insert a number of clean and dry sieves sufficient to encompass the expected range of particle sizes and weigh each sieve to the nearest 0.001 g on an analytical balance. Nest them together in order of diminishing openings with the coarsest sieve on top.

10.3 Insert the stacked sieves into the top of the adapter and apply suction gradually with the rate of flow of air through the sieves adjusted so that no appreciable straining of the finest sieve occurs. After suction has been adjusted, start the vibrator.

10.4 For very fine powders (with most particles ranging in size from 37 to 74 μm) weigh a 1-g sample to the nearest 0.001 g on a clean, dry watch glass. For less fine powders, weigh a 5 to 10-g sample to the nearest 0.001 g into either a beaker or flask. To these samples may be added about 10 percent by weight of either carbon black or graphite having a particle size smaller than the openings in the finest sieve used. The addition of carbon black or graphite facilitates the sieving of fine powders which otherwise might agglomerate and clog the sieve (Note 4).

10.5 Transfer the weighed sample slowly and carefully to the top sieve, preferably in portions. If either carbon black or graphite has been added, first stir or shake the sample to disperse the carbon black or graphite. As the sample is added to the sieve, tap the edge of the sieve and lightly brush the sample onto the sieve to facilitate passage through it.

10.6 When no more powder passes through the top sieve, halt vibration and vacuum, remove that sieve and carefully brush any powder adhering to its under side onto the next succeeding sieve. If any particles are present on the sieve just removed, reweigh it and its contents to the nearest 0.001 g and, by subtraction, determine the weight of those particles.

10.7 Again apply suction and start the vibrator, in that order, and repeat the foregoing procedure for each of the sieves in the stack. The tared sieves with the residues on them obviously may be collected and weighed without interruption at the conclusion of the sieving operation.

11. Report

11.1 The report shall include the following:

11.1.1 Percentage of material retained on each sieve,

11.1.2 If required, the total cumulative percentage of material retained on each sieve,

11.1.3 If desired, an analysis of particle size distribution, described in Section 4 of these methods, and

11.1.4 Presence of any particles of unusual size or character.

Method D

12. Apparatus

12.1 Sieves, vacuum system, vibrator, balance, and accessories as described in Section 9. Also, if required, a laboratory balance of 500-g capacity, sensitive to 0.1 g.

13. Procedure

13.1 Select and weigh a single sieve having openings sufficiently large to permit passage of all particles of acceptable size. Insert it into the top of the adapter and apply suction gradually. Adjust the rate of flow of air through the sieve so that no appreciable straining of the sieve occurs.

13.2 Start the vibrator.

13.3 Weigh a 100-g sample of the powder to the nearest 0.1 g in a beaker of appropriate size and, if necessary, add 1 g of powdered carbon black or graphite to it (Note 4). Pour the sample slowly, in portions, onto the sieve. As the sample is poured onto the sieve, tap the edge of the sieve and lightly brush the sample onto the sieve to facilitate passage through it.

13.4 After all of the sample has been transferred to the sieve and no more particles will pass through it, halt vibration and vacuum and carefully examine the sieve. If particles are present on it, weigh the sieve and its contents on an analytical balance to the nearest 0.001 g and, by subtracting the tare weight of the sieve, determine the weight of the particles. Also, examine the particles closely, note their number and size, and determine whether the particles were originally coarse or are agglomerates.

14. Report

14.1 The report shall include the following:

14.1.1 Percentage of material retained on the sieve(s),

14.1.2 If only a few particles are retained on the coarse sieve, report their number, and

14.1.3 Description of the particles retained on the coarse sieve.

Designation: D 1922 – 67

Standard Method of Test for
PROPAGATION TEAR RESISTANCE OF PLASTIC FILM AND THIN SHEETING[1]

This Standard is issued under the fixed designation D 1922; the number immediately following the designation indicates the year of original adoption, or in the case of revision, the year of last revision. A number in parentheses indicates the year of last reapproval.

NOTE—Section 7 was changed editorially in October 1969.

1. Scope

1.1 This method[2] covers the determination of the average force in grams per specimen required to propagate tearing through a specified length of plastic film or nonrigid sheeting. Two specimens are cited, a rectangular type and one with a constant radius testing length. The latter shall be the preferred or referee specimen.

1.2 Because of (*1*) difficulties in selecting uniformly identical specimens, (*2*) the varying degree of orientation in some plastic films, and (*3*) the difficulty found in testing highly extensible or highly oriented materials, or both, the reproducibility of the test results may be variable and, in some cases, not good or misleading. Provisions are made in the method to compensate for oblique directional tearing which may be found with some materials.

1.3 The testing apparatus used is that described in ASTM Method D 689, Test for Internal Tearing Resistance of Paper,[3] or equivalent, as described in Section 4.

NOTE 1—The values stated in U.S. customary units are to be regarded as the standard. The metric equivalents of U.S. customary units may be approximate.

2. Summary of Method

2.1 The force in grams required to propagate tearing across a film or sheeting specimen is measured using a precisely calibrated pendulum device. Acting by gravity, the pendulum swings through an arc, tearing the specimen from a precut slit. The specimen is held on one side by the pendulum and on the other side by a stationary member. The loss in energy by the pendulum is indicated by a pointer. The scale indication is a function of the force required to tear the specimen.

3. Significance

3.1 This method is of value in ranking relative tearing resistance of various plastic films and thin sheeting of comparable thickness. Experience has shown the test to have its best reliability on relatively less extensible films and sheeting. Variable elongation and oblique tearing effects on the more extensible films preclude its use as a precise production control tool for these types of plastics. This method should be used for specification acceptance testing only after it has been demonstrated that the data for the particular ma-

---

[1] This method is under the jurisdiction of ASTM Committee D-20 on Plastics. A list of committee members may be found in the ASTM Yearbook. This standard is the direct responsibility of subcommittee D-20.10 on Mechanical Properties.

Current edition effective April 11, 1967. Originally issued 1961. Replaces D 1922 – 61 T.

[2] This method has been adapted from TAPPI Standard Method T414m-49 and ASTM Method D689, of Test for Internal Tearing Resistance of Paper. In testing certain plastic films, problems of reproducibility and interpretation of results are encountered which require special treatment to make the method of most value. It is revised here specifically for use with plastic film and thin sheeting. For more complete explanation of certain aspects of the equipment, its calibration and adjustment, refer to TAPPI Standard Method T414m-49.

The following additional references may be of interest in connection with this method:

Painter, E. V., Chu, C. C., and Morgan, H. M., "Testing Textiles on the Elmendorf Tear Tester," *Textile Research Journal*, TRJOA, Vol XX, No. 6, June 1950, pp. 410–417.

Elmendorf, A., "Strength Test for Paper," *Paper*, PPEXA, Vol 26, April 21, 1920, p. 302.

[3] *Annual Book of ASTM Standards*, Part 15.

terial are acceptably reproducible. This method should be used for service evaluation only after its usefulness for the particular application has been demonstrated with a number of different films.

3.2 The method has been widely used as one index of the tearing resistance of plastic film and thin sheeting used in packaging applications. While it may not always be possible to correlate film tearing data with its other mechanical or toughness properties, the apparatus of this method provides a controlled means for tearing specimens at straining rates approximating some of those found in actual packaging service.

3.3 Due to orientation during their manufacture, plastic films and sheeting frequently show marked anisotropy in their resistance to tearing. This is further complicated by the fact that some films elongate greatly during tearing, even at the relatively rapid rates of loading encountered in this test. The degree of this elongation is dependent in turn on film orientation and the inherent mechanical properties of the polymer from which it is made. These factors make tear resistance of some films reproducible between sets of specimens to ±5 percent of the mean value, while others may show no better reproducibility than ±50 percent.

3.4 Data obtained by this method may supplement that from ASTM Method D 1004, of Test for Tearing Resistance of Plastic Film and Sheeting,[4] wherein the specimen is strained at a rate of 50 mm (2 in.) per min. However, specimen geometry and testing speed of the two methods are dissimilar. The rate of tearing in this method, while varying as a function of resistance to tear, is in the range of from 7.6 to 46 m (300 to 1800 in.)/min.

3.5 There is not a direct, linear relationship between tearing force and specimen thickness. Data from this method are expressed as tearing force in grams, with specimen thickness also reported. But sets of data from specimens of dissimilar thickness are usually not comparable. Therefore, only data at the same thickness can be compared.

4. Apparatus

4.1 *Pendulum Impulse Type Testing Apparatus*[5] consisting of the following:

4.1.1 *Stationary Clamp.*

4.1.2 *Movable Clamp*, carried on a pendulum, preferably formed by a sector of a wheel or circle, free to swing on a ball bearing or other substantially frictionless bearing.

4.1.3 *Stop Catch*, for holding the pendulum in a raised position and for releasing it instantaneously.

4.1.4 *Pointer and Scale*, for registering the maximum arc through which the pendulum swings when released. The pendulum shall carry a circumferential scale, graduated from 0 to 100 percent of the machine capacity so as to read against the pointer the average force in grams required to tear a specimen 43 mm (1.7 in.) (Note 2). With the pendulum in its initial position ready for test, separate the two clamps by an interval of 2.54 mm (0.10 in.). So align them that the specimen clamped in them lies in a plane perpendicular to the plane of oscillation of the pendulum with the edges of the jaws gripping the specimen in a horizontal line, a perpendicular to which through the axis of suspension of the pendulum is 102 mm (4 in.) in length and makes an angle of 27.5 deg with the plane of the film specimen. The clamping surface in each jaw shall be at least 25.4 mm (1 in.) in width and at least 12.7 mm (0.5 in.) in depth.

NOTE 2—In this method the work done in tearing through a fixed length of a specimen is measured. The pendulum scale is arbitrarily calibrated to indicate the work done in tearing the specimen a total tearing distance of 86 mm (3.4 in.) as a percentage of the capacity of the machine. The total tearing distance is 86 mm (3.4 in.), because in order to tear a given length of a specimen the tearing force has to be applied through twice that distance.

4.2 *Template, Die, or Shear-Type Cutter*,[6] for cutting specimens.

4.3 *Razor Blades*, single-edged, for cutting specimens where a template is used.

4.4 *Thickness-Measuring Device*—A suitable micrometer, or other thickness gage, reading to 0.0025 mm (0.0001 in.) for measuring the thickness of test specimens. The pressure exerted by the gage on the specimen being measured shall be between 1.6 and 1.9

---

[4] *Annual Book of ASTM Standards*, Part 27.

[5] Equipment available from the Thwing-Albert Instrument Co., Philadelphia, Pa. 19144, meets the requirements for this apparatus.

[6] The TA63 Sample Cutter, Catalog No. 98, from the Thwing-Albert Instrument Co., Philadelphia, Pa. 19144, has been found satisfactory for cutting specimens.

kgf/cm^2 (23 and 27 psi) as prescribed in Method C of ASTM Methods D 374, Test for Thickness of Solid Electrical Insulation.[4]

5. Test Specimens

5.1 Test specimens shall be cut, as shown in Fig. 1, to form a constant-radius testing length. This shall be the preferred or referee specimen type since its geometry automatically compensates for the problem of oblique tearing (Notes 3 and 4). Alternatively, specimens shall be cut to form a rectangle 76 mm (3 in.) or more in width by 63 mm (2.5 in.) in length and plainly marked to denote intended direction of tear. The 63-mm specimen dimension shall be the direction of tear. Two sets of specimens shall be cut from each sample so that their sides are parallel to (*1*) the machine direction, and (*2*) the transverse direction, respectively, of the material being tested. Enough specimens shall be cut in each direction to provide a minimum of ten tear strength determinations.

NOTE 3—Specimens having constant-radius testing lengths are designed to correct for oblique directional tearing encountered in certain anisotropic, elastomeric films and nonrigid sheeting. For purposes of specimen selection, oblique tearing is defined as tearing in a curved or straight line that deviates more than 9.5 mm ($^3/_8$ in.) from the vertical line of intended tear.

NOTE 4—Certain film and sheeting specimens showing oblique tearing may yield data of poor reproducibility because the axis of maximum orientation varies as much as 30 deg from the nominal machine direction. When this is suspected, the sample may be examined by crossed Polaroid plates to determine this direction of maximum orientation and the specimens cut along the axis of anisotropy for testing parallel and normal to it.

5.2 Where a metal template is used, the film or sheeting shall be placed on a hard surface. The template shall be held over it and the specimens cut out using a single-edged razor blade.

5.3 When the specimen is cut out, a slit 20 mm (0.8 in.) deep may be made at the center of the edge perpendicular to the direction to be tested. This leaves exactly 43 mm (1.7 in.) of tearing length between the end of the slit and the opposite edge of the specimen. This slit may be cut into the specimen after it has been placed in the testing apparatus.

NOTE 5—The pendulum apparatus may be fitted with a sharp spring-loaded knife to make this slit in the specimen after it has been clamped in the apparatus. The action of the knife must be such as to make a clean slit exactly 20 mm (0.8 in.) into the specimen from the edge.

6. Adjustment and Calibration of Apparatus

6.1 To check the pendulum swing for freedom from excess friction, level the apparatus and draw a pencil line on the base or stop mechanism 25.4 mm (1 in.) to the right of the edge of the sector stop. With the sector raised to its initial position and the pointer set against its stop, on releasing the sector and holding the stop down, the sector should make at least 20 complete oscillations before the edge of the sector that engages with the stop no longer passes to the left of the pencil line. Otherwise, oil and adjust the bearing.

6.2 Check the pointer friction as follows: Set the pointer at zero reading on the scale before releasing the sector, and after release see that the pointer is not pushed more than three scale divisions beyond zero. A reading of more than three divisions indicates excessive pointer friction and the pointer should be removed the bearing wiped clean, and a trace of oil or petroleum jelly applied. When the pointer friction has been reduced, finally adjust the pointer stop.

6.3 To check the pointer for its zero point, level the apparatus so that, with the sector free, the line on the sector indicating the vertical point of suspension coincides with a corresponding point on the base of the apparatus, usually placed on the stop mechanism. After leveling, operate the apparatus several times with nothing in the jaws, the movable jaw being closed, to ascertain whether the pointer registers zero with no load. If zero is not registered, adjust the position of the pointer stop by means of the pointer stop thumb screw until a zero reading is obtained.

6.4 To calculate the apparatus, first mark the center of gravity of the weight (including means of attaching) by a punched dot on the face of the weight. Then clamp a known weight in grams, W,[7] to the radial edge of the sector beneath the jaws.

6.5 Raise and set the sector as for tearing a specimen and, by means of a suitable scale,

[7] Elmendorf calibration check weights are available from the Thwing-Albert Instrument Co., Philadelphia, Pa. 19144. Use of these weights will permit direct calibration of the apparatus in a shorter time.

measure the height in centimeters, *h*, of the center of gravity of the weight above the surface upon which the apparatus rests. Then release the sector, allow it to swing, and note the pointer reading. Without touching the pointer, raise the sector until the edge of the pointer just meets with its stop, in which position again determine the height in centimeters, *H*, of the center of gravity of the weight above the surface.

6.6 The work done is $W(H - h)$ gram-centimeters. The pointer reading noted above should be the same as that calculated as follows:

$$W(H - h)/137.6$$

6.7 Five weights from 75 to 400 g form a suitable range for calibration of the apparatus, one or more being clamped on the edge of the sector in different positions. Calculate the work done in raising each and add together.

6.8 Make a record of deviations of the pointer from the calculated readings and make corresponding corrections in the test results at the proper points on the scale.

6.9 It is unnecessary to repeat the calibration of the instrument provided it is kept in adjustment and no parts become changed or worn.

7. Conditioning

7.1 *Conditioning*—Condition the test specimens at 23 ± 2 C (73.4 ± 3.6 F) and 50 ± 5 percent relative humidity for not less than 40 h prior to test in accordance with Procedure A of ASTM Methods D 618, Conditioning Plastics and Electrical Insulating Materials for Testing,[4] for those tests where conditioning is required. In cases of disagreement, the tolerances shall be ±1 C (±1.8 F) and ±2 percent relative humidity.

7.2 *Test Conditions*—Conduct tests in the Standard Laboratory Atmosphere of 23 ± 2 C (73.4 ± 3.6 F) and 50 ± 5 percent relative humidity, unless otherwise specified in the test methods or in this specification. In cases of disagreement, the tolerances shall be ±1 C (±1.8 F) and ±2 percent relative humidity.

8. Procedure

8.1 Test not less than ten specimens in each of the principal film or sheeting directions. Measure and record the thickness of each specimen as the average of three readings across its center in the direction in which it is to be torn. Read the thickness to a precision of 0.0025 mm (0.0001 in.) or better except for sheeting greater than 0.25 mm (10 mils) thickness, which read to a precision of 0.025 mm (0.001 in.) or better.

8.2 With the pendulum in its raised position, place the specimen midway in the clamps so that its upper edge is parallel to the top of the clamps and the initial slit (if it was made when the specimen was cut) is at the bottom of and between the clamps at right angles to their top.

8.3 Slit the firmly clamped specimen with the sharp spring-loaded knife if it has not been slit during cutting. Lay the upper, testing portion of the specimen over in the direction of the pendulum pivot.

NOTE 6—The work done in tearing a specimen includes a certain amount of work to bend continuously the film or sheeting as it is torn, to provide for the rubbing of the torn edges of the specimen together, and to lift the specimen against the force of gravity. Consequently, it is necessary to specify certain empirical requirements for both the apparatus and the method to keep the additional work not used for tearing to approximately a definite quantity.

8.4 Release the sector stop and tear the specimen. As the sector completes its return swing, catch it with the thumb and forefinger of the left hand, being careful not to disturb the position of the pointer.

8.5 Examine the specimen. If it tore through the constant-radius section within an approximate angle of 60 deg on either side of the vertical line of intended tear, record the pointer reading to the nearest 0.5 unit. If the line of tear was more than approximately 60 deg from the vertical, reject the reading and test an extra specimen in its place. If rectangular specimens are tested, reject all specimens that tear obliquely more than 9.5 mm (3/8 in.) from the vertical line of intended tear. Test extra specimens to replace those rejected. When oblique tearing is frequent, the test may be performed along and normal to the axis of maximum orientation (see Note 4) instead of along machine and transverse directions.

NOTE 7—In addition to tearing in a curved or oblique direction, some specimens may elongate along the line of tear to such an extent that the actual tearing length may be considerably more than the standard 43-mm (1.7-in.) dimension. As the degree or length of this elongation cannot be measured, the data cannot be corrected for its effect. However, when this has occurred, a note should be included in the report of data. This elongating tendency of certain films may cause poorer reproducibility.

NOTE 8—The maximum accuracy of the pendulum apparatus lies in the scale range between 20 and 60. When thin specimens are being tested, it may be advisable to test enough specimens sandwiched together to produce a scale reading between 20 and 60. However, certain specimens in the same sandwich may tear obliquely in opposite directions, which may lead to falsely high results. When this tearing behavior is encountered, single specimens must be tested, even though scale readings may be in the range below 20. If tearing loads are in excess of 60, the augmenting weight attachment may be used to double the capacity of the apparatus. For thin film, it is recommended that single specimens and a lower capacity tester be used rather than several specimens and a higher capacity machine.

9. Calculations

9.1 Calculate the tearing resistance of the specimen in grams, R, as follows:

$$R = (S \times C)/n$$

where:

S = corrected scale reading,
C = machine capacity, g, and
n = number of sheets torn in each test.

A direct proportionality may not always exist between tearing force and specimen thickness. Therefore, this method provides for reporting data in grams of force required to propagate tearing with specimen thickness reported separately.

9.2 Calculate the arithmetic mean, $\bar{X}$, tearing resistance in each principal direction of the film or sheeting.

9.3 Calculate the standard deviation of the tearing resistance in each principal direction to two significant figures as follows:

$$s = \sqrt{(\Sigma X^2 - n\bar{X}^2)/(n - 1)}$$

where:

s = estimated standard deviation,
X = value of a single observation,
n = number of observations, and
$\bar{X}$ = arithmetic mean of the set of observations.

10. Report

10.1 The report shall include the following:

10.1.1 Complete identification of the sample tested including source, manufacturer's name and code number, method of fabrication, roll or lot number, and date received or made,

10.1.2 Type and direction of specimens tested: rectangular or constant radius, parallel or normal to the machine direction of the film. If tests were performed with reference to an axis of maximum orientation that did not coincide with the machine or transverse direction of the film, the report should also include the location of this axis relative to the latter directions,

10.1.3 Number of specimens tested at one time, and the number tested in each principal direction of the film,

10.1.4 Average, maximum, and minimum values for specimen thickness and for machine and transverse tearing resistance (if data are obtained from specimens in both principal directions), expressed in grams to the nearest whole number,

10.1.5 Standard deviation from the average(s) of the tearing resistance in the machine and transverse directions, if both directions are tested, and

10.1.6 Capacity of the tester.

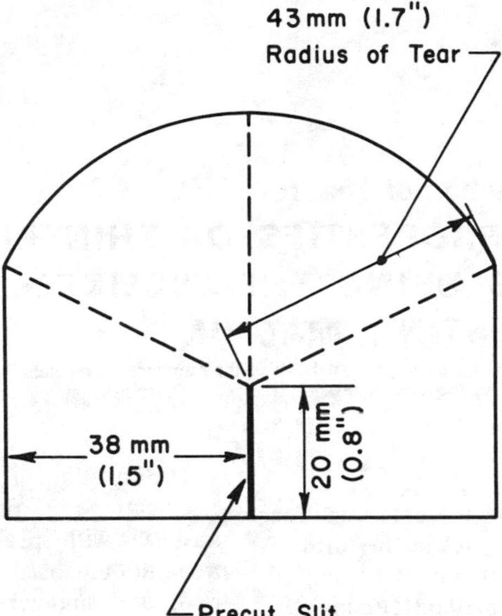

FIG. 1 Constant-Radius Test Specimen for Tear Resistance Test.

Designation: D 1923 – 67

Standard Method of Test for

TENSILE PROPERTIES OF THIN PLASTIC SHEETING USING THE INCLINED PLANE TESTING MACHINE[1]

This Standard is issued under the fixed designation D 1923; the number immediately following the designation indicates the year of original adoption or, in the case of revision, the year of last revision. A number in parentheses indicates the year of last reapproval.

1. Scope

1.1 This method[2] covers the determination of tensile properties of plastics in the form of thin sheeting less than 1.0 mm (0.04 in.) in thickness (Note 1). This method employs an inclined plane weighing head to measure the load applied to break the specimen. The angle of inclination of the tilting table on which the loaded carriage moves to apply the load to the specimen is changed to give a constant rate of load application to the specimen. The rate of grip separation varies, being dependent on the elongation characteristics of the specimen under test (Note 2).

NOTE 1—Tensile properties of plastics 1.0 mm (0.04 in.) or greater in thickness shall be determined in accordance with ASTM Method D 638, Test for Tensile Properties of Plastics,[3] if rigid, and ASTM Method D 412, Tension Testing of Vulcanized Rubber,[3] if nonrigid or semirigid. If the sample quantity is limited, ASTM Method D 1708, Test for Tensile Properties of Plastics by Use of Microtensile Specimens,[3] may be used.

NOTE 2—ASTM Methods D 882, Test for Tensile Properties of Thin Plastic Sheeting[3] describes two types of tension tests for thin plastic sheeting differing in manner of load application:

Method A. Static Weighing—Constant-Rate-of-Grip Separation Test, and

Method B. Pendulum Weighing—Constant-Rate-of-Powered-Grip Motion Test.

Methods D 882 also include a procedure for tensile modulus of elasticity at one strain rate. This procedure is the recommended modulus measurement technique.

NOTE 3—The values stated in U.S. customary units are to be regarded as the standard. The metric equivalents of U.S. customary units may be approximate.

2. Significance

2.1 Tensile properties determined by this method are of value for the identification and characterization of materials for control and specification purposes. Tensile properties may vary with specimen thickness, method of preparation, speed of testing, type of grips used, and manner of measuring extension. Consequently, where precise comparative results are desired, these factors must be carefully controlled.

2.2 Since both method and rate of loading vary between this method and the methods described in Methods D 882, results from these methods cannot be directly compared. Method A of Method D 882 is preferred and shall be used for referee purposes in cases where test results are challenged, unless otherwise indicated in particular materials specifications.

2.3 Tensile properties may be utilized to provide data for research and development and engineering design as well as quality control and specification. However, data from such tests cannot be considered significant for applications differing widely from the load-time scale of the test employed.

2.4 Materials that fail by tearing give data which cannot be compared with those from normal failure.

---

[1] This method is under the jurisdiction of ASTM Committee D-20 on Plastics. A list of committee members may be found in the ASTM Yearbook. This standard is the direct responsibility of Subcommittee D-20.10 on Mechanical Properties.

Current edition effective April 11, 1967. Originally issued 1961. Replaces D 1923 – 61 T.

[2] This test method was formerly Method C of the Methods of Test for Tensile Properties of Thin Plastic Sheets and Films (ASTM Designation: D 882 – 56 T).

[3] *Annual Book of ASTM Standards*, Part 27.

3. Definitions

3.1 Definitions of terms and symbols relating to tension testing of plastics appear in the Appendix to Method D 638.

4. Apparatus

4.1 *Grip*—See 4.1 of Methods D 882.

4.2 *Thickness Gage*—See 4.2 of Methods D 882.

4.3 *Width-Measuring Devices*—See 4.3 of Methods D 882.

4.4 *Specimen Cutter*—See 4.4 of Methods D 882.

4.5 *Conditioning Apparatus*—See 4.5 of Methods D 882.

4.6 *Extension Indicators*—See 4.6 of Methods D 882.

4.7 *Testing Machine*—A film testing machine of the inclined-plane or tilting-table type. It shall be equipped with a device for recording the tensile load carried by the specimen and the amount of separation of the grips during the test; these measuring systems shall be accurate to ±2 percent. Both the stressing and the recording mechanism should be essentially free from inertia lag at the specified rate of stressing, and frictional drag should be reduced to a minimum. The machine shall be equipped with a device for varying the loading rate between approximately 4.5 and 68 kg (10 and 150 lb)/min. Provision for variation of tilting speed is recommended.

5. Test Specimens

5.1 See 5.1 to 5.6 of Methods D 882.

6. Conditioning

6.1 *Conditioning*—Condition the test specimens at 23 ± 2 C (73.4 ± 3.6 F) and 50 ± 5 percent relative humidity for not less than 40 h prior to test in accordance with Procedure A of ASTM Methods D 618, Conditioning Plastics and Electrical Insulating Materials for Testing,[3] for those tests where conditioning is required. In cases of disagreement, the tolerances shall be ±1 C (±1.8 F) and ±2 percent relative humidity.

6.2 *Test Conditions*—Conduct tests in the Standard Laboratory Atmosphere of 23 ± 2 C (73.4 ± 3.6 F) and 50 ± 5 percent relative humidity, unless otherwise specified in the test methods or in this specification. In cases of disagreement, the tolerances shall be ±1 C (±1.8 F) and ±2 percent relative humidity.

7. Number of Test Specimens

7.1 See Section 7 of Methods D 882.

8. Speed of Testing

8.1 The rate of stressing shall be held between the limits of 700 and 2100 kgf/cm^2 (10,000 and 30,000 psi)/min in all tests. The rate of stressing shall be adjusted, depending on the cross-section of the specimen, to approximately 1400 kgf/sq cm^2 (20,000 psi)/min.

8.2 The following table indicates examples of specimen dimensions and rates of loading to obtain a rate of stressing of 1400 kgf/cm^2 (20,000 psi)/min (Note 4):

Nominal Film Thickness		Width of Specimen		Carriage Weight		Rate of Loading	
mm	in.	mm	in.	kg	lb	kg/ min	lb/ min
0.01	0.0005	25.4	1.00	4.5	10	4.5	10
0.08	0.003	12.7	0.50	4.5	10	13.6	30
0.13	0.005	12.7	0.50	4.5	10	22.7	50
0.19	0.0075	12.7	0.50	11.3	25	34.0	75
0.25	0.01	12.7	0.50	11.3	25	45.4	100
0.32	0.0125	12.7	0.50	22.7	50	56.7	125
0.38	0.015	12.7	0.50	22.7	50	68.0	150
0.51	0.02	9.5	0.38	22.7	50	68.0	150
1.0	0.04	4.8	0.19	22.7	50	68.0	150

NOTE 4—It is possible to use a number of combinations of carriage weights and tilting speeds to obtain a given rate of loading. However, due to variation in momentum developed by the various weights during the course of the test, considerable difference among sets of a given test sample will arise as the weights are changed. Consequently, only values obtained under identical machine operating conditions can be used for comparison of materials.

9. Procedure

9.1 Select a specimen width that will produce failure within the upper two thirds of the load range selected. A few trial runs may be necessary to make a proper selection.

9.2 Measure the cross-sectional area of the specimen at several points along its length. Measure the width to an accuracy of 0.25 mm (0.010 in.) or better. Measure the thickness to an accuracy of 0.0025 mm (0.0001 in.) or better for films less than 0.25 mm (0.01 in.) in thickness, and to an accuracy of 1 percent or better for films greater than 0.25 mm (0.01

in.) but less than 1.0 mm (0.04 in.) in thickness. Calculate the minimum cross-sectional area.

9.3 The initial grip separation shall be at least 50 mm (2 in.) for films having total elongation at break of less than 100 percent, and at least 25 mm (1 in.) for films having total elongation at break of more than 100 percent.

NOTE 5—Since slippage is a potential problem in these tests, as great an initial distance between grips as possible should be employed.

9.4 Adjust the rate of loading to give a rate of stressing of approximately 1400 kgf/cm^2 (20,000 psi/min (see Section 8). Balance, zero, and calibrate the load weighing and recording system.

9.5 When it is desired to measure a test section other than the total length between the grips, mark the ends of the desired test section with a soft, fine wax crayon or with ink. Do not scratch these marks onto the surface since such scratches may act as stress raisers and cause premature specimen failure. Extensometers may be used if available; in this case, the test section will be defined by the contact points of the extensometer.

NOTE 6—Measurement of a specific test section is necessary with some materials having high elongation. As the specimen elongates, the accompanying reduction in area results in a loosening of material at the inside edge of the grips. This reduction and loosening moves back into the grips as further elongation and reduction in area takes place. In effect this causes problems similar to grip slippage, that is, exaggerates measured extension.

9.6 Place the test specimen in the grips of the testing machine, taking care to align the long axis of the specimen with an imaginary line joining the points of attachment of the grips to the machine. Tighten the grips evenly and firmly to the degree necessary to minimize slipping of the specimen during test.

9.7 Start the machine and record load versus extension.

9.7.1 When the total length between the grips is used as the test area, record load versus grip separation.

9.7.2 When a specific test area has been marked on the specimen, follow the displacement of the edge boundary lines with respect to each other with dividers or some other suitable device. If a load-extension curve is desired, plot various extensions versus corresponding loads sustained, as measured by the load indicator.

9.7.3 When an extensometer is used, record load versus extension of the test area measured by the extensometer.

10. Calculations

10.1 See 10.1, 10.2, 10.3, 10.4, 10.5, 10.6, 10.9, and 10.10 of Methods D 882.

11. Report

11.1 The report shall include the following:

11.1.1 All items included in Methods D 882 with the exception of 11.1.5 and 11.1.19.

11.1.2 In addition to these values, the rate of stressing as defined in 8.1 shall be included.

ASTM Designation: D 1924 – 70

Recommended Practice for
DETERMINING RESISTANCE OF SYNTHETIC POLYMERIC MATERIALS TO FUNGI[1]

This Recommended Practice is issued under the fixed designation D 1924; the number immediately following the designation indicates the year of original adoption or, in the case of revision, the year of last revision. A number in parentheses indicates the year of last reapproval.

1. Scope

1.1 This recommended practice covers determination of the effect of fungi on the properties of synthetic polymeric materials in the form of molded and fabricated articles, tubes, rods, sheets, and film materials. Changes in optical, mechanical, and electrical properties may be determined by the applicable ASTM methods.

NOTE 1—The values stated in U.S. customary units are to be regarded as the standard. The metric equivalents of U.S. customary units may be approximate.

2. Summary of Method

2.1 The procedure described herein consists of (*1*) selection of suitable specimens for determination of pertinent properties, (*2*) inoculation of the specimens with suitable organisms, (*3*) exposure of inoculated specimens under conditions favorable to growth, (*4*) examination and rating for visual growth, and (*5*) removal of the specimens and observations or testing, either before cleaning or after cleaning and reconditioning.

NOTE 2—Since the procedure involves handling and working with fungi, it is recommended that personnel trained in microbiology perform the portion of the procedure involving handling of organisms and inoculated specimens.

3. Significance

3.1 The resin portion of these materials is usually fungus-resistant in that it does not serve as a carbon source for the growth of fungi. It is generally the other components, such as plasticizers, cellulosics, lubricants, stabilizers, and colorants, that are responsible for fungus attack on plastic materials. It is important to establish the resistance to microbial attack under conditions favorable for such attack, namely, a temperature of 2 to 38 C (35 to 100 F) and a relative humidity of 60 to 100 percent.

3.2 The effects to be expected are as follows:

3.2.1 Surface attack, discoloration, loss of transmission (optical).

3.2.2 Removal of susceptible plasticizers, modifiers, and lubricants, resulting in increased modulus (stiffness), changes in weight, dimensions, and other physical properties, and deterioration of electrical properties such as insulation resistance, dielectric constant, power factor, and dielectric strength.

3.3 Often the changes in electrical properties are due principally to surface growth and its associated moisture and to pH changes caused by excreted metabolic products. Other effects include preferential growths caused by nonuniform dispersion of plasticizers, lubricants, and other processing additives. Attack on these materials often leaves ionized conducting paths. Pronounced physical changes are observed on products in film form or as coatings, where the ratio of surface to volume is high, and where nutrient materials such as plasticizers and lubricants continue to diffuse to the surface as they are utilized by the organisms.

3.4 Since attack by organisms involves a large element of chance due to local accelerations and inhibitions, the order of reproducibility may be rather low. To assure that esti-

[1] This recommended practice is under the jurisdiction of ASTM Committee G-3 on Deterioration of Nonmetallic Materials. A list of committee members may be found in the ASTM Yearbook.

Current edition effective April 13, 1970. Originally issued 1961. Replaces D 1924 – 63.

mates of behavior are not too optimistic, the greatest observed degree of deterioration should be reported.

3.5 Conditioning of the specimens, such as exposure to leaching, weathering, heat treatment, etc., may have significant effects on the resistance to fungi. Determination of these effects is not covered in this recommended practice.

4. Apparatus

4.1 *Glassware*—Glass vessels are suitable for holding specimens when laid flat. Depending on the size of the specimens, the following are suggested:

4.1.1 For specimens up to 76 mm (3 in.) in diameter, 152-mm (6-in.) covered Petri dishes.

4.1.2 For 76-mm (3-in.) and larger specimens, such as tensile and stiffness strips, large Petri dishes, trays of borosilicate glass, or baking dishes up to 408 by 508 mm (16 by 20 in.) in size, covered with squares of window glass.

4.2 *Incubator*—Incubating equipment for all test methods shall maintain a temperature of 28 to 30 C (82.4 to 86 F) and a relative humidity not less than 85 percent. Automatic recording of wet- and dry-bulb temperature is recommended.

5. Reagents and Materials

5.1 *Purity of Reagents*—Reagent grade chemicals or equivalent, as specified in ASTM Methods E 200, for Preparation, Standardization, and Storage of Standard Solutions for Chemical Analysis,[2] shall be used in all tests.

5.2 *Purity of Water*—Unless otherwise indicated, references to water shall be understood to mean distilled water or water of equal purity.

5.3 *Nutrient-Salts Agar*[3]—Prepare this medium by dissolving in 1 liter of water the designated amounts of the following reagents:

Potassium dihydrogen orthophosphate (KH_2PO_4)	0.7 g
Potassium monohydrogen orthophosphate (K_2HPO_4)	0.7 g
Magnesium sulfate ($MgSO_4 \cdot 7H_2O$)	0.7 g
Ammonium nitrate (NH_4NO_3)	1.0 g
Sodium chloride (NaCl)	0.005 g
Ferrous sulfate ($FeSO_4 \cdot 7H_2O$)	0.002 g
Zinc sulfate ($ZnSO_4 \cdot 7H_2O$)	0.002 g
Manganous sulfate ($MnSO_4 \cdot 7H_2O$)	0.001 g
Agar	15.0 g

Sterilize the test medium by autoclaving at 121 C (250 F) for 20 min. Adjust the pH of the medium by the addition of 0.01 *N* NaOH solution so that after sterilization the pH is between 6.0 and 6.5. Prepare sufficient medium for the required tests.

5.4 *Mixed Fungus Spore Suspension:*

NOTE 3—Since a number of other organisms may be of specific interest for certain final assemblies or components, such other pure cultures of organisms may be used if agreed upon by the purchaser and the manufacturer of the plastics. Reference (**1**) illustrates such a choice.

5.4.1 Use the following test fungi in preparing the cultures:

Fungi	ATCC No.[a]	QM No.[b]
Aspergillus niger	9642	386
Penicillium funiculosum	9644	391
Chaetomium globosum	6205	459
Trichoderma sp	9645	365
Pullularia pullulans	9348	279c

[a] American Type Culture Collection, 2112 M St., N. W., Washington 7, D. C.
[b] Quartermaster Culture Collection, QMR & E Command, Natick, Mass.

Maintain cultures of these fungi separately on an appropriate medium such as potato dextrose agar. The stock cultures may be kept for not more than 4 months at approximately 3 to 10 C (37.4 to 50 F). Use subcultures incubated at 28 to 30 C (82.4 to 86 F) for 7 to 20 days in preparing the spore suspension.

5.4.2 Prepare a spore suspension of each of the six fungi by pouring into one subculture of each fungus a sterile 10-ml portion of water or of a sterile solution containing 0.05 g/liter of a nontoxic wetting agent such as sodium dioctyl sulfosuccinate. Use a sterile platinum or nichrome inoculating wire to scrape gently the surface growth from the culture of the test organism.

5.4.3 Pour the spore charge into a sterile 125-ml glass-stoppered Erlenmeyer flask containing 45 ml of sterile water and 10 to 15 solid glass beads, 5 mm in diameter. Shake the flask vigorously to liberate the spores from the fruiting bodies and to break the spore clumps.

5.4.4 Filter the shaken suspension through

[2] *Annual Book of ASTM Standards*, Part 30.
[3] Agar and nutrient-salts agar are obtainable from biological laboratory supply sources.

a thin layer of sterile glass wool in a glass funnel into a sterile flask in order to remove mycelial fragments.

5.4.5 Centrifuge the filtered spore suspension aseptically, and discard the supernatant liquid. Resuspend the residue in 50 ml of sterile water and centrifuge.

5.4.6 Wash the spores obtained from each of the fungi in this manner three times. Dilute the final washed residue with sterile nutrient-salts solution (Note 4) in such a manner that the resultant spore suspension shall contain 1,000,000 ± 200,000 spores/ml as determined with a counting chamber.

5.4.7 Repeat this operation for each organism used in the test and blend equal volumes of the resultant spore suspension to obtain the final mixed spore suspension.

NOTE 4—Nutrient salts solution is identical with the composition for nutrient salts agar given in 5.3 except that the agar is omitted.

5.4.8 The spore suspension may be prepared fresh each day or may be held in the refrigerator at 3 to 10 C (37.4 to 50 F) for not more than 4 days.

6. Viability Control

6.1 With each daily group of tests place each of three pieces of sterilized filter paper, 25.4 mm (1 in.) square, on hardened nutrient-salts agar in separate Petri dishes. Inoculate these with the spore suspension by spraying the suspension from a sterilized atomizer[4] so that the entire surface is moistened with the spore suspension. Incubate these at 28 to 30 C (82.4 to 86 F) at a relative humidity not less than 85 percent and examine them after 14 days' incubation. There shall be copious growth on all three of the filter paper control specimens. Absence of such growth requires repetition of the test.

7. Test Specimens

7.1 The simplest specimen may be a 51 by 51-mm (2 by 2-in.) piece, a 51-mm (2-in.) diameter piece, or a piece (rod or tubing) at least 76 mm (3 in.) long cut from the material to be tested. Completely fabricated parts or sections cut from fabricated parts may be used as test specimens. On such specimens, observation of effect is limited to appearance, density of growth, optical reflection or transmission, or manual evaluation of change in physical properties such as stiffness.

7.2 Film-forming materials such as coatings may be tested in the form of films at least 51 by 25.4 mm (2 by 1 in.) in size. Such films may be prepared by casting on glass and stripping after cure, or by impregnating (completely covering) filter paper or ignited glass fabric.

7.3 For visual evaluation, three specimens shall be inoculated. If the specimen is different on two sides, three specimens of each, face up and face down, shall be tested.

NOTE 5—In devising a test program intended to reveal quantitative changes occurring during and after fungal attack, an adequate number of specimens should be evaluated to establish a valid value for the original property. If five replicate specimens are required to establish a tensile strength of a film material, the same number of specimens shall be removed and tested for each exposure period. It is to be expected that values of physical properties at various stages of fungal attack will be variable; the values indicating the greatest degradation are the most significant (see 3.4). The *ASTM Manual on Quality Control of Materials, STP 15-C*, may be used as a guide.

8. Procedure

8.1 *Inoculation*—Pour sufficient nutrient-salts agar into suitable sterile dishes (see 4.1) to provide a solidified agar layer from 3.2 to 6.4 mm ($^1/_8$ to $^1/_4$ in.) in depth. After the agar is solidified, place the specimens on the surface of the agar. Inoculate the surface, including the surface of the test specimens, with the composite spore suspension by spraying the suspension from a sterilized atomizer[4] with 112-kN/m^2 (16-psi) air pressure so that the entire surface is moistened with the spore suspension.

8.2 *Incubation*—Cover the inoculated test specimens and incubate at 28 to 30 C (82.4 to 86 F) and not less than 85 percent relative humidity for a minimum of 21 days, recording the growth each week.

NOTE 6—Covered dishes containing nutrient agar are considered to have the desired humidity. Covers on large dishes may be sealed with masking tape.

8.3 *Observation for Visible Effects*—If the test is for visible effects only, remove the three specimens from the incubator and judge

[4] DeVilbiss No. 154 atomizer or equivalent has been found satisfactory for this purpose.

them as follows:

Observed Growth on Specimens	Rating
None	0
Traces of growth (less than 10 percent)	1
Light growth (10 to 30 percent)	2
Medium growth (30 to 60 percent)	3
Heavy growth (60 percent to complete coverage)	4

NOTE 7—Traces of growth may be defined as scattered, sparse fungus growth such as might develop from (*1*) a mass of spores in the original inoculum, or (*2*) extraneous contamination such as fingermarks, insect feces, etc. Continuous cobwebby growth extending over the entire specimen, even though not obscuring the specimen, should be rated as 2.

NOTE 8—Considerable physical change in plastics may occur without much visual growth, hence some measure of change in physical property selected from those cited in the Appendix is recommended.

8.4 *Effect on Physical, Optical, or Electrical Properties*—Wash the specimens free of growth, immerse in an aqueous solution of mercuric chloride (1+1000) for 5 min, rinse in tap water, air dry overnight at room temperature, and recondition at the standard laboratory conditions defined in ASTM Methods D 618, Conditioning Plastics and Electrical Insulating Materials for Testing,[5] 23 ± 1.1 C (73.4 ± 2 F) and 50 ± 2 percent relative humidity, and test according to the respective methods used on control specimens (see the Appendix).

NOTE 9—For certain electrical tests, such as insulation resistance and arc resistance, specimens may be tested in the unwashed, humidified condition. Test values will be affected by surface growth and its associated moisture.

9. Report

9.1 The report shall include the following:

9.1.1 Organisms or organism used,

9.1.2 Time of incubation (if progressive),

9.1.3 Visual rating of fungus growth according to 8.3, and

9.1.4 Tabulation of progressive change in physical, optical, or electrical property against time of incubation. Give the number of observations, the mean, and the maximum observed change.

[5] *Annual Book of ASTM Standards*, Part 27.

REFERENCES

(**1**) Bagdon, V. J., Military Specification Mil-P-43018(CE), "Plastic Sheets: Polyethylene Terephthalate, Drafting, Coated," June 13, 1961.

(**2**) Baskin, A. D., and Kaplan, A. M., "Mildew Resistance of Vinyl-Coated Fabrics," *Applied Microbiology*, APMBA, Vol 4, No. 6, November 1956.

(**3**) Berk, S., "Effect of Fungus Growth on Plasticized Polyvinyl Chloride Films," *ASTM Bulletin*, ASTBA, No. 168, September 1950, p. 53 (TP 181).

(**4**) Berk, S., Ebert, H., and Teitell, L., "Utilization of Plasticizers and Related Organic Compounds by Fungi," *Industrial and Engineering Chemistry*, IECHA, Vol 49, No. 7, July 1957, pp. 1115–1124.

(**5**) Brown, A. E., "Problem of Fungal Growth on Synthetic Resins, Plastics, and Plasticizers," *Modern Plastics*, MOPLA, Vol 23, 1946, p. 189.

(**6**) Ross, S. H., "Biocides for a Strippable Vinyl Plastic Barrier Material," Report PB-151-119, U. S. Department of Commerce, Office of Technical Services.

APPENDIX

A1. Test Methods for Evaluation of Effect of Fungi on Synthetic Polymeric Materials

A1.1 For evaluation of the effect of fungi on mechanical, optical, and electrical properties, the following ASTM and other test methods are recommended.

Property	Methods
Tensile strength	D 638, D 882, D 1708[a]
Stiffness	D 747[a] TAPPI Test Method T 451-M-45 Federal Specification Textile Test Methods CCC-T-191b Method 5204 (Clark Stiffness Test) Federal Specification Textile Test Methods CCC-T-191b Method 5206 (Cantilever Bend Method)
Hardness	D 785[a]
Optical transmission	D 791[a]
Haze	D 1003[a]
Water vapor transmission	E 96[a]
Dielectric strength	D 149[a]
Dielectric constant-power factor	D 150[a]
Insulation resistance	D 257[a]
Arc resistance	D 495[a]

[a] These designations refer to the following ASTM methods:

D 149, Tests for Dielectric Breakdown Voltage and Dielectric Strength of Electrical Insulating Materials at Commercial Power Frequencies[5]

D 150, Tests for A-C Loss Characteristics and Dielectric Constant (Permittivity) of Solid Electrical Insulating Materials[5]

D 257, Tests for D-C Resistance or Conductance of Insulating Materials[5]

D 495, Test for High-Voltage, Low-Current Arc Resistance of Solid Electrical Insulating Materials[5]

D 638, Test for Tensile Properties of Plastics[5]

D 747, Test for Stiffness of Plastics by Means of a Cantilever Beam[5]

D 785, Test for Rockwell Hardness of Plastics and Electrical Insulating Materials[5]

D 791, Test for Luminous Reflectance, Transmittance, and Color of Materials[5]

D 1003, Test for Haze and Luminous Transmittance of Transparent Plastics[5]

D 1708, Test for Tensile Properties of Plastics by Use of Microtensile Specimens[5]

E 96, Test for Water Vapor Transmission of Materials in Sheet Form[2]

ASTM Designation: D 1925 - 70

Standard Method of Test for

YELLOWNESS INDEX OF PLASTICS[1]

This Standard is issued under the fixed designation D 1925; the number immediately following the designation indicates the year of original adoption or, in the case of revision, the year of last revision. A number in parentheses indicates the year of last reapproval.

1. Scope

1.1 This method is intended primarily for determining the degree of yellowness (or change of degree of yellowness) under daylight illumination of homogeneous, nonfluorescent, nearly colorless transparent or nearly white translucent or opaque plastics. It is applicable to transmittance of transparent and translucent plastics and to reflectance of opaque plastics (Note 2). It is based upon tristimulus values calculated from data obtained on the Hardy-G.E.-type spectrophotometer,[2] but other apparatus is satisfactory if equivalent results are obtained.

NOTE 1—The values stated in U.S. customary units are to be regarded as the standard. The metric equivalents of U.S. customary units may be approximate.

NOTE 2—This method has not been demonstrated for the determination of transmitted yellowness of plastics having a luminous transmittance below 25 percent, and it has not been demonstrated for the determination of reflected yellowness of translucent plastics.

2. Significance

2.1 Yellowness index obtained by this method correlates reasonably well with the magnitude of yellowness perceived under daylight illumination.

2.2 Yellowness index of transparent and translucent plastics is a function of thickness. Comparison should be made only between specimens of comparable thickness.

2.3 For control work, tristimulus colorimeters are useful so long as their inaccuracies and differences from this primary method are known.

2.4 This method achieves its greatest accuracy in the determination of differences in yellowness index of sample versus a control of similar material and colorant composition. Change of yellowness index determined by this method has proved useful in evaluation of degradation of plastics under exposure to heat, light or other environment.

3. Definitions

3.1 *yellowness*—deviation in chroma from whiteness or water-whiteness in the dominant wavelength range from 570 to 580 nm.

NOTE 3—A definition of a method of obtaining dominant wavelength may be found in the literature.[3]

3.2 *yellowness index* (*YI*)—the magnitude of yellowness relative to magnesium oxide for CIE Source C. Yellowness index is expressed as follows:

$$YI = [100(1.28X_{CIE} - 1.06Z_{CIE})]/Y_{CIE}$$

where:

X_{CIE}, Y_{CIE}, and Z_{CIE} = tristimulus values (Note 4) of the specimen relative to Source C.

NOTE 4—By this test method, positive (+) yellowness index describes the presence and magnitude of yellowness; a specimen with a negative (−) yellowness index will appear bluish.

3.3 *change in yellowness index* (ΔYI)—the difference between an initial value, YI_0, and YI determined after a prescribed treatment of the plastic.

$$\Delta YI = YI - YI_0$$

[1] This method is under the jurisdiction of the ASTM Committee D-20 on Plastics. A list of committee members may be found in the ASTM Yearbook. This standard is the direct responsibility of Subcommittee D-20.40 on Optical Properties.

Current edition effective Jan. 22, 1970. Originally issued 1962. Replaces D 1925 – 63 T.

[2] This spectrophotometer is described in Recommended Practice E 308.

[3] Hardy, A. C., *Handbook of Colorimetry*, Technology Press, Cambridge, Mass.

NOTE 5—By this calculation, positive (+) ΔYI indicates increased yellowness and negative (−) ΔYI indicates decreased yellowness or increased blueness.

4. Apparatus

4.1 *Spectrophotometer, Recording*,[2] conforming to the requirements of ASTM Recommended Practice E 308 for Spectrophotometry and Description of Color in CIE 1931 System.[4] Other apparatus is satisfactory if equivalent results are obtained.

5. Reference Standards

5.1 *Primary Standard*—The primary standard for reflectance measurement is a layer of freshly prepared magnesium oxide prepared in accordance with ASTM Recommended Practice E 259 for Preparation of Reference White Reflectance Standards.[4] The primary standard for transmittance measurement is air.

5.2 *Instrument Standard*—Because of the difficulty of preparing a primary reflectance standard, magnesium carbonate, barium sulfate, or calibrated pieces of white structural glass known as Vitrolite may be used as instrument standards.

6. Test Specimen

6.1 This test procedure does not cover specimen preparation techniques.

6.2 Opaque specimens shall have at least one plane surface; transparent specimens shall have two surfaces that are essentially plane and parallel. Specimens not having plane surfaces may be compared on a relative basis if of the same shape and if similarly positioned for measurement.

7. Conditioning

7.1 *Conditioning*—Condition the test specimens at 23 ± 2 C (73.4 ± 3.6 F) and 50 ± 5 percent relative humidity for not less than 40 h prior to test in accordance with Procedure A of ASTM Methods D 618, Conditioning Plastics and Electrical Insulating Materials for Testing,[5] for those tests where conditioning is required. In cases of disagreement, the tolerances shall be 1 C (1.8 F) and ±3 percent relative humidity.

7.2 *Test Conditions*—Conduct tests in the Standard Laboratory Atmosphere of 23 ± 2 C (73.4 ± 3.6 F) and 50 ± 5 percent relative humidity, unless otherwise specified in the test methods or in this specification. In cases of disagreements, the tolerances shall be 1 C (1.8 F) and ±2 percent relative humidity.

8. Procedure

8.1 Calibrate and operate the spectrophotometer in accordance with Recommended Practice E 308. Calibrate and operate other instruments in accordance with instructions supplied by the manufacturer.

8.2 Obtain spectral transmittance data relative to air. For measurement of transmittance of translucent specimens, place freshly prepared matched magnesium oxide standards at the specimen and reference ports at the rear of the sphere. The interior of the sphere should be freshly coated with magnesium oxide and in good condition.

NOTE 6—Magnesium oxide standards may be considered matched if on interchanging them the percent reflectance is altered by no more than 1 percent at any wavelength between 400 and 700 nm.

NOTE 7—The energy transmitted by plastic specimens varies widely in angular distribution about the normal to the surface, depending upon the ability of the specimens to diffuse the incident light. Thus the amount of energy that strikes the surface at the specimen port may vary from nearly 100 percent of the energy transmitted to a percentage determined by the relative areas of specimen port and sphere wall. Furthermore, the diffuse energy has a spectral composition such that the portion striking the surface at the specimen port is yellower than that which is more widely diffused.

NOTE 8—Matched magnesium carbonate blocks or matched Vitrolite tiles are not recommended in the measurement of translucent samples. Matched barium sulfate blocks with freshly scraped surfaces have been demonstrated to give results comparable to magnesium oxide, within the accuracy of this method.

8.3 Obtain spectral directional reflectance relative to magnesium oxide. Exclude the specular component in reflectance measurement. Back the test specimens with a white standard to obtain spectral directional reflectance.

9. Calculation

9.1 Calculate the tristimulus values for Source C by numerical integration (see Rec-

[4] *Annual Book of ASTM Standards*, Part 30.
[5] *Annual Book of ASTM Standards*, Part 27.

ommended Practice E 308) from recorded spectral data or by automatic integration[5] during spectrophotometer operation.

9.2 Calculate the magnitude and direction of the yellowness index from the following equation:

$$YI = [100(1.28X_{CIE} - 1.06Z_{CIE})]/Y_{CIE}$$

9.3 For direct calculation of the yellowness index from filter photometer readings obtained with a tristimulus colorimeter see the equations in A1.4 of the Appendix. These equations eliminate the need for calculating the tristimulus values.

9.4 If desired, claculate the magnitude and direction of change in yellowness index from the following equation:

$$\Delta YI = YI - YI_0$$

10. Report

10.1 The report shall include the following:

10.1.1 Complete identification of the material tested,

10.1.2 Specimen thickness,

10.1.3 Magnitude and direction (sign) of the yellowness index, and

10.1.4 Identification of the instrument used by the manufacturer's model and serial number.

11. Precision

11.1 The rank-difference coefficient of correlation with subjective ranking for a series of 34 transparent, 38 opaque, and 34 translucent test specimens was 97.7, 99.5, and 98.7 percent, respectively.

NOTE 9—Because of the Tyndall effect in light transmitted by translucent plastics, visual estimation of yellowness should be made with the specimen illuminated from behind by uniform diffuse illumination, nondirectional with respect to a line normal to the surface. Line-of-sight of the observer should also be normal to the viewed surface. The viewed surface should have a very low level of illumination falling upon it.

11.2 Limited tests conducted with this procedure indicate reproducibility among laboratories to be within ±0.3 units, for measurement of yellowness index of transparent and opaque plastics, and ±2.6 units for measurement of transmitted yellowness of translucent plastics, with 95 percent confidence.

11.3 For translucent plastics, rank correlation of instrumental measurement among four laboratories was in every case better than 98.4 percent.

[6] Davidson, H. R., and Imm, L. W., "Continuous, Automatic Tristimulus Integrator for Use with the Recording Spectrometer," *Journal of the Optical Society of America*, JOSAA, Vol 39, 1949, pp. 942–944.

APPENDIX

A.1 DERIVATION OF EQUATIONS FOR CALCULATION OF YELLOWNESS INDEX FROM PHOTOELECTRIC TRISTIMULUS COLORIMETER MEASUREMENTS

A1.1 By the definition given in 3.2, yellowness index has been defined as:

$$YI = [100(1.28X_{CIE} - 1.06Z_{CIE})]/Y_{CIE}$$

where:

X_{CIE}, Y_{CIE}, and Z_{CIE} = tristimulus values (CIE Source C) obtained by integration from spectrophotometric data as described in 9.1.

A1.2 The equations giving calculated tristimulus values from the filter photometer readings are as follows:

$$X_{CIE} = 0.7832\,A_of_x + 0.197\,Z_of_z$$
$$X_{CIE} = 1.0000\,Y_of_y$$
$$Z_{CIE} = 1.18103\,Z_of_z$$

where:

A_o, Y_o, and Z_o = instrumental filter (amber, green, and blue reflectance) values relative to an instrument reference standard, and

f_x, f_y, and f_z = ratios of the reflectances of the instrument reference standard to magnesium oxide for each filter.

A1.3 Substituting these expressions for X_{CIE}, Y_{CIE}, and Z_{CIE} in the equation for yellowness index (A1.1),

$$YI = \frac{100[1.28(0.7832A_of_x + 0.197Z_of_z) - 1.06(1.18103Z_of_z)}{1.0000Y_of_y}$$

$$= \frac{100(1.002A_of_x + 0.2524Z_of_z - 1.2524Z_of_z)}{Y_of_y}$$

$$= 100(1.002A_of_x - 0.999Z_of_z)/y_of_y$$

A1.4 Thus, within the precision of the method, for *reflectance* measurements,

$$YI = 100(A_of_x - Z_of_z)/Y_of_y$$

For *transmittance* measurements ($f_x = f_y = f_z = 1$), this equation reduces to:

$$YI = 100(A_o - Z_o)/Y_o$$

These equations permit calculation of the yellowness index from tristimulus filter colorimeter readings without the labor of calculating approximate tristimulus values.

Designation: D 1929 – 68

Standard Method of Test for

IGNITION PROPERTIES OF PLASTICS[1]

This Standard is issued under the fixed designation D 1929; the number immediately following the designation indicates the year of original adoption or, in the case of revision, the year of last revision. A number in parentheses indicates the year of last reapproval.

1. Scope

1.1 This method[2] covers a laboratory determination of the self-ignition and flash-ignition temperatures of plastics using a hot-air ignition furnace.

NOTE 1—The values stated in U.S. customary units are to be regarded as the standard. The metric equivalents of U.S. customary units may be approximate.

2. Significance

2.1 Tests made under conditions herein prescribed can be of considerable value in comparing the relative ignition characteristics of different materials. Values obtained represent the lowest ambient air temperature that will cause ignition of the material under the conditions of this test. Test values are expected to rank materials according to ignition susceptibility under actual use conditions.

NOTE 2—Round-robin results from three laboratories indicate an ignition temperature range of 5 to 20 C (9 to 36 F) at the 250 to 350 C (482 to 662 F) level and 20 to 45 C (38 to 81 F) at the 350 to 500 C (662 to 932 F) level.

2.2 This test is not intended to be the sole criterion for fire hazard. In addition to ignition temperatures, fire hazard includes such other factors as burning rate or flame spread, intensity of burning, fuel contribution, products of combustion, and others.

3. Definitions

3.1 *flash-ignition temperature*—the lowest initial temperature of air passing around the specimen at which a sufficient amount of combustible gas is evolved to be ignited by a small external pilot flame.

3.2 *self-ignition temperature*—the lowest initial temperature of air passing around the specimen at which, in the absence of an ignition source, the self-heating properties of the specimen lead to ignition or ignition occurs of itself, as indicated by an explosion, flame, or sustained glow.

3.3 *self-ignition by temporary glow*—In some cases slow decomposition and carbonization of the plastic results only in glow of short duration at various points in the specimen without general ignition actually taking place. This is a special case of self-ignition temperature, defined as "self-ignition by temporary glow."

4. Apparatus

4.1 The apparatus[3] shall be a hot-air ignition furnace as shown in Fig. 1 and shall consist primarily of the following parts:

4.1.1 *Furnace Tube*—A vertical tube with an inside diameter of 100 mm (4 in.) and a length of 210 to 250 mm ($8^1/_2$ to 10 in.), made of a ceramic that will withstand 750 C (1382 F), and with an opening at the bottom fitted with a plug for the removal of accumulated residue.

4.1.2 *Inner Ceramic Tube*—A ceramic tube with inside diameter of 76 mm (3 in.), length

[1] This method is under the jurisdication of ASTM Committee D-20 on Plastics. A list of committee members may be found in the ASTM Yearbook. This standard is the direct responsibility of Subcommittee D-20.30 on Thermal Properties.

Current edition effective Sept. 9, 1968. Originally issued 1962. Replaces D 1929 – 62 T.

[2] The following reference may be of interest in connection with this method:

Setchkin, N. P., "A Method and Apparatus for Determining the Ignition Characteristics of Plastics," *Journal of Research*, Nat. Bureau Standards, Vol 43, No. 6, Dec., 1949, (RP 2052) p. 591.

[3] The apparatus described is commercially available as Model CS-88 from Custom Scientific Instruments, Inc., 13 Wing Drive, Whippany, N. J.

of 210 to 250 mm ($8\frac{1}{2}$ to 10 in.), and thickness of about 3.2 mm (0.125 in.); placed inside the furnace tube and positioned 20 mm ($\frac{3}{4}$ in.) above the furnace floor on three small spacer blocks. The top shall be covered by a disk of heat-resistant material with a 25.4-mm (1-in.) diameter opening which is used to insert thermocouple leads, for observation, and for passage of smoke and gases. The pilot flame shall be located immediately above the opening.

4.1.3 *Air Source*—An outside air source to admit clean air tangentially near the top of the annular space between the ceramic tubes through a copper tube at a steady and controllable rate. Air shall be heated and circulated in the space between the two tubes and enter the inner furnace tube at the bottom. Air shall be metered by a rotameter or other suitable device; refer to air calibration curves (Fig. 2) for proper furnace air velocities.

4.1.4 *Heating Unit*—An electrical heating unit made of 50 turns of No. 16 B & S wire.[4] The wires, contained within an asbestos sleeve, shall be wound around the furnace tube, and shall be embedded in cement.[5]

4.1.5 *Insulation*, consisting of a layer of asbestos wool approximately 64 mm ($2\frac{1}{2}$ in.) thick, and covered by a sheet iron jacket.

4.1.6 *Pilot Flame*, consisting of 1.6-mm ($\frac{1}{16}$-in.) inside diameter copper tubing attached to a gas supply and placed horizontally 6.4 mm ($\frac{1}{4}$ in.) above the top surface of the divided disk. The pilot flame shall be adjusted to 19 mm ($\frac{3}{4}$ in.) in length and centered above the opening in the disk.

4.1.7 *Specimen Support and Holder*—A convenient specimen holder, measuring 38 mm ($1\frac{1}{2}$ in.) in diameter by 13 mm ($\frac{1}{2}$ in.) in depth, is a $\frac{1}{2}$-oz metal container of approximately 0.2-mm (5-mil) thick steel.[6] One half of the container shall be used as a specimen holder and shall be held in a ring of 1.6-mm ($\frac{1}{16}$-in.) stainless steel welding rod. The ring shall be welded to a length of the same type rod extending through the cover of the furnace as shown in Fig. 1. The specimen holder shall be located 180 to 190 mm (7 to $7\frac{1}{2}$ in.) down from the top of the furnace.

4.1.8 *Thermocouples*—Chromel-alumel or iron-constantan (0.5-mm or 0.020-in.) thermocouples for temperature measurement. These shall be conveniently connected to a multiple-point recorder and each thermocouple temperature shall be recorded at least every 15 s. Thermocouple No. 1 (T_1) measures the temperature of the specimen. It should be located as near the center of the specimen as possible when the specimen is in place in the furnace. Thermocouple No. 2 (T_2) measures the temperature of the air traveling past the specimen. It shall be located slightly below and to the side of the specimen holder. Thermocouple No. 3 (T_3) measures the temperature of the heating coil. Thermocouple No. 1 is also used for measuring initial air temperature in constant-temperature runs before insertion of the specimen.

NOTE 3—The desired air temperature in the inner tube may be maintained by controlling the electric current supplied to the heating coils through an autotransformer, variable transformer, or equivalent, connected in series with the heating coils. Current adjustment may be made by reference to thermocouple T_3 in the furnace heating coil. This thermocouple is used in preference to the inner-tube thermocouples because of faster response. Constant furnace temperature may also be maintained conveniently by use of an automatic controller.

5. Test Specimens

5.1 Thermoplastic materials may be tested in pellet form normally supplied for molding. Where only sheet samples are available for thermosetting materials, 20 by 20-mm ($\frac{3}{4}$ by $\frac{3}{4}$-in.) squares of the available sheet or film shall be bound together with fine wire. A specimen weight of 3 $\pm$ 0.5 g is required.

6. Conditioning

6.1 *Conditioning*—Condition the test specimens at 23 $\pm$ 2 C (73.4 $\pm$ 3.6 F) and 50 $\pm$ 5 percent relative humidity for not less than 40 h prior to test in accordance with Procedure A of ASTM Methods D 618, Conditioning Plastics and Electrical Insulating Materials for Testing,[7] for those tests where conditioning is required. In cases of disagreement, the tolerances shall be $\pm$1 C ($\pm$1.8 F) and $\pm$2 percent

---

[4] Nichrome V alloy wire made by the Driver Harris Co., Harrison, N. J., or equivalent, has been found satisfactory for the purpose.

[5] Alundum cement made by the Norton Co., Worchester, Mass., or equivalent, has been found satisfactory for this purpose.

[6] A metal salve container, Style 100, made by the Buckeye Stamping Co., Columbus, Ohio, or equivalent, has been found satisfactory for this purpose.

[7] *Annual Book of ASTM Standards*, Part 27.

relative humidity.

6.2 *Test Conditions*—Conduct tests in the Standard Laboratory Atmosphere of 23 ± 2 C (73.4 ± 3.6 F) and 50 ± 5 percent relative humidity, unless otherwise specified in the test methods or in this specification. In cases of disagreement, the tolerances shall be ±1 C (±1.8 F) and ±2 percent relative humidity.

7. Procedure A

7.1 *First Approximation of Flash-Ignition Temperature* (*Effect of Air Flow Rate*):

7.1.1 *Low Air Flow Determination*—Raise the cup to the cover opening and place the specimen in the furance (Note 4). Set the air flow at 152.4 cm/s (5 ft/min) (Note 5). Adjust the transformer controlling current to the furnace coils to provide a rise in temperature T_2 of approximately 600 C (1080 F)/h (±10 percent). Light the gas pilot flame and place it across the hole in the top of the furnace. Note the air temperature (T_2) at which the combustible gases are ignited. This point is evidenced by a rapid rise in the specimen temperature (T_1). This is an approximation of the flash-ignition temperature.

NOTE 4—The furnace must be cooled to 50 C (122 F) or lower before each rising-temperature run is started.

NOTE 5—At all times the air flow shall be adjusted to the actual rate through the full section of the inner furnace tube. (The actual velocity past the specimen surface is unknown, but the effect of variable air supply through the furnace is resolved by controlling the rate through the unrestricted, and thus measurable, portion of the tube.) Refer to the air calibration curves in Fig. 2 for proper settings at various furnace temperatures. An air flow rate within ±10 percent of the nominal value is suitable.

7.1.2 *Medium Air Flow Determination*—Repeat 7.1.1 but with an air setting of 50.8 mm/s (10 ft/min) (Notes 4 and 5).

7.1.3 *High Air Flow Determination*—Repeat 7.1.1 but with an air setting at 101.6 mm/s (20 ft/min) (Notes 4 and 5).

7.2 *First Approximation of Self-Ignition Temperature* (*Effect of Air Flow Rate*):

7.2.1 Repeat 6.1.1 but without the pilot flame. Note the recorded air temperature (T_2) at which the specimen flames, explodes, or glows.

7.2.2 Repeat 7.1.2 but without the pilot flame. Note the recorded air temperature (T_2) at which the specimen flames, explodes, or glows.

7.2.3 Repeat 7.1.3 but without the pilot flame. Note the recorded air temperature (T_2) at which the specimen flames, explodes, or glows.

7.3 *Second Approximation of Flash-Ignition Temperature*—Choose the air setting from 7.1 that gives the lowest flash temperature, and repeat the appropriate determination, 7.1.1, 7.1.2, or 7.1.3, using a temperature rise of 300 C (540 F)/h (±10 percent).

7.4 *Second Approximation of Self-Ignition Temperature*—Choose the air setting from 7.2 that gives the lowest self-ignition temperature, and repeat the appropriate determination, 7.1.1, 7.1.2, or 7.1.3, using a temperature rise of 300 C (540 F)/h (±10 percent).

7.5 *Constant-Temperature Tests to Determine Minimum Ignition Temperatures:*

NOTE 6—The air temperature of ignition in the constant-temperature determinations is taken with the specimen thermocouple (T_1) before the specimen is inserted. The air temperature thermocouple (T_2) in these runs simply records whether the furnace is running under constant conditions during the time of the test.

7.5.1 *Minimum Flash-Ignition Temperature*—Start the furnace with the air setting used in 7.3. Adjust the transformer setting until the initial air temperature (T_1) stays constant as indicated by the recorded temperature readings for a 15-min period. The initial temperature should be maintained not more than 10 C (18 F) below the flash temperature found in 7.3. Place the specimen in the furnace, ignite the pilot flame, and watch for ignition of gases from the specimen. If ignition occurs, repeat this run with temperature T_1 maintained at a 10 C (18 F) lower setting. Repeat at successively lower temperatures until there is no ignition in 30 min. When temperature T_1 is reached at which no ignition occurs, it is suggested that a second run be made to ensure that this is truly below the self-ignition temperature. Report the lowest air temperature (T_1) setting at which ignition occurred as the minimum flash-ignition temperature.

7.5.2 *Minimum Self-Ignition Temperature* (Notes 7 and 8)—Repeat 7.1 but without the pilot flame. Start with an air temperature 10 C lower than the ignition temperature found in 7.4.

NOTE 7—If no ignition point was found in 7.2 or 7.4, this procedure for minimum self-ignition temperature should be started with a constant air temperature of about 100 C (180 F) above the flash-ignition temperature found in 7.3. This is because some plastics, for example, polystyrene, boil away before self-ignition takes place during the rising-temperature test. However, a self-ignition point can be found in the constant-temperature test if the initial temperature is started high enough. The air flow rate selected for this test should be the same as used in 7.3.

NOTE 8—If a sample does not ignite at 750 C (1382 F) at an air-flow rate of 10 ft/min, reference should be made to ASTM Method E 136, for Determining Noncombustibility of Elementary Materials[8] for the purpose of obtaining a "noncombustible" rating for the test material.

8. Procedure B (Short Method)

8.1 *Minimum Flash-Ignition Temperature:*

8.1.1 Set the air flow rate to provide a velocity of 5 ft/min at 400 C in the test chamber of the furnace. Adjust the current to the heating coil until the initial air temperature, T_2, remains constant at 400 C for 15 min.

NOTE 9—The temperature of 400 C is used where no prior knowledge of the probable ignition temperature range is available. Other starting temperatures may be selected if information about the material indicates a better choice.

8.1.2 Locate thermocouple T_1 centrally in the specimen holder intimately surrounded by the test material and lower the unit into the furnace. Start a timer, ignite the gas pilot flame and watch for ignition. Flash-ignition will be evidenced by a flash or mild explosion of combustible gases which may be followed by continuous burning of the specimen. If the specimen burns, by flaming or glowing, a rapid rise will be observed in the temperature at thermocouple T_1 above that at T_2.

8.1.3 If at the end of 5 min ignition has or has not occurred, lower or raise the temperature (T_2) 50 C as required and repeat the test with a fresh specimen. When the minimum ignition temperature has been bracketed, tests are begun 10 C below the lowest ignition temperature observed and repeated, dropping the temperature in 10 C intervals until the temperature is reached at which there is no ignition during 13 min. A repeat run may be desirable at this temperature using an air velocity of 10 ft/min to verify the use of 5 ft/min as optimum.

NOTE 10—Ordinarily, increasing the air velocity above 5 ft/min does not reduce the minimum ignition temperature in this test. However, if a repeat run at the next higher velocity results in ignition, it will be necessary to determine, by repeat tests at lower temperatures, the optimum air velocity for the material.

8.1.4 The lowest air temperature (T_2) at which a flash is observed is recorded as the minimum flash-ignition temperature.

8.2 *Minimum Self-Ignition Temperature:*

8.2.1 Follow the same procedure as in 8.1 but *without* the gas pilot flame.

8.2.2 Self-ignition will be evidenced by flaming or glowing of the specimen. It may be difficult, with some materials, to detect self-ignition visually when burning is by glowing rather than flaming. In such cases, the rapid rise in temperature at thermocouple T_1 above that at T_2 is the more reliable reference.

8.2.3 The lowest air temperature (T_2) at which the specimen burns is recorded as the minimum self-ignition temperature.

9. Report

9.1 The report shall include the following:

9.1.1 Designation of material, including name of manufacturer, composition, and state of subdivision (granules, sheet, etc.),

9.1.2 Air velocities used. If air velocity is not critical, this should be noted,

9.1.3 Flash-ignition temperature,

9.1.4 Self-ignition temperature, and

9.1.5 Visual observations (melting, bubbling, smoking, etc.).

[8] *Annual Book of ASTM Standards*, Part 14.

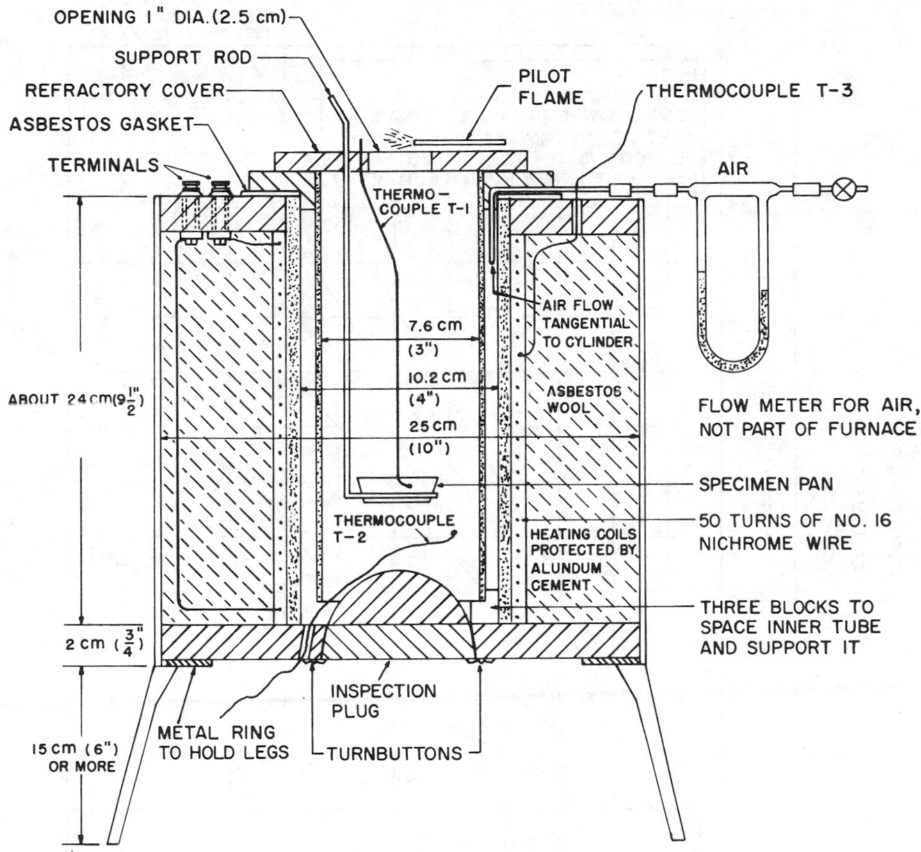

FIG. 1 Cross Section of Hot-Air Ignition Furnace Assembly.

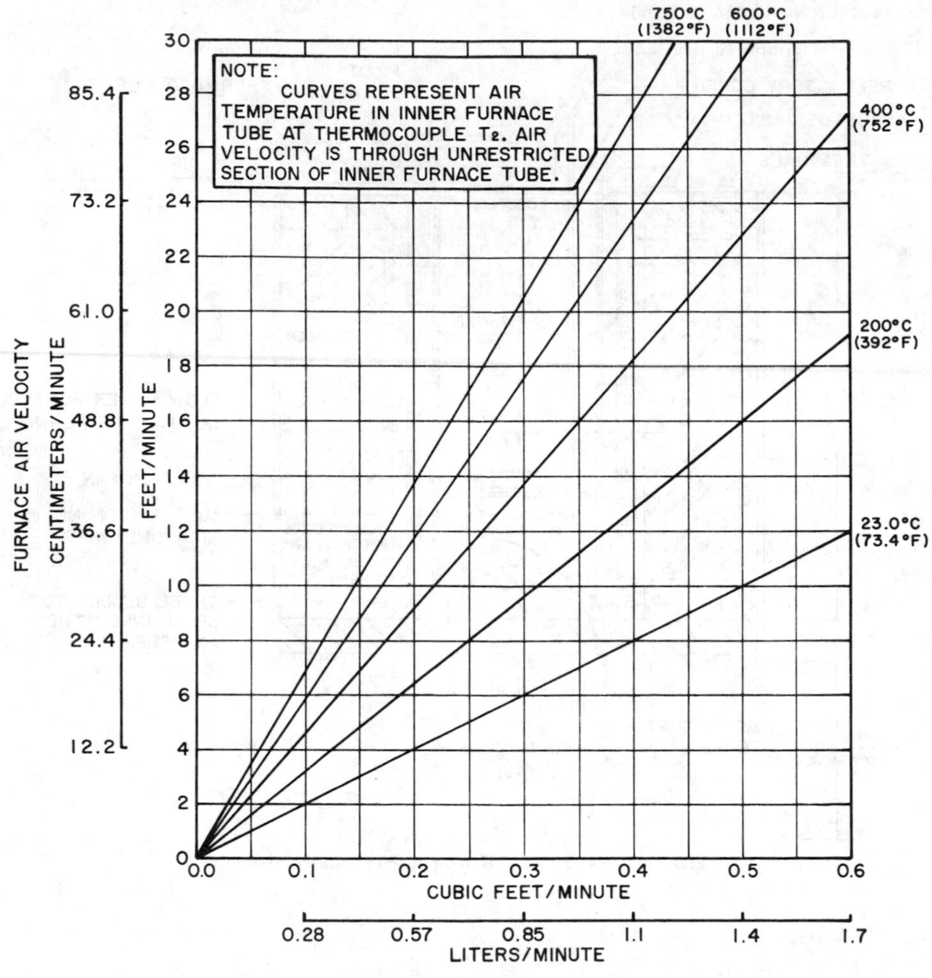

FIG. 2 Air Calibration Curves for Hot-Air Ignition Furnace.

Designation: D 1938 – 67

Standard Method of Test for

RESISTANCE TO TEAR PROPAGATION IN PLASTIC FILM AND THIN SHEETING BY A SINGLE-TEAR METHOD[1]

This Standard is issued under the fixed designation D 1938; the number immediately following the designation indicates the year of original adoption or, in the case of revision, the year of last revision. A number in parentheses indicates the year of last reapproval.

NOTE—Editorial change was made in Section 6 in October 1969.

1. Scope

1.1 This method covers the determination of the force in grams (ounces or pounds) necessary to propagate a tear in plastic film and thin sheeting (thickness of 1 mm (0.04 in.) or less) by a single-tear method.

NOTE 1—The values stated in U.S. customary units are to be regarded as the standard. The metric equivalents of U.S. customary units may be approximate.

2. Summary of Method

2.1 The force in grams (ounces or pounds) to propagate a tear across a film or sheeting specimen is measured using a constant-rate-of-grip separation machine as described in Method A of ASTM Methods D 882, Test for Tensile Properties of Thin Plastic Sheeting.[2] The force in grams (ounces or pounds) necessary to propagate the tear is interpreted from the load-time chart.

3. Significance

3.1 This method is of value in rating the tear propagation force of various plastic films and thin sheeting of comparable thickness. The tear propagation resistance in highly extensible film or sheeting is distinguished from the tear propagation resistance in little or non-extensible film or sheeting in Figs. 2 and 3 in 8.1 and 8.2, respectively.

3.2 This method should be used for specification acceptance testing only after it has been demonstrated that the data for the particular material are acceptably reproducible.

3.3 The data of this method furnish information for ranking the tear propagation resistance of plastic films and sheeting of similar composition. Actual use performance may not necessarily correlate with data from this method. Sets of data from specimens of dissimilar thickness are usually not comparable.

4. Apparatus

4.1 *Film Testing Machine*, with a weighing head that can measure the load applied to tear the specimen. It should be equipped with a device for recording the load carried by the specimen and the amount of separation of the grips during the test. The testing machine shall be essentially free from inertia lag at the specified rate of testing and shall indicate the load with an accuracy of ± 2 percent of the indicated value or better. The accuracy of the testing machine shall be verified in accordance with ASTM Methods E 4, Verification of Testing Machines.[2] A device shall be included to control the grip separation rate at 250 mm (10 in.) ± 5 percent/min.

4.2 *Thickness Measuring Devices* shall be in accord with ASTM Methods D 374, Test for Thickness of Solid Electrical Insulation,[2] or a method of equivalent accuracy.

4.3 *Cutter*—The cutter shall be a sharp razor blade or the equivalent.

---

[1] This method is under the jurisdiction of ASTM Committee D-20 on Plastics. A list of committee members may be found in the ASTM Yearbook. This standard is the direct responsibility of Subcommittee D-20.10 on Mechanical Properties.

Current edition effective Dec. 12, 1967. Originally issued 1962. Replaces D 1938 – 62 T.

[2] *Annual Book of ASTM Standards*, Part 27.

4.4 *Conditioning Apparatus* shall be in accordance with Procedure A of ASTM Methods D 618, Conditioning Plastics and Electrical Insulating Materials for Testing.[2]

5. Test Specimens

5.1 The specimens shall be of the single-tear type and shall consist of strips 75 mm (3 in.) long by 25 mm (1 in.) wide and shall have a clean longitudinal slit 50 mm (2 in.) ±2 percent long cut with a sharp razor blade (Fig. 1) or the equivalent.

NOTE 2—The thickness of the test specimens shall be uniform to within 5 percent of the thickness over the length of the unslit portion of the specimen.

5.2 The thickness of the specimen below the slit (see Fig. 1) shall be measured in several places and recorded in millimeters to the nearest 0.0025 mm (0.0001 in.).

5.3 Enough specimens shall be cut to provide a minimum of five tear propagation force determinations each in the machine direction and in the transverse direction of the material being tested.

6. Conditioning

6.1 *Conditioning*—Condition the test specimens at 23 ± 2 C (73.4 ± 3.6 F) and 50 ± 5 percent relative humidity for not less than 40 h prior to test in accordance with Procedure A of ASTM Methods D 618, Conditioning Plastics and Electrical Insulating Materials for Testing,[2] for those tests where conditioning is required. In cases of disagreement, the tolerances shall be ±1 C (±1.8 F) and ±2 percent relative humidity.

6.2 *Test Conditions*—Conduct tests in the Standard Laboratory Atmosphere of 23 ± 2 C (73.4 ± 3.6 F) and 50 ± 5 percent relative humidity, unless otherwise specified in the test methods or in this specification. In cases of disagreement, the tolerances shall be ±1 C (±1.8 F) and ±2 percent relative humidity.

7. Procedure

7.1 Secure tongue *A* (Fig. 1) in one grip and tongue *B* in the other grip of the constant-rate-of-grip separation testing machine, using an initial grip separation of 50 mm (2 in.). Align the specimen so that its major axis coincides with an imaginary line joining the centers of the grips.

7.2 Using a grip separation speed of 250 mm (10 in.)/min, start the machine, and record the load necessary to propagate the tear through the entire unslit 25-mm (1-in.) portion.

7.3 Test not less than five specimens in each of the principal film or sheeting directions.

8. Calculation

8.1 For thin films and sheeting that have load-time charts characterized by Fig. 2, the average tear propagation force in grams (ounces or pounds) is obtained by averaging the load indicated on the chart over the time period, disregarding the intial and final portions of the curve. This can be done with an integrator or a planimeter. In some cases, a fairly accurate estimate can be made by eye.

8.2 For thin films and sheeting that have load-time charts characterized by Fig. 3, the initial force to continue the propagation of the slit and the maximum force attained are obtained from the chart and reported in grams (ounces or pounds). The initial force may be more readily detected by placing a dot approximately 3 mm ($^1/_8$ in.) in diameter at the base of the razor blade slit with a china marking wax pencil. As the load is applied to the sample, the dot area is observed. When the load is just sufficient to begin the extension of the slit, a "blip" is introduced on the chart (see Fig. 3) by pushing the appropriate button on the recorder or the equivalent to mark this point. The maximum load is the highest reading on the chart as indicated. Both the initial load and the maximum load are reported in grams (ounces or pounds).

8.3 For each series of tests, the median of all values obtained shall be selected to three significant figures and reported as the median value of the particular property.

8.4 Calculate the estimated standard deviation as follows and report to two significant figures:

$$s = \sqrt{(\Sigma X^2 - n\bar{X}^2)/(n - 1)}$$

where:

s = estimated standard deviation,
X = value of a single observation,
n = number of observations, and
$\bar{X}$ = arithmetic means of the set of observations.

9. Report

9.1 The report shall include the following:

9.1.1 Complete identification of the material tested, including type, source, manufacturer's code number, form, principal dimensions, previous history, orientation of samples with respect to principal directions of the material, etc.,

9.1.2 Average thickness of test specimens,

9.1.3 Number of samples tested,

9.1.4 Date of test, and

9.1.5 Median of the five average tear-propagation determinations in grams (ounces or pounds) for materials described in 8.1; and the median of the five initial tear-propagation forces and the medain of the five maximum tear-propagation forces in grams (ounces or pounds) for materials described in 8.2. In each case, the standard deviation of the above sets of data shall be reported. In the cases where the specimens tear to one side, the report should so state, and the values obtained should be reported.

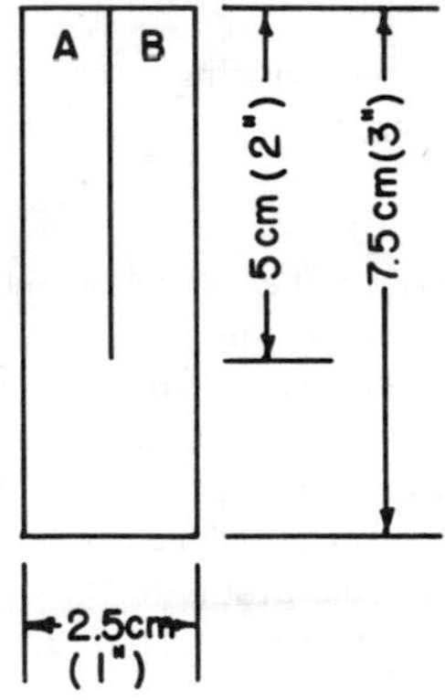

FIG. 1 Single-Tear Specimen.

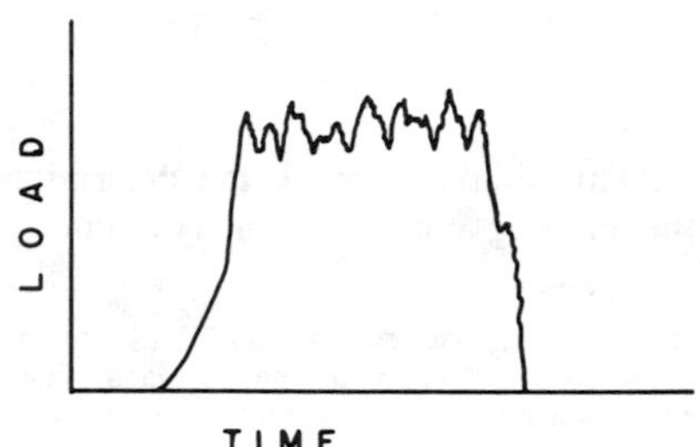

FIG. 2 Load-Time Chart for Low-Extensible Film.

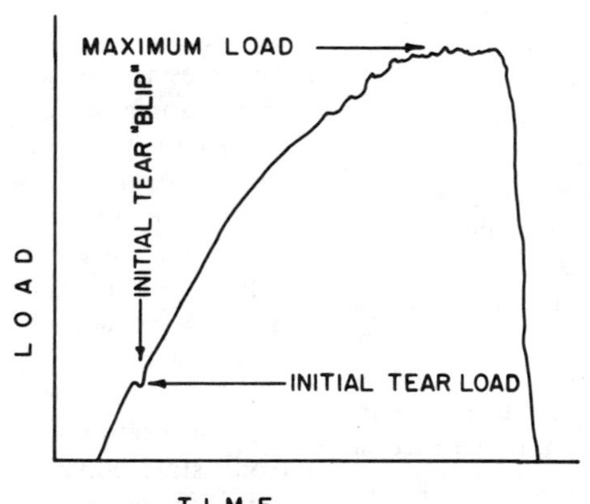

FIG. 3 Load-Time Chart for Highly Extensible Film.

Designation: D 2117 – 64

American National Standard K65.41-1967
American National Standards Institute

Standard Method of Test for
MELTING POINT OF SEMICRYSTALLINE POLYMERS[1]

This Standard is issued under the fixed designation D 2117; the number immediately following the designation indicates the year of original adoption or, in the case of revision, the year of last revision. A number in parentheses indicates the year of last reapproval.

1. Scope

1.1 This method covers the determination of the melting point of semicrystalline polymers (Note 2).

NOTE 1—The values stated in U.S. customary units are to be regarded as the standard. The metric equivalents of U.S. customary units may be approximate.

NOTE 2—Another procedure for the melting point of nylon resins may be found in ASTM Specification D 789, for Nylon Injection Molding and Extrustion Materials.[2]

1.2 The method is applicable to semicrystalline polymers in granular or sheet form, 0.04 mm (1.6 mils) or less in thickness. The normal operating temperature range is 30 to 230 C. A maximum operating temperature of 350 C can be obtained with special equipment (see 6.2).

2. Summary of Method

2.1 This method consists of heating the sample by a hot stage unit mounted under a microscope and viewing it between crossed polars. When the crystalline material melts, the characteristic double refraction from the crystalline aggregates disappears. The point at which the double refraction or birefringence completely disappears is taken as the melting point of the polymer. The effect of exposure to heat or air, or both, on the melting point can be detected by the use of this method.

3. Significance

3.1 This test is useful for determining the melting point of semicrystalline polymers where a high degree of precision is desired (see Section 12).

3.2 The test is suitable for specification acceptance, manufacturing control, development, or research.

4. Definition

4.1 *semicrystalline polymers*—those polymers capable of forming at least two-dimensional intermolecular order as prepared or as developed by suitable thermal treatments. The amount and degree of perfection of the ordering or crystallization depends on the spatial relationship of the atoms in the molecules, as well as the thermal or other treatment.

5. Interferences

5.1 An appreciable increase in particle size above that specified will result in a large melting range and obscure the sharp end point. The presence of anisotropic crystals, for example, some inorganic substances, will reduce the accuracy of the melting point due to the persistence of their double refraction after the greater part of the sample has melted.

5.2 In some cases heat and air or heat alone will cause changes in the melting point. Where it has been shown that there is an effect of heat and air on the melting point, provision is made in 10.3 for an inert gas blanket. In the case of heat alone below the melting point (solid state polymerization) or at the melting point (further polymerization or degradation), the effect of heat should be determined by remelting a sample that has already been run.

[1] This method is under the jurisdiction of ASTM Committee D-20 on Plastics. A list of committee members may be found in the ASTM Yearbook. This standard is the direct responsibility of Subcommittee D-20.10 on Mechanical Properties.

Current edition effective Aug. 31, 1964. Originally issued 1962. Replaces D 2117 – 62 T.

[2] *Annual Book of ASTM Standards*, Part 26.

If the two determinations check within ±1 C, the procedure as outlined may be followed. If the material is affected by heat alone, below or at the melting point, the melting point of the material cannot be determined by this method.

6. Apparatus

6.1 *Microscope*—A compound microscope with magnifications from 50× to 100× and at least a 6-mm working distance. A cap analyzer and disk polarizer to fit the microscope are required. A polarizing microscope with built-in analyzer and capable of withstanding the temperature and having the required working distance gives better visibility and sensitivity than a cap analyzer. A low-power condenser should be employed.

6.2 *Cooling Jacket*—A jacket to keep the microscope objective cool when used above 230 C. (A typical design is shown in Fig. 1.)

6.3 *Micro Hot Stage*[3]—An insulated metal block, which can be mounted slightly above the microscope stage. The block must be: (*1*) provided with a hole for light passage, (*2*) constructed to provide a chamber for blanketing sample in an inert gas and containing a heat baffle and a glass cover, (*3*) electrically heated with sufficient controls for adjustment of heating and cooling rate, and (*4*) provided with an opening for insertion of the thermometer near the light hole (Fig. 2).

6.4 *Temperature Controller*[4]—An electrical temperature controller or variable transformer capable of increasing the temperature rapidly in the hot stage to near the melting point and then by manual adjustment increasing it at 0.5 ± 0.1 C/min through the critical temperature range.

6.5 *Cooling Block (Optional)*[3]—A metal block helps to lower the temperature rapidly after each determination (Fig. 3).

6.6 *Thermometers*[3]—Two calibrated thermometers for ranges from +30 to 230 C and +60 to 350 C.

6.7 *Thermometer Magnifier*[3]—A conventional thermometer magnifier to fit the thermometers.

6.8 *Test Reagent Set*[5]—A set of test reagents, precision 0.5 C, or reagent grade chemicals for checking the calibrated thermometers.

6.9 *Glass Microscope Slides*, 25 by 38 by 1 mm (1.0 by 1.5 by 0.04 in.), and No. 1 cover glasses, 18 mm (0.7 in.) in diameter by 0.15 mm (0.006 in.) thick.

6.10 *Nitrogen Supply* for attachment to hot stage when needed to prevent oxidation in hot air.

6.11 *Hypodermic Punch* as shown in Fig. 4.

7. Calibration

7.1 Each thermometer has an engraved ice point scale to permit checking for secular changes that may occur in all mercury-in-glass thermometers. See ASTM E 77, for Verification and Calibration of Liquid-in-Glass Thermometers[6] for procedure. The thermometers should also be calibrated periodically with the reagent test set available with the hot stage or with reagent grade chemicals, using the rate of temperature rise selected for the tester. Recommended chemicals include 1-menthol melting at 42 to 43 C, hydroxyquinoline at 75 to 76 C, acetanilide at 113 to 114 C, succinic acid at 189 to 190 C, and phenolphthalein at 261 to 262 C.[5]

8. Sample

8.1 *Powdered Samples*—Place a 2 to 3-mg portion on a clean slide and cover with a No. 1 cover glass. Heat the slide, sample, and cover on a hot plate to slightly above the softening point of the polymer so that a thin film can be formed (Note 3). By slight pressure on the cover glass, form a thin film 0.01 to 0.04 mm (0.4 to 1.6 mils), and allow it to cool slowly by turning off the hot plate power to promote crystallization. Transfer the slide with specimen and cover glass to the hot-stage unit. Cover with heat baffle and the large cover glass that is supplied with the hot stage unit. If the procedure of nitrogen blanketing described in 10.3 does not produce the specified reproducibility in the results (±1 C), the melting point of the material cannot be determined by this method (Note 4).

NOTE 3—Sometimes it is necessary to test a

[3] The hot stage, cooling block, thermometers, and thermometer magnifier may be obtained from Arthur H. Thomas Co., Philadelphia, Pa., 19105, or W. J. Hacker & Co., West Caldwell, N. J.

[4] A suitable variable transformer may be obtained from General Radio Co., West Concord, Mass.

[5] These substances may be obtained from Eastman Kodak Co., Rochester, N. Y.

[6] *Annual Book of ASTM Standards*, Part 30.

polymer above its melting point (not merely above its softening point) to develop birefringence upon cooling. Cooling rates as low as 1 C/h may be required for some polymers.

NOTE 4—The melting point of some samples may change with the thermal treatment or exposure to air. Observation of melting on the untreated sample for comparison with a film as prepared in 8.1 above may show a significant effect. See 9.3 for procedure if a change occurs.

8.2 *Molded or Pelleted Samples*—Cut the samples with a microtome or razor blade and hypodermic punch to appropriate size, approximately 1.6 mm ($^1/_{16}$ in.) in diameter, and 0.04 mm ($^1/_{64}$ in.) in thickness, and place on a slide with cover glass for testing.

8.3 *Film or Sheet Samples*—On films with thickness of 0.01 to 0.04 mm (0.4 to 1.6 mils), cut the samples with a diameter of about 1.6 mm ($^1/_{16}$ in.). For films with thickness above 0.04 mm (1.6 mils), cut the samples with a microtome or razor blade to a diameter of about 1.6 mm ($^1/_{16}$ in.) and press thin film as in 8.1.

9. Conditioning

9.1 *Conditioning*—Condition the test specimens at 23 ± 2 C (73.4 ± 3.6 F) and 50 ± 5 percent relative humidity for not less than 40 h prior to test in accordance with Procedure A of ASTM Methods D 618, Conditioning Plastics and Electrical Insulating Materials for Testing,[7] for those tests where conditioning is required. In cases of disagreement, the tolerances shall be ±1 C (±1.8 F) and ±2 percent relative humidity.

9.2 *Test Conditions*—Conduct tests in the Standard Laboratory Atmosphere of 23 ± 2 C (73.4 ± 3.6 F) and 50 ± 5 percent relative humidity, unless otherwise specified in the test methods or in this specification. In cases of disagreement, the tolerances shall be ±1 C (±1.8 F) and ±2 percent relative humidity.

10. Procedure

10.1 Assemble the apparatus with hot stage, analyzer, and polarizer on the microscope, and connect the temperature controller and nitrogen supply, if needed.

10.2 Adjust the light source to maximum light intensity with bright field.

10.3 For specimens that are degraded by heat and air, adjust the gas inlet for hot stage so that a very slow stream of nitrogen blankets the stage. Maintain a small positive pressure by feeding the gas stream through a water bubbler after the hot stage.

10.4 Locate the specimen in the field of view and focus the microscope. Rotate the analyzer to obtain a dark field. Crystalline material will appear bright or multicolored on a dark field.

NOTE 5—Cut film specimens may show residual strains on the edge of the specimen.

10.5 Adjust the electrical controller to bring the temperature rapidly to about 10 C below the sample melting point (as determined by prior test). Then adjust the controller so that the temperature rises at the rate of 0.5 ± 0.1 C/min.

NOTE 6—The scale equivalency of the temperature controller is obtained by plotting the settings versus the thermometer readings. A controller with a fine scale equivalency of 0.1 C for each division or better is ample. When even greater precision is required, a 0.1 C/min rate should be used. In this case, the rapid heating should be stopped at 3 to 4 C below the melting point.

10.6 Observe the temperature to the nearest 0.1 C (or the tolerance as required by the specific material) at which the sample melts. This is the point at which all double refraction disappears, leaving a totally dark field. Disregard any crystals that may be visible on the extreme edges of the field.

NOTE 7—The melting point range will vary significantly for various polymers. Some polymers start melting from room temperature and have a broad melting point range. The lower limit is dificult to establish but the upper limit is unmistakable and is taken as the true melting point.

NOTE 8—Better distinction in some cases may be obtained with a compound microscope equipped with a 550 full wave, first order, red retardation plate which will produce a pink background. When the sample melts, the mosaic pattern disappears, leaving a pink background.

10.7 Turn the electricity off and remove the large glass cover, heat baffle, and sample slide. Place the cooling block on the hot stage unit if rapid cooling is desired.

10.8 Repeat thc procedure with another specimen of the polymer.

11. Report

11.1 The report shall include the following:

11.1.1 Complete identification and descrip-

[7] *Annual Book of ASTM Standards*, Part 27.

tion of the material tested, including source and manufacturer's code,

11.1.2 Results of duplicated determinations and the average of the final melting points. Record heating rate if outside the 0.5 ± 0.1 C range (Note 9), and

11.1.3 Cooling rate during preparation of powdered samples (8.1).

NOTE 9—In some cases it may be helpful to record the effect of reheating the sample on the observed melting point or to record the initial melting point along with the final melting point, particularly if 8.2 was used for specimen preparation.

12. Precision

12.1 The following criteria should be used for judging the acceptability of results (95 percent probability):

12.1.1 *Repeatability*—Duplicate results by the same operator should not be considered suspect unless they differ by more than 1.0 C.

12.1.2 *Reproducibility*—The results submitted by each of two laboratories should not be considered suspect unless the two results differ by more than 2.0 C.

REFERENCES

(**1**) Kofler, L., and Kofler, A., *Thermo-Mikro-Methoden fur Kennzichnung Organischer Stoffe und Stoffgemische*, Universitaetsverlag Wagner, Innsbruck, Austria (1954).

(**2**) McCrone, W. C., Jr., *Fusion Methods in Chemical Microscopy*, Interscience Publishers, New York, N. Y. (1957).

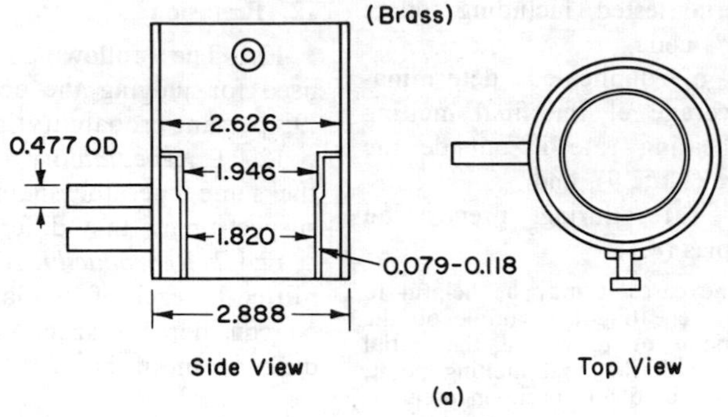

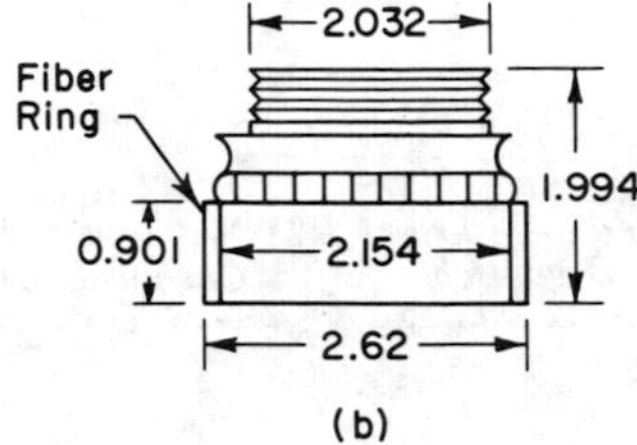

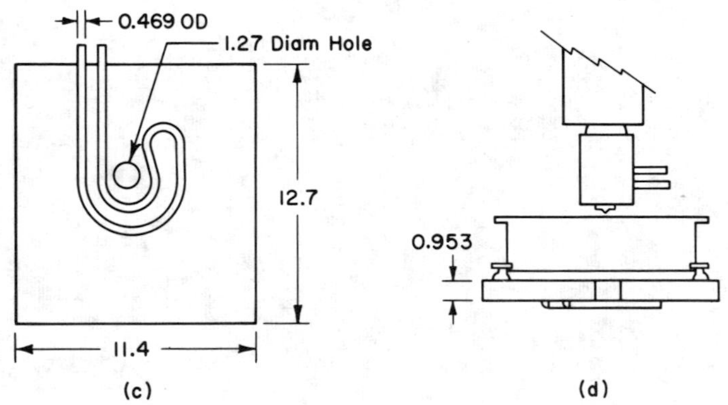

Dimensions in Centimeters

(*a*) Cooling jacket for objective.
(*b*) Adapter for attaching cooling jacket.
(*c*) Cooling plate for hot stage.
(*d*) Assembly.

FIG. 1 Cooling Equipment for Microscope.

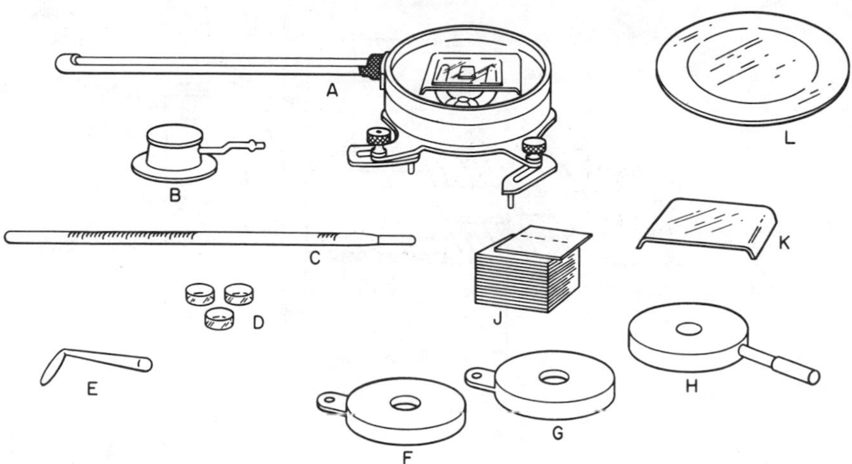

A. Complete hot stage without cover glass.
B. Vacuum sublimation chamber (glass).
C. Thermometer (2) calibrated.
D. Sublimation dishes (glass).
E. Fork lifter.
F. Sublimation block (small).
G. Sublimation block (large).
H. Cooling block.
J. Micro slides.
K. Heat baffle (bridge).
L. Glass cover for stage.

FIG. 2 Hot Stage with Accessories.

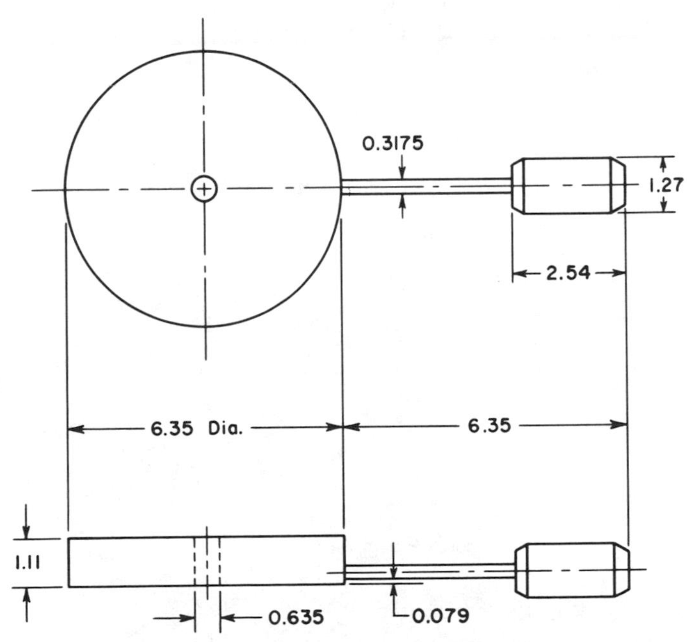

Brass Chromium Plated
Dimensions in Centimeters

FIG. 3. Cooling Block for Hot Stage.

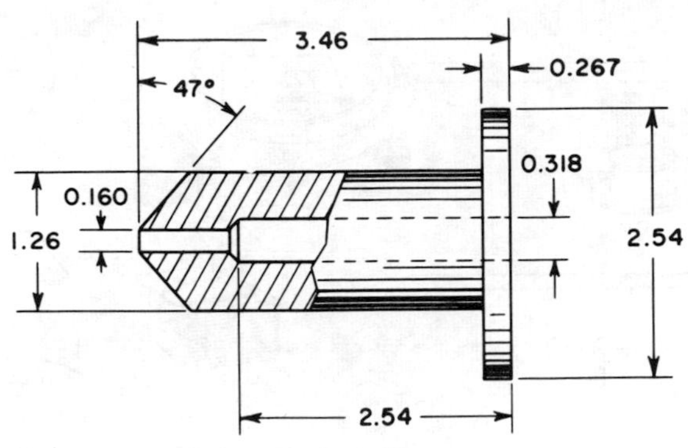

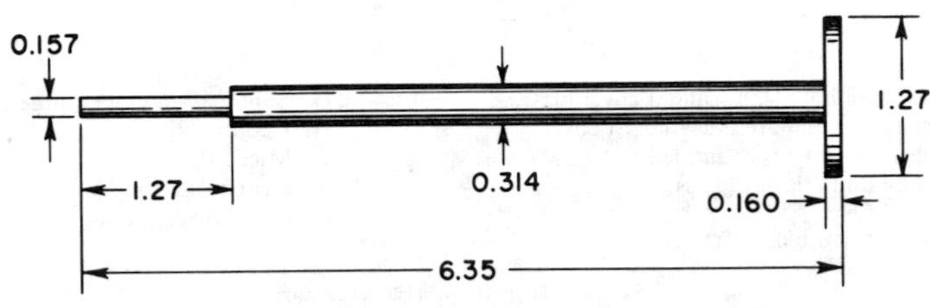

Dimensions in Centimeters

FIG. 4 Hypodermic Punch.

Designation: D 2124 – 70

Standard Method for
ANALYSIS OF COMPONENTS IN POLY(VINYL CHLORIDE) COMPOUNDS USING AN INFRARED SPECTROPHOTOMETRIC TECHNIQUE[1]

This Standard is issued under the fixed designation D 2124; the number immediately following the designation indicates the year of original adoption or, in the case of revision, the year of last revision. A number in parentheses indicates the year of last reapproval.

1. Scope

1.1 This method provides for the identification of certain resins, plasticizers, stabilizers, and fillers in poly(vinyl chloride) (PVC) compounds by an infrared spectrophotometric technique. In many cases, individual components may be measured quantitatively. Complementary procedures, such as chromatographic and other separations, may be used to separate specific components and extend the applications of the method. Other instrumental methods, such as optical emission or X-ray spectroscopic methods, may yield complementary information which may allow more complete or, in some cases, easier measurement of the components. The resin components covered in this method are listed in the Appendix.

1.2 PVC formulations are too varied to be covered adequately by a single method. Using the following method many compounds may be separated into resins, plasticizers, stabilizers, and fillers. A number of components can be quantitatively measured. Many more can be identified and their concentrations estimated. By the use of prepared standards one may determine the usefulness and accuracy of the method for specific PVC formulations. The method is applicable for the resin components listed in the appendix and for other components having similar chemical compositions and solubility characteristics. The method may lead to error in cases where the nature of the components is not known.

NOTE 1—The values stated in U.S. customary units are to be regarded as the standard. The metric equivalents of U.S. customary units may be approximate.

2. Summary of Method

2.1 PVC compound is solvent-extracted in order to separate the plasticizer from the compound. The resin is dissolved from the remaining compound and the inorganic fillers and stabilizers separated by centrifuging. By this technique, the compound is separated into (*1*) plasticizers, (*2*) resin, and (*3*) inorganic stabilizers and fillers. Each may be individually analyzed by an infrared technique to identify and measure the components.

3. Significance

3.1 PVC compounds are used in a wide variety of products and hence they are formulated to provide a wide range of physical properties. The physical properties required in a compound depend upon the product in which it is used. These properties are largely determined by the type, quantity, and quality of the compounding ingredients. The analytical method described below makes use of infrared spectrophotometry for the qualitative or quantitative determination, or both, of many of these ingredients in PVC compounds. The method may be used for a variety of applications including process control, raw material acceptance, product evaluation, and determination of changes in composition resulting from environmental testing.

3.2 The method is directly applicable for only those components listed in the Appendix

---

[1] This method is under the jurisdiction of ASTM Committee D-20 on Plastics. A list of members may be found in the ASTM Yearbook. This standard is the direct responsibility of Subcommittee D-20.70 on Analytical Methods.

Current edition effective April 13, 1970. Originally issued 1962. Replaces D 2124 – 67.

and for those components which are known to be similar in chemical composition and in solubility characteristics to the chemicals listed in the Appendix.

4. Definitions

4.1 For definitions related to the material on infrared spectroscopy, refer to ASTM Definitions E 131, Terms and Symbols Relating to Molecular Spectroscopy.[2]

5. Apparatus

5.1 *Initial Sample Preparation*—Use any of the following apparatus, depending on shape and size of sample, for reducing solid samples to small particle sizes:

5.1.1 *Pencil Sharpener* or grater and a cold box or container capable of holding at least the temperature of solid carbon dioxide.

5.1.2 *Grinding Wheel*, coarse.

5.1.3 *Microtome*.

5.1.4 *Grinding or Cutting Mills*, commercial, for example, a Wiley mill (for samples larger than 1 g).

5.2 *Soxhlet Extraction Apparatus:*

5.2.1 For 0.5 and 1.0-g samples, use an extraction apparatus with a 150-ml flask and a 27 by 100-mm thimble.

5.2.2 For 0.2-g samples, use an extraction apparatus with a 30-ml flask and a 10 by 50-mm thimble.

5.3 *Mold and Press for KBr Pellets*—A mold assembly capable of pelletizing a 12.7-mm ($^1/_2$-in.) minimum diameter pellet under vacuum and a press capable of exerting pressures of at least 140 MN/m^2 (20,000 psi) are required to press clear KBr pellets.

5.4 *Infrared Spectrophotometer*—The spectral region from 4000 cm^{-1} to 650 cm^{-1} (2.5 to 15 μm) is used. Refer to ASTM Recommended Practice E 275, for Describing and Measuring Performance of Spectrophotometers,[3] with particular emphasis on Sections 4 and 14 relating to resolution and spectral slit width measurements. An ultimate resolving power (**1**)[4] of at least 1.5 cm^{-1} at 850 cm^{-1} (0.02 μm at 12 μm) is satisfactory. The suitability of the instrument should be proven in the user's laboratory. Demountable cells, 1.0-mm liquid cells, and a KBr pellet holder are the accessories used.

6. Reagents

6.1 Reagent grade chemicals or equivalent, as specified in ASTM Methods E 200, for Preparation, Standardization, and Storage of Standard Solutions for Chemical Analysis[3], shall be used in all tests.

6.2 *Alumina*, absorption.

6.3 *Carbon Disulfide* (CS_2).

6.4 *Ether, anhydrous*.

6.5 *Potassium Bromide*, (KBr) infrared quality.

6.6 *Tetrachloroethane*, technical.

6.7 *Tetrahydrofuran* (stabilized with 0.1 percent hydroquinone).

7. Component Separations

7.1 *Initial Sample Preparation*—Any method that will increase the surface area of a sample sufficiently to permit complete plasticizer extraction in a reasonable time is satisfactory. PVC compounds as received are usually in the form of powders, granules, slabs, or offshaped pieces. Powders may be used directly. Thin sheets, 0.02 to 0.05 mm thick, molded from individual granules may be used. Granules may be pressed into slabs. Slabs or appropriately shaped pieces may be treated by one of the following techniques:

7.1.1 Buffing on a coarse grinding wheel,

7.1.2 Cooling the sample with solid carbon dioxide and grinding the brittle sample in a clean pencil sharpener or on a grater or clean file, or

7.1.3 Shaving thin slices from the sample with a microtome.

7.2 *Plasticizer Extraction*—Weight to $\pm$0.2 mg approximately 1 g of fine particle size sample into a 27 by 100-mm paper extraction thimble. Place the thimble in a jacketed Soxhlet apparatus fitted with a tared 150-ml flask, and extract with 120 ml of ethyl ether for 6 h (Note 2). Remove the tared 150-ml flask, containing the ethyl ether and the extracted plasticizer, from the jacketed Soxhlet apparatus and gently heat to boil off the ethyl ether. Place the flask in an evacuated desiccator for a minimum of 1 h to remove the last traces of ethyl ether. Weigh to $\pm$0.2 mg the

[2] *Annual Book of ASTM Standards*, Part 32.

[3] *Annual Book of ASTM Standards*, Part 30.

[4] The boldface numbers in parentheses refer to the list of references at the end of this method.

flask containing the extracted plasticizers. Calculate the percentage of plasticizers in the PVC sample as follows:

$$\text{Plasticizers, percent} = \frac{\text{weight of extracted plasticizers} \times 100}{\text{weight of PVC sample}}$$

7.2.1 Keep the plasticizers for infrared identification or determination (8.4).

NOTE 2—Organometallic or organic stabilizer, if present, may partially or wholly separate with either the plasticizer or resin components and should be considered when examining these compounds.

7.3 *Separation of Stabilizers and Fillers*—Empty the resin, stabilizers, and fillers remaining in the extraction thimble into a 50-ml beaker. Add 20 ml of tetrachloroethane and heat the sample gently until the resin has dissolved. Wash the contents of the beaker quantitatively into a tared 50-ml centrifuge tube with 20 ml of tetrahydrofuran (which has been previously passed through a 150 by 12.7-mm (6 by ½-in.) diameter alumina absorption column to remove hydroquinone), swirl to mix, and centrifuge for 30 min. Decant the resin solution and reserve for infrared analysis. Wash the residue remaining in the tared centrifuge tube with 20 ml of tetrahydrofuran and centrifuge again for 30 min. Decant the solution containing the remaining resin. Repeat the operation. Dry the tared centrifuge tube containing the stabilizer and filler in an oven at 110 C for 1 h, cool, weigh, and calculate the percentage of inorganic stabilizer and filler as follows:

$$\text{Inorganic stabilizer and filler, percent} = \frac{\text{weight of stabilizer and filler} \times 100}{\text{weight of PVC sample}}$$

7.3.1 Keep the stabilizer and filler for infrared analysis. Usually carbon black and color pigments are included in this portion.

7.4 *Resin*—Calculate the percentage of resin by difference (100 minus the total percent of plasticizers, stabilizers and fillers).

8. Infrared Analysis of Extracted Plasticizers

8.1 The extracted plasticizers may be run on the infrared spectrophotometer as liquid films for identification or in CS_2 solution for quantitative determinations.

8.2 *Identification of Plasticizers*—Most plasticizers for PVC compounds are liquid at room temperature. A few secondary plasticizers may be solid but would be suspended or dissolved in primary plasticizers. A demountable cell with NaCl windows and a 0.025-mm spacer usually suffices to give a strong plasticizer spectrum (**1**). Scan the spectrum from 4000 cm^{-1} to 650 cm^{-1} (2.5 to 15 μm). By reference to a collection of plasticizer spectra the plasticizers in the sample may be identified (**3, 4, 6, 7, 8, 9, 12, 17**). With experience, rough estimates of concentrations may be made to enable preparation of matching standards for quantitative analysis.

8.3 *Quantitative Analysis of Plasticizers*—The variety of plasticizers and their possible combinations in PVC compounds is extensive. It is impossible to specify a single procedure that determines quantitatively all plasticizers with equal precision and accuracy. The following procedure is useful for a number of plasticizers and their combinations, particularly if dioctyl phthalate or tricresyl is the primary plasticizer. The user should decide whether the efficiency, precision, and accuracy of the procedure is satisfactory for a specific combination of plasticizers to be analyzed.

8.3.1 Weigh 60 ±0.2 mg of extracted plasticizer (from 7.2) into a 25-ml Erlenmeyer flask equipped with a glass stopper; add 20.00 ml of CS_2 to dissolve the plasticizers. Take care to avoid loss of solvent by keeping the Erlenmeyer flask stoppered when possible. Run the resultant 3.00-mg/ml plasticizer solution on the infrared spectrophotometer (**2**) in a 1.0-mm liquid cell. Run a compensating 1.0-mm liquid cell or a variable path cell suitably adjusted, filled with CS_2, in the reference beam. After proper cleaning and drying of the sample cell, run an equivalent blank of CS_2 in the sample cell versus the reference cell.

8.3.2 A chart presentation on absorbance versus frequency (wavelength) paper of 20 cm/100 cm^{-1} for frequency and 18 cm for zero to 1.0 absorbance range is satisfactory. However, other chart presentations may be used. For dioctyl phthalate and tricresyl phosphate the spectra ranges from 1800 to 1650 cm^{-1} (5.4 to 6.1 μm) and from 1345 to 1090 cm^{-1} (7.4 to 9.2 μm) are useful. Dioctyl phthalate bands at 1725 cm^{-1} (5.80 μm), 1270 cm^{-1} (7.87 μm), and 1121 cm^{-1} (8.92 μm), and tricresyl phosphate band at 1191 cm^{-1} (8.4 μm) are satisfactory. The dioctyl phthalate band chosen will depend, in part, on secondary plasticizer interferences. Choice of bands for other plasticizers is left to the discretion of the user. At the analytical band frequency (wavelength) chosen, absorbances for the sample spectrum ($A_s + A_b$) and the blank spectrum (A_b) are measured. Net absorbance due to sample component A_s is $(A_s + A_b) - A_b$.

8.3.3 Prepare plasticizer standards by dissolving the pure plasticizers of interest in CS_2 to give a series of standard solutions covering the 3.0 to 0.5-mg/ml range for each plasticizer. Run these standard plasticizer solutions under conditions identical to those under which the samples are run to obtain the net absorbances of the components at a series of concentrations. Plot Beer's law curves of net absorbances versus concentrations in milligrams per milliliter for each component. All quantitative manipulations shall be in accordance with ASTM Recommended Practices E 168, for General Techniques of Infrared Quantitative Analysis.[3]

8.3.4 Use the net absorbance of a specific plasticizer in conjunction with the appropriate Beer's law curve to determine the concentration in milligrams per milliliter.

8.3.5 Calculate the percentage of plasticizer in the PVC compound as follows:

$$\text{Specific plasticizer, percent} = (AB/3W) \times 100$$

where:

A = concentration of plasticizer, mg/ml,
B = total weight of extracted plasticizers, mg, and
W = weight of PVC sample, mg.

9. Direct Infrared Determination of Plasticizers

9.1 The use of this procedure usually presupposes that a complete formulation analysis is not required and that the plasticizers to be determined are known.

9.2 Weigh 0.25 g (±0.25 mg) of fine particle size sample into a 10 by 50-mm extraction thimble. Place the thimble in a micro Soxhlet extraction apparatus. Extract for 6 h with 20 ml of CS_2. Transfer the CS_2 containing the extracted plasticizers to a 25-ml volumetric flask, dilute to the mark with CS_2, and mix thoroughly. Run this solution on the infrared spectrophotometer. Instrumental conditions and techniques, preparation of standards, and Beer's law curves are the same as those specified in 8.2.

9.3 *Calculation*—Calculate the percentage of a specific plasticizer in the PVC compound as follows:

$$\text{Specific plasticizer, percent} = (25A/W) \times 100$$

where:

A = concentration of plasticizer, mg/ml, (from curve), and
W = weight of PVC sample, mg.

10. Infrared Analysis of Stabilizers and Fillers

10.1 *Identification*—The stabilizers and fillers, as a dry powder after separation from the resin, may be identified by running on the infrared spectrophotometer as a Nujol mull or a KBr pellet. Prepare the Nujol mull by adding a few milligrams of powder to a drop of Nujol in a small mortar and mixing; run the resultant mull as a film between two NaCl plates held in a demountable cell mount. Prepare the KBr pellet by adding approximately 1 mg of powder to 600 mg of dry KBr powder and mixing 1 min in a vibrator mixer. Place the mixture in a 12.7-mm (½-in.) diameter mold assembly, hold under vacuum for 3 min, and press for 3 min at a minimum pressure of 140 MN/m^2 (20,000 psi) while still under vacuum. Higher pressures will produce more stable pellets. Place the resultant pellet in a holder and run it on the infrared spectrophotometer (**1**). By comparison to reference spectra (**3, 4, 6, 10, 13, 14, 15, 17**) the stabilizer and filler components, in many cases, may be identified.

10.2 *Quantitative Analysis of Stabilizers and Fillers*—Analyze these components by the KBr pellet technique. Weigh 1 mg of stabilizer and filler powder and add to 600 mg of dry KBr powder. Prepare KBr pellets as described in 10.1, place in a holder in the sample beam, and run on an infrared spectrophotometer (**2**). A chart presentation of absorbance versus frequency (wavelength) paper of 20 cm/100 cm^{-1} and 18 cm for zero to 1.0 absorbance range is satisfactory. Other chart presentations may be used at the discretion of the user. Base line techniques, in accordance with Recommended Practices E 168, are used to determine absorbances of the bands of interest, and Beer's law curves of net absorbance versus percentage of component in total stabilizers and fillers are plotted. The net absorbances are those which would result if the stabilizers and fillers were exactly 1 mg in 600 mg of KBr powder. The percentage of component in the PVC sample may be calculated from the weight of stabilizers and fillers in the PVC sample previously determined. Standard samples for preparation of Beer's law curves are prepared by mixing the pure compounds of interest in appropriate amounts to give a set of matched standards. The following bands are usable in many cases: basic lead carbonate, 1410 cm^{-1} (7.09 μm); calcined clay, 1075 cm^{-1} (9.30 μm); calcium carbonate, 877 cm^{-1} (11.4 μm); antimony oxide, 741 cm^{-1} (13.5 μm); basic lead sulfate, 1130 cm^{-1}, (8.85 μm); dibasic lead phthalate, 1535 cm^{-1} (6.51 μm).

11. Infrared Analysis of PVC Resin

11.1 *Identification of Resin*—The resin obtained during component separations is in a tetrachloroethane-tetrahydrofuran solution. Evaporate a few milliliters of the solution a few drops at a time on a microscope slide. Gentle heating will accelerate drying. When the resultant film is dry, peel it from the microscope slide. Dry the film in a vacuum desiccator or vacuum oven to reduce solvent spectral interferences. It is advisable to prepare a number of films from each sample in order to obtain one of suitable quality and thickness. Mount the film in the infrared spectrophotometer (**1**) and record its spectrum from 4000 cm^{-1} (2.5 μm) to 650 cm^{-1} (15 μm).

11.2 The PVC may be identified by its over-all infrared spectrum (**3, 4, 5, 6, 11, 16**). If the resin is a copolymer of vinyl chloride and vinyl acetate, a carbonyl band will be present at 1742 cm^{-1} (5.74 μm), and if the amount of acetate is greater than approximately 5 percent, a band attributed to the acetate group is present at 1020 cm^{-1} (9.80 μm). Take care in the interpretation of carbonyl bands in the spectrum of the resin since these may also arise from the following:

11.2.1 Copolymers other than acetate (for example acrylate),

11.2.2 Incomplete extraction of certain polymeric ester plasticizers,

11.2.3 Oxidation of the resin, and

11.2.4 Esterification of the resin by certain compounding ingredients.

11.3 Usually the carbonyl bands due to residual polymeric plasticizers and to oxidation are at lower frequencies than those due to copolymers or to esterification of the resin.

12. Precision and Accuracy

12.1 The precision and accuracy quoted are for PVC formulations comprised of components listed in the Appendix. The precision and accuracy should also apply to PVC formulations that contain plasticizers, fillers, and stabilizers similar in both chemical composition and solubility characteristics to those listed in the Appendix. The precision and accuracy may not be applicable in cases where the nature of the components is not known.

12.2 *Plasticizer Analysis*—The multilaboratory accuracy of the plasticizer analysis is given by a precision of ±0.75 percent (3S%) as defined in ASTM Recommended Practice E 177, for Use of the Terms Precision and Accuracy as Applied to Measurement of a Property of a Material,[3] and a bias of −0.314 percent. In relative percent the coefficient of variation is ±0.79 percent (R1S%).

12.3 *Filler and Stabilizer Analysis*—The multilaboratory accuracy of the filler and stabilizer analysis is given by a precision of ±0.45 percent (3S%) as defined by Recommended Practice E 177 and a bias of −0.10 percent. In relative percent the coefficient of variation is ±1.49 percent (R1S%).

12.4 *Resin Analysis*—The multilaboratory accuracy of the resin analysis is given by a precision of 0.60 percent (3S%) as defined by Recommended Practice E 177 and a bias of +0.38 percent. In relative percent the coefficient of variation is ±0.51 percent (R1S%).

12.5 The basic round-robin data from which these figures were calculated have been filed at ASTM Headquarters as RR22: D-20.

13. Report

13.1 The report shall include the following:

13.1.1 Description of the material tested, that is, the name, color, manufacturer, and other pertinent data,

13.1.2 Description of sample preparation,

13.1.3 Description of spectrophotometer used,

13.1.4 Statement of sections of method used in analysis,

13.1.5 Statement of sections of method modified,

13.1.6 Statement of additional methods used in analysis,

13.1.7 Indication of possible interferences in analytical determinations, and

13.1.8 Indication of precision and accuracy.

References

(**1**) "Proposed Methods for Evaluation of Spectrophotometers," *ASTM Proceedings*, Vol 58, ASTEA, 1959, Part III (IR Spectral Resolution) Resolution C, pp. 472–494.

(**2**) "Proposed Methods for Evaluation of Spectrophotometers," *ASTM Proceedings*, Vol 58, ASTEA, 1959, Part III (IR Spectral Resolution) Resolution B, pp. 472–494.

(**3**) ASTM Formula Cards and ASTM-Wyandotte Infrared Cards, Am. Soc. Testing Mats., 1916 Race St., Philadelphia, Pa. 19103.

(**4**) Burley, R. A., and Bennett, W., J., "Spectroscopic Analysis of Poly(Vinyl Chloride) Compounds," *Applied Spectroscopy*, APSPA, Vol 14, 1960, p. 32.

(**5**) *Infrared Spectroscopy*, March 1961, Chicago Society for Paint Technology, 1350 South Kostner Ave., Chicago, Ill. 60623.

(**6**) Documentation of Molecular Spectroscopy, Spex Industries, Inc., Scotch Plains, N. J.

(**7**) Haslam, J., and Soppett, W. W., "The Determination of Chlorine in Resins Obtained from Poly(Vinyl Chloride) Compositions," *Journal of the Society of Chemical Industry*, JSCIA, Vol 67, 1948, p. 33.

(**8**) Haslam, J., and Newlands, G., "The Examination of Compositions Prepared from Poly(Vinyl Chloride) and Related Polymers," *Journal of the Society of Chemical Industry*, JSCIA, Vol 69, 1950, p. 103.

(**9**) Haslam, J., Soppett, W. W., and Willis, H. A., "The Analytical Examination of Plasticizers Obtained from Poly(Vinyl Chloride) Compositions," *Journal of Applied Chemistry*, JACHA, Vol I, 1951, p. 112.

(**10**) Haslam, J., and Squirrell, D. C. M., "The Examination of Poly(Vinyl Chloride) Compositions Containing Polypropylene Adipate," *Analyst*, ALSTA, Vol 80, 1955, p. 871.

(**11**) Hummel, D., "Kunststoffe-, Lack- und Gummi-Analyse. Chemische and IR Methoden," Carl Hanser, Munich, 1958.

(**12**) Hunt, J. M., Wishert, M. P., and Bonahm, L. C., "Infrared Absorption Spectra of Minerals and Other Inorganic Compounds," *Analytical Chemistry*, ANCHA, Vol 22, 1950, p. 1478.

(**13**) Kagarise, R. E., and Weinberger, L. A., "Infrared Spectra of Plastics and Resins", OTS Publication PB 11438, U. S. Department of Commerce.

(**14**) Kendall, D. N., Hampton, R. R., Hausdorff, H., and Pristera, F., "Catalog of Infrared Spectra of Plasticizers," *Applied Spectroscopy*, APSPA, Vol 7, 1953, p. 179.

(**15**) Lawson, K. E., *Infrared Absorption of Inorganic Substances*, Reinhold, N. Y., 1961.

(**16**) Miller, F. A., Carlson, G. L., Bentley, F. F., and Jones, W. H., *Spectrochimica Acta*, SPACA, Vol 16, 1960, p. 135.

(**17**) Miller, F. A., and Wilkins, C. H., "Infrared Spectra and Characteristic Frequencies of Inorganic Ions," *Analytical Chemistry*, ANCHA, Vol 24, 1952, p. 1253.

(**18**) Nyquist, R. A., "Infrared Spectra of Plastics and Resins," Dow Chemical Co., Midland, Mich., 1960.

(**19**) Sadtler Standard Spectra, Sadtler Research Laboratories, Philadelphia, Pa.

(**20**) Tryon, M., and Horowitz, E., "Infrared Spectrophotometry," Chap. VIII, pp. 291–333, *Analytical Chemistry of Polymers, Part 2*, ed. G. M. Kline, Interscience Publishers, 1962.

APPENDIX

A1. Components Covered by This Method

A1.1 Resin

A1.1.1 PVC.

A1.2 Plasticizers

A1.2.1 Di (2-ethyl hexyl) phthalate (DOP).

A1.2.2 Epoxidized soy bean oil.

A1.2.3 Tricresyl phosphate (TCP).

A1.3 Stabilizers

A1.3.1 Basic lead carbonate.

A1.3.2 Tribasic lead sulfate.

A1.3.3 Bibasic lead phthalate.

A1.4 Fillers

A1.4.1 Clay.

A1.4.2 Calcium carbonate.

A1.4.3 Antimony trioxide.

Designation: D 2151 - 68

Standard Method of Test for
STAINING OF POLY(VINYL CHLORIDE) COMPOSITIONS BY RUBBER COMPOUNDING INGREDIENTS[1]

This Standard is issued under the fixed designation D 2151; the number immediately following the designation indicates the year of original adoption or, in the case of revision, the year of last revision. A number in parentheses indicates the year of last reapproval.

1. Scope

1.1 This method measures the tendency for staining of a poly(vinyl chloride) composition to occur due to migration into the plastic of a staining antioxidant (or other compounding additive) from rubber with which it is in intimate contact.

1.2 This method may be used to determine relative staining resistance of vinyl compounds by testing them against a single standard rubber composition (Note 2). It may also be used to compare relative staining tendencies of two or more rubber compositions by testing them against a single standard poly(vinyl chloride) composition.

NOTE 1—The values stated in U.S. customary units are to be regarded as the standard. The metric equivalents of U.S. customary units may be approximate.

NOTE 2—The preparation of suggested standard compositions is described in the Appendix. This procedure may also be applicable to determine stain resistance of plastic materials other than poly(vinyl chloride) in contact with rubber.

2. Significance

2.1 Compounding ingredients used in the manufacture of rubber can migrate into some vinyl compositions in contact with the rubber. These materials can then cause staining of the vinyl either immediately, or under the action of ultraviolet light. This latter behavior is especially likely when certain antioxidants migrate into the vinyl. Although they may be relatively colorless in their pure state, the action of light can transform these antioxidants into chromophoric agents.

NOTE 3—Vinyl compositions containing lead, cadmium, mercury, or antimony compounds may stain when in contact with rubber due to the formation of colored sulfides. This type of stain is due to chemical interaction between ingredients in the rubber and the vinyl compounds. Sulfide stain tendencies of vinyl compositions can be determined by ASTM Method D 1712, Test for Resistance of Plastics to Sulfide Staining.[2]

3. Apparatus

3.1 *For Test Specimen Preparation*—A mill, a press, and other auxiliary equipment for the preparation of test specimens.

3.2 *Oven*—An oven of the forced-air or convection type, having the following characteristics:

3.2.1 The oven shall be controlled by an accurate thermoregulator, maintaining set point within ± 0.5 C or better.

3.2.2 The oven shall be equipped with a calibrated ASTM thermometer, and the proper stem correction shall be applied to the temperature measurement.

3.2.3 The oven shall be maintained at the temperature of test for at least 1 h prior to insertion of the test specimens.

3.2.4 Uniformity of temperature over the area on which the specimens are to be aged shall be determined by means of thermocouple readings prior to the start of the test.

---

[1] This method is under the jurisdiction of ASTM Committee D-20 on Plastics. A list of committee members may be found in the ASTM Yearbook. This standard is the direct responsibility of Subcommittee D-20.50 on Permanence Properties.

Current edition effective Sept. 9, 1968. Originally issued 1963. Replaces D 2151 – 63 T.

3.3 *Weights*—Weights (for example, lead shot in a suitable flat-bottom container) to exert a pressure of 10 g/cm^2 on the test specimens.

3.4 *Light Source*—An S-1 sunlamp as described in ASTM Recommended Practice D 795, for Exposure of Plastics to S-1 Mercury Arc Lamp[2] is the preferred light source. Other intense light sources such as those described in ASTM Recommended Practice E 188, for Operating Enclosed Carbon-Arc Type Apparatus for Light Exposure of Nonmetallic Materials,[3] and ASTM Recommended Practice E 42, for Operating Light- and Water-Exposure Apparatus (Carbon-Arc Type) for Exposure of Nonmetallic Materials,[2] may be used, provided it is shown that they provide results comparable to that of the sunlamp.

4. Test Specimens

4.1 The test specimen of the desired rubber composition should preferably be approximately 25 by 13 mm (1.0 by 0.5 in.) (Note 4). A useful thickness is 2 mm (0.075 in.). The preparation of suitable rubber compositions is described in the Appendix. The composition to be specified should be designated A, B, or C, depending upon whether the antioxidant is known to be nonstaining, semi-staining, or staining, respectively.

NOTE 4—The minimum size specimen is the same as that used in ASTM Methods D 925, Test for Contact and Migration Stain of Vulcanized Rubber in Contact with Organic Finishes.[4]

4.2 The specimen of the vinyl composition under test should be at least 50 by 50 mm (2 by 2 in.) and in all cases larger than the rubber specimen. A useful thickness is 2 mm (0.075 in.). The preparation of one suitable vinyl composition is described in the Appendix.

5. Conditioning

5.1 None required.

NOTE 5—In test to date, no effect of normal atmospheric variations during processing or storage of these compositions has been observed.

6. Procedure

6.1 Place the rubber test specimen on top of the vinyl test specimen. Place the required weight on top of the rubber test specimen.

6.2 Place the test assembly in the oven for 20 h at 70 C.

6.3 Separate the test assembly, examine the vinyl specimen, and note any signs of discoloration in the area that had been in contact with the rubber. Designate the degree of stain as none, slight, moderate, or severe.

6.4 Simultaneously test a vinyl control specimen without rubber.

6.5 Expose the vinyl samples from 6.3 and 6.4 to the desired light source (see 3.4) for 4 h in accordance with the ASTM standard applicable for the particular light source to be used. The side of the specimen that had been in contact with the rubber shall face the light source.

6.6 Examine the area of the specimen that had been in contact with the rubber and note the type and severity of discoloration, if any. If visual evaluation is used, designate the degree of stain as none, slight, moderate, or severe.

7. Report

7.1 The report shall include the following:

7.1.1 Complete identification of rubber and plastic composition.

7.1.2 Time and temperature of exposure,

7.1.3 Extent of discoloration, if any, immediately after oven aging,

7.1.4 Type of light source, and

7.1.5 Extent of discoloration, if any, after light aging.

[2] *Annual Book of ASTM Standards*, Part 27.

[3] *Annual Book of ASTM Standards*, Part 30.

[4] *Annual Book of ASTM Standards*, Part 28.

APPENDIX

A1. SUGGESTED RUBBER AND VINYL COMPOSITIONS FOR USE IN STAINING TESTS

A1.1 Rubber Composition

A1.1.1 A typical rubber composition for use in staining tests can be prepared by using the procedure described in Section 19 of ASTM Methods D 15, for Sample Preparation for Physical Testing of Rubber Products,[4] with the following recipe:

Ingredient	Parts
GRS-1502	100
Sulfur	2.75
Accelerator[a]	1.5
Accelerator[b]	0.3
Stearic acid	1.5
Zinc oxide	5.0
EPC black (nonstaining)	50
Processing oil[c]	5
Antioxidant	1.5

[a] such as *N*-cyclohexyl benzothiazole sulfonamide.
[b] such as tetramethylthiuram disulfide.
[c] such as naphthenic-type processing oil.

The processing oil should be added in the step indicated in 19.1.4 of Methods D 15, after all the pigment is added. Use a cure of 30 min at 150 C (302 F).

A1.2 Antioxidants

A1.2.1 Staining tendencies of the above will depend upon the specific antioxidant chosen. The following antioxidants are suggested:

A1.2.1.1 *Nonstraining*—2,2′ - methylene bis(6-*tert*-butyl-4-methylphenol).

A1.2.1.2 *Semi-staining*—Alkylated diphenyl-amine.

A1.2.1.3 *Staining*—Phenyl-beta-naphthylamine.

A1.3 Vinyl Composition

A1.3.1 A typical poly(vinyl chloride) composition for use in evaluating the staining tendencies of various rubber compositions is the following:

Ingredient	Parts
PVC resin GP5-00000 (Specification D 1755)[a]	100
2-Ethyl hexyl phthalate (type II of Specifications D 1249)[a]	50
Dibutyl tin dilaurate	2

[a] These designations refer to the following specifications of the American Society for Testing and Materials:

D 1249, Spec. for Primary Octyl Phthalate Ester Plasticizers, *Annual Book of ASTM Standards*, Part 26.

D 1755, Spec. for Poly(Vinyl Chloride) Resins, *Annual Book of ASTM Standards*, Part 26.

Mix in a stainless steel pan or beaker. Mill to form a sheet on rolls set at 166 ± 6 C (330 ± 10 F) for 3 ± 0.5 min. Mold a convenient flat specimen, such as 150 by 150 by 2 mm (6 by 6 by 0.075 in.).

Designation: D 2176 - 69

Standard Method of Test for
FOLDING ENDURANCE OF PAPER BY THE M. I. T. TESTER[1]

This Standard is issued under the fixed designation D 2176; the number immediately following the designation indicates the year of original adoption or, in the case of revision, the year of last revision. A number in parentheses indicates the year of last reapproval.

1. Scope

1.1 This method describes the use of the M.I.T.-type folding apparatus for determining folding endurance of paper. The M.I.T. tester can be adjusted for papers of any thickness; however, if the outer fibrous layers of paper thicker than about 0.25 mm (0.01 in.) rupture during the first few folds, the test loses its significance. The procedure for the Schopper-type apparatus is given in ASTM Method D 643, Test for Folding Endurance of Paper by the Schopper Tester.[2]

2. Apparatus

2.1 *Folding Tester*, consisting of:

2.1.1 A spring-loading clamping jaw constrained to move without rotation in a direction perpendicular to the axis of rotation of the folding head specified below and having its clamping surfaces in the plane of this axis. The load shall be applied through a spring attached to the jaw assembly, which is easily adjustable to provide any desired tension on the specimen from 0.5 to 1.5 kg. The deflection of the spring when loaded shall be at least 17 mm (0.67 in.)/kg.

2.1.2 An oscillating folding head supporting two smooth, cylindrical folding surfaces parallel to, and symmetrically placed with respect to, the axis of rotation. The portion of the axis of rotation shall be midway between the common tangent planes of the two folding surfaces. The folding head shall be provided with a clamping jaw with its nearest edge 9.5 mm (0.375 in.) beyond the axis of rotation. The rotary oscillating movement of the head shall be such as to fold the paper through an angle of 135 ± 2 deg both to the right and to the left of the position of the unfolded specimen. Each of the two folding surfaces shall have a radius of curvature of 0.38 ± 0.015 mm (0.015 ± 0.001 in.) and a width of not less than 19 mm (0.75 in.). The distance separating the folding surfaces shall be greater than the uncompressed thickness of the paper being tested but by no more than 0.25 mm (0.01 in.). Various size folding heads are required for testing different thicknesses of paper. Heads available will accommodate thicknesses from 0 to 0.01, 0.01 to 0.02, 0.02 to 0.03, 0.03 to 0.04, and 0.04 to 0.05 in.

2.1.3 *Motor*—A power-driven device for imparting a rotary oscillating motion of 175 ± 25 periods per min to the folding clamp.

2.1.4 *Counter*, for registering the number of double folds required to break the specimen.

2.2 *Strip Cutter*, to cut 15-mm wide parallel strips to within 0.02 mm with clean edges.

2.3 *Exhaust Fan*, centrifugal, mounted so that its inlet, at least 5 cm in diameter, is next to the oscillating head and the specimen being tested, thus drawing the conditioning room air rapidly over both specimen and head. Its capacity shall be great enough to prevent the temperature of the head from rising more than 0.5 C.

3. Adjustment and Calibration

3.1 Make sure that the folding edges are free from rust or dirt and that the counter

[1] This method is under the jurisdiction of ASTM Committee D-6 on Paper and Paper Products. A list of members may be found in the ASTM Yearbook.

Current edition effective Oct. 3, 1969. Formerly part of Methods D 643. Replaces D 2176 - 63 T.

[2] *Annual Book of ASTM Standards*, Part 15.

operates properly.

3.2 Measure the plunger friction by determining the additional load required to move the plunger perceptibly when displaced under a load of 1.0 kg or the load tension used in testing. The friction should not be greater than 25 g.

3.3 Measure the change in tension due to eccentricity of rotation of folding edges as follows: Place a test strip of strong paper of the proper thickness, cut in the machine direction, in the tester as for making a folding test, and apply a tension of 1 kg or that to be used for the testing. Rotate the folding head slowly throughout the entire folding cycle and measure the maximum change in displacement of the plunger to an accuracy of 0.1 mm (0.004 in.). This should not be greater than that produced by a weight of 35 g. The curvature of the folding edges can be measured by making casts, magnifying them in profile and comparing them to true circles. A folding tester in steady use should be adjusted and calibrated at intervals of not more than 1 month. If not in steady use, then immediately before a standard test.

4. Test Specimen

4.1 The sample to be tested shall be obtained in accordance with ASTM Method D 585, Sampling Paper and Paperboard.[2] From each sample shall be obtained, so as to be representative of the test unit, at least ten specimens in each principal direction of the paper cut accurately to a width of 15 ± 0.02 mm and a length of 13, preferably 15 cm. Strips shall be selected that are free of wrinkles or blemishes not inherent in the paper, and care shall be taken that the area where the flexing takes place does not contain any portion of a watermark. The long edges of the specimens shall be clean cut and parallel.

5. Conditioning

5.1 Condition and test the paper in an atmosphere in accordance with ASTM Method D 618, Conditioning Paper and Paper Products for Testing.[2] Handle a specimen by the ends and do not touch it with the hands in the region in which it is to be folded.

NOTE 1—As folding endurance is very sensitive to the moisture content of the specimen, it is important to observe strictly the requirements for preconditioning from the dry side, for conditioning, and for conditions during testing.

6. Procedure

6.1 Turn the oscillating folding head so that the opening is vertical. Place a weight on the top of the plunger equivalent to the tension desired on the specimen, normally 1 kg, tap the plunger sideways to eliminate friction, check and set the load indicator, and remove the weight. Depress and lock the plunger in position corresponding to the load. Without touching the part of the strip to be folded, clamp the specimen firmly and squarely in the jaws, with the surface of the specimen lying wholly within one plane, that is, flat, and with the sides not touching the oscillating jaw mounting-plate.

6.2 Unscrew the plunger lock to apply the specified tension to the test strip. If the reading of the load indicator then changes, reclamp the specimen under the proper tension. Whenever possible use a tension of 1 kg, but if this does not give good results (that is, between 10 and 1000 double folds), use a greater or lesser tension.

NOTE 2—The number of the folds will vary by about the cube of the applied tension.

6.3 Set the counter to zero and fold the strip at a uniform rate of 175 ± 25 double folds per min until it breaks.

6.4 Record the number of double folds made before fracture.

7. Report

7.1 Report the average results as M.I.T. folding endurance (double folds) to two significant figures and state clearly if a tension other than 1 kg was used. Include the number of tests, and the maximum and minimum results for each of the principle directions of the paper. Tests made on strips having their length in the machine direction are designated as being in the "machine direction."

8. Precision

8.1 Precision has not yet been definitely determined. The coefficient of variation has been indicated[3,4] as being in the range of about 30 percent for different instruments.

[3] Snyder, L. W., and Carson, F. T., "A Study of the M.I.T. Paper Folding Tester," *Paper Trade Journal* PTJOA, Vol 96, No. 22, June 1, 1933, pp. 40–44.

[4] Reitz, L. K., and Sillay, F. J., *Paper Trade Journal* PTJOA, Vol 126, No. 17, April 22, 1948, p. 54.

Designation: D 2188 - 63 T

Tentative Recommended Practice for

STATISTICAL DESIGN IN INTERLABORATORY TESTING OF PLASTICS[1]

This Tentative Recommended Practice has been approved by the sponsoring committee and accepted by the Society in accordance with established procedures, for use pending adoption. Suggestions for revisions should be addressed to the Society at 1916 Race St., Philadelphia, Pa. 19103.

This Tentative has been submitted to the Society for adoption, but balloting was not complete by press time.

1. Scope

1.1 The purpose of this recommended practice is to guide the task group chairman in planning interlaboratory (round-robin) tests. A round robin may be conducted in order to investigate factors in a test method so that a method may be developed that has minimum variability. Or, a round robin may be conducted with an established test method in order to assess the variability of the method. A round robin may be conducted with an established test method in order to set specifications for a particular material. This recommended practice is intended to be broad enough to give the task group chairman guidance in setting up a round robin for any of these purposes.

1.2 A complete explanation of the concept of the statistical application to interlaboratory testing of plastics is not intended, nor do we intend to outline specific round-robin plans, but rather to present some of the philosophy and the usefulness of statistical analysis as a means of appraising quantitatively the various sources of error in a test method or the amount of error in an established test method for a particular material. For details in setting up a particular round robin, the services of a competent statistician may be helpful.

1.3 The sections appear in the following order:

2. Purpose of Interlaboratory Test

2.1 A clear statement of the purpose of the interlaboratory test should be prepared in writing and should reflect one of the purposes described in Section 1 (that is, method development, method variability, material specification). The following information should also be included:

2.1.1 The class of material or type of test method being studied,

2.1.2 The quality characteristics to be measured,

2.1.3 The conditions whose effects are being studied and those which are not being studied,

2.1.4 The conditions under which the tests are to be made,

2.1.5 The objectives to be reached, such as the hypotheses to be tested or the components of variance to be estimated, or both,

2.1.6 A description of the test method and

---

[1] This recommended practice is under the jurisdiction of ASTM Committee D-20 on Plastics. A list of committee members may be found in the ASTM Yearbook. This tentative is the direct responsibility of Subcommittee D-20.13 on Statistical Techniques.

Effective Sept. 30, 1963.

the number of measurements to be made,

2.1.7 A list of any constant errors that the method may have,

2.1.8 An outline of the method for analyzing the results, and

2.1.9 The significance level.

NOTE—The time to seek statistical advice is when the interlaboratory test is being planned. This will ensure that the desired information can be obtained from the test data.

3. Error

3.1 Design of the experiment shall be so formulated that known systematic error is kept to the minimum possible amount. A systematic error is an error that persists throughout a series of measurements of the same type and whose effect cannot be eliminated by the process of averaging the results (**2**).[2] In this respect it is different from random error, which is an error that tends to zero in the process of averaging a large number of measurements. Systematic differences among experimental units or systematic changes in external conditions cause systematic error. Randomization is one procedure whereby the effect of systematic error may be minimized. Systematic errors are likely to occur when all of the units receiving one treatment are collected in a single group and cannot respond independently. In order to conclude that a treatment has caused a change when comparing the results with the results to be expected in the absence of the treatment, it is necessary to include a control series of units that have not received the treatment but are tested under the same conditions as the treated units, and that are not systematically different from them. Examples of conditions that cause systematic error are changes in relative humidity that affect materials being tested, temperature fluctuations, attitudes of test personnel when comparing an old test method with a new one in which more careful attention is devoted to the latter, and other fluctuations of unknown origin.

3.2 Planning must be done prior to the actual running of tests to assure an estimate of the experimental error. There may be little or no information available at the start about the nature of the distribution of the desired property or the magnitude and stability of the components of variance involved in a universe (population) sample to be tested. It may be desirable to run preliminary tests in a single laboratory by a single operator in order to assess the magnitude of the experimental error for use in future interlaboratory tests. Errors may be attributable to the equipment used for testing and whether or not it be in calibration. Random fluctuations that arise from the measurement process must also be considered and appropriate allowance made if necessary.

3.3 In using the preliminary estimate of experimental error we must assume that the measurements included in its calculation are a random sample from the population of such measurements.

3.4 A sample statistic is said to be an unbiased estimate of the population parameter if the average of the statistic over a large number of samples approaches the parameter value. The sample variance, s^2, is an unbiased estimate of the population variance, σ^2. The sample standard deviation, s, is not an unbiased estimate of the population standard deviation, σ. For data that follow the normal distribution, factors for converting s to an unbiased estimate of σ have been tabulated (**9**, p. 405).

4. Statistical Significance

4.1 Before discussing statistical significance it is necessary to understand that there are two types of errors that can be made in drawing conclusions about data:

4.1.1 Type I error is rejecting a hypothesis that is actually true, and

4.1.2 Type II error is accepting as true a hypothesis that is actually false.

4.2 The meaning of Type I and Type II errors may be summarized in tabular form as follows:

Real Situation	Decision	
	Accept	Reject
Hypothesis is true	no error	Type I error
Hypothesis is false	Type II error	no error

4.3 A statistical significance test is a means of evaluating the hypothesis in question. Such a test is made with a predetermined risk of

[2] The boldface numbers in parentheses refer to the list of references at the end of this recommended practice.

Type I error. This risk is the significance level. It is customary to work at a level corresponding to a 95 percent confidence coefficient, which gives a 5 percent chance of a Type I error. Type II error depends on the details of the interlaboratory test program; for example, sample size, value of unknown mean, as well as level of significance. When the avoidance of Type II errors is very important, one may be willing to accept a larger risk for Type I errors.

5. Treatment of Variables

5.1 The experimental design is determined by considering those effects which are to be studied. All variables, either controlled or uncontrolled, that are known or suspected to affect the test results, should be listed (for example, temperature of test, type of material, machine direction of extruder used to produce extruded sheet, laboratory, test operator, type of test machine, etc.). Each variable may be treated in one of four ways, namely, (*1*) varied systematically, (*2*) held constant, (*3*) blocked, or (*4*) randomized. In designing an experimental program, the levels of most important variables are normally either arranged systematically within the design, or else held constant. The required combination of variables or treatments are then carried out according to a randomized schedule. Those suspected to be of little or no importance are more frequently treated by blocking or randomization. Details on blocking may be obtained from various statistics texts (**3,5,6,7,8**). The specific pattern of systematic variation, blocking, and randomization comprises the experimental design. A decision must be made on what variables are to be included and at what levels for the design.

5.2 Wherever possible, the test program within each laboratory shall be so designed that information about day-to-day variability, or run-to-run variability, or about the effects of other systematic variables that are not usually or uniformly controlled in routine work, can be obtained by proper statistical analysis.

6. Size of Experimental Program

6.1 The size of the experimental program is determined by the number of factors, the number of levels of each factor to be studied, and the number of replicates to be run. The number of replicates is determined by the estimated size of the experimental error compared to the precision desired and whether or not estimates of variances are desired. A good reference for determining the approximate number of replicates is given in Ref (**21**).

6.2 The precision of a variance estimate is determined solely by the number of degrees of freedom available for that estimate. These precisions may be determined from percentiles of the chi square over degrees of freedom (**9,** p. 80 and Table A-6b) distributions. Some examples are given in Table 1.

6.3 When the number of factors and their levels and the number of replicates needed to estimate the precision are listed, it is not unusual to find that the experiment is too large to be practical. Also, the experimental program should not be so complicated that the analysis of the data will be almost impossible when one of the participating laboratories does not submit its test results. The interlaboratory test program is derived from a series of compromises between obtaining the desired information and arrangement of a practical size of experimental design.

7. Sampling of Materials

7.1 Where feasible, sampling shall be treated as a controlled variable. Consideration shall be given to sheet-to-sheet or unit-to-unit variability, location of sheets or units in press or mold during curing, and other factors that may increase random error by lack of control.

7.2 Inferences and conclusions are based on the assumption that the sampling has been random. If a sample is not random, bias may be introduced into the test. Systematic biases may be guarded against by randomizing the order in which the tests are performed on the different treatments in a replication. Tables of random numbers are available for assignment of operators to test equipment, distribution, and many other uses. When a test program has to be repeated, the process of randomizing must be a new one and cannot be copied from the previous one, which is no longer random.

7.3 In some cases it is advantageous to follow a systematic pattern in the allocation of the specimens to the laboratories. For example, if the specimens are pieces cut from plastic sheets, it is often advisable to allocate to each laboratory equal numbers of specimens from each sheet. In such cases, only the specimens within each sheet are randomized, rather than the totality of specimens from all sheets. Wherever possible, the method of partial randomization shall be adopted.

7.4 It is desirable to have the specimens allocated and distributed from a single place, except when methods of preparation of test specimens are being studied. From each material enough specimens shall be allocated to provide the required test material for all participating laboratories and a sufficient number of additional specimens for replacement of lost or spoiled specimens. Each specimen shall be properly labeled by means of a code symbol, and the identification of the specimens by means of their code symbols shall be properly recorded and kept for future reference.

8. Types of Experimental Designs

8.1 Enough preliminary work and planning shall be done to fulfill the following requirements:

8.1.1 The test method must be applicable to all materials to be included in the interlaboratory tests,

8.1.2 The personnel in each participating laboratory who are assigned to the testing work for the interlaboratory study shall have acquired complete understanding, preferably by preliminary testing, of the experimental procedure,

8.1.3 The value of standard deviation (and coefficient of variation) shall be determined in a preliminary way in a single laboratory to evaluate its dependence upon the material to be tested and to assess the magnitude of the experimental error for use in future interlaboratory tests, and

8.1.4 The description of the method (or methods) shall be so formulated that known systematic errors are eliminated insofar as possible.

8.2 Statistical design of interlaboratory tests is used in order to obtain mathematical consistency between data acquisition and data analysis. The models recommended are described in Refs (**3**) to (**8**). The user of statistical methods will find much useful information on statistical analysis in Ref (**9**). The basis for selection of experiment design depends upon the individual situation for each investigation. If it is apparent that interactions between factors may be important and it is desired to evaluate these, a factorial design shall be used. Designs that do not allow for calculation of interaction include Latin square, Youden square, lattice squares, and balanced incomplete blocks. Certain designs have been selected for this recommended practice as described in 8.3 to 8.8.

8.3 *Factorial Designs:*

8.3.1 The experimental design in which all factors at all levels are taken in all combinations constitutes a complete or full factorial. The method of subdivision of variance, which is a powerful tool in statistics, may be used to advantage in a factorial design. A complete factorial design is appropriately used where four or less factors are to be studied. More than four factors become unwieldy, particularly where several levels of each factor are being studied. The advantages of a complete factorial design are:

8.3.1.1 All interactions of the effect of one factor on the influence of another factor may be determined and evaluated,

8.3.1.2 Each level of each factor is combined with each level of every other factor in all possible combinations,

8.3.1.3 It contains a large degree of hidden replication which helps to reduce testing error, and

8.3.1.4 When experimental error is large or when high-order interactions are expected, or both, there is no other satisfactory design.

8.3.2 When there are no interactions, the factorial design gives the maximum efficiency in the estimation of effects. The existence of a large interaction means that the effect of one factor is markedly dependent on the level of the other, and when quoting the effect of one factor it is necessary to specify the level of the other. When an interaction is quite large, the corresponding main effects cease to have much meaning.

8.4 *Fractional Factorials*—Fractional factorials are a device to reduce the size of a full factorial though at the expense of confounding main effects with high-order interactions, which may or may not be important. These incomplete factorials require that the experimental error not be large and that any interactions beyond second order be negligible. Confounding is the process by which relatively unimportant comparisons are intentionally confused for the purpose of assessing the more important comparisons with greater precision. The smallest fractional factorial design that permits a degree of confounding is one of eight observations. In the fractional factorial designs, most of the degrees of freedom are often used up in estimating main effects and certain two-factor interactions, leaving few degrees of freedom for obtaining an independent estimate of the error variance. When no prior information exists on the experimental error, it is necessary to use the variance supplied by the experiment itself in order to assess the significance of the effects. With so few degrees of freedom available for error, the test of significance will not be sensitive (**3**). Complete models for fractional factorial designs for factors at two levels are given in Ref (**10**). For information on the usefulness of these see Ref (**11**).

8.5 *Incomplete Block Designs*—Incomplete block designs are useful when it is reasonably certain that no interaction exists and it is desired to screen all variables to find those which have major effects. The method of incomplete blocks, although applicable in theory, is usually impractical for factorial designs because it would require an excessive number of replicates. In a balanced incomplete block design every pair of treatments occurs together in the same number of blocks of a design. Every pair of treatments will be found to occur only once in the same block. The above design ensures that the same standard error may be used for comparing every pair of treatments. These designs have the desirable property that all differences between treatments are estimated with the same precision. Tables for balanced incomplete block designs are available (**3,5**).

8.6 *Youden Square*—A modification of the incomplete block design is the Youden square. It satisfies the requirement that every pair of treatments occur together in the same block the same number of times. The Youden square is an incomplete Latin square in which one column is missing. This does not mean that a Latin square with one column missing is a Youden square, since this may destroy the balance. Fewer replications are required for a Youden square than for a balanced incomplete block design with the same size of error variance. The balanced incomplete block designs make it possible to intercompare a large number of items without the necessity of maintaining conditions constant longer than is required to compare a small number of these items in one block at a time.

8.7 *Latin Square*—If the purpose of the interlaboratory test is to scout a great many factors in a preliminary way, a large amount of work can be saved by sacrificing the information on interactions and concentrating on main effects. This involves somewhat of a risk, since the interaction effects are confused with the main effects. The Latin square or the Graeco-Latin square designs are useful for this purpose. The Latin square permits several factors (other than the ones being studied) to vary during the experiment, and yet it excludes the principle component of their variation from influencing the experiment (**12**). A Latin square is a square array of elements such that each element appears only once in each row and each column.

8.8 *Nested Designs*—Nested sampling design is a method of random sub-sampling in which the laboratories taking part in the study are considered as a small random sample of a large number of laboratories. Single tests are made on each sample on different days by the same analyst for replication and by different analysts. With the nested design it is possible to estimate separately the variability of replicate tests, the variability of the analysts within the same laboratory, and the variability among laboratories. The results of the complete analysis of variance on such a test are the same as can be shown graphically by means of the control chart method. Nested designs will not distinguish between constant bias and laboratory interactions.

8.9 Ref (**22**) and ASTM Recommended Practice D 1749, for Interlaboratory Evaluation of Test Methods Used with Paper and Paper Products,[3] describe a new approach to round-robin studies of a test method. The requirements may be more than needed in many round-robin tests on plastics. But when the variables to be studied are limited to labo-

[3] *Annual Book of ASTM Standards*, Part 15.

ratories and materials, the approach is very useful.

9. Analysis of Variance

9.1 After a round-robin has been conducted and the data have been gathered, it is always true that the data contain variability from low to high values of the measurement involved. If this were not the case, the round robin would not have been conducted. The purpose may have been to explain this variability according to factors in the test method or according to differences between laboratories or between materials, or the purpose may have been to assess this variability and to see whether the variability is the same for different laboratories or different materials. These questions can be answered with the statistical analysis known as the "analysis variance." This is a technique of dividing the total amount of variation into components and testing these components in order to determine whether some of the variation has been caused by the various factors in the round-robin design. In the analysis of variance, therefore, we deal with elements called variances or mean squares. These are indexes of variation. In general, the square root of a variance or a mean square is called a standard deviation. The standard deviation is the element typically used to characterize the reliability of a test.

9.2 If the experimental design includes some sort of replication so that there is more than one measurement associated with each combination of levels of the variables considered in the design, a variance may be calculated for each combination of levels. One of the assumptions of the analysis of variance is that these variances come from the same populations of variances (that is, are homogeneous). It is also assumed that the data are samples from normal populations.

9.3 Reference (**12**) presents a thorough treatment of a computational method in several analyses of various examples. Once the author's notation is mastered, this reference can be used as a "cook book" for calculating the analysis. Reference (**4**) gives a good treatment of the theory of analysis of variance for the uninitiated. Other guides to analysis of variance are also available (**5,6,7,8,9,18**).

10. Graphical Analyses of Data

10.1 Under some conditions more information can be obtained by graphical methods than by calculation of mean and standard deviation of the data. For samples of 20 or more observations, the graphical method will give reliable results in predicting significance in the difference in means of two distributions. A comparison of the graphical method with analysis of variance is given in Refs (**13**) and (**14**). In an article on the use of probability paper in the study of experimental data, a measure of dispersion (geometric dispersion g) is proposed that may be more useful than standard deviation in connection with skewed distributions (**15**). Many significant contributions to the literature have been published on graphical analysis of data and the simplicity and advantages of these procedures (**16**). Other useful applications of graphical analysis are also available (**17**).

References

(**1**) Youden, W. J., "Comparative Tests in a Single Laboratory." *ASTM Bulletin*, ASTBA, No. 166, May 1950, pp. 48–51. (TP 110–113).

(**2**) Youden, W. J., "The Sample, the Procedure, and the Laboratory," *Analytical Chemistry*, ANCHA, Vol. 32, No. 13, 1960, p. 23A.

(**3**) Davies, O. L., *The Design and Analysis of Industrial Experiments*, Hafner Publishing Co., New York, N. Y., 1956.

(**4**) Davies, O. L., *Statistical Methods in Research and Production, with Special Reference to the Chemical Industry*, 3rd Ed., Hafner Publishing Co., New York, N. Y., 1957.

(**5**) Cochran, W. G., and Cox, G. M., *Experimental Designs*, 2nd Ed., John Wiley & Sons, Inc., New York, N. Y., 1957.

(**6**) Kempthorne, O., *Design and Analysis of Experiments*, John Wiley & Sons, Inc., New York, N. Y., 1952.

(**7**) Connor, W. S., *Experimental Designs in Industry*, John Wiley & Sons, Inc., New York, N. Y., 1958.

(**8**) Cox, D. R., *Planning of Experiments*, John Wiley & Sons, Inc., New York, N. Y., 1958.

(**9**) Dixon, W. J., and Massey, F. J., Jr., *Introduction to Statistical Analysis*, 2nd Ed., McGraw-Hill Book Co., Inc., New York, N. Y., 1957.

(**10**) "Fractional Factorial Experiment Designs for Factors at Two Levels," Nat. Bureau Standards, Applied Mathematics Series 48, for sale by Superintendent of Documents, Washington 25, D. C.

(11) Zelen, M., and Connor, W. S., *Industrial Quality Control*, IQCOA, Vol XVI, March 1959, pp. 14–17.

(12) Brownlee, K. A., *Statistical Theory and Methodology in Science and Engineering*, John Wiley & Sons, Inc., New York, N. Y., 1960.

(13) Youden W. J., "Circumstances Alter Cases," *Industrial and Engineering Chemistry*, IECHA, Vol 12, 1958, p. 77A.

(14) Schmidt, P. L., "Graphical Methods of Statistical Analysis and Tests for Significance," *Symposium on Statistical Methods for the Detergent Laboratories, ASTM STP 139*, 1952, Am. Soc. Testing Mats.

(15) Ferrell, E. B., "Plotting Experimental Data on Probability Paper," *Industrial Quality Control*, IQCOA, Vol XV, No. 1, July 1958, pp. 12–15.

(16) Youden, W. J., "Graphical Analysis of Interlaboratory Test Results," *Industrial Quality Control*, IQCOA, Vol XV, May 1959, pp. 24–28.

(17) Quenoille, M. H., *Associated Measurements*, Chapters 1 to 3, Academic Press, Inc., New York, N. Y., 1952.

(18) Freeman, H. A., *Industrial Statistics*, John Wiley & Sons, Inc., New York, N. Y., 1942.

(19) Mandel, J., and Stiehler, R. D., "Sensitivity—A Criterion for Comparison of Methods of Test," *Journal of Research of the National Bureau of Standards*, JRNBA, Vol 53, No. 3, September 1954, p. 155 (*RP* 2527).

(20) Lashof, T. W., Mandel, John, and Worthington, V., "Use of the Sensitivity Criterion for the Comparison of the Bekk and Sheffield Smoothness Testers," *Tappi*, TAPPA, Vol. 39, No. 7, 1956, p. 532.

(21) Harris, M., Horvitz, D. G., and Mood, A. M., "On the Determination of Sample Sizes in Designing Experiments," *American Statistical Association Journal*, JSTNA, Vol 43, September 1948, pp. 391–402; also Vol 46, 1951, p. 515.

(22) *ASTM Manual for Conducting an Interlaboratory Study of a Test Method, ASTM STP 335*, 1963, Am. Soc. Testing Mats.

APPENDIXES

A1. Tests for Homogeneity of Data

A1.1 A test for homogeneity of several estimated variances is the Bartlett test. The hypothesis to be tested is that the variances of K normally distributed populations are equal.

$$\sigma_1^2 = \sigma_2^2 = \cdots = \sigma_i^2 = \cdots = \sigma_k^2 = \sigma^2$$

The test is based on the chi square distribution:

$$\chi^2 \cong 2.3026/c(f \log s^2 - \Sigma_{i=1}^{k} f_i \log s_i^2)$$

where:

f_i = degrees of freedom for the ith estimated variance,

i = $1, 2, \cdots, k$,

f = $\Sigma_{i=1}^{k} f_i$,

s^2 = $\Sigma_{i=1}^{k} f_i s^2 / \Sigma_{i=1}^{k} f_i$

s_i^2 = the ith estimated variance,

c = $1 + [1/3(k - 1)][\Sigma_{i=1}^{k}(1/f_i) - (1/f)]$, and

χ^2 has $k - 1$ degrees of freedom.

A1.2 In an interlaboratory test program it sometimes is apparent that one participating laboratory will have results with extreme individuals that are different from those expected. In this situation the Cochran test for homogeneity of variances can be used to indicate if some of the samples give rise to very large errors (**9**). The within-laboratory variance for each sample at each laboratory is examined for homogeneity with a pooled variance for acceptable laboratories. Let $s_1^2, s_2^2, \cdots s_k^2$ represent the pooled within-laboratory variances; and let

$$g = \text{largest } s^2/(s_1^2 + s_2^2 + \cdots + s_k^2)$$

be the ratio used in this test. The tables for critical values of g for segregation of laboratories or samples that have extensive within-laboratory variances are contained in Ref (**9**).

A2. Sensitivity Criterion for Test Methods (19)

A2.1 "In many methods, particularly when dealing with polymeric materials, the measured quantity, M, and the desired quantity, Q, are not linearly related." "Whether or not a quantity Q can be defined, and whatever the relation may be between a characteristic Q and the measured quantity M, the criterion defined as sensitivity can effectively be used for evaluating and comparing methods of test" (**19**).

A2.2 Consider two alternative methods, A and B, for the measurement of a property, Q. Since both Method A and B are a measure of Q, they are functionally related to each other. The curve in Fig. 1 represents such a relationship. As can be seen from the figure, determination of the sensitivity ratio requires a knowledge of the slope, $\Delta A/\Delta B$, and of the standard deviations, σ_A and σ_B.

A2.3 Steps in the comparison of two methods or procedures by means of the sensitivity criterion are as follows:

A2.3.1 Determine the relationship between the results of the two methods so that the slope of this relationship can be evaluated at each point (that is, at each level of the measured property). If no relationship can be found, the two methods are not measuring the same property, and the sensitivity cannot be used for comparing them.

A2.3.2 For each method, determine the standard deviation and how it varies with the magnitude of the measurement. Remember that in order to obtain a good estimate of standard deviation, considerable replication is required.

A2.3.2.1 The nature of the standard deviation occurring in the definition of sensitivity depends on the nature of the comparison between the two methods. In a within-one-laboratory study, the interest is centered on the intrinsic merit of each instrument. Therefore, the standard deviation used in the evaluation of sensitivity must be a measure of instrumental fluctuations only. Consequently, all errors that are not related to the instrument, such as variability in machine and cross-direction, sheet-to-sheet variability, and systematic operator biases, are eliminated. The remaining error is used in the estimation of the standard deviation.

A2.3.2.2 If two methods are compared on an interlaboratory basis, as when deciding which shall be chosen as the standard method, the standard deviations to be used are the total standard deviations (or standard errors) obtained from the interlaboratory study.

A2.3.3 Compute the ratio of sensitivities in accordance with the formulas given in Fig. 1. Method A is more sensitive wherever the ratio is appreciably greater than unity.

A3. Glossary

analysis of variance—statistical procedure for evaluation of the effects of the systematic variables incorporated in an experiment. In interlaboratory tests, analysis of variance is particularly useful for the evaluation of laboratory-to-laboratory variations in the testing of a variety of materials.

bias—constant error present throughout a set of measurements during a test and therefore not eliminated by any process of averaging.

coefficient of variation—standard deviation divided by the average value. The percentage coefficient of variation (which is the coefficient of variation multiplied by 100) is more commonly used.

components of variance—measures of the main effects and interactions computed from the mean squares of an analysis of variance.

confidence interval—interval estimate of a population parameter computed so that the statement "the population parameter lies in this interval" will be true, on the average, in a stated proportion of the times such statements are made.

confidence level (*or coefficient*)—stated proportion of the times the confidence interval is expected to include the population parameter.

confidence limits—the two statistics that define a confidence interval.

degrees of freedom—a number associated with a mean square, and indicative of the effective number of values on which the evaluation of the mean square is based. The larger the number of degrees of freedom corresponding to residual error, the more reliable is the estimate of random error which is given by the corresponding mean square.

estimate—the particular value, or values, of a parameter for a given sample computed by an estimation procedure.

estimation—procedure for making a statistical inference about the numerical values of one or more unknown population parameters from the observed values in a sample.

factor—any feature of the experimental conditions which may be assigned at will from one trial to another, for example, type of material, laboratory, etc.

inference—logical conclusion from given data.

interaction effect or interaction—variability in the results caused by the lack of additivity of the effects of two or more systematic variables. Of particular interest in interlaboratory studies is the interaction between laboratories and materials, which represents the variation from laboratory-to-laboratory of the differences between materials. Interaction effects can sometimes be eliminated by changing the scales in which either the measurements or systematic variables, or both, are expressed.

interval estimate—estimate of a parameter given by two statistics, defining the end points of an interval.

level—series of values for a factor.

level of a systematic variable—one of the forms or magnitudes that a systematic variable takes in a given experiment.

main effect—main effect of a variation in one systematic variable, averaged over all the levels of all other systematic variables which appear in the experiment.

mean square—a quantitative result yielded by an analysis of variance for the characterization of a main effect, an interaction, or residual error. Mean squares are not in themselves measures of main effect and interactions, but from them such measures can be derived.

null hypothesis—the hypothesis of no difference between expected values of the test measurement at different levels of the round-robin factors.

orthogonality—that property of the design of an experiment which ensures that the different classes of effects shall be capable of direct and separate estimation without any entanglement. The sums of squares of all the effects are then independent and additive.

parameter—as used in statistics, a constant (usually unknown) defining some property of the frequency distribution of a population, for example, a population median or a population standard deviation.

population (*or universe*)—hypothetical collection of all possible test specimens that could be prepared in the specified way from the material under consideration.

random error—deviation of a measured value from the corresponding population average.

randomization—process of ordering a number of

objects (or code symbols representing the objects) with complete avoidance of systematic choice, conscious or subconscious, by the experimenter. Tables of random numbers are available for the purpose.

range—difference between the least and greatest values of the variables in a frequency distribution.

regression methods—statistical procedures dealing with the study of the association or the relationship between two or more variables.

replication—complete set of all the different experimental trials.

residual error—residual variability of a set of measurements after the effects of all main effects and interactions that are considered of importance have been segregated. The residual error is generally considered as an estimate of random error.

sample—specimens selected from the population for test purposes.

sample average (*arithmetic mean or mean*)—sum of all the observed values in a sample divided by the sample size. It is a point estimate of the population mean.

sample median—middle value when all observed values in a sample are arranged in order of magnitude if an odd number of samples are tested. If the sample size is even, it is the average of the two middlemost values. It is a point estimate of the population median, or 50 percent point.

sample standard deviation—square root of the sample variance. It is a point estimate of the population standard deviation, a measure of the "spread" of the frequency distribution of a population.

sample variance—sum of the squares of the differences between each observed value and the sample average divided by the sample size minus 1. It is a point estimate of the population variance.

significance level—stated probability (risk) that a given test of significance will reject the hypothesis that a specified effect is not present when the hypothesis is true (Type I error).

significant—(*statistically significant*)—an effect (difference between populations) is said to be present if the value of a test-statistic is significant, that is, lies outside of predetermined limits.

statistic—summary value calculated from the observed values in a sample.

systematic effect—change in value of a measured quantity due to a change in a systematic variable.

systematic error—systematic effect that is ignored either knowingly or unknowingly.

systematic variable—experimental factor that can be controlled (temperature, pressure, order of test, etc.), or independently measured (hours of sunshine, thickness of tension specimen). Systematic variables may be qualitative (such as a design feature in a testing apparatus or instrument, an operator, a specific instrument or testing apparatus of a given type, or a qualitative difference in operating technique) or quantitative (such as temperature, pressure, or duration).

test of significance—statistical procedure for ascertaining whether a particular observed effect could have been the result of mere chance.

NOTE—While a finding of significance points to the likelihood that the observed effect is real, a finding of non-significance is never complete proof of the absence of an effect.

test-statistic—a function of the observed values in a sample that is used in a test of significance.

tolerance interval—an interval computed so that it will include at least a stated percentage of the population with a stated probability.

tolerance level—the stated probability that the tolerance interval includes at least the stated percentage of the population. (It is not the same as a confidence level, but the term confidence level is frequently associated with tolerance intervals.)

tolerance limits—the two statistics that define a tolerance interval. (One value may be $-\infty$ or $+\infty$.)

treatments—combinations of levels of factors chosen for the experimental design.

TABLE 1 Examples of Determination of Precision

Degrees of Freedom	Percent Confidence Level	Confidence Limits for Estimation of
4	90	$0.422\ s^2 < \sigma^2 < 5.62\ s^2$
	95	$0.358\ s^2 < \sigma^2 < 8.26\ s^2$
9	90	$0.532\ s^2 < \sigma^2 < 2.71\ s^2$
	95	$0.474\ s^2 < \sigma^2 < 3.33\ s^2$
19	90	$0.629\ s^2 < \sigma^2 < 1.88\ s^2$

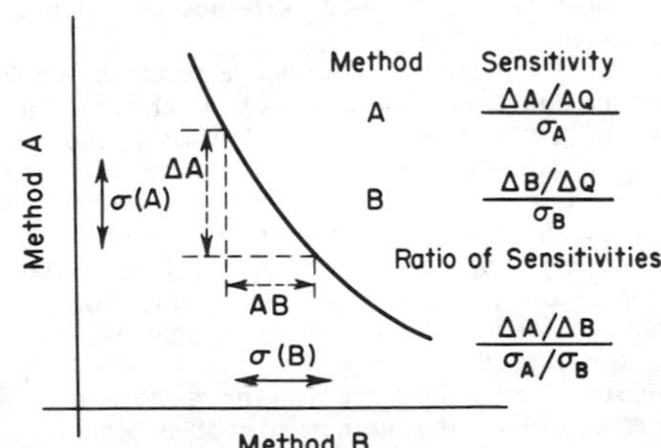

FIG. 1 Comparison of Two Methods for Measurement of Property *Q*.

Designation: D 2236 – 69

Standard Method of Test for DYNAMIC MECHANICAL PROPERTIES OF PLASTICS BY MEANS OF A TORSIONAL PENDULUM[1]

This Standard is issued under the fixed designation D 2236; the number immediately following the designation indicates the year of original adoption or, in the case of revision, the year of last revision. A number in parentheses indicates the year of last reapproval.

1. Scope

1.1 This method covers determination of the elastic and nonelastic components of the complex shear modulus of plastics of logarithmic decrement less than or equal to 1. The method is known to be valid over a range of frequencies from at least 0.1 to 10.0 cP.

1.2 The method is intended for materials that have tensile moduli in the range from 3.5×10^2 to 1.4×10^4 MN/m^2 (5×10^4 to 1×10^6 psi), as determined by ASTM Method D 638, Test for Tensile Properties of Plastics.[2]

1.3 The scope of the method does not include tests in a region where the logarithmic decrement of a material is greater than a value of 1. This limitation is not a function of the technique, but will depend on the mass of the particular specimen and the apparatus inertial member and gripping arrangement. The test has been shown to be valid and useful where the logarithmic decrement is greater than 1 and for materials of tensile modulus up to 3×10^4 MN/m^2 (5×10^6 psi).

NOTE 1—The values stated in U.S. customary units are to be regarded as the standard. The metric equivalents of U.S. customary units may be approximate.

2. Summary of Method

2.1 This method is intended to provide a means of determining the elastic and loss modulus of plastics over a range of temperatures by a free decay technique at about 1 Hz (cycle per second). The specimen is an integral part of an inertial system, thus determining the frequency and rate of decay of the system from which moduli may be determined. These moduli are a function of temperature and frequency in plastics and change rapidly at particular temperatures in the torsion pendulum analysis. The temperatures of rapid moduli change are normally referred to as transition temperatures.

3. Significance

3.1 The elastic component, G', and nonelastic component, G'', of the complex shear modulus, G^*, are two of the fundamental constants of any viscoelastic material. These two constants largely determine the dynamic mechanical performance of the material in shear provided that the maximum strain is relatively low. They are defined by the equation: $G^* = G' + iG''$. G^* is referred to as the complex shear modulus, G' the elastic shear modulus, and G'' the loss modulus of a material. The ratio G''/G' is a measure of the capacity of a viscoelastic material to damp mechanical vibrations.

4. Apparatus

4.1 The purpose of the apparatus is to hold a plastic specimen, either rectangular or circular in cross section, so that the specimen acts as the elastic element in a torsional pendulum.

---

[1] This method is under the jurisdiction of ASTM Committee D-20 on Plastics. A list of committee members may be found in the ASTM Yearbook. This tentative is the direct responsibility of Subcommittee D-20.10 on Mechanical Properties.

Current edition effective Dec. 19, 1969. Originally issued 1964. Replaces D 2236 – 64 T.

4.2 The apparatus shall consist of the following (see Fig. 1):

4.2.1 *Fixed Clamp*—Upper clamp (see 4.4), rigidly fixed, with space below to accommodate specimens anywhere from 25 to 150 mm (1 to 6 in.) long between clamps.

4.2.2 *Inertia Member Clamp*—Lower clamp mounted rigidly to the inertia member.

4.2.3 *Inertia Member*—May be appropriately either a disk or a rod. The inertia member shall be equipped so that oscillations can be counted to determine the frequency of the pendulum. The inertia member shall be suitably equipped to measure the amplitude of the oscillations. The moment of inertia of the inertia member must be known. The total weight of the inertia member or net tension on the upper support shall be such that it will exert a static tensile stress on the plastic specimen of not more than 0.7 MN/m^2 (100 psi).

4.2.4 *Weights*—The rod may have weights attached toward its end in order to increase the moment of inertia for a given total weight of member, and these weights may conveniently be capable of adjustment along the rod in order to vary the moment of inertia of the member.

4.3 The assembly shall hold the specimen along a vertical axis which is common also to the two clamps so that the inertia member will rotate in a horizontal plane.

4.4 Alternatively, the fixed clamp may be in the lower position; in this case, the upper clamp shall be rigidly mounted to the inertia member which shall be attached in such a way as to have a known or negligible torsional effect (see Arrangement B of Fig. 1).

5. Test Specimen

5.1 This method is intended for use with specimens of the following sizes:

5.1.1 *Rectangular Specimens:* 0.38 to 2.5 mm (0.015 to 0.10 in.) in thickness, 2.5 to 15 mm (0.10 to 0.60 in.) in width, and 25 to 150 mm (1 to 6 in.) in length (exclusive of material in the grips).

5.1.2 *Cylindrical Specimens:* less than 7.6 mm (0.30 in.) in radius and 25 to 150 mm (1 to 6 in.) in length (exclusive of material in the grips).

5.2 The specimen shall be of such length that at least 6 mm ($^1/_4$ in.) is held in each grip. Variation in thickness throughout the length of the specimen between grips shall not exceed ±3 percent of the average thickness.

NOTE 2—Due to the numerous types of torsional pendulums and their varying inertia moments, it would be difficult to recommend a typical specimen. However, in many cases, a specimen of 0.75 by 9.4 by 50 mm (0.03 by 0.38 by 2.0 in.) was found to be usable and convenient. Dynamic results can be influenced by specimen thickness if the thickness is not appropriate to the modulus of the material.

6. Conditioning

6.1 *Conditioning*—Condition the test specimens at 23 ± 2 C (73.4 ± 3.6 F) and 50 ± 5 percent relative humidity for not less than 40 h prior to test in accordance with Procedure A of ASTM Methods D 618, Conditioning Plastics and Electrical Insulating Materials for Testing,[2] for those tests where conditioning is required. In cases of disagreement, the tolerances shall be ±1 C (±1.8 F) and ±2 percent relative humidity.

6.2 *Test Conditions*—Conduct tests in the Standard Laboratory Atmosphere of 23 ± 2 C (73.4 ± 3.6 F) and 50 ± 5 percent relative humidity, unless otherwise specified in the test methods or in this specification. In cases of disagreement, the tolerances shall be ±1 C (±1.8 F) and ±2 percent relative humidity.

7. Procedure

7.1 Measure the length, width, and thickness with an accuracy consistent with determining G' with an accuracy of ±5 percent. Measure the length and diameter of cylindrical specimens.

7.2 Cause an initial angular displacement of not more than 2.5 deg/cm of specimen length. It is important that the inertia member rotate in a horizontal plane. This requires in the first place that the inertia member itself be evenly balanced right and left of the clamp. This requires also that the horizontal position of the member not be disturbed in setting the member in motion.

7.3 The test frequency shall be from 0.1 to 10.0 Hz. It is expected that in this test the dynamic properties will be constant for a given frequency of testing, even though strain rates will be different for different dimensions of specimens.

---

[2] *Annual Book of ASTM Standards*, Part 27.

7.4 Determine the time for ten cycles or the maximum number practical. Record the average frequency to the nearest 0.050 Hz or at least three significant figures.

7.5 Determine the amplitude, A, of 6 successive cycles or the maximum number practical and record the average value of A_n/A_{n+1}, the ratio of successive amplitudes.

7.6 Measure and record the temperature at test (Notes 2, 3, and 4). Indicate accuracy and precision of measurement of the sample temperature.

NOTE 3—Preferably tests conducted over a temperature range should be performed in incremental steps or at a rate slow enough to allow equilibration of temperature throughout the entire specimen. The time to equilibrium will depend on the mass of the particular specimen and gripping arrangement.

NOTE 4—The temperature and frequency characteristics of plastics are well known.[3] The rate of heating (or cooling) a specimen is a frequency effect, so it must be treated as such and reported. In addition, the rate of change of mechanical properties increases rapidly in the transition region so that many testers have found it advisable to decrease the rate of change of temperature and to decrease the test temperature intervals in this region to allow for more accurate determinations. Such changes in rate should also be reported. Temperature intervals from 5 C in the glassy region to 2 C in the transition region have been found to be convenient.

8. Calculations

8.1 In the calculations, use the average values from the data for frequency, length, width, and ratio of amplitudes. Use the average value of t^3 in calculations. For values of μ, see Table 1.

8.2 Calculate the logarithmic decrement, Δ, as follows:

$$\Delta = \ln(A_1/A_2)$$

where A_1 and A_2 = the amplitudes (in degrees of rotation) of two successive oscillations.

8.3 Calculate the elastic shear modulus, G', as follows:

NOTE 5—The equations given are approximations of the type equation:

$$G' = (64\pi^2 ILf^2/\mu bt^3)[1 + (\Delta^2/4\pi^2)]$$

This equation is necessary for very high damping ($\Delta > 1$). The reason the term $(1 + (\Delta^2/4\pi^2)$ is omitted is due to the limit in the scope to damping as described in 1.3.

NOTE 6—The effects of gravity and the tensile stress on the specimen are neglected. References for correction factors for high tensile stress are given in the literature.[3,4]

8.3.1 For specimens of rectangular cross section:

8.3.1.1 For G' in dynes per square centimeter and dimensions in centimeters (Note 5):

$$G' = 64\pi^2 ILf^2/\mu bt^3$$

8.3.1.2 For G' in pounds per square inch and dimensions in inches:

$$G' = (ILf^2/\mu bt^3) \times 5.591 \times 10^{-4}$$

8.3.2 For specimens of circular cross section:

8.3.2.1 For G' in dynes per square centimeter and dimensions in centimeters:

$$G' = 8\pi ILf^2/r^4$$

8.3.2.2 For G' in pounds per square inch and dimensions in inches:

$$G' = (ILf^2/r^4) \times 2.22 \times 10^{-5}$$

where:

I = moment of inertia of the inertia member, $g \cdot cm^2$,
f = frequency of oscillation, Hz,
L = length of specimen,
b = width of specimen,
t = thickness of specimen,
r = radius of specimen, and
μ = shape factor for rectangular cross sections; values are given in Table 1.

8.4 Calculate the loss modulus, G'', as follows (Note 5):

$$G'' = (\Delta/\pi)G'$$

9. Report

9.1 The report shall include the following:

9.1.1 Complete identification of the material, including name, stock or code number, date made, form, etc.,

9.1.2 Dimensions of the test specimen and the shape factor, μ,

9.1.3 Frequency and temperature of test,

9.1.4 Average rate of temperature rise (where applicable) or elapsed time at incremental temperature,

9.1.5 Details of conditioning the specimen prior to test,

9.1.6 Table of data and results,

9.1.7 Number of specimens tested,

9.1.8 Average values and standard devia-

[3] Nielsen, L. E., *Mechanical Properties of Polymers*, Reinhold Publishing Corp., New York, N. Y., 1962.

[4] Timoshenko, S., *Strength of Materials*, Part I, D. Van Nostrand Co., Inc., New York, N. Y., 1955.

tion of G', Δ, and G'', reported to two significant figures,

9.1.9 Tensile load on specimen, and

9.1.10 Date of test.

10. Precision[5]

10.1 The precision of this technique was determined from the results of round robins in which 15 laboratories participated. Interlaboratory precision was: for G' in the glassy region, ±7 percent; for G' in the glass transition region, ±30 percent; for G'' and Δ above the glass transition temperature, ±20 percent; and for G'' and Δ below the glass transition temperature, ±10 percent. Utilizing G', G'', or Δ, the glass transition temperature could be determined to within ±3 C. The values for intralaboratory precision were about half those for interlaboratory precision.

NOTE 7—The glass transition temperature was determined by the point of inflection of the log G' versus temperature curve or the maximum of the G'' or Δ versus temperature curve associated with the glass transition.

[5] The round-robin data are on hand at ASTM Headquarters in Research Reports file No. RR 43:D20.

TABLE 1 Values of Shape Factor, μ

Ratio of Specimen Width to Thickness	Shape Factor, μ	Ratio of Specimen Width to Thickness	Shape Factor, μ
1.00	2.249	3.50	4.373
1.20	2.658	4.00	4.493
1.40	2.990	4.50	4.586
1.60	3.250	5.00	4.662
1.80	3.479	6.00	4.773
2.00	3.659	7.00	4.853
2.25	3.842	8.00	4.913
2.50	3.990	10.00	4.997
2.75	4.111	20.00	5.165
3.00	4.213	40.00	5.232

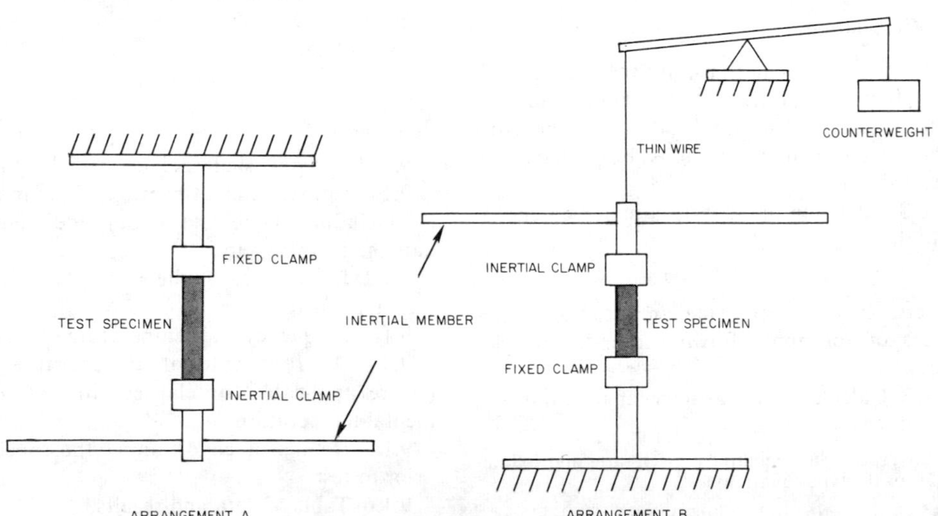

FIG. 1 Schematic Diagrams of Torsional Pendulum Apparatus.

Designation: D 2238 - 68

Standard Method of Test for
ABSORBANCE OF POLYETHYLENE DUE TO METHYL GROUPS AT 1378 CM^{-1} [1]

This Standard is issued under the fixed designation D 2238; the number immediately following the designation indicates the year of original adoption or, in the case of revision, the year of last revision. A number in parentheses indicates the year of last reapproval.

1. Scope

1.1 These methods cover measurement by infrared absorption spectrophotometry of the 1378.4-cm^{-1} (7.255-μm) band in polyethylene due to methyl groups. Two methods are covered:

1.1.1 *Method A* uses compensation with a standard sample film or wedge of known methyl content.

1.1.2 *Method B* uses compensation with a wedge of polymethylene or a polyethylene of known low methyl content.

1.2 These methods are applicable to polyethylenes of Types I (density 0.910 to 0.925 g/cm^3), II (density 0.926 to 0.940), and III (density 0.941 to 0.965).

NOTE 1—For determination of density, see ASTM Specifications D 1248, for Polyethylene Plastics Molding and Extrusion Materials.[2]

NOTE 2—In cases of Type III polyethylene with densities greater than 0.950 g/cm^3, different results are obtained with the two methods.

2. Significance

2.1 When interpreted with the aid of appropriate calibration data, either method can be used to compare the total methyl contents of polyethylenes made by similar processes. Such information can be interpreted in terms of specific alkyl groups with the aid of data on infrared absorption at certain other wavelengths (**3**).[3]

NOTE 3—The accuracy of determination of the concentration of total alkyl groups depends on knowing the concentrations of methyl and ethyl branches present, since these branches have anomalously high absorptivities per group at 1378 cm^{-1} (7.25 μm).

2.2 Knowledge of total methyl groups in polyethylene, when combined with data on molecular weight and on reactive end groups such as vinyl, can lead to assignment of end-group structures and can shed light upon polymerization mechanisms.

2.3 Data on total methyl groups in polyethylene can be correlated qualitatively with certain polymer properties such as melting point, density, stiffness, and other mechanical properties that are closely dependent on the degree of crystallinity of the polymer.

2.4 These methods are especially suitable for research. They have not been tested for use in manufacturing control.

3. Interferences

3.1 Compensation minimizes interference from methylene group absorption bands at 1368 cm^{-1} (7.31 μm) and 1352 cm^{-1} (7.39 μm) with the 1378-cm^{-1} (7.255 μm) methyl deformation band.

3.2 In Method A residual absorption is often present at 1352 cm^{-1} after compensation, but this band is believed not to contribute appreciable interference in the measurement of the methyl peak at 1378.4 cm^{-1} in samples with very low methyl content.

4. Apparatus

4.1 *Infrared Spectrophotometer*, double-beam, with rock salt prism, and spectral resolution as defined by Condition C in Part III (Spectral Resolution) of the Proposed Meth-

[1] These methods are under the jurisdiction of ASTM Committee D-20 on Plastics. A list of committee members may be found in the ASTM Yearbook. This standard is the direct responsibility of Subcommittee D-20.70 on Analytical Methods.

Current edition effective Sept. 9, 1968. Originally issued 1964. Replaces D 2238 - 64 T.

[2] *Annual Book of ASTM Standards*, Part 26.

[3] The boldface numbers in parentheses refer to the list of references at the end of these methods.

ods for Evaluation of Spectrophotometers.[4,5]

4.2 *Compression-Molding Press*, small, with platens capable of being heated to 170 C.[6]

4.3 *Metal Plates* approximately 150 by 150 by 0.5 mm with smooth surfaces.

4.4 *Brass Shims* approximately 75 by 75 mm or larger with an aperture in the center at least 25 by 38 mm in a series of at least five thicknesses from 0.1 to 0.5 mm.

4.5 *Micrometer Calipers* with thimble graduations of 0.001 mm.[7]

4.6 *Mounts* for film specimens with aperture at least 6 by 27 mm.

5. Sampling

5.1 The polyethylene shall be sampled in accordance with ASTM Recommended Practice D 1898, for Sampling of Plastics.[8]

6. Preparation of Apparatus

6.1 The precision obtained using this method depends very markedly upon the condition of the spectrophotometer. Instrument performance should be at least equal to that cited in the manufacturer's specifications for the new instrument. Resolution should be checked to assure conformance with 4.1. The linearity of the photometric system should be measured; linearity should not deviate from absolute by more than 4 percent of the transmittance range of interest. Frequency or wavelength in the 1430 to 1250-cm^{-1} (7 to 8-μm) region should be calibrated.

NOTE 4—For wavelength calibration, it is helpful to record the spectrum of water vapor upon the spectra of the samples.

7. Conditioning

7.1 *Conditioning*—Condition the test specimens at 23 ± 2 C (73.4 ± 3.6 F) and 50 ± 5 percent relative humidity for not less than 40 h prior to test in accordance with Procedure A of ASTM Methods D 618, Conditioning Plastics and Electrical Insulating Materials for Testing,[8] for those tests where conditioning is required. In cases of disagreement, the tolerances shall be ±1 C (±1.8 F) and ±2 percent relative humidity.

7.2 *Test Conditions*—Conduct tests in the Standard Laboratory Atmosphere of 23 ± 2 C (73.4 ± 3.6 F) and 50 ± 5 percent relative humidity, unless otherwise specified in the test methods or in this specification. In cases of disagreements, the tolerances shall be ±1 C (±1.8 F) and ±2 percent relative humidity.

METHOD A. MEASUREMENT OF THE ABSORBANCE AT 1378 CM^{-1} (7.25 μm) BY A STANDARD SAMPLE COMPENSATION METHOD

8. Materials

8.1 *Aluminum Foil.*

8.2 *Crushed Ice.*

8.3 *Reference Wedge or Films*, prepared as described in 9.2.1.

9. Calibration and Standardization

9.1 *Calibration of Reference Polymer by a Self-Compensation Method*—Mold a 0.5 mm film of annealed high-density polyethylene, as well as a series of thinner, shock-cooled films of the same polymer over a range of thickness from 0.1 to 0.4 mm (Note 5). Measure a series of difference spectra, with the annealed film in the sample beam of the spectrophotometer and each shock-cooled film, in turn, in the reference beam. From a graph of absorptivity of the CH_3 band maximum at about 1378 cm^{-1} (7.25 μm) as a function of absorptivity at 1304 cm^{-1} (7.67 μm), obtain a corrected value of absorptivity at 1378 cm^{-1} (7.25 μm) as well as the slope of the graph.

NOTE 5—The polyethylene used for preparation of reference films should have very low methyl group content, preferably less than 0.3 for each 100 carbon atoms. Essentially linear Type III polyethylene with density approximately 0.96 g/cm^3 has been found satisfactory for this purpose (Note 1).

9.2 *Procedure:*

9.2.1 From the reference polyethylene, mold three or four shock-cooled films about 0.5 mm in thickness and a number of films with thicknesses varying from 0.1 to 0.4 mm. The films shall be smooth and free of voids.

---

[4] *Proceedings*, Am. Soc. Testing Mats., Vol 58, 1958, p. 472.

[5] The Perkin-Elmer Model 21 Spectrophotometer, Beckman IR-4 and IR-7, Hilger H-800, and Grubb-Parsons GS-2A Spectrophotometers are capable of this degree of resolution and have been found satisfactory for this purpose.

[6] Hydraulic presses that have been found satisfactory for this purpose are made by Pasadena Hydraulics, Inc., Pasadena, Calif., and Fred S. Carver, Inc., Summit, N. J.

[7] Brown and Sharpe micrometer No. 223RS has been found satisfactory for this purpose.

[8] *Annual Book of ASTM Standards*, Part 27.

The shock-cooled films may be prepared in the following way: Place the desired brass shim on the aluminum foil on top of one of the metal plates. Place sufficient polymer in the aperture of the shim to fill completely this aperture after pressing. Cover the preparation with a second aluminum foil and metal plate. Heat the press to 170 C. Insert the mold assembly between the press platens. Preheat for 15 s, then apply pressure slowly until after 30 s the pressure has reached 30,000 lb. Hold the preparation at this pressure for an additional 30 s. Release the pressure, grasp the assembly with pliers, and quickly plunge it into a bucket containing a slurry of ice and water. Carefully remove thc film and dry it with a cloth or tissue.

9.2.2 Anneal several 0.5-mm shock-cooled films by a suitable press or oven technique to obtain an increase in density at 23 C of at least 0.020 g/cm^3.

9.2.3 Mount each film on a suitable holder. Measure the thickness in millimeters at three places in the aperture and record the average thickness on the sample holder. Measure the density of small clippings made close to but not in the aperture of the holder. Measure the density according to ASTM Method D 1505, Test for Density of Plastics by the Density Gradient Technique.[8]

9.2.4 Scan the spectrum between 11 and 13 μm and reject any film showing interference fringes.

9.2.5 Measure spectra in the range from 1430 to 1250 cm^{-1} (7 to 8 μm) and record the absorbance of the 1368-cm^{-1} (7.31-μm) band on each sample as follows: Place an annealed sample in the sample beam of the spectrophotometer. Place a shock-cooled film in the reference beam. Set the spectrophotometer to achieve the resolution specified in 5.1 (Notes 6, 7). Set the speed at 0.1 μm/min or 17 cm^{-1}/min. Set the gain to produce overshoot of 1 division (1 percent of full scale) at 1368 cm^{-1} (7.31 μm). Adjust the electrical balance so that there is no drift. Adjust the balance control so that no part of the spectrum between 1430 and 1250 cm^{-1} (7 and 8 μm) has more than 90 percent transmittance. Adjust the 0 percent transmittance. Adjust the abscissa scale to at least 75 mm/200 cm^{-1} (1.1 μm).

NOTE 6—The 1378-cm^{-1} (7.25-μm) methyl band has a half-width less than 6 cm^{-1} (0.03 μm) and is thus very sensitive to slit width in prism instruments. Close control of slit width is essential for precise measurements.

NOTE 7—For the Perkin-Elmer Model 21 Spectrophotometer, the following settings are generally satisfactory: Response 1 (electrical and mechanical), Suppression 2, Resolution 3 (960 program), electrical balance to give no drift. For the Beckman Model IR-4 or IR-7 Spectrophotometer, slits 0.40 mm, gain to give single beam/double beam ratio about 1 to 1368 cm^{-1} (7.31 μm), Response 2 s.

9.2.6 Run the spectrum from 1430 to 1250 cm^{-1} (7 to 8 μm) (see Fig. 1).

9.2.7 Return to 1430 cm^{-1} (7 μm) at full speed without releasing the drum or paper lock. Record a spectrum of water vapor on the paper below the polyethylene trace.

NOTE 8—For the Perkin-Elmer Model 21 Spectrophotometer, record the water vapor spectrum with a slit program of 927.

9.2.8 Repeat 9.2.5, 9.2.6, and 9.2.7 for the remaining shock-cooled films.

9.3 *Calculations:*

9.3.1 Draw a base line to each curve from 1396 to 1330 cm^{-1} (7.17 to 7.52 μm). Draw a second base line from 1330 to 1270 cm^{-1} (7.52 to 7.87 μm) on each spectrum.

9.3.2 Measure the absorbance of the methyl band at 1378.4 cm^{-1} (7.255 μm). Measure the absorbance at 1304 cm^{-1} (7.67 μm).

NOTE 9—Type I polyethylene may show weak bands between 7.3 and 8.0 μm which interfere slightly in this measurement.

9.3.3 Calculate the following quantities for each pair of films:

$$\alpha = A_{1378}/(d_s t_s - d_r t_r)$$

$$\beta = A_{1304}/(d_s t_s - d_r t_r)$$

where:

A_{1378} = absorbance at 1378 cm^{-1} (7.25 μm),
A_{1304} = absorbance at 1304 cm^{-1} (7.67 μm),
d_s = density of sample film, g/cm^3,
d_r = density of reference film, g/cm^3,
t_s = thickness of sample film, cm, and
t_r = thickness of reference film, cm.

9.3.4 Plot the quantity α as ordinate against β as abscissa on graph paper for each pair of films.

9.3.5 Draw the best straight line through the points on the graph (see Fig. 2). The intercept on the ordinate is the absorptivity, K'_{1378}, in square centimeters per gram, due to the methyl band at 1378.4 cm^{-1} (7.255 μm). Let

R, the slope of the line, $= \Delta\alpha/\Delta\beta$. A value of 0.5 for R at 1378.4 cm^{-1} is recommended.

NOTE 10—Provided that the 1378.4-cm^{-1} (7.255-μm) band is normal, the proportionality factor R for the correction is the same for all polyethylenes.

NOTE 11—The center of the 1378-cm^{-1} (7.25-μm) methyl band is normally found between 1379.3 and 1378.0 cm^{-1} (7.250 and 7.257 μm). Variation in position of the band center within this range will give a value of R between about 0.40 and 0.52, in direct proportionality.

9.3.6 For each reference film, correct the measured absorbance at 1378.4 cm^{-1}, A_{1378}, for absorption by amorphous bands at the same frequency by use of the equation

$$A_{1378} \text{ (due to methyl groups)} = A_{1378} - R \times A_{1304}$$

Record R and the corresponding A_{1378} (due to methyl groups) on each reference film mount.

9.4 *Factors:*

9.4.1 Factors, f_{1378}, in methyl groups for each 100 carbon atoms, in grams per square centimeter, are used to convert absorptivity data to methyl groups for each 100 carbon atoms by the following relationship:

$$\text{Methyl groups (calculated as methyl in alkyl groups} > C_3) = f_{1378\,(>C_3)} \times K'_{1378}$$

9.4.2 The conversion factors, f_{1378}, must be determined for each spectrophotometer.

NOTE 12—In the calculations, the Beer-Lambert law is assumed to apply, that is, the absorptivity at 1378 cm^{-1} (7.25 μm) is proportional to the concentration of methyl groups.

NOTE 13—The conversion factor, f_{1378}, has been derived from measurements of homopolymers of 1-olefins. In the laboratory of origin, using a Perkin-Elmer Model 21 Infrared Spectrophotometer, $f_{1378\,(>C_3)}$ has been determined to be equal to 0.110. In the absence of suitable polymers for calibration, this factor may be adopted for use with other instruments by means of standard reference samples. For example, *n*-cetane (*n*-hexadecane) of certified purity (NBS or equivalent) is scanned in an 0.025-mm cell as directed in the procedure. K'_{1378} for cetane is then calculated, and f_{1378} is determined as follows:

$$f_{1378} = (114/K'_{1378} \text{ for cetane}) \times 0.110$$

where 114 is the value of K'_{1378} for cetane in the reference instrument.

10. Procedure

10.1 Mold the polyethylene sample into shock-cooled films about 0.3 mm thick, as directed in 9.2.1.

10.2 Mount the films, measure thickness and density, and look for interference fringes, as directed in 9.2.4 and 9.2.5.

10.3 Place the sample film in the sample beam of the spectrophotometer and in the other beam a reference film chosen so that A_{1304} is less than 0.05 (Note 14). Set the spectrophotometer to the operating conditions listed in 9.2.6. Record the spectrum between 1430 and 1250 cm^{-1} (7 and 8 μm).

NOTE 14—Instead of flat reference films, a wedge molded from the reference polymer may be used if desired. Such a wedge is typically 100 mm long, tapering from 0.4 to 0.2 mm, and is mounted in a holder which allows continuous transverse adjustment of its position in the reference beam. Trial scans are performed to establish a wedge position that gives $A_{1304} < 0.05$, before recording the final spectrum.

11. Calculation

11.1 Draw base lines and measure absorbances as directed in 9.3.1 and 9.3.2.

11.2 Correct the 1378.4-cm^{-1} (7.255-μm) absorbance for incomplete compensation by subtracting $R \times A_{1304}$. At 1378.4 cm^{-1} (7.255 μm), $R = 0.5$ (see 9.3.5).

11.3 Add the corrected value of the observed 1378.4 cm^{-1} (7.255-μm) absorbance to the absorbance of the methyl band in the reference film (from 9.3.6) (Note 15). The result is the corrected absorbance of the sample film. Divide this value by the density in grams per cubic centimeter and by the thickness in centimeters to obtain the absorptivity, K'_{1378}, for the sample.

NOTE 15—If a wedge is used as reference, use as the absorbance of the methyl band in the reference the wedge correction, $d_w t_w K'_w$,

where:

d_w = density of wedge, g/cm^3,
t_w = wedge thickness, cm at center of beam, and
K'_w = absorptivity of the reference polymer, cm^2/g (9.3.5).

12. Report

12.1 The report shall include the following:

12.1.1 Complete identification of the material tested, including name, manufacturer, lot code number, and physical form when sampled,

12.1.2 Date of the test,

12.1.3 Method used,

12.1.4 Densities of the films used in the measurements,

12.1.5 Absorptivity values at 1378.4 cm^{-1} (7.255 μm),

12.1.6 Total methyl groups (calculated as methyl in alkyl groups greater than C_3), and

12.1.7 Any unusual absorption bands noted in the spectra.

13. Precision and Accuracy

13.1 In a single laboratory, the measurement of the absorptivity of the 1378.4-cm^{-1} (7.255-μm) band is precise to 0.1 g/cm^2 or 4 percent of K'_{1378}, whichever is greater.

13.2 In an interlaboratory study by six laboratories, the coefficient of variability was 4 percent for two samples of Type III polyethylene, 2 percent for a sample of Type II, and 15 percent for a sample of Type I.

METHOD B. MEASUREMENT OF THE ABSORBANCE AT 1378 CM^{-1} (7.25 μM) BY A WEDGE COMPENSATION METHOD

14. Apparatus

14.1 See Section 4.

14.2 *Wedge Holder*—Framework which can be mounted transversely in the reference beam of the spectrophotometer and adjusted continuously to place any thickness of the wedge in the beam.

15. Materials

15.1 *Aluminum Foil.*

15.2 *Wedge:*

15.2.1 *Preparation*—A wedge is prepared from the reference polymer by pressing in a special mold (7). Typical wedge dimensions are 100 mm long, tapering from 0.4 to 0.2 mm.

15.2.2 *Calibration*—Measure the thickness of the wedge at small intervals (about 3 mm apart) and mark the thickness on the holder.

NOTE 16—If it is suspected that the reference polymer contains significant methyl group absorptivity at 1378 cm^{-1} (7.25 μm), it should be determined using the principle of self-compensation (9.1) but with the base line and wedge adjustment of Method B. An annealed film of reference polymer should therefore be prepared as the "sample film" and the procedure of 16.1 through 16.5 and the calculations of 17.1 and 17.2 used, modifying the calculation of 17.3 to read

$$K'_{1378} = A_{1378}/(d_s t_s - d_r t_r),$$

where subscripts s and r refer to sample and reference films respectively.

16. Procedure

16.1 Mold the polyethylene sample into films having the following thicknesses:

16.1.1 *Type I and II*—Thickness about 0.10 to 0.15 mm.

16.1.2 *Type III*—Thickness about 0.30 to 0.40 mm.

Use a procedure similar to that in 9.2.1 except, after the pressure is released on the press, shut off the heat and leave the preparation in the press. When the platen temperature has fallen below the melting point of the polymer, remove the preparation and separate the film.

16.2 Mount the film, measure its thickness, scan the spectrum, and reject any film showing interferences fringes, as directed in 9.2.4 and 9.2.5.

NOTE 17—For the Perkin-Elmer Model 21 and the Beckman Model IR-4 or IR-7 Spectrophotometer, use the settings given in Note 7. For the Hilger H-800 Spectrophotometer, the following settings have been used: slits 0.1 mm at 3 μm and program 40, speed 132 min/μm, gain and damping adjusted to provide optimum servo response for the particular sample. For the Grubb-Parsons GS-2A Spectrophotometer, the following settings have been used: slits program 10, speed 4 min/μm, gain and damping adjusted to optimum as above.

16.3 Place the sample film in the sample beam of the spectrophotometer and the wedge (the thinnest part) in the reference beam.

16.4 Adjust the wedge in progressive steps with manual scanning until the desired degree of compensation is achieved. This is achieved by alternately setting the wavelength to 1368 cm^{-1} (7.31 μm) and to 1400 cm^{-1} (7.14 μm). When the absorbance is the same at these wavelengths, the wedge is correctly placed.

16.5 Record the spectrum of the 1378-cm^{-1} (7.25-μm) band from 1343 to 1400 cm^{-1} (7.45 to 7.14 μm).

17. Calculations

17.1 Draw a baseline tangent to the recorded curve over the interval from 1343 to 1400 cm^{-1} (7.45 to 7.14 μm) (see Fig. 3).

17.2 Measure the absorbance at the center of the methyl band, A_{1378} (Note 18).

NOTE 18—If a wedge correction is necessary, add to the absorbance of the 1378-cm^{-1} (7.25-μm) methyl band the wedge correction, $t_w K'_w$,

where:

t_w = wedge thickness, cm, and
K'_w = absorptivity of the reference polymer, 8.1, 8.2, and 8.3.1 to 8.3.5).

17.3 Calculate the absorptivity at 1378 cm^{-1} (7.25 μm). K'_{1378}, in square centimeters per gram as follows:

$$K'_{1378} = A_{1378}/(t \times d)$$

where:

A_{1378} = absorbance at 1378 cm^{-1} (7.25 μm) from 17.2,
t = thickness of the sample film, cm, and
d = density of the polymer, g/cm^3.

17.4 Calculate the total methyl content (calculated as methyls in alkyl groups > C_3) as follows:

$$\text{Methyl groups per 100 carbon atoms} = K'_{1378} \times f_{1378}$$

where:

K'_{1378} = absorptivity, as calculated in 17.3, and
f_{1378} = conversion factor, which must be determined for each spectrophotometer.

NOTE 19—An absolute calibration of an infrared method for methyl group determination is described in Reference (**5**). This method involves the preparation of branched deuterated polymethylenes. The calibration obtained in this way is expressed by the equation:

$$\text{Number of methyl groups (calculated as alkyl groups} > C_3\text{) per 100 carbon atoms} = 0.0765\ K'_{1378}$$

Instead of the calibration with the branched deuterated polymethylenes a value of 0.0765 may be adopted for use with other instruments by means of standard samples. For example, *n*-cetane (*n*-hexadecane) of certified purity (NBS or equivalent) is scanned in an 0.075-mm cell as directed in the procedure. K'_{1378} for *n*-cetane is then calculated, and f_{1378} is determined as follows:

$$f_{1378} = [120/(K'_{1378} \text{ for } n\text{-cetane})] \times 0.0765$$

where 120 is the value of K'_{1378} for *n*-cetane in the reference instrument.

18. Report

18.1 See 12.1.1 to 12.1.3 and 12.1.5 to 12.1.7. In addition, report whether methyl group absorptivity at 1378 cm^{-1} has been determined and corrected for (Note 16).

19. Precision and Accuracy

19.1 Precision in one laboratory is estimated to be about ±2 percent relative for methyl content = 3/100 carbon atoms and up to ±5 percent relative for methyl content = 0.3/100 carbon atoms.

19.2 In an interlaboratory study by ten laboratories, the coefficient of variability was 30 percent for one sample of Type III polyethylene, 13 percent for another sample of Type III, 16 percent for a sample of Type II, and 18 percent for a sample of Type I.

REFERENCES

(**1**) Bryant, W. M. D., and Voter, R. C., "The Molecular Structure of Polyethylene. II. Determination of Short-Chain Branching," *Journal of the American Chemical Society*, JACSA, Vol 75, 1953, pp. 6113–6118.

(**2**) Reding, F. P., and Lovell, C. M., "The Effect of Various Branch Lengths on the Crystallinity of Polyethylenic Resins," *Journal of Polymer Science*, JPSCA, Vol 21, 1956, pp. 157–159.

(**3**) Silas, R. S., "Bond Intensities of Methyl and Olefin Groups in Polyethylenes," Presented at Symposium on Molecular Structure and Spectroscopy, Ohio State University, Columbus, Ohio, June 11–15, 1956; abstract in *Spectrochimica Acta*, SPACA, Vol 8, 1956, p. 313.

(**4**) Boyd, D. R. J., Voter, R. C., and Bryant, W. M. D., "Identification and Measurement of Alkyl Groups in Branched Polyethylenes by Infrared Spectrometry," paper presented at 132nd Meeting, Am. Chemical Soc., New York, N. Y., Sept. 9, 1957.

(**5**) Willbourn, A. H., "Polymethylene and the Structure of Polyethylene: Study of Short-Chain Branching, Its Nature and Effects," *Journal of Polymer Science*, JPSCA, Vol 34, 1959, pp. 569–597.

(**6**) Oetgen, R. A., Kao, Chao-lan, and Randall, H. M., "The Infrared Prism Spectrograph as a Precision Instrument," *Review of Scientific Instruments*, RSINA, Vol 13, 1942, p. 515.

(**7**) Harvey, M. C., and Peters, L. L., "Polymer Wedge for the Infrared Determination of Methyl Groups in Polyethylene," *Analytical Chemistry*, ANCHA, Vol 32, 1960, p. 1725.

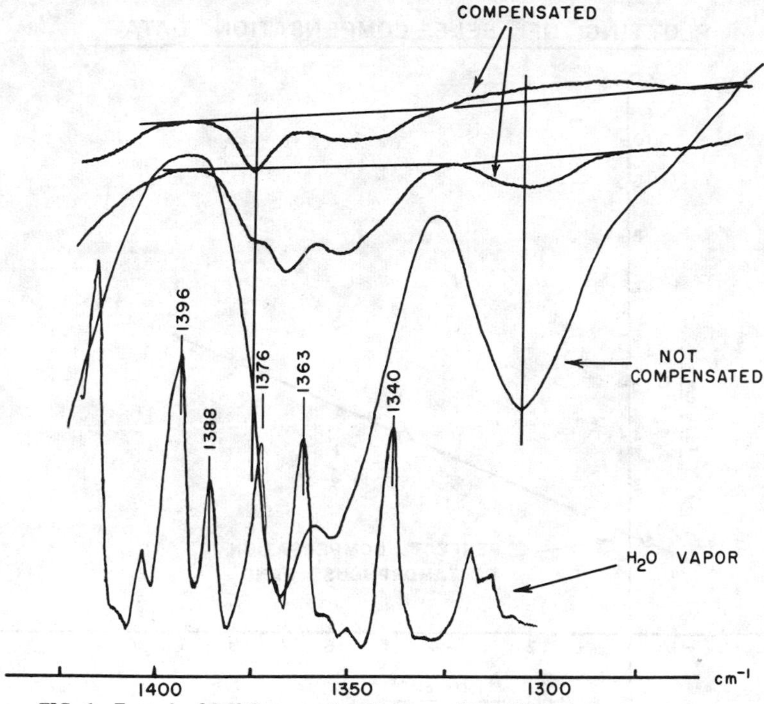

FIG. 1 Example of Self-Compensation Spectrum of Type III Polyethylene (Method A).

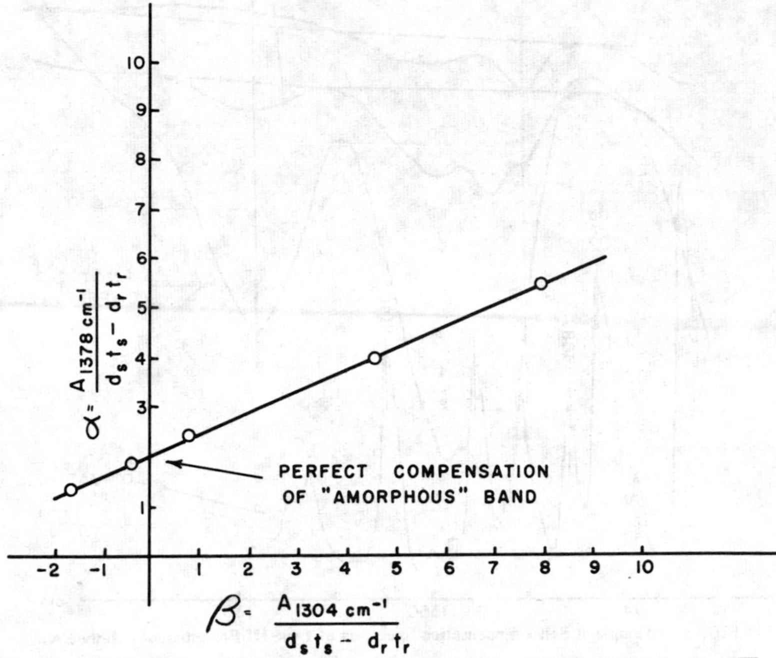

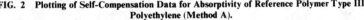

FIG. 2 Plotting of Self-Compensation Data for Absorptivity of Reference Polymer Type III Polyethylene (Method A).

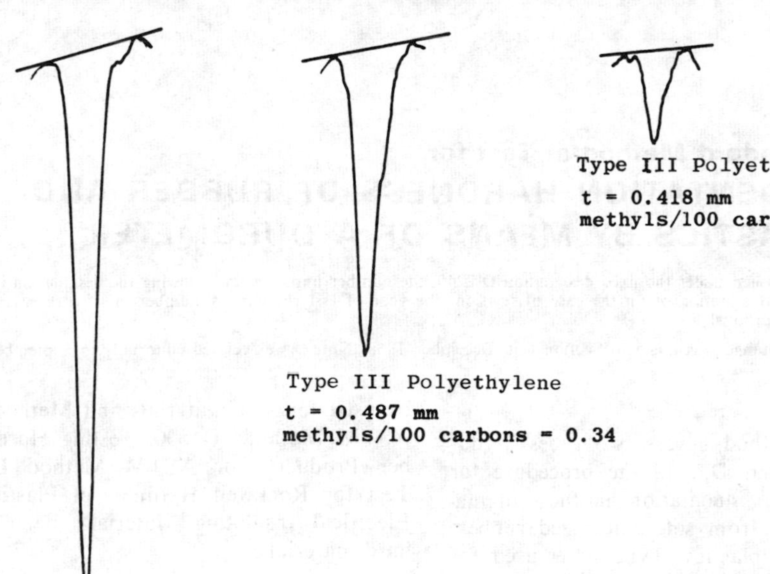

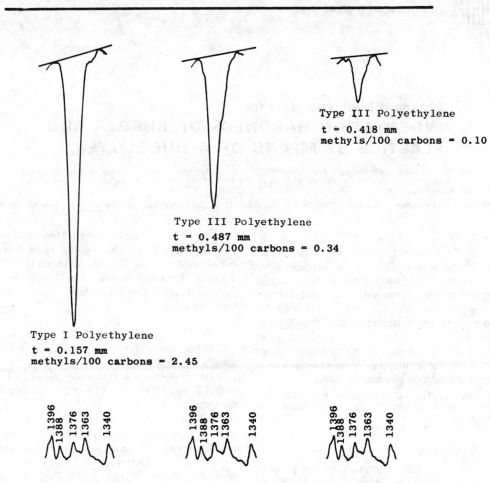

FIG. 3 Examples of Measuring Absorbance at 1378 cm^{-1} (7.25 μm) (Method B).

Designation: D 2240 – 68

Standard Method of Test for
INDENTATION HARDNESS OF RUBBER AND PLASTICS BY MEANS OF A DUROMETER[1]

This Standard is issued under the fixed designation D 2240; the number immediately following the designation indicates the year of original adoption or, in the case of revision, the year of last revision. A number in parentheses indicates the year of last reapproval.

NOTE—Editorial change was made to Section 6 in December 1968. Note 3 was revised editorially in September 1969.

1. Scope

1.1 This method covers two types of durometers, A and D, and the procedure for determining the indentation hardness of materials ranging from soft vulcanized rubbers to some rigid plastics. Type A is used for measuring the softer materials, and Type D for the harder materials (see Note 8, Section 7). This method permits measurements either of initial indentations or of indentations after specified periods of time, or both.

NOTE 1—The values stated in U.S. customary units are to be regarded as the standard. The metric equivalent of U.S. customary units given in the standard may be approximate.

2. Significance

2.1 This method is based on the penetration of a specified indentor forced into the material under specified conditions. The indentation hardness is inversely related to the penetration and is dependent on the elastic modulus and viscoelastic behavior of the material. The shape of the indentor and the force applied to it influence the results obtained so that there may be no simple relationship between the results obtained with one type of durometer and those obtained with either another type of durometer or another instrument for measuring hardness. This method is an empirical test intended primarily for control purposes. No simple relationship exists between indentation hardness determined by this method and any fundamental property of the material tested. For specification purposes, it is recommended that ASTM Method D 1415, Test for International Hardness of Vulcanized Rubbers,[2] be used for soft materials and Method A of ASTM Methods D 530, Testing Hard Rubber Products,[2] or ASTM Method D 785, Test for Rockwell Hardness of Plastics and Electrical Insulating Materials[3] be used for hard materials.

3. Apparatus

3.1 The durometer shall consist of the following components:

3.1.1 *Presser Foot* with a hole between 2.5 and 3.2 mm (0.10 and 0.13 in.) in diameter, centered at least 6 mm (0.25 in.) from any edge of the foot.

3.1.2 *Indentor* formed from hardened steel rod between 1.15 and 1.40 mm (0.045 and 0.055 in.) in diameter to the shape and dimensions shown in Fig. 1 for Type A durometers or Fig. 2 for Type D durometers.

3.1.3 *Indicating Device* on which the amount of extension of the point of indentor may be read in terms of graduations ranging from zero for full extension of 2.46 to 2.54 mm (0.097 to 0.100 in.) (Note 2) to 100 for zero extension obtained by placing presser foot and indentor in firm contact with a flat piece of glass.

NOTE 2—Type A Shore Durometers serial numbers 1 through 16,300 and 16,351 through 16,900 and Type A-2 Shore Durometers numbers 1 through 8077 do not meet the requirement of 2.46 to 2.54 mm (0.097 to 0.100 in.) extension of the

[1] This method is under the joint jurisdiction of ASTM Committee D-11 on Rubber and Rubber-Like Materials and Committee D-20 on Plastics. A list of members may be found in the ASTM Yearbook.

Current edition effective May 7, 1968. Originally issued 1964. Replaces D 2240 – 64 T.

[2] *Annual Book of ASTM Standards*, Part 28.

[3] *Annual Book of ASTM Standards*, Part 29.

indentor at zero reading. These durometers will give readings which are low by amounts ranging from 3 units at 30 hardness to 1 unit at 90 hardness.

3.1.4 *Calibrated Spring* for applying force to the indentor in accordance with one of the following equations:

$$\text{Force, g} = 56 + 7.66\, H_A \quad (1)$$

where H_A is the hardness reading on a Type A durometer, and

$$\text{Force, g} = 45.36\, H_D \quad (2)$$

where H_D is the hardness reading on a Type D durometer.

4. Test Specimen

4.1 The test specimen shall be at least 6 mm (0.25 in.) in thickness unless it is known that identical results are obtained with a thinner specimen (Note 3). A specimen may be composed of thinner pieces to obtain the necessary thickness, but determinations made on such specimens may not agree with those made on one-piece specimens because the surfaces between plies may not be in complete contact. The lateral dimensions of the specimen shall be sufficient to permit measurements at least 12 mm (0.5 in.) from any edge unless it is known that identical results are obtained when measurements are made at a lesser distance from an edge (Note 3). The surface of the specimen shall be flat over sufficient area to permit the presser foot to contact the specimen over an area having a radius of at least 6 mm (0.25 in.) from the indentor point. A suitable hardness determination cannot be made on a rounded, uneven, or rough surface.

NOTE 3—The minimum requirement for the thickness of the specimen is dependent on the extent of penetration of the indentor into the specimen; that is, thinner specimens may be used for materials having hardness values at the upper end of the scale. The minimum distance from edge at which measurements may be made likewise decreases as the hardness increases. For materials having hardness values above 50 Type D durometer, the thickness of the specimen should be at least 3 mm (0.12 in.) and measurements should not be made closer than 6 mm (0.25 in.) to any edge.

5. Calibration

5.1 The spring can be calibrated by supporting the durometer in a vertical position and resting the point of the indentor on a small spacer at the center of one pan of a chemical balance as shown in Fig. 3 in order to prevent interference between presser foot and pan (Note 4). The spacer shall have a small cylindrical stem approximately 2.5 mm (0.1 in.) in height and 1.25 mm (0.05 in.) in diameter, and shall be slightly cupped on top to accommodate the indentor point. Balance the weight of the spacer by a tare on the opposite pan of the balance. Add weights to the opposite pan to balance the force on the indentor at various scale readings. The measured force shall equal the force calculated by either Eq 1 within ±8 g or Eq 2 within ±45 g.

NOTE 4—Instruments specifically designed for calibration of durometers may be used. Zwick & Co., Control Equipment 7501, can be used for calibration as it is capable of measuring or applying a force on the point of the indentor within 0.4 g for a Type A durometer and within 2.0 g for a Type D durometer. Zwick Control Equipment 7501 with serial numbers higher than WA-20301 are satisfactory for this work. Instruments with lower serial numbers must be modified.[4]

6. Conditioning

6.1 Tests shall be made at 23 ± 2 C (73.4 ± 3.6 F) if the temperature of test is not specified. When tests are made at other temperatures, it is recommended that they be made at one or more of the standard temperatures given in ASTM Recommended Practice D 1349, for Standard Test Temperatures for Rubber and Rubber-Like Materials[2] or Procedure A of ASTM Methods D 618, Conditioning Plastics and Electrical Insulating Materials for Testing.[3] The durometer and specimens shall be conditioned at the temperature of test for at least 1 h before test for materials whose hardness is not dependent on the relative humidity (Note 5). For materials whose hardness is dependent on the relative humidity, the specimens shall be conditioned in accordance with Procedure A of Methods D 618 and tested at the same conditions.

NOTE 5—When a durometer is moved from a chamber below room temperature to a higher temperature, the durometer shall be placed in a suitable desiccator or airtight container immediately

[4] The standardization of one type of instrument is described by Lewis Larrick, "The Standardization of Durometers,". *Rubber Chemistry and Technology*, RCTEA, Vol 13, 1940, p. 969.

upon removal and allowed to remain there until the temperature of the durometer is above the dew point of the air in the new environment.

7. Procedure

7.1 Place the specimen on a hard, horizontal surface. Hold the durometer in a vertical position with the point of the indentor at least 12 mm (0.5 in.) from any edge of the specimen, unless it is known that identical results are obtained when measurements are made with the indentor at a lesser distance. Apply the presser foot to the specimen as rapidly as possible without shock, keeping the foot parallel to the surface of the specimen. Apply just sufficient pressure to obtain firm contact between presser foot and specimen.

NOTE 6—Better reproducibility may be obtained by using either a durometer stand or a weight centered on the axis of the indentor or both to apply the presser foot to the specimen. Recommended weights are 1 kg for the Type A durometer and 5 kg for the Type D durometer.

7.2 Unless otherwise specified, read the scale within 1 s after the presser foot is in firm contact with the specimen, unless the durometer has a maximum indicator, in which case the maximum reading is taken. If a reading after a time interval is specified, hold the presser foot in contact with the specimen without change in position or pressure and read the scale after the period specified.

NOTE 7—Durometers having only a maximum indicator cannot be used to obtain hardness values at various time intervals, nor in testing vinyl plastics which require the reading to be taken at 15 s.

7.3 Make five measurements of hardness at different positions on the specimen at least 6 mm (0.25 in.) apart and determine the median value or the arithmetic mean.

NOTE 8—It is recommended that measurements be made with the Type D durometer when values above 90 are obtained with the Type A durometer and that measurements be made with the Type A durometer when values less than 20 are obtained with the Type D durometer.

8. Report

8.1 The report shall include the following:

8.1.1 Complete identification of material tested,

8.1.2 Description of specimen, including thickness and number of pieces plied,

8.1.3 Temperature of test if other than 23 C, and relative humidity when hardness of material is dependent on humidity,

8.1.4 Type and manufacture of durometer,

8.1.5 Indentation hardness time interval at which reading was taken, and data designating basis, median or mean value (Note 9).

8.1.6 Date of test.

NOTE 9—Readings may be reported in the form: A/45/15 where A is the type of durometer, 45 the reading, and 15 the time in seconds that the pressure foot is in firm contact with the specimen. Similarly, D/60/1 indicates a reading of 60 on the Type D durometer obtained either within 1 s or from a maximum indicator.

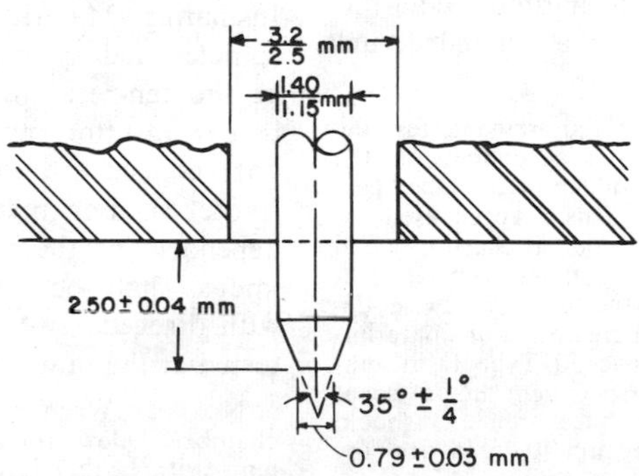

FIG. 1 Indentor for Type A Durometer.

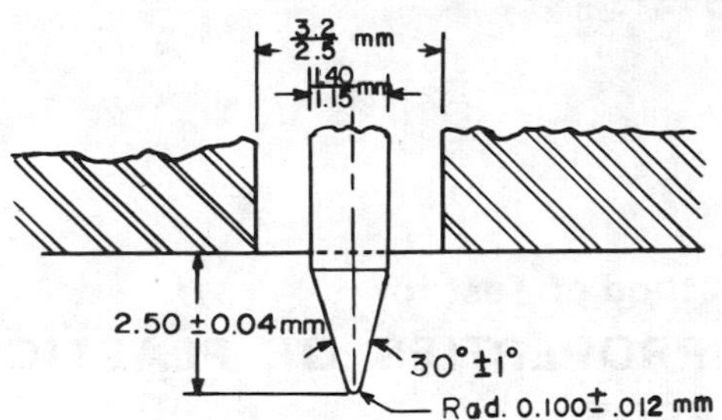

FIG. 2 Indentor for Type D Durometer.

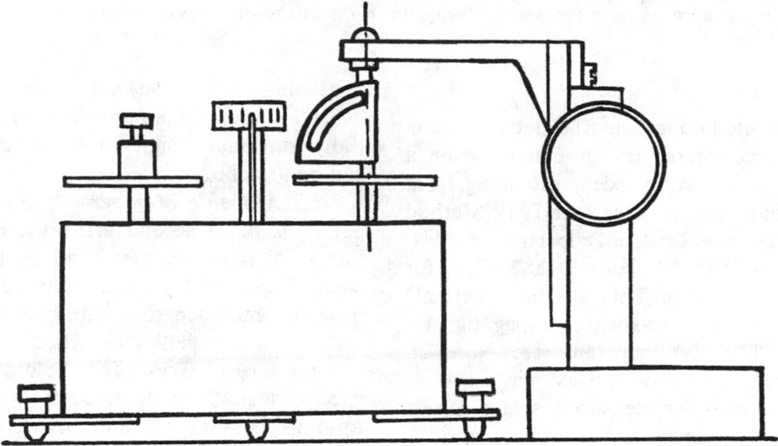

FIG. 3 Apparatus for Calibration of Durometer Spring.

ASTM Designation: D 2289 - 69

Standard Method of Test for
TENSILE PROPERTIES OF PLASTICS AT HIGH SPEEDS[1]

This Standard is issued under the fixed designation D 2289; the number immediately following the designation indicates the year of original adoption or, in the case of revision, the year of last revision. A number in parentheses indicates the year of last reapproval.

1. Scope

1.1 This method covers the determination of the tensile properties of plastics over a broad range of testing speeds extending from conventional speeds (see ASTM Method D 638, Test for Tensile Properties of Plastics,[2] and ASTM Methods D 882, Test for Tensile Properties of Thin Plastic Sheeting[2]) to those at which stress wave propagation effects may become important (see Note 2). This method normally utilizes grip displacement as a means for measuring strain. Direct strain measurement requires use of optical, photographic or strain gage systems.

NOTE 1—The values stated in U.S. customary units are to be regarded as the standard. The metric equivalents of U.S. customary units may be approximate.

NOTE 2—When the speed of test approaches the velocity of sound in a material, the distribution of load throughout the length of the test specimen is not always uniform. This may result in nonuniform straining of the specimen.

2. Significance

2.1 This method is designed to provide the user with a measure of rate sensitivity of a material by generating tensile data over a broad, usually logarithmic, time scale. Data on the response of a material to increases in strain rate may indicate its susceptibility to failure or reduction in energy absorption. Results thus provide an indication of engineering performance in applications where the load-time scale is the same as that specified in this method.

3. Apparatus

3.1 *Testing Machine*, usually hydraulic or pneumatic, of the uniform grip velocity type consisting of the following components:

3.1.1 *Fixed Member*—A fixed or essentially stationary member to which one grip can be attached.

3.1.2 *Movable Member*—A movable member to impel a second grip at a uniform velocity. A slack adapter shall be used in the case of piston type machines to allow the movable member to attain the desired test speed before impelling the grip.

3.1.3 *Grips*—Grips for holding the test specimen shall be constructed with faces as rigid as possible to minimize any effects of extension. The grips shall be aligned with the fixed and movable members. The axis of the test specimen shall coincide with the center line of the grip assembly.

3.1.4 *Drive Mechanism*—A drive mechanism for imparting to the movable member a uniform velocity with respect to the stationary member.

NOTE 3—Grip velocity is to be considered sufficiently uniform if it remains within ±20 percent of the specified speed throughout a single test. It is desirable to measure speed directly such as by a deflection versus time curve.

3.1.5 *Load Sensing System*—A load cell of sufficiently high natural resonance frequency shall be used together with a calibrating network for adjusting load sensitivity.

NOTE 4—Determine the natural frequency response of the load cell system including the load cell, extension rod, and grips. This can be deter-

---

[1] This method is under the jurisdiction of ASTM Committee D-20 on Plastics. A list of committee members may be found in the ASTM Yearbook. This standard is the direct responsibility of Subcommittee D-20.10 on Mechanical Properties.

Current edition effective Feb. 20, 1969. Originally issued 1964. Replaces D 2289 - 64 T.

[2] *Annual Book of ASTM Standards*, Part 27.

mined through utilization of the information in the Appendix.

3.1.6 *Grip Displacement Measurement System*—A means of monitoring the displacement of the moving grip or movable member of the test machine can be accomplished through the use of a suitable type transducer attached directly to the system. Photographic or optical systems can also be utilized. The accuracy should be within the sensitivity of the read-out apparatus used in the system; that is, within ±5 percent.

3.1.7 *Display and Recording Apparatus*—Any suitable means for displaying and recording the data developed from the load and displacement sensing systems can be used, provided its response characteristics are capable of presenting the data sensed with minimal distortion. Convenient apparatus includes a cathode ray oscilloscope used used together with a recording camera.

4. Test Specimens

4.1 Standard test specimens specified in ASTM Method D 1822, Test for Tensile Impact Energy to Break Plastics and Electrical Insulating Materials,[2] Method D 638, or Method D 882 may be used. (In the case of Method D 1822, the Type L specimen is preferred.) Care in preparation of these specimens should be taken as described in these methods. For thin film, a 12.7-mm (0.5-in.) strip has been found to be suitable for this method. The type of specimen used should be specified in the reporting of the data.

5. Conditioning

5.1 *Conditioning*—Condition the test specimens at 23 ± 2 C (73.4 ± 3.6 F) and 50 ± 5 percent relative humidity for not less than 40 h prior to test in accordance with Procedure A of ASTM Methods D 618, Conditioning Plastics and Electrical Insulating Materials for Testing,[2] for those tests where conditioning is required. In cases of disagreement, the tolerances shall be ±1 C (±1.8 F) and ±2 percent relative humidity.

5.2 *Test Conditions*—Conduct tests in the Standard Laboratory Atmosphere of 23 ± 2 C (73.4 ± 3.6 F) and 50 ± 5 percent relative humidity, unless otherwise specified in the test methods or in this specification. In cases of disagreement, the tolerances shall be ±1 C (±1.8 F) and ±2 percent relative humidity.

6. Number of Test Specimens

6.1 A minimum of five specimens should be tested at each speed specified with appropriate statistics developed to determine if additional tests should be conducted to provide more meaningful data.

NOTE 5—In the case of anisotropic materials it may be useful to test at least five specimens normal to, and at least five specimens parallel with, the principal axis of anisotropy at each speed specified.

6.2 Where dumbbell-type specimens are used, only breaks in the constant cross section are acceptable. Specimens that break at some obvious flaw shall be discarded and retests made, unless such flaws constitute a variable to be studied.

7. Speed of Testing

7.1 Speed of testing, which is the relative rate of motion of the grips during the test, shall be dictated by the specification for the material being tested, or by agreement between those concerned. When the speed of testing is not specified, tests at each of three speeds shall be used: 2.5, 25, and 250 m/min (100, 1000, 10,000 in./min).

NOTE 6—The number of test speeds shall be dictated by the viscoelastic response of the material being studied. Some materials exhibit a rapid change in deformation behavior at particular test speeds. When this is observed or suspected, data may be required at additional speeds to describe this behavior fully.

8. Procedure

8.1 Measure the width and thickness of the specimen to the nearest 0.025 mm (0.001 in.) at several points along the constant cross-sectional area portion. For film specimens 0.25 mm (0.010 in.) or less, thickness should be measured and reported to the nearest 0.0025 mm (0.0001 in.). Record the minimum values of cross-sectional area so determined.

8.2 Place specimens in the grips allowing for the specified gage length, taking care to properly align the axis of the specimen with the center line of the grips. Tighten the grips in such a way as to provide uniform clamping pressure and minimize slippage during testing. When testing film, if grip breaks are predominant, a reduction in gripping pressure should be tried. The gripping pressure selected should be that which produces maximum tensile stress and minimum slippage.

8.3 Set the speed of the equipment to the desired value.

8.4 Adjust the slack adapter, allowing slack for sufficient free travel so that the movable member can attain operating velocity before deforming the specimen.

8.5 Conduct test.

9. Calculations (See Notes 7 and 8)

9.1 *Tensile Strength*—Calculate the tensile strength by dividing the maximum load in kilograms-force (or pounds) by the original cross-sectional area of the specimen in square centimeters (or square inches) and report the value to three significant figures. When a yield stress or a breaking stress less than the maximum is present, it may also be desirable to calculate the corresponding tensile stress at yield or break and report the values to three significant figures. When the maximum stress occurs at the yield point it shall be designated tensile strength at yield. When the maximum stress occurs at break, it shall be designated tensile strength at break.

NOTE 7—If mechanical ringing, attributable to the test system rather than to the test specimen, is observed in any portion of the load-elongation curve, these oscillations should be smoothed out for purposes of the calculations. This is done by averaging the load at each peak and its subsequent valley and placing a point at this average load on the elongation scale at the position corresponding to that of the peak. A smooth curve is then drawn connecting these points. Since this averaging procedure may in some instances be a source of error, it is suggested that ringing be kept to a minimum wherever possible (see Appendix A1).

9.2 *Percentage Elongation*—For strip specimens calculate the percentage elongation by dividing the extension at the moment maximum load is attained by the original gage length and multiply by 100. Extension is the difference between the extended length and the original length. When a yield or breaking load less than the maximum is present and applicable, it may also be desirable to calculate both the percentage elongation at yield and the percentage elongation at break. For specimens other than strip specimens, values shall be reported in terms of grip displacement at maximum load, as well as grip displacement at yield and break, where applicable.

9.3 *Work*—Measure the area under the data curve to the maximum load, multiply by the appropriate factors for load and elongation, divide by the cross-sectional area of the specimen and report the normalized values as $kgf \cdot cm/cm^2$ ($ft \cdot lb/in.^2$). When a yield stress or breaking stress less than the maximum is present, the corresponding normalized work-to-yield and work-to-break values shall be reported. These values are comparable only for specimens of identical geometry.

9.4 For each series of tests calculate the arithmetic mean of all values obtained to three significant figures and report it as the "average value" for the particular property in question.

9.5 Calculate the estimated standard deviation as follows and report it to two significant figures:

$$s = \sqrt{(X^2 - n\bar{X}^2)/(n - 1)}$$

where:

s = estimated standard deviation,
X = value of single observation,
n = number of observations, and
$\bar{X}$ = arithmetic mean of the set of observations

NOTE 8—When displacement is measured by a linear output device, readout sensitivity should be selected so as to obtain an initial slope of load versus displacement approaching 45 deg in order to provide maximum reliability of the displacement value.

10. Precision

10.1 For a number of materials regardless of test speed the pooled coefficient of variation for molded specimens in round-robin work has been found to be 6 percent for tensile strength at yield, 4 to 8 percent for tensile stress at break, 18 to 24 percent for elongation at yield and 29 to 34 percent for elongation at break. Comparable values for film were 6 to 7 percent, 4 to 5 percent, 29 to 47 percent, and 14 to 24 percent respectively.

NOTE 9—The precision of the elongation measurement at yield can be improved by minimizing end effects (such as slippage) and when displacement is measured by means of a linear output device by adjusting the sensitivity of the displacement axis to obtain an initial slope of load versus deformation approaching 45 deg. Variability in elongation values at fail can be attributed to the mechanism of fracture as well as to end effects, but probably can be improved by using a direct measure of strain in place of grip separation. Since data are obtained over several decades of test time, however, the trends clearly delineate material response.

11. Report

11.1 The report shall include the following:

11.1.1 Complete identification of the material tested, including type, source, manufacturer's code numbers, form and previous history,

11.1.2 Type of test specimens and dimensions,

11.1.3 Method of preparing test specimens,

11.1.4 Conditioning procedure used,

11.1.5 Atmospheric conditions in test room,

11.1.6 Number of specimens tested,

11.1.7 Tensile strength, percentage elongation and work values and the corresponding values at yield and break, where applicable, for each test speed, and

11.1.8 Plots of the data above versus the logarithm of test speed.

NOTE 10—It may also be desirable in some instances to plot these properties versus log timc to yield and log time to break.

APPENDIX

A1. LOAD CELL RESPONSE AND RINGING

A1.1 Load Cell Response

A1.1.1 Reasonable fidelity of recording of a dynamic signal suggests that the natural frequency of the sensing system be at least twice the frequency of the signal to be recorded or, in terms of time, that the period of the recording system be less than $1/2$ the rise time of the signal to be measured **(1)**, **(2)**, **(3)**.[3] This is a more or less arbitrary guide to the selection of a load cell for general use. The application of this criterion to high-speed testing is as follows:

A1.1.2 Suppose a testing machine operates at a velocity of 24,000 cm/min or 400 cm/s and a specimen reaches maximum stress at 0.08 cm elongation. This would take place in 1/5000 s. It would be desirable that the force transducer (and associated attachments such as a grip) recording the signal have a period of 1/10,000 s or, in other words, a ringing frequency of 10,000 Hz. The frequency of a gage system is given by the formula.

$$f = \frac{1}{2\pi}\sqrt{K/M}$$

where:

f = frequency, s^{-1},

$K = \dfrac{\text{force on gage}}{\text{displacement of end of gage in dynes/cm}}$

and

M = the mass of moving part of the gage and attachments (dyne/980^2).

The ringing frequency of the gage system can be obtained experimentally by tapping the grip with a plastic hammer and recording the oscillatory trace on an oscilloscope.

A1.1.3 Besides frequency, the fidelity of recording depends also on damping. Ideally, the gage should be damped to a given percentage of its critical value. Filtration or other suitable means may be used to accomplish this (Note A1). The requirement for damping is not as important as the requirement for frequency, however. The usual situation for force gages is that they are underdamped. This results in overshoot and an oscillation about the value being measured. This situation is in general not severe and can be handled as suggested in the test method.

NOTE A1—Suggested literature values for load cell damping have ranged from 0.20 to 0.71 of its critical value. Additional experimental work is indicated before a more definite recommendation can be made in this regard.

A1.2 Ringing in Specimen

A1.2.1 Ringing may also take place in the specimen. In such a case, the frequency is determined by the length of the specimen and the velocity of sound in the specimen. The velocity of sound in a solid bar, V, depends in turn on the modulus of elasticity, E, and the density, d. The equation is:

$$V = \sqrt{E/d}$$

where:

V = expressed in cm/s,

E = dynes/cm^2, and

d = g/cm^3 (dyne · s^2/980 cm^4).

A value for the velocity of sound in a typical plastic is $2(10^5)$ cm/s. The time, t, in seconds, required for a stress wave to travel from end to end of a 5-cm specimen and back again is:

$$t = 10/(2 \times 10^5) = 5/10^5$$

This corresponds to a frequency of 20,000 Hz. The response of a 15,000-cycle gage to this frequency is relatively low, about 20 percent **(1)**.[3] The situation is further mitigated by the fact that damping in plastics is rather large. The two effects, that of high frequency and high damping, probably are sufficient to reduce the effects of specimen ringing to tolerable values for most plastics. In cases of

[3] The boldface numbers in parentheses refer to the list of references appended to this method.

materials with low damping, specimen ringing may impose a definite limitation on testing at high rates. The limitation is more deep seated than just the matter of recording. The ringing represents actual fluctuations in stress along the length of the specimen. The effects would be less severe with shorter specimens.

A1.3 Ringing Due to Bounce

A1.3.1 When a slack adaptor is used to allow the piston to attain operating velocity before the specimen is impelled, a ringing effect is sometimes observed (especially at the higher speeds) because at impact the movable grip may move faster than the piston or driving member. To minimize this noise an elastomeric buffer between metal mating surfaces in the ball and socket type slack adaptor has been shown to be rather effective. Also a slack adaptor containing two matched conical sections may be useful to produce "seizing" or "locking" (and thus prevent bouncing) during the test.

References

(**1**) Den Hartog, J. P., *Mechanical Vibrations*, 4th Ed., McGraw-Hill Co., New York, N. Y. 1956.

(**2**) Miller, Arthur, "The Right Angle," *Factors Affecting Accuracy of Oscillographic Recordings*, The Sanborn Co., Waltham 54, Mass., May, 1962.

(**3**) Grimminger, H., *Applied Polymer Symposia No. 1, 13/41*, John Wiley & Sons, New York, N. Y., 1965.

TABLE A1 Round-Robin Analysis of Data Test Speed Versus Pooled Coefficient of Variation[a]

For Tensile Properties of Two Molded Plastics (poly(vinyl chloride) and polycarbonate) $n = 20$ (4 labs, 5 each)

Test Speed, in./min	Break Strength, percent	Yield Strength, percent	Work To Break, percent	Elongation, percent	
				At Yield, percent	At Break, percent
50	8	6	. . .	29	24
5 000	4	6	. . .	34	18

For Tensile Properties of Three Films (polyester, polypropylene, and polystyrene).

Test Speed, in./min	Break Strength, percent	Yield Strength, percent	Work To Break, percent	At Yield, percent	At Break, percent
10	6	5	15	29	16
100	7	4	24	33	17
1 000	6	4	23	47	14
10 000	7	5	16	43	24
$n = 30$					
$n = 20$					

Pooled Coefficient of Variation

$$= \frac{\sqrt{\dfrac{(n_1 - 1)S_1^2 + \cdots + (n_i - 1)S_i^2 + \cdots + (n_m - 1)S_m^2}{(n_1 - 1) + \cdots + (n_i - 1) + \cdots + (n_m - 1)}}}{\dfrac{\sum_1^m \bar{X}_i}{m}} \times 100$$

where:

$S_1, S_2 \cdots S_m$ etc = standard deviation of test results at one test speed,

n_m = number of specimens tested in determining $\bar{X}_m$ and S_m,

$\bar{X}_i$ = average material property, and

m = number of materials at noted speed used for calculations.

[a] Supporting data for this method have been filed at ASTM Headquarters, as RR 58: D-20.

Designation: D 2292 - 68

Recommended Practice for
COMPRESSION MOLDING OF SPECIMENS OF STYRENE-BUTADIENE MOLDING AND EXTRUSION MATERIALS[1]

This Recommended Practice is issued under the fixed designation D 2292; the number immediately following the designation indicates the year of original adoption or, in the case of revision, the year of last revision. A number in parentheses indicates the year of last reapproval.

1. Scope

1.1 This recommended practice covers the compression molding of plaques or test specimens of styrene-butadiene molding and extrusion materials as defined in ASTM Specification D 1892, for Styrene-Butadiene Molding and Extrusion Materials.[2]

NOTE 1—The values stated in U.S. customary units are to be regarded as the standard. The metric eqiuvalents of U.S. customary units may be approximate.

2. Significance

2.1 The method by which samples are molded influences the mechanical properties of the specimen. The Izod impact test values for compression-molded specimens were found to be lower, but were less subject to variation than were values for injection-molded specimens. There is evidence of relationship between the properties of compression-molded specimens and those of extruded specimens.

3. Apparatus

3.1 *Molds:*

3.1.1 *Molding Chase*—A "picture-frame" compression-molding chase having a blanked-out area of suitable size (Note 2) and capable of producing a plaque 3.175 ± 0.13 mm (0.125 ± 0.005 in.) thick. Dimensions are shown in Fig. 1 of ASTM Method D 1693, Test for Environmental Stress-Cracking of Type I Ethylene Plastics[2] for a molding chase to produce 152.4 by 152.4-mm (6 by 6-in.) plaques.

NOTE 2—A 152.4 by 152.4-mm (6 by 6-in.) blanked-out section has been found satisfactory.

3.1.2 *Molding Plates*—Two polished chromium-plated ferrotype plates (dimensions shown in Fig. 1 of Method D 1693), such as used in photography, at least 1.0 mm (0.040 in.) thick, and of adequate surface area to cover the molding chase.

3.1.3 *Cavity Mold*—A cavity-type flash mold capable of producing a plaque of the specified dimensions may be used. Alternative test specimens may be molded to their finished dimensions, exclusive of the notch for impact strength, in a multi-cavity compression-molding chase.

3.2 *Press*—The hydraulic press shall be such that the molding pressure on the specimen can be maintained at the value, and within the tolerances, prescribed.

3.3 *Heating Systems*—Any convenient method of heating the press platens or molds may be used, provided that the heat source is constant enough to maintain the mold temperature within ±3 C (±5 F).

4. Conditioning

4.1 *Conditioning*—Condition the test specimens at 23 ± 2 C (73.4 ± 3.6 F) and 50 ± 5 percent relative humidity for not less than 40 h prior to test in accordance with Procedure A of ASTM Methods D 618, Conditioning Plastics and Electrical Insulating Materials

[1] This recommended practice is under the jurisdiction of ASTM Committee D-20 on Plastics. A list of committee members may be found in the ASTM Yearbook. This recommended practice is the direct responsibility of Subcommittee D-20.90 on Specimen Preparation.

Current edition effective Sept. 9, 1968. Originally issued 1964. Replaces D 2292 - 64 T.

[2] *Annual Book of ASTM Standards*, Part 26.

for Testing,[3] for those tests where conditioning is required. In cases of disagreement, the tolerances shall be ±1 C (±1.8 F) and ±2 percent relative humidity.

4.2 *Conditions*—Conduct tests in the Standard Laboratory Atmosphere of 23 ± 2 C (73.4 ± 3.6 F) and 50 ± 5 percent relative humidity, unless otherwise specified in the test methods or in this specification. In cases of disagreement, the tolerances shall be ±1 C (±1.8 F) and ±2 percent relative humidity.

5. Procedure

5.1 Prepare plaques from which test specimens are to be cut (Note 3) as follows: Clean two polished ferrotype molding plates with solvent and dry. Place the molding chase on top of one of the clean ferrotype molding plates. Introduce into the molding chase a suitable quantity of plastic powder or granules to fill the blanked-out area completely when molded. A slight excess of material is desirable. Place the other molding plate on top of the granules. If the cavity-type mold is used, only one molding plate is required to cover the top of the mold. Insert this assembly between the platens of a compression-molding press previously heated to 175 to 180 C (350 F) (Note 4) as measured by a pyrometer on the press platens, ferrotype plates, and granules. After heating the material until it fluxes (approximately 5 min), apply sufficient pressure to form a smooth, void-free plaque of the required dimensions (Note 5). Shortly thereafter, discontinue the heat supply and cool the press to 66 C (150 F) at a reasonably uniform rate of 7 to 10 C per min. Remove the assembly from the press at 66 C (150 F) or below.

NOTE 3—Test specimens can be rough cut from the plaque with a band saw and then machined with a milling cutter to standard dimensions.

NOTE 4—If acceptable plaques cannot be molded at this temperature, a suitable temperature should be used and this should be reported.

NOTE 5—This is a flash-type molding operation, and excess material should flow out between the chase and the ferrotype plates. Pressures within the range of 70 to 140 kgf/cm^2 (1000 to 2000 psi) are satisfactory. The pressure, however, should be adjusted to avoid splashing material from the mold.

5.2 If cavity-type molds are used (3.1.3), clean them and proceed in accordance with 4.1. Double pressing may be necessary if the initial pressing is not of satisfactory quality. In that event, cut the plaque into strips and place in the mold together with sufficient powder or granules to assure that the required dimensions are maintained. Prepare the second pressing in accordance with the foregoing instructions.

---

[3] *Annual Book of ASTM Standards*, Part 27.

ASTM **Designation: D 2299 - 68**

Recommended Practice for
DETERMINING RELATIVE STAIN RESISTANCE OF PLASTICS[1]

This Recommended Practice is issued under the fixed designation D 2299; the number immediately following the designation indicates the year of original adoption or, in the case of revision, the year of last revision. A number in parentheses indicates the year of last reapproval.

1. Scope

1.1 This recommended practice covers the determination of the susceptibility to staining of plastic compositions on incidental contact with miscellaneous staining materials.

1.2 This recommended practice does not apply when the plastic compositions are intended to be used in intimate and continuous contact with the staining material.

2. Significance

2.1 Polymeric materials may be stained by many substances. Staining tendencies may be affected by additives and compounding ingredients used in the manufacture of various plastic compounds. This recommended practice determines the susceptibility of these polymeric materials and plastic compounds to staining, under a given set of controlled conditions.

3. Apparatus

3.1 *Closed Glass Containers*, for low-viscosity liquids.

3.2 *Applicator*, for applying viscous liquids and pastes, for example, a doctor blade.

3.3 *Oven*, Type I or Type II, meeting requirements of Grade B, as specified in accordance with ASTM Specifications E 145, for Gravity-Convection and Forced-Ventilation Ovens.[2]

4. Reagents

4.1 *Staining Reagents* may be found among foods, cosmetics, beverages, pharmaceuticals, cleaning agents, solvents, etc. Examples of these are asphalt, coffee, ink, lipstick, mustard, shoe polish, tobacco juice, tea, and wax crayon.

5. Test Specimens

5.1 The test specimens shall have a flat, smooth surface, large enough to permit a test area 25 mm square for visual examination. For instrumental evaluation the stained area must be large enough to meet the instrumental requirements.

5.2 Wet-rub thermosetting decorative laminates, prior to application of staining materials, with Grade FF or equivalent grade of pumice just sufficiently to remove the surface gloss, and then wash with a mild soap or detergent.

5.3 Should the staining procedure of 7.1.1 cause delamination of the specimen, use 7.1.2.

6. Conditioning

6.1 *Conditioning*—Condition the test specimens at 23 ± 2 C (73.4 ± 3.6 F) and 50 ± 5 percent relative humidity for not less than 40 h prior to test in accordance with Procedure A of ASTM Methods D 618, Conditioning Plastics and Electrical Insulating Materials for Testing,[3] for those tests where conditioning is required. In cases of disagreement, the tolerances shall be ±1 C (±1.8 F) and ±2 percent relative humidity.

6.2 *Test Conditions*—Conduct tests in the Standard Laboratory Atmosphere of 23 ± 2 C (73.4 ± 3.6 F) and 50 ± 5 percent relative

[1] This recommended practice is under the jurisdiction of ASTM Committee D-20 on Plastics. A list of committee members may be found in the ASTM Yearbook. This recommended practice is the direct responsibility of Subcommittee D-20.50 on Permanence Properties.

Current edition effective Sept. 9, 1968. Originally issued 1964. Replaces D 2299 - 64 T.

[2] *Annual Book of ASTM Standards*, Part 30.

[3] *Annual Book of ASTM Standards*, Part 27.

humidity, unless otherwise specified in the test methods or in this specification. In cases of disagreement, the tolerances shall be ±1 C (±1.8 F) and ±2 percent relative humidity.

7. Procedure

7.1 *Application of Staining Material:*

7.1.1 For low-viscosity liquids, immerse the specimen in the staining material in a glass container, and close tightly.

NOTE 1—In the case of volatile reagents, there may be a hazard connected with heating tightly closed jars. It is recommended that, for such materials, the jar and staining agent be warmed to oven temperature prior to closing tightly.

7.1.2 For viscous liquids and pastes, place the specimen under test on a flat surface, and apply a thin coating of staining material, approximately 0.1 mm, using the applicator.

7.1.3 For solid staining materials, such as wax crayon or lipstick, apply a uniform, opaque coating to the test area.

7.2 Place the specimens prepared in accordance with 7.1, in an oven at 50 ± 2 C for 16 h. Place specimens prepared in accordance with 7.1.2 and 7.1.3 in a horizontal position in the oven. Other temperatures may be used, if desirable in meeting specific requirements.

7.3 After the oven exposure period, remove any excess staining material from the surface.

NOTE 2—Unless otherwise specified, the excess staining material may be removed with a soft dry cloth or a spatula.

7.4 Observe the residual staining on the test specimen, and compare its appearance to that of an unstained sample. The evaluation of degree of staining may be visual or instrumental. If desired, a satisfactory scale for rating the degree of staining may be mutually agreed upon by the buyer and the seller.

NOTE 3—The following system has been used to evaluate the effect of staining reagents on thermosetting decorative laminates:

(*1*) *Unaffected*—No color change and no appreciable change in surface texture.

(*2*) *Superficial*—Applies to stains that are easily removed by a light application of a mild abrasive.

(*3*) *Considerable*—Applies to stains that are not easily removed or that result in etching.

8. Report

8.1 The report shall include the following:

8.1.1 Complete identification of polymeric material or plastic composition under test,

8.1.2 Identification of staining materials used,

8.1.3 Method of measuring the susceptibility to staining and the extent of staining, and

8.1.4 Any deviation from the recommended procedures.

Designation: D 2365 – 68

Standard Method of
CALCULATION OF NEUTRON DOSE TO POLYMERIC MATERIALS AND APPLICATION OF THRESHOLD-FOIL MEASUREMENTS[1]

This Standard is issued under the fixed designation D 2365; the number immediately following the designation indicates the year of original adoption or, in the case of revision, the year of last revision. A number in parentheses indicates the year of last reapproval.

1. Scope

1.1 In irradiation of polymeric materials in a reactor, energy deposition (and therefore changes in the material) will occur both with gamma rays and neutrons. Energy deposition (or dose of the material) from gamma rays can be experimentally determined and will be the subject of other ASTM publications. The energy deposition, or dose, in polymeric materials from neutrons is the basic interest of this method and is accomplished by means of a calculation, with a knowledge of the chemical composition of the material and of the neutron spectrum and integrated flux exposure. The knowledge of the neutron spectrum and integrated fluxes, which can be obtained from the threshold-foil measurements, is sufficient for such a calculation and is the basis of the method presented here. This method is general and can be adapted by the user to his particular type of organic material. The method of making the necessary neutron-flux measurements is also discussed.

1.2 Only calculation of the first collision dose[2] will be considered here. If the sample is large compared to a removal mean free path for the neutron beam, corrections must be made to convert the first collision dose into the actual dose. These corrections can be made accurately by means of Monte Carlo calculations, but can often be estimated with less accuracy by simpler analytical means, or measured in very large specimens by placing detectors at strategic locations throughout the specimen. Usually the dividing line for most substances is a thickness of a few centimeters of material for commonly encountered neutron spectra. If the sample thickness is less than 1 cm, the first collision dose will approximate the actual dose within a few percentage points.

1.3 The procedures appear in the following sections:

2. Calculation of Neutron Dose

2.1 *General Information:*

2.1.1 The basic relationship governing the dose to materials from neutron irradiation is

$$D(E)_{\mathrm{MeV/g}} = E_{\mathrm{MeV}} \, \Sigma f_i Q_i \sigma_i \qquad (1)$$

[1] This method is under the joint jurisdiction of ASTM Committees D-9 on Electrical Insulating Materials and D-20 on Plastics. A list of committee members may be found in the ASTM Yearbook. This standard is the direct responsibility of Subcommittee D-20.20 on Radiation Methods.

Current edition effective Nov. 6, 1968. Originally issued 1965. Replaces D 2365 – 65 T.

[2] For a definition of first collision dose, see *NBS Handbook*, No. 63, pp. 7 and 9.

where:

$D(E)$ = dose in MeV to material per unit neutron flux of energy E,

E = neutron energy, MeV,

Q_i = number of atoms of the ith element in 1 g of the sample,

σ_i = elastic scattering cross section for the ith element for neutrons of energy E, cm^2

f_i = average fraction of energy lost by the neutron to the atom, and is defined as follows:

$$f_i = 2mM_i/(m + M_i)^2 \quad (2)$$

where:

m = 1.009 = atomic weight of a neutron, and

M_i = atomic weight of ith constituent atom.

In Eq 2, it is assumed that the scattering is isotropic in the center-of-mass system, and that the neutrons will make only elastic collisions, both of which are reasonably accurate assumptions up to energies of 3 or 4 MeV. Above these energies, nonelastic collisions may become significant and the scattering may also become anisotropic. While Eq 1 is still reasonably accurate as long as σ refers to the elastic cross section, for many elements only the total cross section has been measured and values of the elastic cross section are not available. Therefore, in such cases, one is forced to use the total cross section in place of the elastic cross section and, as a result, some error is introduced. Most of the neutron spectra in reactors begin to fall off rapidly in intensity at just the energies where the nonelastic cross sections begin to become appreciable. Fortunately, then, if total cross sections are used, the errors made will not be large.

2.1.2 When neutrons are captured either at thermal or higher energies, energetic gamma rays are often produced. These gamma rays by definition do not contribute to the first collision dose. For large-size samples, their contribution to the dose must be calculated separately. (It may be further noted that perhaps as much as 10 percent of the first collision dose may not result in ionization, but may cause molecular excitations of one kind or another.)

2.2 *Calculations for a Known Neutron Spectrum:*

2.2.1 Equation 1 may be rewritten as follows:

$$D(E)_{\text{rads}} = E_{\text{meV}}\, \Sigma f_i Q_i \sigma_i \times 1.6 \times 10^{-8} \quad (3)$$

where:

$$Q_i = (F_i \times 6.02 \times 10^{23})/M_i \quad (4)$$

1.6×10^{-8} = conversion from MeV/g to rads, rad/MeV/g,

F_i = fraction of ith element by weight, and

M_i = atomic mass of the element.

or, combining Eqs 2 and 4, one finally obtains

$$D(E)_{\text{rads}} = 1.9 \times 10^{16}\, E_{\text{meV}} \sum_i F_i \sigma_i(E) m/(m + M_i)^2 \quad (5)$$

where σ_i is the cross section in square centimeters. The dose is given in rads where a rad is defined as being 100 ergs absorbed per gram of the material.

2.2.2 The neutron flux energy spectrum $\varphi(E)dE$, can be normalized such that

$$\int_0^\infty \varphi(E)dE = 1 \quad (6)$$

Using Eq 6, one may then write

$$\Phi = k \int_0^\infty \varphi(E)dE \quad (7)$$

where Φ is the integral neutron flux and k is simply a constant of proportionality numerically equal to the total number of neutrons per square centimeter per second of all energies. The total first collision dose in rads delivered to the sample is now

$$D_{\text{tot}} = k \int_0^\infty D(E)\varphi(E)dE \quad (8a)$$

$$D_{\text{tot}} = \Phi \int_0^\infty D(E)dE \quad (8b)$$

Using this equation, the total dose may be calculated if the neutron spectrum is known and the integral flux can be measured in some manner. For the integral neutron flux above some threshold energy we can write

$$\Phi_{E_t} = k \int_{E_t}^\infty \varphi(E)dE \quad (9)$$

where E_t is the threshold energy. Therefore, combining Eqs 7 and 9 we have

$$\Phi = \Phi_{E_t}\left[\int_0^\infty \varphi(E)dE \Big/ \int_{E_t}^\infty \varphi(E)dE\right] \quad (10)$$

The total first collision dose can now also be given as follows:

$$D_{\text{tot}} = \Phi_{E_t}\left[\int_0^\infty D(E)\varphi(E)dE \Big/ \int_{E_t}^\infty \varphi(E)dE\right] \quad (11)$$

Thus, in the case of a known neutron energy spectrum, the dose will be proportional to the neutron flux measured above any threshold energy.

2.2.3 Again, Eq 11 may be used only if the spectrum is known. This method of calculating total dose has been used many times in the past and is known as the single threshold monitor method. Only one threshold detector (usually radioactivation of sulfur) is used. The spectrum is commonly measured by the nuclear-track technique or perhaps known theoretically; the integrals in Eq 11 are calculated, and, from then on, the total dose is simply proportional to the neutron flux above the threshold energy.

2.3 *Calculations for an Unknown Neutron Spectrum:*

2.3.1 Most of the time, the spectrum will not be known; hence, the method given in 2.2 for calculating the dose will not be usable.

2.3.2 Several experimental methods have been used for characterization of the neutron flux in radiation environments. Some of these are nuclear-emulsion, electronic, and foil techniques, and also the more exotic types such as the chopper and time-of-flight techniques. The objective of a neutron-flux measurement will largely influence the optimum method chosen and the procedures used in the measurements. Obviously, it is certainly not feasible, nor necessary, to use a standard method exclusively for all neutron measurements, since economy and the amount of accuracy desired play important roles in the choice of a particular method. The foil method (**1–4**)[3] is the most widely used and most versatile means presently known where a moderate degree of accuracy is required in the measurement of the intensity and energy distribution of incident neutrons for general experimental conditions. This method and the recommendations for its use as described later are applicable to the specific case of making neutron-flux measurements sufficient for use in calculating first collision dose to organic materials.

2.3.3 In order to calculate the dose for an unknown spectrum, using the threshold detectors ^{239}Pu, ^{237}Np, ^{238}U, and ^{32}S, which are described later, Eq 8 may be rewritten as follows:

$$\begin{aligned} D_{\text{tot}} = & \; k\int_0^{0.010} D(E)\varphi(E)dE \\ & + k\int_{0.010}^{0.60} D(E)\varphi(E)dE \\ & + k\int_{0.60}^{1.5} D(E)\varphi(E)dE \\ & + k\int_{1.5}^{3.0} D(E)\varphi(E)dE \\ & + k\int_{3.0}^{\infty} D(E)\varphi(E)dE \end{aligned} \tag{12}$$

where 0.010, 0.60, 1.5, and 3.0 MeV are the threshold energies above which the various threshold foils will measure the integrated neutron flux. This simply breaks the total integration region on energy into five contiguous intervals. Now, from Eq 9,

$$\Phi_{\text{Pu}} = k\int_{0.010}^{\infty} \varphi(E)dE \tag{13}$$

and

$$\Phi_{\text{Np}} = k\int_{0.60}^{\infty} \varphi(E)dE \tag{14}$$

Subtracting Eq 14 from Eq 13,

$$\Phi_{\text{Pu}} - \Phi_{\text{Np}} = k\int_{0.010}^{0.60} \varphi(E)dE \tag{15}$$

and, in similar fashion,

$$\Phi_{\text{Np}} - \Phi_{\text{U}} = k\int_{0.60}^{1.5} \varphi(E)dE \tag{16}$$

$$\Phi_{\text{U}} - \Phi_{\text{S}} = k\int_{1.5}^{3.0} \varphi(E)dE \tag{17}$$

$$\Phi_{\text{S}} = k\int_{3.0}^{\infty} \varphi(E)dE \tag{18}$$

2.3.4 The first term in Eq 12 is very small and will be assumed hereafter to be negligible. Included in this assumption is the neglect of any dose due to thermal neutron capture. As stated previously, the gamma-ray component of thermal capture is not part of the first collision dose. The proton or alpha part of thermal capture is essentially part of the first collision dose and will be considered separately (see 2.4). Equations 15 through 18 may now be combined with Eq 12, as follows:

$$\begin{aligned} D_{\text{tot}} = & \; (\Phi_{\text{Pu}} - \Phi_{\text{Np}}) \frac{\int_{0.010}^{0.60} D(E)\varphi(E)dE}{\int_{0.010}^{0.60} \varphi(E)dE} \\ & + (\Phi_{\text{Np}} - \Phi_{\text{U}}) \frac{\int_{0.60}^{1.5} D(E)\varphi(E)dE}{\int_{0.60}^{1.5} \varphi(E)dE} \\ & + (\Phi_{\text{U}} - \Phi_{S}) \frac{\int_{1.5}^{3.0} D(E)\varphi(E)dE}{\int_{1.5}^{3.0} \varphi(E)dE} \\ & + \Phi_{S} \frac{\int_{3.0}^{\infty} D(E)\varphi(E)dE}{\int_{3.0}^{\infty} \varphi(E)dE} \end{aligned} \tag{19}$$

2.3.5 This equation for the dose still has the disadvantage that the spectrum must be known. In order to remove this requirement, it may now be assumed that in each of the four intervals of the spectrum in Eq 19, the variation in $D(E)$ is sufficiently small within the interval such that it can be removed from inside the integral. With this assumption, the following equation for the dose results:

[3] The boldface numbers in parentheses refer to the list of references appended to this method.

$$D_{tot} = (\Phi_{Pu} - \Phi_{Np})\overline{D_1(E)} + (\Phi_{Np} - \Phi_U)\overline{D_2(E)} + (\Phi_U - \Phi_S)\overline{D_3(E)} + \Phi_S\overline{D_4(E)} \quad (20)$$

where the D's are the average values so obtained. The D's may be written as follows:

$$\overline{D_1(E)} = \Sigma C_1{}^i \overline{F_i} \quad (21a)$$

$$\overline{D_2(E)} = \Sigma C_2{}^i \overline{F_i} \quad (21b)$$

$$\overline{D_3(E)} = \Sigma C_3{}^i \overline{F_i} \quad (21c)$$

$$\overline{D_4(E)} = \Sigma C_4{}^i \overline{F_i} \quad (21d)$$

where i refers to the ith element of the sample and the C's are the average value of all the terms of Eq 5 other than the F's in this notation. The final equation for the total dose is as follows:

$$D_{tot} = (\Phi_{Pu} - \Phi_{Np})\Sigma C_1{}^i \overline{F_i} + (\Phi_{Np} - \Phi_U)\Sigma_i C_2{}^i \overline{F_i} + (\Phi_U - \Phi_S)\Sigma_i C_3{}^i \overline{F_i} + \Phi_S \Sigma_i C_4{}^i \overline{F_i} \quad (22)$$

An exact expression for C_y is (where $y = 1$, 2, 3, or 4):

$$C_y = \frac{1.9 \times 10^{16} \int_{E_{t_i}}^{E_{t_j}} E_{MeV}\sigma_i(E)dE}{(1 + M_i)^2} \quad (23)$$

2.3.6 In Table 1, the constants $\overline{C}_1$, $\overline{C}_2$, $\overline{C}_3$, and $\overline{C}_4$ have been evaluated for the nine elements that comprise most organic materials. In this table, the assumption was made that the spectrum was uniform within each energy interval, and zero above 5 MeV.

2.3.7 Table 2 is the same as Table 1 except that a dE/E spectrum was assumed. It should be noted that the C's for all energy intervals, except the 10-keV to 0.6-MeV interval, are not very different. Most reactor spectra will probably give values for the C's that lie somewhere between the values in Tables 1 and 2, depending on the degree of neutron moderation.

2.3.8 An example will now be given, showing how to calculate the coefficients of Eq 22 in a particular case. In this example, it is assumed that the substance is tissue having composition by weight H = 0.10, N = 0.04, O = 0.74, and C = 0.12. Table 3 shows the calculation of the coefficients of Eq 22 for tissue, using the uniform neutron spectrum data of Table 1. For this case Eq 22 becomes

$$D = [1.17(\Phi_{Pu} - \Phi_{Np}) + 2.30(\Phi_{Np} - \Phi_U) + 3.12(\Phi_U - \Phi_S) + 4.13\Phi_S] \times 10^{-9} \text{ rads} \quad (24)$$

Table 4 shows the same calculation using the dE/E spectrum of Table 2. The equation for the dose is then

$$D = [0.63(\Phi_{Pu} - \Phi_{Np}) + 2.23(\Phi_{Np} - \Phi_U) + 3.07(\Phi_U - \Phi_S) + 4.04\Phi_S] \times 10^{-9} \text{ rads} \quad (25)$$

Again, it is seen that only the first term is very different for the two spectra.

2.3.9 Sayeg (**4**) calculated this particular case using actual integration over the Godiva reactor spectrum rather than averages used here. His coefficients differ from the ones obtained in Eq 24 by 2 to 10 percent. However, these are often compensating and introduce a much smaller error in total dose.

2.4 *Calculation of Dose from Thermal Neutron-Capture Reactions:*

2.4.1 When a thermal neutron is captured by a nucleus, a gamma ray in most cases is subsequently emitted. This gamma ray has a mean free path in most organic substances of the order of centimeters and therefore does not contribute to the first collision dosage. In the case of two of the elements listed in Tables 1 and 2, namely nitrogen and boron, a thermal capture results in the emission, respectively, of a 0.6-MeV proton and a 2.3-MeV alpha particle. These charged particles have a very short range and may therefore be considered to be part of the first collision dosage.

2.4.2 This part of the dose may be computed from the following equations:

$$D_N = 0.75 \times 10^{-9} F_N \Phi_t \text{ rads} \quad (26)$$

$$D_B = 1.55 \times 10^{-6} F_B \Phi_t \text{ rads} \quad (27)$$

where:

F_N = nitrogen fraction by weight,
F_B = boron fraction by weight, and
Φ_t = thermal activation flux.

2.4.3 It should be noted that, for a substance containing any appreciable amount of boron, the sample size may have to be very thin to prevent self-shielding effects, since the thermal cross section is so large (755 barns).

2.4.4 The thermal activation flux, Φ_t, is measured by the cadmium difference method with gold (see ASTM Method E 262, for Measuring Thermal Neutron Flux by Radioactivation Techniques.)[4] A further small contribution to the dose may come from epicadmium neutrons. Ordinarily, this contribution will be less than 25 percent of the dose computed from Eqs 26 and 27.

3. Measurement of Integrated Neutron Fluxes with Threshold Foils

3.1 *General Information:*

3.1.1 The energy dependence of the neutron flux is important in calculating the dose to materials and for relating any given radiation-induced phenomena to the incident field. The importance of neutron energies in calculation of dose to organic materials has already been discussed in great detail. There are several other neutron variables which should be considered in radiation effects experiments.

3.1.2 The neutron flux intensity, which may or may not be controlled, depends on the reactor capability and the location of the point of measurement with respect to the core. Flux measurements at one point should not, in general, be extrapolated to another point, but all positions of interest should be measured. The angular incidence of the neutron is a factor which may vary between the rather extreme limits of direct-beam incidence to isotropic incidence. This is further complicated, in the case of an anisotropic incidence, by the orientation of the foil with respect to the direction of the neutrons, varying from perpendicular to parallel incidence of the neutron beam. The influence of angular incidence is often overlooked in the interpretation of foil measurements, particularly for cases where relatively thick, flux-perturbing foils are used. Thin foils, which are defined as nonflux-perturbing foils, should be used when possible.

3.1.3 When reporting measurements of the neutron environment of a reactor, the source and associated facilities should either be adequately described or appropriately referenced. For example, in a reactor system, this description should include:

3.1.3.1 The type, kind, size, and configuration of core,

3.1.3.2 The methods and amounts of reflection or moderation, or both,

3.1.3.3 The method of control,

3.1.3.4 The type and amount of shielding,

3.1.3.5 The geometric relation between source and point(s) of interest in flux specification,

3.1.3.6 The description of the nonnuclear environment, including temperature and humidity, which may affect the use of specific foil detectors, and

3.1.3.7 The associated material, equipment, or surroundings, which may influence the flux being measured.

3.1.4 Flux measurements with foils are made by relating the reaction rate in the foil material to the flux producing the reaction. The reaction rate may be obtained by a measurement of the activity produced in the foil at some time following exposure to the flux. The relation between the activity and flux for a thin foil may be expressed as follows:

$$A_s = N\sigma\Phi \tag{28}$$

where:

A_s = saturation activity in disintegrations per second of a foil, on removal from the flux, Φ, after being exposed for an infinite time,

Φ = nv, the integral (Note 1) neutron flux, neutrons/cm^2-s,

n = neutron density, neutrons/cm^3,

v = neutron velocity, cm/s,

N = number of atoms in the foil subject to the reaction, and

σ = cross section for the reaction, cm^2/atom.

Integral fluxes may be defined as follows:

$$\Phi = nv = \int_{E_1}^{E_2} \varphi(E)dE \tag{29}$$

where:

$\varphi(E)$ = neutron flux per unit energy interval in units of n/cm^2-s-unit energy, and

E_1 and E_2 = lower and upper energy limits considered.

NOTE 1—The integral flux is defined as being integrated with respect to energy and not time.

3.1.5 Many nuclear reactions of the type (n, charged particle) or (n, fission) will exhibit a threshold energy, E_t, below which the reaction does not occur and above which it will occur. The cross section for the reaction then rises from zero at the threshold and varies as a function of the neutron energy

[4] *Annual Book of ASTM Standards*, Part 30.

above the threshold. The threshold energy and cross section as a function of neutron energy are peculiar to each particular reaction. Materials that exhibit a threshold energy for some nuclear reaction represent a means of measuring the neutron flux above that energy. By using a series of these threshold detectors and taking the differences in measured neutron fluxes above various energies, the fluxes in several energy increments can be obtained.

3.1.6 Equation 28 is rigorous; however, the actual quantities measured to obtain the activity, A_s, require additional explanation. A foil is never exposed for an infinite length of time and its activity is rarely measured with 100 percent efficiency at the instant of removal from the incident flux. Therefore, A_s is related to the measured number of counts, C, obtained in a count time interval $(t_2 - t_1)$ at times t_1 and t_2 after completion of the exposure. For a monoisotopic foil that decays to a stable daughter and is exposed for a time T, the relation is

$$A_s = \lambda C/(1 - e^{-\lambda T})\ (e^{-\lambda t_1} - e^{-\lambda t_2})\epsilon \quad (30)$$

where:

λ = $\ln 2/T_{1/2}$, the decay constant,
$T_{1/2}$ = half-life of the activated isotope formed in the reaction, and
ϵ = efficiency for the measurement of the true activity.

3.1.7 Equation 30, although rigorous in that it contains a correction for decay during the counting interval, is not often used except when counting times comprise a significant part of the half-life. An approximate equation more often used is

$$A_s = C/(1 - e^{-\lambda T})\ (e^{-\lambda t})\Delta t\epsilon \quad (31)$$

where:

t = mean time from the end of the exposure to the count time, and
Δt = counting interval, $(t_2 - t_1)$, s.

The number of atoms in the foil subject to the reaction is

$$N = WN_O I/M \quad (32)$$

where:

W = foil weight, g,
N_o = Avogadro's number, atoms/mole,
I = isotopic purity in fractional weight, and
M = isotopic weight, g/mole.

3.1.8 Combining Eqs 28, 31, and 32, and rearranging, gives for the flux.

$$\Phi = nv = MC/\Delta t W N_O I\sigma(1 - e^{-\lambda T})\ (e^{-\lambda t})\epsilon \quad (33)$$

Equation 33 is accurate if each factor is precisely known. Experimental measurements of certain of these factors are, however, subject to variation and must be evaluated.

3.1.9 For purposes of the present discussion, the factors in Eq 33 may be classified into two categories, basic and experimental. The basic factors are defined as those which are constant for a particular foil. The experimental factors are defined as those which must be determined for a given set of irradiation conditions. Each of these basic and experimental factors is discussed in terms pertinent to flux measurements.

3.1.10 The isotopic weight, foil weight, Avogadro's number, isotopic purity, and the decay constant are all independently measured or tabulated quantities. These quantities, although indirectly influencing the foil measurement, may be considered constant for a given reaction and are specific properties of a particular foil; therefore, these quantities are not discussed further. The cross section is perhaps the most influential factor of the basic quantities. The cross section is a measure of the probability that a given reaction will occur. This quantity varies with the energy of the incident neutron. Thus, it is a constant only for monoenergetic sources. For materials used in neutron detection, the variable true cross section is replaced with an effective cross section, which has come to be more meaningful in flux measurements.

3.1.11 Time, although not always a controllable factor, may usually be measured with more precision than the other factors. Time is, however, related to the measured activity of foils in a manner needing further explanation. The exponential terms $(1 - e^{-\lambda T})$ and $(e^{-\lambda t})$ in Eq 31 represent the buildup of radioactive atoms in a foil during exposure, with subsequent decay following exposure. The buildup term $(1 - e^{-\lambda T})$ in a simple foil exposure of duration T at a constant source intensity, or power level, is readily computed. It often results, however, that before the foil activity is determined, a series of exposures and decays have occurred at different power levels. Thus, the buildup and decay factors become more complicated, requiring a series computation. The series may be expressed as follows:

$$\sum_i P_i(1 - e^{-\lambda T^i})e^{-\lambda t^i}$$

where the P_i are the relative power levels for the corresponding exposure times T_i, and the t_i are the decay times taken from the end of each exposure to the time the foil activity is determined. This expression assumes linearity of power level and induced activity; it also assumes instantaneous changes in power level. If such is not the case, the induced activity becomes subject to variables of power-level changes and linearity. These variables may be estimated, however, by independent experiments measuring flux power-level linearity and by the adequate recording of irradiation operation. If the exposure time is fixed and is much greater than one or two half-lives of the activated foil, only the latter part of the exposure flux is measured.

3.1.12 Several reactions, for example, (n,λ), (n,p), and (n,α), are possible in the same monoisotopic foil. Each of these reactions may contribute to the measured activity. Foils with more than one isotope or element and which have either desirable or undesirable impurities may further complicate activation data analysis. The separation of the desired activity from extraneous activities is readily made by decay data analysis, provided the half-lives are suitably different and desirable activity levels are obtained. Foil-activation analysis is greatly simplified, however, by the use of extremely pure materials. For example, if a foil measurement is made by utilization of the ^{32}S $(n,p)^{32}P$ reaction, it is desirable to minimize impurities, such as chlorine and phosphorus, which also yield ^{32}P by the (n,α) and (n,λ) reactions, respectively.

3.1.13 In the case of fission foils, analysis of the activity includes many radioisotopes, making a single decay constant impossible. In this case, an experimentally determined decay curve must be used in neutron-flux determination. In the case of all foils, the actual cross section must be replaced with an effective cross section which will be the average value above some effective threshold energy for a given neutron spectrum. It is emphasized that the effective thresholds and cross sections are not applicable to all neutron energy distributions. The activation response in each of these reactions is

$$R_t = \int_0^\infty \sigma(E)\varphi(E)dE \qquad (34)$$

where:

$\sigma(E)$ = cross section as a function of energy, and

$\varphi(E)$ = flux distribution as a function of energy.

3.1.14 In practice, the true cross section, which varies with neutron energy, is replaced with a constant effective cross section above an effective threshold energy using the following relation:

$$\sigma_{\text{eff}}\int_{E_t}^\infty \varphi(E)dE = \int_0^\infty \sigma(E)\phi(E)dE \qquad (35)$$

The effective cross section, σ_{eff}, defined by Eq 35 depends on the specification of the effective threshold energy, E_t, which, in turn, identifies the fast-flux integral,

$$\int_{E_t}^\infty \varphi(E)dE = \Phi_{\text{eff}}$$

The effective threshold energy is often specified in a particular spectral distribution such as a fission distribution. This approximation assumes that the cross section above energy E_t is a constant and that below E_t it is zero.

3.2 *Experimental and Calibration Procedures:*

3.2.1 The threshold foil method employs foils of ^{239}Pu, ^{237}Np, ^{238}U, and ^{32}S (Note 2) for evaluation of the fast neutron flux above various threshold energies. In order to use these materials as threshold detectors, effective cross sections and thresholds must be determined by the relationship given in Eq 35.

NOTE 2—^{32}S is actually used in the form of pellet rather than a foil, but is commonly referred to as a foil.

3.2.2 Figures 1 through 4 show plots of the effective cross section versus threshold energy for a Watt (**6**) fission spectrum, the Godiva Reactor (**7**) spectrum, and several assumed extreme spectra. The values recommended for general use with the foils are shown in these plots. Table 5 lists the nuclear reactions involved, plus the recommended effective threshold energies and cross sections; and gives the approximate time-integrated flux range that can be conveniently detected by this method.

3.2.3 This technique of using threshold foils permits the measurement of the neutron flux above each of the listed effective threshold energies. The difference in flux values, as measured by the various foil materials, gives the neutron flux within the various energy increments between their threshold energies.

3.2.4 *Fission-Foil Calibration Technique:*

3.2.4.1 The method presently used for calibration of the fission foils as threshold detectors is irradiation of bare ^{239}Pu in a known thermal neutron flux. Bare plutonium has a high (746 barns) (Note 3) thermal-neutron-fission cross section and exhibits the threshold and effective cross section indicated in Table 5 only when covered with 1.65 g/cm^2 of boron-10. Thermal-neutron-irradiated bare ^{239}Pu is then used to determine the decay rate and to calibrate the fission-foil nuclear counters. The decay curve of ^{239}Pu, over a period of 30 h after irradiation, has been determined to be identical, to within normal counting statistical variations (±2 to 3 percent), whether fissioned by thermal neutrons or fast neutrons. There are small differences, however, in the decay-curve shapes of ^{237}Np and ^{238}U with respect to the plutonium curve. A comparison of these decay curves, normalized to 2 h after exposure, is shown in Fig. 5. It is not known at what point in time one should normalize the decay curves to obtain a proper comparison. However, it will be noted that regardless of the normalization point, the maximum difference between 2 and 30 h would be about 10 to 15 percent. Variations of this order are often obtained by differences employed in the nuclear counting detection system.

NOTE 3—In actual practice, this thermal neutron capture cross section must be corrected for maxwellian distribution for a nonstandard temperature of the thermal source. This correction is usually from 5 to 10 percent (**10**).

3.2.4.2 Where possible the foils should be calibrated independently of each other by use of a well-known fast neutron spectrum such as the Godiva Reactor. This would allow independent experimental determination of all three decay curves and their relationship to the neutron flux above the various threshold energies. However, this method will include the uncertainty of the original neutron spectrum measurements and will still be of the same order of magnitude (approximately ±10 to 20 percent) as before. Where a fast neutron source of known spectrum is not available, it is recommended that the thermal-neutron-fissioned ^{239}Pu decay curve be used with all three fission foils (Note 4). Combining Eqs 32 and 33, and making modification for the experimentally determined decay factors and counter calibration, we have for each fission foil,

$$\Phi = nv = CK/\Delta t N \sigma_{\text{eff}} T(dF) \tag{36}$$

where $T(dF)$ is the counting rate as a function of time after irradiation and K is the counter calibration factor (incorporating the counter efficiency), both being determined with the bare ^{239}Pu foil irradiated in a known thermal neutron flux or for each foil on a source of fast neutrons with a known spectrum. All other terms are as previously defined. The fission foil decay curves are shown in Fig. 5.

NOTE 4—Calibration with equivalent foils (**3,9**) have been successfully used to determine unmoderated fission spectra. These calibration foils are mixtures of ^{239}Pu and ^{235}U.

3.2.4.3 Since the ^{239}Pu cross section for thermal neutrons is high, there are perturbing effects to be considered when performing the calibration. In exposing a material of high cross section to thermal neutrons, a "neutron sink" occurs at the position of the foil. That is, the exposure flux at the position of the foil, when the foil is not in place, is larger than when the foil is in position. One effect is called "flux depression" and is a result of the following phenomenon. Neutrons in passing through a material of high cross section, have a high probability of being absorbed. Hence, these neutrons do not have the opportunity to be scattered back (as with the absence of foil) to produce the approximate isotropic thermal-neutron-flux distribution. A second effect which causes a smaller flux activating the foil, is the self-shielding of the foil itself. The inner layers of the foil will see a smaller flux due to the absorption of neutrons in the outer layers. The composite effect of flux depression and self-shielding can be severe for thick foils. Thermal-neutron calibration of ^{239}Pu foils of 35 mg/cm^2 will exhibit an effect of about 20 percent. Hurst (**3**) has proposed a method of evaluation for the composite effect. This method involves placing two very thin ^{197}Au foils on each side of the calibration ^{239}Pu foil. This "sandwich" is then exposed to thermal neutrons. Another exposure is made with a thin ^{197}Au foil, with no fission foil present, at the same position and exposed for the same time. Comparison of the induced ^{198}Au activities per unit mass of ^{197}Au with the fission foil present and with the foil not present, will give an index of the total effect of flux depression and self-shielding. For a $^3/_4$-in. diameter ^{239}Pu foil weighing 100 mg (approximately 1 mil thick or 35 mg/cm^2), these effects are found to be approximately 20 percent. Another method, somewhat more tedious, has been employed by Sayeg and others (**5**). This method involves the exposure of different thicknesses of ^{239}Pu to the same integral flux at the same rate and extrapolating to zero foil thickness. The results are in agreement with those of Hurst and his coworkers.

3.2.4.4 When exposing fission foils in the unknown fast-neutron source, they should be surrounded by boron-10. An approximate thickness of 1.65 g/cm^3 of boron-10 should be used. This is necessary in the case of ^{239}Pu to obtain an effective threshold energy by shielding out the thermal and low-energy neutrons. A thermal-neutron shield is required for ^{237}Np to minimize difficulties from the competing neutron-capture reactions. The use of shields other than boron-10 are being studied for this purpose; however, it is convenient to expose the ^{237}Np in the same boron shield with the ^{239}Pu. It is necessary to expose the ^{238}U foils inside the boron containers because of trace amounts of ^{235}U, which has a high thermal-neutron fission cross section. Uranium-238 containing less than 200 ppm ^{235}U should be used for these measurements.

3.2.4.5 A boron container designed for use with fission foils is described by Hurst and his co-workers (**3**). A 30-mil cadmium cover should be used inside the boron container to absorb any neutrons that might be thermalized by the boron. Because of health-hazard considerations, the fissionable materials themselves should be encapsulated in 10-mil thick copper dishes. A convenient amount of material for each foil for most measurement situations would be ^{239}Pu, 0.25 g; ^{237}Np, 0.50 g; and ^{238}U, 2.0 g.

3.2.5 *Sulfur-Foil Calibration Technique:*

3.2.5.1 The nuclear counter for sulfur foil analysis may be calibrated by irradiating a sulfur foil with 14-MeV neutrons from a Cockcroft-Walton accelerator (**5**), or a simulated pellet of ^{31}P may be irradiated with thermal neutrons (**2**). The activation cross section of ^{32}S by 14-MeV neutrons is 248 millibarns, and the thermal neutron activation cross section of ^{31}P is 200 millibarns. The induced ^{32}P activity is corrected for a change in cross section from the reference energy to the 3.0-MeV effective threshold of sulfur, assuming a ^{235}U fission neutron spectrum.

3.2.5.2 Since beta counting must be used in the case of sulfur foils (the product radioisotope ^{32}P does not emit a gamma ray), there exists a problem of self-absorption of the beta particles. The most convenient method of handling this problem is by the use of infinitely thick sulfur pellets. Sulfur compacted into pellet form (density of about 1.5 g/cm^3) to a thickness of $^1/_4$ in. is infinitely thick to the beta particles. The diameter of pellet used is not critical, but must be maintained uniform to obtain consistent results. Diameters in the range of $^3/_4$ to $1\,^1/_2$ in. are commonly used. Consideration must also be given to the purity of the sulfur used to avoid any undesirable activities from contaminating materials. In particular, the sulfur should contain less than 100 ppm of phosphorus and chlorine.

3.2.5.3 Applying Eq 33 to infinitely thick sulfur pellets of a given diameter the integral neutron flux becomes

$$\Phi = nv = CK/\Delta t\sigma_{\text{eff}}(1 - e^{-\lambda T})\,(e^{-\lambda t}) \quad (37)$$

Again, K is the counter calibration factor, incorporating counter efficiency, a determined with the calibration pellet of sulfur or ^{31}P.

3.3 *Instrumentation and Counting Analysis Procedures:*

3.3.1 The choice of nuclear counting instrumentation used to measure the amount of activity induced in the various foils recommended as neutron detectors will depend on what equipment may be available and the volume of measurements to be made. Any method may be used, but the many practical considerations make one or two methods more desirable. Gamma counting should be used whenever possible in order to discriminate against natural low-energy radiations in the fission foils and to minimize self-absorption. The NaI(Tl) crystals are recommended for all gamma counting. In order to avoid corrections for changes in foil diameter the crystal diameter should be large compared to the foil diameter. For most foils in use at the present time, a 3-in. diameter crystal is sufficient. Considerations of over-all efficiencies have led some experimenters to use larger crystals (up to 5 in. in diameter) and double counters in which the foil is inserted between two crystals. The use of these more elaborate systems will depend on the particular applications and the flux levels that one expects to encounter.

3.3.2 The selection of a proper discrimination level is extemely important to the success of the foil system, especially the three materials ^{239}Pu, ^{237}Np, and ^{238}U. Because of the natural activity of these samples and because of possible (n,γ) reactions, a discrimination level of 1.2 MeV is recommended (Note 5). Although ^{237}Np cannot be counted at a bias level much lower than 1.2 MeV, some experimenters have counted ^{239}Pu and ^{238}U at bias levels as low as 0.5 MeV in order to increase their sensitivity. This lower bias level may be of value in some special circumstances, but is not recommended for general use.

NOTE 5—Caution should be exercised in counting very-high-activity foils, which may result in amplifier overloading due to the pulses which are under the bias level.

3.3.3 The ^{237}Np foil presents a special problem because of the abundant 0.3-MeV natural gamma peak from the ^{233}Pa daughter. This gamma ray is present in such a quantity that serious coincidence-buildup problems occur, unless it is reduced by an absorber. At least $^1/_2$ in. of lead is recommended as an absorber when counting ^{237}Np. This absorber transmits a large percentage of the gamma rays over 1.2 MeV and is recommended for counting all fission foils.

3.3.4 Sulfur is the only foil commonly used that does not emit a gamma ray; plastic scintillators have proved desirable for beta-counting the sulfur foil. When scintillation counting is used for all foils, one may choose the electronic components of the system in such a manner that all foils may be counted on the same equipment, changing only the detector itself. If, however, a laboratory wishes to use existing proportional counting equipment for beta-counting the sulfur foil, this would be quite satisfactory. Geiger counters are not recommended for general use in beta-counting the sulfur foils because of the generally wide range of count rates encountered.

3.3.5 The counting of the sulfur foil is straightforward and requires no special bias levels or absorbers. However, a delay of 24 h is recommended between irradiation and counting in order to allow the ^{31}Si, ^{34}S (n,α) ^{31}S, to die out. A check should also be made of the half-life to ascertain that no contaminants are present (for example, ^{35}S).

3.3.6 The standardization of the counting equipment for foil analysis is the same as for any counting procedure. The discriminator should be checked periodically for both zero and slope. The energy calibration should be checked at the same time. Various standards are in use but some appear to have definite advantages. ^{60}Co is recommended for standardizing counting equipment for fission foils. The 1.17 and 1.28-MeV gamma peaks make it extremely sensitive at the 1.2-MeV bias. Equipment used for counting the sulfur foil is best standardized with either a simulated ^{23}P standard or a natural uranium standard, whichever is more convenient. Gamma-ray standards of ^{60}Co (1.17 and 1.33 MeV) will be found useful in general standardization and discriminator energy calibration.

3.3.7 Table 6 summarizes the recommended parameters for counting of the foils used in the threshold detection system.

3.3.8 All foils should be counted several times over a period of time. This procedure will allow a check to be made of the half-life or decay constant, and will add considerably to the reliability of the data.

REFERENCES

(1) Hurst, G. S., Harter, J. A., Hensley, P. N., Mills, W. A., Slater, M., and Reinhardt, P. W., "Techniques of Measuring Neutron Spectra with Threshold Detectors-Tissue Dose, Determinations," *Review of Scientific Instruments*, Vol 27, 1956, p. 153.

(2) Reinhardt, P. W., and Davis, F. J., "Improvements in the Threshold Detector Method of

Fast Neutron Dosimetry," *Health Physics*, Vol 1, 1958, p. 169.

(3) Hurst, G. S., and Ritchie, R. H., "Radiation Accidents: Dosimetric Aspects of Neutron and Gamma Ray Exposures," Oak Ridge National Laboratory Report ORNL-2748, Part A, 1959.

(4) Sayeg, J. A., "The Godiva II Critical Assembly; Dosimetry for Biological Research," Los Alamos Scientific Laboratory Report LA-2435, April, 1960.

(5) Sayeg, J. A., Harris, P. S., and Larkins, J. H., "Experimental Determination of Fast and Thermal Neutron Tissue Dose" Los Alamos Scientific Laboratory Report LA-2174, January, 1958.

(6) Watt, B. E., "Energy Spectrum of Neutrons From Thermal Fission of U-235." *The Physical Review*, Vol 87, 1952, p. 1037.

(7) Frye, G. L., Jr., Gammel, J. H., and Rosen, L., "Energy Spectrum of Neutrons from Thermal Neutron Fission of U-235 and from an Untamped Multiplying Assembly of U-235," Los Alamos Scientific Laboratory Report LA-1670, May, 1954.

(8) Buckalew, W. H., and Humpherys, K. C., "Foil Activation Method of Neutron Flux and Spectra Measurements in Reactor Environments," Sandia Corp. (to be published).

(9) Sayeg, J. A., Ballinger, E. R., and Harris, P. S., "Dosimetry for the Godiva II Critical Assembly-Neutron Flux and Tissue Dose Measurements," Los Alamos Scientific Laboratory Report LA-2310, March, 1959.

(10) Westcott, C. H., "Effective Cross Section Values for Well-Moderated Thermal Reactor Spectra," Chalk River Research Report CRRP-787, 1958 Revision.

TABLE 1 Value of Coefficients to be Used in Calculating Dose From Eq 22 (Assumed Spectrum Uniform in Each Energy Interval and Zero Above 5 MeV)

NOTE—All values in this table are to be multiplied by 10^{-9}.

Energy Interval, MeV ...	0.010 to 0.60	0.60 to 1.5	1.5 to 3.0	>3.0
Elements	Coefficients			
	C_1'	C_2'	C_3'	C_4'
H	10.7	20.6	28.9	36.2
O	0.114	0.273	0.213	0.535
N	0.077	0.176	0.319	0.584
C	0.128	0.299	0.479	0.799
F	0.089	0.176	0.294	0.401
Cl	0.010	0.038	0.099	0.172
Si	0.030	0.078	0.152	0.215
B	0.154	0.351	0.598	0.830
S	0.014	0.046	0.114	0.205

TABLE 2 Value of Coefficients to be Used in Calculating Dose From Eq 22 (Assumed Spectrum dE/E Within Each Energy Interval and Zero Above 5 MeV).

NOTE—All values in this table are to be multiplied by 10^{-9}

Energy Interval, MeV ...	0.010 to 0.60	0.60 to 1.5	1.5 to 3.0	>3.0
Elements	Coefficients			
	C_1'	C_2'	C_3'	C_4'
H	5.84	20.0	28.4	35.4
O	0.048	0.255	0.217	0.524
N	0.042	0.162	0.313	0.572
C	0.065	0.288	0.463	0.782
F	0.047	0.168	0.288	0.392
Cl	0.005	0.035	0.095	0.168
Si	0.013	0.075	0.106	0.210
B	0.079	0.321	0.581	0.812
S	0.008	0.043	0.107	0.201

TABLE 3 Calculation of Coefficients for Tissue (Uniform Spectrum Within Each Interval)

	$C_1'F_i$	$C_2'F_i$	$C_3'F_i$	$C_4'F_i$
H	1.07	2.06	2.89	3.62
N	0.003	0.007	0.013	0.023
O	0.084	0.202	0.158	0.396
C	0.015	0.036	0.057	0.096
	$C_1F = 1.17 \times 10^{-9}$	$C_2F = 2.30 \times 10^{-9}$	$C_3F = 3.12 \times 10^{-9}$	$C_4F = 4.13 \times 10^{-9}$

TABLE 4 Calculation of Coefficients for Tissue (*dE/E* Spectrum With Each Interval)

	$C_1'F_i$	$C_2'F_i$	$C_3'F_i$	$C_4'F_i$
H	0.584	2.00	2.84	3.54
N	0.002	0.006	0.013	0.023
O	0.035	0.189	0.161	0.387
C	0.008	0.035	0.056	0.094
	$C_1F = 0.629 \times 10^{-9}$	$C_2F = 2.23 \times 10^{-9}$	$C_3F = 3.07 \times 10^{-9}$	$C_4F = 4.04 \times 10^{-9}$

TABLE 5 Threshold-Foil Data

Reaction	Effective Energy Threshold, E_t	Effective Cross Section, barns	Normal Range of Measurements nvt
^{239}Pu (n,f) fission products[a]	10 keV	1.7	10^{10} to 10^{14}
^{237}Np (n,f) fission products	0.6 MeV	1.6	10^{10} to 10^{14}
^{238}U (n,f) fission products	1.5 MeV	0.55	10^{11} to 10^{14}
^{32}S (n,p) ^{32}P	3.0 MeV	0.30	10^{10} to 10^{14}

[a] Values for ^{239}Pu apply only with shields of 1.65 g/cm of B-10.

TABLE 6 Recommended Parameters for Counting Foils

Foil	Bias, MeV	Absorber, in.	Standard	Detector
^{239}Pu	1.2	0.50 Pb	Co^{60}	NaI(T1)
^{237}Np	1.2	0.50 Pb	Co^{60}	NaI(T1)
^{238}U	1.2	0.50 Pb	Co^{60}	NaI(T1)
S	above noise	none	uranium	plastic scintillator

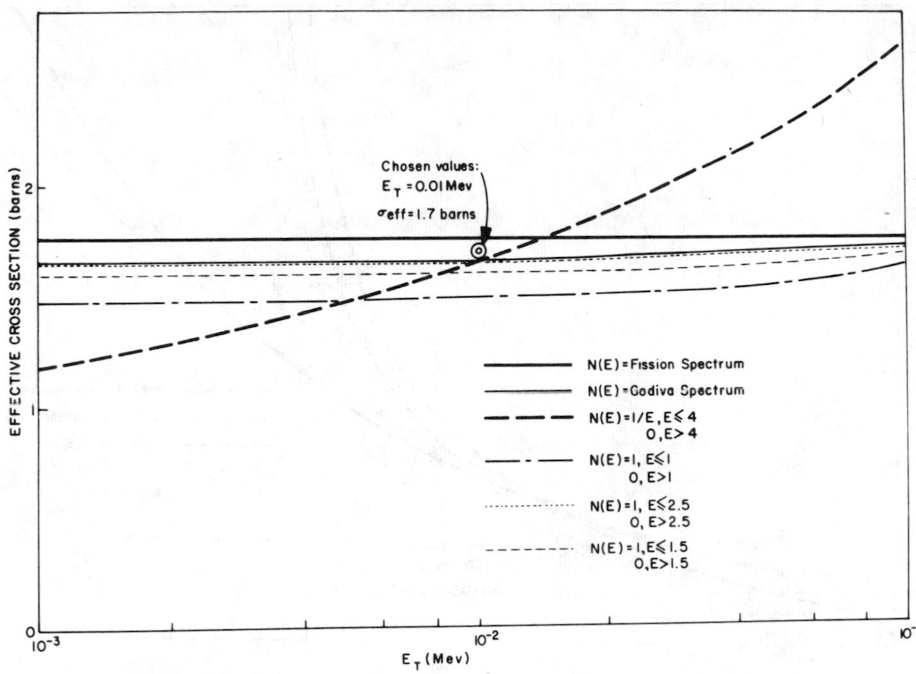

FIG. 1 Effective Cross Section versus Threshold Energy for Various Spectra-Plutonium-239 (Covered with 1.65 g/cm^2 B-10).

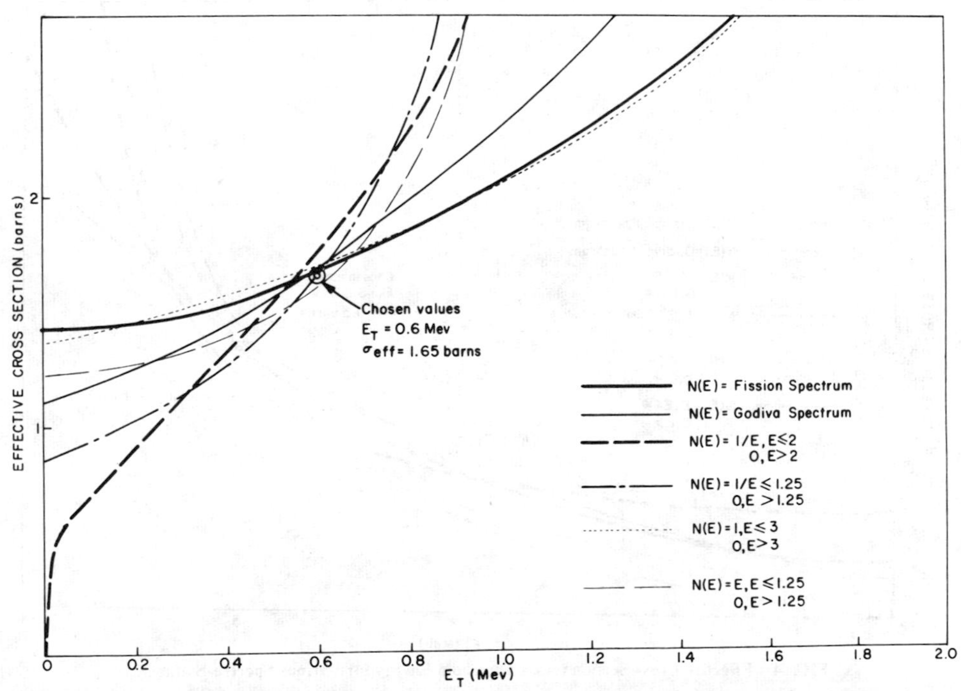

FIG. 2 Effective Cross Section versus Threshold Energy for Various Spectra-Neptunium-237.

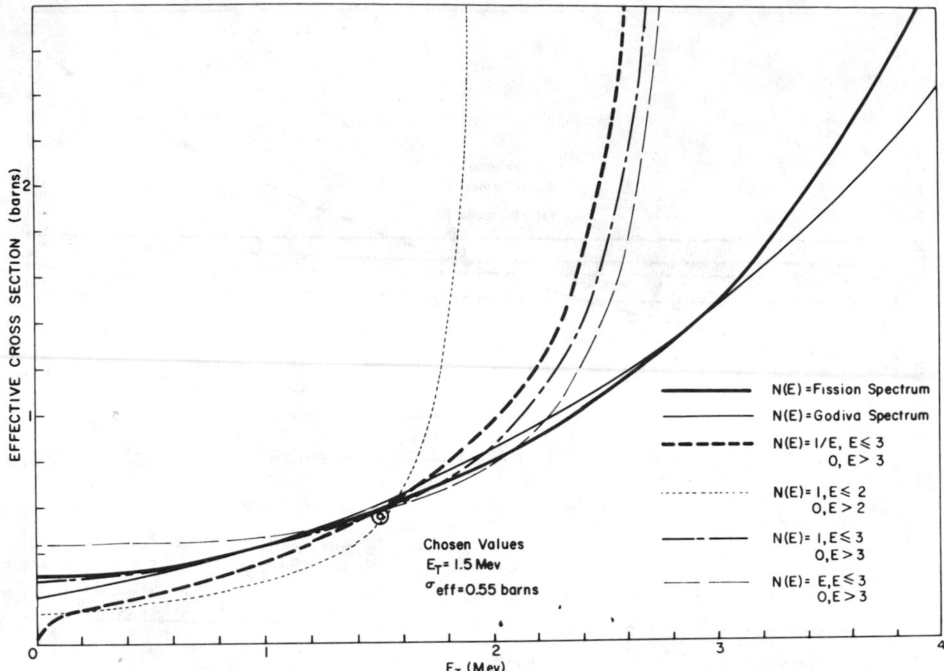

FIG. 3 Effective Cross Section versus Threshold Energy for Various Spectra-Uranium-238.

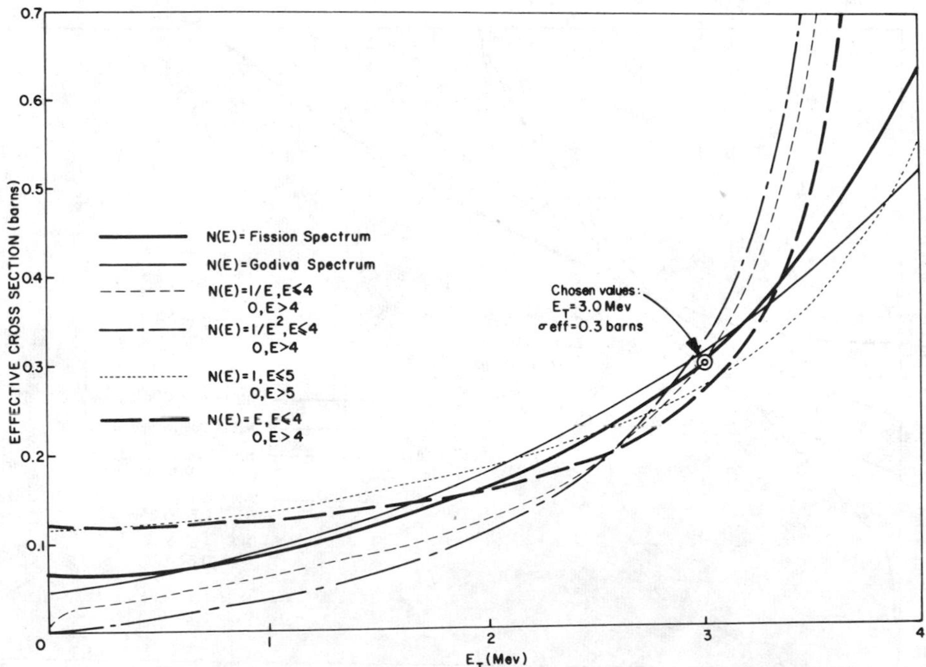

FIG. 4 Effective Cross Section versus Threshold Energy for Various Spectra-Sulfur-32.

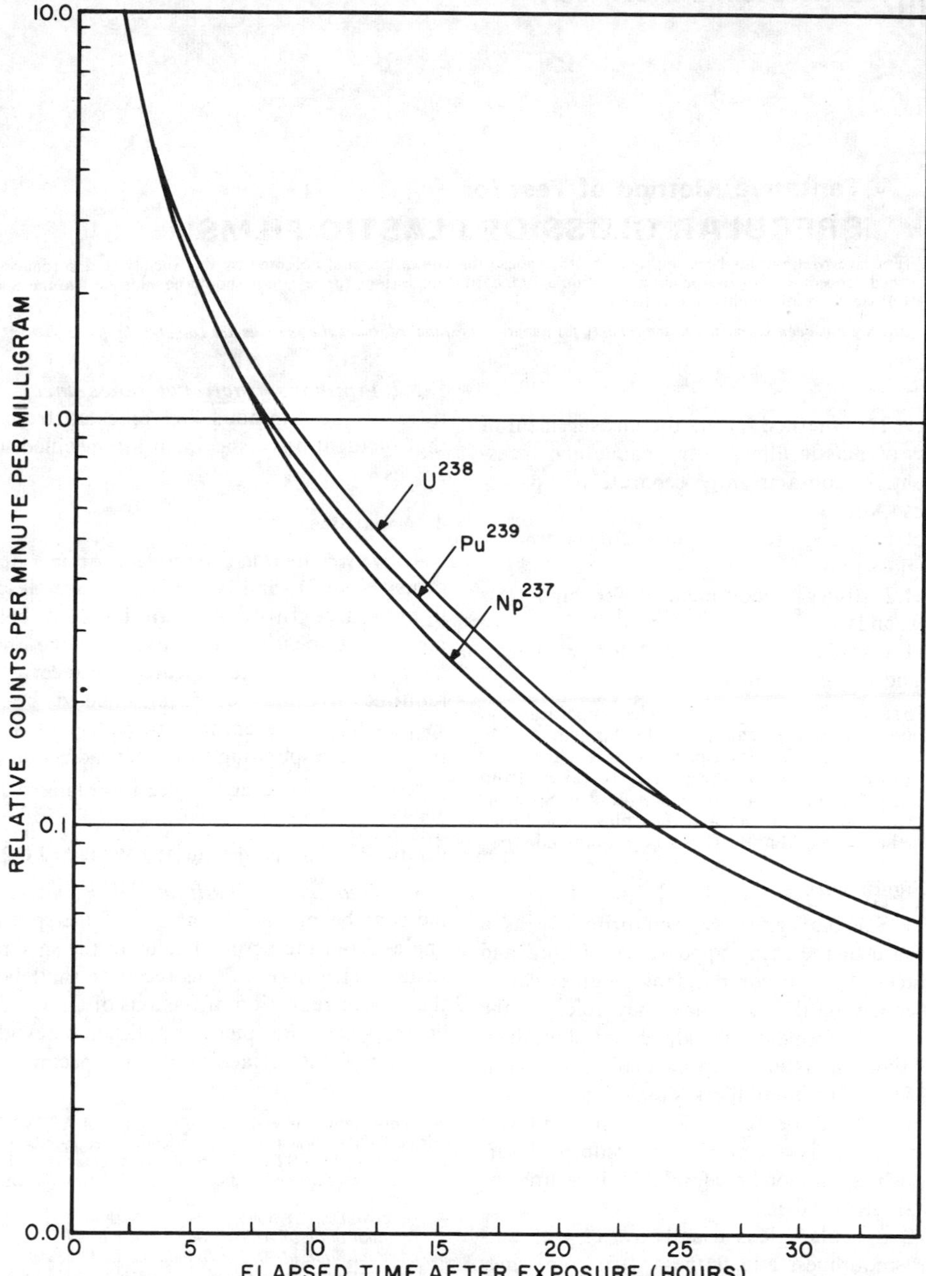

FIG. 5 Gross Fission Product Decay as a Function of Elapsed Time After an Exposure of 10 min. or Less.

ASTM Designation: D 2457 - 65 T

Tentative Method of Test for
SPECULAR GLOSS OF PLASTIC FILMS[1]

This Tentative Method has been approved by the sponsoring committee and accepted by the Society in accordance with established procedures, for use pending adoption as standard. Suggestions for revisions should be addressed to the Society at 1916 Race St:, Philadelphia, Pa. 19103.

This tentative has been submitted to the Society for adoption as standard, but balloting was not complete by press time.

1. Scope

1.1 This method covers the measurement of gloss of plastic films, both opaque and transparent. It contains three separate test procedures (Note 1):

1.1.1 60-deg, recommended for intermediate-gloss films,

1.1.2 20-deg, recommended for high-gloss films, and

1.1.3 45-deg, recommended for intermediate and low-gloss films.

NOTE 1—The 60-deg and 20-deg apparatus and method of measurement duplicate those in ASTM Method D 523, Test for Specular Gloss[2]; those for the 45-deg procedure are similarly taken from ASTM Method C 346, Test for 45-deg Specular Gloss of Ceramic Materials.[3] For plastics in forms other than films, Method D 523 is recommended.

2. Significance

2.1 Specular gloss is used primarily as a measure of the shiny appearance of films and surfaces. Precise comparisons of gloss values are meaningful only when they refer to the same measurement procedure and same general type of material. In particular, gloss values for transparent films should not be compared with those for opaque films, and vice versa. Gloss is a complex attribute of a surface which cannot be completely measured by any single number.[4]

2.2 Specular gloss usually varies with surface smoothness and flatness. It is sometimes used for comparative measurements of these surface properties.

3. Definitions

3.1 *specular gloss*—the relative luminous fractional reflectance[5] of a specimen at the specular direction.

3.2 *luminous fractional reflectance*[5]—the ratio of the luminous flux reflected from, to that incident on, a specimen for specified solid angles.

4. Apparatus

4.1 *Instrumental Components*—Each apparatus (Note 2) shall consist of an incandescent light source furnishing an incident beam, means for locating the surface of the specimen, and a receptor located to receive the required pyramid of rays reflected by the specimen. The receptor shall be a photosensitive device responding to visible radiation.

NOTE 2—The 60 and 20-deg procedures require apparatus identical to that specified in ASTM Method D 523. The 45-deg procedure requires apparatus like that specified in ASTM Method C 346.

4.2 *Geometric Conditions*—The axis of the incident beam shall be at one of the specified angles from the perpendicular to the specimen surface. The axis of the receptor shall be at the mirror reflection of the axis of the incident beam. With a flat piece of polished black glass or other front-surface mirror in specimen po-

---

[1] This method is under the jurisdiction of ASTM Committee D-20 on Plastics. A list of committee members may be found in the ASTM Yearbook. This tentative is the direct responsibility of Subcommittee D-20.40 on Optical Properties.

Effective Dec. 21, 1965.

[2] *Annual Book of ASTM Standards*, Part 27.

[3] *Annual Book of ASTM Standards*, Part 13.

[4] Hunter, R. S., "Gloss Evaluation of Materials," *ASTM Bulletin*, ASTBA, No. 186, December 1952, p. 48. (A study of the history of gloss methods in the American Society for Testing and Materials and other societies; describes the background in the choice of geometry of these methods; contains photographs depicting gloss characteristics of a variety of materials.)

[5] See ASTM Recommended Practice E 179, for Selection of Geometric Conditions for Measurement of Reflectance and Transmittance, *Annual Book of ASTM Standards*, Part 30.

sition, an image of the source shall be formed at the center of the receptor field stop (receptor window). The length of the illuminated area of the specimen shall be equal to not more than one third of the distance from the center of this area to the receptor field stop. The angular dimensions and tolerances of the geometry of the source and receptor shall be as indicated in Table 1. The angular dimensions of the receptor field stop are measured from the center of the test surface. The angular dimensions of the source field stop are most easily measured by the specimen-to-window angular size of the mirror image of the source formed in the receptor field stop. (See Fig. 1 for a generalized illustration of the dimensions.) The tolerances are chosen so that errors of no more than one gloss unit at any point on the scale will result from errors in the source and receptor aperture.[5]

4.3 *Vignetting*—There shall be no vignetting of rays that lie within the field angles specified in 4.2.

4.4 *Spectral Conditions*—Results should not differ significantly from those obtained with a source-filter-photocell combination that is spectrally corrected to yield CIE luminous efficiency with CIE Source C. Since specular reflection is, in general, spectrally nonselective, spectral corrections need be applied only to highly chromatic, low-gloss specimens upon agreement of users of this method.

4.5 *Measurement Mechanism*—The receptor-measurement mechanism shall give a numerical indication that is proportional to the light flux passing the receptor field stop within ±1 percent of full-scale reading.

5. Reference Standards

5.1 *Primary Working Standards* may be highly polished, plane, black glass surfaces. Polished black glass with a refractive index of 1.567 shall be assigned a specular gloss value of 100 for 60 and 20 deg. Polished black glass of refractive index 1.540 shall be assigned a 45-deg specular gloss value of 54.5. For the usual variation of refractive index of black glass, a change in index of 0.001 changes the 45-deg gloss reading by 0.14.

NOTE 3—On the 45 and 60-deg scales, a perfect mirror measures 1000.

5.2 *Secondary Working Standards* of ceramic tile, glass, porcelain enamel, or other materials having hard, flat, and uniform surfaces may be calibrated from the primary standard on a glossmeter determined to be in strict conformance with the requirements prescribed in 4.2.

6. Preparation and Selection of Test Specimens

6.1 This method does not cover preparation techniques. Whenever a test for gloss requires the preparation of a test specimen, report the technique of specimen preparation.

6.2 Specimen surfaces shall have good planarity, since surface warpage, waviness, or curvature may seriously affect test results. The directions of machine marks, or similar texture effects, shall be parallel to the plane of the axes of the two beams, unless otherwise specified.

6.3 Surface test areas shall be kept free of soil and abrasion. Gloss is due chiefly to reflection at the surface; therefore, anything that changes the surface physically or chemically is likely to affect gloss.

7. Mounting Films for Measurement

7.1 Any nonrigid film must be mounted in a device that will hold it flat, but will not stretch it while it is measured. Three different filmholding devices have each proved satisfactory for at least some types of films:

7.1.1 *Vacuum Plate* (see Fig. 2) is required for stiff films. Connect the vacuum plate by rubber tube to a vacuum pump or vacuum line. With thin, soft films it is sometimes necessary to use a valve and pressure gage and to limit the vacuum so as to keep from collapsing the soft film into the pores of the ground plate.

7.1.2 *Flat Plate* with two-side pressure-sensitive tape (see Fig. 3). Make sure each specimen is pulled smooth, but not stretched before holding it by the two strips of adhesive tape. Replace the tape whenever it loses its adhesiveness.

7.1.3 *Telescoping Ring or Hoop* (see Fig. 4)—To mount the specimen in the telescoping ring, lay the flexible film over the base (male) section and drop the top over the base. Push down carefully, taking care to pull the test film taut without stretching it. Measure the taut area.

8. Procedure

8.1 Operate the glossmeter in accordance with the manufacturer's instructions.

8.2 Calibrate the instrument at the start and completion of every period of glossmeter operation and during the operation at sufficiently frequent intervals to assure that the instrument response is practically constant. If at any time an instrument fails to repeat readings of the standard to within 2 percent of the prior setting, the intervening results should be rejected. To calibrate, adjust the instrument to read correctly the gloss of a highly polished standard, and then read the gloss of a standard having poorer image-forming characteristics. If the instrument reading for the second standard does not agree within 1 percent of its assigned value, do not use the instrument without readjustment, preferably by the manufacturer.

NOTE 4—Correct readings on black-glass and intermediate standards do not guarantee instrument conformity to specification requirements. In addition to measurements with gloss standards, dimensional checks for conformity to the geometric requirements of 4.2 should be made.

8.3 Clear plastic films frequently measure more than 100 on the 60 and 20-deg scales, and sometimes on the 45-deg scale. This can happen whenever specular reflection at *both* surfaces of a more or less clear plastic film contribute to specular reflectance. If an instrument has a scale reading only to 100, measure such specimens by taking advantage of the fact that the measurement scales are linear. Adjust the instrument to read some fraction (such as one half) the value of the standard. Then check with an intermediate standard to see that it reads the same fraction of its assigned gloss value. Divide by the same fraction all results obtained with the scale so adjusted.

8.4 Measure at least three portions of specimen surface to obtain an indication of uniformity. Measure parallel to the machined direction as prescribed in 6.2.

NOTE 5—It is often desirable to compare these readings with readings taken across the machine direction. Difference in the readings will relate to the prominence of the machine marks.

9. Report

9.1 The report shall include the following:

9.1.1 Type of specimen, its gloss reading, the geometry (60, 20, or 45-deg), and the specimen holder employed,

9.1.2 If any of the gloss readings differ by more than 10 percent from the average, all gloss readings for that specimen,

9.1.3 Where preparation of the test specimen has been necessary, description or identification of the method of preparation,

9.1.4 Identification of the glossmeter by the manufacturer's name and model designation, and

9.1.5 Identification of the working standard or standards of gloss used.

NOTE 6: *Diffuse Correction*—It can be said that the light reflected by a specimen may be divided into one part reflected specularly in the direction of mirror reflection (associated with gloss) and another part reflected diffusely in all directions (associated with lightness on the white-gray-black scale). According to this picture, a gloss reading always needs to be diminished to compensate for that amount of the measured light attributable to diffuse reflectance. Although it is seldom possible in practice to analyze reflected light according to this picture and say exactly what part is diffuse and what part is specular, it is nevertheless frequent practice where gloss values of light and dark surfaces are being compared to "correct" (diminish) specular gloss settings for diffuse reflectance. If diffuse corrections are desired as additional information, measure 45-deg, 0-deg luminous directional reflectances of specimens in accordance with ASTM Method E 97, Test for 45-deg, 0-deg Directional Reflectance of Opaque Specimens by Filter Photometry.[6] Multiply reflectance values in percentage by the following factors[7] for diffuse corrections in gloss units:

60-deg	0.021
20-deg	0.013
45-deg	0.055

10. Precision

10.1 Readings obtained on the same instrument should be repeatable to within 1 percent of the magnitude of the readings. Readings obtained on different instruments should be reproducible to within 3 percent of the magnitude of the readings.

[6] *Annual Book of ASTM Standards*, Part 30.
[7] Taken from Methods D 523 and C 346.

TABLE 1 Angular Dimensions and Tolerances of Geometry of Source and Receptor Field Stops

Geometry, deg	Incidence Angle, deg	Source Field Stop: In Plane of Measurement, deg	Source Field Stop: Perpendicular to Plane of Measurement, deg	Receptor Field Stop: In Plane of Measurement, deg	Receptor Field Stop: Perpendicular to Plane of Measurement, deg
60	60 ± 0.1	0.75 ± 0.25	3.0 max	4.4 ± 0.1	11.7 ± 0.2
20	20 ± 0.1	0.75 ± 0.25	3.0 max	1.80 ± 0.05	3.6 ± 0.1
45	45 ± 0.1	1.4 ± 0.4	3.0 ± 1.0	8.0 ± 0.1	10.0 ± 0.2

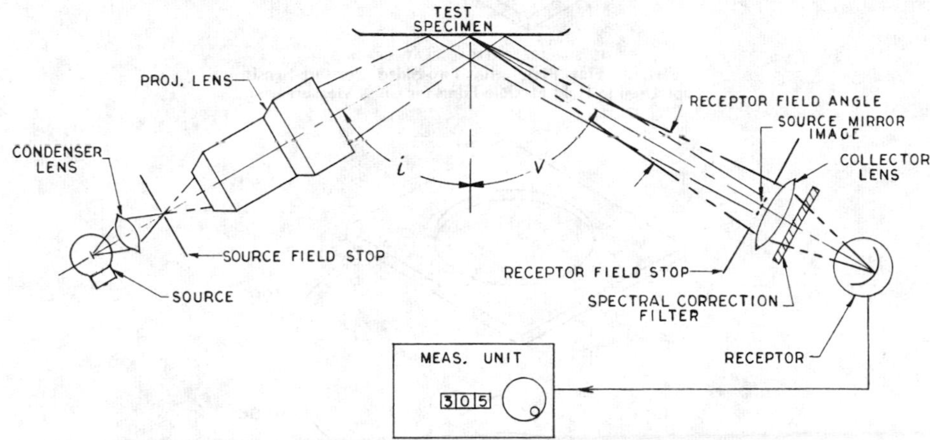

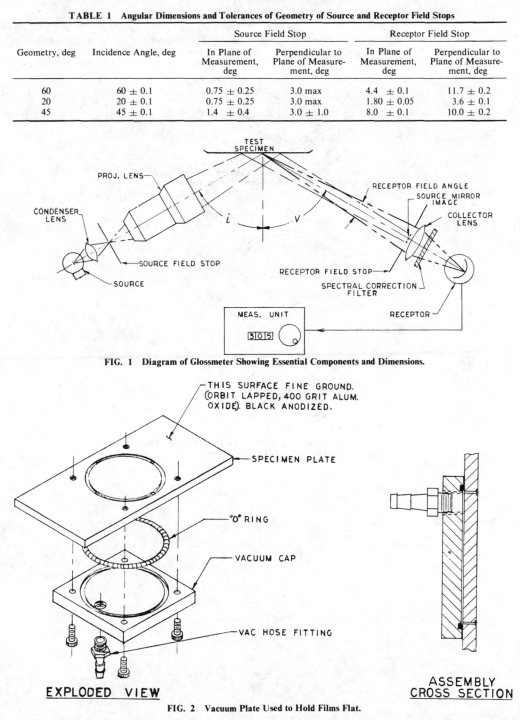

FIG. 1 Diagram of Glossmeter Showing Essential Components and Dimensions.

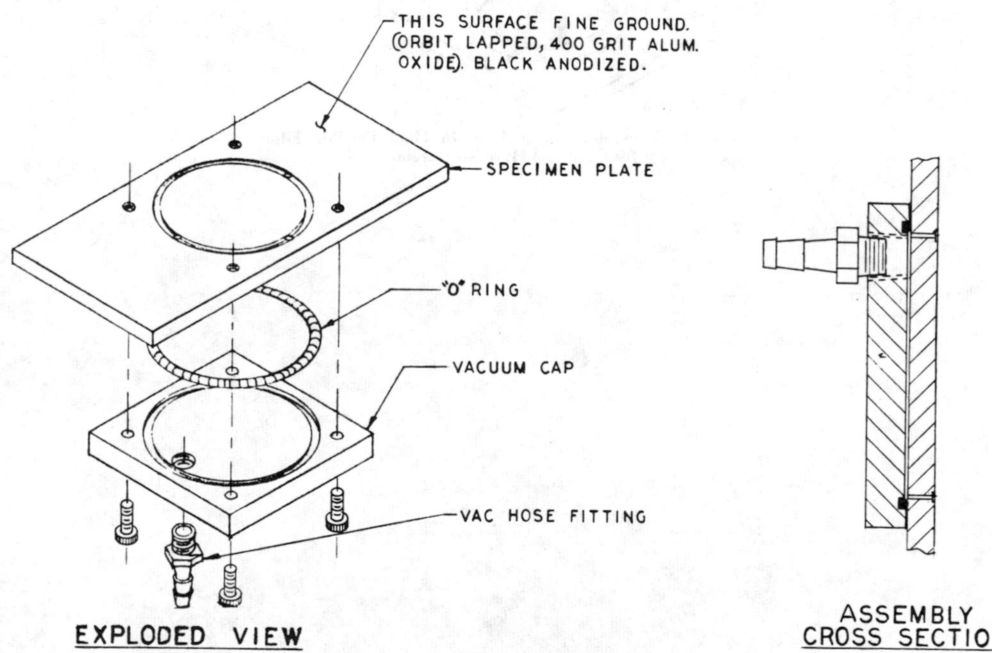

FIG. 2 Vacuum Plate Used to Hold Films Flat.

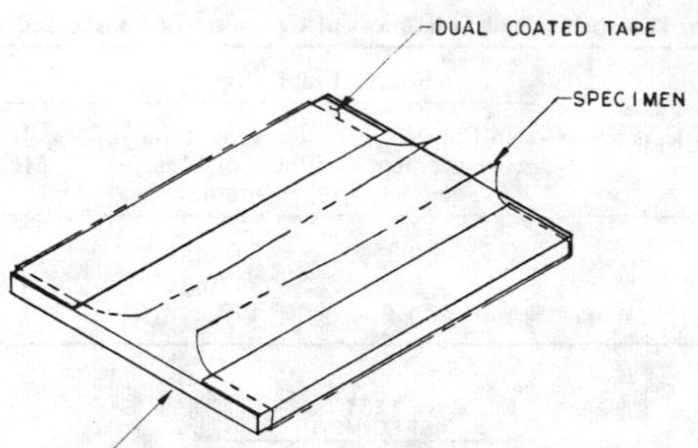

FIG. 3 Flat Plate with Two-Sided Pressure-Sensitive Tape Used to Hold Flexible Films for Gloss Measurement.

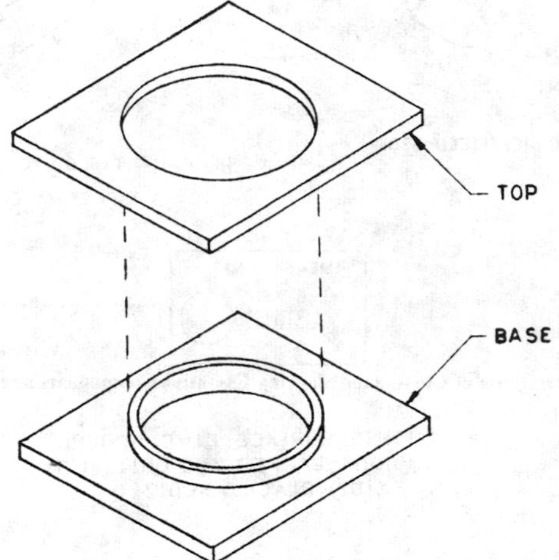

FIG. 4 Telescoping Ring Used to Hold Flexible Films for Gloss and Haze Measurements.

ASTM Designation: D 2565 – 66 T

Tentative Recommended Practice for
OPERATING XENON ARC-TYPE (WATER-COOLED) LIGHT- AND WATER-EXPOSURE APPARATUS FOR EXPOSURE OF PLASTICS[1]

This Tentative Recommended Practice has been approved by the sponsoring committee and accepted by the Society in accordance with established procedures, for use pending adoption. Suggestions for revisions should be addressed to the Society at 1916 Race St., Philadelphia, Pa. 19103.

This tentative has been submitted to the Society for adoption, but balloting was not complete by press time.

1. Scope

1.1 This recommended practice covers procedures applicable when ASTM Recommended Practice E 239, Operating Light- and Water-Exposure Apparatus (Water-Cooled Xenon-Arc Type) for Exposure of Nonmetallic Materials,[2] is employed for exposure of plastics. Reference is made to sample preparation and evaluation of test results.

1.2 Two procedures are provided:

1.2.1 *Procedure A*—Normal operating procedure for comparative evaluation within a series exposed simultaneously in one instrument.

1.2.2 *Procedure B*—Procedure for use in comparison of results among instruments.

2. Significance

2.1 Xenon arcs have been shown to have a spectral energy distribution, when properly filtered, which closely simulates the spectral distribution of sunlight at the surface of the earth.[3,4,5] In comparison to sunlight exposure this test is not necessarily an accelerated test.

2.2 Since the emitted energy from the xenon lamps decays with time and since the other parameters of temperature and water do not represent a specific known climatic condition, the results of laboratory exposure are not necessarily intended to correlate with data obtained by outdoor weathering. There may be no positive correlation of exposure results between the xenon arc and other laboratory weathering devices.

2.3 Procedure A may be used for obtaining empirical results for laboratory comparison. To assure reproducible results for sets of specimens exposed at the same time within any one instrument or even on similar instruments in any location for equivalent exposure time Procedure B must be followed.

3. Test Specimens

3.1 The size and shape of specimens to be exposed will be determined by the specifications of the particular test method used to evaluate the effects of the exposure on the specimens; the test method shall be determined by the parties concerned.

3.2 Specimens the dimensions of which are less than one half the height of the specimen rack must be located equidistant from the center horizontal plane of the arc to assure uniform radiation. Incident energy over the sample rack may vary as much as 20 percent due to the geometry of the field of view. Therefore samples should be periodically rotated.

---

[1] This recommended practice is under the jurisdiction of ASTM Committee D-20 on Plastics. A list of committee members may be found in the ASTM Yearbook. This tentative is the direct responsibility of Subcommittee D-20.50 on Permanence Properties.

Effective Sept. 21, 1966.

[2] *Annual Book of ASTM Standards*, Part 30.

[3] Searle, N. Z., Giesecke, P., Kinmonth, R., and Hirt, R., "Ultraviolet Spectral Distributions and Aging Characteristics of Xenon Arcs and Filters," *Applied Optics*, APOPA, Vol 3, No. 8, August 1964, pp. 923–27.

[4] Boettner, E. A., and Miedler, L. J., "Simulating the Solar Spectrum with a Filtered High Pressure Xenon Lamp," *Applied Optics*, APOPA, Vol 2, No. 1, January 1963, pp. 105–108.

[5] Hirt, R. C., Schmitt, R. G., Searle, N. Z., and Sullivan, A. P., "Ultraviolet Spectral Energy Distributions of Natural Sunlight and Accelerated Test Light Sources," *Journal of the Optical Society of America*, JOSAA, Vol 50, No. 7, 1960, pp.706–13.

NOTE 1—Specimens to be exposed for several hundred hours may be relocated periodically to compensate for the variation in intensity in the vertical plane. The rotation of specimen position in the rack may be in four steps, employed in the same sequence, from upper to lower specimen row and inverted in both specimen rows. The effect of this variation in intensity on the specimen may depend upon the material as well as upon the lamp-to-specimen distance.

3.3 Specimens to be compared should be mounted with the exposed surface in the same vertical plane at equal distances from the lamp.

3.4 Replicate specimens shall be used as unexposed controls when nondestructive original measurements cannot be made on the specimens to be exposed.

NOTE 2—The face of a specimen should not be masked for the purpose of showing the effects of various exposure times in one panel. Misleading results may be obtained by this method since the masked portion is still exposed to temperature and humidity cycles.

4. Apparatus

4.1 The apparatus employed shall utilize a water-cooled xenon-arc lamp as the source of radiation and shall be one of the following four general types or their equivalent:

4.1.1 *Type A*—Water-cooled xenon-arc apparatus, 2500 W, 508-mm (20-in.) diameter specimen rack, automatic control of temperature and cycles, manual control of humidity.

4.1.2 *Type AH*—Same as Type A, except with automatic instead of manual control of humidity.

4.1.3 *Type B*—Water-cooled xenon arc apparatus, 6000 W, 959-mm (37$^3/_4$-in.) diameter specimen rack, automatic control of temperature and cycles, humidity not controlled.

4.1.4 *Type BH*—Same as Type B, except with automatic humidity control.

4.2 The photodegradation of most plastics is due primarily to the ultraviolet portion of the solar spectrum. Since discharge lamps have a progressive drop in radiation output with continued use and since high intensity radiation will produce changes in the transmission characteristics of the filters, provision shall be made to minimize the effect of changes in the energy incident on the sample by pre-exposing the filters. The incident ultraviolet energy in particular is most significantly affected. Emitted energy from the xenon lamp decreases rapidly during the first 100 h followed by a continuous gradual decrease. Glass filters as they are used become less transparent to ultraviolet radiation due to solarization.

4.3 The emitted ultraviolet energy is best controlled by using preaged burners and filters. Borosilicate glass filters[6] should be used because they cut off light below 2750 Å. The solar spectrum at sea level contains insignificant amounts of radiation below 2750 Å.

4.4 A monitoring device consisting of a barrier layer photocell and an interference filter with peak transmittance at 328 nm is suitable for determining the ultraviolet radiation output.[7]

4.5 Deposition on the outer quartz walls of the burner or walls of borosilicate glass filters will reduce incident radiation. The surfaces should be cleaned with a nonabrasive cleanser[8] and a soft cloth followed by flushing with water. Contamination may be minimized by the use of a corrosion-resistant centrifugal pump to circulate cooling water to the lamp.

5. Procedure

5.1 *General:*

5.1.1 Two procedures are presented. Procedure A may be used for routine testing. A control must be used for each series exposed simultaneously because the radiation intensity is not rigorously controlled. In Procedure B, a referee method, the exposure conditions are rigorously controlled and improved reproducibility among similar instruments and different laboratories can be obtained.

5.1.2 Measure the testing temperature by a black panel thermometer as described in 3.8 of Recommended Practice E 239. Regulate the temperature on the basis of the black panel thermometer at an equilibrium temperature measured during the dry portion of the cycle as agreed upon by the parties concerned with the test.

5.1.3 The light-water spray-humidity cycle that may be used is limited only by the type of instrument. The cycle will be determined by

[6] Corning glass No. 7740 (Pyrex) has been found suitable.

[7] Allen, E., "A Monitoring Device for Ultraviolet Energy in the Atlas Xenon Weatherometer," *Applied Optics*, APOPA, Vol 4, No. 7, July 1965, pp. 835–38.

[8] A nonabrasive earth cleanser such as Bon Ami has been found suitable.

the parties concerned with the test. The specimen spray water shall contain less than 20 ppm total solids. To eliminate silica deposits on the specimen it may be necessary to maintain less than 1 ppm of SiO_2. This may require distillation or the use of a four-bed demineralizer. All plumbing from the still or demineralizer to the spray orifice shall be fabricated of an inert material such as aluminum, stainless steel, tin, or suitable unplasticized plastics.

5.2 *Procedure A:*

5.2.1 If a monitor is available to check the ultraviolet energy output, then use no lamp assembly on Type B instruments (4.1.3) that cannot maintain a minimum of 30 $\mu W/cm^2/3$ nm bandpass at 330 nm at sample distance. Use no lamp assembly on Type A instruments (4.1.1) that cannot maintain a minimum of 50 $\mu W/cm^2/3$ nm bandpass at 330 nm at sample distance. It is recommended that the lamp assembly be periodically monitored.

5.2.2 If no monitor is available, and therefore the output is not known, operate the instrument as follows:

5.2.2.1 Use only a burner and outer filter that have been preaged 100 h and have had less than 1000 h of service.

5.2.2.2 At 250-h intervals, replace the inner filter with a preaged filter (20 h).

5.2.2.3 When a newly preaged lamp assembly is installed, set the wattage on the 6000-W unit at 5750 W and on the 2500-W unit at 2250 W. At every 250-h interval, increase the wattage by 250 W.

5.2.2.4 Mount specimens for exposure in accordance with Section 3.

5.2.2.5 Only results from specimens exposed simultaneously may be compared. It is recommended that a standard control specimen accompany each series. Results from exposure at different times or on different instruments are not expected to be comparable.

5.3 *Procedure B:*

5.3.1 Based on round-robin results,[9] this procedure has been shown capable of giving reproducible results at any time within any one instrument or even among similar instruments.

5.3.2 Conduct the test as follows:

5.3.2.1 Start the test with a new preaged burner (100 h).

5.3.2.2 Start the test with a new preaged inner filter (20 h).

5.3.2.3 Start the test with a new preaged outer filter (100 h).

5.3.2.4 Keep the wattage input setting throughout the tests at 2500 W for Type A and 6000 W for Type B instruments.

5.3.2.5 Replace the inner filter every 250 h.

5.3.2.6 Since the intensity from the lamp varies over its 1000-h prescribed life, all polymer test specimens *must* be started simultaneously when the new preaged lamp assembly is installed. The radiation emitted from the lamp decays sufficiently during 1000 h of use to prevent assurance of reproducibility of results for samples exposed to equivalent intervals at different periods of the operation of the lamp.

5.3.2.7 Specimens for exposure shall be so mounted vertically as to be contained within 75 mm (3 in.) of the center horizontal plane of the lamp for the 2500-W unit and 150 mm (6 in.) for the 6000-W unit.

NOTE 3—All indicated numbers should be ±10 percent in value.

6. Report

6.1 The report shall include the following:

6.1.1 Procedure used,

6.1.2 Type and model of exposure device,

6.1.3 Monitoring device used, if any,

6.1.4 Ages of xenon burner(s) and filters used during test,

6.1.5 Ellapsed exposure time,

6.1.6 Relocation procedure used for exposure of specimens,

6.1.7 Light-water-humidity cycle used,

6.1.8 Operating black panel temperature,

6.1.9 Operating relative humidity if controlled,

6.1.10 Spray water used,

6.1.11 Spray nozzle if other than F-80, and

6.1.12 Appearance of exposed test specimens.

---

[9] Supporting data giving results of cooperative tests have been filed at ASTM Headquarters as RR42: D 20.

ASTM Designation: D 2568 – 70

Recommended Practice for
CALCULATION OF ABSORBED DOSE FROM GAMMA RADIATION[1]

This Recommended Practice is issued under the fixed designation D 2568; the number immediately following the designation indicates the year of original adoption or, in the case of revision, the year of last revision. A number in parentheses indicates the year of last reapproval.

1. Scope

1.1 This recommended practice presents a technique for calculating the dose or energy absorbed in a material from knowledge of the gamma radiation field and the composition of the material (**1**).[2] The procedure is also applicable for X-radiation and electrons provided the effective energy of the beam is known and falls within the range from 0.10 to 3.0 MeV. From the absorbed dose in one material, the absorbed dose in any other material exposed to the same radiation field may be calculated. The procedure is restricted to use within the above mentioned energy range and with materials composed of the elements: hydrogen (H), carbon (C), nitrogen (N), oxygen (O), fluorine (F), silicon (Si), sulfur (S), chlorine (Cl), and phosphorus (P).

2. Significance

2.1 Exposure is a measure of the radiation field to which a material is exposed, whereas the absorbed dose is the energy imparted to the irradiated material (**2**). Therefore, the absorbed dose is a more meaningful parameter for use in relating the effects of radiation on materials. In order to obtain a true measure of absorbed dose, charged particle equilibrium must be attained (see Appendix A1). The unit of exposure is the roentgen. This unit describes an electromagnetic radiation field in terms of the ionization which the radiation produces in air.

2.2 Absorbed dose, on the other hand, is expressed solely in terms of energy absorbed by the irradiated material, regardless of the nature of the radiation field. The unit of absorbed dose is the "rad." When different materials are exposed to the same radiation field, they may absorb different amounts of energy.

NOTE 1: **Caution**—If the radiation is attenuated by a significant thickness of absorber the energy of the effective radiation will have been degraded, and it may be necessary to correct for this. For example, if a pool-type ^{60}Co source is used, and there is 10 cm of water between the source and the specimen, the radiation dose to the specimen will be in error by perhaps 20 percent if calculated on the basis of the energy of the virgin ^{60}Co gammas, and if the material contains an appreciable fraction of high Z elements (for example, polymers with chlorine or sulfur).

3. Definitions

3.1 For definitions of terms used in this recommended practice refer to ASTM Definitions E 170, Terms Relating to Dosimetry.[3]

4. Calculation of Absorbed Dose per Unit Exposure

4.1 Table 1 gives values of D_i (absorbed dose in rads per unit exposure in roentgens) for photon energies between 0.1 and 3.0 MeV, for the elements covered by this recommended practice (see Appendix A2 for derivation of D_i values). The values of D_i are valid only under conditions of charged particle equilibrium (**2**) (see Appendix A1 for explanation.

[1] This recommended practice is under the joint jurisdiction of ASTM Committee D-9 on Electrical Insulating Materials and Committee D-20 on Plastics. A list of committee members may be found in the ASTM Yearbook. This standard is the direct responsibility of Subcommittee D-20.20 on Radiation Methods.

Current edition effective Feb. 27, 1970. Originally issued 1966. Replaces D 2568 – 68.

[2] The boldface numbers in parentheses refer to the list of references appended to this recommended practice.

[3] *Annual Book of ASTM Standards*, Part 30.

4.2 The absorbed dose per unit exposure for any material of known composition may be calculated using the following procedure:

4.2.1 Determine the weight fraction of the elements in the material.

4.2.2 From Table 1, obtain D_i values for the elements in the material at the incident photon energy.

4.2.3 Substitute these values and their respective weight fractions in the following equation, and solve for D_m:

$$D_m = \Sigma_i f_i D_i$$

where:

D_m = absorbed dose per unit exposure in the material,

f_i = weight fraction of element i in the material, and

D_i = absorbed dose per unit exposure for element i.

4.2.4 Multiply the value obtained for D_m by the exposure dose in roentgens (obtained by test) to obtain the absorbed dose in the material.

4.3 For example, suppose it is required to calculate the absorbed dose in a polytetrafluoroethylene (PTFE) film, corresponding to an exposure of 10^6 roentgens of 1 MeV photons, the latter having been determined using an air ionization chamber with sufficient wall thickness for "charged particle equilibrium" (**2**). Reference to Appendix A1 indicates the necessity of using an absorber either 0.50-cm thick material with an electron density of 3.3×10^{23} or 0.22 cm of material having an electron density of 7.6×10^{23}.

4.3.1 Polytetrafluoroethylene (PTFE) has the empirical formula $(CF_2)_n$ (disregarding chain ends, unsaturation, and impurities); hence, the weight fractions of C and F are 0.24 and 0.76, respectively.

4.3.2 Table 1 gives $D_C = 0.866 \pm 0.004$ rads/roentgen and $D_F = 0.822 \pm 0.004$ rads/roentgen for 1-MeV photons.

4.3.3 Substituting these values in the equation gives

$$D_{CF_2} = (0.24 \times 0.866) + (0.76 \times 0.822) = 0.83 \text{ rads/roentgen}$$

4.3.4 Therefore, an exposure of 10^6 roentgens gives an absorbed dose in this material of 0.83×10^6 rads. For a photon energy in the range between 0.5 and 1.5 MeV, this equation may be simplified to

$$D_m = 0.85 f_H - 0.05 f_F - 0.04 f_{Cl} - 0.03 f_P + 0.87$$

4.4 Table 2 lists the results of these calculations for some of the commonly used materials at two energy levels.

5. Calculation of Absorbed Dose in One Material from That in Another Material

5.1 The data in Table 1 are also applicable for the comparison of absorbed dose by different media without reference to the exposure, provided the latter is held constant. From Eq 2 in Appendix A2, it is seen that for constant exposure, the ratio between the D_i's for any two media equals the ratio between their energy absorption coefficients, $(\mu_{en})_m$; hence, the ratio equals the ratio between the absorbed doses. When chemical dosimetry is employed, the measured chemical change may be converted directly to the absorbed dose in rads. By utilizing the ratio of D_m values, one can calculate the absorbed dose in any material from a knowledge of the absorbed dose in a chemical dosimeter.

5.2 For example, if one measures an absorbed dose in the Fricke Dosimeter (see ASTM Method D 1671, Test for Absorbed Gamma Radiation Dose in the Fricke Dosimeter[4]) to be 5.0×10^5 rads for a 1-h exposure in a ^{60}Co irradiator and it is desired to determine the absorbed dose in a specimen of polyethylene in the same irradiator for 1 h, one follows the procedure outlined in 4.2.

5.2.1 The Fricke Dosimeter has weight fractions of H, O, and S of 0.11, 0.88, and 0.013, respectively. The empirical formula for polyethylene is $(CH_2)_n$ and the weight fraction of C and H are 0.86 and 0.14, respectively.

5.2.2 From Table 1, $D_H = 1.72$, $D_O = 0.869$, $D_S = 0.869$, and $D_C = 0.866$ for 1.0-MeV photon energy.

5.2.3 Substituting these values in Eq 3 of Appendix A2, one obtains:

$$D_{CH_2} = (0.86 \times 0.866) + (0.14 \times 1.72) = 0.986 \approx 0.99$$

and

$$D_{Fricke} = (0.11 \times 1.72 + (0.88 \times 0.869) + (0.013 \times 0.869) = 0.965 \approx 0.97$$

[4] *Annual Book of ASTM Standards*, Part 29.

The ratio $D_{CH_2}/D_{Fricke} = 0.99/0.97 = 1.02$; therefore, the absorbed dose in polyethylene is equal to

$$1.02 \times 5.0 \times 10^5 \text{ rads} = 5.10 \times 10^5 \text{ rads}$$

5.3 The information in Table 2 may also be used to convert the dose abosrbed in any material listed to the dose absorbed by any other material listed provided the energy of the incident radiation is 0.1 MeV or between 0.5 and 3 MeV. For example, if one has determined the absorbed dose in polyethylene irradiated in a ^{60}Co irradiator, found it to be 5.10 $\times 10^5$ rads, and wishes to determine the dose that will be absorbed by a specimen of poly(vinyl chloride) (PVC) in the same irradiator for the same time, all that is necessary is to multiply the known absorbed dose by the ratio of the D_m's for the material involved:

$$D_{PVC} = 0.89;\ D_{CH_2} = 0.99;$$

therefore,

$$(D_{PVC}/D_{CH_2}) \times 5.10 \times 10^5 \text{ rads}$$
$$= (0.89/0.99) \times 5.10 \times 10^5$$
$$= 4.58 \times 10^5 \text{ rads absorbed by the poly(vinyl chloride) specimen.}$$

6. Limitations

6.1 Since the energy distribution through the specimen being irradiated will vary and is a function of the specimen thickness, density, and the energy of the incident radiation, it is necessary to decide how much variation in dose one is willing to tolerate as the radiation penetrates the specimen. The most commonly used irradiation facilities have radiation sources in the energy range 0.5 to 1.5 MeV. If one arbitrarily sets a limit of 10 percent for the difference between the dose absorbed by the front and rear of the specimen (10 percent attenuation through the specimen), then the specimen thickness is limited to 1.1 cm for 0.5-MeV and 1.8 cm for 1.5-MeV radiation, assuming no buildup and a specimen of electron density 3.3×10^{23} electrons/cm^3 (electron density of water at 25 C) (see Note A1 in Appendix A1).

6.2 Figure A3 is a plot of energy versus thickness of sample for 10 and 25 percent attenuation through a specimen of unit density and 3.3×10^{23} electrons/g (that is, H_2O). The curves in Fig. A3 will shift to the left for higher electron density material and to the right for lower electron density material (Note 2). The thickness for 10 or 25 percent attenuation in the specimen will be the value obtained from Fig. A3 divided by the ratio of the electron density of the specimen to 3.3×10^{23}. Since the curves are calculated on the basis of attenuation only and buildup in the thicker specimens is neglected, the curves represent a maximum attenuation for a given energy and thickness.

NOTE 2—All figures are valid only for materials composed of elements having atomic numbers less than 18.

REFERENCES

(1) Cheek, C. H., and Linnenbom, V. J., "Calculation of Absorbed Dose," *Power Apparatus and Systems*, Am. Inst. Electrical Engrs., Vol 79, December 1960, pp. 1004–1016.

(2) ICRU Report 10B, *Physical Aspects of Irradiation*, Handbook 85, National Bureau of Standards, U. S. Government Printing Office, Washington, D. C., 20402, 1962.

(3) Fano, U., "Gamma Ray Attenuation," *Nucleonics*, Vol 11, No. 8, 1953, pp. 8–12.

(4) Grodstein, G. W., "X Ray Attenuation Coefficients from 10 keV to 100 MeV, *Circular 583*, Nat. Bureau Standards, U. S. Government Printing Office, 1957.

TABLE 1 Absorbed Dose per Unit Exposure Rads/Roentgen

$D_i = 0.869\ [(\mu_{en})_i]/(\mu_{en})_{air}$

Photon Energy, MeV	H	C	N	O	F	Si	S	Cl	P
0.10	1.53	0.802	0.835	0.873	0.896	1.67	2.25	2.56	1.88
0.15	1.67	0.852	0.862	0.872	0.845	1.06	1.22	1.26	1.09
0.20	1.71	0.866	0.866	0.879	0.833	0.947	1.00	1.00	0.947
0.30	1.72	0.869	0.872	0.872	0.823	0.893	0.909	0.950	0.875
0.40	1.72	0.866	0.869	0.869	0.822	0.878	0.884	0.855	0.851
0.50	1.72	0.869	0.869	0.869	0.822	0.866	0.878	0.845	0.843
0.60	1.72	0.869	0.869	0.869	0.825	0.869	0.872	0.839	0.843
0.80	1.72	0.866	0.869	0.869	0.823	0.866	0.863	0.833	0.843
1.0	1.72	0.866	0.869	0.869	0.822	0.863	0.869	0.828	0.837
1.5	1.72	0.869	0.869	0.869	0.821	0.869	0.866	0.824	0.835
2.0	1.72	0.869	0.869	0.869	0.824	0.873	0.873	0.835	0.846
3.0	1.68	0.865	0.869	0.873	0.830	0.890	0.890	0.865	0.865

TABLE 2 Absorbed Dose as a Function of Material Composition and Photon Energy

Material	Empirical Formula	Absorbed Dose, rads/roentgen	
		1 MeV[a]	0.1 MeV
Polystyrene	$(CH)_n$	0.94	0.86
Polyethylene	$(CH_2)_n$	0.99	0.90
Polyamide (ε-amino caproic acid)	$(C_6H_{11}ON)_n$	0.95	0.89
Polydimethyl siloxane	$(C_2H_6OSi)_n$	0.93	1.21
Ethylene polysulfide	$(C_2H_4S_4)_n$	0.90	2.01
Vinylidene chloride copolymer	$(C_4H_5Cl_3)_n$	0.87	2.00
Polytetrafluoroethylene	$(CF_2)_n$	0.83	0.87
Polychlorotrifluoroethylene	$(C_2F_3Cl)_n$	0.83	1.38
Poly(vinyl chloride)	$(C_2H_3Cl)_n$	0.89	1.82
Poly(vinylidene chloride)	$(C_2H_2Cl_2)_n$	0.86	2.10
Poly(vinyl pyrolidone)	$(C_6H_9NO)_n$	0.94	0.87
Poly(vinyl carbazole)	$(C_{14}H_{11}N)_n$	0.92	0.85
Poly(vinyl acetate)	$(C_4H_6O_2)_n$	0.93	0.88
Poly(methyl methacrylate)	$(C_5H_8O_2)_n$	0.94	0.88
Tributyl phosphate	$(C_4H_9)_3PO_4$	0.95	1.02
Fricke Dosimeter	...	0.96	0.96

[a] These data may also be used for ^{60}Co and ^{137}Cs gamma rays and X rays with an effective energy of 0.7 to 3 MeV.

APPENDIXES

A1. Charged Particle Equilibrium Thickness

A1.1 Whenever a material is irradiated with X or gamma rays, there is initially (in the first absorber) a buildup of energy deposition as the radiation penetrates the material. After some finite thickness the radiation energy absorption decreases. The thickness necessary to reach the maximum energy deposition is commonly called the "charged particle equilibrium" (**2**) thickness and is a function of the radiation energy and the electron density of the material being irradiated. Figure A1 is a typical plot of energy deposition as a function of thickness. Whenever one is irradiating a specimen from all sides, it is necessary to surround the specimen with an absorber in order to ensure charged particle equilibrium throughout the specimen. However, if the specimen is being irradiated unidirectionally, it is only necessary to interpose an absorber in front and behind the specimen. Figure A2 is a plot of absorber thickness as a function of energy for material of electron density 3.3 × 10^{23} electrons/cm^3. Figure A3 is a plot of thickness of water for a given attenuation (**3**).

NOTE A1—Electron density, in electrons per cubic centimeter; of any material can be calculated from the following equation:

$$\text{electron density} = \rho \times N \times (1/M) \times \Sigma_i A_i$$

where:

ρ = density, g/cm^3,
N = Avogadro's number, molecules/g-mol,
M = g-molecular weight, and
$\Sigma_i A_i$ = sum of electrons for all atoms in the molecule.

The curve in Fig. A2 will shift to the left as the electron density of the absorber increases above this value and to the right for materials of lesser electron density. The thickness to use is that obtained from Fig. A2 divided by the ratio of the electron density of the absorber to 3.3×10^{23} electrons/cm^3.

A1.2 For example, it is required to irradiate polytetrafluoroethylene (PTFE) film with 1-MeV photons. Referring to Fig. A2, it is noted that 0.5 cm of material of 3.3×10^{23} electrons/cm^3 is needed to assure charged particle equilibrium. Therefore, this thickness of material must surround the film.

A2. Background Information for Calculation of Absorbed Dose per Unit Exposure

A2.1 The efficiency of energy absorption by a material exposed to gamma radiation is determined by its energy absorption coefficient, $(\mu_{en})_m$. This may be determined from the mass absorption coefficient by including only the fraction of the photon energy actually absorbed by the material; corrections are therefore required for fluorescence, scattering, annihilation radiation, and Bremsstrahlung losses (**3,4**). When μ_{en} is expressed in square centimeters per gram, the value for the material being irradiated, whether a compound or a mixture, may be closely approximated by the relation

$$(\bar{\mu}_{en})_m = \Sigma_i f_i (\mu_{en})_i \quad (1)$$

where f_i and $(\mu_{en})_i$ are, respectively, the weight fraction and the energy-absorption coefficient of constituent i.

A2.2 An exposure in air of 1 roentgen results in an absorption of 86.9 ± 0.4 ergs/g of air or 0.869 rads (see Note A2) whenever charged particle equilibrium exists at the point of interest. For the same exposure, the absorbed dose received by the material being irradiated is given by the equation

$$D_m = D_{air}(\mu_{en})_m/(\mu_{en})_{air} = 0.869\ (\mu_{en})_m/(\mu_{en})_{air} \quad (2)$$

where:

D_m = absorbed dose in the material corresponding to an exposure of 1 roentgen,

D_{air} = absorbed dose in air corresponding to an exposure of 1 roentgen,

$(\mu_{en})_m$ = energy absorption coefficient for the material, and

$(\mu_{en})_{air}$ = energy absorption coefficient for air.

Substitution of Eq 1 into Eq 2 gives

$$D_m = 0.869\ [\Sigma_i f_i (\mu_{en})_i/(\bar{\mu}_{en})_{air}] = \Sigma_i f_i D_i \quad (3)$$

where:

$$D_i = 0.869\ [(\mu_{en})_i/(\mu_{en})_{air}] \quad (4)$$

NOTE A2—The value of 86.9 ergs/g of air is based on a value of 33.7 ± 0.2 eV required for one ion pair in air (**2**). In previous years the energy per ion pair has been assumed to be 34.0 eV or 32 eV, giving values for 1 roentgen of 87.7 and 83 ergs/g of air since the roentgen is defined as the energy required to produce one electrostatic unit of ions. Changes in the energy per ion pair also change the values of the conversion factors contained herein.

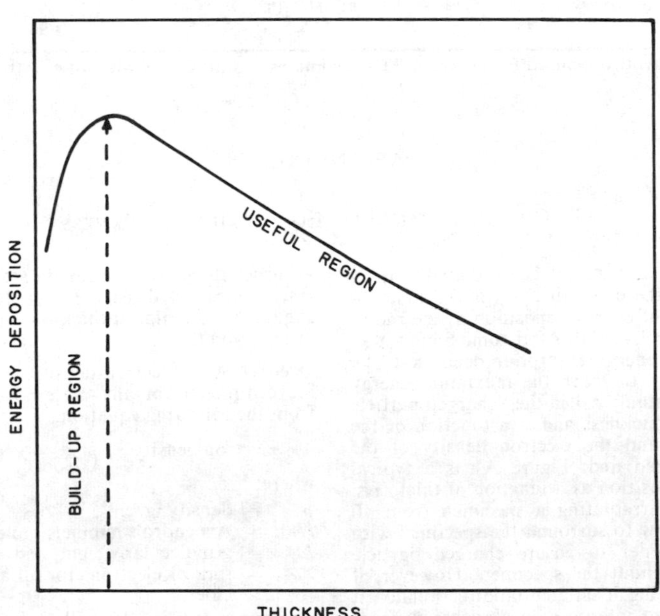

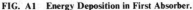

FIG. A1 Energy Deposition in First Absorber.

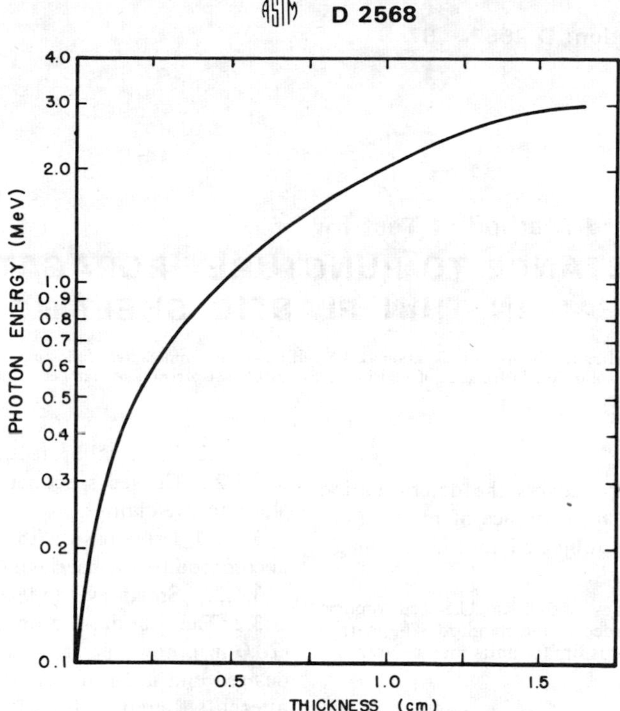

FIG. A2 Absorber Thickness for Charged Particle Equilibrium as a Function of Energy for Water (Material of 3.3×10^{23} electrons/cm^3).

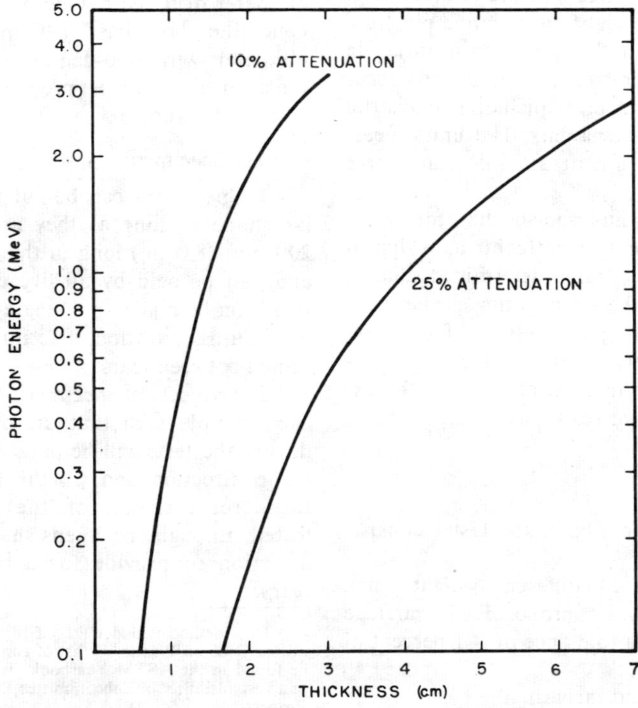

FIG. A3 Thickness of Water (1 g/cm^3 and 3.3×10^{23} electrons/g) for a Given Attenuation.

Designation: D 2582 – 67

Standard Method of Test for
RESISTANCE TO PUNCTURE-PROPAGATION OF TEAR IN THIN PLASTIC SHEETING[1]

This Standard is issued under the fixed designation D 2582; the number immediately following the designation indicates the year of original adoption or, in the case of revision, the year of last revision. A number in parentheses indicates the year of last reapproval.

1. Scope

1.1 This method covers the determination of the dynamic tear resistance of plastic film and thin sheeting subjected to end-use snagging type hazards.

NOTE—The values stated in U.S. customary units are to be regarded as the standard. The metric equivalents of U.S. customary units may be approximate.

2. Significance

2.1 The puncture-propagation of tear test measures the resistance of a material to snagging, or more precisely, to dynamic puncture and propagation of that puncture resulting in a tear. Failures due to snagging hazards occur in a variety of end uses, including industrial bags, liners, and tarpaulins. The units measured by this instrument are kilograms-force (tear resistance).

2.2 Experience has shown that for many materials puncture does not contribute significantly to the force value determined, due to the sharpness of the propagating probe used. However, comparing the results of prepunctured test specimens with normal nonpunctured specimens will accurately show the extent of any puncture resistance in the reported result.

3. Apparatus

3.1 As shown in Fig. 1, the tester consists of:

3.1.1 Carriages of different weight, each with a sharp-pointed probe. Each carriage weight shall have a tolerance of ±1 percent of the weight desired.

3.1.2 A test stand including:

3.1.2.1 Carriage release mechanism,

3.1.2.2 Scale calibrated in millimeters,

3.1.2.3 Curved specimen holder with a tear slot and five clamps,

3.1.2.4 Drop base with a guide channel to accommodate the carriage wheels, and

3.1.2.5 Spirit level to level the base.

3.2 The standard drop height is 508 mm (20.0 in.); however, it can be varied to suit the desired end use rate condition. See 8.3 for an alternative method for calculating tear resistance for heights other than 508 mm.

3.3 The probe is a 3.18-mm (0.125-in.) diameter drill rod having one end a truncated cone, the short base 0.40 mm (0.016 in.) in diameter, with a 30-deg included angle so that most of the tear is propagated against the body of the rod.

4. Test Specimen

4.1 Specimens can be cut in any rectangular shape so long as they are approximately 200 mm (8.0 in.) long in the direction of tear and can be held by all five clamps. Multiple tears can be made on a single sheet provided a minimum separation of 25 mm (1 in.) is maintained between tears.

4.2 Two sets of specimens shall be cut from each sample such that the direction of tear, during the test, will be parallel to (*1*) the machine direction and (*2*) the transverse direction, respectively, of the material being tested. Enough specimens shall be cut in each direction to provide for a minimum of five tears.

---

[1] This method is under the jurisdiction of ASTM Committee D-20 on Plastics. A list of committee members may be found in the ASTM Yearbook. This standard is the direct responsibility of Subcommittee D-20.10 on Mechanical Properties.

Effective June 12, 1967.

5. Adjustment and Calibration of Apparatus

5.1 Level the base of the tester by centering the spirit level bubble by adjustment of legs.

5.2 Check "sharpness" of probes by visual observation under a magnifying glass. If the short base end is not 0.40 mm (0.016 in.) in diameter or any burrs, nicks, or distortions are noted, or both, replace the probe.

5.3 Check the length of the probes by inserting each carriage in turn in the guide channel and lowering to the alignment mark on the curved specimen holder. The point of the probe should be aligned with this mark.

5.4 Measure the vertical drop height from the specimen holder mark, located near the tear slot, to the horizontal mark on the carriage release mechanism. Adjust the selected height to the nearest 2 mm (0.078 in.). The standard drop height is 508 ± 2 mm.

5.5 Check the alignment of the specimen holder receiving slot by lowering a carriage with its probe extending into the slot, up and down the slot length. The probe should be centered, that is, not touching either edge of the slot.

6. Conditioning

6.1 *Conditioning*—Condition the test specimens at 23 ± 2 C (73.4 ± 3.6 F) and 50 ± 5 percent relative humidity for not less than 40 h prior to test in accordance with Procedure A of ASTM Methods D 618, Conditioning Plastics and Electrical Insulating Materials for Testing,[2] for those tests where conditioning is required. In cases of disagreement, the tolerances shall be ±1 C (±1.8 F) and ±2 percent relative humidity.

6.2 *Test Conditions*—Conduct tests in the Standard Laboratory Atmosphere of 23 ± 2 C (73.4 ± 3.6 F) and 50 ± 5 percent relative humidity, unless otherwise specified in the test methods or in this specification. In cases of disagreements, the tolerances shall be ±1 C (±1.8 F) and ±2 percent relative humidity.

7. Procedure

7.1 Measure and record the thickness of each sample tested. Read the thickness to 0.0025 mm (0.0001 in.) or better, except for sheeting greater than 0.25 mm (0.010 in.), which shall be read to a precision of 0.025 mm (0.001 in.) or better.

7.2 Secure the specimen in the holder by placing it under clamps and setting the clamp lever to the down position. The specimen should drape against the holder contour. Stiffer materials should be loosely held adjacent to the holder. Each clamp should apply sufficient pressure to prevent any specimen slippage.

7.3 By trial and error, select the carriage that produces a minimum tear length of 40 mm and does not bottom-out against drop base. Lower selected carriage until the probe point touches but does not indent the specimen. Adjust the tear length indicating rod to "0" on the scale located on the guide channel.

7.4 Place the selected carriage in the release mechanism.

7.5 Cock the release mechanism on the left side and release the carriage by pushing the button on the front of the release mechanism.

7.6 Read the tear length to the nearest 0.5 mm.

7.7 Raise the carriage by pulling the handle on the left side of the guide channel.

7.8 After the carriage has stopped its upward movement, relocate it in the release mechanism by pushing the carriage up in the guide channel by hand. (When the handle is released, the arm should return to the bottom of the guide channel. However, check before releasing the carriage.)

7.9 Re-cock the release mechanism.

7.10 Release the clamps and relocate the specimen for the next tear test cycle. Take care not to relocate the specimen so that the tears are too close to one another, thereby influencing the tear results.

7.11 Make a minimum of five determinations in each direction for each sample.

8. Calculations

8.1 Calculate the tear resistance, F, in kilograms-force, as follows:

$$F = [(W \times H)/L] + W$$

where:

W = weight of carriage, kg,
H = height of carriage before release, mm, and

[2] *Annual Book of ASTM Standards*, Part 27.

L = length of tear, mm.

See Appendix A1 for a derivation of the equation for tear resistance.

8.2 To determine the tear resistance employing a standard drop height of 508 ± 2 mm, use Table 1, normalized factors for each carriage weight, (See A1.1.5 of Appendix A1, for an explanation of normalization.)

8.3 To determine tear resistance employing a nonstandard drop height or carriage weight, other than those listed above, use the equation in A1.1.3 of Appendix A1.

8.3.1 For inter- and intralaboratory data comparisons, the same drop height and weight carriage must be used.

8.4 Calculate the average tear length, L, in both directions to the nearest 0.1 mm.

8.5 Calculate the tear resistance, F, in each direction to the nearest 0.01 kgf.

8.6 Calculate the standard deviation (estimated) for each direction tested as follows:

$$s = \sqrt{(\Sigma X^2 - n\bar{X}^2)/(n - 1)}$$

where:

s = estimated standard deviation,
X = value of single observation,
n = number of observations, and
$\bar{X}$ = arithmetic mean of the set of observations.

9. Report

9.1 The report shall include the following:

9.1.1 Complete identification of the sample tested,

9.1.2 Average tear resistance for the direction tested,

9.1.3 Drop height selected, if nonstandard,

9.1.4 Carriage used, if nonstandard,

9.1.5 Number of specimens tested, if greater or less than five,

9.1.6 Average thickness, in mils,

9.1.7 Type of tears produced, for example, "V" or slit, and

9.1.8 Standard deviation for each direction tested, reported to two significant figures.

10. Precision

10.1 Based on round-robin tests[3], the single-laboratory standard deviation is 4.6 percent relative, and the multilaboratory standard deviation is 7.0 percent relative for the average of five determinations.

[3] The results of this round robin coupled with a statistical analysis entitled "Results of Round Robin on PPT Tear," ASTM Committee D-20, Section I-M, Tear Strength, are on file at ASTM Headquarters.

TABLE 1 Normalized Factors for Each Carriage Weight

Carriage No.	Carriage Weight, kg ± 1 percent	Factors
No. 1	0.1134	$F = 5.761/L + 1.190$
No. 2	0.2268	$F = 11.521/L + 0.936$
No. 3	0.3402	$F = 17.282/L + 0.709$
No. 4	0.4536	$F = 23.043/L + 0.454$
No. 5	0.6804	$F = 34.564/L - 0.029$
No. 6	0.9072	$F = 46.086/L - 0.482$

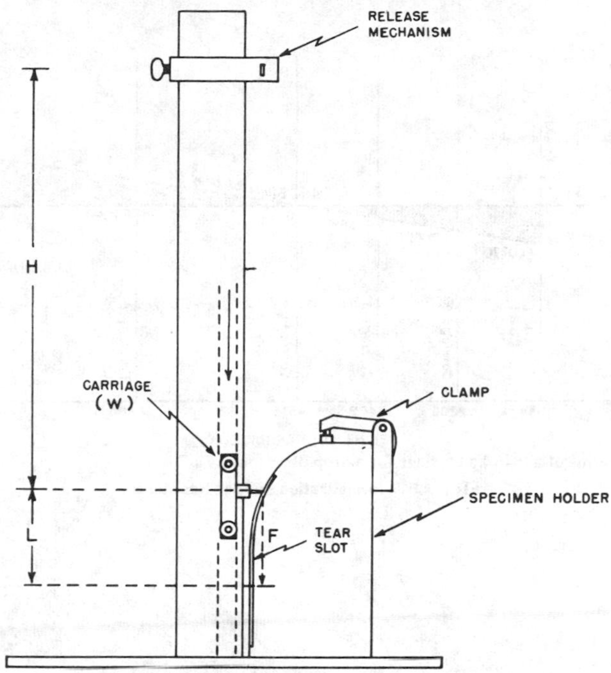

H—drop height, mm
L—tear length, mm
F—force required to produce *L*, kgf
W—weight of carriage, kg

FIG. 1 Puncture-Propagation of Tear Tester.

APPENDIX

A1. Derivation of Equation for Tear Resistance

A1.1 The equation for determining average tear resistance (effective force to tear) has been derived for a drop height of 508 mm (20.0 in.) as follows:

A1.1.1 Potential energy of carriage before release $= W(H + L) =$ work done on sample in bringing carriage to a stop, where: $W =$ weight of the carriage, $H =$ height of carriage before release, and $L =$ length of tear.

A1.1.2 Work done on sample = effective force exerted by probe in propagating tear (= tear resistance) $\times$ length of tear $= F \times L$, where: $F =$ effective force exerted by probe and $L =$ length of tear.

A1.1.3 Equating the two, $W(H + L) = F \times L$, so that: tear resistance $F = (W \times H)/L + W$.

NOTE A1—The above equation is to be used for nonstandard drop height tests.

A1.1.4 The following approximations were made in deriving this equation: tear resistance was assumed proportional to deceleration of the carriage, and the angle at which the probe punctured the sample was neglected.

A1.1.5 In order to compare results of tests performed using different weight carriages, at the standard drop height of 508 mm, an adjustment has been incorporated in the calculation to compensate for the small increase in tear resistance found when the carriage weight is increased (see Fig. A1). To prevent reporting significantly different values due to the use of different carriages, these test results are normalized in terms of the 0.46-kg (1-lb) carriage by: (*1*) plotting tear resistance (*X*-axis) versus weight of carriage (*Y*-axis) for a variety of different type flexible materials, (*2*) determining the slope of best fit from the slope determined by testing each material at two or more carriage weights, (*3*) drawing an *X*-axis through the *Y*-axis for the 0.46-kg carriage at the point of interception of the slope, (*4*) adding a force (in kilograms) increment to the calculated tear resistance where the slope intercepted a Y-axis for a lighter weight carriage below the X-axis and subtracting when it intercepted a Y-axis for a heavy weight carriage above the X-axis.

For tests conducted at nonstandard drop heights, use the equation in A1.1.3. Results can be compared only if the same carriage weight and drop height are employed.

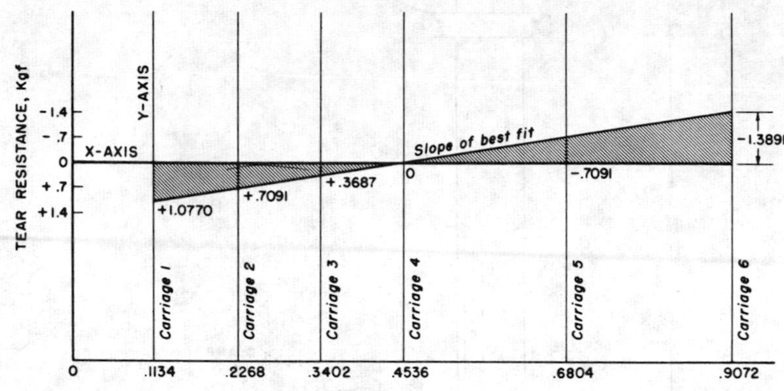

NOTE—Graph is in terms of a 0.46-kg carriage for a drop of 508 mm.

FIG. A1 Normalization of Tear Resistance.

Designation: D 2583 - 67

Standard Method of Test for
INDENTATION HARDNESS OF PLASTICS BY MEANS OF A BARCOL IMPRESSOR[1]

This Standard is issued under the fixed designation D 2583; the number immediately following the designation indicates the year of original adoption or, in the case of revision, the year of last revision. A number in parentheses indicates the year of last reapproval.

1. Scope

1.1 This method covers the determination of indentation hardness of both reinforced and nonreinforced rigid plastics using a Barcol Impressor, Model No. 934-1.

NOTE 1—The values stated in U.S. customary units are to be regarded as the standard. The metric equivalents of U.S. customary units may be approximate.

NOTE 2—Another model, No. 935, is for use on soft plastics but is not included in this method.

2. Significance

2.1 The Barcol impressor is portable and therefore suitable for testing the hardness of fabricated parts and individual test specimens for production control purposes.

3. Apparatus (Figs. 1 and 2)

3.1 *Indentor*—The indentor shall consist of a hardened steel truncated cone having an angle of 26 deg with a flat tip of 0.157 mm (0.0062 in.) in diameter. It shall fit into a hollow spindle and be held down by a spring-loaded plunger. See Fig. 2.

3.2 *Indicating Device*—The indicating dial shall have 100 divisions, each representing a depth of 0.0076-mm (0.0003-in.) penetration. The higher the reading the harder the material.

4. Test Specimens

4.1 The testing area shall be smooth and free from mechanical damage.

4.2 *Dimensions*—Test specimens shall be at least 1.5 mm ($^1/_{16}$ in.) thick and large enough to ensure a minimum distance of 3 mm ($^1/_8$ in.) in any direction from the indentor point to the edge of the specimen.

5. Calibration

5.1 With the plunger upper guide backed out until it just engages the spring, place the impressor on a glass surface and press down until the penetrated point is forced all the way back into the lower plunger guide. The indicator should now read 100. If it does not, loosen the lock-nut and turn the lower plunger guide in or out to obtain a 100 reading. Next, read the "hard" aluminum alloy disk supplied by the manufacturer of the impressor and, if necessary, adjust so that the reading is within the range marked on the disk. Then do the same with the "soft" disk. If these readings cannot be obtained, subsequent measurements are not valid.

6. Conditioning

6.1 *Conditioning*—Condition the test specimens at 23 ± 2 C (73.4 ± 3.6 F) and 50 ± 5 percent relative humidity for not less than 40 h prior to test in accordance with Procedure A of ASTM Methods D 618, Conditioning Plastics and Electrical Insulating Materials for Testing,[2] for those tests where conditioning is required. In cases of disagreement, the tolerances shall be ±1 C (±1.8 F) and ±2 percent relative humidity.

6.2 *Test Conditions*—Conduct tests in the Standard Laboratory Atmosphere of 23 ± 2 C (73.4 ± 3.6 F) and 50 ± 5 percent relative humidity, unless otherwise specified in

[1] This method is under the jurisdiction of ASTM Committee D-20 on Plastics. A list of committee members may be found in the ASTM Yearbook. This standard is the direct responsibility of Subcommittee D-20.10 on Mechanical Properties.

Effective April 23, 1967.

[2] *Annual Book of ASTM Standards*, Part 27.

the test methods or in this specification. In cases of disagreements, the tolerances shall be ±1 C (±1.8 F) and ±2 percent relative humidity.

7. Procedure

7.1 Support the test specimens by a hard, firm surface if they are likely to bend or deform under the pressure of the indentor (Note 3). The indentor must be perpendicular to the surface being tested. Grasp the impressor firmly between leg and point sleeve and set both on the surface to be tested. For small specimens this may require the impressor leg to be shimmed by the thickness of the specimen. Quickly apply by hand sufficient pressure (4 to 7 kg, or 10 to 15 lb) on the housing to ensure firm contact with the test specimen and record the highest dial reading (Note 4). Take care to avoid sliding or scraping while the indentor is in contact with the surface being tested.

NOTE 3—Curved surfaces may be more difficult to support. When the load is applied, bending and spring action in the specimen should be avoided.

NOTE 4—Drift-in readings from the maximum may occur in some materials. This can be nonlinear with time.

7.2 Impressions should not be made within 3 mm (1/8 in.) of the edge of the specimen or of other impressions.

8. Number of Readings

8.1 Application of the Barcol impressor to reinforced plastic (nonhomogeneous) materials will produce greater variation in hardness readings than on nonreinforced (homogeneous) materials. This greater variation may be caused mainly by the difference in hardness between resin and filler materials in contact with the small diameter indenter. There is less variation in hardness readings on harder materials in the range of 50 Barcol and higher and considerably more variation in the readings of softer materials. On homogeneous materials, five readings are needed to maintain a variance-of-average of 0.28 at a 60 Barcol reading; for the same variance-of-average at 30 Barcol, eight readings are needed. On reinforced plastics, in order to maintain a variance-of-average of 0.78 at 60 Barcol, ten readings are needed; and 29 readings are needed for the same variance at the 30 Barcol level (Table 1).

NOTE 5—These findings were obtained with a round robin conducted in a workshop with all participants present. Eight plastic materials of different hardness were evaluated with six different Barcol (934-1) impressors. The report is filed with ASTM Headquarters.

9. Report

9.1 The report shall include the following:

9.1.1 Identification of material tested,

9.1.2 Conditioning of specimen,

9.1.3 Model number of impressor,

9.1.4 Number of readings taken,

9.1.5 Average of hardness values rounded off to the nearest whole scale reading, and

9.1.6 Date of test.

TABLE 1 Recommended Sample Sizes to Equalize the Variance of the Average

Hardness, M-934 Scale	Reading Variance	Coefficient of Variation, percent	Variance of Average	Minimum No. of Readings
Homogeneous Material				
20	2.47	2.6	0.27	9
30	2.20	1.7	0.28	8
40	1.93	1.3	0.27	7
50	1.66	1.1	0.28	6
60	1.39	0.9	0.28	5
70	1.12	0.8	0.28	4
80	0.85	0.7	0.28	3
Nonhomogeneous Material (Reinforced Plastics)				
30	22.4	2.9	0.77	29
40	17.2	2.2	0.78	22
50	12.0	1.7	0.75	16
60	7.8	1.5	0.78	10
70	3.6	1.2	0.75	5

FIG. 1 Barcol Impressor.

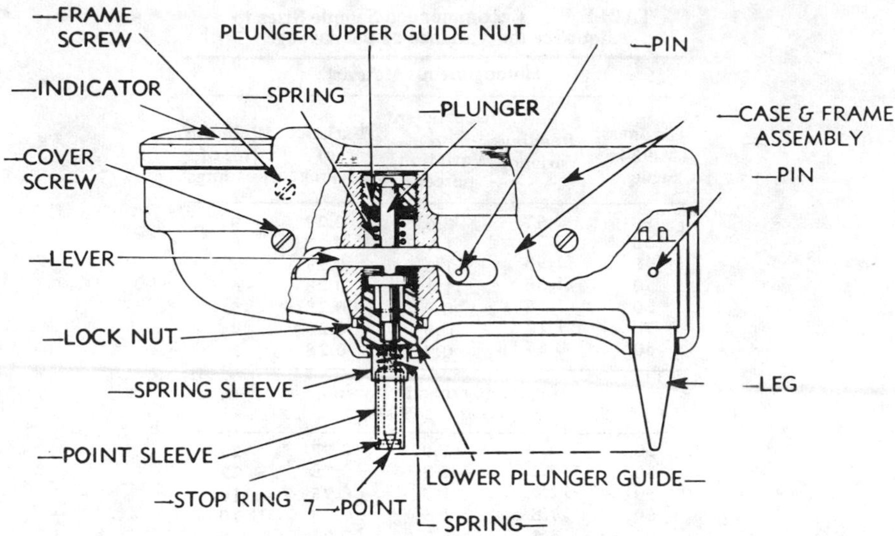

FIG. 2 Diagram of Barcol Impressor.

Designation: D 2648 - 67 T

Tentative Recommended Practice for MEASURING TIME-TO-FAILURE BY RUPTURE OF PLASTICS UNDER TENSION IN VARIOUS ENVIRONMENTS[1]

This Tentative Recommended Practice has been approved by the sponsoring committee and accepted by the Society in accordance with established procedures, for use pending adoption. Suggestions for revisions should be addressed to the Society at 1916 Race St., Philadelphia, Pa. 19103.

This tentative has been submitted to the Society for adoption, but balloting was not complete by press time.

1. Scope

1.1 This recommended practice[2] covers a procedure for measuring the time-to-failure by rupture of plastics in specific environments and under a constant tensile load.

1.2 Environmental stress-rupture tests should be limited to suitable combinations of loads and temperature in each environment to attain reasonable failure times.

2. Summary

2.1 The recommended practice consists of applying a range of fixed loads in tension to specimens in specific environments at constant temperature and measuring the time-to-failure.

3. Significance

3.1 Experimental data from stress-rupture life tests are of importance in comparing the ability of materials to support continuously applied loads for long periods of time. Various environments acting on the surface of the stressed specimen are known to reduce drastically the time a specimen can support a load continuously.

4. Apparatus

4.1 *Testing Machine*, capable of applying *only* a constant tensile load, such as by dead weight, on the specimen. The stress on the specimen should be known within 5 percent.

4.2 *Grips*—The grips and gripping technique should be designed to minimize eccentric loading of the specimen. It is recommended that a swivel or universal joint be used with each grip.

5. Reagents

5.1 *Purity of Reagents*—Reagent grade chemicals or equivalent, as specified in ASTM Methods E 200, for Preparation, Standardization, and Storage of Standard Solutions for Chemical Analysis,[3] shall be used in all tests. Other grades may be used provided it is first ascertained that the reagent is of sufficiently high purity to permit its use without introducing unknown variables in the determination.

5.2 *Purity of Water*—Unless otherwise indicated, references to water shall be understood to mean distilled water or water of equal purity.

5.3 *Specified Reagents*—Should this recommended practice be referenced in a material specification, the specific reagent to be used shall be as stipulated in the specification.

5.4 *Standard Reagents*—A list of standard reagents is also available in ASTM Method D 543, Test for Resistance of Plastics to Chemical Reagents.[4]

6. Safety Precautions

6.1 Safety precautions should be taken to avoid personal contact, to eliminate toxic

---

[1] This recommended practice is under the jurisdiction of ASTM Committee D-20 on Plastics. A list of committee members may be found in the ASTM Yearbook. This tentative is the direct responsibility of Subcommittee D-20.50 on Permanence Properties.

Effective March 23, 1967.

[2] See Carey, R. H., "Creep and Stress Rupture Behavior of Polyethylene Resins," *Industrial and Engineering Chemistry*, IECHA, Vol 50, 1958, pp. 1045–1048.

[3] *Annual Book of ASTM Standards*, Part 30.

[4] *Annual Book of ASTM Standards*, Part 27.

vapors, and to guard against explosion hazards in accordance with the hazardous nature of the particular reagents being used.

7. Test Specimens

7.1 Test specimens may have either circular, square, or rectangular cross-sections and may be made by casting, injection, or compression molding, or machining from sheets, plates, slabs, or similar material. The specimens given in ASTM Method D 638, Test for Tensile Properties of Plastics,[4] or in ASTM Method D 1822, Test for Tensile Impact Energy to Break Plastics and Electrical Insulating Materials[4] are generally used.

7.1.1 The specimen should have a cross-sectional area which is uniform to within ± 0.5 percent throughout the gage length. For round specimens, nominal diameters of 13 mm (1/2 in.), 10 mm, (3/8 in.), 6 mm (1/4 in.), and a gage length of not less than 50 mm (2 in.) are recommended. Longer gage lengths and the 13-mm (1/2-in.) diameter are preferred. Specimens should be straight and free from tool marks and scratches.

7.2 Specimens prepared from sheet or plate should be cut with their axes parallel to each other. When testing materials that may be anisotropic, duplicate sets of specimens should be prepared and tested: one set parallel to the principal axis of anisotropy and another set normal to that direction.

7.3 The number of specimens will be dependent on the number of stress levels used to develop a time-to-failure versus stress curve and the replication at each stress level.

7.4 When this recommended practice is made part of a material specification, specimen geometry and fabricating technique shall be stipulated.

7.5 Usually different materials can best be compared by using specimens essentially free of anisotropy.

8. Conditioning

8.1 Conditioning of test specimens is complicated by changes in sample properties, dimensions, and other factors that influence environmental time to failure.

8.2 Typical alternative preconditioning procedures are as follows:

8.2.1 Subject the specimens to the test temperature and relative humidity to be used in the test for a period of time sufficient to obtain dimensional stability before starting test, or

8.2.2 Subject all specimens to the test temperature to be used in the given series of tests for a period of time sufficient to obtain dimensional stability, or

8.2.3 Start the test 48 h after the specimen has attained the test temperature.

9. Procedure

9.1 Set the test temperature to the selected value and maintain the test space to ±2 C or less. Small variations in temperature can produce large changes in time to failure.

9.1.1 Automatic recording of temperature in the immediate vicinity of the test specimens is recommended. Alternatively, frequent (hourly) observations of temperature by ASTM thermometers or calibrated thermocouples may be recorded.

9.1.2 Standard laboratory test conditions of 23 ± 2 C and 50 ± 5 percent relative humidity are recommended for general use when applicable.

9.2 The water content of moisture-sensitive specimens should be known. For materials that exhibit variations in environmental behavior due to small moisture changes, the relative humidity must be known and rigorously controlled.

NOTE 1—If the specimens are surrounded by an environment that excludes moisture, control of humidity during the test is unnecessary.

9.3 The specimens should be placed in the grips so that the axis of loading of the apparatus is maintained. This practice will reduce flexural forces which can change the stress distribution greatly. Specimens should be handled with clean gloves at all times to avoid contamination.

9.3.1 The specimens should not be excessively tightened in the grips. When excessive clamp or grip pressure is used, fracture or failure of the specimen will occur in the grip instead of in the gage length of the sample.

NOTE 2—To reduce slipping, fine emery paper may be used between the grip face and the specimen. Use of fresh emery paper for each specimen decreases the possibility of contamination. Pins in the tabs of the specimen can also minimize slippage.

9.4 For the selected stress level, calculate the proper load using the initial cross-sectional area of the specimen.

9.4.1 For conventional plots of log time-to-failure as a function of stress, a minimum of three stress levels should be used for each failure curve at each temperature.

9.4.2 Six stress levels are recommended for more satisfactory evaluation at each temperature. Start at a stress value large enough to produce failure in less than 10 h. Decrease stress levels by appropriate increments until failure times of approximately 1000 h are attained.

9.5 The test load should be applied to the specimen gently but quickly with no vibration. Readings of the temperature, relative humidity, and load should be recorded. The time-to-failure is usually recorded automatically to eliminate frequent visual observations.

9.6 The environmental agent (Note 3) should be applied to the entire gage length of the specimen immediately after loading. If the environmental agent is volatile or the time-to-failure very long, the specimen should be covered, to retard evaporation, without affecting the applied load. Volatile agents should be replenished periodically.

NOTE 3—For liquid environmental agents a cotton swab, film, or other device can be wrapped or sealed completely around the gage length of the specimen, and the liquid agent applied to saturate the swab. Testing in a horizontal position is recommended to avoid contamination of grips.

10. Plotting Results

10.1 A plot with stress as abscissa and logarithm of time-to-failure as ordinate is suitable for characterization. At constant temperature this plot has been found to exhibit one or more linear relations for many environmental treatments.

11. Report

11.1 The report shall include the following:

11.1.1 Complete identification of the material tested, including type, source, manufacturer's code numbers, form, previous history, etc.,

11.1.2 Method of preparing test specimens,

11.1.3 Dimensions of test specimens,

11.1.4 Preconditioning used,

11.1.5 Type of testing equipment,

11.1.6 Test temperature, average and range,

11.1.7 Relative humidity in test enclosure,

11.1.8 Reagent used and method of application, and

11.1.9 Curves showing test data as described in Section 10.

NOTE 4—Mode of failure should be described for each linear portion of the curve.

Designation: D 2843 – 70

Standard Method for
MEASURING THE DENSITY OF SMOKE FROM THE BURNING OR DECOMPOSITION OF PLASTICS[1]

This Standard is issued under the fixed designation D 2843; the number immediately following the designation indicates the year of original adoption or, in the case of revision, the year of last revision. A number in parentheses indicates the year of last reapproval.

1. Scope

1.1 This method covers a laboratory procedure for measuring and observing the relative amounts of smoke produced by the burning or decomposition of plastics. It is intended to be used for measuring the smoke producing characteristics of plastics under controlled conditions of combustion or decomposition. Correlation with other fire conditions is not necessarily implied. The measurements are made in terms of the loss of light transmission through a collected volume of smoke produced under controlled, standardized conditions. The apparatus is constructed so that the flame and smoke can be observed during the test.[2]

NOTE 1—The values stated in U.S. customary units are to be regarded as the standard. The metric equivalents of U.S. customary units may be approximate.

2. Significance

2.1 Tests made on a material under conditions herein prescribed can be of considerable value in comparing the relative smoke generating characteristics of plastics.

2.2 This method serves to determine the extent to which plastic materials are likely to smoke under conditions of active burning and decomposition in the presence of flame.

NOTE 2—The visual and instrumental observations from this test compare well with the visual observations of the smoke generated by plastic materials when added to a freely burning, large outdoor fire.[3]

2.3 The usefulness of this test procedure is in its ability to measure the amount of smoke produced in a simple, direct, and meaningful manner under the specified conditions. The degree of obscuration of vision by smoke generated by combustibles can be substantially affected by changes in quantity and form of material, humidity, draft, temperature, and the supply of oxygen.

3. Summary of Method

3.1 The test specimen is exposed to flame for the duration of the test and the smoke is substantially trapped in the chamber in which combustion occurs. A 25 by 25 by 6-mm (1 by 1 by ¼-in.) specimen is placed on a supporting metal screen and burned in a laboratory test chamber (Fig. 1) under active flame conditions using a propane burner operating at a pressure of 2.8 kgf/cm^2 (40 psi). The 300 by 300 by 790-mm (12 by 12 by 31-in.) test chamber is instrumented with a light source, a photoelectric cell, and a meter to measure light absorption horizontally across the 300-mm (12-in.) light beam path. The chamber is closed during the 4-min test period except for the 25-mm (1-in.) high ventilation openings around the bottom.

3.2 The light absorption data are plotted versus time. A typical plot is shown in Fig. 2.

---

[1] This method is under the jurisdiction of ASTM Committee D-20 on Plastics. A list of committee members may be found in the ASTM Yearbook. This standard is the direct responsibility of Subcommittee D-20.30 on Thermal Properties.

Effective Jan. 30, 1970.

[2] Anonymous, "A Method of Measuring Smoke Density," *NFPA Quarterly*, QNFPA, Vol 57, January 1964, p. 276. Reprint NFPA Q57-9. NFPA, 60 Batterymarch St., Boston, Mass. 02110.

[3] Rarig, F. J., and Bartosic, A. J., "Evaluation of the XP2 Smoke Density Chamber," *Symposium on Fire Test Methods—Restraint & Smoke*, *ASTM STP 422*, Am. Soc. Testing Mats., 1966.

Two indexes are used to rate the material: maximum smoke produced, and the smoke density rating.

4. Apparatus

4.1 The smoke chamber shall be constructed essentially as shown in Fig. 1.[4]

4.1.1 *Chamber:*

4.1.1.1. The chamber shall consist of a 14-gage (B & S) 300 by 300 by 790-mm (12 by 12 by 31-in.) aluminum box to which is hinged a heat-resistant glass glazed door. This box shall be mounted on a 350 by 400 by 57-mm (14 by 16 by $2^1/_4$-in.) base which houses the controls. Dependent upon the materials tested, the metal may require protection from corrosion.

4.1.1.2 The chamber shall be sealed except for 25 by 230-mm (1 by 9-in.) openings on the four sides of the bottom of the chamber. A 1700-liters/min (60-ft^3/min) blower shall be mounted on one side of the chamber. The inlet duct to the blower shall be equipped with a close fitting damper. The outlet of the blower shall be connected through a duct to the laboratory exhaust system. If the chamber is in a ventilated hood, no connection to the lab exhaust system through a duct is needed.

4.1.1.3 The two sides adjacent to the door shall be fitted with 70-mm ($2^3/_4$-in.) diameter smoke-tight glazed areas centered 480 mm ($19^3/_4$ in.) above the base. At these locations and outside the chamber, boxes containing the optical equipment and additional controls shall be attached.

4.1.1.4 A removable white plastic plate shall be attached to the back of the chamber. There shall be a 90 by 150-mm ($3^1/_2$ by 6-in.) clear area centered about 480 mm above the bottom of the chamber through which is seen an illuminated white-on-red exit sign. The white background permits observation of the flame, smoke, and burning characteristics of the material. The viewing of the exit sign helps to correlate visibility and measured values.

4.1.2 *Specimen Holder:*

4.1.2.1 The specimen shall be supported on a 64-mm ($2^1/_2$-in.) square of 6 by 6-mm, 0.9-mm gage ($^1/_4$ by $^1/_4$-in., 0.035-in. gage) stainless steel wire cloth 220 mm ($8^3/_4$ in.) above the base and equidistant from all sides of the chamber. This screen shall lie in a stainless steel bezel supported by a rod through the right side of the chamber. From the same rod, a similar bezel shall be located 76 mm (3 in.) below, and it shall support a square of asbestos paper which catches any particles that may drip from the specimen during the test. By rotating the specimen holder rod, the burning specimen can be quenched in a shallow pan of water positioned below the specimen holder.

4.1.3 *Ignition System:*

4.1.3.1 The specimen shall be ignited by a propane flame from a burner operating at a pressure of 2.8 kgf/cm^2 (40 psi). The fuel (Note 3) shall be mixed with air which has been propelled through the burner by the venturi effect of the propane as it passes from a 0.13-mm (0.005-in.) diameter orifice (Note 4), and the burner shall be assembled as shown in the exploded view of the burner in Fig. 3. The burner must be designed to provide adequate outside air.

NOTE 3—Commercial grade 85.0 percent minimum, gross heating value 23,000 cal/liter (2590 Btu/ft^3) propane meets the requirements.

NOTE 4—Since the orifice provides the metering effect proportionate to the supply pressure, care must be taken that the orifice is the only means of fuel egress.

4.1.3.2 The burner shall be capable of being positioned quickly under the specimen so that the axis of the burner falls on a line passing through a point 8 mm /$^3/_{10}$ in.) above the base at one back corner of the chamber extending diagonally across the chamber and sloping upward at 45 deg with the base. The exit opening of the burner shall be 260 mm ($10^1/_4$ in.) from the reference point at the rear of the chamber.

4.1.3.3 A duct at least 150 mm (6 in.) outside of the chamber shall provide the air piped to the burner.

4.1.3.4 Propane pressure shall be adjustable and preferably automatically regulated. Propane pressure shall be indicated by means of a Bourdon tube gage.

4.1.4 *Photometric System:*

4.1.4.1 A light source, a barrier-layer photoelectric cell, and a temperature compen-

[4] The smoke chamber made by George Eysenbach, 315 Western Avenue, Bristol, Pa. 19007, meets the requirements of this method. Detailed drawings of the smoke chamber are also available at a nominal cost from the American Society for Testing and Materials, 1916 Race St., Philadelphia, Pa. 19103.

sated meter shall be used to measure the proportion of a light beam which penetrates a 300-mm (12-in.) path through the smoke. The light path shall be arranged horizontally as shown in Fig. 4.

4.1.4.2 The light source shall be mounted in a box (*4 B1* in Fig. 1) extending from the left side of the chamber at the mean height of 480 mm (19$^3/_4$ in.) above the base. The light source shall be a compact filament microscope lamp No. 1493[5] operated at 5.8 V and a spherical reflector, with power supplied by a voltage regulating transformer. A 60 to 65-mm (2 $^1/_2$-in.) focal length lens shall focus a spot of light on the photocell in the right instrument panel.

4.1.4.3 Another box containing the photometer (*4 B2* in Fig. 1) shall be attached to the right side of the chamber. The barrier-layer photoelectric cell shall have standard observer spectral response. An egg-crate grid in front of the photocell shall be used to protect the cell from stray light. The grid shall be finished in dull black and have openings at least twice as deep as they are wide. The current produced by the photocell is indicated in terms of percent light absorption on a meter. The photocell linearity decreases as the temperature increases; compensations shall therefore be made.

NOTE 5—Photocell manufacturers recommend operating the photocell at temperatures not exceeding 50 deg C.

4.1.4.4 The meter shall have two ranges. The range change shall be accomplished by shunting the meter to one tenth of its sensitivity. When smoke accumulates to absorb 90 percent of the light beam, a momentary switch shall be depressed returning the meter to its basic sensitivity. By doing this the meter scale now reads from 90 to 100 percent instead of 0 to 100 percent.

4.1.5 *Timing Device*—A clock to indicate 15-s intervals shall be used. If the time intervals are audibly marked it will be convenient for the operator to record his observations. A clutch shall be used to reset the clock at the start of a test. The block shall be coupled to the burner-positioning device and it shall start when the burner is swung into test position.

4.1.6 *Planimeter*—A planimeter or other suitable means shall be used for measuring the area under the light-absorption curve.

5. Test Specimen

5.1 The standard specimen shall be 25.4 ± 0.3 by 25.4 ± 0.3 by 6.2 ± 0.3 mm (1 ± 0.01 by 1 ± 0.01 by $^1/_4$ ± 0.01 in.). Thicknesses other than 6.2 mm ($^1/_4$ in.) may be used and their size must be reported with the smoke density values (Note 6). Material thinner than 6.2 mm ($^1/_4$ in.) may be tested either in its normal use thickness or by stacking and forming a composite specimen approximately 6.2 mm ($^1/_4$ in.) thick. Material thicker than 6.2 mm ($^1/_4$ in.) may be tested either in its normal use thickness or by machining the material down to a thickness of 6.2 mm ($^1/_4$ in.).

NOTE 6—If specimens other than the standard specimen are to be used, cooperating laboratories should agree upon preparation procedures and dimensions of the specimen. The results in such cases may vary from the results obtained with the standard specimen.

5.2 The specimens shall be sanded, machined, or die cut in a manner that produces a cut surface that is free from projecting fibers, chips, and ridges.

5.3 The test sample shall consist of three specimens.

6. Conditioning

6.1 *Conditioning*—Condition the test specimens at 23 ± 2 C (73.4 ± 3.6 F) and 50 ± 5 percent relative humidity for not less than 40 h prior to test in accordance with Procedure A of ASTM Methods D 618, Conditioning Plastics and Electrical Insulating Materials for Testing,[6] for those tests where conditioning is required. In cases of disagreement, the tolerances shall be ±1 C (±1.8 F) and ±2 percent relative humidity.

6.2 *Test Conditions*—Conduct tests in the Standard Laboratory Atmosphere of 23 ± 2 C (73.4 ± 3.6 F) and 50 ± 5 percent relative humidity, unless otherwise specified in the test methods or in this specification. In cases of disagreement, the tolerances shall be ±1 C (±1.8 F) and ±2 percent relative humidity.

6.3 Tests shall be conducted in a hood that has a window for observing the test.

[5] The microscope lamps No. 1493 are manufactured by General Electric Company, Westinghouse, and others.

[6] *Annual Book of ASTM Standards*, Part 27.

7. Procedure

7.1 Turn on the photometer lamp, exit sign, and exhaust blower.

7.2 Turn on the propane, ignite the burner, and adjust the propane pressure to 2.8 kgf/cm^2 (40 psi). **Caution**—Do not fail to light the burner immediately.

7.3 Set the temperature compensation.

7.4 Adjust the lamp control to zero percent light absorption.

7.5 Lay the test specimen flat on the screen in such a position that the burner flame will be directly under the specimen when the burner is swung into position.

7.6 Set the timer to zero.

7.7 Shut off the exhaust blower, close the smoke chamber door, and immediately position the burner under the specimen and start the timer.

7.8 If in a hood, shut off the hood fan and close the hood door to within 50 mm (2 in.) of the bottom of the hood.

7.9 Record the percent light absorbed at 15-s intervals for 4 min.

7.10 Record observations during the conduct of the test. Include the time it takes for the sample to burst into flame, the time for flame extinguishment or specimen consumption, the obscuration of the exit sign by smoke accumulation, and any general or unusual burning characteristics noted such as melting, dripping, foaming, or charring.

7.11 Upon completion of the test, turn on the exhaust blower to ventilate the combusion products from the chamber.

NOTE 7—It should be noted that for some materials the products of burning may be toxic, and care should be taken to guard the operator from the effects of these gases. The ventilating fan in the hood should be turned on and the damper opened immediately after the test is completed before opening the hood door in order to remove any irritating products of the test. The exhaust fan is turned off and the hood damper closed during the test to prevent back draft.

7.12 Open the door and clean the combustion deposits from the photometer, exit sign, and door glass with detergent and water. Burn off any material remaining on the screen or replace the screen and asbestos square for the next test.

7.13 Run all tests in triplicate.

7.14 At the beginning of each series or at least once a day, check the light absorption of the meter against a calibrated neutral filter of approximately 50 percent absorption.[7] Check the 100 percent absorption point against an opaque plate.

8. Optional Procedures

8.1 The output of the photocell may be recorded versus time on an appropriate graphic recorder.[8]

8.2 With a suitably sensitive meter, more than one decade change may be used to separate readings in the very dense smoke range.

9. Treatment of Data

9.1 Average the readings at 15-s intervals of light absorption for the three specimens in each group. Plot the average light absorption against time on linear paper. Fig. 2 is a sample curve.

9.2 Read the maximum smoke density as the highest point on the curve.

9.3 Determine the total smoke produced by measuring the area under the curve. The smoke density rating represents the total amount of smoke present in the chamber for the 4-min time interval. Measure the total smoke produced by the area under the curve of light absorption versus time, divided by the total area of the graph, 0–4 min, 0–100 percent light absorption, times 100.

NOTE 8: *Example*—In the light absorption-time plot in Fig. 2, the plot has been made using 10 mm (0.39 in.) equal to 10 percent as the ordinate and 10 mm (0.39 in.) equal to 0.25 min as the absicca. The graph area for 4 min is found to be 16,000 mm^2 (24.80 in.^{2}). The area under the curve is found to be 12,610 mm^2 (19.55 in.^{2}). The smoke density rating is then computed as follows:

Smoke density rating, percent

$$= (12610/16000 \times 100 = 78.8$$
(dimensions in millimeters)

$$= (19.55/24.80) \times 100 = 78.8$$
(dimensions in inches)

10. Report

10.1 The report shall include the following:

10.1.1 Identification of the material,

[7] Jena Glass Neutral Filter NG-5 from Fish-Schurman Corporation, 70 Portman Road, New Rochelle, N. Y. 10802, and No. 96 Wratten Neutral Density Filter (Neutral Density 0.03) 3 in. by 3 in. cemented in glass and recalibrated for density from Eastman Kodak Stores, Inc., 3202 Race St., Philadelphia, Pa. 19104, have proven suitable.

[8] The Honeywell Electronic 19 Recorder Model 19301-01-01 from Honeywell, Inc., Industrial Park, King of Prussia, Pa., has proven suitable.

10.1.2 Dimensions of the specimen,

10.1.3 Readings of light absorption at 15-s intervals for each test and average,

10.1.4 Plots of average light absorption versus time,

10.1.5 Maximum smoke density in percent light absorption,

10.1.6 Area in percent under the light absorption-time curve (smoke density rating),

10.1.7 Observations on behavior of material,

10.1.8 Observations on obscurement of exit sign, and

10.1.9 The details of any departure from the specifications of the method for testing.

11. Precision

11.1 The following criteria should be used in judging acceptability of smoke density rating data.[9]

11.2 *Repeatability*—Two individual results (not averages) determined by a single operator in one laboratory should not be considered suspect (at 95 percent confidence level) unless they differ by more than 18 percent absolute.

11.3 *Reproducibility*—Two results from different laboratories (based on the average of three tests) should not be considered suspect (at 95 percent confidence level) unless they differ by more than 15 percent absolute.

[9] The basis for judging significance is the summary of data from the second round robin performed by five laboratories now filed under Research Reports File Number RR 77: D-20 at the American Society for Testing and Materials, 1916 Race St., Philadelphia, Pa. 19103.

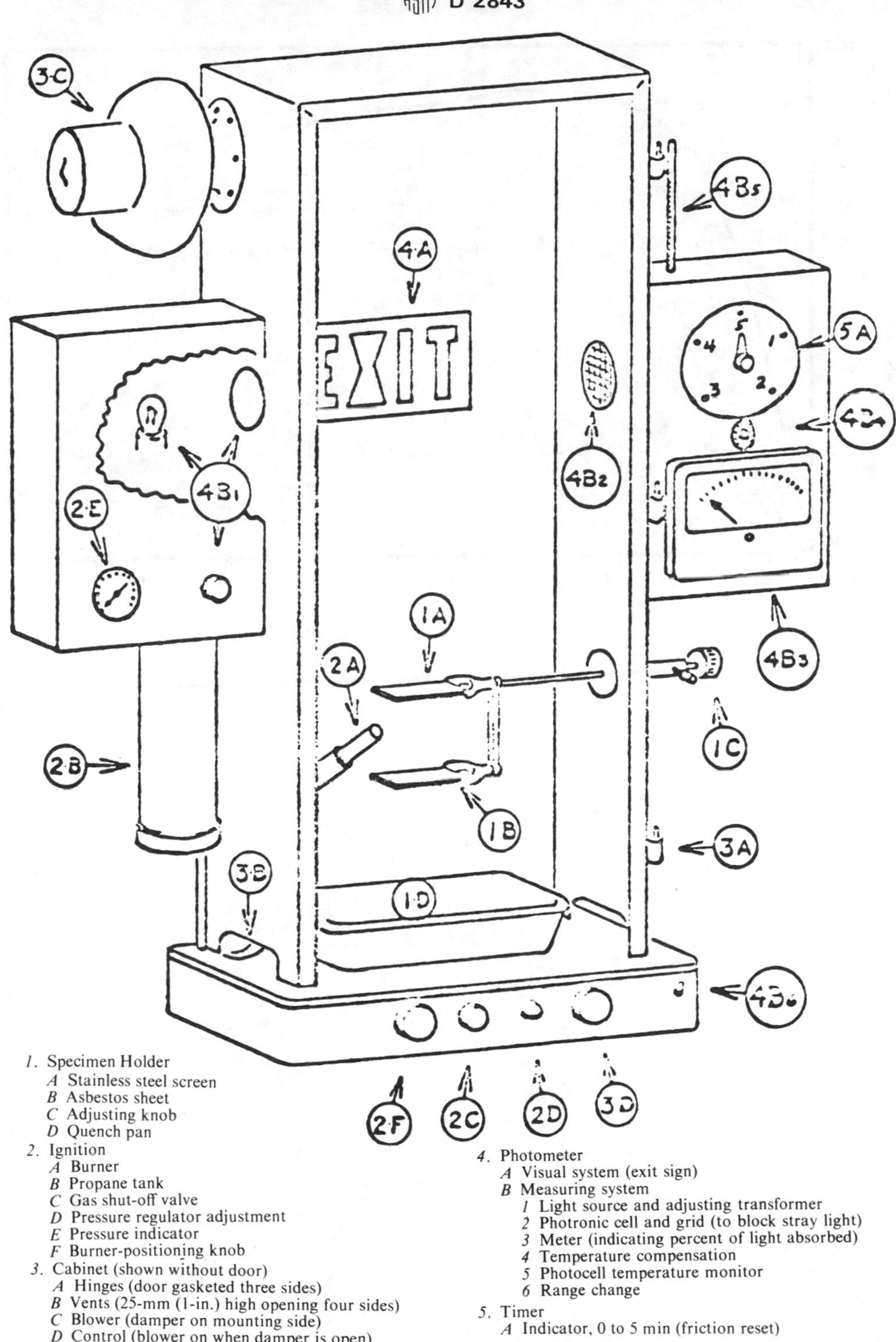

1. Specimen Holder
   - *A* Stainless steel screen
   - *B* Asbestos sheet
   - *C* Adjusting knob
   - *D* Quench pan
2. Ignition
   - *A* Burner
   - *B* Propane tank
   - *C* Gas shut-off valve
   - *D* Pressure regulator adjustment
   - *E* Pressure indicator
   - *F* Burner-positioning knob
3. Cabinet (shown without door)
   - *A* Hinges (door gasketed three sides)
   - *B* Vents (25-mm (1-in.) high opening four sides)
   - *C* Blower (damper on mounting side)
   - *D* Control (blower on when damper is open)
4. Photometer
   - *A* Visual system (exit sign)
   - *B* Measuring system
     - *1* Light source and adjusting transformer
     - *2* Photronic cell and grid (to block stray light)
     - *3* Meter (indicating percent of light absorbed)
     - *4* Temperature compensation
     - *5* Photocell temperature monitor
     - *6* Range change
5. Timer
   - *A* Indicator, 0 to 5 min (friction reset)

FIG. 1 Schematic Diagram of Smoke Chamber.

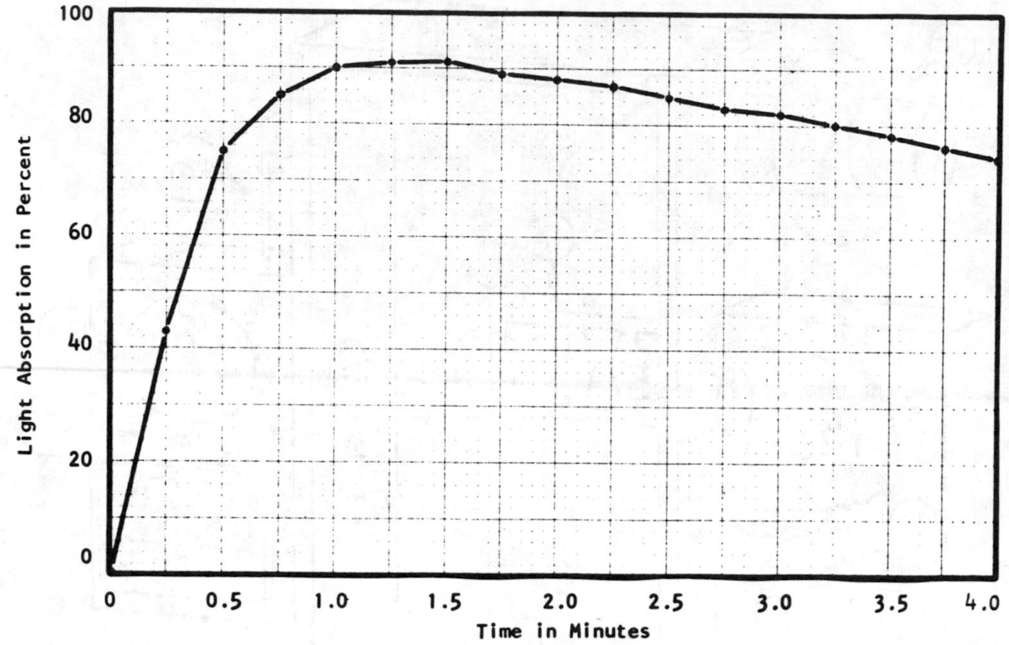

FIG. 2 Light Absorption versus Time.

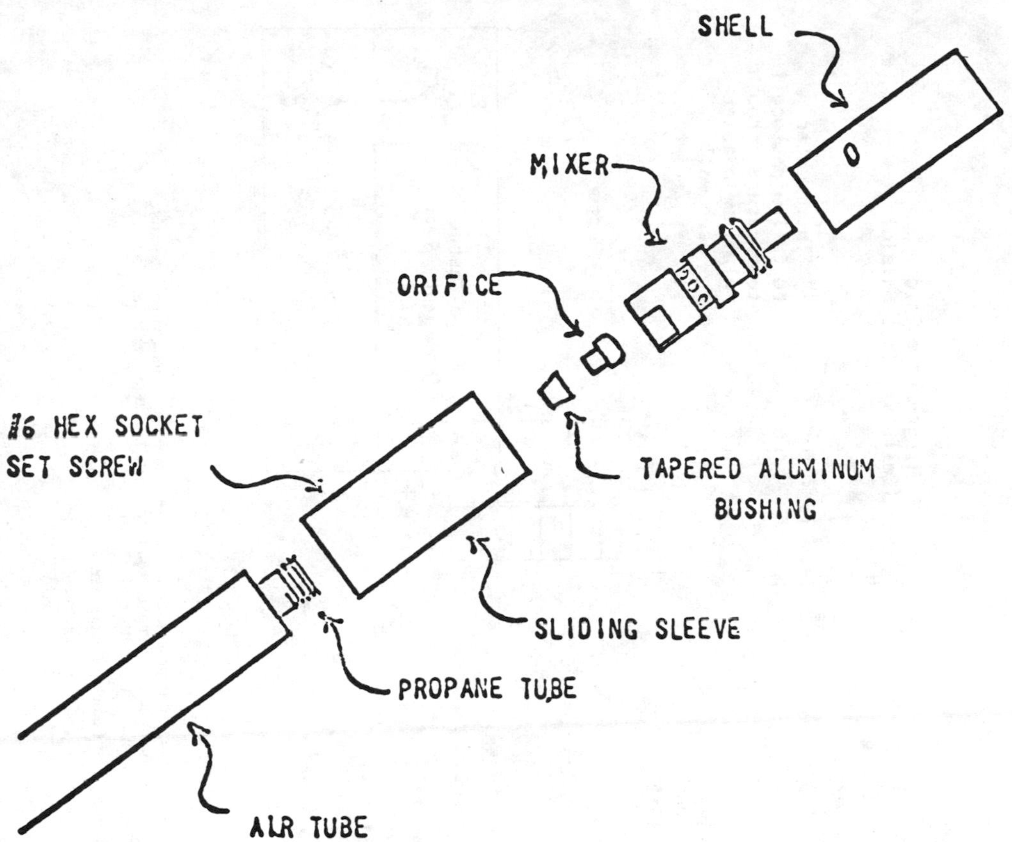

FIG. 3 Exploded View of Burner.

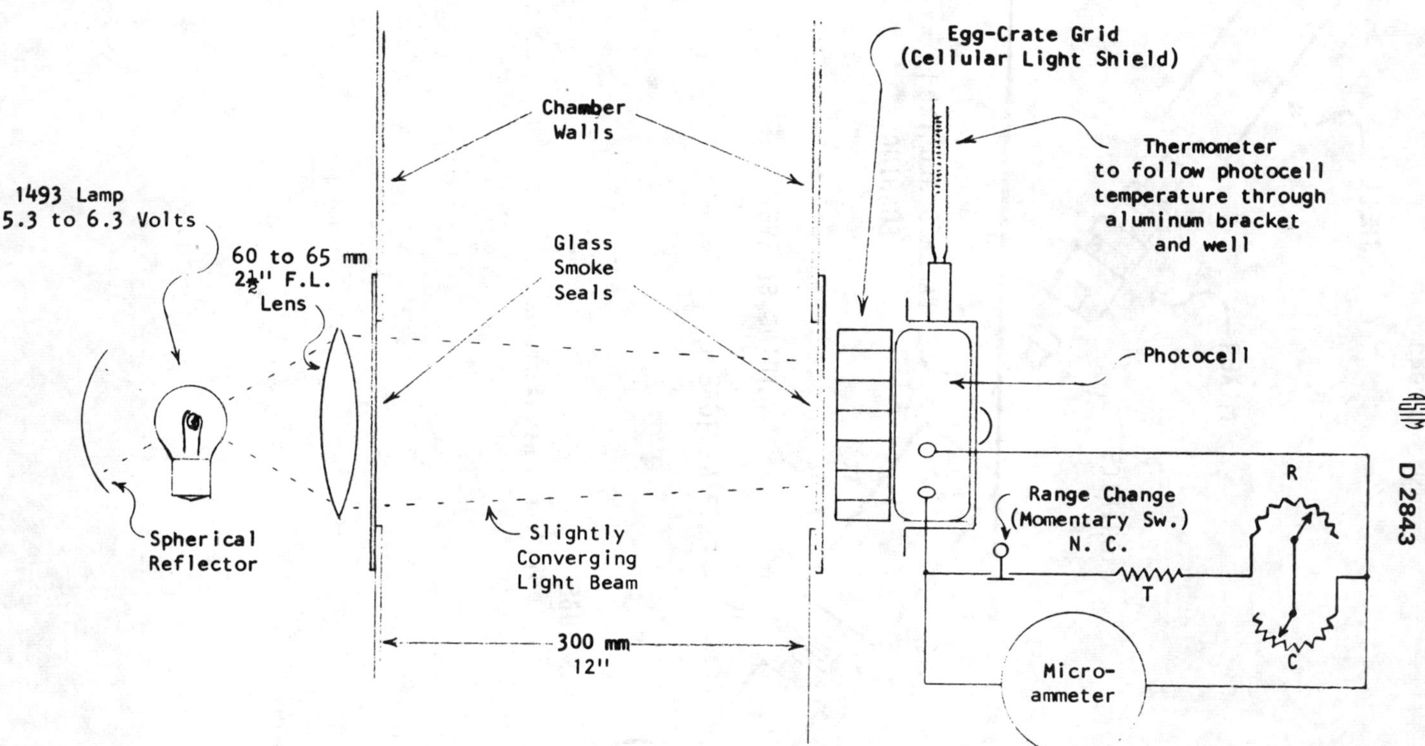

T = Temperature-sensitive winding in or on meter case to increase in resistance in proportion to increase in meter resistance with temperature.
R = Potentiometer with calibrated scale to reduce resistance in proportion to decrease in photocell output with rise in temperature.
C = Potentiometer to calibrate total resistance of shunt to change meter sensitivity exactly by 10:1 ratio.

FIG. 4 Smoke Density Test Chamber Photometer.

Designation: D 2857 – 70

Standard Method of Test for
DILUTE-SOLUTION VISCOSITY OF POLYMERS[1]

This Standard is issued under the fixed designation D 2857; the number immediately following the designation indicates the year of original adoption or, in the case of revision, the year of last revision. A number in parentheses indicates the year of last reapproval.

1. Scope

1.1 This method covers the determination of the dilute-solution viscosity of polymers. It is applicable to all polymers that dissolve completely without chemical reaction or degradation to form solutions that are stable with time at a temperature between ambient and approximately 150 C. Results of the test are usually expressed as viscosity ratio, logarithmic viscosity number, or limiting viscosity number and directions for determining these quantities are given.

NOTE 1—The following two sets of nomenclature are commonly used to describe the dilute-solution viscosity of polymers:

Recommended Term[2]	Former Term[3]
viscosity ratio	relative viscosity
viscosity number	reduced viscosity
logarithmic viscosity number	inherent viscosity
limiting viscosity number	intrinsic viscosity

In this method, only the recommended terms are used.

2. Significance

2.1 The determination of dilute-solution viscosity provides one item of information towards the molecular characterization of polymers. When viscosity data are used in conjunction with other molecular parameters, the properties of polymers depending on their molecular structure may be better predicted.

2.2 With certain restrictions, satisfactory correlations can be obtained between dilute-solution viscosity and molecular parameters such as molecular weight or chain length. The most limiting restrictions which must be observed are as follows:

2.2.1 It must be known that the polymers used to establish the correlations and those to which they are applied do not consist of or contain branched species. Basically a measure of molecular size and not molecular weight, the dilute-solution viscosity can be correlated appropriately with molecular weight or chain length only if there is a unique relationship between the mass and the size of the dissolved polymer molecules. This is the case for linear, but not for most branched, polymers. Caution should be exercised in applying empirical viscosity-molecular weight relationships to polymers of unknown or uncertain structure since unsuspected branching may exist.

2.2.2 For reasons similar to those outlined in 2.2.1, it must be required that the polymers to which the correlations are applied have the same chemical composition as those used in establishing the relationships.

2.3 For polymers meeting the restrictions of 2.2, empirical relationships can be developed between the dilute-solution viscosity of a polymer and its hydrodynamic volume or average chain dimension (radius of gyration or end-to-end distance). Such relationships depend upon any variables influencing this molecular size of the dissolved polymer. The most important of these variables are solvent type and temperature. Thus, the solution viscosity of a given polymer specimen depends on the choice of these variables, and they must always be specified with the viscosity for complete identification.

2.4 The further empirical interpretation of viscosity data in terms of molecular weight

[1] This specification is under the jurisdiction of ASTM Committee D-20 on Plastics. A list of members may be found in the ASTM Yearbook. This standard is the direct responsibility of Subcommittee D-20.70 on Analytical Methods.

Effective April 17, 1970.

[2] International Union of Pure and Applied Chemistry, *Journal of Polymer Science*, JPSCA, Vol 8, 1952, p. 257.

[3] Cragg, L. H., *Journal of Colloid Science*, JCSCA, Vol 1, 1946, p. 261.

rather than size requires control of the variables described in 2.2.

2.5 The solution viscosity of a polymer of sufficiently high molecular weight may depend on rate of shear in the viscometer, and the viscosity of a polyelectrolyte (polymer containing ionizable chemical groupings) will depend on the composition and ionic strength of the solvent. Special precautions beyond the scope of this method are required when measuring such polymers.

2.6 Finally, the viscosity of polymer solutions may be affected drastically by the presence of recognized or unrecognized additives in the sample, including but not limited to colorants, fillers, or low-molecular-weight species.

3. Apparatus

3.1 *Volumetric Flasks*[4], 100-ml or other size found convenient.

3.2 *Transfer Pipets*[4], sizes between 1 and 25 ml, as required. Transfer pipets for use with polymer solutions should have about 2 mm cut from their lower tips to permit more rapid transfer of the solution to the viscometer.

3.3 *Constant-Temperature Bath*, capable of maintaining ± 0.01 C at the desired operating temperature (usually between 25 and 135 C). Less stringent temperature control (±0.02 C) is satisfactory upon demonstration that the precision of the results is not affected.

3.4 *Viscometer*, glass capillary type, as described in ASTM Specification D 2515, for Kinematic Glass Viscometers,[5] subject to the limitations of Appendixes A-C of that method. Efflux time for the solvent and temperature used shall be greater than 200 s (except that efflux time for semi-micro viscometers shall be greater than 80 s), to eliminate the need for kinetic-energy corrections.

NOTE 2—Two types of viscometers are commonly used: one is a constant-volume device of simple construction, recommended for use where solution viscosity is to be measured at a single concentration, as for determination of the viscosity number or logarithmic viscosity number. It may also serve for the determination of the limiting viscosity number through measurement of several solutions having different concentrations.

The second type viscometer, commonly called a dilution viscometer, is a time-saving device for the determination of limiting viscosity number since it does not require constant liquid volume for operation. Several concentrations of a polymer solution can be tested by adding a known quantity of the solvent at the test temperature directly to the viscometer, mixing, measuring the viscosity, and then making the next dilution. The viscosity of the pure solvent must be measured separately.

An alternative procedure is to start with the minimum volume of the pure solvent, then add aliquots of a concentrated stock solution to the viscometer to obtain values of the viscosity ratio at successively higher concentrations. The choice of procedures is dictated by the range of volumes with which the viscometer will operate and the range of concentrations desired for test.

3.5 *Timer*, graduated in divisions of 0.1 s or less, as described in Section 4(*e*) of ASTM Method D 445, Test for Viscosity of Transparent and Opaque Liquids (Kinematic and Dynamic Viscosities).[5]

3.6 *Thermometer*, suitable for the specified test temperature and conforming to the specifications of ASTM Specification E-1, for ASTM Thermometers,[6] Kinematic Viscosity Thermometers ASTM 110F (for use at 135 C) and 118F (for use at 30 C).

3.7 *Fritted Glass Filter Funnel*, coarse grade[7], or equivalent.

4. Materials

4.1 *Solvents* as required or as recommended in the Appendix.

4.2 *Heat-Transfer Liquid* for constant-temperature bath.

NOTE 3—The following materials have been used as heat transfer liquids: (*1*) silicone oil[8], (*2*) mineral oil, (*3*) peanut oil, (*4*) water, (*5*) water-miscible liquid such as glycerol or ethylene glycol. The material selected must not discolor or smoke on prolonged exposure at the test temperature; in some cases discoloring may be inhibited by the use of an antioxidant. The use of water or a water-miscible liquid facilitates cleaning glassware used in the test.

4.3 *Nitrogen*, for purging.

5. Preparation of Sample

5.1 The sample need not be predried or conditioned unless previous specific knowledge, for example that the material is known to be hydroscopic, makes such pretreatment desirable. (See Appendix.)

---

[4] Glassware should conform to the standards of accuracy in National Bureau of Standards Circular No. C602, "Testing of Glass Volumetric Apparatus."

[5] *Annual Book of ASTM Standards*, Part 17.

[6] *Annual Book of ASTM Standards*, Part 30.

[7] A suitable filter type is Corning Glass No. 36060.

[8] Dow Corning Corp., Midland, Mich.; Union Carbide Corp., Linde Silicones Div., New York, N. Y.; General Electric Co., Silicone Products Dept., Waterford, N. Y.

5.2 If it is known that the sample dissolves only slowly in the selected solvent, pretreatment to reduce the particle size of the sample may be advisable.

NOTE 4—Some samples can be pulverized conveniently in a rotary cutting mill with a 20-mesh screen at the outlet of its pulverizing chamber. The Wiley mill, widely available from scientific supply houses, has been found satisfactory. Care must be taken to avoid overheating the sample during pulverization, which might lead to thermal degradation. Low-melting polymers, or hard, tough samples, often can be satisfactorily pulverized only at very low temperature, for example in the presence of dry ice or liquid nitrogen. (See Appendix.)

6. Procedure

6.1 Weigh the sample (see the Appendix for size and tolerance) in a tared 100-ml volumetric flask (or weigh and transfer quantitatively to the flask). If the sample is known to oxidize easily in the subsequent dissolving step, the flask may be purged with nitrogen.

NOTE 5—Other sizes of volumetric flasks may be used, depending on viscometer size and amount of sample available. If another size flask is used, adjust the sample weights and solvent and solution volumes accordingly.

NOTE 6—For greatest reliability of results and in the absence of specific information (see Appendix), select the sample size on the basis of experience to give a viscosity ratio near 1.5. If several concentrations of a solution of a single sample are to be used (Note 7), select them so that the viscosity ratio falls in the range from 1.2 to 2.0.

NOTE 7—Preparation of a single solution may often suffice, either for determining the viscosity ratio or logarithmic viscosity number, or as a stock solution for use in a dilution viscometer to determine the limiting viscosity number. If more than one solution concentration is desired, weigh a series of specimens (often four) into separate flasks, selecting the specimen weights to give the desired solution concentration. Duplicate specimens may be prepared if desired.

6.2 Add approximately 50 ml of solvent to each specimen flask, purge with nitrogen if necessary, and shake the flask once every 10 min until solution is complete (alternatively, use a laboratory shaker). Elevated temperature may enhance solution rate (see Appendix) but should be used with caution since some polymers and solvents have limited high-temperature stability. If solution is carried out at an elevated temperature, subject a flask of pure solvent to the same conditions as the polymer solution.

NOTE 8: **Caution**—Complete solution of all of the specimen is essential. When solution appears complete, examine the flask with care to be sure that no undissolved material, gel particles, or foreign matter is present.

6.3 Place the volumetric flasks containing the solution(s) and the pure solvent in the constant-temperature bath maintained at the test temperature. Add solvent to the solution flask(s) to a total volume of about 99 ml, using solvent maintained at the bath temperature. Mix the solution(s) thoroughly, and replace in the bath. After temperature equilibrium has been achieved (10 to 30 min) complete the dilution to the 100-ml mark by adding solvent maintained at the test temperature, using a transfer pipet. Mix the contents of the flask(s) thoroughly.

NOTE 9: **Caution**—Be sure that the solution is uniformly mixed. If the test temperature is above ambient, avoid cooling the flask excessively while mixing.

6.4 Where necessary to prevent oxidation, purge the viscometer with a slow stream of nitrogen. With the viscometer permanently positioned in the constant-temperature bath at the required temperature, transfer a suitable amount of solution into the viscometer using a suitably modified transfer pipet (3.2). The amount varies with the type of viscometer used (Note 2). If necessary, filter the solution through a fritted-glass filter or equivalent into the viscometer taking care not to lose solvent in the process (pressure filtration). Filtration is usually desirable.

NOTE 10—If the solution is to be handled at elevated temperatures, the transfer pipet may be fitted with a suitable heating mantle to retard precipitation of polymer from the solution during the transfer.

6.5 After temperature equilibration has been achieved (a minimum of 10 min), bring the liquid level in the viscometer above the upper graduation mark by means of gentle air (or, preferably, nitrogen) pressure applied to the arm opposite the capillary. Allow the solution to drain down through the capillary. Start the timer exactly as the meniscus passes the upper graduation mark, and stop it exactly as the meniscus passes the lower mark.

6.6 Determine the efflux time (step 6.5) at least three times each for the solution and for the pure solvent. Three consecutive readings should agree to within 0.1 s or 0.1 percent of their mean, whichever is greater. Larger variations may result from foreign material in the viscometer or from inadequate temperature control, and require repetition of the experiment after their cause is located and corrected.

7. Calculations

7.1 *Viscosity Ratio*—calculate the viscosity ratio from the average efflux time for the solvent, t_0, and the average efflux time for the solution, t, as follows:

$$\text{Viscosity Ratio} = t/t_0$$

NOTE 11—Strictly, the viscosity ratio is defined as η/η_0 where η and η_0 are the viscosities of the solution and solvent, respectively, and are related to the corresponding efflux times by:

$$\eta = Ct\rho - E\rho/t^2$$

$$\eta_0 = Ct_0\rho_0 - E\rho_0/t_0^2$$

where C and E are constants for the particular viscometer used. The equation in 7.1 follows if the second term in these relations, a kinetic-energy correction, is negligible and the respective solvent and solution densities, ρ_0 and ρ, are substantially equal. This kinetic energy correction is negligible for the recommended viscometers and efflux times (see 3.4)[9].

7.2 *Logarithmic Viscosity Number*—calculate the logarithmic viscosity ratio for each solution concentration as follows:

$$\text{Logarithmic Viscosity Number} = \ln(\eta/\eta_0)/c$$

where $\ln(\eta/\eta_0)$ is the natural logarithm of the viscosity ratio and c is the solution concentration in grams per milliliter (grams per milliliter of solution). The units of logarithmic viscosity number are, therefore, ml/g.

7.3 *Limiting Viscosity Number*—Plot the logarithmic viscosity number versus concentration, for several solution concentrations, on rectilinear graph paper. Draw the best straight line through the points and extrapolate it to zero concentration. The limiting viscosity number, $[\eta]$, is the intercept of the line at zero concentration. The units of the limiting viscosity number are ml/g.

NOTE 12—The viscosity number, $(\eta - \eta_0)/\eta_0 c$ may be calculated and plotted versus concentration on the same graph with the logarithmic viscosity number. The two lines should extrapolate to the same point (the limiting viscosity number) at $c = 0$; plotting both functions may serve to fix the limiting viscosity number with greater accuracy. If the limitation on viscosity ratio stated in Note 6 is observed, the extrapolation lines should be accurately straight.

NOTE 13—For some polymer-solvent systems, it can be demonstrated that the slopes of the lines of viscosity number and logarithmic viscosity number versus concentration are closely similar for all samples normally encountered. For such systems, two of which are described in the Appendix, the limiting viscosity number can be approximated from data obtained at a single concentration by one of the formulas tabulated by Billmeyer[10].

8. Precision and Accuracy

8.1 For most polymer-solvent systems, the logarithmic and limiting viscosity numbers can be determined with a standard deviation for reproducibility of about 1.0 ml/g. Since the viscosity numbers are available only from these measurements, accuracy of their values has no meaning.

9. Report

9.1 The report shall include the following:

9.1.1 Complete identification of the sample tested.

9.1.2 Conditioning procedure, if any.

9.1.3 One or more of the following:

9.1.3.1 The viscosity ratio, given to one significant figure beyond the decimal point, followed by the concentration of the test solution in g/ml.

9.1.3.2 The logarithmic viscosity number in ml/g, carried to one significant figure beyond the decimal point, followed by the concentration of the test solution in g/ml.

9.1.3.3 The limiting viscosity number in ml/g, carried to the decimal point.

9.1.4 The solvent employed and the test temperature.

NOTE 14—ISO Draft Recommendation No. 1628, Directives For The Standardization of Methods For The Determination of The Dilute Solution Viscosity of Polymers, differs from this method in several respects. These are:

(1) Application of kinetic energy corrections are different such that errors in viscosity ratios may be as high as 3 percent depending on flow times and viscometer type.

(2) The recommended test temperature is 25 ± 0.05 C.

It is therefore recommended that caution be exercised when comparing data obtained by both methods.

---

[9] Cannon, M. R., Manning, R. E., and Bell, J. D., *Analytical Chemistry*, ANCHA, Vol 32, 1960, p. 355.

[10] Billmeyer, F. W., Jr., *Journal of Polymer Science*, JPSCA, Vol 4, 1949, p. 83.

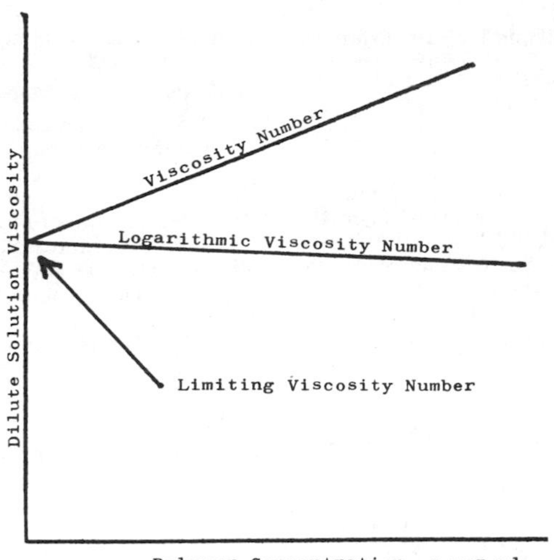

FIG. 1 Example of Plot to Determine Limiting Viscosity Number.

APPENDIX

A1. Details of Materials and Procedures for Important Polymer Types

A1.1 Polyamide

A1.1.1 Recommended solvents are formic acid (90 ± 0.2 percent in water) or *m*-cresol, melting point 11 to 12 C.

A1.1.2 Recommended dissolving conditions are: for formic acid, 30 C; for *m*-cresol, 2 h max at 95 to 100 C or 8 h at 50 C.

A1.1.3 The recommended test temperature is 30 C.

A1.1.4 The recommended solution concentration is 0.0050 ± 0.00002 g/ml.

A1.2 Polycarbonate

A1.2.1 Recommended solvents are methylene chloride or purified grade and dry *p*-dioxane.

A1.2.2 Recommended dissolving temperatures are: for methylene chloride, 30 C; for *p*-dioxane, 60 C.

A1.2.3 If pigments or other fillers are present, the resin may be dissolved in methylene chloride and the solution filtered. The methylene chloride may be evaporated from the resulting clear solution to leave the resin in the form of a thin film. After this has been dried for several hours at 125 C, samples may be taken for viscosity measurement.

A1.2.4 The recommended test temperature is 30 C.

A1.2.5 The recommended solution concentration is 0.0040 ± 0.0002 g/ml, or by convenient dilution from 0.010 ± 0.00002 g/ml.

A1.3 Polyethylene

A1.3.1 *Scope*—Measurement of the dilute-solution viscosity of ethylene polymers presents problems not ordinarily encountered in viscometry, since these polymers are not soluble in any known liquid at room temperature, and some of the higher-density materials are insoluble below 100 C. Accordingly, extreme care must be taken in all operations, such as transferring solution to the viscometer, to avoid loss of polymer from the solution by precipitation. In addition, some ethylene polymers have significant amounts of gel; this test has little significance unless the sample is completely soluble. With these limitations, the method is applicable to ethylene polymers of any density, the usual range being from 0.91 to 0.97 g/cm^3.

A1.3.2 *Apparatus*—In addition to the apparatus described in Section 3, it is convenient to have available an oven maintained at 140 ± 5 C.

NOTE A1—A viscometer especially designed for high temperature operation[11] may be found useful.

A1.3.3 *Solvent*—The recommended solvent is decahydronaphthalene, practical grade, purified and redistilled as follows:

A1.3.3.1 Purify the solvent by percolation through 100 to 200-mesh silica gel, commercial grade, to remove naphthalenes, tetrahydronaphthalene, peroxides, and other oxy compounds.

A1.3.3.2 Distill in accordance with ASTM Method D 86, Test for Distillation of Petroleum Products:[5]

[11] Schulken, R. M., and Sparks, M. L., *Journal of Polymer Science*, JPSCA, Vol 26, 1957, p. 227.

Initial boiling point (760 mm Hg)	190 C min
10 percent	191 C min
20 percent	192 C min
80 percent	194 C max
90 percent	195 C max
Dry point	196 C max

A1.3.3.3 Immediately after distillation, add 0.05 to 0.1 weight percent of an antioxidant. Phenyl-β-naphthylamine and 4-methyl-2,6-ditertiary-butylphenol have been found satisfactory and render the solvent and solutions of most ethylene polymers resistant to oxidation at 135 C for many hours.

A1.3.3.4 Other recommended solvents are 1,2,3,4-tetrahydronaphthalene and xylene (mixed isomers); but see 2.3.

A1.3.4 *Procedure:*

A1.3.4.1 The recommended dissolving conditions are 140 ± 5 C for 2 to 6 h.

NOTE A2: **Caution**—Dissolution should not be prolonged past 6 h because of the possibility of significant oxidation of the polymer. Extreme care should be taken to ascertain whether solution is complete, see Note 8.

A1.3.4.2 The recommended test temperature is 135 C.

A1.3.4.3 The recommended solution concentration is 0.0020 ± 0.00002 g/ml, or convenient dilution from 0.0040 ± 0.00002 g/ml.

A1.4 Poly(Ethylene Terephthalate)

A1.4.1 Recommended solvents are *o*-chlorophenol (melting point 0 C) or a mixture of 1 part trifluoroacetic acid and 3 parts dichloromethane. Solvents must contain less than 0.15 percent water.

NOTE A3: **Caution**—Since *o*-Chlorophenol is toxic, appropriate measures must be taken to avoid breathing its vapors or contact of the liquid with skin or eyes.

A1.4.2 The samples must be dried to a water content of 0.03 percent or less. Drying may be carried out at 100 C for about 3 h under a pressure of 1 mm Hg or less.

A1.4.3 Recommended dissolving conditions are 2 to 3 h at 90 to 100 C.

A1.4.4 The recommended test temperature is 30 C.

A1.4.5 The recommended solution concentration is 0.0050 ± 0.00002 g/ml.

A1.5 Poly(Methyl Methacrylate)

A1.5.1 Recommended solvent is 1,2-dichloroethane (ethylene dichloride), reagent grade.

A1.5.2 Pretreatment of the sample should not include cutting or grinding in such a way as to cause shear degradation of the polymer.

A1.5.3 Recommended dissolving conditions are 24 h at 30 C.

A1.5.4 The recommended test temperature is 30 C.

A1.5.5 The recommended solution concentration is 0.0020 ± 0.00002 g/ml; but the recommendations of Note 6 should be construed to supersede this.

NOTE A4—The limiting viscosity number may be estimated from data obtained at a single concentration by use of the following equation[10]:

$$[\eta] = (2^{1/2}/c)[(\eta/\eta_0) - 1 - \ln(\eta/\eta_0)]^{1/2}$$

A1.6 Poly(Vinyl Chloride)

A1.6.1 The recommended solvent is cyclohexanone, boiling point 155 to 156 C (760 mm Hg).

A1.6.2 Samples should be dried to a moisture content below 0.1 percent.

A1.6.3 The recommended dissolving temperature is 85 ± 10 C.

A1.6.4 The recommended test temperature is 30 C.

A1.6.5 The recommended solution concentration is 0.0020 ± 0.00002 g/ml.

NOTE A5—The limiting viscosity number may be estimated from data obtained at a single concentration by use of the following equation[10]:

$$[\eta] = 1/4\ [(\eta - \eta_0)/\eta_0 c] + 3/4\ [\ln(\eta/\eta_0)/c]$$

A1.6.6 *Precision:*

A1.6.6.1 *Repeatability*—Duplicate results by the same analyst should not be considered suspect unless they differ by more than 1 percent.

A1.6.6.2 *Reproducibility*—The average result of two determinations reported by one laboratory should not be considered suspect unless it differs from that of another laboratory by more than 2 percent.

ASTM Designation: D 2863 - 70

Standard Method of Test for
FLAMMABILITY OF PLASTICS USING THE OXYGEN INDEX METHOD[1]

This Standard is issued under the fixed designation D 2863; the number immediately following the designation indicates the year of original adoption or, in the case of revision, the year of last revision. A number in parentheses indicates the year of last reapproval.

1. Scope

1.1 This method describes a procedure for determining the relative flammability of plastics by measuring the minimum concentration of oxygen in a slowly rising mixture of oxygen and nitrogen that will just support combustion. This method is presently limited to the use of physically self-supporting plastic test specimens.

NOTE 1—This method has been found applicable for testing plastics that are not physically self-supporting and other materials. However, the accuracy of the method has not been determined for these materials, or for specimen geometries and test conditions outside those recommended herein.

2. Significance

2.1 This method provides a means for comparing the relative flammability of physically self-supporting plastics. It may also be used to study the effect of changes in material on the flammability. Correlation with flammability under actual use conditions is not necessarily implied.

3. Definition

3.1 *oxygen index*—the minimum concentration of oxygen, expressed as volume percent, in a mixture of oxygen and nitrogen that will just support combustion of a material under the conditions of this method.

4. Principle of Method

4.1 The minimum concentration of oxygen in a slowly rising mixture of oxygen and nitrogen that will just support combustion is measured under equilibrium conditions of candle-like burning. The balance between the heat from the combustion of the specimen and the heat lost to the surroundings establishes the equilibrium. This point is approached from both sides of the critical oxygen concentration in order to establish the oxygen index.

5. Apparatus

5.1 *Test Column*, consisting of a heat-resistant glass tube of −75 mm minimum inside diameter and 450-mm minimum height. The bottom of the column or the base to which the tube is attached shall contain noncombustible material to mix and distribute evenly the gas mixture entering at this base. Glass beads 3 to 5 mm in diameter in a bed 80 to 100 mm deep have been found suitable (an example is shown in Fig. 1).

NOTE 2—A column with a 95-mm inside diameter and 210 mm high with a restricted upper opening (diameter = 50 mm) has been found to give equivalent results.

NOTE 3—It is helpful to place a wire screen above the noncombustible material to catch falling fragments and aid in keeping the base of the column clean.

5.2 *Specimen Holder*—Any small holding device that will support the specimen at its base and hold it vertically in the center of the column is acceptable. A typical arrangement, shown in Fig. 1, consists of a laboratory thermometer clamp inserted into the end of a glass tube held in place by the glass beads.

5.3 *Gas Supply*—Commercial grade (or better) oxygen and nitrogen shall be used. If an air supply is used with oxygen or nitrogen, it must be clean and dry.

[1] This method is under the jurisdiction of ASTM Committee D-20 on Plastics. A list of members may be found in the ASTM Yearbook. This standard is the direct responsibility of Subcommittee D-20.30 on Thermal Properties.

Effective May 8, 1970.

5.4 *Flow Measurement and Control Devices*—Suitable flow measurement and control devices shall be available in each line which will allow monitoring the volumetric flow of each gas into the column within 1 percent. After the flow is measured in each line, the lines should be joined to allow the gases to mix before being fed into the column.

NOTE 4—One satisfactory flow control system consists of calibrated jeweled orifices (**3**),[2] precision pressure regulating devices, and precision gas gages. An equally satisfactory system consists of needle valves and rotameters. Some manometer-orifice systems have not been found to be sufficiently accurate.

5.5 *Ignition Source*—The igniter should be a tube with a hydrogen, propane, or natural gas flame at the end that can be inserted into the open end of the column to ignite the test specimen. A suitable flame may be from 6 to 12 mm long.

5.6 *Timer*—A suitable timer capable of indicating at least 10 min and accurate to 5 s shall be used.

5.7 *Soot, Fumes, and Heat Removal*—To ensure the removal of toxic fumes, soot, heat, and other possible noxious products, the column shall be installed in a hood or other facilities providing adequate exhaust.

NOTE 5—If soot-generating specimens are being tested, the glass column becomes coated on the inside with soot and should be cleaned as often as necessary for good visibility.

6. Test Specimens

6.1 At least ten specimens 70 to 150 mm long by 6.5 ± 0.5 mm wide and 3.0 ± 0.5 mm thick shall be cut from the materials being tested.

NOTE 6—If other than standard size specimens are used, differences in oxygen index may occur.

6.1.1 The specimens shall be tested in the as-received condition unless otherwise agreed upon.

NOTE 7—Moisture content of some materials has been shown to affect the oxygen index.

6.1.2 The edges of the specimens shall be relatively smooth and free from fuzz or burrs of material left from machining.

7. Procedure

7.1 Calibrate the flow measuring system using a water-sealed rotating drum meter (wet test meter) in accordance with ASTM Method D 1071, for Measurement of Gaseous Fuel Samples,[3] or by equivalent calibration devices. It is recommended that this calibration be repeated at least every 6 months.

NOTE 8—One step in the calibration should be to check carefully for leaks at all joints.

7.2 Clamp the specimen in the holder vertically in the approximate center of the column with the top of the specimen at least 100 mm below the top of the open column.

NOTE 9—If a restricted opening column is used (see Note 2), the top of the specimen should be at least 40 mm below the opening.

7.3 Select the desired initial concentration of oxygen based on past experience with similar materials. If there is no experience with the material, light a specimen in the air and note the burning. If the specimen burns rapidly, start at a concentration of about 18 percent, but if the specimen goes out, select a concentration of about 25 percent or higher depending on the difficulty of ignition and time of burning.

7.4 Set the flow valves so that the desired initial concentration of oxygen is flowing through the column. The gas flow rate in the column shall be 4 ±1 cm/s as calculated at standard temperature (0 C) and pressure (760 mm Hg) from the total flow of gas in cm^3/s, divided by the area of the column in cm^2.

7.5 Allow the gas to flow for 30 s to purge the system.

7.6 Ignite the top of the specimen with the ignition flame so that the specimen is well lit and the entire top is burning. Remove the ignition flame and start the time.

7.6.1 The concentration of oxygen is too high and must be reduced if:

7.6.1.1 The specimen burns 3 min or longer, OR

7.6.1.2 The specimen burns 50 mm.

7.6.2 The concentration of oxygen must be raised if the specimen is extinguished before burning 3 min or 50 mm.

7.7 Adjust the oxygen concentration, insert a new specimen, or if the previous specimen is long enough, turn it end for end or cut off the burned end, then purge and re-ignite.

---

[2] The boldface numbers in parentheses refer to the list of references at the end of this method.

[3] *Annual Book of ASTM Standards*, Part 19.

NOTE 10—Do not adjust the oxygen concentration after igniting the specimen.

7.8 Continue repeating 7.5 through 7.7 until the limiting concentration of oxygen is determined. This is the lowest oxygen concentration which will meet the conditions of 7.6.1. At the next lower oxygen concentration possible with the equipment, the specimen should extinguish as defined in 7.6.2.

7.9 For a material having consistent burning characteristics, the difference in oxygen concentration between burning as defined in 7.6.1 and extinguishing as defined in 7.6.2 will be reproducible within 0.1 to 0.3 percent depending on the sensitivity of the flow measuring equipment and upon the particular oxygen concentration involved. Some materials, however, exhibit erratic burning characteristics because of inhomogeneity, char fòrmation, dripping, bending, etc., which cause less reproducible results. In such cases, the limiting concentration should be determined by a statistical testing method.[4]

7.10 Perform the test at least three times by starting at a slightly different flow rate still within the 3 to 5 cm/s limits and again performing steps 7.4 through 7.8.

8. Calculations

8.1 Calculate the oxygen index, n, of the material as follows:

$$n\text{, percent} = (100 \times O_2)/(O_2 + N_2)$$

where O_2 is the volumetric flow of oxygen, cm^3/s, at the limiting concentration determined in 7.8, and N_2 is the corresponding volumetric flow rate of nitrogen, cm^3/s.

NOTE 11—If air is used and either oxygen or nitrogen is added as required, calculate n assuming that air contains 20.9 percent oxygen as follows:

$$n\text{, percent} = (100 \times O_2) + (20.9 \times A)/(O_2 + N_2 + A)$$

where A is the volumetric flow rate of air, cm^3/s, and O_2 and N_2 are the volumetric flow rates of oxygen and nitrogen added to the mixture. One of these will be zero depending on which gas is added.

9. Report

9.1 The report shall include the following:

9.1.1 Description of the material tested including as much as is known about the type, source, manufacturer's code number, form and principal dimensions, and previous history.

9.1.2 Test specimen dimensions,

9.1.3 Average oxygen index value,

9.1.4 Individual oxygen index values found for each of the tests, and

9.1.5 Description of any unusual behavior such as charring, dripping, bending, etc.

10. Precision

10.1 Based on the results of statistically designed round-robin testing program (**4, 5**) in which 18 laboratories checked 5 materials, the standard deviation of the mean of 3 replicates (for comparing laboratory-to-laboratory) was as follows:

10.1.1 For materials with an oxygen index below 21 percent, the standard deviation was below 0.4.

10.1.2 For materials with an oxygen index above 21 percent, the standard deviation ranged from 0.7 to 1.4. The higher value was for a material that exhibits the erratic behavior noted in 7.9.

10.2 The standard deviation within a laboratory ranged from 0.1 for clean burning materials to 1.0 for erratic materials.

[4] Such a statistical method as the Bruceton Staircase Method may be used. See Dixon, W. J., and Massey, F. J., Jr., *Introduction to Statistical Analysis* (2nd Edition), 1957, Chapter 19, McGraw-Hill Book Co., Inc., New York, N.Y. or Natrella, Mary, "Experimental Statistics," Section 10-4, *National Bureau of Standards Handbook 91*, 1963. Other procedures, such as using ten specimens at each oxygen concentration tried, have also proved successful.

REFERENCES

(**1**) Fenimore, C. P., and Martin, F. J. "Candle-type Test for Flammability of Polymers," *Modern Plastics*, MOPLA, Vol 43, November 1966, p. 141.

(**2**) Goldblum, K. B., "Oxygen Index: Key to Precise Flammability Ratings," *Society of Plastics Engineers Journal*, SPEJA, Vol 25, February 1969, p. 50.

(**3**) Andersen, J. W., and Friedman, R., "An Accurate Gas Metering System for Laminar Flow Studies," *Review of Scientific Instruments*, RSINA, Vol 20, 1949, p. 61.

(**4**) Isaacs, J. L., "The Development, Standardization and Utilization of the Oxygen Index Flammability Test," *General Electric TIS Report 69-MAL-13*, August 1969, Louisville, Ky.

(**5**) Supporting data for this method have been filed at ASTM Headquarters as RR 102: D-20.

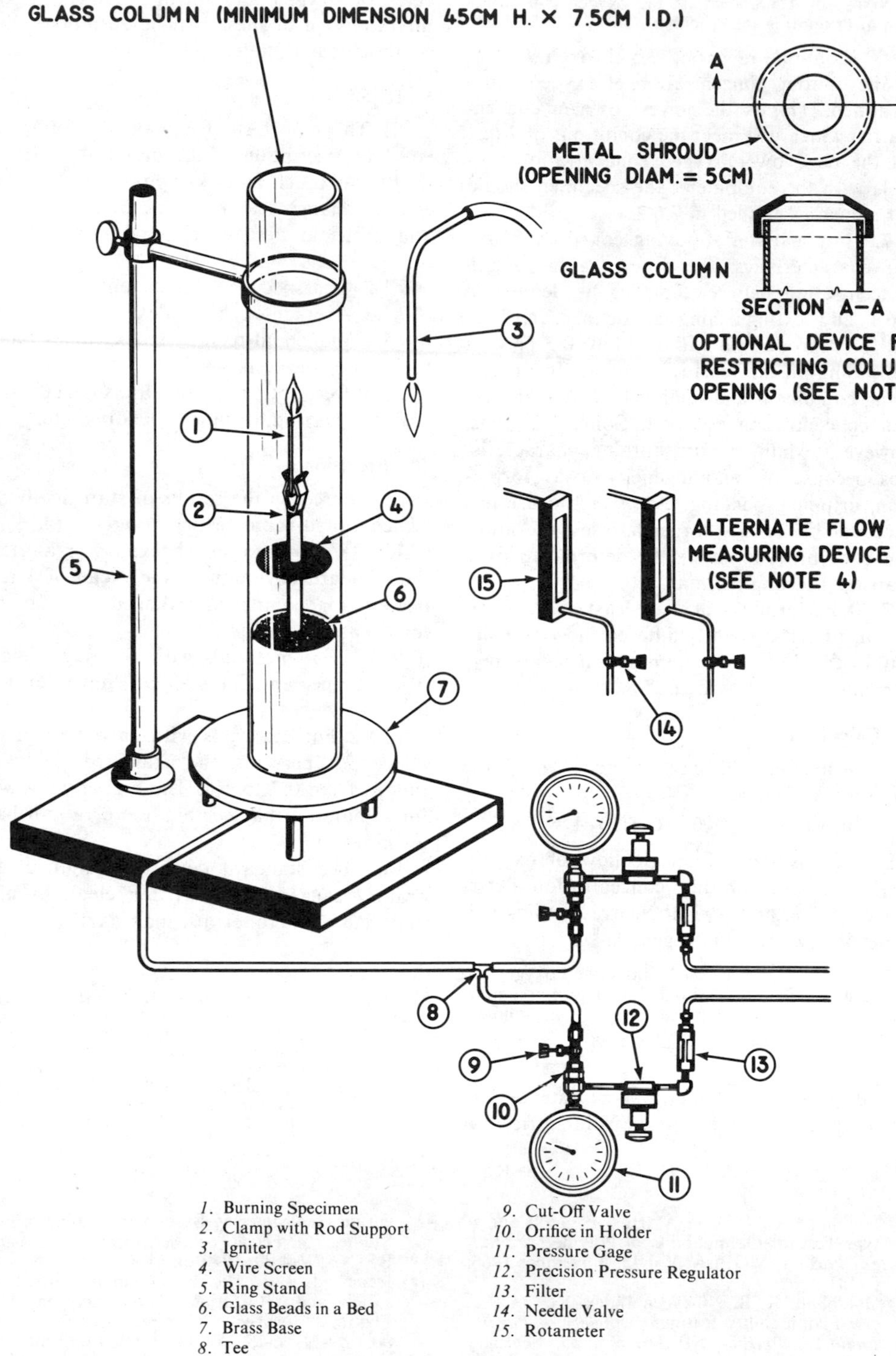

1. Burning Specimen
2. Clamp with Rod Support
3. Igniter
4. Wire Screen
5. Ring Stand
6. Glass Beads in a Bed
7. Brass Base
8. Tee
9. Cut-Off Valve
10. Orifice in Holder
11. Pressure Gage
12. Precision Pressure Regulator
13. Filter
14. Needle Valve
15. Rotameter

FIG. 1 Typical Equipment Layout.

Designation: E 4 – 64

American National Standard Z115.1-1966
American National Standards Institute
American Association State
Highway Officials Standard
A.A.S.H.O. NO.: T 67

Standard Methods of
VERIFICATION OF TESTING MACHINES[1]

This Standard is issued under the fixed designation E 4; the number immediately following the designation indicates the year of original adoption or, in the case of revision, the year of last revision. A number in parentheses indicates the year of last reapproval.

1. Scope

1.1 These methods cover four procedures for the verification by means of standard calibrating devices of testing machines that are designed to measure loads, as follows:

A. Verification by standard weights,

B. Verification by standardized proving levers,

C. Verification by equal-arm balances and standard weights, and

D. Verification by elastic calibration devices.

NOTE 1—The values stated in U.S. customary units are to be regarded as the standard. The metric equivalents of U.S. customary units may be approximate.

NOTE 2—These methods are not intended to be complete purchase specifications for testing machines. Purchase specifications usually state the maximum permissible errors from zero to capacity load for each scale range of the machine.

2. Definitions

2.1 *testing machine*—a mechanical device for applying a load (force) to a specimen.

NOTE 3—Usually the magnitude of the load can be changed at the will of the operator. Many testing machines are arranged to measure the load, but this is not always the case, especially with impact machines and machines for testing ductility.

2.2 *load*—in the case of testing machines, a force measured in units of pound-force or kilogram-force. The pound-force is that force which acting on a 1-lb mass will give to it an acceleration of 980.665 cm/s^2. The kilogram-force is that force which acting on a 1-kg mass will give to it an acceleration of 980.665 cm/s^2. (This internationally accepted standard acceleration converts to 32.174 ft/s^2.)

2.3 *accurate*—A testing machine is said to be accurate if the indicated load is within the specified permissible variation from the actual load.

NOTE 4—In these methods the word "accurate" applied to a testing machine is used without numerical values, for example, "An accurate testing machine was used for the investigation."

The *accuracy* of a testing machine should not be confused with *sensitivity*. For example, a testing machine might be very sensitive; that is, it might indicate quickly and definitely small changes in load, but, nevertheless, be very inaccurate. On the other hand, the accuracy of the results is in general limited by the sensitivity.

2.4 *error*—in the case of a testing machine, the value obtained by subtracting the value indicated by the calibrating device from the value indicated by the testing machine.

NOTE 5—The word "error" shall be used with numerical values, for example, "At a load of 30,000 lb (13,600 kg) the error of the testing machine was +15 lb (6.8 kg)."

2.5 *percentage of error*—in the case of a testing machine, the ratio, expressed as a percentage, of the error to the correct value of the quantity measured.

2.6 *correction*—in the case of a testing machine, the value obtained by subtracting the indicated value from the correct value of the quantity measured.

NOTE 6—The correction has the same magnitude as the error, but the opposite sign. It is recommended that, except for special cases, no corrections be used on machines tested and found to have errors within the permissible variations given in these methods.

2.7 *permissible variation* (*or tolerance*)—in

[1] These methods are under the jurisdiction of ASTM Committee E-1 on Methods of Testing. A list of members may be found in the ASTM Yearbook.

Current edition effective Aug. 31, 1964. Originally issured 1923, to replace in part Method E 1. Replaces E 4–61 T.

the case of testing machines, the maximum allowable error in the value of the quantity indicated.

NOTE 7—It is convenient to express permissible variation in terms of percentage of error. The numerical value of the permissible variation for a testing machine is so stated hereafter in these methods.

2.8 *capacity range*—in the case of testing machines, the range of loads for which it is designed. Some testing machines have more than one capacity range.

2.9 *loading range*—in the case of testing machines, the range of indicated loads for which the testing machine gives results within the permissible variations specified.

3. Calibration Devices

3.1 Verify testing machines, using calibration devices that comply with the requirements of ASTM Methods E 74, Verification of Calibration Devices for Verifying Testing Machines.[2]

4. Advantages and Limitations of Methods

4.1 *Verification by Standard Weights*—Verification by the direct application of standard weights to the weighing mechanism of the testing machine, where practicable, is the most accurate method. Its limitations are; (*1*) the small range of load which can be covered, (*2*) the nonportability of any large amount of standard weights, and (*3*) its nonapplicability to horizontal testing machines or to vertical testing machines the weighing mechanisms of which are not designed to be actuated by a downward force.

4.2 *Verification by Proving Levers*—The second method of verification involves the use of a pair of standardized proving levers. The limitations of this second method are the same as for the verification by standard weights. The range of load practicable with proving levers is much greater than that practicable with standard weights, but the range of load for proving levers is not great enough to cover the capacity load of many large testing machines.

4.3 *Verification by Equal-Arm Balance and Standard Weights*—The third method of verification of testing machines involves measurement of the load by means of an equal-arm balance and standard weights. This method is limited to a still smaller range of loads than either of the foregoing methods, and is generally applicable only to certain types of hardness testing machines in which the load is applied through an internal lever system.

4.4 *Verification by Elastic Calibration Devices*—The fourth method of verification of testing machines involves measurement of the elastic strain or deflection under load of a ring, loop, tension or compression bar, or other elastic device. The elastic calibration device is free from the limitations referred to in 4.1 and 4.2.

5. Gravity Corrections

5.1 In the verification of testing machines under Procedures A, B, and C where loads are applied by standard weights either directly or through lever or balance-arm systems, correct the load for the local value of gravity and for air buoyancy. Calculate the load applied by a standard weight from the following equation:

$$P = M\,[(g/980.665) - 0.00014]$$

where:

P = load, lbf or kgf, force,

M = apparent mass of weights as compared to brass standards in normal air, lb or kg, and

g = local value of gravity, cm/s^2.

NOTE 8—Apparent mass as compared to brass standards in normal air is the usual basis on which standard weights are certified in the United States.

NOTE 9—Local values of gravity for various cities are available in publications of the U.S. Coast and Geodetic Survey and in handbooks. The local value of gravity known to be correct to 0.1 cm/s^2 or better should be used.

6. Method of Applying Load

6.1 In the verification of a testing machine, approach the loads by increasing the load from a lower load.

NOTE 10—For any testing machine the errors observed at corresponding loads taken first by increasing the load to any given test load and then by decreasing the load to that test load, may not agree. This disagreement is especially likely in types of machines in which the load-indicating device is actuated by a Bourdon pressure tube, by a hydraulic indicator or a steam-engine indicator, or by any other load-indicating device depending on the elastic properties of a material. It is also likely to occur in a testing machine in which the load is measured

[2] *Annual Book of ASTM Standards*, Part 30.

by measuring the pressure in a hydraulic jack.

Testing machines are usually used under increasing loads, but if a testing machine is to be used under decreasing loads, it should be calibrated under decreasing loads as well as under increasing loads.

7. Selection of Test Loads

7.1 For any loading range, verify the testing machine by at least five test loads (except for testing machines designed to measure only a smaller number of definite loads, such as certain hardness testing machines). The difference between any two successive test loads shall not exceed one third of the difference between the maximum and minimum test loads. If it is desired to establish the lower limit of a loading range lower than 10 percent of the capacity of the range, verify the lower limit by applying five approximately equal test loads, none of which shall differ from the lowest by more than 5 percent.

8. Eccentric Loading

8.1 For the purpose of determining the loading range of a testing machine, apply all calibration loads so that the resultant load is as nearly along the axis of a testing machine as is possible.

NOTE 11—The effect of eccentric load on the accuracy of a testing machine may be determined by calibration readings taken with proving levers or an elastic calibration device placed so that the resultant load is applied at definite distances from the axis of the machine, and the loading range determined for a series of eccentricities. In the case of testing machines in which the load reading depends on the hydrostatic pressure in a cylinder fitted with a piston, the effect of eccentricity of loading is most serious when the piston is at the most extreme outward position.

A. VERIFICATION BY STANDARD WEIGHTS

9. Procedure

9.1 Place standard metal weights of suitable design, finish, and adjustment on the weighing platform of the testing machine or on trays or other supports suspended from the load-measuring mechanism in place of the specimen. Use weights certified within less than five years to be correct within a limit of error of 0.1 percent. Apply the weights in increments and remove in the reverse order. Arrange the weights symmetrically with respect to the weighing platform, so that the center of gravity of the load lies in the vertical line through the center of the platform. Record the applied load and the indicated load for each test load applied, and the error and the percentage of error calculated from these data.

NOTE 12—The method of verification by direct application of standard weights can be used only on vertical testing machines in which the pressure on the weighing table, hydraulic support, or other weighing device is downward. The total load is limited by the size of the platform and the number of weights available. Fifty-lb (22.7 kg) weights are usually convenient to use. This method of verification is confined to small testing machines and is rarely used above one or two thousand pounds.

NOTE 13—In connection with the required limit of error of 0.1 percent it may be noted that, in addition to the National Bureau of Standards, many of the states, some counties, and some universities have departments or offices of weights and measures equipped and staffed to certify weights up to 50 lb (22.7 kg) within the class "C" tolerances of the Bureau of Standards. These are closer tolerances than the requirement of a limit of error of 0.1 percent.

B. VERIFICATION BY STANDARDIZED PROVING LEVERS

10. Levers

10.1 The common arrangement of proving levers for verification of testing machines is shown in Fig. 1. The two levers rest on supports on the weighing platform of the testing machine. These supports shall move easily in a horizontal direction, which ensures that the forces at each of the knife edges, or other bearings, shall be very nearly vertical. The inner knife edges in each lever shall bear against a suitable block in the crosshead of the testing machine. Suspend weight trays or hangers from each of the outer knife edges, and load these trays or hangers with standard weights. Use weights certified within less than 5 years to meet a tolerance of 0.02 percent accuracy. The increment of load put on the testing machine by the standard weights is the load exerted by the standard weights multiplied by the lever ratio m/n, Fig. 1.

NOTE 14—In testing machines having a hydraulic support, or other load-weighing mechanism in the crosshead with the load during a test pressing upward against the crosshead instead of downward to the table, the lever ratio is $(m - n)/n$.

11. Procedure

11.1 Place the proving levers in the testing

machine to be verified so that the resultant load line coincides with the vertical line through the center of the weighing platform. Bring both levers as near to a horizontal position as is feasible, after applying each increment of load, by means of the movable head of the testing machine. Balance the testing machine with the levers in place and the weight trays empty. Apply (or remove) standard weights in increments, half an increment in (or from) each tray. Place the weights symmetrically on the weight trays with the center of gravity of the weights as nearly over the center of the tray as is feasible. Record the applied load and the indicated load for each test load applied, and the error and the percentage of error calculated from these data.

NOTE 15—The use of standardized proving levers on horizontal testing machines involves the use of bell-crank levers. Such levers require special methods of determination of lever ratio, which have not yet been codified in ASTM standards. Proving levers for vertical testing machines are now available up to 50,000-lb (22,700-kg) capacity.

C. VERIFICATION BY EQUAL-ARM BALANCE AND STANDARD WEIGHTS

12. Procedure

12.1 Position the balance so that the indenter of the testing machine being calibrated bears against a block centered on one pan of the equal-arm balance, the balance being in its equilibrium position when the indenter is in that portion of its travel normally occupied when making an impression. Place standard weights complying with the requirements of Section 9 on the opposite pan to balance the load exerted by the indenter.

NOTE 16—This method may be used for the verification of testing machines other than hardness-testing machines by positioning the load-applying member of the testing machine in the same way that the indenter of a hardness-testing machine is positioned.

12.2 Since the permissible travel of the indenter of a hardness testing machine is usually very small, do not allow the balance to oscillate or swing. Instead, maintain the balance in its equilibrium position through the use of an indicator such as an electric contact, which shall be arranged to indicate when the reaction of the indenter load is sufficient to lift the pan containing the standard weights.

12.3 Using combinations of fractional weights, determine both the maximum value of the dead-weight load that can be lifted by the testing machine indenter load during each of ten successive trials, and the minimum value that cannot be lifted during any one of ten successive trials. Take the correct value of the indenting load as the average of these two values. The difference between the two values shall not exceed 0.5 percent of the average value.

D. VERIFICATION BY ELASTIC-CALIBRATION DEVICE

13. Definition of Elastic Calibration Device

13.1 An elastic calibration device for use in verifying the load readings of a testing machine consists of an elastic member or members to which loads may be applied, combined with a mechanism or device for indicating the magnitude (or a quantity proportional to the magnitude) of deformation under load.

14. Temperature Equalization

14.1 When using an elastic calibration device to verify the readings of a testing machine, place the device near to, or preferably in, the testing machine a sufficient length of time before the test so that the device and the testing machine is at very nearly the same temperature.

15. Procedure

15.1 Place the elastic device in the testing machine so that its center line coincides with the center line of the heads of the testing machine. Balance the machine with no load on the elastic device and take a zero reading of the device. Then apply test loads at suitable increments. Record the indicated load of the testing machinc and the applied load computed from the readings of the elastic device and the error and the percentage of error calculated from these data.

CALCULATION AND REPORT

16. Basis of Verification

16.1 The percentage of error for loads within the loading range of the testing machine shall not exceed ± 1.0.

NOTE 17—This means that the report of the verification of a testing machine will state within what loading range it may be used, rather than reporting a blanket acceptance or rejection of the machine.

16.2 In establishing the lower limit of a loading range below 10 percent of the capacity of the range where five applications of load are required, the algebraic difference between the highest and lowest percentage of error shall not exceed 1.0.

NOTE 18—This means that to establish the lower limit of a loading range less than 10 percent of the capacity of that range, the errors for the series of five readings shall not only not exceed 1 percent but also no two errors shall differ by more than 1.0 percent. If the minimum error in this series is −1.0 percent the maximum error cannot exceed 0.0 percent; if the minimum error is −0.5 percent the maximum error cannot exceed +0.5 percent. If the minimum error is 0.0 percent the maximum error cannot exceed +1.0 percent; etc.

16.3 In no case shall the loading range be stated as including loads below the value which is 100 times the smallest change of load which can be estimated on the load-indicating apparatus of the testing machine.

NOTE 19—This means that in a testing machine which has graduations so spaced that estimations can be made to $^1/_{10}$ division, the loading range could not extend downward to a load less than that corresponding to 10 divisions. If the graduations on the load-indicating scale can be estimated only to 2 divisions the loading range could not extend downward below the load corresponding to 200 divisions. On most machines the smallest load that can be measured is somewhere between the two examples cited.

16.4 In no case shall the loading range be stated as including loads outside the range of loads applied during the verification test.

17. Corrections

17.1 The indicated load of a testing machine shall not be corrected either by calculation or by the use of a calibration diagram to obtain values within the required permissible variation.

18. Time Interval Between Verifications

18.1 It is recommended that testing machines, when in constant use, be verified at intervals of 12 months and, when used intermittently, at intervals of 2 or 3 years. Verify testing machines, however, immediately after making repairs or adjustments of the weighing mechanism, after the testing machine has been moved (this does not apply to portable testing machines), and whenever there is reason to doubt the accuracy of the results, without regard to the time interval since the last verification.

19. Report

19.1 Prepare a clear and complete report of each verification of a testing machine. State the method of verification used, and give the serial numbers and the names of manufacturers of all apparatus employed in carrying out the verification. State how, by whom, and when the calibration of the apparatus used in verifying the testing machinc was made, the loading range of the calibration apparatus, and the loading range of the testing machine.

20. Certificate

20.1 A certificate giving the manufacturer's serial number and a brief description of the testing machine, the manufacturer's name, the date of verification, and the loading range shall be signed by the person in responsible charge of the verification.

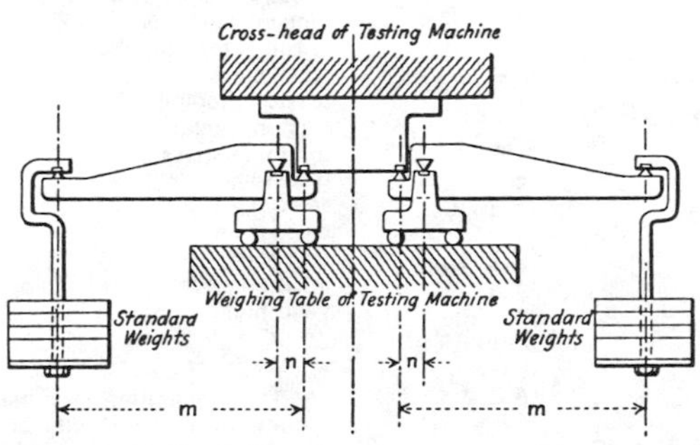

FIG. 1 Proving Levers.

American National Standard Z92.1-1967
American National Standards Institute

Standard Definitions of
TERMS RELATING TO METHODS OF MECHANICAL TESTING[1]

This Standard is issued under the fixed designation E 6; the number immediately following the designation indicates the year of original adoption or, in the case of revision, the year of last revision. A number in parentheses indicates the year of last reapproval.

NOTE—Editorial change was made to the note in Section 1 in May 1968.

1. Scope

1.1 These definitions cover the principal terms relating to methods of mechanical testing of solids (Note). The general definitions are restricted and interpreted, when necessary, to make them particularly applicable and practicable for use in standards requiring or relating to mechanical tests. These definitions are published to encourage uniformity of terminology in product specifications.

NOTE—For the purpose of these Definitions E 6, a solid is defined as a material that undergoes permanent deformation only when subjected to shearing stresses in excess of some finite value (yield stress) characteristic of the material.

2. Index of Terms

2.1 The definitions appear in the following sections:

A. GENERAL

Term	Section
mechanical properties	3
mechanical testing	4
strain	5
stress	6
stress-strain diagram	7

B. TENSION, COMPRESSION, SHEAR, AND TORSION TESTING

Term	Section
angular strain	5.4
axial strain	5.4
bearing load	8
biaxial stress	6.6, Note 2
breaking load	9
chord modulus	20, Note 1(*d*)
compression modulus	20.1
compressive strength	10
compressive stress	6.2.2
constraint	11
ductility	12
elastic constants	13
elastic limit	14
elongation	16
extensometer	15
fracture stress	6.7
gage length	17
initial tangent modulus	20, Note 1(*a*)
least count	18
linear strain	5.1
macrostrain	5.7
malleability	12, Note 2
mechanical hysteresis	19
mechanical properties	3
mechanical testing	4
microstrain	5.8
modulus of elasticity	20
modulus of rigidity	20.2
modulus of rupture in bending	21
modulus of rupture in torsion	22
multiaxial stress	6.6, Note 4
necking	23
nominal stress	6.1
normal stress	6.2
Poisson's ratio	24
principal stress (normal)	6.6
proportional limit	25
reduction of area	26
secant modulus	20, Note 1(*c*)
set	27
shear fracture	28
shear modulus	20.2
shear strain	5.5
shear strength	29
shear stress	6.3
slenderness ratio	30
strain	5
stress	6
stress-strain diagram	7
tangent modulus	20, Note 1(*b*)
tensile strength	31
tensile stress	6.2.1
tension modulus	20.1
torsion modulus	20.2
torsional stress	6.4
transverse strain	5.3
triaxial stress	6.6, Note 3
true strain	5.6
true stress	6.5
uniaxial stress	6.6, Note 1
yield point	32

[1] These definitions are under the jurisdiction of ASTM Committee E-1 on Methods of Testing. A list of members may be found in the ASTM Yearbook.

Current edition effective Sept. 20, 1966. Originally issued 1923. Replaces E 6 – 65a.

yield strength	33
Young's modulus	20.1

C. Hardness Testing

Term	Section
hardness	34
hardness tests	35
indentation hardness	34.1
Brinell hardness number	34.2
Brinell hardness test	35.1
Knoop hardness number	34.3
Knoop hardness test	35.2
Rockwell hardness number	34.4
Rockwell hardness test	35.3
Rockwell superficial hardness number	34.5
Rockwell superficial hardness test	35.4
Vickers hardness number	34.6
Vickers hardness test	35.5

D. Bend Testing

Term	Section
angle of bend	36
bend test	37
free bend	38
guided bend	39
mandrel	40
pin	40
radius of bend	41
semi-guided bend	42
wrap-around bend	43

E. Creep Testing

Term	Section
creep	44
creep recovery	45
creep rupture strength	46
creep strength	47
initial stress	48
instantaneous recovery	49
instantaneous strain	50
instantaneous stress	51
primary creep	44.1
rate of creep	52
secondary creep	44.2
stress relaxation	53
stress-rupture strength	54
tertiary creep	44.3

F. Bearing (Pin) Tests

Term	Section
bearing area	55
bearing load	56
bearing strain	57
bearing strength	58
bearing stress	59
bearing yield strength	60
edge distance	61
edge distance ratio	62

G. Fatigue Testing

Note—The following terms on fatigue testing are defined in ASTM Definitions E 206, Terms Relating to Fatigue Testing and the Statistical Analysis of Fatigue Data.[2]

constant life fatigue diagram
cycle ratio
fatigue
fatigue life
fatigue life for *p* percent survival
fatigue life, median
fatigue limit
fatigue limit for *p* percent survival
fatigue notch factor
fatigue notch sensitivity
fatigue strength at *N* cycles
fatigue strength for *p* percent survival at *N* cycles
maximum stress
mean stress
median fatigue life
median fatigue strength at *N* cycles
minimum stress
nominal stress (E 6, Section 6.1)
range of stress
S-N curve for 50 percent survival
S-N curve for *p* percent survival
S-N diagram
"static fatigue"
steady component of stress
stress amplitude
stress cycle
stress cycles endured
stress ratio
stress-rupture strength
theoretical stress concentration factor
variable component of stress

A. General Definitions

3. **mechanical properties**—those properties of a material that are associated with elastic and inelastic reaction when force is applied, or that involve the relationship between stress and strain.

Note—These properties have often been referred to as "physical properties," but the term "mechanical properties" is much to be preferred.

4. **mechanical testing**—the determination of mechanical properties.

5. **strain**—the unit change, due to force, in the size or shape of a body referred to its original size or shape. Strain is a nondimensional quantity, but it is frequently expressed in inches per inch, centimeters per centimeter, etc.

Note 1—In this standard, "original" refers to dimensions or shape of cross section of specimens at the beginning of testing.

Note 2—In the usual tension, compression, or torsion test, it is customary to measure only one component of strain and to refer to this as "the strain." In a tension or compression test, this is usually the axial component.

Note 3—Linear thermal expansion, sometimes called "thermal strain," and changes due to the effect of moisture are not to be considered strain in mechanical testing.

5.1 *linear* (*tensile or compressive*) *strain*—the change per unit length due to force in an original linear dimension.

Note—An increase in length is considered positive.

[2] *Annual Book of ASTM Standards*, Part 31.

5.2 *axial strain*—linear strain in a plane parallel to the longitudinal axis of the specimen.

5.3 *transverse strain*—linear strain in a plane perpendicular to the axis of the specimen.

NOTE—Transverse strain may differ with direction in anisotropic materials.

5.4 *angular strain*—use *shear strain.*

5.5 *shear strain*—the tangent of the angular change, due to force, between two lines originally perpendicular to each other through a point in a body.

5.6 *true strain*—in a body subjected to axial force, the natural logarithm of the ratio of the gage length at the moment of observation to the original gage length.

5.7 *macrostrain*—the mean strain over any finite gage length of measurement large in comparison with interatomic distances.

NOTE—Macrostrain can be measured by several methods, including electrical-resistance strain gages and mechanical or optical extensometers. Elastic macrostrain can be measured by X-ray diffraction.

5.8 *microstrain*—the strain over a gage length comparable to interatomic distances.

NOTE—These are the strains being averaged by the macrostrain measurement. Microstrain is not measurable by existing techniques. Variance of the microstrain distribution can, however, be measured by X-ray diffraction.

6. **stress** $[FL^{-2}]$—the intensity at a point in a body of the internal forces or components of force that act on a given plane through the point. Stress is expressed in force per unit of area (pounds-force per square inch, kilograms-force per square millimeter, etc.).

NOTE—As used in tension, compression, or shear tests prescribed in product specifications, stress is calculated on the basis of the original dimensions of the cross section of the specimen.

6.1 *nominal stress, S* $[FL^{-2}]$—the stress at a point calculated on the net cross section by simple elastic theory without taking into account the effect on the stress produced by geometric discontinuities such as holes, grooves, fillets, etc.

6.2 *normal stress* $[FL^{-2}]$—the stress component perpendicular to a plane on which the forces act. Normal stress may be either:

6.2.1 *tensile stress* $[FL^{-2}]$—normal stress due to forces directed away from the plane on which they act, or

6.2.2 *compressive stress* $[FL^{-2}]$—normal stress due to forces directed toward the plane on which they act.

NOTE—Tensile stresses are positive and compressive stresses are negative.

6.3 *shear stress* $[FL^{-2}]$—the stress component tangential to the plane on which the forces act.

6.4 *torsional stress* $[FL^{-2}]$—the shear stress on a transverse cross section resulting from a twisting action.

6.5 *true stress* $[FL^{-2}]$—the axial stress in a tension or compression test, calculated on the basis of the instantaneous cross-sectional area instead of the original area.

6.6 *principal stress* (*normal*) $[FL^{-2}]$—the maximum or minimum value of the normal stress at a point in a plane considered with respect to all possible orientations of the considered plane. On such principal planes the shear stress is zero.

NOTE—There are three principal stresses on three mutually perpendicular planes. The state of stress at a point may be:

(*1*) *Uniaxial* $[FL^{-2}]$—A state of stress in which two of the three principal stresses are zero,

(*2*) *Biaxial* $[FL^{-2}]$—A state of stress in which only one of the three principal stresses is zero, or

(*3*) *Triaxial* $[FL^{-2}]$—A state of stress in which none of the principal stresses is zero.

(*4*) *Multiaxial* $[FL^{-2}]$—Biaxial or triaxial.

6.7 *Fracture Stress* $[FL^{-2}]$—The true normal stress on the minimum cross-sectional area at the beginning of fracture.

NOTE—This term usually applies to tension tests of unnotched specimens.

7.1 **stress-strain diagram**—a diagram in which corresponding values of stress and strain are plotted against each other. Values of stress are usually plotted as ordinates (vertically) and values of strain as abscissas (horizontally).

B. Tension, Compression, Shear, and Torsion Testing

8. **bearing load** [F]—a compressive load on an interface.

9. **breaking load** [F]—the load at which fracture occurs.

NOTE—When used in connection with tension tests of thin materials or materials of small diameter for which it is often difficult to distinguish between the breaking load and the maximum load developed, the latter is considered to be the breaking load.

10. **compressive strength** [FL^{-2}]—the maximum compressive stress which a material is capable of sustaining. Compressive strength is calculated from the maximum load during a compression test and the original cross-sectional area of the specimen.

NOTE—In the case of a material which fails in compression by a shattering fracture, the compressive strength has a very definite value. In the case of materials which do not fail in compression by a shattering fracture, the value obtained for compressive strength is an arbitrary value depending upon the degree of distortion which is regarded as indicating complete failure of the material.

11. **constraint**—any restriction to the deformation of a body.

12. **ductility**—the ability of a material to deform plastically before fracturing.

NOTE 1—Ductility is usually evaluated by (*1*) comparing values of elongation or reduction of area from the tension test, (*2*) the depth of cup in a cupping test, or (*3*) the radius or angle of bend in a bend test.

NOTE 2—Malleability is the ability to deform plastically under repetitive compressive forces.

13. **elastic constants**—See **modulus of elasticity** and **Poisson's ratio.**

14. **elastic limit** [FL^{-2}]—the greatest stress which a material is capable of sustaining without any permanent strain remaining upon complete release of the stress.

NOTE—Due to practical considerations in determining the elastic limit, measurements of strain, using a small load rather than zero load, are usually taken as the initial and final reference.

15. **extensometer**—a device for measuring linear strain.

NOTE—Devices for measuring decreases in length are sometimes called "compressometers."

16. **elongation**—the increase in gage length (Note 1) of a tension test specimen, usually expressed as percentage of the original gage length.

NOTE 1—The increase in gage length may be determined either *at* or *after* fracture, as specified for the material under test.

NOTE 2—The term elongation, when applied to metals, generally means measurement after fracture; when applied to plastics and elastomers, measurement at fracture. Such interpretation is usually applicable to values of elongation reported in the literature when no further qualification is given.

NOTE 3—In reporting values of elongation the gage length shall be stated.

17. **gage length** [L]—the original length of that portion of the specimen over which strain or change of length is determined.

18. **least count**—the smallest change in indication that can customarily be determined and reported.

NOTE—In machines with close graduations it may be the value of a graduation interval; with open graduations or with magnifiers for reading, it may be an estimated fraction, rarely as fine as one tenth, of a graduated interval; and with verniers it is customarily the difference between the scale and vernier graduation measured in terms of scale units. If the indicating mechanism includes a stepped detent, the detent action may determine the least count.

19. **mechanical hysteresis**—the energy absorbed in a complete cycle of loading and unloading.

NOTE—A complete cycle of loading and unloading includes any stress cycle regardless of the mean stress or range of stress.

20. **modulus of elasticity** [FL^{-2}]—the ratio of stress to corresponding strain below the proportional limit.

20.1 *tension or compression*, *E* [FL^{-2}]—Young's modulus (modulus in tension or modulus in compression).

20.2 *shear or torsion*, *G* [FL^{-2}]—commonly designated as modulus of rigidity, shear modulus, or torsional modulus.

NOTE 1—The stress-strain relations of many materials do not conform to Hooke's law throughout the elastic range, but deviate therefrom even at stresses well below the elastic limit. For such materials the slope of either the tangent to the stress-strain curve at the origin or at a low stress, the secant drawn from the origin to any specified point on the stress-strain curve, or the chord connecting any two specified points on the stress-strain curve is usually taken to be the "modulus of elasticity." In these cases the modulus should be designated as the "tangent modulus," the "secant modulus," or the "chord modulus," and the point or points on the stress-strain curve described. Thus, for materials where the stress-strain relationship is curvilinear rather than linear, one of the four following terms may be used:

(*a*) *initial tangent modulus* [FL^{-2}]—the slope of the stress-strain curve at the origin.

(*b*) *tangent modulus* [FL^{-2}]—the slope of the stress-strain curve at any specified stress or strain.

(*c*) *secant modulus* [FL^{-2}]—the slope of the secant drawn from the origin to any specified point on the stress-strain curve.

(*d*) *chord modulus* [FL^{-2}]—the slope of the chord drawn between any two specified points on the stress-strain curve.

NOTE 2—Modulus of elasticity, like stress, is expressed in force per unit of area (pounds per square inch, etc.).

21. **modulus of rupture in bending** [FL^{-2}]—the value of maximum tensile or compres-

sive stress (whichever causes failure) in the extreme fiber of a beam loaded to failure in bending computed from the flexure equation:

$$S_b = Mc/I$$

where:

M = maximum bending moment, computed from the maximum load and the original moment arm,

c = initial distance from the neutral axis to the extreme fiber where failure occurs, and

I = initial moment of inertia of the cross section about the neutral axis.

NOTE 1—When the proportional limit in either tension or compression is exceeded, the modulus of rupture in bending is greater than the actual maximum tensile or compressive stress in the extreme fiber, exclusive of the effect of stress concentration near points of load application.

NOTE 2—If the criterion for failure is other than rupture or attaining the first maximum of load, it should be so stated.

22. **modulus of rupture in torsion** [FL^{-2}]—the value of maximum shear stress in the extreme fiber of a member of circular cross section loaded to failure in torsion computed from the equation:

$$S_s = Tr/J$$

where:

T = maximum twisting moment,

r = original outer radius, and

J = polar moment of inertia of the original cross section.

NOTE 1—When the proportional limit in shear is exceeded, the modulus of rupture in torsion is greater than the actual maximum shear stress in the extreme fiber, exclusive of the effect of stress concentration near points of application of torque.

NOTE 2—If the criterion for failure is other than fracture or attaining the first maximum of twisting moment, it should be so stated.

23. **necking**—the localized reduction of the cross-sectional area of a specimen which may occur during stretching.

24. **Poisson's ratio**—the absolute value of the ratio of transverse strain to the corresponding axial strain resulting from uniformly distributed axial stress below the proportional limit of the material.

NOTE 1—Above the proportional limit, the ratio of transverse strain to axial strain will depend on the average stress and on the stress range for which it is measured and, hence should not be regarded as Poisson's ratio. If this ratio is reported, nevertheless, as a value of "Poisson's ratio" for stresses beyond the proportional limit, the range of stress should be stated.

NOTE 2—Poisson's ratio will have more than one value if the material is not isotropic.

25. **proportional limit** [FL^{-2}]—the greatest stress which a material is capable of sustaining without any deviation from proportionality of stress to strain (Hooke's law).

NOTE—Many experiments have shown that values observed for the proportional limit vary greatly with the sensitivity and accuracy of the testing equipment, eccentricity of loading, the scale to which the stress-strain diagram is plotted, and other factors. When determination of proportional limit is required, the procedure and the sensitivity of the test equipment should be specified.

26. **reduction of area**—the difference between the original cross sectional area of a tension test specimen and the area of its smallest cross section (Note 1). The reduction of area is usually expressed as a percentage of the original cross-sectional area of the specimen.

NOTE 1—The smallest cross section may be measured *at* or *after* fracture as specified for the material under test.

NOTE 2—The term reduction of area when applied to metals generally means measurement after fracture; when applied to plastics and elastomers, measurement at fracture. Such interpretation is usually applicable to values for reduction of area reported in the literature when no further qualification is given.

27. **set**—strain remaining after complete release of the load producing the deformation.

NOTE 1—Due to practical considerations, such as distortion in the specimen and slack in the strain indicating system, measurements of strain at a small load rather than zero load are often taken.

NOTE 2—Set is often referred to as permanent set if it shows no further change with time. Time elapsing between removal of load and final reading of set should be stated.

28. **shear fracture**—a mode of fracture in crystalline materials resulting from translation along slip planes which are preferentially oriented in the direction of the shearing stress.

29. **shear strength** [FL^{-2}]—the maximum shear stress which a material is capable of sustaining. Shear strength is calculated from the maximum load during a shear or torsion test and is based on the original dimensions of the cross section of the specimen.

30. **slenderness ratio**—the effective unsupported length of a uniform column divided by the least radius of gyration of the cross-sectional area.

31. **tensile strength** [FL^{-2}]—the maximum tensile stress which a material is capable of sustaining. Tensile strength is calculated from the maximum load during a tension test carried to rupture and the original cross-sectional area of the specimen.

32. **yield point** [FL^{-2}]—the first stress in a material, less than the maximum attainable stress, at which an increase in strain occurs without an increase in stress.

NOTE—It should be noted that only materials that exhibit the unique phenomenon of yielding have a "yield point."

33. **yield strength** [FL^{-2}]—the stress at which a material exhibits a specified limiting deviation from the proportionality of stress to strain. The deviation is expressed in terms of strain.

NOTE—It is customary to determine yield strength by:

(*a*) *Offset Method* (usually a strain of 0.2 percent is specified),

(*b*) *Total-Extension-Under-Load Method* (usually a strain of 0.5 percent is specified although other values of strain may be used). An equation for calculating strain is:

$$x = [Y/E + \text{specified offset (strain)}] \times 100$$

where:

x = limiting strain, in percent,
Y = specified yield strength, in psi, and
E = nominal modulus of elasticity of the material, in psi.

(This strain, x, is multiplied by the gage length to determine the total extension under load).

Whenever yield strength is specified, the method of test must be stated along with the percent offset or the total strain under load. The values obtained by the two methods may differ.

C. HARDNESS TESTING

3.4 **hardness**—the resistance of a material to deformation, particularly permanent deformation, indentation, or scratching.

NOTE—Different methods of evaluating hardness give different ratings because they are measuring somewhat different quantities and characteristics of the material. There is no absolute scale for hardness, therefore to express hardness quantitatively each type of test has its own scale of arbitrarily defined hardness.

34.1 **indentation hardness**—the hardness as evaluated from measurements of area or depth of the indentation made by pressing a specified indenter into the surface of a material under specified static loading conditions.

34.2 **Brinell hardness number,** HB [FL^{-2}]—a number related to the applied load and to the surface area of the permanent impression made by a ball indenter computed from the equation:

$$HB = 2P/\pi D(D - \sqrt{D^2 - d^2})$$

where:

P = applied load, kgf,
D = diameter of ball, mm, and
d = mean diameter of the impression, mm.

NOTE—The Brinell hardness number followed by the symbol HB without any suffix numbers denotes the following test conditions:

Ball diameter	10 mm
Load	3000 kgf
Duration of loading	10 to 15 s

For other conditions, the hardness number and symbol HB is supplemented by numbers indicating the test conditions in the following order: diameter of ball, load, and duration of loading.

34.3 **Knoop hardness number,** HK [FL^{-2}]—A number related to the applied load and to the projected area of the permanent impression made by a rhombic-based pyramidal diamond indenter having included edge angles of 172 deg 30 min and 130 deg 0 min computed from the equation:

$$HK = P/0.07028d^2$$

where:

P = applied load, kgf, and
d = long diagonal of the impression, mm.

In reporting Knoop hardness numbers, the test load is stated.

34.4 **Rockwell hardness number,** HR—a number derived from the net increase in the depth of impression as the load on a penetrator is increased from a fixed minor load to a major load and then returned to the minor load.

NOTE 1—Penetrators for the Rockwell hardness test include a diamond sphero-conical penetrator having an included angle of 120 deg with a spherical tip having a radius of 0.20 mm and steel balls of several specified diameters.

NOTE 2—Rockwell hardness numbers are always quoted with a scale symbol representing the penetrator, load, and dial used.

34.5 **Rockwell superficial hardness number—See Rockwell hardness number.**

34.5 **Vickers hardness number,** HV [FL^{-2}]

—a number related to the applied load and the surface area of the permanent impression made by a square-based pyramidal diamond indenter having included face angles of 136 deg, computed from the equation:

$$HV = 2P \sin(\alpha/2)/d^2 = 1.8544P/d^2$$

where:

P = applied load, kgf,
d = mean diagonal of the impression, mm, and
a = face angle of diamond = 136 deg.

NOTE—The Vickers pyramid hardness number is followed by the symbol HV with a suffix number denoting the load and a second suffix number indicating the duration of loading when the latter differs from the normal loading time, which is 10 to 15 s.

35. **Hardness Tests**

35.1 **Brinell hardness test**—an indentation hardness test using calibrated machines to force a hard ball, under specified conditions into the surface of the material under test and to measure the diameter of the resulting impression after removal of the load.

35.2 **Knoop hardness test**—an indentation hardness test using calibrated machines to force a rhombic-based pyramidal diamond indenter having specified edge angles, under specified conditions, into the surface of the material under test and to measure the long diagonal after removal of the load.

35.3 **Rockwell hardness test**—an indentation hardness test using a calibrated machine to force a diamond spheroconical penetrator (diamond penetrator) or a hard steel ball under specified conditions into the surface of the material under test in two operations, and to measure the difference in depth of the impression under the specified conditions of minor and major loads.

35.4 **Rockwell superficial hardness test**—same as the Rockwell hardness test except that smaller minor and major loads are used.

35.5 **Vickers hardness test**—an indentation hardness test using calibrated machines to force a square-based pyramidal diamond indenter having specified face angles, under a predetermined load, into the surface of the material under test and to measure the diagonals of the resulting impression after removal of the load.

D. BEND TESTING

36. **angle of bend**—the angle between the two legs of the specimen after bending is completed.

NOTE—The angle of bend is measured before release of the bending force, unless otherwise specified.

37. **bend test**—a test for ductility (Note 1) performed by bending or folding, usually by steadily applied forces but in some instances by blows, a specimen having a cross section substantially uniform over a length several times as great as the largest dimension of the cross section.

NOTE 1—The ductility is usually judged by whether or not the specimen cracks under the specified conditions of the test.

NOTE 2—There are four general types of bend tests according to the manner in which the forces are applied to the specimen to make the bend. These are as follows:

1. Free Bend
2. Guided Bend
3. Semi-Guided Bend
4. Wrap-Around Bend

38. **free bend**—the bend obtained by applying forces to the ends of a specimen without the application of force at the point of maximum bending.

NOTE—In making a free bend lateral forces first are applied to produce a small amount of bending at two points. The two bends, each a suitable distance from the center, are both in the same direction.

39. **guided bend**—the bend obtained by use of a plunger to force the specimen into a die in order to produce the desired contour of the outside and inside surfaces of the specimen.

40. **pin or mandrel**—the plunger or tool used in making semi-guided, guided, or wrap-around bend tests to apply the bending force to the inside surface of the bend.

NOTE—In free bends or semi-guided bends to an angle of 180 deg a shim or block of the proper thickness may be placed between the legs of the specimen as bending is completed. This shim or block is also referred to as a pin or mandrel.

41. **radius of bend**—the radius of the cylindrical surface of the pin or mandrel that comes in contact with the inside surface of the bend during bending. In the case of free or semi-guided bends to 180 deg in which a

shim or block is used, the radius of bend is one half the thickness of the shim or block.

42. **semi-guided bend**—the bend obtained by applying a force directly to the specimen in the portion that is to be bent.

NOTE—The specimen is either held at one end or forced around a pin or rounded edge, or is supported near the ends and bent by a force applied on the side of the specimen opposite the supports and midway between them. In some instances, the bend is started in this manner and finished in the manner of the free bend.

43. **wrap-around bend**—the bend obtained when a specimen is wrapped in a closed helix around a cylindrical mandrel.

E. Creep Testing

44. **creep**—the time-dependent part of strain resulting from force.

NOTE—Instantaneous strains are excluded.

44.1 *primary creep*—creep occurring at a diminishing rate.

44.2 *secondary creep*—creep occurring at a constant rate.

44.3 *tertiary creep*—creep occurring at an accelerating rate.

45. **creep recovery**—the time-dependent portion of the decrease in strain following unloading of a specimen.

NOTE—Elastic strains and thermal expansion are excluded.

46. **creep rupture strength** $[FL^{-2}]$—the stress that will cause fracture in a creep test in a specified time and environment.

NOTE—This is sometimes referred to as the *stress-rupture strength*. In glass technology this is termed "static fatigue."

47. **creep strength** $[FL^{-2}]$—the stress that causes a given creep in a given time.

NOTE—The duration of time, the temperature, and other environmental conditions (if important) should be stated.

48. **initial stress**—the stress occurring immediately upon straining a specimen in a relaxation test.

NOTE—Since it is nearly impossible to obtain stress readings at the instant of straining, the stress at a specified small increment of time after straining is a more reproducible value.

49. **instantaneous recovery**—the decrease in strain occurring immediately upon unloading a specimen before any creep recovery takes place.

NOTE—Since it is nearly impossible to obtain strain readings at the instant of loading, the strain at a specified increment of time after loading is a more reproducible value.

50. **instantaneous strain**—the strain occurring immediately upon loading a creep specimen before any creep occurs.

NOTE—Since it is nearly impossible to obtain strain readings at the instant of unloading, the strain at a specified increment of time after unloading is a more reproducible value.

51. **instantaneous stress**—Use **initial stress.**

52. **rate of creep**—the slope of the creep-time curve at a given time.

53. **stress relaxation**—the time-dependent decrease in stress in a constrained specimen.

54. **stress-rupture strength** $[FL^{-2}]$—See **creep rupture strength.**

F. Bearing (Pin) Tests

55. **bearing area** $[L^2]$—the product of the pin diameter and specimen thickness.

56. **bearing load**—See Section 8.

57. **bearing strain**—the ratio of the bearing deformation of the bearing hole, in the direction of the applied force, to the pin diameter.

58. **bearing strength** $[FL^{-2}]$—the maximum bearing stress which a material is capable of sustaining.

59. **bearing stress** $[FL^{-2}]$—the force per unit of bearing area.

60. **bearing yield strength** $[FL^{-2}]$—the bearing stress at which a material exhibits a specified limiting deviation from the proportionality of bearing stress to bearing strain.

61. **edge distance** [L]—the distance from the edge of a bearing specimen to the center of the hole in the direction of applied force.

62. **edge distance ratio**—the ratio of the edge distance to the pin diameter.

G. Fatigue Testing

NOTE—The terms of fatigue testing listed in Section 2 under the Index of Terms are defined in ASTM Definitions E 206, Terms Relating to Fatigue Testing and the Statistical Analysis of Fatigue Data.[2]

APPENDIX

A1. Symbols and Abbreviations

A1.1 The following symbols and abbreviations are frequently used instead of or along with the terms covered by these definitions. For stress, the use of S with appropriate lower case subscripts is preferred for general purposes; for mathematical analysis the use of Greek symbols is generally preferred.

A	Area of cross section, or
A	Stress ratio
HB	Brinell hardness number
C	Cycle ratio
c	Distance from centroid to outermost fiber
D	Diameter
d	Diameter or diagonal
DPH	Diamond pyramid hardness (use VHN, Vickers hardness number)
Δ	Increment
δ	Deviation
E	Modulus of elasticity in tension or compression
	Strain
ϵF	Force
ft · lbf	A unit of work[3]
G	Modulus of elasticity in shear
γ	Shear-strain
I	Moment of inertia
in · lbf	A unit of work[3]
J	Polar moment of inertia
K_f	Fatigue notch factor
K_t	Theoretical stress concentration factor
HK	Knoop hardness number
ksi	Thousands of pounds-force per square inch of kips-force per square inch[3]
lbf · ft	A unit of torque[3]
lbf · in.	A unit of torque[3]
L	Length
M	Bending moment
μ(mu)	Poisson's ratio[4]
N	Fatigue life (number of cycles)
n	Number of stress cycles endured
P	Concentrated load
psi	Pounds-force per square inch[3]
q	Fatigue notch sensitivity
R	Stress ratio
r	Radius
HR	Rockwell hardness number (requires scale designation)
S	Nominal stress
S(or σ)	Normal stress[5]
S_a	Alternating stress amplitude
S_c or σ_c	Compressive stress
S_f	Fatigue limit
S_m	Mean stress
S_{max}	Maximum stress
S_{min}	Minimum stress
S_N	Fatigue strength at N cycles
S_r	Range of stress
S_s, or γ(tau)	Shear stress
S_u	Tensile strength
S_t, or σ_t	Tensile stress
S_{ty}	Tensile yield strength
S_{cy}	Compressive yield strength
σ or S	Normal stress, nominal stress
σ_c or S_c	Compressive stress
σ_t or S_t	Tensile stress
T	Temperature, torque, or twisting moment
t	Time
θ	Angle of twist per unit length
γ(tau) or S_s	Shear stress
HV	Vickers hardness number
W	Work or energy
w	Load per unit distance or per unit area
wL	Total distrubuted load for a given length
wA	Total distributed load for a given area
Y	Yield strength
Z	Section modulus[6]

[3] Because it is general engineering practice in the USA to use "pound" rather than the technically correct term "pound-force," the symbol "lb" rather than "lbf" is used in some standards. Similarly, in the abbreviations "ksi" and "psi" the "pound-force" is the unit involved.

[4] ν (nu) is preferred in applied mechanics.

[5] σ is preferred in applied mechanics but should not be used when statistical treatments are involved.

[6] Many handbooks use S for section modulus, but Z is preferred since S is so widely used for normal or nominal stress.

ASTM Designation: E 29 – 67

Recommended Practice for
INDICATING WHICH PLACES OF FIGURES ARE TO BE CONSIDERED SIGNIFICANT IN SPECIFIED LIMITING VALUES[1]

This Recommended Practice is issued under the fixed designation E 29; the number immediately following the designation indicates the year of original adoption or, in the case of revision, the year of last revision. A number in parentheses indicates the year of last reapproval.

1. Scope

1.1 This recommended practice is intended to assist the various technical committees in the use of uniform methods of indicating the number of places of figures which are to be considered significant in specified limiting values, for example, specified maximum values and specified minimum values. Its aim is to outline methods which should aid in clarifying the intended meaning of specified limiting values with which observed values or calculated values obtained from tests are compared in determining conformance with specifications. Reference to this practice is valid only when a choice of method has been indicated, that is, either *absolute method or rounding-off method*.

1.2 This recommended practice is intended to be used in determining conformance with specifications when the applicable ASTM specifications or standards make direct reference to this practice.

1.3 This recommended practice describes two commonly accepted methods of rounding data, identified as the Absolute Method and the Rounding-Off Method. In the application of this practice to a specific material or materials it is essential to specify which method is intended to apply. In the absence of such specification reference to this recommended practice, which expresses no preference as to which method should apply, would be meaningless. The choice of method is arbitrary, depending upon the current practice of the particular branch of industry or technology concerned, and should therefore be specified in the prime publication.

2. Expression of Numerical Requirements

2.1 The unqualified statement of a numerical limit, such as "2.50 in. max", cannot, in view of different established practices and customs, be regarded as carrying a definite operational meaning concerning the number of places of figures to be retained in an observed or a calculated value for purposes of determining conformance with specifications.

2.2 *Rounding-Off Method*—In some fields, specified limiting values of 2.5 in. max, 2.50 in. max, 2.500 in. max are taken to imply that, for the purposes of determining conformance with specifications, an observed value or a calculated value should be rounded off to the nearest 0.1 in., 0.01 in., 0.001 in., respectively, and then compared with the specified limiting value. This will be referred to as the *rounding-off method*.

2.3 *Absolute Method*—In other fields, specified limiting values of 2.5 in. max, 2.50 in. max, and 2.500 in. max are all taken to imply the same absolute limit of exactly two and a half inches and for purposes of determining conformance with specifications, an observed value or a calculated value is to be compared directly with the specified value. Thus, any deviation, however small, outside the specified limiting value signifies nonconformance with the specifications. This will be referred to as the *absolute method*.

---

[1] This recommended practice is under the jurisdiction of ASTM Committee E-11 on Statistical Methods. A list of members may be found in the ASTM Yearbook.

Current edition effective Nov. 1, 1967. Originally issued 1940. Replaces E 29 – 60 T.

3. Rounding-Off Method

3.1 *Where Applicable*—The rounding-off method applies where it is the intent that a limited number of places of figures in an observed value or a calculated value are to be considered significant for purposes of determining conformance with specifications.

3.2 *How Applied*—With the rounding-off method, an observed value or a calculated value should be rounded off by the procedure prescribed in 2.2 to the nearest unit in the designated place of figures stated in the standard, as, for example, "to the nearest 100 psi," "to the nearest 10 ohms," "to the nearest 0.1 percent," etc. The rounded-off value should then be compared with the specified value, and conformance or nonconformance with the specification based on this comparison.

3.3 *How Expressed*—This intent may be expressed in the standard in one of the following forms:

3.3.1 If the rounding-off method is to apply to all specified limits in the standard, and if all figures expressed in the limiting value are to be considered significant, this may be indicated by including the following statement in the standard:

The following applies to all specified limits in this standard: For purposes of determining conformance with these specifications, an observed value or a calculated value shall be rounded off "to the nearest unit" in the last right-hand place of figures used in expressing the limiting value, in accordance with the rounding-off method of ASTM Recommended Practice E 29, for Indicating Which Places of Figures Are to Be Considered Significant in Specified Limiting Values.

3.3.2 If the rounding-off method is to apply only to the specified limits for certain selected requirements, this may be indicated by including the following statement in the standard:

The following applies to specified limits for requirements on (tensile strength), (elongation), and (. . . . .) given in . . . . , (applicable section number and title) and (. . . . .) of this standard: For purposes of determining conformance with these specifications, an observed value or a calculated value shall be rounded off to the nearest (1000 psi) for (tensile strength), to the nearest (1 percent) for (elongation), and to the nearest (. . . . . .) for (. . . . . .), in accordance with the rounding-off method of ASTM Recommended Practice E 29, for Indicating Which Places of Figures Are to Be Considered Significant in Specified Limiting Values.

3.3.3 If the rounding-off method is to apply to all specified limits in a table, this may be indicated by a note in the manner shown in the following examples:

3.3.3.1 *Example 1*—Same significant places for all items:

	Chemical Composition, percent
Copper	4.5 ± 0.5
Iron	1.0 max
Silicon	2.5 ± 0.5
Other constituents (magnesium + zinc + manganese)	0.5 max
Aluminum	remainder

NOTE—For purposes of determining conformance with these specifications, an observed value or a calculated value shall be rounded off to the nearest 0.1 percent, in accordance with the rounding-off method of ASTM Recommended Practice E 29, for Indicating Which Places of Figures Are to Be Considered Significant in Specified Limiting Values.

3.3.3.2 *Example 2*—Significant places not the same for all items; similar requirements:

	Chemical Composition, percent	
	min	max
Nickel	57	. . .
Chromium	14	18
Manganese	. . .	3
Silicon	. . .	0.40
Carbon	. .	0.25
Sulfur	. . .	0.03
Iron	remainder	

NOTE—For purposes of determining conformance with these specifications, an observed value or a calculated value shall be rounded off "to the nearest unit" in the last right-hand place of figures used in expressing the limiting value, in accordance with the rounding-off method of ASTM Recommended Practice E 29, for Indicating Which Places of Figures Are to Be Considered Significant in Specified Limiting Values.

3.3.3.3 *Example 3*—Significant places not the same for all items; dissimilar requirements:

	Tensile Requirements
Tensile strength, psi	60 000 to 72 000
Yield point, min, psi	33 000
Elongation in 2 in., min, percent	22

NOTE—For purposes of determination of conformance with these specifications, an observed value or a calculated value shall be rounded off to the nearest 1000 psi for tensile strength and yield point and to the nearest 1 percent for elongation, in accordance with the rounding-off method of ASTM Recommended Practice E 29 for Indicating Which Places of Figures Are to Be Considered Significant in Specified Limiting Values.

3.4 *Rounding-Off Procedure*—The actual rounding-off procedure[2] shall be as follows:

3.4.1 When the figure next beyond the last place to be retained is less than 5, retain unchanged the figure in the last place retained.

3.4.2 When the figure next beyond the last place to be retained is greater than 5, increase by 1 the figure in the last place retained.

3.4.3 When the figure next beyond the last place to be retained is 5, and there are no figures beyond this 5, or only zeros, increase by 1 the figure in the last place retained if it is odd, leave the figure unchanged if it is even. Increase by 1 the figure in the last place retained, if there are figures beyond this 5.

3.4.4 This rounding-off procedure may be restated simply as follows: When rounding off a number to one having a specified number of significant places, choose that which is nearest. If two choices are possible, as when the digits dropped are exactly a 5 or a 5 followed only by zeros, choose the one ending in an even digit. Table 1 gives examples of applying this rounding-off procedure.

3.5 The rounded-off value should be obtained in one step by direct rounding off of the most precise value available and not in two or more steps of successive roundings. For example: 89,490 psi rounded off to the nearest 1000 psi is at once 89,000; it would be incorrect to round off first to the nearest 100, giving 89,500 and then to the nearest 1000, giving 90,000.

3.6 *Special Case, Rounding Off to the Nearest 50, 5, 0.5, 0.05 etc.*—If in special cases it is desired to specify rounding off to the nearest 50, 5, 0.5, 0.05, etc., this may be done by so indicating in the standard. In order to round off to the nearest 50, 5, 0.5, 0.05, etc., double the observed or calculated value, round off to the nearest 100, 10, 1.0, 0.10, etc., in accordance with the procedure in 3.4, and divide by 2. For example, in rounding off 6025 to the nearest 50, 6025 is doubled giving 12,050 which becomes 12,000 when rounded off to the nearest 100 (3.4.3). When 12,000 is divided by 2, the resulting number, 6000, is the rounded-off value of 6025. In rounding off 6075 to the nearest 50, 6075 is doubled giving 12,150 which becomes 12,200 when rounded off to the nearest 100 (3.4.3). When 12,200 is divided by 2, the resulting number, 6100, is the rounded-off value of 6075.

4. Absolute Method

4.1 *Where Applicable*—The absolute method applies where it is the intent that all digits in an observed value or a calculated value are to be considered significant for purposes of determining conformance with specifications. Under these conditions, the specified limits are referred to as absolute limits.

4.2 *How Applied*—With the absolute method, an observed value or a calculated value is not to be rounded off, but is to be compared directly with the specified limiting value. Conformance or nonconformance with the specification is based on this comparison.

4.3 *How Expressed*—This intent may be expressed in the standard in one of the following forms:

4.3.1 If the absolute method is to apply to all specified limits in the standard, this may be indicated by including the following sentence in the standard:

> For purposes of determining conformance with these specifications, all specified limits in this standard are absolute limits, as defined in ASTM Recommended Practice E 29, for Indicating Which Places of Figures Are to Be Considered Significant in Specified Limiting Values.

4.3.2 If the absolute method is to apply to all specified limits of some general type in the standard (such as dimensional tolerance limits), this may be indicated by including the following sentence in the standard:

> For purposes of determining conformance with these specifications, all specified (dimensional tolerance) limits are absolute limits, as defined in ASTM Recommended Practice E 29, for Indicating Which Places of Figures Are to Be Considered Significant in Specified Limiting Values.

[2] The rounding-off procedure given in this recommended practice is the same as the one given in the USA Standard Rules for Rounding Off Numerical Values (USAS Z25.1-1940) and in the *ASTM Manual on Quality Control of Materials, STP 15-C*, Part 2, Section 7 (1951).

4.3.3 If the absolute method is to apply to all specified limits given in a table, this may be indicated by including a footnote with the table as follows:

Width, in.	Width Tolerances[a] Plus and Minus, in.	
	0.032 in. and under in Thickness	Over 0.032 in. in Thickness
2 and under	0.005	0.010
Over 2 to 8 incl	0.008	0.013
Over 8 to 14 incl	0.010	0.015
Over 14 to 20 incl	0.013	0.018

[a] Tolerance limits specified are absolute limits as defined in ASTM Recommended Practice E 29, for Indicating Which Places of Figures Are to Be Considered Significant in Specified Limiting Values.

TABLE 1 Examples of Rounding off

Specified Limit	Observed Value or Calculated Value	To Be Rounded Off to Nearest	Rounded-Off Value to be Used for Purposes of Determining Conformance	Conforms with Specified Limit
Tensile strength, 60 000 psi, min	59 940	100 psi	59 900	no
	59 950	100 psi	60 000	yes
	59 960	100 psi	60 000	yes
Nickel, 57 percent, min	56.4	1 percent	56	no
	56.5	1 percent	56	no
	56.6	1 percent	57	yes
Water extract conductivity, 40 micromhos/cm, max	40.4	1 micromho/cm	40	yes
	40.5	1 micromho/cm	40	yes
	40.6	1 micromho/cm	41	no
Sodium bicarbonate 0.5 percent, max	0.54	0.1 percent	0.5	yes
	0.55	0.1 percent	0.6	no
	0.56	0.1 percent	0.6	no

Designation: E 41 – 63

American National Standard Z110.1-1964
American National Standards Institute

Standard Definitions of TERMS RELATING TO CONDITIONING[1]

This Standard is issued under the fixed designation E 41; the number immediately following the designation indicates the year of original adoption or, in the case of revision, the year of last revision. A number in parentheses indicates the year of last reapproval.

Introduction

These definitions pertain to the conditioning of materials for test purposes. Unless otherwise specified, they are intended to apply to all cases where combinations of atmospheric influences are an essential part of the testing of materials.

air conditioning—the simultaneous control of all, or at least the first three, of those factors affecting both the physical and chemical conditions of the atmosphere within any structure. These factors include temperature, humidity, motion, distribution, dust, bacteria, odor, and toxic gases.

air, dry—air containing no water vapor.

air, saturated—a mixture of dry air and water vapor in which the latter is at its maximum concentration for the prevailing temperature and pressure.

atmospheric pressure—the pressure due to the weight of the atmosphere. It is the pressure indicated by a barometer. Standard atmospheric pressure is a pressure of 76 cm Hg having a density of 13.5951 g/cm^3, under standard gravity of 980.665 cm/s^2.

atmosphere, standard—air maintained at a specified temperature, relative humidity, and standard atmospheric pressure. (See Specification E 171.)[2]

condition, standard—the condition reached by a specimen when it is in temperature and moisture equilibrium with a standard atmosphere.

conditioning—the exposure of a material to the influence of a prescribed atmosphere for a stipulated period of time or until a stipulated relation is reached between material and atmosphere.

dehumidify—to reduce, by any process, the quantity of water vapor within a given space.

dew point—the temperature to which water vapor must be reduced to obtain saturation vapor pressure, that is, 100 percent relative humidity.

NOTE—As air is cooled, the amount of water vapor that it can hold decreases. If air is cooled sufficiently, the actual water vapor pressure becomes equal to the saturation water-vapor pressure, and any further cooling beyond this point will normally result in the condensation of moisture.

humidify—to increase, by any process, the quantity of water vapor within a given space.

humidistat—a regulatory device, activated by changes in humidity, used for the automatic control of relative humidity.

humidity—the condition of the atmosphere in respect to water vapor. (See also **humidity, absolute; humidity, relative.**)

humidity, absolute—the weight of water vapor present in a unit volume of air, for example, grains per cubic foot, or grams per cubic meter.

NOTE—The amount of water vapor is also reported in terms of weight per unit weight of dry air, for example, grains per pound of dry air. This value differs from values calculated on a volume basis and should not be referred to as absolute humidity. It is designated as **humidity ratio, specific humidity,** or

[1] These definitions are under the jurisdiction of ASTM Committee E-1 on Methods of Testing. A list of members may be found in the ASTM Yearbook.

Current edition effective Sept. 30, 1963. Originally issued 1942. Replaces E 41 – 60 T.

[2] ASTM Specification E 171, for Standard Atmospheres for Conditioning and Testing Materials which appears in the *Annual Book of ASTM Standards*, Part 30.

moisture content, which also see.

humidity ratio—in a mixture of water vapor and air, the weight of water vapor per unit weight of dry air. (Also called Specific Humidity.)

humidity, relative—the ratio of the actual pressure of existing water vapor to the maximum possible (saturation) pressure of water vapor in the atmosphere at the same temperature, expressed as a percentage.

humidity, specific—see **humidity ratio.**

hygrometer—an instrument that indicates directly or indirectly the relative humidity of the air.

moisture content—the moisture present in a material, as determined by definite prescribed methods, expressed as a percentage of the weight of the sample on either of the following bases:

(*1*) Original weight (Note 1).

(*2*) Moisture-free weight (Note 2).

NOTE 1—This is variously referred to as moisture content or moisture "as is" or "as received."

NOTE 2—This is also referred to as Moisture Regain (frequently contracted to "regain") or moisture content on the "oven-dry," "moisture-free," or "dry" basis.

moisture equilibrium—the condition reached by a sample when the net difference between the amount of moisture absorbed and the amount desorbed, as shown by a change in weight, shows no trend and becomes insignificant.

moisture regain—the moisture in a material determined under prescribed conditions and expressed as a percentage of the weight of the moisture-free specimen.

NOTE—Moisture regain may result from either absorption or desorption, and differs from moisture content only in the basis used for calculation.

preconditioning—any preliminary exposure of a material to the influence of specified atmospheric conditions for the purpose of favorably approaching equilibrium with a prescribed atmosphere.

pressure, saturation—the pressure, for a pure substance at any given temperature, at which vapor and liquid, or vapor and solid, coexist in stable equilibrium.

pressure, vapor—the pressure exerted by a vapor.

NOTE—If a vapor is kept in confinement over its source so that the vapor can accumulate, the temperature being held constant, the vapor pressure approaches a fixed limit called the maximum, or saturated, vapor pressure, dependent only on the temperature and the liquid.

pressure, water vapor—the component of atmospheric pressure caused by the presence of water vapor, usually expressed in millimeters or inches of mercury.

psychrometer—a variety of hygrometer comprising a dry-bulb temperature indicator and a wet-bulb temperature indicator that is cooled to the wet bulb temperature by the spontaneous evaporation of moisture.

psychrometer, sling—a psychrometer containing independent, matched dry- and wet-bulb thermometers suitably mounted for swinging through the atmosphere, simultaneously indicating dry- and wet-bulb temperature.

room temperature—a temperature in the range of 20 to 30 C (68 to 85 F).

NOTE—The term "room temperature" is usually applied to an atmosphere of unspecified relative humidity.

saturation—the condition of coexistence in stable equilibrium of a vapor and a liquid or a vapor and solid phase of the same substance at the same temperature.

saturation, degree of—the ratio of the weight of water vapor associated with a pound of dry air to the weight of water vapor associated with a pound of dry air saturated at the same temperature.

standard laboratory atmosphere—an atmosphere, the temperature and relative humidity of which is specified, with tolerances on each. (See Specification E 171.)[2]

temperature—the thermal state of matter as measured on a definite scale.

temperature, dew point—see **dew point.**

temperature, dry-bulb—the temperature of the air as indicated by an accurate thermometer, corrected for radiation if significant.

temperature, wet-bulb—wet bulb temperature (without qualification) is the temperature indicated by a wet-bulb psychrometer constructed and used according to specifications. (See Method E 337.)[3]

[3] ASTM Method E 337, Determining Relative Humidity by Wet- and Dry-Bulb Psychrometer, which appears in the *Annual Book of ASTM Standards*, Part 30.

vapor—the gaseous form of substances that are normally in the solid or liquid state, and that can be changed to these states either by increasing the pressure or decreasing the temperature.

Designation: E 42 - 69

Recommended Practice for

OPERATING LIGHT- AND WATER-EXPOSURE APPARATUS (CARBON-ARC TYPE) FOR EXPOSURE OF NONMETALLIC MATERIALS[1]

This Recommended Practice is issued under the fixed designation E 42; the number immediately following the designation indicates the year of original adoption or, in the case of revision, the year of last revision. A number in parentheses indicates the year of last reapproval.

1. Scope

1.1 This recommended practice covers the basic principles and standard operating procedure for light- and water-exposure apparatus employing a carbon-arc type light source.

1.2 This recommended practice does not specify the exposure conditions best suited for the material to be tested, but it is limited to the method of obtaining, measuring, and controlling the conditions and procedures of the exposure. Sample preparation, test conditions, and evaluation of results are covered in ASTM methods or specifications for specific materials.

NOTE 1—The values stated in U.S. customary units are to be regarded as the standard. The metric equivalents of U.S. customary units may be approximate.

NOTE 2—See the following for recommendations on specific materials:

ASTM Recommended Practice D 822, for Operating Light- and Water-Exposure Apparatus (Carbon-Arc Type) for Testing Paint, Varnish, Lacquer, and Related Products.[2]

ASTM Recommended Practice D 529, for Accelerated Weathering Test of Bituminous Materials.[3]

ASTM Recommended Practice D 1499, for Operating Light- and Water-Exposure Apparatus (Carbon-Arc Type) for Exposure of Plastics.[4]

ASTM Recommended Practice D 750, for Operating Light- and Weather-Exposure Apparatus (Carbon-Arc Type) for Artificial Weather Testing of Rubber Compounds.[5]

2. Significance

2.1 Several types of apparatus with different exposure conditions are available for use. No single operating procedure for light- and water-exposure apparatus can be specified as a direct simulation of natural exposure. This recommended practice does not imply expressly or otherwise an accelerated weathering test.

2.2 Since the natural environment varies with respect to time, geography, and topography, it may be expected that the effects of natural exposure will vary accordingly. All materials are not affected equally by the same environment. Results obtained by the use of this recommended practice shall not be represented as equivalent to those of any natural weathering test until the degree of quantitative correlation has been established for the material in question.

2.3 Variations in results may be expected among instruments of different types or when operating conditions among similar type instruments vary within the accepted limits of this recommended practice. Therefore, no reference shall be made to results from use of this recommended practice unless accompanied by Section 5 or unless otherwise specified in a referenced procedure.

3. Apparatus

3.1 The apparatus employed shall utilize one or two carbon-arc lamps as the source of radiation, and shall be one of the following twelve general types, or their equivalent. In the following commercial descriptions of the

[1] This recommended practice is under the jurisdiction of ASTM Committee G-3 on Deterioration of Nonmetallic Materials. A list of members may be found in the ASTM Yearbook.

Current edition effective March 21, 1969. Originally issued 1942. Replaces E 42 - 65.

[2] *Annual Book of ASTM Standards*, Part 21.

[3] *Annual Book of ASTM Standards*, Part 11.

[4] *Annual Book of ASTM Standards*, Part 27.

[5] *Annual Book of ASTM Standards*, Part 28.

twelve types, the term "cycle" is defined as the time intervals of light and water spray that are specified differently according to the different testing materials (see 4.6 and 4.7).

3.1.1 *Type A*—Single enclosed carbon-arc lamp apparatus,[6] 30-in. (762-mm) diameter specimen drum, automatic control of temperature and cycle with 1 rpm of the specimen drum, and no automatic control of humidity.

3.1.2 *Type AH*—Same as Type A, except with automatic control of humidity.

3.1.3 *Type B*—Single enclosed carbon-arc lamp apparatus,[6] 30-in. diameter specimen drum, automatic control of temperature and cycle with 3 rph of the specimen drum, and no automatic control of humidity.

3.1.4 *Type C*—Single enclosed carbon-arc lamp apparatus,[6] 30-in. diameter specimen drum, no automatic control of temperature or humidity, and cycle controlled by 3 rph of the specimen drum.

3.1.5 *Type D*—Twin enclosed carbon-arc lamp apparatus,[6] 30-in. diameter specimen drum, automatic control of temperature and cycle with 1 rpm of the specimen drum, and no automatic control of humidity.

3.1.6 *Type DH*—Same as Type D, except with automatic control of humidity.

3.1.7 *Type E*—Single open-flame sunshine carbon-arc lamp apparatus,[6] $37^3/_4$-in. (959-mm) diameter specimen rack, automatic control of temperature and cycle with 1 rpm of the specimen rack, and no automatic control of humidity.

3.1.8 *Type EH*—Same as Type E, except with automatic control of humidity.

3.1.9 *Type F*—Single open-flame sunshine carbon-arc lamp apparatus,[6] $37^3/_4$-in. diameter specimen rack with vertical specimen mounting, automatic control of cycle with 0.5 rph of the specimen rack, and no automatic control of temperature or humidity.

3.1.10 *Type G*—Single open-flame sunshine carbon-arc lamp apparatus,[6] $37^3/_4$-in. effective diameter specimen rack with angular specimen mounting, automatic control of cycle with 0.5 rph of the specimen rack, and no automatic control of temperature or humidity.

3.1.11 *Type H*—Single enclosed carbon-arc lamp apparatus,[6] 20-in. (508-mm) diameter specimen rack, automatic control of temperature and cycle with 1 rpm of the specimen rack, and manual regulation of humidity.

3.1.12 *Type HH*—Same as Type H, except with automatic control of humidity.

3.2 The apparatus shall consist of a suitable frame within which is located a test chamber, and necessary compartments for housing control and regulating equipment.

3.3 Provision shall be made for mounting or supporting the test specimens in a circular rack or drum which is rotated around the arc or arcs, providing uniform distribution of the radiation on all specimens.

3.4 Adequate ventilation shall be provided in the test chamber to prevent contamination of the specimens from products of combustion of the arc.

3.5 The apparatus shall include equipment necessary for measuring and controlling the following:

3.5.1 Arc current,

3.5.2 Arc voltage,

3.5.3 Black-panel temperature (Note 3),

3.5.4 Water-spray temperature,

3.5.5 Operating schedule or cycle,

3.5.6 Exposure time, and

3.5.7 Relative humidity (Types AH, DH, EH, and HH only).

3.6 Types AH, DH, EH, and HH are additionally equipped with a thermostatically actuated, electrically operated vaporizing unit for adding moisture to the air as it passes through the conditioning chamber in the base section of the apparatus prior to its entry into the test chamber. Relative humidity of the air in the test chamber for the purpose of this recommended practice is calculated from the readings of wet- and dry-bulb thermometers, either indicating or recording, whose sensing portion is located in the air stream at its point of exit from the test chamber.

3.7 The black-panel thermometer unit shall consist of a 20-gage stainless-steel panel, $2^3/_4$ by $5^7/_8$ in. (70 by 149 mm), to which is mechanically fastened a stainless-steel bimetallic dial-type thermometer. This thermometer shall have a stem $^5/_{32}$ in. (3.9 mm) in diameter with a $1^3/_4$-in. (44.4-mm) dial. The sensitive portion extending $1^1/_2$ in. (38 mm) from the end of the stem shall be located in the center of the panel $2^1/_2$ in. (64 mm) from the top and $1^7/_8$ in. (48 mm) from the bottom of the panel.

[6] Available from the Atlas Electric Devices Co., 4114 N. Ravenswood Ave., Chicago 13, Ill.

The face of the panel with the thermometer stem attached shall be finished with two coats of a baked-on black enamel selected for its resistance to light and water. Control of the black-panel temperature shall be accomplished preferably by a continuous flow of air over the specimen at a controlled temperature, but an on-off flow of room-temperature air is permissible (Note 3).

NOTE 3—Types B, C, F, and G apparatus may require supplementary ventilation to meet this requirement.

3.8 Detail requirements and operating conditions of the twelve types of apparatus are given in Table 1 and Figs. 1 to 5.

4. Procedure

4.1 Prepare specimens of a suitable size and shape for mounting in the drum or rack of the apparatus in accordance with the detailed requirements specified for the material to be tested.

4.2 Mount the test specimens, except those whose shape or other physical characteristics make it impractical, vertically both above and below the horizontal center line of the source of radiation, except Type G, where the angle of mounting conforms to the equipment. In order to provide uniform exposure conditions over their surface, change the position of the test specimens daily. Rotate the specimen position in the drum or rack in four steps, employed in the same sequence, from upper to lower specimen row and inverted in both specimen rows. If the time of exposure is less than 4 days (the time required for the four position steps in the exposure cycle), all specimens in a given series should be positioned at the same level, either upper or lower (provided that sufficient space is available).

4.3 Where physical characteristics do not permit suspension of specimens in a vertical position, expose then horizontally on a rack $6^1/_2$ in. (165 mm) below the horizontal center of the source or sources of radiation. Mount the specimens on a circular horizontal rack equipped with turntables, so that each specimen is rotated on its own axis as all of the specimens are rotated around the source or sources of radiation.

4.4 Temperature measurement and control shall be based on the black-panel thermometer unit. Support the panel with the thermometer attached in the specimen drum or rack in the same manner as the test specimens so that it will be subjected to the same influences. If possible, read black-panel temperatures through the window in the test chamber without opening the door on Types A, AH, B, D, DH, E, EH, H, and HH apparatus, allowing sufficient time after starting up or after wetting, due to water spray, to reach equilibrium. On Types C, F, and G apparatus, read the temperature, through the curtains or by removing the top, at the point just prior to the entry of the black panel into the specimen spray. Unless other temperatures and tolerances are specified in the applicable ASTM method or detailed material specifications, the black-panel temperature shall be 63 ± 5 C (145 ± 9 F) during the light-on-without-water-spray period of the test cycle.

4.5 The water from the specimen spray shall strike the test specimens in the form of a fine spray equally distributed over the test specimens. Unless otherwise specified in the applicable ASTM method or detailed material specification, the water pressure, number, and type of nozzles shall be in accordance with the detailed requirements for the various types of apparatus as indicated in Figs. 1 to 5. The pH of the water shall be 6.0 to 8.0 and the water shall not leave an objectionable deposit or stain on the specimens after continued exposure in the apparatus. The temperature of the water shall be 16 ± 5 C (60.0 ± 9 F), and recirculation shall not be permitted unless the recirculated water meets the above requirements. For certain test materials or for water systems fabricated of certain materials (for example, aluminum), the use of deionized or distilled water may be advisable.

4.6 Unless otherwise mutually agreed upon or specified by the applicable ASTM or detailed material specification, operate Types A, D, and E, and Types AH, DH, EH, H, and HH with the humidifier off, with a cycle cam that provides 102 min light followed by 18 min light and spray. In these types of apparatus the specimen spray is operated intermittently; and the specimens, when the spray is on, pass through the spray once in each minute or revolution of the drum or rack. In Types B, C, F, and G apparatus allow the water spray to be on continuously, and the interval and duration of the period during which the specimen is in the spray is a function of the rotation of the specimen drum or rack.

NOTE 4—This operating schedule is intended to combine light- and water-spray exposure in a way that permits operation of the various types of carbon-arc-type apparatus that are in common use on approximately the same schedule, in order to promote better correlation between laboratories. For maximum uniformity, a schedule of approximately 20 h, 5 days per week, is recommended, a 2-day rest period being included as a desirable part of the schedule. This schedule is not necessarily to be considered as preferable to any other schedule that combines the effects of light alone, light and water, water alone, and rest periods in a sequence that produces more desirable results for specific types of materials.

4.7 When mutually agreed upon or called for by the applicable ASTM method or detailed material specification, operate Types AH, DH, EH, H, and HH apparatus with automatic control of humidity, unless otherwise specified or agreed upon, on a cycle of 102 min of light only followed by 18 min light with spray repeating for a total of 18 h. Follow the 18-h period by 6 h without light or spray. During the 18-h period of light and spray, the black-panel temperature, except when the specimen spray is on, shall be 63 ± 5 C (145 ± 9 F), and the relative humidity of the air shall be 50 ± 5 percent. During the 6-h period of darkness without spray, the black-panel temperature shall be 24 ± 2 C (75 ± 5 F), and the relative humidity of the air in the test chamber shall be 95 ± 4 percent. In the description of this cycle (see 5.1.1), include the number of 18-h light periods and 6-h dark periods.

4.8 Replace filters (globes and flat panes) after 2000 h of use, or when pronounced discoloration or milkiness develops, whichever occurs first. Clean filters each day by washing with detergent and water. It is recommended that filters be replaced on a rotating replacement schedule in order to provide more uniformity over long periods of exposure. Suggested schedules for Types D, DH, E, EH, F, and G apparatus are replacement of one half of the globes or filter panes each 1000 h of operation.

4.9 Unless otherwise specified in the applicable ASTM method or detailed material specifications, operate Types E, F, and G apparatus with the filters in place and with the carbon electrodes specified in Table 1 and Footnote *c*. If operated without filters or with other types of carbon electrodes, state this in the report of test results.

5. Report

5.1 The report shall include the following:

5.1.1 Type and model of exposure device,

5.1.2 Type of light source,

5.1.3 Age of filters,

5.1.4 Flux density at sample location (watts per square meter), (Note 5),

5.1.5 Spectral irradiance at sample location (watts per square meter, micrometer),

5.1.6 Elapsed exposure time,

5.1.7 Light/dark-water-humidity cycle employed,

5.1.8 Operating black-panel temperature,

5.1.9 Operating relative humidity,

5.1.10 Type of spray water,

5.1.11 Type of spray nozzle, and

5.1.12 Specimen relocation procedure.

NOTE 5—When direct measurement of flux density and spectral irradiance can not be made, data supplied by the manufacturer shall be substituted.

TABLE 1 Detail Requirements and Operating Conditions of Light- and Water-Exposure Apparatus

	Type[a]							
	A, AH	B	C	D, DH	E, EH	F	G	H, HH
Line voltage, V:								
208 to 250	X	X	X	X	X	X	X	X
Arc voltage, V:								
120 to 145	X	X	X	X				X
48 to 52					X	X	X	
Arc current, A:								
15 to 17, ac	X	X	X	X				X
58 to 62, ac					X	X	X	
12 to 14, dc	X	X	X	X				X
58 to 62, dc					X	X	X	
Carbon electrodes, upper, in. (mm):								
½ by 12 (12.7 by 304.8) neutral cored or solid[b]	X	X	X	X				X
⅞ (22.2) by 12 copper-coated sunshine[c]					X	X	X	
Carbon electrodes, lower, in. (mm):								
½ by 4 (101.6) neutral cored or solid[b]	X	X	X	X				X
½ by 12 copper-coated sunshine[c]					X	X	X	
Filter, globe, or flat pane for removing undesired wavelengths of radiation and preventing byproducts of combustion of the arc from contaminating the specimens:								
Globe of optical, heat-resistant glass with cut-off at 2750 A, with an increase in transmission to 91 percent at 3700 A[d]	X	X	X	X				X
Flat panes of optical, heat-resistant glass with cut-off at 2550 A, with an increase in transmission to 91 percent at 3600 A[e]					X	X	X	
Diameter of specimen rack or drum, in (mm):								
20 (508)								X
30 (762)	X	X	X	X				
37¾ (959)					X	X		
33¾ (857) at top and bottom, 39 (991) in cente							X	
Speed of rotation of specimen drum or rack:								
1 rpm	X			X	X			X
3 rph		X	X					
0.5 rph						X	X	
Automatic arc feed:								
Solenoid-operated	X	X	X	X				X
Motor-operated					X	X	X	
Spray (see figure indicated for arrangement, location, and capacity)	Fig. 1	Fig. 1	Fig. 1	Fig. 1	Fig. 2	Fig. 3	Fig. 4	Fig. 5

[a] "X" in column indicates application to that type of apparatus.
[b] No. 70 Solid Carbon Electrodes and No. 20 Cored Carbon Electrodes,[6] or equivalent.
[c] No. 22 Copper-Coated Sunshine Carbon Electrodes and No. 13 Copper-Coated Sunshine Carbon Electrodes,[6] or equivalent.
[d] No. 9200-PX Globe,[6] or equivalent.
[e] No. EX-555-22 Corex D Filter Panes,[6] or equivalent.

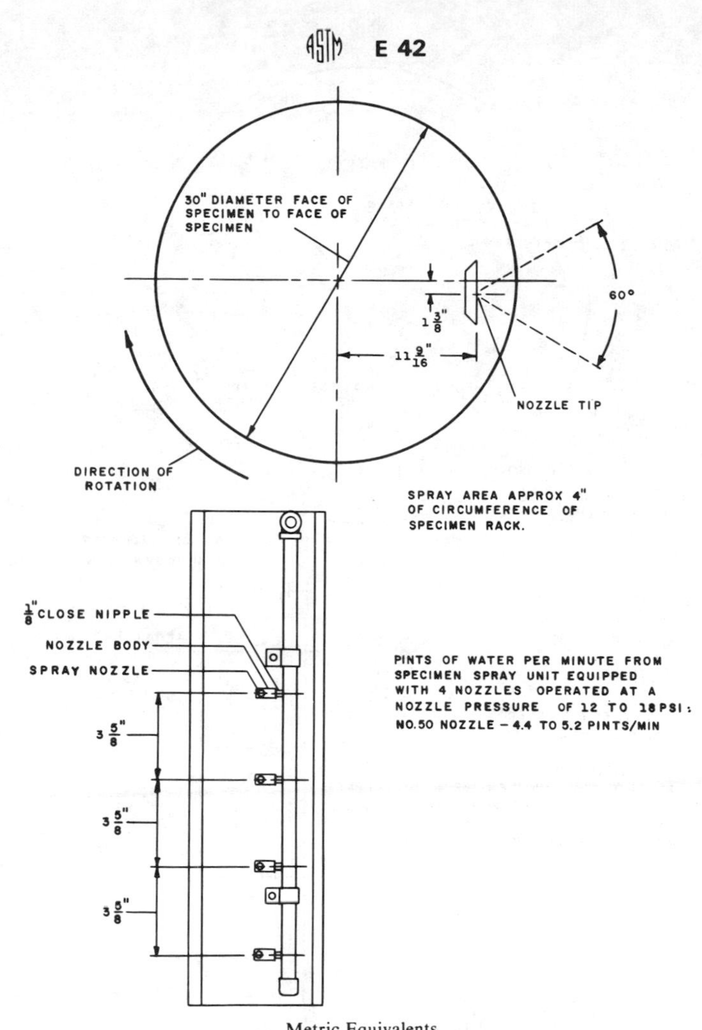

Metric Equivalents

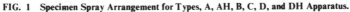

in.	1/8	1 3/8	3 5/8	4	11 9/16	30	12 to 18 psi	4.4 to 5.2 pt
mm	3.3	34.9	92.1	101.6	293.2	762	0.84 to 1.3 kgf/cm^2	2.5 to 2.9 dm^3

FIG. 1 Specimen Spray Arrangement for Types, A, AH, B, C, D, and DH Apparatus.

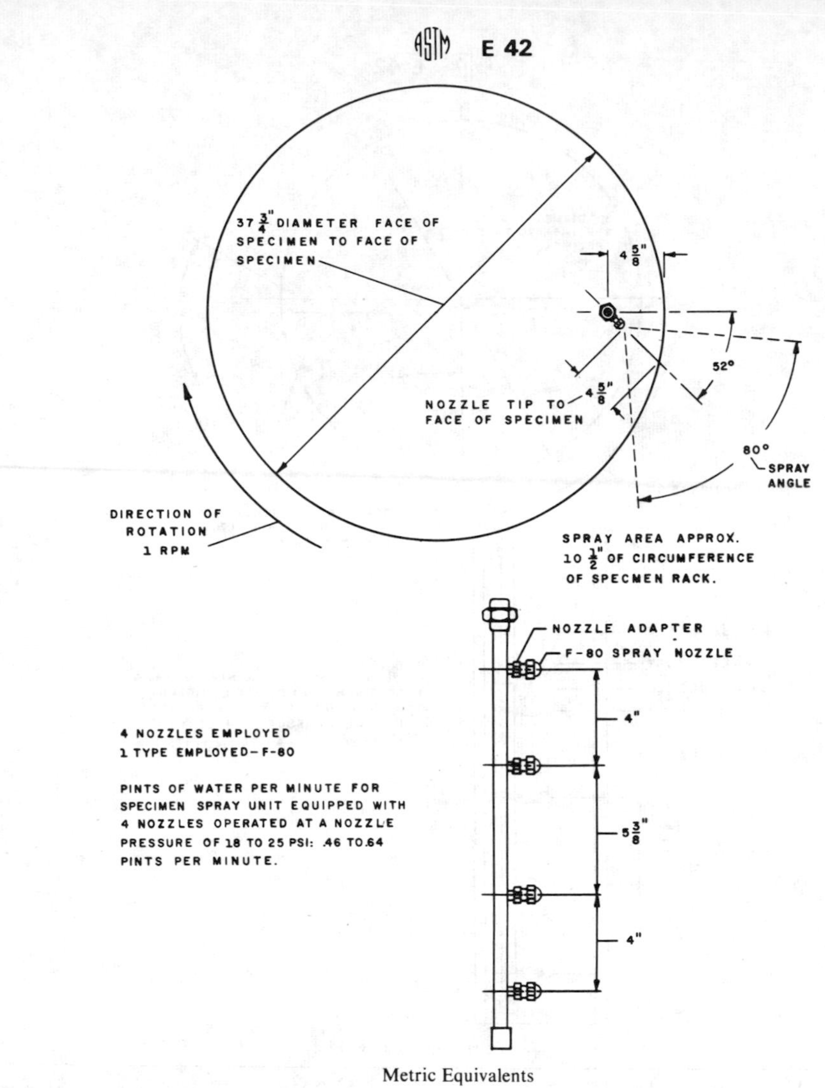

Metric Equivalents

in.	4	4 5/8	5 3/8	10 1/2	37 3/4	18 to 25 psi	0.46 to 0.64 pt
mm	101.6	117.5	136.5	266.7	959	1.3 to 1.8 kgf/cm^2	0.26 to 0.36 dm^3

FIG. 2 Specimen Spray Arrangement for Types E and EH Single Open-Flame Sunshine Carbon-Arc Lamp Apparatus.

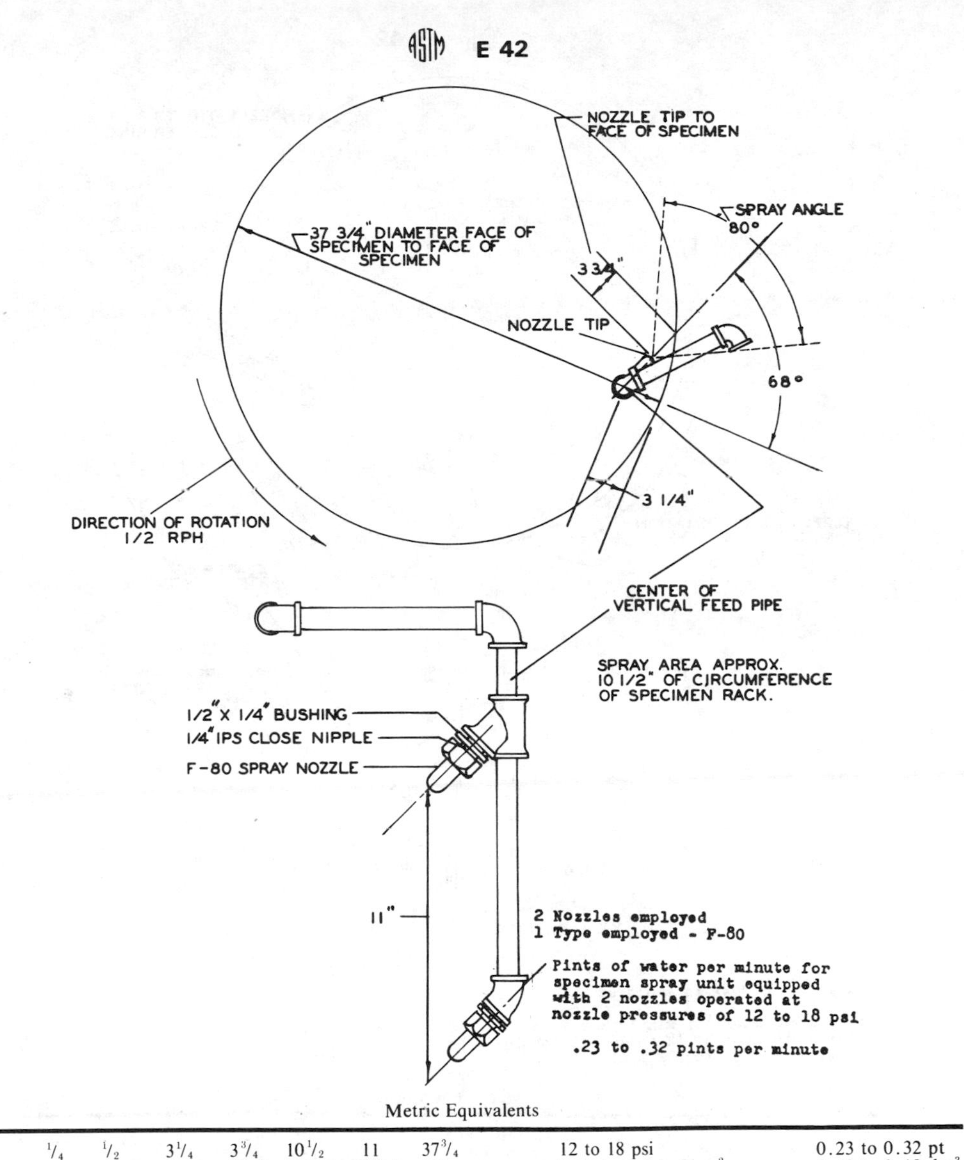

Metric Equivalents

in.	1/4	1/2	3 1/4	3 3/4	10 1/2	11	37 3/4	12 to 18 psi	0.23 to 0.32 pt
min.	6.4	12.7	82.6	92.3	266.7	279.4	959	0.84 to 1.3 kgf/cm^2	0.13 to 0.18 dm^3

FIG. 3 Specimen Spray Arrangement for Type F Single Open-Flame Sunshine Carbon-Arc Lamp Apparatus with Vertical Specimen Mounting.

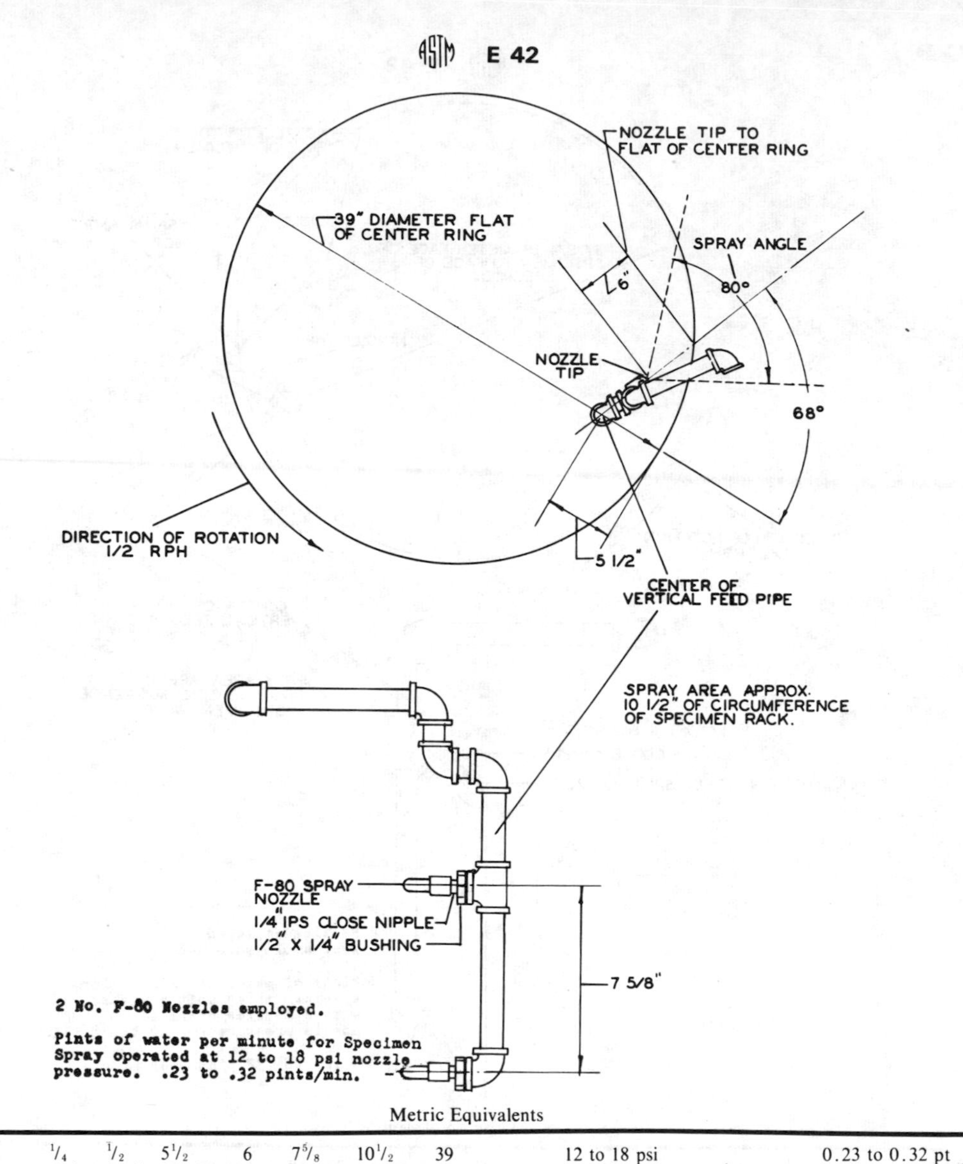

Metric Equivalents

in.	1/4	1/2	5 1/2	6	7 5/8	10 1/2	39	12 to 18 psi	0.23 to 0.32 pt
mm.	6.4	12.7	139.7	152.4	193.7	266.7	991	0.84 to 1.3 kgf/cm^2	0.13 to 0.18 dm^3

FIG. 4 Specimen Spray Arrangement for Type G Single Open-Flame Sunshine Carbon-Arc Lamp Apparatus with Angular Specimen Mounting.

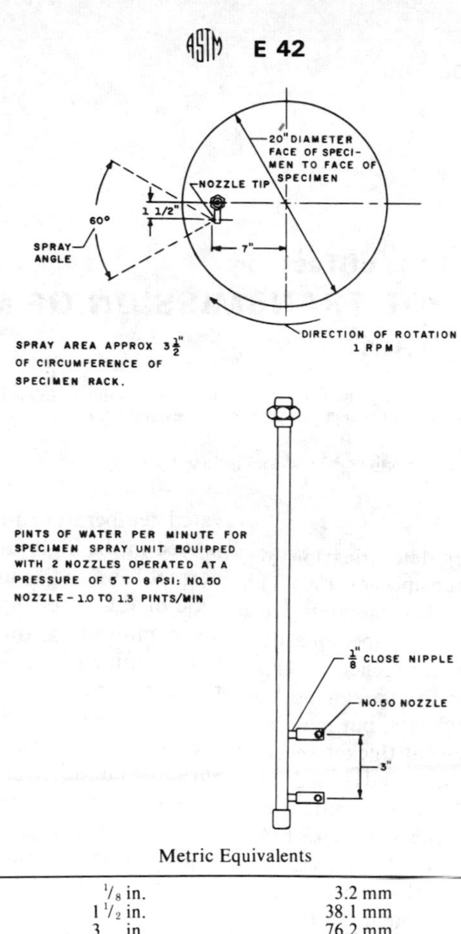

Metric Equivalents

$^1/_8$ in.	3.2 mm
1 $^1/_2$ in.	38.1 mm
3 in.	76.2 mm
3 $^1/_2$ in.	88.9 mm
20 in.	508 mm
5 to 8 psi	0.35 to 0.56 kgf/cm^2
1.0 to 1.3 pt	0.56 to 0.73 dm^3

FIG. 5 Specimen Spray Arrangement for Types H and HH Single Enclosed Carbon-Arc Lamp Apparatus.

Designation: E 96 – 66

Standard Methods of Test for
WATER VAPOR TRANSMISSION OF MATERIALS IN SHEET FORM[1]

This Standard is issued under the fixed designation E 96; the number immediately following the designation indicates the year of original adoption or, in the case of revision, the year of last revision. A number in parentheses indicates the year of last reapproval.

NOTE—Editorial changes were made in Sections 3.3 and 15.3 in June 1967.

1. Scope

1.1 These methods cover determination of the rate of water vapor transmission (WVT) of materials in sheet form. The methods are applicable to materials such as paper, plastic films, and sheet materials in general. The methods are most suitable for specimens $^1/_8$ in. (3.18 mm) or less in thickness, but may be used with caution for somewhat thicker specimens. For specimens of materials used in building construction and similar materials greater than $^1/_8$ in. in thickness, see ASTM Methods C 355, Test for Water Vapor Transmission of Thick Materials.[2] Six procedures are provided for the measurement of transmission under different test conditions, as follows:

1.1.1 *Procedure A*—For use when the materials to be tested are employed in the low range of humidities.

1.1.2 *Procedure B*—For use when the materials to be tested are employed in the high range of humidities, but will not normally be wetted.

1.1.3 *Procedure BW*—For use when materials to be tested may in service be wetted on one surface but under conditions where the hydraulic head is relatively unimportant and moisture transfer is governed by capillary and water vapor diffusion forces.

1.1.4 *Procedure C*—Conducted at an elevated temperature for use with materials employed in the low range of humidities, and intended to shorten the time of testing of highly impermeable materials. This procedure eliminates, in most cases, the need for refrigeration.

1.1.5 *Procedure D*—Conducted at an elevated temperature for use with materials employed in the high range of humidities, but not normally wetted, and intended to shorten the time of testing of highly impermeable materials employed in this range. This procedure also eliminates, in most cases, the need for refrigeration.

1.1.6 *Procedure E*—For use in measuring the WVT at an elevated temperature, with a very low humidity on one side of the sheet and a high humidity on the other side.

NOTE 1—The values stated in U.S. customary units are to be regarded as the standard. The metric equivalents of U.S. customary units may be approximate.

2. Summary of Methods

2.1 The material to be tested is fastened over the mouth of a dish, which contains either a desiccant or water. The assembly is placed in an atmosphere of constant temperature and humidity, and the weight gain or loss of the assembly is used to calculate the rate of water vapor movement through the sheet material under the conditions prescribed in Table 1.

3. Definitions

3.1 *rate of water vapor transmission, WVT*—of a body between two specified parallel surfaces—the time rate of water vapor flow normal to the surfaces, under steady conditions, through unit area, under the conditions of

[1] These methods are under the jurisdiction of ASTM Committee E-1 on Methods of Testing. A list of members may be found in the ASTM Yearbook.

Current edition effective Sept. 20, 1966. Originally issued 1953. Replaces E 96 – 63.

[2] *Annual Book of ASTM Standards*, Part 14.

test. An accepted unit of WVT is 1 g/24 h · m.^{2} The test conditions must be stated.

3.2 *water vapor permeance*—of a body between two specified parallel surfaces—the ratio of its WVT to the vapor pressure difference between the two surfaces. An accepted unit of permeance is a metric perm, or 1 g/24 h · m^2 · mm Hg. Since the permeance of a specimen is generally a function of relative humidity and to a lesser extent temperature, the test conditions must be stated.

3.3 *water vapor permeability*—the product of its permeance and thickness. This property may vary with conditions of exposure. An accepted unit of permeability is a metric perm-centimeter, or 1 g/24 h · m^2 · mm Hg · cm of thickness. Since the permeability of most materials is a function of relative humidity and to a lesser extent temperature, the test conditions must be stated.

NOTE 2—The definition does not imply that permeance is inversely proportional to thickness when the material is not fully homogeneous.

3.4 The following U.S. customary units are also widely used:

For WVT: 1 grain/h · ft^2.

For permeance: 1 perm, or 1 grain/h · ft^2 in. Hg.

For permeability: 1 perm-inch, or 1 grain/h · ft^2 in. Hg · in. of thickness.

Metric-U.S. customary conversion factors are stated in Table 2.

4. Significance

4.1 The purpose of these tests is to obtain, by means of simple apparatus, reliable values for the rate of water vapor transmission of sheet materials, expressed in suitable units. Correlation of test values with any given use must be determined by experience.

4.2 In any application, the vapor transmission of a sheet is determined by its permeance under the given conditions. With many materials the permeance of a sheet increases markedly when the relative humidity on either or both sides is raised. In contrast, the effect of a moderate change of temperature is quite small. Consequently, Procedures A and C, each with an average relative humidity of 25 percent, yield about the same permeance. Procedures B and D, with 75 percent average relative humidity, yield a different pair of similar values. The broad range of relative humidity provided by these two pairs of procedures (sometimes referred to as the desiccant and water methods, respectively) indicates in a practical way the permeance of a given sheet in many applications.

4.3 Procedure E provides the highest WVT, but the permeance generally falls between the values obtained by the desiccant and water methods. Although it has special applications, it is generally not as revealing as the water method for severe relative humidity exposures.

4.4 In Procedure BW, the water cup is inverted to wet the specimen, but otherwise it is the same as Procedure B. For some specimens, capillary transfer will be increased. It is intended specifically for those applications where wetting of one surface over extended periods is anticipated.

4.5 The report of a test result as a permeance value (rather than WVT) largely eliminates a fault in the test space conditions allowed in the tolerances stated in 5.2, provided the actual vapor pressure in the space is properly determined and used in the calculation as required in 15.2. Also, the permeance value of a sheet is a rational basis for evaluating its performance and comparing various sheets for a given application. Therefore, values of vapor permeance both in specifications and in reporting test results are to be encouraged.

4.6 The calculation of permeability (permeance × thickness) is also logical when the sheet is unquestionably homogeneous, with smooth surfaces and uniform thickness. The value so obtained can be used for calculating the permeance of a different thickness of like material in the same exposure. In Table 2 permeability units and metric-U.S. customary conversion factors are shown, but no report of the permeability is required.

5. Apparatus

5.1 *Test Dishes*—Open-mouth dishes shall be of such size or shape that they can be accommodated readily on the pan of an analytical balance. The dishes shall be constructed of a noncorroding, nonpermeable material, and shall be as light as is consistent with the necessary rigidity. The area of the opening shall be as large as practical and at least 30 cm^2. The area of the desiccant or water in the dish shall be at least equal to the exposed area of the specimen (see Procedures A, B, BW, C, D, and E, Sections 9 to 14). There shall be no obstructions within the dish that would restrict the flow of water vapor between the specimen area and the water or desiccant in the dish. For examples of suitable dishes and methods of sealing the specimen to the test dishes, see the Appendix.

5.2 *Test Chambers:*

5.2.1 The room or cabinet where the assembled test dishes are to be placed shall have controlled temperature and humidity which are measured frequently or preferably recorded continuously. The air conditions at any test location shall be one of those shown in Table 1 and shall be held constant within temperature limits of ±0.6 C (1 F) and relative humidity limits of ±2 percemt.

NOTE 3—These tolerances permit a range of vapor pressure up to 11 percent of nominal (in Procedures A and B) when temperature and relative humidity peaks occur together. This variation can be accepted only when control cycles are short and the mean vapor pressure is measured with due regard for such variation.

5.2.2 Air shall be continuously circulated throughout the chamber with a velocity sufficient to maintain uniform conditions at all test locations. Its velocity over the specimen in meters per minute shall be (numerically) not less than five times the permeance of the specimen, in metric perms (fpm = 10 times perms), but not over 150 m (500 ft)/min. Suitable racks shall be provided on which to place the test dishes within the test chamber.

NOTE 4—If the dishes are to be placed on the racks in an inverted position, care must be taken that the specified air velocity over the surfaces of the specimens is maintained.

5.3 *Balance and Weights*—The sensitivity of the balance shall be not less than 1 percent of the weight change during the period when a steady state is considered to exist. The weights used shall be accurate to 1 percent of the weight change during the steady-state period. A light wire sling may be substituted for the usual pan to accommodate a larger and heavier load.

5.4 *Weighing Covers*—Weighing covers for the test dishes shall be provided if the dishes have to be removed from the test room or cabinet for weighing (see the Appendix).

5.5 *Template*—A template may be used for defining the test area and effecting the wax seal (see the Appendix).

6. Materials

6.1 *Sealant*—The sealant used for attaching the specimen to the dish, in order to be suitable for this purpose, must be highly resistant to the passage of water vapor (and water). It must not lose weight to, or gain weight from, the atmosphere in an amount, over the required period of time, that would affect the test result by more than 2 percent. It must not affect the vapor pressure in a water-filled dish. Molten wax or asphalt is required for specimens having a permeance of less than 2.5 metric perms. The sealant shall be so applied that the test area of the specimen is well-defined and the same on both sides, and that no leakage of water vapor can occur at or through the edges of the specimen.

6.2 *Desiccant (for Procedures A, C, and E)*—A desiccant having a high affinity for water vapor and a high drying efficiency, that is, giving a low water vapor pressure after absorbing a large amount of water. The desiccant shall remain essentially unchanged in physical condition and exert no chemical or physical action, other than dehydration effects, on test specimens with which it is in contact. A suitable desiccant is anhydrous calcium chloride in the form of small lumps that will pass a No. 8 (2.38-mm) sieve,[3] and free of fines that will pass a No. 30 (595-μm) sieve.[4] It shall be dried at 200 C (392 F) before use and the moisture gain during the test shall be limited to 10 percent of its starting weight. If $CaCl_2$ will react chemically on the specimen, an absorbing desiccant such as silica-gel, activated at 200 C, may be used but the moisture gain by this desiccant during the test shall be limited to 4 percent.

7. Sampling

7.1 The material shall be sampled in ac-

[3] Detailed requirements for these sieves are given in ASTM Specification E 11, for Wire-Cloth Sieves for Testing Purposes which appears in the *Annual Book of ASTM Standards*, Part 30.

[4] Mobilwax Grade Nos. 2300, 2305, and 2310 of the Mobil Oil Corp. or their equivalent, have been found satisfactory for this purpose.

cordance with standard methods of sampling applicable to the material under test. The sample shall be selected so as to represent the material to be tested, and shall be of uniform thickness. If the material is of nonsymmetrical construction, the two faces shall be designated by distinguishing marks (for example, on a one-side-coated sample, "I" for the coated side and "II" for the uncoated side).

8. Test Specimens

8.1 Test specimens shall be representative of the sample. When the faces of a specimen are indistinguishable, at least three specimens shall be tested by the same method. When the material is designed for use in one position only, at least three specimens shall be tested by the same method with the vapor flow in the designated direction. When the material is in ready-to-use form and may be used with either face toward the vapor source, at least four specimens shall be tested by the same method, two being tested with the vapor flow in each direction and so reported. Great care shall be taken not to contaminate the test area of the specimen.

NOTE 5—When testing a low-permeance material that may be expected to lose or gain weight throughout the test (because of its evaporation or oxidation), it may be advisable to provide an additional specimen or "dummy" tested exactly like the other except that no desiccant or water is put in the dish.

8.2 The thickness of each specimen shall be measured at the center of each quadrant to the nearest 0.025 mm (0.001 in.) and the results averaged.

9. Procedure A. Desiccant Method at 23 C (73.4 F)

9.1 Fill the dish with desiccant to within 6 mm of the specimen. Leave enough space so that shaking of the dish, which must be done at each weighing with the dish in the upright position, will mix the desiccant. The depth of the desiccant shall be at least 12 mm. Seal the specimen to the opening of the dish in such a manner that leakage of water vapor at and through the edges is prevented. (See the Appendix for examples of suitable methods of sealing.)

9.2 Weight the assembly on the analytical balance. Then place the assembly on a rack in a test chamber, as described in 5.2, and locate so that the conditioned air circulates over the exposed surface of the specimen with the specified velocity. The specimen may be placed in either the upright position or inverted so that the desiccant is in direct contact with the specimen. This latter position is preferred for specimens having a high rate of water vapor transmission, but care must be taken to ensure that the seal is not broken or the surface of the specimen damaged.

9.3 Make successive weighings of the assembly at suitable intervals until a constant rate of gain is attained. Weighing should be accomplished without removal of the test dishes from the controlled atmosphere; if removal is necessary, the specimen must be tightly covered, and the weighing made immediately after removal with the assembly returned to the test chamber immediately after each weighing. Plot the results as prescribed in 15.1. Terminate the test or change the desiccant before the water added to the desiccant exceeds the limits specified in 6.2.

10. Procedure B. Water Method at 23 C (73.4 F)

10.1 In this method it is intended to provide 100 percent relative humidity in the dish but to avoid direct wetting of the surface facing the water. This requires that the test be carried out with the dish in the upright position and that a certain distance be maintained between water and specimen. As a result, there will exist a layer of air between the surface of the water and the under surface of the specimen through which water vapor will diffuse at a rate that cannot be measured conveniently. The influence of this air layer will become apparent only on specimens of a high rate of water vapor transmission; in this case the relative humidity on the under side of the specimen will be less than 100 percent; and the measured WVT will be slightly too low.

10.2 Place distilled water in the dish to bring the level of the water to such a height that the distance between the surface of the water and the under surface of the specimen is 20 ± 5 mm. A depth of 5 mm of water will adequately take care of losses by evaporation through the most permeable specimens. The depth shall be not less than 3 mm to ensure coverage of the dish bottom throughout the test. The possibility of splashing may be reduced by the use of a noncorroding grid in the dish to break the water surface. This grid shall be at least 6 mm below the specimen and it shall not reduce the water surface by more than 10 percent.

NOTE 6—For any of the water methods, baking the empty dish and promptly coating its mouth with sealant before assembly is recommended. The water may be added most conveniently after the specimen is attached, through a small sealable hole in the dish above the water line.

10.2 Attach the specimen to the dish, and proceed with the weighing as prescribed for Procedure A in 9.2 and 9.3, except place the assembly on the test racks in the upright position. Small variations in the temperature to which the dish and specimen might be exposed can result in condensation of water vapor on the underside of the specimen. Such condensation can significantly affect the measured water vapor transfer rate of some materials and is to be avoided. For this reason weighing should be carried out within the test chamber. If removal is necessary, the specimen must be tightly covered. The dishes shall not be exposed to a temperature that differs more than 3 C from the cabinet temperature to minimize condensation on the specimen. Plot results as prescribed in 15.1.

11. Procedure BW. Inverted Water Method at 23 C (73.4 F)

11.1 This is similar to Procedure B except that the dish is inverted, so that the water is in contact with the specimen surface at all times during the test.

11.2 Place distilled water in the dish to a depth of not less than 5 mm and not greater than 10 mm.

11.3 Attach the specimen to the dish and proceed with the weighing as prescribed for Procedure A in 9.2 and 9.3, placing the dish on the test racks in an inverted position. The dish must be sufficiently level so that water covers the inner surface of the specimen despite any distortion of the specimen due to the weight of the water. With highly permeant specimens it is especially important to locate the test dish so that air circulates over the exposed surface at the specified velocity. The test dishes may be placed on the balance in the upright position for weighing, but the period during which the wetted surface of the specimen is not covered with water must be kept to a minimum.

12. Procedure C. Desiccant Method at 32.2 C (90 F)

12.1 Follow Procedure A (Section 9), except place the assembly in a chamber maintained at 32.2 C (90 F) as prescribed in 5.2.2.

13. Procedure D. Water Method at 32.2 C (90 F)

13.1 Follow Procedure B (Section 10), except place the assembly in a chamber maintained at 32.2 C (90 F) as prescribed in 5.2.2.

14. Procedure E. Desiccant Method at 37.8 C (100 F)

14.1 Follow Procedure A except place the assembly in a chamber maintained at 37.8 C (100 F) and 90 percent relative humidity, as prescribed in 5.2.2.

NOTE 7—The use of weighing covers is recommended, particularly with this procedure, since weighings will have to be made in a room whose humidity and temperature will differ from the conditions in the cabinet.

15. Plotting and Calculation

15.1 Plot the results of successive weighing against elapsed time and draw a smooth curve through the plotted points. Judgment is required here and therefore numerous points are helpful. When a straight line adequately fits the plot of four properly-spaced, successive points (with due allowance for weighing errors), a nominally steady state exists, and the slope of the straight line is the rate of vapor transmission for the test area.

NOTE 8—The time required to reach the steady state will depend upon the permeance, thickness, hygroscopicity, and initial moisture content of the specimen. For specimens of very low permeance, buoyancy effects due to changes in barometric pressure may cause a greater change in measured weight between weighings than that due to moisture transfer. Long periods of test are required to obtain significant results in these instances.

15.2 Calculate the water vapor transmission (WVT) and permeance as follows:

$$\text{Water vapor transmission (WVT)} = (g \times 24)/(t \times a)$$

where:

g = weight gain or loss, g,
t = time, during which gain or loss, g, was observed, h,
a = exposed area of specimen, m^2, and
WVT = rate of water vapor transmission, $g/m^2 \cdot 24$ h.

$$\text{Permeance} = \text{WVT}/\Delta P = \text{WVT}/[S(R_1 - R_2)]$$

where:

ΔP = vapor pressure difference, mm Hg,
S = saturation vapor pressure, at test temperature, mm Hg,
R_1 = relative humidity at the source,
R_2 = relative humidity at the sink, and
Permeance = WVT/mm Hg expressed in metric perms.

In the controlled chamber the relative humidity and temperature are the average values actually measured during the test and (unless continuously recorded) these measurements shall be made as frequently as the weight measurements. In the dish the relative humidity is nominally 0 percent for the desiccant and 100 percent for the water. These values are usually within 3 percent relative humidity of the actual relative humidity for specimens below 2.5 metric perms when the required conditions are maintained (no more than 10 percent moisture in $CaCl_2$ and no more than 25-mm air space above water).

15.3 When the test specimen is homogeneous (not laminated) and thick relative to pore dimensions, the average water vapor permeability, in metric perm-centimeters, may be calculated as follows:

$$\text{Average permeability} = \text{permeance (metric perms)} \times \text{thickness (centimeters)}$$

For thin specimens the accuracy of the results will depend greatly on the precision with which the average thickness is determined.

15.4 Other units and conversion factors are given in Table 2.

16. Report

16.1 The report shall include the following:

16.1.1 A description of the sheet tested, including identification of side I and side II. If indistinguishable, so state.

16.1.2 Thickness of the specimen tested and exposed area.

16.1.3 The following actual conditions of the test:

16.1.3.1 Procedure, whether A, B, BW, C, D, or E,

16.1.3.2 Measured temperature,

16.1.3.3 Measured relative humidity in the test spaces, and

16.1.3.4 State side of sheet, I or II, exposed to the test space.

16.1.4 WVT of each specimen in stated units as tested.

16.1.5 Average WVT of all specimens tested in each direction of vapor flow.

16.1.6 Permeance of each specimen in stated units under the test conditions.

16.1.7 The average permeance of all specimens tested in each direction of vapor flow.

17. Precision

17.1 Results obtained by any one procedure on several specimens from the same sample may differ as much as 10 percent from their average. Two significant figures in water vapor transmission or permeance will, therefore, usually suffice to characterize the sample. Nevertheless, careful attention to all aspects of the procedure is required in order to obtain test results of acceptable precision.

TABLE 1 Test Conditions

Procedure[a]	Temperature	Relative Humidities on the Two Sides of Sheet: In Dish, percent	Relative Humidities on the Two Sides of Sheet: Outside Dish, percent
A	23.0 C (73.4 F)	0	50
B	23.0 C (73.4 F)	100	50
BW	23.0 C (73.4 F)	100[b]	50
C	32.2 C (90 F)	0	50
D	32.2 C (90 F)	100	50
E	37.8 C (100 F)	0	90

[a] Procedures A and C are desiccant methods; Procedures B, BW, and D are water methods.
[b] Sheet is wet.

TABLE 2 Other Units and Conversion Factors

Multiply	by	To Obtain (for the same test conditions)[a]
	WVT	
g/24 h·m²	0.0598	grains/h·ft²
Grains/h·ft²	16.7	g/24 h·m²
g/24 h·100 in²[b]	15.5	g/24 h·m²
	Permeance[c]	
Metric perms	1.52	perms
Perms	0.659	metric perms
	Permeability[d]	
Metric perm-centimeters	0.598	perm-inches
Perm-inches	1.67	metric perm-centimeters

[a] Data obtained under one test condition cannot be converted to another test condition by these factors.
[b] This unit is not recommended for general use since it contains metric and U.S. customary terms and has no corresponding units of permeance.
[c] To calculate permeance, divide the WVT by the test vapor pressure difference, in millimeters of mercury in the metric or in inches of mercury in the U.S. customary system.
[d] To calculate permeability, multiply the permeance by the specimen thickness.

APPENDIX

A1. Test Dishes and Sealing Methods for Water Vapor Transmission Measurements

A1.1 Several designs for dishes with supporting rings and flanges are shown in Fig. A1. Various modifications of these designs may be made provided that the principle of prevention of edge leakage by means of a complete seal is retained. The dishes may be constructed of any rigid, impermeable, corrosion-resistant material, provided that they can be accommodated on the available analytical balance. A lightweight metal, such as aluminum or one of its alloys, is generally used for larger size dishes. In some cases, when an aluminum dish is employed and moisture is allowed to condense on its surface there may be appreciable oxidation of the aluminum with a resulting gain in weight. Any gain in weight will ordinarily depend on the previous history of the dish and the cleanness of the surface. An empty dish carried through the test procedure as a control will help to determine whether any error may be expected from this cause. When aluminum dishes are used for the water methods, a pressure may develop inside the assembly during a test due to corrosion. This can cause seal failure or otherwise affect the result. Where this is a problem, it can be overcome by providing a protective coating inside the dish of baked-on epoxy resin or similar material. Dishes with flanges or rings that project from the inner walls of the dish are to be avoided, as such projections influence the diffusion of the water vapor. The depth of the dish for the water procedures is such that there is a 20 ± 5-mm distance between the water surface and the under surface of the specimen, with a water depth of about 5 mm.

A1.2 For the desiccant-in-dish procedures, the dishes need not be as deep as those required for the water-in-dish procedures. The desiccant is either in direct contact with the under surface of the specimen or within 6 mm of it, and a minimum depth of only 12 mm of desiccant is required.

A1.3 The dishes shown in Fig A1 require a molten seal.

A1.3.1 An ideal sealing material has the following properties:

A1.3.1.1 Impermeability to water in either vapor or liquid form,

A1.3.1.2 No gain or loss of weight from or to the test chamber (evaporation, oxidation, hygroscopicity, and water solubility being hazards),

A1.3.1.3 Good adhesion to any specimen and to the dish (even when wet),

A1.3.1.4 Complete conformity to a rough surface.

A1.3.1.5 Compatibility with the specimen and no excessive penetration into it,

A1.3.1.6 Strength, pliability, or both, and

A1.3.1.7 Easy handleability (including desirable

viscosity and thermal conductivity of molten sealant).

A1.3.2 Satisfactory sealants possess these properties in varying degrees and the choice is a compromise, with more tolerance in items at the beginning of this list for the sake of those at the latter part of the list when the requirements of 6.1 are met. Molten wax or asphalt is required for permeance tests below 2.5 metric perms. Tests to determine sealant behavior should include:

A1.3.2.1 An impervious specimen (metal) normally sealed to the dish and so tested, and

A1.3.2.2 The seal normally assembled to an empty dish with no specimen and so tested.

A1.3.3 The following materials are recommended for general use when the test specimen will not be affected by the temperature of the sealant:

A1.3.3.1 A mixture of 60 percent microcrystalline wax[4] and 40 percent refined crystaline paraffin wax.

A1.3.3.2 Asphalt, 77 to 88 C softening point, Apply by pouring.

A1.3.3.3 Beeswax and resin (equal weights). A temperature of 135 C is desirable for brush application. Pour at lower temperature.

A1.4 A template such as is shown in Fig. A2 is usually used for defining the test area and effecting the wax seal. It consists of a circular metal dish $^1/_8$ in. (3.18 mm) or more in thickness, with the edge beveled to an angle of about 45 deg. The diameter of the bottom (smaller) face of the template is approximately equal to, but not greater than, the diameter of the effective opening of the dish in contact with the specimen. Small guides may be attached to the template to center it automatically on the test specimen. A small hole through the template to admit air, and petrolatum applied to the beveled edge of the template, facilitate its removal after sealing the test specimen to the dish. In use, the template is placed over the test specimen and, when it is carefully centered with the dish opening, molten wax is flowed into the annular space surrounding the beveled edge of the template. As soon as the wax has solidified, the template is removed from the sheet with a twisting motion. The outside flange of the dish should be high enough to extend over the top of the specimen, thus allowing the wax to completely envelop the edge.

A1.5 Gasketed types of seals are also in use on appropriately designed dishes. These simplify the mounting of the specimen, but must be used with caution, since the possibility of edge leakage is greater with gasketed seals than with wax seals. Gasketed seals are not permitted for the measurement of permeance less than 2.5 metric perms.

A1.6 A suitable weighing cover, for use particularly in connection with Procedure E, consists of a circular disk of aluminum $^1/_{32}$ to $^3/_{32}$ in. in thickness provided with a suitable knob in the center for lifting. The cover fits over the test specimen when assembled, and makes contact with the inside beveled surface of the wax seal at, or just above, the plane of the specimen. The cover is free of sharp edges which might remove the wax, and is numbered or otherwise identified to facilitate its exclusive use with the same dish.

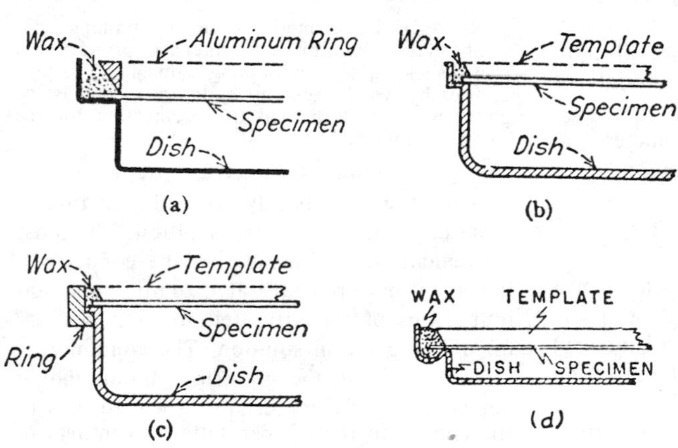

FIG. A1 Several Types of Test Dishes.

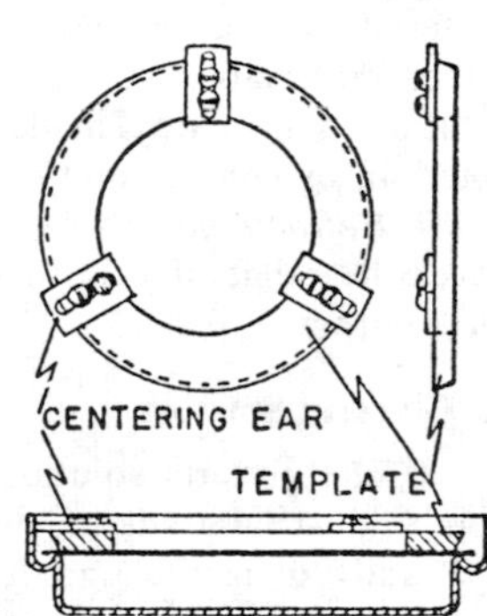

FIG. A2 Template Suitable for Use in Making the Wax Seals on Test Dishes.

Designation: E 104 – 51

Recommended Practice for

MAINTAINING CONSTANT RELATIVE HUMIDITY BY MEANS OF AQUEOUS SOLUTIONS[1]

This Recommended Practice is issued under the fixed designation E 104; the number immediately following the designation indicates the year of original adoption or, in the case of revision, the year of last revision. A number in parentheses indicates the year of last reapproval.

NOTE—Editorial revision of Section 3.3 was made in February 1963.

1. Scope

1.1 This recommended practice describes three methods for obtaining constant relative humidities ranging from 30 to 98 percent at temperatures ranging from 0 to 70 C in relatively small containers.

METHOD A—AQUEOUS GLYCERIN SOLUTIONS

2. Apparatus

2.1 *Container*—The container should be airtight and of glass or copper or any other material not acted on by copper sulfate. Other materials may be used if the glycerin solution is in a tray made of a material not acted on by copper sulfate.

2.2 *Refractometer*—A refractometer to cover the range of 1.33 to 1.47 (sodium) with an accuracy of 0.0003.

3. Glycerin Solution

3.1 The glycerin solution shall consist of a good industrial grade of glycerin ("high gravity" and "dynamite" grades are satisfactory) in distilled water. Calculate the concentration in terms of the refractive index at 25 C for the desired relative humidity at any temperature between 0 and 70 C as follows:

$$(R_1 + A)^2 = (100 + A)^2 + A^2 = (H + A)^2$$
$$R_1 = 715.3\,(R - 1.3333)$$

where:

A = $25.60 - 0.1950T + 0.0008\,T^2$,
H = relative humidity in percent,
R = refractive index of the glycerin solution, and
T = temperature of the solution in degrees Celsius.

This will give the desired relative humidity with an accuracy of ±0.2 percent at a constant temperature of 25 C. At other constant temperatures, the error, if any, may increase with the deviation of the temperature from 25 C. The relative humidity values at 0, 25, 50, and 70 C for a number of refractive index values are given in Table 1. Obtain refractive index for intermediate values of relative humidity and temperature by plotting curves from the values in the table or by calculating from the above formula.

NOTE 1—These methods for maintaining constant relative humidity are intended primarily to provide testing or conditioning ambients. The procedures by which these methods are used must be given in the test method or specification for the particular material.

3.2 Add about 0.1 percent by weight of copper sulfate to the glycerin solution to prevent fungus growth in the solution. The most convenient way of measuring the copper sulfate is to mix up a saturated solution and add four drops of the saturated solution per 100 ml of the glycerin solution. The container or the tray holding the glycerin solution should not be made of a material that will react with the copper in the copper sulfate. Removal of the copper may be followed by fungus growth, which will cause lowering of the humidity value of the glycerin solution.

3.3 Loss of water through evapoation when the container is open or by absorption by the material being conditioned will reduce the

[1] This recommended practice is under the jurisdiction of ASTM Committee E-1 on Methods of Testing. A list of members may be found in the ASTM Yearbook.

Current edition effective 1951. Originally issued 1949 as D 1041. Replaces E 104 – 49 T. Redesignated E 104 in 1954.

humidity value of the solution. The rate of loss with the container open is quite low and is negligible for the normal time the container would be open for loading and unloading (Note 2). A material being conditioned that will absorb a large amount of water may seriously reduce the humidity unless proper precautions are taken. For example, a loss of 0.26 ml of water per cubic inch of a 96 percent solution will reduce the humidity by 0.5 percent relative humidity. If it is estimated that the reduction in humidity will be greater than desired, one or both of two things may be done: the loading may be reduced below that suggested in 8.1.5 or the depth of the solution may be increased, or both.

NOTE 2—A 96 percent solution in an open container in a still atmosphere of 50 percent relative humidity at 25 C will lose water at the rate of approximately 0.01 ml/h $\cdot$in.^{2} of solution surface area. This rate will reduce the relative humidity value of a 96 percent solution having a depth of 1 in. by 0.5 percent relative humidity in 26 h.

METHOD B—AQUEOUS SULFURIC-ACID SOLUTIONS

4. Apparatus

4.1 *Container*—The container should be airtight and of a material not acted on by the sulfuric acid solution. Other materials may be used if the sulfuric acid solution is in a tray made of a material not acted on by the sulfuric acid solution.

4.2 *Hydrometers*—More than one hydrometer may be necessary to cover the range of densities for the range of humidities concerned. The hydrometers should have a minimum scale division of 0.001.

5. Sulfuric Acid Solution

5.1 Table 2 gives the density of sulfuric acid solutions for various relative humidities at various temperatures. Obtain the density for intermediate values of humidity and temperature by plotting curves from the values in the table. It should be noted that hydrometers measure the specific gravity, not the density. It may be necessary to apply a conversion factor to the hydrometer reading in order to obtain the density at 25 C. The magnitude of the conversion factor will depend upon the base temperature of hydrometer. Care should be taken that the solution is at 25 C when measuring the density.

5.2 Loss of water through evaporation when the container is open or by absorption by the material being conditioned will reduce the humidity value of the solution approximately the same as for Method A (see 3.3).

METHOD C—AQUEOUS SALT SOLUTION

6. Apparatus

6.1 *Container*—The container may be airtight or vented by a small hole and of a material not acted on by the salt solution to be used. Other materials may be used if the salt solution is in a glass tray in the bottom of the container.

7. Salt Solution

7.1 Saturated solutions of certain salts maintain definite relative humidities at constant temperature. These solutions should have a surplus of the salts to ensure saturation at all times. Some of the salts with the relative humidities maintained by their solution are listed in Table 3. The values were taken from the *International Critical Tables and Handbooks*. Humidity values for other salts may be obtained from the same sources.

7.2 The values of humidity over salt solutions do not have the reliability of those of Methods A or B since data are not available for comparison between different investigators, while both those of Methods A and B represent the average of the most reliable values of a number of investigators. In addition, the humidity values for most of the salts are available to only the nearest 1 percent.

8. Precautions

8.1 The following precautions should be observed in the use of any of the three Methods A, B, or C for obtaining constant relative humidity.

8.1.1 *Container*—Although an airtight container is called for by both Methods A and B, it may be desirable to have a vent under certain conditions of test or with some kinds of containers. (Changes in pressure may produce undesirable cracks in some types of containers.) The vent should be as small as practical and, as there will be a continual loss of vapor through the vent, check the concentration of the solution periodically and adjust if necessary in this case. The container should be small in order that the temperature throughout the container will differ little from that of the solution. The volume of the air space per unit area of surface of solution should be low. Ten cubic inches or less per square inch is advisable unless a larger volume is necessary because of the size of the device to be conditioned. Creepage distance over the surface between the solution and the material being conditioned should be long enough to prevent the solution creeping to the material being conditioned. Creeping is more likely to occur with some of the salts than with either glycerin or sulfuric-acid solutions.

8.1.2 *Temperature Fluctuations*—Avoid temperature fluctuations. Best results are obtained in a controlled temperature room where the average temperature is constant and the fluctuations are of relatively short duration. Cover the container to shield from drafts. Drafts may cause temperature difference inside the container. Changing temperature causes a temperature difference between that of the solution and the air above it. Changes in the solution temperature, as a rule, lag behind that of the air in the container. This results in a low humidity with rising temperature and a high humidity with falling temperature. If a controlled temperature room is not available, fair results may be obtained by placing the container in a location having the minimum change in temperature and thermally insulating the container with at least 1 in. of glass wool, or the equivalent. Reducing the volume of air space in the container per unit area of solution surface will also reduce the effect of changing temperature. A glass desiccator covered with a corrugated paper box will stand short time (30 min or less) fluctuations of temperature of ±1 C without changing the relative humidity over ±0.1 percent. Where larger fluctuations or long time fluctuations are encountered, heat insulates the container. It is estimated that a heat-insulated container will stand fluctuations of temperature of ±3 C without changing the relative humidity over ±0.1 percent.

8.1.3 *Temperatures Above Room Temperatures*—Operating at temperatures above or below room temperature is not as satisfactory as at room temperature, because of the greater possibility of the air in the container not being at the solution temperature and not being the same throughout the container. For example, with a solution for a relative humidity of 96 percent, a spot having a temperature 0.3 C higher than that of the solution would have a relative humidity of 94 percent, while that having a temperature 0.3 C lower would have a relative humidity of 98 percent. However, with proper care, humidities at temperatures above room temperature may be obtained by heating the container in an oven. Thermally insulate the container as described in 8.1.2 and adjust the oven air circulation so as to have as nearly uniform temperature throughout the container as possible. Load the container while at room temperature.

8.1.4 *Temperatures Below Room Temperature*—Follow the same precautions as for temperatures above room temperatures except reduce the temperature of the container (open) below the conditioning temperature before loading.

8.1.5 *Loading*—Avoid overloading as this may decrease the rate of rise of the humidity in the container to such an extent that an unreasonably long time may be required for the humidity to reach a steady state. The limit of loading cannot very well be specified as this depends upon the amount of moisture the material will absorb; this differs for different materials. A general rule is that the over-all area of the material should be less than the surface area of the solution.

TABLE 1 Relative Humidity Over Glycerin Solutions

Refractive Index at 25 C	Relative Humidity, percent			
	0 C	25 C	50 C	70 C
1.3463	97.7	98.0	98.2	98.4
1.3560	95.6	96.0	96.4	96.7
1.3602	94.5	95.0	95.5	95.8
1.3773	89.2	90.0	90.7	91.2
3.3905	84.0	85.0	85.9	86.6
1.4015	78.8	80.0	81.1	81.8
1.4109	73.7	75.0	76.2	77.0
1.4191	68.6	70.0	71.3	72.2
1.4264	63.4	65.0	66.4	67.3
1.4329	58.4	60.0	61.4	62.5
1.4387	53.3	55.0	56.5	57.6
1.4440	48.3	50.0	51.5	52.6
1.4486	43.3	45.0	46.6	47.7
1.4529	38.3	40.0	41.6	42.7

TABLE 2 Relative Humidity Over Sulfuric Acid Solutions

Density at 25 C, g/cm^3	Relative Humidity, percent			
	0 C	25 C	50 C	75 C
0.9971	100.0	100.0	100.0	100.0
1.0300	98.4	98.5	98.5	98.6
1.0641	95.9	96.1	96.3	96.5
1.0997	92.4	92.9	93.4	93.8
1.1368	87.8	88.5	89.3	90.0
1.1755	81.7	82.9	84.0	85.0
1.2164	73.8	75.6	77.2	78.6
1.2567	64.6	66.8	68.9	70.8
1.2993	54.2	56.8	59.3	61.6
1.3439	44.0	46.8	49.5	52.0
1.3912	33.6	36.8	39.9	42.8
1.4412	23.5	26.8	30.0	33.0
1.4944	14.6	17.2	20.0	22.8
1.5489	7.8	9.8	12.0	14.2
1.6059	3.9	5.2	6.7	8.3
1.6644	1.6	2.3	3.2	4.4
1.7224	0.5	0.8	1.2	1.8

TABLE 3 Relative Humidity Over Saturated-Salt Solutions

Salt	Temperature, deg C	Relative Humidity, percent
$MgCl_2 \cdot 6H_2O$	20	33
	30	32
	40	32
$CaCl_2 \cdot 6H_2O$	5	39.8
	10	38
	18.5	35
	20	32.3
	24.5	31
$K_2CO_3 \cdot 2H_2O$	18.5	44
	20	42
	24.5	43
$Na_2Cr_2O_7 \cdot 2H_2O$	20	52
$Ca(NO_3)_2 \cdot 4H_2O$	18.5	56
	24.5	51
$NaNO_2$	20	66
NaCl	20	76
	30	75
	40	75
NH_4Cl	20	79.2
	25	79.3
	30	79.5
$(NH_4)_2SO_4$	20	81
	25	81.1
	30	81.1
KCl	20	86
	30	84
	40	83
$ZnSO_4 \cdot 7H_2O$	5	94.7
	20	90
$Na_2C_4H_4O_6 \cdot 2H_2O$	20	91
	30	91
	40	91
$NH_4H_2PO_4$	20	93.1
	25	93
	30	92.9
K_2SO_4	20	96.5
	30	96.5
	40	96.5
$CaSO_4 \cdot 5H_2O$	20	98
$Pb(NO_3)_2$	20	98
	103.5	88.4

American National Standard Z110.2-1964
American National Standards Institute

Standard Specification for

STANDARD ATMOSPHERES FOR CONDITIONING AND TESTING MATERIALS[1]

This Standard is issued under the fixed designation E 171; the number immediately following the designation indicates the year of original adoption or, in the case of revision, the year of last revision. A number in parentheses indicates the year of last reapproval.

NOTE—Editorial changes in Table 1 were made in August, 1966.

1. Scope

1.1 This specification defines the standard atmospheres for normal conditioning and testing of materials, at approximately ambient conditions.

NOTE 1—See ASTM Definitions E 41, Terms Relating to Conditioning[2] and ASTM Method E 337, Determining Relative Humidity by Wet- and Dry-Bulb Psychrometer.[2]

2. Standard Atmospheres (see Note 2)

2.1 Unless otherwise specified, conditioning and testing of materials known to be sensitive to variations in temperature or relative humidity shall be carried out in an atmosphere having a temperature of 23 ± 2 C (73.4 ± 3.6 F) and a relative humidity of 50 ± 5 percent. If closer tolerances are required, ±1 C (1.8 F) and ±2 percent relative humidity may be specified. For details of conditioning requirements for specific materials, consult Table 1.

NOTE 2: *International Standard*—ISO/ATCO, the Coordination Committee on Atmospheric Conditioning for Testing, of the International Organization for Standardization (ISO), recommends that no standard atmospheres other than the following be used for testing materials that are used in conditions approximating ambient atmospheres:

Temperature[a]	Temperature Tolerance	Relative Humidity[a]	Relative Humidity Tolerance
20 C	normal ±2 C (if closer tolerance is required, ±1 C may be specified)	65	normal ±5 percent (if closer tolerance is required, ±2 percent may be specified)
27 C		65	
23 C		50	

[a] These requirements for temperature and relative humidity have been adopted by ISO/TC 61 on Plastics.

3. Standard Temperatures

3.1 Unless otherwise specified, the standard test temperature shall be as specified in Section 2. For details of standard test temperatures for specific materials, consult Table 2.

4. Room Temperature

4.1 A temperature in the range of 20 to 30 C (68 to 86 F) shall be called room temperature.

[1] This specification is under the jurisdiction of ASTM Committee E-1 on Methods of Testing. A list of members may be found in the ASTM Yearbook.

Current edition effective Sept. 30, 1963. Originally issued 1960. Replaces E 171 – 60 T.

[2] *Annual Book of ASTM Standards*, Part 30.

TABLE 1 Standard Atmospheres

Material	Temperature	Relative Humidity, percent	ASTM Method[a]
Adhesives	23 ± 2 C (73.4 ± 3.6 F)	50 ± 5	D 618
Cements, hydraulic:			
Room and dry materials	20 to 27.5 C (68 to 81.5 F)	50 min	C 109
Mixing water and moist storage	23 ± 1.7 C (73.4 ± 3 F)	90 min	C 109
Electrical insulating materials	23 ± 2 C (73.4 ± 3.6 F) (± 1 C (± 1.8 F) alternate)	50 ± 5 (±2 alternate)	D 618
Paint, varnish, lacquer, and related products	25 C (77 F)	50	[b]
Paper and paper products	23 ± 2 C (73.4 ± 3.6 F)	50 ± 2	D 685 D 641
Plastics	23 ± 2 C (73.4 ± 3.6 F) (± 1 C (± 1.8 F) alternate)	50 ± 5 (±2 alternate)	D 618
Sandwich constructions	23 ± 1 C (73.4 ± 2 F)	50 ± 2	...
Textiles	70 ± 2 F (21 ± 1.1 C)	65 ± 2	D 1776
Tire cords: cotton	75 ± 2 F (23.9 ± 1.1 C)	55 to 60	D 179
Tire cords: man-made fibers	75 ± 2 F (23.9 ± 1.1 C)	55 ± 2	D 885
Wood	20 ± 3 C (68 ± 6 F)	65 ± 1	D 1037

[a] These designations refer to the following ASTM methods:
C 109, Test for Compressive Strength of Hydraulic Cement Mortars (Using 2-in. Cube Specimens)[3]
D 179, Tolerances and Tests for Cotton Tire Cords and Cord Fabrics[4]
D 618, Methods of Conditioning Plastics and Electrical Insulating Materials for Testing[5]
D 641, Method of Conditioning Paperboard, Fiberboard, and Paperboard Containers for Testing[6]
D 685, Method of Conditioning Paper and Paper Products for Testing[6]
D 885, Testing and Tolerances for Tire Cords from Man-Made Fibers[4]
D 1037, Methods of Evaluating the Properties of Wood-Base Fiber and Particle Panel Materials[7]
D 1776, Method of Conditioning Textiles and Textile Products for Testing[5]
[b] Tolerances specified in individual ASTM standards.

[3] *Annual Book of ASTM Standards*, Part 9.
[4] *Annual Book of ASTM Standards*, Part 24.
[5] *Annual Book of ASTM Standards*, Part 29.
[6] *Annual Book of ASTM Standards*, Part 15.
[7] *Annual Book of ASTM Standards*, Part 16.

TABLE 2 Standard Temperatures

Material	Temperature	ASTM Method[a]
Plastics, electrical insulating materials, adhesives	23 ± 2 C (73.4 ± 3.6 F)	D 618
Rubber	23 ± 2 C (73.4 ± 3.6 F)	D 1349
Paint, varnish, lacquer, and related products	25 C (77 F)	[b]

[a] These designations refer to the following ASTM methods:
D 618, Methods of Conditioning Plastics and Electrical Insulating Materials for Testing[5]
D 1349, Rec. Practice for Standard Test Temperatures for Rubber and Rubber-Like Materials[8]
[b] Tolerances specified in ASTM standards.

[8] *Annual Book of ASTM Standards*, Part 28.

Designation: E 177 - 68 T

Tentative Recommended Practice for
USE OF THE TERMS PRECISION AND ACCURACY AS APPLIED TO MEASUREMENT OF A PROPERTY OF A MATERIAL[1]

This Tentative Recommended Practice has been approved by the sponsoring committee and accepted by the Society in accordance with established procedures, for use pending adoption. Suggestions for revisions should be addressed to the Society at 1916 Race St., Philadelphia, Pa. 19103.

1. Scope

1.1 The purpose of this recommended practice is to outline some general concepts regarding the terms "precision" and "accuracy," to provide some standard usages for ASTM committees in reference to precision and accuracy, and to illustrate some important features of the experimental determination of precision.

General Concepts

2. Test Method

2.1 A test method includes requirements for test apparatus and a well-defined procedure for using it to measure a physical property of a material.

3. Measurement Process

3.1 *General Characteristics*—A measurement process is generated by the repeated application of a test method. Three general characteristics of a measurement process may be mentioned:

3.1.1 *Realization of a Test Method*—If a test method specifies the use of a certain kind of test apparatus, a measurement process following the test method will involve a particular version or versions of such test apparatus. It will also involve a specific operator or operators who are needed to carry out preparation of specimens, measurements, etc., as called for by the test method.

3.1.2 *Realization of a System of Causes*—A system of causes is a collection of factors that may cause variability of measurement such as test apparatus, operator, test specimen, and day of test as well as others more difficult to identify. Some causes may be explicitly involved in the test method, but the system of causes realized in a measurement process is the collection of those present which do, in fact, affect the measurements produced. This realization of a system of causes defines the statistical universe of individual measurements from the measurement process and produces a particular statistical behavior in that universe.

Note 1—Variability of measurements attributable to a single cause such as "operators" is usually considered to be the effect of *differences* among individuals representing either a limited class of specific individuals (such as "the particular operators now employed by Company X") or a broad class of indefinite size and constitution (such as "any and all qualified operators"). Change in reference from one class to the other implies a change in reference from one system of causes to another and therefore may imply a change in degree of variability.

3.1.3 *Capability of Statistical Control*—In a measurement process it is necessary to require the capability of statistical control in the sense of Part 3 of the *ASTM Manual on Quality Control of Materials*.[2] Capability of control means that either the measurements are the product of an identifiable statistical universe or an orderly array of such universes,

[1] This recommended practice is under the jurisdiction of ASTM Committee E-11 on Statistical Methods. A list of members may be found in the ASTM Yearbook.

Current edition effective Feb. 14, 1968. Originally issued 1961. Replaces E 177 - 61 T.

[2] *ASTM Manual on Quality Control of Materials, ASTM STP 15 C*, is available from ASTM Headquarters.

or if not, the physical causes preventing such identification may themselves be identified and, if desired, isolated and suppressed.

3.2 *Material, Level, and Test Specimens*—The level (or levels) of a material is an essential part of the system of causes which deserves special mention. In experimental situations, levels of material may be deliberately selected, if that is possible, or portions of material may be chosen more or less at random with respect to level (Note 2). Except as noted, the term "level" in this practice will refer to the *accepted reference level* of a property of a material. The term "accepted reference level" is used here to cover two distinct but closely related ideas: it may refer to an authorized and accepted value for the measurement of the property concerned in a special portion of material set aside for future use in calibration, registration, etc. (that is, a standard); or it may refer to a value of the measurement of a test specimen that might be taken from any given standard portion of material by a controlled measurement process based on an authorized and accepted test method (that is, the actual or hypothetical measurement yielded by a suitable standard method). In some instances, as in bulk materials, the accepted reference level for a quantity of material may be derived from an accepted calculation using measurements of different portions of material, in accordance with a standard procedure.

NOTE 2—In statistically designed experiments the word "level" is often used to describe a given choice with respect to some other cause of the system (usually then referred to as a "factor") when the measurements are to represent specified or randomly chosen combinations of various "factor levels."

4. Statistical Characteristics of a Measurement Process

4.1 *Concepts of Statistical Characteristics*—In order to define the statistical characteristics of a measurement process, the latter must be of the nature described in Section 3. That is, the statistical characteristics of a process can be assessed and used only when it can be verified that this particular process is generated by the specified method with the assigned system of causes and is capable of a state of statistical control. Two statistical characteristics of a measurement process are of interest here: *precision* and *accuracy*. These concepts have been derived from the ordinary meanings of the adjectives "precise" and "accurate," which are sometimes regarded as absolute and nearly synonymous in indicating conformity to fact, standard or truth. In this sense a measurement might, for instance, be described as "precise" or "imprecise" but not as "more precise" or "less precise" than another measurement. Other considerations arise in applying the terms "precise" and "accurate" to a set of measurements regarded as a sample from a statistical universe, a sample consisting usually of a number of different measured values, or to the statistical universe itself. A measurement process can be described as "precise" only insofar as it meets applicable statistical tests or standards of "precision" as judged from such sample measurements. On the other hand, two different measurement processes can be compared by statistical analysis of sample measurements from both. Such comparisons naturally lead to descriptions of one measurement process or method as "more precise" than the other. The words "precise" and "accurate," as well as "precision" and "accuracy," have thus acquired a relative or comparative sense when applied to the statistical properties of a measurement process. *Precision* of a measurement process refers to the *degree of mutual agreement between individual measurements* from the process, while *accuracy* refers to the *degree of agreement of such measurements with an accepted reference level* of the property in the material measured. Accuracy is therefore used in this practice as an inducive term.

NOTE 3—Accuracy has sometimes been used to refer to the degree of agreement of the average of an indefinitely large number of individual measurements with an accepted reference level. This differs from the above definition in that the difference between such an average and the level depends less and less upon precision as the number of measurements is increased and reflects more and more only whatever systematic error may be common to all measurements (4.2.2).

4.2 *Statements of Statistical Characteristics*—The discussion in 4.1 is intended only to provide a conceptual basis for use of the terms precision and accuracy. It is another matter to formulate specific statements regarding these statistical characteristics of measurement processes.

4.2.1 *Precision*—Precision may be stated in terms of an *index* of precision of the form $\pm a$, where a is some positive number. The numerical value of a in any such index of precision will be smaller the more closely bunched are the individual measurements of a process. However, any such index must have a clearly understandable interpretation regarding variability of measurements. The value (that is, $\pm a$) of an index of precision may be different in different processes because of different methods or systems of causes. These differences in value may arise even when there is statistical control in the different processes since the actual system of causes may differ from process to process in ways not readily recognized. If two indexes of precision refer to two different processes based on the same method or intended to measure the same physical property and both have the same kind of interpretation as to variability, the process having the smaller value of a is said to be more precise or to have higher precision. Thus, these indexes of precision are really direct measures of imprecision. This inversion of reference has been firmly established by custom. The value of the selected index of precision of a process may be referred to simply as its precision.

4.2.2 *Accuracy*—There is no single form of statement of accuracy which can be universally recommended. However, since the concept of accuracy described in 4.1 embraces not only the concept of precision but also the idea of a consistent deviation from the reference level (*systematic error*), it sometimes *may* be appropriate to characterize accuracy by an *index of precision* and a *correction* to be applied in a specified manner to each measurement so as to reduce or, if possible, eliminate systematic error. On the other hand such corrections are often considered inherent in the method of test and may be better incorporated in the description of the test method itself rather than associated with the accuracy of a measurement process. Furthermore, possibly misleading inferences as to the accuracy of particular measurement processes may be induced by relying on a standard correction as a means of eliminating systematic error (see 4.3.3).

4.2.2.1 The sources of systematic error lie largely in the system of causes associated with the process; as, for instance, treatment or preparation of test specimens prior to measurement, calibration of instruments, local environmental conditions such as humidity, temperature, etc., and interpretation of instructions for making measurements. If the indexes of precision of two processes have the same significance and are numerically equal, but if the systematic errors are different, the process of smaller systematic error may be said to be the more accurate or to have higher accuracy. However, appropriate corrections in each case would render them equivalent in accuracy. If the systematic errors of two measurement processes are both negligible, the more precise process of the two may be said to be the more accurate or to have higher accuracy. In other cases no simple comparison may be possible.

4.3 *Changes in Precision and Systematic Error of a Process:*

4.3.1 *Changes Associated with Changes in Level*—The value of an index of precision or the amount of systematic error may change with changes in the accepted reference level of the physical property concerned. Such changes in level occur normally in conjunction with deliberate change of the specific material tested. Even though precision does not involve the level as explicitly as does systematic error, it may also change with changes in level. This change may be due to the particular index of precision used, or it may be inherent in the test method or both. In either case, the statement of precision or systematic error of a process may require qualification or explanation in terms of levels. The suitability of a particular form of statement of precision depends in part on its degree of independence of the level. For instance, an index of precision in terms of the unit of measurement may differ in this respect from one in terms of percentage of level. It is not possible to say that either one is, in general, more independent of the level than the other, but in any given case the degree of independence may be markedly different. In addition, in some cases, there may be a simple mathematical function, like the logarithm or the square root, which can be applied to individual measurements independently of the level with the result that the index of precision or the systematic error expressed in terms of the transformed measurement is found to be stable over a wide range of levels.

4.3.2 *Changes in Precision Associated with Other Changes in Cause System*—The precision of a process depends on the cause system to which it relates. For instance, single-laboratory precision will usually be less than multi-laboratory precision (see 9.1.1). Precision especially must be qualified or explained in order to make it meaningful in terms of cause systems. This qualification should, if possible, be directly incorporated in any statement of precision, even if only in abbreviated form. Variations in precision resulting from serious failure to identify the actual cause system are, of course, not so much to be taken into account as eliminated.

NOTE 4—Failure to identify the actual cause system is often manifested in lack of statistical control.

4.3.3 *Changes in Systematic Error Associated with Other Changes in Cause System*—Systematic error has the quality of being free of random fluctuations. However, it may be fixed, it may change predictably as known changes take place within the scope of the specified cause system, or it may change for reasons difficult to identify. The last is particularly true of systematic errors associated with individual laboratories, sets of test apparatus, and operators. These errors cannot always be relied on to remain fixed for an indefinite time. For instance, it is quite possible that a certain amount of systematic error in a measurement process will pass unobserved until some change associated with a previously unidentified cause takes place. This change may lead to a change in the amount of systematic error and thereby to discovery of its particular cause. If the cause system is changed by increasing the number of laboratories, sets of test apparatus, operators, or test specimens, an error previously regarded as systematic might more fairly be interpreted as only one of a large number of deviations representing a statistical universe (Note 1). Thus, variability associated with a cause modified to include more individuals of a given kind might be attributable to imprecision rather than systematic error, and precision and systematic error (or errors) should be redetermined accordingly. However, in this case, measurements from certain individual laboratories, sets of test apparatus, operators, or test specimens may still deviate so widely from the rest as to suggest the presence of systematic error requiring correction. Such systematic error is not, of course, to be associated with imprecision.

4.4 *Experimental Determination of the Precision of a Measurement Process*—The meaning and usefulness of a statement of the precision of a measurement process depends upon its being able to be verified or determined operationally (that is, by experiment), upon its appropriateness to the test method and system of causes concerned, and upon its being well suited to the purpose of its future use.

4.4.1 *Requirements for Experimental Determination*—Experimental determination requires not only use of the test method but also use of additional plans and procedures, all constituting an experimental *method.* This experimental method has then to be realized in the form of an experimental process, and any given determination of precision may then be regarded as a "measurement" arising from this experimental process. The experimental process for determining the precision of a measurement process has a precision of its own which is determined by *sampling error* and possibly systematic error due to inadequacies of statistical design or irregularities in the conduct of the experiment. The experimental method should be capable of yielding the precision of the basic measurement process without being grossly distorted by imprecision or systematic error in the experimental process. The basic measurement process must have a known system of causes and must be capable of statistical control in order that the experiment be meaningful. It is necessary, therefore, to design the experiment to show whether or not the measurement process is in a state of statistical control and, if not, to provide an opportunity for taking such steps as are needed to bring the measurement process into that state. Finally, the precision of the experimental process should be estimable by means of the experiment itself and the effects of systematic error, of course, kept to a minimum.

4.4.2 *Associating Precision or Systematic*

Error of a Measurement Process with a Method and Class of Comparable Cause Systems—Experimental determination of precision may be attempted in the hope of associating the results with the specified method and any cause system comparable to that involved in the experimental process. That is to say, it may be inferred that other measurement processes in which the same test method and a comparable cause system are realized should have about the same precision. The confidence with which such inferences may be made is highly dependent on the completeness in definition of the test method and experimental method, on the completeness of knowledge of the system of causes, the degree to which the test method and a comparable system of causes are realized in other processes, and the existence of the capability of statistical control in all such processes. In particular there must be enough valid experimental evidence regarding variability among laboratories, operators, sets of test apparatus, test specimens, and days of test to avoid attributing either far too much or far too little variability to the test method when realized in a controlled process under a supposedly comparable system of causes (see 4.3.2). In contrast with precision the accuracy of a particular measurement process may depend on highly individual and perhaps changeable systematic errors characteristic of certain laboratories, sets of test apparatus, operators, or even test specimens. There is no reason to associate that type of systematic error with the test method or, more specifically, with any other measurement process which has a comparable system of causes (Note 5). On the other hand, systematic errors dependent on observable environmental factors or on reference level could be expected to involve deeper scientific knowledge than one could reasonably hope to gain from an experiment designed to serve the purposes of standardization of technology.

NOTE 5—Caution must be taken especially not to incorporate correctible consistent deviations between laboratories into a precision statement about a test method.

PRECISION (OR ACCURACY) AS USED IN INDIVIDUAL ASTM STANDARDS

5. Choice of Form of Statement

5.1 The form of statement of precision (or accuracy) is best chosen according to the purpose the statement is to serve in the future. Among such purposes might be:

5.1.1 To characterize a test method for purposes of comparing its precision (or accuracy) with other test methods used for the same or similar purposes,

5.1.2 To provide a basis for checking the precision (or accuracy) of any particular measurement process or for comparing the precision (or accuracy) of two or more processes. A supporting experimental method, including plans and procedures for taking the measurement, may be required in order to apply the statement effectively, or

5.1.3 To assist in deciding which among a set of measurements should be considered valid in case of doubt.

5.2 A second consideration is whether the numerical values to be presented in the statement can be expected to remain fixed over the range of levels of material (see 4.3.1) and over the range of expected environmental conditions.

6. Scope of Statements

6.1 The recommended forms of statements in Section 10 are intended primarily to suggest standard terms and a manner of reference, not to limit the variety of forms which might be used to express precision quantitatively. If, however, the indexes of precision shown in Section 10 are used in ASTM standards, it is suggested that the terms and forms of statement follow those given here. It will then be appropriate to reference this recommended practice.

6.2 In presenting an index of precision, the particular kind used should be described as well as the system of causes, including such factors as specimens of material, operators, sets of apparatus, laboratories, days, etc. Confidence limits for an index of precision are *not* recommended as a substitute for a definite stated value, but they may be of supplementary interest. The correction (or corrections) given in a statement of accuracy should be intended as a standard one for the test method although its actual value may vary in a known way with the level of material and selected measurable environmental factors. If an index of precision is used in a statement of accuracy, it should pertain to the corrected measurements.

7. Evidence of Experimental Determination

7.1 In some standards it may be useful to present a summary of the experimental method by which precision has been determined and even of the data obtained by the experimental process, presented in such a way as to support the statement of precision. Such a summary should indicate the sampling error associated with the experimental determination of precision.

8. Notation

8.1 The following notation is useful in discussing statements of precision and accuracy:

x_R	= accepted reference level of the property of material to be measured.
X	= a measurement produced by a measurement process at the level x_R.
μ_P	= process at the level x_R.
σ_P	= standard deviation of a measurement process at the level x_R.
$\mu_P - x_R$	= systematic error of a measurement process at the level x_R.
$\delta = x_R - \mu_P$	= correction for systematic error of a measurement process at the level x_R.
$X_c = X + \delta$	= corrected measurement at the level x_R

8.2 An idealized version of some relationships among the above symbols is shown in Fig. 1, in which the area under the curve between any two values of X represents the relative frequency of measurements in that interval. In practice the correction δ and standard deviation σ_P can only be estimated from measurements gathered by an experimental process (see 4.4).

9. Modifiers Relating Precision to a System of Causes

9.1 The following modifiers of the word "precision" are suggested as among those which may be appropriate in ASTM standards:

9.1.1 *"Single-" and "Multi-"*—The prefix "single-" (as in "single-operator") may be used to indicate the contribution of (any) one individual of the kind the prefix modifies. This individual, such as an operator, set of test apparatus, etc., would ordinarily be presumed to meet certain general implicit requirements as to degree of skill, excellence, judgment, etc. Furthermore, when it is important to do so, the context of the statement of precision should make clear whether "single-" refers only to any individual of a limited class or may be imputed to any individual of a broad class (Note 1). On the other hand the prefix "multi-" (as in "multioperator") may be used to indicate that differences among individuals representing a specific cause have contributed to the variability described by the precision; the individuals characterized in this way should be able to be interpreted as any of a broad class unless otherwise indicated. Terms to which the modifiers "multi-" or "single-" could be applied include:

Specimen
Operator (or analyst, etc.)
Apparatus
Laboratory
Day (of test)

"Multi-" would rarely be a suitable modifier of "Material."

9.1.2 *Omitted Causes*—At times it may be convenient to omit mention of some causes because their presence in the system of causes can be readily inferred from the context. Care must be taken, however, that there is such an unambiguous inference in case of omission. For instance, even the term "single-laboratory-operator-material precision" is somewhat ambiguous without further qualification because it is not clear whether the precision applies to measurements made in relatively rapid succession on a single day or to measurements made, say, over a period of several days or weeks, nor whether the measurements are repetitions on one or made on different specimens from any single portion of material.

NOTE 6—Some ASTM standards contain the terms "repeatability" and "reproducibility" to describe systems of causes for indexes of precision. Usually "repeatability" has meant single-laboratory-operator-material precision, and "reproducibility" multilaboratory-operator, single material precision, but variety of usage does not permit straightforward standardization of these terms. If the terms are specially defined for particular applications, it is advisable to adhere to the following principle: "repeatability" should refer to the precision associated with the most restricted system of causes it is reasonable and useful to consider; in other words, the precision that might be expected in a series of "repeated" measurements on the "same" material under essentially the "same" conditions. "Reproducibility" should refer to precision associated with a broader system of causes of variability not accounted for in the corresponding definition of repeatability, especially operators or places of test (however, see Note 5). It is obvious that when these two terms are used in a standard, ambiguity may result unless the terms are clearly defined in the standard.

9.1.3 *Examples:*
Single-operator
Single-analyst
Single-laboratory-operator-material-day
Single-laboratory
Multispecimen
Multilaboratory

10. Indexes of Precision

10.1 *Basis*—The usual basis of indexes of precision is the standard deviation σ_P of the statistical distribution of measurements involved. It leads directly to those indexes of precision shown in Table 1.

10.2 *Interpretation of the Indexes of Precision:*

10.2.1 *Standard Deviation*—See 10.1.

10.2.2 *Two-Sigma Limits*—In practice about 5 percent of all numbers from a statistical universe can be expected to differ from their average value by more than 2 σ_P. Occasionally, this percentage might exceed 5 percent.

10.2.3 *Three-Sigma Limits*—In practice less than 1 percent of all numbers from a statistical universe can be expected to differ from their average value by more than 3 σ_P. The actual percentage is usually negligible, although the percentage could conceivably exceed $^1/_2$ percent.

10.2.4 *Difference Two-Sigma Limits*—In practice less than 5 percent of all random pairs of numbers from a statistical universe can be expected to differ by more than $2\sqrt{2}\ \sigma_P$. Occasionally this percentage might exceed 5 percent.

10.2.5 *Difference Three-Sigma Limits*—In practice less than 1 percent of all random pairs of numbers from a statistical universe can be expected to differ by more than $2\sqrt{2}\ \sigma_P$. The actual percentage is usually negligible, although the percentage could conceivably exceed $^1/_2$ percent.

10.3 *Indexes in Percent*—Precision may be indicated as a percentage of the accepted reference level x_R. A term for such an index in percent may be obtained by adding the words "in percent" after the corresponding term in Column 1 in Table 1. As an abbreviation, the percent sign may be added just before the closing parenthesis of the corresponding abbreviation in Column 2 in Table 1, as in "(1S%)." The value of the index of precision in percent is found by multiplying the corresponding quantity in Column 3 of Table 1 by $100/x_R$.

NOTE 7—The index of precision in percent is based upon the quantity 100 σ_P/x_R. If the process average coincides with the accepted reference level, that quantity becomes the theoretical coefficient of variation 100 σ_P/μ_P.

11. Statements of Accuracy

11.1 *Basis*—In general the usefulness of a statement of accuracy is open to question (see 4.2.2). However, if such a statement is included in an ASTM standard or practice, it may include an index of precision and a standard correction (or corrections) to be applied to individual measurements to reduce systematic error associated with test method, test environment, and level of material. There is no universally best way of stating a standard correction. Some possible general forms, however, are as follows:

11.1.1 The negative of a reasonable estimate of a constant systematic error, possibly supplemented by upper and lower bounds on this error,

11.1.2 An amount dependent on the length of time since the last calibration of test apparatus and needed to correct for a systematic but tolerable expected drift until a time specified for the next calibration,

11.1.3 A step-by-step description of a procedure for applying successive corrections for different reasons, and

11.1.4 A calibration formula or graph which might depend upon measured values of environmental factors as well as, possibly, the raw measurement itself.

11.2 *Index of Accuracy in Percent*—If the raw measurement X is reexpressed as a percent of level (that is, 100 X/x_R), the index of accuracy is changed by converting the index of precision as in 10.3, and any correction is converted by the same procedure of multiplying by $100/x_R$.

12. Use of Maximum Values

12.1 Although an index of precision might be chosen so that it does not change over the range of levels of interest, other indexes may not be constant over the same range. For instance, if an index of precision in terms of percent is constant over the range, an index of precision expressed directly in the unit of measurement cannot be constant. Of course, it may be hard to find any index that is constant. In this case if changes in the index of precision are small or moderate over the range, it may be desirable to state the maximum value of that index over the range. This procedure would then be indicated by adding the abbreviation "max" after the values stated. A similar procedure might possibly be followed for a varying correction for systematic error, but it would rarely be appropriate to do so in a practice since the danger of misinterpretation may be high. If changes in the index of precision are substantial over the useful range of levels in the dimensions of the original measurements, it may be advisable to attempt to reduce these changes by transforming the original measurements as discussed in 4.3.1. Otherwise, the relation between level and precision (or correction for systematic error) should be stated as explicitly as possible.

13. Examples

13.1 *Relationship of Precision and Accuracy*—The following comments on Fig. 2 illustrate the meaning of the terms precision, systematic error, accuracy and their relationship relative to the level x_R. (It may be assumed, for purposes of the example, that the level k_R has been established by use of a standard method.)

13.1.1 The single-laboratory precision of each method is higher than the corresponding multilaboratory precision.

13.1.2 The single-laboratory precision of Method A and that of Method B are about equal.

13.1.3 The multilaboratory precision of Method B is higher than that of Method A.

13.1.4 The single-laboratory systematic error is the same as the multilaboratory systematic error with both methods.

13.1.5 The systematic error of Method A is larger than that of Method B.

13.1.6 The single-laboratory accuracy of Method B is higher than that of Method A.

13.1.7 The multilaboratory accuracy of Method B is higher than that of Method A. If, however, the systematic errors were interchanged but not the precisions, a comparison of accuracy is more difficult to express.

13.2 *Statements of Precision and Accuracy*—The statements below are not complete insofar as describing the system of causes is concerned. Statements in ASTM standards would require further explanation.

13.2.1 The single-specimen precision is ± 3.5 mg (1S) as defined in ASTM Recommended Practice E 177.

13.2.2 The single-laboratory precision is ±0.1 percent (3S%) max, as defined in ASTM Recommended Practice E 177, over the range from 0 to 20 dB.

13.2.3 *Accuracy*—The multispecimen precision of the method is ±2.5 percent (3S%) as defined in ASTM Recommended Practice E 177, with a correction of 0.6 (60 − T) percent for any temperature T (deg F) between 60 and 80 F.

Experimental Determination of Precision and Accuracy

14. General

14.1 *Nature of Experimental Procedures to Determine Precision*—It is usually best, in determining precision, to start with a modest pilot study and proceed, when it has been shown that the measurement process involved in that study is capable of statistical control, to a series of experiments as needed, each experiment involving only one test location. A single all-purpose experiment should be avoided, although it will often be useful ultimately to conduct an interlaboratory test involving a large number of measurements (**6**).[3]

14.2 *Nature of Experimental Procedure to Determine Accuracy*—So far as accuracy is concerned, it is more important to detect and correct error, including systematic error, than to determine its magnitude for future reference. Because of the variety of ways this might be done there is no single procedure that can be recommended for the experimental determination of accuracy, although an informative discussion appears in Ref (**4**).

14.3 *Preliminary Considerations*—In laying plans for experimental determination of

[3] The boldface numbers in parentheses refer to the list of references at the end of this recommended practice.

precision, consideration should be given to securing the services of a well-qualified statistician to advise as to design and analysis of the experiment. Supervision of experiments should be planned to assure that, with normal care, the results will represent the effects to be later identified with them. At the same time, of course, it is desirable to avoid having such an abnormal degree of supervision or of other external influence unique to the experiment that experimental results would be misleading with regard to future use of an index of precision determined from them. If more than one location is involved, uniformity of supervision and standardization of recording and reporting deserve prior discussion and agreement among participating locations.

EXAMPLES

15. General

15.1 These examples illustrate the experimental determination of the precision of a measurement process. A measurement process (see 3.1) includes: the realization of a test method, the realization of a system of causes, and the capability of statistical control.

15.2 As noted in 4.4, experimental determination requires not only use of the test method but also use of additional plans and procedures, all constituting an experimental *method*. The plans and procedures include:

15.2.1 Plans to identify certain known causes of variability of measurements (such as test sets, operators, days, etc.) and to measure their effect on the variability (see 3.1), and

15.2.2 Randomization procedures that tend to balance unknown or uncontrolled causes of variability and permit more accurate estimates of the amount of observed variability attributable to each of the known causes.

15.3 The actual use of an experimental method results in an experimental process, which is used to determine the precision of a measurement process. If the residuals, measurements minus effects of known causes of variability, observed by means of a measurement process are in control (in the sense of Part 3 of the *ASTM Manual on Quality Control of Materials*) several indexes of precision can be computed from a standard deviation of the distribution of the observed residuals.

15.4 Measurement processes differ in many respects, but there is one particular feature of a test method that greatly affects the type of experimental method to be used. That feature is: whether or not the test changes or destroys the test specimen. If repeat measurements can be made on the same part of a test specimen, the experimental determination of the precision of the measurement process is relatively easy. If repeat measurements must be made on different specimens, the experimental method must provide a means of estimating and removing the variability introduced by the difference among specimens.

16. Repeat Measurements on a Single Specimen (Example 1)

16.1 Example 1 illustrates the case where repeat measurements can be made on the same area of a test specimen:

Example 1—Experimental determination of single-operator-micrometer-day precision of micrometer measurements.

16.1.1 *Experimental Method:*

16.1.1.1 *Micrometer*—One that is graduated in thousandths of an inch, chosen at random from those available which are judged to meet applicable standards and to typify the micrometers used for measurements of the kind contemplated.

16.1.1.2 *Test Specimen*—A steel cube approximately 1 in. on a side.

16.1.1.3 *Test Method*—Retract the spindle of the micrometer and reinsert the cube before each measurement.

16.1.1.4 *Operator*—Chosen at random from those in the laboratory who are judged to meet applicable standards and to typify the kind of operator who would ordinarily make micrometer measurements of the kind contemplated.

16.1.1.5 *Experiment*—Twenty readings on one day.

16.1.2 *Experimental Process:*

16.1.2.1 *Micrometer*—One was chosen at random from the half dozen available for use. (Numbers from 1 to 6 were assigned to the micrometers; the random number table was used to choose No. 5.)

16.1.2.2 *Test Specimen*—A 1-in. steel cube was marked on two opposite faces with circles within which all measurements were made.

16.1.2.3 *Operator*—Since only two operators were available, one was chosen by the toss of a coin.

16.1.2.4 *Measurement*—Using the prescribed test method, the operator made 20 measurements of the cube. The last digit was estimated.

16.1.3 *The Data:*

16.1.3.1 *Observed Values*—Table 2 lists the 20 observed values in the order of measurement.

16.1.3.2 *Analysis of Data*—The following statistics were computed from the observed values:

Average, $\bar{X}_P = 1.00403$ in.,
Standard deviation, $s_P = 0.00024$ in.

$\bar{x}_P$ is an estimate of μ_P; s_P is an estimate of σ_P.

16.1.3.3 *Statistical Control*—All observed values fall within the limits, $\bar{x}_P \pm 3\ s_P$. (The use of control charts is illustrated in Example 2.)

16.1.4 *Estimates of Indexes of Precision:*

16.1.4.1 *Individual Measurements:*

± 0.00048 in. (2S),
± 0.00072 in. (3S).

16.1.4.2 *Difference Between Two Measurements:*

± 0.00068 in. (D2S),
± 0.00102 in. (D3S).

16.1.4.3 *Percent*—If the steel cube were a *standard* 1-in. cube, $x_R = 1.0$. The indexes for individual measurements would be:

± 0.048 (2S%),
± 0.072 (3S%).

16.1.5 *Estimate of Systematic Error:*

16.1.5.1 *Individual Measurements*—If the steel cube were a *standard* 1-in. cube, 1.00403 − 1.00000 = 0.00403 would be an estimate of systematic error. It would ordinarily be quite unusual for $\bar{x}_P$ to be farther than $2\ s_P/\sqrt{n}$, or about 0.0001 in this example, from the reference level. Thus the value 0.00403 cannot safely be attributed to sampling variability alone.

16.1.5.2 *Correction*—The estimate of δ is 1.00000 − 1.00403 = −0.00403. Further investigation would be required to determine whether the cube is oversized or whether the micrometer is improperly calibrated. If the error is in the micrometer, the correction would have to be applied, for the sake of accuracy, to all measurements until the micrometer is recalibrated.

16.1.6 *Statements of Precision:*

16.1.6.1 *Individual Measurements*—The single-operator-micrometer-day precision is ± 0.0005 in. (2S) as defined in ASTM Recommended Practice E 177.

16.1.6.2 *Percent*—Several experiments using specimens of different sizes would be necessary to find out if the precision changes with the reference levels.

16.1.6.3 These indexes of precision apply to the operator and micrometer chosen for the experiment. A more general statement would require several experiments on different days with different operators and micrometers.

17. Repeat Measurements on Different Specimens (Example 2)

17.1 Example 2 illustrates the case in which specimens are destroyed by the test. Thus repeat measurements must be made on different specimens.

Example 2—Experimental determination of single-operator-test set precision of tensile strength tests on a strip of phosphor bronze.

17.1.1 *Experimental Method:*

17.1.1.1 *Test Method*—Tension tests are prescribed in ASTM Methods E 8, Tension Testing of Metallic Materials,[4] using test specimens in accordance with the dimensions shown in Fig. 7 of those methods of test. A tensile strength value is the ratio of the maximum load value to the minimum cross-sectional area of a specimen.

17.1.1.2 *Test Set*—A 60,000-lb Baldwin-Tate-Emory testing machine with Templin jaws.

17.1.1.3 *Operator*—Chosen at random from those trained to use the testing machine.

17.1.1.4 *Experiments*—Two experiments of the same general plan run on different days.

17.1.1.5 *Order of Test* (*within experiments*) —Specimens randomly ordered for measurement and testing, using a random number table and the numbers on the specimens.

17.1.1.6 *Material*—ASTM Specification B 103, for Phosphor Bronze Plate, Sheet, Strip, and Rolled Bar,[5] Alloy A phosphor bronze strip with a nominal thickness of 0.040 in. and

[4] *Annual Book of ASTM Standards*, Part 31.
[5] *Annual Book of ASTM Standards*, Part 5.

a nominal tensile strenth of 89.0 ksi.

17.1.1.7 *Sample*—One length chosen at random from the several lengths supplied.

17.1.1.8 *Method of Preparation of Specimens*—The conventional cross-milling technique, using a spiral fluted form cutter.

17.1.2 *Experimental Process:*

17.1.2.1 *Material*—The phosphor bronze strip was manufactured from a single 2.5-in. thick ingot by cold-rolling and annealing. The material was supplied in 17 lengths which had been numbered from 1 to 17 by the manufacturer in the order of production.

17.1.2.2 *Test Specimens*—The length of material that was chosen at random from the 17 lengths supplied was divided crosswise in 10 "columns." Each column was divided into 8 rows, or specimen blanks. Each specimen blank was numbered to show the column and row from which it came, for example 02 04. The specimen blanks required milling to transform the blanks into the type of specimens needed for tension tests. The 80 specimen blanks were assigned to milling lots of 10 specimens each at random, with the constraint that each column and each row be represented in each milling lot. The milling lots, which had been numbered according to the order of their formation, were milled in random order.

17.1.2.3 *Measurement of Specimens*—After milling, a point was located on each specimen by a procedure designed to find the point of minimum thickness. The thickness and width at that point were measured with a micrometer graduated in ten-thousandths of an inch. The computed area is essentially the minimum cross-sectional area of the specimen. (The nominal value is 0.02 in.)

17.1.2.4 *Test Sets*—Two 60,000-lb Baldwin-Tate-Emory testing machines were used. Calibration measurements indicated that both testing machines had load scales accurate to ± 0.06 percent (about 1 lb since the nominal maximum load value is 1780 lb).

17.1.2.5 *Assignment of Test Specimens*—Four of the 8 milling lots were assigned at random (with respect to order of milling) to Experiment I, carried out on the first day. Five specimens from each of these 4 milling lots were assigned to Test Set 1 and the remaining 20 specimens to Test Set 2, with the constraint that 2 specimens from a column of material would be assigned to each test set. The remaining 4 milling lots were assigned to Experiment II, carried out on a later day. The specimens were assigned to the 2 test sets at random as outlined previously.

17.1.2.6 *Tension Tests*—The order of testing the specimens was randomized for each test set within each experiment. During test, the speed of test (rate of jaw separation) was constant and the same for all specimens. The maximum load values were read from the testing machine scale to $\pm ^1/_2$ lb.

17.1.2.7 *Operator*—All measurements and tests were made by one operator.

17.1.3 *The Data:*

17.1.3.1 *Tensile Strength Values*—The maximum load value observed for each specimen was divided by the estimate of the minimum cross-sectional area for that specimen. The resulting tensile strength values were rounded to the nearest 0.1 ksi and are shown in Tables 3 and 4.

17.1.3.2 *Analysis of Data*—The following averages were computed for each of the two experiments:

Milling Lot,
Column,
Test Set,
Over-all.

The over-all average was carried out to 4 decimal places (one more place than is warranted) in order that the sum of the residuals be zero for each experiment. To remove variation caused by differences between milling lots, columns, and test sets, residual values were computed as follows:

Residual = observed value-milling lot average-column average − test set average + 2 (over-all average)

These values are shown in Tables 3 and 4. The standard deviation of residuals is

$$\sqrt{\text{sum of (residuals)}^2}/\sqrt{\text{degrees of freedom}};$$

there are 27 degrees of freedom for each experiment. The computed values are 0.67 ksi for Experiment I and 0.65 ksi for Experiment II. The residual values were plotted in order of test as shown in Fig. 3. The variation of the residual values about zero appears to be less than expected on the basis of a standard deviation value of 0.66 ksi. All the 80 values are within limits based on twice the estimated standard deviation. The standard deviation of residuals may be used as an estimate of precision if there are no significant interactions due to differences in material. An analysis of variance procedure was used to investigate this question. The results are shown in Table 5. The test sets × columns interaction is significantly large at the 5 percent level of significance. The reason for the large interaction is that, for one column of material, the average tensile strength value for Test Set 2 is slightly smaller than for Test Set 1 while for all other columns the tensile strength values for Test Set 2 are equal to or greater than for Test Set 1. Combining the residual sum of squares with the sums of squares of the interactions that are not significantly greater than the residual gives a "pooled" residual mean square equal to 0.3663 $(ksi)^2$, based on 53 degrees of freedom. The resulting estimate of standard deviation, 0.61 ksi, is relatively free of variation attributable to differences in material as well as difference between test sets.

17.1.3.3 *Statistical Control*—All the residual values fall within the limits 0 ± 3 (0.61) ksi, as shown in Fig. 3. Thus, the measurement process is in control for each test separately. The analysis in Table 5 shows that the test sets are significantly different. Before multitest set precision could be estimated, the calibration of the test sets would have to be checked to determine what part of the difference between the test sets is systematic error that can be removed.

17.1.4 *Statements of Precision:*

17.1.4.1 *Individual Measurements*—The single-operator-test set-day precision of ASTM Method E 8, as applied to ASTM Specification B 103 Alloy A phosphor bronze strip with a nominal thickness of 0.040 in., is ±1.22 ksi (2S) as defined in ASTM Recommended Practice E 177.

17.1.4.2 *Percent*—The single-operator-test set-day precision of test method E 8, as applied to ASTM Specification B 103 Alloy A phosphor bronze strip with a nominal tensile strength of 89.0 ksi, is ±1.37 percent (2S%) as defined in ASTM Recommended Practice E 177.

NOTE 8—For a more universally useful statement of precision experimental determination of precision for other nominal values of tensile strength would be required as well as experiments using different operators and test sets.

REFERENCES

(1) Brownlee, K. A., *Statistical Theory and Methodology In Science and Engineering*, John Wiley & Sons, New York, 1965 (2nd Ed.).

(2) Cox, D. R., *Planning of Experiments*, John Wiley & Sons, New York, 1958.

(3) Ducan, A. J., *Quality Control and Industrial Statistics*, Richard D. Irwin, Inc., 1965 (3rd Ed.).

(4) Mandel, John, *The Statistical Analysis of Experiments*, Interscience Publishers, New York 1964.

(5) Murphy, R. B., *Materials Research & Standards*, MTRSA, Vol I, No. 4, April, 1961, pp. 264–267.

(6) Youden, W. J., *Materials Research & Standards*, MTRSA, Vol I, No. 4, April, 1961, pp. 268–270.

(7) *A Guide for Fatigue Testing and the Statistical Analysis of Fatigue Data, ASTM STP 91 A*, Am. Soc. Testing Mats., 1963.

(8) *Manual for Conducting an Interlaboratory Study of a Test Method, ASTM STP 335*, Am. Soc. Testing Mats, 1963.

(9) *Manual on Quality Control of Materials, ASTM STP 15 C*, Am. Soc. Testing Mats., 1951.

TABLE 1 Indexes of Precision

Col. 1	Col. 2	Col. 3
Term	Reference Abbreviation	Notation
Standard deviation	(S)	σ_P
Two-sigma limits	(2S)	$\pm 2\sigma_P$
Three-sigma limits	(3S)	$\pm 3\sigma_P$
Difference two-sigma limits	(D2S)	$\pm 2\sqrt{2\sigma}$
Difference three-sigma limits	(D3S)	$\pm 3\sqrt{2\sigma_P}$

TABLE 2 Observed Values of Dimension of Cube, in

1.0040	1.0042	1.0042	1.0041
1.0038	1.0043	1.0044	1.0043
1.0037	1.0040	1.0038	1.0040
1.0045	1.0039	1.0041	1.0036
1.0039	1.0041	1.0037	1.0040

TABLE 3 Experiment I

NOTE—Order of test was recorded but not shown

Milling Lots	Columns of Material 1	2	3	4	5	6	7	8	9	10	Milling Lot Averages
	Tensile Strength, ksi										
5	88.8$^{(1)a}$	87.5$^{(1)}$	89.0$^{(1)}$	88.1$^{(2)}$	89.6$^{(2)}$	89.3$^{(2)}$	90.4$^{(2)}$	88.2$^{(1)}$	89.1$^{(1)}$	89.7$^{(2)}$	88.97
3	86.9$^{(1)}$	88.1$^{(2)}$	88.9$^{(1)}$	88.2$^{(2)}$	87.4$^{(1)}$	88.6$^{(2)}$	88.3$^{(1)}$	88.8$^{(2)}$	88.8$^{(2)}$	88.7$^{(1)}$	88.37
4	88.2$^{(2)}$	87.5$^{(1)}$	83.3$^{(2)}$	87.4$^{(1)}$	87.9$^{(1)}$	89.4$^{(1)}$	87.4$^{(1)}$	87.5$^{(2)}$	90.4$^{(2)}$	88.4$^{(2)}$	88.34
7	89.7$^{(2)}$	89.2$^{(2)}$	89.8$^{(2)}$	89.7$^{(1)}$	88.0$^{(2)}$	88.1$^{(1)}$	89.8$^{(2)}$	88.1$^{(1)}$	90.5$^{(1)}$	89.2$^{(1)}$	89.21
Column Avg	88.40	88.075	89.25	88.35	88.225	88.85	88.975	88.15	89.95	89.00	88.7225 (over-all average)

Test Set 1 Average = 88.40; Test Set 2 Average = 89.045

RESIDUALS

Milling Lots	1	2	3	4	5	6	7	8	9	10
	Test Set 1									
5	+0.475(19)^{b}	−0.500(7)	−0.175(18)					+0.125(17)	−0.775(6)	
3	−0.825(4)		+0.325(2)		−0.150(11)		0.000(14)			+0.375(3)
4		+0.130(1)		−0.245(5)	+0.380(15)	+1.255(12)	−0.870(13)			
7				+1.185(16)		−0.915(8)		−0.215(20)	+0.385(10)	+0.035(9)
	Test Set 2									
5				−0.820(8)	+0.805(9)	−0.120(6)	+0.855(18)			+0.130(7)
3		+0.055(4)		−0.120(3)		−0.220(12)		+0.680(13)	−0.120(14)	
4	−0.140(11)		+0.110(5)					−0.590(2)	+0.510(15)	−0.540(1)
7	+0.490(17)	+0.315(19)	−0.260(16)		−1.035(20)		+0.015(10)			

[a] Test Set numbers are shown as superscripts.
[b] Order of test is shown in parentheses.

TABLE 4 Experiment II

NOTE—Order of test was recorded, but not shown.

Milling Lots	Columns of Material 1	2	3	4	5	6	7	8	9	10	Milling Lot Averages
	Tensile Strength, ksi										
1	88.6^{2} [a]	87.0^{1}	87.9^{1}	88.0^{2}	88.1^{2}	87.1^{1}	89.9^{2}	87.0^{1}	88.9^{2}	88.5^{1}	88.10
8	88.8^{1}	90.0^{2}	88.4^{2}	88.5^{1}	88.6^{1}	89.9^{2}	88.3^{1}	88.8^{2}	89.5^{1}	88.6^{2}	88.94
2	88.3^{1}	88.6^{2}	88.6^{2}	88.0^{1}	87.4^{2}	90.1^{2}	87.8^{1}	87.4^{1}	90.7^{2}	87.9^{1}	88.48
6	88.6^{2}	88.2^{1}	87.1^{1}	89.2^{2}	88.4^{1}	88.6^{1}	90.7^{2}	88.4^{1}	89.5^{2}	89.4^{2}	88.81
Column Avg	88.575	88.45	88.00	88.425	88.125	88.925	89.175	87.90	89.65	88.60	88.5825 (over-all average)

Test Set 1 Average = 88.12; Test Set 2 Average = 89.045.

RESIDUALS

Milling Lots	1	2	3	4	5	6	7	8	9	10
	Test Set 1									
1		−0.505(15)	+0.845(2)			−0.880(12)		+0.045(14)		+0.845(1)
8	+0.330(20) [b]			+0.180(17)	+0.580(16)		−0.770(7)		−0.045(8)	
2	+0.290(5)			+0.140(3)			−0.810(13)	+0.065(4)		−0.135(11)
6		−0.015(6)	−0.665(18)		+0.510(19)	−0.090(9)			+0.085(10)	
	Test Set 2									
1	+0.045(3)			−0.405(11)	−0.005(2)		+0.745(15)		−0.730(14)	
8		+0.730(17)	−0.420(7)			+0.155(18)		+0.080(6)		−0.820(16)
2		−0.210(13)	+0.240(12)		−1.085(4)	+0.815(1)			+0.690(5)	
6	−0.665(19)			+0.085(9)			+0.835(10)	−0.190(8)		+0.110(20)

[a] Test Set numbers are shown as superscripts.

[b] Order of test is shown in parentheses.

TABLE 5 Analysis of Variance

	Degrees of Freedom	Sum of Squares	Mean Square	Ratio
Known Causes of Variability:				
Experiments	1	0.3920	0.3920	
Test sets	1	12.3245	12.3245	13.51[a]
Columns of material	9	19.5545	2.1727	
Milling lots within experiments	6	9.9236	1.6539	
Interactions:				
Test sets × column	9	8.2105	0.9123	2.53[b]
Experiments × column	9	3.8230	0.4248	1.18
Experiments × test sets	1	0.3920	0.3920	1.09
Test sets × milling lot within experiments	6	1.8674	0.3112	0.86
Residual	37	13.3320	0.3603	
Total	79	69.8195		
"Pooled" residual (see text)	53	19.4144	0.3663 $\sqrt{0.3663} = 0.61$	

[a] Significant at the 1 percent level.
[b] Significant at the 5 percent level.

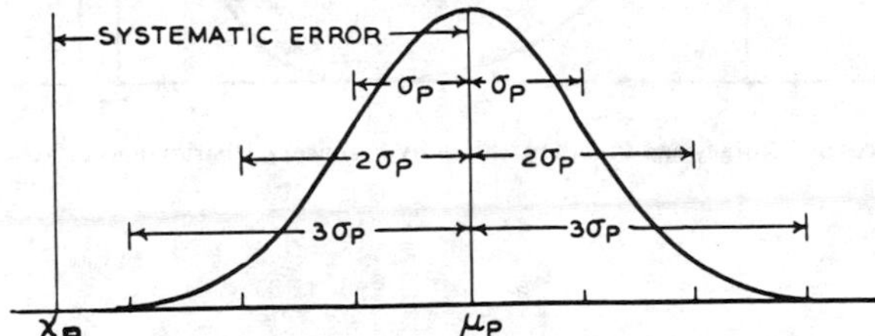

FIG. 1 Frequency Distribution of a Measurement.

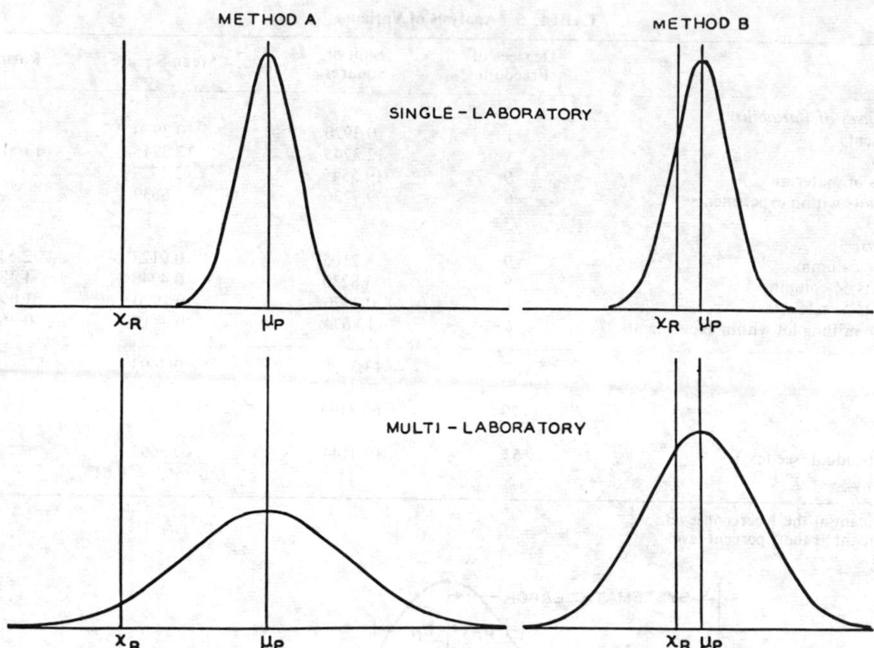

FIG. 2 Differences in Accuracy and Precision Shown by Frequency Distributions of Measurements.

TENSILE STRENGTH—PHOSPHOR BRONZE STRIP

EXPERIMENT I EXPERIMENT II

TEST SET 1

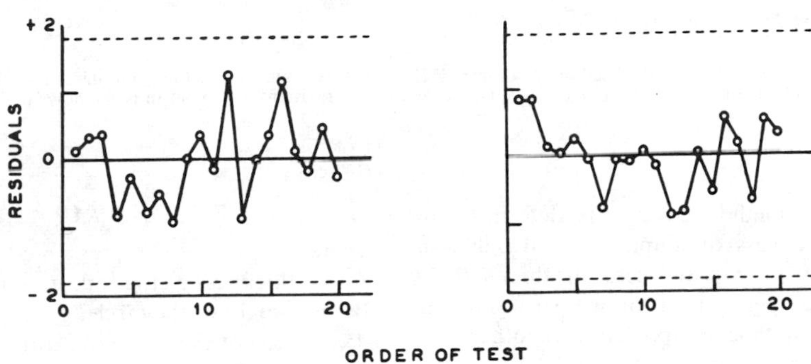

TEST SET 2

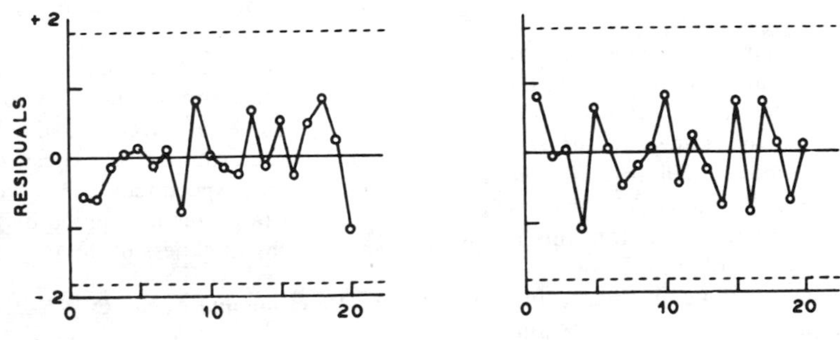

ORDER OF TEST

FIG. 3 Control Charts for Residuals.

Designation: E 252 – 67

American National Standard Z168.8-1969
American National Standards Institute

Standard Method of Test for
THICKNESS OF THIN FOIL AND FILM BY WEIGHING[1]

This Standard is issued under the fixed designation E 252; the number immediately following the designation indicates the year of original adoption or, in the case of revision, the year of last revision. A number in parentheses indicates the year of last reapproval.

1. Scope

1.1 This method covers the determination of the thickness of aluminum foil 0.002 in. (5.1×10^{-3} cm) and less in thickness by weighing a specimen of known area and density. The method is applicable to other foils and films as indicated in Appendix A3.

NOTE—The values stated in U.S. customary units are to be regarded as the standard. The metric equivalents of U.S. customary units may be approximate.

2. Apparatus

2.1 *Precision Blanking Press* to cut foil circles that are 8.000 ± 0.008 in.^{2} (51.616 ± 0.051 cm^2) in area or 3.1915 ± 0.0015 in. (8.106 ± 0.004 cm) in diameter. Other size specimens may be used with the recognition that the accuracy stated in 5.2 is no longer applicable. See Appendix A1 for the selection of other specimen sizes and the resulting change in accuracy of the method.

2.2 *Balance* capable of measuring to the nearest 0.1 mg (equivalent to 0.28 μin. (7.1×10^{-7} cm) of thickness for the 8-in.^{2} (51.616-cm^2) circle).

3. Procedure

3.1 Blank an 8.000 ± 0.008-in.^{2} (51.616 ± 0.051-cm^2) circle representative of the foil, swab with acetone to ensure a surface free from soil, and weigh the clean, dry specimen to the nearest 0.1 mg. Use a suitable solvent to remove any coating known to exceed 0.005 mg/ft^2 (4645 mg/cm^2) of surface area.

4. Calculations

4.1 Determine the thickness from the relationship:

$$T = W/AD$$

where:

T = thickness of the foil, in. (or cm),
W = weight of the circle, g,
A = area of the circle, in.^{2} (or cm^2), and
D = density of the foil, g/in.^{3} (or g/cm^3).

4.2 *Density of Aluminum Alloys:*

4.2.1 Calculate the density of aluminum foil from the nominal chemical composition of the alloy by the method described in Appendix A2. The densities of foil alloys determined in this manner are accurate to within ±0.3 percent.

4.2.2 Table 1 lists densities computed for some of the common alloys. A column headed "Mils/g for 8-in.^{2} Area" is added for convenience in determining thickness of the 8-in.^{2} (51.616-cm^2) specimens. The weight of the specimen in grams multiplied by this factor is equal to the thickness of the foil in mils.

5. Precision and Accuracy

5.1 Following the procedure outlined in this method, repeated weighings of the same specimen on different balances should result in agreement within 1 mg which is equivalent to a thickness of 3 μin. (7.62×10^{-6} cm). It is outside of the scope of this method to describe maintenance and calibration procedures for balances, but disagreement larger than 1 mg warrants attention to maintenance or recalibration of the balance.

---

[1] This method is under the jurisdiction of ASTM Committee E-1 on Methods of Testing. A list of members may be found in the ASTM Yearbook.

Current edition effective Sept. 8, 1967. Originally issued 1964. Replaces E 252 – 66 T.

TABLE 1 Density of Aluminum Alloys at 20 C Applicable to the Determination of Foil Thickness by the Weight Method

Alloy	Density, g/in.^{3}	mils/g for 8-in.^{2} Area
1100	44.44	2.813
1145	44.35	2.818
1180	44.25	2.825
1188	44.25	2.825
1199	44.23	2.826
2024	45.58	2.742
3003	44.82	2.789
3004	44.56	2.805
5005	44.20	2.828
5040	44.54	2.806
5050	44.05	2.838
5052	43.85	2.851
5082	43.36	2.883
5086	43.62	2.866
5154	43.62	2.866
6071	44.26	2.824

APPENDIX

A1. Specimen Size and Shape and Its Effect on Accuracy

A1.1 General

A1.1.1 Specimens of sizes and shapes other than the 8-in.^{2} (51.616-cm^2) circle may be used provided consideration is given to controllable factors affecting the accuracy of the method. Specifically, the area of the specimen shall be known and controlled to an accuracy of ± 0.1 percent and the minimum weight of specimen shall be 70 mg. Specimens ranging in size from 8 to 32 in.^{2} (52 to 206 cm^2) are convenient to work with and can be prepared to meet the aforementioned requirements.

A1.2 Source of Error

A1.2.1 Inherent errors in determining thickness by the weight method result from the limits on the accuracy of the density value assigned to the alloy, the accuracy with which a specimen can be cut and its area determined, and the accuracy of weighing. Much time could be devoted to a discussion of refinement of errors but it shall suffice here to draw on experience as a guide for determining the accuracy of the method.

A1.3 Error from Uncertainty of the Density of the Specimen (E_D)

A1.3.1 The density of aluminum foil alloys shall be those listed in Table 1 or shall be determined by the method described in Appendix A2. Values so arrived at are accurate to ± 0.3 percent of the true density. The error imposed by uncertainty of the density then is $E_D = \pm 0.3$ percent of the thickness determined.

A1.4 Error from Control of the Area of the Specimen (E_A)

A1.4.1 A precision blanking press can cut a specimen whose area is known and reproducible to an accuracy of ± 0.1 percent. If d is the specific diameter required to provide the area used in the thickness computation, then the error in area resulting from a small error, Δd, in the diameter is 200 $\Delta d/d$ percent. It follows then that to maintain an area accurate to ± 0.1 percent, the tolerance on the diameter of the blanked circle shall be ± 0.0005 times the circle diameter. The fact that the tolerance on diameter loosens in direct proportion to the diameter is a factor to consider in selecting the specimen size to use in the method. Compliance with this tolerance limits the area error to $E_A = \pm$ 0.1 percent of the thickness determined.

A1.5 Error from Weighing the Specimen (E_W)

A1.5.1 The accuracy of weighing a foil specimen has been found to be 0.7 mg. This imposes a maximum error on the method of $\pm 0.07/TAD$ percent of the thickness determined. Since D, density of the foil, is fixed, it is seen that the magnitude of the weighing error is a function of the thickness, T, of the foil and the area, A, of the specimen. The area, A, is a controllable factor in the method, and the importance of selecting a large area to minimize the over-all percentage error in the method for thin foil is apparent from a few simple calculations. The product TAD is the weight of the specimen in grams, so to prevent the weighing error from introducing errors in excess of ±1.0 percent, it is necessary that the weight of the minimum thickness specimen be larger than 70 mg. The maximum error in the method due to weighing then is $E_W = \pm 0.07/TAD$ percent of the thickness determined.

A1.6 Maximum Error of the Method

A1.6.1 If E_D, E_A, and E_W represent the errors in percentage of thickness determined as imposed by the limits of accuracy of density, area, and weighing, respectively, then the maximum error of the method is $E_D + E_A + E_W$ percent of the thickness determined. Since these errors at a given test location are normally in the nature of a bias rather than random error, the accuracy of the method is best described in terms of this maximum error. The maximum error of the method in percent is as follows:

$$E_D + E_A + E_W = [0.4 + (0.07/TAD)]$$

where TAD is the weight of the specimen in grams.

A2. Calculating the Density of Aluminum Alloys

A2.1 Calculation

A2.1.1 The density of an aluminum alloy for use in determining the thickness of foil by the weighing method may be calculated from the conventional density equation as follows:

$$D = \frac{100}{[(\%Al/D_{Al}) + (\%Cu/D_{Cu}) + (\%Fe/D_{Fe}) + \text{etc.}]}$$

where:

%Al, %Cu, %Fe = weight percent of each element in the alloy (%Al = 100 − total percent of alloying elements), and

D_{Al}, D_{Cu}, D_{Fe} = density of each element.

The form shown in Table A1 is convenient for making such calculations. A sample calculation is shown in italics for a typical 5052 alloy.

A2.2 Accuracy

A2.2.1 The accuracy of the density arrived at by this method is ±0.3 percent of the determined value for the common foil alloys and ±0.5 percent for highly alloyed compositions such as 2024.

A3. Use of Method for Polyethylene Film

A3.1 Scope

A3.1.1 This method is applicable to a wide range of films and foils. As an example of the slight modifications required for adaptation to other materials the procedure recommended for polyethylene film of 0.002 in. (5.1×10^{-3} cm) and less in thickness is described.

A3.2 Apparatus

A3.2.1 A hand striking die is a convenient way to cut the specimen. The precision blanking press with the polyethylene film on a piece of paper will also produce good results, but any other method capable of the desired precision may be used.

A3.3 Test Specimen

A3.3.1 Because of the lower density of polyethylene film, a minimum specimen of 16.000 ± 0.016 in.^{2} (103.232 ± 0.103 cm^2) is recommended; other specimen sizes may be used as discussed in Appendix A1. If there is evidence of surface contamination, the specimen may be wiped clean with a dry cloth or tissue, but no solvent should be used.

A3.4 Calculations

A3.4.1 The density of polyethylene film may be determined directly by ASTM Method D 1505, Test for Density of Plastics by the Density-Gradient Technique.[2] In many cases the film density will be known within the accuracy needed for this method (about 0.5 percent) from specifications or previous experience. Polyethylene density is usually reported in grams per cubic centimeter. To convert this to grams per cubic inch as needed in this calculation, multiply by 16.39.

[2] *Annual Book of ASTM Standards*, Part 30.

TABLE A1 Density Calculations of Aluminum Alloys at 20 C

Alloy *5052*

Element	1/Density (cm^3/g)	Weight Percent Present	1/Density × Weight Percent Present
Cu	0.1116	(*0.02*)	(*0.002*)
Fe	0.1271	(*0.23*)	(*0.029*)
Si	0.4292	(*0.09*)	(*0.039*)
Mn	0.1346	(*0.03*)	(*0.004*)
Mg	0.5522[a]	(*2.50*)	(*1.380*)
Zn	0.1401	(*0.03*)	(*0.004*)
Ni	0.1123	(*0.01*)	(*0.001*)
Cr	0.1391	(*0.25*)	(*0.035*)
Ti	0.2219	(*0.01*)	(*0.002*)
Pb	0.0882		
V	0.1639		
B	0.4274		
Be	0.5411		
Zr	0.1541		
Ga	0.1693		
Bi	0.1020		
Sn	0.1371		
	Total A	(*3.17*)	(*1.496*)
Al	0.3705	(*96.83*)	(*35.876*)
		Total B	(*37.372*)

Calculated Density
= 100/Total B = (*2.676*) g/cm^3
g/cm^3 × 13.388 = (*43.85*) $g/in.^3$
g/cm^3 × 0.036127 = (*0.0967*) $lb/in.^3$

[a] The magnesium density used in this calculation is 4.2 percent greater than the density of pure magnesium. This corrects for metallurgical formations that normally occur in alloys containing magnesium.

RELATED MATERIAL

Any documents included in this section that are marked "proposed" are published for information only. They have received the approval of the sponsoring technical committee for publication as "proposed," but have not been officially accepted by the Society. Comments are solicited and should be addressed to the American Society for Testing and Materials, 1916 Race St., Philadelphia, Pa. 19103.

LIST BY SUBJECTS, PART 26

1970 ANNUAL BOOK OF ASTM STANDARDS, PART 26

PLASTICS—SPECIFICATIONS; METHODS OF TESTING PIPE, FILM, REINFORCED AND CELLULAR PLASTICS

Since the standards in this book are arranged in numeric sequence, no page numbers are given in this list by subjects.

PLASTICS

Thermoplastic Materials

Specifications for:

D 2133 – 66	Acetal Resin Injection Molding and Extrusion Materials
D 1788 – 68	Acrylonitrile - Butadiene - Styrene (ABS) Plastics, Rigid
*D 706 – 63	Cellulose Acetate Molding and Extrusion Compounds
*D 707 – 63	Cellulose Acetate Butyrate Molding and Extrusion Compounds
D 1562 – 70	Cellulose Propionate Molding and Extrusion Compounds
D 2647 – 67 T	Crosslinkable Ethylene Plastics
*D 787 – 70	Ethyl Cellulose Molding and Extrusion Compounds
D 2116 – 66	FEP-Fluorocarbon Molding and Extrusion Materials
*D 788 – 69a	Methacrylate Molding and Extrusion Compounds
D 789 – 66	Nylon Injection Molding and Extrusion Materials
D 2531 – 68	Phenoxy Plastics Molding and Extrusion Materials
D 2581 – 69	Polybutylene Plastics
D 2473 – 66	Polycarbonate Plastic Molding, Extrusion, and Casting Materials
D 1430 – 69	Polychlorotrifluoroethylene Molding (PCTFE) Plastics
D 1248 – 70	Polyethylene Molding and Extrusion Materials
*D 703 – 67	Polystyrene Molding and Extrusion Materials
D 1047 – 67	*Poly(Vinyl Chloride) Jacket Compound for Electrical Insulated Wire and Cable (see Part 28)*
D 1755 – 66	Poly(Vinyl Chloride) Resins
D 1784 – 69	Poly(Vinyl Chloride) Compounds and Chlorinated Poly(Vinyl Chloride) Compounds, Rigid
D 2146 – 69	Propylene Plastic Molding and Extrusion Materials
D 1431 – 67	Styrene - Acrylonitrile Copolymer Molding and Extrusion Materials
D 1892 – 68	Styrene - Butadiene Molding and Extrusion Materials
D 1457 – 69	TFE-Fluorocarbon Resin Molding and Extrusion Materials
D 1710 – 66	TFE-Fluorocarbon Rod
D 1277 – 59 (1965)	*Thermoplastic Compounds for Automotive and Aeronautical Applications, Nonrigid (see Part 28)*
D 2219 – 68	*Vinyl Chloride Plastic Insulation for Wire and Cable, 60 C Operation (see Part 28)*
D 2220 – 68	*Vinyl Chloride Plastic Insulation for Wire and Cable, 75 C Operation (see Part 28)*
D 2474 – 69	Vinyl Chloride Copolymer Resins
D 2287 – 66	Vinyl Chloride Polymer and Copolymer Molding and Extrusion Compounds, Nonrigid
D 729 – 57 (1965)	Vinylidene Chloride Molding Compounds

Methods of Test for:

D 2659 – 67	Column Crush Properties of Blown Thermoplastic Containers
D 2463 – 65 T	Drop Impact Resistance of Polyethylene Blow-Molded Containers
D 1693 – 70	Environmental Stress-Cracking of Type I Ethylene Plastics
D 2561 – 67 T	Environmental Stress Crack Resistance of Blow-Molded Polyethylene Containers

---

* Approved as **American National Standard** by the American National Standards Institute.

Methods of Test for:

D 2552 – 69 T	Environmental Stress Rupture of Type III Polyethylenes Under Constant Tensile Load
D 2342 – 68	Lot-toLot Stabilization Uniformity of Propylene Plastics
D 2222 – 66	Methanol Extract of Vinyl Chloride Resins
D 2837 – 69	Obtaining Hydrostatic Design Basis for Thermoplastic Pipe Materials
D 1705 – 61 (1968)	Particle Size Analysis of Powdered Polymers and Copolymers of Vinyl Chloride
D 1928 – 68	Preparation of Compression-Molded Polyethylene Test Samples
D 1939 – 67	Quality of Extruded Acrylonitrile - Butadiene - Styrene (ABS) Pipe by Acetic Acid Immersion
D 793 – 49 (1965)	Short-Time Stability at Elevated Temperatures of Plastics Containing Chlorine
D 2134 – 66	Softening of Organic Coatings by Plastic Compositions
D 2741 – 68	Susceptibility of Polyethylene Bottles to Soot Accumulation
D 2445 – 65 T	Thermal Oxidative Stability of Propylene Plastics
D 1823 – 66	Viscosity, Apparent, of Plastisols and Organosols at High Shear Rates by Castor-Severs Viscometer
D 1824 – 66	Viscosity, Apparent, of Plastisols and Organosols at Low Shear Rates by Brookfield Viscometer
D 2288 – 69	Weight Loss of Plasticizers on Heating

Recommended Practices for:

2684 – 68 T	Determining Permeability of Thermoplastic Containers
D 2115 – 67	Oven Heat Stability of Poly(Vinyl Chloride) Compositions
D 2538 – 69	Fusion Test of Poly(Vinyl Chloride) (PVC) Resins Using a Torque Rheometer
D 2383 – 69	Plasticizer Compatibility in Poly(Vinyl Chloride) (PVC) Compounds Under Humid Conditions
D 2396 – 69	Powder-Mix Test of Poly(Vinyl Chloride) (PVC) Resins Using a Torque Rheometer
D 2839 – 69	Use of a Melt Index Strand For Determining the Density of Polyethylene

Thermosetting Materials

Specifications for:

D 1636 – 70 T	Allyl Molding Compounds
D 1763 – 67	Epoxy Resins
D 704 – 62 (1969)	Melamine - Formaldehyde Molding Compounds
D 700 – 68	Phenolic Molding Materials
D 1201 – 62 (1969)	Polyester Molding Compounds
D 705 – 62 (1969)	Urea - Formaldehyde Molding Compounds

Methods of Test for:

D 2471 – 68	Gel Time and Peak Exothermic Temperature of Reacting Thermosetting Plastic Compositions
D 2566 – 69	Linear Shrinkage of Thermosetting Casting Systems During Cure
D 2730 – 68	Sag Flow of Highly Viscous Materials
D 2393 – 68	Viscosity of Epoxy Compounds and Related Components

Thermoplastic Pipe and Fittings

Specifications for:

D 2680 – 70	Acrylonitrile-Butadiene-Styrene (ABS) Composite Pipe
D 2661 – 68	Acrylonitrile-Butadiene-Styrene (ABS) Plastic Drain, Waste, and Vent Pipe and Fittings
D 1527 – 69	Acrylonitrile-Butadiene-Styrene (ABS) Plastic Pipe, Schedules 40 and 80
D 2468 – 69	Acrylonitrile-Butadiene-Styrene (ABS) Plastic Pipe Fittings, Socket-Type, Schedule 40
D 2469 – 69	Acrylonitrile-Butadiene-Styrene (ABS) Plastic Pipe Fittings, Socket-Type, Schedule 80
D 2282 – 69a	Acrylonitrile-Butadiene-Styrene (ABS) Plastic Pipe (SDR-PR and Class T)
D 2465 – 69	Acrylonitrile-Butadiene-Styrene (ABS) Plastic Pipe Fittings, Threaded, Schedule 80
D 2750 – 70	Acrylonitrile-Butadiene-Styrene (ABS) Plastic Utilities Conduit and Fittings
D 2751 – 69	Acrylonitrile-Butadiene-Styrene (ABS) Sewer Pipe and Fittings
D 2672 – 68a	Bell-End Poly(Vinyl Chloride) (PVC) Pipe
D 2610 – 68	Butt Fusion Polyethylene (PE) Plastic Pipe Fittings, Schedule 40
D 2611 – 68	Butt Fusion Polyethylene (PE) Plastic Pipe Fittings, Schedule 80
D 2446 – 68	Cellulose Acetate Butyrate (CAB) Plastic Pipe (SDR-PR) and Tubing
D 1503 – 68	Cellulose Acetate Butyrate (CAB) Plastic Pipe, Schedule 40
D 2846 – 69 T	Chlorinated Poly(Vinyl Chloride) (CPVC) Plastic Hot Water Distribution Systems

Specifications for:

D 2836 – 69 T	Filled Poly(Vinyl Chloride) (PVC) Sewer Pipe
D 2609 – 68	Plastic Insert Fittings for Polyethylene (PE) Plastic Pipe
D 2662 – 68	Polybutylene (PB) Plastic Pipe (SDR-PR)
D 2666 – 68 T	Polybutylene (PB) Plastic Tubing
D 2104 – 68	Polyethylene (PE) Plastic Pipe, Schedule 40
*D 2239 – 68	Polyethylene (PE) Plastic Pipe (SDR-PR)
D 2447 – 68	Polyethylene (PE) Plastic Pipe, Schedules 40 and 80 Based on Outside Diameter
D 2737 – 68a T	Polyethylene (PE) Plastic Tubing
D 2665 – 68	Poly(Vinyl Chloride) (PVC) Plastic Drain, Waste and Vent Pipe and Fittings
D 2466 – 69	Poly(Vinyl Chloride) (PVC) Plastic Pipe Fittings, Socket-Type, Schedule 40
D 2467 – 69	Poly(Vinyl Chloride) (PVC) Plastic Pipe Fittings, Socket-Type, Schedule 80
D 1785 – 68	Poly(Vinyl Chloride) (PVC) Plastic Pipe, Schedules 40, 80, and 120
*D 2241 – 69	Poly(Vinyl Chloride) (PVC) Plastic Pipe (SDR-PR and Class T)
D 2464 – 68	Poly(Vinyl Chloride) (PVC) Plastic Pipe Fittings, Threaded, Schedule 80
D 2740 – 68	Poly(Vinyl Chloride) (PVC) Plastic Tubing
D 2729 – 68	Poly(Vinyl Chloride) (PVC) Sewer Pipe and Fittings
D 2683 – 68 T	Socket-Type Polyethylene Fittings for SDR 11.0 Polyethylene Pipe
D 2235 – 67	Solvent Cement for Acrylonitrile-Butadiene-Styrene (ABS) Plastic Pipe and Fittings
D 2560 – 67	Solvent Cements for Cellulose Acetate Butyrate (CAB) Plastic Pipe, Tubing, and Fittings
D 2564 – 67	Solvent Cements for Poly(Vinyl Chloride) (PVC) Plastic Pipe and Fittings
D 2852 – 69 T	Styrene-Rubber Plastic Drain and Building Sewer Pipe and Fittings
D 2513 – 68	Thermoplastic Gas Pressure Pipe, Tubing, and Fittings

Methods of Test for:

D 1180 – 57 (1961)	*Bursting Strength of Round Rigid Plastic Tubing (see Part 27)*
D 2122 – 67	Dimensions of Thermoplastic Pipe and Fittings, Determining
D 2412 – 68	External Loading Properties of Plastic Pipe by Parallel-Plate Loading
D 2444 – 70	Impact Resistance of Thermoplastic Pipe and Fittings by Means of a Tup (Falling Weight)
D 2837 – 69	Obtaining Hydrostatic Design Basis for Thermoplastic Pipe Materials
D 2152 – 67	Quality of Extruded Poly(Vinyl Chloride) Pipe by Acetone Immersion
D 1599 – 69	Short-Time Rupture Strength of Plastic Pipe, Tubing, and Fittings
D 1598 – 67	Time-to-Failure of Plastic Pipe Under Long-Term Hydrostatic Pressure

Recommended Practices for:

D 2153 – 67	Calculating Stress in Plastic Pipe Under Internal Pressure
D 2657 – 67	Heat Joining of Thermoplastic Pipe and Fittings
D 2855 – 70	Making Solvent Cemented Joints with Poly(Vinyl Chloride) (PVC) Pipe and Fittings
D 2321 – 67	Underground Installation of Flexible Thermoplastic Sewer Pipe
D 2774 – 69 T	Underground Installation of Thermoplastic Pressure Piping

Definitions of Terms Relating to:

D 2749 – 68	Plastic Pipe Fittings

Reinforced Plastics and Pipe

Specifications for:

D 709 – 67	*Laminated Thermosetting Materials (see Part 29)*
D 1532 – 68	*Polyester Glass-Mat Sheet Laminate (see Part 29)*
D 2853 – 70	Reinforced Olefin Polymers for Injection Molding or Extrusion
D 2848 – 69	Reinforced Polycarbonate Injection Molding and Extrusion Materials
D 2517 – 67	Reinforced Thermosetting Plastic Gas Pressure Pipe and Fittings
D 2310 – 64 T	Reinforced Thermosetting Plastic Pipe
D 1694 – 67	Threads for Reinforced Thermosetting Plastic Pipe
D 2660 – 70 T	Woven Glass Fabric, Cleaned and After-Finished with Acrylic-Silane Type Finishes, for Plastic Laminates
D 2408 – 67	Woven Glass Fabric, Cleaned and After-Finished with Amino-Silane Type Finishes, for Plastic Laminates
D 2410 – 67	Woven Glass Fabric, Cleaned and After-Finished with Chrome Complexes, for Plastic Laminates
D 2409 – 67	Woven Glass Fabric, Cleaned and After-Finished with Vinyl-Silane Type Finishes, for Plastic Laminates
D 2150 – 63 T	Woven Roving Glass Fabric for Polyester Glass Laminates

Methods of Test for:

D 2587 – 68	Acetone Extraction and Ignition of Strands, Yarns, and Roving for Reinforced Plastics
D 2344 – 67	Apparent Horizontal Shear Strength of Reinforced Plastics by Short-Beam Method
D 1602 – 60 (1969)	Bearing Load of Corrugated Plastics Panels
C 581 – 68	*Chemical Resistance of Thermosetting Resins Used in Glass Fiber Reinforced Structures (see Part 12)*
D 2143 – 69	Cyclic Pressure Strength of Reinforced, Thermosetting Plastic Pipe
D 1494 – 60 (1969)	Diffuse Light Transmission Factor of Reinforced Plastics Panels
D 2586 – 68	Hydrostatic Compressive Strength of Glass-Reinforced Plastic Cylinders
D 2584 – 68	Ignition Loss of Cured Reinforced Resins
D 2733 – 68 T	Interlaminar Shear Strength of Structural Reinforced Plastics at Elevated Temperatures
D 2105 – 67	Longitudinal Tensile Properties of Reinforced Thermosetting Plastic Pipe and Tube
D 2585 – 68	Preparation and Tension Testing of Filament-Wound Pressure Vessels
D 229 – 68	*Sheet and Plate Materials, Rigid, Used for Electrical Insulation (see Part 27)*
D 2290 – 69	Tensile Strength, Apparent, of Ring or Tubular Plastics by Split Disk Method
D 1502 – 60 (1969)	Transverse Load of Corrugated Reinforced Plastic Panels
D 2343 – 67	Tensile Properties of Glass Fiber Strands, Yarns, and Rovings Used in Reinforced Plastics
D 2734 – 68 T	Void Content of Reinforced Plastics

Recommended Practices for:

D 2562 – 70	Classifying Visual Defects in Compression and Transfer Molded Parts
D 2563 – 70	Classifying Visual Defects in Glass-Reinforced Laminates and Parts Made Therefrom
D 2291 – 67	Fabrication of Ring Test Specimens for Reinforced Plastics

Film and Sheeting

Specifications for:

*D 1463 – 63	Biaxially Oriented Styrene Plastics Sheet
D 702 – 68	Cast Methacrylate Plastic Sheets, Rods, Tubes, and Shapes
D 2411 – 69	Cellulose Acetate Butyrate Film and Sheeting for Fabricating and Forming
*D 786 – 49 (1965)	Cellulose Acetate Plastic Sheets
*D 701 – 49 (1961)	Cellulose Nitrate (Pyroxylin) Plastic Sheets, Rods, and Tubes
D 1547 – 61 (1969)	Extruded Acrylic Plastic Sheet
D 2673 – 69	Oriented Polypropylene Film
D 1202 – 68	*Plasticized Cellulose Acetate Sheet and Film for Primary Insulation (see Part 29)*
D 2103 – 67	Polyethylene Film and Sheeting
D 1927 – 67	Poly(Vinyl Chloride) Plastic Sheet, Rigid
D 2123 – 67	Poly(Vinyl Chloride-Vinyl Acetate) Plastic Sheet, Rigid
D 2530 – 69	Propylene, Nonoriented, Plastic Film
D 1593 – 61 (1969)	Vinyl Chloride Plastic Sheeting, Nonrigid

Methods of Test for:

D 2141 – 68	Adhesion Ratio of Polyethylene Film
D 1790 – 62	*Brittleness Temperature of Plastic Film by Impact (see Part 27)*
**D 1894 – 63*	*Coefficients of Friction of Plastic Film (see Part 27)*
D 2765 – 68	Degree of Crosslinking in Crosslinked Ethylene Plastics as Determined by Solvent Extraction
**D 1433 – 58 (1966)*	*Flammability of Flexible Thin Plastic Sheeting (see Part 27)*
D 568 – 68	*Flammability of Flexible Plastics (see Part 27)*
**D 635 – 68*	*Flammability of Self-Supporting Plastics (see Part 27)*
D 1692 – 68	*Flammability of Plastics Sheeting and Cellular Plastics (see Part 27)*
*D 1604 – 63 (1969)	Flatness of Plastics Sheet or Tubing, Measuring
D 1434 – 66	*Gas Transmission Rate of Plastic Film and Sheeting (see Part 27)*
D 1709 – 67	Impact Resistance of Polyethylene Film by the Free Falling Dart Method
D 1504 – 61	Orientation Release Stress of Plastic Sheeting
D 1922 – 67	*Propagation Tear Resistance of Plastic Film and Thin Sheeting (see Part 27)*
D 1239 – 55 (1966)	*Resistance of Plastic Films to Extraction by Chemicals (see Part 27)*
D 2582 – 67	*Resistance to Puncture-Propagation of Tear in Thin Plastic Sheeting (see Part 27)*
D 2838 – 69	Shrink Tension and Orientation Release Stress of Plastic Film and Thin Sheeting
D 2457 – 65 T	Specular Gloss of Plastic Fibers (see Part 27)

Methods of Test for:

D 637 – 50 (1965) *Surface Irregularities of Flat Transparent Plastic Sheets* (*see Part 27*)
D 1938 – 67 *Tear Propagation, Resistance to, in Plastic Film and Thin Sheeting by a Single-Tear Method* (*see Part 27*)
D 1004 – 66 *Tear Resistance of Plastic Film and Sheeting* (*see Part 27*)
D 1637 – 61 *Tensile Heat Distortion Temperature of Plastic Sheeting* (*see Part 27*)
D 882 – 67 *Tensile Properties of Thin Plastic Sheeting* (*see Part 27*)
D 1923 – 67 *Tensile Properties of Thin Plastic Sheeting Using the Inclined Plane Testing Machine* (*see Part 27*)
E 252 – 67 *Thickness of Thin Foil and Film by Weighing* (*see Part 27*)
D 1746 – 70 *Transparency of Plastic Sheeting* (*see Part 27*)
D 2732 – 68 T Unrestrained Linear Shrinkage of Plastic Film and Sheeting of 0.76 mm (0.030 in.) in Thickness or Less
D 1181 – 56 (1966) *Warpage of Sheet Plastics* (*see Part 27*)
*D 1789 – 65 Welding Performance for Poly(Vinyl Chloride) Structures, Evaluation of
D 2578 – 67 Wetting Tension of Polyethylene and Polypropylene Films

Cellular Plastics

Specifications for:

D 1564 – 64 T *Flexible Urethane Foam, Including Tests for* (*see Part 28*)
D 1565 – 66 *Slab Flexible Foams Made from Polymers or Copolymers of Vinyl Chloride, Including Tests for* (*see Part 28*)
D 1786 – 66 Toluenediisocyanate and Poly(Oxypropylene Glycol)
D 2341 – 65 T Urethane Foam, Rigid

Methods of Test for:

*D 1621 – 64 Compressive Strength of Rigid Cellular Plastics
*D 1622 – 63 Density, Apparent, of Rigid Cellular Plastics
*D 1673 – 61 (1967) Dielectric Constant and Dissipation Factor of Expanded Cellular Plastics Used for Electrical Insulation
D 2735 – 68 T Effect of Cyclic Immersion of Syntactic Foam at Pressure
D 1692 – 68 *Flammability of Plastic Sheeting and Cellular Plastics* (*see Part 27*)
D 2736 – 68 T Hydrostatic Compressive Strength of Syntactic Foam
D 2856 – 70 Measuring the Open Cell Content of Rigid Cellular Plastics by the Air Pycnometer
D 2237 – 64 T Rate-of-Rise (Volume Increase) Properties of Urethane Foaming Systems
D 2126 – 66 Resistance of Rigid Cellular Plastics to Simulated Service Conditions
*D 1623 – 64 Tensile Properties of Rigid Cellular Plastics
D 2849 – 69 Testing Urethane Foam Polyol Raw Materials
D 2326 – 70 Thermal Conductivity of Cellular Plastics by Means of a Probe
D 1638 – 70 Urethane Foam Raw Materials
D 2362 – 65 T Volatile Content, Total, of Expandable Polystyrene Materials
D 2842 – 69 Water Absorption of Rigid Cellular Plastics

Syntactic Foams

Methods of Test for:

D 2840 – 69 Average True Particle Density of Hollow Microspheres
D 2841 – 69 Sampling for Hollow Microspheres

Compounding Ingredients, Chemicals, and Reinforcements

Specifications for:

D 1249 – 69 Primary Octyl Phthalate Ester Plasticizers

Methods of Test for:

D 1045 – 58 (1967) *Sampling and Testing Plasticizers Used in Plastics* (*see Part 27*)
D 1786 – 66 Toluenediisocyanate and Poly(Oxypropylene Glycol)
D 1638 – 70 Urethane Foam Raw Materials

METRIC PRACTICE GUIDE

Standard:

E 380 – 70 *Metric Practice Guide* (*Excerpts*) (*see Related Materials section*)

Designation: E 380 – 70

Excerpts from Standard METRIC PRACTICE GUIDE*

(A Guide to the Use of SI—the International System of Units)

This Standard is issued under the fixed designation E 380; the number immediately following the designation indicates the year of original adoption or, in the case of revision, the year of last revision. A number in parentheses indicates the year of last reapproval.

Following are excerpts from the Metric Practice Guide, which is available as a separate publication. Also included are tables and conversion factors considered to be of greatest general interest.

FOREWORD

To continue to serve the best interests of the public through science and industry the American Society for Testing and Materials is actively cooperating with other standardization organizations in the development of simpler and more universal metrology practices. In some industries both here and abroad, the U.S. customary units are gradually being replaced by those of a modernized metric system known as the International System of Units (SI). Recognizing this trend, the Society considers it important to prepare for broader use of SI units through coexistence of these two major systems. This policy involves no change in standard dimensions, tolerances, or performance specifications.

1. Scope

1.1 This Metric Practice Guide deals with the conversion, from one system of units to another, of quantities that are in general use, and includes the units most frequently used in the various fields of science and industry. The conversion factors given are from U.S. customary units to those of the International System of Units which is officially abbreviated as SI in all languages.

1.2 The recommendations contained herein are based upon the following premises, which are believed to represent the broadest base for general agreement among proponents of the major metrology systems:

1.2.1 That for most scientific and technical work the International System of Units is generally superior to other systems; and that this system is more widely accepted than any other as the common language in which scientific and technical data should be expressed. This is particularly true for the fields of electrical science and technology since the common electrical units (ampere, volt, ohm, etc.) are SI units.

1.2.2 That various U.S. customary units, particularly the inch and the pound, will continue to be used for some time.

1.2.3 That unit usage can and should be simplified; that one means toward such simplification is the elimination of obsolete and unneeded units; and that another is a better understanding of the rational links between SI units and units of other systems.

* This guide is under the jurisdiction of the ASTM Committee on Metric Practice. A list of members may be found in the ASTM Yearbook.

Current edition accepted March 19, 1970. Originally issued 1964 without designation as *ASTM Metric Practice Guide*, revised 1966. Adopted as standard 1968. Replaces E 380 – 68.

4. Rules for Conversion and Rounding

4.1 *General:*

4.1.1 Unit conversions should be handled with careful regard to the implied correspondence between the accuracy of the data and the given number of digits.

4.1.2 In all conversions, the number of significant digits retained should be such that accuracy is neither sacrificed nor exaggerated.

4.1.3 The most accurate equivalents are obtained by multiplying the specified quantity by the conversion factor exactly as given in Appendix A3 and then rounding in accordance with 4.2 to the appropriate number of significant digits (see 4.3). For example, to convert 11.4 ft to meters: 11.4 × 0.3048 = 3.474 72, which rounds to 3.47 m (see 4.3.6).

4.1.4 There is less assurance of accuracy when the equivalent is obtained by first rounding the conversion factor to the same number of significant digits as in the specified quantity, performing the multiplication, and then rounding the product in accordance with 4.3. This procedure may provide approximate values but it is not recommended for tolerances or limiting values. For example, again converting 11.4 ft to meters: 11.4 × 0.305 = 3.4770 which rounds to 3.48 m (see 4.3.6). This compares with the value of 3.47 m obtained by the more exact method of 4.1.3.

4.1.5 Unless greater accuracy is justifiable, the equivalents should be rounded in accordance with the applicable values in Table 1.

4.1.6 Converted values should be rounded to the minimum number of significant digits that will maintain the required accuracy, as discussed above. In certain cases slight deviation from this practice to make use of convenient or whole numbers may be feasible, in which case the word "approximate" must be used following the conversion. For example,

1 ⁷/₈ in.	= 47.625 mm	exact
	= 47.6 mm	normal rounding
	= 48 mm (approx)	rounded to whole number
	= 47.5 mm (approx)	rounded to preferred number

Other examples:

Determine the breaking load of a specimen 20 in. (approx 500 mm) long.

Take individual specimens about 1 yd (1 m) long.

Cut a speciman approximately 4 by 6 in. (100 by 150 mm).

4.1.6.1 A quantity stated as "not more than" is not "nominal." For example, a specimen *at least* 4 in. wide requires a width of at least 101.6 mm.

4.1.7 The practical aspects of measuring must be considered when using SI equivalents. If a scale divided into ¹/₁₆ths of an inch is suitable for making the original measurements, a metric scale having divisions of 1 mm is obviously suitable for measuring the length and width in SI units, and the equivalents should not be reported closer than the nearest millimeter. Similarly, a gage or caliper graduated in divisions of 0.02 mm is comparable to one graduated in divisions of 0.001 in., and equivalents should be similarly indicated. Analogous situations exist in mass, force, and other measurements.

TABLE 6 Pressure and Stress Equivalents—Pounds-Force per Square Inch and Thousand Pounds-Force per Square Inch to Meganewtons per Square Meter

NOTE 1—This table may be used to obtain SI equivalents of values expressed in psi or ksi. SI values are usually expressed in kN/m² when original value is in psi and in MN/m² when original value is in ksi.

NOTE 2—This table may be extended to values below 1 or above 100 psi (ksi) by manipulation of the decimal point and addition as illustrated in 4.4.1.4.

Conversion Relationships: 1 in. = 0.0254 m (exactly)
1 lbf = 4.448 221 615 260 5 N (exactly)

psi ksi	0	1	2	3	4	5	6	7	8	9
					kN/m² MN/m²					
0	0.0000	6.8948	13.7895	20.6843	27.5790	34.4738	41.3685	48.2633	55.1581	62.0528
10	68.9476	75.8423	82.7371	89.6318	96.5266	103.4214	110.3161	117.2109	124.1056	131.0004
20	137.8951	144.7899	151.6847	158.5794	165.4742	172.3689	179.2637	186.1584	193.0532	199.9480
30	206.8427	213.7375	220.6322	227.5270	234.4217	241.3165	248.2113	255.1060	262.0008	268.8955
40	275.7903	282.6850	289.5798	296.4746	303.3693	310.2641	317.1588	324.0536	330.9483	337.8431
50	344.7379	351.6326	358.5274	365.4221	372.3169	379.2116	386.1064	393.0012	399.8959	406.7907
60	413.6854	420.5802	427.4749	434.3697	441.2645	448.1592	455.0540	461.9487	468.8435	475.7382
70	482.6330	489.5278	496.4225	503.3173	510.2120	517.1068	524.0015	530.8963	537.7911	544.6858
80	551.5806	558.4753	565.3701	572.2648	579.1596	586.0544	592.9491	599.8439	606.7386	613.6334
90	620.5281	627.4229	634.3177	641.2124	648.1072	655.0019	661.8967	668.7914	675.6862	682.5810
100	689.4757									

TABLE 7 Pressure and Stress Equivalents Metric Engineering to SI Units

NOTE 1—This table may be used for obtaining SI equivalents of quantities expressed in kgf/cm² by multiplying the given values by 10^{-2}, that is, by moving the decimal point two places to the left.

NOTE 2 This table may be extended to values below 1 or above 100 kgf/cm² by manipulation of the decimal point and addition as illustrated in 4.4.1.4.

Conversion Relationships: 1 mm = 0.001 m (exactly)
1 kgf = 9.80665 N (exactly)

kgf/mm²	0	1	2	3	4	5	6	7	8	9
	MN/m²									
0		9.8066	19.6133	29.4200	39.2266	49.0332	58.8399	68.6466	78.4532	88.2598
10	98.0665	107.8731	117.6798	127.4864	137.2931	147.0998	156.9064	166.7130	176.5197	186.3264
20	196.1330	205.9396	215.7463	225.5530	235.3596	245.1662	254.9729	264.7796	274.5862	284.3928
30	294.1995	304.0062	313.8128	323.6194	333.4261	343.2328	353.0394	362.8460	372.6527	382.4594
40	392.2660	402.0726	411.8793	421.6860	431.4926	441.2992	451.1059	460.9126	470.7192	480.5258
50	490.3325	500.1392	509.9458	519.7524	529.5591	539.3658	549.1724	558.9790	568.7857	578.5924
60	588.3990	598.2056	608.0123	617.8190	627.6256	637.4322	647.2389	657.0456	666.8522	676.6588
70	686.4655	696.2722	706.0788	715.8854	725.6921	735.4988	745.3054	755.1120	764.9187	774.7254
80	784.5320	794.3386	804.1453	813.9520	823.7586	833.5652	843.3719	853.1786	862.9852	872.7918
90	882.5985	892.4052	902.2118	912.0184	921.8251	931.6318	941.4384	951.2450	961.0517	970.8584
100	980.6650									

TABLE A

LENGTH—INCHES AND MILLIMETERS—EQUIVALENTS OF DECIMAL AND BINARY FRACTIONS OF AN INCH IN MILLIMETERS

From 1/64 to 1 Inch

½'s	¼'s	8ths	16ths	32ds	64ths	Milli-meters	Decimals of an inch	Inch	½'s	¼'s	8ths	16ths	32ds	64ths	Milli-meters	Decimals of an inch
					1	= 0.397	0.015625							33	=13.097	0.515625
				1	2	= .794	.03125						17	34	=13.494	.53125
					3	= 1.191	.046875							35	=13.891	.546875
			1	2	4	= 1.588	.0625					9	18	36	=14.288	.5625
					5	= 1.984	.078125							37	=14.684	.578125
				3	6	= 2.381	.09375						19	38	=15.081	.59375
					7	= 2.778	.109375							39	=15.478	.609375
		1	2	4	8	= 3.175	.1250				5	10	20	40	=15.875	.62500
					9	= 3.572	.140625							41	=16.272	.640625
				5	10	= 3.969	.15625						21	42	=16.669	.65625
					11	= 4.366	.171875							43	=17.066	.671875
			3	6	12	= 4.762	.1875					11	22	44	=17.462	.6875
					13	= 5.159	.203125							45	=17.859	.703125
				7	14	= 5.556	.21875						23	46	=18.256	.71875
					15	= 5.953	.234375							47	=18.653	.734375
	1	2	4	8	16	= 6.350	.2500			3	6	12	24	48	=19.050	.75
					17	= 6.747	.265625							49	=19.447	.765625
				9	18	= 7.144	.28125						25	50	=19.844	.78125
					19	= 7.541	.296875							51	=20.241	.796875
			5	10	20	= 7.938	.3125					13	26	52	=20.638	.8125
					21	= 8.334	.328125							53	=21.034	.828125
				11	22	= 8.731	.34375						27	54	=21.431	.84375
					23	= 9.128	.359375							55	=21.828	.859375
		3	6	12	24	= 9.525	.3750				7	14	28	56	=22.225	.8750
					25	= 9.922	.390625							57	=22.622	.890625
				13	26	=10.319	.40625						29	58	=23.019	.90625
					27	=10.716	.421875							59	=23.416	.921875
			7	14	28	=11.112	.4375					15	30	60	=23.812	.9375
					29	=11.509	.453125							61	=24.209	.953125
				15	30	=11.906	.46875						31	62	=24.606	.96875
					31	=12.303	.484375							63	=25.003	.984375
1	2	4	8	16	32	=12.700	.5000	1	2	4	8	16	32	64	=25.400	1.0000

TABLE B

LENGTH—INCHES TO MILLIMETERS

[Reduction Factor: 1 inch = 25.4 millimeters]

Inches	Millimeters	Inches	Millimeters	Inches	Millimeters	Inches	Millimeters	Inches	Millimeters	Inches	Millimeters	Inches	Millimeters	Inches	Millimeters	Inches	Millimeters	Inches	Millimeters
0.00		1.00	25.400	2.00	50.800	3.00	76.200	4.00	101.600	5.00	127.000	6.00	152.400	7.00	177.800	8.00	203.200	9.00	228.600
.01	0.254	1.01	25.654	2.01	51.054	3.01	76.454	4.01	101.854	5.01	127.254	6.01	152.654	7.01	178.054	8.01	203.454	9.01	228.854
.02	.508	1.02	25.908	2.02	51.308	3.02	76.708	4.02	102.108	5.02	127.508	6.02	152.908	7.02	178.308	8.02	203.708	9.02	229.108
.03	.762	1.03	26.162	2.03	51.562	3.03	76.962	4.03	102.362	5.03	127.762	6.03	153.162	7.03	178.562	8.03	203.962	9.03	229.362
.04	1.016	1.04	26.416	2.04	51.816	3.04	77.216	4.04	102.616	5.04	128.016	6.04	153.416	7.04	178.816	8.04	204.216	9.04	229.616
.05	1.270	1.05	26.670	2.05	52.070	3.05	77.470	4.05	102.870	5.05	128.270	6.05	153.670	7.05	179.070	8.05	204.470	9.05	229.870
.06	1.524	1.06	26.924	2.06	52.324	3.06	77.724	4.06	103.124	5.06	128.524	6.06	153.924	7.06	179.324	8.06	204.724	9.06	230.124
.07	1.778	1.07	27.178	2.07	52.578	3.07	77.978	4.07	103.378	5.07	128.778	6.07	154.178	7.07	179.578	8.07	204.978	9.07	230.378
.08	2.032	1.08	27.432	2.08	52.832	3.08	78.232	4.08	103.632	5.08	129.032	6.08	154.432	7.08	179.832	8.08	205.232	9.08	230.632
.09	2.286	1.09	27.686	2.09	53.086	3.09	78.486	4.09	103.886	5.09	129.286	6.09	154.686	7.09	180.086	8.09	205.486	9.09	230.886
0.10	2.540	1.10	27.940	2.10	53.340	3.10	78.740	4.10	104.140	5.10	129.540	6.10	154.940	7.10	180.340	8.10	205.740	9.10	231.140
.11	2.794	1.11	28.194	2.11	53.594	3.11	78.994	4.11	104.394	5.11	129.794	6.11	155.194	7.11	180.594	8.11	205.994	9.11	231.394
.12	3.048	1.12	28.448	2.12	53.848	3.12	79.248	4.12	104.648	5.12	130.048	6.12	155.448	7.12	180.848	8.12	206.248	9.12	231.648
.13	3.302	1.13	28.702	2.13	54.102	3.13	79.502	4.13	104.902	5.13	130.302	6.13	155.702	7.13	181.102	8.13	206.502	9.13	231.902
.14	3.556	1.14	28.956	2.14	54.356	3.14	79.756	4.14	105.156	5.14	130.556	6.14	155.956	7.14	181.356	8.14	206.756	9.14	232.156
.15	3.810	1.15	29.210	2.15	54.610	3.15	80.010	4.15	105.410	5.15	130.810	6.15	156.210	7.15	181.610	8.15	207.010	9.15	232.410
.16	4.064	1.16	29.464	2.16	54.864	3.16	80.264	4.16	105.664	5.16	131.064	6.16	156.464	7.16	181.864	8.16	207.264	9.16	232.664
.17	4.318	1.17	29.718	2.17	55.118	3.17	80.518	4.17	105.918	5.17	131.318	6.17	156.718	7.17	182.118	8.17	207.518	9.17	232.918
.18	4.572	1.18	29.972	2.18	55.372	3.18	80.772	4.18	106.172	5.18	131.572	6.18	156.972	7.18	182.372	8.18	207.772	9.18	233.172
.19	4.826	1.19	30.226	2.19	55.626	3.19	81.026	4.19	106.426	5.19	131.826	6.19	157.226	7.19	182.626	8.19	208.026	9.19	233.426
0.20	5.080	1.20	30.480	2.20	55.880	3.20	81.280	4.20	106.680	5.20	132.080	6.20	157.480	7.20	182.880	8.20	208.280	9.20	233.680
.21	5.334	1.21	30.734	2.21	56.134	3.21	81.534	4.21	106.934	5.21	132.334	6.21	157.734	7.21	183.134	8.21	208.534	9.21	233.934
.22	5.588	1.22	30.988	2.22	56.388	3.22	81.788	4.22	107.188	5.22	132.588	6.22	157.988	7.22	183.388	8.22	208.788	9.22	234.188
.23	5.842	1.23	31.242	2.23	56.642	3.23	82.042	4.23	107.442	5.23	132.842	6.23	158.242	7.23	183.642	8.23	209.042	9.23	234.442
.24	6.096	1.24	31.496	2.24	56.896	3.24	82.296	4.24	107.696	5.24	133.096	6.24	158.496	7.24	183.896	8.24	209.296	9.24	234.696
.25	6.350	1.25	31.750	2.25	57.150	3.25	82.550	4.25	107.950	5.25	133.350	6.25	158.750	7.25	184.150	8.25	209.550	9.25	234.950
.26	6.604	1.26	32.004	2.26	57.404	3.26	82.804	4.26	108.204	5.26	133.604	6.26	159.004	7.26	184.404	8.26	209.804	9.26	235.204
.27	6.858	1.27	32.258	2.27	57.658	3.27	83.058	4.27	108.458	5.27	133.858	6.27	159.258	7.27	184.658	8.27	210.058	9.27	235.458
.28	7.112	1.28	32.512	2.28	57.912	3.28	83.312	4.28	108.712	5.28	134.112	6.28	159.512	7.28	184.912	8.28	210.312	9.28	235.712
.29	7.366	1.29	32.766	2.29	58.166	3.29	83.566	4.29	108.966	5.29	134.366	6.29	159.766	7.29	185.166	8.29	210.566	9.29	235.966
0.30	7.620	1.30	33.020	2.30	58.420	3.30	83.820	4.30	109.220	5.30	134.620	6.30	160.020	7.30	185.420	8.30	210.820	9.30	236.220
.31	7.874	1.31	33.274	2.31	58.674	3.31	84.074	4.31	109.474	5.31	134.874	6.31	160.274	7.31	185.674	8.31	211.074	9.31	236.474
.32	8.128	1.32	33.528	2.32	58.928	3.32	84.328	4.32	109.728	5.32	135.128	6.32	160.528	7.32	185.928	8.32	211.328	9.32	236.728
.33	8.382	1.33	33.782	2.33	59.182	3.33	84.582	4.33	109.982	5.33	135.382	6.33	160.782	7.33	186.182	8.33	211.582	9.33	236.982
.34	8.636	1.34	34.036	2.34	59.436	3.34	84.836	4.34	110.236	5.34	135.636	6.34	161.036	7.34	186.436	8.34	211.836	9.34	237.236
.35	8.890	1.35	34.290	2.35	59.690	3.35	85.090	4.35	110.490	5.35	135.890	6.35	161.290	7.35	186.690	8.35	212.090	9.35	237.490
.36	9.144	1.36	34.544	2.36	59.944	3.36	85.344	4.36	110.744	5.36	136.144	6.36	161.544	7.36	186.944	8.36	212.344	9.36	237.744
.37	9.398	1.37	34.798	2.37	60.198	3.37	85.598	4.37	110.998	5.37	136.398	6.37	161.798	7.37	187.198	8.37	212.598	9.37	237.998
.38	9.652	1.38	35.052	2.38	60.452	3.38	85.852	4.38	111.252	5.38	136.652	6.38	162.052	7.38	187.452	8.38	212.852	9.38	238.252
.39	9.906	1.39	35.306	2.39	60.706	3.39	86.106	4.39	111.506	5.39	136.906	6.39	162.306	7.39	187.706	8.39	213.106	9.39	238.506
0.40	10.160	1.40	35.560	2.40	60.960	3.40	86.360	4.40	111.760	5.40	137.160	6.40	162.560	7.40	187.960	8.40	213.360	9.40	238.760
.41	10.414	1.41	35.814	2.41	61.214	3.41	86.614	4.41	112.014	5.41	137.414	6.41	162.814	7.41	188.214	8.41	213.614	9.41	239.014
.42	10.668	1.42	36.068	2.42	61.468	3.42	86.868	4.42	112.268	5.42	137.668	6.42	163.068	7.42	188.468	8.42	213.868	9.42	239.268
.43	10.922	1.43	36.322	2.43	61.722	3.43	87.122	4.43	112.522	5.43	137.922	6.43	163.322	7.43	188.722	8.43	214.122	9.43	239.522
.44	11.176	1.44	36.576	2.44	61.976	3.44	87.376	4.44	112.776	5.44	138.176	6.44	163.576	7.44	188.976	8.44	214.376	9.44	239.776
.45	11.430	1.45	36.830	2.45	62.230	3.45	87.630	4.45	113.030	5.45	138.430	6.45	163.830	7.45	189.230	8.45	214.630	9.45	240.030
.46	11.684	1.46	37.084	2.46	62.484	3.46	87.884	4.46	113.284	5.46	138.684	6.46	164.084	7.46	189.484	8.46	214.884	9.46	240.284
.47	11.938	1.47	37.338	2.47	62.738	3.47	88.138	4.47	113.538	5.47	138.938	6.47	164.338	7.47	189.738	8.47	215.138	9.47	240.538
.48	12.192	1.48	37.592	2.48	62.992	3.48	88.392	4.48	113.792	5.48	139.192	6.48	164.592	7.48	189.992	8.48	215.392	9.48	240.792
.49	12.446	1.49	37.846	2.49	63.246	3.49	88.646	4.49	114.046	5.49	139.446	6.49	164.846	7.49	190.246	8.49	215.646	9.49	241.046

0.50	12.700	1.50	38.100	2.50	63.500	3.50	88.900	4.50	114.300	5.50	139.700	6.50	165.100	7.50	190.500	8.50	215.900	9.50	241.300
.51	12.954	1.51	38.354	2.51	63.754	3.51	89.154	4.51	114.554	5.51	139.954	6.51	165.354	7.51	190.754	8.51	216.154	9.51	241.554
.52	13.208	1.52	38.608	2.52	64.008	3.52	89.408	4.52	114.808	5.52	140.208	6.52	165.608	7.52	191.008	8.52	216.408	9.52	241.808
.53	13.462	1.53	38.862	2.53	64.262	3.53	89.662	4.53	115.062	5.53	140.462	6.53	165.862	7.53	191.262	8.53	216.662	9.53	242.062
.54	13.716	1.54	39.116	2.54	64.516	3.54	89.916	4.54	115.316	5.54	140.716	6.54	166.116	7.54	191.516	8.54	216.916	9.54	242.316
.55	13.970	1.55	39.370	2.55	64.770	3.55	90.170	4.55	115.570	5.55	140.970	6.55	166.370	7.55	191.770	8.55	217.170	9.55	242.570
.56	14.224	1.56	39.624	2.56	65.024	3.56	90.424	4.56	115.824	5.56	141.224	6.56	166.624	7.56	192.024	8.56	217.424	9.56	242.824
.57	14.478	1.57	39.878	2.57	65.278	3.57	90.678	4.57	116.078	5.57	141.478	6.57	166.878	7.57	192.278	8.57	217.678	9.57	243.078
.58	14.732	1.58	40.132	2.58	65.532	3.58	90.932	4.58	116.332	5.58	141.732	6.58	167.132	7.58	192.532	8.58	217.932	9.58	243.332
.59	14.986	1.59	40.386	2.59	65.786	3.59	91.186	4.59	116.586	5.59	141.986	6.59	167.386	7.59	192.786	8.59	218.186	9.59	243.586
0.60	15.240	1.60	40.640	2.60	66.040	3.60	91.440	4.60	116.840	5.60	142.240	6.60	167.640	7.60	193.040	8.60	218.440	9.60	243.840
.61	15.494	1.61	40.894	2.61	66.294	3.61	91.694	4.61	117.094	5.61	142.494	6.61	167.894	7.61	193.294	8.61	218.694	9.61	244.094
.62	15.748	1.62	41.148	2.62	66.548	3.62	91.948	4.62	117.348	5.62	142.748	6.62	168.148	7.62	193.548	8.62	218.948	9.62	244.348
.63	16.002	1.63	41.402	2.63	66.802	3.63	92.202	4.63	117.602	5.63	143.002	6.63	168.402	7.63	193.302	8.63	219.202	9.63	244.602
.64	16.256	1.64	41.656	2.64	67.056	3.64	92.456	4.64	117.856	5.64	143.256	6.64	168.656	7.64	194.056	8.64	219.456	9.64	244.856
.65	16.510	1.65	41.910	2.65	67.310	3.65	92.710	4.65	118.110	5.65	143.510	6.65	168.910	7.65	194.310	8.65	219.710	9.65	245.110
.66	16.764	1.66	42.164	2.66	67.564	3.66	92.964	4.66	118.364	5.66	143.764	6.66	169.164	7.66	194.564	8.66	219.964	9.66	245.364
.67	17.018	1.67	42.418	2.67	67.818	3.67	93.218	4.67	118.618	5.67	144.018	6.67	169.418	7.67	194.818	8.67	220.218	9.67	245.618
.68	17.272	1.68	42.672	2.68	68.072	3.68	93.472	4.68	118.872	5.68	144.272	6.68	169.672	7.68	195.072	8.68	220.472	9.68	245.872
.69	17.526	1.69	42.926	2.69	68.326	3.69	93.726	4.69	119.126	5.69	144.526	6.69	169.926	7.69	195.326	8.69	220.726	9.69	246.126
0.70	17.730	1.70	43.180	2.70	68.580	3.70	93.980	4.70	119.380	5.70	144.780	6.70	170.180	7.70	195.580	8.70	220.980	9.70	246.380
.71	18.034	1.71	43.434	2.71	68.834	3.71	94.234	4.71	119.634	5.71	145.034	6.71	170.434	7.71	195.834	8.71	221.234	9.71	246.634
.72	18.288	1.72	43.688	2.72	69.088	3.72	94.488	4.72	119.888	5.72	145.288	6.72	170.688	7.72	196.088	8.72	221.488	9.72	246.888
.73	18.542	1.73	43.942	2.73	69.342	3.73	94.742	4.73	120.142	5.73	145.542	6.73	170.942	7.73	196.342	8.73	221.742	9.73	247.142
.74	18.796	1.74	44.196	2.74	69.596	3.74	94.996	4.74	120.396	5.74	145.796	6.74	171.196	7.74	196.596	8.74	221.996	9.74	247.396
.75	19.050	1.75	44.450	2.75	69.850	3.75	95.250	4.75	120.650	5.75	146.050	6.75	171.450	7.75	196.850	8.75	222.250	9.75	247.650
.76	19.304	1.76	44.704	2.76	70.104	3.76	95.504	4.76	120.904	5.76	146.304	6.76	171.704	7.76	197.104	8.76	222.504	9.76	247.904
.77	19.558	1.77	44.958	2.77	70.358	3.77	95.758	4.77	121.158	5.77	146.558	6.77	171.958	7.77	197.358	8.77	222.758	9.77	248.158
.78	19.812	1.78	45.212	2.78	70.612	3.78	96.012	4.78	121.412	5.78	146.812	6.78	172.212	7.78	197.612	8.78	223.012	9.78	248.412
.79	20.066	1.79	45.466	2.79	70.866	3.79	96.266	4.79	121.666	5.79	147.066	6.79	172.466	7.79	197.866	8.79	223.266	9.79	248.666
0.80	20.320	1.80	45.720	2.80	71.120	3.80	96.520	4.80	121.920	5.80	147.320	6.80	172.720	7.80	198.120	8.80	223.520	9.80	248.920
.81	20.574	1.81	45.974	2.81	71.374	3.81	96.774	4.81	122.174	5.81	147.574	6.81	172.974	7.81	198.374	8.81	223.774	9.81	249.174
.82	20.828	1.82	46.228	2.82	71.628	3.82	97.028	4.82	122.428	5.82	147.828	6.82	173.228	7.82	198.628	8.82	224.028	9.82	249.428
.83	21.082	1.83	46.482	2.83	71.882	3.83	97.282	4.83	122.682	5.83	148.082	6.83	173.482	7.83	198.882	8.83	224.282	9.83	249.682
.84	21.336	1.84	46.736	2.84	72.136	3.84	97.536	4.84	122.936	5.84	148.336	6.84	173.736	7.84	199.136	8.84	224.536	9.84	249.936
.85	21.590	1.85	46.990	2.85	72.390	3.85	97.790	4.85	123.190	5.85	148.590	6.85	173.990	7.85	199.390	8.85	224.790	9.85	250.190
.86	21.844	1.86	47.244	2.86	72.644	3.86	98.044	4.86	123.444	5.86	148.844	6.86	174.244	7.86	199.644	8.86	225.044	9.86	250.444
.87	22.098	1.87	47.498	2.87	72.898	3.87	98.298	4.87	123.698	5.87	149.098	6.87	174.498	7.87	199.898	8.87	225.298	9.87	250.698
.88	22.352	1.88	47.752	2.88	73.152	3.88	98.552	4.88	123.952	5.88	149.352	6.88	174.752	7.88	200.152	8.88	225.552	9.88	250.952
.89	22.606	1.89	48.006	2.89	73.406	3.89	98.806	4.89	124.206	5.89	149.606	6.89	175.006	7.89	200.406	8.89	225.806	9.89	251.206
0.90	22.860	1.90	48.260	2.90	73.660	3.90	99.060	4.90	124.460	5.90	149.860	6.90	175.260	7.90	200.660	8.90	226.060	9.90	251.460
.91	23.114	1.91	48.514	2.91	73.914	3.91	99.314	4.91	124.714	5.91	150.114	6.91	175.514	7.91	200.914	8.91	226.314	9.91	251.714
.92	23.368	1.92	48.768	2.92	74.168	3.92	99.568	4.92	124.968	5.92	150.368	6.92	175.768	7.92	201.168	8.92	226.568	9.92	251.968
.93	23.622	1.93	49.022	2.93	74.422	3.93	99.822	4.93	125.222	5.93	150.622	6.93	176.022	7.93	201.422	8.93	226.822	9.93	252.222
.94	23.876	1.94	49.276	2.94	74.676	3.94	100.076	4.94	125.476	5.94	150.876	6.94	176.276	7.94	201.676	8.94	227.076	9.94	252.476
.95	24.130	1.95	49.530	2.95	74.930	3.95	100.330	4.95	125.730	5.95	151.130	6.95	176.530	7.95	201.930	8.95	227.330	9.95	252.730
.96	24.384	1.96	49.784	2.96	75.184	3.96	100.584	4.96	125.984	5.96	151.384	6.96	176.784	7.96	202.184	8.96	227.584	9.96	252.984
.97	24.638	1.97	50.038	2.97	75.438	3.97	100.838	4.97	126.238	5.97	151.638	6.97	177.038	7.97	202.438	8.97	227.838	9.97	253.238
.98	24.892	1.98	50.292	2.98	75.692	3.98	101.092	4.98	126.492	5.98	151.892	6.98	177.292	7.98	202.692	8.98	228.092	9.98	253.492
.99	25.146	1.99	50.546	2.99	75.946	3.99	101.346	4.99	126.746	5.99	152.146	6.99	177.546	7.99	202.946	8.99	228.346	9.99	253.746

SELECTED CONVERSION FACTORS

To convert from	to	multiply by
Btu (International Table) · in./s · ft^2 · deg F (k, thermal conductivity)	watt/meter-kelvin	$5.192\ 204 \times 10^{2}$
board foot	meter^3	$2.359\ 737 \times 10^{-3}$
circular mil	meter^2	$5.067\ 075 \times 10^{-10}$
degree Fahrenheit	degree Celsius	$t_C = (t_F - 32)/1.8$
foot	meter	$3.048\ 000^* \times 10^{-1}$
foot^2	meter^2	$9.290\ 304^* \times 10^{-2}$
foot^3	meter^3	$2.831\ 685 \times 10^{-2}$
gallon (U.S. liquid)	meter^3	$3.785\ 412 \times 10^{-3}$
foot-pound	kilogram-meter	0.138 255
inch^2	meter^2	$6.451\ 600^* \times 10^{-4}$
inch^3	meter^3	$1.638\ 706 \times 10^{-5}$
ounce (U.S. fluid)	meter^3	$2.957\ 353 \times 10^{-5}$
ounce-mass (avoirdupois)	kilogram	$2.834\ 952 \times 10^{-2}$
ounce (avoirdupois)/gallon (U.S. liquid)	kilogram/meter^3	7.489 152
ounce-mass/ft^2	kilogram/meter^2	0.305 152
pound-force (lbf avoirdupois)	kilogram-force	$4.535\ 924 \times 10^{-1}$
pound-force (lbf avoirdupois)	newton	4.448 222
pound-mass (lbm avoirdupois)	kilogram	$4.535\ 924 \times 10^{-1}$
pound-force/inch^2 (psi)	kilogram-force/mm^2	$7.030\ 696 \times 10^{-4}$
pound-force/inch^2 (psi)	newton/meter^2	$6.894\ 757 \times 10^{3}$
pound-mass/inch^3	kilogram/meter^3	$2.767\ 990 \times 10^{4}$
pound-mass/foot^3	kilogram/meter^3	$1.601\ 846 \times 10$
ton (short, 2000 lbm)	kilogram	$9.071\ 847 \times 10^{2}$
yard	meter	$9.144\ 000^* \times 10^{-1}$
yard^2	meter^2	$8.361\ 274 \times 10^{-1}$
yard^3	meter^3	$7.645\ 549 \times 10^{-1}$

* Exact

ISO RECOMMENDATIONS ON PLASTICS PREPARED BY ISO/TC 61[a]

ISO Recommendations	Title	Price	Corresponding ASTM Method[d]
R59-1958	Determination of the Percentage of Acetone Soluble Matter in Phenolic Mouldings	$0.60	D 494
R60-1958	Determination of Apparent Density of Moulding Material that Can be Poured from a Specified Funnel	0.60	D 1895,[b] Method A
R61-1958	Determination of Apparent Density of Moulding Material that Cannot be Poured from a Specified Funnel	0.60	D 1895,[b] Method C
R62-1958	Determination of Water Absorption	0.60	D 570[b]
R62/A1-1965	Amendment 1 to ISO Recommendation R62-1958	(Included in R62)	
R75-1958	Determination of Temperature of Deflection Under Load	1.20	D 648[b]
R117-1959	Determination of Boiling Water Absorption	0.60	D 570,[b] Section 6(*f*)
R118-1959	Determination of Methanol-Soluble Matter in Polystyrene	0.60	. . .
R119-1959	Determination of Free Phenols in Phenol-formaldehyde Mouldings	0.60	. . .
R120-1959	Determination of Free Ammonia and Ammonium Compounds in Phenolformaldehyde Mouldings	0.60	. . .
R171-1961	Determination of Bulk Factor of Moulding Materials	0.60	D 1895[b]
R172-1961	Detection of Free Ammonia in Phenol-formaldehyde Moulding (Qualitative Method)	0.60	. . .
R173-1961	Determination of the Percentage of Styrene in Polystyrene with Wijs Solution	0.60	. . .
R174-1961	Determination of Viscosity Number of Poly(Vinyl Chloride) Resin in Solution	1.20	D 1243
R175-1961	Determination of the Resistance of Plastics to Chemical Substances	1.20	D 543,[b] Procedure I
R176-1961	Determination of the Loss of Plasticizers from Plastics by the Activated Carbon Method	1.20	D 1203
R177-1961	Determination of Migration of Plasticizers from Plastics	1.20	
R178-1961	Determination of Flexural Properties of Rigid Plastics	1.50	D 790[b]
R179-1961	Determination of the Charpy Impact Resistance of Rigid Plastics (Charpy Impact Flexural Test)	2.10	D 256
R180-1961	Determination of the Izod Impact Resistance of Rigid Plastics (Izod Impact Flexural Tests)	1.80	D 256[b]
R181-1961	Determination of Incandescence Resistance of Rigid Self-extinguishing Thermosetting Plastics	1.20	D 757[b]
R182-1961	Determination of the Thermal Stability of Poly(Vinyl Chloride) and related Copolymers and Their Compounds by the Congo Red Method	1.20	D 793[c]
R183-1961	Determination of the Bleeding of Colourants from Plastics	0.60	. . .
R194-1961	List of Equivalent Terms Used in the Plastics Industry (English, French, Russian)	9.90	. . .
R194/P1-1963	Appendix P1: Corresponding Polish Terms	1.00	. . .
R194/CS-1964	Appendix CS: Corresponding Czech Terms	1.00	. . .
R194/C-1964	Appendix D: Corresponding German Terms	1.00	. . .
R194/NL-1968	Appendix NL: Corresponding Dutch Terms	. . .	. . .
R291-1963	Standard Atmospheres for Conditioning and Testing	0.60	D 618[b]
R292-1963	Determination of the Melt Flow Index of Polyethylene and Polyethylene Compounds	1.50	D 1238[b]
R293-1963	Compression Moulding Test Specimens of Thermoplastic Materials	1.50	D 2292
R294-1963	Injection Moulding Test Specimens of Thermoplastic Materials	1.50	D 1897
R295-1963	Compression Moulding Test Specimens of Thermosetting Materials	1.20	D 796
R305-1963	Determination of the Thermal Stability of Poly(Vinyl Chloride) and Related Copolymers and Their Compounds by the Discoloration Method	1.80	. . .
R306-1963	Determination of the Vicat Softening Point of Thermoplastics	1.80	D 1525[b]
R306-1968	Determination of Vicat Softening Temperature of Thermoplastics (2nd Edition)	3.60	. . .
R307-1963	Determination of the Viscosity Number of Polyamide Resins in Dilute Solution	1.80	. . .
R308-1963	Determination of the Acetone Soluble Matter (Resin Content of Material in the Unmoulded State) of Phenolic Moulding Materials	1.20	. . .

ISO Recommendations	Title	Price	Corresponding ASTM Method[d]
R462-1965	Recommended Practice for the Determination of Change of Mechanical Properties After Contact with Chemical Substances	2.40	D 543[b]
R458-1965	Determination of Stiffness in Torsion as a Function of Temperature	3.60	D 1043[b]
R472-1966	Plastics: Definition of Terms	1.20	D 883
R472-1968	Plastics: Definition of Terms (2nd Edition)	...	D 883
R483-1966	Plastics: Methods for Maintaining Constant Relative Humidity in Small Enclosures by Means of Aqueous Solutions	3.60	E 104[b]
R489-1966	Determination of the Refractive Index of Transparent Plastics	2.40	D 542[b]
R527-1966	Plastics, Determination of Tensile Properties	4.80	D 638[b]
R537-1966	Testing of Plastics with the Torsion Pendulum	4.80	D 2236[b]
R584-1967	Determination of the Maximum Temperature and the Rate of Increase of Temperature During the Setting of Unsaturated Polyester Resins	1.20	D 1043
R585-1967	Determination of the Moisture Content of Nonplasticized Cellulose Acetate	1.20	...
R599-1967	Determination of the Percentage of Extractable Materials in Polyamides	2.40	...
R600-1967	Determination of the Viscosity Ratio of Polyamides in Concentrated Solution	3.60	D 789[b]
R604-1967	Determination of Compressive Properties of Plastics	3.60	D 695[b]
R800-1968	Basis for Specification for Phenolic Moulding Materials	2.40	...
R844-1968	Compression Test of Rigid Cellular Plastics	1.80	D 1621
R845-1968	Determination of Apparent Density of Cellular Plastics	1.80	D 1895
R846-1968	Recommended Practice for the Evaluation of the Resistance of Plastics to Fungi by Visual Examination	3.60	D 1924[c]
R868-1968	Determination of Hardness of Plastics by Means of a Durometer (Shore Hardness)	2.80	D 2240[b]
R869-1968	Preparation of Specimens for Optical Tests on Plastics Materials *Moulding Method*	2.00	...
R870-1968	Preparation of Specimens for Optical Tests on Plastics Materials *Casting Method*	2.00	...
R871-1968	Determination of Temperature Evolution of Flammable Gases from Plastics	2.40	D 1929[c]
R872-1968	Determination of Unplasticized Cellulose Acetate	2.80	...
R877-1968	Resistance of Plastics to Color Change Upon Exposure to Daylight	5.60	E 187[c]
R878-1968	Resistance of Plastics to Color Change Upon Exposure to Light of the Carbon Arc	3.60	E 42, D 1499[c]
R879-1968	Resistance of Plastics to Color Change Upon Exposure to Light of a Xenon Lamp	...	D 2565[c]
R899-1968	Determination of Tensile Creep of Plastics	...	D 674[c]
R922-1969	Determination of Soluble Matter in Crystalline Polypropylene by Boiling in *n*-Heptane	...	...
R960-1969	Determination of Water Content in Polyamides	5.60	...
R974-1969	Determination of the Brittleness Temperature by Impact	4.40	D 746
R1043-1969	Abbreviations (Symbols) for Plastics	1.60	...
R1060-1969	Designation of Poly(Vinyl Chloride) Resins	3.20	...
R1061-1969	Determination of Free Acidity of Unplasticized Cellulose Acetate	2.40	...
R1068-1969	PVC Resins: Determination of the Compacted Apparent Bulk Density	2.80	...
R1110-1969	Accelerated Conditioning of Test Specimens of Polyamide 66, 610, and 6	2.40	...
R1133-1969	Determination of the Melt Flow Rate of Thermal Plastics	3.60	D 1238
R1137-1969	Determination of the Behavior of Plastics in a Ventilated Tubular Oven	3.60	...
R1147-1969	Aqueous Dispersions of Polymer and Copolymer: Freeze-Thaw Cycle Stability Test	2.00	...
R1148-1969	Polymer and Copolymer Water Dispersion: Determination of pH	2.00	...
R1157-1970	Determination of Viscosity Number and Viscosity Ratio of Cellulose Acetate in Dilute Solution	2.80	...
R1158-1970	Determination of Chlorine in Vinyl Chloride Polymers and Copolymers	3.60	...

ISO Recommendations	Title	Price	Corresponding ASTM Method[d]
R1159-1970	Determination of Vinyl Acetate in Vinyl Chloride-Vinyl Acetate Copolymers	3.20	...
R1172-1970	Determination of the Loss on Ignition of Textile Glass Reinforced Plastics	2.80	...
R1183-1970	Method for Determining the Density and Relative Density (Specific Gravity) of Plastics, Excluding Cellular Plastics	...	...
R1184-1970	Determination of Tensile Properties of Film	...	...

[a] These publications of the International Organization for Standardization are available from the American National Standards Institute, 1430 Broadway, New York, N. Y. 10018.

[b] These methods are essentially equivalent.

[c] These methods are similar.

[d] These designations refer to the following methods of the American Society for Testing and Materials:

D 256, Tests for Impact Resistance of Plastics and Electrical Insulating Materials
D 494, Test for Acetone Extraction of Phenolic Molded or Laminated Products
D 542, Tests for Index of Refraction of Transparent Organic Plastics
D 543, Test for Resistance of Plastics to Chemical Reagents
D 570, Test for Water Absorption of Plastics
D 618, Conditioning Plastics and Electrical Insulating Materials for Testing
D 638, Test for Tensile Properties of Plastics
D 648, Test for Deflection Temperature of Plastics Under Load
D 674, Testing Long-Time Creep and Stress-Relaxation of Plastics Under Tension or Compression Loads at Various Temperatures
D 695, Test for Compressive Properties of Rigid Plastics
D 757, Test for Flammability of Plastics, Self-Extinguishing Type
D 789, Nylon Injection Molding and Extrusion Materials
D 790, Test for Flexural Properties of Plastics
D 793, Test for Short-Time Stability at Elevated Temperatures of Plastics Containing Chlorine
D 796, Molding Test Specimens of Phenolic Molding Materials
D 883, Nomenclature Relating to Plastics
D 1043, Test for Stiffness Properties of Plastics as a Function of Temperature by Means of a Torsion Test
D 1203, Tests for Loss of Plasticizer from Plastics (Activated Carbon Methods)
D 1238, Measuring Flow Rates of Thermoplastics by Extrusion Plastometer
D 1243, Test for Dilute Solution Viscosity of Vinyl Chloride Polymers
D 1499, Operating Light- and Water-Exposure Apparatus (Carbon-Arc Type) for Exposure of Plastics
D 1525, Test for Vicat Softening Point of Plastics
D 1621, Test for Compressive Strength of Rigid Cellular Plastics
D 1895, Tests for Apparent Density, Bulk Factor, and Pourability of Plastic Materials
D 1897, Injection Molding of Specimens of Thermoplastic Molding and Extrusion Materials
D 1924, Determining Resistance of Synthetic Polymeric Materials to Fungi
D 1929, Test for Ignition Properties of Plastics
D 2236, Test for Dynamic Mechanical Properties of Plastics by Means of a Torsional Pendulum
D 2240, Test for Indentation Hardness of Rubber and Plastics by Means of a Durometer
D 2292, Compression Molding of Specimens of Styrene-Butadiene Molding and Extrusion Materials
D 2565, Operating Xenon Arc-Type (Water-Cooled) Light- and Water-Exposure Apparatus for Exposure of Plastics
E 42, Operating Light- and Water-Exposure Apparatus (Carbon-Arc Type) for Exposure of Nonmetallic Materials
E 104, Maintaining Constant Relative Humidity by Means of Aqueous Solutions
E 187, Conducting Natural Light (Sunlight and Daylight) Exposure Under Glass

RELATED ASTM PUBLICATIONS

STP 460 Composite Materials: Testing and Design (1969) $31.00

STP 386 Plastics in Surgical Implants (1965) $5.00

STP 414 Resinography of Cellular Plastics (1967) $7.50

STP 375 Simulated Service Testing in the Plastic Industry (1965) $3.00

INDEX OF 1970 ANNUAL BOOK OF ASTM STANDARDS, PART 26 AND **PART 27**

STANDARDS AND TENTATIVES ON PLASTICS—SPECIFICATIONS; METHODS OF TESTING PIPE, FILM, REINFORCED AND CELLULAR PLASTICS; GENERAL METHODS OF TESTING; NOMENCLATURE

This index covers those standards and tentatives appearing in both Parts 26 and 27. The respective volume numbers are indicated in each case. With the standards arranged in numeric sequence of serial designations, the reference given is to the ASTM designation rather than the page number. Proposed methods and specifications referred to in this index will be found in the "Related Materials" section of this part. A combined Index, covering all standards and tentatives appearing in all parts of the *Annual Book of ASTM Standards,* is furnished separately.

A

G

M

APPLICATION FOR 1970 ASTM ORGANIZATIONAL MEMBERSHIP

Extracts from Charter and Bylaws and Dues Schedule on REVERSE SIDE of this application

Application is made for membership in the American Society for Testing and Materials in the class of (check one):

☐ **SUSTAINING MEMBER** ☐ **INDUSTRIAL MEMBER** ☐ **INSTITUTIONAL MEMBER**

and if elected the applicant agrees to be governed by the Charter and Bylaws of the Society, and to further its objectives as laid down therein. (This membership includes: opportunities for participation of the Official Representative of the membership in the general and District activities of the Society, and for him [or others in the member organization whom he designates] to make application for participation on its technical committees; subscription(s) to **Materials Research & Standards**; parts of the **Book of Standards**; member discounts on publications; and other benefits and privileges.)

FORMAT FOR LISTING THE ASTM MEMBERSHIP

Please complete the grid below. PRINT or TYPE. Leave a space between words. Use no punctuation marks or special characters, except the ampersand (&). Use initials and abbreviations where necessary.

Line 1:—List the PARENT ORGANIZATION.
Line 2:—If applicable, list the FACILITY which will HOLD THE MEMBERSHIP.
Line 3:—Beginning in 8th position, list LAST NAME, FIRST NAME and MIDDLE INITIAL of the OFFICIAL REPRESENTATIVE who will exercise membership privileges.
Line 4:—TITLE of the Official Representative and DEPARTMENT
Line 5:—STREET ADDRESS or P. O. BOX for the MEMBER FACILITY
Line 6:—CITY (leave 1 space) STATE (leave 2 spaces) ZIP CODE

1
2
3
4
5
6

Major Product. . . . .
(Or service performed by firm or organization)

Representative's Major Field of Interest. . . . .
Examples: chemical research/civil engineering

Specific Application of Effort. . . . .
Examples: polymer plastics/construction

ASTM MEMBERSHIP POLICY — The development of organization memberships and management understanding are vital to the Society's ability to serve the economy effectively in the future. The Society hopes that major organizations will help support its activities with one or more Sustaining Memberships. ASTM also offers inexpensive Industrial and Institutional memberships, providing substantial publications and participating benefits, to help small organizations or subdivisions benefit from affiliation with ASTM.

Each company should be concerned with the number and the classes of membership it maintains. Memberships should be located as needed to take optimum advantage of ASTM's activities. Also, the combination of memberships selected should tend to keep the organization's annual subscription to the Society in line with regard to the benefits it derives from ASTM (in terms of standards, committee participants, etc.) and with the organization's stature in the economy.

Therefore, except for those organizations the Society recognizes as "sister" societies and the agencies of the United States government (each of which may participate on ASTM's Technical Committees with or without a membership in the Society), each organization or major subdivision desiring to obtain publications of the Society at membership prices, or to share in the activities of the Society, should maintain at least one membership in the Society in the name of the organization. The organization's membership may be supplemented by Personal and Associate Memberships as desired.

Name . . . .
Signature of Person Authorized to Apply for Membership

Position . . . .

Address . . . .

SEE REVERSE SIDE FOR DUES AND BENEFITS OF MEMBERSHIP

MAIL TO: American Society for Testing and Materials,
1916 Race St., Philadelphia, Pa. 19103

Application form on reverse side

INFORMATION ON ORGANIZATIONAL MEMBERSHIP FOR 1970

Extract From Charter: The corporation is formed for the promotion of knowledge of the materials of engineering, and the standardization of specifications and the methods of testing.

Extract From Bylaws: ARTICLE 1. Members and Their Election

Sec. 1. . . . . . The rights of membership of Institutional, Industrial, and Sustaining Members shall be exercised by the individual who is designated as the official representative of that membership.

Sec. 3. An Institutional Member shall be a public library, educational institution, a nonprofit professional, scientific or technical society, government department or agency at the federal, state, city, county or township level, or separate divisions thereof, meeting the qualifications established by the Board of Directors for this classification.

Sec. 4. An Industrial Member shall be a plant, firm, corporation, partnership, or other business enterprise, or separate divisions thereof, trade association, or research institute meeting the qualifications established by the Board of Directors for this classification.

Sec. 5. A Sustaining Member shall be a person, plant, firm, corporation, society, department of government or other organization, or separate divisions thereof, electing to give greater support to the Society's activities through the payment of larger dues.

SCHEDULE of DUES and BENEFITS of MEMBERSHIP for 1970

Class of Organizational Membership	SUSTAINING	INDUSTRIAL	INSTITUTIONAL
Annual Dues[1]	$500 min.[2]	$125	$35
Participants Allowed on ASTM Technical Committees	No Limit	No Limit	No Limit
33-part **ASTM Annual Book of Standards**			
Free Parts Allowed (Maximum)[3]	33 Parts	6 Parts	2 Parts
Sets or Parts Available at Discount[4]	No Limit	No Limit	No Limit
Other Publications (Listed with Member Prices) Available at Discount[4]	No Limit	No Limit	No Limit
ASTM Year Book (Abridged Copy to New Members)	1 Free	1 Free	1 Free
Annual **ASTM Proceedings**	$12.00	$12.00	$12.00
Annual Subscriptions to:			
Materials Research and Standards[5]	6 Free[6]	1 Free	1 Free
ASTM News Bulletin[5]	See Limit	See Limit	See Limit
Journal of Materials	$12.00	$12.00	$12.00

[1] Membership year shall commence on the first day of January. Dues are payable in advance. The dues payment retroactive to 1 January must be received to complete a membership. An election to membership shall be voided if payment is not made within three months of notification of election to membership.

[2] Organizations wishing to provide increased support for the operations of the Society may supplement the minimum Sustaining Membership dues by an additional amount in the form of annual payments in such amounts and for such periods as the organization may choose or in the form of a lump sum in anticipation of annual payments of such additional dues.

[3] FREE PARTS of the Book of ASTM Standards are for shipment only to the Official Representative. They must be requested on the copy of the request form provided with the notification of election, and each year thereafter.

[4] Discount on sets or parts of the Book of ASTM Standards approx. 45% off list price. (Discount on other ASTM publications listed with member prices, approx. 20%.) QUANTITY prices are available on request. The member DISCOUNT is allowed ON REQUEST, if the current DUES payment has been received, if the MEMBERSHIP NUMBER is identified, and calls for shipment and billing to, the MEMBERSHIP ADDRESS. An appropriate HANDLING and SHIPPING charge is applied to each order, EXCEPT for the annual pre-publication order for sets or parts of the Book of ASTM Standards on which the PAYMENT in advance ACCOMPANIES the ASTM ORDER FORM (supplied to the Official Representative when the annual dues payment is received).

[5] The ASTM News Bulletin is combined with Materials Research & Standards for distribution to the Official Representative and those designated to receive the allowed subscriptions of MR&S. The ASTM News Bulletin is distributed to the additional individuals the organization designates as its representatives in ASTM's Committee activities. Of the Annual Dues, $5.00 is allocated for each annual subscription to Materials Research & Standards, and $1.00 for each annual subscription to the ASTM News Bulletin for every person designated as eligible to receive copies of one or both magazines, including representatives to the respective committees. (Not deductible.)

[6] ONE subscription to the Official Representative. FIVE subscriptions on request—identify recipients.

NOTE: Benefits of membership and prices of publications are subject to change without notice.

SELECTION OF AN OFFICIAL REPRESENTATIVE: Careful attention should be given the selection of the OFFICIAL Representative of an organizational membership. He should be qualified to exercise the rights and privileges of membership to enhance the overall interests of the organization. This would involve balloting on highly technical matters, and might include participation on technical committees, and selection of other participants for committee work. The Official Representative of an Organization's membership in the Society may be changed as needed.

VOTING RIGHTS: The Official Representative of each organizational membership shall be privileged to cast one ballot in Society matters.

APPLICATION FOR 1970 ASTM INDIVIDUAL MEMBERSHIP

Extracts from Charter and Bylaws and Dues Schedule on REVERSE SIDE of this application
The undersigned hereby applies for membership in the American Society for Testing and Materials in the class of (check one):

☐ **PERSONAL MEMBER** ☐ **SENIOR ASSOCIATE MEMBER ‡** ☐ **ASSOCIATE MEMBER**

and if elected the applicant agrees to be governed by the Charter and Bylaws of the Society, and to further its objectives as laid down therein. (This membership includes: opportunities for the participation of the member in the general and District activities of the Society, and for him to make application for participation on its technical committees; subscription to **Materials Research & Standards**; member discounts on publications; and other benefits and privileges.)

FORMAT FOR LISTING THE ASTM MEMBERSHIP

Please complete the grid below. PRINT or TYPE. Leave a space between words. Use no punctuation marks or special characters, except the ampersand (&). Use initials and abbreviations where necessary.

Line 1:—List your LAST NAME, FIRST NAME and MIDDLE INITIAL
Line 2:—Beginning in 8th position, your TITLE and DEPARTMENT
Line 3:—List your PARENT ORGANIZATION (or self-employment)
Line 4:—If applicable, list the FACILTY in which you are EMPLOYED.
Line 5:—COMPANY street address or P. O. BOX
Line 6:—CITY (leave 1 space) STATE (leave 2 spaces) ZIP CODE
Line 7:—ADDRESS FOR MAIL (if OTHER than above)
Line 8:—CITY (leave 1 space) STATE (leave 2 spaces) ZIP CODE
Line 9:—LEAVE BLANK (for use by the Society)

1
2
3
4
5
6
7
8
9

Major Product. . . . .
(Or service performed by your organization or you, if self-employed)

Major Field of Interest. . . . .
Examples: chemical research/civil engineering

Specific Application of Effort. . . . .
Examples: polymer plastics/construction

Date of Birth: Month. . . . . .Day. . . . . .Year. . . . .

ASTM MEMBERSHIP POLICY — The development of organization memberships and management understanding are vital to the Society's ability to serve the economy effectively in the future. The Society hopes that major organizations will help support its activities with one or more Sustaining Memberships. ASTM also offers inexpensive Industrial and Institutional memberships, providing substantial publications and participating benefits, to help small organizations or subdivisions benefit from affiliation with ASTM.

Each company should be concerned with the number and the classes of membership it maintains. Memberships should be located as needed to take optimum advantage of ASTM's activities. Also, the combination of memberships selected should tend to keep the organization's annual subscription to the Society in line with regard to the benefits it derives from ASTM (in terms of standards, committee participants, etc.) and with the organization's stature in the economy.

Therefore, except for those organizations the Society recognizes as "sister" societies and the agencies of the United States government (each of which may participate on ASTM's Technical Committees with or without a membership in the Society), each organization or major subdivision desiring to obtain publications of the Society at membership prices, or to share in the activities of the Society, should maintain at least one membership in the Society in the name of the organization. The organization's membership may be supplemented by Personal and Associate Memberships as desired.

Applicant's Signature. . . . .

‡Please attach an outline of your prior ASTM affiliation.

SEE REVERSE SIDE FOR DUES AND BENEFITS OF MEMBERSHIP

MAIL TO: American Society for Testing and Materials,
1916 Race St., Philadelphia, Pa. 19103

Application form on reverse side

INFORMATION ON INDIVIDUAL MEMBERSHIP FOR 1970

Extract From Charter:

The corporation is formed for the promotion of knowledge of the materials of engineering, and the standardization of specifications and the methods of testing.

Extract From Bylaws:

ARTICLE 1. Members and Their Election

Sec. 2. A Personal Member shall be a person meeting the qualifications established by the Board of Directors for this classification. ("The basic qualification . . . shall be that he is interested in the Society's field of activities.")

Sec. 6. An Associate Member shall be a person less than thirty years of age. He shall have the same rights and privileges as a Personal Member, except that he shall not be eligible for office. An Associate Member shall not remain in this category beyond the end of the calendar year in which his thirtieth birthday occurs.

Sec. 8. A Senior Associate Member shall be a person who is fully retired from his regular employment, who has been a member of the Society and/or a participant in an activity of the Society for a total of ten years or more, not necessarily continuously, and who elects to continue his association with the Society through the payment of reduced dues. He shall have the same rights and privileges as a Personal Member, except as may be limited by the Board of Directors for this classification.

SCHEDULE of DUES and BENEFITS of MEMBERSHIP for 1970

Class of Individual Membership	PERSONAL	ASSOCIATE	SENIOR ASSOCIATE
Annual Dues[1]	$20	$10	$10
Member's Participation on ASTM Technical Committees	No Limit	No Limit	No Limit
33-part **ASTM Annual Book of Standards**			
Free Parts Allowed[2]	1 Part	None	None
Parts Available at Discount (Maximum)[3]	2 Parts	2 Parts	2 Parts
Other Publications (Listed with Member Prices)			
Available at Discount[3]	No Limit	No Limit	No Limit
ASTM Year Book (Abridged Copy to New Members)	1 Free	1 Free	1 Free
Annual **ASTM Proceedings**	$12.00	$12.00	$12.00
Annual Subscriptions to:			
Materials Research & Standards[4]	1 Free	1 Free	1 Free
ASTM News Bulletin[4]	1 Free	1 Free	1 Free
Journal of Materials	$12.00	$12.00	$12.00
Registration (badge and program) at ASTM National Meetings	...	...	Free

[1] Membership year shall commence on the first day of January. Dues are payable in advance. The dues payment retroactive to 1 January must be received to complete a membership. An election to membership shall be voided if payment is not made within three months of notification of election to membership.

[2] The FREE PART of the Book of ASTM Standards must be requested on the copy of the request form provided with the notification of election and each year thereafter.

[3] Discount on the two PARTS of the Book of ASTM Standards, approx. 45% off list price. (Discount on other ASTM publications listed with member prices, approx. 20%.) The two PARTS can be obtained at member prices ONLY if ordered on, and paid for with the ASTM ORDER FORM (supplied to the member when the annual dues payment is received). The member DISCOUNT is allowed ON REQUEST, if the current DUES payment has been received, if the MEMBERSHIP NUMBER is identified, and calls for shipment and billing to, the MEMBERSHIP ADDRESS. An appropriate HANDLING and SHIPPING charge is applied to each order, EXCEPT for the annual pre-publication order for parts of the Book of ASTM Standards on which the PAYMENT in advance accompanies the ASTM ORDER FORM.

[4] The ASTM News Bulletin is combined with Materials Research & Standards for distribution to the individual members of the Society. Of the Annual Dues, $5.00 is allocated for each annual subscription to Materials Research & Standards, and $1.00 for each annual subscription to the ASTM News Bulletin. (Not deductible.)

NOTE: Benefits of membership and prices of publications are subject to change without notice.

VOTING RIGHTS: Each member shall be privileged to cast one ballot in Society matters, and may apply for participation in any of the Technical Committee activities of the Society. Individual memberships in the Society are non-transferable.